ISBN 978-1-5278-0002-1
PIBN 10900825

CIRCULATES EVERYWHERE IN CANADA
Also in Great Britain, United States, West Indies, South Africa and Australia.

HARDWARE AND METAL

A Weekly Newspaper Devoted to the Hardware, Metal, Heating and Plumbing Trades in Canada.

Office of Publication, 10 Front Street East, Toronto.

VOL. XIX. MONTREAL, TORONTO, WINNIPEG, AUGUST 3, 1907 NO. 31.

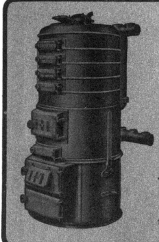

CHICAGO SPRING BUTTS

TRADE CSB CO MARK

A PRODUCT OF RECOGNIZED SUPERIORITY

CHICAGO SPRING BUTT

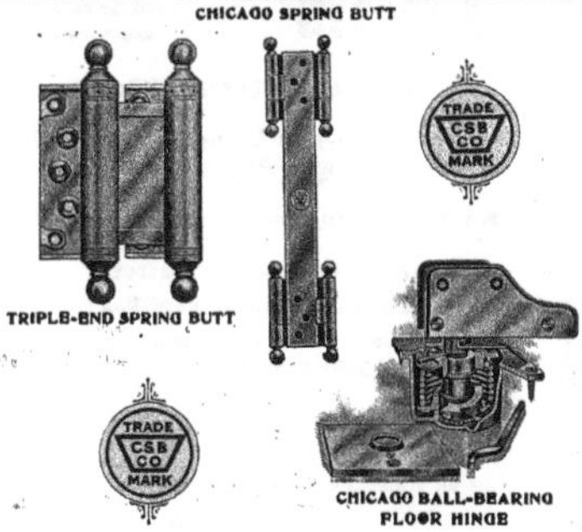

TRIPLE-END SPRING BUTT

CHICAGO BALL-BEARING FLOOR HINGE

Chicago Spring Butt Company

CHICAGO . . NEW YORK

CATALOGUES ON REQUEST

Pig Iron

Jarrow Brand
Clarence "
Lackenby "
Cleveland "
Gartsherrie "
Carnbroe "
Glengarnock "
ETC.

If you are in the market, send us particulars of your requirements, and we shall be glad to quote you favourable prices.

B. & S. H. THOMPSON & CO.
LIMITED
381 and 383 St. Paul Street
MONTREAL

Other Tools are very Good Tools, but **"YANKEE TOOLS"** ARE BETTER

The NEWEST, CLEVEREST and QUICKEST SELLING TOOLS of the KIND.

SCREW DRIVERS

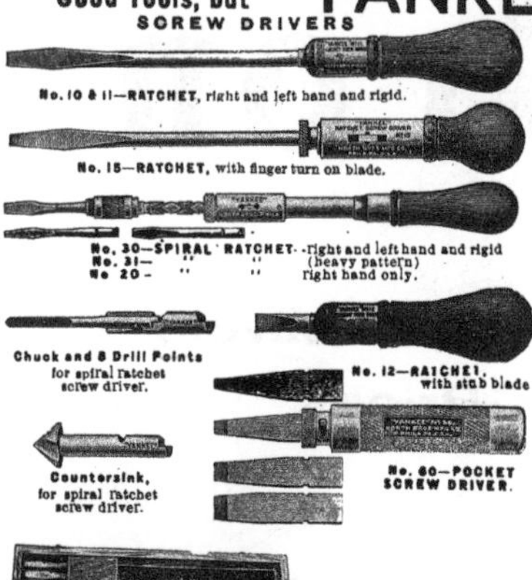

No. 10 & 11—RATCHET, right and left hand and rigid.

No. 15—RATCHET, with finger turn on blade.

No. 30—SPIRAL RATCHET—right and left hand and rigid
No. 31— " " (heavy pattern)
No. 20— " " right hand only.

Chuck and 8 Drill Points for spiral ratchet screw driver.

No. 12—RATCHET, with stub blade

Countersink, for spiral ratchet screw driver.

No. 60—POCKET SCREW DRIVER.

DRILLS

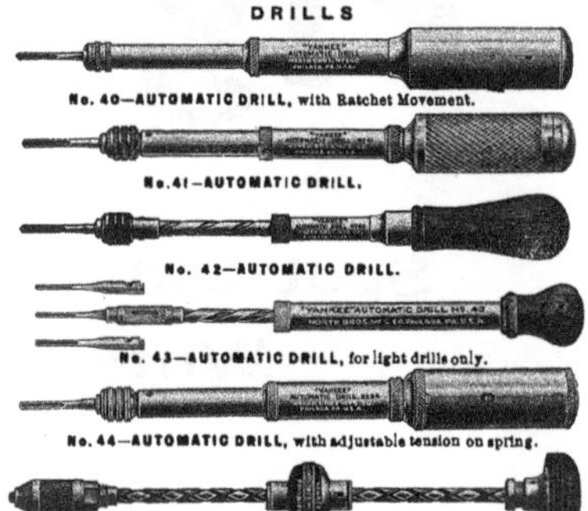

No. 40—AUTOMATIC DRILL, with Ratchet Movement.

No. 41—AUTOMATIC DRILL.

No. 42—AUTOMATIC DRILL.

No. 43—AUTOMATIC DRILL, for light drills only.

No. 44—AUTOMATIC DRILL, with adjustable tension on spring.

No 50—RECIPROCATING DRILL, for wood or metals

SOLD BY LEADING JOBBERS

SEND FOR OUR NEW **"YANKEE"** TOOL BOOK

NORTH BROS. MFG. CO.

PHILADELPHIA, Pa.

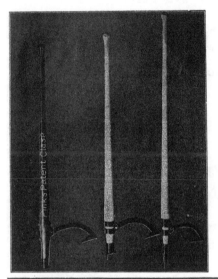

No Woman Should Do Work a Machine Can Do

THE BUCKEYE

Power Washing Machines

Save Women's Lives

An Article Every Dealer Should Have In His Store

TO-DAY

DO IT NOW

Write for PRICES

WE GUARANTEE

all parts of the Buckeye Power Washer against breakage for one year by replacing free of charge.

The Buckeye Motor

is the only motor constructed on scientific principles—the same as a steam engine—only one valve in its entire construction, and that is on the outside of cylinder, and is operated the same as an engine valve.

THE CYLINDER HAS NO VALVES INSIDE, NOTHING BUT A PLAIN PISTON HEAD.

All parts of the Buckeye Motor are made of brass to prevent rusting. Our motor will operate on any pressure from 18 pounds to 250.

The motor does not extend over edge of the tub, being only eleven inches over all, and is absolutely POSITIVE in its action.

Our tubs are of standard make, large lid, being extensions for wringer attachment which will admit of the opening of lid clear of any wringer on the market.

THE BUCKEYE POWER WASHER

Showing Motor on its Base Plate and the Method of Transmitting the Power Direct to the Dasher Inside the Tub which Agitates the Clothes.

Now, Mr. Dealer, if you want a good selling article—one that works itself as well as sells itself—investigate the merits of the Buckeye Power Washer. Every housewife who has running water in her home is a prospective purchaser, and it requires but for her to see the machine in operation and have its merits explained to her to make her enthusiastic about owning one.

Exclusive Canadian Distributors

LEWIS BROS., LIMITED

Shelf and Heavy Hardware

Offices and Warerooms: 14 to 28 Bleury Street

MONTREAL

And at *Ottawa* *Toronto* *Winnipeg* *Calgary* *Vancouver*

Consumers Cordage Co., LIMITED

Manufacture every variety of

CORDAGE — PACKING

BINDER TWINE

LATHYARN — SMALL TWINES

From Manilla, Sisal, Italian, Russian, Jute

Tallow Laid Transmission Rope, 3 or 4 Strand

Oil Well Drilling Cables, Lariat Rope, 3 or 4 Strand,
Halyards, Deep Sea Line, Snap Halters, Snap Plow Lines,
Clothes Lines, Hide Cord, Pulp and Paper Cord,
Tarred Rope, Marline, Houseline

Wrapping, Rolling, Seaming, Sewing and Grape Twines

Careful Execution **SPECIAL ORDERS** Prompt Shipment

Mills: MONTREAL, QUE., and HALIFAX, N.S.

BRANCHES:

F. H. ANDREWS & SON, Quebec. CONSUMERS CORDAGE CO., Limited, St. John, N.B. MacGOWAN & CO., Vancouver
TEES & PERSSE, Limited, Winnipeg, Calgary and Edmonton. GEO. WOOD, London, England.
MERRICK, ANDERSON & CO., Winnipeg, Binder Twine Distributors for Northwest.

Patented 1900. Improved 1905.

Davidson's "Empire" Stove Pipe

Improved Patent

Neatly packed in crates of 25.
Secures lowest possible freight rate.

Deep Swage Prevents Telescoping.
Coated to Prevent Rust.

Sold as low as ordinary Stove Pipes.

SIMPLE and EASY to put together.
No Profane Language Necessary.
The only tools required are a pair of hands.
Pipes being cut out with dies ensure absolute fit and conformity.

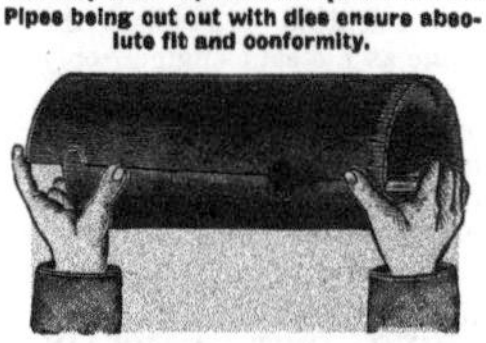

Miners' Camp Kettles

Strong and substantially made in extra heavy tin.

Nineteen sizes, from ⅛ quart to 29 quarts.

Wash Tubs—Galvanized

With Wringer Attachment

Nos.	1	2	3
Top measurement, inches	21½	22¾	25
Bottom " "	17⅞	18¼	20
Height " "	9½	10½	11

The THOS. DAVIDSON MFG. CO., Ltd., Montreal and Winnipeg

H. S HOWLAND, SONS & CO. LIMITED

HARDWARE MERCHANTS

Only Wholesale — 138-140 WEST FRONT STREET, TORONTO. — Wholesale Only

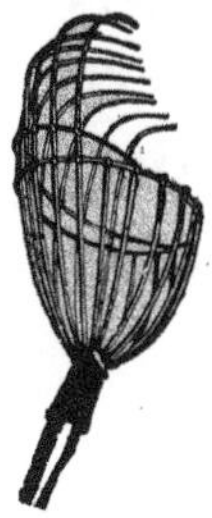

Fruit Pickers
Tinned Wire, 6 in. diameter, 12 in. long

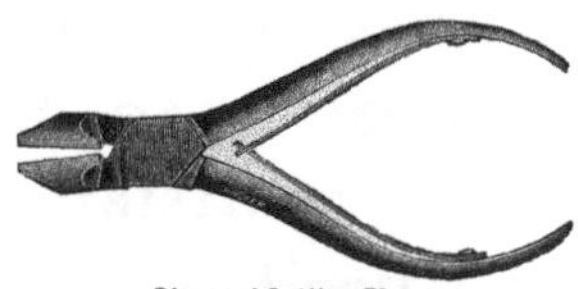

Diagonal Cutting Plyers
For Fruit Packers

Barrel Header

Little Star Apple Parer

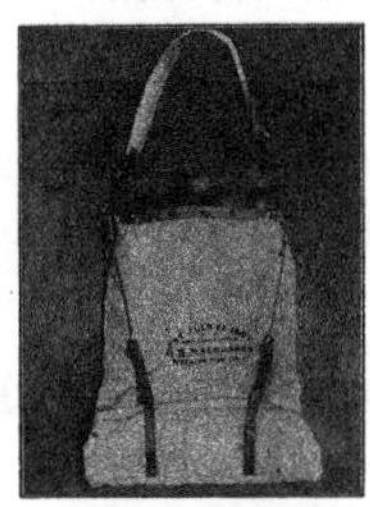

Canvas Apple Sacks

Hudson Apple Parer

Lightning Apple Parer

Barrel Hatchets

Ice Tongs
Cast Steel, 12, 16, 24 in.

H. S. HOWLAND, SONS & CO., LIMITED

Opposite Union Station

GRAHAM WIRE NAILS ARE THE BEST

Are you receiving our monthly illustrated circular? IF NOT WRITE FOR IT.

Our Prices are Right — We Ship Promptly

A New Automatic Pistol— the "Webley"— from the shops of Scott and Webley, Limited, makers of the small arms used in the British Army.

8 shots; weight with magazine, 20 oz.; length, 6¼ in.; depth, 4½ in. Penetration at 20 yards: 7 one-inch pine boards.

The "Webley" pistol combines all the advantages found in the various automatic pistols at present on the market. In addition, however, its construction is simple, it has greater strength, and is lighter and less bulky.

An efficient safety is provided, permitting the pistol to be carried at full cock without danger of discharge. This safety is always in engagement. Before firing it is necessary to raise the safety lever on the side of the pistol until the word safe is covered.

The advantages of automatic pistols are now generally recognized. They are easier to manipulate, can be fired more rapidly, and with greater precision and hitting-force than ordinary revolvers.

This arm uses Kynoch smokeless cartridges adapted to 7.65 M.M. Browning and .32 calibre automatic pistols. Order them as 7.65 M.M. Browning or .32 automatic pistol cartridges.

Because it is British-made and enjoys the preference, it is lower in price than other pistols on the market.

Ask for prices. Speak to our travellers about it.

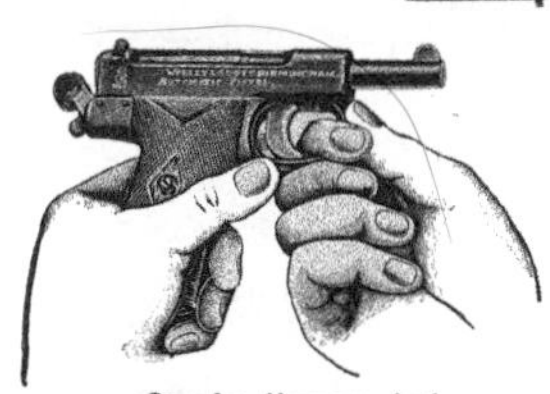

Can be dismounted

and put together again in a few minutes

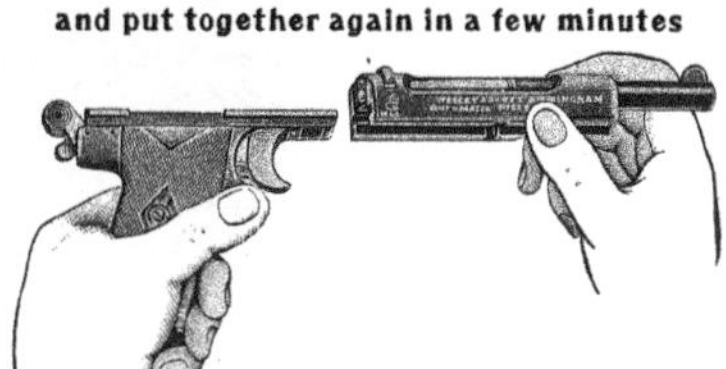

without the use of any tools.

The Cartridge.

Caverhill Learmont & Co.

MONTREAL

WINNIPEG, OTTAWA, QUEBEC, FRASERVILLE

Persons addressing advertisers will kindly mention having seen their advertisement in this paper.

The Dunn Hollow Concrete Block Machine

Many dealers are increasing their trade in Cement by pushing sales of the Dunn Machine. Some are also making a good profit by renting machines to parties building. Machine is compact, portable, simply operated and moderate in price.

WRITE US FOR CATALOGUE H.

The James Stewart Mfg. Co., Limited, Woodstock, Ont.

Past Due Accounts.

We can assist you to convert these into **Working Capital.**

Intelligent and up-to-date attention given to all claims entrusted to us.

Let us assist you.

The Beardwood Agency

Claims Collections and Commercial Reports

313 New York Life Bldg., MONTREAL

MITCHELL'S SELF-TESTING SAFETY CODE

REVISED EDITION

A New and Scientific System for the Transmission of Messages by Telegraph or Cable, so arranged that one word may convey several meanings, such as Article, Quality, Price Quantity, Terms, etc

By other codes three and four words are required to express what this system will convey in one word, thus costing three and four times as much as the same message if sent by this system.

EXAMPLES

Code Words

LUDSELPIEL MEBDIKOLAF WOCUMZEXOI—Ship by Cunard Steamer from your port 28th June, 150 pieces Bevelled Plate Glass, 1½" bevel, 18½ x 24".

WULYBSABNI—Steel sheets 27 g. 24 x 101", 10,000 packages.

WUPUHBABHE—Wire Nails 5d. Buy 500 kegs

WIWULKAFDE—Rosin, "I" quality, quote cost and freight shipment 1st half of next month, 200 barrels

XASYSJABBA—Delivery Waggon Platform Springs 2 to 3,000 lbs., increase our order to ten

Try these with your codes An 8-page prospectus with full particulars sent free on request.

The Mitchell Self-Testing Safety Code Co.

8 COLBORNE ST. - TORONTO, CANADA

Galt "Sure-grip" Shingles

are simply wonders TO SELL, TO LAY AND TO MAKE FRIENDS. Once used and you would never use any other—your customers wouldn't let you.

"Sure-Grips" make a really handsome roof, as well as the most weatherproof. Try them—that's all we ask.

"GALT" CORRUGATED SHEETS are best English galvanized stock, pressed straight and true. Our prices are right, and we can ship promptly.

Consult us before purchasing CEILINGS, SIDINGS, ROOFINGS, CORNICES, SKYLIGHTS, VENTILATORS, FINIALS, FIRE-PROOF WINDOWS, EXPANDED METAL LATH, ETC.

The "GALT KIND" is the line to push; we protect and assist our customers.

THE GALT ART METAL CO., Limited

GALT, - ONT.

Jardine Hand Drills

New Designs

Get samples of these machines for your show room. They will interest your customers and prove a good investment.

It pays to sell the best Tools.

A. B. JARDINE & CO.

HESPELER - - - ONTARIO

...FULL STOCK...

Salt Glazed Vitrified

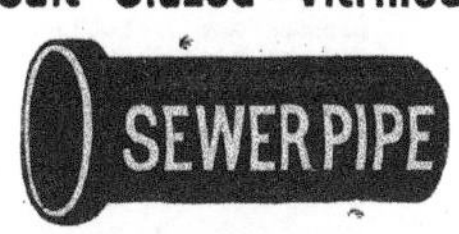

Double Strength Culvert Pipe a Specialty.

THE CANADIAN SEWER PIPE CO.

HAMILTON, ONT. TORONTO, ONT. ST. JOHNS, QUE.

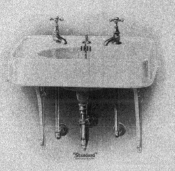

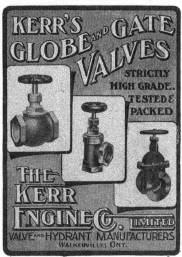

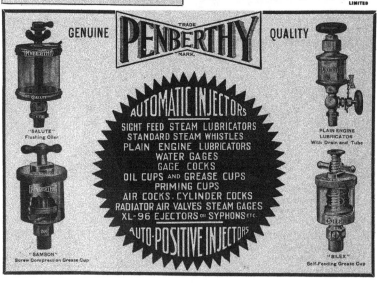

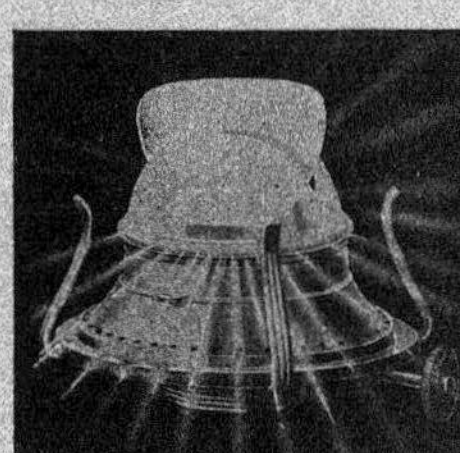

"Alumino" and "Electric"

OIL HEATERS

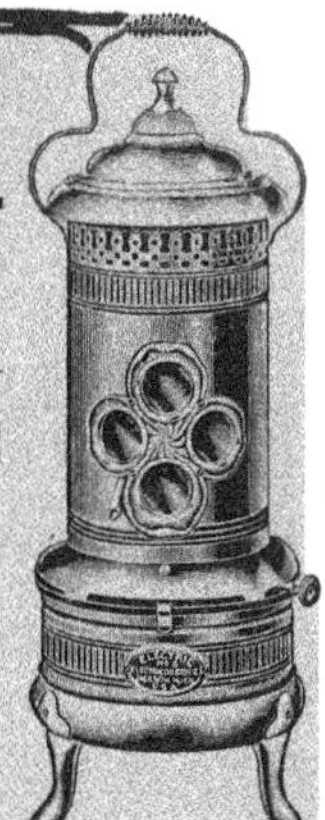

You know the burner is the very soul of the oil heater. If the burner is not perfect the heater is utterly unreliable, unsatisfactory.

"Alumino" and "Electric" Oil Heaters are equipped with our Grand Safety Burner with its solid flame spreader; its liberal air passages, which give thorough circulation and therefore perfect combustion. See below.

No odor. No smoke. No danger. No trouble. And a furnace for heat.

The only oil heaters which radiate heat like a base burner, from the sides and bottom as well as the top—heat the whole room.

Sell on sight and stay sold. Never come back for repairs nor is it ever necessary to send a man out to rewick and explain and operate.

No other oil heater possesses all these merits, sells so quickly, or is so profitable to the dealer.

Sectional view showing construction grand "SAFETY BURNER".

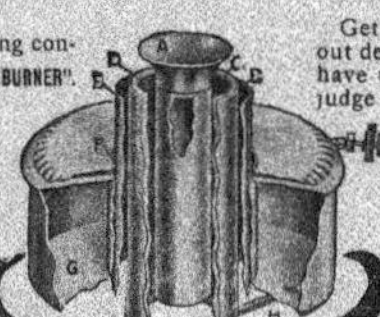

A—Flame Spreader; B—Air space outside of Wick; C—Air space inside of Wick Tube; D—Wick; E—Outside Casing to Burner; F—Air space between Fount and Outside Casing; G—Fount for oil, entirely separate from Burner; H—Pipe Feed carrying oil from Fount to Burner.

Get your order in without delay; somebody will have to wait, if we can judge from the demand.

Lewis Bros., Ltd.

Montreal Winnipeg · Toronto Vancouver · Ottawa Calgary

Emerson & Fisher, Ltd.

ST. JOHN, N.B.

THE BURMAN "Push-Spring" Toilet

THE NEW HAIR CLIPPER

Elegance with True Mechanical Design.
Ease of Action with Accurate Fitting.
Interchangeable Parts. Precise Machining.
Detachable Plates.

MADE IN FOUR SIZES

No. 0 to cut 1-16 inch
No. 1 to cut ⅛ inch
No. 2 to cut ¼ inch
No. 3 to cut 5-16 inch

FROM ALL JOBBERS. A stock of machines and parts is held in Canada.

BURMAN & SONS, Ltd., Lee Bank Works, BIRMINGHAM, ENGLAND

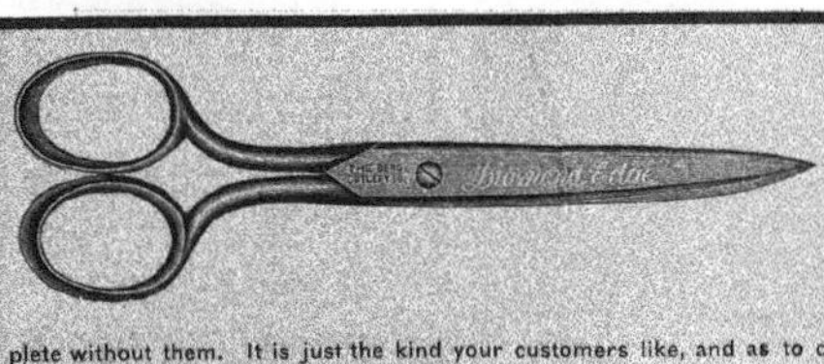

When ordering your stock of Cutlery, specify the

A★1

brand of **The R. Berg Cutlery Co.** There is quite a demand for these goods, as they have proved to be of the **very best quality and finish** which will satisfy even the most particular customers.

Our "Diamond Edge" Scissors are to-day a favorite in the market, and your stock is not complete without them. It is just the kind your customers like, and as to quality, well, we guarantee it to be second to none.

Ask your jobber for details and prices, or write to

F. W. LAMPLOUGH & CO., 9 De Bresoles Street, MONTREAL

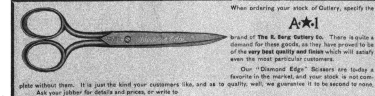

PRESSED ZINC ORNAMENTS

We are offering the trade spendid values in all kinds of architectural pressed zinc ornaments, including:

CABLE ORNAMENTS, CAPITALS,
BRACKETS, CARLANDS, ETC.

We can make up special ornaments to order, and invite your correspondence on the subject.

"WE WORK WITH THE TRADE"

The Metal Shingle and Siding Co., Limited

Montreal **Preston, Ont.** **Toronto**

SALES AGENTS:

ST. JOHN, N.B. — Emerson & Fisher, Limited
QUEBEC — J. A. Bernard
WINNIPEG — Clare & Brockest
CALGARY — Ellis & Grogan
VANCOUVER — McLennan, McFeely & Co.

Canadian National

EXHIBITION

1907—TORONTO, ONT.—1907

Saturday, Aug. 24th to **Monday, Sept. 9th**

$400,000 In New Buildings $400,000
In Premiums $45,000 In Premiums
$40,000 For Attractions $40,000

Historical Art Loan in New Art Gallery
Extra Fine Exhibit of Pure Bred Live Stock

Five Acres of Industrial Exhibits

Grand Musical Festival
Brilliant Military Spectacle
Highest Class Special Features
Battle of Badajos (Nightly)

Largest Fire-Proof Grand Stand in America

Six Tickets for One Dollar	Reduced Rates on all Lines of Travel	Great Dog and Cat Shows

The First Week is the Best Week for Comfort

TRADE MARK

FOR PLATE

Will Call on You

Our Canadian representative is calling on the trade from Montreal to Vancouver. His trip will occupy about three months. He will be pleased to show you our complete line of Silverware, Cutlery, Flasks, etc.
Kindly address the Montreal office, making an appointment.

The House of Hutton

Established in England in 1800

Sole Canadian Agent
W. J. GRANT, Lindsay Bldg., MONTREAL

TRADE MARK

FOR CUTLERY

The Howland Pump Oiler

FOR SALE BY

J. H. Ashdown Hdw Co.,	Winnipeg
Geo. D. Wood & Co.,	"
Miller-Morse Hdw. Co.,	"
Jas. Robinson & Co.,	"
Lewis Bros., Ltd.,	Montreal

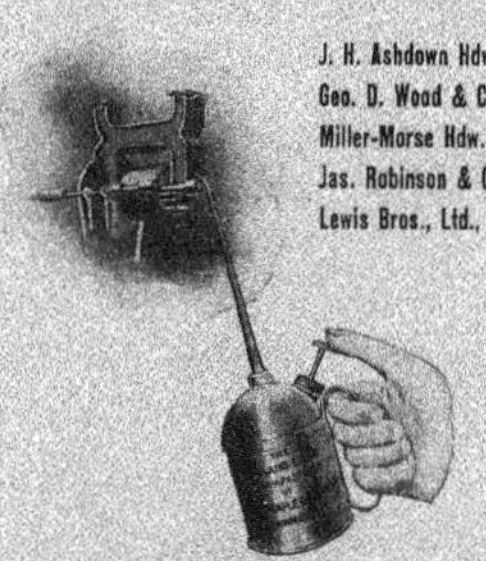

A slight pressure of the thumb regulates the flow.

Maple City Manufacturing Co.

Monmouth, Ill.

PROFITABLE LINES

You are handling the best manufactured in

Cutlery, Electric Plate and Solid Nickel Silver Flatware

if the goods bear our stamps. The trade mark guarantees.

QUALITY. FINISH. DISPATCH.

The McGlashan, Clarke Co., Limited

NIAGARA FALLS, - ONT.

MR. J. MACKAY ROSE, 117 Youville Square, Montreal, Que. MR. N. F. GUNDY, 47 Hayter St., Toronto, Ont.
MR. DAVID PHILIP, 291 Portage Ave., Winnipeg, Man.

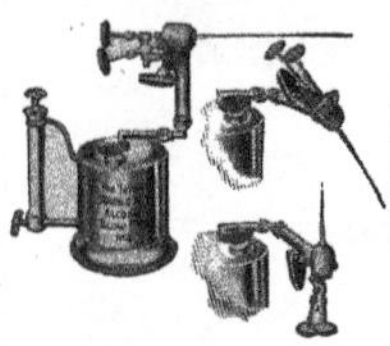

HOTTEST ON EARTH

The No. 8 Alcohol Blow Pipe produces nearly 3000 F. Will do lead burning on storage batteries. Produces a needle blue flame pointed. Satisfaction guaranteed or money refunded. Jobbers sell at factory price, **$10.40 net.**

THE TURNER BRASS WORKS

53 MICHIGAN STREET, - CHICAGO, U.S.A.

REGISTERED TRADE MARK
GRANTED 1682

World's Best Cutlery

You make no mistake in specifying goods made by

JOSEPH RODGERS & SONS, Limited

when ordering cutlery. A reputation of over 200 years at stake, insures the quality of every article.

Write for catalogues from Sole Canadian Agents

JAMES HUTTON & COMPANY, MONTREAL

High-grade Bathroom Fixtures

ARE MADE IN CANADA BY

The Carriage Mountings Co., Limited, Toronto

NO ACID OR GRIT

will be found in "Majestic" polishes The brilliant and lasting lustre imparted by "Majestic" polishes is not susceptible to atmospheric conditions and weather changes.

"Majestic" polishes are unequalled for brass, copper, nickel and all finished metallic surfaces.

Write for sample and prices. 71

MAJESTIC POLISHES, Limited

575 Yonge Street - - Toronto, Canada

Manufacturers' Agents

ALEXANDER GIBB

Manufacturer's Agent and Metal Broker,
13 St. John Street, Montreal

Representing Canadian, British and American Manufacturers. Correspondence invited from firms wishing to be represented.

CHARLES H. FOX

Vancouver, Canada

MANUFACTURERS' AGENT

Selling to the Hardware Jobbers Only.

Representing Canadian, British and American Manufacturers, correspondence invited. Cable Address, Chasfox Vancouver, P O.Box 1105.

H. W. MITCHELL

WINNIPEG, MAN.

Correspondence is solicited from manufacturers wishing a representative in Winnipeg.
Travellers call regularly on the wholesale and retail trade in Western Canada.
Highest references and financial responsibility.

THE WAGGONER EXTENSION LADDER

is the lightest, handiest, strongest and safest ladder made. Re-inforced with steel wire. Patent safety lock. Made in all lengths from 20-ft., 24-ft., 28-ft., etc., up to 100 ft. if desired. Every painter, tinner, stacker and farmer needs one
Write for circulars and prices.

The Waggoner Ladder Co., Ltd.

LONDON, - - CANADA

E. H. Briggs Co., Ltd., Western Representative
Winnipeg, Man.

JENKINS & HARDY
Assignees, Chartered Accountants, Estate and Fire Insurance Agents, 15½ Toronto St., Toronto.
465 Temple Building, Montreal.

Covert Mfg. Co.

TROY, N.Y.

Harness Snaps, Chain, Rope and Web Goods, etc. For sale by Jobbers at Manufacturers' prices.

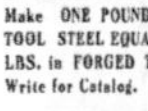

TOOL HOLDERS

For LATHES and PLANERS.
All kinds—All sizes.
ARMSTRONG BROS. TOOL CO.,
106 N. Francisco Ave., Chicago, U. S. A.

Make ONE POUND of TOOL STEEL EQUAL 10 LBS. in FORGED Tools. Write for Catalog.

GALVANIZING AND TINNING

THE CANADA METAL CO.
TORONTO, ONTARIO.

Forwell Foundry Co.

BERLIN, ONT.

Manufacturers of

SOIL PIPE, FITTINGS, and CAST IRON SINKS

Ask Jobbers for "F. F. CO." Brand

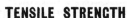

Drummers' Snack on Annual Frolic

Alton Welcomes Travelers on Their 6th Jamboree—An Immortal Outing and a Record-breaking Snack—A Mammoth Concert, An All-night Serenade, a Costume Procession and an Afternoon of Sports Were the Features—Mass Meeting of Drummers' Snack Club Held and New Officers Elected.

(From the Canadian Grocer.)

Laugh, and the world laughs with you :
Weep, and you weep alone.
For the dear old earth must borrow its mirth,
It has trouble enough of its own.

Now this is a true saying, and worthy of all acceptation, for in it lies the full, perfect and sufficient reason for the existence of the Drummers' Snack.

"To be great, one must have a history," says Monmouth sagaciously ; "to have a history, one must have lived," he concludes a trifle obviously ; and thus it is with the Drummers' Snack. They are great, for they have a history (not necessarily that it follows) ; they have a history because they have lived. Individually they have seen the world, and human nature has become to them as an open book. Collectively they are united, knit together in a bond of fellowship that shall endure with the sun. Of history they have no lack. The founding of their club was a performance unattended by civic authorities, destitute of hackneyed and stereotyped speeches, unnoticed by the world, lacking even the conventional corner-stone laid by a worshipful hand. But in the very modesty of its inception lies its present strength. Through its initial obscurity it has attained to the recognition that it enjoys to-day.

Seventeen years ago the Algie family, then in the general store and woolen mills business at Alton, Ont., made a practice of annually entertaining, at a two days outing held in the vicinity of the town, all the travelers who passed through Alton during the year. The boys had nothing to do but come. The Algie family did the rest. As the years went by, and more travelers were added to the list the function became too large and unwieldy for the Algie's to handle, generous and open-homed as they were. The outing had by this time become an event to which a large proportion of the traveling world eagerly looked forward, and its popularity was proved by the clamour that arose at the first hint of abandonment. A club was formed and officers appointed to take charge of the outing, and the Drummer's Snack was fairly launched upon its career. To Billy Colville fell the honor of being its first president, and should one seek a reason for the club's phenomenal growth and prestige since the election of its first chief officer, it is but necessary to glance at the stamp of men who yearly since that date have been selected to watch over and direct its affairs. After Billy Colville came Jas. Cooper, Col. E. E. W. Moore, Bob Keys, Charlie Smith, and Jack Charles, and this year Mike Mal— but of that hereafter.

A Trainload of Smiles.

The sixth annual outing went to Alton on July 26. It returned 36 hours later. There is the bare statement. Here are the details.

On Friday evening the C.P.R. took on a trainload of assorted smiles at Toronto, picked up small quantities of similar freight at wayside stations dotted along the track, and three hours later disgorged the lot onto the Alton platform.

The noise began at once. There was no waiting. Sol Walters' voice was heard raised in hysterical enquiry for a short cut to the cyclone cellar. He was suppressed, and persuaded under pressure to fall in line with the procession. The Alton citizen's band, presided over by Cap. Alberton and Dr. Algie, headed by a profusely flagged and be-ribboned carriage, wherein Harry Coffen and

Drummers' Snack—The baseball teams and some of their admirers.

Reeve Willoughby sat in state, and followed by a bevy of little girls, costumed in white dresses slashed with broad red ribbons, broke into a burst of melody and swung up the road. What

Drummers' Snack—Will J. White and the Occasional Offender.

does it matter what tune they played? Every man in the procession sang a different version of it, whatever it was. The net result, if not musically perfect, was at least harmoniously discordant. All Alton was at its doors, every ear was attuned to intercept those dulcet strains; every eye was levelled at the procession. The latter, proudly conscious of an exalted position, redoubled its efforts to give them something worth listening to. In this manner came the Drummer's Snack to Alton.

The Concert.

Long before eight o'clock. the hour set for the big concert, the mammoth

Drummers' Snack—Mike Malone, the new president.

tent was stretching itself to accommodate the audience. The tent itself is important enough to deserve a passing mention. Just an ordinary tent with no conceit about it at all, made of canvas and built with an eye to solidity rather than beauty. Many tender memories are associated with it by the members of the Drummers' Snack. Here is the home of the club, herein have the concerts been held from the beginning; herein not infrequently have slumbered many dozen of the elect. Staunchly it has withstood the elements for many years, and more exacting test than this has successfully weathered the sonorous snores of Mr. Walters and Jack Charles.

For once the expected came to pass. A general idea appeared to have circulated among the audience that something out of the ordinary was going to happen. But was anybody gifted with imagination sufficient to foresee the net result? If so, they were unfortunate in cheating themselves of the delights of anticipation. This is the programme presented by the Drummers' Snack for the critical appreciation of an audience prepared to be satisfied with anything:

Drummers' Snack—John W. Charles, retiring president.

1. Song, "The Wanderer."—Ford.
   Chas. Leslie.
2. Contralto Solo, "'Twas the Rose." —DeKoven.
   Miss Olive Belyea.
3. Baritone Solo, "In the Days of Old." —Nevin.
   Donald C. McGregor.
4. Soprano Solo, "Sing, Sweet Bird." Shafera.
   Miss Bertha Crawford.
5. Humorous Song, "Insanity."— Will J. White.
6. Duet, "Good-Night, Beloved."— Pensuti.
   Miss Belyea and Donald C. McGregor.

Intermission, Selections by Alton Band.

The Piece de Resistance.

Followed the chief event of the evening. This was the production by the Hamilton branch of the Snack, of a pathetically-humorous plantation sketch, called "The Darkey's Dream," written especially for the club by their old and well-tried friend John B. Nelligan, and produced in Canada for the first time.

The four principal parts were taken by Bay Hill, Court Thompson. Col. John Stoneman, and Hy. Eckstien. but though they were conspicuously prominent, they didn't make an ounce more noise than the following promising bunch of amateur mummers:—Charley

Drummers' Snack—The vice-president in kilts.

Smith, Vernon Smith, George Smye, Neil Pufford, Peter Smith Bob Smith, Jas. Smith, Jack Smye, John Hooper, Ed. Zimmerman. To attempt to make any distinction among these last ten would be an abortive task, similar to splitting hairs. Let them be classed, as they would themselves desire—equally—neither with flattery nor derogatory remarks; only this must be said, they worked hard, they never wearied, they did their best, and it was a good best.

The Artists.

Bay Hill, the Hamilton Dockstader, Court Thompson, a disciple of Munion, by the great horn spoon, Col. John Stoneman, a master of the histrionic art, and Hy. Eckstien, who like Napo-

Drummers' Snack—Will J. White and Billy Colville.

leon, does things without talking, earned individually in the highest degree the thunders of applause that again and again swept like a wave over the audience. John Nelligan must have been a proud man as he watched and listened to the way his work was produced, and heard the manner in which the audience received it. Songs, cake-walks and dances were liberally interspersed throughout the performance, and it is a safe bet that most of the actors surprised themselves as much as they pleased the audience. The production of this play places us in a position flatly to contradict the rumour that the stage is going to the dogs.

Drummers' Snack—A bunch of baseball humorists.

Of the other artists who took part in the concert, special mention is almost impossible, everyone of them was vociferously encored, some more than once. The professional talent, represented by Miss Olive Belyea, Miss Bertha Crawford, Miss Hazel Bell, accompanist, Donald McGregor and Will J. White, gave their services for the evening freely and unconditionally. Will White, who was billed for only one song, gave nearly half-a-dozen. Charles Bodley was to have acted as accompanist, but upon his non-appearance, Miss Bell, who is herself a soprano of high note (top c, anyway) offered to officiate in his stead. The duet by Donald McGregor and Miss Belyea seemed somehow to catch the hearts of the audience, and a burst of clapping and cheering broke out as the last note died away. "God Save the King" was sung in unison by the whole audience standing before the gathering broke up, and the old canvas roof fairly crackled with the strains.

The Night's Work.

All the articles attaching to their persons that could be conveniently got rid of, such as wives, sweethearts, sisters, children, friends and other impediments, were now escorted by the members to the station and carefully returned to Orangeville, whence they re-appeared next day in time for the games, sports and other frivolities that characterized the passing of that Jewish sabbath.

But Friday night will not soon be forgotten in the annuals of the Snack Club. The man into whose mind entered even a thought of slumber was a traitor to the cause, and branded as such. The irrepressible Mr. Walters, gathering around him a choice band of dime novel desperadoes, and arming his little company with torches and instruments of music, descended upon Alton, as a young typhoon descends upon the luckless mariner. Barred doors and closed windows greeted them on every side. The inhabitants, having ascertained by annual and bitter experience approximately what they might expect by attempting to parley with the enemy, have become at last wise in their generation, and upon such occasions as this, retire silently and with what dignity they can command at ten o'clock sharp. Sol, his faithful little band at his heels, took a brief review of the situation and commenced operations with a comprehensive serenade of the entire town. No quarter was given, though in some cases as much as fifty cents was offered to anyone who would remove the midnight Lotharios. The architecture of Alton was thoroughly analyzed, overhauled and replaced as nearly as possible in the same position. One or two of the band becoming hilariously hysterical were promptly resucitated by Mr. Walters, who displayed a truly Napoleonic fortitude in the face of many discomfitures and rebuffs. About four a.m. the party attacked Bill Algie's house and announced their intention of remaining there for the night. Mr. Algie politely showed them into the kitchen, where about a score of their less energetic comrades-in-arms were already slumbering in peace.

The Snack in Session.

At 10 o'clock on Saturday a mass meeting of the club was held in the tent. Considering that it was the morning after the night before the attendance was very creditable. President John W. Charles occupied the chair, on his right and left hand, respectively, being F. C. Hunt, the Toronto secretary, and Robert Algie, of Alton, the home secretary.

M. P. Malone, the Hamilton secretary, read the minutes of the last meeting of the Snack, held several weeks ago in Hamilton to make arrangements for the present outing, They were confirmed. The president theh informed the meeting that under ordinary circumstances the next item would be the treasurer's report, but the present circumstances being far from ordinary, owing to the absence of the treasurer, Bill Irwin, and his presence by proxy in the shape of a mass of incomprehensible figures, he found himself somewhat at a loss for the correct method of procedure. Having spent the whole morning trying to decipher them, he was now prepared to take the figures for granted coincident upon the accession of the meeting. The meeting was graciously pleased to access.

Followed a pause, which Mike Malone filled in a very timely manner by a short recitative descriptive of the way in which the expression "Sure Mike" had first been brought to his notice. It was a good story, but Mr. Malone made a mistake in trying to throw it at the meeting from the floor of the house. The house wouldn't stand for it and Mike had to take the platform.

Drummers' Snack—Part of the costume parade.

Robert Algie, who during the introduction of this extraneous matter had preserved a gravity, but ill according with his customary smile, now rose, and, with some hesitation informed the meeting of a partial victory over the mathematical mess of Bill Irwin, whose

Drummers' Snack—The winner of the Melagama tea race.

character and probable future, he declared himself willing, but unable to describe.

"As far as I can learn," said Mr. Algie, "the financial status of the club is on a very satisfactory basis. There is a handsome balance on the right side, our concert receipts last night having been considerable." Toronto secretary, F. C. Hunt, confirmed this estimate with a statement of the exact amount taken at the concert, giving the president an opportunity to rise and remind the members that the spirit of the Snack was to have a good time not to amass large profits financially.

A Vote of Thanks.

"While I'm talking," said President Charles, "I want to ask everyone here to pay tribute to Donald McGregor and Will White, and all the Hamilton men and lady artists, and to Charlie Leslie. We owe a big debt of gratitude also to that illustrious Dockstader, Bay Hill, and his enthusiastic helpers. His performance was unique in character and daring in enterprise." A vote of thanks to all these members, coupled with the name of John B. Nelligan, was proposed, seconded and unanimously adopted, while the canvas shook to the tune of "They are jolly good fellows."

Donald McGregor, Will White, Bay Hill, Bill Algie, John Nelligan and Col. Stoneman all replied to this evidence of appreciation according to their several temperaments. Some replied wittily, some modestly, some seriously and some humorously, but all enthusiastically and all characteristically. Will White with an air of great sacrifice, said he'd sooner miss a dog-fight than a Snack; that you couldn't bribe him to stay away, and that he'd be there next year as usual, ready to help in any way he could.

Bay Hill, who doesn't relish speechmaking, said that his feelings coincided in a remarkable manner with those of Mr. White, and if it pleased the chair he would like to let it go at that. It didn't please the chair, so Bay added that the feature of the whole Snack which had appealed to him most was the harmonious and frictionless manner in which Hamilton and Toronto had worked together for a common interest, that it had been a great pleasure for him personally to do what he had done, and that he attributed the unique success of this entertainment to the members all having a definite object to work towards. Here he quoted some remarks of Bill Algie, who was promptly called upon to confirm or deny them. Mr. Algie started to speak of himself in a deprecatory manner, but was called to order with great sternness by the chair. Being driven to land on this

Drummers' Snack—F. C. Hunt, vice-president.

tack, Mr. Algie put out on another one and referred with the deepest sympathy and respect to the memory of their old comrade and brother, Samuel E. Ryan, who was a prominent member of the Drummers' Snack Club and a favorite with everybody. Mr. Algie displayed to the members a handsome memorial, bound in black morocco and engrossed in gold lettering, "Condolence of the Drummers' Snack Club," whose preparation had been in the hands of Bob Algie. The memorial read as follows:

MRS. RYAN.

The members of the Drummers' Snack Club desire to place a tiny forget-me-not on the grave of your late husband and to express to you their united sympathy in your bereavement.

Our comrade and brother, Samuel E. Ryan, was a loved and loving husband, an affectionate father, a good citizen, and an honest man.

No nobler or grander epitaph can be inscribed on the greatest monument.

A life well lived ends well at any time. Your husband lived, he loved, he was loved. This fills the vase of joy. The longest life contains no more.

His old comrades in our social club will sadly miss his presence, which was always cordially welcomed, either as an officer, tried and trusted, or as a social companion, on the road, in the holiday camp, or at the festive board.

The gentle, loving hand of time will partly heal the wounds made by the grim sergeant, and we realize that "words are but empty barren sounds" which are utterly inadequate to express sympathy when death calls a dear one.

In the windowless Palace of rest all must in time sleep, and we who are daily traveling towards the end of life's journey extend our best wishes to you and yours.

Signed on behalf of the Club.

JNO. W. CHARLES,
President.

ROBERT ALGIE,
Secretary.

A vote of thanks to those responsible for this memorial was proposed and carried. Mr. Algie's concluding remarks were characteristic of the man's great heart and generosity. "I thank you all for coming to see us once again in the old place," he said, "and you all know that as long as there is a square foot of ground in my house, or a crust of bread in the larder, no one of you will ever have to go hungry. (Cheers). In concluding I must confess that this morning I stole a ham from the cyclone cellar to feed my 30 or 40 guests with. Having no defence I throw myself unreservedly upon the mercy of the court." Discharged.

Drummers' Snack—Sol Walters and the Comfort Soap Race second prize winner.

A Few Remarks.

Jno. Nelligan, who was next called upon, spoke feelingly of the spirit of universal kindness, which was the key-note of the Snack. "The air of Alton is full of kindness," he said. "I'll try and write you something even better than "The Darkey's Dream' next year, and I think I can do it, because the object for which I write will be an inspiration in themselves."

Col. Stoneman, M. P. Malone and Hy. Eckstein each said a few words, and Court Thompson was called out. Court proceeded in a dignified manner to the rostrum where he assumed the old Dr. Munion attitude. His listeners knew that one of the great truths of Dr. Munion was imminent, but there was no time for escape. "I've heard some of you people talking of shake-downs in Dr. Algie's house," he drawled, "Well," he continued with stern impressiveness, "I stayed out all night, and Dr. Munion says there is no punishment too great for those who neglect the afflicted."

The president called on the members for any criticisms or suggestions. A few were offered—tentatively—and—accepted—gracefully. A photograph was then taken of the meeting in session. The election of officers was next taken up, previous to which President Charles, in the course of a short address, spoke highly of the help rendered him by the members of the executive. "If I'd had an executive made to order," he said, "it couldn't have been improved upon."

Drummers' Snack—The Baby Show—The winner is in the go-cart.

The New President.

Col. Stoneman proposed M. P. Malone for the new president. Bob Keys seconded it and there being no dissenting voice, Mike was duly boosted to the platform. In his address of thanks, or welcome, or whatever it was, he remarked candidly:

"It is no cinch to be president of the Snack Club. It means hard work and lots of it. But I think that to be president of such a bunch as I see below me is well worth a little trouble. I know you all. If I didn't I wouldn't take the office. But you'll all turn out and work for the Snack of 1908 as unselfishly, as generously, and as enthusiastically as you did for the Snack of 1907. Another thing, I'm following a hard man—a popular man, a man who made last year the most enthusiastic president we ever had. I refer to Jack Charles, and I don't expect to beat him. But with your help we'll give the Snack of 1908 as little cause as possible to blush with shame by comparison with that of this year. Gentlemen, I thank you for the honor you have done me."

On the proposal of Bill Colville, an enthusiastic vote of thanks was tendered J. W. Charles for having made the Snack of 1907 the greatest in the history of the club.

Drummers' Snack—Members and friends in Wm. Algie's grounds.

For Next Year's Snack.

The following officers were then elected:

Vice-president (an office created for the first time after some discussion)—F. C. Hunt, Toronto.

Hamilton secretary—Bay Hill.

Toronto secretary—Robert Asher.

Home secretary (elected by acclamation)—Bob Algie. Mr. Algie also assumes the duties of treasurer. He does the work anyway and might as well get the title.

Auditors—Peter Smith, Hamilton, and Will Meen, Toronto.

Hamilton executive—George Smye, Hy. Eckstein, Bob Smith, Court Thompson, Col. John Stoneman, Jas. Hooper and Chas. Smith.

Toronto executive—J. Wildfong, W. Meen, Sol. Walters, D. McGregor, W. J. White, N. Oakley, Jno. W. Charles, G. Campbell, T. Gloster, F. C. Hunt, Robt. Keys.

The following were created honorary members—Jno. B. Nelligan, Ben Arthur, Chas. Leslie, B. McIntosh, of the Central Business College, Toronto; Alex. Earle, of Creemore, and Donald McGregor.

Amos Mason and Ross MacKenzie were made assistants to the secretary. Before the meeting closed, Bob Asher was presented with Bill Colville's prize for having sold the most buttons. Bob sold 32, and said it was pie. Snack buttons cost $1 each and entitle the owner to full membership in the club.

As the final wind-up to the meeting, William Algie recited a poem by Robt. W. Service, author of "Songs of a Sour Dough." The poem, which was called "The Cremation of Sam M'Ghee," was of a humorous persuasion, and made a great hit.

The Procession.

Lunch was served to all who had appetites, by the ladies of the Presbyterian Church, in the Science Hall. One paid 25 cents and ate all one could. The fare was extraordinarily good value for the money, but this may have been on account of the ladies relying on the well-known generosity of all travelers to come forward at a slightly advanced price over the one asked. Their expectations, if they entertained any, were gratified many-fold. The boys dived into their pockets and wouldn't take any change.

After lunch, came an event from which the writer, for one, has barely recovered. From the hill-top floated the strains of sad music, and down the street there swung a procession, so marvellous in composition, so unique in character, that one regrets the paucity of the English language as a medium of description. Most of the spectators are now happily reported to be convalescent, but many of them suffered severely. A glance at the accompanying photograph will give the unprivileged outsider a faint, vague, and shadowy idea of the reality. Observe the illustrous Mr. Sol Walters, consider that he was by no means the most picturesque of that outrageous company, and in time you may train your mind to comprehend in a dim way why Galbraith, the photographer, is taking a forced holiday from business. To his credit be it said that he flinched not from the call of duty, but with clenched teeth and ashy countenance exposed his plate and staggered back to the cyclone cellar.

When the kilted processionists reached the sport's field, a baseball game was

Drummers' Snack—William Algie—The Snack Club's friend.

immediately arranged between a picked team from Toronto and Hamilton. It lasted most of the afternoon, and considering the strict impartiality with which Col. Stoneman and Bay Hill handed out their decisions, was remarkably devoid of friction. Jack Charles, by lashing out a magnificent home-run in the last inning, secured the fruits of victory for the Toronto IX. The score was 13 to 11. Each member got a box of Old Abe cigars and Jack Charles, as captain, a tiny silver cup on a mahogany stand. The winners declared that the referees had conducted the game in an able manner.

The teams were as follows:—

Toronto :—Jack Charles, captain, Bob Asher, F. C. Hunt, Tommy Gloster, Jack Wildfong, Bill Meen, Pud Oakley, A. R. Fraser.

Hamilton :—Bob Smith, captain, Geo. Smye, Hy. Eckstien, Billy Miles, Fat Gufford, Court Thompson, Bill Wark, Slim Arthurs, Ed. Nally.

Meanwhile the games and sports were in full progress. The Sport's Committee consisted of Wm. Colville, chairman; Col. Stoneman, referee; R. Asher and P. Smith, entry clerks. The committee were as follows :—William Irwin, Court Thompson, William Meen, E. F. Clarke, Bob Keys, Guy Long and Sol Walters.

The Sports.

All the sports and competitions were pulled off in fine style, and reflect great credit on the management. The following list of the various events with their winners, wherever it was possible to ascertain who were the winners, fairly represents the large scale upon which the whole programme was conducted. Competitors in most of the events were almost too numerous to admit of perfect handling. Sol Walters, smitten with the idea that he would like to help the committee, being one of them himself, succeeded in making himself recognized to the extent of being on nine separate occasions forcibly removed from the grounds. But nothing could daunt that indomitable spirit, and Sol was never far away from the centre of action. When last seen he was making for Orangeville, in his wake an infuriated mob of fellow-competitors, who with arms raised to heaven were calling the very firmaments to witness that through the strategical ingenuity (only they didn't put it like that) of the man they were pursuing, they had been wrongfully defrauded of their just rights.

It appears that Sol had stolen a march on them in one of the races, by rushing out of the crowd into the course when the judges were looking at other people's sisters and finishing a yard or two ahead of the legitimate winner. So let us leave him, with best wishes for his success if he was captured, and condolences for his pursuers if he escaped.

Baby show—1st prize-winner may be seen in the go-cart in the photograph; 2nd prize, Clara Gracy; 3rd prize, Wm. Edward Algie; 4th, Dorothy Cockerene; 5th, Baby Stevens.

Past Presidents' race—1st, Robert Keys; 2nd, Wm. Colville.

Tug of war—Winning team, Jack Burnett, R. L. McKenzie, William Algie, Joe, Dodds, John McLaughlin and Wm. Alexander.

Ladies' Comfort Soap race—2nd, Miss Hall; 3rd, Miss Smith; 4th, Miss Alexander; 5th, Miss Ada Saunders; 6th, Miss Oakley; 7th, Mrs. Smith; 9th, Miss Saunders; 10th, Miss Neeley; 11th, Miss Rodwi.

Artists' race—1st, Hy. Eckstein; 2nd, Ed. Nally; 3rd, F. C. Hunt.

Musical race—1st, Wm. Collins; 2nd, R. A. Scott; 3rd, Jim. Algie.

Ladies' Blue Ribbon hockey broom match—Winning team, Miss Saunders, Miss Eddie Saunders, Mrs. Wilson, the Misses Mason and Miss Campbell.

Needle race—1st, B. McIntosh; 2nd, J. H. Thurston; 3rd, F. C. Hunt.

Three-deaded race—1st, J. H. Thurston, A. R. Fraser and R. M. Thurston; 2nd, Walter Scott and company; 3rd, Bob Smith, Fred. C. Hunt and Jack Wildfong,

Early-call race—1st, Ed. Nally; 2nd, F. C. Hunt; 3rd, R. M. Thurston.

Frog-race—1st, Ed. Nally; 2nd, A. R. Fraser.

Smoking race—1st, Billy Mill; 2nd, Sol. Walters; 3rd, F. C. Hunt.

Sack race—1st, A. R. Fraser; 2nd, Bob Smith; 3rd, Walter Scott.

Fat man's race—1st, Billy Mill; 2nd, Jack Charles; 3rd, Jack Wildfong.

Kicking the football—1st, Billy Meen; 2nd, B. McIntosh.

Married men's race—1st, Walter Scott 2nd, Billy Meen; 3rd, O. E. Wallace.

Unmarried men's race—1st, B. McIntosh; 2nd, A. R. Fraser.

Consolation race—1st, F. J. White, of Melagama Tea; 2nd, F. Oakley; 3rd, Mel Tufford.

F.J. White had a Melagama Tea race for ladies. The names of the winners did not transpire, but a picture of the first lady home is shown elsewhere.

On the sports field a big tent was erected, wherein was dispensed ice cream fruit and cooling drinks. It was well patronized.

Prizes for the various events were the gifts of the following travelers and

Drummers' Snack—A fair spectator.

firms: Tug-of-war, J. W. Charles; baseball match, Havana Cigar Co., cigars and silver cup; ladies' Comfort Soap race, silver salver and other prizes donated by the Comfort Soap Co., Aprons and soap supplied free. Little maidens' Comfort Soap race—12 rings given by the Comfort Soap Co. Blue Ribbon Tea hockey broom race, one pound of one dollar tea to each member of the winning team, donated by George F. Campbell. Boys' Comfort Soap race, twelve prizes given by the Comfort Soap Co. Three-headed race (a new one), three case pipes given to the first team by W. H. Steele, tobacconist, Toronto. Smoking race, "Mary Ann" cigars, donated by Manners & Son, of London, through their representative, Ed. Bingham. Football kicking contest, Sol. Walters.

Envoie.

So let us take our leave of the Drummers' Snack. With a royal send-off that came straight from the hearts of the people of Alton, with a last cheer sent ringing back along the rails from the men and women in the vanishing train, a memorable outing was at an end, and another frolic added to the credit of the Snack.

And if, as Stevenson asserted, it is the duty of all men to be happy, surely this organization is fulfilling that duty in the highest degree. A club of men formed in the interests of clean mirth, of honorable enjoyment and of mutual service to one another, cannot well fail of success, and if the future of the Drummers' Snack Club is to be measured even proportionately by its past and present work, in after years a great body of men is destined to arise, whose motto, "Laugh and grow fat," will echo through the trackless wastes of the Siberian desert, and tickle the risibilities of the humble Esquimaux in the bleak solitude of his frozen home.

Short Snacks.

The president wore kilts. He's Scotch anyway.

* * *

As usual the star feature of the games was Bill Colville's Comfort Soap race.

* * *

Fred Hunt, the vice-president, came in third in more races than anyone.

* * *

Jack Charles cannot bé beaten as a chairman. He knows how to keep things moving.

* * *

Many of the boys were inquiring for Walter Armstrong, editor of The Canadian Grocer, who was away on his vacation.

* * *

What would a Snack be without Sol. He looked immense in kilts and used a broom for a sporan.

The Snack resembled one huge, happy family picnic. The verdict returned was "The best ever!" There will be no appeal.

* * *

Will White, former member of the executive, now living in Vermillion, Alberta, was wired the regrets of the club at his absence.

* * *

Jack Wildfong, of Gordon Mackay, a director of the Commercial Travelers' Association, was on hand early. It was Jack's first visit, but he says he'll never miss a snack again.

* * *

The bonfire and freworks display which were held after the concert made a fit ending to a glorious night. As a home secretary, Bob Algie has them all skinned. He forgets nothing.

* * *

Some members of the defeated baseball team were unkind enough to say that the Colonel's knowledge of the game antedated the new rules. And yet his decisions were perfectly satisfactory to the winners. Incomprehensible!

* * *

A telegram was received by Bob Algie from George Nicholson, familiarly known as Big Nick. Nick was in Vancouver, B.C., and wrote "Sorry not with you. Success is my wishes to all."

* * *

Mr. and Mrs. F. J. White, of Melagama Tea, were interested spectators of the games. Bill Meen, who travels for Melagama, was very much to the fore in the races. He won several firsts.

* * *

Mike Malone, the new president, has his work cut out for him for next year's snack. This one was so good. But everyone knows Mike, and when the time comes he'll deliver the goods.

* * *

Nowhere is there anybody just like Bert Menzies. During the Friday night serenade he kept the town alive and sleepless by pounding the kettle drum. He made up in energy what he lacked in artistic feeling.

Drummers' Snack—The members in session.

Bill Colville, as a games starter and general sports director, holds the belt. He is energy personified.

* * *

John B. Nelligan surpassed himself in "The Darkey's Dream." But he says he can write something better for next year's snack.

The Blue Ribbon Tea hockey-broom match, under the able direction of George Campbell was an unqualified success. Many brooms were broken, but no skulls. The single ladies won fittingly by a single goal.

* * *

Col. E. E. W. Moore, whose kindly presence was so greatly missed at the snack, telegraphed to Jack Charles:

Drummers' Snack—Bay Hill, the Hamilton Dockstader.

"Sorry, cannot be with you. Best wishes for record snack." That his wish came true is now ancient history.

* * *

Bay Hill, the Hamilton Dockstader, was largely responsible for the success of the concert. Bay was made up as an old nigger, and acted, as well as looked, the part. With Bay Hill and Bob Asher as the Hamilton and Toronto secretaries, the success of the next snack is already assured. They are untireable.

IT PAYS TO ADVERTISE.

A theatrical manager was holding forth on the value of publicity the other day and pointed his moral with this :

"When the teacher was absent from the school room, Billy, the mischievous boy of the class, wrote on the blackboard :

" 'Billy Jones can hug the girls better thqn any boy in the school.'

" Upon her return the teacher called him up to her desk.

" 'William, did you write that ?' she asked, pointing to the blackboard.

" 'Yes, ma'am,' said Billy.

" 'Well, you may stay after school,' said she, 'as punishment.'

" The other pupils waited for Billy to come out, and then they began guying him.

" 'Got a licking, didn't you ? '

" 'Nope,' said Billy.

" 'Got jawed ? '

" 'Nope.'

" 'What did she do ?' they asked.

" 'Shan't tell,' said Billy. 'but it pays to advertise.' "

What To Do This Month

All men are good sports at heart. They are always on the trail, whether it be for money, for wide reputation, for amiable life-partners, or for maskinonge, for deer, for foxes, or rabbits. It depends largely upon their appropriation how much fun they will get out of the sport. It is certain that only money can get money, and often it requires money to win a buxsom maiden, and money coupled with a longing for the wild to be able to land big game in the woods and in the river.

Little pressure is needed to effect the sale of a shotgun or sectional fishing rod to a good sport. He wants to have them as much as the store-keeper wants to sell them. A good deal of diplomacy and patience is essential to sell a gun to a man who is not quite certain whether he wants it or not. He is the indifferent sport. There are no bad sports.

It is up to the hardware merchants this month to make the indifferent sports good ones and convince them that they need guns and other autumn sporting requisites in order to spend what the Londoner calls "a decent time" in the fall months. It will be well for the merchants not to ignore those who have always bought guns and who never would be without them. They must be looked after.

* * *

To wholesalers the latter part of August and September has come to be an almost proverbially bad time to book orders. The travelers come in to the cities at the week-end thoroughly disgusted with their week's luck. "All the merchants," they say, "are trotting around to the fairs, and there's no catching them." Obviously, if there are no fish in the brook there will be few bites. Such times however, can be profitably spent by retail clerks in re-arranging stocks for busier months.

* * *

Every year the jobbers are emphasizing more and more the importance, the necessity to the retailer of getting his orders booked early. This fall undoubtedly will be a bumper season for hunters, and those merchants who are going to derive the greatest amount of business from the game-seekers will book their orders now for all lines of sporting goods for the fall and winter, such as guns, ammunition, skates, hockey sticks, pucks, snowshoes, moccasins, and curlers' supplies. The farmer must not be forgotten and a good stock of cow-ties, halters, blankets, sleigh bells, etc., should be ordered during the next two or three weeks in order to insure prompt shipment. Let not the retailer be responsible for delays in shipments.

* * *

As for the display of fall and winter goods, a model window of sporting will be illustrated, described, and criticized in the next issue of this paper on the "Window Display" page.

Cow chains can be effectively displayed in the front window by hanging them from the ceiling and having them meet to form different diagrams in the centre of the window, making them as the centre work for some display of cutlery or mechanics' tools. A good way of displaying horse blankets and robes is by suspending them from the ceiling above the counters. It is easy to show sleigh bells to good effect as they have in their burnished surfaces an attractiveness independent of their arrangement. Merchants will do well to display them in the front window as they are the most attractive of all the fall and winter stock.

* * *

Carry a full line of fall and winter goods and, besides pushing their sale in your store, accentuate them in your fall advertising in the local papers. Don't be foolish enough to allow yourself to think that money expended in judicious advertising is foolishly spent. A wise advertising campaign will increase your business to an extent you never dreamed of. In this connection the New York Evening Journal is constantly trying to educate businessmen on the value of advertising and for that purpose is publishing daily hints which deserve consideration by all who succeed in business. Here are a few valuable sentences from one of that paper's recent sermons on advertising: "The Non-advertiser Pays the Other Man's Advertising Bills.—Does the man who doesn't advertise ever stop to think that he is the one who really pays the advertising bills of the man who booms his business? The advertiser himself isn't out, for he gets back four-fold in increased business. The customer doesn't pay for the ads, for he can buy better goods, from a larger assortment, at lower prices, from the man who is constantly turning over and renewing his stock, than from the one whose goods are carried over season after season. It's the non-advertiser who pays for his lack of initiative and enterprise in decreased sales and dwindling profits. The profits are all going to the other fellow to pay for more advertising."

* * *

Those who carry a stock of stoves and heating supplies should already be thinking of ways and means to increase their sales in these during the coming autumn. When you begin to display your stoves don't keep the same stoves standing always in the same places. Change them around once in a while. Put the big baseburner where you had the steel range and let the oak and gasoline stove change places. Then get out the stove blacking and brush and allow your assistant to limber up his arm. People will think you have some new stoves in till you tell them you haven't. Don't forget to make the show window sell stoves for you, either. For a while yet the window is too valuable for other things to give it up to stoves, still if you can get a small gasoline stove in one corner or work a camp stove in effectively with other camping goods, do it. In the fall, when stoves are the thing to push hard, let the window do its share by making new and attractive displays of stoves from time to time.

* * *

What are you doing to clear off the balance of this season's stock of fishing tackles? The sumemr will soon be over and you will want your money to pay for fall goods which will soon be arriving. You cannot afford to devote valuable space in your store to stock which will not move again for another year. I have heard of one dealer who made a silent salesman display of fishing tackle last week and as an inducement to sportsmen and holiday-seekers, offered a couple of fly baits as a bonus with every dollar's worth of tackle sold; the experiment worked well and his sales in this class of goods were largely increased. You may not use the same plan, but the method is immaterial as long as you get busy in this regard and are not left with a heavy stock of fishing tackle to carry through the winter months.

* * *

Are you satisfied with your seasons sale of sporting goods? If not, what have you done to create a demand for the goods you have been handling? The amount of your sales in sporting goods will depend largely on creating a demand for sporting requisites in the town in which you live, and the sooner this fact is realized by some hardware dealers, the better it will be for their business. One of the most effective methods that can be used by the trade in small towns to arouse interest in baseball, lacrosse, football, or any other sport, is for the merchant to have his son, clerk, or any of his employes join a local team, and thus be instrumental in helping to keep the enthusiasm of the local players at the highest pitch. The next thing to be done by the dealer after the team has been organized, is to sell them uniforms at the lowest price possible, or even at actual cost, as afterwards he will realize enough on the rest of the club's trade to make up for the lost profit on the uniforms. After the team has been supplied with uniforms, bats, balls, etc., and the grounds selected, it would be good policy on the part of the dealer in the case of a baseball club, to erect a backstop, and set forth on it in large letters that his house is the headquarters for all kinds of baseball and sporting supplies. A window display of goods a couple of weeks before the season for each individual sport commences will also arouse interest and put the sporting members of the community in the notion of buying. Your display will also advertise the fact that you have a complete line of sporting goods, and will capture for you much local business.

Business Management

By HOWARD R. WELLINGTON.

The Management of a Retail Store.

In introducing a subject of this nature a few general remarks might not be out of place as to the recognized value of method and system in every business, not only as to the recording of transactions from a bookkeeper's point of view, but in every department of the business. The young man who lacks "order" will, nineteen times out of twenty, fail to succeed. The merchant should cultivate systematic habits in his own work. If it is known among his clerks and employes that the same hour each morning will find him in his store or office, this very fact will infuse punctuality throughout the house. The work should be systematically distributed among the clerks and each article of goods should have its special place, convenient of access and ready to handle. The clerks should be so thoroughly familiar with their stock that all reasonable questions can be answered promptly.

Buying.

The old maxim "Goods well bought are half sold" is truer to-day than ever before. Nearly every mercantile failure finds about half the stock so old and shopworn and dusty and dead as to be almost unsalable, even under the hammer. An elaborate system of stock records may be installed from which the merchant may know at any period just what stock he carries of each line of goods, but as a usual thing neither merchant nor clerk, especially in a small store where all the work is done by two or three, has time to post up the sales of goods. A practicable method would be to have an indexed book arranged alphabetically, in which might be entered the goods required from time to time to assort the stock depleted by sales. These goods may then be ordered by mail each week or through traveling salesmen, if they call regularly.

Of course in a large store it is advisable to keep a complete record of all goods bought and sold, also prices, both cost and selling, and this may be done by means of stock cards arranged alphabetically and ruled in some convenient way, such as the following:

better impression on the mind of a customer than attractive display of the stock, keeping to the front such articles as are likely to attract trade; remanants and "dead stock" instead of being put out of reach on the top shelf are carefully sorted, specially marked down, and so displayed and advertised as to be easily worked off. Always bear in mind that "the first loss is the smallest loss" on goods which have been carried in stock too long.

The Location of Your Business.

In the retail trade the location is a very important matter. Trade centres around centres. People do not like to go out of their way to buy things. An inferior or small store in a good location is preferable in every case to a fine large store poorly located. Jut a little around the corner is not as good as on the corner. Two or three steps up is not as good as on a level with the sidewalk. A rent of $400 on the main street may appear high as compared with $40 a month three or four doors from the main street, but the rent is only ten times as great and there may be twenty-five shoppers on the main street for every one who passes up a side street. A Boston dealer made money enough in a little three-angled room scarcely big enough for three customers at once, but located on a famous corner, to build one of the largest and finest business blocks for which the city is noted.

Selling.

Endeavor to impress upon your salespeople the absolute necessity of cultivating a memory for faces and names. A customer who is addressed by name feels a certain personal interest in the store where purchases are made, and it goes a long way toward making a sale. It is not the quantity, but the quality of speech, which tells. The successful salesman knows how to talk, what to talk about, and, more especially, when to stop talking. To say the right thing at the right time is not nearly so difficult as to say nothing at the right time. The seller should only talk enough to keep the buyer talking.

VELVET — (BLACK SILK)

Date.	From Whom Purchased or Total Sales.	Bought.	Sold.	Balance.	Cost.	Selling.	% Profit.
Mar. 25	J. S. & Co.	10 yds.	....	....	$1.00	$1.50	50
" 28	Sales	....	5	5	....	....	....

This card would be filed alphabetically, the purchases posted up from the purchase invoices and the sales from the sales files for the day, week or month, whichever is thought advisable.

The Character of Your Business.

Strive to maintain, as far as possible, a reputation for sterling values and moderate prices. The customer, who leaves your store feeling that he has secured an article which is good value, is the best asset a merchant has in building up a business. Nothing will create a

The faculty of holding trade, or of selling repeatedly to the same people, is the highest attribute in the qualifications of a successful salesman.

Understand thoroughly what you are trying to sell. Know your goods, believe in yourself, and you are sure to inspire confidence in the buyer. There is an old maxim:

"When you buy, keep one eye on the goods, the other on the seller; when you sell, keep both eyes on the buyer."—Dry Goods Review.

HARDWARE AND METAL

Established - - - - - - 1888

The MacLean Publishing Co.
Limited

JOHN BAYNE MACLEAN - President

Publishers of Trade Newspapers which circulate in the Provinces of British Columbia, Alberta, Saskatchewan, Manitoba, Ontario, Quebec, Nova Scotia, New Brunswick, P.E. Island and Newfoundland.

OFFICES:

MONTREAL, - - - - - - 232 McGill Street
Telephone Main 1255
TORONTO - - - - - - 10 Front Street East
Telephones Main 2701 and 2702
WINNIPEG, - - - - 511 Union Bank Building
Telephone 3726
LONDON, ENG. - - - - - 88 Fleet Street, E.C.
J. Meredith McKim
Telephone, Central 12960

BRANCHES:

CHICAGO, ILL. - - - - - 1001 Teutonic Bldg
J. Roland Kay
ST. JOHN, N B - - - - No. 7 Market Wharf
VANCOUVER, B.C. - - - - Geo. S. B. Perry
PARIS, FRANCE - Agence Havas, 8 Place de la Bourse
MANCHESTER, ENG. - - - - 92 Market Street
ZURICH, SWITZERLAND - - - - Louis Wolf
Orell Fussli & Co.

Subscription, Canada and United States, $2.00
Great Britain, 8s. 6d., elsewhere - 12s

Published every Saturday.

Cable Address { Adscript, London / Adscript, Canada

WESTERN POSTAL SERVICE.

Complaints of inadequate postal service are still heard from all parts of Western Canada and lately the Winnipeg Free Press devoted an entire page to the extracts from complaints from subscribers as to late delivery of their papers. Every other paper circulating in the west could tell a similar story and western businessmen are suffering serious losses through delays in the delivery of their correspondence.

At the recent convention of the Associated Boards of Trade of Western Canada this grievance came up for discussion and a strong resolution was passed calling the attention of the department to the serious state of affairs now existing. In the course of the discussion F. T. Fisher, ex-secretary of the Edmonton Board of Trade, told how the trouble had been dealt with in Edmonton and his remarks pointed out a remedy. One year ago Edmonton had probably the worst postal service of any town in the west and the businessmen were loud in their complaints. But complaints that are not specific seldom accomplish much and this is the reason why the west has been complaining so long and getting so little redress. The Edmonton Board of Trade took the matter up in a businesslike way and soon accomplished results. The secretary requested the businessmen of the town to give him the envelopes of all delayed letters received. Soon he had a vast collection of envelopes, the post marks on which showed long and vexatious delays in delivery. These were forwarded to the department authorities and very soon an inspector was in Edmonton looking into the matter and investigating the causes of all these delays. Edmonton businessmen established conclusively the fact that they had a serious grievance and to a great extent that grievance has been redressed.

The Edmonton plan might, with advantage, be followed in every town in the west and specific complaints of this kind could not possibly be overlooked. If the readers of this paper would go to the secretaries of their local board of trade and induce them to collect evidence of inefficient postal service according to the simple, direct method followed by Mr. Fisher in Edmonton, there would soon be in the hands of the postal authorities a mass of evidence that would compel instant action.

EDMONTON IN THE CENT BELT.

Startling news comes over the wire from far off Edmonton. If information received at this office is correct the Alberta capital is now within the limits of the "Cent Belt" and the inhabitants of that once progressive, ambitious, hustling and truly western city, if they fail to resent in truly western fashion the shameful innovation introduced by a newcomer from Charlottetown, P.E.I., will be no better than their benighted friends whom they have left behind in the "effete east" where it is said that occasionally the collection plate on Sundays receives more coppers than silver.

As every person knows it has for years been the proud boast of the Western Provinces that nothing of less value than the five cent piece was accepted by the western people. Newspapers have sold for five cents and until recently ninety-eight-cent and thirty-seven-cent articles have been quite unknown in the territory west of the Great Lakes. The genuine westerner has despised the red copper and for years the east has been referred to in derision as the "Cent Belt" of Canada. A man offering a red copper on the streets of Winnipeg or in any of the western towns for a newspaper, the price of which was plainly marked one cent, would incur the scorn and derision of the newsboy and everyone would know that he had just got off the train from the east. Two years ago the opening in Winnipeg of the branch of a big Toronto departmental store had the effect of introducing coppers to the people of the Manitoba capital, and people living farther west have been sorry to find that the "Cent Belt" was creeping towards them. The startling news now comes from Edmonton that a Mr. Sentner who hails from Charlottetown, P.E.I., is determined to introduce red cents in that city. The punsters might say that his name would imply his fondness for that particular kind of currency that is so much despised in the Western Provinces. Consternation reigns in the Alberta capital. Are the people in that city to be no better than the old bogey easterners who live in Toronto or are they to be even as the weaklings who live in Winnipeg? Is Mr. Sentner to be allowed so to degrade Edmonton from her proud position? Is Edmonton to be included within the "Cent Belt?" What will they say in Calgary where the red coppers are still despised?

Perish the thought that Edmonton should submit to this outrage. Four or five years ago a peddlar is said have visited Nelson, B.C., and to have offered coppers as change to the people of that thriving city. It was more than flesh and blood could stand, and the story goes that he was given a bath in the Kootenay and driven back to the east where he belonged. Since that day no merchant has been so rash as to attempt the introduction of coppers in Nelson, and it is not to be expected that the people of Edmonton will submit to that which Calgary and Nelson would never have allowed. Mr. Sentner is a rash man and we predict for him a speedy exit from the Alberta capital.

ADOPT NEW IDEAS.

A good suggestion made by an exchange is that every merchant should have a new idea book. Many new ideas come to you in crude form as time goes by. Have a place to put them down. You can develop them and work them out to a practical conclusion as you get to them.

Remember that many new ideas are old ones made over. You may see what some merchant is doing in another part of the country and by working it over you can apply it to your local conditions. It works just as effectively with your people as if you picked it red hot from the sun itself. Constant studying of your problems will bring ideas. What your competitors are doing will furnish ideas for you. Never be backward about selecting the good things in your competitor's methods and work them into your own program as time goes by. If your own store makes the right impression upon your community you must first have it right within itself. Your store staff and you yourself must believe earnestly in yourselves. The foundation of your progress should be your belief that you are giving your trade a square deal.

To go farther, you should have faith all of the time that you are giving the same quality for less money in many lines than can be purchased at other stores or of any competitors, no matter where they may be located. If you are a progressive merchant, interested in your business and mindful of your opportunities, you are confident that you are doing that.

Mere boast and loud talk does not accomplish this. The work of your store from day to day, the number of satisfied customers you make, tells the story. And as each day and each week go by in your talks with customers, in your talks with clerks and in your talks to the general public through your advertising, keep your faith in yourself, your values, and your business well to the front.

Believing so thoroughly in yourself, you will find that time brings to you many people who have learned to believe as you do.

FIRE INSURANCE RATES.

If you are paying a high rate of insurance, what are you doing to reduce the rate? Insurance is a steady expense and its cost is based on a schedule which charges something for every deficiency in building construction for every lack of precaution against fire that you can be charged with.

Are you known as a careless man in these particulars? Do you allow waste paper and any other inflammable material to collect in the basement, around stairways, or in the rear of the building?

Have you failed to keep the chimneys in good condition? Have you endeavored to so manage things that the stove pipe will not have to pass through so many walls, floors, or partitions?

Just where do you stand with the insurance companies? What is the schedule of items fixing the status of your fire risk? You can get this list through your insurance broker. Get it and see where you can improve your position and reduce your rate.

Make a few changes in the construction of the building, in the arrangement of stock, and in the habits of yourself and clerks around the store, and when you are done you will have pulled the rate down.

Always be an enthusiastic advocate and supporter of a fireman's organization in your town. Be in favor of good fire equipment and enough of an appropriation to have it well taken care of.

ADVENT OF MOLDING MACHINE.

No feature of the recent foundry conventions at Philadelphia attracted more attention than the exhibit of molding machines. The booths where the various devices for turning out castings were being demonstrated were constantly surrounded by curious and interested crowds of foundrymen, asking questions and keenly watching the operations of the machines.

Occasionally some expressions of skepticism were heard, but the generally expressed opinions were that it is only a question of time—and a short time, too—when these machines will be in general use. The devices themselves were their own strongest advocates, showing how one man and a machine would do the work of several molders, giving promise that each shop will be able to turn out more work at a decreased cost.

The fear most often expressed was that the union molders will make a fight against the general introduction of the machine, but it is to be hoped that such will not be the case. History has shown the futility of all such struggles against the evolution of manufacture. It has been demonstrated over and over again that machinery, instead of lowering wages, results in the final betterment of conditions, both as to wages and as to the hours that men have to work, and journeymen molders will be standing in their own way if they fight the introduction of successful machinery. If they know their own interests they will welcome the day when molding machines have been brought to perfection, and placed in general use.

CONCRETE SCHOOLS.

It has been suggested by a number of recent writers in the American industrial press that separate departments for the education of those engaged or preparing to engage in concrete architecture be established in the colleges and universities throughout the country. That such a suggestion is highly plausible is obvious. We cannot have too much or too highly technical education for our architects. After all the science of architecture is perhaps the most important of the sciences of our day. To have our nation solidly established we must have our houses, our stores, and our factories built well. If our buildings are not durable, neither is our nation.

Skilful concrete architecture means economy of material. J. F. Haskin, in the Detroit Free Press (June 12) makes this startling assertion: "The enormous growth of the use of reinforced concrete in building construction wastes more lumber than any other one line of business. As much timber is used in the boards into which the concrete is poured as would be used in an ordinary brick building. This lumber is usually an almost total loss." If such a waste of material is always the case in the erection of concrete buildings, it is very needful that some remedy be looked for. The remedy may be found in the establishing of concrete departments in our national schools.

EMPLOY SYSTEM IN SHIPPING

In the majority of hardware stores money is being lost every day through the merchant's neglect of freight bills and all matters pertaining to the proper transportation of merchandise; for this reason freight matters should receive the most careful attention.

There is good money to be made for the time spent in figuring over freight bills and making out claims for overcharges. It is an easy matter to obtain from your agent and your jobbers the correct freight rates on each class of goods you handle, and this information will enable you to speedily check over the rates charged on your freight bill to see that some railway clerk has not given the shipment too high a rate. You will see the necessity for doing this, when you learn that a railway clerk who makes an error which costs the company money, always has to stand the loss himself. It is only natural then, when there is any doubt on the part of the clerk as to exactly what classification your goods should come under, that you may be charged the highest rate.

Those merchants who use system in shipping will hardly credit the fact that some retailers return their goods without advising the jobber or manufacturer that the goods are being returned. Many wholesale houses are forced to enter these in an "unknown book" until weeks afterwards the matter crops up in settling the account. It is only right that the retail merchant should advise the concern to whom the shipment is returned as soon as the goods start for the depot.

Do not allow yourself or your draymen to receipt for goods that have not been received. Check the number of packages received against the express receipt or freight bill carefully.

If packages are in damaged condition when received, have your local freight agent endorse "in bad order" on the freight bill. Then have him come to your store and check the goods over with you after the cases are open. Make a bill against the railroad company for the lost or damaged goods. If unable to get redress from the railroad company, ask your wholesaler to help you.

All railroads have what is termed a minimum charge, which is the lowest amount they will charge, no matter how small the shipment. Bear this in mind, and it will save you many dollars in a year; when sending in a special order for some small thing, glance around your store and take a squint through your order book and you will invariably find that there is something else you will need before long that can be ordered with that shipment.

HARDWARE TRADE GOSSIP

Ontario.

A. E. Harding, harness dealer, Kingston, Ont., has assigned to J. Lemmon.

B. B. Sriven, Otterville, Ont., is advertising a tinsmithing business for sale.

J. H. Henderson, hardware merchant, Niagara Falls, Ont., spent Wednesday of last week in Toronto.

The death occurred recently of Jacob Hose, a partner of the hardware firm of Hose & Canniff, Kenora, Ont.

Q. Adams, of the hardware firm of J. Q. Adams & Co., Ravensworth, Ont., was in Toronto last Saturday.

S. J. Frame, with the firm of Jenkins & Hardy, Toronto, has left the Queen City on a couple of weeks' holidays.

Among hardwaremen noticed in Toronto on Friday was W. M. Knight, Maple, and Geo. Peaker, Brampton, Ont.

F. R. Reade, buyer for the Kennedy Hardware Co., Toronto, is spending a couple of weeks holidaying in Muskoka.

May Bros., Toronto Junction, have taken possession of Thos. Hoar's old stand and are conducting a general hardware business there.

K. A. Cameron, with the wholesale hardware firm of H. S. Howland, Sons & Co., Toronto, left that city last Saturday for a two weeks' vacation.

Frank E. Hallitt, of the firm of Hallitt & Bradley, hardware merchants, plumbers, etc., Oshawa, was married on July 18th to Miss Phee Hezzlewood, of the same town.

David Mann, Toronto, has bought out the Pinkerton Hardware Co., and has opened out a general hardware business at 766 Bathurst St., in that city. Mr. Mann will carry a full line of hardware, cutlery and tinware.

Edward Hawes, of Hawes & Co., Toronto, dealers in stove and shoe polishes, is spending two or three months in England, among old friends. Just at present he is enjoying a few weeks in London. He will return about the middle of August.

F. C. Baker, customs house clerk, for Rice Lewis & Son, Toronto, has left for the north to spend a couple of weeks' vacation in Magnetewan district.

Mr. Smith, of the hardware firm of Smith and Schaeffer, Bolton, Ont., was in Toronto on Wednesday.

The condition of Harry Wilson, traveler for H. S. Howland, Sons & Co., wholesale hardware merchants, Toronto, is somewhat improved. For the past four weeks Mr. Wilson has been confined to a private hospital in the Queen City, suffering from an attack of typhoid fever.

Quebec.

Lecuyer & Daniel, Montreal, plumbers, have dissolved partnership.

R. J. H. Douglas, of Caverhill, Learmont & Co., Montreal, is away on a holiday.

W. H. Martin, who travels for Caverhill, Learmont & Co., Montreal, in the Sudbury district, was in Montreal last week.

H. Sylvestre, hardware merchant, Montreal, has returned from a successful fishing expedition in the Laurentian Mountains.

H. W. Aird, of the Canada Paint Co., Montreal, is spending a three weeks' vacation with his family at Prout's Neck, Maine.

T. H. Jordan, the managing director of A. C. Leslie & Co., Montreal, is spending a few weeks' holidays at Old Orchard, Maine.

J. N. Walker, a city traveler for Caverhill, Learmont & Co., and one of the oldest travelers in Montreal, has gone west for a holiday.

A. W. Benedict, in charge of the sales department for the Standard Paint Co., Montreal, is making an extended tour through the Maritime Provinces.

MR. GEO. CAVERHILL, Montreal.

R. W. Leonard, of Coniagas Mines, Cobalt, spent a few days in Montreal last week. He speaks hopefully of the strike situation and of future prospects of development.

Amongst those who called in Montreal last week were: A. Lemieux, Henryville; O. B. H. Maille, Longueil; H. E. Crepin, Chateauquay; N. C. Polson, Kingston; Mr. Corriveau, St. Sebastien.

The travelers for Starke, Seybold & Co., Montreal, were the guests of the company at the Corona Hotel, Thursday evening, July 25. William Starke was master of ceremonies and a very enjoyable evening was spent.

K. M. Ireland, of Dalton & Sons, Kingston, was married recently, and passed through Montreal, looking up old friends, on his way home from a trip down the Saguenay. Hardware and Metal joins in wishing him a happy married life.

Joseph Ridge, general manager of John Round & Son, silversmiths, Sheffield, Eng., spent last week in Montreal at the Canadian offices and left on his way west through Canada. He was accompanied to Toronto by J. R. S. McLernon, the Montreal manager.

A. D. Leblanc, a representative of Lewis Bros., Montreal, for a number of years past, is severing his connection with that firm, and is going into business for himself. He will start up a sporting goods store in Montreal. J. A. Rochette, of the sales department of Lewis Bros., will succeed him on the road.

Western Canada.

Trenholme Bros., of High River, Alta. have sold their hardware business to Ballantyne Bros.

G. E. Kingsbury, harness dealer, North Battleford, has assigned to W. H. Jarman.

The Winnipeg Galvanizing & Mfg. Co. have bought the works of the Red River Metal Co., of Winnipeg. Louis Sheldinq is the manager of the new company.

The hardware firm of Frame & Miller, Virden, Man., have sold out. The business is now being carried on by Miller Bros., who are carrying a full line of hardware and have already enlarged the premises to double their former size.

The Moose Jaw Evening Times of July 26, says: R. J. Gay, representing the Hardware and Metal Journal, one of the McLean publications, was in the city to-day collecting material for an article on conditions of the trade in Western Canada. Mr. Gay said that nowhere along the line had he seen the crops looking better than in the Moose Jaw district. He is going through to the coast.

A MAN OF PARTS.

Last week George Caverhill, of Caverhill, Learmont & Co., Montreal, was appointed a director of the Montreal Street Railway. This is but another instance showing the demand for Mr. Caverhill as a director of corporations. His business capacity and influence are always called into requisition and they always prove their value. To show the powerful influence Mr. Caverhill wields in commercial circles, it might be interesting to know that he is president of the Montreal Board of Trade, vice-president of the Montreal Loan & Mortgage Co., a director of the Dominion Iron & Steel Co., chairman of the board of directors of the R. & O. Navigation Co., a director of the Canadian Colored Cotton Mills Co., of the Royal Victoria Life Insurance Co., and a chairman of the Canadian Board of the London & Lancashire Fire Insurance Co. Mr. Caverhill is also a member of all the known clubs in Montreal.

IMPROVED FILE.

An invention by H. Getaz, Schenectady, N.Y., refers to that class of files in which the teeth are composed of a series of cutting blades clamped together in an angular relation and adapted to be readily sharpened when dull. The object is to improve the files, especially in the matter of providing for the deflection of the blades in an effective manner.

Markets and Correspondence

(For detailed prices see Current Market Quotations, page 66.)

MARKETS IN BRIEF.

Montreal.

Antimony—Weak.
Copper—Weak.
Iron—Firmer.
Lead—Firm.
Linseed Oil—Decline of 2 cents.
Paris Green—Firm with heavy demand.
Turpentine—Further decline of 3 cents.

Toronto.

Linseed Oil—Decline of 2 cents.
Copper—Decline of 1 cent.
Old Material—Lower prices.

MONTREAL HARDWARE MARKET.

Montreal, August 2.—Considerable activity is noticeable in local circles. The fall trade is commencing to open up and in another week or so will be in full swing. More activity than is usual at this period of the year is now experienced. Manufacturers are working at a greater pressure than is customary, not so much because their order books are all cleared up, but because they are sowewhat uneasy regarding the future. The rate at which they are producing material now would make it easy to accumulate a stock which is not at all wished for this season. Jobbers are not quite as liberal in their orders and even now when the manufacturers of lines which have been scarce this last winter and spring, such as green wire cloth, screws, nuts, and bolts, are working hard, retailers have to wait for their supplies. The fact that the employes of the manufacturers and the jobbing houses are so busy is striking evidence of the fact that trade is in a very healthy condition. Prices on lines are very firm with no immediate prospects of advances. Supplies are quite adequate to the demand and freight conditions are in a more satisfactory condition. This fall, however, it is expected there will be a shortage of shed room for heavy freight.

Screws—Stocks are being gradually repleted. The manufacturers are getting under way and another month or so will be in a position to cope with the demand. Prices are firm and unchanged, the recent advances being well maintained.

Building Paper—An unusually large amount of building operations is being carried through this year, and the demand for all lines of building paper is strong. The manufacturers are busy, and supplies are coming in promptly. Tarred felt is still quoted at 2.25 per 100 pounds.

Green Wire Cloth—The situation is unchanged. Prices are well maintained, but stocks are still short. Now that the demand has somewhat diminished the factories will be given a chance to catch up.

Fire Brick—The fall season will soon open again and it is expected a big business will be done. Last season made a splendid showing in the amount of business transacted.

Cement—Prices are firm and unchanged and a strong demand exists owing to the increased call for this in architecture.

Mechanics' Tools—The demand for all lines has been well maintained throughout the summer season. Prices are unchanged.

Builders' Hardware—This season will probably be a record-breaker in the amount of business done in all lines of builders' material. This business has been the chief feature of the hardware trade this year, especially in metropolitan districts.

Sporting Goods—Now is the time for dealers in these goods to get in their supplies and be in perfect readiness for this fall's business which undoubtedly will be a bumper season. Guns and ammunition, and hockey and curling supplies should be well looked after.

The travelers for the various jobbing houses have been in the warehouses and are again on the road, and fall business will in a few days be in full swing.

TORONTO HARDWARE MARKETS

Toronto, Aug. 2.—There is very little activity in the local hardware markets this week and evidently the usual midsummer calm has settled upon the trade. However, jobbers do not figure on being busy at this time of the year and they are quite satisfied with the more or less steady stream of sorting orders which, even in this quiet time, continues to flow to the large jobbing houses here from all parts of the porvince. Though the weather was very backward during the earlier portion of the season, and trade was retarded as a consequence, this fact simply served to prolong business farther ineo the summer months; the early trade was not lost, but simply distributed over a wider period of time. Now that local jobbers have reviewed their season's business, they invariably report that the aggregate business of this season has broken all records.

Screws—The demand for these, as for most hardware commodities, has fallen off considerably during the last two weeks, consequently the manufacturers have had an oportunity to at least partially catch up with the trade. Considering how far the makers were behind a few weeks ago they have done good work in repleting their stocks and supplies are now much better.

Nails, etc.—Supplies of these are once more well up with the trade, though the best selling sizes continue in strong demand. Supplies of wire are also much improved, principally owing to the fact that the demand has slackened off and the factories are pushing the manufacture to be ready for the fall trade, which usually commences about September.

Green Wire Cloth—So far as jobbers are concerned, the season for this is over and some local houses did not even get one-third of the amount they required. Fortunately the season has not been a bad one for flies, so that retailers and consumers have not been half as much inconvenienced by the lack of this article as they surely would have been had the spring been hot.

Builders' Hardware—This has been a record-breaking year for the sale of builders' hardware and mechanics' tools. Even now when the hardware trade is quiet, a strong demand prevails for this class of goods, and the unparallelled amount of building being done all over the country will cause an equally strong demand all through the fall months.

Sporting Goods—Baseball, lacrosse and football requisites have had a large sale this year, and a good average business is still being done in these goods. Fishing tackle continues to be the strongest seller, however, though the trade in tourists' supplies and general summer resort goods is very satisfactory. It is noticeable that each year steel fishing rods are coming more and more into popular favor among sportsmen.

Glass—Trade is quiet in window glass at present and prices remain unchanged. Plate glass, however, has taken a slight rise in price owing to an advance in the European market; it is now quoted at 30 per cent. off list.

Binder Twine—Trade has greatly improved during the last two weeks and jobbers are greatly elated at the prospects of a far better season than they expected. Car shipments have reached their destinations and a good business in repeat orders is being done.

MONTREAL METAL MARKETS.

Montreal, August 2.—Local market conditions are somewhat improved. The dullness characteristic of the summer months was of short duration this year. The mills have opened up earlier this year than formerly. A good tonnage is moving, the cargoes which arrived ten days or a fortnight ago being readily distributed.

Much uncertainty has ruled in the American market owing to the unsettled condition of prices on many lines. On some lines, however, orders are well booked up with the American mills, prices are firm, and a large tonnage is contracted for ahead. For the past five weeks there has been absolutely no demand for material in the United States.

The English market is still subject to wide fluctuations. The chief factor in causing this lack of stability is speculation. Merchants in America have been receiving continually such contradictory reports of conditions as are supposed to rule in England that now the find it more satisfactory to ignore them. This present unstable condition has maintained itself for such a period of time that it might assume serious proportions. Copper is firm one day and the next day falls flat. So with iron.

Heavy inquiries for fall stocks are commencing to arrive in the local market, and it will not be long before the fall trade is in full swing.

Pig Iron.—There has been a little more enquiry for iron in the United States, incited probably by the uncertain conditions existing at Lake Superior. The dullness which has been reigning in the local market has ceased and orders for fall stocks are arriving in increased numbers. Scotch iron is plentiful. Prices are firm and unchanged. Middlesboro No. 1, $21.50; No. 2, $20.50; Summerlee, $25.50.

Ingot Copper.—Is weak, corresponding to American and foreign conditions. In some cases, however, even in United States a strengthening is noticeable. How long this firmness will last is hard to state. Prices are unchanged and easy.

Ingot Tin.—No further lessening has been made in premiums offered for early delivery and conditions are easier. Up to last week business in the United States was good, but since then a general dullness has reigned. Lamb and Flag, and Straits, are quoted at $44.

Antimony.—The market is much easier. Not much demand exists. Cookson's is still quoted at 18c.

Sheet Zinc.—Is quiet. Prices are unchanged.

Zinc Spelter.—The situation is unchanged. The tendency in the American market is downward. Domestic is quoted at $6.50.

Old Materials.—Very little is doing in the local market. It will not be long before business is enlivened. It will be interesting to note that the mines have set the price of copper this year at or about 22 cents, a cent or two lower than last year's price. Prices are weak. No change is made this week. Machinery cast scrap, $17; heavy red brass, $15.50; heavy copper, $17.

Lead.—The demand is steady. Pig lead is still quoted at $5.45.

TORONTO METAL MARKETS.

Toronto, Aug. 2—There is little doing in metals on the local market, a seasonable dullness having settled upon the trade. Jobbers are quite elated over the record-breaking amount of business that has been transacted this year, and are confidently looking forward to a heavy autumn business as soon as trade revives again, which it no doubt will before the end of the present month. Meanwhile, few trqnsactions of any magnitude are reported, but a fair amount of orders continue to arrive for the various metals in sorting quantities.

Antimony—Since the decline of last week there has been no change in local quotations, the price remaining easy at the reduced figures. The American market shows no encouraging features, sellers are again urging for orders, and the market is weaker on spot and futures with nothing doing. In spite of the unusually heavy shipments of antimony ore and metal from the foreign ports to the United States, the European market has continued to shade prices and this uncertainty has practically demoralized the American market. During the past five months the market has been dull, but indications seem to favor an improvement in the consumption of antimony regulus, and naturally the execution of new orders should stimulate the market. While a few still look for further reductions in prices, on the other hand there is good reason to believe that the American dealers would raise their prices if the European market would assume a stronger tone.

Pig Iron—The market continues extremely dull locally, with practically nothing doing.

Tin—The market is quiet and very few enquiries are coming in. Prices remain steady as last quoted.

Lead—The price is fairly steady locally, considering the fact that both the British and American markets are at present very weak.

Copper—The local price has declined one cent, and casting ingot is now quoted here at 22c. The London market is rather quiet at present, and as for the American market, the copper situation is dull, and there is litele buying in evidence. However, it would seem that stocks in the hands of consumers are pretty low, and developments in the buying line in the near future would not be surprising. There has been a decrease in copper production since the first of the year, and it would not be surprising in the least, if the output of the country's mines in 1907 fell below the production in 1906. For this reason many dealers across the line do not expect a further cut from the present figures.

Old Materials —Prices continue to drop, and the local market is exceedingly quiet. It is not expected that business will pick up to any extent for two or three weeks yet. Heavy copper and wire has dropped three cents and is now quoted at 15c. Light copper and heavy red brass have each taken a further decline of one cent and the price of both is now 14c.; heavy yellow brass has also dropped one cent, and is now quoted at 11c.; tea lead and scrap zinc have both declined ½c. during the week and both are now quoted at 3¾c

LONDON METAL MARKETS.

London, July 30—Cleveland warrants are quoted at 57s. 4½d., and Glasgow standards at 56s. 4d., making prices as compared with last week, on Cleveland warrants, 1s. 1½d. lower, and on Glasgow standards, 1s. 5d. lower.

Tin—Spot tin opened firm at £182, futures at £180 15s., and after sales of 250 tons spot and 450 tons futures, closed easy at £181 15s. for spot and £180 10s. for futures, making price as compared with last week £2 5s. lower on spot, and £1 lower on futures.

Copper—Spot copper opened weak at £88 15s., futures at £84, and after sales of 300 tons spot and 900 tons futures, closed firm at £89 for spot and £84 15s. for futures, making price as compared with last week £4 10s. lower on spot, and £2 10s. lower on futures.

Spelter—The market closed at £23 5s., making price as compared with last week, 12s. 6d. lower.

Lead—The market closed at £19 7s. 6d., making price as compared with last week, £1 7s. 6d. lower.

U.S. METAL MARKETS.

Cleveland, O., Aug. 1.—The Iron Trade Review to-day says:

There is a marked increase in inquiry for pig iron in eastern markets, and a buying movement of some proportions has been started in New York, but in the central west the pig iron markets are listless, although prices show little change. It is understood that the Corporation is willing to take from 10,000 to 15,000 tons of Bessemer pig iron at $22 valley furnace. There is a scarcity of Bessemer, but an ample supply of basic is obtainable.

Although the buying of steel rails is not general, contracts for 50,000 tons have been placed in Chicago. Railroads are showing more of a tendency to buy, and the market in track material is stronger, especially in spikes and bolts.

Lower prices for scrap are having effect on bar iron, and the quotation of 1.60c Pittsburg has been shaded. The present tendency is toward firmness. One concern, which has been well known for its tendency to name low prices for bar iron, has been placed in the hands of a receiver.

U.S. IRON AND METAL MARKETS.

New York, Aug. 1—The Iron Age today says: "The real test of the situation at the ore mines and at the shipping docks will come as soon as the Oliver Iron Mining Co. starts resumption seriously. This is expected at a very early date. The attitude of the local authorities and of the residents in the mining regions has been strongly adverse to the strikers, and the conviction has been growing that the trouble will soon be ended.

The strike has had little effect upon the markets, it being generally understood that it does not influence the early supply.

The Steel Corporation has bought 11,000 tons of Bessemer pig, practically all of the floating supply in the Pittsburgh district, for August delivery, but has not purchased any iron in the eastern markets.

On the whole, deliveries of pig iron are very well taken, thus indicating that current consumption is still very heavy. It is only rarely that complaint is heard by melters that they are getting too much pig iron and fear an embargo.

There is a good deal of delay in putting through contracts for structural material, many of which have been pending for a considerable time. This is apparently due to jockeying for lower prices, since concessions are being freely made by fabricators on erected work.

The rail trade is very quiet, the only large transaction being reported from Chicago, where a new traction system has bought 25,000 tons, after negotiations protracted through efforts to finance the new road.

The leading metals continue to show weakness. In small quantities electrolytic copper has sold at 20c. to-day. No details as to reported transactions on a very large scale can be obtained. Spelter is weaker, and a further reduction in the price of lead by the American Smelting & Refining Co. is regarded as imminent.

CORRESPONDENCE

Travelers, hardware merchants and clerks are requested to forward correspondence regarding the doings of the trade and the industrial gossip of their town and district. Addressed envelopes, stationery, etc., will be supplied to regular correspondents on request. Write the Editor for information.

HALIFAX HAPPENINGS.

Halifax, July 29—Mid-summer dullness in the hardware trade is now on, and there is very little business moving. Prices throughout the list are pretty steady. There was a fairly good demand for haying tools this season, but the rush is now over. The trade in builders' supplies has been exceptionally good, and the sales of glass, paints and oils, have been quite heavy. The new price lists have not yet come to hand.

* * *

Thomas R. Gue, president of the Acadia Powder Co., and president of the Dominion Electrical Co., died in this city on Sunday. He was one of the wealthiest citizens of Halifax, and practically owned both of the above companies. He was born in New York in 1843, but came to Canada thirty years ago, and was for some time secretary of the Hamilton Powder Co., of Montreal. He established the Acadia Powder Co., which plant at Waverly, N.S., is one of the largest of the kind in Canada. The plant of the Dominion Electrical Works, of which he was the head, is near Windsor Junction, and the business has grown rapidly of late. This company manufactures electrical appliances for mining. The late Mr. Gue was considered an expert authority on mining matters.

* * *

The Dominion Iron & Steel Co has been awarded a contract for 43,000 tons of steel rails for the National Transcontinental Railway. The contract price is said to be $1,353,000. The first half of the order is to be delivered in November of this year, and the remainder in 1908. The company's output from the rail mill is said to be the principal asset, and with the orders already on hand, the additional one just received insures a profitable period for a considerable time to come.

* * *

The Board of Trade, of Amherst, has appointed a strong committee with H. J. Logan, M.P., chairman, and A. P. Atherton, secretary, to carry out appropriate connection with the opening of the Chignecto Power Plant, on the 31st inst.

* * *

The City Board of Works has decided to ask for tenders for three thousand water meters, four hundred of lead water pipe, and a supply of valves The meters will be installed in the houses in the city as soon after the tenders are accepted as possible.

* * *

Alexander McDougall, superintendent of the Boston Woven Hose and Rubber Co., of Cambridge, Mass., spent several days in the city last week.

Mr. Donly, Canada's commercial agent in Mexico, left last week for Western Canada. He will visit the principal cities en route.

* * *

Charles Hargraves, manager of the Springhill Collieries, accompanied by his wife and daughter, have left on a two months tour of Europe.

* * *

L. J. Fader, chief engineer of the Halifax Cold Storage Co., left to-day for Port Hawkesbury, where he will superintend the erection of a cold storage plant.

* * *

At a recent meeting of the Water and Light Committee, of Moncton, the tender of the Canada Iron & Foundry Co, of Londonderry, N.S., for water pipe was accepted. The tenders were: R. D. Wood & Co., $13,935.00; D. J. Stewart & Co., $12,776.25; Singleton, Dunn & Co., $12,384.00; Canadian Iron & Foundry Co., $12,592.50; Summer & Co., $12,894.00.

* * *

A telegram was received in Halifax last week from Winnipeg announcing that Prof. E. Brydon-Jack had been appointed to the chair of civil engineering in the University of Manitoba. It is understood that Mr. Brydon-Jack has wired his acceptance of the position, and that his resignation has been tendered to the board of governors of Dalhousie College, in which institution he has for two years been in charge of the department of civil engineering.

SCENIC ST. JOHN.

St. John, N.B., July 2.—The hardware business in most lines is rather dull at present. The trade in builders' material however, continues brisk.

It is stated that a local company, with a large capitalization, has been formed with the intention of manufacturing a patented dress skirt fastener, and that the property now being used by the Mowry Lock Nut Company, which is to remove to Sydney, shortly, has been purchased, for carrying on the industry. Thos. A. Linton, of Linton & Sinclair, is said to be the principal stockholder. Mr. Linton, however, was not ready to discuss the plans of the new company.

In a fire on Mill street on Friday last ten families were burned out; part of the stock of John McGoldrick, junk and metal dealer, was destroyed; three stores, occupied by N. S. Springer for his grocery and feed business, were damaged considerably, and slight damage done to E. J. Carpenter's meat store. The loss was $10,000, and is partly covered by insurance.

Keith & Plummer and Horace R. Nixon, whose general stores were destroyed in the Hartland fire, have decided to rebuild at once. Ziba Orser has bought out the building and stock of T. J. Hurley, and has taken possession of the store. He will fit it up especially for the hardware trade. A new building is being erected for the Hartland monument works. No new evidence leading to the solution of the fire mystery is forthcoming, but the matter will be thoroughly investigated.

Log operations in the St. John river have amounted to more than 35,000,000 feet during the season, which has now been practically brought to a close, making it the best which lumbermen have had in this district in many years.

H. H. Dryden, Sussex, has secured property from G. W. Fowler, M.P., on Court street, Sussex, and will at once commence the erection of a large tinware factory. The factory will be served by a railway siding and its location is one of the finest manufacturing sites in the town.

Matthew Lodge, of Moncton, returned home last week from a visit to England. Mr. Lodge is secretary of the New Brunswick Petroleum Company and his visit to the Old Country was to interest capitalists in the enterprise with a view to developing the Westmoreland oil fields. It is understood that the visit was very successful and that quite extensive operations will soon be undertaken.

Mr. Lodge was accompanied by J. McLaren Henderson and J. A. Henderson, of London, who are representing English interests. F. W. Summer, president of the company; O. P. Boggs, of Pittsburg, and the Englishmen visited the oil wells at Memramcook a few days ago. It is understood that Mr. Boggs, who was at the wells last year looking over the proposition, has been engaged by the company to manage the work of development.

John E. Moore, W. H. Barnaby and Col. H. H. McLean returned Saturday from Ottawa, where they interviewed Sir Wilfrid Laurier in regard to west side dredging. Mr. Moore reported that the dredging matter will be gone into by the premier when ministers absent from Ottawa return to the capital this week.

The wharf building work on the west side is progressing favorably. Clark & Adams, contractors for the new wharf, have two cribs of 200 feet in length built to a height of 20 feet, and the other two cribs well under way. One crib was moved into deeper water on Friday. The building of the warehouse on the D. C. Clark wharf will be commenced this week. Several large gangs of men are engaged by the city in rebuilding Union street, which fell away on account of the dredging operations. C.P.R. and street railway gangs are engaged in laying tracks, and the west side presents a busy scene at present.

KINGSTON KINETOSCOPE.

Kingston, July 29.—The local hardwaremen here report business as being exceptionally brisk in all lines at present. So much building going on seems to be the cause of this and from all appearances it is likely to continue this way for quite a while, as a large number of building permits are being taken out and the buildings under way are being rushed to completion; many are the plumbers and tinsmiths who are working overtime owing to the rush in this line of business. The painters and carpenters are also very busy still, as a result of the late spring we had, along with the numerous new buildings which are being erected. There has been a great demand for all lines of sporting goods as we have had very fine weather the past month and a large number are taking their summer vacation. This, with the various Sunday School and private picnics which are being held, adds considerably to the sporting trade.

* * *

Contracts have been awarded for improvements to St. George's Hall and for the building of a sexton's residence to the following: Masonry, Henry W. Watts; carpentering, W. J. Chapman; heating and lighting, Taylor & Hamilton; painting, etc., T. McMahon & Co. The work is to be completed by October.

* * *

The main corridors and stairways of the city buildings are now going under needed repairs and are at least assuming a respectable look. The walls are being painted green, while the woodwork is to be grained an oak hue. It will be a fortnight before the painters are through.

* * *

A meeting of the creditors of George Sears, hardware merchant of this city, who recently made an assignment, was held on Thursday afternoon. J. B. Walkem was appointed assignee and A. J. Macdonnell, J. McDonald Mowatt and W. B. Dalton were named inspectors. The meeting was adjourned till July 30th. If Mr. Sears is able to arrange certain matters, the claims may be paid in full, otherwise about fifty cents on the dollar will be the result.

* * *

A happy time was held on Wednesday evening, when the employes of McKelvey & Birch's celebrated the re-opening of the new building, erected in the place of the one recently destroyed by fire. About fifty of their friends gathered to take part in the house-warming, which proved to be one of the best and most enjoyable parties of the season. The fine polished hardwood floor downstairs furnished ample space for the dancers, who tripped about to the strains of suitable music from Watson & Salisbury's orchestra. Upstairs, those who preferred cards, found tables daintily arranged for their accommodation. During the evening refreshments were served and it was early in the morning before the merry-makers were able to tear themselves away from the happy scene. The store is about completed, there being just some of the finishing-up jobs being done now and the electricians are still at work, but they expect to get moved into their new apartments in the course of a week and will have one of the finest fitted up stores for miles around.

* * *

Work on the new cement platform at the Kingston and Pembroke railway station here, commenced several weeks ago, was concluded this week. The work was done by Mr. Holder, contractor, of this city. The new platform and steps gives the station a splendid appearance. Concrete cement is becoming a very favorite material here for the construction of buildings and residences. R. G. Armstrong has opened a new store on Montreal Street, built of concrete blocks, which he is manufacturing in his new building. He also handles farm implements and has now space enough to show them to good advantage.

* * *

Another fine brick building is rapidly rising into space on the corner of Johnson and Bagot Streets for P. S. Mahood, merchant, of this city. The contracts for this were awarded as follows: Masonry, Free & Morley; carpentering, Hunter & Harold; plumbing and heating, Simmons Bros.; painting, Robinson Bros. G. R. Thomlinson has the contract for electric wiring and fittings. Mr. Mahood expects to be able to move into his new residence in the fall.

* * *

The largest shipment of binder twine ever sent out from any factory in Ontario was loaded here recently to be taken westward, the consignment consisted of eight carloads, the output of the Kingston Penitentiary plant. It was being forwarded to Alberta. A few days later nine carloads were sent to the same destination, making the enviable record of seventeen carloads in one week. The twine goes to the Farmers' Association in Alberta. Last year the association tried samples and were so well pleased that this year they filled their entire needs from the prison. The prices paid were: 550-foot, mixed, 10c per lb.; 600-foot, mixed, 11c; 600-foot, pure, 11½c. The seventeen carloads were shipped on three steamers from this city.

* * *

The Lake Ontario park power house, owned by the Kingston Street Railway Co., was burned to the ground last week. The damage will inconvenience the company greatly, but will not interfere with the lighting of the park, as power can be switched from the city current and Edison machine will be brought into use. The cause of the blaze remains a mystery as it happened about four o'clock in the morning. Lake Ontario Park is one of the cool spots visited by the people of this city, the street railway having charge of the grounds.

BOOMING BELLEVILLE.

Belleville, July 30—All conditions pertaining to the hardware, plumbing and steam-heating trade in this city were never in a more flourishing condition than at present. Such is the opinion expressed this week by the leading local dealers. Shops and factories are running full blast with a full staff of workmen. Some places have even had to refuse big contracts owing to the lack of help.

* * *

The Belleville Hardware Co., of which Sir Mackenzie Bowell is president, find it difficult to keep up with orders. The firm makes a specialty of all kinds of locks, and they are gaining a great reputation for this class of goods A great many expert mechanics are given employment at these works. A great many people in this city believe it was the starting of these works a few years ago that has been mainly responsible for Belleville's prosperity in recent years. This company started in a very small way, and by strict attention to business, good management on the part of the officers and square dealing, it has now become one of the leading industries of the city, and it is constantly growing. W. C. Springer is general manager, and Mr. H. C. Hunt, secretary.

* * *

The Mac Machine Co., is another important industry which deserves special mention in your representative hardware journal. Under the able management of Harry Bennett, it has established a business known from one end of the country to the other. This firm makes a specialty of all kinds of mining machinery, and there is a demand for their goods even from far Alaska. Nearly every mine in this part of the country is installed with mining apparatus from the Mac Machine Co., whose mining drills have a splendid reputation.

* * *

Marsh & Hawthorn's Foundry must not be over-looked, when one is speaking of Belleville's iron industries. This is one of the oldest foundries, having been established about 60 years ago by the late James Brown. For over half a century it was known far and wide as Brown's Foundry, from which hundreds of boilers and engines were sent forth, many of which are in existence and doing duty to-day. This firm makes a specialty of boilers and steam hoists, and every week big shipments are made throughout the country. Mr. Marsh, the manager, is about 38 years of age, and is one of the most popular residents of the city.

* * *

The steel work on the new hotel Quinte has been completed, and reflects credit on the Hamilton Bridge Co., the contractors for the work. Mr. Maitland Roy, and old Belleville boy, now engaged with the Hamilton Bridge Co., was here last week, and in company with the gas committee of the council put in plans and specifications for the new gas plant and also submitted tenders.

* * *

Mr. Bennen, representing Morrison's Brass Works, of Toronto, was a caller on the trade here this week. He thinks Belleville is a plumhing city, and predicts for her a brilliant future.

* * *

Over 350 men are now employed on the erection of the Lehigh Portland Cement Co.'s plant, which will be when completed, the largest of its kind in Canada. The main buildings are to be built entirely of steel and cement, over 25,000 barrels of the latter being necessary.

PETERBOROUGH PARAGRAPHS.

Peterborough, July 30.—An agitation is on foot among the hardware merchants of Peterborough to close shop on Thursday afternoons during the months of July and August. Last summer a movement was made in this direction but it did not materialize. The Thursday half holiday movement is gaining ground here and this summer practically all the dry goods stores, clothing establishments, jewelry shops, and several other lines of business close on Thursday afternoons during the summer. Next year it is hoped that the hardwaremen will get into line and also have the half holiday.

* * *

The firm of F. J. R. MacPherson & Co., hardware merchants and dealers in electrical supplies, has been given the contract of making numerous improvements at the local customs house. Among other things, the electrical wiring is being overhauled and a more up-to-date system being installed.

* * *

With scarcely one exception, the hardware firms in Peterborough do not advertise. This is somewhat remarkable in a progressive city like Peterborough, but a number of local men who are in the trade state that advertising is not necessary. "We are doing all the business we can handle now," said one of them, "so what is the use of advertising." If this is true in connection with all the local hardware establishments, then they do not need to advertise, but if a dealer has the goods and wants to dispose of them more rapidly than he may be doing, judicious newspaper advertising would no doubt accomplish his purpose. Peterborough has three live daily papers and advertising in them would no doubt pay.

* * *

Two local hardwaremen, Ernest M., and Ray P. Best, of the Best Stove Company, have recently obtained considerable repute as walkers. A short time ago they walked from Bests' summer cottage at Chemong to Peterborough, a distance of 6½ miles, in one hour and seventeen minutes.

* * *

Damage to the extent of $38,000 was done by fire to the mills of the Cavendish Lumber Company at Lakefield on Saturday, July 27. A large amount of lumber was burned, but the loss is covered by insurance.

* * *

There is some talk of a company being organized in Peterborough to establish a plant and carry on the manufacture of bits and augurs. This city already has an industry engaged in the manufacture of these goods, the Rapid Tool Company. This company has not yet been running a year, but it is getting along splendidly and is working overtime to keep up with orders. The goods turned out are said to be of superior quality and are finding a ready market.

WIDE AWAKE WOODSTOCK.

Woodstock, July 29.—Hardware men in Woodstock report that during the summer months, business has been quite up to the average. There have been a number of new buildings, private residences and extensions to industrial establishments; the new Y.M.C.A., Carnegie Library, a new hotel, etc., are in process of construction, and these all furnish sales. "I have no reason to kick" is the prevailing tone of the dealers.

* * *

Quite an extensive business is done in this district, in the sale of wire fence. No more the old-time rail fences are going up, and what remain are gradually being used as a fire wood. We know of a number of farmers, who, each spring in their maple sugar camps, burn up cord after cord of old rails, replacing the fences as they are torn down by modern wire fences. The barbarous barbed wire is just as much " out " as the rail fence, and not a small amount of it is being removed, and the wire buried deep in the earth, where it will never more be seen or heard of. The smooth wire fence is undoubtedly the best kind made. There are two establishments in Woodstock, which annually turn out a good many miles of fence. Many farmers also have the apparatus for making their own wire fence, and weaving the stays. This is cheaper, and it is possible, they say, to construct just as good an article as they may purchase, already made.

* * *

Local dealers are this season observing, during July and August, each Wednesday afternoon as a half holiday. No programs of amusement are arranged, but merchants and businessmen go out fishing or amuse themselves quietly in some way.

* * *

Something was said last week about finding messages in packages of oranges, etc. I know of a man who found something of more value than a mere address. He is a boot and shoe repairer in this city, L. E. Edwards, by name. One day a few months ago, when he was emptying out a package of small nails, he noticed in the pile a yellow glitter. He investigated, and found a twenty-five cent gold piece. The date was an old one, and the piece is worth considerably more than the face value. How it got there is a mystery. Mr. Edwards couldn't find out. He has only one explanation. That is, that some person, probably when leaning over looking into a bin of nails, dropped the gold piece in, and then was unable to find it again. Since he made the discovery, he has carefully watched all packages, but no such luck has come his way again.

* * *

Where are the flies. So far this year, they have been conspicuous by their almost entire absence. Doubtless hardwaremen appreciate their forbearance, but all the same, they notice a difference in the sael of screen windows and doors. "If there are no flies to keep out, what is the use of providing anything to keep them out" is the way people argue, and then they spend the money somewhere else, and the hardwareman goes without that sale. However, as green wire cloth is abnormally scarce this year, it certainly is a godsend that flies are not as plentiful as usual.

* * *

Most men agree that a business may be carried on more satisfactorily on a cash basis than a credit. But the difficulty is, to get them to put the principle into practice. They are afraid of offending some customer who has carried an account, perhaps for years, and always pays up in the end, by insisting that he pay cash with every purchase. They look on the dark side, and see their business gradually dwindling away, until at last the sheriff takes possession. If they looked on the bright side, they would see a business system simplified by cash sales, no accounts to be sent out, no bad debts to weigh against the year's balance, no embarrassing moments while refusing credit to a customer, for all would be treated alike. They should recognize that a dealer would be better off with a smaller business, and on a cash basis, than with a large business, and collections indefinite.

INKLINGS FROM INGERSOLL.

Ingersoll, July 31.—Few towns of similar size can boast of larger or better equipped industries than those to be found in Ingersoll. The citizens of this town are fully alive to the importance of manufactories, and the Board of Trade and the council are constantly on the alert to augment those already so well established whenever an oportunity presents itself. The industry which pays its employes good wages and provides steady employment is an asset, the value of which it is difficult to determine. A sign of industrial expansion is to be seen in a large factory, on Thames St., which is just nearing completion and which will be occupied by Mitchell & Co., hearse builders. The new factory will give employment to more hands than are at present employed by the firm, and of course the merchants generally will be benefited to a larger extent. All the industries in the town are running to their full capacity and harmony reigns supreme between employes and employers.

* * *

"This has been the biggest year in the history of our store," said a hardware dealer to your correspondent today. Continuing, the dealer explained that the large increase in the volume of business was in a great measuer due to the prosperity which the farming community has enjoyed for the past six or seven years. While Canada as a whole has been unusually prosperous for the period mentioned, no class of people have been favored to a greater extent than the formers. As a result of the profitable seasons, which have come with such regularity, the farmers are now spending their money freely. Many are erecting large up-to-date barns while others are building handsome modern residences, which in many instances compare favorably with the finest to be found in towns and cities. This state of affairs, of course, indicates that the hardware dealers and the plumbers are reaping their harvest. The hardware merchants are a unit in stating that this has been a most busy season. Contrary to expectations, a great deal of building has been going on in the town, while many citizens have greatly improved their premises with fresh coats of paint. It was feared earlier in the season that the increased cost of paint materials would interfere to a marked extent with trade, but people seem to have made up their mind to go on with the work, and little or no difference can be noted.

* * *

The season is fast approaching when the hardware dealer will be giving attention to the stove trade. This is a very important part of the business, and one that is full of possibilities if properly handled. What applies to other retail business, applies with equal force to the hardware business, and if it pays the grocer to make attractive displays, it should also benefit the stove dealer to do likewise. At least this is the argument of a local dealer who had "a most successful season" last year. When people are looking for a stove, they, as a rule, want a variety as well as does the lady who enters the dry goods store. Inspecting and question-asking is a big part of purchasing nowadays, and if the dealer has the variety that holds the attention of the prospective purchaser, a sale is often made, largely on this account. Opinions in regard to stoves differe as well as in regard to other things, and it is well that the dealer should have a good assortment from which to choose. A large and attractive display of stoves at the proper season creates the impression that the dealer has anticipated a busy season, and soe how or other people like to patronize the busy store.

* * *

Nothing is heard here about the hardware dealers cutting prices. They are all well established and apparently get along with that friendliness which indicates that they have all they can do to attend to their own business without resorting to price-chopping.

* * *

Although proclaimed as a "public holiday," Thursday was in reality Ingersoll's civic holiday. The feature of the day was an excursion to Port Stanley which was largely attended. A rather peculiar state of affairs led to the day being proclaimed a "public holiday" instead of civic holiday. A petition submitted to the council requested that the day be observed as civic holiday, and the request was granted. Some time previous to this three lodges of the town had arranged to run an excursion out of town, and August 1 was the day set. As soon as it was learned that civic holiday would come on this date it became known that the railway companies have entered into an agreement not to run excursions on a civic holiday, consequently complications arose at once. A little tact, however, saved the day and let both parties out. Mayor Coleridge had not yet issued his proclamation, and instead of proclaiming the day civic holiday, it was announced as a "public holiday" and the Grand Trunk promptly announced reduced fares, which gave the occasion the same effect as civic holiday.

* * *

H. Walter Knight, for the past seven years manager of the factory here of the St. Charles Condensing Co., has severed his connection with that firm. Knight's resignation came as a great surprise to the people of Ingersoll and the farmers of this district. It is understood that he has accepted the position of general manager of a new concern. The Canadian Condensed Milk Co., which, until a location for its factory has been decided upon, will have its headquarters in Ingersoll.

* * *

There are doubtless many readers of Hardware and Metal who are interested in educational matters, who will also be interested in the action of the property committee of the public school at the last meeting of the board of education. In a report the committee recommended that the manual training and domestic science schools be discontinued and that the services of a special singing and music teacher be dispensed with. They also suggested that to relieve the congested condition at the central school that the kindergarten be suspended for the fall term. After much spirited discussion the report was referred to the property and teachers' committees and the action of the board is awaited with much interest by all citizens.

GO-AHEAD GALT.

Galt, July 31.—There is more activity in hardware circles in Galt at the present time than there has been for some years. The activity is due mainly to the building operations under progress and also the brightening up the town is receiving in preparation for the old boys' reunion on August 7, 8 and 9. Mr. W. J. McMurtry, Main street hardware merchant, remarked to-day that he had sold more paint during the past few months than would otherwise have been disposed of in a year. The reunion will be a big event, and all old boys and their friends from the furthermost ends of the earth are cordially invited.

The Galt Art Metal Company are experiencing a very busy season. They have completed the contracts awarded them for the four new Normal shools to be erected by the Government and are now busy on a number of other contracts. They have recently added a machine to their plant to be used in making steel lath. The machine was built here at a cost of $10,000, including patents. A full description of the machine will appear in the Galt column shortly.

The Clare Bros.' stove and furnace foundry in Preston is working overtime, and there is some talk of enlarging the premises on account of the rapidly increasing business.

The first machine for the Canadian Brass Manufacturing Company was installed last week. The building has been completed and it is expected that operations will be commenced in earnest in about a month.

The Down Draft Furnace Company intends putting a new line of stoves on the market, but as yet no information is available.

The Maple Leaf Motor Company, which recently moved its plant from London to Galt, is making good in every sense of the word. The plant is being worked overtime, and the prospects are that the concern will grow to large proportions.

The Galt Board of Trade has several new industries under consideration at present, but as the day of large bonusing is passing, it is doubtful if they will locate here.

The Galt Brass Manufacturing Company, composed of local men, is doing a large business. It is expected that the present quarters will be enlarged in the fall, or new quarters obtained.

CHAT FROM CHATHAM.

Chatham, July 30—For this season of the year trade has been exceptionally good. The majority of the seasonable lines are reported pretty nearly cleaned out. Notwithstanding the late start, the seasonable trade has been very good, much better, in fact, than the dealers anticipated. Plumbing and galvanized work, especially heating work, are rather brisk at present.

* * *

Building operations have been slow this year. The high price of lumber and other materials is unquestionably to blame for this; people who would otherwise have built say they simply can't afford to. Building this year has fallen far short of previous years, particularly 1905. But, notwithstanding this tendency, it is a noticeable fact that a great many householders are having baths put in, installing modern conveniences and improved heating apparatus. All of which goes to show that people are better off, and are giving more attention than hitherto to sanitary improvements.

* * *

C. A. Thomas, of the Sherwin-William's Paint Co. was in the city this week, calling on the local agents, Westman Bros.

* * *

Wallaceburg is beginning to discuss the waterworks question. The Messrs. Winters, surveyors and engineers, are engaged in surveying between the sugar factory and the chenal ecarte, with a view to locating the best route for bringing "Sny" water to the factory. Citizens are urging that the town join hands with the sugar company in bringing in the water and securing an up-to-date waterworks system.

* * *

A charter has been granted by the Provincial Government to the Blenheim & South Kent Telephone Co., Blenheim, capital, $10,000. The provisional directors are: Geo. Taylor, merchant, Jas. Rutherford, miller; Harry Drane, Andrew Denholm and Chas. Henry Echlin, manufacturers, all of Blenheim; Wilber James Huffman, farmer, and Neil Watson, manufacturer, both of Harwich, and Thomas Letson Pardo, of Raleigh. At the organization meeting held at Blenheim last Friday afternoon, by-laws were adopted and the following officers elected: President, Dr. C. B. Langford; vice-president, Neil Watson; secretary-treasurer, Geo. Taylor; directors, Messrs. Paedo, Huffman, Rutherford, Denholm and Dr. Hanks. The intention is to operate rural telephones through South Kent, preparations being already under way and franchises secured.

* * *

Geo. Watt, of Jas. Watt & Sons, was a Toronto visitor last week. He has accepted a position with the Dominion Radiator Co. of that city, and will leave on Aug. 15th to take up his new duties.

* * *

Work on the C.W. & L.E. extension to Lake Erie is progressing rapidly, grading being now complete to Cedar Springs. Last Thursday evening the first electric car crossed the Grand Trunk track, the city route along Raleigh street having been finished in a temporary fashion. The new car will be used in connection with the construction work. The laying of the ties is progressing rapidly. The C.W. & L.E. is, through the continuance of the injunction procured by the Robertson, of Inches estate, prevented from building its permanent road on the western side of the highway; but to overcome this difficulty, the city council on Monday evening granted the railway permission to lay temporary tracks in the centre of the street.

* * *

Tilbury merchants are discarding electric lights and substituting natural gas. It is stated that D. W. Kett, proprietor of the electric lighting plant, intends entering action against the village for damages, claiming that his lighting franchise was exclusive and that the village had no right to grant the gas franchise. The councillors express no uneasiness.

* * *

Word has been received that the Weber Gas Engine Co., of Kansas City, intend sending a representative here to look over the ground in reference to establishing a plant in the Maple City.

* * *

Mayor Stone and Chairman Austin, of the finance committee, last week instructed City Solicitor Lewis to enter an appeal against the decision of Magistrate Houston, in the Brody transient trader case. The magistrate, it will be remembered, dismissed Brody on the ground that a $100 license fee was prohibitive. The appeal has now been entered, and will come before the divisional court at Toronto some time in Sept. The transient question was the theme of a rather warm tilt between Mayor Stone and Ald. Stephens at Monday night's council meeting, the latter objecting to the action of the mayor in appealing the case without first laying the matter before the council. The council, however, finally endorsed the mayor's action by an almost unanimous vote.

* * *

Word was received last week that the Canadian Wolverine Co. had been granted a provincial charter. L. A., H. C. and C. W. Cornelius, of Grand Rapids, Mich., and Robert Gray and W. H. Westman, of Chatham, are named as provisional directors, the capital stock being $100,000, in 1,000 shares of $100 each. It is expected that the Grand Rapids people will be here shortly to arrange for the organization meeting. The company are authorized to manubuy, sell and deal in plumbers' and waterworks' supplies, brass, metal and wooden goods and wares, machinery and tools of all kinds. This list affords the concern a pretty wide scope for expansion.

* * *

An important business change took place to-day in the well-known firm of James Watt & Sons, plumbers and fitters of this city, the firm being dissolved by mutual consent. The business will be continued by James and John B. Watt, Thomas Watt retiring from the partnership.

* * *

J. W. Fleming & Sons, of Blenheim, had a large delivery of "Success" manure spreaders a week ago Saturday, fifteen machines being taken home by purchasers from Raleigh, Harwich and Tilbury. The firm had a parade of the "delivery" that afternoon, which attracted much attention. The delivery is believed to be the largest ever held in Blenheim.

* * *

An automobile garage is the latest for Chatham. Wm. Haven and T. B. Tarsney, of Detroit, are in the city, and have announced their intention of opening an auto livery here. They will also go into the repairing business. Their location has not yet been decided upon, but they state that the concern will be inaugurated in the near future.

LEAFY LONDON.

London, Ont., July 31.—The second day's session of the Master House Painters' and Decorators' Association opened with 150 delegates present. The forenoon was taken up with the reading of papers dealing with various phases of the trade, followed by interesting discussions.

"The Successful Master Painter of the Twentieth Century" was the first paper, which was read by Mr. Graves, of this city. In outlining the different qualifications which were necessary to make a success as a painter, Mr. Graves first called attention to the fact "that a man must be adapted to the trade," that is, he must naturally have no fear of dizzy heights, as the business necessitated a great deal of work up in the air at different heights. He must also have a good understanding of different colors, so as to detect the slightest shade in hues, and above all, he must have a liking for the trade or success would ever come to him.

The next paper read was "The Masterter Painter in His Relations to the Architect," by J. W. Knott, of Toronto. The speaker deplored the fact that the architects as a general rule were not explicit enough in making specifications for painting, and this generally caused a great deal of annoyance and misunderstanding among the members of the trade.

Different delegates spoke of misunderstandings they had experienced in this respect, and thought an attempt should be made to improve the conditions as they existed.

After a great deal of discussion a committee was appointed to confer with the Canadian Society of Architects with a view to having specifications made plainer.

E. J. Linnington, of Toronto, followed with a paper on "Pure Shellac." The history of shellac was dealt with. Its source, process of manufacture and uses were the more important points touched upon. The subject was dealt with in a very able manner, the delegates displaying intense interest throughout. A number of formulas were given for making shellac varnish, which would be suitable for verious kinds of work.

In the afternoon the delegates enjoyed a pleasant time at Springbank park.

The most interesting discussion of the convention took place on the closing day, Thursday, when G. W. Freeman read a paper on "Best Means of Organization." At the outset he explained that he had been requested to formulate a plan for the best means of organizing a Canadian organization apart from the international association. He thought that every loyal Canadian should be allowed to use his own discretion as to whether he should join the international or not. As the matter stands now every master painter is compelled to join the international as the Canadian association is merely a branch. He thought that if the proper measures were taken a Canadian organization could be formed which would be just as successful as the international. He did not altogether advocate secession from the international, but said that the matter of joining that association should be optional with every master painter in Canada. Although the international had been their protection and but for it they would not be here to-day, he predicted that the time was not far distant when the master painters of Canada would be an exclusively Canadian organization.

The paper proved a startler to the delegates, none apparently being prepared for such a radical proposition as made by Mr. Freeman.

Mr. Phinnemore, of Toronto, was first to express disagreement He thought to attempt the formation of a Canadian society at present would mean disorganization throughout Canada. He advised considering the matter for some time and then perhaps the way might seem clear to secede from the international.

President Brookes thought that if they went at it with a will a Canadian organization could be formed.

Others spoke on similar lines, some suggesting that it would be better to allow matters to remain as they are while others thought it better to lay the matter before the next convention.

An interesting feature of the session was a practical demonstration by W. E. Wall, of Somerville, Mass., of the art of graining in imitation of various woods.

The following officers were elected for the ensuing term: President, H. R. Reynolds, Guelph; vice-president, R. Booth, London; secretary-treasurer, A. M. McKenzie, Hamilton; executive board, Benjamin Goodfellow, Galt; John Reede, Toronto; W. R. Talbot, Winnipeg; W. T. Mossop, London, and Wm. Laessen, Windsor.

The president, vice-president and secretary-treasurer were appointed as delegates to the New Orleans convention next February.

Stewart N. Hughes, secretary-treasurer of the Canadian Association, has been elected president of the International Association. After a vote of thanks and three cheers for the newly elected officers the convention was dismissed.

* * *

Mr. Donald, the Chicago man who is interested in the proposed new plough works for Chelsea Green, a southeast suburb of the city, has purchased from Col. Gartshore, of the McClary Co., two acres of land upon which to erect the works. The new industry will employ a large number of men, the great majority of whom will be skilled mechanics, and it will be a big addition to the south end's factories. Contracts have been let for the buildings, and it is understood that work will begin at once.

* * *

General quiet pervades the hardware trade. Travelers are all off on their holidays.

* * *

Mention was made in this correspondence a few weeks ago of the fact that owing to pressure of business the Dennis Wire & Iron Co. were contemplating the erection of an addition to their present premises or building a new factory. This afternoon Ernest Dennis gave out something definite in connection with the matter. The company, he says, has secured the property to the east of their present works and will at once commence the erection of an addition 40x80 feet, and two storeys high, to be used chiefly by the steel structural department. Some $13,000 will be spent on the building, and additional plant this year, which will enable the company to add 25 workmen to their staff.

* * *

J. C. McConnell, advertising manager for the McClary Mfg. Co., leaves next week on a two weeks' trip to the Maritime Provinces.

SASKATOON SAYINGS.

Saskatoon, July 26.—Several carloads of brick and steel have arrived for the various buildings and are being unloaded at the C.N.R. station.

* * *

The elevator addition to the Saskatoon flour mill is now completed. The whole building looks neat in its coat of red paint and is rendered fireproof by a covering of metal siding.

* * *

S. A. Clark, one of the pioneer hardwaremen of Saskatoon, has disposed of his business. The new firm will take over the business on August 1 and will be known as the S. A. Clark, Limited.

J. F. Cairns is showing a neatly trimmed window of graniteware in various colors. To show the goods to better advantage stepladders are used and on these the smaller articles are displayed.

* * *

The plumbing contract for the new Windsor Hotel has been secured by M. Isbister & Son. The same firm will also install a steam heating system both in the Windsor and in the new addition to the Royal Hotel. In both cases Florence steam boilers are specified, which will be supplied by the Pease-Waldon Co., Winnipeg.

* * *

W. P. Landon & Co., plumbers and metal workers are building an addition in the rear of their workshop in which a reserve stock will be kept.

LEGALITY OF ESTABLISHED RESALE PRICES

Paper read by George Puchta before the recent convention of the National Supply and Machinery Dealers' Association, Cincinnati, Ohio.

We are all trying to do a legitimate, honest business, and to make it such we must have a fair profit. How can that be done if goods are sold at cost or less? The business man who knows what he is doing will not likely commence this practice. It is too dangerous. It therefore follows that it must be instigated by the competitor who does not know, and which we will call unintelligent competition, and which causes most of the trouble and should be eradicated.

Not in a radical manner, but by showing him the error of his ways and making him intelligent competition, and right here is where our associations can do much to improve conditions. How many have we in business that do not know the percentage of cost to do business, that keep no accurate accounts, but if they did, would surprise themselves, and show them how much profit it is really necessary to have before they make anything. How many salesmen they have that are not producers: How often they permit salesmen to make prices on their goods, besides many others.

Now this brings us to the present condition where we are selling some lines so low that we would be better off if we did not sell them at all, and as one of the remedies, the resale price is suggested where prices are demoralized, and is in operation in some lines with good results. The resale price is an agreement between the manufacturer and dealer whereby the manufacturer specifies a minimum price at which he will permit his make of goods to be sold by the dealer.

Resale Prices on Rebate Plan.

The resale price on the rebate plan is an agreement between the manufacturer and dealer whereby the manufacturer specifies a minimum price at which he will permit his goods to be sold by the dealer, and if the dealer has kept this agreement for an agreed period to the satisfaction of the manufacturer he receives a rebate or commission, which in some cases is the dealer's entire profit.

The resale price on the rebate plan has produced the best results. It is an individual agreement between the manufacturer and the dealer, and as long as it remains an individual agreement between the two parties the legality of it is not questioned. When two or more manufacturers collectively, or two or more dealers collectively act, then the legality is questioned.

When advised that I was expected to prepare a paper on this subject, I secured twelve different resale price agreements in effect in various lines, and gave them to a reliable attorney for an opinion, and also prepared these questions, and received the answers affixed:

First. Do these agreements violate any laws? Ans.—No.

Second. Are these contracts binding on the parties signing them? Ans.—Yes.

I also quote from this attorney's report the following:

Important Legal Opinion.

"If the articles manufactured are protected by patents, the manufacturers have the right to make any terms, prices, conditions or restrictions they see fit.

"The patent statute grants an exclusive monopoly which can not be cut down by the rule against restraint, for that would be to grant a monopoly by law and then proceed to take it away by law."

If two or more manufacturers, who produce goods of similar class or character, should combine either orally or in writing, or by a secret understanding to put up prices limit output, or do other things against public policy or obnoxious to the law, it would be in restraint of trade to an extent that would be unreasonable and illegal.

But if a single manufacturer of an unpatented article should make an agreement with his customers to give them a rebate, commission or discount if they should refuse to sell his product at less than a certain price, it would not be in restraint of trade, as there would be nothing to prevent others engaging in the business, or the manufacturers of other articles from selling their products to any one willing to buy.

The attorney further states: "I have examined a large number of decisions and have used the language of the courts in the foregoing very largely."

Resale Agreement in Force.

There are many resale price agreements in effect in many lines of trade with benefit to all concerned.

First. The manufacturer in many cases has found that unrestrained competition means serious loss to him, because the dealer, the selling end of the manufacturer could not make a profit, and he was thereby forced to sell his own goods direct as best he could, at a largely increased selling expense.

Second. The dealer, who is conceded to be a necessity, cannot exist without a profit any more than the manufacturer's salesman can exist without his salary; consequently, when a line of goods becomes unprofitable, his exis-

tence demands a change to goods that are profitable, or forces him to become a manufacturer, which in many cases is a detriment to both manufacturer and dealer.

Third. The consumer is not injured as long as fair profits are made, and he is protected against unreasonable charges, because when profits become unreasonable that in itself invites additional competition, which is soon forthcoming and adjusts the price.

Proof Of Popularity.

The resale price has been adopted in some lines that we as supply and machinery dealers handle with excellent results, and while it may not be advisable to adopt it in all, it could be in lines that are sold at little or no profit with much benefit to manufacturer and dealer alike, and should be encouraged. To show its popularity, in a recent letter sent out by a manufacturer, asking the dealers what they thought of resale price agreements, and if they would support same, answers were as follows:

87.5 per cent. yes, with a few qualifications.

9.7 per cent. no.

2.8 per cent. did not answer.

But when a manufacturer arranges a resale price for his goods, it should be the solemn duty of the dealer to uphold that agreement to the letter and not in any manner violate same, and thereby make the profit the manufacturer is willing to allow him. The dealer must have the courage and honor to lose an order rather than violate his agreements.

Pays to Uphold Agreement.

What encouragement is it to a manufacturer if he allows the dealer, say, 15 per cent. differential, and the dealer in many ways gives part of it away, secretly or otherwise? Every time that it is done and discovered, and it usually is discovered, it invariably leads to a reduced differential, because the dealers violation of the agreement is sufficient evidence to the manufacturer that the profit allowed is more than the dealer wants, and consequently it is reduced, and the dealer alone is responsible.

I believe many manufacturers are willing to establish resale prices on their goods if the dealers will be fair, but fair play and a square deal are essential on the part of both manufacturer and dealer. Let us prove to the world that every member of these three associations holds sacred any agreement he may make, and if this be done many of the goods sold to-day at little or no profit can be made profitable (thereby improving his condition as well as his competitor), and while the resale price will not cure all ailments and is not a cure-all, it can be effectively applied in many lines with benefit and profit to all and injury to none, and these associations here assembled can do much to bring same about.

Files clogged with tin or lead will be cleaned by a few seconds in strong nitric acid. For iron filings use blue vitriol, rinse in water and dip in nitric. For copper or brass, use nitric several times; for zinc, dilute sulphuric acid. After any of these treatments rinse the files in water, brush vigorously and dry in sawdust or by burning alcohol on the file.

MANITOBA HARDWARE AND METAL MARKETS

Corrected by telegraph up to 12 a.m. Friday Aug. 2. Room 511, Union Bank Bldg, Winnipeg, Man.

There are no changes of any account in prices this week, values continuing steady. A good average business is being transacted for this time of the year. Buyers still continue to show a disposition to refrain from making large purchases until they can form some accurate idea concerning the amount of this season's western crop.

ROPE—Sisal, 11c. per lb., and pure manila, 15¾c.

LANTERNS—Cold blast, per dozen, $7; coppered, $9; dash, $9.

WIRE—Barbed wire, 100 lbs., $3.22½; plain galvanized, 6, 7 and 8, $3.70; No. 9, $3.25; No. 10, $3.70; No. 11, $3.80; No. 12, $3.45; No. 13, $3.55; No. 14, $4; No. 15, $4.25; No. 16, $4.40; plain twist, $3.45; staples, $3.50; oiled annealed wire, 10, $2.90; 11, $2.96; 12, $3.04; 13, $3.14; 14, $3.24; 15, $3.39; annealed wires (unoiled), 10c. less; soft copper wire, base, 36c.; brass spring wire, base, 30c.

POULTRY NETTING—The discount is now 47½ per cent. from list price, instead of 50 and 5 as formerly.

HORSESHOES—Iron, No. 0 to No. 1, $4.65; No. 2 and larger, $4.40; snowshoes, No. 0 to No. 1, $4.90; No. 2 and larger, $4.65; steel, No. 0 to No. 1, $5; No. 2 and larger, $4.75.

HORSENAILS—No. 10 and larger, 22c; No. 9, 24c.; No. 8, 24c.; No. 7, 26c.; No. 6, 28c.; No. 5, 30c.; No. 4, 36c. per lb. Discounts: "C" brand, 40, 10, 10 and 7½ p.c.; "M.R.M." cold forged process, 50 and 5 p.c. Add 15c. per box. Capewell brand, quotations on application.

WIRE NAILS.—$3 f.o.b. Winnipeg, and $2.55 f.o.b. Fort William.

CUT NAILS—Now $3.20 per keg.

PRESSED SPIKES — ¼ x 5 and 6, $4.75; 5-6 x 5, 6 and 7, $4.40; ⅜ x 6, 7 and 8, $4.25; 7-16 x 7 and 9, $4.15; ½ x 8, 9, 10 and 12, $4.05; ⅝ x 10 and 12, $3.90. All other lengths 25c. extra net.

SCREWS — Flat head, iron, bright, 85 and 10; round head, iron, 80; flat head, brass, 75 and 10; round head, brass, 70 and 10; coach, 70.

NUTS AND BOLTS — Bolts, carriage, ⅜ or smaller, 60 p.c.; bolts, carriage, 7-16 and up, 50; bolts, machine, ⅜ and under, 50 and 5; bolts, machine, 7-16 and over, 50; bolts, tire, 65; bolt ends, 55; sleigh shoe bolts, 65 and 10; machine screws, 70; plough bolts, 55; square nuts, cases, 3; square nuts, small lots, 2½; hex nuts, cases, 3; hex nuts, small lots, 2½ p.c. Stove bolts, 70 and 10 p.c.

RIVETS — Iron, 60 and 10 p.c.; copper, No. 7, 43c.; No. 8, 42½c.; No. 9, 45½c.; copper, No. 10, 47c.; copper, No. 12, 50½c.; assorted, No. 8, 44½c., and No. 10, 48c.

COIL CHAIN — ¼-in., $7.25; 5-16, $5.75; ⅜, $5.25; 7-16, $5; ½, $4.75; 9-16, $4.70; ⅝, $4.65; ¾, $4.65.

SHOVELS—List has advanced $1 per dozen on all spades, shovels and scoops.

HARVEST TOOLS—60 and 5 p.c.

AXE HANDLES—Turned, s.g. hickory, doz., $3.15; No. 1, $1.90; No. 2, $1.60; octagon extra, $2.30; No. 1, $1.60.

AXES—Bench axes, 40; broad axes, 25 p.c. discount off list; Royal Oak, per doz., $6.25; Maple Leaf, $8.25; Model, $8.50; Black Prince, $7.25; Black Diamond, $9.25; Standard flint edge, $8.75; Copper King, $8.25; Columbian, $9.50; handled axes, North Star, $7.75; Black Prince, $9.25; Standard flint edge, $10.75; Copper King, $11 per dozen.

CHURNS—45 and 5; list as follows: No. 0, $9; No. 1, $9; No. 2, $10; No. 3, $11; No. 4, $13; No. 5, $16.

AUGER BITS—"Irwin" bits, 47½ per cent., and other lines 70 per cent.

BLOCKS—Steel blocks, 35; wood, 55.

FITTINGS — Wrought couplings, 60; nipples, 65 and 10; T.'s and elbows, 10; malleable bushings, 50; malleable unions, 55 p.c.

HINGES—Light "T" and strap, 65.

HOOKS — Brush hooks, heavy, per doz., $8.75; grass hooks, $1.70.

STOVE PIPES—6-in., per 100 feet length, $9; 7-in., $9.75.

TINWARE, ETC.—Pressed, retinned, 70 and 10; pressed, plain, 75 and 2½; pieced, 30; japanned ware, 37½; enamelled ware, Famous, 50; Imperial, 50 and 10; Imperial, one coat, 60; Premier, 50; Colonial, 50 and 10; Royal, 60; Victoria, 45; White, 45; Diamond, 50; Granite, 60 p.c.

GALVANIZED WARE — Pails, 37½ per cent.; other galvanized lines, 30 per cent.

CORDAGE — Rope sisal, 7-16 and larger, basis, $11.25; Manilla, 7-16 and larger, basis, $16.25; Lathyarn, $11.25; cotton rope, per lb., 21c.

SOLDER—Quoted at 27c. per pound. Block tin is quoted at 45c. per pound.

WRINGERS—Royal Canadian, $36; B.B., $40.75 per dozen.

FILES—Arcade, 75; Black Diamond, 60; Nicholson's, 62½ p.c.

LOCKS—Peterboro and Gurney, 40 per cent.

BUILDING PAPER—Anchor, plain, 66c.; tarred, 69c.; Victoria, plain, 71c.; tarred, 84c.; No. 1 Cyclone, tarred, 84c.; No. 1 Cyclone, plain, 66c.; No. 2 Joliette, tarred, 69c.; No. 2 Joliette plain, 51c.; No. 2 Sunrise, plain, 56c.

AMMUNITION, ETC. — Cartridges, rim fire, 50 and 5; central fire, 33½ p.c.; military, 10 p.c. advance. Loaded shells: 12 gauge, black, $16.50; chilled, 12 gauge, $17.50; soft, 10 gauge, $19.50; chilled, 10 gauge, $20.50. Shot: ordinary, per 100 lbs., $7.75; chilled, $8.10. Powder: F.F., keg, Hamilton, $4.75; F.F.G., Dupont's, $5.

REVOLVERS — The Iver Johnson revolvers have been advanced in price, the basis for revolver with hammer being $5.30 and for the hammerless $5.95.

IRON AND STEEL—Bar iron basis, $2.70. Swedish iron basis, $4.95; sleigh shoe steel, $2.75; spring steel, $3.25; machinery steel, $3.50; tool steel, Black Diamond, 100 lbs., $9.50; Jessop, $13.

SHEET ZINC—$8.50 for cask lots, and $9 for broken lots.

CORRUGATED IRON AND ROOFING, ETC.—Corrugate iron 28 gauge painted $3, galvanized $4.10; 26 gauge $3.35 and $4.35. Pressed standing seamed roofing 28 gauge painted $3.10, galvanized $4.20; 26 gauge $3.45 and $4.45. Crimped roofing 28 gauge painted $3.20, galvanized, $4.30; 26 gauge $3.55 and $4.55.

PIG LEAD—Average price is $6.

COPPER—Planished copper, 44c. per lb.; plain, 39c.

IRON PIPE AND FITTINGS—Black pipe, ¼-in., $2.65; ⅜, $2.80; ½, $3.50; ¾, $4.40; 1, $6.35; 1¼, $8.65; 1½, $10.40; 2, $13.85; 2½, $19; 3, $25. Galvanized iron pipe, ⅜-in., $3.75; ½, $4.35; ¾, $5.65; 1, $8.10; 1¼, $11; 1½, $13.25; 2-inch, $17.65. Nipples, 70 and 10 per cent.; unions, couplings, bushings and plugs, 60 per cent.

GALVANIZED IRON — Apollo, 16-gauge, $4.15; 18 and 20, $4.40; 22 and 24, $4.65; 26, $4.65; 28, $4.50; 30 gauge or 10¾-oz., $5.20; Queen's Head, 20 $4.60,; 24 and 26, $4.90; 28, $5.15.

LEAD PIPE—Market is firm at $7.80.

TIN PLATES—IC charcoal, 20 x 28, box, $10; IX charcoal, 20 x 28, $12; XXI charcoal, 20 x 28, $14.

TERNE PLATES—Quoted at $9.50.

CANADA PLATES — 18 x 21, 18 x 24, $3.50; 20 x 28, $3.80; full polished, $4.30.

LUBRICATING OILS—600W, cylinders, 80c.; capital cylinders, 55c. and 50c.; solar red engine, 30c.; Atlantic red engine, 29c.; heavy castor, 28c.; medium castor, 27c.; ready harvester, 28c.; standard hand separator oil, 35c.; standard gas engine oil, 35c. per gallon.

PETROLEUM AND GASOLENE — Silver Star, in bbls., per gal., 20c.; Sunlight, in bbls., per gal., 22c.; per case, $2.35; Eocene, in bbls., per gal., 24c.; per case, $2.50; Pennoline, in bbls., per gal., 24c.; Crystal Spray, 23c.; Silver Light, 21c.; engine gasoline in barrels, gal., 27c.; f.o.b. Winnipeg, in cases, $2.75.

GOOD CHEER LINE AT WINNIPEG EXHIBITION

The Winnipeg branch of the James Stewart Mfg. Co., of Woodstock, made a fine display of their well-known line of Good Cheer stoves and furnaces.

Their new, first-class steel range, the "Sunray Good Cheer," was given a prominent place in the exhibit and the interest in it, shown by both the general public and the many visiting stove dealers, evidences the attractiveness of this new line. The "Sunray" is supplied with either a 18 or 20-inch oven, 4 and 6-hole top, on nickel-plated feet or substantial cast iron base, and in the usual combinations of square or with reservoir, high shelf, high closet, and with either cast or coil waterfront.

The construction throughout is of the most substantial nature, everything possible to place it in the front rank of first-class steel ranges having been done. In addition to the usual protection under oven bottoms of steel and asbestos, an additional bottom is provided, assuring the maximum radiation of heat with the oven through oven bottom and making the range a perfectly safe one to set on the floor without the protective zinc or galvanized sheet usually required.

The fire-box linings are all of cast iron, the fireback being of a pattern which the company claim to be the most lasting of iron backs, as supplied for coal-burning ranges. All gates and linings are removable without disturbing waterfront. Front and end draft doors and sieves assure ample air to the fire, while the convenience of the end broiler door is well known.

Another line which will be welcomed by the trade is their new series of Good Cheer furnaces for anthracite or bituminous coal, coke or wood. This line is the result of several years' designing, construction and testing, the aim of the company being to produce a warm air furnace capable of generating the maximum of heat from the fuel consumed, and of a construction which will stand the wear and tear of many years' service. There has been a most decided call for a furnace of its qualities by furnace men; the trouble having been that despite the first-class work of installation which they do, the average furnace will often disappoint in both operating and radiating capacity and in lasting qualities, and as his workmanship reputation is apt to suffer and further furnace contracts consequently influenced adversely through no fault of his, the dealer asks for a furnace as good as his own work—and here it is. In the first place, it is most easily set up. The radiator comes from factory ready mounted and is of a height (20 inches) which will permit of its being readily taken through the average cellar door opening and down crooked stairways. In the average good quality furnace there is a similarity up to the top of the firepot, and there ends, the radiators differing materially. This is the vital point of a furnace, as it determines just what amount of heat is generated from the fuel consumed.

"Good Cheer" Stoves, Ranges and Furnaces at the Winnipeg Exhibition.

The "Good Cheer," which having all the good constructive points as to roomy ash-pit, sectional fire-pot with cup joints, and heavy radiating flanges, also has a radiator which is not equalled by any other in the market. This part alone of the 20-inch fire-pot size weighs 400 lbs. No steel is anywhere exposed to direct action of the fire, being only used for outside jacket of radiator.

A series of heavy cast iron tubes surround the fire-pot and in conjunction with the outside steel jacket, form a flue, through which all gases and products of combustion must pass completely around furnace in one direction before passing into chimney. While this means that the series of tubes are heated from both sides and therefore made most powerful heat radiators, all necessity of diving flues or divided drafts is avoided, and the result is a furnace the flues of which cannot possibly choke up and which can be readily cleaned in five minutes with the fire going at the same time.

It has been demonstrated satisfactorily that the air admitted through feed door of the average furnace is not nearly sufficient to create complete combustion of the gases and heat products arising from the fire, and as a consequence, much heat energy passes to the chimney and into the outside air unconsumed. This is the secret of the wastefulness of the average furnace.

In the "Good Cheer" the top of firepot is surrounded by an air blast ring, which is furnished with air through tubes opening on either side of furnace. This ring distributes the necessary oxygen evenly over surface of fire and provides the only medium yet devised for obtaining the maximum of heat from the fuel consumed. While this feature is of service in the use of any kind of fuel, it is invaluable in our Canadian Northwest, where bituminous coals are so largely used.

In other lines this company is also right up to the mark, as shown by their display of Oak Heaters, the well-known Good Cheer art base burner, the smaller size of which can now be supplied in the highly artistic form as shown in their No. 55. Hot blast heaters, cast iron ranges, coal and wood cooks and a most complete line of small steel range cooks with or without high closets or reservoirs.

A. E. Karges, the company's manager of their Winnipeg branch, was in constant attendance welcoming the many out-of-town customers who called, and Charles E. Stewart, president and manager of the company, was also present making the acquaintance of the trade.

LETTER BOX.

Correspondence on matters of interest to the hardware trade is solicited. Manufacturers, jobbers, retailers and clerks are urged to express their opinions on matters under discussion.

Any questions asked will be promptly answered. Do you want to buy anything, want some shelving, a silent salesman, any special line of goods, anything in connection with the hardware trade? Ask us. We'll supply the necessary information.

Interest on Deposits and Loans.

Editor Hardware and Metal:

Sir —Your editorial about bank buildings and interests was very timely taken up. Why should the bankers be allowed to form a combine in restraint of trade, in only giving depositors three per cent. on deposits, then turning round and loaning out funds at seven, eight and ten per cent.? Take for instance the case of the Home Bank, which previously to entering the bankers' association paid three and a half per cent., and made more money than they are able to do at the present time. In order to secure the privileges of clearing house they had to reduce their rate to three per cent.

A number of loan companys are paying four per cent. Of course depositors are not protected to the same extent as in chartered banks, but, still they come under supervision of the government, which forced several weak loan companys to the wall.

It looks as though the banks are keeping too much of the interests the depositor should get, and putting it into handsome buildings. DEPOSITOR.

Toronto, July 23.

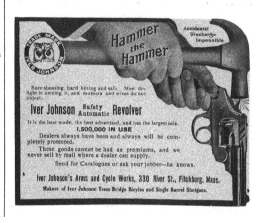

BUILDING AND INDUSTRIAL NEWS

For additional items see the correspondence pages. The Editor solicits information from any authoritative source regarding building and industrial news of any sort, the formation or incorporation of companies, establishment or enlargement of mills, factories or foundries, railway or mining news.

Industrial Development.

A large boiler works will be established at Halifax, N.S.

Valuable deposits of ochre have been found near Saskatoon, Man.

Fire destroyed five blocks at Victoria recently; damage, $3,250,000.

Smith, Runciman & Co., Toronto, will erect a warehouse to cost $45,000.

The stock of the Fruitland Brick & Supply Co., Fruitland, Ont., is for sale.

The H. H. Dryden Co., Sussex, N.B., will considerably enlarge their tinware plant.

The assets of the Good Roads Machinery Co., Hamilton, are advertised for sale.

The Massey-Harris Co. will erect a large factory at Brantford, Ont., to cost $20,000.

The Bloomfield Packing Co. will erect a large canning factory at Hillier, Ont., costing $25,000.

A firm manufacturing heaters, stoves and furnaces is anxious to obtain a site in Sarnia, Ont.

The Canadian Bag Co., Winnipeg, will erect a new bag factory in that city, to cost $10,000.

The International Turpentine Co. will install a plant in New Westminster, B.C., to cost $30,000.

The Canada Woodenware Co. have been offered strong inducements to locate in Chatham, N.B.

The Canadian Shovel & Tool Co., Hamilton, Ont., will erect an addition to their plant to cost $1,000.

The Toronto Bolt & Forging Co., Swansea, will erect a plant at their present works, to cost $30,000.

The British Columbia Electric Co. suffered damage to the extent of $1,000 in the recent fire at Victoria.

The pumping plant at Well 5 of Fort William's waterworks system was destroyed by fire; loss, $12,000.

A new steam heating plant and ventilating system is to be installed in St. Joseph's Hospital, Guelph, Ont.

The Deleware Seamless Tube Co., Reading, Pa., may build a large plant at Sarnia, Ont., to cost $200,000.

A company has been formed for making cast iron culverts and will establish a plant in Port Arthur, Ont.

The representatives of the Ross Oscillating Pump are in Calgary looking for a purchaser for the Canadian rights.

The plant of the Midland Engine Works is again in operation. The capacity of the molding room has been doubled.

The North Arm Lumber Co., Vancouver, has obtained a very suitable mill site and will at once erect a mill at a cost of $150,000. The machinery has been ordered and the mill will be in running order about the beginning of the year.

The huge steam engine in the Ogilvie Flour Mills, Winnipeg, has been replaced by an electric motor of 1,200 horse-power.

The saw mill and woodworking factory belonging to J. H. Simonson, at Spragues Mills, N.B., was destroyed by fire; loss, $10,000.

The Eells Lime Co., at Rockport, N.B., and the Rockport Ice Co. suffered heavily by fire recently. The damage was estimated at $75,000.

The Union Lumber Co. has been incorporated in Edmonton to supply lumber to union men at a little above cost. The capital is $250,000.

The first installment of machinery for the new pulp mill of the Gordon Paper Co., Dryden, Ont., is on the ground and installation will commence immediately.

Fox & Co., mantel and woodwork manufacturers, will erect a factory in Windsor, Ont., to cost $40,000 if the city will give them exemption from taxation and free light and water.

The Dominion Car & Foundry Co., Montreal, will erect a large addition to their plant, to cost $2,000,000. A cast steel plant will cost $1,000,000 and a malleable iron plant and three steel car works will be erected.

The fifteenth annual meeting of the shareholders of the Page Wire Fence Co. was held recently. The year's report showed an enormous increase, the volume of sales being the largest in the history of the company. The officers were all re-elected as follows: President, Walter Clement; vice-president, N. L. Clement, secretary-treasurer, Merton Church.

Companies Incorporated.

Weston Tool and Noveliits, Limited, manufacture and sell tools, cutlery, etc. Provisional directors, C. M. M. Colquhoun, H. E. Erwin, J. C. Webster, all of Toronto.

O'Keeffe-Sanford, Limited, Toronto; capital, $40,000. To manufacture and deal in mantels, tiles and fire place fittings. Provisional directors, W. Webb, M. Whalen, T. Main, B. F. Bennet, all of Toronto.

Onaping Iron Ore Co., Toronto; capital, $200,000. To carry on in all its branches mining, milling and reducing of iron ores. Provisional directors, F. Denton, A. R. Cochrane, G. J. Valin, all of Toronto.

General Industries Construction Co., Toronto; capital, $100,000. To carry on a general construction business. Provisional directors, J. A. Patterson, G. F. McFarland, A. McKenzie, W. H. Templeton, all of Toronto.

The Dominion Wheel Co., Lindsay, Ont.; capital, $40,000. To manufacture and deal in all kinds of wheels and turned goods. Provisional directors, J. D. Flavelle, W. Mc. Flavelle, J. Carew, and T. Stewart, all of Lindsay.

The Crown Oil Refining Co., Hamilton, Ont.; capital, $40,000. To manufacture and deal in oil refining apparatus and refiners' supplies. Provisional directors, E. Hull, W. Perkins, J. A. Hull, T. Barnes, G. F. Hull, all of Hamilton.

Guelph Oil Clothing Co., Guelph; capital, $50,000. To manufacture and deal in oil clothing, tents, awnings, etc. Provisional directors, G. McPherson, J. T. McPherson, R. E. McPherson, G. A. McPherson, all of Guelph.

The Tecumseh and Walkerville Oil and Gas Co., Walkerville, Ont.; capital, $40,000. To carry on an oil and gas business. Provisional directors, J. Dugal, P. Dugal, H. A. Walker, H. C. Walker, R. J. Collotou, all of Walkerville.

William Hamilton Co., Peterborough, Ont.; capital, $300,000. To deal in machinery, contractors' and builders' supplies and carry on the business of machinists and engineers. Provisional directors, W. G. Ferguson, W. S. Davidson, G. L. Hay, R. M. Glover, J. D. Clarke, all of Peterborough.

The New Liskeard Clock Co., New Liskeard, Nip.; capital, $40000; To manufacture clocks, novelties, dies, tools and sheet metal work and to carry on the business of machinists and engineers. Provisional directors S. D. Briden, J. Armstrong, M. McLeod, D. McKelvie, J. Redpath, all of New Liskeard.

Building Notes.

A large armory is to be erected at Fernie, B.C.

Tenders are invited for an armory for Brandon, Man.

The Bank of Montreal will erect a block in Sudbury, Ont.

A new church will be erected at Lorette, Que., costing $100,000.

A fish hatchery will be erected at Newcastle, N.B., to cost $5,000.

A mission school will be erected at Kitamaat, B.C., to cost $25,000.

The Canadian Bank of Commerce are erecting a branch at Elkhom, Man.

Municipal Undertakings.

The new jail at Sydney, N.S., will cost $13,550.

Gracefield, Que., will install a waterworks system.

Halifax, N.S., will instal water meters at a cost of $50,000.

The land title offices at Battleford, Sask., will cost $31,000.

Whitewater, Man., will install a municipal telephone system.

A municipal waterworks system will be installed at South Vancouver, B.C.

The contract for the municipal lighting system for Battleford, Sask., has

CONDENSED OR "WANT" ADVERTISEMENTS.

Advertisements under this heading 2c. a word first insertion; 1c. a word each subsequent insertion.

Contractions count as one word, but five figures (as $1,000) are allowed as one word.

Cash remittances to cover cost must accompany all advertisements. In no case can this rule be overlooked. Advertisements received without remittance cannot be acknowledged.

Where replies come to our care to be forwarded, five cents must be added to cost to cover postage, etc.

AGENT WANTED.

AGENT wanted to push an advertised line of Welsh tinplates; write at first to "B.B.," care HARDWARE AND METAL, 88 Fleet St., E.C., London, Eng. [tf]

BUSINESS CHANCES.

MINING hardware, stoves and tinware business for sale; sales $4,000 monthly; buildings and lot, $3,500; present stock, $8,000; proprietor's health failing; a bonanza. Box 643, HARDWARE AND METAL, Toronto. [32]

EXPERIENCED business man wishes to meet capable energetic young man with a thorough knowledge of hardware and stoves and with two or three thousand dollars in cash; advertiser would join him in purchasing some desirable business and provide seven or eight thousand capital. Address Box 644, HARDWARE AND METAL, Toronto.

FOR SALE — Well established hardware, tinshop, implement and undertaking business, also good lumber yard, well fenced, with lumber and lime sheds in good condition; we will sell above altogether, or divide same to suit purchaser; proprietors are retiring from business in Manitoba, and therefore wish for immediate sale. Apply to Eakins & Griffin, Shoal Lake, Man. [33]

WANTED—Partner to take half interest in one of the best hardware propositions in Algoma; plumber preferred. Box 636, HARDWARE AND METAL, Toronto. [32]

HARDWARE, tinware, stove and plumbing business in manufacturing town in the Niagara Peninsula; no competition; $250,000 factory and waterworks will be completed this summer; stock about $3,000; death of owner reason for selling. Box 85, Thorold, Ont. [tf]

HARDWARE, Stove and Tinware Business for sale in live Western Ontario Village; first class chance; good reasons for selling. Address Box 640, HARDWARE AND METAL.

FOR SALE—A good hardware business in Western Ontario; stock about $6,000. For further reference apply The Hobbs Hardware Co., Limited, London, Ont. [32]

HARDWARE and Tin Business for Sale in good Western Ontario town of 3,000; stock about $3,500; good reasons for selling. Address, Box 647 HARDWARE AND METAL, Toronto. (34)

FOR SALE.

FOR SALE — First-class set of tinsmith's tools second-hand but almost as good as new; includes an 8-foot iron brick almost new. Apply Pease Waldon Co., Winnipeg. [tf]

FOR SALE, CHEAP—About thirteen kegs cut nails, sizes two to five inches. Box 642, HARDWARE AND METAL, Toronto. [31]

SITUATIONS VACANT.

TINSMITHS WANTED — First-class tinsmiths wanted for points west of Winnipeg; must be good mechanics capable of taking charge of a metal department; thorough knowledge of furnace work necessary. Pease Waldon Co., Winnipeg, Man. [tf]

WANTED — A salesman familar with plumbing supplies and gas ranges, to represent a Canadian manufacturer direct to the wholesale and retail trade; must have acquaintance with the trade; best of references will be expected; give full information concerning experience, acquaintance with the trade, salary expected; good position for the right person. Only those who can fill the above qualifications need apply Box 639, HARDWARE AND METAL, Toronto.

WANTED—Hardware clerk, experienced, who can keep stock, and is willing to do so; sober and active; state age, experience and salary expected at start. Hose & Canniff, Kenora, Ont. [32]

SITUATIONS WANTED.

SITUATION wanted as master mechanic or chief engineer by man of 22 years' experience as a mechanic; can give A1 reference as to ability; strictly temperate. Box A, HARDWARE AND METAL, Toronto. [32]

GERMAN (31), 14 years' commercial experience desires situation; perfect knowledge of tools, hardware, fittings of every description, wooden goods, bar iron, steel, metals, also bookkeeping, shorthand (English and German) typewriting, storekeeping; at liberty on about October 1st. Address 0783 care of Messrs. Deacon's, Leadenhall Street, London, E.C.

COMMERCIAL gentleman with nine years' trade connection with ironmongers, architects and public institutions in Great Britain desires position as representative of a Canadian manufacturing firm. Box X, HARDWARE AND METAL, Montreal.

HARDWARE Salesman or Clerk, seven years' experience, desires situation. Abstainer, best of references; position in the West preferred. A. Tilley, Brantford, Ont. (31)

WANTED (by Englishman) position in Hardware store, experience in all branches, bookkeeper, would invest $400 to $500. Address, Box 646 HARDWARE AND METAL, Toronto. (31)

WANTED.

OFFICE space wanted by manufacturer in Toronto; state location and terms. Box 645, HARDWARE AND METAL, Toronto. [32]

THE WANT AD.

The want ad. has grown from a little used force in business life, into one of the great necessities of the present day.

Business men nowadays turn to the "want ad" as a matter of course for a hundred small services.

The want ad. gets work for workers and workers for work.

It gets clerks for employers and finds employers for clerks. It brings together buyer and seller, and enables them to do business, though they may be thousands of miles apart.

The "want ad." is the great force in the small affairs and incidents of daily life.

been awarded to J. Stewart & Co., Winnipeg, agents for the Canadian Westinghouse Co.

The council of the town of Morden, Man., have awarded the contracts for the municipal electric lighting plant. The steam plant will be installed by the Robb Engineering Co., Amherst, N.S., and the electric portion by the Allis-Chalmers-Bullock, Limited.

Railroad Construction.

The C.P.R. will erect a coal handling plant at Fort William, Ont., to cost $1,250,000.

Track-laying will be commenced on the Mackenzie & Mann line at Garneau Junction, Que.

The new coal chute for the C.P.R. at Lethbridge is completed and giving complete satisfaction.

The Canadian Northern line from St. Jerome to Montford, Que., will be in operation this fall.

The C.P.R. will immediately take steps to double-track their line from Brandon to Winnipeg.

Saskatoon and Goose Lake, Man., have petitioned Mackenzie & Mann for a branch line between those towns.

The Sydney and Glace Bay Railway, Sydney, N.S., will erect their central power station at Dominion No. 4.

SIGHT FOR SHOOTING AT NIGHT.

The development of accuracy in shooting at night has received a double impetus in the British service. A new sight has been adopted, and is being manufactured and issued with all possible dispatch. This sight, a telescopic pattern, is defined by electric light for night work. It has been extensively tried under all conditions and has proved a signal success. Quite recently, also, a greatly improved thirty-six-inch searchlight has supplanted the regulation twenty-four-inch light, which is capable of defining an enemy at over double the distance of its predecessor. Owing to the excellent training afforded by the gunnery branch, naval gunners can now with the aid of their latest sight and searchlight depend on disabling the enemy at from 3,000 to 4,000 yards on a favorable night.

ELECTRICAL SHOW.

A big electrical show will be held in the Drill Hall, at Montreal, from Sept. 2 to 14 next. Space has already been reserved by prominent firms, including the Canadian General Electric, the Allis-Chalmers-Bullock, the Canadian Westinghouse Co., the Bell Telephone Co., the Montreal Light, Heat & Power Co., the Crocker-Wheeler Co., of New York City, the Wagner Electric Co., of St. Louis, and the Packard Electric Co., of St. Catharines. Without doubt this display will be very attractive and interesting. The show will be essentially of an electrical character, and various exhibitions will be given of the uses to which electricity can be put. The convention of the Electrical Association, will be held in Montreal, on Sept. 11, 12, 13.

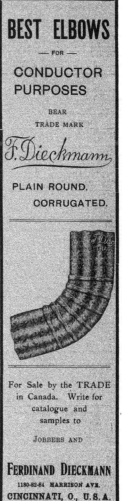

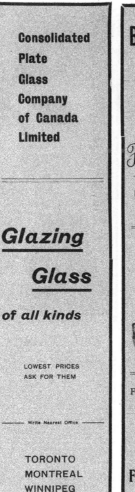

News of the Paint Trade

ISLAND CITY PAINT WORKS.

Since the large extensions were made in their works last year P. D. Dods & Co., manufacturers of paints and varnishes, Montreal, have transacted a large volume of business, and the demand for their brands of paint are steadily increasing, so much that even now with the enlarged plant they find it difficult to meet the demand. The best evidences of progress are growth and expansion, for they go hand in hand. From the day that "Island City" brands were first put on the market the popularity and demand for them has steadily grown, and the factories have steadily expanded. Ten years ago when this paint manufactory was in its infancy its dimensions were insignificant, 30 feet frontage and 30 feet depth. To-day it occupies a frontage of 250 feet and a depth of 130 feet.

The strongest factor in the upbuilding of an industry is convenience of location. Proximity to railroads and steamships means economy of time and economy in expenses. Few firms could be more conveniently situated than P. D. Dods & Co. A railroad siding runs past their front doors, and but a block distant is the Lachine Canal, thus affording to the company excellent facilities for transportation.

The Island City Paint Co. claim to have the largest and most complete pulp-color and dry-color plant in Canada. All the latest improvements have been adopted in their process both for economizing manual help and for perfecting the products. Very little pulp-color is now being made by this company as a small demand exists for it, so that a large percentage of the pulp-color is manufactured into dry-colors, for which there is always a steady demand.

P. D. Dods & Co. claim for themselves the discovery of a paint which is so impervious to rain and other forms of moisture as to prevent any iron or steel material to which the paint is applied, from rusting. This paint has proved a great success and has become popular amongst structural steel manufacturers. It is known by the brand "Anti-Rust" paint.

The company has realized the value of branch establishments and have them extending across the Dominion—one in Toronto at 145 Church Street, recently partially destroyed by fire; another in Winnipeg at 328 Smith Street, and their Pacific coast branch in Vancouver. The Henry Darling Co. have their agency in Vancouver.

PAINT AND OIL GOSSIP.

Alexandre Cooker and Aldemars Cooker, St. Poly-Carpe, Que., have been registered to carry on a painting business. The new firm will be known as A. Cooker & Sons.

T. W. Milo, Kingston, Ont., who has been running a painting and livery business, has sold the livery part of his establishment to A. McIlquhan.

Frank Ruelens and Robert Lansloot, Montreal, (Maisonneuve) have been registered to carry on business as painters under the firm name of F. Ruelens & Co.

PAINT ENDURANCE

There is a lot of paint on the market which looks all right when it's first put on. But the beauty of

NEW ERA PAINT

is that IT LASTS. It looks fine when it's first put on, and it KEEPS ON LOOKING FINE. That's why hardwaremen find it profitable. It pleases customers—same thing with all the paints and varnishes made by

STANDARD PAINT AND VARNISH WORKS CO., LIMITED
WINDSOR - ONTARIO

IT IS OFTEN EASY TO PERSUADE

your customer that the best is the cheapest.

JAMIESON'S
Floor Enamel

while more expensive than ordinary floor paint, is so much better for the floor that it should not be difficult to induce your customer to try it.

It Dries Hard
Is Never Sticky
Gives a Varnish Gloss
Stands Repeated Washing
and Wears Like Iron

You can make more profit on **Floor Enamel** than you can on floor paint.

Write for show cards and prices to

R. C. JAMIESON & CO.
LIMITED
16-30 Nazareth St., MONTREAL.

HEADQUARTERS
FOR
Window Glass,
Paints, Oils,
Fine Varnishes

SANDERSON PEARCY & CO., Ltd.
Toronto, Canada

STOVE PIPE

SCHEIP'S PATENT NESTED
25 lengths in crate

We also manufacture a full line of Cooper's Elbows.

The Best on the Market.

This is the only perfect working Knock Down Pipe made, will not telescope.

E. T. WRIGHT & CO., HAMILTON, ONT.

GLAZIERS' DIAMONDS OF PROVED WORTH,
Having a Century's Reputation for Reliability.
MADE BY
A. SHAW & SON, London, Eng.
CANADIAN AGENT
GODFREY S. PELTON, - 201 Coristine Building, Montreal

Paint and Oil Markets

TORONTO.

Toronto, August 2.—Business continues dull throughout the paint and oil markets here, though there has been a slight increase in the volume of trade during the past two or three days. Orders continue to arrive for paints and oils in sorting quantities, but, generally speaking, the market is quiet. However, as many members of jobbing staffs, as well as heads of departments, are still away on vacations, there is plenty of business to keep the reduced staffs of the local supply houses busy.

White Lead—The demand is good for the end of July, when jobbers do not count on disposing of much. Prices continue firm and show no disposition to change. Genuine pure white lead is quoted at $7.65, and No. 1 is held at $7.25.

Red Lead—The market continues a trifle dull, though a fair business is being done in sorting quantities. Present quotations are as follows: Genuine, in casks of 500 lbs., $6.25; ditto, in kegs of 100 lbs., $6.75; No. 1, in casks of 500 lbs., $5; ditto, in kegs of 100 lbs., $5.50.

Paris Green—This commodity at present furnishes the most interesting feature of the market. The demand for Paris green, which almost invariably falls off about the second or third week in July, is still strong and a supply of pound packages is no longer available. Prices remain unchanged, but the best selling cask goods of Paris green are also becoming scarce and only the fact that the demand is likely to slacken before long keeps the price from advancing. We still quote this commodity at the following prices: Canadian Paris green, 29½c. base; English Paris green, 30½c.

Petroleum—The demand is steady and will not increase until the nights begin to grow longer. Present prices are: Prime white, 13c.; water white, 14½c.; Pratt's astral, 18c.

Shellac—The market is dull, though the quantity sold for this time of the year is good: Pure orange, in barrels, $2.70; white, $2.82½ per barrel; No. 1, (orange), $2.50.

Turpentine—There has been no change in the price since the drop of last week and turpentine still remains a little dull. The price has now reached a trifle lower figure than was at any time during the month of July last year. In the month of August last year the prices began to advance and during the month an advance of 4c. per gal. took place. There is no certainty that this will be repeated this year, but as the price is now lower than was anticipated at the beginning of the season, it would not be surprising if the market should take a firmer tone again this August. In rosins there has not been much change during the week. The extra fine grades have suffered a slight decline, while the cheap grades have advanced at the point of production. The local market is still quiet and the following prices are still quoted: Single barrels 80c.; two barrels, and upwards, 79c. f.o.b. point of shipment, net 30 days;, less than barrels, 5c. advance. Terms: 2 per cent., 30 days.

Linseed Oil—The market in Great Britain a week ago seemed to indicate some considerable decline, but during the past few days prices have again advanced and if any great demand should spring up in the next two weeks prices may stiffen, as stocks are light, and it will yet be six weeks or possibly two months before Canadian crushers have new seed to work on. However, just at present the local market is weak and a two cent drop has taken place since last week. We now quote: Raw, 1 to 3 barrels, 65c; 4 barrels and over, 64c.

Add three cents to this price for boiled oil f.o.b. Toronto, Hamilton, London and Guelph, 30 days.

For additional prices see current market quotations at the back of the paper.

MONTREAL.

Montreal, August 2.—The feature of the past week in local paint and oil circles has been the multitude of sorting orders received for Paris green, especially from the Eastern Provinces. Ontario appears to have had its quota, and it looks as if all the Paris green in the hands of the makers will be required for the eastern districts.

The demand for general mixed paints has been very well maintained, and although the weather has been unsettled, a great deal of work is being done. Industrial paints are experiencing a strong demand. The condition of the trade on the whole is quite satisfactory. Transportation facilities are steadily improving, although there are even yet glaring instances of delay and incapacity.

Turpentine—This continues at the lowest point in price that has been touched for some years, and consumers are somewhat inclined to recover, although the outlook is very uncertain. Dealers report good trade. Owing to a controversy which is supposed to be in progress in the south, much uncertainty exists, and a further drop of 3 cents has occurred in the prices. Single barrels are now quoted at 80 cents.

Linseed Oil—Continues to weaken and prices have been marked down 2 cents. We now quote: Raw, 1 to 4 bbls., 62c; 5 to 9, 61c; boiled, 1 to 4 bbls., 65c; 5 to 9 bbls., 64c.

Ground White Lead—A moderate demand prevails. Prices are unchanged: Government standard, $7.50; No. 1, $7; No. 2, $6.75; No. 3, $6.35.

Dry White Zinc—Situation is unchanged. Demand is moderate with adequate supplies: V.M. Red Seal, 7½c; Red Seal, 7c; French V.M., 6c; Lehigh, 5c.

White Zinc Ground in Oil—The call for this is steady. Prices are firm: Pure, 8½c; No. 1, 7c; No. 2, 5⅜c.

Putty—Manufacturers are busily engaged endeavoring to keep pace with the demand. No change is made in the prices: Pure linseed oil, $1.85 bulk; in barrels, $1.60; in 25-lb. irons, $1.90; in tins, $2; bladder putty, in barrels, $1.85.

Red Lead—A fair demand exists. We continue to quote: Genuine red lead, in casks, $6.25; in 100-lb. kegs, $6.50; in less quantities at $7.25 per 100 lbs. No. 1 red lead, casks, $6; kegs, $6.25, and smaller quantities, $7.

Paris Green—The volume of business being done in this is steadily increasing. Eastern districts are taking a large percentage of the shipments. Prices are firm and unchanged.

Gum Shellac—No change has occurred in the situation: Fine orange, 60c per lb.; medium orange, 55c per lb.; white (bleached), 65c.

Shellac Varnish—The demand for all lines of shellac and varnish is steady. Prices are firm: Pure white bleached shellac, $2.80; pure orange, $2.60; No. 1 orange, $2.40.

BRUSHES IN CARTONS.

Scrub brushes, shoe brushes and stove brushes put in half-dozen cartons make a very neat way for placing brushes on the shelves, besides helping to make good window displays. The United Factories are placing a large number of their lines in this more convenient way to handle brushes of all kinds.

STANDARD SPECIFICATION FOR JAPAN.

Robert Job, chemist of the Pennsylvania railroad, after a series of tests and analyses of japan, has drawn up the following standard specifications:

(1) When equal parts by weight of japan and pure turpentine are thoroughly mixed and poured over a slab of glass, which is then placed nearly vertical at a temperature of 100 degrees Fahr., with free access to air, but not exposed to draught, the coating should be hard and dry, neither brittle nor sticky in not exceeding 12 hours. (2) When thoroughly mixed with ordinary pure linseed oil at the ordinary temperature in proportions of 5 per cent. weight of japan to 95 per cent. weight of raw linseed oil, no curdling should result, nor any observable separation or settling on standing. (3) When the above mixture is flowed over a slab of glass, which is then placed nearly vertical at a temperature of 100 degrees Fahr., with free access to air, but not exposed to draught, the coating shall dry throughout, neither brittle nor sticky, in not exceeding two hours. (4) When 5 c.c. of the japan are poured into 95 c.c. of pure turpentine ot the ordinary temperature and thoroughly shaken, a clear solution shall result without residue on standing an hour. (5) After evaporation of the turpentine the solid residue must be tough and hard, and must not "dust" when scratched with a knife. (6) Benzine or a mineral oil of any kind will not be permitted.

MASTER PAINTERS' CONVENTION.

The sessions of the convention of the Master Painters' and Decorators' Association of Canada, held at London, Ont. last week, were very well attended. Many interesting and educative papers were read and the keen discussion which followed each showed the deep interest taken in the subjects by those present.

The convention concluded its sessions on Friday and officers for the year were elected as follows: President, H. R. Reynolds, Guelph; vice-president, Ald. Richard Booth, London; secretary-treasurer, Alex. McKenzie, Hamilton. The delegates to the international convention are Messrs. Reynolds, Booth and McKenzie.

The choosing of the place for the next convention was left with the executive.

FACTS ABOUT GASOLINE.

In view of the scare which the large number of recent gasoline explosions have caused among Canadian sailing men regarding the use of gasoline on yachts, a few facts concerning the conditions under which gasoline will explode will be interesting and perhaps reassuring just at this time.

A gasoline tank rarely explodes. It cannot unless it contains gasoline vapor and air in explosive proportions, which latter condition is almost never present. It does not explode because it contains too little air or too much gasoline. Explosive gasoline vapor consists of ten per cent. gasoline and ninety per cent. air. The accidental formation of this vapor might not occur in years as the proportions must be more or less exact. Even if a tank of gasoline were to burst from heat applied to its exterior, the confined heavy gas would not explode if in contact with flame or fire, but would burn instead. A tank of gasoline with no vent could do considerable damage were it to burst and throw burning oil and flaming gas about, but 1,000 gallons of gasoline in a vessel's bilges would not be so dangerous from explosion as a hundredth of that amount. The larger quantity would burn rapidly, while the smaller would be sufficient, if mixed with the proper amount of air, to demolish utterly almost any boat.

REMOVING PAINT FROM IRON TANK.

To remove paint from an iron tank, take lime and mix with common lye into a thick paste by the addition of water, says Engineering Review, and apply over the surface of the metal with a mason's trowel to a thickness of about ¼ in. After allowing the mixture to remain a short time, wash off with a hose and most of the old paint will be entirely removed, the remainder being easily scraped off with a scraper. If the tank contains several coats of dried paint, two or three applications will be necessary before the entire surface is clean.

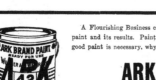

CURRENT MARKET QUOTATIONS.

August 2, 1907

These prices are for such qualities and quantities as are usually ordered by retail dealers on the usual terms of credit, the lowest figures being for larger quantities and prompt pay. Large cash buyers can frequently make purchases at better prices. The Editor is anxious to be informed at once of any apparent errors in this list, as the desire is to make it perfectly accurate.

—

METALS.

ANTIMONY.

Cookson's ....per lb.	0 17
Hallett's	0 16½

BOILER PLATES AND TUBES.

Plates, ¼ to ½ inch, per 100 lb.	2 40	2 50
Heads, per 100 lb	2 60	2 75
Tank plates 3-16 inch	2 60	2 10
Tubes per 100 feet, 1½ inch ..	8 25	8 50
" " 2 " ..	9 10	9 10
" " 2½ " ..	10 50	11 00
" " 3 " ..	12 00	12 50
" " 3½ " ..	15 00	16 00
" " 4 " ..	19 25	20 00

BOILER AND T.K. PITTS.

Plain tinned and Spun, 25 per cent. off list.

BABBIT METAL.

Canada Metal Company—Imperial, genuine 60c; Imperial Tough, 60c; White Brass, 50c; Metallic, 35c; Harris Heavy Pressure, 25c; Hercules, 25c; White Bronze, 15c; Star Frictionless, 14c; Alluminoid, 10c.; No. 4, 9c. per lb.

James Robertson Co.—Extra and genuine Monarch, 60c.; Crown Monarch, 50c; No 1 Monarch, 40c.; King 30c; Fleur-de-lis, 20c; Thurber, 15c; Philadelphia, 12c; Canadian, 10c; hardware, No. 1, 15c; No. 2, 12c.; No. 3, 10c. per lb.

BRASS.

Rod and Sheet, 14 to 30 gauge, 25 p c. advance.

Sheets, 12 to 14 in.	0 30
Tubing, base, per lb 5-16 to 2 in	0 33
Tubing, ½ to 3-inch, iron pipe size..	0 31
" 1 to 3-inch, seamless..	0 36

Copper tubing, 6 cents extra.

COPPER.

		Per 100 lb.
Casting ingot	23 00	23 50
Cut lengths, round, bars, ½ to 2 in...		35 00
Plain sheets, 14 oz..........		36 00
Plain, 16 oz., 14x48 and 14x60		35 00
Tinned copper sheet, base		38 00
Planished base............		43 00
Braziers' (in sheets), 4x6 ft., 25 to 30 lb. each, per lb., base..	0 34	0 35

BLACK SHEETS.

	Montreal.	Toronto
8 to 10 gauge	2 70	2 75
12 gauge..................	2 70	2 75
14 "	2 50	2 60
17 "	2 50	2 60
18 "	2 50	2 60
20 "	2 50	2 60
22 "	2 55	2 65
24 "	2 55	2 70
26 "	2 65	2 85
28 "	2 70	3 00

CANADA PLATES.

Ordinary, 52 sheets	2 75	3 05
All bright "	3 75	4 05

Galvanized—	Dom.Crown.	Ordinary.
18x24x52	4 45	4 35
60..................	4 70	4 60
20x28x80	8 90	8 70
... .	9 40	9 20

GALVANIZED SHEETS.

B W. gauge	Queen's Head	Fleur-de-Lis	Gordon Crown	Colborne Crown Gorbal's Best
16-20 ..	3 95	3 80	3 95	....
22-24 ..	4 20	4 05	4 00	4 05
26........	4 45	4 30	4 40	4 30
28..... ..	4 70	4 55	4 60	4 55

Less than case lots 10 to 25c. extra.

Apollo Brand.

24 guage, American	3 85
26 " "	4 10
28 " "	4 55
10¾ oz.	4 85

25c. less for 1,000 lb. lots.

IRON AND STEEL.

	Montreal.	Toronto.
Middlesboro, No. 1 pig iron..	21 50	24 50
Middlesboro, No. 3 pig iron	20 50	23 50
Summerlee, " ..	25 50	24 50
" special " .	24 50	....
" soft " ..	24 00	....
Carron...	26 00	....
Carron Special	24 50	....
Carron Soft	24 00	....
Clarence, No. 3	21 50	23 50
Glengarnock, No. 1	....	27 00
Midland, Londonderry and Hamilton, off the market but quoted nominally at ..	....	26 00
Radnor, charcoal iron	32 00	34 00
Common bar, per 100 lb......	2 20	2 30
Forged iron "	2 45	..
Refined " "	2 60	2 70
Horseshoe iron "	2 60	2 70
Hoop steel, 1½ to 3 in. base...	....	2 80
Sleigh shoe steel "	2 25	2 30
Tire steel	2 40	2 50
Best sheet cast steel........	...	0 12
B. K. Morton "Alpha" high speed.		0 65
Colonial black diamond......	0 08	0 09
Sanderson's	0 08	0 45
Jessop's standard	0 14	0 14
" ark high speed	0 60	0 60
" Leonard brand	0 08	0 08
Jonas & Colver's tool steel....	0 10	0 20
Jowett & Sons B.P.L. tool steel	0 10½	0 11

INGOT TIN.

Lamb and Flag and Straits— 56 and 28-lb. ingots, 100 lb.	$44 00	$45 00

TINPLATES.

Charcoal Plates—Bright

M.L.S., Famous (equal Bradley)	Per box.
I C, 14 x 20 base	$6 50
IX, 14 x 20 "	8 00
IXX, 14 x 20 base	9 50

Raven and Vulture Grades—

I C, 14 x 20 base	5 00
I X "	6 00
I X X "	7 00
I X X X "	8 00

"Dominion Crown Best"—Double Coated, Tissued.

		Per box.
I C, 14 x 20 base..........	5 50	5 75
I X, 14 x 20 "	6 50	6 75
IXX x 20 "	7 50	7 75

"Allaway's Best"—Standard Quality.

I C, 14 x 20 base..........	4 65	5 00
I X, 14 x 20 "	5 40	5 75
I XX, 14 x 20 "	6 15	6 50

Bright Cokes.

Bessemer Steel—

I C., 14 x 20 base	4 25	4 35
20x28, double box	8 50	8 70

Charcoal Plates—Terne

Dean or J. G Grade—

I C., 20x28, 112 sheets	7 25	8 00
IX., Terne Tin	...	9 50

Charcoal Tin Boiler Plates.

Cookley Grade—

X X, 14x56, 50 sheet bxs. / " 14x60, " / " 14x65, "	7 50

Tinned Sheets.

72x30 up to 24 gauge......	8 50
" " 26 "	9 00

LEAD.

Imported Pig, per 100 lb......		5 50
Bar, "	5 75	6 00
Sheets, 2½ lb. sq. ft., by roll		6 50
Sheets, 3 to 6 lb. "		6 25

Cut sheets ½c. per lb., extra.

SHEET ZINC.

5-cwt. casks	7 75	8 00
Part casks....................	8 00	8 25

ZINC SPELTER.

Foreign, per 100 lb	6 75	7 00
Domestic	6 50	6 75

COLD ROLLED SHAFTING.

9-16 to 11-16 inch..............	0 06
¾ to 1 7-16 "	0 05½
1 7-16 to 3 "	0 05

30 per cent.

OLD MATERIAL.

Dealers buying prices:

	Montreal	Toronto
Heavy copper and wire, lb.	0 17	0 16½
Light copper	0 16	0 15
Heavy red brass....	0 15½	0 15
" yellow brass	0 12	0 12
Light brass................	0 08½	0 08
Tea lead	0 03¾	0 04
Heavy lead................	0 04	0 04¼
Scrap zinc	0 03¾	0 04
No. 1 wrought iron	15 50	11 50
" 2 " "	6 00	6 00
Machinery cast scrap ...	17 00	16 50
Stove plate................	13 00	12 00
Malleable and steel	8 00	8 00
Old rubbers	0 10½	0 10
Country mixed rags, 100 lbs.	1 00	1 25

—

PLUMBING AND HEATING

BRASS GOODS, VALVES, ETC.

Standard Compression work, 57½ per cent.
Cushion work, discount 40 per cent.
Fuller work, 70 per cent.
Flatway stop and stop and waste cocks, 60 per cent.; roundway, 55 per cent.
J.M.T. Globe, Angle and Check Valves, 45; Standard, 55 per cent
Kerr standard globes, angles and checks, special, 42½ per cent.; standard, 47½ p c.
Kerr Jenkins' disc, copper-alloy disc and heavy standard valves, 40 per cent.
Kerr steam radiator valves, 60 p c., and quick-opening hot-water radiator valves, 60 p c.
Kerr brass, Weber's straightway valves, 40; straightway valves, I B & M, 60.
J. M. T. Radiator Valves 50; Standard, 60; Patent Quick-Opening Valves, 65 p.c.
Jenkins' Valves—Quotations on application to Jenkins' Bros., Montreal.

No 1 compression bath cock......net	2 00
No. 4 " " ...... "	1 90
No 7 Fuller's .... "	2 25
No. 4½, " .... "	2 35
Patent Compression Cushion, basin cock, hot and cold, per dcz., ...	$16.20
Patent Compression Cushion, bath cock, No. 2208	25

Square head brass cocks, 50; iron, 60 p.c.
Thompson Smoke-test Machine $25.00

BOILERS—COPPER RANGE.

Copper, 30 gallon, $33, 15 per cent.

BOILERS—GALVANIZED IRON RANGE.

30-gallon, Standard, $5; Extra heavy, $7.75

BATH TUBS.

Steel clad copper lined, 15 per cent.

CAST IRON SINKS.

16x24, $1; 18x30, $1; 18x36, $1.30.

ENAMELED BATHS, ETC.

List issued by the Standard Ideal Company Jan. 3, 1907, shows an advance of 10 per cent. over previous quotations.

ENAMELED CLOSETS AND URINALS

Discount 15 per cent.

HEATING APPARATUS.

Stoves and Ranges—40 to 70 per cent.
Furnaces—45 per cent.
Registers—70 per cent.
Hot Water Boilers—50 per cent.
Hot Water Radiators—50 to 55 p.c.
Steam Radiators—50 to 55 per cent.
Wall Radiators and Specials—50 to 55 p.c.

LEAD PIPE

Lead Pipe, 7c. per pound, net list.
Lead waste, 8c. per pound, net list.
Caulking lead, 6½c. per pound.
Traps and bends, 40 per cent.

IRON PIPE.

Size (per 100 ft.)	Black.		Galvanized
¼ inch........	2 35	¼ inch........	3 20
⅜ "	2 35	⅜ "	3 20
½ "	2 90	½ "	3 75
¾ "	3 90	¾ "	5 00
1 "	5 80	1 "	7 25
1¼ "	7 65	1¼ "	9 90
1½ "	9 20	1½ "	11 90
2 "	12 25	2 "	15 80
2½ "	20 10	2½ "	26 00
3 "	26 75	3 "	34 00
3½ "	34 25	3½ "	42 75
4 "	39 00	4 "	48 60

Malleable Fittings—Canadian discount 30 per cent.; American discount 25 per cent.
Cast Iron Fittings 57½; Standard bushings 57½; headers, 57½; flanged unions 57½, malleable bushings 55; nipples, 70 and 10; malleable lipped unions, 55 and 5 p.c.

SOIL PIPE AND FITTINGS

Medium and Extra heavy pipe and fittings, up to 6 inch, 60 and 10 to 70 per cent.
7 and 8-in. pipe, 40 and 5 per cent.
Light pipe, 50 p.c.; fittings, 50 and 10 p.c.

OAKUM.

Plumbers ....per 100 lb... ..	4 50	5 00

STOCKS AND DIES.

American discount 25 per cent.

SOLDERING IRONS.

½-lb. to 1½ .... per lb.	0 45½	0 48
2-lb. or over .... "	0 42½	0 46

SOLDER.

Per lb.	Montreal	Toronto
Bar, half-and-half, guaranteed	0 25	0 26
Wiping........................	0 22	0 23

—

PAINTS, OILS AND GLAS

BRUSHES

Paint and household, 70 per cent.

CHEMICALS.

	In casks per lb.
Sulphate of copper (bluestone or blue vitrol)	0 09
Litharge, ground	0 06
" flaked	0 06½
Green copperas (green vitrol)........	0 01
Sugar of lead........	0 08
Lump olive..........	0 01¾

COLORS IN OIL.

Venetian red, 1-lb. tins pure.	....	0 08
Chrome yellow "	....	0 15
Golden ochre "	....	0 10
French " "	....	0 08
Marine black "	....	0 04½
Chrome green "	....	0 09
French permanent green "	....	0 13
Signwriters' black "	....	0 15

GLUE.

Domestic sheet 0 10 0 10½
French medal.. .... 0 12 0 12½

PARIS GREEN.

	Berger's	Canadian
600-lb. cask	....	0 27½
250-lb. drums	0 28	0 27¾
100-lb. "	0 29½	....
50-lb. "	0 29½	0 28½
1-lb. pkgs, 100 in box	0 30½	0 29½
½-lb. "	0 32½	0 31½
1-lb. tins, 100 in box	0 31½	0 30½
½-lb. tins	0 33½	0 32

PARIS WHITE.

In bbls 0 90

PIGMENTS.

Orange mineral, casks........ .... 0 08
" " 100-lb. kegs 0 08½

PREPARED PAINTS.

Barn (in bbls.).............. 0 65 0 90
Sherwin-Williams paints, ... 1 55
Canada Paint Co.'s pure...... .... 1 40
Standard P. & V. Co.'s "New Era." 1 30
Benj. Moore Co.'s "Ark" B'd 1 25
" British Navy deck 1 50
Brandram-Henderson's "English" 1 45
Ramsay's paints, Pure, per gal. 1 20
" Thistle, " 1 10
Martin-Senour's 100 p.c. pure, ... 1 55
Senour's Floor Paints......... .... 1 35
Jamieson's "Crown and Anchor" 1 20
Jamieson's floor enamel...... .... 1 50
"Island City" paint............ .... 1 25
Sanderson, Pearcy's, pure 1 20
Robertson's pure paints...... .... 1 20

PUTTY.

Bulk in bbls......................... 1 60
Bladders in bbls.................... 1 85
25-lb. tins.......................... 1 90
Bladders in bulk or tins less than 100 lb. 2 00
Bulk in 100-lb. irons............... 1 80

SHINGLE STAINS.

In 5 gallon lots 0 85 0 90

SHELLAC.

White, bleached 2 65
Fine orange....... .... 2 60
Medium orange.............. .... 2 55

TURPENTINE AND OIL.

Prime white petroleum 0 13
Water white " 0 14½
Pratt's astral " 0 18
Castor oil....... .... 0 08 0 10
Gasoline 0 22½
Benzine, per gal 0 17 0 20
Turpentine, single barrels 0 79 0 80
Linseed Oil, " raw.. 0 63 0 65
" " boiled 0 66 0 68

WHITE LEAD GROUND IN OIL. Per 100 lbs

Canadian pure 7 15 7 50
No. 1 Canadian 6 80 7 15
Munro's Select Flake White.. .... 7 65
Elephant and Decorators' Pure 7 65
Monarch 7 40
Standard Decorator's 7 15
Essex Genuine 6 80
Brandram's B. B. Genuine... .... 8 70
"Anchor," pure 7 40
Ramsay's Pure Lead 7 00
Ramsay's Exterior 6 65
"Crown and Anchor," pure... .... 7 25
Sanderson Pearcy's 7 40
Robertson's C.P., lead........ .... 7 20

RED DRY LEAD.

Genuine, 560 lb. casks, per cwt 6 25
Genuine, 100 lb. kegs, " 6 50
No. 1, 560 lb. casks, per cwt 6 00
No. 1, 100 lb. kegs, per cwt 6 25

WINDOW GLASS

Size United inches.	Star	Double Diamond
Under 26	$4 25	$6 25
26 to 40	4 65	6 75
41 to 50	5 10	7 50
51 to 60	5 35	8 50
61 to 70	5 75	9 75
71 to 80	6 25	11 00
81 to 85	7 00	12 50
86 to 90	....	15 00
91 to 95	....	17 50
96 to 100	....	20 50
101 to 105	....	24 00
100 to 110	....	27 50

Discount—16-oz., 25 per cent.; 21-oz. 30 per cent. per 100 feet. Broken boxes 50 per cent.

WHITING.

Plain, in bbls 0 70
Gilders bolted in bands........ .... 0 90

WHITE DRY ZINC.

Extra Red Seal, V.M... ... 0 07½ 0 08

WHITE ZINC IN OIL.

Pure, in 25-lb. irons 8 08½
No. 1, " " 8 07
No. 2, " " 0 05½

VARNISHES.

Per gal cans.

Carriage, No. 1 .... 1 50
Pale durable body........... .... 3 50
" hard rubbing 3 00
Finest elastic gearing 3 00
Elastic oak 1 50
Furniture, polishing 2 00
Furniture, extra.............. .... 1 15
" " No. 1........ .... 0 90
" " union....... .... 0 80
Light oil finish................ .... 1 40
Gold size japan 1 80
Brown japan 0 95
No. 1 brown japan 8 95
Baking black japan........... .... 3 35
No. 1 black japan 0 90
Benzine black japan 0 70
Crystal Damar................. .... 2 80
No. 1 " 2 50
Pure asphaltum................ .... 1 40
Oilcloth....................... .... 1 50
Lightning dryer 0 70
Elastilite varnish, 1 gal. can, each.. 2 00
Granitine floor varnish, per gal 2 50
Maple Leaf coach enamels; size 1, 1 20
Sherwin-Williams' kopal varnish, gal., 2 50
Canada Paint Co's sun varnish 2 00
"Kyanize" Interior Finish.... .... 2 40
"Flint-Lac," coach. 1 80
B H. Co's "Gold Medal," in cases 2 00
Jamieson's Copaline, per gal. 2 00

BUILDERS' HARDWARE.

BELLS.

Brass hand bells, 60 per cent.
Nickel, 55 per cent.
Gongs, Sargeant's door bells.. 5 50 8 00
American, house bells, per lb. 0 35 0 40
Peterboro' door bells, 37½ and 10 off new list.

BUILDING PAPER, ETC.

Tarred Felt, per 100 lb. 2 25
Ready roofing, 2-ply, not under 45 lb. per roll 1 00
Ready roofing, 3-ply, not under 65 lb., per roll........................... 1 25
Carpet Felt ....per ton 60 00
Heavy Straw Sheathing.......per ton 40 00
Dry Surprise...................... 0 45
Dry Sheathing...... per roll, 400 sq. ft. 0 40
Tar " ...... " 400 " 0 50
Dry Fibre.......... " 400 " 0 55
Tarred Fibre ...... " 400 " 0 65
O. K. & I. X. L.... " 400 " 0 70
Resin-sized .... " 400 " 0 45
Oiled Sheathing.... " 600 " 1 00
Oiled " .... " 400 " 0 70
Roof Coating, in barrels......per gal. 0 17
Roof " small packages " 0 25
Refined Tar...............per barrel 3 00
Coal Tar .... " 4 00
Coal Tar, less than barrels...per gal. 0 15
Roofing Pitch....... per 100 lb. 0 80 0 90
Slater's felt ....per roll 0 70
Heavy Straw Sheathing f. o. b. St. John and Halifax................ 42 50

BUTTS.

Wrought Brass, net revised list.
Wrought Iron, 70 per cent.
Cast Iron Loose Pin, 60 per cent.
Wrought Steel Fast Joint, and Loose Pin, 70 per cent.

CEMENT AND FIREBRICK.

Canadian Portland............. 2 00 2 10
Belgium..... .... 1 60 1 90
White Bros. English 1 80 2 05
"Lafarge" cement in wood 2 40
"Lehigh" cement, in wood 2 54
"Lehigh" cement, cotton sacks. 2 39
"Lehigh" cement, paper sacks 2 31
Fire brick, Scotch per 1,000 .... 27 00 30 00
" English.............. 17 00 21 00
" American, low 23 0 25 00
" " high...... 27 50 35 00
Fire clay (Scotch), net ton 4 95

Paving Blocks per 1,000.

Blue metallic, 9"x4½"x3", ex wharf .. 35 00
Stable pavers, 12"x6"x2", ex wharf .. 50 00
Stable pavers, 9"x4½"x3", ex wharf .. 36 00

DOOR SETS.

Peterboro, 37½ and 10 per cent.

DOOR SPRINGS

Torrey's Rod.........per doz. 1 75
Coil, 9 to 11 in........ " 0 95 1 65
English .... " 2 00 4 00
Chicago and Reliance Coil 25 per cent.

ESCUTCHEONS.

Discount 50 and 10 per cent., new list
Peterboro, 37½ and 10 per cent.

ESCUTCHEON PINS.

Iron, discount 40 per cent.
Brass, 45 per cent.

HINGES.

Blind, discount 50 per cent.
Heavy T and strap, 4-in., per lb. net.. 0 06
" " 5-in., " 0 05¾
" " 6-in., " 0 05½
" " 8-in., " 0 05¼
" " 10-in. and larger.. 0 05
Light T and strap, discount 65 p.c.
Screw hook and hinge—
under 12 in.... per 100 lb. 4 65
over 12 in........ " 3 65
Crate hinges and back flaps, 65 and 5 p.c.
Hinge hasps 65 per cent.

SPRING HINGES.

Spring, per gro., No. 5. $17.50 No. 10, $18; No. 20, $10 80; No. 120, $20; No. 51, $10; No 50, $27 50
Chicago Spring Butts and Blanks, 12½ per cent.
Triple End Spring Butts, 30 and 10 per cent.
Chicago Floor Hinges, 37½ and , off.
Garden City Fire House Hinges, 12½ p.c.
"Chief" floor hinge, 50 p c.

CAST IRON HOOKS.

Bird cage............. per doz. 0 50 1 10
Clothes line, No. 61.. " 0 00 0 70
Harness............... " 0 60 12 00
Hat and coat .... per gro. 1 10 10 00
Chandelier ....per doz. 0 50 1 00
Wrought hooks and staples—
2 x 5 .... per gross 2 65
5-16 x 5 .... " 3 30
Bright wire hooks, 60 p c.
Bright steel gate hooks and staples, 40 p.c.
Cresent hat and coat wire, 60 per cent.
Screw, bright wire, 65 per cent.

KNOBS.

Door, japanned and N.P., doz 1 50 2 50
Bronze, Berlin....... per doz. 2 75 3 25
Bronze, Genuine .... " 6 00 9 00
Shutter, porcelain, F. & L. screw .... per gross 1 30 2 00
White door knobs....per doz. 2 00
Peterboro knobs, 37½ and 10 per cent.
Porcelain, mineral and jet knobs, net list.

KEYS.

Lock, Canadian 40 to 40 and 10 per cent.

LOCKS.

Peterboro, 37½ and 10 per cent
Russell & Erwin, steel rim $2.50 per doz
Eagle cabinet locks, discount 30 per cent
American padlocks, all steel, 10 to 15 per cent.; all brass or bronze, 10 to 25 per cent.

SAND AND EMERY PAPER.

B. & A. sand, discount, 35 per cent
Emory, discount 35 per cent
Garnet (Burton's) 5 to 10 per cent. advance

SASH WEIGHTS.

Sectional..........per 100 lb. 2 00 2 25
Solid .... " 1 50 1 75

SASH CORD.

Per lb....................... .... 0 31

BLIND AND BED STAPLES.

All sizes per lb............... 0 07½ 0 10

WROUGHT STAPLES.

Galvanized 2 75
Plain 2 50
Coopers', discount 45 per cent.
Poultry netting staples, discount 40 per cent.
Bright spear point, 75 per cent. discount.

TOOLS AND HANDLES.

ADZES.

Discount 22½ per cent.

AUGERS.

Gilmour's, discount 60 per cent. off list.

AXES.

Single bit, per doz. 6 00 9 00
Double bit, " 10 00 11 00
Bench Axes, 40 per cent.
Broad Axes. 25 per cent.
Hunters' Axes............... 5 50 6 00
Boys' Axes 6 25 7 00
Splitting Axes 7 00 12 00
Handled Axes 7 00 9 00
Red Ridge, boys', handled.. .... 5 75
" hunters 5 25

BITS.

Irwin's auger, discount 47½ per cent.
Gilmour's auger, discount 60 per cent.
Rockford auger, discount 50 and 10 per cent.
Jennings' Gen. auger, net list.
Gilmour's car, 47½ per cent.
Clark's expansive, 40 per cent.
Clark's gimlet, per doz 0 65
Diamond, Shell, per doz...... .... 1 00
Nail and Spike, per gross..... .... 2 25

BUTCHERS CLEAVERS

German............per doz. 7 00 9 00
American......... " 12 00 18 00

CHALK.

Carpenters' Colored, per gross 0 45 0 75
White lump..........per cwt. 0 60 0 65

CHISELS.

Warnock's, discount 70 and 5 per cent.
P. S. & W. Extra, discount, 70 per cent.

CROSSCUT SAW HANDLES.

S. & D., No. 3 ....per pair 0 13
S. & D., " 5 .... " 0 11½
S. & D., " 6 .... " 0 18
Boynton pattern.............. " 0 20

CROWBARS.

3½c. to 4c. per lb.

DRAW KNIVES.

Coach and Wagon, discount 75 and 5 per cent.
Carpenters' discount 75 per cent.

DRILLS.

Millar's Falls, hand and breast. net list.
North Bros., each set, 50c.

DRILL BITS.

Morse, discount 37½ to 40 per cent.
Standard, discount 50 and 5 to 55 per cent.

FILES AND RASPS.

Great Western .... 75 per cent.
Arcade .... 75 "
Kearney & Foot .... 75 "
Disston's....... .... 75 "
American .... 75 "
J. Barton Smith .... 75 "
McClellan .... 75 "
Eagle .... 75 "
Nicholson.... .... 66⅔ "
Globe .... 75 "
Black Diamond, 60, 10 and 5 p.c.
Jowitt's, English list, 27½ per cent.

GAUGES.

Stanley's, discount 50 to 60 per cent.
Winn's, Nos. 26 to 33 ....each 1 65 2 40

HANDLES.

Second growth ash fork, hoe, rake and shovel handles, 40 p c.
Extra ash fork hoe, rake and shovel handles, 45 p c
No. 1 and 2 ash fork, hoe, rake and shovel handles, 50 p.c
White ash whiffletrees and neckyokes, 35 p.c.
All other ash goods, 40 p c
All hickory, maple and oak goods, excepting carriage and express whiffletrees, 40 p c.
Hickory, maple, oak carriage and express whiffletrees, 45 p.c.

HAMMERS.

Maydole's, discount 5 to 10 per cent.
Canadian, discount 25 to 27½ per cent.
Magnetic tack ....per doz. 1 10 1 20
Canadian sledge .... per lb. 0 07 0 08½
Canadian ball pean, per lb. .. 0 22 0 25

HATCHETS.

Canadian, discount 40 to 42½ per cent.
Shingle, Red Ridge 1, per doz.......... 4 40
" " 2, " 4 80
Barrel Underhill...................... 5 05

HOES.

Mortar, 50 and 10 per cent.

MALLETS.

Tinsmiths'............per doz. 1 25 1 50
Carpenters', hickory, " 1 25 3 75
Lignum Vitae........ " 3 85 5 00
Caulking, each 0 60 2 00

MATTOCKS.

Canadian............ per doz. 5 50 6 00

MEAT CUTTERS.

German, 15 per cent.
American discount, 33⅓ per cent.

PICKS.

Per dozen 6 00 9 00

PLANES.

Wood bench, Canadian, 40, American, 25
Wood, fancy, 37½ to 40 per cent.
Stanley planes, $1.55 to $3 60, net list prices.

PLANE IRONS.

English ....per doz. 2 00 5 00
Stanley, 2⅝ inch, single 24c., double 39c.

PLIERS AND NIPPERS.

Button's genuine, 37½ to 40 per cent.
Button's imitation....per doz. 5 00 9 00
Berg's wire fencing.... ...... 1 72 5 50

PUNCHES.

Saddler's ....per doz. 1 00 1 85
Conductor's.......... " 3 00 15 00
Tinners', solid..........per set 0 72
" hollow......per inch 1 00

RIVET SETS.

Canadian, discount 35 to 37½ per cent.

RULES.

Boxwood, discount 70 per cent.
Ivory, discount 20 to 25 per cent

SAWS.

Atkins, hand and crosscut, 25 per cent.
Disston's Hand, discount 12½ per cent
Disston's Crosscut ....per foot 0 35 0 55
Hack, complete..........each 0 75 2 75
" frame only........each 0 50 1 25

S. & D. solid tooth circular shingle, concave and band, 50 per cent; mill and ice, drag, 30 per cent; cross-cut, 35 per cent., hand saws, butcher, 35 per cent; buck, New Century, $6 25; buck No. 1 Maple Leaf, $5 25; buck, Happy Medium, $4.25; buck, Watch Spring, $4.25; buck, common frame, $4.00.

Spear & Jackson's saws—Hand or rip, 26 in., $12 75; 24 in., $11 25, panel 18 in., $8 25; 20 in., $9; tenon, 10 in., $9.90; 12 in., $10 90; 14 in., $11 50.

SAW SETS.

Lincoln and Whiting 4 75
Hand Sets, Perfect.................. 4 00
X-Cut Sets, " 7 50
Maple Leaf and Premiums saw sets, 40 off.
S. & D. saw swages, 40 off.

SCREW DRIVERS.

Sargent's ....per doz. 0 65 1 00
North Bros., No. 30 ..per doz. 16 80

SHOVELS AND SPADES.

Canadian, discount 45 per cent.

SQUARES.

Iron, discount 20 per cent.
Steel, discount 65 and 10 per cent.
Try and Bevel, discount 50 to 52½ per cent.

TAPE LINES.

English, ass skin .....per doz. 2 75 5 00
English, Patent Leather 5 50 9 75
Chesterman's.............each 0 90 2 85
" steel ....each 0 80 8 00
Berg's, each.................. 0 75 2 50

TROWELS.

Disston's, discount 10 per cent.
S. & D., discount 35 per cent.
Berg's, brick, 924x11 .... 4 00
" pointing, 924x5.......... .... 2 10

FARM AND GARDEN GOODS

BELLS.

American cow bells, 63⅓ per cent.
Canadian, discount 45 and 50 per cent.
American, farm bells, each .. 1 35 3 00

BULL RINGS.

Copper, $2.00 for 2½-inch

CATTLE LEADERS

Nos. 32 and 33 .. ...per gross 7 50 8 50

BARN DOOR HANGERS

		doz. pairs
Stearns wood track	4 50	6 00
Zenith.............	....	9 00
Atlas, steel covered	5 00	6 00
Perfect	8 00	11 00
New Milo, flexible	....	6 50
Steel, track, 1 x 3-16 in(100 ft)	....	3 25
" " 1¼ x 3-16 in(100 ft)	....	4 75
Double strap hangers, doz. sets....		6 40
Standard jointed hangers, "	....	6 40
Steel King hangers "	....	6 25
Storm King and safety hangers	....	7 00
" " rail...........	....	4 25

Chicago Friction, Oscillating and Big Twin Hangers, 5 per cent.

HARVEST TOOLS.

50 and 10 per cent.
S. & D. lawn rakes, Dunn's, 40 off.
" sidewalk and stable scrapers, 40 off.

HAY KNIVES.

Net list.

HEAD HALTERS.

Jute, Rope, ⅝-inch....per gross 9 00
" " ⅝ " .... " 10 00
" " ⅞ " .... " 12 00
Leather, 1-inch ......per doz 4 00
Leather, 1½ " ...... " 5 20
Web................. " 2 45

HOES.

Garden, 50 and 10 per cent.
Planter.............. per doz. 4 00 4 50

LAWN MOWERS

Low wheel 12, 14 and 16-inch $2 30
9-inch wheel, 12-inch 2 85
" " 14 " 3 10
" " 16 " 3 12½
High wheel, 12 " 4 05
" " 14 " 4 50
" " 16 " 4 75

SCYTHES.

Per doz. net.................. 6 25 9 25

SCYTHE SNATHS.

Canadian, discount 40 per cent.

SNAPS

Harness, German, discount 25 per cent.
Lock, Andrews' 4 50 11 00

STABLE FITTINGS.

Warden King, 35 per cent
Dennis Wire & Iron Co., 33⅓ p c.

WOOD HAY RAKES.

40 and 10 per cent.

HEAVY GOODS, NAILS, ETC.

ANVILS.

Wright's, 80-lb. and over.............. 0 10½
Hay Budden, 80-lb. and over 0 09½
Brook's, 80-lb. and over 0 11½
Taylor-Forbes, prospectors 0 05
Columbia Hardware Co., per lb. 0 09½

VISES.

Wright's...................... .... 0 13½
Berg's, per lb...... .... 0 12½
Brook's....................... .... 0 12½
Pipe Vise, Hinge, No. 1...... .... 3 50
" " " No. 2...... .... 5 50
Saw Vise..... .. .. 4 50 5 00
Blacksmiths' (discount) 60 per cent.
" parallel (discount) 45 per cent.

BOLTS AND NUTS

	Per cent.
Carriage Bolts, common ($1 list	
" " ⅜ and smaller..	60, 10 and 10
" " 7-16 and up	55 and 5
" " Norway Iron ($3 list)	50
Machine Bolts, ⅜ and less	60 and 10
Machine Bolts, 7-16 and up....	55 and 5
Plough Bolts	55 and 10
Blank Bolts......	55
Bolt Ends...................	55
Sleigh Shoe Bolts, ⅜ and less ..	60 and 10
" " " 7-16 and larger	50 and 5
Coach Screws, conepoint......	70 and 5

Nuts, square, all sizes, 4c. per cent. off
Nuts, hexagon, all sizes, 4¼c. per cent. off.
Stove Rods per lb., 5½ to 6c.
Stove Bolts, 75 per cent.

CHAIN.

Proof coil, per 100 lb. ¼ in., $6 00; 5-16 in., $4.85; ⅜ in., $4.25; 7-16 in., $4 00; ½ in., $3 75; 9-16 in. $3 70; ⅝ in., $3.65; ¾ in., $3.60; ⅞ in., $3.45; 1 in. $3.40.

Halter, kennel and post chains, 40 to 40 and 5 per cent; Cow ties, 40 per cent; Tie out chains 65 per cent.; Stall fixtures, 35 per cent; Trace chain, 45 per cent.; Jack chain iron, 50 per cent.; Jack chain, brass, 50 per cent.

HORSE NAILS

M R M. cold forged process, list May 15, 1907, 50 and 5 per cent.
"C" brand, 57½ per cent. off list.
Capewell brand, quotations on application.

HORSESHOES.

M R.M. brand: iron, light and medium No. 1 and smaller, $3 90; No. 2 and larger, $3 65; snow pattern No 1 and smaller $4.15, No. 2 and larger, $3 90; "X L." new light steel, No. 1 and smaller, $4 25; No 2 and larger, $4; "X L" featherweight steel, No. 0 to 4, $5 60; toe-weight, all sizes, $6 85. F o b Montreal Extras for packing.

Belleville brand: No. 0 and 1, light and medium iron $3 90; snow, $4 15; light steel, $4.25; No. 2 and larger, light and medium iron, $3 65; snow, $3.90; light steel, $4. F o.b. Belleville Two per cent., 30 days.

HORSE WEIGHTS.

Taylor-Forbes, 4½c. per lb.

NAILS.	Cut.	Wire.
2d...........................	4 00	3 50
3d...........................	3 15	3 20
4 and 5d....	2 90	2 90
6 and 7d	2 80	2 80
8 and 9d	2 65	2 65
10 and 12d	2 60	2 60
16 and 20d...................	2 55	2 55
30, 40, 50 and 60d (base)	2 50	2 50

F.o.b. Montreal. Cut nails, Toronto 20c. higher.
Miscellaneous wire nails, discount 75 per cent
Coopers' nails, discount 40 per cent.

PRESSED SPIKES.

Pressed spikes, ⅝ diameter, per 100 lbs $3.15

RIVETS AND BURRS.

Iron Rivets, black and tinned, 60, 10 and 10.
Iron Burrs, discount 60 and 10 and 10 p.c.
Copper Rivets, usual proportion burrs, 15 p.c.
Copper Burrs only, net list.
Extras on Coppered Rivets, ½-lb. packages 1c. per lb.; ¼-lb. packages 2c. lb.
Tinned Rivets, net extra, 4c. per lb.

SCREWS.

Wood, F. H., bright and steel, 85 and 10 p c.
" R. H., bright, 80 and 10 per cent.
" F. H., brass, 75 and 10 per cent.
" R. H., " 70 and 10 per cent.
" F. H., bronze, 70 and 10 per cent.
" R. H., " 65 and 10 per cent.
Drive Screws, dis. 87½ per cent.
Bench, wood ....per doz. 3 25 4 00
" iron .... " 4 25 5 00
Set, case hardened, dis. 60 per cent.
Square Cap, dis. 50 and 5 per cent.
Hexagon Cap, dis. 45 per cent.

MACHINE SCREWS.

Flat head, iron and brass, 35 per cent.
Felister head, iron, discount 30 per cent.
" " brass, discount 25 per cent.

TACKS, BRADS, ETC.

Carpet tacks, blued 75 p.c.; tinned, 80 and 10; (in kegs), 40; cut tacks, blued, in dozens only, 75; ½ weights, 60; Swedes cut tacks, blued and tinned, bulk, 75 dozens, 75; Swedes, upholsterers', bulk, 85 and 12½; brush, blued and tinned, bulk, 70; Swedes, gimp, blued, tinned and japanned, 75 and 12½; zinc tacks, 35; leather carpet tacks, 40; copper tacks, 25; copper nails 30, trunk nails, black, 65; trunk nails, tinned and blued, 65; clout nails, blued and tinned 65; chair nails, 35; patent brads, 40; fine finishing, 40; lining tacks, in papers, 10; lining tacks, in bulk, 15; lining tacks, solid heads, in bulk, 75; saddle nails, in papers, 10; saddle nails, in bulk, 15; tufting buttons, 22 line in dozens only, 60; zinc glaziers' points, 5, double pointed tacks, papers, 90 and 10; double pointed tacks, bulk, 40; clinch and duck rivets, 45; cheese box tacks, 85 and 5 trunk tacks, 80 and 10.

WROUGHT IRON WASHERS.

Canadian make, discount 40 per cent.

SPORTING GOODS.

CARTRIDGES.

"Dominion" Rim Fire Cartridges and C.B. caps, 50 and 2½ per cent.; Rim Fire B B. Round Caps 80 and 2½ per cent; Centre Fire, Pistol and Rifle Cartridges 30 p c.; Centre Fire Sporting and Military Cartridges, 20 and 5 p c.; Rim Fire, Shot Cartridges, 50 and 7½ p c.; Centre Fire, Shot Cartridges, 30 p.c.; Primers, 25 p c.

LOADED SHELLS.

"Crown" Black Powder, 15 and 10 p.c; "Sovereign" Empire Bulk Smokeless Powder, 30 and 10 p.c.; "Regal" Ballistite Dense smokeless Powder, 30 and 5 p.c; "Imperial" Empire or Ballistite Powder, 30 and 10 p.c.

EMPTY SHELLS.

Paper Shells, 25 and 5; Brass Shells, 5 and 5 p.c.

Wads.	per lb.
Best thick brown or grey felt wads, in ½-lb. bags	$0 70
Best thick white card wads, in boxes of 500 each, 12 and smaller gauges	29
Best thick white card wads, in boxes of 500 each, 10 gauge............	0 35
Thin card wads, in boxes of 1,000 each, 12 and smaller gauges	0 20
Thin card wads, in boxes of 1,000 each, 10gauge........................	0 25
Chemically prepared black edge grey cloth wads, in boxes of 250 each—	Per M
11 and smaller gauge	0 60
9 and 10 gauges.................	0 70
and 8 "	0 90
5 and 6 "	1 10
Superior chemically prepared pink edge, best white cloth wads in boxes of 250 each—	
11 and smaller gauge	1 15
9 and 10 gauges	1 40
7 and 8 "	1 65
6 and 9 "	1 90

SHOT.

Ordinary drop shot, AAA to dust $7.50 per 100 lbs. Discount 5 per cent; cash discount, 2 per cent, 30 days; net extras as follows subject to cash discount only; Chilled, 40 c.; buck and seal, 80c.; no. 28 ball, $1 20 per 100 lbs; bags less than 25 lbs., ½c. per lb; F.O.B. Montreal, Toronto, Hamilton, London, St. John and Halifax, and freight equalized thereon.

TRAPS (steel.)

Game, Newhouse, discount 30 and 10 per cent.
Game, Hawley & Norton, 50, 10 & 5 per cent
Game, Victor, 70 per cent.
Game, Oneida Jump (B. & L.) 40 & 2½ p.
Game, steel, 60 and 5 per cent.

SKATES.

Skates, discount 37½ per cent.
Empire hockey sticks, per doz .. 3 00

CUTLERY AND SILVERWARE.

RAZORS.

		per doz.
Elliot's	4 00	18 00
Boker's	7 50	11 00
" King Cutter	13 50	18 50
Wade & Butcher's	3 60	10 00
Lewis Bros. "Klean Kutter"	8 50	10 50
Henckel's	7 50	20 00
Berg's	7 50	20 00

Clauss Razors and Strops, 50 and 10 per cent

KNIVES.

Farrlers-Stacey Bros., doz 3 50

PLATED GOODS

Hollowware, 40 per cent. discount.
Flatware, staples, 40 and 10, fancy, 40 and 5.
Hutton's "Cross Arrow" flatware, 47½; "Singalese" and "Alaska" Nevada silver flatware, 42 p c.

SHEARS.

Clauss, nickel, discount 60 per cent.
Clauss, Japan, discount 67½ per cent.
Clauss, tailors, discount 40 per cent.
Seymour's, discount 50 and 10 per cent.
Berg's 6 00 12 00

HOUSE FURNISHINGS.

APPLE PARERS.

Hudson, per doz., net 5 75

BIRD CAGES.

Brass and Japanned, 40 and 10 p. c.

COPPER AND NICKEL WARE.

Copper boilers, kettles, teapots, etc. 30 p c.
Copper pitts, 30 per cent.

KITCHEN ENAMELED WARE.

White ware, 75 per cent.
London and Princess, 50 per cent.
Canada, Diamond, Premier, 50 and 10 p c.
Pearl, Imperial, Crescent and granite steel, 50 and 10 per cent.
Premier steel ware, 40 per cent.
Star decorated steel and white, 25 per cent.
Japanned ware, discount 45 per cent.
Hollow ware, tinned cast, 35 per cent. off.

KITCHEN SUNDRIES.

Can openers, per doz.	0 40	0 75
Mincing knives per doz	0 50	0 81
Duplex mouse traps, per doz	....	0 65
Potato mashers, wire, per doz.	0 60	0 70
" wood "	0 50	0 60
Vegetable slicers, per doz	....	2 25
Universal meat chopper No. 1	....	1 15
Enterprise chopper, each	....	1 30
Spiders and fry pans, 50 per cent.		
Star Al chopper 5 to 32	1 35	4 10
" " 100 to 103	1 35	2 00
Kitchen hooks, bright	....	0 60

LAMP WICKS.

Discount, 60 per cent.

LEMON SQUEEZERS.

Porcelain lined	per doz.	2 20	5 60
Galvanized	"	1 87	3 85
King, wood	"	2 75	2 90
King, glass	"	4 00	4 50
All glass	"	0 50	0 90

METAL POLISH.

Tandem metal polish paste 6 00

PICTURE NAILS.

Porcelain head	per gross	1 35	1 50
Brass head		0 40	1 00

Tin and gilt, picture wire, 75 per cent

SAD IRONS.

Mrs. Potts, No. 55, polished ....per set	0 90
" No. 50, nickle-plated, "	0 95
" handles, japaned, per gross	9 25
" " nickled, "	9 75
Common, plain	4 25
" plated	4 50
Asbestos, per set	1 50

TINWARE.

CONDUCTOR PIPE.

2-in. plain or corrugated, per 100 feet, $3.30; 3 in., $4.40; 4 in., $5.80; 5 in., $7.45; 6 in., $9.90.

FAUCETS.

Common, cork-lined, discount 35 per cent.

EAVETROUGHS.

10-inch .... per 100 ft. 3 30

FACTORY MILK CANS.

Discount off revised list, 35 per cent.
Milk can trimmings, discount 25 per cent.
Creamery Cans, 45 per cent

LANTERNS.

No. 2 or 4 Plain Cold Blast	per doz.	6 50
Lift Tubular and Hinge Plain,	"	4 75
No. 0, safety	"	4 00

Better quality at higher prices.
Japanning, 50c. per doz. extra.
Prism globes, per doz., $1.20.

OILERS.

Kemp's Tornado and McClary's Model galvanized oil can, with pump, 5 gallon, per dozen 10 92
Davidson oilers, discount 40 per cent.
Zinc and tin, discount 50 per cent.
Coppered oilers, 20 per cent. off.
Brass oilers, 50 per cent. off.
Malleable, discount 25 per cent

PAILS (GALVANIZED).

Dufferin pattern pails, 45 per cent.
Flaring pattern, discount 45 per cent.
Galvanized washtubs 40 per cent.

PIECED WARE.

Discount 35 per cent off list, June, 1899.
10-qt. flaring sap buckets, discount 35 per cent.
6, 10 and 14-qt. flaring pails dis 35 per cent.
Copper bottom tea kettles and boilers, 30 p.c.
Coal hods, 40 per cent.

STAMPED WARE.

Plain, 75 and 12½ per cent. off revised list.
Retinned, 72½ per cent. revised list.

SAP SPOUTS

Bronzed iron with hooks ....per 1,000
Eureka tinned steel, hooks 8 00

STOVEPIPES.

5 and 6 inch, per 100 lengths	7 64	7 91
7 inch. " "	...	8 18

Nestable, discount 40 per cent.

STOVEPIPE ELBOWS

5 and 6-inch, common	per doz.	1 32
7-inch	"	1 48

Polished, 15c. per dozen extra.

THERMOMETERS

Tin case and dairy, 75 to 75 and 10 per cent.

TINNERS' SNIPS

Per doz 3 00 15
Clauss, discount 35 per cent

TINNERS' TRIMMINGS.

Discount, 45 per cent.

WIRE.

ANNEALED CUT HAY BAILING WIRE.

No. 12 and 13, $4; No. 13½, $4 10; No 14, $4.20; No. 15, $4.50; in lengths 6' to 11', 25 per cent.; other lengths 20c. per 100 lbs extra; if eye or loop on end add 25c. per 100 lbs. to the above.

BRIGHT WIRE GOODS

Discount 60 per cent.

CLOTHES LINE WIRE.

7 wire solid line, No. 17, $4.90; No. 18, $3.00; No. 19, $2 70; [illegible] wire solid line, No. 17, $4 45; No. 18, $3.10; No 19, $2 80. All prices per 1000 ft. measure; 6 strand, No. 18, $2.60; No. 19, $2 90. F.o.b. Hamilton, Toronto, Montreal

COILED SPRING WIRE

High Carbon, No. 9, $2 95; No. 11, $3.50; No. 12, $3.20.

COPPER AND BRASS WIRE.

Discount 37½ per cent.

FINE STEEL WIRE.

Discount 25 per cent List of extras In 100-lb. lots: No. 17, $5 — No. 18, $5 50 — No. 19, $6 — No. 20, $6 65 — No. 21 $7 — No. 22, $7.30 — No 23, $7.65 — No 24, $8 — No. 25, $9 — No. 26, $9.50 — No. 27, $10 — No. 28, $11 — No. 29, $12 — No. 30, $13 — No. 31, $14 — No. 32, $15 — No. 33, $16 — No. 34, $17. Extras net—tinned wire, Nos. 17-25, $2 — Nos. 26-31, $4 — Nos. 32-34, $6. Coppered, 75c.—oiling, 10c.—in 25-lb. bundles, 15c.—in 5 and 10-lb. bundles, 25c.—in 1-lb. hanks, 25c. —in ½-lb. hanks, 38c.—in ¼-lb. hanks, 50c.—packed in casks or cases, 15c.—bagging or papering, 10c

FENCE STAPLES.

Bright .... 2 80 Galvanized ... 32

HAY WIRE IN COILS.

No 13, $2.70; No. 14, $2 80; No. 15, $2.95; f.o b., Montreal.

GALVANIZED WIRE.

Per 100 lb.—Nos. 4 and 5, $3.95 — Nos. 6, 7, 8, $3 35 — No. 9, $2.85 — No. 10, $3 40 — No. 11, $3.45 — No. 12, $3.00 — No. 13, $3.10 — No. 14, $3.95 — No 15, $4.30 — No. 16, $4.30 from stock. Base sizes, Nos. 6 to 9, $2.35 f.o.b. Cleveland. Extras for cutting.

LIGHT STRAIGHTENED WIRE.

Gauge No.	Over 20 in. per 100 lbs.	10 to 20 in.	5 to 10 in.
0 to 5	$0.50	$0.75	$1 25
6 to 9	0 75	1.25	2 00
10 to 11	1.00	1 75	2 50
12 to 14	1 50	2.25	3.50
15 to 16	2.00	3.00	4.50

SMOOTH STEEL WIRE.

No. 0-9 gauge, $2 40; No. 10 gauge, 6c. extra; No. 11 gauge, 12c extra; No. 12 gauge, 20c. extra; No. 13 gauge, 30c. extra; No 14 gauge 40c. extra; No. 15 gauge, 55c. extra; No 16 gauge, 70c. extra. Add 60c. for coppering and $2 for tinning.

Extra net per 100 lb.—Oiled wire 10c., spring wire $1.25, bright soft drawn 15c., charcoal (extra quality) $1.25, packed in casks or cases 15c., bagging and papering 10c., 50 and 100-lb. bundles 10c., in 25-lb. bundles 15c., in 5 and 10-lb. bundles 25c., in 1-lb hanks, 50c., in ½-lb hanks 75c., in ¼-lb. hanks $1

POULTRY NETTING.

2-in. mesh, 19 w g., 50 p.c. Other sizes, 45 p c.

WIRE CLOTH

Painted Screen, in 100-ft. rolls, $1.72½, per 100 sq. ft.; in 50-ft. rolls $1 77½, per 100 sq ft.

WIRE FENCING.

Galvanized barb 2 95
Galvanized, plain twist 3 30
Galvanized barb, f.o.b. Cleveland $2 70 for small lots and $2.60 for carlots

WOODENWARE.

CHURNS.

No. 0, $9; No. 1, $9; No. 2, $10; No. 3, $11; No. 4, $13; No. 5, $16; f.o.b. Toronto Hamilton, London and St. Marys, 30 and 30 per cent; f.o.b. Ottawa, Kingston and Montreal, 40 and 15 per cent. discount.

CLOTHES REELS

Davis Clothes Reels, dis. 40 per cent.

FIBRE WARE

Star pails, per doz		$ 3 00
0 Tubs,	"	14 00
1 "	"	12 00
2 "	"	10 00
3 "	"	8 50

LADDERS, EXTENSION.

3 to 6 feet, 12c. per foot; 7 to 10 ft., 13c.
Waggoner Extension Ladders, dis. 40 per cent.

MOPS AND IRONING BOARDS.

"Best" mops 1 25
"900" mops 1 25
Folding ironing boards 12 00 16 50

REFRIGERATORS

Discount, 40 per cent.

SCREEN DOORS.

Common doors, 2 or 3 panel, walnut stained, 4-in. style .... per doz. 7 25
Common doors, 2 or 3 panel, grained only, 4-in., style .... per doz. 7 55
Common doors, 2 or 3 panel, light stair per doz. 9 55

WASHING MACHINES.

Round, re-acting per doz.	60 00
Square "	63 00
Eclipse, per doz	54 00
Dowswell "	39 00
New Century, per doz	75 00
Daisy	54 00
Stephenson	74 00

WRINGERS.

Royal Canadian, 11 in., per doz.	35 00
Royal American 11 in	35 00
Eze 10 in., per doz	36 75

MISCELLANEOUS

AXLE GREASE.

Ordinary, per gross	6 00	7 00
Best quality	10 00	12 00

BELTING.

Extra, 60 per cent.
Standard, 60 and 10 per cent.
No. 1, not wider than 6 in., 60, 10 and 10 p.c.
Agricultural, not wider than 4 in., 75 per cent
Lace leather, per side, 75c.; cut laces, 80c.

BOOT CALKS.

Small and medium, ball .... per M 4 25
Small heel .... " 4 50

CARPET STRETCHERS.

American	per doz.	1 00	1 50
Bullard's	"	....	6 50

CASTORS.

Bed, new list, discount 55 to 57½ per cent.
Plate, discount 55½ to 57½ per cent.

PINE TAR.

½ pint in tins .... per gross ... 7 80
" " 9 60

PULLEYS.

Hothouse	per doz.	0 55	1 00
Axle	"	0 22	0 33
Screw	"	0 22	1 00
Awning	"	0 35	2 50

PUMPS.

Canadian cistern	1 40	2 00
Canadian pitcher spout	1 80	3 16

Berg's wing pump, 75 per cent.

ROPE AND TWINE.

Sisal		0 10½
Pure Manilla		0 15
"British" Manilla		0 12
Cotton, 3-16 inch and larger	0 21	0 23
" 5-32 inch	0 25	0 27
" ⅛ inch	0 25	0 28
Russia Deep Sea		0 16
Jute		0 09
Lath Yarn, single		0 10
" double		0 10½
Sisal bed cord, 48 feet .... per doz.		0 65
" " 60 feet .... "		0 80
" " 72 feet .... "		0 95

Twine.

Bag, Russian twine, per lb.	....	0 27
Wrapping, cotton, 3-ply	....	0 25
" " 4-ply	....	0 29
Mattress twine per lb.	0 33	0 45
Staging " "	0 27	0 35

BINDER TWINE.

500 feet, sisal	0 09½
500 " standard	0 09½
550 " " manilla	0 10½
600 " " "	0 12½
650 " " "	0 13½

Car lots, ½c. less; 5-ton lots, ¼c. less. Central delivery.

SCALES.

Gurney Standard, 35; Champion, 45 p c.
Burrow, Stewart & Milne — Imperial Standard, 35; Weigh Beams, 35; Champion Scales, 45
Fairbanks Standard, 30; Dominion, 50 Richelieu, 50.
Warren new Standard, 35; Champion, 45 Weigh Beams, 30.

STONES—OIL AND SCYTHE.

Washita	per lb.	0 25	0
Hindostan	"	0 06	0 10
" slip	"	0 18	0 20
" Axe	"	....	0 10
Deer Creek	"	....	0 10
Deerlick	"	....	0 25
" Axe	"	....	0 15
Lily white	"	....	0 42
Arkansas	"	....	1 50
Water-of-Ayr	"	....	0 10
Scythe	per gross	3 50	5 00
Grind, 40 to 200 lb., per ton		20 00	22 00
" under 40 lb., "		....	24 00
" 200 lb. and over		....	28 00

INDEX TO ADVERTISERS.

CLASSIFIED LIST OF ADVERTISEMENTS.

Alabastine.
Alabastine Co , Limited, Paris, Ont

Auditors.
Davenport, Pickup & Co., Winnipeg.

Automobile Accessories.
Canada Cycle & Motor Co., Toronto Junction
Carriage Mountings Co , Ltd.. Toronto.

Awnings.
Tobin Tent and Awning Co , Ottawa

Babbitt Metal.
Canada Metal Co., Toronto.
Canadian Fairbanks Co., Montreal.
Frothingham & Workman, Montreal.
Robertson, Jas. Co., Montreal.

Bar Urns.
Buffalo Mfg. Co., Buffalo, N.Y.

Bath Room Fittings.
Buffalo Mfg. Co., Buffalo, N.Y.
Carriage Mountings Co , Ltd , Toronto.
Forsyth Mfg Co., Buffalo, N.Y.
Ontario Metal Novelty Co., Toronto

Belting, Hose, etc.
Gutta Percha and Rubber Mfg. Toronto.
Sadler & Haworth. Toronto.

Bicycles and Accessories.
Johnson's, Iver, Arms and Cycle Works Fitchburg, Mass

Binder Twine.
Consumers Cordage Co , Montreal.

Bolts.
Toronto & Belleville Rolling Mills, Belleville.

Box Strap.
J. N. Warminton, Montreal.

Brass Goods.
Frothingham & Workman, Montreal.
Glauber Brass Mfg. Co., Cleveland, Ohio.
Kerr Engine Co., Walkerville, Ont.
Lewis, Rice, & Son., Toronto.
Morrison, Jas., Brass Mfg. Co., Toronto.
Mueller Mfg Co., Decatur, Ill.
Penberthy Injector Co., Windsor, Ont.
Taylor-Forbes Co., Guelph, Ont.

Bronze Powders.
Canadian Bronze Powder Works, Montreal.

Brushes.
United Factories, Toronto.

Cans.
Acme Can Works, Montreal.

Builders' Tools and Supplies.
Covert Mfg. Co., West Troy, N.Y.
Frothingham & Workman Co., Montreal.
Howland, H. S. Sons & Co., Toronto.
Hyde, F., & Co., Montreal.
Lewis Bros. & Co., Montreal.
Lewis, Rice, & Son, Toronto.
Lockerby & McComb, Montreal.
Lufkin Rule Co., Saginaw, Mich.
Newman & Sons, Birmingham.
North Bros. Mfg. Co., Philadelphia, Pa.
Stanley Rule & Level Co., New Britain.
Stanley Works, New Britain, Conn.
Stephens, G. F., Winnipeg.
Taylor-Forbes Co., Guelph, Ont.

Carriage Accessories.
Carriage Mountings Co., Ltd., Toronto.
Covert Mfg. Co., West Troy, N.Y.

Carriage Springs and Axles.
Guelph Spring and Axle Co., Guelph.

Carpet Beateas.
Ontario Metal Novelty Co., Toronto.

Cartridges.
Dominion Cartridge Co., Montreal.

Cattle and Trace Chains.
Greening, B., Wire Co., Hamilton.

Chafing Dishes.
Buffalo Mfg Co., Buffalo, N.Y.

Churns.
Dowswell Mfg. Co., Hamilton.

Clippers—All Kinds.
American Shearer Mfg. Co., Nashua, N.H

Clothes Reels and Lines.
Hamilton Cotton Co., Hamilton, Ont.

Clutch Nails.
J. N. Warminton, Montreal.

Congo Roofing
Buchanan Foster Co., Philadelphia, Pa.

Cordage.
Consumers' Cordage Co., Montreal.
Hamilton Cotton Co., Hamilton.

Cork Screws.
Erie Specialty Co., Erie, Pa.

Cow Ties
Greening, B., Wire Co., Hamilton

Cuspidors.
Buffalo Mfg Co., Buffalo, N.Y.

Cut Glass.
Phillips, Geo., & Co., Montreal.

Cutlery—Razors, Scissors, etc.
Birkett, Thos., & Son Co., Ottawa.
Clauss Shear Co., Toronto
Dorken Bros. & Co., Montreal.
Frothingham & Workman Montreal.
Heinisch's, R., Sons Co., Newark, N.J.
Howland, H. S. Sons & Co., Toronto.
Hutton, Wm., & Sons, Ltd., London, Eng.
Lamplough, F. W., & Co., Montreal.
Phillips, Geo., & Co., Montreal.
Round, John, & Son, Montreal.

Electric Fixtures.
Canadian General Electric Co., Toronto.
Morrison James, Mfg. Co., Toronto.
Munderloh & Co., Montreal.

Electro Cabinets.
Cameron & Campbell Toronto.

Enameled Ware
Kemp Mfg. Co., Toronto.

Engines, Supplies, etc.
Kerr Engine Co., Walkerville, Ont.

Eavetroughs
Wheeler & Bain, Toronto

Fencing—Woven Wire
Dominion Wire Mfg. Co., Montreal.
Owen Sound Wire Fence Co., Owen Sound.
Banwell Hoxie Wire Fence Co., Hamilton.

Files and Rasps.
Barnett Co., G. & H., Philadelphia, Pa.
Nicholson File Co., Port Hope

Firearms and Ammunition.
Hamilton Rifle Co., Plymouth, Mich.
Harrington & Richardson Arms Co., Worcester, Mass.
Johnson's, Iver, Arms and Cycle Works Fitchburg, Mass.

Fishing Tackle.
Enterprise Mfg. Co., Akron, Ohio

Food Choppers
Enterprise Mfg. Co., Philadelphia, Pa.
Lamplough, F. W., & Co., Montreal.
Shirreff Mfg. Co., Brockville, Ont.

Furnaces.
Pease Foundry Co , Toronto.

Galvanizing.
Canada Metal Co., Toronto.
Dominion Wire Mfg. Co., Montreal.
Montreal Rolling Mills Co., Montreal.
Ontario Wind Engine & Pump Co., Toronto.

Glass Ornamental
Hobbs Mfg. Co., London
Consolidated Plate Glass Co., Toronto

Glaziers' Diamonds.
Gibsone, J. B , Montreal.
Pelton, Godfrey S.
Sharratt & Newth, London, Eng.
Shaw, A., & Son, London, Eng.

Handles.
Still, J H., Mfg. Co.

Harvest Tools.
Maple Leaf Harvest Tool Co , Tillsonburg Ont.

Hockey Sticks
Still, J. H Mfg. Co., St. Thomas.

Hoop Iron.
Frothingham & Workman, Montreal.
Montreal Rolling Mills Co , Montreal.
J. N. Warminton, Montreal.

Horse Blankets.
Heney, E. N., & Co., Montreal.

Horseshoes and Nails.
Canada Horse Nail Co., Montreal.
Montreal Rolling Mills, Montreal.
Capewell Horse Nail Co., Toronto
Toronto & Belleville Rolling Mills Belleville.

Hot Water Boilers and Radiators.
Cluff, R J., & Co Toronto.
Pease Foundry Co. Toronto.
Taylor-Forbes Co., Guelph.

Ice Cream Freezers.
Dana Mfg. Co., Cincinnati, Ohio.
North Bros. Mfg. Co., Philadelphia, Pa.

Ice Cutting Tools.
Erie Specialty Co., Erie, Pa.
North Bros. Mfg. Co., Philadelphia, Pa.

Injectors—Automatic.
Morrison, Jas., Brass Mfg. Co., Toronto.
Penberthy Injector Co., Windsor, Ont.

Iron Pipe.
Montreal Rolling Mills, Montreal.

Iron Pumps.
Lamplough, F. W., & Co., Montreal.
McDougall, R., Co., Galt, Ont.

Lanterns.
Kemp Mfg. Co., Toronto.
Ontario Lantern Co., Hamilton, Ont.
Wright, E. T., & Co., Hamilton.

Lawn Mowers.
Birkett, Thos., & Son Co., Ottawa.
Frothingham & Workman, Montreal.
Maxwell, D., & Sons, St. Mary's, Ont.
Taylor, Forbes Co., Guelph.

Lawn Mower Grinders
Root Bros. & Co., Plymouth, Ohio.

Ledgers—Loose Leaf.
Business Systems, Toronto.
Copeland-Chatterson Co , Toronto.
Crain, Rolla L., Co., Ottawa.
Universal Systems, Toronto.

Lithographing.
London Printing & Lithographing Co., London, Ont.

Locks, Knobs, Escutcheons, etc.
Peterborough Lock Mfg. Co., Peterborough, Ont.
National Hardware Co., Orillia, Ont.

Lumbermen's Supplies.
Pink, Thos., & Co., Pembroke Ont.

Lye
Gillett, E. W., & Co., Toronto

Manufacturers' Agents.
Fox, C. H., Vancouver.
McIntosh, H. F., & Co., Toronto.
Gibb, Alexander, Montreal.
Scott, Bathgate & Co., Winnipeg.

Metals.
Canada Iron Furnace Co., Midland, Ont.
Canada Metal Co., Toronto.
Eadie, H. G., Montreal.
Frothingham & Workman, Montreal.
Gibb, Alexander, Montreal.
Kemp Mfg Co., Toronto
Leslie, A. C., & Co., Montreal.
Lysaght, John, Bristol, Eng.
Nova Scotia Steel and Coal Co., New Glasgow, N.S.
Robertson, Jas., Co., Montreal
Roper, J. H., Montreal.
Samuel, Benjamin & Co., Toronto.
Stairs, Son & Morrow, Halifax, N.S.
Thompson, B. & S. H. & Co. Montreal.

Metal Lath.
Galt Art Metal Co , Galt.
Metallic Roofing Co., Toronto.
Metal Shingle & Siding Co., Preston, Ont.

Metal Polish, Emery Cloth, etc.
Oakey, John, & Sons, London, Eng.

Nails Wire
Dominion Wire Mfg. Co., Montreal.

Oil Tanks.
Bowser, S. F., & Co., Toronto.

Ornamental Iron and Wire.
Dennis Wire & Iron Co , London, Ont.

Packing.
Gutta Percha & Rubber Co Toronto,

Paints, Oils, Varnishes, Glass.
Blanchite Process Paint Co., Toronto.
Brandram-Henderson, Montreal
Canada Paint Co., Montreal.
Canadian Oil Co., Toronto
Consolidated Plate Glass Co., Toronto.
Dods, P. D., & Co., Montreal
Imperial Varnish and Color Co., Toronto
Jamieson, R. C., & Co., Montreal.
Lucas John & Co., New York
McArthur, Corneille & Co., Montreal.
McCaskill, Dougall & Co., Montreal.
Moore, Benjamin, & Co Toronto.
Ottawa Paint Works, Ottawa
Queen City Oil Co., Toronto.
Ramsay & Son, Montreal
Sanderson Pearcy & Co., Toronto
Sherwin-Williams Co., Montreal.
Standard Paint Co , Montreal
Standard Paint and Varnish Works Windsor, Ont.
Stephens & Co , Winnipeg.
Martin-Senour Co., Montreal
Winnipeg Paint & Glass Co., Winnipeg

Perforated Sheet Metals.
Greening, B., Wire Co., Hamilton.

Plumbers' Tools and Supplies.
Canadian Fairbanks Co., Montreal.
Cluff, R. J., & Co., Toronto
Frothingham & Workman, Montreal.
Glauber Brass Co., Cleveland, Ohio.
Jardine, A. B., & Co , Hespeler, Ont.
Jenkins Bros., Boston, Mass.
Kerr Engine Co , Walkerville, Ont.
Lewis, Rice, & Son, Toronto.
Merrell Mfg. Co., Toledo, Ohio.
Montreal Rolling Mills Montreal.
Morrison, Jas., Brass Mfg. Co., Toronto.
Mueller, H., Mfg. Co , Decatur, Ill.
Oshawa Steam & Gas Fitting Co., Oshaw
Robertson Jas., Co. Montreal.
Robertson, Jas., Co., Limited, Toronto
Somerville, Limited, Toronto
Stairs, Son & Morrow, Halifax, N.S.
Standard Ideal Sanitary Co., Port Hope,
Standard Sanitary Co., Pittsburg.
Stephens, G F , & Co., Winnipeg, Man.
Turner Brass Works, Chicago.
Vickery, Orlando, Toronto.

Polishes.
Majestic Polishes, Toronto

Portland Cement.
International Portland Cement Co. Ottawa, Ont.
Hanover Portland Cement Co., Hanover, Ont.
Hyde, F., & Co., Montreal.
Thompson B. & S. H. & Co., Montreal.

Poultry Netting.
Greening, B., Wire Co., Hamilton, Ont.

Printing.
London Printing & Lithographing Co., London, Ont.

Razors.
Clauss Shear Co., Toronto.

Refrigerators.
Fabien, C. P., Montreal

Registers
Pease Foundry Co., Toronto.

Roofing Supplies.
Brantford Roofing Co., Brantford.
Barrett Mfg. Co., New York
F. W. Bird, East Walpole, Mass.
Buchanan Foster Co , Philadelphia, Pa.
McArthur, Alex., & Co., Montreal
Metal Shingle & Siding Co., Preston, Ont.
Metallic Roofing Co., Toronto.
Paterson Mfg. Co., Toronto & Montreal.
Wheeler and Bain, Toronto

Saws.
Atkins, E. C., & Co., Indianapolis, Ind
Shurly & Dietrich, Galt, Ont.
Spear & Jackson, Sheffield, Eng.

Scales.
Canadian Fairbanks Co., Montreal.
Frothingham & Workman, Montreal.

Screw Cabinets.
Cameron & Campbell, Toronto.

Screws, Nuts, Bolts.
Dominion Wire Mfg. Co , Montreal.
Montreal Rolling Mills Co., Montreal.

Soil Pipe
McFarlane, Walter, Glasgow

Sewer Pipes.
Canadian Sewer Pipe Co., Hamilton
Hyde, F., & Co., Montreal.

Shelf Boxes.
Cameron & Campbell, Toronto.

Shears, Scissors.
Clauss Shear Co., Toronto.

Shovels and Spades.
Eclipse Mfg. Co., Ottawa
Frothingham & Workman, Montreal
Peterboro Shovel & Tool Co., Peterboro

Silverware.
Hutton, Wm , & Sons, Ltd., London, Eng
McGlashan, Clarke Co., Niagara Falls, Ont.
Phillips, Geo., & Co., Montreal.
Round, John, & Son, Sheffield, Eng.

Skates.
Canada Cycle & Motor Co., Toronto.
McFarlane, Walter, Glasgow.

Sprayers
Cavers Bros., Galt

Spring Hinges, etc.
Chicago Spring Butt Co., Chicago, Ill.

Stable Fittings
Dennis Wire & Iron Co., London

Steel Rails.
Nova Scotia Steel & Coal Co., New Glasgow, N.S.

Stove Pipe.
Chown, Edwin, and Son, Kingston

Stoves, Tinware, Furnaces
Canadian Heating & Ventilating Co. Owen Sound.
Copp, W. J., Son & Co., Fort William
Davidson, Thos., Mfg. Co., Montreal
Down Draft Furnace Co , Galt
Guelph Stove Co., Guelph.
Gurney Foundry Co., Toronto.
Harris, J. W., Co., Montreal.
Howard, Wm , Toronto
Kemp Mafg Co. Toronto.
McClary Mfg. Co. London.
Merrick Anderson, Winnipeg
Pease Foundry Co., Toronto.
Smart, James, Mfg Co., Brockville
Stewart, Jas., Mfg. Co., Woodstock, Ont.
Taylor-Forbes Co., Guelph, Ont.
Wright, E. T., & Co., Hamilton.

Tacks.
Montreal Rolling Mills Co., Montreal.
Ontario Tack Co., Hamilton.

Tents.
Tobin Tent and Awning Co., Ottawa

Tin Plate.
American Sheet & Tin Plate Co , Pittsburg, Pa
Baglan Bay Tin Plate Co , Briton Ferry South Wales
Lysaght, John, Bristol, Newport and Montreal

Turpentine
Defiance Mfg. Co., Toronto

Ventilators.
Harris, J. W , Co., Montreal.
Pearson, Geo. D., Montreal.

Wall Paper
Staunton Limited Toronto

Wall Paper Cleaner.
Gilbert, Frank U. S., Cleveland

Washing Machines, etc
Dowswell Mfg. Co., Hamilton, Ont
The Shultz Bros Co., Brantford.
Taylor Forbes Co., Guelph, Ont.

Water Filters.
Buffalo Mfg. Co., Buffalo, N.Y.

Wheelbarrows
London Foundry Co., London Ont.
Schultz Bros. Co., Ltd., The Brantford.

Wholesale Hardware
Birkett, Thos., & Sons Co., Ottawa.
Caverhill, Learmont & Co., Montreal.
Frothingham & Workman, Montreal.
Hobbs Hardware Co., London.
Howland, H. S., Sons & Co., Toronto.
Lamplough, F. W., & Co., Montreal.
Lewis Bros. & Co., Montreal.
Lewis, Rice, & Son, Toronto.

Window and Sidewalk Prisms
Hobbs Mfg. Co., London, Ont.

Wire, Wire Rope, Cow Ties, Fencing Tools, etc
Banwell-Hoxie Fence Co., Hamilton
Dennis Wire and Iron Co., London, Ont.
Dominion Wire Mafg. Co., Montreal
Greening, B., Wire Co., Hamilton.
Owen Sound Wire Fence Co., Owen Sound
Montreal Rolling Mills Co., Montreal.
Western Wire & Nail Co., London, Ont

Wrapping Papers.
Canada Paper Co., Toronto.
McArthur, Alex., & Co , Montreal.
Stairs, Son & Morrow, Halifax, N.S.

Wringers
Connor, J. H. & Son, Ottawa, Ont

A CHAIN of REASONS
IN APPEARANCE the most attractive, in use the smoothest and most flexible and in construction the strongest of their kind on the market are the
AMERICAN WELDLESS HALTER CHAINS
By means of our patented Lock Ring you can adjust the loop to any size and it locks fast. No slipping is possible, in spite of any pull or strain. The Snap is the most durable and effective made, never sticking on account of rust.
MADE IN ALL SIZES FROM DOG LEADS UP
ONEIDA COMMUNITY Limited
NIAGARA FALLS Ontario

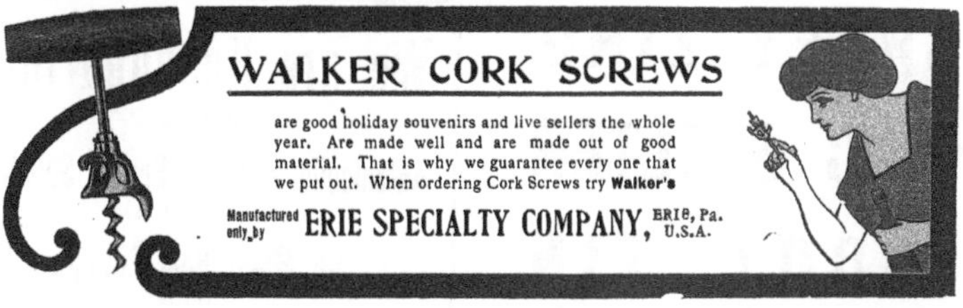
WALKER CORK SCREWS
are good holiday souvenirs and live sellers the whole year. Are made well and are made out of good material. That is why we guarantee every one that we put out. When ordering Cork Screws try Walker's
Manufactured only by ERIE SPECIALTY COMPANY, ERIE, Pa. U.S.A.

LUFKIN MEASURING TAPES
Steel, Metallic, Linen, Pocket, Ass Skin, Pat. Leather, Bend Leather, Etc
ARE THE BEST AND MOST POPULAR TAPES IN THE WORLD. YOUR STOCK IS NOT COMPLETE WITHOUT THEM.
LUFKIN RULE CO., Saginaw, Mich, U.S.A.
Canadian Factory - - Windsor, Ontario
London Office and Warehouse—24 and 26 Holborn. New York City Branch—280 Broadway.
For sale by ALL PROMINENT CANADIAN HARDWARE JOBBERS.

CIRCULATES EVERYWHERE IN CANADA
Also in Great Britain, United States, West Indies, South Africa and Australia.

HARDWARE AND METAL

A Weekly Newspaper Devoted to the Hardware, Metal, Heating and Plumbing Trades in Canada.

Office of Publication, 10 Front Street East, Toronto.

VOL. XIX. MONTREAL, TORONTO, WINNIPEG, AUGUST 10, 1907 **NO. 32.**

See Classified List of Advertisements on Page 66.

GEM FOOD CHOPPER

A modern household utensil that is needed in every kitchen. Carefully made, nicely tinned, self-cleaning, self-sharpening. It has steel cutters, and is made in four sizes.

No. 16	**capacity**	**1 lb.**	**per minute**
No. 20	**"**	**2 lbs.**	**"**
No. 22	**"**	**2½ lbs.**	**"**
No. 24	**"**	**3 lbs.**	**"**

Gem Food Choppers chop all kinds of Raw or Cooked Meat, Fish, Clams, Oysters, Vegetables of all kinds, Fruit, Bread, Crackers or Cheese, and many other articles used in making substantial dishes and dainty desserts. It is more useful than a chopping bowl and a great deal handier. It may be easily clamped to the kitchen table and is ready for use whenever needed.

RICE LEWIS & SON

LIMITED

TORONTO.

If you have to do a

CREDIT BUSINESS

why not make it as easy and convenient as possible?

THE CRAIN MONTHLY ACCOUNT SYSTEM

is designed especially for the retail trade. It will cut your bookkeeping in half; accounts are always ready.

We manufacture and devise systems for any kind of business.

Write us for Catalogue "E," which will give you all information.

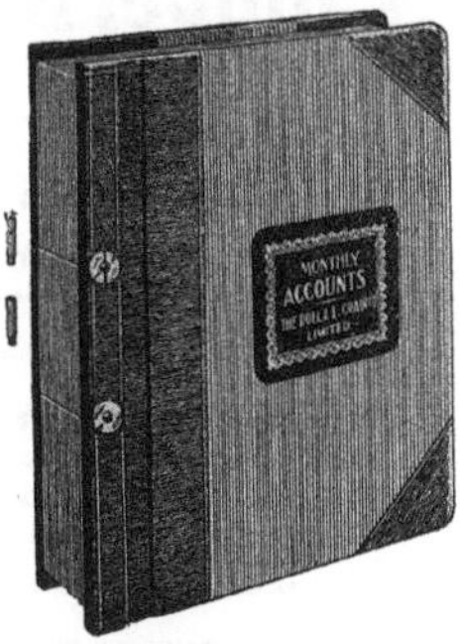

THE ROLLA L. CRAIN CO., Limited
OTTAWA, CANADA

Branches—TORONTO, 18 Toronto Street; MONTREAL, 28 Alliance Bldg., WINNIPEG, 11 Nanton Block.

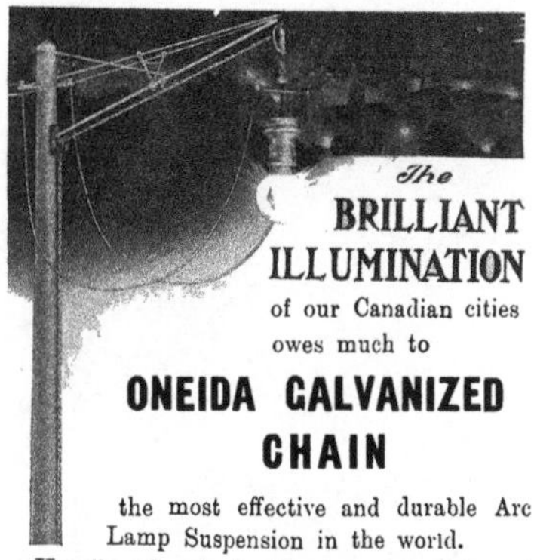

The **BRILLIANT ILLUMINATION** of our Canadian cities owes much to

ONEIDA GALVANIZED CHAIN

the most effective and durable Arc Lamp Suspension in the world.

Heavily galvanized—therefore absolutely rust proof. Impervious to ice and sleet. Uniform in strength. Perfectly flexible. ***Will outwear cord, cable or rope many times over.***

MILLIONS OF FEET IN USE.

ONEIDA COMMUNITY, Ltd, Niagara Falls, Ont.

Freezes two flavors of Ice Cream or an Ice or Sherbet and Ice Cream at one and same time, in one freezer.

SOMETHING ENTIRELY NEW —NEVER DONE BEFORE.

THE can is divided in two parts by a vertical partition. One flavor is placed in one, the other in opposite side of can, in each of which the scrapers, etc., operate. The can cannot be turned as in old style freezers, because of the partition. It is therefore swung to and fro by a rocking motion of the crank. This is a more comfortable and much easier movement. It is very much less tiresome than turning a crank. One can sit back in a chair with freezer alongside and rock the crank to and fro without discomfort or undue exertion and even read at same time. Although two flavors are frozen at once and with less exertion, no more time is required than in old style freezers.

The **Pails have electric welded wire hoops**, which are guaranteed not to break or fall off; Cans are of heavy tinplate with **drawn steel bottoms** that are guaranteed not to fall out or break and do not leak, the strongest and most durable Freezer Can made; the **Automatic Twin Scrapers** by their positive action insure perfect scraping of frozen particles from side of can. All inside parts are heavily coated with pure block tin, and outside parts all thoroughly galvanized.

SOLD BY LEADING JOBBERS

ICE CREAM FREEZERS THAT SELL

Send for Catalog.

THEY ARE	THEY EXCEL IN
Well Advertised	Easy Running
In Demand	Quick Freezing
Easily Sold	Economy
Satisfactory in Use	Convenience
Of Known Reputation	Practical Results

North Bros. Mfg. Co.
Philadelphia, Pa.

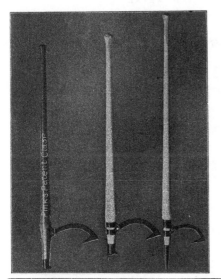

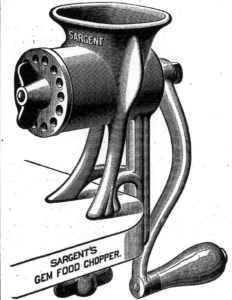

BUSINESS SUCCESS

Mr. Dealer, if you could take out a policy assuring commercial success in your line you certainly would do so. The next best thing is to place your orders for

Carpenters' Tools and Builders' Hardware

with a specialty house—a house which guarantees quality and prompt deliveries. We handle

Stanley's, Union Hardware Co. and Sarjent's Iron and Wood Bottom Planes.
Disston's, Atkins', and Shurly & Dietrich—High Grade Hand, Rip and Panel Saws.
Sarjent's Front, Store Door and Inside Locks.
Stanley's Butt Hinges—all kinds, all finishes.
Also headquarters for Wood Sash Pegs, sizes ⅞, 1¼, 1⅝-in. in one-bushel bags.

Remember our prices are right and we deliver like lightning

THOS. BIRKETT & SON CO., Ottawa, Canada

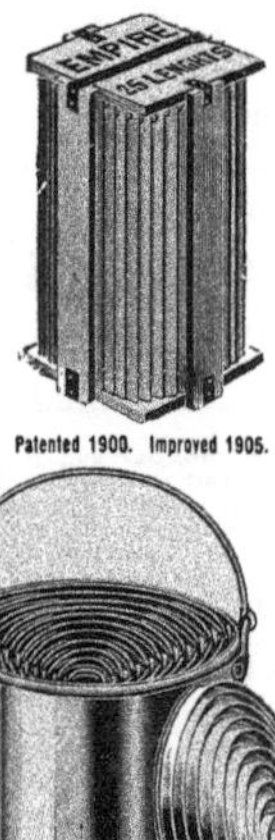

Patented 1900. Improved 1905.

Davidson's "Empire" Stove Pipe

Improved Patent

Neatly packed in crates of 25.
Secures lowest possible freight rate.

Deep Swage Prevents Telescoping.
Coated to Prevent Rust.

Sold as low as ordinary Stove Pipes.

SIMPLE and EASY to put together.
No Profane Language Necessary.
The only tools required are a pair of hands.
Pipes being cut out with dies ensure absolute fit and conformity.

Wash Tubs—Galvanized

With Wringer Attachment

Nos.	1	2	3
Top measurement, inches	21⅛	22¾	25
Bottom " "	17⅞	18¼	20
Height " "	9½	10½	11

Miners' Camp Kettles

Strong and substantially made in extra heavy tin.

Nineteen sizes, from ⅛ quart to 29 quarts.

The THOS. DAVIDSON MFG. CO., Ltd., Montreal and Winnipeg

When there is work to be done the tools that will do it *quickest* are "*Reece*" and "*Derby*" screw plate sets.

They will also do it *best.*

They make *perfect* threads at a *single cut.* Threads could not be cut more quickly than that nor better than perfect.

"*Reece*" and "*Derby*" screw plate sets are finely finished tools, put up in neat hardwood varnished boxes.

Sets are made up of both part and full assortments of sizes, adapted to every demand.

We keep in stock all the best selling sets and sell lots of them.

If you want to handle screw plate sets that will sell, handle "*Reece*" and "*Derby.*"

Caverhill Learmont & Co

MONTREAL

WINNIPEG, OTTAWA, QUEBEC, FRASERVILLE

Canada Leads the World in the Manufacture of

HIGH-CLASS SAWS

No. 81

The best and cheapest Hand Saw on the market, quality, temper and finish considered. Every saw warranted. Carved applewood handle. Patented, wave finish blade.

No. 1 Maple Leaf Lance Tooth Cross-Cut Saw, and the No. 0 Racer are tempered under our secret process.

No. 0 Narrow Racer Cross-Cut Saw

If you want saws that will sell and give your customer satisfaction, you will stock the Maple Leaf Saws.

Manufactured by **SHURLY & DIETRICH, - GALT, Ontario**

THREE BUSINESS BOOMERS

Maxwell Lawn Mower **Two Household Favorites**

Durable, Clean-cutting, and Beautifully Finished.

They are both "Self-Sellers." Keep them in stock.

David Maxwell & Sons, St. Mary's, Ont.

SIMPLE

DURABLE

RELIABLE

STYLE F.

THE LEADER

A Name Familiar to Butter Makers in Connection with the

Highest Grade Churns

ITS STRONG POINTS:

Stiff Steel Frame, with malleable iron fittings, Combination Bow Lev r and Pedal Drive. Bicycle Ball Bearings. Bolted Trunnions. Cream Breakers. Barrel easily drained and conveniently detatchable for cleaning.

DEALERS: When you handle this line you handle QUALITY.

The Dowswell Manufacturing Co., Limited

W. L. HALDIMAND & SON, Montreal — EASTERN AGENTS

HAMILTON — ONTARIO

GALVANIZED FENCE HOOK FOR FASTENING WOODEN PICKET OR WIRE FENCES

WIRE NAILS, COILED SPRING, BARB and PLAIN FENCE WIRE, OILED and ANNEALED, CLOTHES LINE WIRE STAPLES, etc.

THE WESTERN WIRE & NAIL CO., Limited, ——LONDON, ONT.

The Peterborough Lock Manufacturing Company, Limited

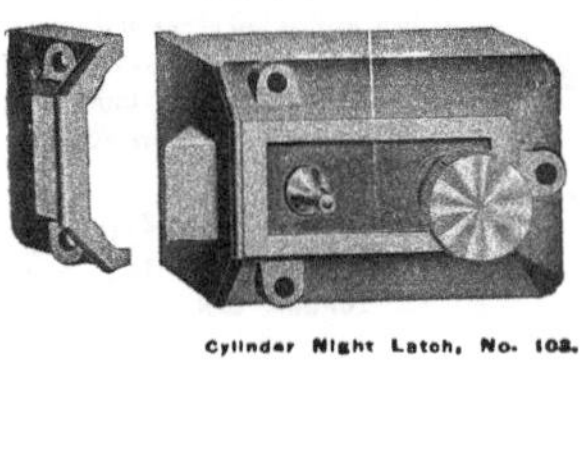

Cylinder Night Latch, No. 103.

Peterborough, Ont.

Manufacturers of all kinds

Rim and Mortise Locks, Inside, Front Door, and Store Door Sets, also full line High-class Builders' Hardware.

Sold by all Leading Jobbers in the Dominion.

Persons addressing advertisers will kindly mention having seen their advertisement in this paper.

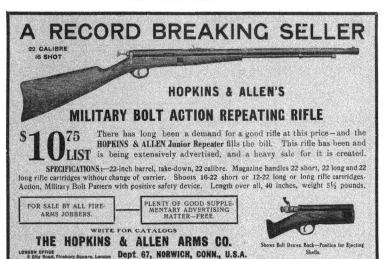

ESTABLISHED 1795

JOHN SHAW & SONS
WOLVERHAMPTON, Limited
WOLVERHAMPTON

Coil Chain
Canada Plates
Tin Plates
Black Sheets
Galvanized Sheets
Bar Iron
Lowest Prices for Import.

J. H. ROPER
CANADIAN REPRESENTATIVE
82 St. Francois Xavier Street, - MONTREAL

MONEY IN THE FENCE

There is money for you in the

"Dillon" Hinge-Stay Field Fence

It is the most ***DURABLE, ECONOMICAL*** and ***SALEABLE*** fence on the market.

It costs you nothing to get our descriptive literature—just **ask for it.**

THE OWEN SOUND WIRE FENCE CO., Limited
OWEN SOUND, ONT.

Sold by
Messrs. Caverhill, Learmont & Co., Montreal
Messrs. Christie Bros. Co., Limited, Winnipeg
The Abercrombie Hardware Co., Vancouver

BEAVER POST-HOLE DIGGER

will please your customer. A pleased customer is an asset to any business. No wood to rot, check or break.

Special Discounts to Trade

CANADA FOUNDRY COMPANY
LIMITED
HEAD OFFICE and WORKS, TORONTO, ONT.

District Offices: Montreal, Halifax, Ottawa, Winnipeg
Vancouver, Rossland

TRADE WITH ENGLAND

Every Canadian who wishes to trade successfully with the Old Country should read

"Commercial Intelligence".

(The address is 168 Fleet St., London, England)

The cost is only 6c. per week. (Annual subscription, including postage, $4.80).

Moreover, regular subscribers are allowed to advertise without charge in the paper. See the rules.

Persons addressing advertisers will kindly mention having seen their advertisement in this paper.

"KANDA" Reg'd. LIQUID METAL POLISH

BRITISH MANUFACTURE.

Absolutely non-inflammable. Has no bad smell.
Leaves no Verdigris. Contains no acid or grit.
Leaves no greasy smear. KANDA polishes all metal.

KANDA & CO., 45 Basinghall St., LONDON, ENGLAND

SOLE AGENTS for Ontario:

THE KENNEDY HARDWARE CO., Ltd.
TORONTO

Expert Bookkeeping

Investigation and auditing of all classes of accounts undertaken. Trading, profit and loss accounts and balance sheets prepared. Sets of books opened and office systems devised in an up-to-date manner. Collection of accounts a specialty.

Davenport, Pickup & Co.
622 McIntyre Block and 422 Ashdown Block
WINNIPEG, MAN.
AND AT BRANDON, MAN.

Square your Accounts

FURNACE SATISFACTION

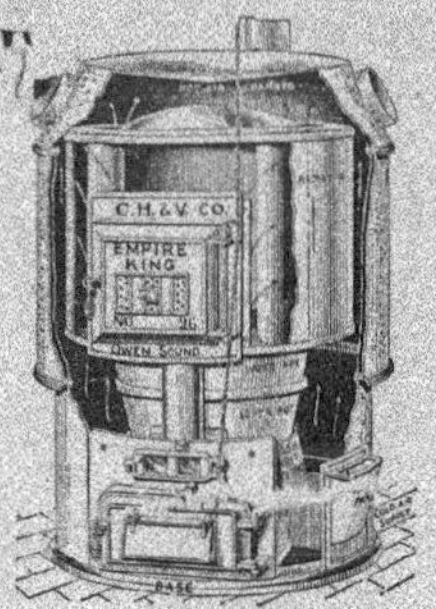

Apart from the Shrewd Dealer's desire to handle **The Best Furnace** the exigencies of business compel all wise Hardwaremen to stock and recommend That Particular Furnace which yields most satisfaction to their customers. It's an easily proven fact that **The Empire King** is known from ocean to ocean as **The Furnace that satisfies.**

Here are some of the reasons for superiority of

The Empire King Furnace

Simplicity of construction. **The Empire King** has few parts — therefore few joints. Furnaces having a large number of parts and complicated flues are trouble-creators and soon become useless.

Low Built. A low cellar ceiling does not eliminate the comforts a first-class heating apparatus affords. **Base ring**—is strong, heavy and solid—a fitting foundation for **The Best Furnace.**

Ask for more reasons why **you should talk the Empire King Furnace.**

The Canadian Heating and Ventilating Co., LIMITED

OWEN SOUND, ONT.

THE CANADA STOVE AND FURNITURE CO., 126 West Craig St., Montreal, Que., Agents for the Province of Quebec.

THE CHRISTIE BROS. CO., Limited, 238 King St., Winnipeg, Man., Western Agents.
The Abercrombie Hardware Co., Vancouver, B.C.

IT'S A LATCH THAT EVERY BARN NEEDS

WHITCOME

STEEL BARN DOOR LATCH

It is convenient; it is strong: it is durable; it is adapted to any thickness of door; and harness, etc., will not catch on it. Fully guaranteed.

Get a sample and see what an excellent Latch it is.

ALBANY HARDWARE SPECIALTY MFG. CO.

Manufacturers of PATENTED SPECIALTIES BY CONTRACT

ALBANY, WISCONSIN, U.S.A.

THERE IS MONEY IN MERIT!

An article which sells readily on its merit is a profitable line for you to handle. It cannot be denied that the

JONES SIDE-WALL REGISTERS

have all the customer-pleasing points—saving your customer's money on a heating system by reducing piping 50 per cent and giving thorough and lasting satisfaction.

WRITE FOR PARTICULARS

The Jones Register Co.

732 King St. West ***Toronto, Canada***

Wrought Steel Registers

Standard Sizes, All Finishes
Perfect Operating Device
Unusually Large Air Space
For Sale by Leading Jobbers

HART & COOLEY CO.

New Britain, Conn., U.S.A.

H. & C. No. 100.

FERROSTEEL REGISTERS

The only house in the world manufacturing

CAST IRON FLOOR
SEMI STEEL FLOOR
WROUGHT STEEL FLOOR
CAST IRON SPECIAL SIDE WALL
WROUGHT STEEL SPECIAL SIDE WALL
FLAT OR CONVEX EXTRA SHALLOW SIDE WALLS
DEFLECTORS, PANCAKES, WAFERS

In all Designs and Finishes.
Mechanism Perfect.
Capacity Greatest.
Finish Unsurpassed.
No reputable line can undersell us.

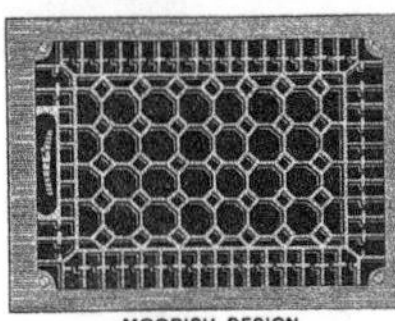

MOORISH DESIGN

FERROSTEEL COMPANY, - - CLEVELAND, OHIO

"Brantford" "Roofing"

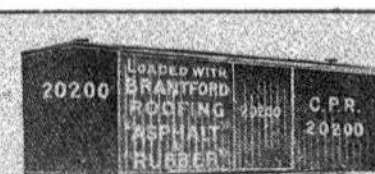

We beg to announce to the Trade throughout

Manitoba, Saskatchewan, Alberta and British Columbia

that our representative, MR. J. S. McGUIRL will call upon you shortly in the interest of

Brantford Roofing Co., Brantford, Canada.

You May Be a Good Salesman

But, is everything in your favor? Are you selling the new Imperial Oxford? A continuous stream of repeat orders tells the story to us.

Removable Nickel
Removable Fire Linings
An Oval Firebox
The "Divided Draft"

All contained by the best looking stove in the market.

It gives a good salesman a chance to spread himself.

Drop us a card, or, better still, tell us the size range you want on your floor. If you get this agency you're lucky.

The Gurney Foundry Company, Limited

Toronto Winnipeg Vancouver Hamilton and London

THE GURNEY-MASSEY CO., LIMITED, MONTREAL, Que.
THE GURNEY STANDARD METAL CO., LIMITED, CALGARY, Alta., and EDMONTON, Alta.

"SUNSHINE" FURNACE

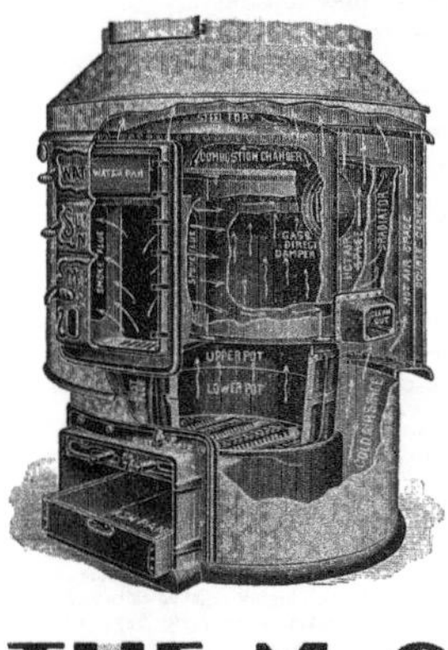

The most successful hot air furnace in Canada.

Built to burn all kinds of fuel successfully.

Large double feed doors admit rough chunks of wood and make it an easy furnace to feed with coal.

All parts exposed to fire and wear are made extra heavy.

Never gives any trouble after being properly installed.

Has more good selling arguments than any other furnace made.

Extensively advertised from Halifax to Vancouver.

If you are dissatisfied with your present line write for catalog and complete information on the "Sunshine."

THE McCLARY MFG. CO.

LONDON, TORONTO, MONTREAL, WINNIPEG, VANCOUVER, ST. JOHN, N.B., HAMILTON, CALGARY

To

EVERY MAN

who uses

Fire Bricks

We handle the best makes of Scotch Fire Bricks, and our prices are right, but there are occasions when you must have more refractory materials, and then please remember that we are the Canadian representatives of the

"HARBISON WALKER REFRACTORIES" COMPANY

and this should indicate that we are the Fire Brick House of Canada.

Others of our exclusive specialties are

Lafarge
Non-Staining Cement

Iron Clad
Portland Cement

United Paving Bricks
Best in the World

Enamelled Bricks
All Colors

Drain Pipes

Plaster, etc.

Francis Hyde & Co.

KING, QUEEN and WELLINGTON STS.

MONTREAL

Smooth finish, perfect threads and first-class material, are the reasons why

"Diamond" brand Fittings

are in such constant demand.

WHOLESALE ONLY

OSHAWA STEAM & GAS FITTING CO., LIMITED

OSHAWA, CANADA

Don't Stop

to ask your neighbor, lift the load yourself with **The Burr Self-Locking Tackle Block.** Saves the labor of two or three men. 600 to 5,000 pounds capacity.

Lewis Bros., Ltd., Montreal
Munro Wire Works, Winnipeg
Vokes Hardware Co., Toronto
F. Hamilton & Co., Hamilton
The Burr Mfg. Co., Cleveland, Ohio

YOU RUN NO RISK

when you handle our Stove and Furnace Cement. We give you a money-back guarantee that

STERNE'S ASBESTOS STOVE AND FURNACE CEMENT

is the very best. We are sure of it, or we wouldn't guarantee it.

Send for particulars anyway.

C. F. STERNE & SONS,

BRANTFORD, - - ONTARIO

Steel Troughs and Tanks

We Manufacture

Steel Tanks, Stock Tanks, Steel Cheese Vats, Thresher's Tanks, Hog Troughs, Water Troughs, Feed Cookers, Grain Boxes, Coal Chutes, Smoke Stacks.

Agents Wanted.

The STEEL TROUGH and MACHINE CO. Ltd., TWEED, ONT.

Forwell Foundry Co.

BERLIN, ONT.

Manufacturers of

SOIL PIPE, FITTINGS, and CAST IRON SINKS

Ask Jobbers for **"F. F. CO." Brand**

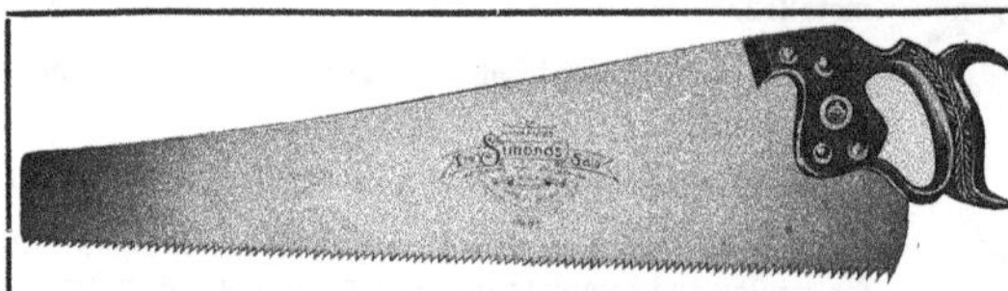

SIMONDS HAND SAWS

are good sellers. Dealers should send for 1907 Catalogue and discounts.

SIMONDS MANUFACTURING CO.

Fitchburg, Mass.

Branches Throughout The United States

SIMONDS No. 8½

This Saw is of the general style and shape which carpenters have been accustomed to using for years past. Has a carved apple handle with highly polished sides and edges, making a very pretty Saw. Holds its cutting edge and saws true. Its high quality and even temper are due to the fact that it is **Made of Simonds Steel.** Covered by the broadest Simonds warranty.

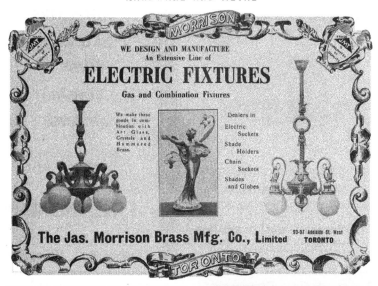

DOMINION WIRE MANUFACTURING CO.

MONTREAL TORONTO LIMITED

Manufacturers of

PLAIN WIRE *All Sizes and for all purposes.*

GALVANIZED WIRE *For Fencing, Telegraph and Telephone purposes.*

Copper and Brass Wire

WIRE NAILS
WOOD SCREWS

Bright Wire Goods
Staples
Steel and Brass Jack Chain
Spring Cotters, Strand Guy Wire, Steel Wire Barrel Hoops

It's the **GLASS CONE**

that makes this new burner a great success

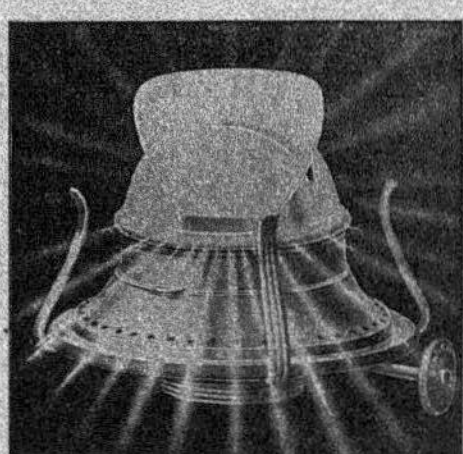

It gives full benefit of flame down to the very wick top.

FOR SALE BY ALL PROMINENT DEALERS.

Manufactured by the

ONTARIO LANTERN AND LAMP CO.
HAMILTON, ONT. LIMITED

The Buffalo Manufacturing Co.

Buffalo, N.Y.

When you get our goods you know you get THE BEST.

We manufacture
Water Filters
Water Coolers
Chafing Dishes
Table Kettles and Stands
Coffee Extractors
Wine Coolers
Nursery Chests
Baking Dishes
Crumb Trays and Scrapers
Tea and Bar Urns
Bathroom Fixtures
Coal Vases and Hods
Candlesticks
Cuspidors
Match Safes, Etc.

All High Grade and exceedingly presentable.

REPRESENTED BY

H. F. McINTOSH & CO.

34 Yonge Street

Write for Catalogue TORONTO, ONT.

PROFIT and REPUTATION ASSURE SUCCESS

Are your profits reasonable and secure? Unless you are absolutely certain on both points you are playing a losing game. You can secure both reputation and profit permanently by stocking the proper kind of

WINDOW GLASS

We do not pretend to manufacture imperishable window glass, but it is an easily proven fact that we manufacture the next best thing.

Have you our catalogue? You ought to have it because it contains suggestions for **Fan Lights, Door Lights, Hall Windows, Side Lights, Etc.,** made in Copper or German Brass.

All orders shipped landed in proper condition because they are properly packed.

LET'S MAIL YOU SOME INTERESTING FACTS.

The Hobbs Manufacturing Co., Limited

LONDON TORONTO WINNIPEG

McCaskill, Dougall & Co.

Manufacturers RAILWAY, CARRIAGE AND BOAT **VARNISHES.**
HIGH-GRADE FURNITURE and HOUSE VARNISHES

MONTREAL.

Above is a reproduction of Show Card 9 in. x 15 in. in colors which we will send free to dealers in exchange for business card.

HENRY T. SEYMOUR SHEAR COMPANY. WIEBUSCH AND HILGER, LIMITED, Sole Agents, NEW YORK.

A Certain Sense of Satisfaction

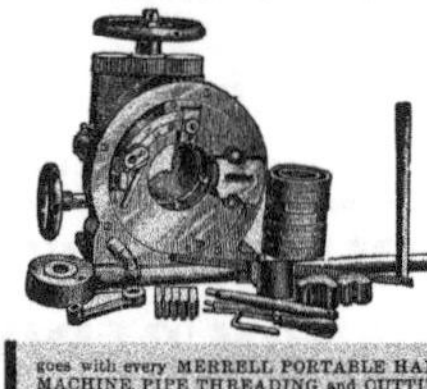

goes with every MERRELL PORTABLE HAND MACHINE, PIPE THREADING and CUTTING MACHINE. We build them that way.

For doing ALL kinds of threading—under ALL conditions—labor trouble and faulty material included—where speed, durability and ease of control count for something, our word to you is to investigate the MERRELL.

Whether it be stationary or portable—hand or power driven—30 DAYS FREE TRIAL must prove conclusively that your choice should be the MERRELL.

This Portable Hand Machine has encased gears, MERRELL standard quick opening and closing die-head, and the latest improved Cutting-off Knife.

Let us tell you more about this machine—The Chasers, Vise and the large range of work covered.

Catalogue for the asking.
Quick shipments.

THE CANADIAN FAIRBANKS CO.

Sole Agents for Canada Limited

MONTREAL, TORONTO, WINNIPEG, VANCOUVER

PRESSED ZINC ORNAMENTS

We are offering the trade spendid Values in all kinds of architectural pressed zinc ornaments, including:

CABLE ORNAMENTS, CAPITALS, BRACKETS, CARLANDS, ETC.

We can make up special ornaments to order, and inVite your correspondence on the subject.

" WE WORK WITH THE TRADE "

The Metal Shingle and Siding Co., Limited

Montreal Preston, Ont. Toronto

SALES AGENTS:

ST. JOHN, N.B. Emerson & Fisher, Limited
QUEBEC J. A. Bernard
WINNIPEG Clare & Brockest
CALGARY Ellis & Grogan
VANCOUVER McLennan, McFeely & Co.

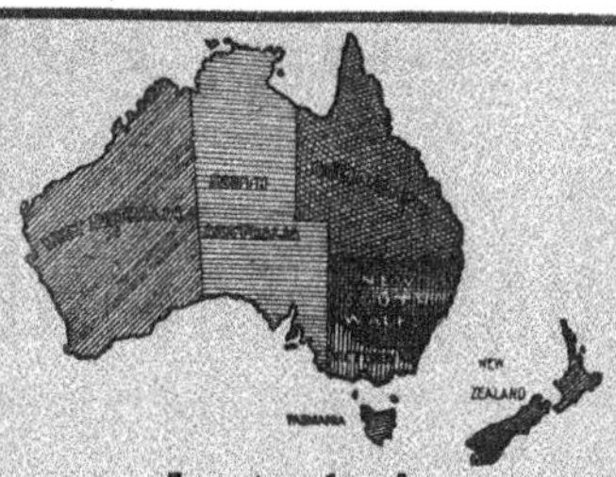

Australasian Hardware and Machinery.

The Organ of the Hardware, Machinery and Kindred trades of the Antipodes.

SUBSCRIPTION $1.25 PER ANNUM,

post free to any part of the world.

PUBLISHING OFFICES:
Melbourne - Fink's Buildings.
Sydney, - - - Post Office Chambers.

BRITISH OFFICES:
London, - - 42 Cannon St., E.C.

CANADIAN AND AMERICAN ENQUIRIES will receive prompt attention if addressed to the **LONDON OFFICE, 42 CANNON STREET, E.C.**

Specimen Copies Free on Application.

TRADE MARK
FOR PLATE

Will Call on You

Our Canadian representative is calling on the trade from Montreal to Vancouver. His trip will occupy about three months. He will be pleased to show you our complete line of Silverware, Cutlery, Flasks, etc.
Kindly address the Montreal office, making an appointment.

The House of Hutton

Established in England in 1800

Sole Canadian Agent
W. J. GRANT, Lindsay Bldg., MONTREAL

TRADE MARK
FOR CUTLERY

The Howland Pump Oiler

FOR SALE BY

J. H. Ashdown Hdw Co.,	Winnipeg
Geo. D. Wood & Co.,	"
Miller-Morse Hdw. Co.,	"
Jas. Robinson & Co.,	"
Lewis Bros., Ltd.,	Montreal

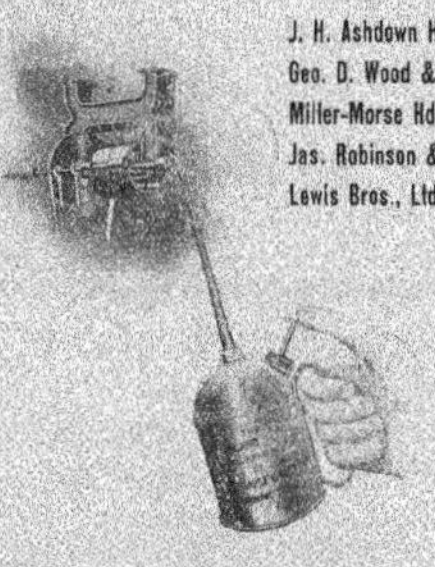

A slight pressure of the thumb regulates the flow.

Maple City Manufacturing Co.

Monmouth, Ill.

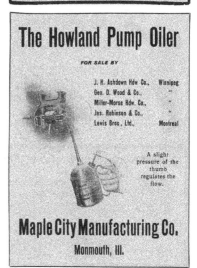

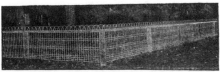

Does Underhand Competition Pay?

Some instances told by C. A. Kiler tend to show that co-operation is more profitable.

Competitors in business generally misunderstand each other, and nearly always each is ready to attribute unholy motives to the other fellow without thoughtful consideration or investigation of any kind.

Is it wiser for the merchant to work in harmony with his competitors, agreeing to sell certain staple goods at a common price, or persevere in the old practice of underselling competitors and thus seek to win trade by the power of lower prices?

Thoughtful people must know that a merchant cannot sell goods at cost, except such merchandise as shop-worn articles, passed styles and finishes, and goods purchased under abnormal conditions from overloaded factories and the like. A checker player frequently gives away one man to his opponent in order to win two or three by the new position he has gained. And everybody must understand that merchants frequently sell staple articles at cost in order to sell many other articles at a profit.

One buying an article of which he knows the value recognizes a bargain when he sees it. But the average customer doesn't know values and must depend on the storekeeper to give him what he wants.

Storekeepers are closer to the people than any other class of men, and they run their stores to meet the requirements of their customers—in other words, they are what their customers force them to be.

Cheap Goods.

People who complain of shoddy goods which they buy generally want cheap goods or live in a community where the demand is greater for cheap goods than it is for goods of a substantial character.

The man who thinks a home merchant is trying to skin him may be unwilling to pay a fair profit on his purchases and therefore courts the treatment of which he complains. He generally is the man who assumes to know what goods cost and goes from one store to another until he gets the kind of a price he wants. Then in the end if he finds he has to pay for unaccounted extras he says with a loud voice that he has been skinned.

If competitors would play fair with each other and let it be known when the sharp man is going the rounds he would find about the same prices at each store and be forced to buy goods at a proper price. He would then be likely to quote prices from catalogue houses or catalogues from out of town stores, but this talk is easily met; most merchants are happy to duplicate goods shown in catalogues at the same price and under the same conditions at which the goods are offered. These conditions are cash in advance and the customers to pay all the freight, drayage, and carting charges.

It is never good for the merchant or the people when staple goods are sold below the staple price at which such goods are sold the country over.

In a certain Illinois city standard 9x12 Wilton rugs are sold for $29, while over the rest of the entire country these rugs bring $35. They are listed at $32 wholesale and must be bought in lots of 100 in order to be sold at the price for which they go in this particular city. This low price came from the jealousy existing between competitors — one thought the other was cutting prices and therefore determined that he would not be undersold.

Price Cutter Sick of His Policy.

Some people think that it is a good thing for them when merchants get to cutting prices down to cost or lower—the price of the rugs, for instance, puts them where many people can buy them. People will pay $29 much quicker than $35. There is no doubt but that more rugs are sold in this city than in any other city of its size in the country, but the dealers are getting sick of handling them. The man responsible for the ruinous price is to-day advertising an inferior rug as "the best domestic rug made—price $35"—which indicates that he is anxious to discourage the sale of the standard rugs at cost. He even is willing to sell an inferior rug at the price the good one should bring in order to make up for the loss he has sustained. Is this good for the people? Is it good for the merchant? Decidedly not.

It may be said that law "caveat emptor"—the buyer takes the risk—still obtains, and people should know good rugs from poor ones. It may be that people should know values, but it is certain that people do not. We read of great bankers and railroad owners paying fortunes for imitations of ancient art, and the best of men do not know values outside of their own particular fields of endeavor.

A furniture dealer recently went into a competing store to ask a favor which readily was granted to him. While there he saw a sideboard like one he had and marked at the same price. On reaching his own store he marked his sideboard $5 cheaper and considered himself a smooth man. When the competitor found out what had been done it was natural for him to mark his own sideboard still lower—both men expect to make up for the loss in other ways.

Should be Fair With Competitors.

How different was the competitor who went into the same furniture store and saw some porch shades which he was also handling, inquiring the prices at which they were being sold, he found that he was lower on one size and higher on another, so he went back to his store and made his prices exactly like those of his competitor. He was wise and his customers are sure of fair treatment always, because by this action he proved himself to be a fair-minded man. By the same token the merchant who marked his sideboard down below the staple price in order to outdo his friendly competitor showed an unfair mind and a disposition which will take advantage of a customer's lack of knowledge. It is better for competitors to get together than seek to ruin each other. It is better because when they are fair to each other they are fair to their customers, and nothing but good can come from such business conduct. If the laborer is worthy of his hire, the merchant certainly is worthy of his reasonable profit. Getting together enables them to make this profit and to feel safe in asking it.

Honest, fair competition really is not competition at all—but legitimate business. The great trouble is that we have come to misunderstand the word competition and confuse it with throat cutting. Competition has become throat cutting and that is not the life but death of trade. Price agreements—if not carried to criminal ends are legitimate and fair to all.

How do You Figure?

It is claimed that one of the chiefest reasons for commercial failure throughout the country, is the lack of merchandising ability. Merchants may buy right but it is another thing to sell right.

Many merchants figure their percentages too low to enable them to cover all the expenses of their business, they allow nothing for their own labor, nor do they count on taxes, depreciation, wear and tear on building and fixtures, heating, stationery and scores of little things that in the aggregate make a fair burden of expense that is not considered in price-making.

Another thing that the merchant frequently neglects is the matter of getting profit sufficient to cover losses on leftovers or shop-worn goods.

It would be well to avoid this loss and there is only one way in the world to do it. Moreover in this day it is easy to abide by the rule. Buy frequently and in small quantities. In this way fresh goods, new and attractive goods are always on display and they will certainly increase your business. Moreover they will help you to get your profit.

From Newsboy to Hardware Jobber

J. West tells in the Hardware Trade how a newsboy developed the qualities of a successful merchant.

Jimmie Connors was a scientific newsboy. That is why he now is a trusted department manager in a downtown hardware firm, and if he ever reaches a higher rung on the ladder of success than the one he is on now he will owe it to the same energy and foresight that enabled him to reduce his first job, that of selling newspapers, to an exact system. Jimmie was born while the silver spoon was on its annual vacation, and his early education was pounded into him by that rough old teacher to whom the poets refer as the best in the pedagogical profession. When he was still in the primary class he was given a bunch of afternoon extras for a textbook, and Jimmie, with an appetite that was bordered on all sides by a first-class vacuum, found that the only way to eliminate this large and unnecessary border was to dispose of his textbooks to whatever passerby cared to take them off his hands at one cent each.

That was the start, and those who know Jimmie predict a finish that will land him well up among the high and mighty. He early learned the importance of little things, and this learning he applied to his first business venture just as he is applying it to his present work.

Long on Brains.

Jimmie was a newsboy only a few years before he developed into a merchant. His stock in trade was the same after he became a merchant as it was while he was a newsboy, but there is a great distinction between newsboys and merchants. Anybody can be a newsboy, but it requires brains to be a merchant, and Jimmie was long on brains.

All merchants are students of human nature, and Jimmie, true to his calling, made it a plan to know the men and women with whom he dealt. He did'nt bother about knowing their names, because that was not as necessary as a thorough knowledge of their natures and tastes. He early learned what the majority of present day newsboys have not yet found out, and that is, that the average man or woman does not like to look for his or her paper on the doorstep in the morning only to find that it has blown over into the next yard, or, what is equally bad, that it was thrown carelessly on the floor of the vestibule, where everybody coming into or leaving the house or flat could walk over it.

He looked upon this practice, so common among newsboys, as a defect in the business, and soon determined to alleviate it. This he did by always making sure that every paper was securely folded and wrapped with a rubber band and then carefully pressed behind the door knob, where it could not fall out or be blown away.

Studied His Customers.

At first Jimmie worked in the downtown district, but he soon picked out a better location in an outlying part of the city, where he worked up a sort of delivering and selling business. That is, he had a number of customers to whose homes he regularly delivered the morning and evening papers, and then, to fill in the day, he had a stand at an important transfer corner where he sold papers to those who bought them from day to day rather than subscribe for any one regularly. Delivering, however, he looked upon as a side line, as it was in selling that he made most of his profits.

Nobody ever sold papers like Jimmie did. It was not his plan to merely hand out the paper and grab the money, like the average newsboy does. He studied his business and his customers. Located, as he was, at a transfer corner, he made it a point to learn in what particular department of the paper each of his customers was interested. One of the men who bought off him every morning was a broker, and Jimmie saw that he invariably turned to the financial page the first thing. When he became sure that this man was interested in stocks and bonds, he turned his paper to that page and handed it to him, neatly folded, as he hurried to the street car.

Gave Broker Financial Page.

At first the broker thought it was merely an accident that his paper was turned to his favorite page, and while he was unconsciously thankful for being thus relieved of the necessity of wrestling with his paper in the wind, or jabbing his fellow passengers in the crowded street car as he turned the pages, he did not associate the convenience with Jimmie Graduallv, though, he noticed that every morning his paper was turned and folded the same way, and then he remarked to one of his friends whom he met on the car, that this certainly was an accommodating newsboy.

But the friend was not surprised. He also was a customer of Jimmie, and he had found out that the newsboy had learned in some way that his interest lay in sports. Every mornibg he had found his paper turned and folded to the sporting page, and he, like the broker, at first had supposed it was merely an accident and not business acumen on Jimmie's part. The broker thought it was the funniest thing he ever heard of, and that day he told all his board of trade friends about it. Having a good income of his own, and appreciating the extra attention that was given him by the newsboy, the broker soon formed the habit of slipping Jimmie a quarter as he hurried on to his office. And the beauty of it was that this quarter came to him as a reward of merit and not as a tip.

Attention Doubled Business.

Jimmie knew that he was giving the broker something he wanted and for which he was willing to pay. He didn't take the trouble to turn and fold papers for nothing. He knew he had a good scheme and he soon found out that he could double and treble his income by that little extra attention to his business

With the women he was the same way. He learned where each of his women customers traded and he figured, and rightly, that the average woman on her way downtown was interested more in department store advertisements than anything else, and with this knowledge as a basis he soon learned where each of his customers did most of her trading.

One handsomely dressed woman was surprised one morning, just as the broker had been surprised, to find her paper turned so that the advertisement of her favorite store confronted her when she got on the car. She, too, thought it was an accident, but when the same thing happened several mornings in succession she began to think. Then she told her friends, and then—what was more important to Jimmie—she gave him a dime with the remark that he was a genius. Jimmie didn't bother about the genius part, but he knew that dime would come handy in his business.

Before long, Jimmie had worked up such a trade that the circulation managers of the various newspapers began to make inquiries about him, and one man even condescended to visit him, None of them had been able to figure out why so many papers went to that corner, and when they found out there was a frantic effort on their part to have his methods applied to other newsboys in other parts of the city. But it was no use.

Won Success in Business.

Jimmie was Jimmie. His individuality was his own, and try as they would the circulation managers could not convince the other newsboys that his system was a good one. In their estimation his plan involved too much work and as the income was without a promise of immediateness or certainty, the idea did not appeal to them.

Jimmie could not long remain a newsboy. One of the men to whom he sold a daily paper became interested in him and finally, when the time was ripe, he made Jimmie an offer that could not well be rejected. This man was in the hardware business, and he figured that any boy that could sell newspapers as Jimmie sold them was fit for better things. His judgment proved correct, and it was not long before the former newsboy was setting a pace in the hardware firm's office that made the other clerks wonder if he was run by electricity. In a few months he was promoted to a foremanship and then to the superintendency of a branch house. Before long he was taken back into the home office as a department manager, and now the head of the firm is beginning to wonder how long it will be before he is crowded out of the business altogether by his hustling and energetic subordinate.

Use one of each kind of gasoline and kerosene stoves you handle for demonstrating. That is, keep some oil in them and light them up for the customer—we do not mean that you should bake biscuits in them all the time.

Window and Interior Display

A MISCELLANEOUS DISPLAY.

The cut herewith reproduced shows a splendid arrangement of a miscellaneous display of hardware, the chief lines being cutlery, sporting goods, mechanics' tools, and plumbing fixtures. This display in the window of the store of L. J. A. Surveyer Montreal, was dressed by H. A. Bernier, who has many times previously proved an eye to artistic workmanship.

The floor of the window is covered with green baize, and arrayed on it are boat fittings and mechanics' tools. Toward the back and middle of the window will be noticed a simply constructed arch beautified by an ingenious arrangement of trolling hooks and a small tiller at the top. On either side of the arch are vertically arranged cutlery—knives, spoons and razors. Underneath the arch are three ice water tanks. At the back of the window is a display of feather dusters. At either side of the window are displayed mechanics' tools, bird cages and plumbing fixtures. Every line of hardware is shown and for such a miscellaneous assortment, the display is an exceedingly creditable one.

One or two suggestions, however, might be offered. The mechanics' tools, such as bits, augurs and chisels, which are so attractively shown, might be displayed to even better advantage, and to the improvement of the general appearance of the window, by having them nearer the centre of the display. Again the floor of the window might have been raised at the back to more effect.

One thing which the window-dresser kept in mind, and which is absolutely essential to an effective display, was cleanliness. Everything in the window, including the plate glass itself was clean and bright.

MISCELLANEOUS HARDWARE DISPLAY BY MONTREAL STORE.

DISPAYS IN SMALL TOWNS.

An American hardware contemporary recently raised the question, 'Does it pay to dress show windows in small cities and towns, and how frequently?" The editor published answers from twenty-six of his readers, who spoke on the subject from personal experience. One gave a decided "No" to the first part of the question, another could not trace any results to his efforts in this direction, and a third did not think it paid in small towns, "as there is no protection from dust, which damages the goods." All the rest were satisfied that window dressing pays even in the smallest towns. Many of them were most emphatic in their testimony. "An attrtctive show window is worth more in my estimation than a two-column ad. in a newspaper." "Our window advertising is the cheapest and best and most profitable part of our business." "Good show windows increase your sales every day," "It pays well to dress windows, even in small towns." And so on. But the thing has to be well done; this is insisted upon by one contributor after another. The windows must be kept clean, and so, too, the goods shown in them. That means frequent overhauling. Change every week is the usual suggestion, although a few are content with fortnightly redressing, and one, like the moon, comes up new only once a month. To facilitate this frequent change, simple, instead of elaborate, displays are recommended by one retailer. "The people will soon learn that you change your window often," writes the dealer, "and will stop and look every time they go along to see what you have in your window this week."

HARDWARE AND METAL

Established 1888

The MacLean Publishing Co.
Limited

JOHN BAYNE MACLEAN - President

Publishers of Trade Newspapers which circulate in the Provinces of British Columbia, Alberta, Saskatchewan, Manitoba, Ontario, Quebec, Nova Scotia, New Brunswick, P E. Island and Newfoundland.

OFFICES:

MONTREAL, ... 232 McGill Street
Telephone Main 1255
TORONTO . . . . . . 10 Front Street East
Telephones Main 2701 and 2702
WINNIPEG, ... 511 Union Bank Building
Telephone 3726
LONDON, ENG . . . . . 88 Fleet Street, E.C.
J. Meredith McKim
Telephone, Central 12960

BRANCHES:

CHICAGO, ILL. - . . . . 1001 Teutonic Bldg
J. Roland Kay
ST. JOHN, N.B. . . . . No. 7 Market Wharf
VANCOUVER, B.C. . . . . Geo. S. B. Perry
PARIS, FRANCE - Agence Havas, 8 Place de la Bourse
MANCHESTER, ENG. . . . . 92 Market Street
ZURICH, SWITZERLAND . . . . Louis Wolf
Orell Fussli & Co.

Subscription, Canada and United States, $2.00
Great Britain, 8s. 6d., elsewhere . 12s

Published every Saturday.

Cable Address { Adscript, London / Adscript, Canada

BUSINESS POLITICS.

To be invariably successful, it is obvious that every deliberation and transaction in a business concern be politic and canny. An Italian proverb says, "If you would be successful, do not be too good" From the austere viewpoint of the moralist, the above proverb is a glaring fallacy and in practice must be severely let alone. It really means that if a merchant is desirous of making a big sale he is justified in being unscrupulous as to the means he shall adopt to effect the sale. It cannot be denied that the tendency in commercial circles is swinging around in this direction.

Far from defending the principles of unscrupulous bargain-driving, we are constrained to put down in type some of those things which characterize the politic business man. Commercial methods are every year undergoing radical changes. The fact that such methods are sensible to the workings of evolutionary law is the best proof that they are working onward and upward, each step bringing them nearer that standard of perfection which should be the goal of all institutions

The first essential to the successful carrying on of a business is persistence and self-reliance. Especially do these two qualities prove invaluable in a newly organized concern It needs courage and persistence to stick to the helm in stormy seasons of financial depression and scanty patronage, when each week sees an ominous increase on the credit side of the ledger and still more ominous decrease on the debit side.

Then is the opportune time to bring into requisition a third essential quality, enterprise Happy the merchant who can in times when destiny is unkind, call upon the power of invention and receive a prompt response. If he can hit on some new scheme for attracting people to his store, and thereby increase his patronage, he is fortunate. Every man in this world has some reserve power. It is seldom called into requisition, and eminently successful is the man who can call out his reserves at the right time It is not those powers possessed by men as evidenced in their daily round of business duties which place them higher or lower in the world's estimation, nor is it so much the aggregate power of their reserve forces, but it is the wisdom and canniness displayed in calling out these reserve forces at the opportune moment, and in thus gaining the day. It is said by our pushing neighbors across the border that we lack enterprise, and the audacity to invest, that our industries are still in the primitive stages of development because of lack of enterprise on the part of the promotors. It is hard to deny the truth of these complaints. Even in the hardware trade there are evidences of a lack of enterprise and the investor's boldness.

SHALL COURTESY BE RESTRICTED?

Doubtless the average merchant would regret the slightest suggestion of placing limitations upon the degree of urbanity to be accorded patrons as a straining after a subject of debate. Unquestionably, too, he would not be disposed to bestow much thought upon the proposal that visitors to the shop be treated with scant courtesy if their purchasing propensities are not of the most desirable sort.

Yet the fact remains that the question is one which is claiming the attention of merchants in this country and abroad. Recent issues of English trade journals have contained numerous references to the part politeness should play in business. And be it said, the comments have been varied, and quite emphatic on one side as on the other. Some have contended that commercial transactions do not call for more than ordinary civility; that a man gets what he pays for; that the merchant is in no way indebted to him for his patronage. It has been argued, also, that too frequently this quality of politeness is overdone, and the clerk or the proprietor who puts himself in the light of one who must affect an extremely affable manner in order to hold his custom thereby detracts from his shop's reputation as a place where value-giving alone is relied upon to hold trade. Now appears a prominent haberdasher with a plea for a change of tactics on the part of his fellow merchants. Says he:—

"You may discuss this matter eternally from the standpoint of propriety, and different men will still be of different minds. What I maintain is based upon practical operations. I have seen many a man enter a shop and buy one collar. The salesman's face would beam. He would draw forth a mirror and aid the purchaser in adjusting his cravat. In fact, in every way the merchant has impressed me as appreciating highly that man's custom when he knew absolutely that his visitor bought only when he felt it unavoidable, and then in the smallest possible quantity.

"Now in England such a man would mighty soon be made cognizant of the fact that no great value was placed upon his custom. If he repeated the offence several times the merchant would make his opinion so apparent the man would go elsewhere the next time.

"I believe that such an attitude toward the chronically small buyer should be encouraged. I believe we could create a new conception of business, one that would make men hesitate to play the 'cheap' act with us when they are extremely free with their shekels in the cigar store and elsewhere. Mind you, I do not advocate a coarse, rough treatment of anyone, yet I believe that by his manner the merchant and his salesmen may indicate their sentiments towards those who make a continual practice of buying the smallest amount of merchandise possible to carry them through the day or week."

With all due deference to this view, we incline to the belief that whatever be the policy pursued abroad, the Canadian people, to speak bluntly, would not stand for it. They are remarkably independent. They realize how keen competition is, especially in the large cities. They are given to resenting distinctions of dollars or class. What they must have is careful, considerate handling. If this is not to be found in one shop they will hie themselves to another. And in the long run, the establishment of such a practice would, we think, merely tend to emphasize the democracy of the department stores, which to-day are cutting deeply into the trade of even the most exclusive shops and where all men are on equal footing.

FOLLOW THE MARKET.

When the manufacturers' and jobbers' price has advanced on a certain article, many retailers neglect to raise their selling price until their entire supply of that particular article has been exhausted and they are forced to buy again at the enhanced prices. A case in point has recently come under our notice. On July 1st the Asbestos Sad Iron Co. found it necessary to advance the minimum retail price of their goods in Canada, on account of the increase in customs charges. The advance was delayed as long as possible by the company in the hope that their goods might be placed under a more favorable classification. When finally the manufacturers were compelled to make the advance, most dealers at once raised the retail price to a figure that would enable them to make a profit commensurate with the increased cost of the goods Some, however, who were fortunate enough to have a large stock on hand, are still retailing them at the old figures. Such a procedure is obviously as unfair to themselves as it is to their competitors. Had the change in the manufacturers' price taken the form of a decline instead of an advance, these same retailers would have been forced to follow the market by selling at reduced profits, or perhaps at an actual loss. It is, therefore, quite fair and legitimate for them to take advantage of this advance in price, for it is a case where the rule quite properly works both ways. By selling at the old figures, they are simply cutting prices against all other retailers who are compelled to re-stock at the advanced rate. In this, as well as all similar cases where the manufacturers advance their prices, we would strongly advise all dealers to adopt the new retail schedule at once.

BRIGHT OUTLOOK FOR BETTER RAILS.

It is seldom that an agitation meets with such immediate success as that which has resulted from the recent exposure of the poor quality of rails which have lately been furnished to the railroads as between the railroads and the railmakers conditions have been completely reversed. Three months ago, says the Scientific American, the railroads were urging the manufacturers to give them a better product; to-day, it is the manufacturers who are urging the railroads to come to a speedy conclusion as to what kind of rail they require, in order that the present stagnation of the steel rail business may be relieved. It was a wise move on the part of the railroads, when they jointly determined to place no more orders with the mills until some better understanding had been arrived at, and a specification drawn that would meet the present conditions. There is evidence that the manufacturers are sincerely anxious to co-operate with the engineers of the railroads; and it is only fair to recognize the fact that the former had already shown a conciliatory spirit before the present falling off of orders began.

We note with satisfaction that more than one of the more important railmaking establishments are preparing to overhaul their plants, and put in such improvements as are necessary to meet the demands for a better rail. A despatch from Pittsburg announces that the Carnegie Steel Company are about to rehabilitate, at a cost of $2,000,000, their famous Edgar Thomson rail-making plant at Braddock. In addition to the installation of engines to be run by the fuel gas from blast furnaces, the improvements include extensive changes in the rolling mills, which, it is believed, will secure that more thorough working of the rail which is universally admitted to be necessary. It is also the intention of the company to build open-hearth furnaces, to enable them to furnish rails of open-hearth steel whenever they may be called for.

COST OF SELLING GOODS AT RETAIL.

The cost of selling must be considered at some time, for it certainly has to be paid. What the cost is remains a mystery to many merchants who have to look after most of the details of their business themselves. The great merchants of the big cities have clerks who figure all these things out for them, according to systems devised by experts for keeping proper account of the business. The merchant who cannot take time to figure out what per cent. it costs him to sell goods had better accept at least 20 per cent, as the figure. If it is much more, it is too much for the keen competition of these days.

Then this percentage must be considered in marking the price on goods. An article that is billed at a dollar thus really costs the store when it is sold at least a doller and twenty cents. If sold for a dollar and a quarter, it may prove a good leader, but it cannot be considered ed a profit-maker. The merchant who figures to make 25 per cent., or 33 1-3 per cent. profit, and marks goods on that basis, "guessing" that that will allow for all expenses and leave him a good profit, will have a lot of "guessing" to do to find out what has become of the profit he thought he was making.

MARITIME BOARD OF TRADE

No business organization in Canada is more worthy the support of its members than the Maritime Board of Trade. It has been justified by its works and what it has done is but an earnest of its possibilities. The year's meeting opens in St. John on Aug. 21 and it is expected to excel that of a year ago at Amherst. The veteran and able secretary, Charles W. Creed, writing last week of that meeting, said: "In my view the meeting at Amherst was the best since the board was formed."

In the same letter Mr. Creed wrote. "The Government at Ottawa took notice of nearly all the resolutions and the greater portion of them were discussed in Parliament. The Nova Scotia Government also fell into line. Technical education is an assured fact and the Government is now looking for a site on which to erect the college. A Bureau of Publicity and Immigration has also been established."

This is all very encouraging and should excite every member of the board and every business man in the Maritime Provinces to put forth an effort to make this year's meeting a new record for activity and attainment.

CANADIAN NATIONAL EXHIBITION.

The Canadian National Exhibition will open at Toronto on August 24 and continue until September 9. To justify its title it must make big strides in advance each year, and the preparations that have been made indicate that it will be more worthy than ever to represent this great Canada of ours.

A number of new processes of manufacture will be demonstrated, and in these the keenest interest will centre. The Process Building has at present a strong educational value, which may be increased materially. The exhibition management recognize the possibilities of this feature, and seek steadily to enlist the co-operation of manufacturers, particularly those who make articles of everyday use, the evolution of which, from the raw material to the finished product, would be interesting to everyone.

It is impossible here to detail the multitude of attractions that will be found at the 1907 exhibition. Suffice to say that they will be better and more numerous than any that have yet been offered.

Hardwaremen will visit Toronto at this time in large numbers. Besides attending the exhibition, they will visit the various Toronto wholesale houses. Thus business may advantageously be combined with pleasure.

When merchants are in the city Hardware and Metal will be glad to receive a call from them. They may have any correspondence addressed here, and a place will be provided where they can do any writing they desire in a quietude not present in any of the hotels at this time.

HARDWARE TRADE GOSSIP

Western Canada.

John A. Thompson, hardware merchant, Elm Creek, Man., has sold his business to C. C. Clough.

The tinsmithing firm of Wells & Cornish, Saskatoon, Sask., have been succeeded by the firm of Elford & Cornish.

Illsey Bros., Red Deer, Alta., have recently secured a nice contract which speaks well for the enterprise and success of this hustling young firm. It is for the installation of a steam heating and ventilating system in the Red Deer public school. The specifications call for the Pease-Waldon system.

Mr. R. E. Walton, Medicine Hat, Alberta, who has been in the contracting business for about fifteen years, is now building a two-story brick store on Main Street and when completed will run a general hardware, tinsmithing, plumbing, heating and gasfitting business. He expects to be in his new building about Oct. 1.

The Great Northern Supply Co., Swift Current, Sask., are putting up a new building 11th Ave., two storey and basement 75x115. They are fitting up the hardware part with the finest fixtures in that part of that country. The northern half will be used for hardware and the southern half for implements. The total cost when completed will be in the neighborhood of $20,000. The fittings for the hardware department were made by the Northwestern Marble Works of Minneapolis.

Quebec.

W. Strachan, Ottawa, was in Montreal this week.

S. M. Howe, of the Dover Mfg. Co., Canal Dover, O., is in Montreal this week.

G. W. Dunn, representing Lewis Bros., Montreal, in Algoma, is spending his vacation on a fishing expedition.

The stock and premises of Denis Bros. hardware merchants, Montreal, were recently damaged to a considerable extent by fire.

The hardware store of E. Garon, St. Gregoire, Que., was gutted by fire last week. The loss was partly covered by insurance.

J. T. Smith, secretary-treasurer for Lewis Bros., Montreal, is spending a few weeks with his family at Trois Pistoles, Que.

L. F. Blue, of the Standard Stamping Co., Maryville, O., was in Montreal last week, also H. Moulden, of the Taylor-Forbes Co., Guelph.

Edward Goodwill, manager of the Thos. Davidson Mfg. Co., Montreal, is spending a month in the Maritime Provinces, combining business with pleasure.

Thos. Blaikie, representing the Dominion Wire Mfg. Co., Montreal, has returned to Montreal after spending two weeks amongst the trade in Ontario, calling in Toronto.

Among the July callers at the London office of Hardware and Metal was Mr. W. H. Evans of the Canada Paint Co., Montreal. Mr. Evans, accompanied by his wife, is making one of his frequent visits to the Old Land, and is keeping his English friendships in good repair. He is fully appreciating the restfulness of country life in England and his holiday is proving a great benefit.

Ontario.

Mr. Binns, hardware merchant, Newmarket, Ont., was in Toronto on Wednesday.

W. L. Allen of W. L. Allen & Co., Cobourg, is holidaying in the Maritime Provinces.

The stock of R. C. Benjamin, painter, Ottawa, was sold recently at fifty-one cents on the dollar.

A. D. Kenelly, of the Toronto office of the McClary Mfg. Co., is at Port Cockburn spending a couple of weeks' vacation.

W. J. Good, Toronto, city traveler for the Canada Paint Co., left that city for a couple of weeks' vacation on Thursday.

Walter Eyer, junior partner of the hardware firm of Jacob Eyer & Son, Richmond Hill, Ont. was a caller in Toronto last Tuesday.

Watt & King, hardware merchants, Toronto, have dissolved partnership. Mr. Watt will continue the business, and the new firm will be known as F. Watt & Co.

A new plumbing firm has started business in Toronto, under the name of Robinson Bros. The firm consists of E. B. Robinson and G. H. Robinson, late of R. T. Robinson, Toronto.

J. J. Little, hardware merchant, has received a government appointment in the shape of the postmastership of Earlscourt, a newly created post office near the city of Toronto.

E. T. Dean, manager of the Toronto office of Stewart & Wood, wholesale paint dealers, has left the Queen City for a fortnight's vacation at Lake of Bays, Muskoka, at which place his daughter has been spending the summer.

A. J. Jackson, Toronto, has taken over the business of Orlando Vickory, and will have the Ontario agency for Beaver Brand enamelware, manufactured by the Amherst Foundry Co., Amherst, N.S. It is rumoured that Mr. Vickory will start a plumbing supply house.

R. G. Kingan, of the Kingan Hardware Co., Peterborough, rode up to Toronto on civic holiday by automobile. He called on several of his hardware friends in the Queen City on Tuesday, and reported that the roads between Peterborough and Toronto were in splendid condition.

On Friday of last week the large block owned and occupied by J. W. Richardson, hardware merchant, North Bay, was completely destroyed by fire. Mr. Richardson's stock of hardware was valued at $25,000, and the loss was covered by insurance to the extent of $14,360. The building was valued at $6,000, two-thirds of which was covered by insurance.

H. W. Lightburn, a member of the hardware firm of John S. George & Co., Nassau, Bahama Islands, has been spending an extended vacation in Toronto for the past six weeks, and expects to embark for home about the end of August. Mr. and Mrs. Lightburn—together with their daughter, Miss E. Lightburn—came to the Queen City for the purpose of spending the summer months with their youngest daughter, Miss Hilda, who will shortly enter her second year at Ontario Ladies' College, Whitby.

A VETERAN TRAVELER.

H. G. Allen, of the Oneida Community, Limited, Niagara Falls, called at the office of Hardware and Metal on Wednesday, Aug. 6. This was Mr. Allen's 74th birthday and his many friends throughout Canada will be glad to know he is still enjoying excellent health.

It is forty years since Mr. Allen made his first trip in Canada for the Oneida Community. At that time traps were the chief line his firm had to sell to the Canadian trade. During the past forty years, however, the Oneida Community have made great progress, have added many new lines, and have established a Canadian factory at Niagara Falls, Ont.

Mr. Allen received many hearty congratulations from his friends in Toronto and Montreal, all of whom look forward with a great deal of pleasure to a call from this genial "knight of the grip."

Mr. Allen will leave in a few days for a trip across the continent to Los Angeles.

A RUST DESTROYER.

Hydrogen ions are said to be responsible for iron rust. This is a new discovery of a Washington government employe. He says that iron rusts because there is electro-chemical action and reaction on its surface. The hydrogen ions are essential to this process, and if they can be destroyed the rust will be prevented. The Washington man, Allerton S. Cushman by name, claims that chromic acid and its salts are death to the hydrogen ions. It kills the "little devils," as he calls them, and thus stops the process of rusting.

If this method proves to be all that its inventor claims for it, he has given to the world a most valuable discovery. Put his preparation in the form of a paint and it will save millions of dollars of damage now done by rust to iron work of all kinds. A demonstration of the efficiency of chromic acid ought not to be difficult. Meantime let no man decry its potency, for is not this an age of wonders in every department of science and art?

COOLING VAT FOR MILK OR CREAM.

A storage and cooling vat for milk or cream has been invented by Z. S. Lawrence, West Shefford, Que. The vat, which forms the object of this improvement, is preferably of large capacity and of such construction that the milk or cream contained therein may be subjected to a slow or gradual cooling or be suddenly chilled, as desired. The invention also provides for a thorough mixing of the contents, and for bringing the same to a uniform consistency before drawing it off.

Markets and Correspondence

(For detailed prices see Current Market Quotations, page 62.)

MARKETS IN BRIEF.

Montreal.

Antimony—Firming.
Copper—Weak.
Lead Pipe—Decline from net list to 5 per cent. off.
Old Material—Weak, with a general decline this week.
Pig Iron—Stronger.
Poultry Netting—Next year's list out. New discount on galvanized poultry netting, 50 and 5 per cent. off.

Toronto.

Lead.—Slight decline.
Tin Ingots—Weaker.
Lead Pipe—Decline from net list to 5 per cent. trade discount.
Old Materials—Still declining.
Poultry Netting—Next year's list is out. New discount 50 and 5 off.

TORONTO HARDWARE MARKETS.

Toronto, Aug. 7.—The local hardware market has been more or less inactive for the last two or three weeks, but business is already commencing to revive and August trade has started out at a good clip. All seasonable goods are selling well and jobbers appear to be quite gratified with the amount of business being done for this time of the year. As last Monday was observed as civic holiday in Toronto, the total business of this week will have to be crowded into the short space of four and a half days; as many members of the jobbing staffs are still away on vacation, to clean off the orders which accumulated over the holiday and catch up with business by the end of the week means some hustling on the part of the present depleted jobbing staffs.

Nails—The demand for nails is not as brisk as it was a couple of weeks ago. Meanwhile manufacturers availed themselves of this opportunity to push the manufacture and supplies are now in a very satisfactory condition.

Screws—Supplies have greatly improved during the past fortnight, but a slight scarcity still exists in those sizes for which the trade has most call. The manufacturers have made great strides during the last few weeks and before long supplies should be quite adequate. Quotations have not changed since the recent advance and prices remain firm at the advanced figures.

Builders' Hardware—The strong demand which has characterized the entire season still prevails. Jobbers have had a bumper season in all kinds of builders' supplies, mechanics' tools and indications are that the demand will continue equally strong all through the coming fall.

Glass—Trade has been rather quiet but should begin to pick up, as many houses and other buildings are nearing the stage when they will be ready for the glazier. There has been no further change in the price of plate glass since last week, the advance being well maintained.

Binder Twine—The season is practically over, though a few sorting orders continue to arrive. Managers of local twine houses report having had a much better season's business than they had expected. It is now safe to say that, compared with other years, they have at least held their own.

Sporting Goods—A good business is still being done in all summer resort requisites. Guns are beginning to sell and by the number of sportsmen who are examining firearms and asking prices it looks as if next month will see a lively trade in guns and ammunition. An advance of five per cent. has been made in Dominion loaded shells and these are now quoted as follows: Crown, fifteen and five per cent. off list; sovereign, thirty and five off; empty brass shells, fifty-five and five off.

Stoves, Ranges and Hollow-ware—Will advance five per cent. on 1st October, 1907. All orders received up to Sept. 30th will be filled at the old prices, but all orders received subsequent to that date will be subject to the five per cent. advance. It is up to retailers to place their orders now and not be caught napping.

Lead Pipe—Has declined from 10, net list to five per cent. trade discount.

Poultry Netting—Next year's list on galvanized poultry netting is out and the new discounts are 50 and 5 off list price.

MONTREAL HARDWARE MARKETS

Montreal, August 9.—Local hardware trade appears to have suffered a slight reaction this week. At any rate, the activity is not so strenuous as it was ten days ago. It is not right that we should always look for a booming business. It would not be good in the long run.

These statements apply only to the wholesale business. Manufacturers are very busy. Orders are piling up on them and it is only by keeping everlastingly at it that they will be able to cope with the demand. The makers of screws, nails, nuts, and bolts are in a better condition, although it will take a few months yet to render conditions entirely satisfactory.

Next year's list on galvanized poultry netting is out. The new discounts are 50 and 5 per cent. off. Standard poultry netting per roll of fifty yards is now listed at $1.50 for 12-inch, $3 for 24-inch, $4.50 for 36-inch, and $7.50 for 60-inch; galvanized after made, $9.50 per roll of 50 yards, 24 inches wide, and $14 for 36-inch.

Fall Goods.—The outlook for sales of sleigh heaters and sleigh bells is bright. Jobbers here are, however, somewhat anxious about this season's business in sporting goods, such as skates, hockey sticks, and curlers' supplies. Cow-ties are experiencing an increasingly strong demand. The demand for meat-choppers is good.

Screws.—While the factories are pretty well caught up on large sizes, there is still a serious shortage on the smaller sizes. Supplies of nuts and bolts are fairly adequate.

Wire Goods.—In view of the heavy demand which will shortly commence for shipments of nails to the northwest it is not likely there will be any surplus with the makers for some time to come. It is probable there will be a heavy demand for annealed hay-baling wire this fall. Although in the west hay crops are light in places, in Quebec it is heavy, and the calls for baling wire commencing about the end of this month will last well on to the close of navigation.

Building Paper.—The situation is unchanged. The demand is steady, makers are busy, and prices are firm. Tarred felt is still quoted at $2.25 per 100 pounds.

Fire Brick.—In another month or so the demand for this will enliven; at present business is quiet.

Cement.—Concrete and cement block architecture is strongly maintaining the demand for this.

Builders' Hardware.—All through the season the call for all lines of building material and interior fittings has been splendidly maintained.

Sporting Goods.—Guns and ammunition, fishing tackle, and campers' outfits are experiencing an increased call this season. Orders for fall and winter goods are well booked up. Some anxiety is expressed by wholesalers for the trade this fall in skaters' outfits.

MONTREAL METAL MARKETS.

Montreal, Aug. 9—Somewhat of a quietness has again settled upon the metal trade. Some apprehension is expressed in local circles for the prospects of trade the coming season. The outlook is not encouraging. The stringency of money due to an unusual large investment of Canadian money in American interests is largely the cause of the present darkness.

American market conditions are weak and uncertain. Strikes and holidays are helping to aggravate present conditions.

The English market is uncertain owing to speculation. It is, however, fairly strong for this season. A German demand seems to have taken the place of the American demand for English iron. Supplies continue to decrease.

Pig Iron—At the reduced prices a better demand in many districts is experienced Since the United States Steel Corporation bought 14,000 tons of Bessemer pig, practically all of the floating supply in the Pittsburg district has gone, and nothing since this big transaction has been available for less than $22.50 or $22.75. In the early part of the season there was in that district about 570,000 tons of pig, now there is less than 200,000 tons. Good Scotch brands of iron are scarce, and prices are strongly maintained. English prices have again advanced 1s. a ton. Locally, little change has occurred. Good supplies are being received, and heavy enquiries are arriving. Prices are firm and unchanged. Middlesboro, No. 1, $21.50; No. 2, $20.50, Summerlee $25.50.

Antimony—Contrary to anticipations prices are becoming firmer. No changes, however, are recorded this week. Cookson's still at 18c.

Ingot Copper—Locally, this is weak. Consumers of copper in the United States refuse to buy, although lower prices have been named, and the producers are anxious sellers. In Canada, as well as in the United States, producers are of the opinion that there will be no pronounced improvement in the copper trade until financial conditions improve. Prices are weak and unchanged.

Ingot Tin—Continues weak. A general quietness prevails in American circles. Premiums are low, and little demand exists. Lamb, and flag and straits are still quoted at $11.

Sheet Zinc—Is weak.

Zinc Spelter—Is weak.

Lead—The demand is moderate.

Old Materials—The market is still very quiet. A general decline in prices is made this week. Heavy copper and wire, 16c; light copper, 15c., heavy red brass, 11c; heavy yellow brass, 10½c.; No. 1 wrought iron, $15.00, stove plate, $12.00.

TORONTO METAL MARKETS.

Toronto, Aug. 9—Very little buying or selling is being done in the local metal markets, and with the exception of old materials, prices continue much the same as last week. Generally speaking, both the British and American metal markets are dull and inactive, and the quietness of the local market is but a natural reflection of the situation in the primary markets of supply. What little business that is being here done in metals is of the sorting order variety, and jobbers do not expect the market to show much activity for a month or six weeks yet. However, as local dealers did an exceptionally heavy business right up till the time when the present mid-summer lull struck the trade, they are not grumbling now, but are content to wait until trade picks up again, which it undoubtedly will some time in September.

Antimony—The local market is dull and inactive, the price remaining fairly steady at figures last quoted. The American market, however, has a better look than at any time in months, and many think the bottom has been reached. But there is very little interest shown yet in the way of purchases.

Pig Iron—There is practically nothing doing on the local market, and prices remain unchanged.

Tin—Very few enquiries are coming in and local prices are very easy; although no general decline has been made still the fact that one large local jobber is quoting ingot tin from $43.50 to $44, shows that the tendency of the market is downward. The London market continues dull and the American market continues to have an easier tendency, but the price of tin there may be said to be holding very well in view of the extreme dullness and the weak and unsettled condition of nearly all commercial and financial markets.

Lead—This metal has undergone a slight decline in the local market during the past week, and imported pig lead is now quoted here at $5.35 instead of $5.50. The demand for lead is limited and the market is extremely dull.

Copper—Since the decline of last week there has been no change in local prices which remain comparatively steady at the reduced figures. Concerning the American market it is noticeable that the consumption of copper has slackened up a bit there, but the decrease is fully compensated by the declining production.

Old Materials — Business continues quiet on the local market, and consequently further reductions in price have taken place during the week. Heavy copper and wire has declined ½c and is now quoted at 16c.; light copper has dropped one cent and is now quoted at 14c.; heavy red brass has lost 1½c., and the market price is now 13½c.; yellow brass and light brass have each declined 1c. and are now quoted at 11c. and 7c. respectively, tea lead is ¼c. lower and is now quoted at 3¾c.

LONDON METAL MARKETS.

London, Aug. 6.—Cleveland warrants are quoted at 57s 9d, and Glasgow standards at 56s 9d, making prices as compared with last week, on Cleveland warrants 4½d higher, and on Glasgow standards 5d higher.

Tin.—Spot tin opened easy at £179, futures at £178 5s, and after sales of 180 tons spot and 280 tons futures, closed easy at £178 10s for spot and £178 for futures, making the price as compared with last week £3 5s lower on spot and £2 10s lower on futures.

Copper.—Spot copper opened weak at £84, futures at £80 15s, after sales of 250 tons spot and 1,100 tons futures closed steady at £84 for spot and £80 15s for futures, making price as compared with last week £5 lower on spot and £4 lower on futures.

Spelter.—The market closed at £22 12s 6d, making price as compared with last week 12s 6d lower.

Lead.—The market closed at £19 2s 6d, making price as compared with last week 5s lower.

U. S. METAL MARKETS.

Cleveland, O., Aug. 8—The Iron Trade Review to-day says: The demand for structural material is strong, and some independent interests have been important factors during the past week, among the orders taken by them being 10,000 tons of steel for the Edison power plant, New York, and 7,000 tons for the Hay Foundry. The American Bridge Co. will furnish the 20,000 tons of steel needed to complete the Pennsylvania terminals, the contract having been originally taken by the firm of Milliken Bros., now bankrupt.

While large premiums are not so frequently paid for delivery of plates, the demand continues strong. A contract for steel for another lake vessel has been placed, and it is expected that orders for several others will be booked shortly. A number of inquiries of good size for steel piling are pending in the lake district, and a traction company has placed an order for steel ties. Orders for 11,000 tons of standard sections for a steam railroad, and 6,000 tons for a traction line, which had been deferred pending the making of financial arrangements, have been placed with the Lackawanna Steel Co.

U.S. IRON AND METAL MARKETS.

New York, Aug. 8.—The Iron Age today says: The July output of coke and anthracite pig iron was somewhat disappointing. According to the returns received by the Iron Age, the production in July was 2,259,682 tons, which is 72,893 tons per day, as compared with 2,234,575 tons in June, at the rate of 74,486 tons per day. The steel works furnaces were chiefly responsible for this, the output having been 1,452,557 tons in July, compared with 1,457,230 tons in June. The industry entered the month of August with a capacity of 514,121 tons per week, as compared with 528,170 tons on July 1.

The strikes in the iron ranges and at the docks are practically ended, and it is estimated that this week 700,000 tons of ore will be sent down from the docks at the head of the lakes. It may take some time, however, before the scattered workmen may be collected to get back to the former record-breaking rate of work.

While the pig iron markets in the east are irregular and weak, but show a moderate amount of activity, the southern and western markets are very dull, because the great majority of sellers refuse to make any concessions whatever. This is notably true of the southern furnaces, concerning whom there is only an occasional whisper of weakness. So far as can be judged, the rate of melting is as large as it has been, but the question, of course, is whether the founders are running along under the momentum of a past boom.

Considering the general impression of absolute stagnation in the steel rail trade, this week's aggregate of sales of about 40,000 tons is not a bad showing. Among the purchases are 10,000 tons for a new road, made in Chicago; 6,000 tons to the Idaho & Northern, for next year; 5,000 tons for the Northern Pacific, for this year, and 8,000 tons to a frog and crossing plant, partly for this year's and partly for next year's delivery.

The mills are reporting an improved demand from the car builders, and the shipyards have again placed material for two boats.

In the bar iron trade interest centres chiefly in the forthcoming wage scale for puddlers and finishers by the Conciliation Board appointed to adjust the differences between the Western Bar Iron mills and the Amalgamated Association. The decision is expected on Saturday.

Chicago reports that in the structural trade mill orders and specifications are again outstripping the rolling capacities. In New York the contract for 10,000 tons for the new Edison power house has been placed.

The reports from the wire, tube and sheet trades continue favorable.

The markets for old material all over the country seem overstocked and prices are lower.

In melting copper do not allow the ends of the ingot, or other form that is being melted, to project above the top of the crucible so that charcoal will not cover them. They are thus greatly oxidized or "burnt." The castings made from "burnt" copper will be weak and full of blowholes.

CORRESPONDENCE

Travelers, hardware merchants and clerks are requested to forward correspondence regarding the doings of the trade and the industrial gossip of their town and district. Addressed envelopes, stationery, etc., will be supplied to regular correspondents on request. Write the Editor for information.

HALIFAX HAPPENINGS.

Halifax, Aug. 6—The opening of the big power plant at the Chignecto coal mines last week, marks an important era in the industrial development of Nova Scotia. It is unique on the American continent in that the power is obtained from coal at the mouth of the pit, and transmitted over 6½ miles to the town of Amherst. Lieut-Governor Fraser turned the switch which set the wheels in motion and many prominent engineers from various parts of Canada were present at the inaugeration. The success of the undertaking was clearly demonstrated.

As a result of the installation of this plant, manufacturers will be able to obtain power at a much less cost than they are now paying.

The employment of electric power has many advantages over the use of steam in point of economy, not only in capital expenditure for its installation, but in cost of running and maintenance. No engine house is required; engines, boilers and steam pipes are done away with, fireman and engineers are not required; the cost of fuel and water is saved. Instead of all these, an induction motor is used, which may be set in any part of a building, and costs from $150 for five horse power to $1,000 for fifty horse power. The usual price of electric power developed hydraulically is from $25 to $30 per horse power per year, while steam power usually may be figured at about $60.

The power house at the mouth of the Chignecto mine is a very substantial brick structure, 60 x 24 feet, and 35 feet high, located about 60 yards from the pit mouth. It is fitted up with a 750-horse power vertical compound engine, high-speed, 300 revolutions, with forced lubrication, and four Robb Engineering company's boilers of 300 horse power each. The generator is from the Westinghouse company, is of 500 kilowatts, of the revolving field type, wound for three phase currents at 11,000 volts. Both the engine and generator have a capacity for 25 per cent. over-load The switchboard consists of the usual instruments, lighting arresters, etc., and is also from the Westinghouse company.

The transmission line is six and a half miles in length in a straight line to Amherst, and consists of three No. 4 copper wires, run on cedar poles and entering the sub-station, Amherst Highlands, near the Amherst Malleable Iron Works.

The Amherst sub-station consists of a brick two-storey building with concrete base. The first storey contains the low pressure or 2,000-volt gear, and the second storey the 11,000-volt gear. Here the current is transformed from 11,000 to 2,000-volts, by three Westinghouse transformers of 150 kilowatts each, the voltage of 2,000 being for distribution to the consumers of Amherst The buildings at Chignecto and Amherst were built by J. N Fage, contractor, Amherst.

At Chignecto the coal is conveyed from the pit mouth by mechanical conveyors to a coal bin at the boiler house, a distance of about thirty-five yards, from which it ascends in chutes to the mechanical stokers, which are of the Jones type Ashes and clinkers are also taken away by means of conveyors.

The people required to man the system are:—An engineer for the day shift and one for the night shift, two firemen for day and one for night, and a cleaner, all at Chignecto Two linemen. always ready for repair work, by day or night. At Amherst, one person, principally for telephone. A superintendent has charge of the entire power system.

The following is an example of the manner in which power may be used in a large factory:—Rhodes, Curry & Co. have taken 200 horse power to supplement the present power in their car works. At different points in the works are placed ten or twelve motors of ten to forty-horse power each, for driving fans, blowers, shafting, cranes, etc., some working on 2,000 and some on 220 volts The company has three large transformers of its own, of forty kilowatts each, which are for the purpose of reducing the pressure from 2,000 to 220 volts for the protection of the workmen in cases where they are likely to come in close proximity to the wires.

The capacity at present is 1,000 horse power; but in order to ensure the effectiveness of the plant under all circumstances, it is intended to duplicate it at an early date, with the exception of the wiring, which is capable of transmitting 3,000 horse power. By duplication, an accident to the machinery could be overcome in a few minutes. The energy which it starts to supply is 600 horse power.

The following are the officers and the directorate of this enterprising company:—President, Senator William Mitchell, Montreal: vice-president, Nat. Curry, Amherst; secretary-treasurer, Walter G. Mitchell, Montreal; manager, David Mitchell, Chignecto, directors, William Ewing Montreal; D. Smith, stock raiser, Montreal; Mr. Dunn, M.P, Ontario.

During the inauguration proceedings, the following telegram from Thos. A. Edison, the famous inventor, was read:

"Orange, N.J., July 31, 1907.

"H. J. Logan, M.P.

"Chairman Board of Trade Committee.

"Permit me to congratulate your Board and Senator Mitchell on the inauguration of the first plant on the American continent for the generating of electricity at the mouth of a coal mine, and the distribution of the same to distant commercial centres It is a bold attempt and I never thought it would be first accomplished in Nova Scotia, where my father was born over one hundred years ago.

(Signed) "THOMAS A. EDISON."

SCENIC ST. JOHN.

St. John, N. B., August 6. — St. John merchants are looking forward with interest to the 21st and 22nd of this month, when the annual meeting of the Maritime Board of Trade will be held here. The whole of the first day and one-half of the second will be devoted to the discussion of issues which are of vital importance to the progress and prosperity of the Maritime Provinces. These discussions will be taken part in by the most active business men from all sections and should prove of deep interest.

The delegates, of whom there are expected to be more than a hundred, will be well looked after by a reception committee from the local board, who are making arrangements for their proper entertainment.

There will probably be an excursion on the river on the afternoon of the second day, and a good time is looked for. W. S. Fisher, the president, is working energetically to make this year's convention one of the best yet held. There will be many subjects of interest presented for discussion.

* * *

It is expected that the winter port business of 1907-08 will be greater than ever before. The Allan line steamers will make weekly sailings, also the C. P. R., Donaldson, Manchester, Head, Furness, South African and West Indian services. About 150 sailings are promised.

H. G. Hunter, civil engineer, who has been engaged on the water extension system here for the past two years, has accepted a position as one of the assistant engineers on the water extension system ae New York, and will leave shortly to take up his new duties.

The corner stone of the new $60,000 Y.M.C.A. building on Chipman Hill, was laid on Wednesday, July 31, by his worship, Mayor Sears. The work of construction is proceeding very satisfactorily.

At a meeting of the civic board of works last week, it was decided that a tender for tar to be used for asphalt sidewalks, be not accepted, as the price $3.50 a barrel, was considered exorbitant. The only tenderer was the Carritte, Patterson Co., who control the tar output of the Maritime Provinces. Under a previous call for tenders they had offered to supply the tar for $2.25 a barrel but would not agree to take half the contract, the St. John Ry. Co. taking the other half. The city fathers, rather than submit to what they termed a hold-up, have decided to try and purchase elsewhere or to put down concrete side walks.

The town of Woodstock, N.B., suffered from a severe fire on Wednesday last when the buildings occupied by the Baird Company, wholesale druggists; Sheasgreen Drug Co., and Troy's restaurant, were burned. The fire started from the crossing of electric light and telephone wires. The Baird Co.s' loss amounted to over $15,000, fairly covered by insurance, but the other concerns' loss was not very great. One of the firemen, Charles McKinney, was killed by coming in contact with a live electric wire.

KINGSTON KINETOSCOPE.

Kingston, Aug. 5.—Business continues rather brisk here still as it has been doing for some time past, especially in all lines of sporting and seasonable goods. The merchants have done an exceptionally good trade in all lines of builders' supplies all spring, and in such lines as paints, oils, glass, etc., they have done remarkably well, but the rush is beginning to quiet down a little now. All our hardware stores keep open each day until six o'clock. The boot and shoe, gents' furnishings and dry goods stores all close at five o'clock each evening during the summer months. The hardware dealers keep the same time here the year round.

* * *

The employes of the Canadian Locomotive Works intend holding a huge picnic at Lake Ontario Park on Saturday, August 10th, the proceeds to be equally divided between the Hotel Dieu and General Hospitals here. Fine weather is being looked forward to and large crowds are expected to be present on this occasion to help make it a grand success as it is for the good of these two charitable institutions.

* * *

The annual meeting of the shareholders of the Kingston, Portsmouth and Cataraqui Electric Railway Company was called to meet at the secretary's office, Ontario Street, on the 31st of July, but was on that date postponed until October 11th, some of the shareholders not being present.

* * *

A meeting of the creditors of George Sears, hardware merchant, this city, was held one day last week in the office of the assignee, J. B. Walkem. It was decided at that meeting to offer the stock for sale by tender; in the meantime, Mr. Sears may make an offer. Some years ago this business was carried on under the name of John Mucklestone & Co., and at that time they did an exceptionally large trade and employed about twenty-seven men, many of whom are now our leading hardware merchants. It seems a shame to see such a good business as this fall through, but for some time it has been steadily decreasing but no advertising whatever was done to help build it up again, and at last an assignment had to be made. The business was under the name of Mucklestone until about seven or eight years ago, when Mr. Sears took it over, the latter having been an employe some twenty-five years before. Subsequently Mr. Sears left for the States, where he remained for some time. On returning again he went into partnership with Mr. Mucklestone and finally took over the business under his own name when Mr. Mucklestone went out of it.

* * *

James Swift, general manager of the Collingwood Shipbuilding Company, Collingwood, formerly of Garden Island, is renewing old acquaintances in this district. "Sunny Jim," as he is known by all his old friends, is receiving a warm welcome. Another visitor to the city over Sunday was Oscar Johnson, invoice clerk for the Queen City Oil Co., Toronto.

* * *

Tenders are being called here by the county clerk for the construction of the abutments and concrete floor for the new steel bridge to be erected at Grass Lake in the Township of Kingston.

* * *

On Tuesday evening, while riding home to supper on a bicycle with a friend, one of our leading hardwaremen met with a slight accident which he is likely to remember for a few days, when near the city buildings he was about to go on the sidewalk when suddenly his front wheel struck the walk, which was considerably higher than he had anticipated, throwing him with great force to the asphalt pavement below. Luckily no bones were broken and with the exception of a few scratches and bruises he escaped further injury and is once more in the ring with the same happy smile as before, looking little the worse for his experience.

* * *

Tenders were called here a short time ago for the rebuilding of the Tete du pont barracks, built here in 1672, the soldiers at present being away at Petawawa. The work was to have been done while they were absent, but a number of times there have been rumors afloat that the Grand Trunk Railway Co. had made offers for this site and would build a freight station there. A rumor of this nature was current lately and either it was this reason or that the contractors' cheques were returned on account of the tenders being too high, that the contracts have not been awarded. New contracts will probably be called for.

* * *

There has been a great demand for all kinds of motor boat supplies here this season, many merchants and residents having purchased gasoline launches, and the local builders have received contracts in quick succession. It seems at present as though the sailing yacht would almost be done away with in the near future, many people with smaller boats are having engines installed, and "put-puts" are to be heard in every direction on the water now. This greatly increases the business of the hardwaremen, who keep on hand a goodly supply of goods needed for that purpose.

* * *

There have been several large wagons about the city the past few days advertising the "Comfort Range." They travel through the country around here taking orders and in some cases taking the farmers' old stoves as part payment. One farmer was rejoicing the other day because a traveler had called on him and offered to take his old stove at a reduction of $20 on the new one. He bought the new stove but they never called back to remove the old one and he now has the new stove with the reduction of the $20 and also his old stove. One of these wagons met with a mishap one evening while crossing the Street Railway Co.'s tracks, one of the rear wheels was wrenched off, the driver being thrown from his high seat and receiving a bad shaking up, but otherwise escaping serious injury.

* * *

Civic Holiday was observed by the people of this city on Monday last and was proclaimed throughout the city as a public holiday. The annual homecoming of the old boys and girls from Toronto was celebrated and on the Saturday evening previous a warm welcome was given them on their arrival by the citizens, with the 14th regimental band playing "Home, Sweet Home." It was estimated that close upon 960 people were on the train, made up of thirteen coaches. A band concert was given in their honor on the market square that evening, and another in Macdonald Park on Sunday evening was attended by several thousand people. Monday was an ideal day for the visitors, who took in the many picnics and excursions held. Among some of the features of the day were baseball and lacrosse in the morning and an excursion among the Thousand Islands in the afternoon, the boat being crowded. The local Catholic Irish societies held a picnic at Lake Ontario Park at which a large number were present. The old boys from Ottawa were here the week before to spend Sunday, when two band concerts were given, and also took in what excursions were being held on Sunday. There were reduced rates on the Grand Trunk on both occasions.

WIDE AWAKE WOODSTOCK.

Woodstock, Aug. 6.—An American concern, which desires to establish a branch factory in Western Ontario for the manufacture of brass goods, is conducting negotiations with the city council and board of trade, and the officials are hopeful of landing the industry, which seems like a good one, for Woodstock.

Hardly a day passes but what one or more letters are received by the civic authorities, asking what inducements would be offered manufacturers to locate here. Most of them are from American promoters. Some of them are of course genuine, and the writers really desire to build a factory somewhere, but very many are the outcome of wildcat schemes, and the problem is to pick out those worth while. Almost every town and city has had some experience with manufactories brought to the municipality by a loan or other inducements, which failed to make good, and left the plant and business to fall back into the hands of the town, sometimes entailing a considerable loss. Eternal vigilance is the price of safety.

* * *

One valuable new industry is all but landed for Woodstock. It is a milk condensing plant which would employ a large number of hands, and have a capacity of one hundred thousand pounds of milk per day. The promoters are the Borden Condensed Milk Company, of New York, who have a large number of branches across the line. They propose to erect another plant in Western Ontario, in addition to the one at Ingersoll. Negotiations are proceeding very satisfactorily and it is probable that a bylaw providing for the concessions will be submitted to the people in a few weeks. The company ask for a free site, free sewer connection, exemption from taxation for all save school purposes, for ten years.

* * *

Active steps are being taken for the erection of a new city hall for Woodstock. It is not before a new one is needed, for the structure at present serving in that capacity has been antiquated for years past. The people will be given an opportunity of voting on the question next January.

* * *

The new building which Jones Bros., cigar manufacturers, are erecting, is nearly completed. It is a handsome three-storey structure. On the street floor will be a cigar retail store and pool rooms, and the remainder of the building will be devoted to manufacturing purposes.

LEAFY LONDON.

London, August 7.

That there is need for real organization and agreement among the retail hardware dealers of London, whereby living prices for goods of all kinds might be maintained, is every day becoming more apparent. As it is, merchants "hoe their own row," and in order to get business do not hesitate to cut prices to a ruinous degree. Much of this price-cutting is in the nature of baits, whereby the unsuspecting housekeeper is allured to a store as a result of the advertising of certain articles at prices away below their actual value. An instance of this was furnished during the past week. One dealer put out the bait in his advertisement in the daily papers that he was offering charcoal irons, the regular selling price of which is $1, at 75 cents each. Then out came another merchant with the announcement that in his store the same iron could be got for 60 cents. Where this sort of thing will end—whether it will ultimately lead to dealers offering to throw in a charcoal iron with every pound of nails, or something as ridiculous, it is hard to say; but it is certain that if dealing in hardware is to continue a profitable business in London something must be done, and that right soon, to bring these price-cutters to their senses.

* * *

One of London's most enterprising and progressive firms is that of Wortman & Ward. In 1879 Wm. H. Wortman started in business with a capital of $350; to-day he is at the head of one of the largest concerns of the kind in the country, with business connections extending to various parts of the world. Among the articles manufactured by this firm are revolving barrel churns, washing machines, iron pumps, pump cylinders of brass and iron, horse hay forks, pea harvesters, wagon skeins and miscellaneous castings. The fame of Wortman & Ward's goods has spread over four continents, for besides doing business in every section of Canada, they find profitable markets in England, Australia, South Africa and New Zealand. Their extensive establishment in the eastern part of the city is up-to-date in every particular, the entire buildings being 160x250 feet. The front section is three storeys in height and 50x250 feet, and the foundry is a one-storey structure, about 100x200 feet. From ninety to one hundred men are constantly employed in the various departments. In the foundry some twenty molders are in use and six tons of metal are melted every day. One cause, and no unimportant one, either, of Wortman & Ward's success is their straightforward business methods, and another is that they keep abreast of the times in their various lines of manufacture. The business is steadily growing, and it would surprise no one to learn before long that it has reached such dimensions as to necessitate an enlargement of the firm's already very large establishment. When that time comes Wortman & Ward will be prepared to meet it.

* * *

The two L. & K. test boring machines manufactured by the Scott Manufacturing Company, of this city, for the Transcontinental Railway, have proven most satisfactory and the general adoption of these machines by those requiring them is considered almost certain. The local firm have filled an order for one of them to the C.P.R., and to-morrow will ship one for use by the Dominion Department of Public Works. The prospects are that there will be more to follow.

* * *

Business is fairly good in hardware circles. Travelers are still off on their holidays, and the result is unusually large mail and telephone orders.

* * *

The employes of the Jackson Manufacturing Company and their friends, numbering about two hundred in all, held their annual picnic at Springbank park yesterday.

MALAYSIAN TIN OUTPUT.

The following table, furnished by Vice-Consul-General G. E. Chamberlin, of Singapore, shows the output of tin from the Federated Malay States for the first four months of the years 1906 and 1907. The figures represent tons of 2,240 pounds.

	1906	1907
Perak	8,076	8,187
Selangor	5,255	5,067
Negri Sembilan	1,500	1,386
Pahang	636	623
Total	15.467	15.263
Decrease		204

It will be noticed that there is a falling off from last year's output of 204 tons. However, this is a slight gain for the months of March and April, as January and February showed a decrease of 272 tons.

WHY RUBBER TIRES GET HOT.

When an automobile is running at high speed the rubber tires are rapidly warmed, and the heat sometimes becomes very great, with resultant injury to the rubber. The cause of this accumulation of heat in the tire is ascribed to the kneading of the rubber, which generates heat faster than it can be radiated away. For this reason manufacturers have found it to be an advantage to have metal parts in the tread, such as the ends of rivets, in contact with the tire, because the metal being a good radiator, helps to carry off the heat.

A RARE FIND.

A Danish philologist browsing in a convent in Constantinople recently discovered a manuscript of the Greek geometrician and practical inventor, Archimedes, who had a monopoly of the hardware trade in screws and levers in Athens, who shouted "Eureka" long before Longfellow's unknown cried "Excelsior," and who mastered the difference between the subjective and the objective sufficiently to announce B.C. that if he only had a place to put his lever he could move the world. Savants are now on tenterhooks to know whether the manuscript is poetry or prose, a new stiff problem in Q. E. D. or a claim to recognition at the Athenian patent office.

NEW BRASS FACTORY.

The Turner Brass Works, of Chicago, have completed their new factory at Sycamore, Ill., and are now occupying it. The main factory building has a floor space of 40,000 square feet, with power and heating plants additional. Improved machinery has been installed and their facilities enlarged and greatly improved. They are located on the main line of the Chicago & Great Western Ry. and the Galena division of the Chicago & Northwestern Ry., with side-tracks to the factory. As soon as completely settled, all orders will be shipped promptly. All correspondence and orders should be sent to the main office, Sycamore, Ill.

TOUGHENING METALS.

In a novel method of toughening metals lately introduced the metal to be treated is placed in a closed retort with a small quantity of mercury, says the London Engineer. The retort is subjected to pressure, and as it is heated to a point below melting temperature, a current of electricity is passed through the metal. Besides increased toughness, greater resistance to sea-water corrosion is imparted. The hardening is especially adapted for iron and steel, but is claimed to be useful for other metals.

LARGEST CHAINS IN THE WORLD.

The heaviest work ever done in the chainmaking industry is the mooring chain made for the two new Cunard steamships Mauretania and Lusitania, now being constructed at Pontypridd, South Wales. The links are made from 4¼ and 5¾-inch iron and weigh from 243 to 336 pounds each. The swivel connection weighs 4,485 pounds, and each shackle weighs 711 pounds. The anchors will weigh twelve tons apiece. The chains are 720 feet long, and the entire mooring will weigh altogether about 200 tons.

MANITOBA HARDWARE AND METAL MARKETS

Corrected by telegraph up to 12 a m. Friday Aug 9 Room 511, Union Bank Bldg, Winnipeg, Man.

Crop reports are more favorable now than for some weeks, and the trade are in more hopeful mood. A good average sorting trade is being done and the volume of business is considered quite satisfactory for an ordinarily quiet month.

ROPE—Sisal, 11c. per lb., and pure manila, 15¾c.

LANTERNS—Cold blast, per dozen, $7; coppered, $9; dash, $9.

WIRE—Barbed wire, 100 lbs., $3.22½; plain galvanized, 6, 7 and 8, $3.70; No. 9, $3.25; No. 10, $3.70; No. 11, $3.80; No. 12, $3.45; No. 13, $3.55; No. 14, $4; No. 15, $4.25; No. 16, $4.40; plain twist, $3.45; staples, $3.50; oiled annealed wire, 10, $2.90; 11, $2.96; 12, $3.04; 13, $3.14; 14, $3.24; 15, $3.39; annealed wires (unoiled), 10c. less; soft copper wire, base, 36c.; brass spring wire, base, 30c.

POULTRY NETTING—The discount is now 47½ per cent. from list price, instead of 50 and 5 as formerly.

HORSESHOES—Iron, No. 0 to No. 1, $4.65; No. 2 and larger, $4.40; snowshoes, No. 0 to No. 1, $4.90; No. 2 and larger, $4.65; steel, No. 0 to No. 1, $5; No. 2 and larger, $4.75.

HORSENAILS—No. 10 and larger, 22c; No. 9, 24c.; No. 8, 24c.; No. 7, 26c.; No. 6, 28c.; No. 5, 30c.; No. 4, 36c. per lb. Discounts: "C" brand, 40, 10, 10 and 7½ p.c.; "M.R.M." cold forged process, 50 and 5 p.c. Add 15c. per box. Capewell brand, quotations on application.

WIRE NAILS.—$3 f.o.b. Winnipeg, and $2.55 f.o.b. Fort William.

CUT NAILS—Now $3.20 per keg.

PRESSED SPIKES — ¼ x 5 and 6, $4.75; 5-6 x 5, 6 and 7, $4.40; ⅜ x 6, 7 and 8, $4.25; 7-16 x 7 and 9, $4.15; ½ x 8, 9, 10 and 12, $4.05; ⅝ x 10 and 12, $3.90. All other lengths 25c. extra net.

SCREWS—Flat head, iron bright, 80, 10, 10 and 10; round head, iron, 80; flat head, brass, 75; round head, brass, 70; coach, 70.

NUTS AND BOLTS — Bolts, carriage, ⅜ or smaller, 60 p.c.; bolts, carriage, 7-16 and up, 50; bolts, machine, ⅜ and under, 50 and 5; bolts, machine, 7-16 and over, 50; bolts, tire, 65; bolt ends, 55; sleigh shoe bolts, 65 and 10; machine screws, 70; plough bolts, 55; square nuts, cases, 3; square nuts, small lots, 2½; hex nuts, cases, 3; hex nuts, small lots, 2½ p.c. Stove bolts, 70 and 10 p.c.

RIVETS — Iron, 60 and 10 p.c.; copper, No. 7, 43c.; No. 8, 42½c.; No. 9, 45½c.; copper, No. 10, 47c.; copper, No. 12, 50½c.; assorted, No. 8, 44½c., and No. 10, 48c.

COIL CHAIN — ¼-in., $7.25; 5-16, $5.75; ⅜, $5.25; 7-16, $5; ½, $4.75; 9-16, $4.70; ⅝, $4.65; ¾, $4.65.

SHOVELS—List has advanced $1 per dozen on all spades, shovels and scoops.

HARVEST TOOLS—60 and 5 p.c.

AXE HANDLES—Turned, s.g. hickroy, doz., $3.15; No. 1, $1.90; No. 2, $1.60; octagon extra, $2.30; No. 1, $1.60.

AXES—Bench axes, 40; broad axes, 25 p.c. discount off list; Royal Oak, per doz., $6.25; Maple Leaf, $8.25; Model, $8.50; Black Prince, $7.25; Black Diamond, $9.25; Standard flint edge, $8.75; Copper King, $8.25; Columbian, $9.50; handled axes, North Star, $7.75; Black Prince, $9.25; Standard flint edge, $10.75; Copper King, $11 per dozen.

CHURNS—45 and 5; list as follows: No. 0, $9; No. 1, $9; No. 2, $10; No. 3, $11; No. 4, $13; No. 5, $16.

AUGER BITS—"Irwin" bits, 47½ per cent., and other lines 70 per cent.

BLOCKS—Steel blocks, 35; wood, 55.

FITTINGS — Wrought couplings, 60; nipples, 65 and 10; T.'s and elbows, 10; malleable bushings, 50; malleable unions, 55 p.c.

HINGES—Light "T" and strap, 65.

HOOKS — Brush hooks, heavy, per doz., $8.75; grass hooks, $1.70.

STOVE PIPES—6-in., per 100 feet length, $9; 7-in., $9.75.

TINWARE, ETC.—Pressed, retinned, 70 and 10; pressed, plain, 75 and 2½; pieced, 30; japanned ware, 37½; enamelled ware, Famous, 50; Imperial, 50 and 10; Imperial, one coat, 60; Premier, 50; Colonial, 50 and 10; Royal, 60; Victoria, 45; White, 45; Diamond, 50; Granite, 60 p.c.

GALVANIZED WARE — Pails, 37½ per cent.; other galvanized lines, 30 per cent.

CORDAGE — Rope sisal, 7-16 and larger, basis, $11.25; Manilla, 7-16 and larger, basis, $16.25; Lathyarn, $11.25; cotton rope, per lb., 21c.

SOLDER—Quoted at 27c. per pound. Block tin is quoted at 45c. per pound.

WRINGERS—Royal Canadian, $36; B.B., $40.75 per dozen.

FILES—Arcade, 75; Black Diamond, 60; Nicholson's, 62½ p.c.

LOCKS—Peterboro and Gurney, 40 per cent.

BUILDING PAPER—Anchor, plain, 66c.; tarred, 69c.; Victoria, plain, 71c.; tarred, 84c.; No. 1 Cyclone, tarred, 84c.; No. 1 Cyclone, plain, 66c.; No. 2 Joliette, tarred, 69c.; No. 2 Joliette plain, 51c.; No. 2 Sunrise, plain, 56c.

AMMUNITION, ETC. — Cartridges, rim fire, 50 and 5; central fire, 33⅓ p.c.; military, 10 p.c. advance. Loaded shells: 12 gauge, black, $16.50; chilled, 12 gauge, $17.50; soft, 10 gauge, $19.50; chilled, 10 gauge, $20.50. Shot: ordinary, per 100 lbs., $7.75; chilled, $8.10. Powder: F.F., keg, Hamilton, $4.75; F.F.G., Dupont's, $5.

REVOLVERS — The Iver Johnson revolvers have been advanced in price, the basis for revolver with hammer being $5.30 and for the hammerless $5.95.

IRON AND STEEL—Bar iron basis, $2.70. Swedish iron basis, $4.95; sleigh shoe steel, $2.75; spring steel, $3.25;

machinery steel, $3.50; tool steel, Black Diamond, 100 lbs., $9.50; Jessop, $13.

SHEET ZINC—$8.50 for cask lots, and $9 for broken lots.

CORRUGATED IRON AND ROOFING, ETC.—Corrugate iron 28 gauge painted $3, galvanized $4.10; 26 gauge $3.35 and $4.35. Pressed standing seamed roofing 28 gauge painted $3.10, galvanized $4.20; 26 gauge $3.45 and $4.45. Crimped roofing 28 gauge painted $3.20, galvanized, $4.30; 26 gauge $3.55 and $4.55.

PIG LEAD—Average price is $6.

COPPER—Planished copper, 44c. per lb.; plain, 39c.

IRON PIPE AND FITTINGS—Black pipe, ¼-in., $2.65; ⅜, $2.80; ½, $3.50; ¾, $4.40; 1, $6.35; 1¼, $8.65; 1½, $10.40; 2, $13.85; 2½, $19; 3, $25. Galvanized iron pipe, ⅜-in., $3.75; ½, $4.35; ¾, $5.65; 1, $8.10; 1¼, $11; 1½, $13.25; 2-inch, $17.65. Nipples, 70 and 10 per cent.; unions, couplings, bushings and plugs, 60 per cent.

GALVANIZED IRON — Apollo, 16-gauge, $4.15; 18 and 20, $4.40; 22 and 24, $4.65; 26, $4.65; 28, $4.50; 30 gauge or 10¾-oz., $5.20; Queen's Head, 20 $4.60,; 24 and 26, $4.90; 28, $5.15.

LEAD PIPE—Market is firm at $7.80.

TIN PLATES—IC charcoal, 20 x 28, box, $10; IX charcoal, 20 x 28, $12; XXI charcoal, 20 x 28, $14.

TERNE PLATES—Quoted at $9.50.

CANADA PLATES — 18 x 21, 18 x 24, $3.50; 20 x 28, $3.80; full polished, $4.30.

LUBRICATING OILS—600W, cylinders, 80c.; capital cylinders, 55c. and 50c.; solar red engine, 30c.; Atlantic red engine, 29c.; heavy castor, 28c.; medium castor, 27c.; ready harvester, 28c.; standard hand separator oil, 35c.; standard gas engine oil, 35c. per gallon.

PETROLEUM AND GASOLENE — Silver Star, in bbls., per gal., 20c.; Sunlight, in bbls., per gal., 22c.; per case, $2.35; Eocene, in bbls., per gal., 24c.; per case, $2.50; Pennoline, in bbls., per gal., 24c.; Crystal Spray, 23c.; Silver Light, 21c.; engine gasoline in barrels, gal., 27c.; f.o.b. Winnipeg, in cases, $2.-75.

PAINTS AND OILS — White lead, pure, $6.50 to $7.50, according to brand; bladder putty, in bbls., 2¼c.; in kegs, 2¾c.; turpentine, barrel lots, Winnipeg, 90c; Calgary, 97c; Lethbridge, 97c; Edmonton, 98c. Less than barrel lots, 5c. per gallon advance. Linseed oil, raw, Winnipeg, 72c.; Calgary, 79c.; Lethbridge, 79c.; Edmonton, 80c.; boiled oil, 3c. per gallon advance on these prices.

WINDOW GLASS — 16-oz. O. G., single, in 50-ft. boxes—16 to 25 united inches, $2.25; 26 to 40, $2.40; 16-oz. O.G., single, in 100-ft. cases—16 to 25 united inches, $4; 26 to 40, $4.52; 41 to 50, $4.75; 50 to 60, $5.25; 61 to 70, $5.75. 21-oz. C.S., double, in 100-ft. cases, 26 to 40 united inches, $7.35; 41 to 50. $8.40; 51 to 60, $9.45; 61 to 70, $10.50; 71 to 80, $11 55; 81 to 90, $17.-30.

INTERESTING FACTS ABOUT ROOFING.

Nearly a quarter of a century has elapsed since the Standard Paint Company began business, manufacturing the P. & B. products. These include P. & B. paint, electrical compounds, P. & B. insulating papers, and insulating tape.

In addition to these, they at that time made a roofing, the base of which was burlap. This, however, in a very short time proved to be useless as a base—something more substantial had to be employed. Their experts were at once put to work to devise a roofing that would withstand every internal strain and all weather changes.

About sixteen years ago, after much experimenting, they finally succeeded in making ruberoid roofing. From that time until this, without any change in its composition or construction, it has withstood the test of time.

The Standard Paint Company have on file in their offices samples of ruberoid roofing which have been on buildings for the last fourteen years, and, although these have had no attention, they are as good to-day as when they were put on.

Owing to the constantly increasing demand for ruberoid roofing, the Standard Paint Company, in addition to their factories in the United States, and at Hamburg, Germany, have had to build factories in Canada. The factories in the United States only supply South America, the West Indies, and the Orient; that in Hamburg the continental and English colonial trade, and the Canadian factory at Montreal, was built to supply the large increasing demand for ruberoid roofing in Canada.

Ruberoid has withstood the severest tests to which it is possible to subject any roofing. Not only has it been used in the most northerly Arctic regions, but at the equator, with equal success. Changes in climate have absolutely no effect on it. Neither is it affected by oxidation, corrosion, acid fumes, or steam arising from the interior of buildings in round houses, factories, fertilis or plants, laundries, etc. Ruberoid is equally suitable for glass factories.

FREIGHT QUESTION IN MONTREAL

Street Railway Company Proposes a Plan to Relieve Congestion.

A plan has been submitted by the Montreal Street Railway whereby the freight congestion which at present exists in the city of Montreal may be relieved. Their proposition is nothing less than an offer to institute a service of freight cars which will enable them to transport merchandise from one section of the city to another.

At the moment merchants in all branches of trade find it difficult to obtain prompt delivery of their goods within the city limits. It is next to impossible to secure carters when they are wanted and it is probable in consequence that any action such as that proposed by the railway company would meet with the hearty approval of business men.

It is proposed to run the freight cars at night in order that the passenger service accorded the public be not interfered with during daylight hours. Since the system owned by the company practically covers the Island of Montreal it would be possible for them to take freight for delivery to any of the suburbs, as well as within the city limits.

If the plan they propose be adopted it is probable that many manufacturers and wholesalers would have sidings built to their shipping entrances. The whole matter is now under discussion in the city council.

Heating and Housefurnishings

SOME MODERN ENGLISH STOVES AND RANGES

From a recent article in the Building News, London, Eng.

There is an unquestionable tendency at the present day to reduce expenditure and this is being reflected in the building industries by the production of one apparatus after another which will affect a household economy by its adoption, either in the matter of service and labor or in the utilization of otherwise

FIG. 1

waste products. It is possible in fire grates and cooking ranges to effect both these economies simultaneously, and to combine with them an economy in first cost also. The devices for doing this which are now upon the market are very many.

So far as ordinary fireplaces are concerned, the first step in advance, from the English standpoint at least, was made when the Pridge-Teale fireback was introduced, now many years ago, and by the attention which was drawn to the subject in the various smoke prevention exhibitions which have been held since from time to time. The general idea has been that of so constructing the fire as to cause perfect combustion of the fuel, thus utilizing for heating purposes both the waste gases and the solid smoke. The utilization of the heat for warming incoming air, although not often adopted on account of its expense, was shortly afterward successfully achieved in what at the time was known as the Galton grate, and in many of the forms of hospital grates which are still in use in England. Then

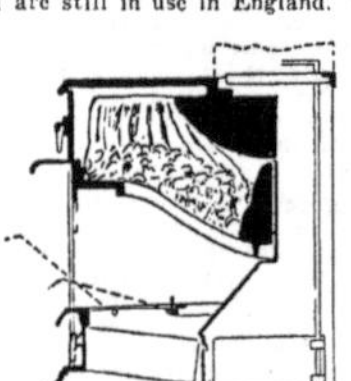

FIG. 2

followed many different contrivances for economizing fuel, most of them based upon the idea of a solid hearth. Grates constructed on this principle, however, were found to choke with ashes and become dull. They are not often seen now, but have developed into "well fires," "heaped fires," and many others which rest upon either a sunk or a raised hearth having an ash pan beneath.

In almost every case the projecting back is used, and sometimes projecting sides also, these being so arranged that the fire is narrower at back than front, so as to radiate the heat into the room. If there is to be perfect combustion, however, it is necessary to fill up the rather sharp angle between the bottom grating and the fireback with a block or pillow of some sort, preferably of fire clay, as done in the modern Venetian grate, as shown at A in Fig. 1. Lime burners know perfectly well that some such contrivance is necessary, and always introduce it into the flare kiln, as otherwise a dead corner results and the lime is imperfectly burnt. Sometimes householders overcome the difficulty by introducing a movable piece of fire brick into this position, but not many of them understand the reason for doing so.

If anthracite coal is to be burned, it is better to replace the block by a grating of curved bars, so as to admit a large supply of air into the midst of the fire. Throughout the greater part of England this hard coal is practically unknown, but there are other districts in which it is used to the exclusion of all other, while even in London there are some householders who prefer to use it in spite of its additional cost. There is no more sure way of curing a smoky chimney than that of burning a smokeless coal.

The introduction of a back block of the nature described has been adopted also in the Glow portable cooking ranges, combined with a roof block, which radiates the heat downward in such a way as to promote combustion of all the coal products, the fuel being arranged on a hearth grating of entirely unusual shape, so as to economize the depth of coal while obtaining a large fire area, as shown in Fig. 2. There are other ranges, doubtless, which do the same sort of thing when needed by means of a hinged bottom, the object in all cases being the same—namely, to obtain the greatest amount of heat with the least expenditure of fuel, and to utilize it for cooking and boiling purposes to the utmost, so that it is comparatively cool gas only which eventually passes up the flue. It is naturally essential in all forms of range that every part should be accessible for cleaning.

Another sitting room contrivance, which is likely to become popular on account of its economy, particularly in flats and small houses where the sitting rooms are placed back to back, is that of a revolving fire pivoted in the centre of the wall, which can be used, say, for the dining room until dinner commences and the room is warm, then turned away from the back of the carver during the dinner hour, so as to face toward the drawing room, which is then heated ready to be entered immediately after dinner. There are certain difficulties connected with this arrangement, but they are not insurmountable, and the advantages are considerable where great economy has to be exercised, while it is always possible to light a fire on each side of the revolving partition, and so heat both rooms simultaneously, if so desired. A movable apparatus such as this might, for example, be a great temptation to children, who would enjoy playing with it, seeing the fire appear and disappear, and calling out to one another in excitement as the changes took place. Whatever releasing and swinging arrangement is adopted it should, therefore, be such as to be under the control of adults only, else deplorable accidents might result.

A somewhat noticeable feature in the recent Building Trades Exhibition in London was a series of exhibits by different firms of appliances for the secondary utilization of heat developed from an ordinary fire grate. This was generally

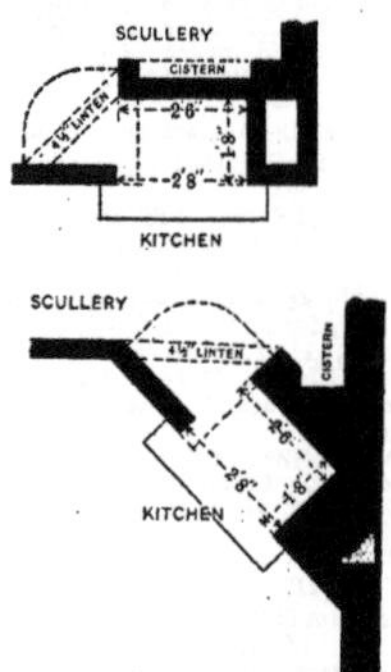

FIG. 3

done by the introduction of boilers of various forms and conveyance therefrom of water in small ball pipes to heat radiators or towel racks in other parts of the house. This is obviously a legitimate thing to do, for, although the boiler must be small, it was generally possible in this way to at least heat one radiator, say in a hall or much used bedroom, and a towel rack placed in a linen cupboard, thereby adding much to the comfort of a house. This can be done much more effectually from a kitchen range, though it requires a really large boiler to supply a series of radiators in addition to the usual baths and sinks. There is this also to bear in mind—that where the water is constantly changing on account of its being drawn off for household use there is constant deposit of fur. This does not take place where the only waste of water occurs through exceedingly slow evaporation, for once all the lime in the water has been deposited this action necessarily ceases until further water is added. There are some boilers made from which the incrustation is easily removed, but in many of the commoner kinds it causes a great amount of trouble and even occasionally leads to burning out.

Another combination arrangement is one now being much employed in artisans' dwellings, by means of which a copper is heated from a small range fitted into a 2 ft. 8 in. opening in a kitchen, it being generally further contrived to supply a bath also. As the range is in one room and the scullery in another, a rather special arrangement is necessary. The house has, in fact, to be built to fit the range, as the range cannot be introduced into any existing house. Fig. 3 shows two ways in which this can be done when the Lanco combination is being used.

TIN MINING IN CORNWALL.

Consul Joseph G. Stephens, of Plymouth, writes as follows on renewed mining activity in that part of England:

Cornish tin mining points to a revival. As long as the Americans delay developing the rich deposits in South Dakota and in the Southern States, so long will the Cornish mines profit through the lack of rivalry. There is no doubt that an enormous quantity of tin still remains to be extracted from the Cornish hills. Many old mines, which ceased working when the bad times came some thirty years ago, were closed, not because they had not an abundant deposit of tin ore, but because the market was flooded with foreign tin, which was then being produced at a price with which the deep and costly Cornish working could not compete. This foreign supply has now greatly diminished, and the world again turns for its supply to Cornwall. Old mines are being re-opened and new claims started

IMPROVED HEATING STOVE.

W. Heuermann, Sedala, Mo., has invented an improved heating stove. There is provision in this stove for a relatively large heating surface, and a long flue or passage for transverse of the heated gaseous products of combustion. The stove proper, occupies a relatively small space. It comprises a combustion chamber and a superposed heating chamber connected and supported together from the top of the downwardly extended flues or pipes, the latter serving practically for the place of legs or other usual form of support.

TO MAKE STEEL RUST-PROOF.

The invention attributed to a Birmingham (Eng.) apothecary of a simple process for rendering steel rust-proof is one of revolutionary importance. The process would seem to consist of an application of phosphorus quite different from that which has produced the strong and beautiful, but seldom used, alloy, phosphor-bronze. It is likely to be more akin to the process by which the ancients made copper hard and capable of taking an edge like steel, a lost art that every metallurgist has dreamed of rediscovering.

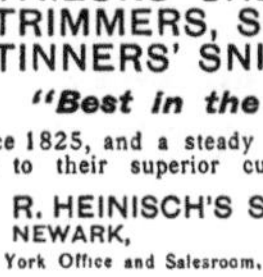

CATALOGUES AND BOOKLETS.

When sending catalogues for review, manufacturers would confer a favor by pointing out the new articles that they contain. It would assist the editor in writing the review.

By mentioning HARDWARE AND METAL to show that the writer is in the trade, a copy of these catalogues or other printed matter will be sent by the firms whose addresses are given.

Elaborate Stove Catalogue.

The Gurney Foundry Co., Toronto, have just issued a magnificent 9 x 11 inch catalogue containing 187 pages replete with illustrations, descriptive of their various styles of stoves, ranges, warm air furnaces, steam boilers, hot water heaters, radiators, general castings, etc. In the centre of the front cover is a magnificent 5x5-inch colored engraving representing a typical foundry scene. Ten moulders are seen carrying away the molten metal from two huge cauldrons into which it falls from glowing furnaces in the background. The dancing light of the red hot metal, which casts a lurid glare on windows, walls and floor and is reflected back on the animated faces and sinewy arms of the moulders, makes the picture extremely realistic. The rest of the cover surrounding this scene is handsomely embossed in black, gold, green and pale green colors, the artistic effect obtained by this color combination leaving nothing to be desired. The first two inner pages of the catalogue contain fine engravings of the company's huge stove and furnace foundry, and boiler and radiator foundry, respectively. Immediately after that the company's telegraphic code is explained and then follow a few pages of general directions for operating stoves, ranges, furnaces, etc. Many pages are devoted to attractive designs and complete descriptions of steel ranges, the Oxford Chancellor steel plate range occupying the premier position, after which are shown various styles of Imperial Oxford ranges. Gas ranges, hot water boilers, radiators, hotel ranges and the various parts of each are all shown and described in an interesting manner. On the closing pages of the catalogue complete directions are given for ordering repairs, which makes it an easy matter for the customer to obtain the repair he wants without the usual confusion and delay. This handsome and comprehensive catalogue, together with a separate price list booklet, may be obtained by writing the company, mentioning this paper.

Attractive I.C.R. Booklets.

For the lover of history and romance, as well as for the tourist and sportsman, Eastern Canada possesses attractions unsurpassed by any other portion of North America. To present a few of the attractions of this beautiful summer land, which, notwithstanding the thousands who travel over the road each year, is yet a veritable terra incognito to the average tourist and sportsman, the passenger department of the Intercolonial Railway has prepared a beautiful and useful series of booklets. For general information this series is far in advance of anything of the kind ever presented to the public by this railway before, and should be in the hands of everyone seeking rest and sport.

The first of these publications is entitled "Tours to Summer Haunts." It is a pamphlet of some hundred pages, printed on coated book paper and generously illustrated with artistic halftone reproductions of scenes incident to the text, with sub-heads in marginal indent, supplementing which is a schedule of tours and excursions from Montreal to the principal points of interest, and facilities offered to votaries of rod and gun, all interesting and instructive reading for even those who have no opportunity of visiting the places described. Then follow "Fishing and Hunting," "The Hunting Grounds of the Micmacs," "A Week in the Canaan Woods," "Big Game of the Southwest Miramichi," and "The Maritime Express," all of which are eminently practical subjects indicated by their respective titles.

"Fishing and Hunting," has the game laws of the different provinces compiled in addition to tabulated information concerning the varieties of game and fish to be found in each locality along the whole line. It is a splendid effort to present desired information in concise form, and meets the requirements admirably.

"The Hunting Grounds of the Micmacs," presents to its readers a short historical reference to the tribe for which it is named, and directs the lovers of rod and gun where to find victims. It is an admirable little handbook, readable and full of information.

"A Week in the Canaan Woods" and "Big Game of the Southwest Miramichi," informs the hunter and tourist of the attractions of the territories indicated by the names.

"The Maritime Express" is really an annotated time table describing the locality of each station, and giving valuable information regarding business, sport and other data necessary or desired by travelers from Montreal to the eastern termini of the road.

The booklets are legibly printed, well illustrated with half-tones, and each has a full four-page map of the country traversed by the railroad. Sportsmen and tourists may obtain these interesting publications by applying to the company mentioning this notice.

New Sporting Catalogue.

A very artistic sporting goods catalogue is now being issued to the trade by Lewis Bros., Montreal, one of Canada's greatest wholesale hardware houses. The covers are of light green stiff paper, the outside front cover containing an illustration of the Montreal home of the firm, and an elaborate contents list. On the inside of the front cover is a message to customers regarding complaints about high charges for gun repairs, and also advice as to shipping of guns to the firm for repairs.

The catalogue consists of 100 pages fully illustrated, containing cuts and prices on all lines of shot guns, rifles and ammunition.

The lines of rifles carried by Lewis Bros., are : the Winchester, the Marlin, the Savage, and Stevens'. Illustrations and prices are given also of hunters' supplies, such as traps, sheath knives, hunting knives, pocket compasses, axes, police goods, skates, hockey sticks, and fishers' supplies are included. Altogether, the catalogue contains interesting information, and is attractively gotten up, and is eminently a credit to its compilers. Any who have not received this catalogue will do well to procure one by applying to the company, mentioning this paper months are as follows:

Convenient Cartridge Chart.

C. Edgar Wood, general sales manager for the Dominion Cartridge Co., Montreal, who has for years sold ammunition to the retail trade—having been associated with one of the large jobbing houses— from practical experience, became decidedly impressed with the fact that the trade needed an educational hanger or chart that would show the exact difference between each cartridge. Heretofore the cartridges have been mounted on board that by reason of their size had to be placed high on the walls, where minute inspection as to size and shape was impracticable. To do away with this difficulty Mr. Wood has gotten up a new chart or hanger with the different sizes and types grouped for immediate inspection, and the new hanger will no doubt prove of decided benefit to all hardware and sporting goods dealers. The Dominion Cartridge Company will take pleasure in forwarding one of these convenient hangers to any dealer applying for it on his business stationery.

Asbestology.

"Asbestology" for July, published by Canadian Asbestos Co., Montreal, contains some interesting information regarding the various asbestic products, with a few lines more or less pungent. This little paper is done on red paper and will be interesting and instructive to all engaged in asbestos business.

Compo-Rubber Roofing.

Merrick Anderson Co., Winnipeg, have recently been appointed agents in Western Canada for the sale of compo-rubber roofing. A superior line manufactured by the Lineolu Waterproof Cloth Co., Bound Brook, N.J. This is a superior line of roofing specially prepared to withstand the effects of heat, rain, snow and cold. Western Canada affords an immense market for roofing materials of superior quality and Merrick Anderson Co. state that their new line is meeting with a favorable reception from the trade.

Wants Catalogues.

F. R. Agnew, Box 1181, Calgary, Alta., writes this office that he would like to receive catalogues from manufacturers of hardware and plumbing supplies.

New Stove Catalogues.

The Thos. Davidson Mfg. Co., Montreal, have issued to the trade two splendidly gotten up catalogues, containing illustrations and price lists of their full line of stoves and ranges. The larger of the two catalogues is bound in heavy brown linen with an illustration of their factories. Inside is a page of instructions on how to put up stoves. This catalogue throughout is replete with illustrations and information. The smaller booklet is enclosed in yellow cover and contains full illustrations of their special line of ranges. All interested in the stove trade will do well to procure a copy of these by applying to the company mentioning this paper.

WIRE NAILS TACKS WIRE

Prompt Shipment

The ONTARIO TACK CO.
Limited.
HAMILTON ONT.

ALL INGOT METALS IN STOCK

A. C. LESLIE & CO. Limited
MONTREAL

"MIDLAND"

BRAND.

Foundry Pig Iron.

Made from carefully selected Lake Superior Ores, with Connellsville Coke as Fuel, "Midland" will rival in quality and grading the very best of the imported brands.

Write for Price to Sales Agents

Drummond, McCall & Co.
MONTREAL, QUE.
or to
Canada Iron Furnace Co.
MIDLAND, ONT. Limited

NOVA SCOTIA STEEL & COAL CO., Limited

NEW GLASGOW, N.S.

Manufacturers of

Ferrona Pig Iron

And SIEMENS-MARTIN
OPEN HEARTH STEEL

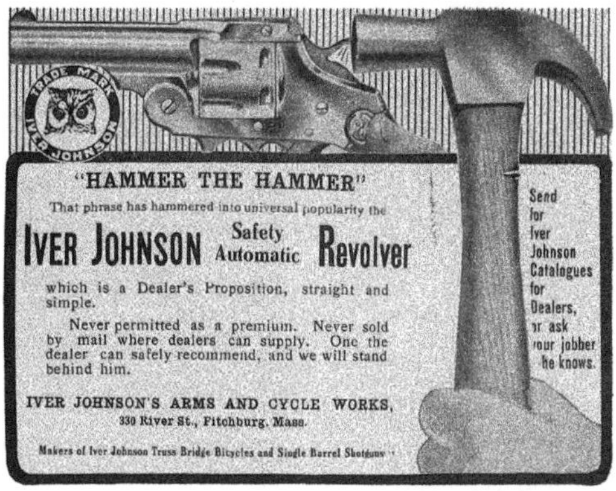

M.R.M. CO.

SHOT

The season for the sale of this line is at hand, and in ordering your supply of Shot it is well to consider the question of quality.

"M.R.M." Shot is made from the very best quality of pig lead. Care is taken in its manufacture to have it

Uniform, Round and True to Size

It has earned for itself the reputation of always being up to standard, and can be relied upon to give satisfaction.

MADE BY

The Montreal Rolling Mills Co.

BUILDING AND INDUSTRIAL NEWS

For additional items see the correspondence pages. The Editor solicits information from any authoritative source regarding building and industrial news of any sort, the formation or incorporation of companies, establishment or enlargement of mills, factories or foundries, railway or mining news.

Companies Incorporated.

The Spanish River Navigation Co., Massey, Ont., capital, $40,000; to build ships and vessels. Provisional directors: J. Errington, J. Sheets, J. S. Lowe, all of Massey, Ont.

The Watford Milling Co., Watford, Ont., capital, $10,000; to carry on a general milling business. Provisional directors: A. Dunlop, G. A. Dunlop and S. Rivers, all of Watford.

Bain & Cubitt, Toronto, capital, $40,000; to manufacture paper boxes, envelopes and printers' machinery. Provisional directors: D. Bain, W. C. Cubitt, J. S. Denison, all of Toronto.

The Cobbler-Sexton Co., Woodstock, Ont., capital, $1,024,000; to carry on mining and milling and reduction. Provisional directors: Wm. A. Hayward, J. C. Hovey, E. L. Greer, J. McClement, I. Draper, all of Woodstock.

The Interlocking Piling and Engineering Co., Limited, Toronto, capital, $200,000; to manufacture and deal in interlocking, piling and piling of all kinds, and carry on the business of engineers. Provisional directors: H. E. Pearce, A. Gate, W. H. Smith, G. Kerwin, M. Irving, all of Toronto.

Industrial Development.

There is a great influx of machinery to the Cobalt regions.

Fire did damage to the extent of $50,000 in Winchester, Ont.

The Doty Engine Co., Goderich, Ont., are erecting a new foundry.

W. C. Edwards' sash and door factory Ottawa, was destroyed by fire.

The Hull Electric Co., Hull, Que., are erecting an addition to their plant.

The Canada Steel Goods Co., Hamilton, will erect a factory to cost $75,000.

A large cement mill will be erected at Owen Sound by some Toronto capitalists.

The Ontario Iron and Steel Co., Welland, will begin smelting operations soon.

The Canadian Cutlery Co., want a loan of $25,000 to erect a plant at Grimsby, Ont.

The Chestnut Canoe Co., Fredericton, N.B., are erecting a factory to cost $13,000.

Fire damaged the Flemming's foundry, St. John, N.B., to the extent of $1,000.

The Mortimer Co., Ottawa, are erecting an addition to their premises, to cost $25,000.

The Western Bag Co., Winnipeg, will erect a warehouse to cost $200,000 in that city.

The flour mill and electric plant of Claresholm, Alta., was destroyed by fire recently.

The pork-packing factory at Aylmer, Ont., was destroyed by fire caused by lightning.

The Rocky Mountain Cement Co. will establish a large cement works at Blairmore, Alta.

The sash and door factory of Henry Henks, Toronto, was destroyed by fire; loss $6,000.

The Dominion Car and Foundry Co., Montreal, will enlarge their works at a cost of $2,000.

The Cobalt Concentrators, Ltd., are asking for tenders for a concentrating mill at Cobalt.

The buildings of the Manitoba Peat Works, Fort Francis, Man., will be rebuilt immediately.

The Dominion Paint Co.'s premises at Hamilton were damaged by fire to the extent of $1,200.

The storage tank of the Imperial Oil Co., at Sarnia, Ont., was destroyed by lightning recently.

Whitman & Barnes Mfg. Co., St. Catharines, Ont., will spend $50,000 in an addition to their factory.

The factory of the Dominion Furniture Co., St. Therese, Que., was destroyed by fire; loss $100,000.

A firm manufacturing stoves for burning soft coal wants a free site and loan of $10,000 from Sarnia, Ont.

The Kensington Furniture Co., Goderich, Ont., are making extensive additions to their furniture factory.

A brick company has been formed by some residents of Lacombe, Alta., and a large industry will be established.

Mackay Bros., North Bay, with a capital of $100,000, will acquire the hardware business of David Purvis.

The pulp and wood mills of the North River Lumber Co., Murray, N.S., were totally destroyed by fire recently.

The firm of J. M. Ross, Sons & Co., St. Catharines, which received a bonus of $20,000 from that city, has failed.

Work has commenced on the factory of the Aluminium and Crown Stopper Co., Toronto. The factory will cost $50,000.

The cement brick works at Radisson, Sask., have started operations and the product has been found satisfactory in every way.

The machinery is being installed in the new buildings of the St. Thomas Canning Co., St. Thomas, and will be running in three weeks.

The Ottawa Steel Casting Co., Ottawa will increase the capacity of their furnace to 8 tons per day. A new addition will be erected and a new plant installed.

The Chapman Double Ball Bearing Co., Toronto, have been awarded the contract for the entire equipment of the new plant of the Standard Valve and Fitting Co., Guelph.

The Don Valley Brick Works, Toronto, have received the contract for the fireproof brick for the new Royal Bank branch in that city.

The International Heating and Lighting Co. have obtained a franchise from Fort Saskatchewan, and will erect a plant there. Machinery has been ordered to cost $100,000.

A unique electric power plant, indeed the first of its kind in America, was opened recently at Chignecto mines, N. S., The power house is situated at the mouth of the mine and the waste screenings are used for fuel. This does away with transportation of fuel and at the same time uses fuel that would otherwise be useless. It is said the power can be obtained at such a low rate that it will be available for every sphere of manufacturing, The power is transmitted to Amherst, N.S., a distance of 7 miles. This style of plant has long been advocated by authorities who have made the production of cheap power a study. It is hoped that this will be the means of bringing new industries to Amherst.

Building Notes.

A town hall will be erected at Rosthern, Sask.

A new high school will be erected at Beachville, Ont.

The new high school at Calgary, Alta. will cost $68,000.

A hotel will be erected at Kenora, Ont. to cost $225,000.

The high school Ottawa, will be enlarged and re-modelled.

J. Murphy will erect a business block in Fort William to cost $15,000.

A. E. Snell, Calgary, Alta., is enlarging his dry goods establishment.

E. R. Wayland, Fort William, Ont., will erect a residence to cost $11,000.

A morgue and ambulance house will be erected in Toronto to cost $30,000.

A police station will be erected in Toronto on Queen street to cost $25,000.

A building will be erected by the Y.M.C.A. in St. John, N.B. to cost $60,000.

A new wing is being erected at the Hotel Dieu Hospital, Campbellton, N.B. to cost $40,000.

The grand stand at the Ottawa exhibition grounds was destroyed by fire, entailing a loss of $60,000.

The Edmonton Steam Laundry Co., Edmonton, Alta., are erecting an addition to their plant to cost $2,000.

A large fire in North Bay recently did damage to the extent of $25,000 to the hardware stock of J. W. Richardson.

The contract for the new Bank of Commerce building, Brantford, has been let to Schultz Bros. It will cost $30,000.

Mining News.

The Silver Queen mine, Cobalt, Ont., will erect a smelter.

A valuable depoist of silver and lead has been discovered near Loon Lake, N.S.

The Montreal Copper Co., Montreal, have recently received large orders from foreign countries.

A mining company is being formed at Sturgeon Lake, Ont., for the purpose of

CONDENSED OR "WANT" ADVERTISEMENTS.

Advertisements under this heading 2c. a word first insertion; 1c. a word each subsequent insertion.

Contractions count as one word, but five figures (as $1,000) are allowed as one word.

Cash remittances to cover cost must accompany all advertisements. In no case can this rule be overlooked. Advertisements received without remittance cannot be acknowledged.

Where replies come to our care to be forwarded, five cents must be added to cost to cover postage, etc.

AGENT WANTED.

AGENT wanted to push an advertised line of Welsh tinplates; write at first to "2.R." care HARDWARE AND METAL, 88 Fleet St., E.C., London, Eng. [tf]

BUSINESS CHANCES.

MINING hardware, stoves and tinware business for sale; sales $4,000 monthly; buildings and lot, $3,500; present stock, $8,000; proprietor's health failing; a bonanza. Box 643, HARDWARE AND METAL, Toronto. [32]

FOR SALE — Well established hardware, tinshop, implement and undertaking business, also good lumber yard, well fenced, with lumber and lime sheds in good condition; we will sell above altogether, or divide same to suit purchaser; proprietors are retiring from business in Manitoba, and therefore wish for immediate sale. Apply to Eakins & Griffin, Shoal Lake, Man. [33]

WANTED—Partner to take half interest in one of the best hardware propositions in Algoma; plumber preferred. Box 636, HARDWARE AND METAL, Toronto. [32]

HARDWARE, tinware, stove and plumbing business in manufacturing town in the Niagara Peninsula; no competition; $250,000 factory and water-marks will be completed this summer; stock about $3,000; death of owner reason for selling. Box 85, Thorold, Ont. [tf]

FOR SALE—A good hardware business in Western Ontario; stock about $6,000. For further reference apply The Hobbs Hardware Co., Limited, London, Ont. [32]

HARDWARE and Tin Business for Sale in good Western Ontario town of 3,000; stock about $3,500; good reasons for selling. Address, Box 647 HARDWARE AND METAL, Toronto. (34)

HARDWARE, Tinware, Stove and Furnace Business for sale, in live Eastern Ontario Village; first class chance for a practical man; English speaking community; stock can be reduced to suit purchaser; can give possession September 15th. 1907; premises for Sale or Rent, Apply to D. Courville, Manxville, Ont. (34)

HARDWARE Business and Tinshop for sale in Saskatchewan; population 1500; stock carried about $14,000 turnover, $45,000 practically all cash business; cash required, $8,000 would rent building; Do not answer without you have the money and mean business; it will pay to investigate this. Box 648 Hardware & Metal, Toronto. (41)

FOR SALE.

FOR SALE — First-class set of tinsmith's tools second-hand but almost as good as new; includes an 8-foot iron brick almost new. Apply Pease Waldon Co., Winnipeg. [tf]

FOR SALE—A quantity of galvanized plain twist wire. Apply to C. B. Miner, Cobden, Ont. (34)

SITUATIONS VACANT.

TINSMITHS WANTED — First-class tinsmiths wanted for points west of Winnipeg; must be good mechanics capable of taking charge of a metal department; thorough knowledge of furnace work necessary. Pease Waldon Co., Winnipeg, Man. [tf]

WANTED—Hardware clerk, experienced, who can keep stock, and is willing to do so; sober and active; state age, experience and salary expected at start. Hose & Canniff, Kenora, Ont. [32]

Brighten Up
Our "Brighten Up" Campaign last year was a winner. This year we have another—even better than the last. It begins September 1st. The features have been specially planned for it—they're strong—they're distinctive—they'll get the business. This campaign is going to produce results.
If you are not one of those fortunate dealers who handle S-W. Products, let us hear from you. Take advantage of the business opportunities our big "Brighten Up" Campaign will bring. We have other campaigns, too, and are constantly helping our agents. We co-operate to mutual advantage.
Get our proposition today. Address
THE SHERWIN-WILLIAMS CO.
LARGEST PAINT AND VARNISH MAKERS IN THE WORLD
Canadian Headquarters and Plant: 639 Centre St. Montreal, Que.
Warehouses: 86 York St., Toronto, and Winnipeg, Man.

SITUATIONS WANTED.

SITUATION wanted as master mechanic or chief engineer by man of 22 years' experience as a mechanic; can give A1 reference as to ability; strictly temperate. Box A, HARDWARE AND METAL, Toronto. [32]

GERMAN (31), 14 years' commercial experience desires situation; perfect knowledge of tools, hardware, fittings of every description, wooden goods, bar iron, steel, metals, also bookkeeping, shorthand (English and German) typewriting, storekeeping; at liberty on about October 1st. Address 0783 care of Messrs. Deacon's, Leadenhall Street, London, E.C.

COMMERCIAL gentleman with nine years' trade connection with ironmongers, architects and public institutions in Great Britain desires position as representative of a Canadian manufacturing firm. Box X, HARDWARE AND METAL, Montreal.

HARDWARE Salesman or Clerk, seven years' experience, desires situation. Abstainer, best of references; position in the West preferred. A. Tilley, Brantford, Ont. (31)

WANTED (by Englishman) position in Hardware store, experience in all branches, bookkeeper, would invest $400 to $500, Address, Box 646 HARDWARE AND METAL, Toronto. (31)

WANTED.

OFFICE space wanted by manufacturer in Toronto; state location and terms. Box 645, HARDWARE AND METAL, Toronto. [32]

THE WANT AD.

The want ad. has grown from a little used force in business life, into one of the great necessities of the present day.

Business men nowadays turn to the "want ad" as a matter of course for a hundred small services.

The want ad. gets work for workers and workers for work.

It gets clerks for employers and finds employers for clerks. It brings together buyer and seller, and enables them to do business, though they may be thousands of miles apart.

The "want ad." is the great force in the small affairs and incidents of daily life.

operating the Wyndego, near Kenora, Ont.

A copper smelter with a capacity of 250 tons of ore per day will be erected at the Whitehorse mine, Yukon, by the Yukon Smelting and Power Co.

Railroad Construction.

The permit has been issued for the new C.N.R. shops at Winnipeg, to cost $200,000.

The Canadian Northern have been asked to build a line from Cobalt to Parry Sound, Ont.

The G.T.R. will erect a large roundhouse at Durand, Mich., to have accommodation for forty engines.

Municipal Undertakings.

A court house will be erected at Saskatoon, Sask.

A provincial jail will be erected at Moosemin, Sask., to cost $50,000.

The ratepayers of Whitewater, Man., have decided to construct a telephone system.

A site has been obtained in Morden, Man. for the new municipal electric light plant.

The ratepayers of Amherstburg, Ont., will spend $2,500 in improving the waterworks system.

HAMMER FIRST KNOWN TOOL.

The hammer, besides being a tool of universal use, is probably the oldest representative of a mechanic's tool kit. The hammer was originally a stone fastened to a handle with thongs and itw as useful as a weapon as a tool. Hammers are represented on the monuments of Egypt 20 centuries before our era. They greatly resemble the hammer now in use, save that there were no claws on the back for the extraction of nails. Claw hammers were invented some time during the middle ages. Illuminated manuscripts of the eleventh century represent carpenters with claw hammers. Hammers are of all sizes, from the dainty instruments used by all the jewelers, which weigh less than half an ounce, to the gigantic 50-ton hammer of shipbuilding establishments, some of which have a falling force of 90 to 100 tons. Every trade has its own hammer and its own way of using it.

CANADA'S OLDEST IRON FURNACE

The St. Maurice Forges, on the right bank of the St. Maurice river, about seven miles above Three Rivers, are the oldest smelting furnaces in Canada, and dispute with those of Principio, in Maryland, the right to be considered the oldest in America. The deposits of bog-ore were known very early to the Jesuits. In 1668 they were examined by the Sieux la Potardien, who reported unfavorably to the Intendant Talon as to their quality and quantity. Frontenac and De Denonville gave a better account of them, and it seems that tests were made before the year 1700. It was not until 1737, however, that a company was found to work them. This company was granted a large tract, including the site where the old forges now stand, and erected furnaces, but exhausted its capital, and in 1740 had to surrender its charter. The Government carried on the works very successfully, as a report of the Colonial Inspector Tranguet shows, and must have extended them, as appears by the erection of the old château that stands on a flat bluff overlooking the river. On an iron plate in its chimney are the official fleurs de lis and the date 1752. Its walls, some two and a half feet thick, withstood the fire that destroyed its woodwork in 1863.

A brook flows through the ravine immediately below the chateau. It furnished water power for the oldest works, remains of which are to be seen near its mouth. The attachments of an old shaft show that a trip hammer were used, and there are other signs of extensive works for making wrought iron. From 250 to 300 men were employed, under directors who had gained their skill in Sweden. Many of the articles made then—notably stoves—still attest the quality of the iron and of the work. Pigs and bars were sent to France. During the war shot and shells were cast. When the English came to take possession, the chateau was occupied by

Canada's Oldest Iron Furnace.

a Desmoiselle Poulin, who threw the keys into the river rather than yield them. Legends of mysterious lights and buried treasure cling to the place. After the conquest the works were leased to private persons, and have passed through several hands before coming into those of the present owners, who use most of the product in the manufacture of car-wheels in Three Rivers.

The original blast-furnace, or cupola—a huge block of granite masonry, thirty feet square at the base—is still used for smelting; the fire has rarely been extinguished, except for repairs, during the past 150 years.

RAZOR BLADE HOLDER.

T. F. Curley, New York, N.Y., has invented an improved holder for razor blades. The object in this improvement is to provide a holder for razor blades used in ordinary and safety razors, and to be held in stropping and honing machines and other devices. The blade holder can be arranged so as to hold a blade securely in position during the use of the razor, machine, or other device, and will allow an interchange of blades of different thicknesses.

News of the Paint Trade

PAINTING THE EIFFEL TOWER.

The Eiffel Tower Company has just had its concession renewed until the year 1914, and is celebrating the event by giving the tower a new coat of paint, thirty tons of liquid paint being required to give it a single coat. Fifty painters, working continuously together, take three months to do one coat. In 1889 three of them fell. The tower had already had eighteen victims in constructing, and everyone working about it went carefully. In the repainting of 1895 seven men fell, burying their bodies deep in the earth. When, after five men had been killed, the directors would have rigged them up with life-saving belts and ropes, says the Oil and Color Trade Journal, they rebelled to a man. The result was that two others fell before the job was finished. In 1900 the directors called for painters who had worked in 1895, to give them preference. They responded in a mass and asked exorbitant wages. When it was finally got into their heads that the selection was being made not for their skill as painters, but as a mere act of humanity, they agreed to work for the usual advance of 30 per cent. on the union scale—on condition of not being required to disgrace themselves with life-saving belts and rigging. The result was five men killed in 1900. Now, last week there fell the first victim of the series of 1907. The actual painting has not been begun. He was a foreman, and he was just climbing about, looking over the job. He fell from a point a trifle below the third platform, struck three times going down, deflecting his fall, and yet his body sank almost five feet into the earth of a flower-bed beside the southeast pile. The third platform is 940 feet from the ground. Every artist in Paris is concerned about the color, on which depends whether the gigantic mass of iron shall be the eye-sore or the glory of the capital. Its first shade, in 1889, was "dead leaf," and the sun lit it up to hazy gold. In 1895, after immense discussion, they repainted it orange. In strong sun the orange showed gleaming copper. Then, for the exposition of 1900, they painted it "sun-color." In bright lights the tower became a thing of glory once again. That was seven years ago. The atmospheric electricity received every hour by such an unprecedented mass of iron reaching up in the air is incalculable. Conducting-tubes a foot and a half in diameter lead down its four piles to 50 feet below the water-bearing stratum of the earth; but the effect on the paint is there for everyone to see. It does not crack off; it simply disappears, leaving the ugly, dingy brown of the oxydising iron.

A recipe for tin varnish is given as follows: Asphalt 10 parts by weight, rosin 5 parts, linseed oil varnish 20 parts, oil of turpentine 8 parts.

Paint and Oil Markets

TORONTO.

Toronto, Aug. 9.—On account of last Monday being Toronto's civic holiday, time for business this week has been shortened by one day, and consequently the large supply houses here have been kept busy during the last couple of days attending to the orders which accumulated over the holiday. In all departments of the paint and oil markets the amount of business being done is satisfactory in every respect for the beginning of August, with prices throughout remaining the same as quoted last week.

White Lead—A steady trade is being carried on for August. Prices are firm and unchanged at the following figures: Genuine pure white lead is quoted at $7.65, and No. 1 is held at $7.25.

Red Lead—The market continues quiet though a fair amount of sorting orders continue to arrive. There are no changes in quotations, which run as follows; Genuine, in casks of 500 lbs., $6,-25; ditto. in kegs of 100 lbs., $6.75; No. 1, in casks of 500 lbs., $5; ditto, in kegs of 100 lbs., $5.50.

Paris Green—The demand has fallen off to a considerable degree during the week and it may be safely said that as far as the local market is concerned the season is practically over. Of course, the cool weather which prevailed during the latter part of last week is partly responsible for the decreased demand and no doubt the present hot spell will have the effect of restoring trade in repeat orders for a few days more at least. The following prices are still quoted; Canadian Paris green, 29½c. base; English Paris green, 30½c.

Petroleum—Little variation is noticeable in the demand and locally prices show no disposition to change from these figures: Prime white, 13c.; water white, 14½c.; Pratt's astral, 18c.

Shellac—For this time of the year the demand is good, with prices unchanged as under: Pure orange, in barrels, $2.-70; white, $2.82½ per barrel; No. 1, (orange), $2.50.

Turpentine — The local prices for spirits of turpentine are unchanged. The market in the southern States has advanced slightly within the past week and while there is a good supply of turpentine apparently on hand, it looks as if there will be a steady advance now in the market for the next two months. Local prices remain at the following figures: Single barrels, 80c.; two barrels and upwards, 79c. f.o.b. point of shipment, net 30 days; less than barrels, 5c. advance. Terms: 2 per cent., 30 days.

Linseed Oil—The linseed oil market is unchanged, futures depending very largely upon the flax crop in the western provinces. At present reports are not very favorable. English oil may be slightly cheaper for futures, as the seed supply from the East Indies is said to be somewhat larger than last year. Trade for the past week in both oil and turpentine has been about normal for this time of the year; the ordinary demand having been as good as usual from the country. Generally speaking, prices remain unchanged as under, but the weakness of the local market is shown by the fact that one firm continues to quote three cents lower all round: Raw, 1 to 3 barrels, 65c.; 4 barrels and over, 64c. Add three cents to this price for boiled oil f.o.b. Toronto, Hamilton, London and Guelph, 30 days.

For additional prices see current market quotations at the back of the paper.

MONTREAL.

Montreal, Aug. 9—The chief feature in local paint and oil circles continues to be the unusually strong call for Paris Green, more especially from the eastern districts. Makers here are experiencing but a weak demand from Ontario dealers, as already they appear to have adequate stocks The manufacturers are kept very busy in an endeavor to make prompt and full shipments.

Mixed paints are still receiving a good call. The backwardness of spring compelled painters to postpone a large fraction of their work to the summer months, thus, for this season of the year an unseasonably strong demand exists. Extensive renovations and the erection of an unusually large number of new buildings in local districts is also largely responsible for the active calls for mixed paints.

No material changes have been made in the prices on any lines.

Turpentine—Local market conditions are unchanged. Prices have reached the minimum, and the outlook is uncertain. Lack of demand in southern markets has, however, caused a drop in American prices of 1½ cents during the past week. The habit of waiting for lower prices by purchasers is probably the cause for this decline. Single barrels are still quoted at 80 cents.

Linseed Oil—No further decline has occurred in prices. We continue to quote: Raw, 1 to 4 bbls., 62c; 5 to 9, 61c.; boiled, 1 to 4 bbls., 65c.; 5 to 9 bbls., 64c.

Ground White Lead — Supplies are quite adequate to the call which at present is moderate. Prices are unchanged. Government standard, $7.50; No. 1, $7; No. 2, $6.75; No. 3 $6.35.

Dry White Zinc — Prices are firm and unchanged, with a moderate demand. V.M. Red Seal, 7½c.; Red Seal, 7c.; French V.M., 6c.; Lehigh 5c.

White Zinc Ground in Oil — Situation is unchanged. Prices are firm. Pure, 8½c.; No. 1, 7c.; No. 2, 5¾c.

Red Lead—Business in this is somewhat slack. The demand is moderate. Prices are firm and unchanged. Genuine red lead, in casks, $6.25; in 100-lb. kegs, $6.50; in less quantities at $7.25 per 100 lbs. No. 1 red lead, casks, $6; kegs, $6.25, and smaller quantities, $7.

Gum Shellac — Prices are unchanged. Fine orange, 60c. per lb.; medium orange, 55c. per lb.; white (bleached), 65c.

Shellac Varnish — For shellac and varnishes the demand is steady. Prices are firm Pure white bleached

shellac, $2.80 ; pure orange, $2.60 ; No. 1 orange $2.40.

Putty—Manufacturers are busily engaged filling the orders which are increasing in number. Prices are firm. Pure linseed oil, $1.85 bulk ; in barrels, $1.60 ; in 25-lb. irons, $1.90 ; in tins, $2 ; bladder putty, in barrels, $1.85.

Paris Green—The business done this year will be a record-breaker. Eastern districts are at present monopolizing the trade.

UNINFLAMABLE PAINT REMOVERS.

Carbon tetra-chloride forms an excellent medium for removing old paint, but is too volatile to be used alone. On the other hand, it has the property of rendering inflammable liquids safe when used in suitable proportion. Recently also it has been made miscible with water, since it forms with sulphonated oils, like Turkey red oil, a gelatinous soap which is perfectly homogeneous, and will mix with water in all proportions. Such a solution containing, for instance, 1 part of the said gelatinous soap and ½ to 1 part of water, when stirred up with 1 to 2 parts of carbon-tetra-chloride and mixed with alkali and spirit, will form a very good paint remover. Another suitable class of remover is obtained by dissolving caustic alkalies in spirit. For instance, a solution containing equal parts of alkali and water is warmed with sufficient soap to form a gelatinous mass, and diluted with strong alcohol. The soap acts on the varnish covering the paint, and exposes the latter to attack by the alkali. A French preparation for the same purpose consist of alkali cellulose, which has been converted into viscose by treatment with carbon disulphide, and dissolving the product in water. The viscose is mixed with alkali, and in this condition will rapidly corrode even the oldest layers of paint, laying the underlying surface bare.—Farben Zeitung.

HOW TO PAINT IRON, ZINC, ETC.

The best time for painting new iron is at the foundry as soon after casting, or being wrought or rolled, as possible, says the Master Painter. Paint it when a dry wind or warm sun will act upon it; do not paint it in the early morning or damp evening. First see that the iron is thoroughly dry and free from rust, and then coat with red lead and linseed oil, a thin coat, just enough to penetrate the pores of the iron. The first coat must dry hard. Follow up with three other coats containing red or white lead in as great proportion as possible.

To paint old iron, burn off all rust and scale, brush with turpentine or paraffin and proceed precisely as with new iron.

For zinc, the first coat should consist of white lead, red lead and turpentine, tempered with varnish. Wash new rolled sheet zinc with a solution of a tablespoonful of hydrochloric or nitric acid to a gallon of water, or scratch the surface with No. 2. glass paper before painting.

Treat galvanized iron the same as zinc, but do not use the acid preparation, nor scratch. Very smooth, bright tin plate must be first dulled or scratched and the first coat should be oilless.

LIQUID FLOOR WAX.

R. C. Jamieson & Co., Montreal, are putting on the market a new line of floor wax. Every householder who has waxed floors to keep in order knows how difficult it is to apply floor wax in stiff paste form. The liquid hard floor wax manufactured by R. C. Jamieson & Co. is put on like oil or varnish with a wide flat brush and the person applying it does not need to kneel. The makers claim that it can be applied in one-tenth the time and with less than one-tenth the trouble and fatigue. Most American and Canadian floor wax is largely composed of beeswax or other soft wax, and consequently dries sticky, collects dust, and allows dirt to grind into it, spoiling the clean appearance and destroying the lustre of the wax. This new wax is claimed to be entirely free from a vegetable wax, and to be hard as stone, drying quite free from stickiness, and to give hard, durable finish.

BRIDLING A BRUSH.

To bridle a tool, one very simple method is to form a loop crossed at commencement, and wind a sufficient number of turns over it, then pass the end through the loop and draw it tightly until the loop is drawn well under the bridling; and then cut off close each end of the string. This is a neat method and does not require tacks.

Before use brushes should be soaked for a short time in water. This tightens the bridling, and prevents it slipping off, which it would otherwise have a tendency to do.

Fine whipcord or fishing line is better than common twine for bridling, says the Decorators' Magazine, and it is a good precaution to place a band of paper round the bristles and tie the bridle over it. It can be slipped out after the tying has been completed. The bridling must not be drawn too tightly, as the bristles will swell when put into water or paint, and tighten up, and if the bridling is too tight the bristles will twist and cripple.

TO REMOVE PAINT SPOTS.

Where a house is being done up paint is not infrequently spilt on doorsteps and it is sometimes found difficult to remove. In that case make a strong solution of potash and wash the steps, simply leaving the solution to soak in. In a short time the paint will become soft and then can be washed off with soap and water. Then use cold water. Paint which has been left for some time will yield to this treatment.

Paints are booming now. There is always a good demand for floor finishes, interior finishes and varnishes and lacquers for furniture, etc. A woman in the window demonstrating a good varnish or finish will bring a lot of business into the store. Paint brushes and similar goods sell with this line.

Holding Trade

Must be accomplished by right methods. Progressive merchandising requires powerful and direct advertising of you and your business. If you would like to secure the co-operation of one of the oldest and largest paint manufacturing concerns in the world in giving you a bigger reputation as a live, up-to-date dealer, increasing your business and helping you

To Make More Money

this season than ever before in your business history all you have to do is to sit down and write to the makers of **Martin-Senour Paint 100% Pure.** Ask us for our special proposition to agents and how we advertise dealers who handle our goods. We'll tell you something to your advantage.

The MARTIN-SENOUR CO. Ltd.,
"Pioneers of Pure Paints"
142-144 Inspector Street, Montreal.
2514-2522 Quarry St. and Archer Ave., Chicago. (Est. 1878.)
The Winnipeg Paint & Glass Co. Ltd., Winnipeg. The Maritime Paint & Varnish Co. Ltd., Halifax, N. S.

BUSINESS INSURANCE

100% Durability

The shrewd dealer takes out a policy assuring lasting commercia success by stocking a paint with ***lasting qualities.*** The business acumen of hardware men accounts for the ever-increasing sales of

ARK BRAND PAINT

because of its durability and lasting lustre. You can preach economy to your customers when recommending Ark Brand Paint. The best paint is always the most economical for your customers to use. To preach economy is sound business policy. Stocking Ark Brand is positive assurance of the paying trade and a lasting reputation.

MURESCO WALL FINISH

is so much superior to the next best Wall Finish that it stands in a class by itself. ***Muresco*** is peerless in ***quality*** and ***durability,*** and the ease with which it is applied makes it the next best thing to ***Ark Brand Paint*** as a trade-builder and profit-maker.

We have interesting particulars for the man who wants more business.

Manufactured by

BENJAMIN MOORE & CO., LIMITED

(Formerly The Francis-Frost Co., Limited)

NEW OFFICES AND WORKS:

TORONTO JUNCTION, CANADA.

New York. Chicago. Cleveland. Carteret, N.J.

BENJAMIN MOORE & CO. FOR WALL & CEILING DECORATIONS MURESCO NEW YORK AND CHICAGO

CURRENT MARKET QUOTATIONS.

August 9, 1907

These prices are for such qualities and quantities as are usually ordered by retail dealers on the usual terms of credit, the lowest figures being for larger quantities and prompt pay. Large cash buyers can frequently make purchases at better prices. The Editor is anxious to be informed at once of any apparent errors in this list, as the desire is to make it perfectly accurate.

—

METALS.

ANTIMONY.

Cookson's .... per lb.	0 17
Hallett's	0 16½

BOILER PLATES AND TUBES.

Plates, ¼ to ½ inch, per 100 lb.	2 40	2 50
Heads, per 100 lb	2 65	2 75
Tank plates 3-16 inch	2 60	2 10
Tubes per 100 feet, 1½ inch ..	8 25	8 50
" " 2 " ..	9 40	9 10
" " 2½ " ..	10 50	11 00
" " 3 " ..	12 00	12 50
" " 3½ " ..	15 00	16 00
" " 4 " ..	19 25	20 00

BOILER AND T.K. PITTS.

Plain tinned and Spun, 25 per cent. off list.

BABBIT METAL.

Canada Metal Company—Imperial,genuine 60c.; Imperial Tough, 66c; White Brass, 50c.; Metallic, 35c.; Harris Heavy Pressure, 25c.; Hercules, 25c; White Bronze, 15c; Star Frictionless, 14c; Alluminoid, 10c; No. 4, 9c per lb.

James Robertson Co.—Extra and genuine Monarch, 60c.; Crown Monarch, 50c; No. 1 Monarch, 40c; King, 30c; Fleur-de-lis, 20c; Thurber, 15c; Philadelphia, 12c., Canadian, 10c; hardware, No. 1, 15c; No. 2, 12c; No. 3, 10c. per lb.

BRASS.

Rod and Sheet, 14 to 30 gauge, 25 p.c. advance.

Sheets, 12 to 14 in.	0 30
Tubing, base, per lb 5-16 to 2 in	0 33
Tubing ¼ to 3-inch, iron pipe size..	0 31
" ½ to 3-inch, seamless..	0 36

Copper tubing, 6 cents extra

COPPER.

	Per 100 lb.	
Casting ingot	23 10	23 50
Cut lengths, round, bars, ½ to 2 in...		35 00
Plain sheets, 14 oz.		36 00
Plain, 16 oz., 14x48 and 14x60		35 00
Tinned copper sheet, base		38 00
Planished base		43 00
Braziers' (in sheets), 4x6 ft., 25 to 30 lb. each, per lb., base..	0 34	0 35

BLACK SHEETS.

	Montreal.	Toronto
8 to 10 gauge	2 70	2 75
12 gauge	2 70	2 75
14 "	2 50	2 60
17 "	2 50	2 60
18 "	2 50	2 60
20 "	2 50	2 60
22 "	2 55	2 65
24 "	2 55	2 70
26 "	2 65	2 85
28 "	2 70	3 00

CANADA PLATES

Ordinary, 52 sheets	2 75	3 05
All bright "	3 75	4 05
Galvanized –	Dom.Crown.	Ordinary.
18x24x52	4 45	4 35
60	4 70	4 60
20x28x80	8 90	8 70
....	9 40	9 20

GALVANIZED SHEETS.

B W gauge	Queen's Head	Fleur-de-Lis	Gordon Crown	Colborne Crown Gorbal's Best
16-20 ..	3 95	3 80	3 95	....
22-24 ..	4 20	4 05	4 00	4 05
26	4 45	4 30	4 40	4 30
28	4 70	4 55	4 60	4 55

Less than case lots 10 to 25c. extra.

Apollo Brand

24 guage, American	3 85
26 " "	4 10
28 " "	4 55
10¾ oz.	4 85

25c less for 1,000 lb. lots.

IRON AND STEEL.

	Montreal.	Toronto.
Middlesboro, No 1 pig iron..	21 50	24 50
Middlesboro, No. 3 pig iron ..	20 50	23 50
Summerlee, " ..	25 50	24 50
" special " ..	24 50	....
" soft " ..	24 00	....
Carron	26 00	....
Carron Special	24 50	....
Carron Soft	24 00	....
Clarence, No. 3	21 50	23 50
Glengarnock, No. 1	....	27 00
Midland, Londonderry and Hamilton off the market but quoted nominally at ..	...	26 00
Radnor, charcoal iron	32 00	34 10
Common bar, per 100 lb.	2 20	2 30
Forged iron "	2 45	
Refined " "	2 60	2 70
Horseshoe iron "	2 60	2 70
Band iron, 1½ to 3 in. base...	....	2 80
Sleigh shoe steel "	2 25	2 30
Tire steel	2 40	2 50
Best sheet steel	...	0 12
Mining cast steel	....	0 08
Warranted cast steel	....	0 14
Annealed cast steel	....	0 15
High speed		0 60
B.P.L. tool steel ...	0 10½	0 11

INGOT TIN.

Lamb and Flag and Straits—56 and 28-lb. ingots, 100 lb. $44 00 $45 00

TINPLATES.

Charcoal Plates—Bright

M.L.S., Famous (equal Bradley)	Per box
I C, 14 x 20 base	$6 50
IX, 14 x 20 "	8 00
IXX, 14 x 20 base	9 50

Raven and Vulture Grades—

I C, 14 x 20 base	5 00
I X "	6 00
I X X "	7 00
I X X X "	8 00

"Dominion Crown Best"—Double Coated, Tissued

		Per box.
I C, 14 x 20 base	5 50	5 75
I X, 14 x 20 "	6 50	6 75
IXX x 20 "	7 50	7 75

"Allaway's Best"—Standard Quality.

I C, 14 x 20 base	4 65	5 00
I X, 14 x 20 "	5 40	5 75
I XX, 14 x 20 "	6 15	6 50

Bright Cokes.

Bessemer Steel—

I.C., 14 x 20 base	4 25	4 35
20x28, double box	8 50	8 70

Charcoal Plates—Terne

Dean or J. G. Grade—

I.C., 20x28, 112 sheets	7 25	8 00
IX., Terne Tin	...	9 50

Charcoal Tin Boiler Plates.

Cookley Grade—

X X, 14x56, 50 sheet bxs. / " 14x60, " / " 14x65, " }	7 50

Tinned Sheets.

72x30 up to 24 gauge........	8 50
" " 26 "	9 00

LEAD.

Imported Pig, per 100 lb.	5 25	5 35
Bar, "	5 75	6 00
Sheets, 2½ lb. sq. ft., by roll ..	....	6 50
Sheets, 3 to 6 lb. "	....	6 25

Cut sheets ½c. per lb., extra.

SHEET ZINC.

5-cwt. casks	7 75	8 00
Part casks	8 00	8 25

ZINC SPELTER.

Foreign, per 100 lb	6 75	7 00
Domestic	6 50	6 75

COLD ROLLED SHAFTING.

9-16 to 11-16 inch	0 06
¾ to 1 7-16 "	0 05½
1 7-16 to 3 "	0 05

30 per cent.

OLD MATERIAL.

Dealers buying prices:

	Montreal	Toronto
Heavy copper and wire, lb.	0 17	0 16½
Light copper	0 16	0 15
Heavy red brass	0 15½	0 15
" yellow brass	0 12	0 12
Light brass	0 08½	0 08
Tea lead	0 03¾	0 04
Heavy lead	0 04	0 04¼
Scrap zinc	0 03¾	0 04
No. 1 wrought iron	15 50	11 50
" 2 " "	6 00	6 00
Machinery cast scrap ...	17 00	16 50
Stove plate	13 00	12 00
Malleable and steel	8 00	8 00
Old rubbers	0 10½	0 10
Country mixed rags, 100 lbs.	1 00	1 25

—

PLUMBING AND HEATING

BRASS GOODS, VALVES, ETC.

Standard Compression work, 57½ per cent.
Cushion work, discount 40 per cent.
Fuller work, 70 per cent.
Flatway stop and stop and waste cocks, 60 per cent; roundway, 55 per cent.
J M T. Globe, Angle and Check Valves, 45; Standard, 55 per cent
Kerr standard globes, angles and checks, special, 42½ per cent; standard, 47½ p.c.
Kerr Jenkins' disc, copper-alloy disc and heavy standard Valves, 40 per cent.
Kerr steam radiator Valves, 60 p.c., and quick-opening hot-water radiator valves, 60 p.c.
Kerr brass, Weber's straightway Valves, 40; straightway Valves, I.B.B.M., 60.
J. M. T. Radiator Valves 50; Standard, 60; Patent Quick-Opening Valves, 65 p.c.
Jenkins' Valves—Quotations on application to Jenkins' Bros., Montreal.

No. 1 compression bath cock ...... net	2 00
No. 4 " " ...... "	1 90
No 7 Fuller's .... "	2 25
No. 4½ " .... "	2 35
Patent Compression Cushion, basin cock, hot and cold, per doz., ...	$16.20
Patent Compression Cushion, bath cock, No. 2208	25

Square head brass cocks, 50; iron, 60 p.c.
Thompson Smoke-test Machine $25.00

BOILERS—COPPER RANGE.

Copper, 30 gallon, $33, 15 per cent.

BOILERS—GALVANIZED IRON RANGE.

30-gallon, Standard, $5; Extra heavy, $7.75

BATH TUBS.

Steel clad copper lined, 15 per cent.

CAST IRON SINKS.

16x24, $1; 18x30, $1; 18x36, $1.30.

ENAMELED BATHS, ETC.

List issued by the Standard Ideal Company Jan. 3, 1907, shows an advance of 10 per cent over previous quotations.

ENAMELED CLOSETS AND URINALS

Discount 15 per cent

HEATING APPARATUS.

Stoves and Ranges—40 to 70 per cent.
Furnaces—45 per cent.
Registers—70 per cent.
Hot Water Boilers—50 per cent
Hot Water Radiators—50 to 55 p.c.
Steam Radiators—50 to 55 per cent.
Wall Radiators and Specials—50 to 55 p.c.

LEAD PIPE

Lead Pipe, 5 p.c. off
Lead waste, 5 p.c. off.
Caulking lead, 6½c. per pound.
Traps and bends, 40 per cent.

IRON PIPE.

Size (per 100 ft.)	Black.		Galvanized
¼ inch	2 35	¼ inch	3 20
⅜ "	2 35	⅜ "	3 20
½ "	2 90	½ "	3 75
¾ "	3 90	¾ "	5 00
1 "	5 60	1 "	7 25
1¼ "	7 65	1¼ "	9 90
1½ "	9 20	1½ "	11 90
2 "	12 25	2 "	16 80
2½ "	20 10	2½ "	26 00
3 "	26 75	3 "	34 00
3½ "	34 25	3½ "	42 75
4 "	39 00	4 "	48 60

Malleable Fittings—Canadian discount 30 per cent.; American discount 25 per cent.
Cast Iron Fittings 57½; Standard bushings 57½.; headers, 57½; flanged unions 57½, malleable bushings 55; nipples, 70 and 10; malleable lipped unions, 55 and 5 p.c.

SOIL PIPE AND FITTINGS

Medium and Extra heavy pipe and fittings, up to 6 inch, 60 and 10 to 70 per cent.
7 and 8-in. pipe, 40 and 5 per cent.
Light pipe, 50 p.c.; fittings, 50 and 10 p.c.

OAKUM.

Plumbers ....per 100 lb. 4 50 5 00

STOCKS AND DIES.

American discount 25 per cent.

SOLDERING IRONS.

½-lb. to 1½ .... per lb.	0 45½	0 48
2-lb. or over .... "	0 42½	0 46

SOLDER.

Per lb.	Montreal	Toronto
Bar, half-and-half, guaranteed	0 25	0 26
Wiping	0 22	0 23

—

PAINTS, OILS AND GLASS.

BRUSHES

Paint and household, 70 per cent..

CHEMICALS.

	In casks per lb.
Sulphate of copper (bluestone or blue vitrol)	0 09
Litharge, ground	0 06
" flaked	0 06½
Green copperas (green vitrol)	0 01
Sugar of lead	0 08
Lump olive	0 01½

COLORS IN OIL.

Venetian red, 1-lb. tins pure.	....	0 09
Chrome yellow "	....	0 15
Golden ochre "	....	0 10
French " "	....	0 08
Marine black "	....	0 04½
Chrome green "	....	0 09
French permanent green "	....	0 13
Signwriters' black "	....	0 15

GLUE.

Domestic sheet	0 10	0 10½
French medal	0 12	0 12½

PARIS GREEN.

	Berger's	Canadian
600-lb. cask	....	0 27½
250-lb. drums	0 28	0 27¾
100-lb. "	0 29½	....
50-lb. "	0 29½	0 28½
1-lb. pkgs, 100 in box	0 30½	0 29½
½-lb. "	0 32½	0 31½
1-lb. tins, 100 in box	0 31½	0 30½
½-lb. tins	0 33½	0 32

PARIS WHITE

In bbls	0 90

PIGMENTS.

Orange mineral, casks	....	0 08
" " 100-lb. kegs	....	0 08½

PREPARED PAINTS.

Barn (in bbls.)	0 65	0 90
Sherwin-Williams paints,	....	1 55
Canada Paint Co.'s pure	....	1 40
Standard P. & V. Co.'s "New Era."		1 30
Benj. Moore Co.'s "Ark" B'd	....	1 25
" British Navy deck	....	1 50
Brandram-Henderson's "English"		1 45
Ramsay's paints, Pure, per gal.	....	1 30
" Thistle, "	....	1 10
Martin-Senour's 100 p.c. pure,	....	1 55
Senour's Floor Paints	....	1 35
Jamieson's "Crown and Anchor"		1 20
Jamieson's floor enamel	....	1 50
"Island City" paint	....	1 25
Sanderson Pearcy's, pure	....	1 20
Robertson's pure paints	....	1 20

PUTTY.

Bulk in bbls	1
Bladders in bbls	1
25-lb. tins	1
Bladders in bulk or tins less than 100 lb	2 60
Bulk in 100-lb. irons	1 86

SHINGLE STAINS.

In 5 gallon lots	0 85	0 90

SHELLAC.

White, bleached	....	2 65
Fine orange	....	2 60
Medium orange	....	2 55

TURPENTINE AND OIL.

Prime white petroleum	....	0 13
Water white "	....	0 14½
Pratt's astral "	....	0 18
Castor oil	0 08	0 10
Gasoline	....	0 22½
Benzine, per gal	0 17	0 20
Turpentine, single barrels	0 79	0 80
Linseed Oil, " raw	0 61	0 64
" " boiled	0 64	0 67

WHITE LEAD GROUND IN OIL. Per 100 lbs

Canadian pure	7 15	7 50
No. 1 Canadian	6 80	7 15
Munro's Select Flake White	....	7 65
Elephant and Decorators' Pure	....	7 65
Monarch	....	7 40
Standard Decorator's	....	7 15
Essex Genuine	....	6 80
Brandram's B. B. Genuine	....	8 70
"Anchor," pure	....	7 40
Ramsay's Pure Lead	....	7 00
Ramsay's Exterior	....	6 65
"Crown and Anchor," pure	....	7 25
Sanderson Pearcy's	....	7 40
Robertson's O.P., lead	....	7 20

RED DRY LEAD.

Genuine, 560 lb. casks, per cwt	....	6 25
Genuine, 100 lb. kegs, "	....	6 50
No. 1, 560 lb. casks, per cwt	....	6 00
No. 1, 100 lb. kegs, per cwt	....	6 25

WINDOW GLASS

Size United inches	Star	Double Diamond
Under 26	$4 25	$6 25
26 to 40	4 65	6 75
41 to 50	5 10	7 50
51 to 60	5 35	8 50
61 to 70	5 75	9 75
71 to 80	6 25	11 00
81 to 85	7 00	12 50
86 to 90	....	15 00
91 to 95	....	17 50
96 to 100	....	20 50
101 to 105	....	24 00
100 to 110	....	27 50

Discount—16-oz., 25 per cent; 21-oz. 30 per cent. per 100 feet. Broken boxes 50 per cent.

WHITING.

Plain, in bbls	....	0 70
Gilders bolted in bands	....	0 90

WHITE DRY ZINC.

Extra Red Seal, V.M	0 07½	0 08

WHITE ZINC IN OIL.

Pure, in 25-lb. irons	....	0 08½
No. 1, " "	....	0 07
No. 2, " "	....	0 05½

VARNISHES.

		Per gal cans
Carriage, No. 1	....	1 50
Pale durable body	....	3 50
" hard rubbing	....	3 00
Finest elastic gearing	....	3 00
Elastic oak	....	1 50
Furniture, polishing	....	2 00
Furniture, extra	....	1 15
" " No. 1	....	0 90
" " union	....	0 80
Light oil finish	....	1 40
Gold size japan	....	1 80
Brown japan	...	0 95
No 1 brown japan	....	0 95
Baking black japan	....	1 35
No. 1 black japan	....	0 90
Benzine black japan	....	0 70
Crystal Damar	....	2 80
No. 1 "	....	2 50
Pure asphaltum	....	1 40
Oilcloth	....	1 50
Lightning dryer	....	0 70
Elastilite varnish, 1 gal. can, each		2 00
Granitine floor Varnish, per gal	....	2 50
Maple Leaf coach enamels; size 1,		1 20
Sherwin-Williams' kopal Varnish, gal.,		2 50
Canada Paint Co's sun Varnish	....	2 00
"Kyanize" Interior Finish	....	2 40
"Flint-Lac," coach	....	1 80
B.H Co's "Gold Medal," in cases		2 00
Jamieson's Copaline, per gal.	....	2 00

BUILDERS' HARDWARE.

BELLS.

Brass hand bells, 60 per cent.

Nickel, 55 per cent.

Gongs, Sargeant's door bells	5 50	8 00
American, house bells, per lb.	0 35	0 40

Peterboro' door bells, 37½ and 10 off new list.

BUILDING PAPER, ETC.

Tarred Felt, per 100 lb.		2 25
Ready roofing, 2-ply, not under 45 lb. per roll		1 00
Ready roofing, 3-ply, not under 65 lb., per roll		1 25
Carpet Felt ....per ton		60 00
Heavy Straw Sheathing ....per ton		40 00
Dry Surprise		0 45
Dry Sheathing ....per roll, 400 sq. ft.		0 40
Tar " " 400 "		0 50
Dry Fibre " 400 "		0 55
Tarred Fibre " 400 "		0 65
O. K. & I. X. L. " 400 "		0 70
Resin-sized " 400 "		0 45
Oiled Sheathing " 600 "		1 00
Oiled " " 400 "		0 70
Roof Coating, in barrels ....per gal.		0 17
Roof " small packages "		0 25
Refined Tar ....per barrel		5 00
Coal Tar "		4 00
Coal Tar, less than barrels ....per gal.		0 15
Roofing Pitch ....per 100 lb.	0 80	0 90
Slater's felt ....per roll		0 70
Heavy Straw Sheathing f. o. b. St. John and Halifax		42 50

BUTTS.

Wrought Brass, net revised list.

Wrought Iron, 70 per cent.

Cast Iron Loose Pin, 60 per cent.

Wrought Steel Fast Joint and Loose Pin, 70 per cent.

CEMENT AND FIREBRICK.

Canadian Portland	2 00	2 10
Belgium	1 60	1 90
White Bros. English	1 80	2 05
"Lafarge" cement in wood		3 40
"Lehigh" cement, in wood		2 54
"Lehigh" cement, cotton sacks		2 39
"Lehigh" cement, paper sacks		2 31
Fire brick, Scotch, per 1,000	27 00	30 00
" English	17 00	21 00
" American, low	23 0	25 00
" " high	27 50	35 00
Fire clay (Scotch), net ton		4 95

Paving Blocks per 1,000.

Blue metallic, 9"x4½"x3", ex wharf	35 00
Stable pavers, 12"x6"x2", ex wharf	50 00
Stable pavers, 9"x4½"x3", ex wharf	36 00

DOOR SETS.

Peterboro, 37½ and 10 per cent.

DOOR SPRINGS.

Torrey's Rod ....per doz.	....	1 75
Coil, 9 to 11 in. "	0 95	1 65
English "	2 00	4 00

Chicago and Reliance Coil 25 per cent

ESCUTCHEONS.

Discount 50 and 10 per cent., new list

Peterboro, 37½ and 10 per cent.

ESCUTCHEON PINS

Iron, discount 40 per cent.

Brass, 45 per cent.

HINGES.

Blind, discount 50 per cent.

Heavy T and strap, 4-in., per lb. net	0 06
" " 5-in., "	0 05¾
" " 6-in., "	0 05½
" " 8-in., "	0 05¼
" " 10-in. and larger	0 05

Light T and strap, discount 65 p.c.

Screw hook and hinge—

under 12 in. ....per 100 lb.	....	4 65
over 12 in. "	....	3 65

Crate hinges and back flaps, 65 and 5 p. c.

Hinge hasps, 65 per cent.

SPRING HINGES.

Spring, per gro., No. 5, $17.50 No. 10, $18; No. 20, $10 80; No. 120, $20; No. 51, $10; No. 50, $27.50

Chicago Spring Butts and Blanks 12½ per cent.

Triple End Spring Butts, 30 and 10 per cent.

Chicago Floor Hinges, 37½ and 5 off.

Garden City Fire House Hinges, 12½ p c.

"Chief" floor hinge, 50 p.c.

CAST IRON HOOKS.

Bird cage ....per doz.	0 50	1 10
Clothes line, No. 61 "	0 00	0 70
Harness "	0 60	12 00
Hat and coat ....per gro.	1 10	10 00
Chandelier ....per doz.	0 50	1 00

Wrought hooks and staples—

¼ x 5 ....per gross	....	2 65
5-16 x 5 "	....	3 30

Bright wire hooks, 60 p c.

Bright steel gate hooks and staples, 40 p.c.

Cresent hat and coat wire, 60 per cent.

Screw, bright wire, 65 per cent.

KNOBS.

Door, japanned and N.P., doz	1 50	2 50
Bronze, Berlin ....per doz.	2 75	3 25
Bronze, Genuine "	6 00	9 00
Shutter, porcelain, F. & L. screw ....per gross	1 30	2 00
White door knobs. ..per doz.	....	2 00

Peterboro knobs, 37½ and 10 per cent.

Porcelain, mineral and jet knobs, net list.

KEYS.

Lock, Canadian 40 to 40 and 10 per cent.

LOCKS.

Peterboro 37½ and 10 per cent

Russell & Erwin, steel rim $2 50 per doz

Eagle cabinet locks, discount 30 per cent

American padlocks, all steel, 10 to 15 per cent.; all brass or bronze, 10 to 25 per cent

SAND AND EMERY PAPER.

B & A. sand, discount, 35 per cent

Emery, discount 35 per cent.

Garnet (Rurton's) 5 to 10 per cent. advance

SASH WEIGHTS.

Sectional ....per 100 lb	2 00	2 25
Solid "	1 50	1 75

SASH CORD.

Per lb.	....	0 31

BLIND AND BED STAPLES.

All sizes per lb.	0 07½	0 10

WROUGHT STAPLES.

Galvanized	....	2 75
Plain	....	2 50

Coopers', discount 45 per cent.

Poultry netting staples, discount 40 per cent.

Bright spear point, 75 per cent. discount.

TOOLS AND HANDLES.

ADZES.

Discount 22½ per cent.

AUGERS.

Gilmour's, discount 60 per cent. off list.

AXES.

Single bit, per doz.	6 00	9 00
Double bit, "	10 00	11 00

Bench Axes, 40 per cent.

Broad Axes, 25 per cent.

Hunters' Axes	5 50	6 00
Boys' Axes	6 25	7 00
Splitting Axes	7 00	12 00
Handled Axes	7 00	9 00
Red Ridge, boys', handled	....	5 75
" hunters	....	5 25

BITS.

Irwin's auger, discount 47½ per cent

Gilmour's auger, discount 60 per cent.

Rockford auger, discount 50 and 10 per cent

Jennings' Gen. auger, net list.

Gilmour's car, 47½ per cent.

Clark's expansive, 40 per cent.

Clark's gimlet, per doz	....	0 65
Diamond, Shell, per doz.	....	1 00
Nail and Spike, per gross	....	2 25

BUTCHERS CLEAVERS

German ....per doz.	7 00	9 00
American "	12 00	18 00

CHALK.

Carpenters' Colored, per gross	0 45	0 75
White lump ....per cwt.	0 60	0 66

CHISELS.

Warnock's, discount 70 and 5 per cent.

P. S. & W. Extra, discount, 70 per cent.

CROSSCUT SAW HANDLES.

S. & D., No. 3 ....per pair	0 13
S. & D., " 5 "	0 11½
S. & D., " 6 "	0 18
Boynton pattern "	0 20

CROWBARS.

3½c to 4c. per lb.

DRAW KNIVES.

Coach and Wagon, discount 75 and 5 per cent.

Carpenters' discount 75 per cent.

DRILLS.

Millar's Falls, hand and breast. net list

North Bros., each set, 50c.

DRILL BITS.

Morse, discount 37½ to 40 per cent.

Standard, discount 50 and 5 to 55 per cent.

FILES AND RASPS.

Great Western	75 per cent.
Arcade	75 "
Kearney & Foot	75 "
Disston's	75 "
American	75 "
J. Barton Smith	75 "
McClellan	75 "
Eagle	75 "
Nicholson	66⅔ "
Globe	75 "

Black Diamond, 60, 10 and 5 p.c.

Jowitt's, English list, 27½ per cent.

GAUGES.

Stanley's discount 50 to 60 per cent.

Winn's, Nos. 26 to 33 ....each	1 65	2 40

HANDLES.

Second growth ash fork, hoe, rake and shovel handles, 40 p c.

Extra ash fork, hoe, rake and shovel handles, 45 p.c.

No. 1 and 2 ash fork, hoe, rake and shovel handles, 50 p.c

White ash whiffletrees and neckyokes, 35 p.c

All other ash goods, 40 p c.

All hickory, maple and oak goods, excepting carriage and express whiffletrees, 40 p c.

Hickory, maple, oak carriage and express whiffletrees, 45 p.c.

HAMMERS.

Maydole's, discount 5 to 10 per cent.

Canadian, discount 25 to 27½ per cent.

Magnetic tack ....per doz.	1 10	1 20
Canadian sledge ....per lb.	0 07	0 08½
Canadian ball pean, per lb.	0 22	0 25

HATCHETS.

Canadian, discount 40 to 42½ per cent.
Shingle, Red Ridge 1, per doz.......... 4 40
" " 2, " 4 80
Barrel Underhill...................... 5 05

HOES

Mortar, 50 and 10 per cent.

MALLETS.

Tinsmiths'............per doz. 1 25 1 50
Carpenters', hickory, " 1 25 3 75
Lignum Vitae........ " 3 85 5 00
Caulking, each 0 60 2 00

MATTOCKS.

Canadian............ per doz. 5 50 6 00

MEAT CUTTERS.

German, 15 per cent
American discount, 33⅓ per cent.

PICKS.

Per dozen 6 00 9 00

PLANES.

Wood bench, Canadian, 40, American, 25
Wood, fancy, 37½ to 40 per cent
Stanley planes, $1.55 to $3 60, net list prices.

PLANE IRONS.

English ....per doz. 2 00 5 00
Stanley, 2⅝ inch, single 24c, double 39c

PLIERS AND NIPPERS.

Button's genuine, 37½ to 40 per cent.
Button's imitation....per doz. 5 00 9 00
Berg's wire fencing 1 72 5 50

PUNCHES.

Saddlers .. ..per doz. 1 00 1 85
Conductor's.......... " 3 00 15 00
Tinners', solid..........per set 0 72
" hollow......per inch 1 00

RIVET SETS.

Canadian, discount 35 to 37½ per cent.

RULES.

Boxwood, discount 70 per cent.
Ivory, discount 20 to 25 per cent.

SAWS.

Atkins, band and crosscut, 25 per cent.
Disston's Hand, discount 12½ per cent
Disston's Crosscut ....per foot 0 35 0 55
Hack, complete..........each 0 75 2 75
" frame only........each 0 50 1 25

S. & D. solid tooth circular shingle, concave and band, 50 per cent; mill and ice, drag, 30 per cent; cross-cut, 35 per cent.; band saws, butcher, 35 per cent.; buck, New Century $6 25; buck No. 1 Maple Leaf, $5.25; buck, Happy Medium, $4.25; buck, Watch Spring, $4.25; buck, common frame, $4.00.

Spear & Jackson's saws—Hand or rip, 26 in., $12 75; 24 in., $11 25; panel, 18 in., $8 25; 20 in., $9; tenon, 10 in., $9.90; 12 in., $10 90; 14 in., $11.50.

SAW SETS.

Lincoln and Whiting 4 75
Hand Sets, Perfect.................... 4 00
X-Cut Sets, " 7 50
Maple Leaf and Premiums saw sets, 40 off.
S. & D saw swages, 40 off.

SCREW DRIVERS.

Sargent's.............per doz. 0 65 1 00
North Bros., No. 30 ..per doz. 16 80

SHOVELS AND SPADES.

Canadian, discount 45 per cent.

SQUARES.

Iron, discount 20 per cent.
Steel, discount 65 and 10 per cent.
Try and Bevel, discount 50 to 52½ per cent.

TAPE LINES.

English, ass skin .....per doz. 2 75 5 00
English, Patent Leather 5 50 9 75
Chesterman's..............each 0 90 2 85
" steel ....each 0 80 8 00
Berg's, each.................... 0 75 2 50

TROWELS.

Disston's, discount 10 per cent.
S. & D., discount 35 per cent
Berg's, brick, 924x11 .... 4 00
" pointing, 924x5.......... .... 2 10

FARM AND GARDEN GOODS

BELLS.

American cow bells, 63⅓ per cent.
Canadian, discount 45 and 50 per cent.
American, farm bells, each .. 1 35 3 00

BULL RINGS.

Copper, $2.00 for 2½-inch

CATTLE LEADERS

Nos. 32 and 33 .. ...per gross 7 50 8 50

BARN DOOR HANGERS

	doz.	pairs.
Stearns wood track	4 50	6 00
Zenith......................	....	9 00
Atlas, steel covered	5 00	6 00
Perfect	8 00	11 00
New Milo, flexible	....	6 50
Steel, track, 1 x 3-16 in(100 ft)	....	3 25
" " 1½ x 3-16 in(100 ft)	....	4 75
Double strap hangers, doz sets....		6 40
Standard jointed hangers, "	....	6 40
Steel King hangers "	....	6 25
Storm King and safety hangers	....	7 00
" " rail................	....	4 25

Chicago Friction, Oscillating and Big Twin Hangers, 5 per cent.

HARVEST TOOLS.

50 and 10 per cent.
S. & D. lawn rakes, Dunn's, 40 off.
" sidewalk and stable scrapers, 40 off.

HAY KNIVES.

Net list.

HEAD HALTERS.

Jute Rope, ⅝-inch....per gross 9 00
" " ½ " .... " 10 00
" " ¾ " .... " 12 00
Leather, 1-inch ......per doz. 4 00
Leather, 1¼ " ...... " 5 20
Web................ " 2 45

HOES.

Garden, 50 and 10 per cent.
Planter.............. per doz. 4 00 4 50

LAWN MOWERS

Low wheel, 12, 14 and 16-inch $2 30
9-inch wheel, 12-inch 2 85
" " 14 " 3 10
" " 16 " 3 12½
High wheel, 12 " 4 05
" " 14 " 4 50
" " 16 " 4 75

SCYTHES.

Per doz. net................ .. 6 25 9 25

SCYTHE SNATHS.

Canadian, discount 40 per cent.

SNAPS.

Harness, German, discount 25 per cent.
Lock, Andrews' 4 50 11 00

STABLE FITTINGS.

Warden King, 35 per cent
Dennis Wire & Iron Co., 33⅓ p.c.

WOOD HAY RAKES.

40 and 10 per cent.

HEAVY GOODS, NAILS, ETC.

ANVILS.

Wright's, 80-lb. and over.............. 0 10½
Hay Budden, 80-lb. and over 0 09½
Brook's, 80-lb. and over 0 11½
Taylor-Forbes, prospectors 0 05
Columbia Hardware Co., per lb 0 09½

VISES.

Wright's...................... .... 0 13½
Berg's, per lb.................... 0 12½
Brook's...................... .... 0 [illegible]
Pipe Vise, Hinge, No. 1...... 3
" " " No. 2...... .... 5 12
Saw Vise .. 4 50 5 60
Blacksmiths' (discount) 60 per cent.
" parallel (discount) 45 per cent.

BOLTS AND NUTS

	Per cent.
Carriage Bolts, common ($1 list	
" " ⅜ and smaller..	60, 10 and 10
" " 7-16 and up	55 and 5
" " Norway Iron ($3 list)	50
Machine Bolts, ⅜ and less	60 and 10
Machine Bolts, 7-16 and up....	55 and 5
Plough Bolts	55 and 10
Blank Bolts..................	55
Bolt Ends..................	55
Sleigh Shoe Bolts, ⅜ and less ..	60 and 10
" " " 7-16 and larger	50 and 5
Coach Screws, conepoint......	70 and 5

Nuts, square, all sizes, 4c. per cent. off
Nuts, hexagon, all sizes, 4¼c. per cent. off
Stove Rods per lb., 5½ to 6c.
Stove Bolts, 75 per cent.

CHAIN.

Proof coil, per 100 lb, ¼ in., $6 00; 5-16 in., $4.85; ⅜ in., $4 25; 7-16 in., $4.00; ½ in., $3 75; 9-16 in., $3 70; ⅝ in., $3 65, ¾ in., $3 60, ⅞ in., $3.45, 1 in., $3.40

Halter, kennel and post chains, 40 to 40 and 5 per cent.; Cow ties, 40 per cent; Tie out chains 65 per cent; Stall fixtures, 35 per cent; Trace chain, 45 per cent.; Jack chain iron, 50 per cent.; Jack chain, brass, 50 per cent.

HORSE NAILS

M R M cold forged process, list May 15, 1907, 50 and 5 per cent.
"C" brand, 57½ per cent. off list.
Capewell brand, quotations on application.

HORSESHOES.

M R M. brand: iron, light and medium No. 1 and smaller, $3 90; No. 2 and larger, $3 65; snow pattern No 1 and smaller $4.15, No 2 and larger, $3 90; "X L" new light steel, No. 1 and smaller, $4 25; No 2 and larger, $4; "X L" featherweight steel, No. 0 to 4, $5 60, toe-weight, all sizes, $6.85. F o.b Montreal. Extras for packing

Belleville brand: No 0 and 1, light and medium iron, $3.90; snow, $4 15; light steel, $4 25; No. 2 and larger, light and medium iron, $3 65; snow, $3 90; light steel, $4. F.o b. Belleville. Two per cent., 30 days.

HORSE WEIGHTS.

Taylor-Forbes, 4½c. per lb.

NAILS.	Cut.	Wire
2d	4 00	4 50
3d	3 15	3 20
4 and 5d..............................	2 90	2 90
6 and 7d	2 80	2 85
8 and 9d	2 65	2 65
10 and 12d	2 60	2 60
16 and 20d..............................	2 55	2 55
30, 40, 50 and 60d (base)	2 50	2 50

F.o.b. Montreal, Cut nails, Toronto 20c. higher.
Miscellaneous wire nails, discount 75 per cent
Coopers' nails, discount 40 per cent.

PRESSED SPIKES

Pressed spikes, ⅝ diameter, per 100 lbs $3.15

RIVETS AND BURRS.

Iron Rivets, black and tinned, 60, 10 and 10.
Iron Burrs, discount 60 and 10 and 10 p c.
Copper Rivets, usual proportion burrs, 15 p
Copper Burrs only, net list.
Extras on Coppered Rivets, ½-lb. packages 1c. per lb.; ¼-lb. packages 2c. lb.
Tinned Rivets, net extra, 4c. per lb.

SCREWS.

Wood, F. H., bright and steel, 85 and 10 p c.
" R. H., bright, 80 and 10 per cent.
" F. H., brass, 75 and 10 per cent.
" R. H., " 70 and 10 per cent.
" F. H., bronze, 70 and 10 per cent.
" R. H., " 65 and 10 per cent.
Drive Screws, dis. 87½ per cent.
Bench, wood ....per doz. 3 25 4 00
" iron .... " 4 25 5 00
Set, case hardened, dis. 60 per cent.
Square Cap, dis. 50 and 5 per cent.
Hexagon Cap, dis. 45 per cent.

MACHINE SCREWS.

Flat head, iron and brass, 35 per cent.
Felister head, iron, discount 30 per cent.
" " brass, discount 25 per cent.

TACKS, BRADS, ETC.

Carpet tacks, blued 75 p.c.; tinned, 80 and 10; (in kegs), 40; cut tacks, blued, in dozens only, 75; ½ weights, 60; Swedes cut tacks, blued and tinned, bulk, 75 dozens, 75; Swedes, upholsterers', bulk, 85 and 12½; brush, blued and tinned, bulk, 70; Swedes, gimp, blued, tinned and japanned, 75 and 12½; zinc tacks, 35; leather carpet tacks, 40; copper tacks, 25; copper nails 30; trunk nails, black, 65; trunk nails, tinned and blued, 65; clout nails, blued and tinned 65; chair nails, 35; patent brads, 40; fine finishing, 40; lining tacks, in papers, 10; lining tacks, in bulk, 15; lining tacks, solid heads, in bulk, 75; saddle nails, in papers, 10; saddle nails, in bulk, 15; tufting buttons, 22 line in dozens only, 60; zinc glaziers' points, 5; double pointed tacks, papers, 90 and 10; double pointed tacks, bulk, 40; clinch and duck rivets, 45; cheese box tacks, 85 and 5 trunk tacks, 80 and 10.

WROUGHT IRON WASHERS.

Canadian make, discount 40 per cent.

SPORTING GOODS.

CARTRIDGES.

"Dominion" Rim Fire Cartridges and C.B. caps, 50 and ½ per cent; Rim Fire B.B Round Caps 60 and 2½ per cent; Centre Fire, Pistol and Rifle Cartridges 30 p c.; Centre Fire Sporting and Military Cartridges, 25 and 5 p c; Rim Fire, Shot Cartridges, 50 and 7½ p.c; Centre Fire, Shot Cartridges, 30 p.c; Primers, 25 p.c.

LOADED SHELLS.

"Crown" Black Powder, 15 and 5 p c.; "Sovereign" Empire Bulk Smokeless Powder, 30 and 5 p.c; "Regal" Ballistite Dense smokeless Powder, 30 and 5 p c; "Imperial" Empire or Ballistite Powder, 30 and 10 p.c.

EMPTY SHELLS.

Paper Shells, 25 and 5; Brass Shells, 5 and 5 p.c.

Wads.	per lb.
Best thick brown or grey felt wads, in ½-lb. bags	$0 70
Best thick white card wads, in boxes of 500 each, 12 and smaller gauges	29
Best thick white card wads, in boxes of 500 each, 10 gauge............	0 35
Thin card wads, in boxes of 1,000 each, 12 and smaller gauges	0 20
Thin card wads, in boxes of 1,000 each, 10gauge...	0 25
Chemically prepared black edge grey cloth wads, in boxes of 250 each—	Per M
11 and smaller gauge	0 60
9 and 10 gauges....................	0 70
7 and 8 "	0 90
5 and 6 "	1 10
Superior chemically prepared pink edge, best white cloth wads in boxes of 250 each—	
11 and smaller gauge	1 15
9 and 10 gauges	1 40
7 and 8 "	1 65
5 and 6 "	1 90

SHOT.

Ordinary drop shot, AAA to dust $7.50 per 100 lbs. Discount 5 per cent; cash discount, 2 per cent, 30 days; net extras as follows subject to cash discount only; Chilled, 40 c; buck and seal, 80c.; no 28 ball, $1 20 per 100 lbs.; bags less than 25 lbs., ½c. per lb.; F.O.B. Montreal, Toronto, Hamilton, London, St. John and Halifax, and freight equalized thereon.

TRAPS (steel.)

Game, Newhouse, discount 30 and 10 per cent.
Game, Hawley & Norton, 50, 10 & 5 per cent
Game, Victor, 70 per cent.
Game, Oneida Jump (B. & L.) 40 & 2½ p.
Game, steel, 60 and 5 per cent.

SKATES.

Skates, discount 37½ per cent.
Empire hockey sticks, per doz .. 3 00

CUTLERY AND SILVERWARE.

RAZORS. per doz.

Elliot's	4 00	18 00
Boker's	7 50	11 00
" King Cutter	13 50	18 50
Wade & Butcher's	3 60	10 00
Lewis Bros. "Klean Kutter"	8 50	10 50
Henckel's	7 50	20 00
Berg's	7 50	20 00

Clauss Razors and Strops, 50 and 10 per cent

KNIVES.

Farriers-Stacey Bros., doz 3 50

PLATED GOODS

Holloware, 40 per cent. discount.
Flatware, staples, 40 and 10, fancy, 40 and 5.
Hutton's "Cross Arrow" flatware, 47½; "Singalese" and "Alaska" Nevada silver flatware, 42 p c.

SHEARS.

Clauss, nickel, discount 60 per cent.
Clauss, Japan, discount 67½ per cent.
Clauss, tailors, discount 40 per cent.
Seymour's, discount 50 and 10 per cent.
Berg's 6 00 12 00

HOUSE FURNISHINGS.

APPLE PARERS.

Hudson, per doz., net 5 75

BIRD CAGES.

Brass and Japanned, 40 and 10 p. c.

COPPER AND NICKEL WARE.

Copper boilers, kettles, teapots, etc. 30 p c.
Copper pitts, 30 per cent.

KITCHEN ENAMELED WARE.

White ware, 75 per cent.
London and Princess, 50 per cent.
Canada, Diamond, Premier, 50 and 10 p.c.
Pearl, Imperial, Crescent and granite steel, 50 and 10 per cent.
Premier steel ware, 40 per cent.
Star decorated steel and white, 25 per cent.
Japanned ware, discount 45 per cent.
Hollow ware, tinned cast, 35 per cent. off.

KITCHEN SUNDRIES.

Can openers, per doz	0 40	0 75
Mincing knives per doz	0 50	0 [illegible]
Duplex mouse traps, per doz	..	0
Potato mashers, wire, per doz	0	0
" wood "	0 [illegible]	[illegible]
Vegetable slicers, per doz	....	[illegible]
Universal meat chopper No. 1	....	1 15
Enterprise chopper, each	...	1 30

Spiders and fry pans, 50 per cent.

Star A1 chopper 5 to 32	1 35	4 10
" " 100 to 103	1 35	2 00
Kitchen hooks, bright	....	0 60

LAMP WICKS.

Discount, 60 per cent.

LEMON SQUEEZERS.

Porcelain lined	per doz.	2 20	5 60
Galvanized	"	1 87	3 85
King, wood	"	2 75	2 90
King, glass	"	4 00	4 50
All glass	"	0 50	0 90

METAL POLISH.

Tandem metal polish paste 6 00

PICTURE NAILS.

Porcelain head ......per gross	1 35	1 50
Brass head	0 40	1 00

Tin and gilt, picture wire, 75 per cent.

SAD IRONS.

Mrs. Potts, No. 55, polished....per set	0 90
" No. 50, nickle-plated, "	0 95
" handles, japaned, per gross	9 25
" " nickled, "	9 75
Common, plain	4 25
" plated	5 50
Asbestos, per set	1 5[illegible]

TINWARE.

CONDUCTOR PIPE.

2-in. plain or corrugated, per 100 feet, 3.30; 3 in., $4.40; 4 in., $5.80; 5 in., $7.45; 6 in., $9.90.

FAUCETS.

Common, cork-lined, discount 35 per cent.

EAVETROUGHS.

10-inch ....per 100 ft. 3 30

FACTORY MILK CANS.

Discount off revised list, 35 per cent.
Milk can trimmings, discount 25 per cent.
Creamery Cans, 45 per cent

LANTERNS.

No. 2 or 4 Plain Cold Blast....per doz.		6 50
Lift Tubular and Hinge Plain,	"	4 75
No. 0, safety	"	4 00

Better quality at higher prices.
Japanning, 50c. per doz. extra.
Prism globes, per doz., $1 20.

OILERS.

Kemp's Tornado and McClary's Model galvanized oil can, with pump, 5 gallon, per dozen 10 92
Davidson oilers, discount 40 per cent.
Zinc and tin, discount 50 per cent.
Coppered oilers, 20 per cent. off.
Brass oilers, 50 per cent. off.
Malleable, discount 25 per cent

PAILS (GALVANIZED).

Dufferin pattern pails, 45 per cent.
Flaring pattern, discount 45 per cent.
Galvanized washtubs 40 per cent.

PIECED WARE.

Discount 35 per cent off list, June, 1899.
10-qt. flaring sap buckets, discount 35 per cent.
6, 10 and 14-qt. flaring pails dis. 35 per cent.
Copper bottom tea kettles and boilers, 30 p.c.
Coal hods, 40 per cent.

STAMPED WARE.

Plain, 75 and 12½ per cent. off revised list.
Retinned, 72½ per cent. revised list.

SAP SPOUTS

Bronzed iron with hooks ....per 1,000
Eureka tinned steel, hooks 8 00

STOVEPIPES.

5 and 6 inch, per 100 lengths	7 64	7 91
7 inch " "	....	8 18

Nestable, discount 40 per cent.

STOVEPIPE ELBOWS

5 and 6-inch, common........per doz.	1 32
7-inch "	1 48

Polished, 15c. per dozen extra.

THERMOMETERS.

Tin case and dairy, 75 to 75 and 10 per cent.

TINNERS' SNIPS.

Per doz........ 3 00 15
Clauss, discount 35 per cent.

TINNERS' TRIMMINGS.

Discount, 45 per cent.

WIRE.

ANNEALED CUT HAY BAILING WIRE.

No. 12 and 13, $4; No 13½, $4 10. No 14, $4.25; No. 15, $4.50; in lengths 6' to 11', 25 per cent.; other lengths 20c per 100 lbs extra; if eye or loop on end add 25c. per 100 lbs. to the above.

BRIGHT WIRE GOODS

Discount 60 per cent.

CLOTHES LINE WIRE.

7 wire solid line, No. 17, $4.90; No. 18, $3.00; No. 19, $2.70; 7 wire solid line, No. 17, $4.45; No. 18, $3.10; No. 19, $2 80. All prices per 1000 ft measure; 6 strand, No 18, $2 60; No. 19, $2 90. F o b. Hamilton, Toronto, Montreal.

COILED SPRING WIRE

High Carbon, No. 9, $2 95; No. 11, $3.50; No. 12, $3.20.

COPPER AND BRASS WIRE.

Discount 37½ per cent.

FINE STEEL WIRE.

Discount 25 per cent. List of extras In 100-lb. lots: No. 17, $5 — No. 18 $5.50 — No. 19, $6 — No. 20, $6.65 — No. 21 $7 — No. 22, $7.30 — No 23, $7.65 — No. 24, $8 — No. 25, $9 — No. 26, $9.50 — No. 27, $10 — No. 28, $11 — No. 29, $12 — No. 30, $13 — No. 31, $14 — No. 32, $15 — No. 33, $16 — No. 34, $17. Extras net—tinned wire, Nos. 17-25, $2 — Nos. 26-31, $4 — Nos. 32-34, $6. Coppered, 75c.—oiling, 10c.—in 25-lb. bundles, 15c.—in 5 and 10-lb. bundles, 25c.—in 1-lb. hanks, 25c.—in ½-lb. hanks, 38c.—in ¼-lb. hanks, 50c.—packed in casks or cases, 15c.—bagging or papering, 10c

FENCE STAPLES.

Bright .... 2 80 Galvanized ... 32

HAY WIRE IN COILS.

No 13, $2 70; No. 14, $2 80; No. 15, $2.95; f.o.b., Montreal.

GALVANIZED WIRE.

Per 100 lb.—Nos. 4 and 5, $3.90 — Nos. 6, 7, 8, $3.35 — No. 9, $2.85 — No. 10, $3.40 — No. 11, $3.45 — No. 12, $3 00 — No. 13, $3 10 — No. 14, $3.95 — No 15, $4.30 — No. 16, $4.30 from stock. Base sizes, Nos. 6 to 9, $2.35 f.o.b. Cleveland. Extras for cutting.

LIGHT STRAIGHTENED WIRE.

Gauge No.	Over 20 in. per 100 lbs.	10 to 20 in.	5 to 10 in.
0 to 5	$0.50	$0 75	$1 25
6 to 9	0.75	1.25	2 00
10 to 11	1.00	1.75	2 50
12 to 14	1 50	2.25	3 50
15 to 16	2.00	3.00	4.50

SMOOTH STEEL WIRE.

No. 0-9 gauge, $2 40; No. 10 gauge, 6c extra; No. 11 gauge, 12c extra; No. 12 gauge, 20c. extra; No. 13 gauge, 30c. extra; No 14 gauge, 40c. extra; No. 15 gauge, 55c. extra; No. 16 gauge, 70c extra. Add 60c. for coppering and $2 for tinning.

Extra net per 100 lb.—Oiled wire 10c., spring wire $1.25, bright soft drawn 15c., charcoal (extra quality) $1.25, packed in casks or cases 15c., bagging and papering 10c., 50 and 100-lb. bundles 10c., in 25-lb. bundles 15c., in 5 and 10-lb. bundles 25c., in 1-lb hanks, 50c., in ½-lb. hanks 75c., in ¼-lb. hanks $1.

POULTRY NETTING.

2-in. mesh, 19 w g., 50 and 5 p.c. off. Other sizes, 50 and 5 p c. off.

WIRE CLOTH.

Painted Screen, in 100-ft rolls, $1.72½, per 100 sq. ft.; in 50-ft. rolls, $1.77½, per 100 sq ft.

WIRE FENCING.

Galvanized barb........ 2 95
Galvanized, plain twist 3 30
Galvanized barb, f.o.b. Cleveland, $2 70 for small lots and $2.60 for carlo's

WOODENWARE.

CHURNS.

No. 0, $9; No. 1, $9; No. 2, $10; No. 3, $11; No. 4, $13; No. 5, $16; f.o b. Toronto Hamilton, London and St. Marys, 30 and 30 per cent.; f o.b Ottawa, Kingston and Montreal, 40 and 15 per cent. discount.

CLOTHES REELS.

Davis Clothes Reels, dis., 40 per cent.

FIBRE WARE.

Star pails, per doz		$ 3 00
0 Tubs,	"	14 00
1 "	"	12 00
2 "	"	10 00
3 "	"	8 50

LADDERS, EXTENSION.

3 to 6 feet, 12c. per foot; 7 to 10 ft., 13c.
Waggoner Extension Ladders, dis. 40 per cent.

MOPS AND IRONING BOARDS.

"Best" mops		1 25
"900" mops		1 25
Folding ironing boards	12 00	16 50

REFRIGERATORS

Discount, 40 per cent.

SCREEN DOORS.

Common doors, 2 or 3 panel, walnut stained, 4-in. style........per doz. 7 25
Common doors, 2 or 3 panel, grained only, 4-in., style ....per doz. 7 55
Common doors, 2 or 3 panel, light stair per doz. 9 55

WASHING MACHINES.

Round, re-acting per doz.	60 00
Square "	63 00
Eclipse, per doz	54 00
Dowswell "	39 00
New Century, per doz	75 00
Daisy	54 00
Stephenson	74 00

WRINGERS.

Royal Canadian, 11 in., per doz.	35 00
Royal American, 11 in.	35 00
Eze 10 in., per doz	36 75

MISCELLANEOUS

AXLE GREASE.

Ordinary, per gross	6 00	7 00
Best quality	10 00	12 00

BELTING.

Extra, 60 per cent.
Standard, 60 and 10 per cent.
No. 1, not wider than 6 in., 60, 10 and 10 p c
Agricultural, not wider than 4 in., 75 per cent
Lace leather, per side, 75c.; cut laces, 80c.

BOOT CALKS.

Small and medium, ball ....per M 4 25
Small heel .... " 4 50

CARPET STRETCHERS.

American ....per doz.	1 00	1 50
Bullard's "	....	6 50

CASTORS.

Bed, new list, discount 55 to 57½ per cent.
Plate, discount 52½ to 57½ per cent.

PINE TAR.

½ pint in tins ....per gross ... 7 80
" " .... " 9 60

PULLEYS.

Hothouse	per doz.	0 55	1 00
Axle	"	0 22	0 33
Screw	"	0 22	1 00
Awning	"	0 35	2 50

PUMPS.

Canadian cistern	1 40	2 00
Canadian pitcher spout	1 80	3 16

Berg's wing pump, 75 per cent.

ROPE AND TWINE.

Sisal		0 10½
Pure Manilla		0 15
"British" Manilla		0 12
Cotton, 3-16 inch and larger	0 [illegible]	0
" 5-32 inch	0 21	0
" ⅛ inch	0 [illegible]	0
Russia Deep Sea		0
Jute		0
Lath Yarn, single		0
" double		0
Sisal bed cord 48 feet ....per doz.		0
" " 60 feet .... "		0 23
" " 72 feet .... "		0 [illegible]

Twine

Bag, Russian twine, per lb.	....	0 27
Wrapping, cotton, 3-ply	....	0 25
" " 4-ply	....	0 29
Mattress twine per lb.	0 33	0 45
Staging " "	0 27	0 35

BINDER TWINE.

500 feet, sisal	0 09½
500 " standard	0 09½
550 " " manilla	0 10½
600 " " "	0 12½
650 " " "	0 13½

Car lots, ½c. less; 5 ton lots, ¼c. less. Central delivery.

SCALES.

Gurney Standard, 35; Champion, 45 p c.
Burrow, Stewart & Milne — Imperial Standard, 35; Weigh Beams, 35; Champion Scales, 45
Fairbanks Standard, 30; Dominion, 50 Richelieu, 50
Warren new Standard, 35; Champion, 45 Weigh Beams, 30.

STONES—OIL AND SCYTHE.

Washita	per lb.	0 25	0
Hindostan	"	0 06	0 10
" slip	"	0 18	0 20
" Axe	"	....	0 10
Deer Creek	"	....	0 10
Deerlick	"	....	0 25
" Axe	"	....	0 15
Lily white	"	....	0 42
Arkansas	"	....	1 50
Water-of-Ayr	"	....	0 10
Scythe	per gross	3 50	5 00
Grind, 40 to 200 lb., per ton	....	20 00	22 00
" under 40 lb., "	....	....	24 00
" 200 lb. and over	....	....	28 00

CLASSIFIED LIST OF ADVERTISEMENTS

Alabastine.
Alabastine Co., Limited, Paris, Ont.

Auditors.
DaVenport, Pickup & Co., Winnipeg.

Automobile Accessories.
Canada Cycle & Motor Co., Toronto Junction
Carriage Mountings Co., Ltd., Toronto.

Awnings.
Tobin Tent and Awning Co., Ottawa

Babbitt Metal.
Canada Metal Co., Toronto.
Canadian Fairbanks Co., Montreal
Frothingham & Workman, Montreal
Robertson, Jas. Co., Montreal.

Bar Urns.
Buffalo Mfg. Co., Buffalo, N.Y.

Bath Room Fittings.
Buffalo Mfg. Co., Buffalo, N.Y.
Carriage Mountings Co., Ltd., Toronto.
Forsyth Mfg. Co., Buffalo, N.Y.
Ontario Metal NoVelty Co., Toronto

Belting, Hose, etc.
Gutta Percha and Rubber Mfg. Toronto.
Sadler & Haworth Toronto.

Bicycles and Accessories.
Johnson's, IVer, Arms and Cycle Works Fitchburg, Mass

Binder Twine.
Consumers Cordage Co., Montreal.

Bolts.
Toronto & Belleville Rolling Mills, BelleVille

Box Strap.
J. N. Warminton, Montreal.

Brass Goods.
Frothingham & Workman, Montreal.
Glauber Brass Mfg. Co., Clevel and, Ohio.
Kerr Engine Co., WalkerVille, Ont.
Lewis, Rice, & Son., Toronto.
Morrison, Jas., Brass Mfg. Co., Toronto.
Mueller Mfg. Co., Decatur, Ill.
Penberthy Injector Co., Windsor, Ont.
Taylor-Forbes Co., Guelph, Ont.

Bronze Powders.
Canadian Bronze Powder Works, Montreal.

Brushes.
United Factories, Toronto.

Cans.
Acme Can Works, Montreal

Builders' Tools and Supplies.
CoVert Mfg. Co., West Troy, N.Y.
Frothingham & Workman Co., Montreal
Howland, H. S. Sons & Co., Toronto.
Hyde, F., & Co., Montreal.
Lewis Bros. & Co., Montreal.
Lewis, Rice, & Son, Toronto.
Lockerby & McComb, Montreal.
Lufkin Rule Co., Saginaw, Mich.
Newman & Sons, Birmingham.
North Bros. Mfg. Co., Philadelphia, Pa
Stanley Rule & Level Co., New Britain.
Stanley Works, New Britain, Conn.
Stephens, G. F., Winnipeg.
Taylor-Forbes Co., Guelph, Ont.

Carriage Accessories.
Carriage Mountings Co., Ltd., Toronto.
CoVert Mfg. Co., West Troy, N.Y.

Carriage Springs and Axles.
Guelph Spring and Axle Co., Guelph.

Carpet Beaters.
Ontario Metal NoVelty Co., Toronto.

Cartridges.
Dominion Cartridge Co., Montreal.

Cattle and Trace Chains.
Greening, B., Wire Co., Hamilton.

Chafing Dishes.
Buffalo Mfg. Co., Buffalo, N.Y.

Churns.
Dowswell Mfg. Co., Hamilton.

Clippers—All Kinds
American Shearer Mfg. Co., Nashua, N.H.

Clothes Reels and Lines.
Hamilton Cotton Co., Hamilton, Ont.

Clutch Nails.
J. N. Warminton, Montreal.

Congo Roofing
Buchanan Foster Co., Philadelphia, Pa.

Cordage.
Consumers' Cordage Co., Montreal.
Hamilton Cotton Co., Hamilton.

Cork Screws.
Erie Specialty Co., Erie, Pa.

Cow Ties
Greening, B., Wire Co., Hamilton

Cuspidors.
Buffalo Mfg. Co., Buffalo, N.Y.

Cut Glass.
Phillips, Geo., & Co., Montreal.

Cutlery—Razors, Scissors, etc.
Birkett, Thos., & Son Co., Ottawa.
Clauss Shear Co., Toronto
Dorken Bros. & Co., Montreal.
Frothingham & Workman, Montreal.
Heinisch's, R., Sons Co., Newark, N.J.
Howland, H. S. Sons & Co., Toronto.
Hutton, Wm., & Sons, Ltd., London, Eng.
Lamplough, F. W., & Co., Montreal.
Phillips, Geo., & Co., Montreal.
Round, John, & Son, Montreal.

Electric Fixtures.
Canadian General Electric Co., Toronto.
Morrison James, Mfg. Co., Toronto.
Munderloh & Co., Montreal.

Electro Cabinets.
Cameron & Campbell Toronto.

Enameled Ware
Kemp Mfg. Co., Toronto.

Engines, Supplies, etc.
Kerr Engine Co., WalkerVille, Ont.

Eavetroughs
Wheeler & Bain, Toronto

Fencing—Woven Wire
Dominion Wire Mfg. Co., Montreal.
Owen Sound Wire Fence Co., Owen Sound.
Banwell Hoxie Wire Fence Co., Hamilton.

Files and Rasps.
Barnett Co., G. & H., Philadelphia, Pa
Nicholson File Co., Port Hope

Firearms and Ammunition.
Hamilton Rifle Co., Plymouth, Mich.
Harrington & Richardson Arms Co., Worcester, Mass.
Johnson's, IVer, Arms and Cycle Works Fitchburg, Mass.

Fishing Tackle.
Enterprise Mfg. Co., Akron, Ohio

Food Choppers
Enterprise Mfg. Co., Philadelphia, Pa.
Lamplough, F. W., & Co., Montreal.
Shirreff Mfg. Co., BrockVille, Ont.

Furnaces.
Pease Foundry Co., Toronto.

Galvanizing.
Canada Metal Co., Toronto.
Dominion Wire Mfg. Co., Montreal.
Montreal Rolling Mills Co., Montreal.
Ontario Wind Engine & Pump Co., Toronto.

Glass Ornamental
Hobbs Mfg. Co., London
Consolidated Plate Glass Co., Toronto

Glaziers' Diamonds.
Gibsone, J. B., Montreal.
Pelton, Godfrey S.
Sharratt & Newth, London, Eng.
Shaw, A., & Son, London, Eng.

Handles.
Still, J. H., Mfg. Co.

Harvest Tools.
Maple Leaf HarVest Tool Co., Tillsonburg, Ont.

Hockey Sticks
Still, J. H. Mfg. Co., St. Thomas.

Hoop Iron.
Frothingham & Workman, Montreal.
Montreal Rolling Mills Co., Montreal.
J. N. Warminton, Montreal.

Horse Blankets.
Heney, E. N., & Co., Montreal.

Horseshoes and Nails.
Canada Horse Nail Co., Montreal.
Montreal Rolling Mills, Montreal.
Capewell Horse Nail Co., Toronto
Toronto & BelleVille Rolling Mills Belleville.

Hot Water Boilers and Radiators.
Cluff, R. J., & Co. Toronto.
Pease Foundry Co. Toronto.
Taylor-Forbes Co., Guelph.

Ice Cream Freezers.
Dana Mfg. Co., Cincinnati, Ohio.
North Bros. Mfg. Co., Philadelphia, Pa.

Ice Cutting Tools.
Erie Specialty Co., Erie, Pa.
North Bros. Mfg. Co., Philadelphia, Pa.

Injectors—Automatic.
Morrison, Jas., Brass Mfg. Co., Toronto.
Penberthy Injector Co., Windsor, Ont.

Iron Pipe.
Montreal Rolling Mills, Montreal.

Iron Pumps.
Lamplough, F. W., & Co., Montreal.
McDougall, R., Co., Galt, Ont.

Lanterns.
Kemp Mfg. Co., Toronto.
Ontario Lantern Co., Hamilton, Ont.
Wright, E. T., & Co., Hamilton.

Lawn Mowers.
Birkett, Thos., & Son Co., Ottawa.
Frothingham & Workman, Montreal.
Maxwell, D., & Sons, St. Mary's, Ont.
Taylor, Forbes Co., Guelph.

Lawn Mower Grinders
Root Bros. & Co., Plymouth, Ohio.

Ledgers—Loose Leaf.
Business Systems, Toronto.
Copeland-Chatterson Co., Toronto.
Crain, Rolla L., Co., Ottawa.
UniVersal Systems, Toronto.

Lithographing.
London Printing & Lithographing Co., London, Ont.

Locks, Knobs, Escutcheons, etc.
Peterborough Lock Mfg. Co., Peterborough, Ont.
National Hardware Co., Orillia, Ont.

Lumbermen's Supplies.
Pink, Thos., & Co., Pembroke Ont.

Lye
Gillett, E. W., & Co., Toron

Manufacturers' Agents.
Fox, C. H., Vancouver
McIntosh, H. F., & Co., Toronto.
Gibb, Alexander, Montreal.
Scott, Bathgate & Co., Winnipeg.

Metals.
Canada Iron Furnace Co., Midland, Ont
Canada Metal Co., Toronto.
Eadie, H. G., Montreal.
Frothingham & Workman Montreal.
Gibb, Alexander Montreal.
Kemp Mfg. Co., Toronto
Leslie, A. C., & Co., Montreal.
Lysaght, John, Bristol, Eng.
Nova Scotia Steel and Coal Co., New Glasgow, N.S.
Robertson, Jas., Co., Montreal
Roper, J. H., Montreal.
Samuel, Benjamin & Co., Toronto.
Stairs, Son & Morrow, Halifax, N.S.
Thompson, B. & S. H. & Co. Montreal

Metal Lath.
Galt Art Metal Co., Galt
Metallic Roofing Co., Toronto.
Metal Shingle & Siding Co., Preston, Ont.

Metal Polish, Emery Cloth, etc.
Oakey, John, & Sons, London, Eng.

Nails Wire
Dominion Wire Mfg Co., Montreal.

Oil Tanks.
Bowser, S. F., & Co., Toronto

Ornamental Iron and Wire.
Dennis Wire & Iron Co., London, Ont.

Packing.
Gutta Percha & Rubber Co Toronto

Paints, Oils, Varnishes, Glass
Blanchite Process Paint Co., Toronto
Brandram-Henderson, Montreal
Canada Paint Co., Montreal.
Canadian Oil Co. Toronto
Consolidated Plate Glass Co., Toronto.
Dods, P. D. & Co., Montreal
Imperial Varnish and Color Co., Toronto
Jamieson, R. C., & Co., Montreal.
Lucas John & Co. New York
McArthur, Corneille & Co., Montreal.
McCaskill, Dougall & Co. Montreal
Moore, Benjamin, & Co. Toronto.
Ottawa Paint Works, Ottawa
Queen City Oil Co., Toronto.
Ramsay & Son, Montreal
Sanderson Pearcy & Co., Toronto
Sherwin-Williams Co., Montreal.
Standard Paint Co., Montreal
Standard Paint and Varnish Works Windsor, Ont.
Stephens & Co., Winnipeg.
Martin-Senour Co., Montreal
Winnipeg Paint & Glass Co., Winnipeg

Perforated Sheet Metals
Greening, B., Wire Co., Hamilton.

Plumbers' Tools and Supplies
Canadian Fairbanks Co., Montreal.
Cluff, R. J., & Co., Toronto
Frothingham & Workman, Montreal
Glauber Brass Co., Cleveland, Ohio.
Jardine, A. B., & Co., Hespeler, Ont.
Jenkins Bros., Boston, Mass.
Kerr Engine Co., Walkerville, Ont
Lewis, Rice, & Son, Toronto.
Merrell Mfg. Co. Toledo, Ohio.
Montreal Rolling Mills Montreal
Morrison, Jas., Brass Mfg. Co., Toronto.
Mueller, H., Mfg. Co., Decatur, Ill
Oshawa Steam & Gas Fitting Co., Oshaw
Robertson Jas., Co Montreal
Robertson, Jas., Co., Limited, Toronto
Somerville, Limited, Toronto
Stairs, Son & Morrow, Halifax, N.S.
Standard Ideal Sanitary Co., Port Hope,
Standard Sanitary Co., Pittsburg.
Stephens, G. F., & Co., Winnipeg, Man.
Turner Brass Works, Chicago
Vokery, Orlando, Toronto.

Polishes.
Majestic Polishes, Toronto

Portland Cement
International Portland Cement Co Ottawa, Ont.
Hanover Portland Cement Co., Hanover, Ont.
Hyde, F., & Co., Montreal.
Thompson B. & S. H. & Co., Montreal

Poultry Netting
Greening, B., Wire Co., Hamilton, Ont

Printing.
London Printing & Lithographing Co., London, Ont.

Razors.
Clauss Shear Co., Toronto.

Refrigerators
Fabien, C. P., Montreal

Registers
Pease Foundry Co., Toronto.

Roofing Supplies
Brantford Roofing Co Brantford.
Barrett Mfg. Co., New York
F. W. Bird, East Walpole, Mass
Buchanan Foster Co., Philadelphia, Pa
McArthur, Alex., & Co., Montreal
Metal Shingle & Siding Co., Preston, Ont
Metallic Roofing Co., Toronto
Paterson Mfg. Co., Toronto & Montreal.
Wheeler and Bain, Toronto

Saws
Atkins, E. C., & Co., Indianapolis, Ind
Shurly & Dietrich, Galt, Ont
Spear & Jackson, Sheffield, Eng.

Scales
Canadian Fairbanks Co., Montreal.
Frothingham & Workman, Montreal

Screw Cabinets.
Cameron & Campbell, Toronto.

Screws, Nuts, Bolts.
Dominion Wire Mfg. Co., Montreal.
Montreal Rolling Mills Co., Montreal.

Soil Pipe
McFarlane, Walter, Glasgow

Sewer Pipes
Canadian Sewer Pipe Co., Hamilton
Hyde, F., & Co., Montreal

Shelf Boxes.
Cameron & Campbell, Toronto

Shears, Scissors.
Clauss Shear Co., Toronto

Shovels and Spades
Eclipse Mfg. Co., Ottawa
Frothingham & Workman, Montreal
Peterboro Shovel & Tool Co., Peterboro.

Silverware.
Hutton, Wm., & Sons, Ltd., London, Eng
McGlashan, Clarke Co., Niagara Falls, Ont
Phillips, Geo., & Co., Montreal
Round, John, & Son, Sheffield, Eng.

Skates.
Canada Cycle & Motor Co., Toronto.
McFarlane, Walter, Glasgow.

Sprayers
Cavers Bros., Galt

Spring Hinges, etc.
Chicago Spring Butt Co., Chicago, Ill

Stable Fittings
Dennis Wire & Iron Co., London

Steel Rails
Nova Scotia Steel & Coal Co., New Glasgow, N.S.

Stove Pipe.
Chown, Edwin and Son, Kingston

Stoves, Tinware, Furnaces
Canadian Heating & Ventilating Co. Owen Sound.
Copp, W. J., Son & Co., Fort William
Davidson, Thos., Mfg. Co., Montreal
Down Draft Furnace Co., Galt
Guelph Stove Co., Guelph.
Gurney Foundry Co., Toronto.
Harris, J. W., Co., Montreal.
Howard Wm., Toronto
Kemp Mnfg. Co. Toronto.
McClary Mfg. Co. London.
Merrick Anderson, Winnipeg
Pease Foundry Co., Toronto.
Smart, James, Mfg. Co., Brockville
Stewart, Jas., Mfg. Co., Woodstock, Ont.
Taylor-Forbes Co., Guelph, Ont.
Wright, E. T., & Co., Hamilton.

Tacks.
Montreal Rolling Mills Co., Montreal.
Ontario Tack Co., Hamilton.

Tents.
Tobin Tent and Awning Co., Ottawa

Tin Plate.
American Sheet & Tin Plate Co., Pittsburg, Pa.
Baglan Bay Tin Plate Co., Briton Ferry South Wales
Lysaght, John, Bristol, Newport and Montreal

Turpentine
Defiance Mfg. Co., Toronto.

Ventilators.
Harris, J. W., Co., Montreal.
Pearson, Geo. D., Montreal.

Wall Paper
Staunton Limited Toronto

Wall Paper Cleaner.
Gilbert, Frank U. S., Cleveland

Washing Machines, etc
Dowswell Mfg. Co., Hamilton, Ont
The Shultz Bros. Co., Brantford.
Taylor Forbes Co., Guelph, Ont.

Water Filters
Buffalo Mfg. Co., Buffalo, N.Y.

Wheelbarrows
London Foundry Co., London Ont
Schultz Bros. Co., Ltd., The Brantford.

Wholesale Hardware
Birkett, Thos., & Sons Co., Ottawa.
Caverhill, Learmont & Co., Montreal.
Frothingham & Workman, Montreal.
Hobbs Hardware Co., London
Howland, H. S., Sons & Co., Toronto.
Lamplough, F. W., & Co., Montreal.
Lewis Bros. & Co., Montreal.
Lewis, Rice, & Son, Toronto

Window and Sidewalk Prisms
Hobbs Mfg. Co., London, Ont.

Wire, Wire Rope, Cow Ties, Fencing Tools etc
Banwell-Hoxie Fence Co., Hamilton
Dennis Wire and Iron Co., London, Ont.
Dominion Wire Mnfg. Co., Montreal
Greening, B., Wire Co. Hamilton.
Owen Sound Wire Fence Co., Owen Sound
Montreal Rolling Mills Co., Montreal.
Western Wire & Nail Co., London, Ont

Wrapping Papers
Canada Paper Co., Toronto.
McArthur, Alex., & Co., Montreal.
Stairs, Son & Morrow, Halifax, N.S.

Wringers
Connor, J. H. & Son, Ottawa, Ont

"Seeing is Believing"

¶ As the Ross Sporting Rifle represents the best in modern rifle manufacture, the highest grade materials and minute accuracy in workmanship, it is to your interest to learn fully about this Canadian Rifle—

¶ The Ross Straight Pull 303 Sporting Model M is one of our leaders to retail at $26. The barrel is the celebrated Ross high pressure, non-fouling, gun-barrel steel — The breech action is dependable.

Our representatives are now demonstrating the merits of this rifle—It places you under no obligation to drop us a line to make sure you will be on their calling list.

Our advertising policy is co-operative.

THE ROSS RIFLE CO., - QUEBEC

THE SUMMER COTTAGES

need something to supersede the old-fashioned way of drawing water.

OUR GILT EDGE PUMPS

are built for the purpose and a few words will convince you that they would be a good thing to have in stock. The real article will sell itself. We have others, of course, and all through it is Quality first.

McDOUGALL PUMPS ARE STANDARD

THE

R. McDOUGALL CO., Ltd.

GALT, CANADA

WORKS ALL THE TIME

Some ventilators will work only in a wind-storm. That spoils their effectiveness

AEOLIAN VENTILATORS

are so skilfully constructed and so nicely balanced that they are driven, not only by the slightest current of air, but by the difference in temperature, within and outside the building.

Montreal, June 1st, 1903,
St. Peter School.

Messrs. The J. W. Harris Co., Limited,
Dear Sirs,

In answer to yours of 3rd inst. in connection with your system of ventilation which you have placed in the Brothers' House, in the School, etc., I beg to state that we are getting the greatest satisfaction we may desire.

Brother Jordanus, Dir.

MANUFACTURED BY

THE J. W. HARRIS CO, LIMITED

CONTRACTORS, MONTREAL

ROBERTSON'S SOLDERS

Bar-King, Strictly Merchant's, Wire (any gauge) Triangular, Oval, Plumbers' and Canners' Solders, Electricians' Solder

ARE STANDARD

Ask Quotations

The James Robertson Co., Limited
144 William St., MONTREAL, QUE.

CHICAGO SPRING BUTTS

TRADE CSB CO MARK

A PRODUCT OF RECOGNIZED SUPERIORITY

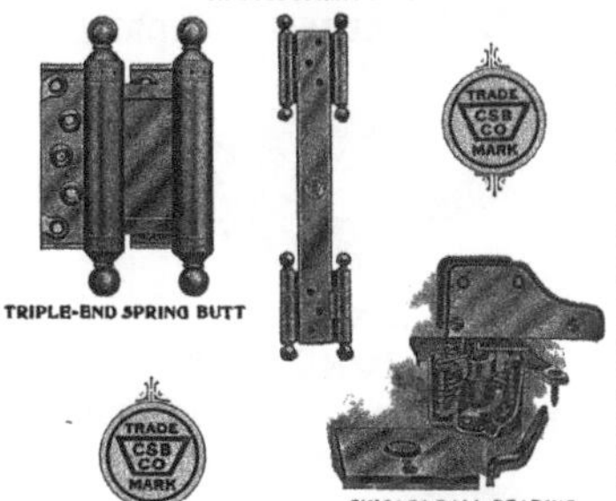

Chicago Spring Butt Company
CHICAGO NEW YORK
CATALOGUES ON REQUEST

WALKER CORK SCREWS

are good holiday souvenirs and live sellers the whole year. Are made well and are made out of good material. That is why we guarantee every one that we put out. When ordering Cork Screws try **Walker's**

Manufactured only by **ERIE SPECIALTY COMPANY,** ERIE, Pa. U.S.A.

LUFKIN MEASURING TAPES

Steel, Metallic, Linen, Pocket, Ass Skin, Pat. Leather, Bend Leather, Etc

ARE THE BEST AND MOST POPULAR TAPES IN THE WORLD.
YOUR STOOK IS NOT COMPLETE WITHOUT THEM.

LUFKIN RULE CO., Saginaw, Mich, U.S.A.

Canadian Factory - - Windsor, Ontario

London Office and Warehouse—24 and 26 Holborn. New York City Branch—[illegible] Broadway.

For sale by ALL PROMINENT CANADIAN HARDWARE JOBBERS.

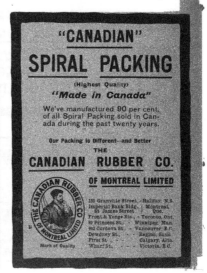

OUR "WANT ADS." get clerks for employers and find employers for clerks.

HARDWARE AND METAL

A Weekly Newspaper Devoted to the Hardware, Metal, Heating and Plumbing Trades in Canada.

Office of Publication, 10 Front Street East, Toronto.

VOL. XIX. MONTREAL, TORONTO, WINNIPEG, AUGUST 17, 1907 **NO 33.**

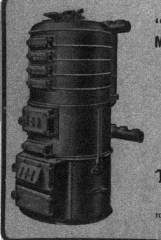

Read "Want Ads." on Page 51

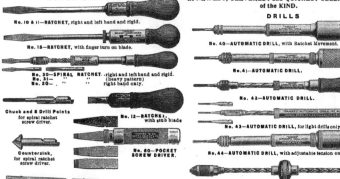

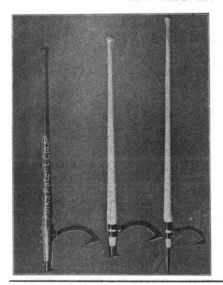

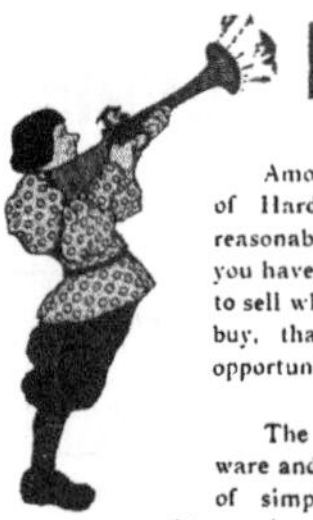

Hits the Mark

Among the thousands of readers of Hardware and Metal it seems reasonable that someone wants what you have to sell, or that someone has to sell what you have tried in vain to buy, that someone is hunting the opportunity you have to offer.

The want ad. columns of Hardware and Metal are the simplest form of simplified advertising. No ad. writer need be employed, no drawing or cut need be made, no knowledge of display type is necessary, for display type will not be used in these columns. All that you have to do is state your wants, state them simply and clearly and send to any of our offices along with remittance to cover cost of advertisement. No accounts opened in this department.

Don't despise the want ad. because of its small size or small cost. Don't forget that practically all the hardware merchants, clerks, manufacturers and travellers read our paper each week.

RATES:—
2c. per word first insertion.
1c. per word subsequent insertions.
5c. additional each insertion for box number.

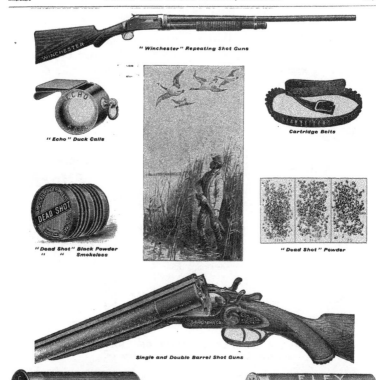

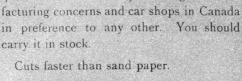

Stanley Tools are Standard

They mark the highest perfection in workmanship, igenuity of design and adaptability.

They are the tools competitors imitate. They know it is useless to endeavor to originate anything better.

Stanley tools are the results of many years of study, testing and improvement.

It pays to handle the best. Your store is known by the quality of its goods. A reputation for handling cheap substitutes for the genuine article will bring the trade of those who HAGGLE over prices instead of studying the merits of the goods they buy.

You don't want that kind of business.

Build up your reputation.

Sell good tools, and those are STANLEY.

Style of No. 9½, 15, 16 and 17 adjustable block planes.

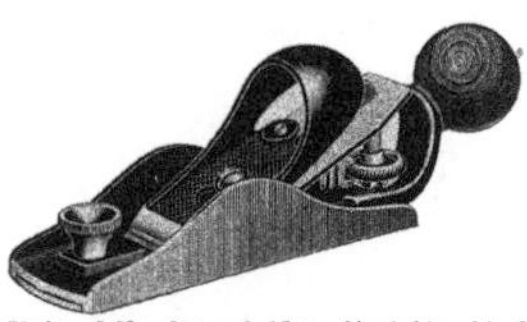

Style of No. 9¾ and 15½ adjustable block planes with rosewood handle.

Style of wood bottom jack, fore and jointer planes.

No. 20 adjustable circular plane. Full nickel plated.

Style of "Bed Rock' jack, fore and jointer planes. The highest perfection in the art of plane making

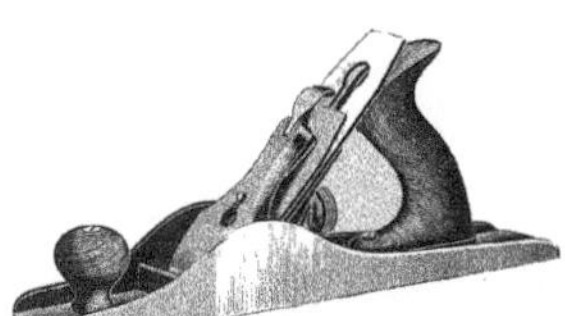

Style of 'Bailey" jack, fore and jointer planes. Plain or corrugated bottoms.

Caverhill Learmont & Co

MONTREAL

Winnipeg, Ottawa, Quebec, Fraserville

Simonds Crescent-Ground Cross-Cut Saws

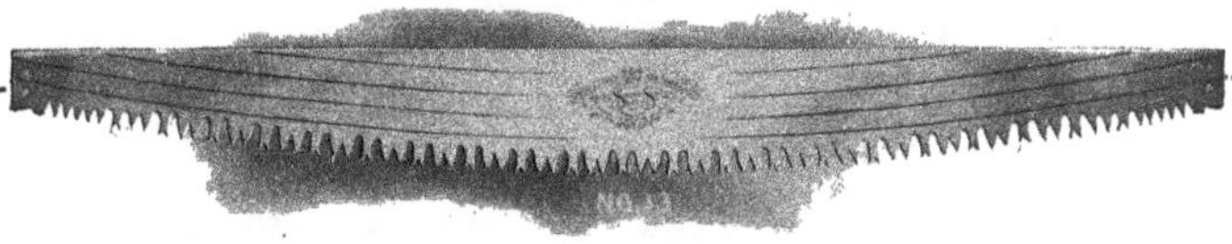

For logging camps where work must be fast and economical there is no other saw made that can give results equal to the Crescent-Ground Saw. **Made of Simonds Steel.** It cuts easy, runs fast and gives good results. This cross-cut saw is sold by most of the leading jobbers in Canada.

Simonds Canada Saw Co., Limited

TORONTO, ONT. MONTREAL, QUE. ST. JOHN, N.B.

PRATT AND WHITNEY MILLING CUTTERS

are made in any size and shape. The standard for accuracy, uniform in temper, and of the best steel.

TAPS, DIES, REAMERS, DRILLS, PUNCHES, GAUGES

PRATT AND WHITNEY SMALL TOOLS

Arrangements made with dealers to carry stock.

THE CANADIAN FAIRBANKS CO., Limited,

CANADIAN SALES AGENTS,

MONTREAL - TORONTO - WINNIPEG - VANCOUVER

Wheelbarrows

All kinds and sizes. The cut reproduced here is just of one of the many, but the kind every contractor should use. The substantial, satisfactory, steel tray Contractor's Barrow.

The London Foundry Co.

LONDON, CAN.

PURITAN
REACTING
WASHING
MACHINE.
PURITAN
ROLLER BEARING
RUNS EASY.

THREE PROFIT PULLERS!

MAXWELL LAWN MOWER

The best for Durability, Finish and Efficiency.

FAVORITE CHURN

Best for the customer and best for you.

David Maxwell & Sons, St. Mary's, Ont.

SIMPLE

DURABLE *RELIABLE*

STYLE F.

THE LEADER

A Name Familiar to Butter Makers in Connection with the

Highest Grade Churns

ITS STRONG POINTS:

Stiff Steel Frame, with malleable iron fittings, Combination Bow Lever and Pedal Drive. Bicycle Ball Bearings. Bolted Trunnions. Cream Breakers. Barrel easily drained and conveniently detatchable for cleaning.

DEALERS: When you handle this line you handle QUALITY.

The Dowswell Manufacturing Co., Limited

W. L. HALDIMAND & SON, Montreal — *EASTERN AGENTS*

HAMILTON *ONTARIO*

GALVANIZED FENCE HOOK FOR FASTENING WOODEN PICKET OR WIRE FENCES

WIRE NAILS, COILED SPRING, BARB and PLAIN FENCE WIRE, OILED and ANNEALED, CLOTHES LINE WIRE STAPLES, etc.

THE WESTERN WIRE & NAIL CO., Limited, —LONDON, ONT

The Peterborough Lock Manufacturing Company, Limited

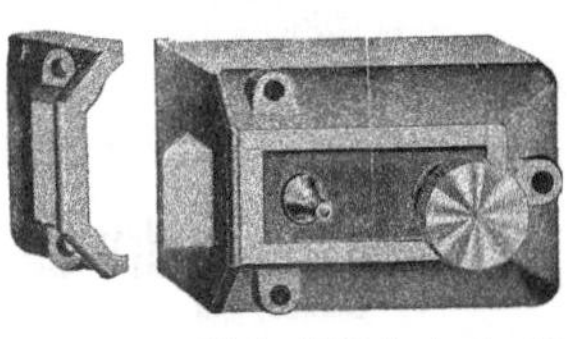

Cylinder Night Latch, No. 108.

Peterborough, Ont.

Manufacturers of all kinds

Rim and Mortise Locks, Inside, Front Door, and Store Door Sets, also full line High-class Builders' Hardware.

Sold by all Leading Jobbers in the Dominion.

Persons addressing advertisers will kindly mention having seen their advertisement in this paper.

Our Monogram is Your Protection

Send us your orders for

Cutlery, Electro Plate and Solid Nickel Silver Flatware

and we guarantee to fill them quickly and satisfactorily. Our goods have a distinguishing quality and finish which pull business.

The McGlashan, Clarke Co., Limited

NIAGARA FALLS, - ONT.

MR. J. MACKAY ROSE, 117 Youville Square, Montreal, Que. MR. N. F. GUNDY, 47 Hayter St., Toronto, Ont.
MR. DAVID PHILIP, 291 Portage Ave., Winnipeg, Man.

THE WAGGONER EXTENSION LADDER

is the lightest, handiest, strongest and safest ladder made. Re-inforced with steel wire. Patent safety lock. Made in all lengths from 20-ft., 24-ft., 28-ft., etc., up to 100 ft. if desired. Every painter, tinner, stacker and farmer needs one

Write for circulars and prices.

The Waggoner Ladder Co., Ltd.

LONDON, - - CANADA

E. H. Briggs Co., Ltd., Western Representative Winnipeg, Man.

Can. Feb. 27, 1889

FACTS

about the best

CONE ROTARY VENTILATOR

on the market

Known everywhere as PEARSON'S

This ventilator is set in motion by the least current of air and will give a perfect draught at all times.

Substantial in make, Ornamental in appearance.

For balky chimneys and ventilating Churches, Schools, Halls, Factories, Steamers, Public and Private Buildings, Stables, Water Closets, &c.

This Ventilator is in use throughout the Dominion, and has proved a great success.

Remember every Ventilator is guaranteed to work satisfactorily.
They are made in sizes from 2½ to 48 inches.
All orders by mail promptly attended to Testimonials on application.

Write to-day for particulars to the owners.:

Geo. D. Pearson & Co., Office and Works: 17 Pontiac Street, MONTREAL
Bell Telephone East 2106

Past Due Accounts.

We have just issued an eight page booklet regarding our Collection Department. If you have not received one write us. It will interest you.

The Beardwood Agency

Claims, Collections and Commercial Reports

313, New York Life Bldg., MONTREAL

Galt "Sure-grip" Shingles

are simply wonders TO SELL, TO LAY AND TO MAKE FRIENDS. Once used and you would never use any other—your customers wouldn't let you.

"Sure-Grips" make a really handsome roof, as well as the most weatherproof. Try them—that's all we ask.

"GALT" CORRUGATED SHEETS are best English galvanized stock, pressed straight and true. Our prices are right, and we can ship promptly.

Consult us before purchasing CEILINGS, SIDINGS, ROOFINGS, CORNICES, SKYLIGHTS, VENTILATORS, FINIALS, FIRE-PROOF WINDOWS, EXPANDED METAL LATH, ETC.

The "GALT KIND" is the line to push; we protect and assist our customers.

THE GALT ART METAL CO., Limited

GALT, - ONT.

Jardine Hand Drills

New Designs

Get samples of these machines for your show room. They will interest your customers and prove a good investment.

It pays to sell the best Tools.

A. B. JARDINE & CO.

HESPELER - - - ONTARIO

...FULL STOCK...

Salt Glazed Vitrified

Double Strength Culvert Pipe a Specialty.

THE CANADIAN SEWER PIPE CO.

HAMILTON, ONT. TORONTO, ONT.
ST. JOHNS, QUE.

There's a Profit Story Here!

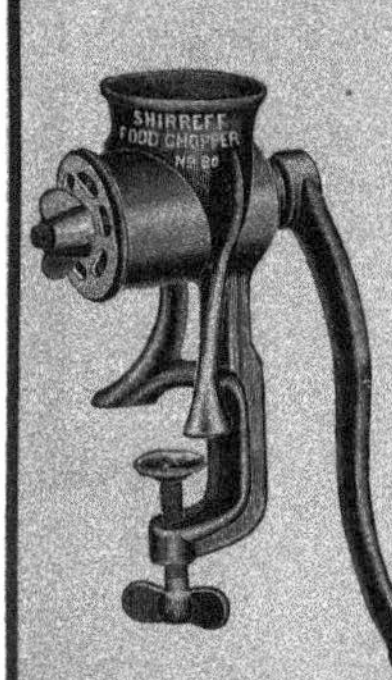

Most shrewd dealers know it. Do you? It's an easily proven fact that our Food Choppers are the best—by actual test—on the Canadian market. Before placing your orders for

FOOD CHOPPERS

you should have our price list and ask for proofs of our claims. We give a money-back guarantee with every machine. All parts are tinned and interchangeable, and **these Choppers have Steel Knives.**

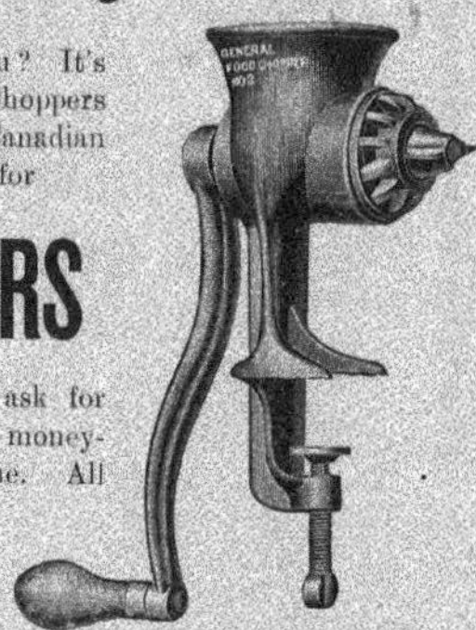

BUY IN CANADA **BUY THE BEST**

Shirreff Mfg. Co., Ltd., - Brockville, Ont.

A RECORD BREAKING SELLER

22 CALIBRE
16 SHOT

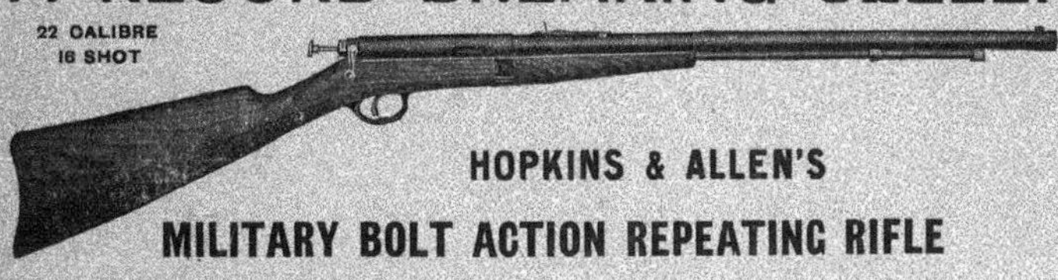

HOPKINS & ALLEN'S

MILITARY BOLT ACTION REPEATING RIFLE

$10.75 LIST

There has long been a demand for a good rifle at this price—and the **HOPKINS & ALLEN Junior Repeater** fills the bill. This rifle has been and is being extensively advertised, and a heavy sale for it is created.

SPECIFICATIONS:—22-inch barrel, take-down, 22 calibre. Magazine handles 22 short, 22 long and 22 long rifle cartridges without change of carrier. Shoots 16-22 short or 12-22 long or long rifle cartridges Action, Military Bolt Pattern with positive safety device. Length over all, 40 inches, weight 5½ pounds.

FOR SALE BY ALL FIRE-ARMS JOBBERS. — PLENTY OF GOOD SUPPLEMENTARY ADVERTISING MATTER—FREE.

Shows Bolt Drawn Back—Position for Ejecting Shells.

WRITE FOR CATALOGS

THE HOPKINS & ALLEN ARMS CO.

Dept. 67, NORWICH, CONN., U.S.A.

LONDON OFFICE
6 City Road, Finsbury Square, London

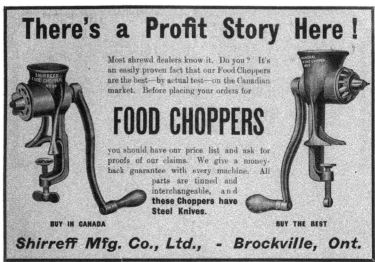

WOULDN'T YOU RATHER SELL

a register which will give your customer entire satisfaction? The

JONES SIDE-WALL REGISTER

is the **register that satisfies.** It saves your customers money by reducing piping 50 per cent. on a heating system, and works to his **entire satisfaction.**

The Jones Register Co.

732 King St. West ***Toronto, Canada***

Wrought Steel Registers

Standard Sizes, All Finishes
Perfect Operating Device
Unusually Large Air Space
For Sale by Leading Jobbers

H. & C. No. 100.

HART & COOLEY CO.

New Britain, Conn., U.S.A.

FERROSTEEL REGISTERS

The only house in the world manufacturing

CAST IRON FLOOR
SEMI STEEL FLOOR
WROUGHT STEEL FLOOR
CAST IRON SPECIAL SIDE WALL
WROUGHT STEEL SPECIAL SIDE WALL
FLAT OR CONVEX EXTRA SHALLOW SIDE WALLS
DEFLECTORS, PANCAKES, WAFERS

In all Designs and Finishes.
Mechanism Perfect.
Capacity Greatest.
Finish Unsurpassed.
No reputable line can undersell us.

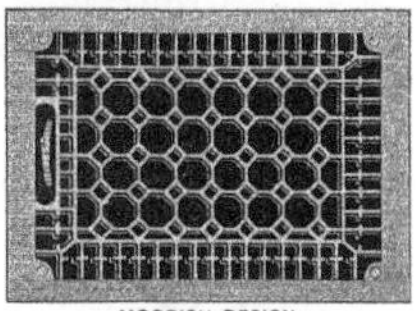

MOORISH DESIGN

FERROSTEEL COMPANY, - - CLEVELAND, OHIO

"Brantford" "Roofing"

We beg to announce to the Trade throughout

Manitoba, Saskatchewan, Alberta and British Columbia

that our representative, MR. J. S. McGUIRL will call upon you shortly in the interest of

Brantford Roofing Co., Brantford, Canada.

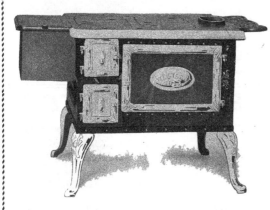

You May Be a Good Salesman

But, is everything in your favor? Are you selling the new Imperial Oxford? A continuous stream of repeat orders tells the story to us.

Removable Nickel
Removable Fire Linings
An Oval Firebox
The "Divided Draft"

All contained by the best looking stove in the market.

It gives a good salesman a chance to spread himself.

Drop us a card, or, better still, tell us the size range you want on your floor. If you get this agency you're lucky.

The Gurney Foundry Company, Limited

Toronto Winnipeg Vancouver Hamilton and London

THE GURNEY-MASSEY CO., LIMITED, MONTREAL, Que.
THE GURNEY STANDARD METAL CO., LIMITED, CALGARY, Alta., and EDMONTON, Alta.

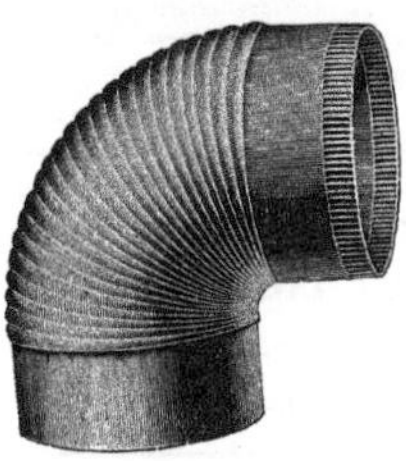

Experienced dealers select McClary's Stove-Piping and Elbows because, first, they sell, and second, it pays to sell them.

"Easy Lock" Stove-Piping is made to meet the requirements for a non-telescoping, easy-fitting, long-lasting pipe that any dealer can conscientiously recommend and any consumer wisely buy. **Nested 25 lengths in each crate.**

The McClary Elbow you see here is the acme of strength and neatness. The seam is grooved and riveted, which ensures firmness, which ensures a perfect fit. Round or bevel. Crates of one dozen each.

Profit as fellow-dealers have done:
Sell the McClary Line

The McClary Mfg. Co.

LONDON, TORONTO, MONTREAL, WINNIPEG, VANCOUVER, ST. JOHN, N.B., HAMILTON, CALGARY

To

EVERY MAN

who uses

Fire Bricks

We handle the best makes of Scotch Fire Bricks, and our prices are right, but there are occasions when you must have more refractory materials, and then please remember that we are the Canadian representatives of the

"HARBISON WALKER REFRACTORIES" COMPANY

and this should indicate that we are the Fire Brick House of Canada.

Others of our exclusive specialties are

Lafarge
Non-Staining Cement

Iron Clad
Portland Cement

United Paving Bricks
Best in the World

Enamelled Bricks
All Colors

Drain Pipes

Plaster, etc.

Francis Hyde & Co.

KING, QUEEN and WELLINGTON STS.

MONTREAL

Smooth finish. perfect threads and first-class material, are the reasons why

"Diamond" brand Fittings

are in such constant demand.

WHOLESALE ONLY

OSHAWA STEAM & GAS FITTING CO., LIMITED

OSHAWA, CANADA

Don't Stop to ask your neighbor, lift the load yourself with **The Burr Self-Locking Tackle Block.** Saves the labor of two or three men. 600 to 5,000 pounds capacity.

Lewis Bros., Ltd., Montreal
Munro Wire Works, Winnipeg
Vokes Hardware Co., Toronto
F. Hamilton & Co., Hamilton
The Burr Mfg. Co., Cleveland, Ohio

DON'T TAKE CHANCES, BECAUSE IT DOESN'T PAY TO DO SO

You can make any and all claims for Furnace Cement, if you have stocked with

Sterne's Asbestos Stove and Furnace Cement

because it is more durable, economical and has more heat-resisting power than any other Cement on the market. ***Every pound guaranteed.***

G. F. STERNE & SONS,

BRANTFORD, - - ONTARIO

Queen City Water White Oil

GIVES PERFECT LIGHT

The most economical high-grade oil ever sold in Canada.

FOR SALE BY ALL DEALERS

SIMONDS HAND SAWS

are good sellers. Dealers should send for 1907 Catalogue and discounts.

SIMONDS MANUFACTURING CO.

Fitchburg, Mass.

Branches Throughout The United States

SIMONDS No. 8½

This Saw is of the general style and shape which carpenters have been accustomed to using for years past. Has a carved apple handle with highly polished sides and edges, making a very pretty Saw. Holds its cutting edge and saws true. Its high quality and even temper are due to the fact that it is **Made of Simonds Steel.** Covered by the broadest Simonds warranty.

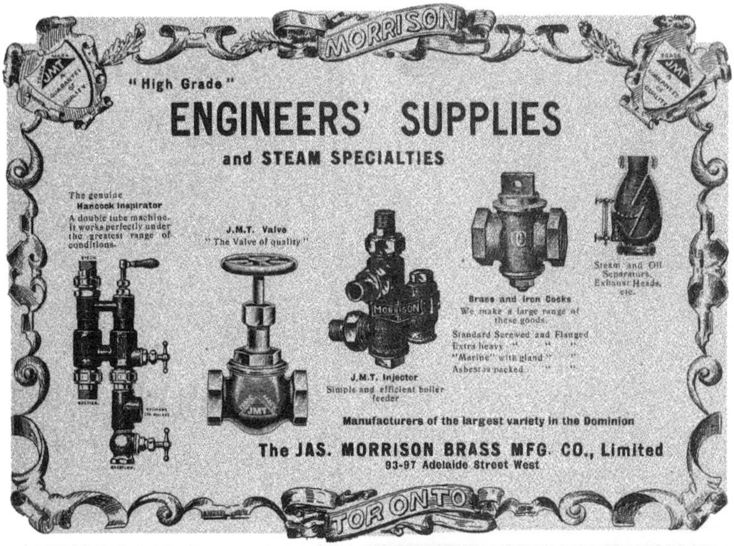
MORRISON
"High Grade"
ENGINEERS' SUPPLIES
and STEAM SPECIALTIES
The genuine
Hancock Inspirator
A double tube machine. It works perfectly under the greatest range of conditions.
J.M.T. Valve
"The Valve of quality"
J.M.T. Injector
Simple and efficient boiler feeder
Brass and Iron Cocks
We make a large range of these goods.
Standard Screwed and Flanged
Extra heavy " " "
"Marine" with gland " "
Asbestos packed " "
Steam and Oil Separators, Exhaust Heads, etc.
Manufacturers of the largest variety in the Dominion
The JAS. MORRISON BRASS MFG. CO., Limited
93-97 Adelaide Street West
TORONTO

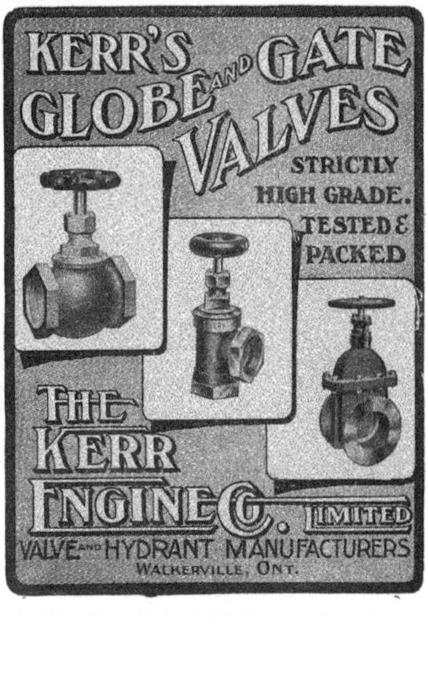
KERR'S GLOBE AND GATE VALVES
STRICTLY HIGH GRADE.
TESTED & PACKED
THE KERR ENGINE CO. LIMITED
VALVE AND HYDRANT MANUFACTURERS
WALKERVILLE, ONT.

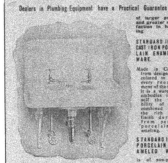

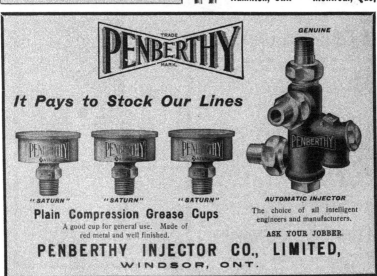

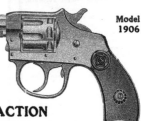

I WILL TALK

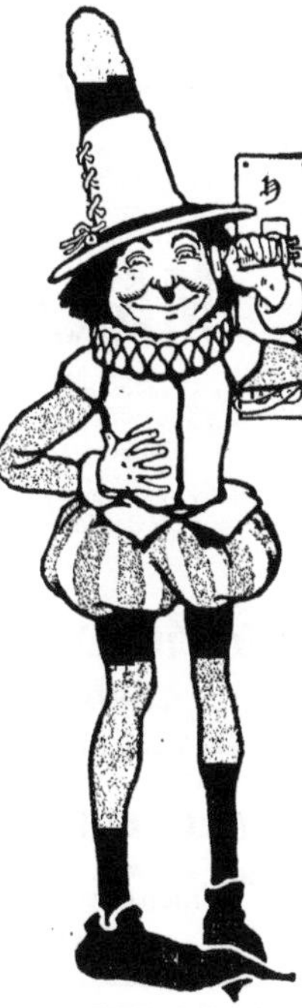
THE WANT AD MAN

to practically every Hardware merchant in Canada from the Atlantic to the Pacific. I cannot do it all in one day, but during the first twenty-four hours I will deliver your message to every Hardware merchant in Ontario. I travel all day Sunday and on Monday morning there will not be a village within the limits of Halifax in the East and Brandon in the West, into which I will not have penetrated.

I cannot go any further East, so I now devote all my energies to the West, and so many new towns are springing up here each week that I haven't as much time as I used to have to enjoy the scenery. But I like talking to hardwaremen, clerks, travellers and manufacturers, especially as they are always glad to see me and hear the news I have to tell them. Tuesday noon I am at Calgary, Wednesday noon at Kamloops, and by Thursday morning I reach Vancouver, having been in all the mining towns and all through the fruit districts of British Columbia.

I have been eighteen years on the road and I have a pretty good connection. I never intrude when a man is busy, but just bide my time, because I know men pay far more attention to what you have to say if you catch them when they have a few moments to spare. So I often creep into their pocket when they are going home at night, and when supper is over Mr. Hardwareman usually finds me. He must be glad to see me, because he listens to what I have to say for an hour or more.

I try to always tell the truth, and men put such confidence in what I say that I would feel very sorry to deceive them even inadvertently. Probably some other week I will tell you about the different classes of people I meet. In the meantime if you want a message delivered to HARDWAREMEN, PLUMBERS, CLERKS, MANUFACTURERS or TRAVELLERS—and want it delivered quickly—I'm your man.

Condensed Advertisements in Hardware and Metal cost 2c. per word for first insertion, 1c. per word for subsequent insertions. Box number 5c. extra. Send money with advertisement. Write or phone our nearest office

Hardware and Metal

MONTREAL **TORONTO** **WINNIPEG**

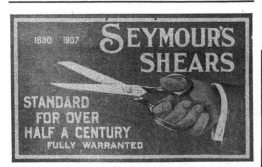

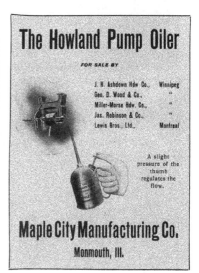

Our Fireproof Windows

ARE WINDPROOF

These windows are easily distinguished from the ordinary kind by their windproof qualities. No other Fireproof Windows on the Canadian market exclude the wind so effectively as ours.

All parts are perfectly uniform, because the whole window is stamped by steam power with steel dies.

Our Fireproof Windows are fully described in our new Catalogue No. 18 and Folder A 103, both of which are sent free for the asking.

Let us figure on your contracts.

"WE WORK WITH THE TRADE."

The Metal Shingle & Siding Co., Limited

Montreal — Cor St. Catherine and Delorimier.
Preston, Ont.
Toronto — 100 Esplanade St. East

SALES AGENTS:

ST. JOHN, N.B. — Emerson & Fisher, Limited
QUEBEC — J. A. Bernard
WINNIPEG — Clare & Brockest
CALGARY — Ellis & Grogan
VANCOUVER — McLennan, McFeely & Co., Limited

The Buffalo Manufacturing Co.

Buffalo, N.Y.

When you get our goods you know you get THE BEST.

We manufacture
Water Filters
Water Coolers
Chafing Dishes
Table Kettles and Stands
Coffee Extractors
Wine Coolers
Nursery Chests
Baking Dishes
Crumb Trays and Scrapers
Tea and Bar Urns
Bathroom Fixtures
Coal Vases and Hods
Candlesticks
Cuspidors
Match Safes, Etc.

All High Grade and exceedingly presentable.

REPRESENTED BY

H. F. McINTOSH & CO.

51 Yonge Street

Write for Catalogue

TORONTO, ONT.

QUICK SALES

and good profits are naturally the outcome of handling

Empire, Imperial and Champion

AXE HANDLES

To keep up the profitable pace see that you are always kept supplied with our

Hammer, Sledge and Pick Handles

Lists and discounts sent free upon request.

J. H. Still Manufacturing Company, Limited

ST. THOMAS, ONTARIO

Cash Better Than Credit System

Pertinent comment by George Brett on the only safe and profitable method of conducting a mercantile business.

Every few years the business world imagines it has discovered some new principle. Not infrequently after it has tried to put the new principle into practice it wakes up to the fact that a mistake has been made.

One of the latest theories in business is that it is possible to build up a big paying business on credit. People are urged to believe that they can live and die on credit. There has been an orgy of credit talk; but there are indications that the world of industry is waking up to the fact that the new theory is not all that it is "cracked up to be." It is coming to realize that the old ideas about credit being a vicious principle on which habitually to do business, are sound.

Chief among the guiding rules of the late Marshall Field were the mottos: "Do business on a cash basis," "Never borrow." If it is absolutely necessary to give credit then "Sell on a shorter time than competitors." Make every inducement to the buyer to pay for his goods as soon as possible.

People often fail to grasp the truth that the success of the gigantic mail order houses is as much due to their doing business on a strictly cash basis as to any other factor. The enormous success of the mammoth stores of Philadelphia, New York, Chicago, etc., largely can be explained by the "small profit, quick return" idea being everlastingly put into operation.

Pay When You Buy.

In stores the most satisfactory way of doing business is to see what you are getting after a careful examination, pay for your purchase, and walk away with your parcel under your arm. Of course, if it is too heavy, you have to ask for van delivery.

This is the way the storekeeper likes things to be done, and if everyone would pay for small purchases and take them away it would enable the retailer to sell his goods on a smaller margin of profit. The stores asked to send a few needles or a box of handkerchiefs by delivery van lose much of their profit on such transactions. They usually do not make a large amount on shoppers who continually ask to have goods sent "C.O.D." Often by the time the shopper has reached home her mind has changed, and when the delivery man trudges up five flights of stairs the lady is "not at home; no instructions have been left." Then, after a vain attempt to deliver the goods the next day and the day after, the goods are put back into stock. This is not doing business on a credit basis; but it is the halfway post to it. The element of delay in getting the money enters into such transactions, with attendant danger to the retailer.

The man who aims to do business on a credit basis is often in an infinitely worse predicament. Not infrequently he loses a great amount of time and trouble in making a sale and then loses the goods as well Who pays for such losses? The honest man or woman who pays up. That is why it pays both the retailer and individual to do business on a cash basis as much as possible. You are not paying for the sins of "dead beats" and chronic debtors when you trade with a man who does business on a cash basis only.

Collect When You Sell.

The beneficial results of doing business on a cash basis are far reaching. The retailer who abstains from giving credit and buying more goods than he can pay for quickly has a better reputation with wholesalers than the one who always has a lot of money outstanding and who continually is buying goods for which he is going to pay when he himself gets his money from a hundred different sources.

If this matter were pursued further it would be seen that the individual himself purposes paying provided he gets some money. The chain of credit has generally a weak link somewhere.

D. A. Kimbark, in a lecture before the University of Chicago, gave the following lucid example, showing how it pays to do business on a cash basis. Suppose that two firms applied for credit on a bill of goods. The first trader has:

Assets.	
Stock on hand	$28,000
Trusted out	14,000
Cash on hand	....
Total	$42,000

Liabilities.	
Due bank	$ 7,500
Merchandise not due	15,000
Merchandise due	4,500
Total	$27,000
Net worth	15,000

The second firm had:

Assets.	
Stock on hand	$5,000
Cash on hand	1,000
Total	$6,000

Liabilities—none.

Cash Business Valued Highly.

It is safe to say that the latter trader would be given credit gladly by any wholesale house, while in the former case, though the man was worth two and a half times as much as the latter, it would be hard for him to get credit. He is carrying too large a stock, living too much in the regions of credit for his good.

The man who does business on a cash basis can secure far better terms than the man who wants a long time to pay his account, with the wholesaler. Some of the bargains secured by the mammoth mail order houses who have plenty of ready money on hand all the time are almost incredible. The small retailer who can pay for goods immediately undoubtedly gets the cream of the bargains in the wholesale world. A hundred credit men know his record. They pass no sleepless nights in thinking whether it is safe to trust such a man, and he is going to get a better deal from the wholesaler than his competitor who sells on credit.

The trader who does business on a cash basis really gets the cream of the buying market. He usually gets those who have money fifty-two weeks in the year.

The merchant who does business on a credit basis does not get such a good class of people.

Specific Date for Payment Best.

The question of giving credit centres at all times into the life of almost everyone. It is almost impossible to make a rule never to extend credit, but when it is necessary there should be a hard and fast understanding about repayment.

A writer for a weekly paper sagely remarked that nearly all bad debts are due to lack of proper investigation and a want of understanding at first. When a man asks for credit he should be asked the least amount that will satisfy him. Suppose a man goes into a grocery store, and, acting upon this rule, the storekeeper finds out that a monthly credit of $25 will be satisfactory. The retailer then tells the man flatly and in strong terms, that having set his own time, the customer must absolutely abide by the terms made. A certain date should always be named—say the first of the month, and the time of payment vividly impressed on the customer's mind. Nine times out of ten he will meet his obligations. But if the date set be indefinite, say between the first and seventh, negligence will almost inevitably ensue.

Cash on Hand Big Help.

The example is applicable in a large degree to individuals. It is not enough to have wealth; but it is also a wise thing to have a certain amount of cash on hand to meet emergencies. How many men turn their cash into a lot of personal effects and unnecessary household goods which could not be sold for one-third of what was paid for them if they had to be converted into ready money? How many people mortgage their future earnings by buying a lot of stuff "on time" which is not half as useful when bought as the money itself would be? The habit of buying things when the ready money has been saved up makes people infinitely wiser judges of values and enables them to make better bargains. Further, those who deal on a cash basis save much needless worry about making payments.

"Never place a mortgage on your holdings" was another motto of Mr. Field He referred to present holdings. If he thought it unwise to mortgage present belongings he would have said it was infinitely more foolish to mortgage future holdings Yet this is exactly what the man does who takes credit.

Effective Hardware Advertising

HOW TO PRODUCE IT

By T. Johnston Stewart.

"A very good test for an advertiser to apply to an advertisement is to ask himself the question, 'What would a mail order house or a department store do with the space which I have at my command?' If the advertiser feels that he has employed the space as it would be employed by one of the great concerns whose success has been built up by advertising, he may rest satisfied with his ad. But one thing is certain—no expert and successful advertiser would dream of paying for space in a newspaper and use it to say, 'Hardware and Stoves for Sale by John Smith,' or 'Come to John Johnson's for your Stoves and Ranges.' In fancy you can see the space filled by the catalogue house or the department store with snappy descriptions of the goods, with plenty of prices to attract and interest the reader, and with cuts—not too big, but of a suitable size for the space.

"It is worth while to spend time and effort in preparing advertising copy, for the mere renting of the space is only a preliminary. The filling it with proper 'pulling' matter is the main thing."

That's sound commercial sense—every word of it, and both you and I know more than a few dealers who might read it a second time with profit. The idea that ordinary business publicity required exceptional talent has been exploded. Shrewd dealers know that the fundamental principle in all advertising is just average, every-day common sense, combined with a salesmanlike knowledge of their wares. Of course, there are a few dealers who advertise immensely better than their fellows. They are to the manner born, as it were. If there is a better way of stating a fact—a more seductive, sales-compelling way, you may be sure that they state it in just that way. These men would make splendid sales-managers for the largest concerns in the Dominion. They serve to prove a favorite contention of the writer's, viz., that the best admen are pretty nearly always practical men, with a thorough knowledge of their public and the goods they have for sale. While this is undeniably true, yet, there are scores of dealers who cling to the old ideas of publicity, fancying that it is enough for the general public to know that they are in the hardware and metal business.

* * *

There is no secret about the extreme success of most successful merchants. They simply recognize advertising as a necessary factor in all twentieth century business—not as a means conceived by publishers for the purpose of adding to their bank accounts. And Mr. Dealer! if trade has been slow, you can depend upon it that the lack of advertising—or the fault of your advertising, explains the reason why. It is not enough that you should buy space in your favorite paper—the shrewd dealer plays no favorites in the advertising game—you must fill up that space with a trade-bringing announcement. Most of my readers have seen the catalogues issued by big mail order establishments. Your chief competitor does not waste space. He probably has paid dearly for his experience and he fills his catalogue up with breezy, snappy descriptive matter. The whole is nicely arranged and price talks—not loudly, but convincingly, throughout. Now, if you read over our first paragraph again you will benefit yourself and understand just why we quoted so much of our contemporary—The American Artisan and Hardware Record.

* * *

The McLeod Hardware Company some time ago mailed us a few samples of very effective advertising. They say: "We aim to make every dollar spent for advertising pay, and it does, as we have found much profit in what others have contributed. We are sending you some efforts of our own, in the hope that they will benefit some others who push the interesting trade."

The McLeod way of advertising must pay. They specialize some particular line in each ad., such as washing machines, tools, builders' supplies, etc., and it is evident that they change their ad. quite frequently. The ad. we reprint and improve somewhat in the set-up, may or may not be the best ad. this firm has sent in, but at all events, it is strong and good enough to clearly portray the McLeod style. That introduction does seem a little too long, but when one reads it over one finds lots of good business sense in it. The McLeod Company simply tell the reasons they have for holding a kitchen utensil sale, and then go on to let the prices talk. The arrangement does not call for any eulogy, neither can one criticize it much. Gentlemen, if you have an original copy of this ad., you will agree with us when we say that prices might have been more effectively brought out. All the ads. you have forwarded are good—the illustrated ones especially so. Of course, you understand that it is practically impossible for us to reproduce an ad. illustrated by manufacturers' cuts. This is so obvious that we need not give any other reason for our choice in this particular case.

New Kitchen Utensils

Our Household and Kitchen Goods section has just been loaded up with new Enameled and Tinware, bought direct from the makers at prices that make it possible for us to save you money on every article you buy from us. Every piece is warranted first quality.

WE DO NOT SELL SECONDS.

Economical housekeepers will find this department of our store an interesting and profitable place toshop. We are glad to have you come in and look, whether you buy or not.

SOME PRICES.

Sauce Pans .. .. .. 15c. to 75c.	Double Boilers .... .. 65c. to $1.50
Tea Pots .... 35c. to 70c.	White Pudding Dishes .... 15c.
Large Tin Dish Pans 35c. to 45c.	Deep Stove Pots .... 80c. and 90c.

KITCHEN CUTLERY.

No need to spoil good table cutlery in the kitchen. A small outlay here will buy a complete outfit of knives, forks and spoons for kitchen use.

Kitchen Tea Spoons .... 20c. doz.	Kitchen Knives .... 75c. to $1 doz.
" Dessert Spoons. 40c. doz.	" Knives and Forks, 60c. to $1 per set of 6
" Table Spoons .. 50c. doz.	" Bread Knives 15c. to 60c. ea.
" " Forks .. 50c. doz.	" Carving Knives 25c. to 60c.
" Sharp'g Stones, 15c. each	" Mounted Grindstones.. $1.25

Our strong suit is High Grade Table Cutlery. It's the reputation for quality that is building our big cutlery business. We're showing two special lines of Celluloid Handle Steel Table Knives at $2.50 and $3.50 per doz. Ask to see them.

The Hardwaremen

THE McLEOD HARDWARE CO., LTD.

233 CHARLOTTE STREET

A Study of Co-operative Stores

Some of the claims of the Co-operatives analyzed by an English editor—Retailers in Great Britain forming societies of their own along co-operative lines.

The pretentious claim that the co-operative stores movement is based on noble ideals no longer carries conviction to sensible people. We all recognize that at the outset the promoters of the co-operative idea were inspired by lofty visions of the brotherhood of man. So much credit is certainly due to them: their intentions were excellent. The co-operative movement is now, however, a matter of history; it has a long record of years behind it; it is to be criticized not by its aims but by its deeds.

That the movement has been a success of a kind is beyond doubt. The enormous figures to which we are treated at the co-operative congresses are eloquent evidence of the fact that a great section of the public, more especially in the north of England and in Scotland, has fastened on to the system with enthusiasm. Dividends have a potent charm for us all, whether we are co-operators or non-co-operators. There is nothing inherently contemptible in dividends, so long as they are earned on sound business lines.

"Idealism" a Delusion.

A distinction, however, must be drawn between "divi" and dividend. The root difference between co-ops. and non-co-ops. is that the former draw their "divi" with unwarranted references to idealism and to the brotherhood of man, whereas the latter draw their profits without deception and without any sugared platitudes. A few leaders of "the movement," and some amiable, but unbusinesslike, divines may still harbor the delusion that co-operative stores have brought us nearer to the millennium. The practical man who faces facts as they are, has long ago rid himself of any of these delusions. The very manner in which the advocates of co-operative stores treat criticism is a proof of their lack of even the rudiments of fraternal charity.

Recently the Weekly Scotsman has been giving considerable space to arguments pro and con the co-operative stores. The correspondence was started in the interests of truth and fair play, and equal opportunities were given to both sides to state their views. The result has been that a co-operative journal has charged the editor of our contemporary with concocting the letters which were published against "co-op." Needless to say, the Weekly Scotsman repudiates this gross charge of falsehood, fraud and wilful imposition. The very fact, however, that the charge was made, is evidence of the state of mind of these men who cry aloud that they are conducting business on noble ideals.

Craving for Dividends.

There is little need for us to inform our readers that the idealism with which co-operators attempt to vest their stores is not justified. Not need we tear a passion to tatters in condemning this claim. Rather is it better to accept the fact that "co-op, divi," based although it is on ignorance of business principles, is favored by a certain section of the public. It follows that the private trader might, where the exigiencies of co-op. competition render it necessary, attempt to adapt his own business to the craving for "divi."

Strip the co-operator of his shams, and we find he is a shrewd business man who takes full advantage of the fact that his customer is not influenced by love for his fellow man, but by love for himself. However much we may be opposed to the "divi" system, the fact that it appeals to a certain class of customers must be taken into account. If they cannot appreciate the direct business method of selling an article at bed rock prices, after allowing sufficient margin for expenses and fair profit, the private trader must of sheer necessity adapt his system to suit them. One method is to change discounts into dividends.

Co-operatives Only Advantage.

The private trader can sell as cheaply as the co-operative store. He can buy better, and he can manage his shop better. So far the competitors meet on fair fighting terms. The only advantage that the co-operative store in reality possesses is its method of returning a part of the purchase money to the customer. Why, some people are now asking, should not private traders combine among themselves, and also arrange to return to their customers a part of the purchase money? The suggestion is being made that private traders should form co-operative societies among themselves, and agree to give their customers dividends instead of discounts. Not only the co-operative store, but also the trading stamp system and the clothing club system—both of which have some strong features of resemblance to the co-operative system—can, it is believed, be most effectually combatted by these means. Retailers are, in short, being invited to fight these competitors with the competitors' own weapons.

The formation of a co-operative private traders' society will be, it is argued, an effective counter-blast to all these forms of competition. The weakness of the trading stamp system and the clothing club system lies in the fact that they represent intervention between tradesman and customer in the interest of some outside third party. Now a co-operative private traders' society, properly conducted, would intervene only in the interest of the members of the society. Already societies of this kind have been formed, and we wish them success so long as they are organized on the right lines, namely, lines of mutual benefit and not of profit to outside third parties. They should be societies in which all private traders are eligible for membership without distinction, provided that they are willing to contribute their share to the expenses. We should regret any proposal to allow only one retailer in each class of trade to join.

A Genuine Co-operative Movement.

A society which singles out certain tradesmen for participation in its benefits is not a whole hearted co-operative society, but merely a society to give one particular tradesman an advantage over another. It does not seem to offer any real or lasting solution to the problem of the competition of the co-op. stores. It is merely adding confusion and making the lot of the private tradesman, regarded as a class, more difficult and more complicated. We trust, therefore, that this new method of meeting co-operative and other competition—a method which is already taking shape in certain districts—will be a genuine co-operative method among all private traders without distinction, and not merely a society which is a clique, and which, while ostensibly fighting the co-operative stores, is in reality tending to crush out other private traders who are not privileged to join the clique.

Where retailers are invited to join societies which are not open to all bona fide private traders without distinction but which are open to only one retailer of a particular trade, we should advise them to refuse the application. The societies which should be joined are those which are conducted on lines similar to the many mutual plate glass assurance societies throughout the kingdom—societies which hold an "open door" to all bona fide traders in the district.

HANDY WRENCH.

An improved wrench has been patented by A. L. Moss, Sandusky, Ohio. As no swinging movement of the handle is required in this improvement, it is evident that the wrench can be used to great advantage for turning bolts, nuts, and the like located in places not readily accessible to an ordinary wrench. The tool may, however, be used as an ordinary wrench. Mr. Moss has invented another wrench, such as shown and described in his application for former Letters Patent of the U. S. Its object is to provide a new and improved tool, more especially designed for turning nuts, screws, and other articles in places not easily accessible by ordinary wrenches.

A CHEAP PRESENT.

Mrs. Greene met a friend, and in the course of conversation, remarked:

"My husband's birthday comes next week."

"Well," asked the other lady, "what gift are you choosing him?"

Mrs. Greene smiled proudly as she replied:

"I've been taking a cigar out of his case every day now for the last four months, and have a hundred now which I'll give to him for a birthday present."

Sporting Goods

MANUFACTURE OF BOWS AND ARROWS.

The growth in the revival on this side of the Atlantic of the ancient sport of archery has brought to the trade's notice one of the most novel forms of expert handicraft in the range of American industries. This is the manufacture of long bows and hunting and target ar-

Fig. 1.

rows by Mr. F. S. Barnes, of Forest Grove, Oregon. The Barnes bows are rapidly being adopted by the most expert archers in all parts of the world, not so much perhaps because of his excellent skill in manufacture—although that is not to be underestimated—as by reason of the marvelous quality of the wood he employs, the famous Oregon mountain yew. The bows used for generations past by European archers have in almost every instance been made of yew wood, usually of Spanish or Italian origin. Some years ago Mr. Barnes, who has been interested in archery practically all his life, discovered the adaptability of the Oregon yew for bow-moking purposes. Indeed, the wood found in the high altitudes of the mountains of our North Pacific coast is much finer grained and supêrior in every way to anything obtainable in the forests of the old world. Mr. Barnes, in addition to drawing upon an important new source of supply, has, after long experiment, developed a new method of seasoning and treating the wood, which is revolutionary in its radical departure from accepted practice abroad, and is claimed to be responsible for the production of long bows that are in many respects a distinct advance over the output of the best known English and continental bow-makers.

COLLAPSIBLE MINNOW TRAP.

We are in receipt of a descriptive booklet from A. J. Algate, Toronto, Can., in connection with a new minnow catcher. The device is known as Algate's Collapsible Minnow-Trap, and, as the name implies, this new trap can be folded into very small space.

Undoubtedly, there is a great need for something of this nature, and many attempts have been made to produce a satisfactory and effective device for the purpose, but their success has been very doubtful. The glass traps have always been a source of trouble and anxiety while the wire outfits are altogether too bulky to ever become popular.

This new device is made of transparent celluloid—thus affording the necessary transparency and serviceableness; it is most simple in its construction and requires but the fraction of a minute to unpack and set. When not in use it folds up and can be carried in the metal case.

It will be seen from Fig. 1, in the accompanying illustrations, that the device is triangular. Three small independent pieces set on springs at either end form the funnels, and as the lower section works free from the others, it is merely necessary to compress same to recover the entire catch at one operation (Fig. 2). Then, too, the triangular formation enables its quick recovery from the water.

Its compactness, as shown in Fig. 3, will appeal very strongly to every fisherman, as every extra square inch and every additional ounce means much on the last half mile of a heart-breaking portage. The trap when folded in case actually measures 16¾x6¾x1¾ in. and weighs twenty-eight ounces.

There is no question but that Mr. Al-

Fig. 2.

gate has produced an article of superior merit—something which will meet the popular approval of the fishing fraternity. It is unique in its simplicity and compactness, and being made of celluloid will undoubtedly stand all kinds of rough usage. The traps are constructed from the best of materials, and will certainly be a valuable addition to the sportsman's kit. They retail in Canada at $3.50 each. Descriptive booklet and prices to the trade may be obtained by addressing A. J. Algate, 98 King Street West, Toronto, Can.

TWO KINDS OF AMMUNITION.

"There are two kinds of cartridges on the market," remarked a manufacturer of ammunition, "one of which is for hunters and the other for those who only think they are hunters. Let me tell you an incident illustrating what I mean: A rich New Yorker who owns a preserve in the Adirondack mountains invited a number of friends up to hunt with him. One of them acted as if he had never seen a gun before. He shot one of the guides in the leg the first morning. The next day he had wretched luck. Bang, bang, went his gun, but he had hit nothing. He was much embarrassed. It seemed, too, that at each one of the misses the guides smiled at one another oddly. Finally his cartridges gave out. He hurried to the nearest guide and demanded more. 'There ain't no more, sir,' the man answered. 'No more? Nonsense. Why, you've got at least a thousand in that box.' The man blushed and stammered: 'Ah, but them ain't for you, sir; they're for another gent. They've got shot in 'em, sir, and I can't let you use 'em.'"

HANGING GLASS SHELF.

An exceedingly convenient and practical window display method is described herewith, showing how a glass shelf may be hung in the window for the display of various kinds of sporting goods The supports for this shelf are made of small chain; nickel or brass may be used or some less attractive article may be covered with cloth or crepe paper. The chains are attached to hooks in the ceiling and at the bottom run through short pieces of pipe, forming supports which should also be enclosed in the covering material. On the two supports thus contrived the shelf is suspended, being, of course, cut the proper length and width and made of good, substantial plate glass for best effect. The advantage of such a shelf is not only the ease with which it may be put up or taken down, but also the fact that it offers scarcely any obstruction to the view and affords a bright background for articles displayed.

A GERMAN EXPERIMENT.

German military authorities are experimenting with a device by which the location of troops using smokeless powder may be easily discovered. By this device it is proposed to survey the landscape through pale red glasses. The flash

Fig. 3.

of smokeless powder appears strong in red light, while ordinary objects are dimmed. By furnishing field glasses with the device in question, which is provided with screens of the proper tint, the position of concealed marksmen can be detected.

HARDWARE TRADE GOSSIP

Quebec.

C. O. Jervas, of St. John, N.B., was in Montreal last week.

Mr. Cuzner, of McDougall & Cuzner, hardware merchants, Ottawa, died last week.

E. W. McCarty, general agent, New York City, has been calling on the trade in Montreal.

J. H. Roper, manufacturers' agent, Montreal, is spending a month in the Maritime Provinces.

E. C. Budge, of the order department of Sherwin-Williams Co., Montreal, is away on his vacation.

J. Fred Booth, son of J. R. Booth the lumber king, Ottawa, spent a few days last week in Montreal.

C. R. Cummings, of the Shelby Steel Tube Co., Pittsburg, Pa., is spending a few weeks in Montreal on business.

L. D. Robertson, representing the Comet Motor Co., Montreal, sailed this week for Europe on the Empress of Ireland.

Mr. Hahn, of the Hahn Brass Co., New Hamburg, Ont., was in Montreal last week representing the interests of his company.

Mr. Newman, manager of the Winnipeg branch of Caverhill, Learmont & Co., Montreal, is spending a short time in the latter city.

F. Wilkinson, secretary of B. & S. H. Thompson & Co., Montreal, is spending a few days this week at the rifle meet near Montreal.

R. E. Thorne, of the Canadian Bronze Powder Works, Montreal, is spending a month in Western Canada combining business with pleasure.

It will be noted with interest by hardware men that a representative of their profession in the person of G. C. Kelley, of Lewis Bros., Montreal, distinguished himself last week at the canoe meet in Montreal, by winning the senior singles championship.

R. B. Coulson, of the Dominion Wire Manufacturing Co., Montreal, who underwent an operation recently, is recovering rapidly, and expects to be able to leave the hospital in a few days. It will be some weeks before he can resume his duties on the road.

Ontario.

Thomas Marshall, hardware merchant, Dunnville, was in Toronto last week.

C. Kerouock, hardware merchant, L'Orignal, Ont., has assigned to G. H. Pharnand.

Brickman & Orman, plumbers, Stratford, Ont., have been succeeded by Brickman & Stoll.

Walter Gainer, of the hardware firm of Garner Bros., Niagara Falls, spent Tuesday in Toronto.

Mr. Russel, of the Georgian Bay Lumber Co., Waubaushene, Ont., was in Toronto last Saturday.

W. J. McClung, plumber, Port Hope, Ont., recently had his premises and stock damaged by fire.

The hardware store and stock of Richardson & Co., North Bay, Ont., was recently destroyed by fire.

J. N. Rowan, head bookkeeper for Frankel Bros., dealers in old metals, has left for a fortnight's vacation. Mr. Frankel will spend a week at Preston and a week at Niagara Falls, Ont.

Fred Donnell, Ravenshoe, and Charles Hall, Trafalgar, Ont., hardware merchants, spent Wednesday in Toronto.

Herbert King, at present with the Russell Hardware Co., Toronto, leaves in a few days to seek his fortune in the west.

R. G. Swalwell, harness dealer, Ripley, Ont., has assigned to R. D. McKenzie. A meeting of creditors will be held on August 19th.

A five hundred light ascetylene gas plant has been purchased by H. W. Brown, hardware merchant, Plattsville, which will be used to illuminate the majority of the business places and many of the private houses in that town.

Gerald Davison, formerly with the Vokes Hardware Co., Toronto, and at present with G. A. Richardson, hardware merchant, Guelph, Ont., will leave shortly for Edmonton, Alta., where he will seek a position in the same line.

H. F. McIntosh & Co., manufacturers' agents and Canadian representatives of the Buffalo Manufacturing Co., makers of cooking and table ware, have moved to new quarters at 51 Yonge street, Toronto, where a full line of samples are on display.

Charles Penfold, of the hardware firm of S. & G. Penfold, Guelph, Ont., was a caller at the Toronto office of Hardware and Metal on Friday. Mr. Penfold was on his way home after spending an enjoyable fortnight's vacation in the Niagara peninsula.

Western Canada.

Hammill & Featherston, hardware merchants, Nanton, Alta., have dissolved partnership.

J. B. Wright, hardware merchant, Midale, Sask., has sold out his hardware business to Mr. Purdy.

Isaac Saunders, Calgary, Alta., has sold his interest in the Calgary Saddlery Co. to W. A. Johnson of the same city.

Fire did considerable damage recently to the stock and premises of Heath & Howard, hardware merchants, Vermilion, Alta.

J. G. Doonan & Co., hardware merchants, Swift Current, Sask., have sold out their business to the Great Northern Supply Co.

The hardware and furniture firm of Stringer & Bennett, Rocanville, Sask., have been succeeded by the firm of Stringer & Thurston.

The Manitoba Rolling Mills Co., Winnipeg, are now manufacturing bar iron in good quantities. Their plant has been in operation for nearly a month.

The hardware firm of Lundy & McLeod, Edmonton, Alta., have admitted W. D. Smith into partnership and the firm will henceforth be styled Lundy, McLeod & Co.

The Vulcan Iron Works, Winnipeg, have the contract to instal a galvanizing plant for the Winnipeg Galvanizing and Manufacturing Company. The plant is to be installed during the present month.

Maritime Provinces.

T. P. Calkin & Co. have established a plumbing, heating and sheet metal business at Kentsville, N.S., under the management of Lewis G. Ellis.

The hardware establishment of A. E. Alexander & Co., Campbellton, N.B. was broken into a fortnight ago by a couple of burglars. Charles Alexander was awakened by the burglar alarm and he and his brother-in-law ran to the store. The burglars tried to make their escape but they were pounced upon and after a struggle, lasting about twenty minutes, were reduced to submission and taken to the lock-up.

FREIGHT RATES COME DOWN.

After some months of careful consideration, the members of the Dominion Board of Railway Commissioners have issued an order which will have a far-reaching effect on the commerce of the Dominion east of the great lakes.

The order remedies the long standing complaint of Ontario and eastern Canada shippers that the railways discriminate in favor of through traffic from the United States to eastern Canada points as compared with traffic originating on this side of the border, and puts into effect the principle of uniform rates for equal distances. As a result, there will be a marked reduction of freight rates to the east from practically all points in western Ontario, the existing unfair discrimination will be done away with, and shippers and the public generally will greatly benefit.

The change will make a difference of several cents in connection with all points. Another feature is that the "owner's risk" condition has been dropped, according to the order, against some 250 items.

VALUES IN WASTE.

These are days when all waste matter is utilized for something or other, but few people realize what a great business it is. Waste paper, as everybody knows, is gathered and sold to be re-made. In the office buildings of Philadelphia all paper taken from the waste basket is gathered into a bin in the cellar. In one of the buildings its monthly sale nets $150, enough to pay the chief engineer. In most of the buildings, however, the engineers are allowed to profit by the sale of the paper, and each generally makes a tidy sum every month.

MILITARY OR CAMP STOVE.

W. B. Kimmel, Boise, Idaho, has invented a stove which is especially designed for military or camping uses. The object of the invention is to provide a stove strong, light, and durable, and which can be packed into a small compass by placing certain parts within other parts. The oven is adapted for cooking of food through chambers for the circulation of hot gases from the fire.

HARDWARE AND METAL

Established 1888

The MacLean Publishing Co.
Limited

JOHN BAYNE MACLEAN . *President*

Publishers of Trade Newspapers which circulate in the Provinces of British Columbia, Alberta, Saskatchewan, Manitoba, Ontario, Quebec, Nova Scotia, New Brunswick, P.E. Island and Newfoundland.

OFFICES:

MONTREAL, ... 232 McGill Street
Telephone Main 1255
TORONTO . . . . . . 10 Front Street East
Telephones Main 2701 and 2702
WINNIPEG, ... 511 Union Bank Building
Telephone 3726
LONDON, ENG. . . . . . 88 Fleet Street, E.C.
J. Meredith McKim
Telephone, Central 12960

BRANCHES:

CHICAGO, ILL. - . . . . 1001 Teutonic Bldg
J. Roland Kay
ST. JOHN, N B. . . . . No. 7 Market Wharf
VANCOUVER, B.C. . . . . Geo. S. B. Perry
PARIS, FRANCE - Agence Havas, 8 Place de la Bourse
MANCHESTER, ENG. . . . . 92 Market Street
ZURICH, SWITZERLAND . . . . Louis Wolf
Orell Fussli & Co.

Subscription, Canada and United States, $2 00
Great Britain, 8s. 6d., elsewhere - 12s

Published every Saturday.

Cable Address { Adscript, London / Adscript, Canada

NEW TERMS OF PAYMENT.

Much laxity has existed in dealing with the recognized terms of payment in use amongst those engaged in the various branches of the hardware trade, and it was inevitable that the advantages taken by many jobbers and retailers would ultimately result in a tightening up of the terms allowing discounts for payments of accounts within a specified time.

The action of the American Hardware Manufacturers' Association at its convention in June at Richmond, Virginia, in adopting a resolution calling for a more stringent observance of the conditions of payment does not come as a surprise, therefore, and, taken in conjunction with the general tightening in the money market, will undoubtedly result in a closer observance of the clearly defined rules under which the average purchases of goods are made. The practice of taking discounts where not really earned will also, in consequence, be checked.

The resolution adopted reads as follows:

"Whereas, the terms of payment of a large proportion of our members provide a cash discount if paid within a stated period, and, whereas, sometimes in the past the cash discount has been allowed when remittances have not been received by the seller within the agreed period; hence some buyers have not regarded it necessary to conform to the terms stipulated, and this laxity has resulted in disadvantage to all. Resolved, that the association urges all members to refuse to allow cash discount unless the remittances are strictly in accordance with the terms of purchase. Resolved, that it is not sufficient excuse for allowing cash discount after the prescribed limit has expired because the goods have not reached destination. Resolved, that it is understood each member will report to the secretary any violation of these resolutions for such action as is deemed advisable, and resolved, that all members are requested to send a copy of these resolutions to their customers."

REGAINED HIS SIGHT.

Charles W. Creed, secretary of the Maritime Board of Trade, is one of Canada's grand old men of business. The Maritime Board has demonstrated its usefulness as an organization of business men for the general advantage of Canada, as Parliament says, and its

CHAS. W. CREED.

success is in no small degree due to the enthusiastic, intelligent and self-sacrificing efforts put forth by Mr. Creed as the secretary and (one might well add) manager.

Writing last week to a member of the MacLean staff, Mr. Creed said: "I entered on my 76th year on 22nd July, and am pleased to say my health is excellent. I have been wearing glasses for the past 25 years, but on Feb. 13th of this year my sight suddenly returned and I can now both read and write without them. This is to me a great comfort."

Mr. Creed is a remarkable man, and readers of Hardware and Metal everywhere will join with this paper in wishing him many more years of health and usefulness.

MAY HAVE SOLVED IT.

A South Dakota firm who have wrestled with the problem of cash vs. credit have formulated a plan by which they hope to increase their business and still be able to accommodate their customers who are short of cash—at the same time protecting themselves and taking advantage of every available opportunity to buy goods right.

Here is their plan: "Notes for small amounts and chattel mortgages for large amounts, both bearing interest. Special 10 per cent. added on all bargain table goods sold on security. Cash takes anything in the store at bargain prices."

There are many merchants who could not adopt this plan and there are others who might do it with profit. That it is right and proper that debtors should give notes bearing interest is unquestionable. It is the basis of all financial operations not immediately settled in cash—and where the buyer asks the merchant to carry him for an indefinite length of time the latter has a perfect right to insist on security.

THE ODD PRICE QUESTION.

Business men are quick and expert in detecting human weaknesses and in making use of them to the full. Twenty-five or thirty years ago all goods were sold at an even price of 5c., 25c., $1.00, $3.00 or $5.00, etc., and odd prices were seldom marked on goods, merchants having apparently a horror of "making change," pennies and coppers being seldom used.

Departmental stores were quick to see the deluding power of marking 25c. goods at 23c., $3.00 goods at $2.98, $5.00 goods at $4.98, and so on. Shoppers, especially women, jumped at the conclusion that 25c. goods were cheap at 23c., and the departmental stores were jammed with bargain hunters, who had rather peculiarly forgotten about the 10c. car fare they were spending to gain 2c. on a bargain.

People dearly love to get a bargain and one of the regular duties the housewife assiduously performs every evening is to search the evening papers for $2.98 hats, 98c. wall paper, etc. It is rather inhuman for the big merchant to impose upon the morbid bargain appetite of the public. However, that is their own business.

The country merchant has never adopted this clever though perhaps precarious device. He has tenaciously clung to the idea of even prices for everything to save time in making change. There are numerous articles in a country store which should in all fairness to the dealer, be marked up 1c., 2c., 3c., or 4c. They would be surprised at the increase in profits if they would dispel the illusion they have been laboring under for so many years, and mark their goods to the cent. We have not the slightest doubt it would pay them and we strongly urge them to do so. The merchant has a perfect right to all he can make.

their worst they begin to mend, we may confidently look for some improvement in the next fortnight. We continue to quote: Lamb, flag and straits at $44.

Sheet Zinc—Is weak.

Lead—Is strengthening and prices are firmer. Supplies are adequate.

Old Materials—The market is still quiet and it expected to continue so for a fortnight yet. Prices are weak and a still further decline is noted on some lines. We quote: Heavy copper and wire, 16c.; light copper, 15c.; heavy red brass, 15c.; light brass, 8c.; No. 1, wrought iron, $15.50; stove plate, $12.00.

TORONTO METAL MARKETS.

Toronto, August 15.—The metal trade continues quiet with no outstanding features in the local market. Business has not yet recovered from the midsummer quietness which has prevailed for the past three or four weeks, and, consequently, Toronto prices are still declining. While no heavy sales are reported, a good sorting-order business is, however, being done and local jobbers are getting as much business as they expect for the middle of August, when the metal markets are usually dull.

Antimony—The local market is extremely dull with practically nothing doing. Antimony prices have taken another drop of two cents during the week and Cookson's is now quoted locally at 15c. and Hallett's at 14½c. There is little doing on the American or London markets, and prices are a shade easier on account of the continued dullness.

Pig Iron—The market here continues very quiet, with prices unchanged. A few enquiries continue to arrive but these are almost exclusively for iron in sorting quantities.

Tin—A sharp decline has taken place during the week and lamb, flag and straits ingot tin is now quoted locally from $42 to $44, as against $44 to $45 last week. Little buying or selling is being done and supplies are quite adequate for present needs. Concerning the American market the improved demand of the last few days shows that some consumers there are beginning to reckon with the decline, and are considering the advisability of taking advantage of present prices. Records show that since the first of the month the London price of spot tin has declined over £12 per ton, and the New York price has declined very nearly three cents per pound.

Lead—Since the decline of last week there has been no further change in lead and local prices for imported pig lead remain very firm at $5.35.

Copper—A further decline of 1½c. took place during the week and casting ingot copper is now quoted here at 22c. Copper decidedly furnishes the most interesting feature in the world's markets at present. A well-versed Boston writer sums up the American market as follows: "The repeated declarations of the big copper producers that there will be no further reduction in prices is having little effect with purchasers, for it was but a short time ago that those same producers announced that under no consideration would there be any cut. They declared that the market was thoroughly sound and there was absolutely no occasion for any reduction in the prices. It is now getting to be a case of another cut or no sales, with the prospect good for a large accumulation of the surplus metal if the present attitude of the producers is adhered to. Already rumours are current that many of the larger companies are beginning to pile up their copper, although naturally there is no official confirmation of this rumour. These large producing interests are, by reason of their strong financial condition, in a position to hold out for a long time against the demands of consumers for further concessions, but it is doubtful if they will consider it a wise policy to do so. A 20-cent copper market may not be a pleasing thing to contemplate, after the long continued period of high prices, but it surely begins to look as if this level—and possibly a lower one—must be reached before any considerable fresh demand for the metal is stimulated."

Old Materials—Are still declining locally and the market continues very weak. Heavy copper and wire has declined 1½c. and is now quoted at 15c., light copper and heavy red brass have each dropped 2c. and are now 13c.; heavy yellow brass has dropped from 12c. to 10½c., and light brass from 8c. to 6½c.; tea lead is now 3¾c. instead of 4c., and heavy lead has lost ¼c. and is now quoted at 4c., scrap zinc is also ¼c. easier and now brings 3¾c.

LONDON METAL MARKETS.

London, Aug. 13.—Cleveland warrants are quoted at 57s. 4½d., and Glasgow standards at 56s. 6d., making prices as compared with last week, on Cleveland warrants 4½d. lower, and on Glasgow standards 3d. lower.

Tin.—Spot tin opened weak at £168 15s., futures at £168 5s., and after sales of 250 tons of spot and 250 tons of futures closed weak at £167 10s. for spot and £167 10s. for futures, making price as compared with last week £11 lower on both spot and futures.

Copper.—Spot copper opened weak at £78 5s., futures at £75 5s., and after sales of 200 tons of spot and 700 tons of futures, closed weak at £76 10s. for spot and £73 10s. for futures, making price as compared with last week £7 10s. lower on spot and £7 5s. lower on futures.

Spelter.—The market closed at £22, making price as compared with last week 12s. 6d. lower.

Lead.—The market closed at £19 15s., making price as compared with last week 12s. 6d. higher.

U.S. METAL MARKETS.

Cleveland, O., Aug. 15.—The Iron Trade Review to-day says:

"Conditions in the iron market have been very quiet during the past week and an anxious eye has been turned in the direction of the New York stock market to see what effect, if any, it will have upon the situation. The last previous flurry came in the height of the buying movement, when new business was being received from every direction, and the halt it occasioned was but momentary. At this time there is practically no new business, as compared with the former occasion, but it is not thought the effect of the heavy stock fluctuations will be, or should be, taken as an indication of still further decline in quotations. With the light new buying the pressure for deliveries continues unabated in most lines, and, as far as can be learned, there have been no cancellations and but very few requests for suspensions.

"In nearly all points of difference between the amalgamated association and the manufacturers of bar iron, the reconciliation boards decided in favor of the employes. Manufacturers are not bound to accept the decision of the board, but will probably do so.

Following a long period of unprecedented demand, fall business in wire and wire nails has opened about a month earlier than usual, and mills, which are practically destitute of stocks, are exceedingly busy. The very encouraging crop reports, recently issued, are favorable for the wire business.

The Illinois and Carnegie steel companies now have on hand orders for 40,000 tons of plates and shapes for boat construction and instruction for shipping nearly 60,000 tons of steel bars and shapes for implement manufacturers. Bridge builders are also placing new orders freely and the general demand for structural material is strong. In New York bids for 15,000 tons for piers have been submitted in addition to 10,000 tons recently contracted and 11,000 tons more are to be awarded later."

U. S. IRON AND METAL MARKETS.

New York, Aug. 15.—The Iron Age to-day says: "Foundrymen, large and small, seem to persist in the policy of keeping out of the market, and even the largest melters are buying only from week to week for prompt deliveries. They are encouraged in their course by constantly lower prices in some sections. In other districts, notably the south, the makers are clinging to the prices which have prevailed for a considerable time, but they are not making any sales, and in some competitive markets are dollars above the asking prices of other producers.

In steel making irons interest centres entirely on basic iron, which is weaker. Aside from one lot of 5,000 tons in the east, no business is reported, but it is more than probable that a buying movement in this branch of the iron trade will set in during the next few weeks in the district east of the Allegheny mountains.

There has been no further buying of steel by the leading interest since last week.

While steel works and rolling mills have assurances of full work for the balance of the year, and while they are crowded now, it is undeniable that on the whole new orders are coming in at a considerably reduced rate, so that the winter may find a slackening of operations necessary.

The deliveries of galvanized sheets are still unsatisfactory, the mills being overcrowded. In black sheets, however, the situation is improving. Thus far the season's business in tin plates has been a disappointment.

There is no sign as yet that the large consumers of copper are tempted to buy, although small lots of electrolytic have sold at 18 cents.

CORRESPONDENCE

Travelers, hardware merchants and clerks are requested to forward correspondence regarding the doings of the trade and the industrial gossip of their town and district. Addressed envelopes, stationery, etc., will be supplied to regular correspondents on request. Write the Editor for information.

HALIFAX HAPPENINGS.

Halifax, August 12.—The annual meeting of the Maritime Board of Trade will be held at St. John, N.B., this year, the opening session being on Wednesday, August 21st. It is expected that the attendance of delegates will be large, as a number of important subjects are to be considered. Secretary C. M. Creed has just issued a circular showing the many important questions to be discussed, some of which are as follows:

The importance of double-tracking the Intercolonial Railway between Halifax and St. John, N.B.

The need of increased transportation facilities, engines, freight cars and passenger cars for the Intercolonial Railway.

The importance of the founding of a School or University of Technology that will have the support of the three Maritime Provinces.

Abrogation Modus Vivendi.

Appointment of a Canadian Atlantic Fisheries Board.

Resolution endorsing recommendation of colonial conference for fast line to the Orient through Canada.

Bonus to steel shipbuilding.

Running rights of the C.P.R. and other railways over the Intercolonial Railway through Nova Scotia to Sydney.

Federal and local subsidies for steamship service between Sydney, ports in Bras d'Or Lakes on Inverness shore and Charlottetown and Summerside.

The necessity of an export duty on rossed pulp wood.

A national banking system for Canada.

The desirability of permanent high roads between counties and provinces.

Encouragement of sheep raising.

Proposed good roads.

Encouragement of trade between Canada and the British West Indies.

Maritime union.

Development of our national resources.

National system of technical education.

Re-affirmation of the tunnel resolution.

SCENIC ST. JOHN.

St. John, N.B., Aug. 12.—Business in the hardware line has been very good ately although it is naturally the quiet eason. The demand for builders' hardware continues brisk and smallwares are meeting with a ready sale.

* * *

Emerson & Fisher's big hardware establishment on Germain Street was invaded last week by the Red Rose Tea travelers from all over Canada, who were meeting here in annual convention. The big, seven-story brick building was examined from top to bottom and the firm was greatly congratulated on the fine appearance of their up-to-date establishment and the methodical way in which they conducted the various departments.

* * *

The New Brunswick Cold Storage Company was organized at a meeting held here on Thursday last and is an amalgamation of the provincial company, with the company now building the warehouse on Main Street, near the I.C.R. pier. The following officers were elected: R. J. Graham, Belleville, Ont., president; George McAvity, St. John, vice-president; L. S. Macoun, Ottawa, secretary-treasurer; H. R. Ross, Sussex, and A. I. Trueman, St. John, directors. The company will proceed with the building of the warehouses, now being erected at a cost of $125,000, and will operate under the provincial charter granted to the New Brunswick company.

* * *

Justice John Barnett, of Hartland, will this week, at the instance of the village fire commissioners, hold an investigation into the origin of the fire of July 15 under the act providing for such cases. Hon. H. A. McKeown, of St. John, will handle the case for the commissioners.

* * *

Not for many years has there been so much water in the river during the summer season. At Fredericton a freshet of nine feet is reported, and the lumbermen have taken full advantage to make a clean-up of logs. The rafting operations this year are likely to be more successful than for many years past. If the lumbermen are happy, however, the farmers are correspondingly gloomy, for the wet weather has destroyed great quantities of hay. If fine weather does not soon come, the farmers will suffer heavy loss. It is likely that the C.P.R. will bring one of their floating elevators here this winter to use in connection with the new wharf on the west side, as there will not be time to erect a conveyer from the elevator.

KINGSTON KINETOSCOPE.

Kingston, Aug. 12.—McKelvey & Birch, hardware merchants, have moved into their new premises on Brock street, built in place of the building recently destroyed by fire, and have now one of the best equipped stores of its kind to be found in these parts. They have added another storey to the old store, making it a four-storey building, whereas the other was only three. The additional storey will be used to keep their reserve stock of enameled ware and all lines of lighter goods, generally handled in the hardware trade. The hardware business has been well represented by this firm for the past forty years and there is hardly a more progressive house than this in the city. They carry a full line of general hardware, both shelf and heavy goods, the quality being always first-class, selected from the best markets in Canada and the States. The store shows good arrangement and everything about it is well planned; they occupy a large floor space, which enables them to show goods to the best advantage. The customer is always given the most courteous atttention by competent salesmen. In connection with their business they have also a well-fitted-up tinshop, as well as a plumbing, gas and steam-fitting department next door to the hardware, where all plumbing is scientifically and carefully done. The tinsmithing department is one of the best, the highest order of work being turned out, and gives employment to a large number of men. The members of the firm are men of sterling worth, and men who rank high in commercial circles.

* * *

Dr. Waugh, dentist, and E. S. Suddard, grocer, both of this city, have purchased a plant for the manufacture of cement blocks for building purposes, and will commence operations at once.

* * *

The Knox Company are going to open one of their 5c., 10c. and 15c. stores in Kingston, in the store recently occupied by George Wills & Co., hatters. They have had carpenters and workmen busy fitting up the store for the past few weeks and will open as soon as the work is completed. Robinson & Copley are doing the carpentering.

* * *

James McGrath, of this city, has been awarded the contract for rebuilding the Roman Catholic church, at Sharbot Lake. This edifice was destroyed by fire last spring. Architect H. P. Smith has reduced the size of the church, as a smaller building would suffice, owing to the first contract price being too high.

* * *

The city engineer's department began the asphalt construction work to-day. A number of crossings are to be laid throughout the city, and quite a lot of patch-work has to be done on walks and crossings, which have been torn up by the gas excavations.

* * *

The people made good use of their screen doors and windows on Sunday, and many of them stayed away from their houses as much as possible, it being one of the hottest days experienced this summer. The thermometer registered about eighty-two degrees in the shade, which is considered awfully hot in Kingston, though many a place would consider itself lucky to have such a moderate temperature. The hot weather should be relished, however, as there won't be a great many warm nights like that of Sunday now, as the autumn is fast approaching.

* * *

George Mills & Co., hatters and furriers, have purchased a store on Princess street, Great alterations have been made and they have now a much larger store than before. James Halliday, electrician, had the contract for lighting and fixtures, and made an excellent job of the work. Besides placing a number of lights in each window, they have placed about forty lights in the doorway, which shows the store up fine from the outside. They have glass cases along the walls on either side of the store, as well as lights along the top of these cases, the length of the store. It is expected, now Mr. Mills has started the outside electrical sign lighting, that other merchants will follow in his footsteps and take advantage of the cheaper rates offered by the city for outside lighting; when one starts, other usually follow.

* * *

J. B. Walkem, assignee for George Sears, hardware merchant, has called tenders for the estate and stock in trade of the above mentioned, consisting of shelf and heavy hardware, paints, oils, etc., and a large assortment of other goods usually carried in a hardware business. The stock is valued at $16,000. Tenders will be received by the assignee until the 23rd of August.

* * *

Arthur Ellis, architect, is having a fine brick house erected in one of the best residential parts of the city. Workmen have been at the building for some time, and Mr. Ellis expects to have it completed in the course of a month, so as to be able to move into it in the fall.

* * *

W. B. Dalton, hardware merchant, left with his wife one day last week for Old Orchard beach, where they will remain during the month of August.

* * *

The sloop Granger arrived here on Thursday with a cargo of cement from Belleville, for John Lemmon & Sons.

* * *

The Grand Trunk Railway Company has no intention of purchasing the Tete du Pont barracks, and the rumor of their purchase has since been contradicted. The work of reconstructing the buildings is to go on. The public works department has not yet returned the revised plans for tenders, but will do so shortly. The cost will be considerably reduced so as to keep within the appropriation. The plans for further additions to Artillery park barracks have arrived for approval by the eastern Ontario officer in command. They provide for alterations to the stable buildings, which will be veneered and otherwise improved.

* * *

The employes of the Canadian Locomotive Works held a picnic at Lake Ontario park on Saturday last in aid of the Hotel Dieu and general hospitals. The day was an ideal one for the outing, and had the committee been given the making, a finer day could not have been secured. The men made an all-day affair of it and people streamed out from early morning until evening. Some of the sports were, football in the morning between the Englishmen and Scotchmen employed at the works, the teams being evenly matched, and the Englishmen winning the game and the prize of one hundred cigars. At 1.30 the other sports started, and there was a contest for everybody—running races, long and high jumps, greasy pole, tug-of-war and other contests which called for show of strength. The main attraction was a baseball match between two city teams for a purse of $50. The jolly day was brought to a close by a dance in the pavilion after the show, the street cars being kept busy until nearly midnight bringing back the crowds. A long list of names of the generous merchants of the city, who were kind enough to offer goods for prizes, amongst which are most of our leading hardwaremen, has been published. This was the first affair of this nature held by the men of the works and it was such a grand success that they intend making it an annual one. A goodly sum was realized.

BOOMING BELLEVILLE.

Belleville, Aug. 15.—No doubt many of the readers of Hardware and Metal will be pleased to know that Belleville is soon to have in operation another large and new industry in the shape of a brass foundry, work on which commenced in real earnest on Monday of this week. The works of the said industry will be situated in the centre of West Belleville, on a splendid site. The building will be of brick and stone and will be 136x40 feet. George A. Bennett of this city is the contractor and builder. The foundry, when completed, will be one of the best equipped of the kind between Toronto and Montreal, the company having plenty of capital behind the undertaking. The ouput will consist of brass fixtures of all kinds. There is said to be a splendid and growind demand for the prospective products of the new factory, and a large number of orders have already been booked.

* * *

The Belleville Rolling Mills can hardly keep up with the demand for horseshoes, 50 kegs being shipped this week to one Hamilton firm. The mills are running night and day with the largest staff in their history.

* * *

The company is capitalized at $100,-000 and the shares have been principally taken up by local capitalists. It is said that H. C. Hunt will be the manager of the concern and his long experience with the Belleville Lock Works well qualifies him for the position. The new building, which will cost about $7,000, is expected to be ready for occupation about November 1 ; in the meantime the company have rented a building and will manufacture the necessary tools and equipment. A large staff will be employed at the works as soon as operations begin. The promoters had excellent inducements to locate in other places but preferred this city. The inducements furnished by the city council were free installation of sewerage, water and gas connection—the company to pay the same rate for water and gas as other consumers.

* * *

The Belleville Tubular Axle Works have had to seek larger quarters owing to increased business.

LEAFY LONDON.

London, August 11.—The destruction by fire on Saturday last of the South-Western Traction Company's car barns with five cars, the machine shop, oil house and winding room, is a serious blow to that concern. The direct loss is covered by insurance, but the company seriously suffers indirectly through the stoppage of traffic over the line. They have been in full operation between this city and St. Thomas for over a year, and have their rails laid close to Port Stanley, but now they find themselves with scarcely any rolling stock or machinery necessary to the running of the road at a season when they should be busiest. However, the men at the head of the company are energetic, enterprising fellows, and it may be taken for granted that as little time as possible will be lost in getting things in running shape again.

* * *

A dozen or so local contractors have sent in tenders to the Grand Trunk Railway Company for the erection of the addition which is to be made to the car works on the east end. The building is to be of brick, and it is understood the tenders vary from $55,000 to $60,000. The addition will mean the doubling of the facilities for the building of freight cars and the employment of at least one hundred more men than at present. The Grand Trunk shops are already among London's chief industries, and when the addition is completed they will mean more than ever to the people of this city. As the tenders were sent to Montreal direct, an early announcement of the successful bidder may be looked for.

* * *

J. E. McConnell, who for a number of years has been manager of the advertising department of the McClary Manufacturing Company, is about to sever his connection with that company, and will hereafter be more intimately connected with the advertising agency of the McConnell-Ferguson Company, whose business has grown to such proportions as to demand Mr. McConnell's entire time and attention. The position of advertising manager for the McClary Company will be taken by A. A. Briggs, who for some years has been connected with the practical printing, advertising and literary branches of several of the leading journals of Toronto.

* * *

Andrew R. Simpson, for the past eleven years manager of the Ontario Spring Bed & Mattress Company, is retiring from the firm to seek a respite from business, in travel and change of scene. With his partner R. C. Williams, Mr. Simpson came to London a little over eleven years ago, and the two established and extended the business situated at 90 York street, until now their goods are sold all over the Dominion and certain lines are sent out of the country. About forty hands are regularly employed, and a large sum is annually paid out in wages. The firm is one of London's growing concerns and there is every reason to believe it will continue to flourish in the future as it has in the past

* * *

Hardware travelers have returned to the road again after their two weeks' vacation. Trade does not seem to have suffered to any serious extent during their absence, business having been quite satisfactory right through.

* * *

The annual picnic of the brass finishers at Port Stanley, Saturday, was a huge success. About 500 employes of Labatt's, the Empire, London Brass and the Mann Company works, took in the trip and enjoyed themselves immensely. The sports provided were good.

CHAT FROM CHATHAM.

Chatham, Aug. 13.—Hardwaremen report the last week or two exceptionally good, business being quite in advance of last year. Merchants are clearing out their summer stocks, preparatory to making way for fall goods, which are now beginning to come in.

* * *

Dresden ratepayers recently defeated bylaws for a new school and street paving.

* * *

Recent press despatches state that negotiations are in progress whereby the Windsor, Essex and Lake Shore Electric Railway may pass into the control of the Cataract Power Co., of Hamilton. The Windsor road holds a franchise for the city of Chatham, and its extension to this point was ultimately looked for. It is stated that the intention is to complete an all electric line from Detroit to Niagara, with an international ferry across the Detroit River.

President Piggott, of this city, when interviewed regarding the matter, stated that the report in the Detroit paper was the first he had heard of it.

The railway is reported to be progressing finely and the power house at Kingsville is rapidly nearing completion.

* * *

Blenheim's tax rate for the current year is 24 mills on the dollar, being one mill less than last year. The town council have granted the Blenheim and South Kent Telephone Co. a franchise to place poles and wires on the streets, if a suitable agreement cannot be made with the Bell Company for the use of their poles.

* * *

The Volunteer Firemen's Convention at Wallaceburg last week was a big gathering, many being present from all parts of Ontario. The convention lasted two days, a third day being taken up with the firemen's tournament. The town was resplendent, the merchants decorating in a manner befitting the occasion.

* * *

Geo. Watt left on Saturday for Toronto, where he has secured a position with the Dominion Radiator Co.

* * *

Negotiations are in progress between the Civic Industrial Committee and certain American capitalists which, it is hoped, may result in the landing of a new industry for the Maple City. The proposed concern will employ between 60 and 70 hands, and Chairman Westman, while discreetly silent as to the precise nature of the enterprise, is sanguine of success.

* * *

Cortland H. Rayment, contractor, has made an assignment for the benefit of creditors, the assignee being Sheriff Gemmill. Sept. 4 is fixed as the date of distribution.

* * *

A party of Towanda, Pa., capitalists, interested in the C. W. & L. E. electric road, were in the city last week inspecting the work on the southern extension. They expressed themselves as much gratified at the progress made. The line was opened to the fair grounds on Monday, Chatham's civic holiday, cars being run to the races.

SASKATOON SAYINGS.

Saskatoon, August 10, 1907.—The hardware store of James Clinkskill was broken into on Thursday night. The thieves gained entrance by a basement window. About $16 was taken from the cashier's drawer, besides stamps and a considerable quantity of goods, including cutlery, knives, razors, hat pins, etc.

* * *

On Thursday night the prize drawing took place at the Cairns store. As ten o'clock came round the big store became crowded with expectant coupon holders. The lucky customers were as follows: First prize, A. Marriott, driving outfit; second, Miss F. Bulmer, malleable steel range; third, W. McNab, lady's costume; fourth, N. J. Anderson, suit; fifth, T. Borgford, goods to the value of $20; sixth, G. McFarlane, dinner set, seventh, J. W. Raynor, Stetson hat, eighth, Mason Bros., pair of shoes; ninth, A. Randall, baseball outfit; tenth, T. Borgford, girl's suit.

* * *

A very successful three day's fair was held this week. Amongst the goods exhibited were to be seen saddlery, harness and leather goods, binder twine, paints, plumbing goods, a furnace, electrical fixtures, etc.

* * *

Contracts for the Flanagan Hotel, now in course of erection, have been awarded as follows: general contract, Shannon Bros.; plumbing, G. G. Taylor; heating, Splayfords'; metallic roofing, cornices and galvanized iron work, M. Isbister & Son.

The Western Plumbing Company have secured the contract for plumbing and heating the Northern Bank building, also plumbing and heating of a double house being erected for the Hoeschen-Wentzler Brewing Company.

* * *

W. F. Moser, of the firm of J. G. Moser & Son, formerly of Blyth, Ont., is now manager of the hardware and plumbing department of the J. F. Cairns departmental store. Mr. Moser reports trade brisk and the firm has recently been awarded the following contracts: Plumbing and heating of the Bank of Commerce, Hanson block, Sutherland block, and Gordon & Sparling block. Plumbing of the Alexander residence, Bulmer residence, Howes residence and F. Cahill's residence.

* * *

John Gunn & Sons, who are at present putting in the concrete work for the Grand Trunk Pacific bridge here, have been awarded another contract at Lethbridge, Alta. This for an immense steel bridge for the Canadian Pacific Railway. Gunn & Sons will build all the abutments and the concrete substructure, which will cost in the neighborhood of $200,000.

NAIL-HOLDING HAMMER.

A tool adapted to be used for driving nails in shingles and lathing, and especially for overhead work, has been invented by H. C. Lyon, Howard Lake, Minn. This hammer is provided with means to contain a quantity of nails, and to deliver them singly at the ball of the tool and hold them in such position in line with the hammer head that they may be partially driven into an object without being handled.

CUTTING-TOOL HOLDER.

F. A. Hummel, New York, N.Y., has invented an improved cutting-tool holder. The instrument has been designed to operate upon a rod, shaft, tube, or the like, held by a chuck or a face-plate and dog, or in any desired manner, at the head centre of a lathe, so as to be rotated. It is intended to be applied to the work and held by hand or other means in a stationary position centered by the lathe and fed up to the work by the tail centre or other means, so that upon rotation of the work the operation will be performed upon it by the stationary cutting tool.

FIREPROOF CHRISTMAS TREE.

An ingenious Christmas tree has recently been invented by F. L. McGahan, Los Angeles, Cal. While the construction may be employed as a Christmas tree, it may be used as an advertising device or a display rack, and when made upon a small scale may be employed as a toy. The tree may be mounted in various ways, and may be lighted by gas, electricity, or candles.

GOOD PROMISE BETTER THAN BAD CHECK.

The salesman is always ready to take your order, and it's easy to buy; but paying sometimes gets to be what Sherman said war was. In paying your bills, if you do so by check, be sure there is enough money in the bank, and don't calculate that Messrs. Jones & Brown will not put your check through for a day or so, and pay another bill with what you have on deposit, hoping for good business or some fellow to pay a bill to tide you over, because it's a known fact that some of these credit men have wonderful intuitive powers, and just about that time the credit man with Jones & Brown is going to rush your check through. And then—well you figured wrong or the bank was wrong. Don't do it—if business is quiet, collections poor, or unlooked-for expenses have depleted your cash balance, tell the truth, and say that in about ten days or two weeks you will remit. Always remember that a good promies is better than a returned check marked "No Funds."

POLISH FOR CLEANING FURNITURE.

Splashes of dirt on polished furniture are removed with soap and water, and the wood is well rubbed with a mixture of equal parts of spirit and oil or spirit and turpentine, applied with a woolen rag. This mixture has both a cleansing and polishing action, the polish being retained for a long time if well rubbed in and the surplus wiped off. Another good preparation for the same purpose is a solution of stearine in oil of turpentine and a little spirit, care being taken not to use so much stearine that white streaks are produced in the mass. When the turpentine and spirit have evaporated, the wood is well rubbed with a woolen rag. This gives an excellent polish that can be renewed by rubbing when dimmed. Furniture with a matt finish can be renovated with a thin solution of white wax in oil of turpentine, or by rubbing it over with linseed oil.

CLEANING PORCELAIN BATH TUBS

To clean a porcelain lined bath tub, use hot water and a rag saturated with gasoline. If the gasoline is objectionable, smear a little vaseline on the dirtiest parts and remove dirt and vaseline at once with rag and hot water. Never scour porcelain tub, nor nick nor scratch its surface in any way.

THIS POINT WAS WELL TAKEN.

A story is told of a poor boy who, while walking along a busy thoroughfare, saw a pin on the pavement before him. Quickly he stayed his steps, and picking up the pin, stuck it safely and securely in his coat. A wealthy man chancing to pass at that time, saw the action, and was much impressed by it; so much so that he took the boy into his bank and finally adopted him. Thirty years passed and the poor boy became a millionaire. Struck with an idea, he one day drew a cheque for 2,000 pounds and gave it to a former schoolmate, who had not prospered in the race of life. "All that I am now I owe to you John," said the millionaire. " But I don't understand," stammered the other. "Simple enough. If I hadn't hated you so at school I should never have picked up that pin to stick into you while we were in class."

SIMPLE METAL POLISH.

Metal polish is one of the easiest things in the world to manufacture, and, as the cost of materials is little or nothing, the profits are very large, advises an exchange. Take any quantity of yellow or blue clay, perfectly free from sand, and allow it to dry. Pulverize by pounding it and run through a flour sieve, or a finer one, if one is obtainable.

To five pounds of this sifted clay add one pound of sifted wood ashes and mix to a very stiff paste with a solution of water and lye, in the proportion of one gallon of water to two heaping tablespoonfuls of lye. Spread this mixture on a flat, planed board and level off the top until it has a uniform thickness of about one inch. When it has become set, mark it off into squares about two inches on a side and cut apart.

To use this polish, scrape off a little on a moist cloth and rub the article to be polished, thereby giving it a glittering, lustrous shine. It is said that one man has made a good living making and selling this polish. If the material is sifted carefully in the making the preparation will not scratch the finest polished surface.

ORIGINAL ADVERTISING.

An original form of advertising comes from Russia where a shopkeeper posted up the following announcement:

"The reason why I have hitherto been able to sell goods so much cheaper than anybody else is that I am a bachelor, and do not need to make a profit for the maintenance of a wife and children.

"It is now my duty of informing the public that this advantage will shortly be withdrawn from them, as I am about to be married. They will, therefore, do well to make their purchases at once at the old rate."

The result was that there was such a run on the shop that in the course of a few days this shopkeeper had made money enough to pay the expense of his wedding on a very lavish scale.—Ex.

ODD STREET PAVEMENT.

The city of Monterey, California, has a street paved with the backbones of whales. The segments of the vetebrae were used as we use stone for street paving. The history connected with this pavement relates that about one hundred years ago a large school of whales was driven ashore and perished. After the flesh had decayed the backbones of these whales were used for paving the streets in place of large sea shells, which are frequently used for the same purpose. The streets in the old Spanish towns of California often served for man and beast, there being no sidewalks.

FOUR WAYS OF DOING BUSINESS.

A "successful manufacturer" is quoted by an American exchange as instancing the four following ways of doing business:

1. The Dishonest Way.—Sell so low that you can't earn a living, and the sheriff will finally wind up your affairs, and your creditors carry your losses.
2. The Misrepresentation Way.—Palm off upon your customers unfair goods and persuade them that they are the best.
3. The Suicidal Way.—Employ workmen at less than living wages; buy the cheapest materials; pare everything down to the lowest notch, and patch results.
4. The Straight Way.—Buy the best; employ skilled labor; thoroughly know your own business and business values. Provide special facilities for the execution of the greatest amount of high grade production at a minimum cost. Avoid extravagant management, expensive methods, and have your dealings with customers who appreciate honest treatment.

JAPS FIND COPPER IN B. C.

The discovery of copper ore in the northern inland of British Columbia by Japs reads like a romance. The Japs were there fishing and diving for abalona shells used in Japanese commerce. They found that the reefs where they were carrying on their operations on Queen Charlotte Sound, were not country rock as supposed, but great reefs of mineral. One of their divers when at the bottom of the sound came across a chunk of pure yellow metal he thought was gold, but which was native copper. The Japs then quit their search for shells and became miners. Besides copper, quantities of iron ore were found. The little brown men said nothing until they secured ample capital to work their claims, which they are now doing. Several Americans, however, are located on properties beside them.

FINE METAL POLISH.

A good polish for fine metals is made by mixing a little vaseline with the ashes of burned out or broken gas mantles. Apply with a rag or finger and polish with a clean rag. The best finish can be obtained by using a soft rag.

WATER CLOSET FLUSHER.

An improved flushing device for water closets, bowls and the like has been invented by L. W. Eggleston, Appleton, Wis. The usual tank and supply tank are employed by the inventor. At the upper end of the pipe is a valve casing in which is a nozzle or injector discharging water into the tank. A plug valve having slidable movement within the casing and movably connected with which is an actuating lever, through whose medium the valve is closed to open the nozzle outlet and again close it. A controlling member is employed for the lever intermediate of which and a co-operating float are other members of special construction. The flushing devices are primarily actuated by the usual pull chain.

MANITOBA HARDWARE AND METAL MARKETS

Corrected by telegraph up to 12 a.m Friday, Aug. 16 Room 511, Union Bank Bldg, Winnipeg, Man.

An average business for the season of the year is reported by the wholesale houses. The trade are waiting for the crop and there will not be much activity until after harvest, when dealers will know what they have to expect. Values are steady.

ROPE—Sisal, 11c. per lb, and pure manila, 15¾c.

LANTERNS—Cold blast, per dozen, $7; coppered, $9; dash, $9.

WIRE—Barbed wire, 100 lbs., $3.22½; plain galvanized, 6, 7 and 8, $3.70; No. 9, $3.25; No. 10, $3.70; No. 11, $3.80; No. 12, $3.45; No. 13, $3.55; No. 14, $4; No. 15, $4.25; No. 16, $4.40; plain twist, $3.45; staples, $3.50; oiled annealed wire, 10, $2.90; 11, $2.96; 12, $3.04; 13, $3.14; 14, $3.24; 15, $3.39; annealed wires (unoiled), 10c. less; soft copper wire, base, 36c.; brass spring wire, base, 30c.

POULTRY NETTING—The discount is now 47½ per cent. from list price, instead of 50 and 5 as formerly.

HORSESHOES—Iron, No. 0 to No. 1, $4.65; No. 2 and larger, $4.40; snowshoes, No. 0 to No. 1, $4.90; No. 2 and larger, $4.65; steel, No. 0 to No. 1, $5; No. 2 and larger, $4.75.

HORSENAILS—No. 10 and larger, 22c, No. 9, 24c.; No. 8, 24c.; No. 7, 26c.; No. 6, 28c.; No. 5, 30c.; No. 4, 36c. per lb. Discounts: "C" brand, 40, 10, 10 and 7½ p.c.; "M.R.M." cold forged process, 50 and 5 p.c. Add 15c. per box. Capewell brand, quotations on application.

WIRE NAILS.—$3 f.o.b. Winnipeg, and $2.55 f.o.b. Fort William.

CUT NAILS—Now $3.20 per keg.

PRESSED SPIKES — ¼ x 5 and 6, $4.75; 5-6 x 5, 6 and 7, $4.40; ⅜ x 6, 7 and 8, $4.25; 7-16 x 7 and 9, $4.15; ½ x 8, 9, 10 and 12, $4.05; ⅝ x 10 and 12, $3.90. All other lengths 25c. extra net.

SCREWS—Flat head, iron bright, 80, 10, 10 and 10; round head, iron, 80; flat head, brass, 75; round head, brass, 70; coach, 70.

NUTS AND BOLTS — Bolts, carriage, ⅜ or smaller, 60 p.c.; bolts, carriage, 7-16 and up, 50; bolts, machine, ⅜ and under, 50 and 5; bolts, machine, 7-16 and over, 50; bolts, tire, 65; bolt ends, 55; sleigh shoe bolts, 65 and 10; machine screws, 70; plough bolts, 55; square nuts, cases, 3; square nuts, small lots, 2½; hex nuts, cases, 3; hex nuts, small lots, 2½ p.c. Stove bolts, 70 and 10 p.c.

RIVETS — Iron, 60 and 10 p.c.; copper, No. 7, 43c.; No. 8, 42½c.; No. 9, 45½c.; copper, No. 10, 47c.; copper, No. 12, 50½c.; assorted, No. 8, 44½c., and No. 10, 48c.

COIL CHAIN — ¼-in., $7.25; 5-16, $5.75; ⅜, $5.25; 7-16, $5; ½, $4.75; 9-16, $4.70; ⅝, $4.65; ¾, $4.65.

SHOVELS—List has advanced $1 per dozen on all spades, shovels and scoops.

HARVEST TOOLS—60 and 5 p.c.

AXE HANDLES—Turned, s.g. hickory, doz., $3.15; No. 1, $1.90; No. 2, $1.60; octagon extra, $2.30; No. 1, $1.60.

AXES—Bench axes, 40; broad axes, 25 p.c. discount off list; Royal Oak, per doz., $6.25; Maple Leaf, $8.25; Model, $8.50; Black Prince, $7.25; Black Diamond, $9.25; Standard flint edge, $8.75; Copper King, $8.25; Columbian, $9.50; handled axes, North Star, $7.75; Black Prince, $9.25; Standard flint edge, $10.75; Copper King, $11 per dozen.

CHURNS—45 and 5; list as follows: No. 0, $9; No. 1, $9; No. 2, $10; No. 3, $11; No. 4, $13; No. 5, $16.

AUGER BITS—"Irwin" bits, 47½ per cent., and other lines 70 per cent.

BLOCKS—Steel blocks, 35; wood, 55.

FITTINGS — Wrought couplings, 60; nipples, 65 and 10; T.'s and elbows, 10; malleable bushings, 50; malleable unions, 55 p.c.

HINGES—Light "T" and strap, 65.

HOOKS — Brush hooks, heavy, per doz., $8.75; grass hooks, $1.70.

STOVE PIPES—6-in., per 100 feet length, $9; 7-in., $9.75.

TINWARE, ETC.—Pressed, retinned, 70 and 10; pressed, plain, 75 and 2½; pieced, 30; japanned ware, 37½; enamelled ware, Famous, 50; Imperial, 50 and 10; Imperial, one coat, 60; Premier, 50; Colonial, 50 and 10; Royal, 60; Victoria, 45; White, 45; Diamond, 50; Granite, 60 p.c.

GALVANIZED WARE — Pails, 37½ per cent.; other galvanized lines, 30 per cent.

CORDAGE — Rope sisal, 7-16 and larger, basis, $11.25; Manilla, 7-16 and larger, basis, $16.25; Lathyarn, $11.25; cotton rope, per lb., 21c.

SOLDER—Quoted at 27c. per pound. Block tin is quoted at 45c. per pound.

WRINGERS—Royal Canadian, $36; B.B., $40.75 per dozen.

FILES—Arcade, 75; Black Diamond, 60; Nicholson's, 62½ p.c.

LOCKS—Peterboro and Gurney, 40 per cent.

BUILDING PAPER—Anchor, plain, 66c.; tarred, 69c.; Victoria, plain, 71c.; tarred, 84c.; No. 1 Cyclone, tarred, 84c.; No. 1 Cyclone, plain, 66c.; No. 2 Joliette, tarred, 69c.; No. 2 Joliette plain, 51c.; No. 2 Sunrise, plain, 56c.

AMMUNITION, ETC. — Cartridges, rim fire, 50 and 5; central fire, 33⅓ p.c.; military, 10 p.c. advance. Loaded shells: 12 gauge, black, $16.50; chilled, 12 gauge, $17.50; soft, 10 gauge, $19.50; chilled, 10 gauge, $20.50. Shot: ordinary, per 100 lbs., $7.75; chilled, $8.10. Powder: F.F., keg, Hamilton, $4.75; F.F.G., Dupont's, $5.

REVOLVERS — The Iver Johnson revolvers have been advanced in price, the basis for revolver with hammer being $5.30 and for the hammerless $5.95.

IRON AND STEEL—Bar iron basis, $2.70. Swedish iron basis, $4.95; sleigh shoe steel, $2.75; spring steel, $3.25; machinery steel, $3.50; tool steel, Black Diamond, 100 lbs., $9.50; Jessop, $13.

SHEET ZINC—$8.50 for cask lots, and $9 for broken lots.

CORRUGATED IRON AND ROOFING, ETC.—Corrugate iron 28 gauge painted $3, galvanized $4.10; 26 gauge $3.35 and $4.35. Pressed standing seamed roofing 28 gauge painted $3.10, galvanized $4.20; 26 gauge $3.45 and $4.45. Crimped roofing 28 gauge painted $3.20, galvanized, $4.30; 26 gauge $3.55 and $4.55.

PIG LEAD—Average price is $6.

COPPER—Planished copper, 44c. per lb.; plain, 39c.

IRON PIPE AND FITTINGS—Black pipe, ¼-in., $2.65; ⅜, $2.80; ½, $3.50; ¾, $4.40; 1, $6.35; 1¼, $8.65; 1½, $10.40; 2, $13.85; 2½, $19; 3, $25. Galvanized iron pipe, ⅜-in., $3.75; ½, $4.35; ¾, $5.65; 1, $8.10; 1¼, $11; 1½, $13.25; 2-inch, $17.65. Nipples, 70 and 10 per cent.; unions, couplings, bushings and plugs, 60 per cent.

GALVANIZED IRON — Apollo, 16-gauge, $4.15; 18 and 20, $4.40; 22 and 24, $4.65; 26, $4.65; 28, $4.50; 30 gauge or 10¾-oz., $5.20; Queen's Head, 20 $4.60,; 24 and 26, $4.90; 28, $5.15.

LEAD PIPE—Market is firm at $7.80.

TIN PLATES—IC charcoal, 20 x 28, box, $10; IX charcoal, 20 x 28, $12; XXI charcoal, 20 x 28, $14.

TERNE PLATES—Quoted at $9.50.

CANADA PLATES — 18 x 21, 18 x 24, $3.50; 20 x 28, $3.80; full polished, $4.30.

LUBRICATING OILS—600W, cylinders, 80c.; capital cylinders, 55c. and 50c.; solar red engine, 30c.; Atlantic red engine, 29c.; heavy castor, 28c.; medium castor, 27c.; ready harvester, 28c.; standard hand separator oil, 35c.; standard gas engine oil, 35c. per gallon.

PETROLEUM AND GASOLENE — Silver Star, in bbls., per gal., 20c.; Sunlight, in bbls., per gal., 22c.; per case, $2.35; Eocene, in bbls., per gal., 24c.; per case, $2.50; Pennoline, in bbls., per gal., 24c.; Crystal Spray, 23c.; Silver Light, 21c.; engine gasoline in barrels, gal., 27c.; f.o.b. Winnipeg, in cases, $2.-75.

PAINTS AND OILS — White lead, pure, $6.50 to $7.50, according to brand; bladder putty, in bbls., 2¼c.; in kegs, 3¾c.; turpentine, barrel lots, Winnipeg, 90c; Calgary, 97c; Lethbridge, 97c; Edmonton, 98c. Less than barrel lots, 5c. per gallon advance. Linseed oil, raw, Winnipeg, 72c.; Calgary, 79c.; Lethbridge, 79c.; Edmonton, 80c.; boiled oil, 3c. per gallon advance on these prices.

WINDOW GLASS — 16-oz. O. G., single, in 50-ft. boxes—16 to 25 united inches, $2.25; 26 to 40, $2.40; 16-oz. O.G., single, in 100-ft. cases—16 to 25 united inches, $4; 26 to 40, $4.52; 41 to 50, $4.75; 50 to 60, $5.25; 61 to 70, $5.75. 21-oz. C.S., double, in 100-ft. cases, 26 to 40 united inches, $7.35; 41 to 50. $8.40; 51 to 60, $9.45; 61 to 70, $10.50; 71 to 80, $11.55; 81 to 90, $17.-30.

SMOKE CONSUMER.

An apparatus for consuming smoke in stoves and furnaces has been invented by C. J. Roux, 12 Rue Doudeauville, Paris, France. The invention is applicable to domestic and industrial heating apparatus of all kinds. By its means complete combustion may be obtained and absolute consumption of smoke, whatever the nature of the fuel may be, as soon as normal conditions have been established, even with the softest coals.

STOVE, FURNACE OR DRUM.

J. H. Hanson, Aitkin, Minn., is the inventor of an important heating improvement, by means of which hot gases are brought into close contact with the outer wall of the stove so as to give opportunity for the wall to absorb the heat from them. An arrangement of disks tends to choke the flow so as to give time for this heat absorption. There is no danger of an actual choking of the draft, as the area of annular spaces surrounding the disks through which the gases pass, is always equal to or more than equal to the area of the stove pipe.

VALUE OF FIELD STORE BUILDING

A valuation of $4,083,400 has been placed on the Marshall Field & Co. retail store, Chicago, as the worth of the building alone, and does not include either the stock or the land. The great structure, running from Randolph to Washington Street in State Street and from 73 to 92 Wabash Avenue, is therefore considered the most valuable mercantile building in Chicago, if not in the world. The assessment valuation is put at a higher rate per foot because of the costly mahogany woodwork and other finishings. The valuation this year is the first that has been made since the entire building has been completed.

BATH CABINET.

An invention referring to cabinets for steam or medicated vapor baths and especially useful as an attachment for and in connection with bath tubs of the usual kind has been successfully worked out by C. W. Groover, Valdosta, Ga. The aim is to provide a cabinet or cover by means of which the ordinary bath tub can be converted into a steam or vapor bath, which is capable of being removed and packed small when not in use, and which the bather can manipulate without assistance.

IMPROVED HASP.

S. B. Phelps, Green Hill, Chester County, Pa., has invented a hasp which is simple in form and so constructed that it will lie upon the interior; the general purpose being to prevent its being tampered with by a dishonest person. It relates to hasps such as used on chest doors, boxes, or in similar constructions. The fact that the entire hasp is within the interior of the chest and not in position to be reached by an intruder, is not only an advantage from the point of utility, but tends to give the chest a neat appearance.

THOUGHT IT A BARGAIN.

A smartly dressed young man was chatting with a prominent financier of a most economical disposition, when the latter suddenly drew attention to the suit of clothes he was then wearing.

"I never pay fancy prices for made-to-order garments," he said. "Now, here's a suit for which I paid $12. Appearances are very deceptive. If I told you I purchased it for $20 or $25 you'd probably believe that to be the truth."

"I would if you told me by telephone," replied the young man.

Heating and Housefurnishings

INCREASING STOVE SALES.

"A special plan for increasing stove sales came to my knowledge since your issue of July 13th, when this matter was presented by several persons," writes Charles D. Chown. "A few years ago an Ottawa firm, who handled bicycles, automobiles, etc., during the summer months, arranged to take up the sale of a first-class line of stoves and ranges, made in Hamilton. They had a large show room that permitted them to make a display of the full range of styles, so they could show a prospective customer a stove or range with the equipment he or she desired.

"They engaged a canvasser, fortified him with circulars such as the manufacturers provide their customers with for distribution at exhibitions and to prospective customers. These circulars bore the name and address of the selling concern. The canvasser did not endeaver to make sales, simply make enquiries as to whether the party called upon needed or would need or knew of any one that would need a stove or range. If he found that they did or would he left circulars of the lines they were likely to be interested in and asked them to call and examine their display.

"The results were such as to justify the expenses incurred, as they sold over 400 stoves that season, which brought them further business the following season. This line of work would be especially desirable where gas, either manufactured or natural, is about to become available for cooking or heating, and would probably result in keeping the trade in gas stoves in the hands of the stove trade, rather than in the hands of the gas companies, which have in many places, in order to get persons to use gas, sold stoves at a margin of profit that made it unprofitable for the regular store trade to handle gas stoves,"

STRUCTURE OF METALS.

Dr. J. A. Ewing, F.R.S., recently gave before a scientific society in England a lecture on the structure of metals. Under the microscope all metals show practically the same structure. A piece of metal is built up of innumerable particles, similar to each other, as so many bricks, or perhaps stones, built into a wall. The separate grains are not quite so regular in shape and size as bricks. They are laid in a perfectly regular formation. This is revealed by the microscopic examination of highly-polished and etched surfaces of the metal. When a metal is strained beyond the elastic limit, what takes place is that slips occur between the layers of brick-bats and the separate grains. Looked at from above these "slip lines" seem like minute cracks, but they are really steps caused by the slipping of one layer on its neighbors, just as cards might slip in a pack. Cards can be slipped only in two directions—sideways and lengthways. But the grains in metal can be slipped in a third direction—across the thickness of the card, owing to the fact that the metal is built up of such innumerable small particles. When a piece of metal is bent backwards and forwards several times the "slip" is repeated, the cohesion of the particles to each other is broken and a crack results. When a piece of metal is strained beyond the elastic point the slip has taken place, and it is obvious that it will be a long time before the particles that have slipped away from each other will cohere as firmly to those that they have at last come next to. The technical name of this phenomenon is "fatigue." Another thing that is constantly observed is also explained by this particular formation of metals, though not referred to by the lecturer. When a punching machine comes down upon a three-sixteenths plate of iron the punch must move slowly. If it moves too rapidly the hole will not be punched, and the punch itself will be smashed. If the punch moves down slowly and deliberately the particles of metal have time to move out of the way, and the hole is punched and no damage is done. In the same way if a cannon ball strikes a brick wall it will smash through the bricks, cutting some of them in half. If a lever is applied the wall breaks at the lines of mortar.

MORE EARLY BUYING.

The Gurney Foundry Company, Toronto, has issued the following circular to the trade:

During the fall months it is almost impossible for us to promptly fill all orders for repair castings called for, and to assist the prompt shipment of such orders we urge that you place a stock order with us for a few standard sets of repairs for such stoves as you have in your vicinity, of our make, that are likely to be called for. If such orders are placed at once, vexatious delays will be saved, and prompt service assured in the busier season.

HOW THE LAMP CHIMNEY WAS DISCOVERED.

The comfortable and convenient lamp chimney of every-day use is to be attributed to a child's restlessness. Argand, a native of Switzerland, a poor man, invented a lamp the wick of which was fitted into a hollow cyclinder, that allowed a current of air to supply oxygen to the interior as well as the exterior of the circular frame. The lamp was a success, but its inventor had never thought of adding a glass chimney, and probably never would have thought of it, had not his little brother been playing in his workroom while Argand was engaged with the burning lamp. The boy had gained possession of an old bottomless flask, and was amusing himself by putting it over various small articles in the room. Suddenly he placed it over the top of the lamp, and the flame instantly responded by shooting with increased brilliance up the narrow neck of the flask. Argand's ready brain at once caught the idea, and his lamp was perfected by the addition of a glass chimney.

GASOLINE STOVES.

This is the time when the gasoline stove season is on in full blast. Wholesalers report that dealers have been large buyers of high grade gasoline stoves for this summer and retailers say that business up to date has been in many cases in excess of that of any previous year.

The remarkably low prices at which these stoves are being placed in dealers' hands this year, notwithstanding the large advances which have taken place in nearly all the materials entering into their construction, make them unusually profitable to handle. They are easy to sell, for the reason that they are so much preferable to the larger coal or wood cook or range in warm weather and manufacturers have brought them to such a high state of perfection that they are very attractive.

Old Timidity Overcome.

The old timidity about gasoline stoves has pretty well disappeared. People have become educated to the fact that they are no more dangerous than any other kind of stoves, says Hardware Trade, and that only when the operator is guilty of gross and inexcusable carelessness is there chance for an accident. Fully ninety per cent. of the "gasoline stove explosions" we read about are not explosions at all, but simply cases in which the operator has been burned. Gasoline is allowed to drop down into the bottom tray and when a good deal of it has collected there a lighted match gets into it, there is a quick flaming up and a loose, flowing sleeve catches fire. Then people say "gasoline stove explosion."

The vapor gasoline stoves are wonderfully clean, neat and convenient. They are about as near a gas stove as any-one could wish for and they have so many strong points that the dealer should have no trouble in making the customer see their advantages.

There is always a chance for a good business on small one and two-burner gasoline stoves for camping outfits, as they are much more convenient for camp cooking that the more romantic camp fire.

Kerosene Stoves Popular.

There are lots of kerosene stoves being sold all the time and they are often preferred by people who cannot overcome their timidity about gasoline. There is a blue flame kerosene stove with a wick that is an enormous seller and that is pronounced by dealers to be the best thing they have ever seen.

Of course, for towns where there is gas, gas ranges are the best thing to buy and to sell. There has been a wonderful increase in the use of these ranges within the last few years and they are becoming more popular every day. Gas water heaters are a very profitable line for dealers to handle in towns where there is gas.

COPPER ALLOYS.

In a recent issue of Metallurgie,' an investigation of the alloys of copper and phosphorus by Messrs. Heyn and Bauer is published, with numerous micro-photographs of the various results obtained by them. The method of producing the alloys is described, and tables and graphic diagrams to indicate the general outcome of the experiments. The alloys were partly fused in the electro furnace and partly in a Roester gas blast furnace. There is a marked entetic line at a temperature of 707 degrees C., and a distinct entetic point with about 8.25 per cent. of phosphorus.

INDUSTRIAL VENTILATION.

In a recent article by Miss Gertrude Beeks, welfare secretary of the National Civic Federation, the subject of industrial ventilation is thus presented:

Systems of ventilation which permit a complete change of air in the workrooms at least every fifteen minutes are already becoming installed in many modern structures. Emplayers are beginning to realize the desirability of going to the expense of installing such systems in old buildings. In one notable case the cost of installation was six thousand dollars, but the reduction thereafter of the percentage of absences because of illness was so great that the employer was compensated for the outlay. He also found that while previously the employes were likely to become stupid the latter part of the afternoon, the new system maintained alertness during the entire day.

It has been found advantageous to install, even in the old cotton mills, exhaust systems for the removel of the lint resultant from the first processes of manufacture. Where it is necessary to humidify the air in the textile industry, employers have found that the cold water spray provides an atmospheric condition much more comfortable for the operatives than the steam spray during the summer season, and that heat is not essential to the successful manufacture of cotton goods. In the foundry, where the pipes which are used in the winter for heating serve to bring in the cool air in the summer, great comfort is afforded the molders.

Galvanized iron pipes carry fresh air to the faces of the men employed in rolling mills. Previously, in very hot weather, the men were frequently overcome, and sometimes it was necessary to shut down the entire mill, the workmen thereby losing their wages and the company the output. Where these ventilating systems have been installed, not a single hour's time has been lost because of the excessive heat. In a watch factory fresh air, which has been forced through sheets of water, is conducted through pipes to the faces of the young women who sit in front of the ovens baking the faces of the watches.

In the case of stationary engineers and firemen, some thoughtful employers have carried the pipes for forced ventilation above the furnace doors to prevent the firemen from baking their faces when "hauling the fires." Others have contributed greatly to their comfort by exhausting the foul, hot air and throwing fresh air into the furnace and boiler rooms. In many places this is seriously needed, because the rooms are located in the sub-cellar or interiors of structures.

Energy in selling goods is right, but is should be backed up by determination to collect the cash as soon as due.

LETTER BOX.

Correspondence on matters of interest to the hardware trade is solicited. Manufacturers, jobbers, retailers and clerks are urged to express their opinions on matters under discussion.

Any questions asked will be promptly answered. Do you want to buy anything, want some shelving, a silent salesman, any special line of goods, anything in connection with the hardware trade? Ask us. We'll supply the necessary information.

Chas. Worrod, hardware merchant, Tottenham, Ont., writes: "Kindly give me the address of the Standard Sewing Machine Co. I want to get the agency."

Ans.—"Write Standard Sewing Machine Co., Hamilton, Ontario."—Editor.

An increasing demand for small, portable oil heaters is noted by wholesalers. These are mighty handy things on cool evenings and mornings and they have many strong points which make them big sellers.

QUEBEC STOVE TRADE.

During the past year the demand for stoves and ranges throughout the province, and eastwards, has steadily increased. Not only has the number of orders increased, but a higher order of stoves is being demanded. This has naturally affected prices. The higher the class of production, the greater will be the cost of manufacture.

Generally speaking, there has been an advance of 5 per cent. over last year's cost. The difficulty in procuring efficient workmen and raw material has been the cause of this advance. It is expected that even a greater demand will prevail during the coming year as new fields have been opened, especially in northern Quebec, in the districts immediately north of Three Rivers and Quebec city. If the Grand Trunk Pacific project is completed, thereby opening up northern Quebec, the new field should prove a fruitful one to stove dealers.

One new manufactory has come into the field this summer, namely, the Thos. Davidson Mfg. Co., Montreal, famous throughout the Dominion for their various lines of hardware manufactures. Everything now from the highest class of range to the smallest teaspoon are manufactured in their extensive factories. The excellence of their lines is not less than the variety. Their works cover a large area, and their advent in the stove field is but a natural outcome of their progress and success in other fields.

Now that the range has become universally popular and is going into the homes of the poor as well as the rich, it was considered advisable by the Thos. Davidson Co., in putting a new line of stoves on the market, to do away with all such needless things as ornaments in order to lessen the cost of manufacture and, therefore, the cost to the user. The most noticeable feature in the "leader" line of ranges and stoves is the absence of ornaments. Plainness is what is aimed at. This permits of putting a better quality of material in necessary parts of the stove.

The Record Foundry Co., of Montreal, a pioneer stove company, has done a record-breaking volume of business during the past year. Their business is steadily expanding and all prospects point to an even greater volume being transacted during the coming year.

FURNACE IMPROVEMENT.

An improvement in furnaces has recently been invented by W. J. Hatcher and J. W. Crim, Johnston, South Carolina. By this construction of furnace, the inventors provide an efficient heating means, requiring but a small amount of fuel and adapted for use in or out of doors For out of door use a bottom or casing in the furnace is not needed.

DAMPER REGULATOR.

A regulator such as is used in connection with boilers and furnaces has recently been invented by J. Scales, New York, N.Y. The object of the invention is to produce a mechanism for automatically controlling the position of the damper in the flue leading from the firebox, in order to reduce the amount of draft when the boiler or furnace becomes too hot or is supplying too great a quantity of steam.

GAS REGULATORS FOR BURNERS.

In a recent invention C. F. Gaffney, New York, N.Y., aims to provide an attachment to a burner, whereby when a vessel or object to be heated is placed over an opening in the stove above the burner, a full head of gas will be automatically supplied to the burner. Upon removal of such vessel or object from over the opening the supply to the burner will be automatically reduced to a greater or lesser extent according to the set adjustment of the device, the supply cock being meanwhile open.

WATER-BACK SHIELD.

Heat radiated by a hot water boiler connected with a water back attached to a range frequently renders a kitchen uncomfortable, especially in summer, and in many cases the backs are removed at such time, and it is often necessary to open the hot water faucet so as to cool the boiler.

A device recently invented by S. M. Stevens, Asheville, N.C., dispenses with such inconvenience and also avoids heating water when not wanted, thereby economizing in fuel.

COOKING STOVE.

In an improvement to a cooking stove jointly worked out by F. Oberbeck, New Athens, and C. T. Taylor, Mount Sterling, Ill., fresh heated air is admitted to the oven, causing the evaporation to take place faster and thereby removing the moisture from the material being cooked and causing such material, particularly bread, to bake much quicker. The number of flues and dampers existing in the common form of cooking stoves is reduced, and the means for providing air circulation through the oven results in thorough, even and healthful cooking of food.

A BIG DITCH THIS.

The present digging equipment on the Panama canal consists of 63 steam shovels, 32 of 95 tons, 28 of 70 tons, and 3 of 45 tons each, while 15 further 95-ton and 7 45-ton steam shovels are to be delivered this year. There are also 184 locomotives in service, 228 steam or pneumatic drills and 73 machine or well drills. While there was a falling off in the rate of digging in May and June, due to the rain season, it is expected that 1,000,000 cubic yards a month will be reached later in the year. The total excavation necessary to dig the canal was figured at 111,280,000 cubic yards. To July the amount taken out was 8,651,802 cubic yards. At 1,000,000 cubic yards a month the channeling would be completed in 1915.

Invite the hot-air spouter to call again in December.

CATALOGUES AND BOOKLETS.

When sending catalogues for review, manufacturers would confer a favor by pointing out the new articles that they contain. It would assist the editor in writing the review.

By mentioning HARDWARE AND METAL to show that the writer is in the trade, a copy of these catalogues or other printed matter will be sent by the firms whose addresses are given.

Handsome Export Catalogue.

Hardware and Metal is in receipt this week of a handsome export catalogue issued by A. Willander, Stockholm, Sweden. This catalogue is printed in English, German, French and Spanish, indicating that export business is being aggressively sought after. The principal line illustrated consists of a variety of "Svea" cooking stoves burning petroleum. No wick is used in these stoves and the flame is free from smoke, smell or soot. The "Svea" stoves burn ordinary petroleum, which during the circulation through the heated burner is transformed into petroleum gas. This gas escapes through a fine opening called the nipple, and burns with a blue atmospheric flame.

The catalogue may be secured by writing A. Willander, Stockholm, Sweden, mentioning this paper.

Welsh Tin-Plate.

The Welsh Tinplate & Metal Stamping Co., of Llanelly, S. Wales, have recently issued their 1907 catalogue of enamelled, tinned, etc. hollowware, a copy of which has been received. This catalogue conforms to the requirements of the trade in regard to size, and all its illustrations are made direct from photographs. Every article shown in the catalogue is made in the company's works at Llanelly, and it is the opinion of these manufacturers that they make a larger variety of enamelled and tinned hollowware, than any other individual makers, either British or continental. The present issue of their catalogue certainly conveys the impression that there is a very large organization at the back of it. The list is copiously indexed and is printed in English, French, German and Spanish. The catalogue is for distribution amongst wholesale houses.

PERMANANT ROAD MAKING.

Many experiments are being made in various parts of Great Britain in permanent road making. In Wolverhampton (twelve miles from Birmingham) the borough engineer has been testing four different kinds of materials in contrast with the old-fashioned ordinary macadam, in order to see which system was the best qualified for the making of a substantial road surface in its resistance to heavy traffic, its imperviousness to wet, and a consequent accumulation of mud and dust. The materials used were what is known as tarmac—two different kinds—and two other processes alike embodying the use of broken granite and broken furnace slag, also soaked in tar. As a binding ingredient, tarmac seems to be gaining great favor.

WIRE NAILS TACKS WIRE

Prompt Shipment

The ONTARIO TACK CO.
Limited
HAMILTON ONT.

CANADA PLATES

Galvanized
Polished
Half Bright

STOCK AND IMPORT.

A. C. LESLIE & CO., Limited
MONTREAL

"MIDLAND"

BRAND.

Foundry Pig Iron.

Made from carefully selected Lake Superior Ores, with Connellsville Coke as Fuel, "Midland" will rival in quality and grading the very best of the imported brands.

Write for Price to Sales Agents

Drummond, McCall & Co.
MONTREAL, QUE.
or to
Canada Iron Furnace Co.
MIDLAND, ONT. Limited

NOVA SCOTIA STEEL & COAL CO., Limited

NEW GLASGOW, N.S.

Manufacturers of

Ferrona Pig Iron

And SIEMENS-MARTIN

OPEN HEARTH STEEL

TINPLATES

All made out of the

BEST SIEMEN'S MARTIN STEEL

Brand	
Best Coke or Charcoal	"Rhyd"
Good Charcoal	"Edith"
Second Charcoal	"Sawel"
Best Charcoal	"Dryw"

We are ready to offer the Canadian Trade very best value and finest quality in Tinplates.

Baglan Bay Tinplate Co., Limited
Briton Ferry, South Wales

Please mention above brands in your indents and inquiries.

OAKEY'S

The original and only Genuine Preparation for Cleaning Cutlery, 6d. and 1s. Canisters

'WELLINGTON' KNIFE POLISH

JOHN OAKEY & SONS, Limited
Manufacturers of
Emery, Black Lead, Emery, Glass and Flint Cloths and Papers, etc.

Wellington Mills, London, England

Agent:
JOHN FORMAN, - 644 Craig Street
MONTREAL.

NICKEL

The Canadian Copper Co.

NICKEL FOR NICKEL STEEL

The Orford Copper Co.

WRITE US FOR PARTICULARS AND PRICES

General Offices
43 EXCHANGE PLACE
NEW YORK CITY

BUILDING AND INDUSTRIAL NEWS

For additional items see the correspondence pages. The Editor solicits information from any authoritative source regarding building and industrial news of any sort, the formation or incorporation of companies, establishment or enlargement of mills, factories or foundries, railway or mining news.

Industrial Development.

The Hawes-Yongol Mfg. Co., New York, are looking for a site in Brantford, Ont.

R. T. Godman, Vancouver, B.C., will erect a mill between Point Atkinson and the Narrows.

The Superior Oil Co., Sault Ste. Marie, Ont., suffered damage by fire to the extent of $1,200.

Work has commenced on the tunnel for the Calgary Power and Transmission Co., Calgary, Alta.

Fire did damage to the building occupied by the Waters Printing Co., Montreal, to the extent of $15,000.

The Adams River Lumber Co., Shuswap, B.C., will erect a large sawmill on Little Shuswap Lake.

The North American Timber Co., St. Paul, Minn., will erect six large sawmills in British Columbia.

The shingle mill of the Valley Shingle Co., at Padden, B.C., was destroyed by fire, entailing a loss of $7,000.

A local firm are contemplating the erection of a plant for the manufacture of wooden pipe at Penticton, B.C.

Pressed Bricks, Limited, are anxious to obtain a site in Strathcona, Alta. They want an exemption from taxes.

The Victor Wood Works Co., Amherst, N.S., have gone into liquidation. The Maritime Heating Co. have also failed.

The saw and shingle mills of the Miller & Seim firm at Hampden, Ont., were destroyed by fire recently; loss, $7,000.

The Jackson Wagon Works, Galt, have been purchased by D. Clement, Ayton, Ont., who will erect a sawmill there.

The Prescott Marble & Granite Works have been purchased by Bowers Bros., marble dealers, of Ogdensburg and Prescott.

The W. P. Demond Upholstering Co. wants to secure a loan of $12,000 and free water, before erecting a plant in Strathroy, Ont.

The Rogers Mfg. Co., Kansas City, will erect a large plant at Strathroy, Ont., and manufacture malleable iron and journal boxes.

The Ryan Storm Canopy Co., Sault Ste. Marie, want to locate in Western Ontario and ask a free site and a loan of $15,000 for two years.

The Cavendish Lumber Co., Lakefield, Ont., suffered loss by fire recently to the extent of $38,000. Nearly 1,500,000 feet of lumber were destroyed.

The barns of the Southwestern Traction Co., London, Ont., were destroyed by fire, with twenty valuable motors. The total loss was about $150,000.

The work of building the plant of the cement works in Welland, Ont., is progressing rapidly. The buildings will be completed by the end of the year.

The Rocky Mountain Cement Co., Toronto, will receive tenders for engines, boilers, mills, etc., to cost $90,300. E. H. Keating is the managing director.

H. New, Hamilton, will form a company and manufacture vitrified brick and other materials. A modern plant will be erected and a high grade product will be turned out.

The Pintsch Gas Co. have secured the contract for lighting the C.P.R. and C.N.R. trains on the Edmonton branches and will erect a large gas plant at Edmonton.

An experiment is going on in Medicine Hat with natural gas. It is proposed to compress it in tanks and sell it for use in place of gasoline. It is claimed to be very much cheaper and easier to handle.

The Eadie-Douglas Co., Montreal, have been appointed Canadian agents for the waterproofing for concrete, plaster, wood, steel and iron manufactured by the Preservating Products Co., New York.

The Uxbridge Organ and Piano Co., Uxbridge, Ont., suffered severely by fire recently. The building in which the greater part of their machinery was kept was destroyed by fire, entailing a loss of $25,000.

An electric power plant may be established at Medicine Hat to operate the C.P.R. works and the pusher engines.

The Canadian Machine Telephone Co. will erect a building in Brantford for their automatic exchange.

The Schaake Machine Co., New Westminster, B.C., have secured the right to manufacture and sell a line of lath machines manufactured on a large scale by the Bolton Lath & Shingle Machinery Co., Minneapolis, Minn.

The electrical exhibition held by the Canadian Electrical Exhibition Co. will take place in Montreal for two weeks commencing Sept. 2. The annual convention of the Canadian Electrical Association will be held on Sept. 11, 12 and 13.

The Beach Mfg. Co., manufacturing the Beach Triple Expansion Engine and Cast Iron Culverts, are anxious to locate in Port Arthur, Ont., and want a free site and twenty-five per cent. of capital subscribed. The proposed capital is $100,000.

The Canadian Association of Stationary Engineers held its annual convention in Guelph on Aug. 13, 14 and 15, when the following officers were elected: President, F. R. Chowen; vice-president, W. G. Waters; treasurer, Geo. Hird; secretary, Geo. Grievson.

A company to be known as the Monitor Manufacturing Co., has been formed in Fredericton, N.B. The company will manufacture the Monitor Acetylene Generator. The capital of the company is $25,000. They have secured a site and will begin operations immediately. Lighting plants of all sizes will be manufactured.

Building Notes.

A sanitarium may be erected at Ninette, Man.

The contract for the building of the new court house at Victoria, B.C., has been awarded to McDonald, Wilson & Snyder of that city. It will cost about $400,000.

A Presbyterian church will be erected at Finch, Ont.

A large school will be erected at Haileybury, Ont.

The new C.P.R. sample room at Vancouver will cost $28,000.

H. Robb, Winnipeg, will erect an apartment block to cost $58,000.

D. Gibb & Son, Vancouver, will erect an apartment house to cost $115,000.

The new Seamen's Institute Building, at St John, N.B. will cost $15,000.

The new medical building at McGill University, Montreal, will cost $500,000.

The Robert Crean Co., Toronto, will erect a warehouse in that city to cost $10,500.

A. E. Carter, Vancouver, will erect an apartment block in that city to cost $40,000.

The trustees of the University of Toronto will erect a residence in that city at a cost of $150,000.

The corner stone of the new high school at Picton, Ont. was laid recently. The building will cost $50,000.

The congregation of St. Alban's Church, New Westminster, B.C., will shortly erect a large church, the plans having been accepted.

Companies Incorporated.

Imperial Rubber Co., Montreal; capital, $20,000; to make and deal in rubber and rubber goods. Incorporators: R. C. McMichael, D. J. Angus, R. O. McMurty, F. G. Bush, all of Montreal.

The Canadian Jack Co., Windsor, Ont.; capital, $25,000; to manufacture and sell a combination lifting jack and farmers' tool. Provisional directors: J. W. Yakey, M. Riddle, H. H. Calkins, all of Windsor.

James L. Burton & Son Lumber Co., Barrie, Ont.; capital, $250,000; to cut and deal in lumber. Provisional directors: J. L. Burton, F. L. Burton, F. C. Lett, A. Alexander, W. H. Walter, all of Barrie.

Mackie Bros., Ltd., North Bay, Ont.; capital, $100,000; to take over and conduct the hardware business of D. Purvis, North Bay. Provisional directors: A. T. Mackie, W. L. Mackie, B. S. Leeak, all of Pembroke, Ont.

Victor Automatic Carriers, Montreal; capital, $20,000; to manufacture and deal in electric motors and dynamos and other power machinery. Incorporators: F. Filteau, F. H. Markey, R. C. Grant, G. G. Hyde, all of Montreal.

The Great Northern Petroleum & Asphaltum Co., Ottawa; capital, $100,000; to prospect for and deal in oil and asphaltum. Incorporators: J. G. Gibson, H. H. Williams, M. C. Edey, W. C. Perkins, R. M. Perkins, all of Ottawa.

The Lachute Graphite Mining Co., Township of Wentworth, Que.; capital, $20,000; to mine, prepare and deal in graphite, plumbago and silver, clay bricks, marbles, and all kinds of artificial stone. Incorporators: A. Guilbault, Lachute; Z. A. Fournier, St. Andrews,

CONDENSED OR "WANT" ADVERTISEMENTS.

Advertisements under this heading 2c. a word first nsertion ; 1c. a word each subsequent insertion.

Contractions count as one word, but five figures (as $1,000) are allowed as one word.

Cash remittances to cover cost **must** accompany all advertisements. **In no case** can this rule be overlooked. Advertisements received without remittance cannot be acknowledged.

Where replies come to our care to be forwarded, five cents must be added to cost to cover postage, etc.

AGENT WANTED.

AGENT wanted to push an advertised line of Welsh tinplates; write at first to "B.B.," care HARDWARE AND METAL, 88 Fleet St., E.C., London, Eng. [tf]

BUSINESS CHANCES.

MINING hardware, stoves and tinware business for sale; sales $4,000 monthly; buildings and lot, $3,500; present stock, $8,000; proprietor's health failing; a bonanza. Box 643, HARDWARE AND METAL, Toronto. [32]

FOR SALE — Well established hardware, tinshop, implement and undertaking business, also good lumber yard, well fenced, with lumber and lime sheds in good condition; we will sell above altogether, or divide same to suit purchaser; proprietors are retiring from business in Manitoba, and therefore wish for immediate sale. Apply to Eakins & Griffin, Shoal Lake, Man. [33]

WANTED—Partner to take half interest in one of the best hardware propositions in Algoma; plumber preferred. Box 636, HARDWARE AND METAL, Toronto. [32]

HARDWARE, tinware, stove and plumbing business in manufacturing town in the Niagara Peninsnla; no competition; $250,000 factory and watermarks will be completed this summer; stock about $3,000; death of owner reason for selling. Box 85, Thorold, Ont. [tf]

FOR SALE—A good hardware business in Western Ontario; stock about $6,000. For further reference apply The Hobbs Hardware Co., Limited, London, Ont. [32]

HARDWARE and Tin Business for Sale in good Western Ontario town of 3,000: stock about $3,500; good reasons for selling. Address, Box 647 HARDWARE AND METAL, Toronto. (34)

HARDWARE, Tinware, Stove and Furnace Business for sale. in live Eastern Ontario Village; first class chance for a practical man; English speaking community; stock can be reduced to suit purchaser; can give possession September 15th. 1907; premises for Sale or Rent, Apply to D. Courville, Maxville, Ont. (35)

HARDWARE Business and Tinshop for sale in Saskatchewan; population 1500; stock carried about $14,000 turnover, $45,000 practically all cash business; cash required, $8,000 would rent building; Do not answer without you have the money and mean business; it will pay to investigate this. Box 648 Hardware & Metal, Toronto. (41)

WELL established tinsmith business in thriving Ontario village, including Eavetroughing, metal roofing, furnaces, pumps, pipefitting, hardware, paints, glass and stoves. Present stock about $1500; could be reduced. Shop and house combined, can be rented, including tools, horse and wagon, for $14.00 per month; possession given at once. Box 682, HARDWARE AND METAL. (33)

FOR SALE.

FOR SALE — First-class set of tinsmith's tools second-hand but almost as good as new; includes an 8-foot iron brick almost new. Apply Pease Waldon Co., Winnipeg. [tf]

FOR SALE—A quantity of galvanized plain twist wire. Apply to C. B. Miner, Cobden, Ont. (34)

TEN gross of "Antisplash" Tap Filters. Made to fit on all ordinary kitchen taps, acts as a filter and prevents water splashing. Supplied with showcards to hold 1 dozen. Price per dozen $1.00; price per gross $11.00. Sample on request. G. T. Cole, Box 460, Owen Sound, Ont. [36]

SITUATIONS VACANT.

TINSMITHS WANTED — First-class tinsmiths wanted for points west of Winnipeg; must be good mechanics capable of taking charge of a metal department; thorough knowledge of furnace work necessary. Pease Waldon Co., Winnipeg, Man. [tf]

WANTED—Hardware clerk, experienced, who can keep stock, and is willing to do so; sober and active; state age, experience and salary expected at start. Hose & Canniff, Kenora, Ont. [32]

WANTED—Hardware clerk, experienced and sober; state age and salary expected at start. Thomas Oliver, Copper Cliff, Ont. [33]

SITUATIONS WANTED.

SITUATION wanted as master mechanic or chief engineer by man of 22 years' experience as a mechanic; can give A1 reference as to ability; strictly temperate. Box A, HARDWARE AND METAL, Toronto. [32]

COMMERCIAL gentleman with nine years' trade connection with ironmongers, architects and public institutions in Great Britain desires position as representative of a Canadian manufacturing firm. Box X, HARDWARE AND METAL, Montreal.

WANTED.

OFFICE space wanted by manufacturer in Toronto; state location and terms. Box 645, HARDWARE AND METAL, Toronto. [32]

THE WANT AD.

The want ad. has grown from a little used force in business life, into one of the great necessities of the present day.

Business men nowadays turn to the "want ad" as a matter of course for a hundred small services.

The want ad. gets work for workers and workers for work.

It gets clerks for employers and finds employers for clerks. It brings together buyer and seller, and enables them to do business, though they may be thousands of miles apart.

The "want ad." is the great force in the small affairs and incidents of daily life.

Que.: J. R. Hyer, A. T. Woeltje, F. B. Kelly, all three of Watertown, N.Y.

The Natural Gas Supplies Co., Montreal; capital, $18,000, to carry on a plumbing and gas fitting business and deal in oil and gas wells. Incorporators: H. G. Eadie, H. P. Douglas, W. L. Bond, E. Chamberland, M. H. Barclay, all of Montreal.

Municipal Undertakings.

A water system is being installed in the Snowshoe Mine, B.C.

The ratepayers of Rapid City voted on a by-law for municipal phones.

New municipal buildings will be erected at Stirling, Alta., to cost $12,000.

A town hall and public library will be erected at Hanover, Ont. at a cost of $23,000.

A by-law has been passed at Welland, Ont., to raise $11,910 for waterworks extension

The ratepayers of Wellington, Ont. will vote on a by-law for $1,000 for sidewalk construction.

The Portage la Prairie council will ask for a grant of $30,000 for the completion of the waterworks system

A Worthington steam pump has been installed at the Point St. Charles station of the Montreal waterworks.

The work on the Trent Valley Canal has been handed over to the Dominion Government by the Ontario Government

Chatham, Ont., has let the contract to the Colonial Engineering Co., Montreal for a complete electric lighting system.

The Pelham, Ont , town council have given permission to two power companies to furnish electricity to that town.

The corporation of Berlin, Ont., have received permission to acquire the interests of the Berlin & Waterloo Street Railway.

There is a by-law before the municipality of Alliston, Ont., to grant the Lloyd & Buchanan Mfg Co a free site and a loan grant of $500. A factory will be erected to cost $10,000

Railroad Construction.

The Canadian Northern will erect a bridge over the Assiniboia river at Winnipeg.

The G.T.R. have commenced the construction of the line from Kitamaat to Kitsalas Canyon, B.C.

The Canadian Northern have placed the contract for two switching locomotives with the American Locomotive Co.

The C.P.R. will erect new roundhouses at Coleridge and Strathcona, Alta.: Swift Current, Sask., and Cranbrook, B.C.

The C.P.R. has let the contract for the line running northwest from Moose Jaw, by way of Lacombe and Edmonton.

The contract for the stores building for the T. & N.O. railroad at North Bay has been let to O'Boyle Bros. Construction Co., North Bay.

The farmers around Goose Lake, Sask., have given the Canadian Northern 52 miles of free right of way to run a line from Saskatoon to Wigins.

The Temiskaming and Northern Ontario Railway commissioners recently decided to purchase ten new locomotives from the Canadian Locomotive Co., of Kingston.

The bridge which the C.P.R. proposes to build at Lethbridge will be one of the most remarkable structures of its kind in the world. For two-thirds of its length the bridge will be over 300 feet high. There will be 22 spans 100 feet long, 44 sixty-seven feet long and one 167 feet long. The steel used in its construction will weigh 10,000 tons and will be supplied by the Canada Bridge Company, Walkerville, Ont.

TIN ORE DISCOVERED IN NOVA SCOTIA.

E. E. Bishop, connected with the well known firm of Austen Bros., is the owner of a one-half interest in the tin ore discovery at Lunenburg, Nova Scotia, which is attracting wide attention. It is the only place in Canada outside the Yukon where tin has been found, and its occurrence at granite suggests the importance of scrutinizing carefully the large area of contacts extending through the centre of Nova Scotia, and also the great granite intrusions at different districts. Valuable discoveries will probably be the result of such attention. Mr. Bishop has associated with him Messrs O'Brien and King, of Toronto, whose names are identified with the important O'Brien mine at Cobalt. The other owner of an interest in the tin prospect is the finder, Keddy Development work for investigation of the deposit is now proceeding, and the progress of the work will be watched with intense interest. Mr Faribault, assisted by Dr. Young of the Geological Survey, have been directed to study the tin occurrence closely, and are now on the ground. The ore is cassiterite and appears at a dyke of pegmatite John Reeves, Benjamin Meister, E. E. Bishop, and Charles Keddy took up the areas under license to search, Oct. 26, 1906

A STOLEN TRADE SECRET.

The manufacture of tinware in England originated in a stolen secret. Few readers need to be informed that tinware is simply thin sheet iron plated with tin by being dipped into the molten metal. In theory it is an easy matter to clean the surface of iron. Dip the iron in a bath of boiling tin and remove it enveloped in the silvery metal to a place of cooling. In practice, however, the process is one of the most difficult of arts. It was discovered in Holland and guarded from publicity with the utmost vigilance for nearly half a century. England tried to discover the secret in vain until James Sherman, a Cornish miner, crossed the channel, insinuated himself surreptitiously into a tin plate manufactory, made himself master of the secret and brought it home.

Are you a consistent member of the home patronage club ? Do you buy lines not found in your store of your home merchants ?

BEST ELBOWS

— FOR —

CONDUCTOR PURPOSES

BEAR TRADE MARK

F. Dieckmann

PLAIN ROUND.
CORRUGATED.

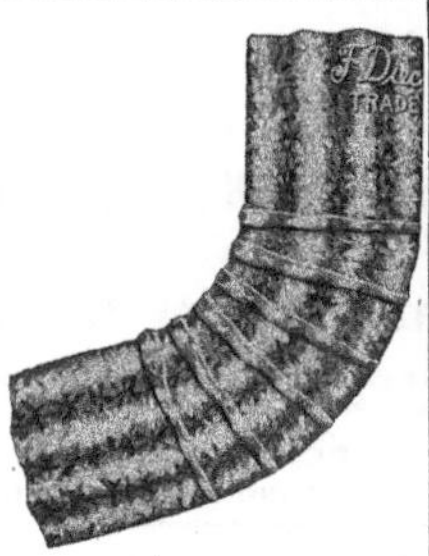

For Sale by the TRADE in Canada. Write for catalogue and samples to

JOBBERS AND

FERDINAND DIECKMANN
1180-82-84 HARRISON AVE.
CINCINNATI, O., U.S.A.

Pertinent Pease Points

From January 1st to August 1st, 1907, many more Pease Heaters have been shipped than during the same months of any previous year.

On August 1st, 1907, The Pease Company had vastly more orders entered for Fall delivery than on the same date in any previous year.

At the beginning of August, 1907, The Pease Company are better equipped than ever before to fill orders promptly; to satisfy every requirement of the Heating Trade and to maintain their pre-eminent position in the Heating business in Canada.

We have a new illustrated price list. Would you like to have one?

Pease-Waldon Co., Ltd.
Winnipeg

Pease Foundry Co., Ltd.
Toronto

News of the Paint Trade

IRON PAINTING.

There are many cases of the rusting of iron. It may be produced by atmosphere action alone, but in a majority of cases galvanism plays a large part in the destruction of the metal. Long experience has shown how rapidly iron nails employed in fastening sheets of lead and copper upon roofs are destroyed, the other, the electro-negative metal, remaining comparatively unaffected. The electrolyte, or exciting fluid, which by acting on the iron and not on the other metal or by acting more upon the former than upon the latter, causes the electric current, in either actual water from rain or snow or the water vapor always present in the atmosphere. The decomposition of the water causes the liberation of oxygen at the positive pole, which is the iron, and this nascent oxygen rapidly combines with the iron. Now, it is claimed that red lead is an excellent material for protecting iron from rust and electrical action. Unfortunately, however, red lead is more electro-negative to iron than either copper or lead, Hence, should moisture by any chance get between the red lead and the iron, the destruction by rust is more rapid than when iron is in contact with copper or lead. This electro-chemical action is at the same time strengthened by the purely chemical action between the red lead and the carbonic acid always present in the air, an action which converts the red lead into ceruse, whereby an additional quantity of nascent oxygen is set free to rust the iron. It is also highly probable that the carbonic acid has an independent action upon the iron, thereby much facilitating its oxidation. It must not be forgotten that every porous place and still more every crack in the paint becomes sooner or later an entrance for water and carbonic acid. A good oil varnish is by far the best protection for iron, but it must, of course, be properly used. Not only must the iron be scrupulously, practically and chemically clean and dry when the varnish is applied, but the covering must be without a flaw. Varnish will not adhere to greasy, rusty, or wet iron, and the contraction of the varnish on drying will cause minute cracks at such places and the iron-destroying gases will find their way through these cracks and get between the iron and the non-adherent varnish.—Building Management.

AUSTRALIAN WHITE LEAD RULING.

The following notification was recently issued from the Commonwealth Government office in London: "A recent importation of 'white lead' was described on the kegs as 'Ground white lead reduced.' Analysis showed that the white lead was adulterated with barium to the extent of about 25 per cent. The

term 'reduced' does not, in this case, sufficiently describe 'adulteration' for the purpose of the Commerce Act, although in the trade it has that meaning. In future, the percentage of barium, or any other adulterant in white lead, must be as clearly and as permanently shown on the covering of the goods as the ordinary trade description applied thereto and delivery is not to be permitted until this condition is fulfilled."

The Melbourne Journal of Commerce says: "It is satisfactory to be able to report anent the recent agitation against alleged adulterated white lead, that an analysis of samples drawn from six different shipments of various brands by the customs department has satisfied the authorities that they were pure, some of them going as high as 96 per cent. carbonate lead. The customs department has, however, notified that it will not be sufficient to state that on mixed lead that it is 'reduced.' The proportion of the mixture must be specified plainly on packages."

PAINT AND OIL MARKETS

MONTREAL.

Montreal, August 16.—Market conditions on the whole are quite satisfactory Important contracts have been placed in Montreal during the past week. These include mixed paints and special contracts for corporations. The activity referred to last week is being well maintained, and a general feeling of satisfaction amongst manufacturers exists in regard to the amount of business being done. No changes of importance, with the exception of linseed oil, have been made.

Linseed Oil—This continues to slowly weaken. Prices this week are marked down 2c. The new prices are: Raw, 1 to 4 barrels, 60c; 5 to 9, 59c.; boiled, 1 to 4 barrels, 63c.; 5 to 9, 62c.

Turpentine—Has taken an upward turn after the long period of depression. Conditions at Savannah are much improved. Local quotations are unchanged: Single barrels, 80c.

Ground White Lead—Is in rather better demand. Local corroding work is fully implied, and grinders find it moving more steadily. Prices are firm: Government standard, $7.50; No. 1, $7; No. 2, $6.75; No. 3, $6.35.

Dry White Zinc—The situation is unchanged; the demand is moderate and prices are firm: V.M. Red Seal, 7½c.; Red Seal, 7c.; French V.M., 6c.; Lehigh, 5c.

White Zinc Ground in Oil—The demand is steady with unchanged prices: Pure, 8½c.; No. 1, 7c.; No. 2, 5⅝c.

Red Lead—The volume of business done at present is somewhat diminished. Prices are firm and unchanged: Genuine red lead, in casks, $6.25; in 100-lb. kegs, $6.50; in less quantities at $7.25 per 100 lbs. No. 1 red lead, casks, $6; kegs, $6.25, and smaller quantities, $7.

Gum Shellac—Prices are firm and unchanged During the summer season little demand is expected, and conditions are up to expectations Fine orange, 60c. per lb; medium orange, 55c. per lb, white (bleached), 65c.

Shellac Varnish — The demand is steady and prices are firm and unchanged. Pure white bleached shellac, $2.80, pure orange, $2.60; No. 1 orange, $2.40.

Putty—Grinders are very busy supplying the demand, which continues strong. Prices are unchanged: Pure linseed oil, $1.85 bulk; in barrels, $1.60; in 25-lb. irons, $1.90; in tins, $2; bladder putty, in barrels, $1.85.

Paris Green—The season is now practically over and stocks are almost exhausted. While it lasted a large volume of business was done in this commodity. The season, however, was of short duration owing to lateness of spring

TORONTO.

Toronto, August 16.—A fair August business is being transacted, but compared with a few weeks ago trade is rather inactive. Jobbers do not expect to be rushed at this season of the year and apparently are satisfied with the amount of mid-August trade that is coming in. The sale of mixed paints is largely over for this season, and the enquiries that are now coming in are mostly for small quantities to enable retailers to sort up their stocks in those colors in which they are low. Despite the continued cool weather of the early summer, this has been a splendid season for the sale of paints and oils; in fact, the managers of local supply houses are quite gratified by the amount of business that has been done.

White Lead—A few enquiries are coming in and trade is all that can be expected for August. Prices remain firm and unchanged as follows: Genuine pure white lead is quoted at $7.65, and No. 1 is held at $7.25.

Red Lead—A fair amount of orders for red lead in sorting quantities continue to arrive. There are no alterations in prices which are still quoted as follows: Genuine, in casks of 500 lbs., $6.25; ditto, in kegs of 100 lbs., $6.75; No. 1, in casks of 500 lbs., $5, ditto, in kegs of 100 lbs., $5.50.

Paris Green—The season is now practically over, though the hot weather which prevailed during the first part of the week had the effect of temporarily stimulating the demand, and even yet a few repeat orders continue to come in from various parts of the province. Prices are unchanged and are quoted as follows: Canadian Paris green, 29½c base; English Paris green, 30½c.

Petroleum—The demand is a trifle stronger but has not yet increased sufficiently to warrant any change in prices. Prime white, 13c.; water white, 14½c.; Pratt's astral, 18c.

Shellac—Trade in shellac is quiet as it usually is at this time in August. There are no changes in prices. Pure orange, in barrels, $2.70; white, $2.82½ per barrel; No. 1, (orange), $2.50.

Turpentine—Local prices are fairly steady, though one firm is quoting one.

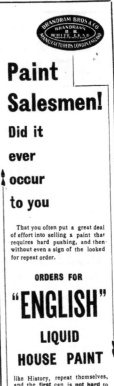

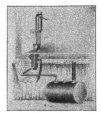

cent lower than the market. On account of the telegraph operators' strike, it is impossible to report the latest developments of the southern market, but the last intelligence received by letter was to the effect that the market there had slightly declined. Last year turpentine began to advance in July and the upward movement of prices continued till September; so far there has been no corresponding advance this year, but lately prices in the south have been fluctuating up and down to such an extent that it is exceedingly difficult to give a reliable forecast of the market. The local demand is good for this time of the year and the following are the prevailing prices: Single barrels, 79c.; two barrels and upwards, 78c. f.o.b. point of shipment, net 30 days; less than barrels, 5c. advance. Terms: 2 per cent., 30 days.

Linseed Oil—The Canadian market has been considerably undermined by large importations of English oil at Montreal, and consequently, a two cent drop has resulted locally, though a few dealers are still quoting last week's figures. We now quote: Raw, 1 to 3 barrels, 63c.; 4 barrels and over, 62c. Add 3c. to this price for boiled oil f.o.b. Toronto, Hamilton, London and Guelph, 30 days.

For additional prices see current market quotations at the back of the paper.

MAKING ZINC-LEAD PIGMENT.

In the manufacture of zinc-lead pigment all grades of ore can be used, varying from those containing no lead up to those containing equal proportions of zinc and lead. It has been found by mixing ores from various districts that an average will contain one-third lead and two-thirds zinc. From such a charge a pigment is produced containing very nearly one-third basic lead sulphate and two-thirds zinc oxide. In preparing the charge, ores high in lead are mixed with ores low in lead, the amount of lead that may be used being calculated from the zinc content as a basis.

The ores, says Wm. F. Gordon, in the Engineering and Mining Journal, are sized and carefully roasted to a proper sulphur content, the amount of sulphur left in the ore depending on the amount needed to combine with the lead to form basic lead sulphate.

After roasting, the ore is taken to the pigment furnaces, and the zinc and lead are volatilized, the zinc going through the reaction as above described, and producing zinc oxide. The lead, which theoretically should also produce an oxide, does not do so, but unites with the sulphur left in the ore and forms basic lead sulphate. I have studied this phenomenon to a considerable extent, and am of the opinion that the lead is also volatilized and, having a greater affinity for sulphur than for oxygen, combines as the sulphate. The reason for so believing is that an ore charge of zinc and lead carbonate may be used and sufficient sulphur in form of iron pyrite be added. It is hardly probable that a chemical combination takes place between the lead as carbonate, and the sulphur in the pyrite before they are volatilized.

After passing from the pigment furnaces the floating zinc and lead is conducted as in the zinc oxide process to the bag room. In this process about 150 square feet of muslin in the bag room is used for every square foot of grate surface in the pigment furnaces.

In concluding his article on this subject Mr. Gordon says:

"The zinc-lead pigment is one of great beauty. Its covering capacity is equal to the same amounts of white lead (carbonate) and zinc oxide. Since neither the zinc oxide nor lead sulphate is affected by atmospheric conditions it is very durable It does not darken by contact with sulphur gases, and therefore holds its color in coal burning cities."

TESTS FOR LUBRICATING OILS.

The viscosity and other characteristics of oils used for lubricating form the subject for frequent experiment, and the use of viscometers is extending. According to a daily paper, the simplest means of testing the viscosity of an oil is to measure the relative time taken for the oil and water to flow through an orifice. Accurate tests of the viscosities of oil as compared with water have shown that sperm oil is one of the least viscous, and rape seed oil is about double that of sperm. Moisture in oil, which is especially objectionable for oil cooled transformers, may be readily detected by immersing a hot wire in the oil when, if there be moisture present, a crackling sound will be heard, and simply a puff of smoke if the oil is dry.

A NEW LUBRICANT.

Bean cake is an important production in Japan and China. It is used chiefly for fertilizing, and to some extent for the feeding of live stock. Bean oil is also an important article of commerce. Whether bean oil is a by-product of the manufacture of bean-cake, or vice versa, is a question that is answered differently by different people. But a consular report from Mukden is authority for the statement that "one of the principal exports of Manchuria is bean oil, whose use at the present time is largely for illuminating and cooking purposes. The manager of the Mitsui Bussan Kaisha at Tieling, however, believes that by combining it with a small percentage of some other oil it will be possible to produce a very satisfactory and cheap lubricant, and states that his company is at present performing experiments to that end."

To remove oil paint from tin goods in the case of fresh paint, rub off with oil of turpentine or petroleum. Otherwise, use hot, saturated solution of potash, hot water afterward. The most powerful means is caustic soda lye.

CURRENT MARKET QUOTATIONS.

August 16, 1907

These prices are for such qualities and quantities as are usually ordered by retail dealers on the usual terms of credit, the lowest figures being for larger quantities and prompt pay. Large cash buyers can frequently make purchases at better prices. The Editor is anxious to be informed at once of any apparent errors in this list, as the desire is to make it perfectly accurate.

METALS.

ANTIMONY.

Cookson's .... per lb.	0 15
Hallett's	0 14½

BOILER PLATES AND TUBES.

Plates, ¼ to ½ inch, per 100 lb.	2 40	2 50
Heads, per 100 lb	2 65	2 75
Tank plates 3-16 inch	2 60	2 70
Tubes per 100 feet, 1½ inch	8 25	8 50
" " 2 "	9 00	9 10
" " 2½ "	10 50	11 00
" " 3 "	12 00	12 50
" " 3½ "	15 00	16 00
" " 4 "	19 25	20 00

BOILER AND T.K. PITTS.

Plain tinned and Spun, 25 per cent. off list.

BABBIT METAL.

Canada Metal Company—Imperial, genuine 60c; Imperial Tough, 56c.; White Brass, 50c.; Metallic, 35c.; Harris Heavy Pressure, 25c.; Hercules, 25c; White Bronze, 15c.; Star Frictionless, 14c; Alluminoid, 10c; No 4, 9c. per lb.

James Robertson Co.—Extra and genuine Monarch, 60c.; Crown Monarch, 50c; No. 1 Monarch, 40c, King, 30c; Fleur-de-lis, 20c; Thurber, 15c; Philadelphia, 12c; Canadian, 10c; hardware, No. 1, 15c; No. 2, 12c; No. 3, 10c per lb.

BRASS.

Rod and Sheet, 14 to 30 gauge, 25 p.c. advance.

Sheets, 12 to 14 in.	0 30
Tubing, base, per lb 5-16 to 2 in	0 33
Tubing, ¼ to 3-inch, iron pipe size	0 31
" ¼ to 3-inch, seamless	0 36

Copper tubing, 6 cents extra

COPPER. Per 100 lb.

Casting ingot	22 50	23 50
Cut lengths, round, bars, ½ to 2 in.		35 00
Plain sheets, 14 oz.		36 00
Plain, 16 oz., 14x48 and 14x60		35 00
Tinned copper sheet, base		38 00
Planished base		43 00
Braziers' (in sheets), 4x6 ft., 25 to 30 lb. each, per lb., base	0 34	0 35

BLACK SHEETS.

	Montreal.	Toronto
to 10 gauge	2 70	2 75
12 gauge	2 70	2 75
14 "	2 50	2 60
17 "	2 50	2 60
18 "	2 50	2 60
20 "	2 50	2 60
22 "	2 55	2 65
24 "	2 55	2 70
26 "	2 65	2 85
28 "	2 70	3 00

CANADA PLATES

Ordinary, 52 sheets	2 75	3 05
All bright "	3 75	4 05

Galvanized—	Dom Crown.	Ordinary.
18x24x52	4 45	4 35
60	4 70	4 60
20x28x80	8 90	8 70
....	9 40	9 20

GALVANIZED SHEETS.

B.W. gauge	Queen's Head	Fleur-de-Lis	Gordon Crown	Colborne Crown Gorbal's Best
16-20	3 95	3 80	3 95	....
22-24	4 20	4 05	4 00	4 05
26	4 45	4 30	4 40	4 30
28	4 70	4 55	4 60	4 55

Less than case lots 10 to 25c. extra.

Apollo Brand.

24 guage, American	3 85
26 " "	4 10
28 " "	4 55
10¾ oz.	4 85

25c. less for 1,000 lb. lots.

IRON AND STEEL.

	Montreal.	Toronto.
Middlesboro, No. 1 pig iron	21 50	24 50
Middlesboro, No 3 pig iron	20 50	23 50
Summerlee, "	25 50	26 50
" special "	24 50	....
" soft "	24 00	....
Carron	26 00	....
Carron Special	24 50	....
Carron Soft	24 00	....
Clarence, No. 3	21 50	23 50
Glengarnock, No. 1	....	27 00
Midland, Londonderry and Hamilton, off the market but quoted nominally at	...	26 00
Radnor, charcoal iron	32 00	34 00
Common bar, per 100 lb.	2 20	2 30
Lowmoor iron		6 50
Angles		2 50
Forged iron "	2 45	..
Refined " "	2 60	2 70
Horseshoe iron "	2 60	2 70
Band iron, No. 10 x ⅜ in.	....	3 60
Sleigh shoe steel "	2 25	2 30
Iron finish steel		2 40
Reeled machinery steel		3 50
Tire steel	2 40	2 50
Best sheet steel	...	0 12
Mining cast steel	....	0 08
Warranted cast steel		0 14
Annealed cast steel	....	0 15
High speed	....	0 60
B.P.L. tool steel	10½	0 11

INGOT TIN.

Lamb and Flag and Straits—56 and 28-lb. ingots, 100 lb.	$42 00	$44 00

TINPLATES.

Charcoal Plates—Bright

M.L.S., Famous (equal Bradley)		Per box
I C,14 x 20 base	...	$6 50
IX, 14 x 20 "	....	8 00
IXX, 14 x 20 base	....	9 50

Raven and Vulture Grades—

I C, 14 x 20 base	5 00
I X "	6 00
I X X "	7 00
I X X X "	8 00

"Dominion Crown Best"—Double Coated, Tissued. Per box.

I C, 14 x 20 base	5 50	5 75
I X, 14 x 20 "	6 50	6 75
IXX x 20 "	7 50	7 75

"Allaway's Best"—Standard Quality.

I C, 14 x 20 base	4 65	5 00
I X, 14 x 20 "	5 40	5 75
I XX, 14 x 20 "	6 15	6 50

Bright Cokes.

Bessemer Steel—

I C., 14 x 20 base	4 25	4 35
20x28, double box	8 50	8 70

Charcoal Plates—Terne

Dean or J. G. Grade—

I.C., 20x28, 112 sheets	7 25	8 00
IX., Terne Tin	...	9 50

Charcoal Tin Boiler Plates.

Cookley Grade—

X X, 14x56, 50 sheet bxs. / " 14x60, " / " 14x65, "	7 50

Tinned Sheets.

72x30 up to 24 gauge	8 50
" " 26 "	9 00

LEAD.

Imported Pig, per 100 lb.	5 25	5 35
Bar, "	5 75	6 00
Sheets, 2½ lb. sq. ft., by roll	....	6 50
Sheets, 3 to 6 lb. "	....	6 25

Cut sheets ½c. per lb., extra.

SHEET ZINC.

5-cwt. casks	7 75	8 00
Part casks	8 00	8 25

ZINC SPELTER.

Foreign, per 100 lb	6 75	7 00
Domestic	6 50	6 75

COLD ROLLED SHAFTING.

9-16 to 11-16 inch	0 06
¾ to 1 7-16 "	0 05½
1 7-16 to 3 "	0 05

30 per cent

OLD MATERIAL

Dealers buying prices:

	Montreal	Toronto
Heavy copper and wire, lb	0 17	0 15
Light copper	0 15	0 13
Heavy red brass	0 15	0 13
" yellow brass	0 12	0 10½
Light brass	0 08	0 06½
Tea lead	0 03¾	0 03¾
Heavy lead	0 04	0 04
Scrap zinc	0 03¾	0 03¾
No 1 wrought iron	15 50	11 50
" 2 " "	6 00	6 00
Machinery cast scrap	17 00	16 50
Stove plate	13 00	12 00
Malleable and steel	8 00	8 00
Old rubbers	0 10½	0 10
Country mixed rags, 100 lbs.	1 00	1 25

PLUMBING AND HEATING

BRASS GOODS, VALVES, ETC.

Standard Compression work, 57½ per cent.

Cushion work, discount 40 per cent.

Fuller work, 70 per cent.

Flatway stop and stop and waste cocks, 60 per cent; roundway, 55 per cent.

J.M.T. Globe, Angle and Check Valves, 45, Standard, 55 per cent

Kerr standard globes, angles and checks, special, 42½ per cent.; standard, 47½ p.c.

Kerr Jenkins' disc, copper-alloy disc and heavy standard Valves, 40 per cent.

Kerr steam radiator valves 60 p.c., and quick-opening hot-water radiator Valves, 60 p.c.

Kerr brass, Weber's straightway Valves, 41; straightway Valves, I.B.B.M., 60.

J.M.T. Radiator Valves 50; Standard, 60; Patent Quick-Opening Valves, 65 p.c.

Jenkins' Valves—Quotations on application to Jenkins' Bros., Montreal.

No 1 compression bath cock ... net	$2 00
No. 4 " " "	1 90
No 7 Fuller's "	2 25
No. 4½ " "	2 35
Patent Compression Cushion, basin cock, hot and cold, per doz.	$16 20
Patent Compression Cushion, bath cock, No. 2208	2 25

Square head brass cocks, 50; iron, 60 p.c.

Thompson Smoke-test Machine 25.00

BOILERS—COPPER RANGE

Copper, 30 gallon, $33, 15 per cent.

BOILERS—GALVANIZED IRON RANGE.

30-gallon, Standard, $5; Extra heavy, $7.75

BATH TUBS.

Steel clad copper lined, 15 per cent.

CAST IRON SINKS.

16x24, $1; 18x30, $1; 18x36, $1.30.

ENAMELED BATHS, ETC.

List issued by the Standard Ideal Company Jan. 3, 1907, shows an advance of 10 per cent. over previous quotations.

ENAMELED CLOSETS AND URINALS

Discount 15 per cent.

HEATING APPARATUS

Stoves and Ranges—40 to 70 per cent.

Furnaces—45 per cent.

Registers—70 per cent

Hot Water Boilers—50 per cent.

Hot Water Radiators—50 to 55 p.c.

Steam Radiators—50 to 55 per cent

Wall Radiators and specials—50 to 55 p.c.

LEAD PIPE

Lead Pipe, 5 p.c. off

Lead waste, 5 p.c. off

Caulking lead, 6½c. per pound.

Traps and bends, 40 per cent.

IRON PIPE.

Size (per 100 ft.)	Black.	Size	Galvanized
¼ inch	2 35	¼ inch	3 20
⅜ "	2 35	⅜ "	3 20
½ "	2 90	½ "	3 75
¾ "	3 90	¾ "	5 00
1 "	5 60	1 "	7 25
1¼ "	7 65	1¼ "	9 90
1½ "	9 20	1½ "	11 90
2 "	12 25	2 "	15 80
2½ "	20 10	2½ "	26 00
3 "	26 75	3 "	34 00
3½ "	34 25	3½ "	42 75
4 "	39 00	4 "	48 60

Malleable Fittings—Canadian discount 30 per cent.; American discount 25 per cent.

Cast Iron Fittings 57½; Standard bushings 57½; headers, 57½; flanged unions 57½, malleable bushings 55; nipples, 70 and 10; malleable lipped unions, 55 and 5 p.c.

SOIL PIPE AND FITTINGS

Medium and Extra heavy pipe and fittings, up to 6 inch, 60 and 10 to 70 per cent.

7 and 8-in. pipe, 40 and 5 per cent.

Light pipe, 50 p.c.; fittings, 50 and 10 p.c.

OAKUM.

Plumbers ....per 100 lb.	4 50	5 00

STOCKS AND DIES.

American discount 25 per cent.

SOLDERING IRONS.

½-lb. to 1½ ...... per lb.	0 45½	0 48
2-lb. or over "	0 42½	0 46

SOLDER. Per lb.

	Montreal	Toronto
Bar, half-and-half, guaranteed	0 25	0 26
Wiping	0 22	0 23

PAINTS, OILS AND GLASS.

BRUSHES

Paint and household, 70 per cent.

CHEMICALS.

	In casks per lb.
Sulphate of copper (bluestone or blue vitrol)	0 09
Litharge, ground	0 06
" flaked	0 06½
Green conperas (green vitrol)	0 01
Sugar of lead	0 08
Lump olive	0 01½

COLORS IN OIL.

Venetian red, 1-lb. tins pure.	0 08
Chrome yellow "	0 15
Golden ochre "	0 17
French " "	0 08
Marine black "	0 04½
Chrome green "	0 09
French permanent green "	0 13
Signwriters' black "	0 15

GLUE.

Domestic sheet	0 10	0 10½
French medal	0 12	0 12½

PARIS GREEN.

	Berger's	Canadian
600-lb. cask	....	0 27½
250-lb. drums	0 28	0 27½
100-lb. "	0 29½	....
50-lb. "	0 29½	0 28½
1-lb. pkgs, 100 in box	0 30½	0 29½
½-lb. "	0 32½	0 31½
1-lb. tins, 100 in box	0 31½	0 30½
½-lb. tins	0 33½	0 32

PARIS WHITE.

In bbls	0 90

PIGMENTS.

Orange mineral, casks	....	0 08
" " 100-lb kegs	....	0 08½

PREPARED PAINTS.

Barn (in bbls.)	0 65	0 90
Sherwin-Williams paints,	....	1 55
Canada Paint Co.'s pure	....	1 40
Standard P. & V. Co.'s "New Era."		1 30
Benj. Moore Co.'s "Ark" B'd	....	1 25
" British Navy deck	....	1 50
Brandram-Henderson's "English"		1 45
Ramsay's paints, Pure, per gal.	....	1 30
" Thistle, "	....	1 10
Martin-Senour's 100 p.c. pure,	....	1 55
Senour's Floor Paints	....	1 35
Jamieson's "Crown and Anchor"		1 20
Jamieson's floor enamel	....	1 50
"Island City" paint	....	1 25
Sanderson Pearcy's, pure	....	1 20
Robertson's pure paints	....	1 20

PUTTY.

Bulk in bbls	1 60
Bladders in bbls	1 85
25-lb. tins	1 90
Bladders in bulk or tins less than 100 lb	2 00
Bulk in 100-lb. irons	1 80

SHINGLE STAINS.

In 5 gallon lots	0 85	0 90

SHELLAC.

White, bleached	....	2 65
Fine orange	....	2 60
Medium orange	....	2 55

TURPENTINE AND OIL.

Prime white petroleum	....	0 13
Water white "	....	0 14½
Pratt's astral "	....	0 18
Castor oil	0 08	0 10
Gasoline	....	0 22½
Benzine, per gal	0 17	0 20
Turpentine, single barrels	0 78	0 79
Linseed Oil, " raw	0 61	0 64
" " boiled	0 64	0 67

WHITE LEAD GROUND IN OIL. Per 100 lbs

Canadian pure	7 15	7 50
No. 1 Canadian	6 80	7 15
Munro's Select Flake White	....	7 65
Elephant and Decorators' Pure	....	7 65
Monarch	....	7 40
Standard Decorator's	....	7 15
Essex Genuine	....	6 80
Brandram's B. B. Genuine	....	8 70
"Anchor," pure	....	7 40
Ramsay's Pure Lead	....	7 00
Ramsay's Exterior	....	6 65
"Crown and Anchor," pure	....	7 25
Sanderson Pearcy's	....	7 40
Robertson's C.P., lead	....	7 20

RED DRY LEAD.

Genuine, 560 lb. casks, per cwt	....	6 25
Genuine, 100 lb. kegs, "	....	6 50
No. 1, 560 lb. casks, per cwt	....	6 00
No. 1, 100 lb. kegs, per cwt	....	6 23

WINDOW GLASS

Size United inches.	Star	Double Diamond
Under 26	$4 25	$6 25
26 to 40	4 65	6 75
41 to 50	5 10	7 50
51 to 60	5 35	8 50
61 to 70	5 75	9 75
71 to 80	6 25	11 00
81 to 85	7 00	12 50
86 to 90	....	15 00
91 to 95	....	17 50
96 to 100	....	20 50
101 to 105	....	24 00
100 to 110	....	27 50

Discount—16-oz., 25 per cent; 21-oz. 30 per cent. per 100 feet. Broken boxes 50 per cent.

WHITING.

Plain, in bbls	....	0 70
Gilders bolted in bands	....	0 90

WHITE DRY ZINC.

Extra Red Seal, V.M.	0 07½	0 08

WHITE ZINC IN OIL.

Pure, in 25-lb. irons	....	0 08½
No. 1, " "	....	0 07
No. 2, " "	....	0 05½

VARNISHES.

		Per gal cans.
Carriage, No. 1	....	1 50
Pale durable body	....	3 50
" hard rubbing	....	3 00
Finest elastic gearing	....	3 00
Elastic oak	....	1 50
Furniture, polishing	....	2 00
Furniture, extra	....	1 15
" " No. 1	....	0 90
" " union	....	0 80
Light oil finish	....	1 40
Gold size japan	....	1 80
Brown japan	....	0 95
No 1 brown japan	....	0 85
Baking black japan	....	1 35
No. 1 black japan	....	0 90
Benzine black japan	....	0 70
Crystal Damar	....	2 80
No. 1 "	....	2 50
Pure asphaltum	....	1 40
Oilcloth	....	1 50
Lightning dryer	....	0 70
Elastilite varnish, 1 gal. can, each		2 00
Granitine floor Varnish, per gal	....	2 50
Maple Leaf coach enamels; size 1,		1 20
Sherwin-Williams' kopal Varnish, gal.,		2 50
Canada Paint Co's sun Varnish	....	2 00
"Kyanize" Interior Finish	....	2 40
"Flint-Lac," coach	....	1 80
B H Co's "Gold Medal," in cases		2 00
Jamieson's Copaline, per gal.	....	2 00

BUILDERS' HARDWARE.

BELLS.

Brass hand bells, 60 per cent.
Nickel, 55 per cent.

Gongs, Sargeant's door bells	5 50	8 00
American, house bells, per lb.	0 35	0 40

Peterboro' door bells, 37½ and 10 off new list.

BUILDING PAPER, ETC.

Tarred Felt, per 100 lb.		2 25
Ready roofing, 2-ply, not under 45 lb. per roll		1 00
Ready roofing, 3-ply, not under 65 lb., per roll		1 25
Carpet Felt	per ton	60 00
Heavy Straw Sheathing	per ton	40 00
Dry Surprise		0 45
Dry Sheathing	per roll, 400 sq. ft.	0 40
Tar "	" 400 "	0 50
Dry Fibre	" 400 "	0 55
Tarred Fibre	" 400 "	0 65
O. K. & I. X. L.	" 400 "	0 70
Resin-sized	" 400 "	0 45
Oiled Sheathing	" 600 "	1 00
Oiled "	" 400 "	0 70
Roof Coating, in barrels	per gal.	0 17
Roof " small packages	"	0 25
Refined Tar	per barrel	5 00
Coal Tar	"	4 00
Coal Tar, less than barrels	per gal.	0 15
Roofing Pitch	per 100 lb. 0 80	0 90
Slater's felt	per roll	0 70
Heavy Straw Sheathing f. o. b. St. John and Halifax		42 50

BUTTS.

Wrought Brass, net revised list.
Wrought Iron, 70 per cent.
Cast Iron Loose Pin, 60 per cent.
Wrought Steel Fast Joint and Loose Pin, 70 per cent.

CEMENT AND FIREBRICK.

Canadian Portland	2 00	2 10
Belgium	1 60	1 90
White Bros. English	1 80	2 05
"Lafarge" cement in wood		3 40
"Lehigh" cement, in wood		2 54
"Lehigh" cement, cotton sacks		2 39
"Lehigh" cement, paper sacks		2 31
Fire brick, Scotch, per 1,000	27 00	30 00
" English	17 00	21 00
" American, low	23 0	25 00
" " high	27 50	35 00
Fire clay (Scotch), net ton		4 95

Paving Blocks per 1,000.

Blue metallic, 9"x4½"x3", ex wharf	35 00
Stable pavers, 12"x6"x2", ex wharf	50 00
Stable pavers, 9"x4½"x3", ex wharf	36 00

DOOR SETS.

Peterboro, 37½ and 10 per cent.

DOOR SPRINGS.

Torrey's Rod .... per doz.	....	1 75
Coil, 9 to 11 in. "	0 95	1 65
English "	2 00	4 00

Chicago and Reliance Coil 25 per cent.

ESCUTCHEONS.

Discount 50 and 10 per cent., new list
Peterboro, 37½ and 10 per cent.

ESCUTCHEON PINS.

Iron, discount 40 per cent.
Brass, 45 per cent.

HINGES.

Blind, discount 50 per cent.

Heavy T and strap, 4-in., per lb. net	0 06
" " 5-in., "	0 05¾
" " 6-in., "	0 05½
" " 8-in., "	0 05¼
" " 10-in. and larger	0 05

Light T and strap, discount 65 p.c.
Screw hook and hinge—

under 12 in. .... per 100 lb.	4 65
over 12 in. "	3 65

Crate hinges and back flaps, 65 and 5 p. c.
Hinge hasps, 65 per cent.

SPRING HINGES.

Spring, per gro., No. 5, $17 50 No. 10, $18; No. 20, $10.80; No. 120, $20; No. 51, $10; No. 50, $27.50
Chicago Spring Butts and Blanks 12½ per cent.
Triple End Spring Butts, 30 and 10 per cent.
Chicago Floor Hinges, 37½ and 5 off.
Garden City Fire House Hinges, 12½ p c.
"Chief" floor hinge, 50 p.c.

CAST IRON HOOKS.

Bird cage .... per doz.	0 50	1 10
Clothes line, No. 61 "	0 00	0 10
Harness "	0 60	12 00
Hat and coat .... per gro.	1 10	10 00
Chandelier .... per doz.	0 50	1 00

Wrought hooks and staples—

¼ x 5 .... per gross	....	2 65
5-16 x 5 "	....	3 30

Bright wire hooks, 60 p c.
Bright steel gate hooks and staples, 40 p.c.
Cresent hat and coat wire, 60 per cent.
Screw, bright wire, 65 per cent.

KNOBS.

Door, japanned and N.P., doz	1 50	2
Bronze, Berlin .... per doz.	2 75	3 50
Bronze, Genuine "	6 00	9 00
Shutter, porcelain, F. & L. screw .... per gross	1 30	2 00
White door knobs .... per doz.	....	2 00

Peterboro knobs, 37½ and 10 per cent.
Porcelain, mineral and jet knobs, net list.

KEYS.

Lock, Canadian 40 to 40 and 10 per cent.

LOCKS.

Peterboro, 37½ and 10 per cent.
Russell & Erwin, steel rim $2.50 per doz
Eagle cabinet locks, discount 30 per cent
American padlocks, all steel, 10 to 15 per cent; all brass or bronze, 10 to 25 per cent

SAND AND EMERY PAPER.

B. & A. sand, discount, 35 per cent
Emery, discount 35 per cent.
Garnet (Burton's) 5 to 10 per cent. advance

SASH WEIGHTS.

Sectional .... per 100 lb.	2 00	2 25
Solid "	1 50	1 75

SASH CORD.

Per lb.	....	0 31

BLIND AND BED STAPLES.

All sizes per lb.	0 07½	0 10

WROUGHT STAPLES.

Galvanized	....	2 75
Plain	....	2 50

Coopers', discount 45 per cent.
Poultry netting staples, discount 40 per cent.
Bright spear point, 75 per cent. discount.

TOOLS AND HANDLES.

ADZES.

Discount 22½ per cent.

AUGERS.

Gilmour's, discount 60 per cent. off list.

AXES.

Single bit, per doz.	6 00	9 10
Double bit, "	10 00	11 00

Bench Axes, 40 per cent.
Broad Axes, 25 per cent.

Hunters' Axes	5 50	6 00
Boys' Axes	6 25	7 00
Splitting Axes	7 00	12 00
Handled Axes	7 00	9 00
Red Ridge, boys', handled	....	5 75
" hunters	....	5 25

BITS.

Irwin's auger, discount 47½ per cent.
Gilmour's auger, discount 60 per cent.
Rockford auger, discount 50 and 10 per cent.
Jennings' Gen. auger, net list.
Gilmour's car, 47½ per cent.
Clark's expansive, 40 per cent.

Clark's gimlet, per doz	....	0 65
Diamond, Shell, per doz.	....	1 00
Nail and Spike, per gross	....	2 25

BUTCHERS CLEAVERS

German .... per doz.	7 00	9 00
American "	12 00	18 00

CHALK.

Carpenters' Colored, per gross	0 45	0 75
White lump .... per cwt.	0 60	0 65

CHISELS.

Warnock's, discount 70 and 5 per cent.
P. S. & W. Extra, discount, 70 per cent.

CROSSCUT SAW HANDLES.

S. & D., No. 3 .... per pair	0 13
S. & D., " 5 "	0 11½
S. & D., " 6 "	0 18
Boynton pattern "	0 20

CROWBARS.

3½c. to 4c. per lb.

DRAW KNIVES.

Coach and Wagon, discount 75 and 5 per cent.
Carpenters' discount 75 per cent.

DRILLS.

Millar's Falls, hand and breast, net list.
North Bros., each set, 50c.

DRILL BITS.

Morse, discount 37½ to 40 per cent.
Standard, discount 50 and 5 to 55 per cent.

FILES AND RASPS.

Great Western	75 per cent.
Arcade	75 "
Kearney & Foot	75 "
Disston's	75 "
American	75 "
J. Barton Smith	75 "
McClellan	75 "
Eagle	75 "
Nicholson	66⅔ "
Globe	75 "

Black Diamond, 60, 10 and 5 p c.
Jowitt's, English list, 27½ per cent.

GAUGES.

Stanley's discount 50 to 60 per cent.

Winn's, Nos. 26 to 33 .... each	1 65	2 40

HANDLES.

Second growth ash fork, hoe, rake and shovel handles, 40 p c.
Extra ash fork, hoe, rake and shovel handles, 45 p c.
No 1 and 2 ash fork, hoe, rake and shovel handles, 50 p c.
White ash whiffletrees and neckyokes, 35 p c.
All other ash goods, 40 p.c.
All hickory, maple and oak goods, excepting carriage and express whiffletrees, 40 p.c.
Hickory, maple, oak carriage and express whiffletrees, 45 p.c.

HAMMERS.

Maydole's, discount 5 to 10 per cent.
Canadian, discount 25 to 27½ per cent.

Magnetic tack .... per doz.	1 10	1 20
Canadian sledge .... per lb.	0 07	0 08½
Canadian ball pean, per lb.	0 22	0 25

HATCHETS.

Canadian, discount 40 to 42½ per cent.

Shingle, Red Ridge 1, per doz.	4 40
" " 2, "	4 80
Barrel Underhill	5 05

HOES.

Mortar, 50 and 10 per cent

MALLETS.

Tinsmiths' ... per doz.	1 25	1 50
Carpenters', hickory, "	1 25	3 75
Lignum Vitae ... "	3 85	5 00
Caulking, each	0 60	2 00

MATTOCKS.

Canadian ... per doz.	5 50	6 00

MEAT CUTTERS.

German, 15 per cent.
American discount, 33⅓ per cent.

PICKS.

Per dozen	6 00	9 00

PLANES.

Wood bench, Canadian, 40, American, 25
Wood, fancy, 37½ to 40 per cent.
Stanley planes, $1 55 to $3 60, net list prices

PLANE IRONS.

English ... per doz.	2 00	5 00

Stanley, 2⅝ inch, single 24c, double 39c.

PLIERS AND NIPPERS.

Button's genuine, 37½ to 40 per cent.

Button's imitation ... per doz.	5 00	9 00
Berg's wire fencing	1 72	5 50

PUNCHES.

Saddlers ... per doz.	1 00	1 85
Conductor's ... "	3 00	15 00
Tinners', solid ... per set	....	0 72
" hollow ... per inch	....	1 00

RIVET SETS.

Canadian, discount 35 to 37½ per cent.

RULES.

Boxwood, discount 70 per cent.
Ivory, discount 20 to 25 per cent

SAWS.

Atkins, hand and crosscut, 25 per cent.
Disston's Hand, discount 12½ per cent

Disston's Crosscut ... per foot	0 35	0 55
Hack, complete ... each	0 75	2 75
" frame only ... each	0 50	1 25

S. & D. solid tooth circular shingle, concave and band, 50 per cent, mill and ice, drag, 30 per cent, cross-cut, 35 per cent, hand saws, butcher, 35 per cent.; buck, New Century, $6.25, buck, No. 1 Maple Leaf, $5 25; buck, Happy Medium, $4 25, buck, Watch Spring, $1.25; buck, common frame, $4.00.

Spear & Jackson's saws—Hand or rip, 26 in., $12 75; 28 in., $14.25; panel, 18 in., $8 25; 20 in., $9; tenon, 10 in., $9.90, 12 in., $10 90; 14 in., $11 50.

SAW SETS.

Lincoln and Whiting	4 75
Hand Sets, Perfect	4 00
X-Cut Sets, "	7 50

Maple Leaf and Premiums saw sets, 40 off.
S & D. saw swages, 40 off.

SCREW DRIVERS

Sargent's ... per doz.	0 65	1 00
North Bros, No. 30 .. per doz.	....	16 80

SHOVELS AND SPADES.

Canadian, discount 45 per cent.

SQUARES.

Iron, discount 20 per cent.
Steel, discount 65 and 10 per cent.
Try and Bevel, discount 50 to 52½ per cent.

TAPE LINES.

English, ass skin ... per doz.	2 75	5 00
English, Patent Leather	5 50	9 75
Chesterman's ... each	0 90	2 85
" steel ... each	0 80	8 00
Berg's, each	0 75	2 50

TROWELS.

Disston's, discount 10 per cent.
S & D., discount 35 per cent.

Berg's, brick, 9¼x11	4 00
" pointing, 9¼x5	2 10

FARM AND GARDEN GOODS

BELLS.

American cow bells, 63⅓ per cent.
Canadian, discount 45 and 50 per cent.

American, farm bells, each	1 35	3 00

BULL RINGS.

Copper, $2.00 for 2¼-inch

CATTLE LEADERS.

Nos. 32 and 33 ... per gross	7 50	8 50

BARN DOOR HANGERS.

	doz.	pairs.
Stearns wood track	4 50	6 00
Zenith	....	9 00
Atlas, steel covered	5 00	6 00
Perfect	8 00	11 00
New Milo, flexible	....	6 50
Steel, track, 1 x 3-16 in (100 ft)	....	3 25
" " 1¼ x 3-16 in (100 ft)	....	4 75
Double strap hangers, doz. sets	....	6 40
Standard jointed hangers, "	....	6 40
Steel King hangers "	....	6 25
Storm King and safety hangers	....	7 00
" " rail	....	4 25

Chicago Friction, Oscillating and Big Twin Hangers, 5 per cent.

HARVEST TOOLS.

50 and 10 per cent.
S. & D. lawn rakes, Dunn's, 40 off.
" sidewalk and stable scrapers, 40 off.

HAY KNIVES.

Net list.

HEAD HALTERS.

Jute Rope, ⅜-inch ... per gross	....	9 00
" " ½ " ... "	....	10 00
" " ¾ " ... "	....	12 00
Leather, 1-inch ... per doz.	....	4 00
Leather, 1¼ " ... "	....	5 20
Web ... "	....	2 45

HOES.

Garden, 50 and 10 per cent.

Planter ... per doz.	4 00	4 50

LAWN MOWERS

Low wheel, 12, 14 and 16-inch	$2 30
9-inch wheel, 12-inch	2 85
" " 14 "	3 00
" " 16 "	3 12½
High wheel, 12 "	4 05
" " 14 "	4 50
" " 16 "	4 75

SCYTHES.

Per doz. net	6 25	9 25

SCYTHE SNATHS.

Canadian, discount 40 per cent.

SNAPS.

Harness, German, discount 25 per cent.

Lock, Andrews'	4 50	11 00

STABLE FITTINGS.

Warden King, 35 per cent
Dennis Wire & Iron Co., 33⅓ p.c.

WOOD HAY RAKES.

40 and 10 per cent

HEAVY GOODS, NAILS, ETC.

ANVILS.

Wright's, 80-lb. and over	0 10½
Hay Budden, 80-lb. and over	0 09½
Brook's, 80-lb. and over	0 11¾
Taylor-Forbes, prospectors	0 05
Columbia Hardware Co., per lb.	0 09½

VISES.

Wright's	....	0 13½
Berg's, per lb.	....	0 12½
Brook's	....	0 12½
Pipe Vise, Hinge, No. 1	....	3 50
" " " No. 2	....	5 50
Saw Vise	4 50	5 00

Blacksmiths' (discount) 60 per cent.
" parallel (discount) 45 per cent.

BOLTS AND NUTS

	Per cent.
Carriage Bolts, common ($1 list	
" " ⅜ and smaller	60, 10 and 10
" " 7-16 and up	55 and 5
" " Norway Iron ($3 list)	50
Machine Bolts, ⅜ and less	60 and 10
Machine Bolts, 7-16 and up	55 and 5
Plough Bolts	55 and 10
Blank Bolts	55
Bolt Ends	55
Sleigh Shoe Bolts, ⅜ and less	60 and 10
" " " 7-16 and larger	50 and 5
Coach Screws, conepoint	70 and 5

Nuts, square, all sizes, 4c. per cent. off
Nuts, hexagon, all sizes, 4¼c. per cent. off.
Stove Rods per lb., 5½ to 6c.
Stove Bolts, 75 per cent.

CHAIN.

Proof coil, per 100 lb., ¼ in., $6 00; 5-16 in., $4.85; ⅜ in., $4 25; 7-16 in., $4.00; ½ in., $3 75; 9-16 in., $3 70; ⅝ in., $3 65; ¾ in., $3 60; ⅞ in., $3.45; 1 in., $3.40.

Halter, kennel and post chains, 40 to 40 and 5 per cent.; Cow ties, 40 per cent; Tie out chains 65 per cent; Stall fixtures, 35 per cent; Trace chain, 45 per cent.; Jack chain iron, 50 per cent; Jack chain, brass, 50 per cent.

HORSE NAILS

M R.M cold forged process, list May 15, 1907, 50 and 5 per cent.
"U" brand, 57½ per cent. off list.
Capewell brand, quotations on application.

HORSESHOES.

M R.M. brand: iron, light and medium No. 1 and smaller, $3 90; No. 2 and larger $3 65; snow pattern No. 1 and smaller, $4 15, No. 2 and larger, $3 90; "X.L." new light steel, No 1 and smaller, $4 25; No 2 and larger, $4; "X.L." featherweight steel, No. 0 to 4, $5.60; toe-weight, all sizes, $6.85. F o b Montreal Extras for packing.

Belleville brand: No. 0 and 1, light and medium iron, $3 90; snow, $4 15; light steel, $4 25; No 2 and larger, light and medium iron, $3.65; snow, $3 90; light steel, $4. F o b. Belleville. Two per cent., 30 days.

Toecalks—Standard No 1 and smaller, $1 50; No 2 and larger, $1 25 Blunt No 1 and smaller, $1 75; No 2 and larger, $1.50 per box Sharp Put up in 25 lb. bxs

HORSE WEIGHTS.

Taylor-Forbes, 4½c. per lb.

NAILS.	Cut.	Wire.
2d	4 00	3 50
3d	3 15	3 20
4 and 5d	2 90	2 90
6 and 7d	2 80	2 80
8 and 9d	2 65	2 65
10 and 12d	2 60	2 60
16 and 20d	2 55	2 55
30, 40, 50 and 60d (base)	2 50	2 50

F.o.b. Montreal. Cut nails, Toronto 20c. higher.
Miscellaneous wire nails, discount 75 per cent
Coopers' nails, discount 40 per cent.

PRESSED SPIKES.

Pressed spikes, ⅝ diameter, per 100 lbs $3 15

RIVETS AND BURRS.

Iron Rivets, black and tinned, 60, 10 and 10.
Iron Burrs, discount 60 and 10 and 10 p.c.
Copper Rivets, usual proportion burrs, 15 p
Copper Burrs only, net list.
Extras on Coppered Rivets, ½-lb. packages 1c. per lb.; ¼-lb. packages 2c. lb
Tinned Rivets, net extra, 4c. per lb.

SCREWS.

Wood, F. H., bright and steel, 85 and 10 p.c.
" R. H., bright, 80 and 10 per cent.
" F. H., brass, 75 and 10 per cent.
" R. H., " 70 and 10 per cent.
" F. H., bronze, 70 and 10 per cent.
" R. H., " 65 and 10 per cent.
Drive Screws, dis. 87½ per cent.

Bench, wood ... per doz.	3 25	4 00
" iron ... "	4 25	5 00

Set, case hardened, dis. 60 per cent.
Square Cap, dis. 50 and 5 per cent.
Hexagon Cap, dis. 45 per cent.

MACHINE SCREWS.

Flat head, iron and brass, 35 per cent.
Felister head, iron, discount 30 per cent.
" " brass, discount 25 per cent.

TACKS, BRADS, ETC.

Carpet tacks, blued 75 p.c.; tinned, 80 and 10; (in kegs), 40; cut tacks, blued, in dozens only, 75; ½ weights, 60; Swedes cut tacks, blued and tinned, bulk, 75 dozens, 75; Swedes, upholsterers', bulk, 85 and 12½; brush, blued and tinned, bulk, 70; Swedes, gimp, blued, tinned and japanned, 75 and 12½; zinc tacks, 35; leather carpet tacks, 40; copper tacks, 25; copper nails 30; trunk nails, black, 65; trunk nails, tinned and blued, 65; clout nails, blued and tinned 65; chair nails, 35; patent brads, 40; fine finishing, 40; lining tacks, in papers, 10; lining tacks, in bulk, 15; lining tacks, solid heads, in bulk, 75; saddle nails, in papers, 10; saddle nails, in bulk, 15; tufting buttons, 22 line in dozens only, 60; zinc glaziers' points, 5; double pointed tacks, papers, 90 and 10; double pointed tacks, bulk, 40, clinch and duck rivets, 45; cheese box tacks, 85 and 5 trunk tacks, 80 and 10.

WROUGHT IRON WASHERS.

Canadian make, discount 40 per cent.

SPORTING GOODS.

CARTRIDGES.

"Dominion" Rim Fire Cartridges and C.B caps, 50 and 7½ per cent; Rim Fire B.B Round Caps, 60 and 2½ per cent.; Centre Fire, Pistol and Rifle Cartridges, 30 p c; Centre Fire Sporting and Military Cartridges, 20 and 5 p c; Rim Fire, Shot Cartridges, 50 and 7½ p.c.; Centre Fire, Shot Cartridges, 30 p c; Primers, 25 p c.

LOADED SHELLS.

"Crown" Black Powder, 15 and 5 p c; "Sovereign" Empire Bulk Smokeless Powder, 30 and 5 p.c; "Regal" Ballistite Dense smokeless Powder, 30 and 5 p.c; "Imperial" Empire or Ballistite Powder, 30 and 10 p.c

EMPTY SHELLS

Paper Shells, 25 and 5; Brass Shells, 55 and 5 p c.

Wads.	per lb.
Best thick brown or grey felt wads, in ½-lb. bags	$0 70
Best thick white card wads, in boxes of 500 each, 12 and smaller gauges	29
Best thick white card wads, in boxes of 500 each, 10 gauge	0 35
Thin card wads, in boxes of 1,000 each, 12 and smaller gauges	0 20
Thin card wads, in boxes of 1,000 each, 10 gauge	0 25
Chemically prepared black edge grey cloth wads, in boxes of 250 each—	Per M
11 and smaller gauge	0 60
9 and 10 gauges	0 70
7 and 8 "	0 90
5 and 6 "	1 10
Superior chemically prepared pink edge, best white cloth wads in boxes of 250 each—	
11 and smaller gauge	1 15
9 and 10 gauges	1 40
7 and 8 "	1 65
5 and 9 "	1 90

SHOT.

Ordinary drop shot, AAA to dust $7.50 per 100 lbs. Discount 5 per cent; cash discount 2 per cent, 30 days; net extras as follows subject to cash discount only; Chilled, 40 c.; buck and seal, 80c; no 28 ball, $1 20 per 100 lbs; bags less than 25 lbs., ½c. per lb; F.O.B. Montreal, Toronto, Hamilton, London, St. John and Halifax, and freight equalized thereon.

TRAPS (steel.)

Game, Newhouse, discount 30 and 10 per cent.
Game, Hawley & Norton, 50, 10 & 5 per cent
Game, Victor, 70 per cent.
Game, Oneida Jump (B. & L.) 40 & 2½ p.
Game, steel, 60 and 5 per cent.

SKATES.

Skates, discount 37½ per cent.
Empire hockey sticks, per doz .. 3 00

CUTLERY AND SILVERWARE.

RAZORS. per doz.

Elliot's 4 00 18 00
Boker's 7 50 11 00
" King Cutter 13 50 18 50
Wade & Butcher's 3 60 10 00
Lewis Bros.' "Klean Kutter" 8 50 10 50
Henckel's 7 50 20 00
Berg's 7 50 20 00
Clauss Razors and Strops, 50 and 10 per cent

KNIVES

Farriers-Stacey Bros., doz 3 50

PLATED GOODS

Hollowware, 40 per cent. discount.
Flatware, staples, 40 and 10, fancy, 40 and 5.
Hutton's "Cross Arrow" flatware, 47½; "Singalese" and "Alaska" Nevada silver flatware, 42 p c.

SHEARS.

Clauss, nickel, discount 60 per cent.
Clauss, Japan, discount 67½ per cent.
Clauss, tailors, discount 40 per cent.
Seymour's, discount 50 and 10 per cent.
Berg's 6 00 12 00

HOUSE FURNISHINGS.

APPLE PARERS.

Hudson, per doz., net 5 75

BIRD CAGES.

Brass and Japanned, 40 and 10 p. c.

COPPER AND NICKEL WARE.

Copper boilers, kettles, teapots, etc. 30 p c.
Copper pitts, 30 per cent.

KITCHEN ENAMELED WARE.

White ware 75 per cent.
London and Princess, 50 per cent
Canada, Diamond, Premier, 50 and 10 p c.
Pearl, Imperial, Crescent and granite steel, 60 and 10 per cent.
Premier steel ware, 40 per cent.
Star decorated steel and white, 25 per cent.
Japanned ware, discount 45 per cent.
Hollow ware, tinned cast, 35 per cent. off.

KITCHEN SUNDRIES.

Can openers, per doz 0 40 0 75
Mincing knives per doz 0 50 0 80
Duplex mouse traps, per doz 0 65
Potato mashers, wire, per doz... 0 60 0 70
" wood " .. 0 50 0 60
Vegetable slicers, per doz 2 25
Universal meat chopper No. 1 .... 1 15
Enterprise chopper, each 1 30
Spiders and fry pans, 50 per cent.
Star A1 chopper 5 to 32 .... 1 35 4 10
" " 100 to 103 .... 1 35 2 00
Kitchen hooks, bright 0 60

LAMP WICKS.

Discount, 60 per cent.

LEMON SQUEEZERS.

Porcelain lined...... per doz. 2 20 5 60
Galvanized .... " 1 87 3 85
King, wood .... " 2 75 2 90
King, glass .... " 4 00 4 50
All glass .... " 0 50 0 90

METAL POLISH.

Tandem metal polish paste 6 00

PICTURE NAILS.

Porcelain head ......per gross 1 35 1 50
Brass head 0 40 1 00
Tin and gilt, picture wire, 75 per cent.

SAD IRONS.

Mrs. Potts, No. 55, polished....per set 0 90
" No 50, nickle-plated, " 0 95
" handles, japaned, per gross 9 25
" " nickled, " 9 75
Common, plain 4 25
" plated 5 50
Asbestos, per set 1 50

TINWARE.

CONDUCTOR PIPE.

2-in. plain or corrugated, per 100 feet, $3.30; 3 in., $4 40; 4 in, $5.80; 5 in, $7.45; 6 in., $9 90.

FAUCETS.

Common, cork-lined, discount 35 per cent.

EAVETROUGHS.

10-inch .... per 100 ft. 3 30

FACTORY MILK CANS.

Discount off revised list, 35 per cent.
Milk can trimmings, discount 25 per cent.
Creamery Cans, 45 per cent

LANTERNS.

No 2 or 4 Plain Cold Blast....per doz. 6 50
Lift Tubular and Hinge Plain, " 4 75
No 0, safety .... " 4 00
Better quality at higher prices.
Japanning, 50c. per doz. extra.
Prism globes, per doz., $1.20.

OILERS.

Kemp's Tornado and McClary's Model galvanized oil can, with pump, 5 gallon, per dozen 10 92
Davidson oilers, discount 40 per cent.
Zinc and tin, discount 50 per cent
Coppered oilers, 20 per cent. off.
Brass oilers, 50 per cent. off.
Malleable, discount 25 per cent

PAILS (GALVANIZED).

Dufferin pattern pails, 45 per cent.
Flaring pattern, discount 45 per cent.
Galvanized washtubs 40 per cent.

PIECED WARE.

Discount 35 per cent off list, June, 1899.
10-qt. flaring sap buckets, discount 35 per cent.
6, 10 and 14-qt. flaring pails dis. 35 per cent.
Copper bottom tea kettles and boilers, 30 p.c.
Coal hods, 40 per cent.

STAMPED WARE.

Plain, 75 and 12½ per cent. off revised list.
Retinned, 72½ per cent. revised list.

SAP SPOUTS.

Bronzed iron with hooks ....per 1,000
Eureka tinned steel, hooks 8 00

STOVEPIPES.

5 and 6 inch, per 100 lengths 7 64 7 91
7 inch .... " " 8 18
Nestable, discount 40 per cent.

STOVEPIPE ELBOWS

5 and 6-inch, common........per doz. 1 32
7-inch .... " 1 48
Polished, 15c. per dozen extra.

THERMOMETERS.

Tin case and dairy, 75 to 75 and 10 per cent.

TINNERS' SNIPS.

Per doz 3 00 15
Clauss, discount 35 per cent.

TINNERS' TRIMMINGS.

Discount, 45 per cent.

WIRE.

ANNEALED CUT HAY BAILING WIRE.

No. 12 and 13, $4; No. 13½, $4 10; No. 14, $4 20; No. 15, $4.50; in lengths 6' to 11', 25 per cent; other lengths 20c. per 100 lbs extra; if eye or loop on end add 25c. per 100 lbs. to the above.

BRIGHT WIRE GOODS

Discount 60 per cent.

CLOTHES LINE WIRE.

7 wire solid line, No. 17, $4.90; No. 18, $3.00; No. 19, $2 70; 6 wire solid line, No. 17, $4.45; No. 18, $3.10; No. 19, $2 80. All prices per 1000 ft measure; 6 strand, No. 18, $2 60; No. 19, $2 90. F.o.b. Hamilton, Toronto, Montreal.

COILED SPRING WIRE.

High Carbon, No. 9, $2.95; No. 11, $3.50; No. 12, $3.20.

COPPER AND BRASS WIRE.

Discount 37½ per cent.

FINE STEEL WIRE.

Discount 25 per cent. List of extras In 100-lb. lots: No. 17, $5 — No. 18, $5.50 — No. 19, $6 — No. 20, $6.65 — No. 21 $7 — No. 22, $7 30 — No. 23, $7.65 — No. 24, $8 — No. 25, $9 — No. 26, $9 50 — No. 27, $10 — No. 28, $11 — No. 29, $12 — No. 30, $13 — No. 31, $14 — No. 32, $15 — No. 33, $16 — No. 34, $17. Extras net — tinned wire, Nos. 17-25, $2 — Nos. 26-31, $4 — Nos. 32-34, $6. Coppered, 75c. — oiling, 10c. — in 25-lb. bundles, 15c. — in 5 and 10-lb. bundles, 25c. — in 1-lb. hanks, 25c. — in ½-lb. hanks, 38c. — in ¼-lb. hanks, 50c. — packed in casks or cases, 15c. — bagging or papering, 10c

FENCE STAPLES.

Bright. .... 2 80 Galvanized 32

HAY WIRE IN COILS.

No 13, $2.70; No. 14, $2 80; No. 15, $2.95; f.o b., Montreal.

GALVANIZED WIRE.

Per 100 lb. — Nos. 4 and 5, $3 90 — Nos. 6, 7, 8, $3.35 — No 9, $2.85 — No. 10, $3 40 — No. 11, $3.45 — No. 12, $3.00 — No. 13, $3 10 — No. 14, $3.95 — No 15, $4.30 — No. 16, $4 30 from stock. Base sizes, Nos. 6 to 9, $2.35 f.o.b. Cleveland. Extras for cutting.

LIGHT STRAIGHTENED WIRE.

Gauge No.	Over 20 in. per 100 lbs.	10 to 20 in.	5 to 10 in.
0 to 5	$0 50	$0 75	$1 25
6 to 9	0.75	1 25	2 00
10 to 11	1.00	1 75	2 50
12 to 14	1 50	2.25	3 50
15 to 16	2.00	3.00	4.50

SMOOTH STEEL WIRE.

No. 0-9 gauge, $2.40; No. 10 gauge, 6c extra; No. 11 gauge, 12c extra; No. 12 gauge, 20c. extra; No. 13 gauge, 30c. extra; No 14 gauge, 40c. extra; No 15 gauge, 55c. extra; No. 16 gauge, 70c. extra. Add 60c. for coppering and $2 for tinning.

Extra net per 100 lb. — Oiled wire 10c., spring wire $1.25, bright soft drawn 15c., charcoal (extra quality) $1.25, packed in casks or cases 15c., bagging and papering 10c, 50 and 100-lb. bundles 10c., in 25-lb. bundles 15c., in 5 and 10-lb. bundles 25c., in 1-lb hanks, 50c., in ½-lb. hanks 75c., in ¼-lb. hanks $1.

POULTRY NETTING.

2-in. mesh, 19 w. g., 50 and 5 p.c. off. Other sizes, 50 and 5 p.c. off.

WIRE CLOTH.

Painted Screen, in 100-ft rolls, $1.72½, per 100 sq. ft.; in 50-ft. rolls, $1.77½, per 100 sq ft

WIRE FENCING.

Galvanized barb 2 95
Galvanized, plain twist 3 30
Galvanized barb, f.o.b. Cleveland, $2.70 for small lots and $2.60 for carlots

WIRE ROPE

Galvanized 1st grade, 6 strands, 24 wires, ⅝, $5; 1 inch $16 80.
Black, 1st grade, 6 strands, 19 wires. ⅝, $5; 1 inch $15 10. Per 100 feet f o.b. Toronto.

WOODENWARE.

CHURNS.

No. 0, $9; No. 1, $9; No. 2, $10; No. 3, $11; No. 4, $13; No. 5, $16.; f o b. Toronto Hamilton, London and St. Marys, 30 and 30 per cent; f.o.b. Ottawa, Kingston and Montreal, 40 and 15 per cent. discount.

CLOTHES REELS.

Davis Clothes Reels. dis. 40 per cent.

FIBRE WARE.

Star pails, per doz .... $ 3 00
0 Tubs, " 14 00
1 " " 12 00
2 " " 10 00
3 " " 8 50

LADDERS, EXTENSION.

3 to 6 feet, 12c. per foot; 7 to 10 ft, 13c.
Waggoner Extension Ladders, dis. 40 per cent.

MOPS AND IRONING BOARDS.

"Best" mops 1 25
"900" mops 1 25
Folding ironing boards 12 00 16 50

REFRIGERATORS

Discount, 40 per cent.

SCREEN DOORS.

Common doors, 2 or 3 panel, walnut stained, 4-in. style..........per doz. 7 25
Common doors, 2 or 3 panel, grained only, 4-in., style ....per doz. 7 55
Common doors, 2 or 3 panel, light stair per doz 9 55

WASHING MACHINES.

Round, re-acting per doz. 60 00
Square " " 63 00
Eclipse, per doz 54 00
Dowswell " 39 00
New Century, per doz 75 00
Daisy 54 00
Stephenson 74 00

WRINGERS.

Royal Canadian, 11 in., per doz. ... 35 00
Royal American, 11 in. .. 35 00
Eze 10 in., per doz 36 75

MISCELLANEOUS

AXLE GREASE.

Ordinary, per gross 6 00 7 00
Best quality 10 00 12 00

BELTING.

Extra, 60 per cent.
Standard, 60 and 10 per cent.
No. 1, not wider than 6 in., 60, 10 and 10 p c
Agricultural, not wider than 4 in., 75 per cent.
Lace leather, per side, 75c.; cut laces, 80c.

BOOT CALKS.

Small and medium, ball ....per M 2 25
Small heel 4 50

CARPET STRETCHERS.

American ....per doz. 1 00 1 50
Bullard's .... " 6 50

CASTORS.

Bed, new list, discount 55 to 57½ per cent.
Plate, discount 52½ to 57½ per cent.

PINE TAR.

½ pint in tins ....per gross ... 7 80
" " .... " 9 60

PULLEYS.

Hothouse ....per doz. 0 55 1 00
Axle .... " 0 22 0 33
Screw .... " 0 22 1 00
Awning .... " 0 35 2 50

PUMPS.

Canadian cistern 1 40 2 00
Canadian pitcher spout ... 1 80 3 16
Berg's wing pump, 75 per cent.

ROPE AND TWINE.

Sisal 0 10½
Pure Manilla 0 15
"British" Manilla 0 12
Cotton, 3-16 inch and larger... ..0 21 0 23
" 5-32 inch 0 25 0 27
" ¼ inch 0 25 0 28
Russia Deep Sea 0 16
Jute 0 09
Lath Yarn, single 0 10
" double 0 10½
Sisal bed cord, 48 feet.......per doz. 0 65
" " 60 feet .... " 0 80
" " 72 feet .... " 0 95

Twine.

Bag, Russian twine, per lb. 0 27
Wrapping, cotton, 3-ply 0 25
" " 4-ply 0 29
Mattress twine per lb. 0 33 0 45
Staging " " 0 27 0 35

BINDER TWINE.

500 feet, sisal 0 09½
500 " standard 0 09½
550 " " manilla 0 10½
600 " " " 0 12½
650 " " " 0 13½

Car lots, ½c less; 5-ton lots, ¼c. less. Central delivery.

SCALES.

Gurney Standard, 35; Champion, 45 p c.
Burrow, Stewart & Milne — Imperial Standard, 35; Weigh Beams, 35; Champion Scales, 45.
Fairbanks Standard, 30; Dominion, 50 Richelieu, 50.
Warren new Standard, 35; Champion, 45 Weigh Beams, 30.

STONES—OIL AND SCYTHE.

Washita ....per lb. 0 25 0
Hindostan .... " 0 06 0 10
" slip .... " 0 18 0 20
" Axe .... " 0 10
Deer Creek .... " 0 10
Deerlick .... " 0 25
" Axe .... " 0 15
Lily white .... " 0 42
Arkansas .... " 1 50
Water-of-Ayr .... " 0 10
Scythe ....per gross 3 50 5 00
Grind, 40 to 200 lb., per ton.... 20 00 22 00
" under 40 lb., " 24 00
" 200 lb. and over 28 00

INDEX TO ADVERTISERS.

CLASSIFIED LIST OF ADVERTISEMENTS.

Alabastine.
Alabastine Co , Limited, Paris, Ont

Auditors.
Davenport, Pickup & Co., Winnipeg.

Automobile Accessories.
Canada Cycle & Motor Co , Toronto Junction
Carriage Mountings Co , Ltd.. Toronto.

Awnings.
Tobin Tent and Awning Co , Ottawa

Babbitt Metal.
Canada Metal Co , Toronto.
Canadian Fairbanks Co., Montreal
Frothingham & Workman, Montreal
Robertson, Jas. Co., Montreal.

Bar Urns.
Buffalo Mfg Co , Buffalo, N Y.

Bath Room Fittings.
Buffalo Mfg Co., Buffalo, N Y
Carriage Mountings Co , Ltd , Toronto.
Forsyth Mfg Co., Buffalo N.Y
Ontario Metal Novelty Co , Toronto

Belting, Hose, etc.
Gutta Percha and Rubber Mfg. Toronto
Sadler & Haworth Toronto.

Bicycles and Accessories.
Johnson's, Iver, Arms and Cycle Works Fitchburg, Mass

Binder Twine.
Consumers Cordage Co , Montreal.

Bolts.
Toronto & Belleville Rolling Mills, Belleville.

Box Strap.
J. N. Warminton, Montreal.

Brass Goods.
Frothingham & Workman, Montreal
Glauber Brass Mfg Co , Cleveland, Ohio.
Kerr Engine Co., Walkerville, Ont
Lewis, Rice, & Son., Toronto
Morrison, Jas , Brass Mfg Co., Toronto.
Mueller Mfg Co., Decatur, Ill.
Penberthy Injector Co., Windsor, Ont.
Taylor-Forbes Co., Guelph, Ont.

Bronze Powders.
Canadian Bronze Powder Works, Montreal.

Brushes.
United Factories, Toronto.

Cans.
Acme Can Works, Montreal.

Builders' Tools and Supplies.
Covert Mfg Co., West Troy, N.Y.
Frothingham & Workman Co., Montreal.
Howland, H. S. Sons & Co., Toronto.
Hyde, F., & Co., Montreal.
Lewis Bros. & Co., Montreal.
Lewis, Rice, & Son, Toronto.
Lockerby & McComb, Montreal.
Lufkin Rule Co., Saginaw, Mich.
Newman & Sons, Birmingham.
North Bros. Mfg. Co., Philadelphia, Pa.
Stanley Rule & LeVel Co., New Britain.
Stanley Works, New Britain, Conn.
Stephens, G. F., Winnipeg.
Taylor-Forbes Co., Guelph, Ont.

Carriage Accessories.
Carriage Mountings Co , Ltd , Toronto
CoVert Mfg. Co., West Troy, N.Y.

Carriage Springs and Axles.
Guelph Spring and Axle Co., Guelph.

Carpet Beateas.
Ontario Metal NoVelty Co., Toronto.

Cartridges.
Dominion Cartridge Co., Montreal.

Cattle and Trace Chains.
Greening, B , Wire Co., Hamilton.

Chafing Dishes.
Buffalo Mfg Co., Buffalo, N.Y.

Churns.
Dowswell Mfg. Co., Hamilton.

Clippers—All Kinds.
American Shearer Mfg. Co ,Nashua,N.H

Clothes Reels and Lines.
Hamilton Cotton Co., Hamilton, Ont.

Clutch Nails.
J. N. Warminton, Montreal.

Congo Roofing
Buchanan Foster Co., Philadelphia, Pa.

Cordage.
Consumers' Cordage Co., Montreal.
Hamilton Cotton Co., Hamilton.

Cork Screws.
Erie Specialty Co., Erie, Pa.

Cow Ties
Greening, B , Wire Co., Hamilton

Cuspidors.
Buffalo Mfg Co , Buffalo, N.Y.

Cut Glass.
Phillips, Geo., & Co., Montreal.

Cutlery—Razors, Scissors, etc.
Birkett, Thos., & Son Co., Ottawa.
Clauss Shear Co , Toronto
Dorken Bros. & Co., Montreal.
Frothingham & Workman, Montreal.
Heinisch's, R., Sons Co., Newark, N.J.
Howland, H. S. Sons & Co., Toronto.
Hutton, Wm., & Sons, Ltd., London, Eng.
Lamplough, F. W., & Co , Montreal.
Phillips, Geo., & Co., Montreal.
Round, John, & Son, Montreal.

Electric Fixtures.
Canadian General Electric Co., Toronto.
Morrison James, Mfg. Co., Toronto.
Munderloh & Co., Montreal.

Electro Cabinets.
Cameron & Campbell Toronto.

Enameled Ware
Kemp Mfg. Co., Toronto.

Engines, Supplies, etc.
Kerr Engine Co., Walkerville, Ont.

Eavetroughs
Wheeler & Bain, Toronto

Fencing—Woven Wire
Dominion Wire Mfg. Co., Montreal
Owen Sound Wire Fence Co., Owen Sound.
Banwell Hoxie Wire Fence Co., Hamilton.

Files and Rasps.
Barnett Co., G. & H., Philadelphia, Pa.
Nicholson File Co., Port Hope

Firearms and Ammunition.
Hamilton Rifle Co , Plymouth, Mich.
Harrington & Richardson Arms Co., Worcester, Mass.
Johnson's, IVer, Arms and Cycle Works Fitchburg, Mass.

Fishing Tackle.
Enterprise Mfg. Co., Akron, Ohio

Food Choppers
Enterprise Mfg. Co., Philadelphia, Pa.
Lamplough, F. W., & Co , Montreal.
Shirreff Mfg. Co., BrockVille, Ont.

Furnaces.
Pease Foundry Co , Toronto.

Galvanizing.
Canada Metal Co., Toronto.
Dominion Wire Mfg. Co , Montreal.
Montreal Rolling Mills Co., Montreal.
Ontario Wind Engine & Pump Co., Toronto.

Glass Ornamental
Hobbs Mfg. Co , London
Consolidated Plate Glass Co , Toronto

Glaziers' Diamonds.
Gibsone, J. B., Montreal.
Pelton, Godfrey S.
Sharratt & Newth, London, Eng.
Shaw, A., & Son, London, Eng.

Handles.
Still, J. H., Mfg. Co.

Harvest Tools.
Maple Leaf Harvest Tool Co., Tillsonburg, Ont.

Hockey Sticks
Still, J. H. Mfg. Co , St. Thomas.

Hoop Iron.
Frothingham & Workman Montreal.
Montreal Rolling Mills Co , Montreal.
J. N. Warminton, Montreal.

Horse Blankets.
Heney, E. N., & Co., Montreal.

Horseshoes and Nails.
Canada Horse Nail Co., Montreal.
Montreal Rolling Mills, Montreal.
Capewell Horse Nail Co., Toronto
Toronto & BelleVille Rolling Mills Belleville.

Hot Water Boilers and Radiators.
Cluff, R J., & Co Toronto
Pease Foundry Co. Toronto.
Taylor-Forbes Co., Guelph.

Ice Cream Freezers.
Dana Mfg. Co., Cincinnati, Ohio.
North Bros. Mfg. Co., Philadelphia, Pa.

Ice Cutting Tools.
Erie Specialty Co., Erie, Pa.
North Bros. Mfg. Co., Philadelphia, Pa.

Injectors—Automatic.
Morrison, Jas., Brass Mfg. Co., Toronto.
Penberthy Injector Co., Windsor, Ont.

Iron Pipe.
Montreal Rolling Mills, Montreal.

Iron Pumps.
Lamplough, F. W., & Co , Montreal.
McDougall, R., Co., Galt, Ont.

Lanterns.
Kemp Mfg. Co., Toronto.
Ontario Lantern Co., Hamilton, Ont.
Wright, E. T., & Co., Hamilton.

Lawn Mowers.
Birkett, Thos., & Son Co., Ottawa.
Frothingham & Workman, Montreal.
Maxwell, D., & Sons, St. Mary's, Ont.
Taylor, Forbes Co., Guelph.

Lawn Mower Grinders
Root Bros. & Co., Plymouth, Ohio.

Ledgers—Loose Leaf.
Business Systems, Toronto.
Copeland-Chatterson Co., Toronto.
Crain, Rolla L., Co., Ottawa.
UniVersal Systems, Toronto.

Lithographing.
London Printing & Lithographing Co., London, Ont.

Locks, Knobs, Escutcheons, etc.
Peterborough Lock Mfg. Co., Peterborough, Ont.
National Hardware Co., Orillia, Ont.

Lumbermen's Supplies.
Pink, Thos., & Co., Pembroke Ont.

Lye
Gillett, E. W., & Co , Toron

Manufacturers' Agents.
Fox, O. H., Vancouver.
McIntosh, H. F., & Co., Toronto.
Gibb, Alexander, Montreal.
Scott, Bathgate & Co., Winnipeg.

Metals.
Canada Iron Furnace Co., Midland, Ont.
Canada Metal Co., Toronto.
Eadie, H. G., Montreal.
Frothingham & Workman, Montreal.
Gibb, Alexander, Montreal.
Kemp Mfg Co., Toronto
Leslie, A. C., & Co., Montreal.
Lysaght, John, Bristol, Eng.
Nova Scotia Steel and Coal Co., New Glasgow, N.S.
Robertson, Jas., Co., Montreal.
Roper, J. H., Montreal.
Samuel, Benjamin & Co., Toronto.
Stairs, Son & Morrow, Halifax, N.S.
Thompson, B., & S. H. & Co. Montreal.

Metal Lath.
Galt Art Metal Co., Galt.
Metallic Roofing Co., Toronto.
Metal Shingle & Siding Co., Preston, Ont.

Metal Polish, Emery Cloth, etc.
Oakey, John, & Sons, London, Eng.

Nails Wire
Dominion Wire Mfg. Co., Montreal.

Oil Tanks
Bowser, S. F., & Co., Toronto.

Ornamental Iron and Wire.
Dennis Wire & Iron Co, London, Ont.

Packing.
Gutta Percha & Rubber Co Toronto.

Paints, Oils, Varnishes, Glass.
Blanchite Process Paint Co., Toronto.
Brandram-Henderson, Montreal
Canada Paint Co., Montreal.
Canadian Oil Co., Toronto.
Consolidated Plate Glass Co., Toronto.
Dods, P. D., & Co., Montreal
Imperial Varnish and Color Co., Toronto.
Jamieson, R. C., & Co., Montreal.
Lucas John & Co., New York
McArthur, Corneille & Co., Montreal.
McCaskill, Dougall & Co., Montreal.
Moore, Benjamin, & Co. Toronto.
Ottawa Paint Works, Ottawa
Queen City Oil Co., Toronto.
Ramsay & Son, Montreal.
Sanderson Pearcy & Co., Toronto
Sherwin-Williams Co., Montreal.
Standard Paint Co., Montreal
Standard Paint and Varnish Works Windsor, Ont.
Stephens & Co., Winnipeg.
Martin-Senour Co., Montreal
Winnipeg Paint & Glass Co., Winnipeg

Perforated Sheet Metals.
Greening, B., Wire Co., Hamilton.

Plumbers' Tools and Supplies
Canadian Fairbanks Co, Montreal.
Cluff, R. J, & Co, Toronto
Frothingham & Workman, Montreal.
Glauber Brass Co., CleVeland, Ohio.
Jardine, A. B., & Co, Hespeler, Ont.
Jenkins Bros, Boston, Mass
Kerr Engine Co, Walkerville, Ont
Lewis, Rice, & Son, Toronto.
Merrell Mfg. Co., Toledo, Ohio.
Montreal Rolling Mills Montreal.
Morrison, Jas., Brass Mfg. Co., Toronto.
Mueller, H., Mfg. Co, Decatur, Ill
Oshawa Steam & Gas Fitting Co., Oshaw
Robertson Jas., Co. Montreal.
Robertson, Jas, Co., Limited, Toronto
SomerVille, Limited, Toronto
Stairs, Son & Morrow, Halifax, N.S.
Standard Ideal Sanitary Co., Port Hope.
Standard Sanitary Co., Pittsburg.
Stephens, G F., & Co., Winnipeg, Man.
Turner Brass Works, Chicago.
Vickery, Orlando, Toronto.

Polishes.
Majestic Polishes, Toronto

Portland Cement.
International Portland Cement Co. Ottawa, Ont
HanoVer Portland Cement Co., HanoVer, Ont.
Hyde, F., & Co., Montreal.
Thompson B. & S. H. & Co., Montreal.

Poultry Netting.
Greening, B., Wire Co., Hamilton, Ont.

Printing.
London Printing & Lithographing Co., London, Ont.

Razors.
Clauss Shear Co., Toronto.

Refrigerators.
Fabien, C. P., Montreal.

Registers
Pease Foundry Co., Toronto.

Roofing Supplies.
Brantford Roofing Co., Brantford.
Barrett Mfg. Co, New York.
F. W. Bird, East Walpole, Mass
Buchanan Foster Co., Philadelphia, Pa.
McArthur, Alex., & Co., Montreal
Metal Shingle & Siding Co, Preston, Ont.
Metallic Roofing Co., Toronto.
Paterson Mfg. Co., Toronto & Montreal.
Wheeler and Bain, Toronto

Saws.
Atkins, E. C., & Co., Indianapolis, Ind
Shurly & Dietrich, Galt, Ont.
Spear & Jackson, Sheffield, Eng.

Scales.
Canadian Fairbanks Co., Montreal.
Frothingham & Workman, Montreal.

Screw Cabinets.
Cameron & Campbell, Toronto.

Screws, Nuts, Bolts.
Dominion Wire Mfg. Co, Montreal.
Montreal Rolling Mills Co., Montreal.

Soil Pipe
McFarlane, Walter, Glasgow

Sewer Pipes.
Canadian Sewer Pipe Co., Hamilton
Hyde, F., & Co., Montreal.

Shelf Boxes.
Cameron & Campbell, Toronto.

Shears, Scissors.
Clauss Shear Co, Toronto

Shovels and Spades
Eclipse Mfg Co, Ottawa
Frothingham & Workman, Montreal
Peterboro Shovel & Tool Co, Peterboro.

Silverware.
Hutton, Wm, & Sons, Ltd, London, Eng
McGlashan, Clarke Co, Niagara Falls, Ont
Phillips, Geo, & Co., Montreal.
Round, John, & Son, Sheffield, Eng.

Skates.
Canada Cycle & Motor Co, Toronto.
McFarlane, Walter, Glasgow.

Sprayers
CaVers Bros., Galt

Spring Hinges, etc.
Chicago Spring Butt Co, Chicago, Ill.

Stable Fittings
Dennis Wire & Iron Co., London

Steel Rails.
NoVa Scotia Steel & Coal Co., New Glasgow, N.S.

Stove Pipe.
Chown, Edwin and Son, Kingston

Stoves, Tinware, Furnaces
Canadian Heating & Ventilating Co. Owen Sound.
Copp, W. J., Son & Co, Fort William
Davidson, Thos., Mfg. Co., Montreal
Down Draft Furnace Co, Galt
Guelph Stove Co., Guelph.
Gurney Foundry Co., Toronto.
Harris, J. W., Co., Montreal.
Howard, Wm, Toronto
Kemp Mnfg. Co. Toronto.
McClary Mfg. Co. London.
Merrick Anderson, Winnipeg
Pease Foundry Co., Toronto.
Smart, James, Mfg. Co., Brockville
Stewart, Jas., Mfg. Co, Woodstock, Ont.
Taylor-Forbes Co., Guelph, Ont.
Wright, E. T., & Co., Hamilton.

Tacks.
Montreal Rolling Mills Co, Montreal
Ontario Tack Co., Hamilton.

Tents.
Tobin Tent and Awning Co., Ottawa

Tin Plate.
American Sheet & Tin Plate Co., Pittsburg, Pa
Baglan Bay Tin Plate Co, Briton Ferry South Wales
Lysaght, John, Bristol, Newport and Montreal

Turpentine
Defiance Mfg Co, Toronto.

Ventilators.
Harris, J. W, Co., Montreal.
Pearson, Geo. D., Montreal.

Wall Paper
Staunton Limited, Toronto.

Wall Paper Cleaner.
Gilbert, Frank U. S., CleVeland

Washing Machines, etc
Dowswell Mfg. Co., Hamilton, Ont.
The Shultz Bros. Co, Brantford.
Taylor-Forbes Co., Guelph, Ont.

Water Filters.
Buffalo Mfg. Co, Buffalo, N.Y.

Wheelbarrows
London Foundry Co, London Ont.
Schultz Bros. Co., Ltd, The Brantford.

Wholesale Hardware
Birkett, Thos., & Sons Co., Ottawa.
Caverhill, Learmont & Co., Montreal.
Frothingham & Workman, Montreal.
Hobbs Hardware Co., London.
Howland, H. S., Sons & Co., Toronto.
Lamplough, F. W., & Co., Montreal.
Lewis Bros. & Co., Montreal.
Lewis, Rice, & Son, Toronto.

Window and Sidewalk Prisms
Hobbs Mfg. Co., London, Ont.

Wire, Wire Rope, Cow Ties, Fencing Tools, etc.
Banwell-Hoxie Fence Co., Hamilton
Dennis Wire and Iron Co., London, Ont.
Dominion Wire Mnfg. Co., Montreal
Greening, B., Wire Co., Hamilton.
Owen Sound Wire Fence Co., Owen Sound
Montreal Rolling Mills Co., Montreal.
Western Wire & Nail Co, London, Ont

Wrapping Papers.
Canada Paper Co., Toronto.
McArthur, Alex., & Co, Montreal.
Stairs, Son & Morrow, Halifax, N.S.

Wringers
Connor, J. H. & Son, Ottawa, Ont

ROBERTSON'S SOLDERS

Bar-King, Strictly Merchant's, Wire (any gauge) Triangular, Oval, Plumbers' and Canners' Solders, Electricians' Solder

ARE STANDARD

Ask Quotations

The James Robertson Co., Limited
144 William St., MONTREAL, QUE.

NIAGARA CHAIN
FOR
PIPE HANGING

Most convenient and neatest as well as the most economical hanger on the market.

Pipe may be hung any distance from ceiling with this hanger.

Permits of easy and exact adjustment.

Write for Samples and Prices

ONEIDA COMMUNITY, Limited
NIAGARA FALLS, ONT.

WALKER CORK SCREWS

are good holiday souvenirs and live sellers the whole year. Are made well and are made out of good material. That is why we guarantee every one that we put out. When ordering Cork Screws try **Walker's**

Manufactured only by **ERIE SPECIALTY COMPANY,** ERIE, Pa. U.S.A.

LUFKIN MEASURING TAPES

Steel, Metallic, Linen, Pocket, Ass Skin,
Pat. Leather, Bend Leather, Etc

ARE THE BEST AND MOST POPULAR TAPES IN THE WORLD.
YOUR STOCK IS NOT COMPLETE WITHOUT THEM.

LUFKIN RULE CO., Saginaw, Mich, U.S.A.

Canadian Factory - - Windsor, Ontario

London Office and Warehouse—24 and 26 Holborn. New York City Branch—[illegible] Broadway.

For sale by ALL PROMINENT CANADIAN HARDWARE JOBBERS.

CIRCULATES EVERYWHERE IN CANADA
Also in Great Britain, United States, West Indies, South Africa and Australia.

HARDWARE AND METAL

A Weekly Newspaper Devoted to the Hardware, Metal, Heating and Plumbing Trades in Canada.

Office of Publication, 10 Front Street East, Toronto.

VOL. XIX. | MONTREAL, TORONTO, WINNIPEG, AUGUST 24, 1907 | NO 34.

Read "Want Ads." on Page 51

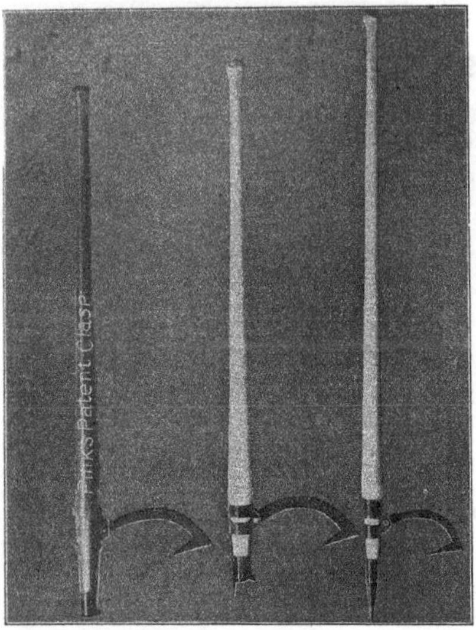

Pink's Lumbering Tools

MADE IN CANADA

Send for Catalogue and Price List

THE STANDARD TOOLS

in every Province of the Dominion, New Zealand, Australia, Etc.

We manufacture all kinds of Lumber Tools

Pink's Patent Open Socket Peaveys.
Pink's Patent Open Socket Cant Dogs.
Pink's Patent Clasp Cant Dogs, all Handled with Split Rock Maple.

These are light and durable tools.

Sold throughout the Dominion, by all Wholesale and Retail Hardware Merchants

MANUFACTURED BY

Long Distance Phone No. 87

THOMAS PINK
Pembroke, Ont., Canada.

Pig Iron

"JARROW" and "GLENCARNOCK."

Agents for Canada,

M. & L. Samuel, Benjamin & Co.

TORONTO

Covertite Roofing

THE ORIGINAL HIGH GRADE OF RUBBER ROOFING

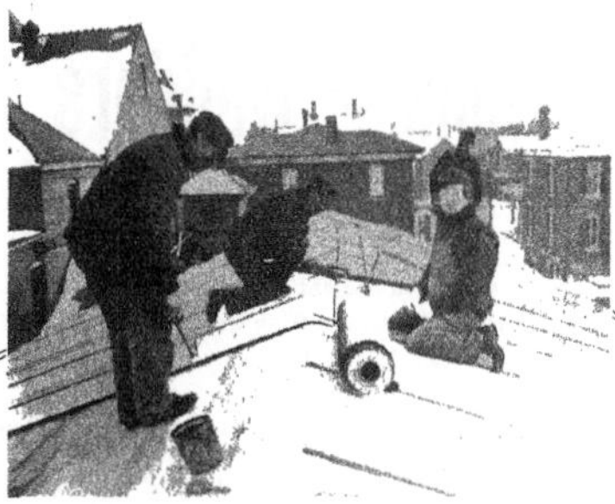

Laying Covertite Roofing in Midwinter.

Mr. Dealer :-

Did you ever stop to consider the responsibility you assume in recommending to a customer a certain kind of Roofing? If the Roofing you recommend gives good service through a sufficient number of years to prove its durability, you are assured of a large patronage as a result of the reputation of these roofs. However, if the Roofing does not give satisfaction, the leakage will be a source of loss and annoyance to those who use it, and you will be certain to suffer a loss of trade and the confidence and respect of your customers.

The fact that a Roofing has kept the water out for a season or two is no assurance of its durability, as it requires at least 5 years' wear to prove that it is worthy of your recommendation.

Owing to the fact that the various brands, good and bad, are so similar in appearance and the representation of some manufacturers to secure trade so questionable, it is necessary for the dealer to be extremely cautious. Do not be deceived by the shrewdly-prepared, worthless guarantee that a number of firms are offering--your best guide is the integrity and reputation of the firm from whom you purchase.

The large trade that we have developed on our various lines is to a great extent due to a truthful representation of the value of each article that we sell, and before adding Covertite Roofing to our list, we have satisfied ourselves fully as to its quality and durability. The fact that our firm name is on every roll of Covertite Roofing makes both dealer and user absolutely secure.

We are confident that all dealers reading the above lines will be interested, and if you will send as your name and address, we will be glad to furnish you with samples, prices and further information.

Yours truly,

EXCLUSIVE CANADIAN DISTRIBUTORS

LEWIS BROTHERS, LIMITED

SHELF AND HEAVY HARDWARE

OFFICE AND WAREROOMS: 14 TO 28 BLEURY STREET, MONTREAL

and at TORONTO, OTTAWA, VANCOUVER, CALGARY, WINNIPEG

BUSINESS SUCCESS

Mr. Dealer, if you could take out a policy assuring commercial success in your line you certainly would do so. The next best thing is to place your orders for

Carpenters' Tools and Builders Hardware

with a specialty house—a house which guarantees quality and prompt deliveries. We handle

Stanley's, Union Hardware Co. and Sarjent's Iron and Wood Bottom Planes.
Disston's, Atkins', and Shurly & Dietrich—High Grade Hand, Rip and Panel Saws.
Sarjent's Front, Store Door and Inside Locks.
Stanley's Butt Hinges—all kinds, all finishes.
Also headquarters for Wood Sash Pegs, sizes ⅞, 1¼, 1⅝-in. in one-bushel bags.

Remember our prices are right and we deliver like lightning

THOS. BIRKETT & SON CO., Ottawa, Canada

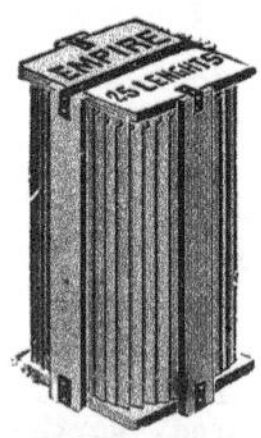

Patented 1900. Improved 1905.

Davidson's "Empire" Stove Pipe

Improved Patent

Neatly packed in crates of 25.
Secures lowest possible freight rate.

Deep Swage Prevents Telescoping.
Coated to Prevent Rust.

Sold as low as ordinary Stove Pipes.

SIMPLE and EASY to put together.
No Profane Language Necessary.
The only tools required are a pair of hands.
Pipes being cut out with dies ensure absolute fit and conformity.

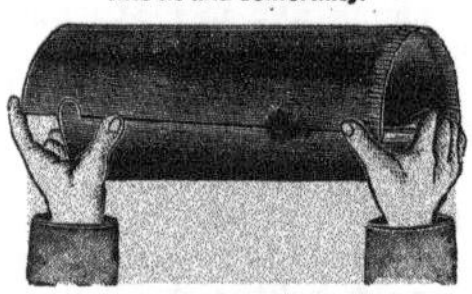

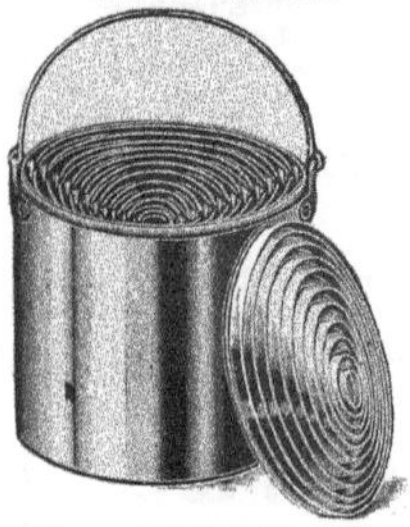

Miners' Camp Kettles

Strong and substantially made in extra heavy tin.

Nineteen sizes, from ⅛ quart to 29 quarts.

Wash Tubs—Galvanized

With Wringer Attachment

Nos.	1	2	3
Top measurement, inches	21½	22¾	25
Bottom " "	17⅞	18¼	20
Height " "	9½	10½	11

The THOS. DAVIDSON MFG. CO., Ltd., Montreal and Winnipeg

H. S HOWLAND, SONS & CO. LIMITED

HARDWARE MERCHANTS

Only Wholesale — 138-140 WEST FRONT STREET, TORONTO. — Wholesale Only

Section of Sample Room—looking East.

Call and see us when in Toronto

Section of Sample Room—looking West.

Our Warehouse is opposite the Union Station

H. S. HOWLAND, SONS & CO., LIMITED

Opposite Union Station

GRAHAM WIRE NAILS ARE THE BEST

Are you receiving our monthly illustrated circular? IF NOT, WRITE FOR IT.

Our Prices are Right — We Ship Prompt

GARNET PAPER

Should be tough, pliable and not liable to crack.

In Sheets, 9 x 11 inches.

H. Barton's Best American have these qualities. It is used by the largest manufacturing concerns and car shops in Canada in preference to any other. You should carry it in stock.

Cuts faster than sand paper.

IN ROLLS

24, 30, 36, 40, 42, 46

INCHES.

Garnet Cloth

In Rolls, 14-inch.

SAND PAPER

Star, in sheets, 8½ x 10½-inch.
Baeder & Adamson, in sheets, 9 x 11-inch
Baeder & Adamson, in Rolls,
24, 30, 36, 40, 42, 48-in.

EMERY CLOTH

Atlas, Improved, in sheets, 8½ x 11½-in.
Baeder & Adamson, in sheets, 9 x 11-in.

WE HAVE A COMPLETE STOCK OF THE ABOVE.

ASK OUR TRAVELLERS TO QUOTE, OR WRITE US.

F. & W. Hardware Montreal

F. & W.

FROTHINGHAM & WORKMAN, Limited

YOU MAY HAMMER

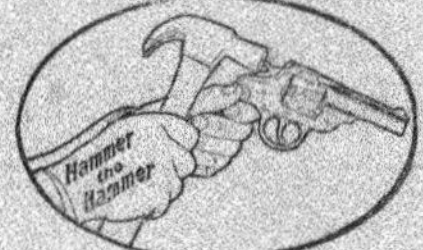

at many things besides **Iver Johnson** revolvers without their going off.

And some of the things that won't go off—your shelves—are inferior revolvers, generally imitations of **Iver Johnson Automatic Safety Revolvers.**

Iver Johnson revolvers sell without effort. They have the call. They are the kind that are advertised. It requires no "hammering" to sell them, and no amount of hammering will discharge them. Accidental discharge is impossible.

We have large stocks of these revolvers in all the different styles.

They are shown in our **Sporting Goods Blue Book** for 1907.

Have you received a copy?

If not, a post card request will bring you one.

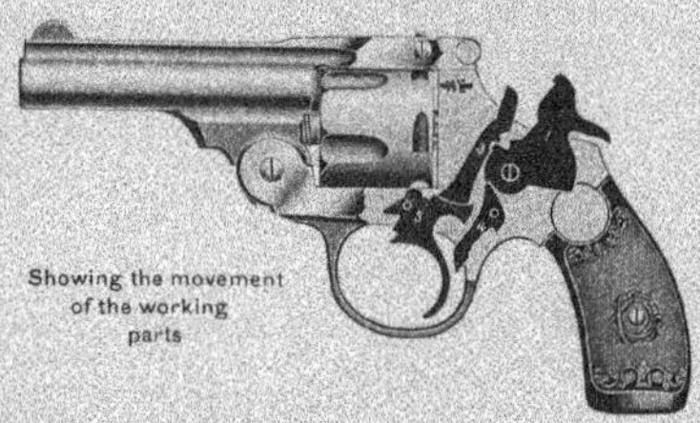

Showing the movement of the working parts

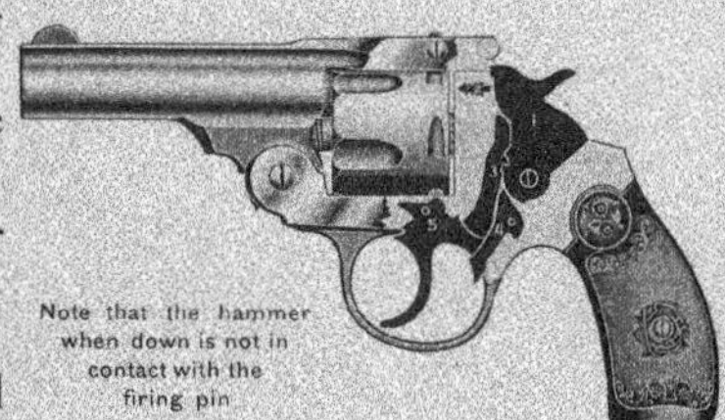

Note that the hammer when down is not in contact with the firing pin

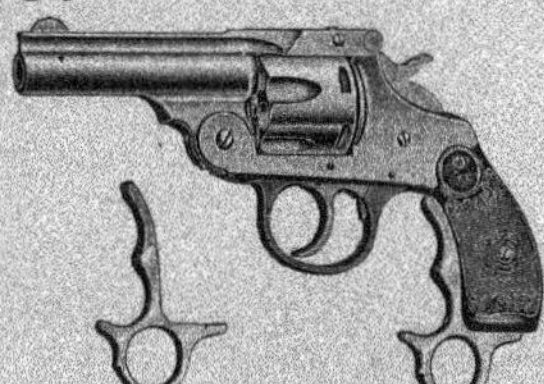

The new grip gives a firm hold on the weapon

The hammerless model made in 32 and 38 S. & W.

Caverhill Learmont & Co

MONTREAL

Winnipeg **Ottawa** **Quebec** **Fraserville**

Simonds
Crescent-Ground Cross-Cut Saws

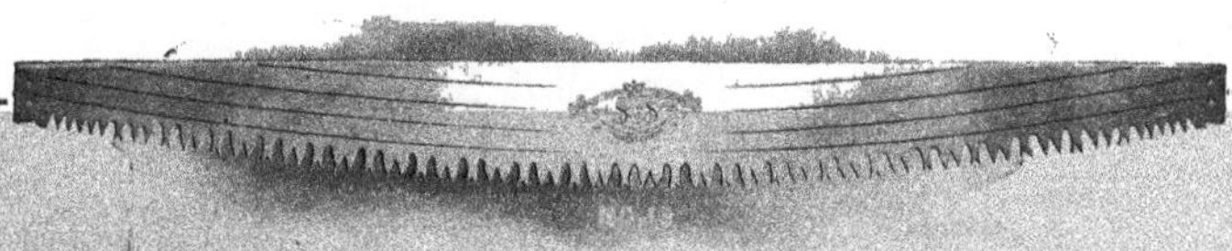

For logging camps where work must be fast and economical there is no other saw made that can give results equal to the Crescent-Ground Saw. **Made of Simonds Steel.** It cuts easy, runs fast and gives good results. This cross-cut saw is sold by most of the leading jobbers in Canada.

Simonds Canada Saw Co., Limited

TORONTO, ONT. MONTREAL, QUE. ST. JOHN, N.B.

Mr. Hardware Dealer ---

We have a proposition : Why not carry a stock of

PRATT AND WHITNEY SMALL TOOLS

and add strength and dignity to your other lines? . . . The quality and accuracy of these tools and the high standard set by the makers in their manufacture, is known the world over. ***Pratt and Whitney Small Tools*** are made, and advertised extensively, in Canada. Discriminating buyers demand and use them. Why not get your share of this business? Attractive terms made to dealers who will carry a stock.

THE CANADIAN FAIRBANKS CO., LTD. TORONTO MONTREAL WINNIPEG VANCOUVER

Exclusive Canadian Selling Agents

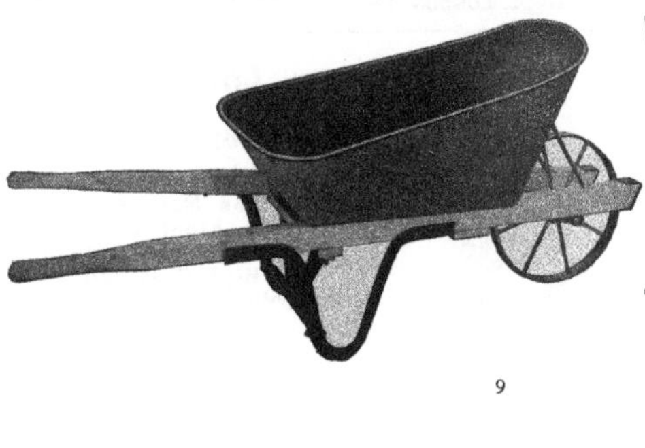

Wheelbarrows

All kinds and sizes. The cut reproduced here is just of one of the many, but the kind every contractor should use. The substantial, satisfactory, steel tray Contractor's Barrow.

The London Foundry Co.

LONDON, CAN.

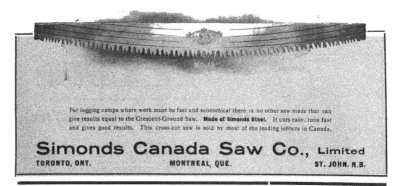

PURITAN
REACTING
WASHING
MACHINE.
ROLLER BEARING
RUNS EASY.
THREE PROFIT PULLERS!
MAXWELL LAWN MOWER
FAVORITE CHURN
The best for Durability, Finish and Efficiency.
Best for the customer and best for you.
David Maxwell & Sons, St. Mary's, Ont.

SIMPLE
DURABLE
RELIABLE
STYLE F.
THE LEADER
A Name Familiar to Butter Makers in Connection with the
Highest Grade Churns
ITS STRONG POINTS:
Stiff Steel Frame, with malleable iron fittings, Combination Bow Lever and Pedal Drive. Bicycle Ball Bearings. Bolted Trunnions. Cream Breakers. Barrel easily drained and conveniently detatchable for cleaning.
DEALERS: When you handle this line you handle QUALITY.
The Dowswell Manufacturing Co., Limited
W. L. HALDIMAND & SON, Montreal
EASTERN AGENTS
HAMILTON
ONTARIO

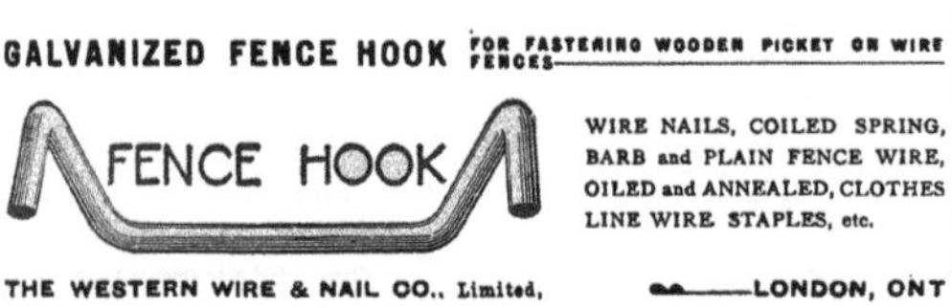
GALVANIZED FENCE HOOK
FOR FASTENING WOODEN PICKET OR WIRE FENCES
FENCE HOOK
WIRE NAILS, COILED SPRING, BARB and PLAIN FENCE WIRE, OILED and ANNEALED, CLOTHES LINE WIRE STAPLES, etc.
THE WESTERN WIRE & NAIL CO., Limited,
LONDON, ONT

PRIEST'S CLIPPERS
BALL BEARINGS
Largest Variety,
Toilet, Hand, Electric Power
ARE THE BEST.
Highest Quality Grooming and Sheep-Shearing Machines.
WE MAKE THEM.
SEND FOR CATALOGUE TO
American Shearer Mfg. Co., Nashua, N.H., USA
Weibusch & Hilger, Limited special New York representatives, 9-15 Murray Street.

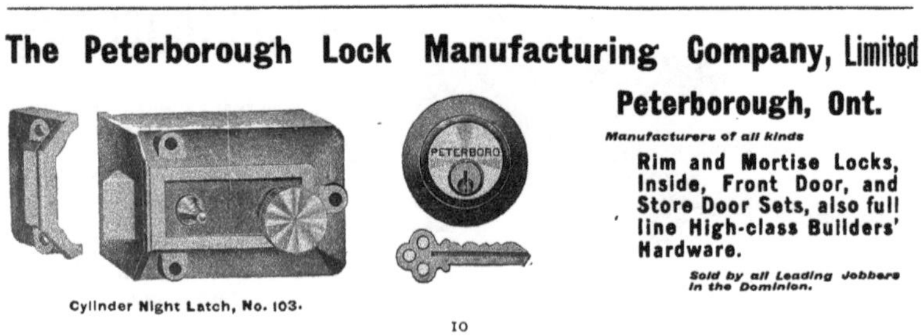
The Peterborough Lock Manufacturing Company, Limited
Peterborough, Ont.
Manufacturers of all kinds
Rim and Mortise Locks, Inside, Front Door, and Store Door Sets, also full line High-class Builders' Hardware.
Sold by all Leading Jobbers in the Dominion.
PETERBORO
Cylinder Night Latch, No. 103.

"KANDA" Reg'd. LIQUID METAL POLISH

BRITISH MANUFACTURE.

Absolutely non-inflammable. Has no bad smell.
Leaves no Verdigris. Contains no acid or grit.
Leaves no greasy smear. KANDA polishes all metal.

KANDA & CO., 45 Basinghall St., LONDON, ENGLAND

SOLE AGENTS for Ontario:

THE KENNEDY HARDWARE CO., Ltd.

TORONTO

BEAVER POST-HOLE DIGGER

will please your customer. A pleased customer is an asset to any business. No wood to rot, check or break.

Special Discounts to Trade

CANADA FOUNDRY COMPANY

LIMITED

HEAD OFFICE and WORKS, TORONTO, ONT.

District - Offices: Montreal, Halifax, Ottawa, Winnipeg, Vancouver, Rossland

Galt "Sure-grip" Shingles

are simply wonders TO SELL, TO LAY AND TO MAKE FRIENDS. Once used and you would never use any other—your customers wouldn't let you.

"Sure-Grips" make a really handsome roof, as well as the most weatherproof. Try them—that's all we ask.

"GALT" CORRUGATED SHEETS are best English galvanized stock, pressed straight and true. Our prices are right, and we can ship promptly.

Consult us before purchasing CEILINGS, SIDINGS, ROOFINGS, CORNICES, SKYLIGHTS, VENTILATORS, FINIALS, FIRE-PROOF WINDOWS, EXPANDED METAL LATH, ETC.

The "GALT KIND" is the line to push; we protect and assist our customers.

THE GALT ART METAL CO., Limited

GALT, ONT.

Buy the Best.

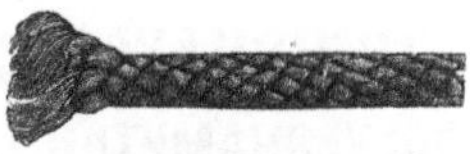

HERCULES

Sash Cord.

Star Brand Cotton Rope
Star Brand Cotton Clothes Lines
Star Brand Cotton Twine

For Sale by all Wholesale Dealers.

Manufacturers' Agents

H. W. MITCHELL
WINNIPEG, MAN.

Correspondence is solicited from manufacturers wishing a representative in Winnipeg. Travellers call regularly on the wholesale and retail trade in Western Canada. Highest references and financial responsibility.

ALEXANDER GIBB

Manufacturers' Agent and Metal Broker, 13 St. John Street, Montreal

Representing Canadian, British and American Manufacturers. Correspondence invited from firms wishing to be represented.

Jardine Hand Drills

New Designs

Get samples of these machines for your show room. They will interest your customers and prove a good investment.

It pays to sell the best Tools.

A. B. JARDINE & CO.
HESPELER - - - ONTARIO

...FULL STOCK...

Salt Glazed Vitrified

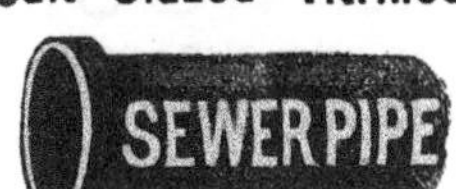

Double Strength Culvert Pipe a Specialty.

THE CANADIAN SEWER PIPE CO.

HAMILTON, ONT. TORONTO, ONT. ST. JOHNS, QUE.

ESTABLISHED 1795

JOHN SHAW & SONS

WOLVERHAMPTON, Limited

WOLVERHAMPTON

ENGLISH GALVANIZED POULTRY NETTING

Advantageous Prices
to Jobbers

J. H. ROPER

CANADIAN REPRESENTATIVE

82 St. Francois Xavier St., - Montreal

The Man on This Fence

shows why the

"DILLON" HINGE STAY FIELD FENCE

is the best for you and your customers. The *hinge* of the *stay prevents* the fence from *sagging*. Take the *weight off* the fence and the *stays spring back*. Your *farmer customers* will buy the "*Dillon*" before any other.

Write and ask us about your profit.

THE OWEN SOUND WIRE FENCE CO., Limited

OWEN SOUND, ONT.

Sold by
Messrs. Caverhill, Learmont & Co., Montreal
Messrs. Christie Bros. Co., Limited, Winnipeg
The Abercrombie Hardware Co., Vancouver

Quick Deliveries **Our Specialty**

Send us your orders for **Cutlery, Electroplate and Solid Nickel Silver Flatware**

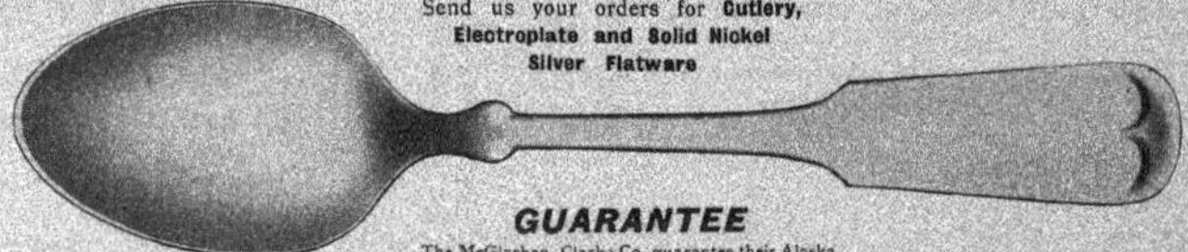

GUARANTEE

The McGlashan, Clarke Co. guarantee their Alaska McG., C. Co. brand of goods to be the same special high grade metal throughout as on the surface. In quality, finish and design they are superior to all others. There being no plating to wear off, they will last a lifetime if properly cared for. Every box contains our guarantee and every piece warranted to give perfect satisfaction.

The McGlashan, Clarke Co., Limited - - Niagara Falls, Ont.

Shear making "up-to-date" is found in

HEINISCH

TAILORS' SHEARS,
TRIMMERS, SCISSORS,
TINNERS' SNIPS, ETC.

"Best in the World"

Since 1825, and a steady increase in sales due to their superior cutting quality.

R. HEINISCH'S SONS CO.
NEWARK, N.J., U.S.A.
New York Office and Salesroom, 155 Chambers St.

TRADE WITH ENGLAND

Every Canadian who wishes to trade successfully with the Old Country should read

"Commercial Intelligence"

(The address is 168 Fleet St., London, England.)

The cost is only 6c. per week. (Annual subscription, including postage, $4.80.)
Moreover, regular subscribers are allowed to advertise without charge in the paper. See the rules.

Persons addressing advertisers will kindly mention having seen their advertisement in this paper.

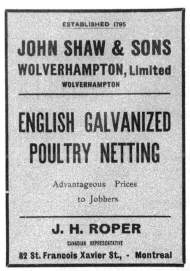

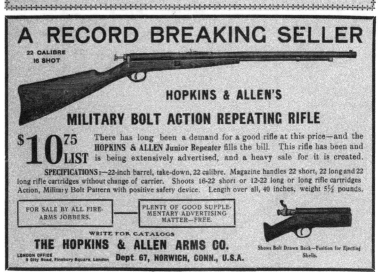

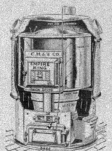

"GOOD CHEER" STEEL RANGES

THE

SUNRAY

A moderately priced
first-class steel range

4-hole top, with 18 inch oven
6 " " 20 "

with cast base or on feet.

The attractiveness of design speaks for itself, and we guarantee the quality A1.

THE JAS. STEWART MFG. CO., LIMITED, WOODSTOCK, ONT.

Western Warehouse, James Street, Winnipeg, Man.

The Points That Sell a

warm air register, are **economy** and **efficiency**. These are the points, easily proven, which make the

JONES SIDE-WALL REGISTER

the most saleable register on the market. **Its construction** makes it the **cheapest to instal** and the **healthiest** and **most satisfying** in daily use.

Write and ask us for the proof!

The Jones Register Co.

732 King St. West ***Toronto, Canada***

Wrought Steel Registers

Standard Sizes, All Finishes
Perfect Operating Device
Unusually Large Air Space
For Sale by Leading Jobbers

HART & COOLEY CO.

New Britain, Conn., U.S.A.

H. & C. No. 100.

FERROSTEEL REGISTERS

The only house in the world manufacturing

CAST IRON FLOOR
SEMI STEEL FLOOR
WROUGHT STEEL FLOOR
CAST IRON SPECIAL SIDE WALL
WROUGHT STEEL SPECIAL SIDE WALL
FLAT OR CONVEX EXTRA SHALLOW SIDE WALLS
DEFLECTORS, PANCAKES, WAFERS

In all Designs and Finishes.
Mechanism Perfect.
Capacity Greatest.
Finish Unsurpassed.
No reputable line can undersell us.

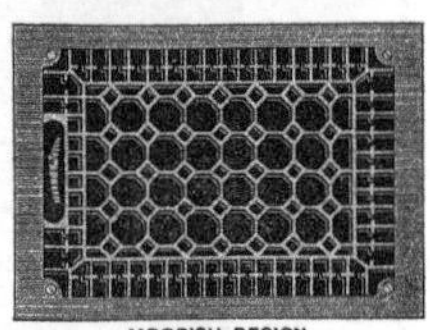

MOORISH DESIGN

FERROSTEEL COMPANY, - - CLEVELAND, OHIO

"Brantford" ASPHALT & RUBBER "Roofing"

The Climate of Canada probably is more severe on Roofing material than any other. Shingles warp and curl, Tin and Iron sweats and rapidly rusts out. Corrugated Iron expands and contracts until it is useless as a protection against the elements. Coal Tar runs and clogs the drain pipes and gutters in Summer and cracks and opens up in the cold of Winter. To meet and overcome these difficulties "Brantford Roofings" are manufactured. For more than twenty-five years this Roofing has been made in the United States, under another name. WRITE FOR PRICES AND SAMPLES.

BRANTFORD ROOFING CO., LIMITED - Brantford, Ont.

You May Be a Good Salesman

But, is everything in your favor? Are you selling the new Imperial Oxford? A continuous stream of repeat orders tells the story to us.

Removable Nickel
Removable Fire Linings
An Oval Firebox
The "Divided Draft"

All contained by the best looking stove in the market.

It gives a good salesman a chance to spread himself.

Drop us a card, or, better still, tell us the size range you want on your floor. If you get this agency you're lucky.

The Gurney Foundry Company, Limited

Toronto Winnipeg Vancouver Hamilton and London

THE GURNEY-MASSEY CO., LIMITED, MONTREAL, Que.
THE GURNEY STANDARD METAL CO., LIMITED, CALGARY, Alta., and EDMONTON, Alta.

STOVE BOARDS

WITH SIZES STAMPED ON BACK AND CORNERS CRIMPED

Crystallized and Lithographed Wood-lined

Square

26 x 26 in. 28 x 28 in.
30 x 30 in. 32 x 32 in.

Oblong

24 x 32 in. 25 x 36 in.
28 x 41 in.

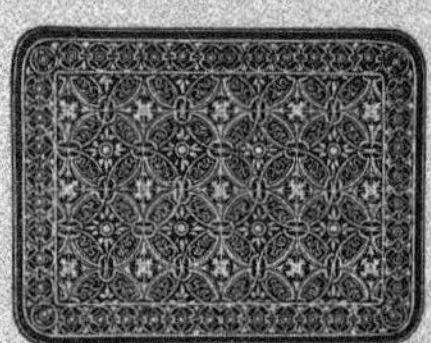

Crystallized and Stencilled Wood-lined

Square

26 x 26 in. 28 x 28 in.
30 x 30 in. 32 x 32 in.
36 x 36 in.

Oblong

24 x 32 in. 25 x 36 in.
28 x 41 in.

NEW AND ATTRACTIVE DESIGNS FOR 1908

We Ship Orders Quickly

The McClary Mfg. Co.

LONDON, TORONTO, MONTREAL, WINNIPEG, VANCOUVER, ST. JOHN, N.B., HAMILTON, CALGARY

I WILL TALK

to practically every Hardware merchant in Canada from the Atlantic to the Pacific. I cannot do it all in one day, but during the first twenty-four hours I will deliver your message to every Hardware merchant in Ontario. I travel all day Sunday and on Monday morning there will not be a village within the limits of Halifax in the East and Brandon in the West, into which I will not have penetrated.

I cannot go any further East, so I now devote all my energies to the West, and so many new towns are springing up here each week that I haven't as much time as I used to have to enjoy the scenery. But I like talking to hardwaremen, clerks, travellers and manufacturers, especially as they are always glad to see me and hear the news I have to tell them. Tuesday noon I am at Calgary, Wednesday noon at Kamloops, and by Thursday morning I reach Vancouver, having been in all the mining towns and all through the fruit districts of British Columbia.

I have been eighteen years on the road and I have a pretty good connection. I never intrude when a man is busy, but just bide my time, because I know men pay far more attention to what you have to say if you catch them when they have a few moments to spare. So I often creep into their pocket when they are going home at night, and when supper is over Mr. Hardwareman usually finds me. He must be glad to see me, because he listens to what I have to say for an hour or more.

I try to always tell the truth, and men put such confidence in what I say that I would feel very sorry to deceive them even inadvertently. Probably some other week I will tell you about the different classes of people I meet. In the meantime if you want a message delivered to HARDWAREMEN, PLUMBERS, CLERKS, MANUFACTURERS or TRAVELLERS—and want it delivered quickly—I'm your man.

THE WANT AD MAN

Condensed Advertisements in Hardware and Metal cost 2c. per word for first insertion, 1c. per word for subsequent insertions. Box number 5c. extra. Send money with advertisement. Write or phone our nearest office

Hardware and Metal

MONTREAL **TORONTO** **WINNIPEG**

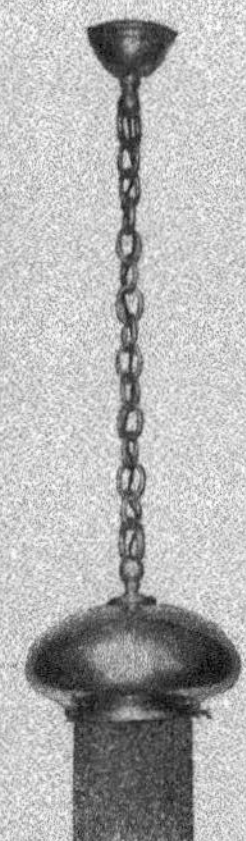

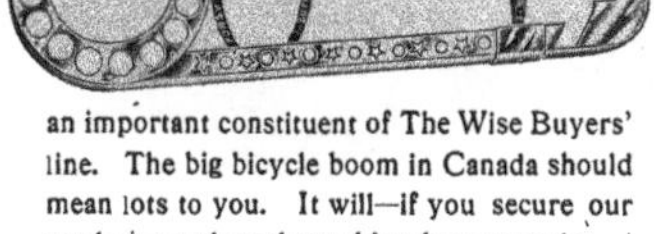

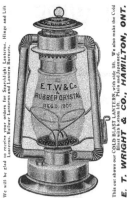

Recommend The Best

Shrewd dealers not only recommend the best window glass—they also stock it. **There's a reason for it.** Once your customer becomes convinced that his interests are yours, you've secured his trade for all time.

While high-grade, almost imperishable,

WINDOW GLASS

is our specialty, we also manufacture **Fan Lights, Door Lights, Hall Windows, Side Lights, etc., made in Copper, or German Silver Bars.**

Our catalogue should be on every dealer's desk. A post card brings it to you. Write to-day for particulars.

The Hobbs Manufacturing Co., Limited

LONDON TORONTO WINNIPEG

McCaskill, Dougall & Co.

Manufacturers RAILWAY, CARRIAGE AND BOAT **VARNISHES.**
HIGH-GRADE FURNITURE and HOUSE VARNISHES

MONTREAL.

Above is a reproduction of Show Card 9 in. x 15 in. in colors which we will send free to dealers in exchange for business card.

HENRY T. SEYMOUR SHEAR COMPANY. WIEBUSCH AND HILGER, LIMITED, Sole Agents, NEW YORK

A Certain Sense of Satisfaction

goes with every MERRELL PORTABLE HAND MACHINE, PIPE THREADING and CUTTING MACHINE. We build them that way.

For doing ALL kinds of threading—under ALL conditions—labor trouble and faulty material included—where speed, durability and ease of control count for something, our word to you is to investigate the MERRELL.

Whether it be stationary or portable—hand or power driven—30 DAYS FREE TRIAL must prove conclusively that your choice should be the MERRELL.

This Portable Hand Machine has encased gears, MERRELL standard quick opening and closing die-head, and the latest improved Cutting-off Knife.

Let us tell you more about this machine—The Chasers, Vise and the large range of work covered.

Catalogue for the asking. Quick shipments.

THE CANADIAN FAIRBANKS CO.
Sole Agents for Canada Limited

MONTREAL, TORONTO, WINNIPEG, VANCOUVER

Your Customer is Annoyed

when the handle of his axe breaks at the critical moment. **He blames you for it.** But he will **never be annoyed** this way if you sell him only

Empire, Imperial and Champion

AXE HANDLES

They are the strong kind. They are made to be used; and it takes a lot of abuse to break 'em. The same qualities of durability are found in every **HAMMER, SLEDGE and PICK HANDLE** shipped from our factory.

Free lists and discounts sent upon request.

J. H. Still Manufacturing Co. Limited

ST. THOMAS, ONTARIO

The Buffalo Manufacturing Co.

Buffalo, N.Y.

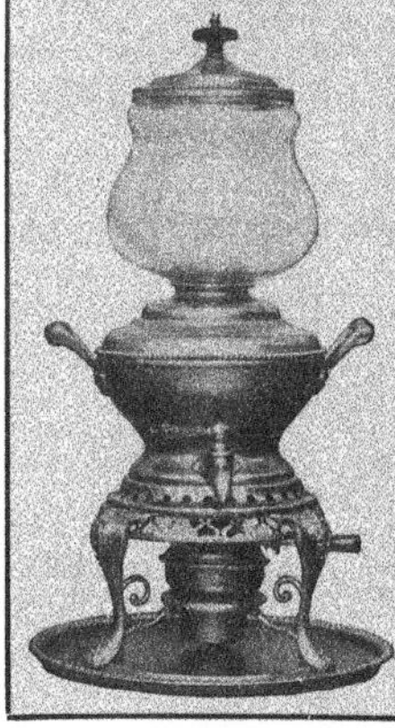

When you get our goods you know you get THE BEST.

We manufacture

Water Filters
Water Coolers
Chafing Dishes
Table Kettles and Stands
Coffee Extractors
Wine Coolers
Nursery Chests
Baking Dishes
Crumb Trays and Scrapers
Tea and Bar Urns
Bathroom Fixtures
Coal Vases and Hods
Candlesticks
Cuspidors
Match Safes, Etc.

All High Grade, and exceedingly presentable.

Represented by

H. F. McINTOSH & CO.

51 Yonge Street,

TORONTO, - ONT.

Write for Catalogue

TRADE MARK

FOR PLATE

Will Call on You

Our Canadian representative is calling on the trade from Montreal to Vancouver. His trip will occupy about three months. He will be pleased to show you our complete line of Silverware, Cutlery, Flasks, etc.

Kindly address the Montreal office, making an appointment.

The House of Hutton

Established in England in 1800

Sole Canadian Agent

W. J. GRANT, Lindsay Bldg., MONTREAL

TRADE MARK

FOR CUTLERY

WE WANT YOUR ORDERS

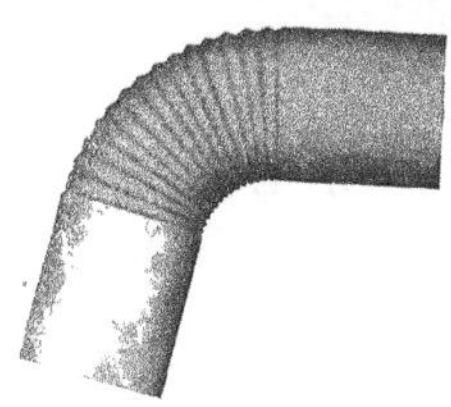

The season for fall repairs is almost here. Are you fortified with a good stock of our

Troughing, Pipe and Elbows?

Our lines are made with greater care, and are, therefore, erected more easily and with less solder than any other. Send us a trial order.

"WE WORK WITH THE TRADE."

The Metal Shingle & Siding Co., Limited

Montreal **Preston, Ont.** **Toronto**

SALES AGENTS:

ST. JOHN, N.B. Emerson & Fisher, Limited
CALGARY Ellis & Grogan
QUEBEC J. A. Bernard
WINNIPEG Clare & Brockest
VANCOUVER McLennan, McFeely & Co., Limited

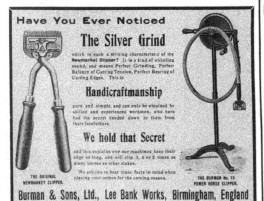

IT'S EASY TO SELL

a rifle which you are sure will not burst and injure your customer. The superior point about

HAMILTON RIFLES

is that they simply **can't burst.** Only really skilled workmen are employed in making them, and only **specially selected** and **tested rolled steel** is used in their manufacture.

No. 15 sells at $2.00
No. 19 sells at $2.50
No. 23 sells at $3.50

WRITE AND ASK US ABOUT THE PROFIT STORY

THE HAMILTON RIFLE CO.
Dept. 71 ***Plymouth, Mich.***

PAGE WHITE FENCES

The Acme style you see above costs only from 16 cents a linear foot. Handsome and durable for lawns, parks, cemeteries, etc. Any height or length. Gates to match from $2.25. Write for catalog, or inspect this ideal fence. From us direct or any Page dealer.

The PAGE Wire Fence Co., Limited Walkerville Toronto Montreal St. John Winnipeg 812

BIRKMYER'S WATERPROOF FLAX-CANVAS

Thousands in use.

Best Waterproof in the country.

Just as cheap as the common ones, but **infinitely better.**

For Horses, Wagons, Feed-bags, Capes, etc., etc.

All styles and sizes in stock.

Samples cheerfully furnished

TOBIN TENT, AWNING and TARPAULIN CO.

Ottawa	*Toronto*	*Montreal*
201 Sparks St.	*125 Simcoe St.*	*28 St. Peter St.*

Persons addressing advertisers kindly mention having seen their advertisement in Hardware and Metal.

TINPLATES

All made out of the

BEST SIEMEN'S MARTIN STEEL

Brand	
Best Coke or Charcoal	"Rhyd"
Good Charcoal	"Edith"
Second Charcoal	"Sawel"
Best Charcoal	"Dryw"

We are ready to offer the Canadian Trade very best value and finest quality in Tinplates.

Baglan Bay Tinplate Co., Limited
Briton Ferry, South Wales

Please mention above brands in your indents and inquiries.

OAKEY'S The original and only Genuine Preparation for Cleaning Cutlery, 6d. and 1s. Canisters

'WELLINGTON' KNIFE POLISH

JOHN OAKEY & SONS, Limited
Manufacturers of
Emery, Black Lead, Emery, Glass and Flint Cloths and Papers, etc.

Wellington Mills, London, England

Agent:
JOHN FORMAN, - 644 Craig Street
MONTREAL.

NICKEL

The Canadian Copper Co.

NICKEL FOR NICKEL STEEL

The Orford Copper Co.

WRITE US FOR PARTICULARS AND PRICES

General Offices
43 EXCHANGE PLATE
NEW YORK CITY

Hardware Convention at Detroit

The editor of Hardware and Metal spent a few days in Detroit last week attending the thirteenth annual convention of the Michigan Retail Hardware Association. During the thirteen years' existence of this organization a record for progressiveness has been built up, the tangible result being an association comprising 677 members enthusiastically paying an annual membership fee of $4.

Michigan's need for a retail hardware association is no greater than Ontario's, but the lesson of organization was early learned and there are now few reputable retail hardware firms in the state outside the organization. And the interest shown in the annual convention was amply proven by the attendance of about 550 of the members at Detroit in spite of the fact that all efforts to secure special rates from the railway companies were futile. Paying full fares going and coming the hardwaremen gathered from every part of the state, many bringing their wives and daughters with them. Fully 75 per cent., however, of the delegates in attendance were young men under 35 or 40 years of age, this giving a vigorous tone to the gathering and making the meetings lively and entertaining.

The object of the editor's visit was to see how our United States cousins conducted their association and learn from their experience lessons which may be of value to the hardwaremen of Canada, who as yet, are comparatively new to association work.

The first fact which impressed itself upon the visitor was the marked interest manifested in the convention by the large number of delegates present. True, some attended the ball game, enticed away by jobbing house salesmen overanxious to entertain customers, but there were few vacant seats left in the large convention hall during any of the sessions, and the delegates listened attentively to the various addresses and discussions.

Pure Paint Legislation.

The subjects demanding the attention of the delegates were much the same as those which would interest a Canadian convention of retail hardwaremen. First, came the subject of pure paint, a topic which has won considerable importance since the praiseworthy efforts of Prof. Ladd, of North Dakota, to check the sale of cheap, adulterated mixed "paints" and white lead put on the market by mail order houses and unscrupulous manufacturers. Representatives of the Paint Manufacturers' Association and National Lead Company attended the Detroit convention and delivered addresses, pointing out that they approved of Prof. Ladd's work and were assisting in his experiments as well as in getting pure paint laws adopted by the various state legislatures. The growing importance of this subject is worthy of note, as the Ontario Retail Hardware Association executive now have the matter of white lead adulteration and marking under consideration.

Another important subject discussed was the ever-present catalogue house competition. The delegates were cheered by the announcement that, owing to the refusal of many large hardware manufacturers to allow their goods to be listed in mail order catalogues two large Chicago mail order concerns had withdrawn from their catalogues the 60 to 80 pages devoted to hardware lines. While the campaign against this form of selling is producing good results, the evil still manifests itself in many ways, however, many delegates telling how consignments of catalogues were sent to express and freight agents for delivery to farmers and others. The railway companies have given instructions to refuse to deliver them, however, and one agent who persisted in doing so nearly lost his job as a result of the activity of the association secretary.

One hardwareman who had been a telegraph operator told of how he had relieved an operator recently to allow the fellow to attend a funeral. In the office he found seventeen undelivered mail order catalogues addressed to customers of his store and enquiry proved that 27 boxes of goods came to the depot in one day, through sales encouraged by the operator. Needless to say, the hardwareman got busy and the operator found it advisable to cut out that line of work.

Parcels Post Legislation.

A hardware jobber notified the convention that he had learned that the catalogue house interests were preparing to have legislation introduced at the next session of Congress providing for the establishment of a parcels post system, enabling local merchants to send parcels up to ten pounds in weight to their customers by the rural mail carriers, etc. The plan is to disarm the opposition of the small retailers by having the law appear to be in their interest and then at the last minute an amendment would be introduced giving the large mail order concerns the same privileges extended to the little fellows. Truly a wily game, but the cat is out of the bag.

Some very effective methods of opposing mail order business were outlined at the convention, reference to which will be made in future issues when some of the addresses delivered will be reproduced.

Various trade evils were discussed, thirteen years of organization having failed to make the state a paradise. Such evils as selling to consumers, etc., were not as much in evidence, however, as in Ontario, and the general feeling seemed to be far more friendly, few stories of price-cutting being heard.

Prominence was given to the subject of mutual fire insurance, one man telling of having received $36 as a refund on a payment of $90 made to mutual hardware companies at old line company rates. Secretary Scott stated that much of the success of the Michigan association was due to the benefits received from mutual insurance.

Secretary Peck, of Wisconsin, and National Vice-president Stebbins, of Minnesota, delivered addresses on mutual fire insurance, the former telling of how the Retail Hardware Association in Illinois had been built up by canvassers for the hardware mutual companies, it being necessary to belong to the hardware association before a mutual policy could be obtained.

Manufacturers' Exhibits.

A feature of the gathering was the large number of exhibits made in the hotel parlors, rooms and corridors by jobbers and manufacturers selling to the hardware trade. The Cadillac Hotel provided the convention hall, rooms for executive, offices, etc., making no charge to the association. They recompensed themselves, however, by charging for space for exhibits.

Another feature of the work in Michigan is the publication of an annual souvenir, the jobbers and manufacturers contributing liberally for advertising space. The $4 fee does not cover all the expenses of the association, as the large amount of work requires a secretary and an assistant secretary, both of whom receive small salaries. The deficiency is made up from the "souvenir" receipts. In Ontario the association has always stood entirely independent, never having asked or received any money from the jobbers or manufacturers, although it is understood the latter intend to provide some entertainment at the next convention. The entertainment at Detroit consisted of an evening at an amusement park, a moonlight and banquet and auto rides for the visiting ladies. The exhibitors also handed out souvenirs, some of which were quite valuable.

The "Question Box" was, of course, one of the most interesting and valuable features of the gathering. The "box" was open on the chairman's table during each session of the convention, the afternoon of the third day being devoted to the discussion. Keen interest was manifested in the short talks by the various members on subjects introduced through the questions asked.

The gathering, on the whole, was a huge success and undoubtedly resulted in much advantage to the trade. Men who get together once a year and talk over methods of business with men in their own line in towns 50 or 100 miles away are undoubtedly going to learn a few things and be more friendly with their nearby competitors on matters of price-cutting, early closing, etc. The money spent by retailers in attending the convention and by those who provided the entertainment will be returned many times over in all probability.

About Private Price Marks

Some Important Secrets of the Cost and also Prices Ciphers of the Big Retail Stores that may be Used by the Country Retailer if Desirable.

Almost every merchant has a secret price mark. It is a business necessity. The greatest point in its favor is that the dealer who marks his articles with a code is not subject to comparisons of price, which the one price dealer always has to complain of.

The majority of dealers have a secret code from which to determine the price. A great many of them so arrange this code that they can tell the cost as well at a glance.

Most frequently the price code is arranged on a word of nine or ten letters, in which word there are no two letters alike. The letters correspond with the numbers from one to ten, like this:

m a n u s c r i p t
1 2 3 4 5 6 7 8 9 0

This is the commonest and easiest way of marking the price. All dealers are familiar with it, but they seldom change their key-word, on account of the difficulty of looking up a new word or the bother of becoming accustomed to the new code.

It is essential that the key be changed often, for when two or three persons, aside from the clerks, know the key it is useless as far as any idea of secrecy is concerned.

Nine Little Price Marks.

Here is a list of nine-letter words which can be used for keys. They are all adapted to the use, they offer great variety, and are difficult to decipher. These being nine lettered words, the o, or last number, can be used as it stands, or any other letter or sign substituted for it.

Thus.

s i g n a t u r e
1 2 3 4 5 6 7 8 9 0

Similar nine letter words are:

cavernous	machinery
blasphemy	manifesto
clerkship	monastery
dangerous	nocturnal
drinkable	numerical
dropsical	obscurity
duplicate	observant
factious	outwardly
fisherman	outspread
gunpowder	prudently
hamstring	pneumatic
harmonize	porcelain
heptacord	subaltern
labyrinth	voluntary
longevity	

"Abruptness" also may be used in this class by substituting O for the final "s."

The advantages of the nine letter word are many, they are simpler to remember, simpler to read (on account of the O), and more inexplicable than the ten letter words. The nine letter word, however, is not used as often as the ten letter word.

List of Ten Letter Words.

The following list of keys of ten letters will aid the dealer in picking out a new price mark:

background	lachrymose
birthplace	manipulate
blacksmith	manuscript
chivalrous	plastering
daughterly	pneumatics
deaconship	profligate
fishmonger	formidable
handsomely	phlegmatic
importable	forgivable

The following eleven letter words, also may be used by dropping the final letter, or using it as a repeater:

Candlestick
Disturbance
Neighborly

The two words, "birthplace" and "blacksmith," in the ten letter list, are well fitted for pricemarks. Being double words and each word containing five letters, they are much easier to remember and far simpler for the clerk to decipher.

The simplicity is seen from the above, the first word ends with the number five and the second begins with number six, the letters being equally divided in the compound word makes the key much simpler.

Using Hyphen for Figure.

But even more suitable is the hyphenated word. There are few, if any, dealers who have discovered and used this class of words for keys. They are ideally adapted to such use.

h a i l s t o n e
1 2 3 4 5 6 7 8 9 0

Other hyphenated words, which are suitable for price marks, are:

back-slide
hail-stone
make-shift
ship-board
shop-lifter
yard-stick

The word "fish monger" also may be used, the same as "shoplifter," by cutting off the final letter.

Five is the most frequent number occurring in price marks. There usually is a five somewhere in the combination, and for this reason the type of words above makes the ideal work key. At a glance the clerk can read the dash or hyphen and it is confusing to the curious customer, who cannot account for a figure in a letter code, unless it represents a figure one or a naught. The five never is thought of by the investigator and easily thought of by the clerk. In those two points lies the beauty of such a word for a price mark.

Letter X for a Repeater.

The figure x, as a rule, is used for a repeater thus, with "manuscript" for the key;

Mstx—$V.

It will be found in key words that the p's and m's predominate and words beginning with those letters are used by the majority of dealers. For that reason it would be advisable to pick out one beginning with a letter other than p or m.

There is some humor in price marks; for instance, "shoplifter," as a constant reminder and warning to the clerk; "yardstick," how ironical it must sound to the "counter jumper." Think of a clerk determining the price of face powder from the key "gunpowder." And how suitable the code "labyrinth" is to the new clerk winding in and out among the letters of the word in a vain effort to find the price of a 5-cent cake of soap.

With the introduction of the spelling reform there has been an innovation in price marks. It opens up a wide field, and articles marked on such a key surely are inexplicable. For instance, among the novelties are the following:

Sell th gudz.
Charg enuf.
Plez be onist.

The keys are novel and intricate, but it is not seriously thought, however, that many of the dealers will adopt this scheme of marking.

Phrases Better than Words.

Some merchants use phrases instead of words. For instance, a big wholesale jewelry house in New York uses the legend, "Now be sharp." A firm in Chicago uses, "Be watchful." Some houses think that phrases have certain advantages over mere words.

Many merchants form keys upon their business names, for instance:

Klein Bros.	C. L. Hornsby.
Born & Smith	Jones Drug.
Yalding Co.	S. & G. Penfold.

There is a personality about such a key that makes it of value to the merchant, and it is, as a rule, hard to decipher.

A simple device, which is preferred by many, is plain figure marking, The real figures are written down as a stock number, in a row.

For instance, in 2468793: $4.80 would be the price mark. Beginning wih the first number and using every other number as a blind, this is read easily and not as liable to mistake as the word method. This may be varied, beginning with the first and skipping every other one, reading backwards, and divers other ways. The merit of this lies in its readability and simplicity, together with the fact that it is usually taken for a stock number, instead of a price mark.

First Figure is Blind.

An even simpler way is to use merely one blind, as, 6,148.

Price would be $1.48.

Any letter or figure may be used for the blind.

One objection to this is that an article marked to sell at less than a dollar often is sold for more, and the scarcity of figures would lead the customer to believe that the price came in two, rather than three figures. This easily may be overcome by using an extra figure, a plus sign, or an x, before the number to make the price look greater.

This method often is used as a cost mark, the selling price being found by adding whatever per cent. profit is to be figured.

A new method of cost marking which would be almost impossible to solve, and yet one which the clerk could handle and learn easily, is the following:

For example, 61483: 6 is the blind; 148 is the cost; 3 is the figure which, multiplied by the cost, will give the selling price.

Curiosities in Cost Mark.

Often a buyer picks up a line of goods at a bargain which he either can afford to sell cheaply as a leader, or which he prefers to get a better profit from. By this system of marking he can get whichever price he wishes and still know the original cost simply by looking at his price mark.

There are some curiosities in price marks, which can be seen once in a while. For instance, a haberdasher in Clark street, in Chicago, uses Stevenson's "dancing man." The marks are curious but almost impracticable.

The solution is that the number correspond to the various parts of Stevenson's famous "dancing man.'

The various limbs must be made with care and even then there are frequent mistakes in the reading. The adoption of such a cipher scarcely is advisable.

Probably the best mark possible never has been used. It is most simple and could be adopted with great safety for it is practicaly unknown. There is no code to it, nothing to remember, nothing to figure out. It is simply this, in putting down a price mark merely use the preceding figure in the numerical scale instead of the real figure.

An article to be priced $17.50 would be marked 0649. What could be more simple? What could be more practical?

Or, by reversing the process the article priced at $17.50 would be marked 2861, according to the following table:

2 3 4 5 6 7 8 9 0 1
1 2 3 4 5 6 7 8 9 0

There are thousands of different price marks in use to-day, and more coming in, as the advisability of an exclusive mark grows upon the dealer. It is a business necessity, as is the frequent changing of the code or cipher for the sake of protection.

R. C .BROWN.

LETTER BOX.

Correspondence on matters of interest to the hardware trade is solicited. Manufacturers, jobbers, retailers and clerks are urged to express their opinions on matters under discussion.

Any questions asked will be promptly answered. Do you want to buy anything, want some shelving, a silent salesman, any special line of goods, anything in connection with the hardware trade? Ask us. We'll supply the necessary information.

L. P. Beaulue, Hawkesbury, Ont., writes: "Could you direct me where I could get a book or periodical that would treat on how to dress windows and make displays in a store where gents' furnishings, boots and shoes, and ready-made clothing are kept?"

Ans.—A monthly periodical called the Merchants' Record and Show Window is the best medium we know of treating on window dressing and store display. This paper is published by the Merchants' Record Co., 315 Dearborn St., Chicago, Ill.—Editor.

Karl Freeman, Bridgetown, N.S., writes: "Can you direct me to a maker of bottling and bottle-washing machinery?"

Ans.—Communicate with J. J. McLaughlin, Limited, manufacturing chemists, 145-155 Sherbourne St., Toronto, and they will give you the desired information.—Editor.

Retail Store Development.

To the Editor :—

In Hardware and Metal of the 17th inst. you call attention to the department store development and give illustrations of the building additions being made in Toronto by the leaders of this class of business in Canada, namely, the T. Eaton Co. and the R. Simpson Co.

These establishments have been steadily growing larger. They must therefore be serving a legitimate purpose in the life of the community. It may therefore be profitable to glance at their methods of doing business as seen by an outsider:

Departmental stores are, as a rule, conducted without giving credit. The Hy. Morgan Co., Montreal, have recently ceased to give credit. They use printers' ink freely in advertising catalogues and display cards. They dispose of seasonable goods before the end of the season for their use, at any old price to clear so as to start the next season as nearly as possible with a fresh stock of the latest designs and patterns. This method accomplishes the removal of stock that would otherwise occupy valuable space and require rehandling several times before the next season. It makes cash available for use. It is the feature that gives the impression that they sell cheap and attracts those who are looking for bargains. And what thrifty wife is not on the lookout for a bargain?

There is no overlapping of lines in one department with those sold in another, as is the case in the different branches of the retail trade.

Their salespeople have short hours (8 or 9) and are consequently alert and in condition to discharge their duties.

If these methods are compared with the system of conducting retail businesses, it will not be wondered at that they grow so rapidly. The reader will commence by this time to wonder why this article is headed Retail Store Development. It is preparatory to outlining what I feel confident must be the way of development for the retail merchant; that would distribute trade more evenly and not require the long hours that the retailer is largely compelled to put in at present, trying to make a success of his business, often at the sacrifice of his health. The late John Macdonald stated in a book he wrote that nine out of ten men failed in business at some period of their life. The writer is one of them.

Rev. Prof. Bland, in a sermon last Sunday in Sherbourne St. Methodist Church, Toronto, on "The Ideal City," stated that business competition stood in the way of many men being Christians; that competition was, in a sense, war. The writer some years ago heard him state that the man that brought goods together which he required, and sold them to him, was as much entitled to his thanks as the merchant had to thank him for his buying from him, and this is so.

Business should be directed towards the doing away as far as possible with undue competition.

In the manufacturing and transportation businesses it has been found desirable to co-operate and amalgamate. I believe it will be found desirable and necessary for smaller concerns to pursue the same course, namely, to cooperate and learn to work together.

This will take time to bring about but it will be the part of wisdom for two of the same class of business serving the same community to seek to cooperate and amalgamate; then seek to amalgamate (in large places) the next congenial business and keep doing so until a similar business to a departmental store is evolved, only with this difference that at the head of each department would be one interested in the welfare of the business with a knowledge of its special requirements. Such a local concern could give its customers a service that would be more satisfactory and quicker than the departmental store. These stores could be scattered over a wider territory where they would be convenient to the homes which require the lines sold. With your kind permission, I will reply to any questions sent you (the editor) during September, bearing on this matter.

"Retail Store Development."

LOW-LIFT PLOW.

D. Halloran, Paris Station, Ont., has invented an improved plow. The object of this inventor is to provide means to enable the plow to make a short turn; to permit it to be readily adjusted to regulate the depth of cut, to regulate the width of cut of the furrow; to enable its being readily lifted out of the ground or placed therein; to enable the rear furrow wheel to be locked in position or to run at will, and to provide means whereby the wheel and pivotal connection of the plow with the main frame may be operated by the main lever.

HARDWARE TRADE GOSSIP

Quebec.

J. F. Berrill, Melbourne, Que., called last week in Montreal.

Harry Moulden, of Taylor, Forbes Co., Guelph, was in Montreal again this week.

J. A. Paquin, St. Eustache, and C. O. Jervas, St. John, were in Montreal this week.

Mr. Bright, of Zip Mfg. Co., Sutton, Que., called on the trade in Montreal this week.

Mr. Fronhoffer, manager of the Orillia Construction Co., was in Montreal last week purchasing supplies.

G. A. Jordan, of Caverhill, Learmont & Co., Montreal, is spending his holidays on the Maine coast.

Mr. Purvis, Calgary, was in Montreal this week, where he met his wife, who has returned from England.

Banks Rucker, of Peck, Stow & Wilcox Co., Southington, Conn., was calling on the trade in Montreal last week.

Ontario.

D. A. Jones, hardware merchant, Beaton, Ont., was in Toronto last Saturday.

The hardware firm of H. R. Manders & Co., Owen Sound, Ont., have gone out of business.

The hardware and tinsmithing business of D. Cowville, Maxville, Ont., is being advertised for sale.

The stock and premises of J. Weber, hardware merchant, Neustadt, Ont., suffered considerable damage by fire last Saturday.

D. A. Husband, of Husband & Son, Wallaceburg, spent last week at Orillia as one of the delegates to the I.O.O.F. grand lodge.

H. T. Eager, manager of the Toronto office of Wood, Vallance & Co., wholesale hardware merchants, Hamilton, is spending a fortnight's vacation in Atlantic City.

Lawrence J. Levy, of United Factories, Limited, has returned to Toronto after a pleasant trip to Detroit and Buffalo. While in the former city Mr. Levy took in the Michigan State Hardware Dealers' convention.

Mr. A. W. Wills, secretary of United Factories, Limited, Toronto, was at Old Orchard, Maine, at the time of the big fire last week, but is reported uninjured, being in the only hotel that was left standing. He expects to arrive home early next week.

A. E. Lech, who for the past year and a half has been managing the financial end of the Nelson Hardware Co.'s business, at Windsor, has taken over the active business management. Wm. Moore, formerly in charge, is planning to go into business on his own account.

J. M. Brown, formerly of the hardware firm of Brown & Mitchell, Brandon, Man., is spending a few days in Toronto and is registered at the Walker House. Fifteen years ago Mr Brown started a hardware business in a tent at Brandon and to-day he takes rank among the most successful business men of the Prairie Province.

Owing to their large and constantly increasing business and to facilitate city deliveries, Rice, Lewis & Son, Toronto, wholesale hardware merchants, are announcing to their customers this week that an office has been opened at their new west-end warehouse, Pacific and Atlantic Avenues, and that orders for iron and steel will be taken and delivery made from that warehouse as well as from King and Victoria Sts.

Western Canada.

Eakins & Griffin, Shoal Lake, Man., have sold their hardware business to D. McDonald & Co.

Fire recently totally destroyed the hardware store and stock of D. McDonald, Shoal Lake, Man.

CAVERHILL, LEARMONT & CO'S FIRE.

Just as we go to press Caverhill, Learmont & Co., Montreal, telegraph that the fire in their warerooms on Friday morning was confined to the top floor of their St. Peter street building.

Surplus stocks are carried in their Colborne street warerooms at Montreal, and they state that no delay whatever will be caused in getting out shipments of orders received by them.

The loss cannot be estimated at time of writing, but the fire was promptly extinguished.

The stock and premises of Gaines & Wylie, hardware merchants, Oxbow, Sask., recently suffered loss by fire.

C. Rudel, of the Canadian Fairbanks Co., Montreal, was in Winnipeg last week in connection with the western business of the company.

A business change of some importance recently took place in Saskatoon, Sask., when the S. A. Clark Co., Ltd., took over the hardware concern formerly conducted by S A. Clark. The new company, the capital of which is $75,000, numbers among the shareholders, Dr. J H. C Willoughby, Gerald Willoughby, A. J. E. Sumner, S. A. Clark and D. M. Leyden. The last-named gentleman is leaving his position with Jas. Clinkskill on September 1st to take over the managership of the new company. The S. A. Clark Co., Ltd , will continue business on the same lines as conducted by the old firm, although it is their intention in time to go largely into the wholesale and jobbing business

BUILDERS' EXHIBIT.

The first exhibition of builders' and contractors' supplies and hardware ever held in Canada, was held at Victoria rink, Montreal, on August 20, 21, 22. The exhibition was a very interesting one comprising every imaginable material used in building, with an equally imposing display of interior equipments for the finished structure, from plumbers' supplies to wall paper The rink throughout was tastefully decorated with flags and bunting, and in the evenings was brilliantly illumined with electric lights and gas lamps. Altogether there was one hundred and ten booths and sixty-five exhibitors, some exhibitors exhibiting in two or more booths.

The exhibition was not only interesting, but was also very useful; not merely to those directly connected with the building trade, but to the general public, as the variety and novelty of the articles on exhibition furnished interesting proof of the great progress being made in the science of architecture Amongst the various lines of material displayed were plumbers' supplies, builders' supplies, wall paper, stained glass, cut stone and imitations of cut stone, mosaic and other patent floorings, electrical supplies, heating appliances, concrete mixers, derricks, metallic work, paints and varnishes.

Amongst the exhibitors at the builders' show were: The Canadian Fairbanks Co., Maison Jean Paquette, International Steel Co., P. D. Dods & Co., Martin-Senour Paint Co., Metal Shingle & Siding Co., Dominion Radiator Co., Alex. McArthur & Co., Dodge Mfg. Co., Francis Hyde Co., Standard Paint Co., Warden-King Co., Gurney-Massey Co., Lockerby & McCoomb.

All the exhibits were tastefully and effectively arranged, and Hardware and Metal hopes to publish in next week's issue illustrations of these various exhibits.

To enliven the exhibition an orchestra was in attendance, with a vaudeville performance at frequent intervals. Altogether the show was interesting and instructive, and should its success be a criterion of future exhibitions, it will prove a great benefit to the architectural world.

ACME CAN WORKS' PICNIC.

Employes of the Acme Can Works, Montreal, recently held their second annual picnic at Isle Gros Bois, and a very enjoyable one it was. Fully four hundred employes and friends were there and they all seemed to enjoy themselves. Jas. B. Campbell, the head of the company, was on hand to help out with a word here and there.

The programme of sports was a long one and every event was keenly contested. The winners:

Boys' race, 6 years and under—R. Page; girls' race, 6 years and under—Mary Welsh; boys' race, 12 years and under—N. Desjardins; girls' race, 12 years and under—Eva Beauchemin; boys' race, 16 years and under—Hy. Laurendeau; girls' race, 16 years and under—Nellie McConnell; married ladies' race—Mrs. T. H. Pratt; young ladies' race—Miss Morris; young men's race—A. Muncaster; sack race—H. Gohn; hop, step and jump—Thos. H. Pratt; putting 32-lb. shot—Nicola Leckas; prize waltz—George Chisholm and Miss Galarneau; committeee race—Edward F. Pratt; married men's race—D. Dubeau; baseball match—Tomato Cans vs. Corn Cans, won by Corn Cans; baby competition, 18 months and under—Winner, Marguerite DeGruchy, age 18 months; baby competition, 18 months and three years—Winner, Doris Beard, age 3 years; three-legged race—H. Cohn and E. Larin; tug-of-war—Married men vs single men (won by married men); broad jump—N. Fortis.

HARDWARE AND METAL

Established 1888

The MacLean Publishing Co.
Limited

JOHN BAYNE MACLEAN . President

Publishers of Trade Newspapers which circulate in the Provinces of British Columbia, Alberta, Saskatchewan, Manitoba, Ontario, Quebec, Nova Scotia, New Brunswick, P.E. Island and Newfoundland.

OFFICES:

MONTREAL, ... 232 McGill Street
Telephone Main 1255
TORONTO . . . . . . 10 Front Street East
Telephones Main 2701 and 2702
WINNIPEG, ... 511 Union Bank Building
Telephone 3726
LONDON, ENG. . . . . . 88 Fleet Street, E.C.
J. Meredith McKim
Telephone, Central 12960

BRANCHES:

CHICAGO, ILL. . . . . . 1001 Teutonic Bldg
J. Roland Kay
ST. JOHN, N.B. . . . . No. 7 Market Wharf
VANCOUVER, B.C. . . . . Geo. S. B. Perry
PARIS, FRANCE . Agence Havas, 8 Place de la Bourse
MANCHESTER, ENG. . . . . 92 Market Street
ZURICH, SWITZERLAND . . . . Louis Wolf
Orell Fussli & Co.

Subscription, Canada and United States, $2.00
Great Britain, 8s. 6d., elsewhere . 12s

Published every Saturday.

Cable Address { Adscript, London
Adscript, Canada

WEST'S BUSINESS SITUATION.

Much nonsense is being talked about the business situation in western Canada, not only in the east, where it might be expected and excused because of lack of knowledge of conditions obaining at present, but also in the west itself, where there is little or no excuse for it. Because the combination of a tight money market and a late and hence somewhat uncertain crop has resulted in checking, at least temporarily, the rapid rise in real estate values, a few calamity howlers will insist on having it that the west is on the verge of commercial disaster. As a matter of fact, the checking of real estate speculation is the best thing that has happened to the west in recent years, and while it is, no doubt, causing some temporary inconvenience, it will very soon be generally recognized as a blessing in disguise.

The outlook is, of course, a little uncertain at the moment, but there is no reason for undue anxiety. The crop is late, but the answer of the west to the calamity howlers, who insist that there will be no crop to harvest, is a call for 21,000 men, and the railways are taking steps to secure the labor required. With favorable weather, there is now little doubt that a fairly good crop will be harvested. If the crop escapes the September frosts there will be a yield in Manitoba slightly below the average, an average yield in Saskatchewan, and the largest yield on record in Alberta. The price is certain to be much higher than last year, and the farmer should have quite as much money from the sale of the 1907 crop as from that of 1906.

This is the hopeful outlook, and it seems the reasonable one. If, however, a good portion of the crop should be destroyed by frost, the west can stand it. Ten years ago the result would have been a business panic, but to-day the prosperity of western Canada rests upon a basis so stable and secure that trouble of this kind could be endured. The west is able to stand a hard year if necessary, and, as a matter of fact, a crop failure has already been pretty generally discounted in advance. The next month will be an anxious time, but there is no excuse for any undue excitement.

THE PERCENTAGE OF PROFIT.

A correspondent signing himself "Percentage" sends an interesting comment on the article under the above heading, published in Hardware and Metal, of July 20. He contends that the questions of profits and commissions were mixed, and writes:

"The tables at the end are not applicable to the question taken up. They are the basis of making from a net list a price list that will be subject to the discounts named. That is, if you wanted to make a price list and allow discounts of 20 per cent., you would add 25 per cent. to the net price, etc.

For example:

Net price, .80; add 25 per cent., making list price $1.00; deduct discount of 20 per cent., is a fifth, or .20, giving net price of .80.

Net price, .80; add 50 per cent., making list price $1.20; deduct discount of 33 1-3 per cent., is a third, or .40, giving net price of .80.

Net price, .80; add 100 per cent., making list price $1.60; deduct discount 50 per cent., is one-half, or .80, giving net price of .80.

"The answer given to the question: 'An article bought for $10.00 and sold for $20.00,' is correct, as he realizes 100 per cent., or 1 cent for every cent invested. Less than that percentage is a decimal of 1, for instance, 25 per cent. is .25 of 1. So that a merchant can and does often sell at even greater than 100 per cent., as he can add any percentage to the cost of goods. They will stand even to 200 or 300 per cent. which it is necessary to do in some lines, such as drugs, etc., where the cost of service is a large item in the conduct of a business.

"The error made in regard to selling a house was in overlooking that the commission is deductable from the profits, and is paid on the price obtained. We will put the figures in another way: A man buys a horse for $50 and agrees to allow a broker 20 per cent. for its sale. What percentage of profit on original cost does he make if broker sells at $75? Answer, 20 per cent.

"Sold at $75. Deduct commission, 20 per cent., $15, leaving $60.

"Cost $50. Add 20 per cent., $10; totalling $60.

"If the question is altered to, What shall I sell a horse at to gain 20 per cent., after paying a broker 20 per cent.?

"Cost of horse, $50. Add 20 per cent., $10; total, $60. To deduct 20 add 25 per cent., as shown above, $15, leaving selling price $75.

"So that if a merchant wants to make a clear profit of any percentage, he must first add the cost of freight and selling expenses and then add the percentage of profit he wishes to make on the goods."

THE CURSE OF CHEAPNESS.

A conference was recently convened in London by the Worshipful Company of Plumbers, the Master Plumbers' Association, to which were invited representatives of the masters of other trade organizations, for the purpose of reviving apprenticeships.

Many diverse expressions of opinion were heard as to the cause of the admittedly poor craftsmanship of the present day. The master plumbers required an apprenticeship of seven years, but some of the members contended that this term could be cut in half with advantage, as most of the first three years was wasted, and ordinarily lads of the present day were not willing to wait so long for an opportunity to earn good wages, when an errand boy could do better than they while acquiring a trade. Another member declared that the poor work of the present day was the curse of cheapness, that as good craftsmen existed to-day as ever, but they were not given an opportunity to exercise their originality in carrying out work.

The first consensus of opinion appeared to be that apprenticeship should be revived of a reduced term, and that the apprentices should be given technical school training in conjunction with practical work. A committee of prominent masters was appointed to formulate plans for carrying out the reforms suggested.

Markets and Correspondence

(For detailed prices see Current Market Quotations, page 62.)

MARKETS IN BRIEF.

Montreal.

Antimony—Weak.
Linseed Oil—Weak.
Pig Iron—Firm.
Sporting Goods—Season opens exceptionally early.
Tin—Strengthening.
Turpentine—Firm.

Toronto.

Antimony—In better demand.
Copper—One-half cent lower.
Tin—Demand increasing.
Linseed Oil—Weak.
Turpentine—Steadier.

TORONTO HARDWARE MARKETS.

Toronto, Aug. 23.—A good August business is being done and, now that the traveling salesmen of local houses are again on the road, orders for all lines of seasonable hardware are arriving in increasing numbers from all parts of the province. As far as western trade is concerned, the amount of business booked for fall delivery is not as large as jobbers hoped it would be. On account of the present stringency of the money market and the consequent tightening of the lines of credit by many eastern houses, western retailers are buying much more conservatively than last year and booked orders for the west are consequently not as large as they otherwise would be. Local jobbers will be kept busy for the next couple of weeks entertaining their numerous retail customers from all parts of the country who usually make it a point to call on their jobbers here during the national exposition. The exposition, together with the reduced railway rates, never fails to bring a large number of retailers to Toronto and by personal attention to these, local jobbers develop a closer acquaintance with their outside customers and incidentally book a considerable amount of business.

Screws—Brass and bright screws in almost all sizes are still very scarce, but the manufacturers are daily gaining ground in catching up with the trade. Jobbers here are doing the best they can in the way of supplying customers by only partially filling all large orders and holding the balance of the order to be filled as the supply of screws becomes greater. Instead of selling five hundred gross to one large concern, they cut that order down to one hundred, and utilize the remaining four hundred gross to meet the demands of smaller customers.

Nails—The demand is good, with supplies quite adequate to meet it. Prices continue firm at former figures.

Wire—During the last few weeks there has been practically no wire selling and the manufacturers have had a chance to get a supply on hand to meet the fall demand, which usually sets in towards the end of September.

Glass—Trade is now much better in glass and orders are being received daily from many retailers who had not been previously booked.

Building Materials and Supplies—All hardware building supplies are in strong demand and jobbers are experiencing a record-breaking trade. Farmers are beginning to enquire about cement and indications are that a hustling autumn trade will be transacted in this material.

Sporting Goods—The ammunition and gun trade is opening up in splendid style. With the close approach of the game season, retailers should instruct their clerks to be on the alert and lose no opportunity to push this trade. Just at present in Toronto, on account of the Ontario Rifle Matches being held here, sporting goods dealers report a good trade in verniers, orthoptics, cleaners, wood-cased ramrods, marksmen's caps, and other rifle-shooting requisites. Prices on the various lines of sporting goods remain unchanged.

Fall Goods—The outlook is bright for a good autumn business. Booked orders for horse blankets, sleigh bells, axes and saws are exceptionally heavy and the additional orders for fall goods arriving from travelers make it an assured fact that this autumn's trade will be much heavier than usual.

MONTREAL HARDWARE MARKETS

Montreal, Aug. 23.—Hardware market conditions are somewhat strengthened this week. This trade, like all others, is subject to fluctuations, which though only slight, are sufficient sometimes to cause a little discomfort and apprehension amongst the merchants. At present all the jobbing houses are exceptionally busy, some finding it necessary to keep part of their staffs employed until late at night.

Carters are still in some cases asking exorbitant rates for delivery, and the merchants throughout the city will welcome the day when the civic authorities give the Street Railway permission to handle freight. The present congestion of freight on the waterfront would be soon lifted, and trouble with cartage charges done away with.

Prices on all lines are very firm and unchanged.

Sporting Goods—The feature of the hardware trade at present is the exceptionally early opening of the sporting goods trade. The retailers are getting in larger stocks this year and the outlook for this season's business is very hopeful.

Poultry Netting—The demand maintains itself splendidly for this season of the year. A diminution is generally looked for now, but so far such anticipations have not been realized. New discounts, 50 and 5 off.

Screws—The factories are getting in a very much improved condition, and they are making strong efforts to replete their stocks for the fall rush. Supplies of raw material are arriving much more freely. The shortage in small sizes is not so acute as formerly.

Wire Goods—Prospects for this season's trade are encouraging, as the hay crop in Quebec is expected to be heavy, thus creating a strong demand for hay baling material.

Building Paper—Trade is steady. Prices are firm and unchanged. The makers are endeavoring to enlarge their stocks for the fall rush, as well as meet the current demand, and are consequently very busy. Raw material supplies at present are hardly adequate.

Builders' Hardware and Mechanics' Tools—The trade in these throughout the year has been splendidly maintained. Supplies at present are adequate to the demand.

Cement—The demand is moderate, with prospects of enlivenment in a fortnight or more. The volume of business transacted in this line during the past season will, in all probability, exceed all previous records.

MONTREAL METAL MARKETS.

Montreal, August 23.—With the exception of pig iron the local market conditions are characterized generally by dullness. These conditions are quite in consonance with those in the American market, which are far from satisfactory or encouraging. Speculation is largely responsible for the depression prevailing in American circles and no doubt the fruit of speculation, money stringency, is doing much towards keeping local conditions in their present unsatisfactory state. The buying public have no confidence in the present prices and they have a suspicious hesitancy in buying up supplies. Absolute necessity is the only thing that can persuade them to buy these days.

The English market is somewhat firmer. Supplies are rather short, and strictly No. 1 qualities are hardly available.

American market prices even now are considerably above English prices and consequently a large fraction of the supplies is being procured in England and Scotland, thus unusually diminishing the stocks there. Moreover, the furnaces in England have not been working satisfactorily, and this accounts for the difficulty experienced by some American and local dealers in procuring supplies.

Pig Iron—English brands are difficult to procure and prices are high. Locally, a heavy demand exists at present for all lines. They are not mere hand-to-mouth orders, but extensive, and such as to guarantee safety for some time to come. Upwards of 10,000 tons have been bought during the past ten days or fortnight. Stagnation characterizes the American situation.

The Metal Worker (August 17) says: "The practical suspension of buying for all deliveries makes it extremely difficult to determine the actual level of market prices. All indications, however, point to further weakness. It is becoming more and more apparent that foundry stocks are by no means depleted. In fact, there is good reason to believe that work has slackened up sufficiently to stretch the supply of iron in melters' yards over a longer period than was anticipated. There is a decided lack of harmony in sellers' views as to the outlook. But it is very evident that, on a firm inquiry for desirable tonnage, prices considerably under those nominally quoted would be brought out. The market is barren of interest in first quarter requirements, for which no sales or inquiries are reported."

Pig Tin—In the local market this has firmed a little, as it has in the English market. In the United States it is still very weak. We continue to quote. Lamb and Flag and Straits, $44.

Ingot Copper—Large lots in the American market are sold at a reduction. Prices are weak. The weakness in financial circles is largely responsible for the depression.

Antimony—Continues weak. In the United States purchasers of large lots can secure this at reduced prices. Local prices are unchanged: Cookson's, 8c.

Old Materials—The market continues weak. Further declines are noted on some lines this week. Heavy copper and wire, 15½c; light copper, 14½c; heavy red brass, 14c; heavy yellow brass, 10½c; light brass, 7c.

TORONTO METAL MARKETS.

Toronto, Aug. 23.—Conditions on the Toronto metal markets have considerably improved during the past week, and it is quite evident that the low prices prevailing for the various metals are beginning to attract buyers.

Antimony—Sales have increased to a considerable degree during the week, the low price being responsible for much of this trade. There has been no further decline in local antimony prices this week; Cooksons is still quoted at 15c, and Hallett's at 14½c. The American market is dull, with very little doing and prices there are again easier.

Pig Iron—Business has improved considerably during the week, and more buyers are now in evidence for this metal. Local prices remain unchanged at figures previously quoted.

Tin—The improved demand of the last few days is an evidence that many local consumers are beginning to reckon with the decline, and not a few consumers placed orders during the week in order to take advantage of present low prices. Lamb, flag and straits ingot tin is still being quoted locally at from $42 to $44, though one large house is quoting slightly lower than these figures. As for the foreign situation, both the British and American markets are dull and, apart from the steady increase in the Australian output, there is little of fresh interest regarding the metal.

Lead—The demand for lead shows a slight increase over last week. Local prices are comparatively steady, imported pig lead being still quoted at $5.35.

Copper—There has been an increased sale of copper locally; casting ingot copper is quoted at 22c, and has attracted quite a number of local buyers at that price during the week. As for the American market, the past week has seen but little change in the copper metal market, although naturally some consumers there have been drawn into the market by the reduction in price and are making small purchases. The big consumers, however, are still far from being satisfied with conditions and are holding off their big orders for a more stable market. The stagnation of the metal market is having its effect on the stock market, so that both the metal users and the stock brokers are anxious for some kind of a peaceable settlement of the controversy, and all efforts are being directed along that line. Meanwhile, the producers are contented; they are harboring their resources for the demand which they know must come sooner or later, and when it does come they want to be prepared to meet every order in the shortest possible time. A few shipments of orders placed months ago have been made during the past few days, but that is about all.

Old Materials—The market continues extremely dull and it will not be surprising if buyers who are now afraid of the market, have to purchase later at higher figures. The price of tea lead has dropped ¼c per pound, and is now quoted at 3¼c. Prices of other scrap metals remain easy at figures last quoted.

LONDON METAL MARKETS.

London, Aug. 21.—Cleveland warrants are quoted at 57s 3d, and Glasgow standards at 56s 6d, making prices as compared with last week, on Cleveland warrants 1½d lower, and on Glasgow standards unchanged.

Tin—Spot tin opened steady at £167 10s, futures at £166 15s and after sales of 80 tons spot and 220 tons futures, closed easy at £166 12s 6d for spot and £166 for futures, making price as compared with last week 18s 6d lower on spot and £1 10s lower on futures.

Copper—Spot copper opened strong at £80, futures at £77 15s, and after sales of 500 tons spot and 800 tons futures, closed easy at £79 for spot, £76 15s for futures, making price as compared with last week £2 10s higher on spot and £3 5s higher on futures.

Spelter—The market closed at £22, making price as compared with last week unchanged.

Lead—The market closed at £19, making price as compared with last week 19s lower.

U. S. METAL MARKET.

Cleveland, O., Aug. 22.—The Iron Trade Review to-day says:

With the background of trade somewhat more sombre, if there be included in the view the financial situation, there are, nevertheless, this week a few indications of increasing activity in the metal markets in some centres, such as have come at times in seasons heretofore at about this time of year. The fear or expectation of lower prices hold back or defer considerable business, coupled with the further view that prices in some lines are held to be high. If the retarding influences of those considerations and of the money market could be eliminated, the current developments seem to indicate that trade in the iron and steel industries might forge ahead again at something like the old pace. Perhaps the most encouraging developments is that the railroad ordering is on the mend. The transportation companies are specifying on their contracts much more heavily than they did a month or six weeks ago and the inference seems warranted, taking into consideration the well-known attitude of the carriers, that the buying is no greater than it must be. In other words, the absolute demands of the railroads are compelling more liberal purchases of material.

The demand for structural material is very good and the outlook is that the orders taken by the leading interest will be about 50,000 tons for the present month. The contract for the steel for three additional piers in New York has been awarded, and in a few days the city will award contracts for the third lot of piers, involving about 10,000 tons of steel. About 12,000 tons will be needed for the Manhattan terminal, New York city. The Panama railway is inquiring for 3,000 tons of rails, but new business in rails is of very small tonnages.

U. S. IRON AND METAL MARKETS.

New York, Aug. 22.—The Iron Age to-day says: Generally speaking the undertone throughout the Iron trade is one of increased nervousness over the future. The plants are all still running under high pressure, and will do so for the balance of this year, in order to fill orders now on the books. This is the season of the year when ordinarily there is a lull, but the contrast between recent experiences of record business and a quiet market may have an undue sentimental effect.

There is a falling off in new business in nearly all directions, but specifications against old contracts continue very heavy and there is no indication in the finished lines of any cessation of work under way.

Throughout the country the buying of pig iron is from hand to mouth and the market has not been really seriously tested, in foundry iron, for some time. But prices are weakening. There is some complaint that certain smelters are exhausting every pretext for cancelling high-priced contracts and that iron is being thrown on the market in this way.

It is admitted that the amount of foundry iron which is being melted has fallen off in quantity, but against that must be taken into account the fact that foreign iron, which was coming in at the rate of 30,000 to 40,000 tons a month, is no longer a factor in the supply.

The Steel Corporation has purchased an additional lot of 5,000 tons of Steel Billets from an eastern Pennsylvania open hearth plant. The Western market is quite bare of steel. In the Chicago district two orders for an aggregate of 11,000 tons of billets still remain unfilled. On the other hand, the Steel Corporation participated last week to the extent of 10,000 tons in a heavy selling movement of sheet bars, in the English market, for delivery during the last quarter, the German Steel Works also taking a considerable tonnage.

The steel rail trade is very quiet. There is pending, however, some very large business for export, which may be closed at an early date.

CORRESPONDENCE

Travelers, hardware merchants and clerks are requested to forward correspondence regarding the doings of the trade and the industrial gossip of their town and district. Addressed envelopes, stationery, etc., will be supplied to regular correspondents on request. Write the Editor for information.

HALIFAX HAPPENINGS.

Halifax, N.S., Aug. 19.—The Hardware trade is very quiet at present. Very few orders are coming to hand. Collections are fairly good. Many of the clerks are now enjoying well earned vacations. Outside of that, in builders' supplies, the trade coming in is very small. The jobbers are now straightening out their stocks and putting things in shape for the autumn business.

The Works Department of Halifax last week opened tenders for two thousand water meters. The tenders have been deferred for a week, and in the meantime the City Engineer will test the samples of the meters submitted. The tenders, which include the couplings, are as folows:

J. McPatridge—Crown meter, ½ or ⅝-inch, $17.08; 3¼-inch, $30.08; delivered in Halifax.

J. McPatridge—A.A. Empire meter, ½ or ⅝, $14.53; ¾-inch $23.07, delivered in Halifax.

J. McPatridge—Nash meter, ½ or ⅝-inch, $11.98; ¾-inch, $17.33; delivered in Halifax.

John McDougall—Worthington type B meter, ½ or ⅝-inch, $9.10; ¾-inch, $15.50; F.O.B., Montreal.

Neptune Meter Co.—Neptune cast iron bottom, ½ or ⅝-inch $9.40; ¾-inch, $13.60, and duty.

Neptune Meter Co.—Resident meter, ½ or ⅝-inch, $10.40; ¾-inch, $15.60, and duty.

H. B. Clarke & Son—Lambert meter, ½ or ⅝-inch, $11.20; ¾-inch, $16.75, delivered in Halifax.

H. B. Clarke & Son—Keystone meter, ½ or ⅝-inch, $10.70; ¾-inch, $15.50, delivered at Halifax.

Standard Meter Co.—Standard, ½ or ⅝-inch, $8.25; ¾-inch, $12.50, F.O.B., New York.

Austen Bros.—Genuine Hersey disc, ½ or ⅝-inch, $11.98; ¾-inch, $17.34, delivered at Halifax.

Austen Bros.—American Meter Co.'s bronze top case, ½ and ⅝-inch, $9.59; ¾, $14.31, delivered at Halifax.

Austen Bros.—Niagara, all galvanized, iron case, ½ or ⅝-inch, $8.47; ¾, $12.63, delivered in Halifax.

W. Stairs, Sons & Morrow—Beggs & Son, London, meter, ½ or ⅝-inch, $15.40; ¾, $16.95, delivered at Halifax.

W. Stairs, Son & Morrow—Kennedy, London, meter, ½ or ⅝-inch, $23.10; ¾, $31.85, delivered in Halifax.

W. Stairs, Son & Morrow, Hersey, ½ or ⅝-inch, $11.95; ¾, $17.75, delivered in Halifax.

W. Stairs, Son & Morrow—Columbia Union meter, ½ or ⅝-inch, $10.60; ¾-inch, $16.20, delivered in Halifax.

Buffalo Meter Co.—½ or ⅝-inch, $9.85; ¾, $14.81, delivered in Halifax.

Everything is humming in industrial life in the town of New Glasgow, and there is a good demand for labor. No man need be idle in the town at present. The new Sutherland Rifle Sight Factory is nearing completion, and the work of installing the machinery is now in progress. The W. P. McNeil Co. are preparing the site for the new iron works on the I.C.R. eastern line, and A. McCulloch & Co., have just awarded a tender for a new brick block. The keel of the first steel ship has been laid in the shipyard of Matheson & Co., James A. Stairs has become associated with the Brown Machine Co., and it is understood that the works will be removed to Trenton, and the capacity of the plant increased.

QUAINT QUEBEC.

Quebec, August 19.—With the advance of the season the local hardware trade is a little quieter this week. The usual calm of the present season has evidently settled upon the trade. However, as local dealers did an exceptionally heavy business this summer they are content to wait until the trade picks up again at the beginning of September. They have done good business in all lines, especially in paints, oils and glass. Stoves and builders' supplies are particularly active this year. During the week a large quantity of agricultural implements were sold.

* * *

The annual excursion of the Quebec Clerks' Association will take place on the 21st of September. Our jolly fellows will go and visit Montreal.

* * *

The Atlantic, Quebec and Western Railway will open the first two sections of their line for traffic in a few days. The road is now completed from New Carlisle to Port Daniel, some 23 miles. Tenders will soon be invited for the supply of ties, rails and construction of bridges, and as these contracts will be awarded this fall, the whole line will be under construction by spring. At Port Daniel the company will tunnel through Hell's Cape, some four hundred feet.

* * *

The growing importance of the passenger trade between Quebec and Scotland is recognized by the Allan Line Steamship Co., which is planning for a vastly improved service between the St. Lawrence and Glasgow for next season. The Allan Line will have on the route four large vessels, giving a weekly service with the St. Lawrence. The two new steamships which will be added to that service are the Grampian and the Hesperian.

F. H. Drolet, proprietor of large machine works in our city, has purchased for the sum of $50,000 an area of ground near the St. Charles River, on which new shops will soon be erected. The works will be 200x300 feet and cost over $500,000.

* * *

The differences which arose last week between the ironworkers on the Quebec bridge and the Phoenix Co. have been adjusted. The strike caused a cessation of work for a couple of days. The differences are said to have taken place because the Phoenix Bridge Co. refused to allow a number of employes, who were leaving for their homes, the railroad fare which they had expected coming to Quebec to work on the bridge.

* * *

The greater part of the hardware travelers are still off on their holidays and will give the retailers a rest for a few weeks. Meanwhile, mail orders are arriving in fair numbers.

* * *

Some hardware retailers are complaining that the greater part of the wholesalers in our city are doing some retail trade and, as they buy in larger quantities, sell their goods at a lower price than it can be done by local dealers. This price-cutting is noticeable in nearly every line. Your correspondent was told that these wholesalers, who persist in slashing the retail prices, buy from manufacturers at 15 per cent. cheaper than can be done by retailers who don't keep a very large stock. This way of doing business creates a remarkable absence of cohesion amongst local hardwaremen.

* * *

O. Picard & Sons, plumbers, have been awarded important contracts for installing heating by hot water at the English Cathedral, as well as the plumbing, heating and electric wiring at the new building of The Daily Telegraph.

* * *

Business in wood working machinery is very active. It is noticeable that Canadian machinery is in very large demand on the Quebec market. Dealers cannot secure all they need because manufacturers are over worked. The same activity prevails in sporting goods. Retailers say that they have great difficulty in having their orders for rifles and cartridges filled from European firms.

* * *

The Clark City Pulp Co., Seven Islands, will open its doors in December next. That new mill is just being completed and will manufacture over 250 tons of pulpwood every day.

* * *

Another party of prospectors has just left for the North Coast of the St. Lawrence Gulf. It is reported that a few weeks ago some copper was found in fairly large quantity in some of the Mingnan Islands. An American specialist from Boston is accompanying the party, in order to assay the ore.

* * *

The Quebec Government will again have the Temiscamingue region explored, near Cobalt, Ontario. New mining lands are said to have been discovered lately.

* * *

The door and sash factory of Langevin & Freres, at Scott Junction, was destroyed by fire Thursday last. The loss is placed at $12,000.

* * *

T. O. Rouleau was appointed manager of the engine-builder society of Chicoutimi.

* * *

Elisee Lemieux, of Beauceville, has just been named director of the Beauceville Foundry Co.

* * *

At the last meeting of the Quebec Provincial Cabinet the principal business transacted was the passing of an order in council to construct and establish an agricultural school at Oka.

* * *

The last stone of the new Bank of Montreal building, in St. Peter St., was placed in position Saturday. That construction will cost over $500,000.

BOOMING BELLEVILLE.

Belleville, August 21.—Local and district hardware dealers as well as plumbers and tinsmiths report business in a first-class state. Every industry is rushed with orders and in many cases the work cannot be promised. Local business was never in a better condition than at present.

* * *

The Ontario Steel Tubular Axle Company, a new concern which started work here recently, has found business growing so large that their old premises were not sufficient to carry on the work and they have just purchased a large site on Front St. This week they have a number of men engaged in tearing down the old buildings on the new site and will erect a large factory. The company is composed entirely of local men and is capitalized at $20,000, half of which is fully paid up. The officers and directors are:

President—Ex-Mayor R. J. Graham.

Vice-president—Byron Hudson, of the Hudson Hardware Company.

Secretary-treasurer — H. Parker Thomas, city assessor.

Manager—Charles Panter.

Board of Directors—R. J. Graham, John McKeown, Dr. MacColl, H. Ackerman.

The work on the new building is expected to be completed this fall.

* * *

Work commenced this week on the erection of the new brass foundry on Coleman St., quite a staff of men being engaged. The new company have already in stock, and at present making a fine supply of tools ready for use when the plant opens. The company's men are working at the Walker Foundry and the tools already manufactured and in stock are said to represent a value of $5,000. P. C. Jones is one of the chief promoters of the company and H. C. Hunt is another, the latter resigning the secretaryship of the Belleville Hardware Company in order to devote all his time to the new organization. He has been succeeded by M. Charles McMullen.

* * *

That this city is progressing rapidly is shown by the assessment now about completed. Mr. Thomas, the assessor, says the city has increased in population during the past year by at least 500, the largest increase in the city's history. There is still a great scarcity of small houses and the assessor told of one house where four families are living, as well as another family in the barn. What some of the newly arrived citizens are going to do the coming winter it is hard to say.

* * *

Part of the steel work for the Government drill hall, now under construction, has arrived and is being placed in position. The steel is from the works of the Dominion Bridge Company, Lachine.

* * *

The gas department are removing a lot of the old iron gas lamps used so long in lighting the city streets. Tenders will shortly be given out for the erection of the new works.

KINGSTON KINETOSCOPE.

Kingston, August 20, 1907.—At present the city lighting plant is making big alterations here owing to the great demand for gas and electric lighting since the cheaper rates went into force. During the past year great improvements have been made, new machinery has been installed, new buildings put up and the list of gas and electric services greatly increased. A new telescope gas tank with a capacity of one hundred and fifty thousand cubic feet is being installed on Barrack Street and will be ready for its final test in a few weeks. The foundation is of concrete, about two feet thick. New electrical machines of the latest design have been put in, in order to carry the increased load. During the past six months, one hundred and forty new gas mains have been put in. Gas consumers are now paying the same price for both lighting and heating gas, viz., one dollar a thousand cubic feet, whereas last year they paid one dollar and a half for lighting and one dollar for cooking. Special rates are now given for exterior and interior, and for other ways in which much of the current is used. The new gas tank will likely be ready about the first of October and when alterations are all completed, Kingston will own one of the best plants in the Dominion and citizens will be receiving the product with very little profit going into the city treasury.

* * *

Plans are now being arranged for a new parsonage for the pastor of Princess Street Methodist Church. The residence will be built on the church property and it is estimated that it will cost in the neighborhood of $3,000. It is expected that work will be commenced within the course of the next few weeks.

* * *

M. Elliott and five workmen left for Toronto the latter part of last week to complete a big contract which Elliott Bros. have in that city.

* * *

It is expected that a new smelting works will be started in this city shortly, as the citizens carried two exemption bylaws on the 28th of May last; one bylaw for the smelting of lead and the other for the smelting of zinc. The board of trade worked hard on these two propositions for a couple of months and were all out doing what they could on polling day. The Stanley Smelting Co., of New York, is the interested concern. The city was asked for a grant of some of the smelter site below Cataraqui bridge, and freedom from taxation for ten years. The lead smelter company agreed in consideration of the above mentioned concession to establish a smelter whose buildings would cost $40,000, and to complete the plant inside of twelve months. They also agreed to employ not less than fifty men. The zinc smelter company's agreements were similar to the above, the only difference being the latter agreed to erect buildings to cost $100,000. The zinc smelter building will be all of concrete. Alderman Craig, chairman of the committee of industries, received a letter recently saying that the plans were not quite completed, but would be finished shortly. They also made the following statement: "We are quite as anxious to make a start as the Kingston people are to see us there." It was pointed out that it would be a great saving to the city, if, instead of sending money out of the country for these products, as is now required, they should retain it, and much outside money would begin to flow into the city. Again, an opening is made for the employment of our own youth and brain right at home, and instead of allowing our young men to wander to other parts, it would give them an opportunity to remain with us and help to build up Kingston and their own country.

* * *

W. B. Dalton, hardware merchant, and wife, were stopping at one of the hotels now burned at Old Orchard Beach, at the time of the great fire.

* * *

The dome of St. George's Cathedral here is receiving a coat of aluminum paint.

* * *

The civic buildings are being greatly improved. The painters are still at work painting the doors and corridors and gardens have now been made with a concrete border about a foot high all around.

* * *

Among the leading concerns of Kingston which is worthy of special mention is the Canadian Locomotive Works, the factory of which is fitted up with all modern appliances and the latest machinery. Employment is given to upwards of six hundred men. New additions have been going on for some time and are now nearing completion. A new power house is being built and new boiler and erecting shops are being put up, together with a new iron foundry, machine and tank shop. These buildings are being built chiefly of concrete so that they will be fireproof. When the alterations have all been completed they will probably employ one thousand men.

* * *

Tenders are being called for the alterations, additions and repairs to the Tete du Pont Barracks, by the Department of Public Works, Ottawa. Tenders will be received until the 9th of September.

* * *

Kingston takes a strong interest in Picton's new Collegiate Institute, the building of which is estimated to cost about $50,000. The structure, which gives ample promise of being one of the finest and most thoroughly modern of Ontario collegiates, is the product of the architectural ingenuity of William Newlands, Kingston. Two contracts, which total nearly ten thousand dollars, have come to the Limestone City. Elliott Bros., Kingston, are the contractors for the plumbing, ventilation and heating, while the cut and red sandstone being used in the new building is supplied by Hugh McBratney, Kingston.

HUSTLING HAMILTON.

Hamilton, Aug. 20.—This is the quiet season in the hardware line in Hamilton but despite that fact the local hardware merchants are not suffering for lack of trade. Despite the fact that business has fallen off a little during the months of July and August, the merchants report a decided increase in business over last year, and they are being kept well on the move. The chief demand with the hardwaremen at present is for building materials. Building operations have become so extensive in Hamilton within the last few years that, no matter how much the trade may fall off in other lines, there is a steady call for builders' supplies. A large percentage of the merchants are taking advantage of the quiet season to get a lot of their winter orders out to the west. It is expected that there will be heavy shipments to that section this fall.

* * *

There is a strong probability that the city will issue contracts this year for a complete new electric pumping service in connection with the municipal waterworks. It is proposed to submit a bylaw to the people in a short time for the expenditure of $30,000 on new pumps of the modern type. The present steam pumps have been in operation ever since the inauguration of the present system, but owing to the rapid growth of the city, they are becoming entirely inadequate for the city's needs. The new pumps will have a capacity of 10,000,000 gallons a day.

* * *

Things look bright for hardware and metal merchants next year. Although the proposition to build the Hamilton, Waterloo and Guelph railway was dropped some time ago because the promotors could not secure a route through Dundurn park, it begins to look now as though the road may be built after all. If such is the case, it will mean that a large additional area of farming country will turn to Hamilton for the market, and, with good railway facilities, a large number of farmers and townspeople, who at present do their trading in other places, will be directed toward Hamilton; the hardware men will reap the benefits in increased sales, as much, and probably more so, than any other branch of trade. John Patterson, one of the former promoters, has taken the matter in hand again and another company is said to also have its eyes on the proposition.

* * *

This has been a banner year for those engaged in the building trades. So far the building permits are over a million dollars more than last year's, and the prospects are that before the building season closes, they will reach the two-million mark and probably more. It is conservatively estimated that the total amount of permits this year will be about $3,000,000. Some of the big works which are still in the course of completion, are the new terminal station for the Dominion Power and Transmission Co., costing $250,000; the Bennett vaudeville theatre, $90,000; the armories for the local regiments, $250,000; Home for Incurables, $30,000; an addition to the blast furnaces of the Hamilton Steel and Iron Co., $150,000, and numerous other works, including additions to the Canadian Westinghouse works, and the International Harvester works. The Hamilton Bridge Works Co. is also making extensive additions to its plant.

* * *

The numerous knights of the grip will be pleased to learn that there is a proposition on foot to build a new Grand Trunk station in Hamilton. A short time ago Superintendent Gillen, of London, intimated that a station might be erected within the next year, and it is rumored that the railway people have secured options on sites along their right of way, near the corner of James and Stuart streets, about three blocks east of the present station.

* * *

Interest is keen at present in the electric lighting question. The advent of the Government power scheme has been the occasion for vigorous preparations on the part of the Cataract Power Co. to compete with it for the control of the electric fluid in Hamilton. Already the local power magnates are making propositions to the city with a view to reducing the cost of street lighting and several officials of the company have been quoted as saying that their corporation was prepared to underbid the Government. Local users of power will watch the outcome of this competition with interest.

LEAFY LONDON.

London, Aug. 21.

Local hardware dealers are looking far a big trade during the Western Fair week. The Hobbs Hardware Company are making preparations on a large scale to receive their customers and the trade generally. The travelers will all be in during the week, and visitors are invited to make themselves at home.

* * *

London has lost one of its most popular young men in the departure of Geo. Scott Murray, for Winnipeg, where he will reside in future. Mr. Murray has been connected with the McClary Manufacturing Company, in this city, for about five years, being employed in the order department, and later as city traveler, in which capacity he will act in the Prairie City. For several years he has been connected with the Hermitage Club, and has held the offices of president and vice-president on several occasions. Recently the members of the club made him the recipient of a handsome traveling bag. Mr. Murray's host of friends in this city, while sorry to see him go, wish him every success in his new field of labor, and predict for him a brilliant future.

* * *

Munro Bros., of Alvinston, have sold out their hardware business to Johns and Miner, who will take possession on Sept. 1st. Mr. Johns was formerly in business in Elmsville, and Mr. Miner comes from Hensall.

* * *

Frank E. Kestle, of Claudehoze, has disposed of his general store, which includes hardware, to S. C. Chown, of the same place. Mr. Kestle is going into business with his uncle, T. J. Kestle, Ilderton.

* * *

To a man in the metal trade, Ald. Samuel Stevely, belongs the credit of settling the barbers' strike in this city, which lasted for two weeks, and which, if prolonged, might have proven a most serious matter. Mr. Stevely is president of the Board of Trade, and in that capacity was the means of bringing the bosses and the men together and effecting a settlement satisfactory to all concerned.

* * *

The representative of a large American flax company was in the city a few days ago, and, accompanied by a local merchant, made a tour of the city for the purpose of looking over London's offerings in the way of a site for a factory, to cost in the neighborhood of $125,000. The visitor, while here, interviewed a member of the manufacturers' committee and discussed the matter with him. He declined to give the name of his firm for publication, but said he had been very favorably impressed with the city's offerings. The establishment of such an industry here would be not only of the greatest benefit to the city, but would be a benefit to the farmers of western Ontario. The visitor promised that we would hear from him again within a few weeks.

* * *

It is stated that a move is on foot to establish a glass factory in London, and that the man behind the guns is Wm. Gray, the well-known traveler and politician. The European firm, for which Mr. Gray travels, has, it is said, come to the conclusion that much of the fancy glassware imported into Canada could be made right here, and Mr. Gray is credited with the intention of having the factory located in London. According to the story, only the finest cut glass would be manufactured, and the factory would give employment to about one hundred men. Mr. Gray, it is understood, sails for Europe shortly to complete the deal. All the workmen who would be employed, would have to be imported, as it is said there are no cut glass workers in this country at present.

* * *

The Scott Machine Company, of this city, has just received an order from an Evansville, Wis., firm for 150 patent Giant Wonder disc plough-sharpening machines, and the local firm will probably act as distributing agents in Canada for the Wisconsin people.

* * *

J. A. Frazell, traveler for the Hobbs Hardware Company, is ill in St. Joseph's hospital.

* * *

C. A. Whitman, vice-president of the Hobbs Hardware Co., is holidaying in Muskoka.

CHAT FROM CHATHAM.

Chatham, August 20.—An important business change took place here last week when the Ark, conducted for many years by H. Macaulay, was sold to Jas. E. Gray, of Gray's China Hall. The change is largely due to the continued ill health of Mr. Macaulay, who has only recently recovered from a severe and prolonged illness, and consequently desires to retire from active business life.

Mr. Gray, the new proprietor, will assume possession in the course of a week or ten days, and the combined businesses will be ultimately conducted at the present Ark stand.

The Ark was opened 17 years ago last April, and during the interval Mr. Macaulay had made many friends in Chatham and built up an extensive trade. The business includes the local agency for the McClary ranges, of London. Mr. Macaulay's friends—and they are legion—will wish him many years of well-earned rest.

* * *

Four new and promising industries are now in sight for the Maple City. Negotiations between one of these and the city council have been concluded, and await only the sanction of the ratepayers. The firm will have a paid-up capital of $30,000, and will expend $40,000 on building and plant, with 75 hands steadily employed and an annual wage roll of $18,000. They ask a loan of $20,000, repayable in 20 annual instalments with interest at 5 per cent. The board of trade has endorsed the proposition.

Negotiations with the other concerns mentioned are still in the initial stages, but Chairman Westman of the industrial committee is busy interviewing delegations, and hopes to land all of the concerns in due course.

* * *

The long promised natural gas purifiers have at last arrived, and henceforth Chathamites may look for odorless gas. The foundations for the purifiers are ready, and they will probably be placed in position in the course of the next few weeks. Though the gas has unquestionably proven a good thing from a financial point of view, resulting in a distant saving to a large number of citizens, the pronounced odor emanating from sundry leaks in the main has occasioned much adverse comment. The odor has, in fact, been one of the striking features of the Maple City for months past.

The Volcanic Oil & Gas Co. now have about 20,000,000 feet of gas at their disposal, which ensures a practically unlimited supply for Chatham, while the danger of the 3-inch supply line being frozen during the cold weather is being eliminated by the laying of a 6-inch pipe, which will be completed before the winter.

At a special meeting of the water commissioners last week the proposition of the Chatham Gas Co. to supply the waterworks plant with natural gas for a period of 5 years at 12 cents net per thousand cubic feet was accepted.

* * *

Chatham's civic tax rate for 1907 is 28 mills on the dollar. This, after strenuous efforts on the part of every committee to cut down expenses, is, all things considered, a rather commendable figure, since this year the city has for the first time to meet the new market, school and electric light debentures. The whole rateable property of the city amounts to $4,895,518. If the cost of running the city does come a little high, there is the consolation that the cheapest is not always the best, and that if people want to live in a city where life is worth while, they will always have to pay a little more for it.

* * *

Samuel Trotter, of McKeough & Trotter, arrived home last week from Edmonton, where he has been superintending the construction of a new dredge.

* * *

The Western Bridge & Equipment Co. have much enhanced the appearance of their building on King St., with the aid of a coat of paint and a number of improvements. They intend building a concrete addition at a cost of $900.

WIDE-AWAKE WOODSTOCK.

Woodstock, August 20.—One night last week the annual meeting of the Retail Merchants' Association was called. There were not enough members present to form a quorum, so nothing was done. It is understood that the present officers will retire, and not accept re-election. The year's work has been most discouraging, and their decision is not to be wondered at. Three or four of the faithful have worked hard for the interests of themselves and their fellow-merchants, but it has not been possible to stir up any active interest among the rank and file of the association, and for several months there has been no meeting, no quorum. It is hard to assign a reason for this indifference. No difficulty is experienced in getting men to join the association and pay up the dollar. But when it comes to attending regularly, it is a different story. Perhaps the real reason is in the jealousy which exists between merchants. It is unfortunate that this feeling should exist, but it does, to a greater or less extent in every town and small city. The solution of the difficulty, is to educate the merchants up to a higher plane, to teach them to look at questions affecting themselves, from the standpoint of how they would affect the whole body of merchants.

* * *

When there are only five hardware stores in a town, it is hard to form an organization. The number is too small. Nevertheless, a branch of the Hardware Association would be very beneficial to the local dealers. It would prevent what price-cutting is carried on and bring the men nearer together for their mutual protection and interests.

* * *

Extensive improvements and alterations are being made to the Sentinel-Review building, which is also the quarters for the magazine, Rod and Gun, in Canada. When the work is completed the building will be one of the finest newspaper buildings in Western Ontario.

MANITOBA HARDWARE AND METAL MARKETS

Corrected by telegraph up to 12 noon Friday. Aug. 23 Room 511, Union Bank Bldg, Winnipeg, Man.

Business conditions continue unchanged. Until the crop is assured buying will be restricted to immediate requirements.

Rope—Sisal, 11c per lb., and pure manila, 15¾c.

Lanterns—Cold blast, per dozen, $7; coppered, $9; dash. $9.

Wire—Barbed wire, 100 lbs.. $3.22½; plain galvanized. 6, 7 and 8, $3.70; No. 9, $3.25; No. 10, $3.70; No. 11, $3.80; No. 12, $3.45; No. 13, $3.55; No. 14, $4; No. 15, $4.25; No. 16, $4.40; plain twist, $3.45; staples, $3.50; oiled annealed wire, No. 10, $2.90; No. 11, $2.96; No. 12, $3.04; No. 13, $3.14; No. 14, $3.24; No. 15, $3.39; annealed wires (unoiled). 10c less; soft copper wire, base, 36c; brass spring wire, base, 30c.

Poultry Netting—The discount is now 47½ per cent. from list price, instead of 50 and 5 as formerly.

Horseshoes—Iron, No. 0 to No. 1, $5.65; No. 2 and larger, $4.40; snowshoes, No. 0 to No. 1, $4.90; No. 2 and larger, $4.65; steel, No. 0 to No. 1, $5; No. 2 and larger, $4.75.

Horsenails—No. 10 and larger, 22c; No. 9, 24c; No. 8, 24c; No. 7, 26c; No. 6, 28c; No. 5, 30c; No. 4, 36c per lb. Discounts: "C" brand, 40, 10, 10 and 7½ p.c.: "M.R.M" cold forged process, 50 and 5 p.c. Add 15c per box. Capewell brand, quotations on application.

Wire Nails—$3 f.o.b. Winnipeg and $2.55 f.o.b. Fort William.

Cut Nails—Now $3.20 per keg.

Pressed Spikes—¼ x 5 and 6, $4.75; 5-6 x 5, 6 and 7, $4.40; ⅜ x 6, 7 and 8, $4.25; 7-16 x 7 and 9, $4.15; ½ x 8, 9, 10 and 12, 4.05; ⅝ x 10 and 12, $3.90. All other lengths 25c extra, net.

Screws—Flat head, iron, bright, 80, 10, 10 and 10; round head, iron, 80; flat head, brass, 75; round head, brass, 70; coach, 70.

Nuts and Bolts—Bolts, carriage, ⅜ or smaller, 60 p.c.; bolts, carriage, 7-16 and up, 50; bolts, machine, ⅜ and under, 50 and 5; bolts, machine, 7-16 and over, 50; bolts, tire, 65; bolt ends, 55; sleigh shoe bolts, 65 and 10; machine screws, 70; plough bolts, 55; square nuts, cases, 3; square nuts, small lots, 2½; hex nuts, cases, 3; hex nuts, small lots, 2½ p.c. Stove bolts, 70 and 10 p.c.

Rivets—Iron, 60 and 10 p.c.; copper, No. 7, 43c; No. 8, 42½c; No. 9, 45½c; copper, No. 10, 47c; copper, No. 12, 50½c, assorted, No. 8, 44½c, and No. 10, 48c.

Coil Chain—¼-in., $7.25; 5-16, $5.75; ⅜, $5.25; 7-16, $5; 9-16, $4.70; ⅝, $4.65; ¾, $4.65.

Shovels—List has advanced $1 per dozen on all spades, shovels and scoops.

Harvest Tools—60 and 5 p.c.

Axe Handles—Turned, s.g. hickory, $3.15; No. 1, $1.90; No. 2, $1.60; octagon extra, $2.30; No. 1, $1.60.

Axes—Bench axes, 40; broad axes, 25 p.c. discount off list; Royal Oak, per doz, $6.25; Maple Leaf, $8.25; Model, $8.50; Black Prince, $7.25; Black Diamond, $9.25; Standard flint edge, $8.75; Copper King, $8.25; Columbian, $9.50; handled axes, North Star, $7.75; Black Prince, $9.25; Standard flint edge, $10.75; Copper King, $11 per dozen.

Churns—45 and 5; list as follows: No. 0, $9; No. 1, $9; No. 2, $10; No. 3, $11; No. 4, $13; No. 5, $16.

Auger Bits—"Irwin" bits, 47½ per cent. and other lines 70 per cent.

Blocks—Steel blocks, 35; wood, 55.

Fittings—Wrought couplings, 60; nipples. 65 and 10K Ts and elbows, 10; malleable bushings, 50; malleable unions, 55 p.c.

Hinges—Light "T" and strap, 65.

Hooks—Brush hooks, heavy, per doz., $8.75; grass hooks, $1.70.

Stove Pipes—6-in., per 100 feet length, $9; 7-in., $9.75.

Tinware, Etc.—Pressed, retinned, 70 and 10; pressed, plain, 75 and 2½; pieced, 30; japanned ware, 37½; enamelled ware, Famous, 50; Imperial, 50 and 10; Imperial, one coat, 60; Premier, 50; Colonial, 50 and 10; Royal, 60; Victoria, 45; White, 45; Diamond, 50; Granite, 60 p.c.

Galvanized Ware—Pails, 37½ per cent.; other galvanized lines, 30 per cent.

Cordage—Rope sisal, 7-16 and larger, basis, $11.25; manila, 7-16 and larger, basis, $16.25; lathyarn, $11.25; cotton rope, per lb., 21c.

Solder—Quoted at 27c per pound. Block tin is quoted at 45c per pound.

Wringers—Royal Canadian, $36; B.B., $40.75 per dozen.

Files—Arcade, 75; Black Diamond, 60; Nicholson's, 62½ p.c.

Locks—Peterboro and Gurney 40 per cent.

Building Paper—Anchor, plain, 66c; tarred, 69c; Victoria, plain, 71c; tarred, 84c; No. 1 Cyclone, tarred, 84c; No. 1 Cyclone, plain, 66c; No. 2, Joliette, tarred, 69c; No. 2 Joliette plain, 51c; No. 2 Sunrise, plain, 56c.

Ammunition, etc.—Cartridges, rim fire, 50 and 5; central fire, 33½ p.c.; military, 10 p.c. advance. Loaded shells: 12 gauge, black, $16.50; chilled, 12 gauge, $17.50; soft, 10 gauge, $19.50; chilled, 10 gauge, $20.50. Shot: ordinary, per 100 lbs., $7.75; chilled, $8.10. Powder: F.F., keg, Hamilton, $4.75; F.F.G., Dupont's, $5.

Revolvers—The Iver Johnson revolvers have been advanced in price, the basis for revolver with hammer being $5.30 and for the hammerless $5.95.

Iron and Steel—Bar iron basis, $2.70. Swedish iron basis, $4.95; sleigh shoe steel, $2.75; spring steel, $3.25; machinery steel, $3.50; tool steel, Black Diamond, 100 lbs., $9.50; Jessop, $13.

Sheet Zinc—$8.50 for cask lots, and $9 for broken lots.

Corrugated Iron and Roofing, etc.—Corrugated iron, 28 gauge, painted, $3; galvanized, $4.10; 26 gauge, $3.35 and $4.35. Pressed standing seamed roofing, 28 gauge, $3.45 and $4.45. Crimped roofing, 28 gauge, painted, $3.20; galvanized, $4.30; 26 gauge, $3.55 and $4.55.

Pig Lead—Average price is $6.

Copper—Planished copper, 44c per lb.; plain, 39c.

Iron Pipe and Fittings—Black pipe, ¼-in., $2.65; ⅜, $2.80; ½, $3.50; ¾, $4.40; 1, $6.35; 1¼, $8.65; 1½, $10.40; 2, $13.85; 2½, $19; 3, $25. Galvanized iron pipe, ⅜-in., $3.75; ½, $4.35; ¾, $6.65; 1, $8.10; 1¼, $11; 1½, $13.25; 2-inch, $17.65. Nipples, 70 and 10 per cent.; unions, couplings, bushings and plugs, 60 per cent.

Galvanized Iron—Apollo, 16-gauge, $4.15; 18 and 20, $4.40; 22 and 24, $4.65; 26, $4.65; 28, $4.50; 30-gauge or 10¾-oz., $5.20; Queen's Head, 20, $4.60; 24 and 26, $4.90; 28, $5.15.

Lead Pipe—Market is firm at $7.80.

Tin Plates—IC. charcoal, 20x28, box, $10; IX charcoal, 20x28, $12; XXI charcoal, 20x28, $14.

Terne Plates—Quoted at $9.50.

Canada Plates—18x21, 18x24, $3.50; 20x28, $3.80; full polished, $4.30.

Lubricating Oils—600W, cylinders, 80c; capital cylinders, 55c and 50c; solar red engine, 30c; Atlantic red engine, 29c; heavy castor, 28c; medium castor, 27c; ready harvester, 28c; standard hand separator oil, 35c; standard gas engine oil, 35c per gal on.

Petroleum and Gasolene—Silver Star, in bbls., per gal., 20c; Sunlight, in bbls., per gal., 22c; per case, $2.35; Eocene, in bbls., per gal., 24c; per case, $2.50; Pennoline, in bbls., per gal., 24c; Crystal Spray, 23c; Silver Light, 21c. Engine gasoline, in barrels, gal., 27c; f.o.b. Winnipeg, in cases, $2.75.

Paints and Oils—White lead, pure, $6.50 to $7.50, according to brand, bladder putty, in bbls., 2¼c; in kegs, 2¾c; turpentine, barrel lots, Winnipeg, 90c; Calgary, 97c; Lethbridge, 97c; Edmonton, 98c. Less than barrel lots, 5c per gallon advance. Linseed oil, raw, Winnipeg, 72c; Calgary, 79c; Lethbridge, 79c; Edmonton, 80c; boiled oil, 3c per gallon advance on these prices.

Window Glass—16-oz., O. G., single, in 50-ft. boxes—16 to 25 united inches, $2.25; 26 to 40, $2.40; 16-oz. O.G. single, in 100-ft. cases—16 to 25 united inches, $4; 26 to 40, $4.52; 41 to 50, $4.75; 50 to 60, $5.25; 61 to 70, $5.75; 21-oz. C.S., double, in 100-ft. cases, 26 to 40 united inches, $7.35; 41 to 50, $8.40; 51 to 60, $9.45; 61 to 70, $10.50; 71 to 80, $11.55; 81 to 90, $17.30.

When estimating on the cost of a stove we should not forget that it costs 20 per cent. to 30 per cent. to do business.

WHERE THEY BEAT US.

The builders of the old world were more ambitious than our own. No such theatre has ever been built in the modern world as the Coliseum with its diameter of 615 feet, its height of 164 feet, and its seats for 100,000 people. No wall has ever been built to equal the great wall of China, which runs 30 feet high and 24 feet thick for 1,200 miles; and the pyramids remain the wonder of the world in the twentieth century as in the first. Ancient Egypt had twelve palaces, each with 3,000 rooms; and the walls of Nineveh ran for 100 miles 100 feet high, and wide enough for three chariots to drive abreast along the top. Who builds so well and on so magnificent a scale today?

Practical Talks on Warm Air Heating

The second of a series of articles by E. H. Roberts

Warm Air Heating Plant for Seven-roomed Cottage.

Whatever may be the opinion of heating experts in regard to the practicability of furnace heat for large houses, it is generally conceded that in cottages like the one here shown there is no system of heating so economical in fuel or so satisfactory in operation as a good warm air furnace, properly installed.

In explanation of this plan, you will notice first that we have indicated side

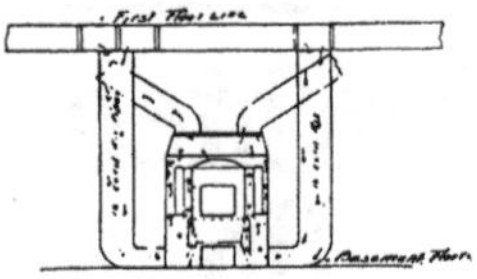

wall registers for all the rooms on the first floor, and with the exception of the one in the kitchen, these are intended to connect with wall stacks leading to rooms directly above. Each side wall register is provided with a valve, so

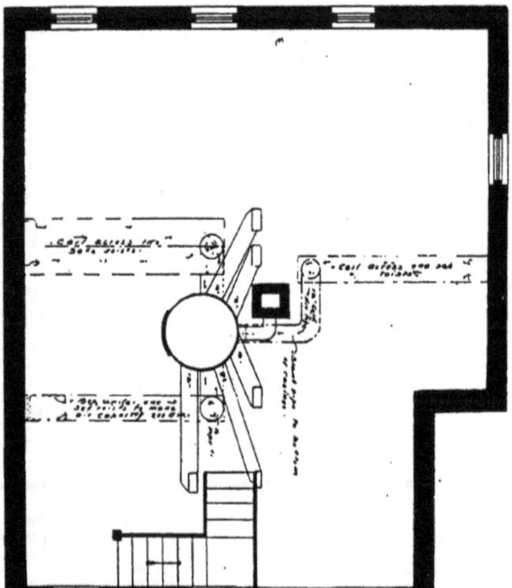

Basement Plan.

that the heat can be proportioned between the two rooms, or closed off entirely from either of them. This arrangement for heating two rooms from a single basement pipe we have found a

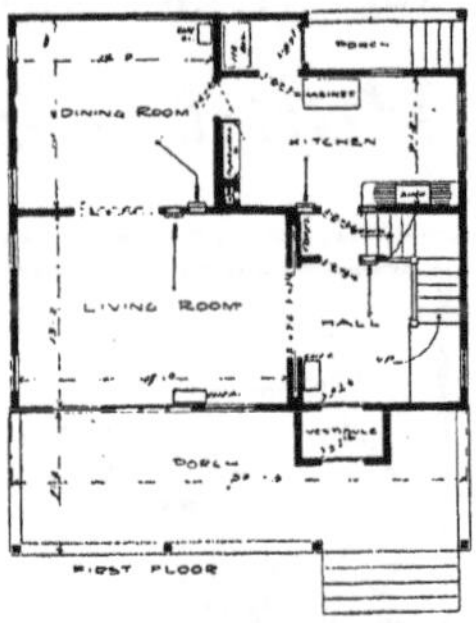

FIRST FLOOR

great improvement over the old method of running independent pipes to each room, as it minimizes the loss of heat from radiation in basement by reducing the number of basement pipes and ensures a very positive circulation of the warm air to every part of the building. We recommend side wall registers in preference to floor registers, not only because they make it possible to heat two rooms satisfactorily with a single basement pipe, but also because they save the cutting of floors and carpets, are more out of the way, and, if of sufficient size, they offer less resistance to the flow of warm air.

One of the most frequent causes of trouble with warm air furnaces is the lack of sufficient cold air supply proper-

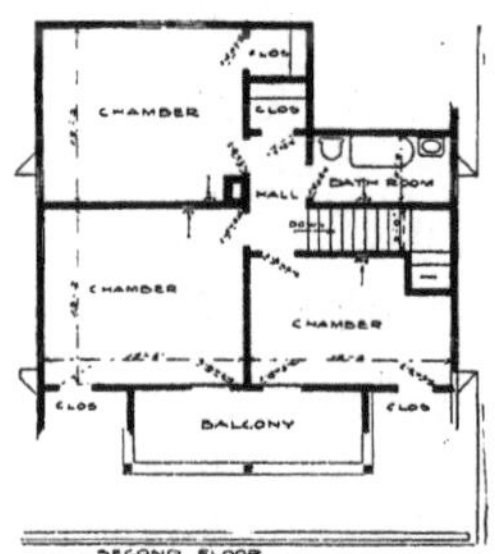

SECOND FLOOR

ly arranged. In this northwestern country, an outside air supply is never satisfactory except in mild weather, so we have indicated on our plans all inside circulation. If desired, an eight or nine-inch fresh air pipe may be run overhead from nearest basement window and connected with furnace casing at back of heater, but too much outside air is apt to interfere with the proper circulation of air through the returns, so it is generally safer to use inside air altogether.

In order to avoid underground ducts, also a pit underneath the furnace, and

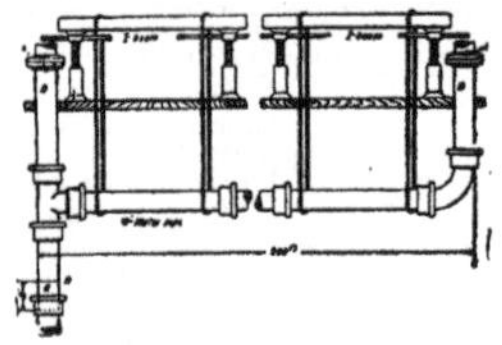

Method Used for Lowering Water Pipe.

at the same time to keep the basement as free from piping as possible, we would advise ceiling across joists as indicated on plans, using galvanized iron 36 inches wide where two sets of joists are enclosed, and the same width iron cut once in two for one set of joists.

At the points indicated by circles, are placed top flange collars and from these vertical or oblique pipes connect with cold air shoes at bottom of furnace casings in the manner shown on detail.

We have shown on plan floor faces for cold air, but side wall faces can be used if desired. We recommend the floor faces, however, as the coldest stratum of air always lies close to the floor, and as a consequence the floor faces take the air at a lower temperature than those placed in the side wall. When the air supply is taken from inside, as in this case, it is very essential that the aggregate area of the cold air duct should be at least equal to the combined areas of the warm air pipes, for while the volume of the warm air is increased by expansion, this is more than counterbalanced by the sluggishness of the cold air. On this particular plan, we have figured on 467 square inches of warm air outlet and a cold air supply of 513 square inches.

You will notice on plan that the furnace is centrally located and that warm air pipes are all comparatively short and have plenty of elevation. We advise taking warm air pipes out of the top of the hood, as shown on detail, rather than off the side, as the pipes seem to fill more quickly this way and the circulation of warm air through them is faster.

We would advise, however, having the hoods made with a slight mansard, as otherwise the air can work up along the sides of casings and get to pipes without being thoroughly heated.

Another important point to be considered when the air is taken in above the base ring, is the necessity for a shield between the firepot and the opening in cold air shoe to prevent the radiation from firepot from creating a counter current of warm air in the cold air pipes. We have shown on detail two shields hung from top of firepot in front of the side returns and there should also be another at the back in front of the rear return.

In conclusion, will say that a warm air furnace installed in the manner we have described ought to give the very best of satisfaction and it will cost no more than a furnace placed according to less modern methods.

CHIMNEYS AND FLUES.

A special committee of the National Fire Protection Association of the United States appointed to consider the question of chimneys and flues, reports in general terms and not in the customary draft clauses for a building code.

The committee thinks that all heat conveying flues should be free from all contact with inflammable materials, should have a free ventilating space surrounding them, should be securely built and supported, should be so placed that they can be readily reached so as to clean them off on the tops, the distance from inflammable materials is to a great extent dependent upon the construction of the flue, the temperature of the heat passing through it and the continous length of time the heat is passing through.

In all chimneys and flues of brick construction, only good hard, well-burnt brick should be used; soft brick should be prohibited; all joints should be struck smooth on inside excepting where the flue is lined with well-burnt clay or terra cotta pipe; no pargeting mortar shall be used on the inside; for bake ovens, low pressure boilers and similar purposes the brickwork shall be at least eight inches thick and lined continuously on the inside with well-burnt clay or terra cotta pipe, and be capped with terra cotta, stone or cast iron; for high-pressure boilers the brickwork shall be not less than twelve inches thick, with the inside four inches of this wall built up of fire-brick laid in fire mortar for a distance of twenty-five feet in any direction from the source of heat; for smelting furnaces or of steam boilers or other apparatus which heat the flues to a high temperature, shall be built with double walls of suitable thickness for the temperature, with an air space between the walls, the inside four inches to be of fire-brick laid in fire mortar for a distance of not less than twenty-five feet in any direction from the source of heat. All other chimney flues shall be lined continuously on the inside with well-burnt clay or terra cotta pipe made smooth on inside from the bottom of the flue or throat of the fireplace if the flue starts from the latter; chimneys not in continual use, or in dwellings from fireplaces or stoves, need not be lined, but must be struck smooth on inside. It is not advisable to have any bends or curves requiring a smaller upward inclination than seventy-five degrees, and all curves and bends are to be deprecated. No flue should be less than eight inches by eight inches. All unused flue holes in chimneys shall be bricked up, or closed with permanent tightly fitting metal covers. Horizontal brick flues should be covered on their tops with neat cement.

COPPER AND THE ELECTRIC AGE.

"The nineteenth century was the age of steel and the twentieth century, in its beginning at least, is the age of electricity. Electricity demands copper in ever-increasing quantities. For the last six decades of the nineteenth century," says the Copper Handbook for 1906, published by Horace J. Stevens, of Houghton, Mich., "the production of the leading metals increased at the rate of almost exactly 6½ per cent., compounded yearly, but in the last decade of the century copper left the other metals behind, and for fifteen years past the average ratio of increase in copper production has been 8 per cent., compounded yearly. This difference of only 1½ per cent. in the ratio of yearly increase may seem small, but like the fable of the horse's shoes, for which the first nail brought a penny, and each nail thereafter was doubled in price, the cumulative results are surprising, and at the end of fifteen years have run into hundreds of millions of dollars. In 1881, a quarter-century past, the world's production of copper was 163,000 long tons, worth an average of 18 cents a pound, giving a total value of $65,000,000 in round figures, while at the present moment the copper mines of the world are producing at the rate of nearly 800,000 tons per annum, worth, with copper at a little more than 20 cents a pound, $1,000,000 for every day in the year, or at the rate of $365,000,000 for 1906—an increase of almost fivefold in the world's copper bill in a quarter of a century."

CATALOGUES AND BOOKLETS.

When sending catalogues for review, manufacturers would confer a favor by pointing out the new articles that they contain. It would assist the editor in writing the review.

By mentioning HARDWARE AND METAL to show that the writer is in the trade, a copy of these catalogues or other printed matter will be sent by the firms whose addresses are given.

Advertising Enterprise.

Women take three million copies of one book and call for more.

Someone has said that the hand that rocks the cradle rules the world and it is the same hand that holds the purse. However that may be, the man with home goods to sell who does not study the ways of women and cater to their desires is far off the straight road to success. A successful retailer must know what women want, and what they are going to want next.

A careful study of home needs lies behind the little book illustrated herewith which has been a tremendous educational force for The Enterprise Manufacturing Co., of Pennsylvania, which has used successive editions as a compliment to their women customers, and as an advertising medium, until over 3,000,000 copies have been distributed, by mail and through the retail hardware dealer. The sixth edition of this book, revised and improved, is now being distributed.

It includes a lot of famous old recipes which all housekeepers value because of their simplicity, economy and goodness, and also a selected collection of newer recipes, all tested and proved right.

It contains in addition descriptions of the household Enterprise meat and food choppers, sausage stuffers and lard presses, coffee mills, fruit, wine and jelly presses, sadirons, etc. When imprinted with the business card of the dealer this book advertises his store for years. Dealers who desire to increase their sales of "Enterprise" specialties should communicate with the advertising department of The Enterprise Manufacturing Co., of Pa., Philadelphia, mentioning this paper.

BRIGHT OUTLOOK IN ALBERTA.

The Department of Agriculture of Alberta recently issued the following estimate of the crops in that province this year, with last year's figures for comparison:

Year.	Acres.	Yield.
	Spring Wheat.	
1906	97,760	2,322,292
1907	162,643	3,600,881
	Winter Wheat.	
1906	43,661	907,421
1907	92,382	2,039,509
	Oats.	
1906	323,923	14,553,522
1907	384,344	13,192,150
	Barley.	
1906	75,678	2,101,887
1907	76,433	2,201,179

BRITISH COMMERCIAL ATTACHES.

A London report from Consul-General R. J. Wynne states that there is a new scheme affecting the sphere of work of the British commercial attaches at Berlin, Vienna, Madrid and St. Petersburg. It is proposed that these gentlemen shall pay periodical visits to the more important trade centres of the United Kingdom. The principal object of these visits will be to enable the commercial attaches to get into touch with those members of the commercial community who are interested in any particular branch of export trade to the countries with which the commercial attaches are officially connected. It is probable that the first of these visits will shortly be undertaken by the British attache for Austria-Hungary, Italy and Greece.

AUSTRALIA'S TIMBER SUPPLY.

There are at the present time nearly six and a half million acres of forest reserves in New South Wales. In South Australia there are nearly two hundred thousand acres of forest reserves and plantations; in Queensland where forest conservation is of recent date, the reserved areas form a total of over three million acres; in Victoria the forest reserves cover a total of four million six hundred and seventy-five thousand five hundred and forty acres out of eleven million seven hundred and ninety-seven thousand acres of forest country, the balance being mostly timber country difficult of access.

ASBESTOS HORSESHOES.

Visitors to the volcano of Kilauea, on the island of Hawaii, generally ride on horseback, and in crossing what is known as the "pit," the horses suffer much from the great heat. The earth is so hot that the hoofs of the horses are not infrequently scorched. As some protection became very necessary, a clever blacksmith in Honolulu has recently devised a very successful method by which asbestos may be used.

The idea is to provide the hoofs of the horses with an asbestos covering much after the fashion of the outer shield of iron-studded leather or canvas over the automobile tires. These hoof shields may be put on and removed at pleasure.

WIRE NAILS TACKS WIRE

Prompt Shipment

The ONTARIO TACK CO.
Limited.
HAMILTON ONT.

Anvils and Vises
Machinery Steel
Wire Rope

IN STOCK

A. C. LESLIE & CO., Limited
MONTREAL

"MIDLAND"

BRAND.

Foundry Pig Iron.

Made from carefully selected Lake Superior Ores, with Connellsville Coke as Fuel, "Midland" will rival in quality and grading the very best of the imported brands.

Write for Price to Sales Agents

Drummond, McCall & Co.
MONTREAL, QUE.

or to

Canada Iron Furnace Co.
MIDLAND, ONT. Limited

NOVA SCOTIA STEEL & COAL CO., Limited

NEW GLASGOW, N.S.

Manufacturers of

Ferrona Pig Iron

And SIEMENS-MARTIN

OPEN HEARTH STEEL

OVER A MILLION AND A HALF

IVER JOHNSON

SAFETY AUTOMATIC REVOLVERS

ALREADY SOLD

Three Things Have Done it:

The superiority of the goods in every regard.

The immense popularity of "HAMMER ***THE HAMMER***" advertising.

The complete protection of the dealer. These goods cannot be used as premiums, and we never sell by mail where dealers can supply.

Send for Dealers' Catalogues, or ask your jobber—he knows

Iver Johnson's Arms and Cycle Works, 330 River Street, Fitchburg, Mass.
Makers of
Iver Johnson Truss Bridge Bicycles and Single Barrel Shotguns

TRADE MARK IVER JOHNSON

M.R.M. CO.

DON'T FORGET

That Iron worked under a hammer makes the best

HORSE SHOES

as in working the material in this way, all impurities are eliminated, the result being a much finer grade of Iron than otherwise obtainable.

"M.R.M." Horse Shoes are made from Iron worked under a powerful steam hammer, and for unifotmity of material, shape and finish, are second to none.

MADE BY

The Montreal Rolling Mills Co.

BUILDING AND INDUSTRIAL NEWS

For additional items see the correspondence pages. The Editor solicits information from any authoritative source regarding building and industrial news of an- sort, the formation or incorporation of companies, establishment or enlargement of mills, factories or foundries, railway or mining news.

Industrial Development.

The Canadian Bag Co. are building a factory at Winnipeg.

The Alberta Biscuit Co. will erect a factory at Edmonton, Alta.

T Lawson will erect a blacksmith shop in Ottawa, to cost $2,000.

The Brandon Mfg. Co. will erect a factory in that city to cost $2,000.

The Hanbury Mfg. Co. will erect a factory at Brandon, to cost $4,500.

Fire did damage to the town of Shoal Lake, Man., to the extent of $25,000.

Fire did damage to the town of Oxbow, Sask., to the extent of $175,000.

Williams & Wilson, Montreal, are erecting an addition to their warehouse.

The planing mill owned by Geo Wood & Son, Dundalk, Ont., was destroyed by fire.

An Illinois concern may erect a factory at Edmonton, for the manufacture of furniture.

An American brass manufacturing concern are anxious to obtain a site in Sarnia, Ont.

A gasoline engine and cream separator manufacturing concern may locate in Chatham, Ont.

The sash and door factory of V. E. Traversy, Montreal, was destroyed by fire; loss, $20,000.

A. I. Isaac's cigar factory at St. John, N.B., was destroyed by fire, entailing a loss of $4,000.

The warehouse of S. Hill & Son, Saskatoon, was razed to the ground by fire recently; loss, $15,000.

The Canadian Pin Co., a new concern, will erect a factory at Woodstock, Ont., for the manufacture of pins.

The Brantford Linen Manufacturing Company are going to transfer their factory to Tillsonburg, Ont.

Work has commenced on the erection of the steel plant of the Coughlan Company, at Vancouver, B.C.

The Great Northern Supply Co., of Swift Current, Sask., are putting up a new block, to cost $20,000.

The Otis-Fensom elevator will build a branch for British Columbia at Vancouver, B.C., to cost $35,000.

The Miramichi Lumber Company are negotiating for the purchasing of the Clark Spool Mill, Newcastle, N.B.

Fire did damage in Janesville, near Ottawa, to the extent of $9,000. The paint shop of A. Proulx suffered.

The warehouse of Shim & Son, Saskatoon, was destroyed by fire, and $15,000 worth of goods was destroyed.

The Ladysmith Lumber Company will erect a saw mill near Nanaimo, B.C., to cut 35,000 feet of lumber per day.

A large American firm, manufacturing agricultural implements, will make their western headquarters at Regina, Sask.

Morrison Bros. have been granted a loan of $1,000, freedom from taxation and free site by the town of Lloydminster, to erect an electric supply system.

The pressed brick plant of L. H. Pruitt, Medicine Hat, is nearing completion, and will soon be in operation.

Hamilton purchasers have acquired the timber limits, mill, and machinery of McManus & McKelvie, New Liskoard, Ont.

J. Davidson, Millbrook, Ont., has purchased the lighting plant and has been awarded the contract for lighting the town.

Gates & Carpenter, of Salt Lake City, were in Haileybury. They propose the erection of a large smelter there, to cost $6,000,000.

A plant capable of delivering 10,000,000 gallons per day will be installed by the Rainier Development Co., at New Westminster.

There is a chance of a large cut glass manufacturing concern being established at London, Ont. Mr. Wm. Gray is at the back of it.

The power situation in Moose Jaw is reaching an alarming condition. The present plant is incapable of keeping up with the demand.

The Canadian branch of the royal mint will cost, when completed, about $500,000. The salaries will run as high as $80,000 per annum.

R. Forbes Co., Hespeler, Ont., has secured a Dominion charter and will manufacture textile goods. The capital of the company is $1,000,000.

W, J. Campbell, Ottawa, will erect a boiler works in that city to cost $20,000. A new bridge will be erected at Lac du Bonnet, to cost $40,000.

The W. C. Edwards Co., Ottawa, will immediately commence the erection of their mill, which was destroyed by fire. The new mill will cost $300,000.

H. Proctor and C. W. Hughes, of Toronto, have formed a contracting firm in Fort William, Ont. They will construct all classes of cement work.

The buildings of the new plant of Jenkins Bros., Montreal, have been completed and the machinery is being installed. They will make valves of all kinds.

Negotiations are going on for the purchase of the Ottawa Electric Railway, the Ottawa Electric Light Co., and the Ottawa Gas Co. American capitalists are interested.

A large iron and steel plant will be constructed at Kootenay, B.C. Construction will commence shortly, and the initial expenditure will be $2,500,000.

The Library Bureau of Canada, Ottawa, will not rebuild their factory on the old site, but will move to another part of the town, where a modern plant will be erected.

The North American Bent Chair Co., Owen Sound, Ont., have awarded the contract to the Dominion Heating and Ventilating Co., Hespeler, Ont., for an extensive shaving system.

The magnificent new transformer building for the Canadian General Electric Co., Peterboro, Ont., has been handed over by the contractors, the Dominion Engineering and Construction Co.

The large plant of the Red Cliff Brick Co., near Medicine Hat, is in full operation. The plant cost $100,000, and will turn out 60,000 bricks a day. This number will be rapidly increased.

The contract for the construction of the power house of the Sydney and Glace Bay Railways, at Dominion No. 4, has been let to Rhodes, Curry & Co., Amherst. The plant and machinery will cost $25,000.

The Smart-Turner Machine Co., Hamilton, are supplying the following with standard duplex pumps: The Hamilton asylum, the Temiscaming & Northern Ontario Railway; the Exeter Canning and Preserving Co., Exeter, Ont.; the Canadian Canners, Leamington, Ont. The Leigh Portland Cement Co., Allantown, Pa., have also ordered three duplex plunger pumps.

Municipal Undertakings.

Massey, Ont., will spend $10,000 in waterworks.

A drill hall will be erected at Sherbrooke, Que.

A city hall will be erected at High River, Alta.

Sudbury, Ont., will spend $10,000 in waterworks extension.

A new pump house and equipment will be installed at Ottawa East.

The village of Hintonburg, Ont., will raise $10,000 for waterworks.

Nelson, B.C., is considering the installation of a municipal telephone system.

The ratepayers of St. Felicien, Que., will vote on a by-law for a waterworks system.

The municipality of Chilliwack, B.C., have voted on a by-law for a municipal telephone system.

The ratepayers of Medicine Hat will vote on a by-law to raise $10,000 for an Isolation Hospital.

A by-law will be submitted to the ratepayers of Hamilton, Ont., to raise $50,000 for waterworks pumps.

Ingersoll, Ont., will purchase the waterworks for $95,000, subject to the passing of the necessary by-law.

The ratepayers of Battleford, Sask., will vote on a by-law to raise $10,000 for a bonus to the Battleford Milling & Elevator Co.

Mining News.

A large two-stand converting plant will be erected at the Boundary Mines, B.C.

The first machine drill was recently started at the workings of the Phoenix Amalgamated.

There is a scheme on foot for the erection of a large smelter at Cobalt. Guggenheims are interested.

The compressor plant of the Cleveland-Cobalt Mine is in operation, and giving satisfaction. It is one of the largest in the district.

Spokane capitalists have formed a company to operate the Sure Copper

CONDENSED OR "WANT" ADVERTISEMENTS.

Advertisements under this heading 2c. a word first insertion; 1c. a word each subsequent insertion.

Contractions count as one word, but five figures (as $1,000) are allowed as one word.

Cash remittances to cover cost **must** accompany all advertisements. **In no case** can this rule be overlooked. Advertisements received without remittance cannot be acknowledged.

Where replies come to our care to be forwarded, five cents must be added to cost to cover postage, etc.

AGENTS WANTED.

AGENT wanted to push an advertised line of Welsh tinplates; write at first to "B.B.," care HARDWARE AND METAL, 88 Fleet St., E.C., London, Eng. [tf]

ONE who has sufficient organization to travel Canada east and west, fine line of Sheffield cutlery, brands already well known in Canada; none but applications from experienced and well-established men can be considered. Box 650, HARDWARE AND METAL, Toronto. [38]

BUSINESS CHANCES.

FOR SALE — Well established hardware, tinshop, implement and undertaking business, also good lumber yard, well fenced, with lumber and lime sheds in good condition; we will sell above altogether, or divide same to suit purchaser; proprietors are retiring from business in Manitoba, and therefore wish for immediate sale. Apply to Eakins & Griffin, Shoal Lake, Man. [33]

HARDWARE, tinware, stove and plumbing business in manufacturing town in the Niagara Peninsula; no competition; $250,000 factory and watermarks will be completed this summer; stock about $3,000; death of owner reason for selling. Box 85, Thorold, Ont. [tf]

WANTED—Partner to take half interest in one of the best hardware propositions in Algoma; plumber preferred. Box 636, HARDWARE AND METAL, Toronto. [32]

FOR SALE—A good hardware business in Western Ontario; stock about $6,000. For further reference apply The Hobbs Hardware Co., Limited, London, Ont. [32]

HARDWARE and Tin Business for Sale in good Western Ontario town of 3,000: stock about $3,500; good reasons for selling. Address, Box 647 HARDWARE AND METAL, Toronto. (34)

HARDWARE, Tinware, Stove and Furnace Business for sale. in live Eastern Ontario Village; first class chance for a practical man; English speaking community; stock can be reduced to suit purchaser; can give possession September 15th. 1907; premises for Sale or Rent, Apply to D. Courville, Maxville, Ont. (35)

HARDWARE Business and Tinshop for sale in Saskatchewan; population 1500; stock carried about $14,000 turnover, $45,000 practically all cash business; cash required, $8,000 would rent building; Do not answer without you have the money and mean business; it will pay to investigate this. Box 648 Hardware & Metal, Toronto. (41)

WELL established tinsmith business in thriving Ontario Village, including Eavetroughing, metal roofing, furnaces, pumps, pipefitting, hardware, paints, glass and stoves. Present stock about $1500; could be reduced. Shop and house combined, can be rented, including tools, horse and wagon, for $14.00 per month; possession given at once. Box 682, HARDWARE AND METAL (33)

FOR SALE.

FOR SALE — First-class set of tinsmith's tools second-hand but almost as good as new; includes an 8-foot iron brick almost new. Apply Pease Waldon Co., Winnipeg. [tf]

FOR SALE—A quantity of galvanized plain twist wire. Apply to C. B. Miner, Cobden, Ont. (34)

TEN gross of "Antisplash" Tap Filters. Made to fit on all ordinary kitchen taps, acts as a filter and prevents water splashing. Supplied with showcards to hold 1 dozen. Price per dozen $1.00; price per gross $11.00. Sample on request. G. T. Cole, Box 460, Owen Sound, Ont. [36]

LIVE AGENT WANTED FOR CUTLERY

Large and Old Established Sheffied Firm

manufacturing cutlery in all its branches,

Table and Dessert Knives and Forks, Carvers, Painters' and Glaziers' Goods, Butcher Knives, Steels, Pen Knives, Pocket Knives, Daggers, Shears, Scissors, Razors, Case Goods, Spoons and Forks, Electroplate in all its Branches

Is anxious to correspond with a First Class Man or Firm, NOT HANDLING GERMAN GOODS, who would be prepared to take up the agency, and push same thoroughly in Canada from coast to coast.

None but persons thoroughly acquainted with the business need apply. Highest references given and required.

Liberal commission to the right party.

Correspondence, in first instance, to "CUTLER," care of Hardware and Metal 10 Front St. E., Toronto

SITUATIONS VACANT.

TINSMITHS WANTED — First-class tinsmiths wanted for points west of Winnipeg; must be good mechanics capable of taking charge of a metal department; thorough knowledge of furnace work necessary. Pease Waldon Co., Winnipeg, Man. [tf]

WANTED general hardware clerk, experienced, sober, good stock-keeper and salesman; steady job for right man; state experience and salary expected. C. Richardson & Son, Box 325, Harrow, Ont. [36]

WANTED—Hardware clerk, experienced and sober; state age and salary expected at start. Thomas Oliver, Copper Cliff, Ont. [33]

SITUATIONS WANTED.

COMMERCIAL gentleman with nine years' trade connection with ironmongers, architects and public institutions in Great Britain desires position as representative of a Canadian manufacturing firm. Box X, HARDWARE AND METAL, Montreal.

WANTED.

OFFICE space wanted by manufacturer in Toronto; state location and terms. Box 645, HARDWARE AND METAL, Toronto. [32]

WANTED—Two store ladders, "Myers" preferred, complete with fixtures. Send full description and price to Illsey Bros., Red Deer, Alta. (36)

THE WANT AD.

The want ad. has grown from a little used force in business life, into one of the great necessities of the present day.

Business men nowadays turn to the "want ad" as a matter of course for a hundred small services.

The want ad. gets work for workers and workers for work.

It gets clerks for employers and finds employers for clerks. It brings together buyer and seller, and enables them to do business, though they may be thousands of miles apart.

The "want ad." is the great force in the small affairs and incidents of daily life.

Group, near Bella Coola, B.C. Machinery and supplies have been purchased, and are being shipped with all haste.

A deal was completed recently by which a syndicate of Nelson men secured control of the Goodenough and Blue Bird properties at Sandon. A fund of $25,000 was subscribed and development will begin immediately. J. A. Whittier will be in charge of the operations and L. Pratt is treasurer.

The large smelter of the Ontario Electric Smelting & Refining Co., at Newark, N.J., will be removed to Ottawa and the capacity doubled. All kinds of ore will be treated by the electric process, which is much cheaper than the old way. The by-products will all be utilized, and these alone will pay for the smelting. The company is capitalized at $2,000,000, and the plant will cost $1,000,000.

Railroad Construction.

The C.P.R. have given $30,000 for a Railroad Y.M.C.A., to be erected at Kenora, Ont.

The contract for the Moose Jaw-Edmonton line of the C.P.R. has been let, the cost being $300,000.

The G.T.R. will spend $150,000 in enlarging and improving their shops and roundhouse at St. Thomas, Ont.

The plans for the new Central Station for the G.T.R. at Ottawa have been approved and work will be commenced immediately. The cost will be a million dollars.

Building Notes.

The new courthouse for Vancouver will cost $400,000.

The new college building to be erected at Point Grey, B.C., will cost $100,000.

Bushnell & Varty will erect an apartment block in Vancouver, to cost $20,000.

The new edifice for the congregation of St. Andrew's Presbyterian Church, Fort William, wil cost $946,500.

Business Systems, Limited, will erect a large building in Toronto for a commercial school, to cost $50,000.

The congregation of Annette Street Methodist Church, Toronto Junction, will build an edifice to cost $50,000.

The British Columbia Permanent Loan & Savings Co. will erect an office building in New Westminster, to cost $40,000.

Companies Incorporated.

The Peterboro Boiler & Radiator Co., has changed its name to Canadian Boiler & Radiator Co.

McEwen Bros., New York, capital $25,000, have secured an Ontario license and will make and deal in steam engines and machinery fittings.

The Wiarton Steamboat Co., Wiarton, Ont.: capital, $20,000; to own and operate steamboats and carry on a wrecking business. Incorporators: S. Rutherford, T. C. Allan, J. A. Acres, W. H. Buchan, all of Wiarton.

The Bottle Exchange Company of Canada, Toronto, capital $20,000; to manufacture and sell milk cans and dairy supplies. Incorporators: J. H. Lock, R. W. Dockeray, E. Grace, A. Anderson, V. E. Vansant, all of Toronto.

E. and T. Fairbanks & Co., Sherbrooke, Que.: capital, $150,000; to manufacture and deal in scales, weighing instruments, and machinery of all kinds. Incorporators: H. N. Turner, J. C. Clark, P. F. Hazen, C. H. H. Turner, all of St. Johnsbury, Vermont, U.S.A.

The Benson Lumber Truck Co., Port Arthur, Ont.: capital, $20,000; to manufacture, sell and dispose of heavy farm and lumber machinery. Incorporators: A. W. Benson, N. O. Werner, E. L. Mattson, all of Mineapolis, Minn., and G. S. Clark, and F. H. Keefer, both of Port Arthur.

CEMENT INDUSTRY OF ONTARIO.

U. S. Consul A. G. Seyfert, is convinced that the cement industry of Ontario, which in its earliest struggle had costly experiments, has at last been put on its feet. Writing to his government from Collingwood, Ont., he says:

"At the present time the plants of the province are turning out 10,000 barres of Portland cement a day, or 3,000,000 barrels a year. Half of this output is from this consular district in the locality of Owen Sound. The coal consumed in the industry throughout the province is all imported from Pennsylvania, and amounts to 250,000 tons annually. The average price is $3.50 per ton. Of the total output of cement in Ontario 25 per cent. is shipped to the Northwest, while the remainder is used by the home market.

"The business on the whole is now giving satisfactory financial returns, for dividends were paid at all the plants during last year. The Canadian demand is ahead of the immediate supply, which means an increase of output at an early date. The price received per barrel at the mill is $1.60 to $1.70, according to quality. The demand for cement has been created in many new directions, and perhaps the most novel in use now is in the construction of reenforced telegraph poles 60 feet high, such as are seen along the Welland canal."

A RECOMMEND.

How often are you asked to write a recommend for someone of your acquaintance seeking some situation? Sometimes you hardly feel like doing it, but you usually comply hoping no harm will result. We heard of a gentleman who dismissed his gardener for dishonesty, but for the sake of his family gave him the following "character": "This certifies that the bearer has been my gardener for over two years, and during that time he succeeded in getting more out of my garden than any man I ever employed."

News of the Paint Trade

MAKING OF PAINT BRUSHES.

The most important factor in making up the cost of a paint brush is the bristles, as the cost of labor is a secondary item. In preparing bristles to make a satisfactory working brush it is necessary first of all, that they be sorted to various lengths and degrees of stiffness and taper. They are then washed and straightened—a most important process—then mixed, and finally prepared for the size required. Bristles prepared in this way are the foundation of a good working tool, but the small and inexperienced maker will simply take the bristles as they come from the cask, jab them into a ferrule and call it a paint brush. How can a retailer recommend such goods to a customer whose trade he wants to hold season after season? Another source of trouble is that firms who know better often turn out brushes with a showy finish, but which have not the real working qualities behind them, and are not the perfect tools they should be. They catch the eye, says an exchange, but do not give solid satisfaction. The best way to protect customers against these things and thus hold their confidence, is to handle the goods of some well-known maker, preferably one who brands his brushes. You would not have a man working for you who wouldn't tell his name. Then why expect to build up your trade by giving customers a brush which a manufacturer is not willing to guarantee by placing his name thereon?

THE SCARCITY OF BRISTLE.

Lack of bristle is disturbing the brush manufacturers to a considerable degree. This product gets scarcer and dearer constantly, and it is hard for the trade to figure out just where its supply for the future will come from. While the trade seems to call for more brushes each succeeding year, still the crop of bristle seems to go backward, rather than forward, nor can any different state of affairs be expected, when the facts are considered. While bristle of a high quality at present seems absolutely indispensable to many qualities of toilet brushes, it is, nevertheless, a serious question whether various fibres will not have to be gradually substituted wherever such is possible. It may be that the future holds discoveries and improvements in the matter of fibre development and preparation which will largely solve the present difficult problem as to bristle supply. The manufacturer of fine brushes certainly will hope for such a state of things to transpire, as his troubles in securing bristle become more and more serious with each year that passes.

A rapid review of the world's leading bristle markets will be of interest at this time. Only last week the English bristle market underwent an advance of five per cent. Chinese, Russian, Polish, and American bristles continue very scarce and firmly held, at the figures prevailing last month. French bristles have recently been advanced materially, as

available supplies are almost exhausted and shrewd buyers are contracting ahead as far as next summer. Chinese merchants continue to withhold their stocks in the hope of securing still higher figures within the near future, and it is becoming more and more difficult to obtain reasonably prompt shipments. Russian and Polish bristles are especially scarce, as the production has been curtailed by labor troubles in Russia, while the demand is still increasing by leaps and bounds. American bristles are being purchased rapidly. Many brush manufacturers are placing orders for far distant deliveries of Chinese bristles and, in many instances, are trying to persuade the importers with whom they have already contracted for fall and winter shipments, to anticipate these deliveries and let them have at least a part of their orders within the near future. Such action goes to show that not a few brush manufacturers are looking for a further advance in the leading markets before long.

All descriptions of hair, including goat and squirrel, are likewise in brisk demand, and prices are very firm. Offerings of fibres are very small, and a strong undertone characterizes the market; there has been no additional advance in the prices of bass, but there is no likelihood of a reduction for some time to come. The output of all fibres has been curtailed somewhat, and as demand continues very active, prices are being very firmly held.

PAINT AND OIL MARKETS

MONTREAL.

Montreal, August 23.—The usual quietness prevalent in all branches of trade in August prevails in paint and oil circles. Makers are busy not so much in filling current orders as in filling a few back orders, and in repleting their stocks for the fall trade. The demand for all seasonable lines is moderate and steady. The factories are receiving prompt delivery of supplies of raw material and merchants are finding little difficulty in securing prompt consignments. As the amount of freight to be handled by the transportation companies has considerably diminished, they are in better position to handle the freight now offered and will be given ample opportunity to prepare themselves for the fall rush, and all commercial men sincerely hope that the railroads will embrace this opportunity to preclude such disastrous delays as have occurred during the past summer. There are no changes in this week's quotations. Prices are on the whole firm.

Linseed Oil.—Considerable weakness exists in the oil market. Local prices are weak and unchanged. Raw, 1 to 4 barrels, 60c.; 5 to 9, 59c.; boiled, 1 to 4 barrels, 63c.; 5 to 9, 62c.

Turpentine—The prices which were advanced in foreign markets last week are well maintained. Local prices are firm and unchanged. Conditions in Savannah are quiet with a slight tinge of uncertainty. Single barrels, 80c.

Ground White Lead—The demand is strengthening slightly. Prices are a little firmer. Local corroding work is active and makers are busy. Government standard, $7.50; No 1, $7; No. 2, $6.75, No. 3, $6.35.

Dry White Zinc—A moderate demand exists. Prices are firm and unchanged. Painters are coming steadily to recognize the superiority of this commodity for finishing touches. V.M. Red Seal, 7½c.; Red Seal, 7c; French V.M., 6c.; Lehigh, 5c.

White Zinc Ground in Oil—Situation is unchanged. Prices are firm and the demand is moderate. Pure, 8½c.; No. 1, 7c.; No. 2, 5⅞c.

Red Lead—The activity in this line has slackened. Prices are firm and unchanged. Genuine red lead, in casks, $6.25; in 100-lb. kegs, $6.50; in less quantities at $7.25 per 100 lbs. No. 1 red lead, casks, $6; kegs, $6.25, and smaller quantities, $7.

Gum Shellac—The situation is unchanged. Demand is quiet and prices are firm. Fine orange, 60c. per lb.; medium orange, 55c. per lb.; white (bleached), 65c.

Shellac Varnish—Prices are firm and unchanged. A seasonable business is being transacted. Pure white bleached shellac, $2.80; pure orange, $2.60; No. 1 orange, $2.40.

Putty—Makers are busy with the demand which is unseasonably strong. Prices are firm. Pure linseed oil, $1.85 in bulk; in barrels, $1.60; in 25-lb. irons, $1.90; in tins, $2; bladder putty, in barrels, $1.85.

Paris Green—The season with the makers is practically over. The demand this year has been strong but of short duration.

TORONTO.

Toronto, August 23.—While business in the local paint and oil markets cannot be said to be brisk, nevertheless, a fair amount of trade is being done for this time in August. Mixed paints are in good average demand and, with the approach of fall, the volume of transactions in glass and putty is increasing daily. Strictly city business is rather quiet this week, but this is the usual condition of affairs the week prior to the opening of the great national exposition, which will be held here from Aug. 26 to Sept. 9. After the excitement of the exhibition is over, retailers will begin to place their full orders and jobbers are confidently looking forward to a brisk September business.

White Lead—A good average demand exists and trade is all that can be expected at this advanced stage of the summer. Prices show no disposition to change and are firm at the following figures: Genuine pure white lead is quoted at $7.65, and No. 1 is held at $7.25.

Window Glass

Come to us for all kinds of glass. Let us quote you on every specification. We have immense stocks in our large new warehouses.

Star Window Glass
Diamond Picture Glass
Double Diamond Glass
Colored Glass
Rolled Glass
Cathedral Glass
Wire Glass
Heavy 26 oz. 32 oz. Glass
and Glass of all kinds.

LET US QUOTE

A. RAMSAY & SON COMPANY

Glass Dealers MONTREAL 10 to 22 Inspector St.

What to Write

Cut 9
FOR PAINT OILS.

If you want to learn how to save from $500 to $5,000 a year on oil expense, just write the following on a postal:

S. F. Bowser & Co., Limited
Toronto, Ontario.

Gentlemen:

Please send us your Catalogue V with complete information and prices regarding your oil tanks which will stop every form of loss. We handle the following oils: ————, ————, ————.

(Signed) ————————

The postal will cost you but one cent, but it will show you the way to save hundreds of dollars. Write to-day.

S. F. BOWSER & CO., Limited

66-68 Fraser Ave. Toronto, Can.

LOSS BY FIRE

is rendered improbable if you sell your customer our

ASBESTOS STOVE LINING

It is the lining par excellence for every kind of **Stove, Grate, Hot Air Furnace**, etc., and has put the old-fashioned **fire brick out of the game.** It can be **applied** in a few minutes by **anyone.**

Put up in 6-lb. packages and in bulk.

A Broken Stove, Furnace, Range or Heater can be repaired quickly and permanently by our

ASBESTOS FIREPROOF FURNACE CEMENT

It is the best cement for mounting and repairing stoves, furnaces, etc., because it adheres firmly to the castings and **sets harder than iron.** Acids and chemicals have no effect upon it and fire cannot melt it. **Moreover, there's money in it for you.** Investigate anyway!

Benjamin Moore & Co., Limited

(Formerly The Francis-Frost Co., Limited)
New Offices and Works: Toronto Junction, Canada
New York, Chicago, Cleveland, Carteret, N. J.

McArthur, Corneille & Co.

MONTREAL

Glue and Gelatine

An extensive assortment, to suit all requirements.
WILL BE PLEASED TO SUBMIT SAMPLES AND PRICES

MANUFACTURERS AND IMPORTERS OF . . .

White Lead, Oils and Colors, Prepared Paints, Window Glass, Varnishes, Etc.

SELLING AGENTS IN CANADA

For the GENUINE **Imperial French Green** of JOHN LUCAS & CO., PHILADELPHIA

And CELEBRATED **English Varnishes** of CHAS. TURNER & SON, LONDON.

Please mention HARDWARE AND METAL when writing.

Red Lead—A good number of enquiries for red lead in sorting quantities continue to arrive and jobbers find no reason to complain of August business in this commodity. Prices are firm and unchanged as follows: Genuine, in casks of 500 lbs., $6.25; ditto, in kegs of 100 lbs., $6.75; No. 1 in casks of 500 lbs., $5; ditto, in kegs of 100 lbs., $5.50.

Paris Green—Last week the demand for Paris green was temporarily stimulated by a few days of very hot weather; this week, however, orders have practically ceased to arrive and the season for this commodity may now be safely said to be over. Prices are unchanged as here quoted: Canadian Paris green, 29½c base; English Paris green, 30¼c.

Petroleum—The lengthening nights are gradually strengthening the demand and in less than a month the heavy autumn trade will set in. Present prices are: Prime White, 13c; water white, 14½c; Pratt's astral, 18c.

Shellac—The demand is rather slack at present, but that is the usual condition of the shellac market at this stage of the season; most jobbers, however, report shellac sales to be much in excess of last year. The following prices still hold good: Pure orange, in barrels, $2.70; white, $2.82½ per barrel; No, 1 (orange), $2.50.

Linseed Oil—The local market is easy and prices are unchanged since last week. The British market is rather weak at present, owing to large shipments of Indian seed which have arrived within the last ten days as well as larger consignments which are known to be on the ocean en route for Britain. Little business in oil is being done on the local market and present prices are: Raw, 1 to 3 barrels, 63c; 4 barrels and over, 62c. Add 3c to this price for boiled oil f.o.b. Toronto, Hamilton, London and Guelph, 30 days.

Turpentine—Prices are unchanged on the local market and business is fair for August. The most interesting feature about the turpentine situation to local importers is the fact that the classification agents of the United States have raised the weight of the minimum car of turpentine from 30,000 to 36,000 pounds. This means that an importer instead of buying 72 barrels must now buy 87, the extra amount necessitating an extra investment of from four to five hundred dollars on each car load. As the classification agents' decisions are backed up by inter-State laws, local importers will simply have to accustom themselves to the new order of things. Local turpentine prices are unchanged: Single barrels, 79c; two barrels and upwards, 78c f.o.b. point of shipment, net 30 days; less than barrels, 5c advance. Terms: 2 per cent., 30 days.

For additional prices see current market quotations at the back of the paper.

L. Cleophas Dumas, painter and trader, Montreal, has made an assignment.

Argall & Riche have been registered to carry on business as painters in Montreal.

PAINTING BRIDGES IN WINTER.

The following article was written by a foreman painter of one of the large railroads of the United States. This gentleman has had experience of many years in painting railroad bridges, under all kinds of conditions, and the subject of this article, "Painting Bridges in Winter," is of special interest at this time:

That experience is a good teacher, no one will deny, and after repeated warnings by different authorities on the increased cost of labor and material and the lack of durability from painting metal in the winter months, let us recall our experience of six years.

In January, 1901, a large railroad system sent a gang of painters into western Iowa to clean and paint fifty steel girder span bridges. With the force at hand, and owing to other conditions of the work, they were twenty months completing the undertaking. During this time they were painting through the winter months of 1901 and 1902, as well as through the summer months of both years, and every precaution was taken not to paint when there was moisture in the air or when frost was on. The paint was applied on girders in February, 1902, when the thermometer registered as low as 4 degrees above zero, and there seems to be no difference now in the general condition of this work, whether painted in summer or winter.

At no time did the work over-run the estimate of cost, although the cost was slightly increased, both for labor and material, in the winter. However, this extra cost was only a trifle, as compared to the care of the structures, and now after more than five years it is impossible to tell whether the structures were painted during the summer or the winter, as the general conditions are the same, and only by looking at the date painted on when finished can you tell when the work was done.

All of these bridges received exactly the same treatment, having first been coated at the factory with boiled linseed oil, which was only a detriment to the surface. The oil bound fast the mill scale, rust, grease and blisters and in the field, without a sand blast, could only be removed by hand scrapers. After the steel was erected it received two coats of paint.

The condition of the work had several advantages, by having a land climate without alkali, sea fogs or gases, which are enemies to paint; yet the bridges have had the disadvantages of being painted, in many cases, six months after erection, together with the coating of linseed oil and the drippings of salt brine from meat refrigerator cars.

After experiences of this kind, which may occur over and over again, it only convinces one that the time to paint is just as soon as the structure is erected and before rust starts. Like filling the purse, "Little and Often" makes a good rule for painting, and the season of the year can not be considered when a structure needs paint.

Growth of Canada's Export Trade.

Value of Annual Exports from Canada:

1886—$85,251,314.00 1896—$118,011,508.00 1906—$256,586,630.00

What is YOUR share of this ever-increasing business?

THE BEST WAY to reap the profits of foreign trade is through the unique foreign trade service of the

AMERICAN EXPORTER

HALF A THOUSAND Manufacturers use this method of reaching the buyers abroad. Among them:

Dominion Cartridge Co., Limited	Montreal	Goold, Shapley & Muir Co.	Brantford
Waterous Engine Works Co.	Brantford	Brantford Carriage Co.	Brantford
Canada Horse Nail Co.	Montreal	Metallic Roofing Co.	Toronto
Sylvester Bros. Mfg. Co.	Lindsay	Page Wire Fence Co., Limited	Walkerville

☞ **ANNUAL HARDWARE NUMBER PUBLISHED AUGUST 1st** ☜

AMERICAN EXPORTER

Established 1877 **Conducted by Export Specialists**

135 William Street, New York City, U.S.A.

Fresh Paint

It's a sure sign of progress wherever it is seen and when it's Martin-Senour 100% pure Paint it's a sign that there is a wide-awake dealer not far away. Wide-awake dealers handle Martin-Senour 100% pure Paint because it's a money-maker, that is one reason. It's a trade builder, that's another reason. The manufacturers do a deal of powerful direct advertising that helps the dealer's trade grow and widens his reputation, that's another reason. Would you like to belong to the wide-awakes? There is money in it and we will tell you how we help you get it if you write us. Better do it today.

The MARTIN-SENOUR CO. Ltd.,

"Pioneers of Pure Paints"

142-144 Inspector Street, Montreal 2514-2522 Quarry St. and Archer Ave., Chicago (Est. 1878)

The WINNIPEG PAINT & GLASS CO. Ltd., Winnipeg. The MARITIME PAINT & VARNISH CO. Ltd., HALIFAX, N. S.

CURRENT MARKET QUOTATIONS.

August 23, 1907

These prices are for such qualities and quantities as are usually ordered by retail dealers on the usual terms of credit, the lowest figures being for larger quantities and prompt pay. Large cash buyers can frequently make purchases at better prices. The Editor is anxious to be informed at once of any apparent errors in this list, as the desire is to make it perfectly accurate.

METALS.

ANTIMONY.

Cookson's ....per lb.	0 15
Hallett's	0 14½

BOILER PLATES AND TUBES.

Plates, ¼ to ½ inch, per 100 lb..	2 40	2 50
Heads, per 100 lb	2 65	2 75
Tank plates 3-16 inch	2 60	2 70
Tubes per 100 feet, 1½ inch ..	8 25	8 50
" " 2 " ..	9 00	9 10
" " 2½ " ..	10 50	11 00
" " 3 " ..	12 00	12 50
" " 3½ " ..	15 00	16 00
" " 4 " ..	19 25	20 00

BOILER AND T.K. PITTS.

Plain tinned and Spun, 25 per cent. off list.

BABBIT METAL.

Canada Metal Company—Imperial, genuine 60c; Imperial Tough, 60c; White Brass, 50c; Metallic, 35c; Harris Heavy Pressure, 25c.; Hercules, 25c; White Bronze, 15c.; Star Frictionless, 14c; Alluminoid, 10c; No. 4, 9c per lb

James Robertson Co—Extra and genuine Monarch, 60c.; Crown Monarch, 50c; No 1 Monarch, 40c; King, 30c; Fleur-de-lis, 20c; Thurber, 15c; Philadelphia, 12c.; Canadian, 10c; hardware, No. 1, 15c; No. 2, 12c; No. 3, 10c. per lb.

BRASS.

Rod and Sheet, 14 to 30 gauge, 25 p c. advance.

Sheets, 12 to 14 in.	0 30
Tubing, base, per lb 5-16 to 2 in	0 33
Tubing, ¼ to 3-inch, iron pipe size..	0 31
" ¼ to 3-inch, seamless..	0 36

Copper tubing, 6 cents extra.

COPPER.

	Per 100 lb	
Casting ingot	22 00	23 00
Cut lengths, round, bars, ½ to 2 in...		35 00
Plain sheets, 14 oz..........................		36 00
Plain, 16 oz., 14x48 and 14x60		35 00
Tinned copper sheet, base		38 00
Planished base.....................		43 00
Braziers' (in sheets), 4x6 ft., 25 to 30 lb. each, per lb., base .	0 34	0 35

BLACK SHEETS.

	Montreal.	Toronto
to 10 gauge	2 70	2 75
12 gauge.........................	2 70	2 75
14 "	2 50	2 60
17 "	2 50	2 60
18 "	2 50	2 60
20 "	2 50	2 60
22 "	2 55	2 65
24 "	2 55	2 70
26 "	2 65	2 85
28 "	2 70	3 00

CANADA PLATES.

Ordinary, 52 sheets	2 75	3 05
All bright "	3 75	4 05
Galvanized—	Dom.Crown.	Ordinary.
18x24x52	4 45	4 35
60..........................	4 70	4 60
20x28x80	8 90	8 70
....	9 40	9 20

GALVANIZED SHEETS.

B.W. gauge	Queen's Head	Fleur-de-Lis	Gordon Crown	Colborne Crown Gorbal's Best
16 - 20 ..	3 95	3 80	3 95	....
22 - 24 ..	4 20	4 05	4 00	4 05
26........	4 45	4 30	4 40	4 30
28........	4 70	4 55	4 60	4 55

Less than case lots 10 to 25c. extra.

Apollo Brand.

24 guage, American	....	3 85
26 " "	....	4 10
28 " "	....	4 55
10¾ oz.	....	4 85

25c. less for 1,000 lb lots.

IRON AND STEEL.

	Montreal.	Toronto.
Middlesboro, No. 1 pig iron..	21 50	24 50
Middlesboro, No. 3 pig iron..	20 50	23 50
Summerlee, " ..	25 50	26 50
" special " .	24 50	....
" soft " ..	24 00	....
Carron...	26 00	....
Carron Special	24 50	....
Carron Soft	24 00	....
Clarence, No. 3	21 50	23 50
Glengarnock, No 1		27 00
Midland, Londonderry and Hamilton, off the market but quoted nominally at		26 00
Radnor, charcoal iron.........	32 00	34 00
Common bar, per 100 lb......	2 20	2 30
Lowmoor iron........................		6 50
Angles..		2 50
Forged iron "	2 45	..
Refined " "	2 60	2 70
Horseshoe iron "	2 60	2 70
Band iron, No. 10 x ⅜ in.		3 60
Sleigh shoe steel "	2 25	2 30
Iron finish steel		2 40
Reeled machinery steel............		3 50
Tire steel	2 40	2 50
Best sheet steel......		0 12
Mining cast steel....		0 08
Warranted cast steel....		0 14
Annealed cast steel....		0 15
High speed...........		0 60
B.P.L. tool steel	10½	0 11

INGOT TIN.

Lamb and Flag and Straits—
56 and 28-lb. ingots, 100 lb. $42 00 $44 00

TINPLATES.

Charcoal Plates—Bright

M.L.S., Famous (equal Bradley)		Per box.
I C,14 x 20 base	...	$6 50
IX, 14 x 20 "	....	8 00
IXX, 14 x 20 base	....	9 50

Raven and Vulture Grades—

I C, 14 x 20 base	....	5 00
I X "	....	6 00
I X X "	....	7 00
I X X X "	....	8 00

"Dominion Crown Best"—Double Coated, Tissued.

		Per box.
I C, 14 x 20 base..........	5 50	5 75
I X, 14 x 20 "	6 50	6 75
IXX x 20 "	7 50	7 75

"Allaway's Best"—Standard Quality.

I C, 14 x 20 base..........	4 65	5 00
I X, 14 x 20 "	5 40	5 75
I XX, 14 x 20 "	6 15	6 50

Bright Cokes.

Bessemer Steel—

I.C., 14 x 20 base	4 25	4 35
20x28, double box	8 50	8 70

Charcoal Plates—Terne.

Dean or J. G. Grade—

I C., 20x28, 112 sheets	7 25	8 00
IX., Terne Tin	...	9 50

Charcoal Tin Boiler Plates.

Cookley Grade—
X X, 14x56, 50 sheet bxs. }
" 14x60, " } 7 50
" 14x65, " }

Tinned Sheets.

72x30 up to 24 gauge...... ...	....	8 50
" " 26 "	....	9 00

LEAD.

Imported Pig, per 100 lb......	5 25	5 35
Bar, ...	5 75	6 00
Sheets, 2½ lb. sq. ft., by roll ..	....	6 50
Sheets, 3 to 6 lb. "	....	6 25

Cut sheets ½c. per lb., extra.

SHEET ZINC.

5-cwt. casks	7 75	8 00
Part casks.......................	8 00	8 25

ZINC SPELTER.

Foreign, per 100 lb	6 75	7 00
Domestic	6 50	6 75

COLD ROLLED SHAFTING.

9-16 to 11-16 inch..............	....	0 06
¾ to 1 7-16 "	....	0 05½
1 7-16 to 3 "	....	0 05

30 per cent.

OLD MATERIAL.

Dealers buying prices:

	Montreal	Toronto
Heavy copper and wire, lb.	0 15½	0 15
Light copper..............	0 14½	0 13
Heavy red brass....	0 14	0 13
" yellow brass	0 10½	0 10½
Light brass................	0 07	0 07½
Tea lead	0 03½	0 03½
Heavy lead................	0 04	0 04
Scrap zinc	0 03¾	0 03¾
No. 1 wrought iron	15 50	11 50
" 2 " "	6 00	6 00
Machinery cast scrap ...	17 00	16 50
Stove plate................	13 00	12 00
Malleable and steel	8 00	8 00
Old rubbers	0 10½	0 10
Country mixed rags, 100 lbs	1 00	1 25

PLUMBING AND HEATING

BRASS GOODS, VALVES, ETC.

Standard Compression work, 57½ per cent.
Cushion work, discount 40 per cent.
Fuller work, 70 per cent.
Flatway stop and stop and waste cocks, 60 per cent; roundway, 55 per cent.
J.M.T. Globe, Angle and Check Valves, 45; Standard, 55 per cent
Kerr standard globes, angles and checks, special, 42½ per cent., standard, 47½ p.c.
Kerr Jenkins' disc, copper-alloy disc and heavy standard valves, 40 per cent.
Kerr steam radiator valves, 60 p c., and quick-opening hot-water radiator valves, 60 p c.
Kerr brass, Weber's straightway valves, 40; straightway valves, I B B.M., 60.
J. M. T. Radiator Valves 50; Standard, 60; Patent Quick-Opening Valves, 65 p c.
Jenkins' Valves—Quotations on application to Jenkins' Bros, Montreal.

No. 1 compression bath cock......net	2 00
No. 4 " " ...... "	1 90
No 7 Fuller's...... .... "	2 25
No. 4½, " .... "	2 35
Patent Compression Cushion, basin cock, hot and cold, per doz., ...	$16.20
Patent Compression Cushion, bath cock, No. 2208	2 25

Square head brass cocks, 50; iron, 60 p c.
Thompson Smoke-test Machine 25.00

BOILERS—COPPER RANGE.

Copper, 30 gallon, $33, 15 per cent.

BOILERS—GALVANIZED IRON RANGE.

30-gallon, Standard, $5; Extra heavy, $7.75

BATH TUBS.

Steel clad copper lined, 15 per cent.

CAST IRON SINKS.

16x24, $1; 18x30, $1; 18x36, $1.30.

ENAMELED BATHS, ETC.

List issued by the Standard Ideal Company Jan. 3, 1907, shows an advance of 10 per cent. over previous quotations.

ENAMELED CLOSETS AND URINALS

Discount 15 per cent.

HEATING APPARATUS.

Stoves and Ranges—40 to 70 per cent.
Furnaces—45 per cent.
Registers—70 per cent
Hot Water Boilers—50 per cent.
Hot Water Radiators—50 to 55 p.c
Steam Radiators—50 to 55 per cent.
Wall Radiators and Specials—50 to 55 p.c.

LEAD PIPE

Lead Pipe, 5 p.c. off
Lead waste, 5 p.c. off
Caulking lead, 6½c. per pound.
Traps and bends, 40 per cent.

IRON PIPE.

Size (per 100 ft.)	Black.		Galvanized
¼ inch........	2 35	¼ inch........	3 20
⅜ "	2 35	⅜ "	3 20
½ "	2 90	½ "	3 75
¾ "	3 90	¾ "	5 00
1 "	5 80	1 "	7 25
1¼ "	7 65	1¼ "	9 90
1½ "	9 20	1½ "	11 90
2 "	12 25	2 "	15 80
2½ "	20 10	2½ "	26 00
3 "	26 75	3 "	34 00
3½ "	34 25	3½ "	42 75
4 "	39 00	4 "	48 60

Malleable Fittings—Canadian discount 30 per cent.; American discount 25 per cent.
Cast Iron Fittings 57½; Standard bushings 57½; headers, 57½; flanged unions 57½, malleable bushings 55; nipples, 70 and 10; malleable lipped unions, 55 and 5 p c.

SOIL PIPE AND FITTINGS

Medium and Extra heavy pipe and fittings, up to 6 inch, 60 and 10 to 70 per cent.
7 and 8-in. pipe, 40 and 5 per cent.
Light pipe, 50 p.c.; fittings, 50 and 10 p c.

OAKUM.

Plumbers ....per 100 lb......	4 50	5 00

STOCKS AND DIES.

American discount 25 per cent.

SOLDERING IRONS.

½-lb. to 1½ .... per lb.	0 45½	0 43
2-lb. or over .... "	0 42½	0 46

SOLDER.

Per lb.	Montreal	Toronto
Bar, half-and-half, guaranteed	0 25	0 26
Wiping........................	0 22	0 23

PAINTS, OILS AND GLASS.

BRUSHES

Paint and household, 70 per cent..

CHEMICALS.

	In casks per lb.
Sulphate of copper (bluestone or blue vitrol)	0 09
Litharge, ground	0 06
" flaked	0 06½
Green coperas (green vitrol)	0 01
Sugar of lead..........	0 08
Lump olive..........	0 01¾

COLORS IN OIL.

Venetian red, 1-lb. tins pure.	....	0 08
Chrome yellow "	....	0 15
Golden ochre "	....	0 10
French " "	....	0 08
Marine black "	....	0 04½
Chrome green "	....	0 09
French permanent green "	....	0 13
Signwriters' black "	....	0 15

GLUE.

Domestic sheet	0 10	0 10½
French medal	0 12	0 12½

PARIS GREEN.

	Berger's	Canadian
600-lb. cask	....	0 27½
250-lb. drums	0 28	0 27½
100-lb. "	0 29½	....
50-lb. "	0 29½	0 28½
1-lb. pkgs, 100 in box	0 30½	0 29½
½-lb. "	0 32½	0 31½
1-lb. tins, 100 in box	0 31½	0 30½
½-lb. tins	0 33½	0 32

PARIS WHITE.

In bbls		0 90

PIGMENTS.

Orange mineral, casks	....	0 08
" " 100-lb. kegs	....	0 08½

PREPARED PAINTS.

Barn (in bbls.)	0 65	0 90
Sherwin-Williams paints	....	1 55
Canada Paint Co.'s pure	....	1 40
Standard P. & V. Co.'s "New Era."		1 30
Benj. Moore Co.'s "Ark" B'd	....	1 25
" British Navy deck	..	1 50
Brandram-Henderson's "English"		1 45
Ramsay's paints, Pure, per gal.	....	1 30
" Thistle, "	....	1 10
Martin-Senour's 100 p.c. pure,	....	1 55
Senour's Floor Paints	....	1 35
Jamieson's "Crown and Anchor"		1 20
Jamieson's floor enamel	....	1 50
"Island City" paint	....	1 25
Sanderson Pearcy's, pure	....	1 20
Robertson's pure paints	....	1 20

PUTTY.

Bulk in bbls	1 60
Bladders in bbls	1 85
25-lb. tins	1 90
Bladders in bulk or tins less than 100 lb	2 00
Bulk in 100-lb. irons	1 80

SHINGLE STAINS.

In 5 gallon lots	0 85	0 90

SHELLAC.

White, bleached	....	2 65
Fine orange	....	2 60
Medium orange	....	2 55

TURPENTINE AND OIL.

Prime white petroleum	....	0 13
Water white "	....	0 14½
Pratt's a tral "	....	0 18
Castor oil	0 08	0 10
Gasoline	....	0 22½
Benzine, per gal	0 17	0 20
Turpentine, single barrels	0 78	0 79
Linseed Oil, " raw	0 59	0 64
" " boiled	0 62	0 67

WHITE LEAD GROUND IN OIL. Per 100 lbs

Canadian pure	7 15	7 50
No. 1 Canadian	6 80	7 15
Munro's Select Flake White	....	7 65
Elephant and Decorators' Pure	....	7 65
Monarch	....	7 40
Standard Decorator's	....	7 15
Essex Genuine	....	6 80
Brandram's B. B. Genuine	....	8 70
"Anchor," pure	....	7 40
Ramsay's Pure Lead	....	7 00
Ramsay's Exterior	....	6 65
"Crown and Anchor," pure	....	7 25
Sanderson Pearcy's	....	7 40
Robertson's C.P., lead	....	7 20

RED DRY LEAD.

Genuine, 560 lb. casks, per cwt	....	6 25
Genuine, 100 lb. kegs, "	....	6 50
No. 1, 560 lb. casks, per cwt	....	6 00
No. 1, 100 lb. kegs, per cwt	....	6 25

WINDOW GLASS

Size United Inches.	Star	Double Diamond
Under 26	$4 25	$6 25
26 to 40	4 65	6 75
41 to 50	5 10	7 50
51 to 60	5 35	8 50
61 to 70	5 75	9 75
71 to 80	6 25	11 00
81 to 85	7 00	12 50
86 to 90	....	15 00
91 to 95	....	17 50
96 to 100	....	20 50
101 to 105	....	24 00
106 to 110	....	27 50

Discount—16-oz., 25 per cent; 21-oz. 30 per cent. per 100 feet. Broken boxes 50 per cent.

WHITING.

Plain, in bbls	....	0 70
Gilders bolted in bands	....	0 90

WHITE DRY ZINC.

Extra Red Seal, V.M.	0 07½	0 08

WHITE ZINC IN OIL.

Pure, in 25-lb. irons	....	0 08½
No. 1, " "	....	0 07
No. 2, " "	....	0 05½

VARNISHES.

	Per gal. cans
Carriage, No. 1	1 50
Pale durable body	3 50
" hard rubbing	3 00
Finest elastic gearing	3 00
Elastic oak	1 50
Furniture, polishing	2 00
Furniture, extra	1 15
" " No. 1	0 90
" " union	0 80
Light oil finish	1 40
Gold size japan	1 80
Brown japan	0 95
No 1 brown japan	0 95
Baking black japan	1 35
No. 1 black japan	0 90
Benzine black japan	0 70
Crystal Damar	2 80
No. 1 "	2 50
Pure asphaltum	1 40
Oilcloth	1 50
Lightning dryer	0 70
Elastilite varnish, 1 gal. can, each	2 00
Granitine floor varnish, per gal	2 50
Maple Leaf coach enamels; size 1,	1 20
Sherwin-Williams kopal varnish, gal.,	2 50
Canada Paint Co's sun varnish	2 00
"Kyanize" Interior Finish	2 40
"Flint-Lac," coach	1 80
B H. Co's "Gold Medal," in cases	2 00
Jamieson's Copaline, per gal.	2 00

BUILDERS' HARDWARE.

BELLS.

Brass hand bells, 60 per cent.
Nickel, 55 per cent.

Gongs, Sargeant's door bells	5 50	8 00
American, house bells, per lb.	0 35	0 40

Peterboro' door bells, 37½ and 10 off new list.

BUILDING PAPER, ETC.

Tarred Felt, per 100 lb.		2 25
Ready roofing, 2-ply, not under 45 lb. per roll		1 00
Ready roofing, 3-ply, not under 65 lb., per roll		1 25
Carpet Felt	per ton	60 00
Heavy Straw Sheathing	per ton	40 00
Dry Surprise		0 45
Dry Sheathing	per roll, 400 sq. ft.	0 40
Tar "	" 400 "	0 50
Dry Fibre	" 400 "	0 55
Tarred Fibre	" 400 "	0 65
O. K. & I. X. L.	" 400 "	0 70
Resin-sized	" 400 "	0 45
Oiled Sheathing	" 600 "	1 00
Oiled "	" 400 "	0 70
Roof Coating, in barrels	per gal.	0 17
Roof " small packages	"	0 25
Refined Tar	per barrel	5 00
Coal Tar	"	4 00
Coal Tar, less than barrels	per gal.	0 15
Roofing Pitch	per 100 lb. 0 80	0 90
Slater's felt	per roll	0 70
Heavy Straw Sheathing f. o. b. St. John and Halifax		42 50

BUTTS.

Wrought Brass, net revised list.
Wrought Iron, 70 per cent.
Cast Iron Loose Pin, 60 per cent.
Wrought Steel Fast Joint and Loose Pin, 70 per cent.

CEMENT AND FIREBRICK.

Canadian Portland	2 00	2 10
Belgium	1 60	1 90
White Bros. English	1 80	2 05
"Lafarge" cement in wood	....	3 40
"Lehigh" cement, in wood	....	2 54
"Lehigh" cement, cotton sacks	....	2 39
"Lehigh" cement, paper sacks	....	2 31
Fire brick, Scotch, per 1,000	27 00	30 00
" English	17 00	21 00
" American, low	23 0	25 00
" " high	27 50	35 00
Fire clay (Scotch), net ton	....	4 95

Paving Blocks per 1,000.

Blue metallic, 9"x4½"x3", ex wharf	35 00
Stable pavers, 12"x8"x2", ex wharf	50 00
Stable pavers, 9"x4½"x3", ex wharf	36 00

DOOR SETS.

Peterboro, 37½ and 10 per cent.

DOOR SPRINGS.

Torrey's Rod per doz.	....	1
Coil, 9 to 11 in. "	0 95	1 75
English "	2 00	4 66

Chicago and Reliance Coil 25 per cent.

ESCUTCHEONS.

Discount 50 and 10 per cent., new list
Peterboro, 37½ and 10 per cent.

ESCUTCHEON PINS.

Iron, discount 40 per cent.
Brass, 45 per cent.

HINGES.

Blind, discount 50 per cent.

Heavy T and strap, 4-in., per lb. net	0 06
" " 5-in., "	0 05¾
" " 6-in., "	0 05½
" " 8-in., "	0 05¼
" " 10-in. and larger	0 05

Light T and strap, discount 65 p.c.
Screw hook and hinge—

under 12 in.... per 100 lb.	....	4 65
over 12 in. "	....	3 65

Crate hinges and back flaps, 65 and 5 p. c.
Hinge hasps, 65 per cent.

SPRING HINGES.

Spring, per gro., No. 5, $17.50 No. 10, $18; No. 20, $10 80; No. 120, $20; No. 51, $10; No. 50, $27 50
Chicago Spring Butts and Blanks 12½ per cent.
Triple End Spring Butts, 30 and 10 per cent.
Chicago Floor Hinges, 37½ and 5 off.
Garden City Fire House Hinges, 12½ p c.
"Chief" floor hinge, 50 p.c.

CAST IRON HOOKS.

Bird cage per doz.	0 50	1 10
Clothes line, No. 61 "	0 00	0 70
Harness "	0 60	12 00
Hat and coat per gro.	1 10	10 00
Chandelier per doz.	0 50	1 00
Wrought hooks and staples—		
¼ x 5 per gross	....	2 65
5-16 x 5 "	....	3 30

Bright wire hooks, 60 p c.
Bright steel gate hooks and staples, 40 p.c.
Cresent hat and coat wire, 60 per cent.
Screw, bright wire, 65 per cent.

KNOBS.

Door, japanned and N.P., doz	1 50	2 50
Bronze, Berlin per doz.	2 75	3 25
Bronze, Genuine "	6 00	9 00
Shutter, porcelain, F. & L. screw per gross	1 30	2 00
White door knobs per doz.	....	2 00

Peterboro knobs, 37½ and 10 per cent.
Porcelain, mineral and jet knobs, net list.

KEYS.

Lock, Canadian 40 to 40 and 10 per cent.

LOCKS.

Peterboro, 37½ and 10 per cent.
Russell & Erwin, steel rim $2.50 per doz
Eagle cabinet locks, discount 30 per cent
American padlocks, all steel, 10 to 15 per cent.; all brass or bronze, 10 to 25 per cent.

SAND AND EMERY PAPER.

B. & A. sand, discount, 35 per cent
Emery, discount 35 per cent.
Garnet (Burton's) 5 to 10 per cent. advance

SASH WEIGHTS.

Sectional per 100 lb.	2 00	2 25
Solid "	1 50	1 75

SASH CORD.

Per lb.	....	0 31

BLIND AND BED STAPLES.

All sizes per lb.	0 07½	0 10

WROUGHT STAPLES.

Galvanized	....	2 75
Plain	....	2 50

Coopers', discount 45 per cent.
Poultry netting staples, discount 40 per cent.
Bright spear point, 75 per cent. discount.

TOOLS AND HANDLES.

ADZES.

Discount 22½ per cent.

AUGERS.

Gilmour's, discount 60 per cent. off list.

AXES.

Single bit, per doz.	6 00	9 00
Double bit, "	10 00	11 00
Bench Axes, 40 per cent.		
Broad Axes, 25 per cent.		
Hunters' Axes	5 50	6 00
Boys' Axes	6 25	7 00
Splitting Axes	7 00	12 00
Handled Axes	7 00	9 00
Red Ridge, boys', handled	....	5 75
" hunters	....	5 25

BITS.

Irwin's auger, discount 47½ per cent.
Gilmour's auger, discount 60 per cent.
Rockford auger, discount 50 and 10 per cent.
Jennings' Gen. auger, net list.
Gilmour's car, 47½ per cent.
Clark's expansive, 40 per cent.

Clark's gimlet, per doz	....	0 65
Diamond, Shell, per doz.	....	1 00
Nail and Spike, per gross	....	2 25

BUTCHERS CLEAVERS

German per doz.	7 00	9 00
American "	12 00	18 00

CHALK.

Carpenters' Colored, per gross	0 45	0 75
White lump per cwt.	0 60	0 65

CHISELS.

Warnock's, discount 70 and 5 per cent.
P. S. & W. Extra, discount, 70 per cent.

CROSSCUT SAW HANDLES

S. & D., No. 3	per pair	0 13
S. & D., " 5	"	0 11½
S. & D., " 6	"	0 18
Boynton pattern	"	0 20

CROWBARS.

3½c. to 4c. per lb.

DRAW KNIVES.

Coach and Wagon, discount 75 and 5 per cent.
Carpenters' discount 75 per cent.

DRILLS.

Millar's Falls, hand and breast. net list.
North Bros., each set, 50c

DRILL BITS.

Morse, discount 37½ to 40 per cent.
Standard, discount 50 and 5 to 55 per cent.

FILES AND RASPS.

Great Western	75 per cent.
Arcade	75 "
Kearney & Foot	75 "
Disston's	75 "
American	75 "
J. Barton Smith	75 "
McClellan	75 "
Eagle	75 "
Nicholson	66⅔ "
Globe	75 "

Black Diamond, 60, 10 and 5 p.c.
Jowitt's, English list, 27½ per cent.

GAUGES.

Stanley's discount 50 to 60 per cent.

Winn's, Nos. 26 to 33 ....each	1 65	2 40

HANDLES.

Second growth ash fork, hoe, rake and shovel handles, 40 p c.
Extra ash fork, hoe, rake and shovel handles, 45 p c
No 1 and 2 ash fork, hoe, rake and shovel handles, 50 p c.
White ash whiffletrees and neckyokes, 35 p c.
All other ash goods, 40 p c.
All hickory, maple and oak goods, excepting carriage and express whiffletrees, 40 p c.
Hickory, maple, oak carriage and express whiffletrees, 45 p.c.

HAMMERS.

Maydole's, discount 5 to 10 per cent.
Canadian, discount 25 to 27½ per cent.

Magnetic tack per doz	1 10	1 20
Canadian sledge per lb.	0 07	0 08½
Canadian ball pean, per lb.	0 22	0 25

HATCHETS.

Canadian, discount 40 to 42½ per cent.
Shingle, Red Ridge 1, per doz.......... 4 40
" " 2, " 4 80
Barrel Underhill...................... 5 05

HOES

Mortar, 50 and 10 per cent.

MALLETS.

Tinsmiths'............per doz. 1 25 1 50
Carpenters', hickory, " 1 25 3 75
Lignum Vitae........ " 3 85 5 00
Caulking, each 0 60 2 00

MATTOCKS.

Canadian............ per doz. 5 50 6 00

MEAT CUTTERS.

German, 15 per cent.
American discount, 33⅓ per cent.

PICKS.

Per dozen 6 00 9 00

PLANES.

Wood bench, Canadian, 40, American, 25.
Wood, fancy, 37½ to 40 per cent.
Stanley planes, $1 55 to $3 60, net list prices.

PLANE IRONS.

English ....per doz. 2 00 5 00
Stanley, 2⅝ inch, single 24c, double 39c

PLIERS AND NIPPERS.

Button's genuine, 37½ to 40 per cent.
Button's imitation....per doz. 5 00 9 00
Berg's wire fencing 1 72 5 50

PUNCHES.

Saddler's .. ..per doz. 1 00 1 85
Conductor's.......... " 3 00 15 00
Tinners', solid..........per set 0 72
" hollow......per inch 1 00

RIVET SETS.

Canadian, discount 35 to 37½ per cent.

RULES.

Boxwood, discount 70 per cent.
Ivory, discount 20 to 25 per cent.

SAWS.

Atkins, hand and crosscut, 25 per cent.
Disston's Hand, discount 12½ per cent
Disston's Crosscut ....per foot 0 35 0 55
Back, complete ....each 0 75 2 75
" frame only.........each 0 50 1 25

S. & D. solid tooth circular shingle, concave and band, 50 per cent; mill and ice, drag, 30 per cent; cross-cut, 35 per cent; band saws, butcher, 35 per cent.; buck, New Century $6 25, buck No. 1 Maple Leaf, $5 25; buck, Happy Medium $4 25; buck, Watch Spring, $1.25; buck, common frame, $4.00.

Spear & Jackson's saws—Hand or rip, 26 in., $12 75, 24 in., $12.25; panel 18 in., $8 25; 20 in., $9; tenon, 10 in., $9.90; 12 in., $10.90; 14 in., $11 50

SAW SETS.

Lincoln and Whiting 4 75
Hand Sets, Perfect 4 00
X-Cut Sets, " 7 50
Maple Leaf and Premiums saw sets, 40 off.
S. & D. saw swages, 40 off

SCREW DRIVERS.

Sargent's ....per doz. 0 65 1 00
North Bros., No. 30 ..per doz. 16 80

SHOVELS AND SPADES.

Canadian, discount 45 per cent.

SQUARES.

Iron, discount 20 per cent.
Steel, discount 65 and 10 per cent.
Try and Bevel, discount 50 to 52½ per cent.

TAPE LINES.

English, ass skin .....per doz. 2 75 5 00
English, Patent Leather..... 5 50 9 75
Chesterman's ....each 0 90 2 85
" steel ....each 0 80 8 00
Berg's, each........ .... 0 75 2 50

TROWELS.

Disston's, discount 10 per cent.
S. & D., discount 35 per cent
Berg's, brick, 924x11 .... 4 00
" pointing, 924x5.......... .... 2 10

FARM AND GARDEN GOODS

BELLS.

American cow bells, 63¾ per cent.
Canadian, discount 45 and 50 per cent.
American, farm bells, each .. 1 35 3 00

BULL RINGS.

Copper, $2.00 for 2½-inch

CATTLE LEADERS

Nos. 32 and 33 .. ...per gross 7 50 8 50

BARN DOOR HANGERS

	doz.	pairs.
Stearns wood track	4 50	6 00
Zenith........................	....	9 00
Atlas, steel covered	5 00	6 00
Perfect	8 00	11 00
New Milo, flexible	....	6 50
Steel, track, 1 x 3-16 in(100 ft)	....	3 25
" " 1¼ x 3-16 in(100 ft)	....	4 75
Double strap hangers, doz. sets...	...	6 40
Standard jointed hangers, "	....	6 40
Steel King hangers "	....	6 25
Storm King and safety hangers	....	7 00
" " rail................	....	4 25

Chicago Friction, Oscillating and Big Twin Hangers, 5 per cent.

HARVEST TOOLS.

50 and 10 per cent.
S. & D. lawn rakes, Dunn's, 40 off.
" sidewalk and stable scrapers, 40 off.

HAY KNIVES.

Net list.

HEAD HALTERS.

Jute Rope, ⅜-inch....per gross 9 00
" " ½ " .... " 10 00
" " ⅝ " .... " 12 00
Leather, 1-inch ......per doz. 4 00
Leather, 1¼ " ...... " 5 20
Web................. " 2 45

HOES.

Garden, 50 and 10 per cent.
Planter.............. per doz 4 00 4 50

LAWN MOWERS

Low wheel, 12, 14 and 16-inch $2 30
9-inch wheel, 12-inch 2 85
" " 14 " 3 10
" " 16 " 3 12½
High wheel, 12 " 4 05
" " 14 " 4 50
" " 16 " 4 75

SCYTHES.

Per doz. net................... 6 25 9 25

SCYTHE SNATHS

Canadian, discount 40 per cent.

SNAPS

Harness, German, discount 25 per cent.
Lock, Andrews' 4 50 11 00

STABLE FITTINGS.

Warden King, 35 per cent
Dennis Wire & Iron Co., 33⅓ p c.

WOOD HAY RAKES.

40 and 10 per cent.

HEAVY GOODS, NAILS, ETC.

ANVILS.

Wright's, 80-lb. and over............... 0 10½
Hay Budden, 80-lb. and over.......... 0 09½
Brook's, 80-lb. and over 0 11½
Taylor-Forbes, prospectors 0 05
Columbia Hardware Co., per lb. 0 09½

VISES.

Wright's...................... .... 0 13½
Berg's, per lb.................................. 0 12½
Brook's........................ .... 0 12½
Pipe Vise, Hinge, No. 1...... .... 3 50
" " " No. 2...... .... 5 50
Saw Vise........................ 4 50 5 00
Blacksmiths' (discount) 60 per cent.
" parallel (discount) 45 per cent.

BOLTS AND NUTS

	Per cent.
Carriage Bolts, common ($1 list	
" " ⅜ and smaller..	60, 10 and 10
" " 7-16 and up....	55 and 5
" " Norway Iron ($3 list)	50
Machine Bolts, ⅜ and less	60 and 10
Machine Bolts, 7-16 and up....	55 and 5
Plough Bolts	55 and 10
Blank Bolts......	55
Bolt Ends......................	55
Sleigh Shoe Bolts, ⅜ and less ..	60 and 10
" " " 7-16 and larger	50 and 5
Coach Screws, conepoint......	70 and 5

Nuts, square, all sizes, 4c. per cent. off
Nuts, hexagon, all sizes, 4¼c. per cent off
Stove Rods per lb., 5½ to 6c.
Stove Bolts, 75 per cent.

CHAIN.

Proof coil, per 100 lb., ¼ in., $6 00; 5-16 in., $4.85; ⅜ in., $4 25; 7-16 in., $4.00; ½ in., $3 75; 9-16 in. $3 70; ⅝ in., $3 65; ¾ in., $3.60; ⅞ in., $3.45; 1 in., $3.40.

Halter, kennel and post chains, 40 to 40 and 5 per cent; Cow ties, 40 per cent; Tie out chains 65 per cent.; Stall fixtures, 35 per cent; Trace chain, 45 per cent.; Jack chain iron, 50 per cent.; Jack chain, brass, 50 per cent.

HORSE NAILS.

M R.M. cold forged process, list May 15, 1907, 50 and 5 per cent.
"C" brand, 57½ per cent. off list.
Capewell brand, quotations on application

HORSESHOES.

M R M brand: iron, light and medium No 1 and smaller, $3 90; No. 2 and larger $3 65; snow pattern No 1 and smaller $4 15, No 2 and larger, $3 90; "X L" new light steel, No 1 and smaller, $4 25; No 2 and larger, $4; "X. L" featherweight steel, No. 0 to 4, $5 60; toe-weight, all sizes, $6 85. F o b Montreal. Extras for packing

Belleville brand: No. 0 and 1, light and medium iron $3 90, snow, $4 15; light steel, $4 25; No 2 and larger, light and medium iron, $3.65; snow, $3 90; light steel, $4. F o b Belleville. Two per cent. 30 days.

Toecalks—Standard No 1 and smaller, $1 50; No. 2 and larger, $1 25 Blunt No. 1 and smaller, $1.75; No 2 and larger, $1 50 per box Sharp Put up in 25 lb. bxs.

HORSE WEIGHTS.

Taylor-Forbes, 4¼c. per lb.

NAILS.	Cut.	Wire.
2d...........................	4 00	3 70
3d...........................	3 15	3 20
4 and 5d........	2 90	2 90
6 and 7d	2 80	2 80
8 and 9d	2 65	2 65
10 and 12d	2 60	2 60
16 and 20d....................	2 55	2 55
30, 40, 50 and 60d (base)	2 50	2 50

F.o.b. Montreal. Cut nails, Toronto 20c. higher.
Miscellaneous wire nails, discount 75 per cent
Coopers' nails, discount 40 per cent.

PRESSED SPIKES

Pressed spikes, ⅝ diameter, per 100 lbs $3.15

RIVETS AND BURRS

Iron Rivets, black and tinned, 60, 10 and 10.
Iron Burrs, discount 60 and 10 and 10 p c.
Copper Rivets, usual proportion burrs, 15 p c.
Copper Burrs only, net list.
Extras on Coppered Rivets, ½-lb. packages 1c. per lb.; ¼-lb. packages 2c. lb
Tinned Rivets, net extra, 4c. per lb.

SCREWS.

Wood, F. H., bright and steel, 85 and 10 p c.
" R. H., bright, 80 and 10 per cent.
" F. H., brass, 75 and 10 per cent.
" R. H., " 70 and 10 per cent.
" F. H., bronze, 70 and 10 per cent.
" R. H., " 65 and 10 per cent.
Drive Screws, dis. 87½ per cent.
Bench, wood ....per doz. 3 25 4 00
" iron .... " 4 25 5 00
Set, case hardened, dis. 60 per cent.
Square Cap, dis. 50 and 5 per cent.
Hexagon Cap, dis 45 per cent.

MACHINE SCREWS.

Flat head, iron and brass, 35 per cent.
Felister head, iron, discount 30 per cent
" " brass, discount 25 per cent.

TACKS, BRADS, ETC.

Carpet tacks, blued 75 p c.; tinned, 80 and 10; (in kegs), 40; cut tacks, blued, in dozens only, 75; ½ weights, 60; Swedes cut tacks, blued and tinned, bulk, 75 dozens, 75; Swedes, upholsterers', bulk, 85 and 12½; brush, blued and tinned, bulk, 70; Swedes, gimp, blued, tinned and japanned, 75 and 12½; zinc tacks, 35; leather carpet tacks, 40; copper tacks, 25; copper nails 30; trunk nails, black, 65; trunk nails, tinned and blued, 65; clout nails, blued and tinned 65; chair nails, 35; patent brads, 40; fine finishing, 40; lining tacks, in papers, 10; lining tacks, in bulk, 15; lining tacks, solid heads, in bulk, 75; saddle nails, in papers, 10; saddle nails, in bulk, 15; tufting buttons, 22 line in dozens only, 60; zinc glaziers' points, 5; double pointed tacks, papers, 90 and 10; double pointed tacks, bulk, 40; clinch and duck rivets, 45; cheese box tacks, 85 and 5 trunk tacks, 80 and 10.

WROUGHT IRON WASHERS.

Canadian make, discount 40 per cent

SPORTING GOODS.

CARTRIDGES.

"Dominion" Rim Fire Cartridges and C.B caps, 50 and 7½ per cent; Rim Fire B B Round Caps 60 and 2½ per cent.; Centre Fire, Pistol and Rifle Cartridges 30 p c; Centre Fire Sporting and Military Cartridges, 20 and 5 p c; Rim Fire, Shot Cartridges, 50 and 7½ p c.; Centre Fire, Shot Cartridges, 30 p.c.; Primers, 25 p.c.

LOADED SHELLS.

"Crown" Black Powder, 15 and 5 p c; "Sovereign" Empire Bulk Smokeless Powder, 30 and 5 p.c.; "Regal" Ballistite Dense smokeless Powder, 30 and 5 p c.; "Imperial" Empire or Ballistite Powder, 30 and 10 p c.

EMPTY SHELLS.

Paper Shells, 25 and 5; Brass Shells, 55 and 5 p c.

Wads.	per lb.
Best thick brown or grey felt wads, in ½-lb. bags	$0 70
Best thick white card wads in boxes of 500 each, 12 and smaller gauges	29
Best thick white card wads, in boxes of 500 each, 10 gauge............	0 35
Thin card wads, in boxes of 1,000 each, 12 and smaller gauges	0 20
Thin card wads, in boxes of 1,000 each, 10 gauge.....................	0 25
Chemically prepared black edge grey cloth wads, in boxes of 250 each—	Per M
11 and smaller gauge	0 60
9 and 10 gauges..................	0 70
7 and 8 "	0 90
5 and 6 "	1 [illegible]
Superior chemically prepared pink edge, best white cloth wads in boxes of 250 each—	
11 and smaller gauge	1 15
9 and 10 gauges	1 40
7 and 8 "	1 65
5 and 9 "	1 90

SHOT.

Ordinary drop shot, AAA to dust $7.50 per 100 lbs. Discount 5 per cent; cash discount, 2 per cent, 30 days; net extras as follows subject to cash discount only; Chilled, 40 c.; buck and seal, 80c.; no 28 ball, $1 20 per 100 lbs.; bags less than 25 lbs., ½c. per lb.; F O.B. Montreal, Toronto, Hamilton, London, St. John and Halifax, and freight equalized thereon.

TRAPS (steel.)

Game, Newhouse, discount 30 and 10 per cent.
Game, Hawley & Norton, 50, 10 & 5 per cent
Game, Victor, 70 per cent.
Game, Oneida Jump (B. & L.) 40 & 2½ p.
Game, steel, 60 and 5 per cent.

SKATES.

Skates, discount 37½ per cent
Empire hockey sticks, per doz .. 3 00

CUTLERY AND SILVERWARE.

RAZORS.

	per doz.	
Elliot's	4 00	18 00
Boker's	7 50	11 00
" King Cutter	13 50	18 50
Wade & Butcher's	3 60	10 00
Lewis Bros. "Klean Kutter"	8 50	10 50
Henckel's	7 50	20 00
Berg's	7 50	20 00

Clauss Razors and Strops, 50 and 10 per cent

KNIVES.

Farriors-Stacey Bros., doz 3 50

PLATED GOODS

Hollowware, 40 per cent. discount.
Flatware, staples, 40 and 10, fancy, 40 and 5.
Hutton's "Cross Arrow" flatware, 47½; "Singalese" and "Alaska" Nevada silver flatware, 42 p c.

SHEARS.

Clauss, nickel, discount 60 per cent.
Clauss, Japan, discount 67½ per cent.
Clauss, tailors, discount 40 per cent.
Seymour's, discount 50 and 10 per cent.
Berg's 6 00 12 00

HOUSE FURNISHINGS.

APPLE PARERS.

Hudson, per doz., net 5 75

BIRD CAGES.

Brass and Japanned, 40 and 10 p. c.

COPPER AND NICKEL WARE.

Copper boilers, kettles, teapots, etc. 30 p.c.
Copper pitts, 30 per cent.

KITCHEN ENAMELED WARE.

White ware, 75 per cent.
London and Princess, 50 per cent
Canada, Diamond, Premier, 50 and 10 p c.
Pearl, Imperial, Crescent and granite steel, 50 and 10 per cent.
Premier steel ware, 40 per cent.
Star decorated steel and white, 25 per cent.
Japanned ware, discount 45 per cent.
Hollow ware, tinned cast, 35 per cent off.

KITCHEN SUNDRIES.

Can openers, per doz.	0 40	0 75
Mincing knives per doz	0 50	0 80
Duplex mouse traps, per doz.	....	0 65
Potato mashers, wire, per doz.	0 60	0 70
" wood "	0 50	0 60
Vegetable slicers, per doz	....	2 25
Universal meat chopper No. 1	....	1 15
Enterprise chopper, each	....	1 30
Spiders and fry pans, 50 per cent.		
Star Al chopper 5 to 32	1 35	4 10
" " 100 to 103	1 35	2 00
Kitchen hooks, bright	....	0 60

LAMP WICKS.

Discount, 60 per cent.

LEMON SQUEEZERS.

Porcelain lined	per doz.	2 20	5 60
Galvanized	"	1 87	3 85
King, wood	"	2 75	2 90
King, glass	"	4 00	4 50
All glass	"	0 50	0 90

METAL POLISH.

Tandem metal polish paste 6 00

PICTURE NAILS.

Porcelain head ......per gross	1 35	1 50
Brass head	0 40	1 00

Tin and gilt, picture wire, 75 per cent.

SAD IRONS.

Mrs. Potts, No. 55, polished....per set	0 90
" No. 50, nickle-plated, "	0 95
" handles, japaned, per gross	9 25
" " nickled, "	9 75
Common, plain	4 25
" plated	4 50
Asbestos, per set	1 50

TINWARE.

CONDUCTOR PIPE.

2-in, plain, or corrugated., per 100 feet, $3 30; 3 in., $4.40; 4 in., $5.80; 5 in., $7.45; 6 in., $9.90.

FAUCETS.

Common, cork-lined, discount 35 per cent.

EAVETROUGHS.

10-inch .... per 100 ft. 3 30

FACTORY MILK CANS.

Discount off revised list, 35 per cent.
Milk can trimmings, discount 25 per cent.
Creamery Cans, 45 per cent

LANTERNS.

No. 2 or 4 Plain Cold Blast....per doz.	6 50
Lift Tubular and Hinge Plain, "	4 75
No 0, safety .... "	4 00

Better quality at higher prices.
Japanning, 50c. per doz extra.
Prism globes, per doz., $1 20.

OILERS.

Kemp's Tornado and McClary's Model galvanized oil can, with pump, 5 gallon, per dozen 10 92
Davidson oilers, discount 40 per cent.
Zinc and tin, discount 50 per cent
Coppered oilers, 20 per cent. off.
Brass oilers, 50 per cent. off.
Malleable, discount 25 per cent

PAILS (GALVANIZED).

Dufferin pattern pails, 45 per cent.
Flaring pattern, discount 45 per cent.
Galvanized washtubs 40 per cent.

PIECED WARE.

Discount 35 per cent off list, June, 1899.
10-qt. flaring sap buckets, discount 35 per cent.
6, 10 and 14-qt. flaring pails dis 35 per cent.
Copper bottom tea kettles and boilers, 30 p.c.
Coal hods, 40 per cent.

STAMPED WARE.

Plain, 75 and 12½ per cent. off revised list.
Retinned, 72½ per cent. revised list.

SAP SPOUTS.

Bronzed iron with hooks ....per 1,000
Eureka tinned steel, hooks 8 00

STOVEPIPES.

5 and 6 inch, per 100 lengths	7 64	7 91
7 inch .... " "	....	8 18

Nestable, discount 40 per cent.

STOVEPIPE ELBOWS

5 and 6-inch, common........per doz.	1 32
7-inch .... "	1 48

Polished, 15c. per dozen extra.

THERMOMETERS.

Tin case and dairy, 75 to 75 and 10 per cent.

TINNERS' SNIPS.

Per doz 3 00 15
Clauss, discount 35 per cent.

TINNERS' TRIMMINGS.

Discount, 45 per cent.

WIRE.

ANNEALED CUT HAY BAILING WIRE.

No. 12 and 13, $4; No. 13½, $4 10; No. 14, $4.25; No. 15, $4 50; in lengths 6' to 11', 25 per cent; other lengths 20c. per 100 lbs extra; if eye or loop on end add 25c. per 100 lbs. to the above.

BRIGHT WIRE GOODS

Discount 60 per cent.

CLOTHES LINE WIRE.

7 wire solid line, No. 17 $4.90; No. 18, $3.00; No. 19, $2 70; 6 wire solid line, No. 17, $4 45; No. 18, $3.10; No. 19, $2 80. All prices per 1000 ft. measure; 6 strand, No. 18, $2 60; No. 19, $2 90. F.o b. Hamilton, Toronto, Montreal

COILED SPRING WIRE

High Carbon, No. 9, $2 95; No. 11, $3.50; No. 12, $3.20.

COPPER AND BRASS WIRE.

Discount 37½ per cent.

FINE STEEL WIRE.

Discount 25 per cent List of extras In 100-lb. lots: No. 17, $5 — No. 18, $5 50 — No. 19, $6 — No. 20, $6.65 — No. 21 $7 — No. 22, $7.30 — No 23, $7.65 — No. 24, $8 — No. 25, $9—No. 26, $9.50—No. 27, $10—No. 28, $11—No. 29, $12—No. 30, $13—No. 31, $14—No. 32, $15—No. 33, $16—No 34, $17. Extras net—tinned wire, Nos. 17-25, $2—Nos. 26-31, $4—Nos. 32-34, $6. Coppered, 75c.—oiling, 10c.—in 25-lb. bundles, 15c.—in 5 and 10-lb. bundles, 25c.—in 1-lb. hanks, 25c.—in ½-lb. hanks, 38c.—in ¼-lb. hanks, 50c.—packed in casks or cases, 15c.—bagging or papering, 10c

FENCE STAPLES.

Bright. .... 2 80 Galvanized.... 32

HAY WIRE IN COILS.

No 13, $2 70; No. 14, $2.80; No. 15, $2.95; f o.b, Montreal.

GALVANIZED WIRE.

Per 100 lb.—Nos. 4 and 5, $3.95 — Nos. 6, 7, 8, $3 35 — No. 9, $2.85 — No. 10, $3.40 — No. 11, $3 45—No. 12, $3 00 —No. 13, $3 10—No. 14, $3.95—No 15, $4.30 —No. 16, $4.30 from stock. Base sizes, Nos. 6 to 9, $2 35 f.o.b. Cleveland. Extras for cutting.

LIGHT STRAIGHTENED WIRE.

Gauge No.	Over 20 in. per 100 lbs.	10 to 20 in.	5 to 10 in.
0 to 5	$0.50	$0 75	$1.25
6 to 9	0.75	1 25	2 00
10 to 11	1 00	1.75	2 50
12 to 14	1 50	2.25	3 50
15 to 16	2.00	3 00	4.50

SMOOTH STEEL WIRE.

No. 0-9 gauge, $2 40; No. 10 gauge, 6c. extra; No. 11 gauge, 12c extra; No. 12 gauge, 20c. extra; No. 13 gauge, 30c. extra; No 14 gauge, 40c. extra; No. 15 gauge, 55c. extra; No 16 gauge, 70c extra. Add 60c. for coppering and $2 for tinning

Extra net per 100 lb.—Oiled wire 10c., spring wire $1.25, bright soft drawn 15c., charcoal (extra quality) $1.25, packed in casks or cases 15c., bagging and papering 10c., 50 and 100-lb. bundles 10c., in 25-lb. bundles 15c., in 5 and 10-lb. bundles 25c., in 1-lb hanks, 50c., in ½-lb. hanks 75c., in ¼-lb. hanks $1.

POULTRY NETTING.

2-in. mesh, 19 w. g., 50 and 5 p.c. off Other sizes, 50 and 5 p c. off.

WIRE CLOTH.

Painted Screen, in 100-ft. rolls, $1.72½, per 100 sq. ft.; in 50-ft. rolls, $1 77½, per 100 sq ft.

WIRE FENCING.

Galvanized barb 2 95
Galvanized, plain twist 3 30
Galvanized barb, f.o.b. Cleveland, $2 70 for small lots and $2.60 for carlots

WIRE ROPE

Galvanized 1st grade, 6 strands, 24 wires, ⅜, $5, 1 inch $16 80
Black, 1st grade, 6 strands, 19 wires. ⅜, $5; 1 inch $15 10. Per 100 feet f o.b Toronto

WOODENWARE.

CHURNS.

No. 0, $9; No. 1, $9; No. 2, $10; No. 3, $11; No. 4, $13; No, 5, $16; f.o b. Toronto Hamilton, London and St. Marys, 30 and 30 per cent.; f o b Ottawa, Kingston and Montreal, 40 and 15 per cent. discount,

CLOTHES REELS.

Davis Clothes Reels, dis. 40 per cent.

FIBRE WARE.

Star pails, per doz		$ 3 00
0 Tubs,	"	14 00
1 "	"	12 00
2 "	"	10 00
3 "	"	8 50

LADDERS, EXTENSION.

3 to 6 feet, 12c. per foot; 7 to 10 ft., 13c.
Waggoner Extension Ladders, dis 40 per cent.

MOPS AND IRONING BOARDS.

"Best" mops		1 25
"900" mops		1 25
Folding ironing boards	12 00	16 50

REFRIGERATORS

Discount, 40 per cent

SCREEN DOORS.

Common doors, 2 or 3 panel, walnut stained, 4-in. style........per doz. 7 25
Common doors, 2 or 3 panel, grained only, 4-in., style ....per doz. 7 65
Common doors, 2 or 3 panel, light stair per doz. 9 55

WASHING MACHINES.

Round, re-acting per doz.	60 00
Square "	63 00
Eclipse, per doz	54 00
Dowswell "	39 00
New Century, per doz	75 00
Daisy	54 00
Stephenson	74 00

WRINGERS.

Royal Canadian, 11 in., per doz.	35 00
Royal American, 11 in.	35 00
Eze 10 in., per doz	36 75

MISCELLANEOUS

AXLE GREASE.

Ordinary, per gross	6 00	7 00
Best quality	10 00	12 00

BELTING.

Extra, 60 per cent.
Standard, 60 and 10 per cent.
No. 1, not wider than 6 in., 60, 10 and 10 p.c.
Agricultural, not wider than 4 in., 75 per cent.
Lace leather, per side, 75c.; cut laces, 80c.

BOOT CALKS.

Small and medium, ball ....per M 2 25
Small heel 4 50

CARPET STRETCHERS.

American	per doz.	1 00	1 50
Bullard's	"	....	6 50

CASTORS.

Bed, new list, discount 55 to 57½ per cent.
Plate, discount 52½ to 57½ per cent.

PINE TAR.

½ pint in tins ....per gross	....	7 80
1 " "	....	9 60

PULLEYS.

Hothouse	per doz.	0 55	1 00
Axle	"	0 22	0 33
Screw	"	0 22	1 00
Awning	"	0 35	2 50

PUMPS.

Canadian cistern	1 40	2 00
Canadian pitcher spout	1 80	3 16

Berg's wing pump, 75 per cent.

ROPE AND TWINE.

Sisal		0 10½
Pure Manilla		0 15
"British" Manilla		0 12
Cotton, 3-16 inch and larger	0 21	0 23
" 5-32 inch	0 25	0 27
" ¼ inch	0 25	0 28
Russia Deep Sea		0 16
Jute		0 09
Lath Yarn, single		0 10
" double		0 10½
Sisal bed cord, 48 feet........per doz.		0 65
" " 60 feet........ "		0 80
" " 72 feet........ "		0 95

Twine.

Bag, Russian twine, per lb.		0 27
Wrapping, cotton, 3-ply		0 25
" " 4-ply		0 29
Mattress twine per lb.	0 33	0 45
Staging " "	0 27	0 35

BINDER TWINE.

500 feet, sisal	0 09½
500 " standard	0 09½
550 " " manilla	0 10½
600 " " "	0 12½
650 " " "	0 13½

Car lots, ½c. less; 5-ton lots, ¼c. less. Central delivery.

SCALES.

Gurney Standard, 35, Champion, 45 p c.
Burrow, Stewart & Milne — Imperial Standard, 35; Weigh Beams, 35; Champion Scales, 45.
Fairbanks Standard, 30; Dominion, 50 Richelieu, 50
Warren new Standard, 35; Champion, 45 Weigh Beams, 30.

STONES—OIL AND SCYTHE.

Washita	per lb.	0 25	0
Hindostan	"	0 06	0 10
" slip	"	0 18	0 20
" Axe	"	....	0 10
Deer Creek	"	....	0 10
Deerlick	"	....	0 25
" Axe	"	....	0 15
Lily white	"	....	0 42
Arkansas	"	....	1 50
Water-of-Ayr	"	....	0 10
Scythe	per gross	3 50	5 00
Grind, 40 to 200 lb., per ton		20 00	22 00
" under 40 lb., "		....	24 00
" 200 lb. and over		....	28 00

INDEX TO ADVERTISERS.

CLASSIFIED LIST OF ADVERTISEMENTS

Alabastine.
Alabastine Co., Limited, Paris, Ont.

Auditors
DaVenport, Pickup & Co., Winnipeg.

Automobile Accessories.
Canada Cycle & Motor Co., Toronto Junction.
Carriage Mountings Co., Ltd., Toronto.

Awnings.
Tobin Tent and Awning Co., Ottawa

Babbitt Metal.
Canada Metal Co., Toronto.
Canadian Fairbanks Co., Montreal
Frothingham & Workman, Montreal.
Robertson, Jas. Co., Montreal.

Bar Urns.
Buffalo Mfg. Co., Buffalo, N.Y.

Bath Room Fittings.
Buffalo Mfg Co . Buffalo, N.Y.
Carriage Mountings Co., Ltd., Toronto.
Forsyth Mfg Co., Buffalo N.Y.
Ontario Metal Novelty Co., Toronto

Belting, Hose, etc.
Gutta Percha and Rubber Mfg. Toronto
Sadler & Haworth Toronto.

Bicycles and Accessories.
Johnson's, IVer, Arms and Cycle Works Fitchburg, Mass

Binder Twine
Consumers Cordage Co., Montreal.

Bolts
Toronto & BelleVille Rolling Mills, BelleVille.

Box Strap.
J. N. Warminton, Montreal.

Brass Goods.
Frothingham & Workman, Montreal.
Glauber Brass Mfg. Co., CleVeland, Ohio.
Kerr Engine Co., WalkerVille, Ont.
Lewis, Rice, & Son., Toronto.
Morrison, Jas., Brass Mfg. Co., Toronto.
Mueller Mfg. Co., Decatur, Ill.
Penberthy Injector Co., Windsor, Ont.
Taylor-Forbes Co., Guelph, Ont.

Bronze Powders.
Canadian Bronze Powder Works, Montreal.

Brushes.
United Factories, Toronto.

Cans.
Acme Can Works, Montreal.

Builders' Tools and Supplies.
CoVert Mfg. Co., West Troy, N.Y.
Frothingham & Workman Co., Montreal.
Howland, H. S. Sons & Co., Toronto.
Hyde, F., & Co., Montreal.
Lewis Bros. & Co., Montreal.
Lewis, Rice, & Son, Toronto.
Lockerby & McComb, Montreal.
Lufkin Rule Co., Saginaw, Mich.
Newman & Sons, Birmingham.
North Bros. Mfg. Co., Philadelphia, Pa.
Stanley Rule & Level Co., New Britain.
Stanley Works, New Britain, Conn.
Stephens, G. F., Winnipeg.
Taylor-Forbes Co., Guelph, Ont.

Carriage Accessories.
Carriage Mountings Co., Ltd., Toronto.
CoVert Mfg. Co., West Troy, N.Y.

Carriage Springs and Axles.
Guelph Spring and Axle Co., Guelph.

Carpet Beateas
Ontario Metal NoVelty Co., Toronto.

Cartridges.
Dominion Cartridge Co., Montreal.

Cattle and Trace Chains.
Greening, B., Wire Co., Hamilton.

Chafing Dishes.
Buffalo Mfg. Co., Buffalo, N.Y.

Churns
Dowswell Mfg. Co., Hamilton.

Clippers—All Kinds
American Shearer Mfg. Co., Nashua, N.H

Clothes Reels and Lines.
Hamilton Cotton Co., Hamilton, Ont.

Clutch Nails.
J. N Warminton, Montreal.

Cong. Roofing
Buchanan Foster Co., Philadelphia, Pa.

Cordage.
Consumers' Cordage Co., Montreal.
Hamilton Cotton Co., Hamilton.

Cork Screws.
Erie Specialty Co., Erie, Pa.

Cow Ties
Greening, B., Wire Co., Hamilton

Cuspidors.
Buffalo Mfg. Co., Buffalo, N.Y.

Cut Glass.
Phillips, Geo., & Co., Montreal.

Cutlery—Razors, Scissors, etc.
Birkett, Thos., & Son Co., Ottawa.
Clauss Shear Co., Toronto
Dorken Bros. & Co., Montreal.
Frothingham & Workman Montreal.
Heinisch's, R., Sons Co., Newark, N.J.
Howland, H. S. Sons & Co., Toronto.
Hutton, Wm., & Sons, Ltd., London, Eng.
Lamplough, F. W., & Co., Montreal.
Phillips, Geo., & Co., Montreal.
Round, John, & Son, Montreal.

Electric Fixtures.
Canadian General Electric Co., Toronto.
Morrison James, Mfg. Co., Toronto.
Munderloh & Co., Montreal.

Electro Cabinets.
Cameron & Campbell Toronto.

Enameled Ware
Kemp Mfg. Co., Toronto.

Engines, Supplies, etc.
Kerr Engine Co., Walkerville, Ont.

Eavetroughs
Wheeler & Bain, Toronto

Fencing—Woven Wire
Dominion Wire Mfg. Co., Montreal.
Owen Sound Wire Fence Co., Owen Sound.
Banwell Hoxie Wire Fence Co., Hamilton.

Files and Rasps.
Barnett Co., G. & H., Philadelphia, Pa.
Nicholson File Co., Port Hope

Firearms and Ammunition.
Hamilton Rifle Co., Plymouth, Mich.
Harrington & Richardson Arms Co., Worcester, Mass.
Johnson's, IVer, Arms and Cycle Works Fitchburg, Mass.

Fishing Tackle.
Enterprise Mfg. Co., Akron, Ohio

Food Choppers
Enterprise Mfg. Co., Philadelphia, Pa.
Lamplough, F. W., & Co., Montreal.
Shirreff Mfg. Co., Brockville, Ont.

Furnaces.
Pease Foundry Co., Toronto.

Galvanizing.
Canada Metal Co., Toronto.
Dominion Wire Mfg Co., Montreal.
Montreal Rolling Mills Co., Montreal.
Ontario Wind Engine & Pump Co., Toronto.

Glass Ornamental
Hobbs Mfg Co., London
Consolidated Plate Glass Co., Toronto

Glaziers' Diamonds.
Gibsone, J. B., Montreal.
Pelton, Godfrey S.
Sharratt & Newth, London, Eng.
Shaw, A., & Son, London, Eng.

Handles.
Still, J H., Mfg. Co.

Harvest Tools.
Maple Leaf HarVest Tool Co., Tillsonburg Ont.

Hockey Sticks
Still, J. H. Mfg. Co., St. Thomas.

Hoop Iron.
Frothingham & Workman Montreal.
Montreal Rolling Mills Co., Montreal.
J. N. Warminton, Montreal.

Horse Blankets.
Heney, E. N., & Co., Montreal.

Horseshoes and Nails.
Canada Horse Nail Co., Montreal.
Montreal Rolling Mills, Montreal.
Capewell Horse Nail Co., Toronto
Toronto & BelleVille Rolling Mills BelleVille.

Hot Water Boilers and Radiators.
Cluff, R. J., & Co Toronto.
Pease Foundry Co. Toronto.
Taylor-Forbes Co., Guelph.

Ice Cream Freezers.
Dana Mfg. Co., Cincinnati, Ohio.
North Bros. Mfg. Co., Philadelphia, Pa.

Ice Cutting Tools.
Erie Specialty Co., Erie, Pa.
North Bros. Mfg. Co., Philadelphia, Pa.

Injectors—Automatic.
Morrison, Jas., Brass Mfg. Co., Toronto.
Penberthy Injector Co., Windsor, Ont.

Iron Pipe.
Montreal Rolling Mills, Montreal.

Iron Pumps.
Lamplough, F W., & Co., Montreal.
McDougall, R., Co., Galt, Ont.

Lanterns.
Kemp Mfg. Co., Toronto.
Ontario Lantern Co., Hamilton, Ont.
Wright, E. T., & Co., Hamilton.

Lawn Mowers.
Birkett, Thos., & Son Co., Ottawa.
Frothingham & Workman, Montreal.
Maxwell, D., & Sons, St. Mary's, Ont.
Taylor. Forbes Co., Guelph.

Lawn Mower Grinders
Root Bros. & Co., Plymouth, Ohio.

Ledgers—Loose Leaf.
Business Systems, Toronto.
Copeland-Chatterson Co., Toronto.
Crain, Rolla L., Co., Ottawa.
Universal Systems, Toronto.

Lithographing.
London Printing & Lithographing Co., London, Ont.

Locks, Knobs, Escutcheons, etc.
Peterborough Lock Mfg. Co., Peterborough, Ont.
National Hardware Co., Orillia, Ont.

Lumbermen's Supplies.
Pink, Thos., & Co., Pembroke Ont.

Lye
Gillett, E. W., & Co., Toron

Manufacturers' Agents.
Fox, C. H., Vancouver.
McIntosh, H. F., & Co , Toronto.
Gibb, Alexander, Montreal.
Scott, Bathgate & Co , Winnipeg.

Metals.
Canada Iron Furnace Co., Midland, Ont.
Canada Metal Co., Toronto.
Eadie, H. G., Montreal.
Frothingham & Workman, Montreal.
Gibb, Alexander, Montreal.
Kemp Mfg. Co., Toronto
Leslie, A. C., & Co , Montreal.
Lysaght, John, Bristol, Eng.
Nova Scotia Steel and Coal Co., New Glasgow, N.S.
Robertson, Jas., Co., Montreal.
Roper, J. H , Montreal.
Samuel, Benjamin & Co., Toronto.
Stairs, Son & Morrow, Halifax, N S.
Thompson, B. & S. H. & Co. Montreal.

Metal Lath.
Galt Art Metal Co., Galt.
Metallic Roofing Co., Toronto.
Metal Shingle & Siding Co., Preston, Ont.

Metal Polish, Emery Cloth, etc.
Oakey, John, & Sons, London, Eng.

Nails Wire
Dominion Wire Mfg. Co., Montreal.

Oil Tanks
Bowser, S. F., & Co., Toronto.

Ornamental Iron and Wire.
Dennis Wire & Iron Co , London, Ont.

Packing.
Gutta Percha & Rubber Co Toronto

Paints, Oils, Varnishes, Glass.
Blanchite Process Paint Co , Toronto.
Brandram-Henderson, Montreal
Canada Paint Co., Montreal.
Canadian Oil Co., Toronto.
Consolidated Plate Glass Co., Toronto.
Dods, P. D., & Co., Montreal
Imperial Varnish and Color Co., Toronto.
Jamieson, R. C., & Co., Montreal.
Lucas John & Co., New York
McArthur, Corneille & Co., Montreal.
McCaskill, Dougall & Co., Montreal.
Moore, Benjamin, & Co. Toronto.
Ottawa Paint Works, Ottawa
Queen City Oil Co., Toronto.
Ramsay & Son, Montreal.
Sanderson Pearcy & Co , Toronto
Sherwin-Williams Co., Montreal.
Standard Paint Co., Montreal
Standard Paint and Varnish Works Windsor, Ont.
Stephens & Co , Winnipeg.
Martin-Senour Co., Montreal
Winnipeg Paint & Glass Co., Winnipeg

Perforated Sheet Metals.
Greening, B., Wire Co., Hamilton.

Plumbers' Tools and Supplies.
Canadian Fairbanks Co., Montreal.
Cluff, R. J., & Co., Toronto
Frothingham & Workman, Montreal.
Glauber Brass Co., Cleveland, Ohio
Jardine, A. B., & Co , Hespeler, Ont.
Jenkins Bros., Boston, Mass
Kerr Engine Co , Walkerville, Ont
Lewis, Rice, & Son, Toronto.
Merrell Mfg. Co , Toledo, Ohio.
Montreal Rolling Mills Montreal.
Morrison, Jas , Brass Mfg. Co., Toronto.
Mueller, H., Mfg. Co., Decatur, Ill.
Oshawa Steam & Gas Fitting Co., Oshaw
Robertson Jas., Co. Montreal.
Robertson, Jas , Co , Limited, Toronto
Somerville, Limited, Toronto
Stairs, Son & Morrow, Halifax, N.S.
Standard Ideal Sanitary Co., Port Hope,
Standard Sanitary Co., Pittsburg.
Stephens, G. F , & Co , Winnipeg, Man.
Turner Brass Works, Chicago.
Vickery, Orlando, Toronto.

Polishes.
Majestic Polishes, Toronto

Portland Cement.
International Portland Cement Co. Ottawa, Ont.
Hanover Portland Cement Co., Hanover, Ont.
Hyde, F., & Co., Montreal.
Thompson B. & S. H. & Co., Montreal.

Poultry Netting.
Greening, B., Wire Co., Hamilton, Ont.

Printing.
London Printing & Lithographing Co., London, Ont.

Razors.
Clauss Shear Co., Toronto.

Refrigerators.
Fabien, C. P., Montreal.

Registers
Pease Foundry Co., Toronto.

Roofing Supplies.
Brantford Roofing Co., Brantford.
Barrett Mfg Co., New York.
F. W. Bird, East Walpole, Mass.
Buchanan Foster Co., Philadelphia, Pa.
McArthur, Alex., & Co., Montreal.
Metal Shingle & Siding Co., Preston, Ont.
Metallic Roofing Co., Toronto.
Paterson Mfg. Co., Toronto & Montreal.
Wheeler and Bain, Toronto

Saws.
Atkins, E. C., & Co., Indianapolis, Ind
Shurly & Dietrich, Galt, Ont.
Spear & Jackson, Sheffield, Eng.

Scales.
Canadian Fairbanks Co., Montreal.
Frothingham & Workman, Montreal.

Screw Cabinets.
Cameron & Campbell, Toronto.

Screws, Nuts, Bolts.
Dominion Wire Mfg. Co., Montreal.
Montreal Rolling Mills Co., Montreal.

Soil Pipe
McFarlane, Walter, Glasgow

Sewer Pipes.
Canadian Sewer Pipe Co., Hamilton
Hyde, F., & Co., Montreal.

Shelf Boxes.
Cameron & Campbell, Toronto.

Shears, Scissors.
Clauss Shear Co., Toronto.

Shovels and Spades.
Eclipse Mfg Co., Ottawa
Frothingham & Workman, Montreal
Peterboro Shovel & Tool Co , Peterboro.

Silverware.
Hutton, Wm., & Sons, Ltd., London, Eng
McGlashan, Clarke Co , Niagara Falls, Ont.
Phillips, Geo., & Co., Montreal.
Round, John, & Son, Sheffield, Eng.

Skates.
Canada Cycle & Motor Co , Toronto.
McFarlane, Walter, Glasgow.

Sprayers
Cavers Bros., Galt

Spring Hinges, etc.
Chicago Spring Butt Co., Chicago, Ill.

Stable Fittings
Dennis Wire & Iron Co., London

Steel Rails.
Nova Scotia Steel & Coal Co., New Glasgow, N.S.

Stove Pipe.
Chown, Edwin, and Son, Kingston

Stoves, Tinware, Furnaces
Canadian Heating & Ventilating Co. Owen Sound.
Copp, W. J., Son & Co , Fort William
Davidson, Thos., Mfg. Co., Montreal.
Down Draft Furnace Co., Galt
Guelph Stove Co., Guelph.
Gurney Foundry Co., Toronto.
Harris, J. W., Co., Montreal.
Howard, Wm , Toronto
Kemp Mufg. Co. Toronto.
McClary Mfg. Co. London.
Merrick Anderson, Winnipeg
Pease Foundry Co., Toronto.
Smart, James, Mfg Co., Brockville
Stewart, Jas., Mfg. Co., Woodstock, Ont.
Taylor-Forbes Co., Guelph, Ont.
Wright, E. T., & Co., Hamilton.

Tacks.
Montreal Rolling Mills Co , Montreal.
Ontario Tack Co., Hamilton.

Tents.
Tobin Tent and Awning Co., Ottawa

Tin Plate.
American Sheet & Tin Plate Co., Pittsburg, Pa
Baglan Bay Tin Plate Co., Briton Ferry South Wales
Lysaght, John, Bristol, Newport and Montreal

Turpentine
Defiance Mfg Co., Toronto.

Ventilators.
Harris, J. W., Co , Montreal.
Pearson, Geo. D., Montreal.

Wall Paper
Staunton Limited, Toronto.

Wall Paper Cleaner.
Gilbert, Frank U. S., Cleveland

Washing Machines, etc
Dowswell Mfg. Co., Hamilton, Ont.
The Shultz Bros. Co , Brantford.
Taylor-Forbes Co., Guelph, Ont.

Water Filters.
Buffalo Mfg. Co., Buffalo, N.Y.

Wheelbarrows
London Foundry Co., London Ont.
Schultz Bros. Co., Ltd., The Brantford.

Wholesale Hardware.
Birkett, Thos., & Sons Co., Ottawa.
Caverhill, Learmont & Co., Montreal.
Frothingham & Workman, Montreal.
Hobbs Hardware Co., London.
Howland, H. S., Sons & Co., Toronto.
Lamplough, F. W., & Co., Montreal.
Lewis Bros. & Co., Montreal.
Lewis, Rice, & Son, Toronto

Window and Sidewalk Prisms.
Hobbs Mfg. Co., London, Ont.

Wire, Wire Rope, Cow Ties, Fencing Tools, etc
Banwell-Hoxie Fence Co., Hamilton
Dennis Wire and Iron Co., London, Ont.
Dominion Wire Mnfg. Co., Montreal
Greening, B., Wire Co., Hamilton.
Owen Sound Wire Fence Co., Owen Sound
Montreal Rolling Mills Co., Montreal.
Western Wire & Nail Co., London, Ont.

Wrapping Papers.
Canada Paper Co., Toronto.
McArthur, Alex., & Co , Montreal.
Stairs, Son & Morrow, Halifax, N.S.

Wringers
Connor, J. H. & Son, Ottawa, Ont

HARDWARE AND METAL

Vol. XXI. Montreal, Toronto, Winnipeg, August 31st, 1907. No. 35

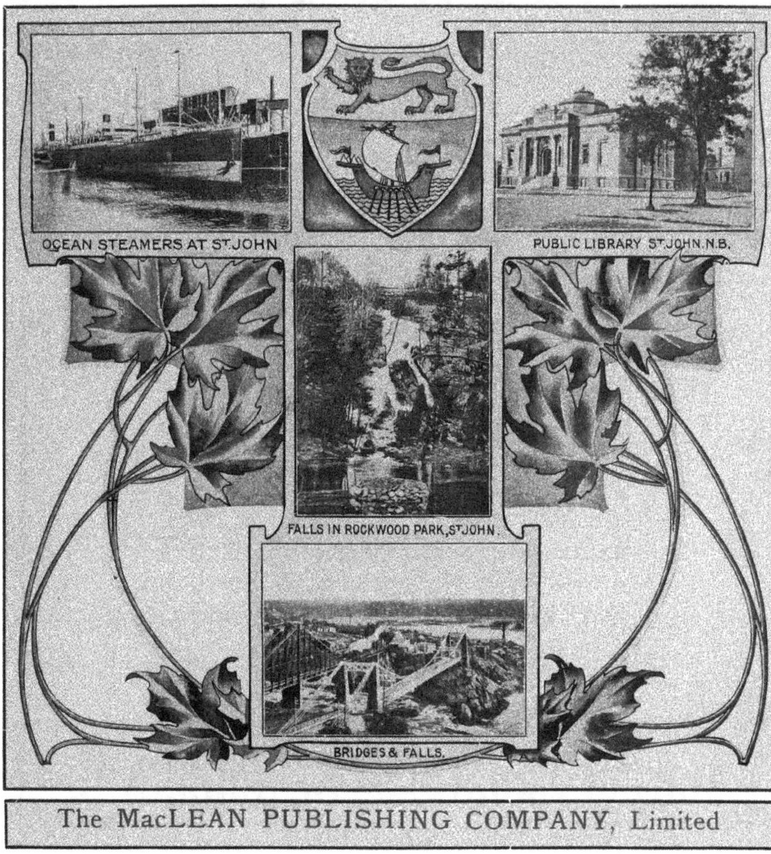

OCEAN STEAMERS AT ST. JOHN

PUBLIC LIBRARY ST. JOHN N.B.

FALLS IN ROCKWOOD PARK, ST. JOHN.

BRIDGES & FALLS.

The MacLEAN PUBLISHING COMPANY, Limited

W. H. Thorne & Co.

LIMITED

HARDWARE

The Largest Hardware Establishment
in the Maritime Provinces

Market Square, - St. John, N.B.

CIRCULATES EVERYWHERE IN CANADA
Also in Great Britain, United States, West Indies, South Africa and Australia.

HARDWARE AND METAL

A Weekly Newspaper Devoted to the Hardware, Metal, Heating and Plumbing Trades in Canada.

Office of Publication, 10 Front Street East, Toronto.

VOL. XIX. | MONTREAL, TORONTO, WINNIPEG, AUGUST 31, 1907 | NO. 35.

Read "Want Ads." on Page 79

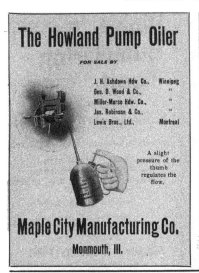

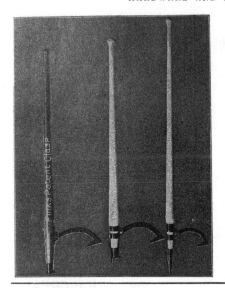

Shears and Scissors

WILL MAKE FRIENDS OF YOUR CUSTOMERS

ARE you really satisfied with your scisssor, shear and cutlery trade? Have you the trade of the dressmakers, tailors and people who continually use good shears?

¶ Whenever one of your customers wants good scissors or shears, have you a pair to recommend which you know will not come back, but which will give absolute satisfaction?

¶ If you have not, *you should secure the Wiss line just as quickly as you possibly can.* Wiss Shears will advertise themselves—you will see your cutlery business grow, not only on account of the extensive advertising we are doing, but because every pair of Wiss Scissors or Shears that you sell will be an advertisement for your house That is the line you should handle.

¶ The Wiss guarantee of absolute satisfaction not only protects the consumer but the dealer as well.

The Wiss cost you no more than ordinary shears—but cost more to make—sell better—and satisfy.

Send to-day for new catalogue and prices.

LEWIS BROS., Ltd.

OTTAWA
TORONTO

Montreal
WINNIPEG

VANCOUVER
CALGARY

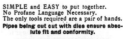

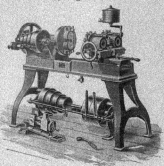

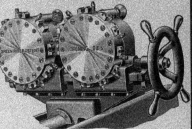

Canada Leads the World in the Manufacture of

HIGH-CLASS SAWS

No. 81

The best and cheapest Hand Saw on the market, quality, temper and finish considered. Every saw warranted. Carved applewood handle. Patented, wave finish blade.

No. 1 Maple Leaf Lance Tooth Cross-Cut Saw, and the No. 0 Racer are tempered under our secret process.

No. 0 Narrow Racer Cross-Cut Saw

If you want saws that will sell and give your customer satisfaction, you will stock the Maple Leaf Saws.

Manufactured by **SHURLY & DIETRICH, - GALT, Ontario**

Quality Tells

in saws more than in almost any other tools a carpenter uses.

The best guarantee of quality in tools is the reputation of the maker. And reputation is a manufacturer's most valuable asset.

Henry Disston & Sons have been making saws since **1840,** and have learned how. They learned how long ago, and their name is now a by-word for quality with carpenters and mechanics.

Your **trade wants quality,** and will pay for it. You cannot do better than offer them **Disston** hand saws. Two of the best sellers are the well known brands, D. 8 and No. 7.

D. 8.

No. 7.

We have them in hand, rip and panel, in all the saleable lengths and numbers of teeth per inch. Let us have your orders for Disston saws.

Caverhill Learmont & Co

MONTREAL

Winnipeg **Ottawa** **Quebec** **Fraserville**

Simonds
Crescent-Ground Cross-Cut Saws

For logging camps where work must be fast and economical there is no other saw made that can give results equal to the Crescent-Ground Saw. **Made of Simonds Steel.** It cuts easy, runs fast and gives good results. This cross-cut saw is sold by most of the leading jobbers in Canada.

Simonds Canada Saw Co., Limited

TORONTO, ONT. MONTREAL, QUE. ST. JOHN, N.B.

The Best Dealers Carry the Best Tools

If you are not stocking the ***PRATT AND WHITNEY SMALL TOOLS,*** made in Canada, a large volume of the small tool business is slipping away from you.

QUALITY ***TAPS - DIES - MILLING CUTTERS***
REAMERS - DRILLS - GAUGES **ACCURACY**
AND SMALL TOOLS GENERALLY

Would you like to consider a proposition that will enable you to handle this line ?

The Canadian Fairbanks Company, Limited,

Exclusive Canadian Selling Agents

MONTREAL - TORONTO - WINNIPEG - VANCOUVER

Wheelbarrows

All kinds and sizes. The cut reproduced here is just of one of the many, but the kind every contractor should use. The substantial, satisfactory, steel tray Contractor's Barrow.

The London Foundry Co.

LONDON, CAN.

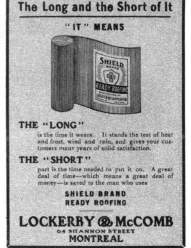

OUR "WANT ADS." get clerks for employers and find employers for clerks.

KANDA

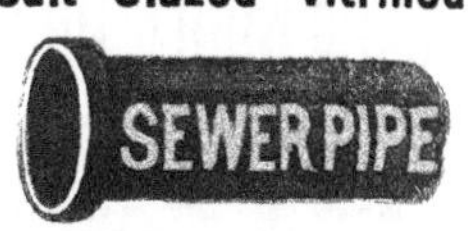
SEWER PIPE

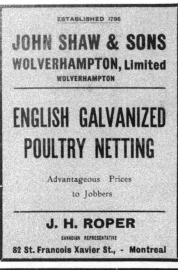

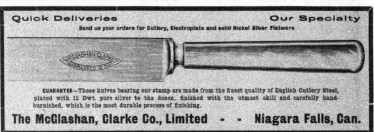

How You can make more Money

Out of Your Business

There are more than a few dealers in Canada who are making more money—lots more than their fellows. One of their reasons for their success is that they make it their business to handle the best Paint Brushes manufactured on this Continent, the

BOECKH BRAND

No practical man, knowing Paint Brush Goodness, but asserts the superiority of Boeckh's Brushes. And they are easy-priced, too, because our foreign representatives are in close touch with the bristle markets in the countries of the Old World where the best bristles come from.

You certainly can make more money by stocking the famous **BOECKH BRAND.**

UNITED FACTORIES, Limited
TORONTO, CANADA

A RECORD BREAKING SELLER

22 CALIBRE
16 SHOT

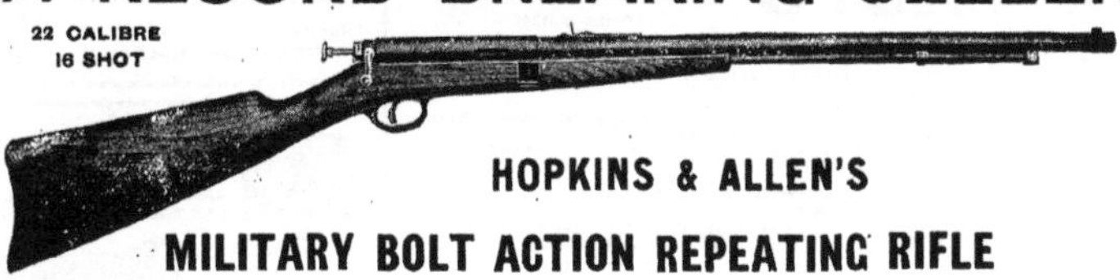

HOPKINS & ALLEN'S

MILITARY BOLT ACTION REPEATING RIFLE

$10.75 LIST There has long been a demand for a good rifle at this price—and the **HOPKINS & ALLEN Junior Repeater** fills the bill. This rifle has been and is being extensively advertised, and a heavy sale for it is created.

SPECIFICATIONS:—22-inch barrel, take-down, 22 calibre. Magazine handles 22 short, 22 long and 22 long rifle cartridges without change of carrier. Shoots 16-22 short or 12-22 long or long rifle cartridges Action, Military Bolt Pattern with positive safety device. Length over all, 40 inches, weight 5½ pounds.

FOR SALE BY ALL FIRE-ARMS JOBBERS.

PLENTY OF GOOD SUPPLEMENTARY ADVERTISING MATTER—FREE.

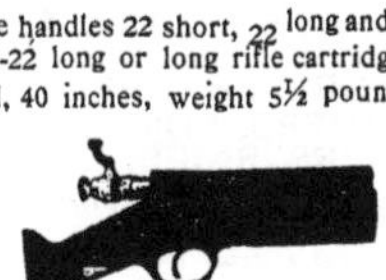

Shows Bolt Drawn Back—Position for Ejecting Shells.

WRITE FOR CATALOGS

THE HOPKINS & ALLEN ARMS CO.

LONDON OFFICE
6 City Road, Finsbury Square, London

Dept. 67, NORWICH, CONN., U.S.A.

A TRADE BRINGER....

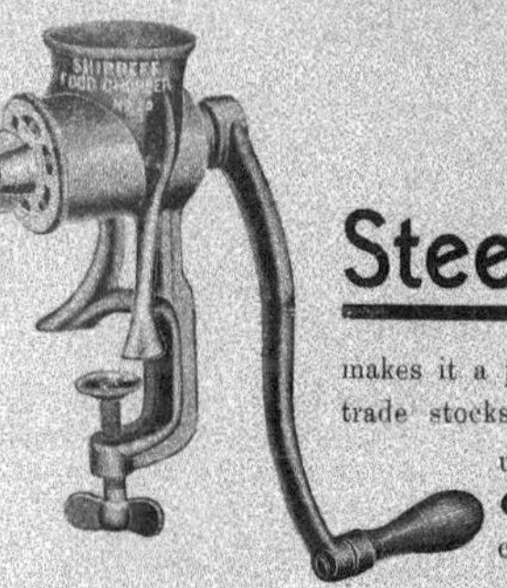

The simplicity and durability of the

Shirreff

Steel Knife Food Cutter

makes it a popular favorite. The Dealer who is out for the staying trade stocks and recommends the ***SHIRREFF*** because it gives utter and lasting satisfaction. The ***Shirreff Steel Cutters*** do not break—they are self-sharpening and cut uniformly.

There's Money in the Shirreff for you

Shirreff Mfg. Co., Ltd., - Brockville, Ont.

Can. Feb. 27, 1889

FACTS

about the best

CONE ROTARY VENTILATOR

on the market

Known everywhere as PEARSON'S

This ventilator is set in motion by the least current of air and will give a perfect draught at all times.

Substantial in make, Ornamental in appearance.

For balky chimneys and Ventilating Churches, Schools, Halls, Factories, Steamers, Public and Private Buildings, Stables, Water Closets, &c.

This Ventilator is in use throughout the Dominion, and has proved a great success.

Remember every Ventilator is guaranteed to work satisfactorily.

They are made in sizes from 2½ to 48 inches.

All orders by mail promptly attended to Testimonials on application.

Write to-day for particulars to the owners :

Geo. D. Pearson & Co., Office and Works: 17 Pontiac Street, MONTREAL
Bell Telephone East 2106

COVERT MFG. CO.

TROY, N.Y.

DERBY SNAP

With Plated Rust Proof and Guarded Springs

"The Latest and Best."

For Sale by all Jobbers at Manufacturers' Price

ARMSTRONG PLANER TOOLS
Are ALWAYS Ready for Use.

7 Sizes. Write for Catalog.

ARMSTRONG BROS. TOOL CO.
116 N. Francisco Ave., Chicago. U.S.A.

ONE
with an assortment of Cutters.
EQUAL A DOZEN FORGED TOOLS.

Shear making "up-to-date" is found in

HEINISCH

TAILORS' SHEARS,
TRIMMERS, SCISSORS,
TINNERS' SNIPS, ETC.

"Best in the World"

Since 1825, and a steady increase in sales due to their superior cutting quality.

R. HEINISCH'S SONS CO.
NEWARK, N.J., U.S.A.

New York Office and Salesroom, 155 Chambers St.

Past Due Accounts.

We have just issued an eight page booklet regarding our Collection Department. If you have not received one write us. It will nterest you.

The Beardwood Agency

Claims, Collections and Commercial Reports

313, New York Life Bldg., MONTREAL

"Good Cheer" Furnaces

No. 100 SERIES

For Anthracite or Butuminous Coal, Wood and Coke

The class of Furnace which means every satisfaction to the dealer in selling and installing and to the householder in operating

The radiating qualities of this furnace are so greatly enhanced by the adoption of the cast iron tube principle therein as to place it in a class entirely by itself, and, in combination with the solidity of construction and air-blast feature, make it a warm air generator of such capacity, durability and economical fuel-consuming qualities as to place it beyond comparison with any other of the moderately-priced furnaces in the market.

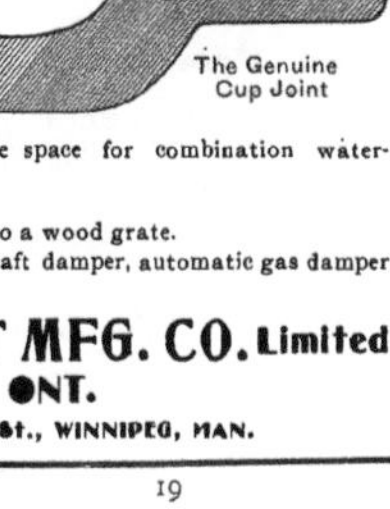

The Genuine Cup Joint

A Few of its Many Good Points

Cast iron tube radiating principle. Tubes surround fire chambers and form most powerful heat producers.

No sheet steel exposed to direct action of fire, being used only in outside jacket of radiator.

All steel set in genuine cup joints, positively gas and dust tight.

Commen-sense, positively unblockable and most easily cleaned flues.

Large combustion chamber, affording ample space for combination water-heating attachment if required.

Large double-feed door with smoke shield.

Roller grates for either hard or soft coal. Also a wood grate.

Roomy ashpit, waterpan, dust flue, direct draft damper, automatic gas damper and **AIR BLAST FUEL ECONOMIZER.**

The JAMES STEWART MFG. CO. Limited

WOODSTOCK, ONT.

Western Branch: Foot of James St., WINNIPEG, MAN.

QUICK

A want ad in **Hardware and Metal** will reach practically every Hardware Merchant in Canada, from Halifax to Vancouver, within four days.

If you are a Clerk and want to secure another position; if you are a Hardware Dealer and want to secure a clerk—in fact, if you want anything in which the Hardwaremen, Clerks, Travellers or Manufacturers are interested,

Use our Condensed Ad column

RATES

2c. per word................1st insertion
1c. " " ......subsequent insertions

HARDWARE & METAL

Montreal Toronto Winnipeg

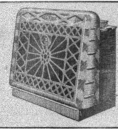

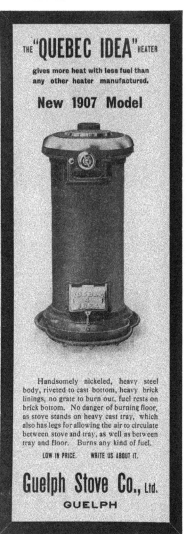

THE POSTED DEALER KNOWS

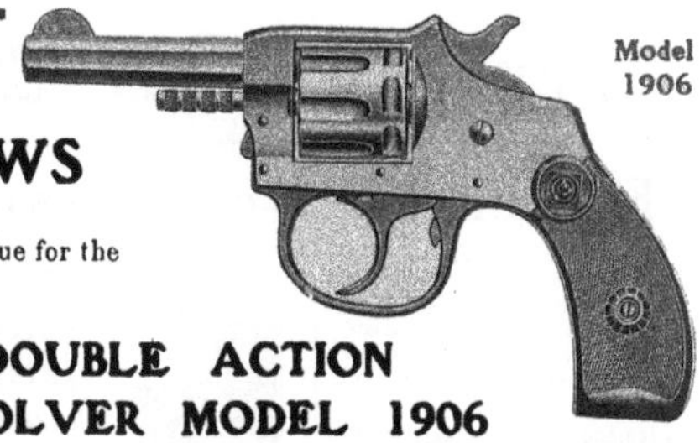

Model 1906

that **H & R** gives greatest revolver value for the money. The new

H & R DOUBLE ACTION REVOLVER MODEL 1906

is no exception to the rule. It is surprisingly good for the price. Surprisingly cheap for so good a revolver. Is a thoroughly well made, durable and serviceable arm. Light in weight, only ten ounces, and small in size, it is particularly adapted for those who desire a safe and efficient revolver at a moderate price.

H & R Greatest Revolver Value for the Money

Specifications 22 Calibre, Seven Shot, Rim Fire, Double Action; 2½ in. Barrel, Finest Nickel Finish. Also made with 4½ in. and 6 in. barrel.

The Celebrated **H & R Hammerless** is always a good seller. Write for Catalog of Revolvers and Single Barrel Shot Guns.

HARRINGTON & RICHARDSON ARMS CO.

314 Park Ave., - Worcester, Mass.

THE LAW SAYS YOU MUST

Do you know that a man is obliged by law to keep a proper set of books?

The law says:

"Any man who fails and has not for five years before his failure kept proper books shall be liable to a fine of $800.00 and one year's imprisonment."

Since you must keep books of some sort, and as the law insists that you keep proper books, why not keep them properly?

It is just as easy to do a thing right as it is to do it wrong.

* * * * *

Business Systems mean absolute accuracy, and with less work than the old-style methods of bookkeeping.

Business Systems are not expensive to put in in the first place—the cost is in proportion to the size of the business, and they enable one clerk to do the work of two.

Business Systems allow your clerks to put their work before you in such shape that you can see at a glance how everything stands.

* * * * *

Drop us a one-cent post card asking for further particulars.

You don't have to buy unless we can suit you.

System the Secret of Success

BUSINESS SYSTEMS LIMITED

84 SPADINA AVENUE
TORONTO - - - CANADA

MY BUSINESS

is to talk each week to hardware merchants, plumbers, clerks, manufacturers and travellers. I find out what they want and then I satisfy those wants.

Sometimes I find a man in Halifax who wants to buy a set of tinner's tools and I may find a purchaser for him in Vancouver.

Frequently hardware merchants buy new show cases and then they want to sell the old ones. Well, no person in their own town wants to buy them and they can't afford to start out on the road, for they can't travel as cheaply as I can. So I find a purchaser for them, it may be in a nearby town or it may be half way across the continent. My charge is the same in each case.

Hardware merchants are always wanting clerks, and clerks are always wanting new positions, and when you know all these people it isn't so hard to introduce them and let them come to terms if they can.

I had some work last week that kept me pretty busy. A hardware merchant in Eastern Ontario wanted to buy some cut nails, so he told me to go into all the hardware stores in Canada and deliver this message :—

> ANY hardwaremen having stock of cut nails they wish to dispose of at reduced prices should write at once to box 501 HARDWARE & METAL, Toronto.

Over sixty dealers left their work at once and wrote offering to sell their stock. A number of others seemed to be interested and will probably write later. I charged him half a dollar for my work. Do you think this was too much?

I am always glad to carry any message no matter what it is. You see, I am a sort of Jack-of-all Trades. I'm an employment bureau and a second-hand store although I sell a good many new articles as well.

I do my work well because I have been going over the ground for over 18 years and know practically every hardware merchant, plumber, stove and tinware dealer in Canada.

Have you a message to be delivered to any of these people?

Do you want it delivered quickly?

Condensed Advertisements in Hardware and Metal cost 2c. per word for first insertion, 1c. per word for subsequent insertions. Box number 5c. extra. Send money with advertisement. Write or phone our nearest office.

Hardware and Metal

MONTREAL　　TORONTO　　WINNIPEG

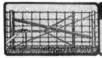

HORSE THE "C" BRAND NAILS

The old reliable "hot-forged" horse shoe nails, as "Made in Canada" by us for over forty years, have no equal in strength at the neck, where the most severe strain is encounted in service.

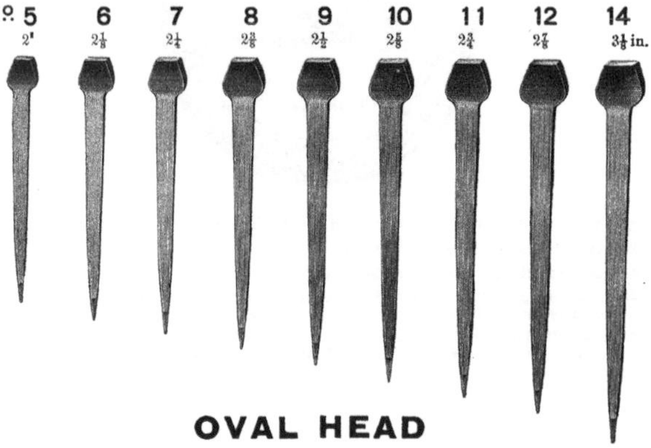

OVAL HEAD

They are made from a special quality of Swedish Charcoal Steel Nail Rods manufactured expressly for our purpose, and is positively the best material known or used in the world by any maker.

Canada Horse Nail Company
MONTREAL

HORSE THE "C" BRAND NAILS

They are made to a graduated taper and have fine hardened bevel points, which enable them to be driven easily into the hardest hoofs without bending or breaking off under the head.

No. 5	6	7	8	9	10	11	12	14
2"	$2\frac{1}{8}$	$2\frac{1}{4}$	$2\frac{3}{8}$	$2\frac{1}{2}$	$2\frac{5}{8}$	$2\frac{3}{4}$	$2\frac{7}{8}$	$3\frac{1}{8}$ in.

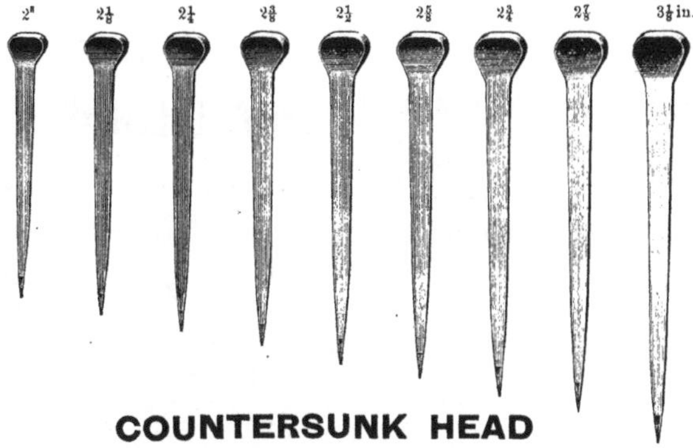

COUNTERSUNK HEAD

We have a fixed scale of prices to the horseshoer, at which all sales are made. Our discounts and terms to merchants afford a liberal profit in their sale. We solicit your enquiries or orders.

Canada Horse Nail Company
MONTREAL

FOR BIG GAME

The Best Hunters of Big Game

make Dominion Ammunition their first choice—because of its reliability, light recoil and high power.

Dominion Cartridges are as sure—accurate and effective as experience and money can make them; and they have a stopping power second to none—hence their popularity with big game hunters.

The hunting season will soon be here with its big demand for Dominion Cartridges.

Will you supply this demand?

There is a cumulative profit to be obtained from selling a first class article like Dominion Cartridges—for they will bring new customers to your store and the old ones back again.

Made in Canada by Canadians to fit all popular rifles—Savage, Stevens, Colt, Winchester, etc., and sold at a "less-the-duty" price.

If your jobber will not supply you—write us.

Dominion Cartridge Co., Limited, Montreal

Kemp's Seamless Savory Roaster

Self-Basting — *Seamless*

Self-Browning — *Sanitary*

MADE IN SHEET IRON AND ENAMEL WARE.

			Per Doz
No. 200	Finest Siemens-Martin Blued Steel, - - - - -	List Price,	$16.00
No. 400	Finest Siemens-Martin Blned Steel, - - - - -	List Price,	33.60
No. 200	Pearl Enameled Ware, - - - - - - - - -	List Price,	40.00

Perfectly smooth and seamless, without a corner or an angle, or a corrugation or fold of any kind—just a plain, smooth surface, easy to clean. This is what makes the ***SAVORY*** an easy seller to the women.

MANUFACTURED BY

The Kemp Manufacturing Co.

Cor. Gerrard and River Sts. — Toronto, Can.

The Kemp Mfg. & Metal Co., Limited
111 Lombard St., WINNIPEG, MAN.

The Kemp Mfg. Co. of Montreal
39 St. Antoine St., MONTREAL, QUE.

Maritime Board of Trade

Splendid Annual Gathering of the Business Men in the Provinces by the Sea—Many Matters of Importance Discussed—Not Jealous of the West, But Determined That the East Will Make Good, Too—Demand for the Improvement of Transportation Facilities—The All Red Line.

The annual meeting of the Maritime Board of Trade was held last week in St. John, on Wednesday, Thursday and Friday, Aug. 21, 22 and 23. Last year's meeting at Amherst was thought to be high-water mark of success for a meeting of the amalgamated boards, but this meeting marks a new record of achievement. The attendance at this, the 13th meeting of the Maritime Board was large and representative and included many leaders of commerce and industry in the provinces. Promptly at 10.30 Wednesday morning President Fisher called the board to order, and addresses of welcome were delivered by President McRobbie, of the St. John Board of Trade; Mayor Sears, of St. John, and President Fisher. The most striking matter of the opening of the meeting was the passing of a resolution endorsing the recommendation of the colonial conference for the All Red Line. This resolution was carried unanimously. The meeting was held in the cosy rooms of the St. John Board of Trade, which had been specially fitted up for the meeting and which were also elaborately decorated with flags and flowers.

His worship, Mayor Sears, after a few preliminary remarks of welcome said he thought in looking over the list of subjects to be discussed that technical education was one of the most important matters for their consideration. He referred to the recent visit of His Excellency the Governor-general to the Elinor Home Farm, and while it might not come under their jurisdiction he thought this was a matter of much importance and that many more such institutions should be founded. He extended on behalf of the city a cordial welcome and trusted their deliberations would result in a great benefit to business men throughout the province. His worship's remarks were heartily applauded.

Programme Committee.

President Fisher then appointed the following committee to arrange the list of subjects for discussion: Rev. A. E. Burke, J. H. McRobbie, M. G. De Wolfe, Hector McDougall, A. M. Bell, F. C. Whitman and W. B. Snowball.

The Delegates.

Following is the list of delegates by provinces:

Nova Scotia.

Annapolis—F. C. Whitman, Geo. E. Corbett.

Amherst—E. B. Elderkin, R. Robertson.

Berwick—D. C. Crosby.

Bridgewater—R. Dawson, G. W. Godard.

Canso—E. C. Whitman.

Digby—H. B. Short, L. Peters.

Halifax—E. A. Saunders, A. H. Whitman, N. McHall, J. E. De Wolfe, W. A. Major, A. Y. Wilson, A. M. Bell, G. S. Campbell and H. O. Smith.

Kentville—R. H. Dodge, C. O. Allen, E. J. Ward, M. G. De Wolfe, J. C. Anderson, E. W. Webster, Robert Harrington.

Kings County—A. E. McMahon, J. A. Kinsman, Geo. R. Pineo, E. Seaman.

Lunenburg—D. Frank Matheson, J. Frank Hall.

Oxford—Geo. F. Bissett.

Truro—C. E. Bentley.

Weymouth—Rev. Chas. R. Cunning.

Yarmouth—E. H. Armstrong, A. W. Eakin.

Sydney, C.B.—H. F. McDougall.

New Brunswick.

Chatham—W. B. Snowball, J. L. Stewart, Jas. Beveridge, W. S. Loggie.

Newcastle—John Morrissey, Geo. F. McWilliam.

Sackville—W. W. Andrews, F. B. Black.

Woodstock—J. P. Maney, J. E. Sheasgreen.

St. John—John H. McRobbie, W. E. Foster, W. H. Throne, T. H. Estabrooks, Jas. Pender, W. M. Jarvis, W. F. Hatheway, J. A. Likely, J. N. Harvey, J. Hunter White, W. F. Burditt, H. B. Schofield, F. W. Daniel, F. E. Dykeman, G. F. Fisher, T. H. Somerville, W. H. Barnaby, Geo. Robertson, M.P.P.; R. O'Brien, John Sealy, Wm. Kirkpatrick, A. H. Wetmore, Wm. Pugsley.

Prince Edward Island.

Alberton—Rev. A. E. Burke, Jas. E. Birch, W. H. Turner.

Charlottetown—Hon. Geo. E. Hughes.

St. John, the Ambitious City of the Atlantic Seaboard—Where the Maritime Board of Trade Met.

SECRETARY'S REPORT.

Chas. M. Creed Presents His Statement of the Year's Work—Advances Made.

The venerable and able Secretary-Treasurer, Chas. M. Creed, presented his report, as follows:

To the President and members of the Maritime Board of Trade:

Gentlemen: Your Secretary-Treasurer begs herewith to hand you his report for the current year.

All the recommendations and resolutions passed at the last annual meeting were forwarded to the Ministers of the various departments of the Federal Government, those concerning local Governments were forwarded to the Premiers of same, copies of the Canadian Grocer, containing the proceedings of the annual meeting were forwarded to the Ministers of the Federal Cabinet, also to the Premiers of the local Governments, to Boards of Trade and to delegates who attended the meeting. On Nov. 8, copies of the Grocer, also all resolutions, were forwarded to the following Maritime Province members in advance of the opening of the Federal Parliament: W. S. Loggie and Dr. J. W. Daniels, New Brunswick; J. J. Hughes Prince Edward Island; Dr. J. B. Black, H. J. Logan, B. B. Law, and A. K. McLean, Nova Scotia, in order that they could become familiar with them.

On Dec. 3, shortly after the opening of the Federal Parliament, copies of the the Grocer and also all resolutions were forwarded to all members and Senators from the Maritime Provinces. The correspondence with the same being, as also with various Boards, too voluminous to read, are now on the table for the inspection of members.

Annual Meeting at Amherst, N.S.

The annual meeting at Amherst, in August last, was a success, the subjects were well debated and nearly all the resolutions passed were brought forward and discussed at the Federal Parliament with good results.

Technical education has had the attention of the local Governments, and the Government of Nova Scotia are now about securing a site for the institution in this Province, and also have established a Bureau of Immigration and Publicity.

Annual Report.

The annual report was not published this year, the Canadian Grocer having such a full report of the proceedings of the annual meeting, it was used instead, at a cost of $25, for 500 copies, it has been found quite satisfactory, has been appreciated, and quite a good sum has been saved thereby.

New Board.

Boards have been formed at Oxford and Weymouth, N.S., and have affiliated. One has also been formed at Shelburne, N.S. It has not yet affiliated, but I hope it may do so during the coming year.

Unaffiliated Boards.

The Boards of Trade not yet affiliated have had every attention paid them during the year, and I am pleased to report the Boards of Bridgwater and Lunenburg, N.S., have affiliated; also the Board of Woodstock, New Brunswick.

Reorganized Boards.

The Boards of Pictou and New Glasgow have been reorganized. Pictou Board has affiliated, and the Board at New Glasgow, it is hoped, will soon follow.

Correspondence With Board.

This has been conducted with vigor. During the year some of the secretaries have promptly responded to notices and letters, while many of them have been very slow, which has been the means of delaying the work, and it makes quite a rush as the annual meeting approaches. It is hoped, however, the secretaries will take notice and be more prompt the coming year.

Notices of Annual Meeting.

The subjects for discussion at the annual meeting were mailed to all Boards on Aug. 10, with notices of the date, place of meeting and railway arrangements for delegates, and also to the delegates from each Board.

In conclusion, during the past year as usual, everything has been done by me for the enlargement and advancement of the Board, at the same time doing all posssible to assist your worthy President, in which I had the able assistance of M. G. De Wolfe, Esq., of Kentville as also that of W. E. Anderson, Esq., Corresponding Secretary, at St. John, N. B.

Accounts for the Year.

The accounts for the year are all prepared, but as several Boards will pay their per capita tax before the close of this day's meeting, they will be presented to-morrow for your approval and audit.

All of which is respectfully submitted.

CHARLES M. CREED,
Secretary-Treasurer.

Following is a list of the Maritime Boards of Trade:

Affiliated Boards:

Nova Scotia.

Annapolis, Amherst, Antigonish, Berwick, Bear River, Bridgewater, Canning, Canso, Chester, Caledonia Digby, Dartmouth, Halifax, Kentville, Kings County, Lockport, Liverpool, Lunenburg, Oxford, Pictou, Truro, Wolfville, Windsor, Yarmouth, Sydney, North Sydney, Weymouth.

New Brunswick.

Chatham, Moncton, St. John, Sackville, St. Stephen, Woodstock, Newcastle.

Prince Edward Island.

Charlottetown, Summerside, Souris Southern Kings (Montague), West Prince (Alberton).

Unaffiliated Boards:

Nova Scotia.

Middleton, New Glasgow, Shelburne, Pugwash, Glace Bay.

New Brunswick.

St. Andrews, Fredericton, Campbellton.

Visits to Industries.

President Fisher, informed the delegates that invitations had been received from the following industrial enterprises, to go through their premises: The Partington Pulp & Paper Co., Andre Cushing & Co.'s sawmill, The Cornwall & York Cotton Mills, T. S. Simms & Co.'s brush factory, and others. He said that letters of regret at their inability to attend had been received from the Boards at Bangor, Portland and the State of Main Board of Trade. Forty-nine Boards were represented in this body.

Facilities would be provided, the President said, to take any members who might desire to go to the west side to view the harbor improvements.

The Secretary read a number of letters of regret at inability to attend from Sir Wilfrid Laurier, the Governors of the Maritime Provinces, Lieut.-Col. J. B. McLean, Secretaries of the Montreal, Toronto and Ottawa Boards of Trade, and the Canadian Manufacturers' Association, and several western newspapers.

PRESIDENT'S ADDRESS.

Review of the Commercial and Industrial Situation in Canada.

President W. S. Fisher then delivered his annual report. He said:

"Gentlemen of the Maritime Board of Trade:

"I would like to add a word to what you have already heard from the Mayor of the city and the President of the St. John Board, to express the pleasure it gives me in common with all our citizens, to welcome to this city and meeting so representative a gathering of the commercial, professional and industrial interests of these Maritime Provinces.

"I also wish to express my appreciation of the honor conferred upon me in electing me as your President; and hope that the same kindly feeling which has prompted this, will also incline you to overlook my shortcomings.

"Boards of Trade have been aptly termed the Business Men's Parliaments; and while they have no power to enact laws or to put into effect such legislation as they may consider desirable, yet their influence in creating public opinion and directing attention to matters along lines of public interest is very considerable and is being more and more recognized.

"The opportunity afforded by such a gathering to exchange ideas concerning the problems that face us all cannot fail but have a stimulating effect, not alone upon each individual present, but also upon the community in which he lives.

"The list of subjects that we will be

called upon to discuss covers a wide range of interesting and live topics, and are such as to demand most careful consideration before decisions are reached, in order that they may stand the test when fairly discussed by the public, in the press and before the legislative bodies to whom they will be submitted and who alone have the power to finally put them into effect.

"The Secretary, in his report, will deal with the work accomplished during the past year. I will, therefore, not attempt to discuss it; but in passing wish to pay a tribute to the zeal and earnestness he has shown in following up the work and in keeping it well before the members and Ministers of both the Federal and Provincial Parliaments.

To the West Indies.

"I also wish to place on record the hearty appreciation of the liberality of Messrs. Pickford & Black, who so kindly sent at their expense delegates from different sections of Canada to the West Indies for the purpose of discussing with the merchants and others there the development of trade between the two countries. That good results will follow there can be no doubt. As this subject is on the agenda, we will hear further particulars at a later stage.

"Our main concern now is with the present and the future. We are here to determine what can be done by mutual co-operation to improve conditions; to find the weak spots and suggest the remedies.

"The turn of the east to share in the great progress and prospects of the country, while somewhat delayed, is surely coming; and as the west fills up attention will be drawn and capital invested to a greater degree in the development of the great natural resources of the section in which we live.

"It behooves us to help this along by being alert to every opportunity; to keep our case well to the front; to prevent our advantages being overlooked; and so hasten the time when through this development our population and wealth may be increased and our young men made to realize that there exists for them in the east as good an opportunity for advancement and progress, as good a prospect for the intelligent man who is willing to put his shoulder to the wheel and work, as there is in any part of this broad Dominion.

Advantage of Difficulties.

"The Governor-General is just completing his tour of the Maritime Provinces and while here has given us many inspiring thoughts and has referred specially to the general prosperity and absence of poverty in evidence on every side.

"Among other things, he drew attention to the Dutch, a people who by their patient, untiring industry have overcome tremendous natural drawbacks.

"This suggests two thoughts: First, that those individuals and countries succeed best who have many difficulties to encounter; and, second, that our natural advantages are very great, greater than perhaps we realize; and that while we have many difficulties to overcome, they are small in comparison with those that have been surmounted by others in the past.

"The lumber and pulp industries have been very prosperous for some years, and there is a much greater recognition of the increasing value of our forests, as well as a disposition through reforestation, protection from fires and, in many other ways, to preserve and increase their value as a source of future wealth.

"A new feature in this trade is the fact that a considerable quantity of spruce lumber is now being shipped to Ontario, owing to its growing scarcity in that Province.

"This fact is suggestive, as showing the possibilities in that connection and the widening market for one of our principal products.

Fruit Farming.

"Fruit farming is attracting more attention each year and there seems no reason why it should not develop enormously, situated as well as we are to cater to the export demand.

"If one might make a suggestion regarding this branch of industry, it would be a co-operative movement on the part of growers and shippers with a view to such a system of grading and packing as will ensure absolute confidence on the part of buyers.

"In the American States of Oregon and Washington, co-operative apple growers' associations exist, and have accomplished a great deal for their members in establishing a uniform standard of quality and, as a result, securing the very highest market prices.

"The establishment of evaporating factories now under way at different points in the apple growing districts, will help this very much by providing a market on the spot for the poorer grades.

Cold Storage.

"Another movement marking a distinct advance is the establishment of cold storage facilities at central shipping points, which will prove of great advantage in handling fruit, fish and other products specially.

Fishing.

"The fishing industry, which employs many thousands of our population, is of great importance, as shown by the statement that the annual yield is valued at fourteen millions of dollars. On the agenda paper several topics under this heading appear, which will, no doubt, be so fully debated as to throw much new light on a topic of such deep interest.

Mining.

"The coal industry of Nova Scotia has grown with great rapidity during the past few years, the output for the past season being in the vicinity of six million tons. Unfortunately, labor and other troubles exist at present in some of the most important centres, which, let us hope, will soon be settled in such a way as to ensure prompt resumption of work and such an agreement for future operations as will be fair and equitable to all interests.

"In New Brunswick the past year or two has seen considerable increase in the quantity and improvement in the quality of coal produced; and it is said the prospects for future development are excellent.

Transportation.

"Canada, forming as it does a long and comparatively narrow strip of country, the question of cheap and rapid transportation is one of the utmost importance. With one complete line of railway from ocean to ocean, which, with its ramifications makes a total of over 10,000 miles, and two other trunk lines in course of construction, the interior development is being well provided for.

"To supplement this excellent work and to complete these chains of communication so as to make it possible to carry on the entire commerce of the country over its own rails and through its own ports with the maximum of despatch and the minimum of expense, and also to secure that share of the all-round-the-world trade in passengers and freight that our geographical position entitles us to, no effort should be spared and no expense considered too great for the country to undertake in improving our ports and making their approaches safe and easy.

"Much has been done to improve the St. Lawrence route, that most important of Canadian highways. Much still remains to be done. Something also has been done in equipping the ports of the Lower Provinces which afford the only entrance or exit for the Dominion for six months in the year. Much more is required. The United States Government spends millions every year in deepening the channels and improving the harbors of Portland, Boston, New York, Philadelphia, Baltimore and New Orleans; and these ports have captured far too large a percentage of our ocean traffic. If we are ever to become independent of them we must do as they have done. It means much for the future of the country and no question is of greater importance. It is not a matter in which the Maritime Provinces alone are interested; the wellbeing of the whole of Canada is involved. Unless this work is done promptly, it will be impossible to capture the rapidly growing traffic of the Northwest and keep it within Canadian channels.

Building of the West.

"We are deeply interested in and immensely proud of the development and prosperity of the West, and have for many years cheerfully contributed by heavy drafts upon our pockets and upon our population to its up-building, looking forward to the time when we would secure our share of the wonderful prosperity that its growth has brought to the whole country. As Sir

Wm. Van Horne so well put the case a few years ago, when referring to the slow development of our ports, he said: 'We have enlarged the hopper and not the spout.' It is not a safe thing that we should continue to be dependent to so great an extent upon the ports of a foreign country. Therefore let us urge that the cause be removed. The case is strong and the matter important.

"Do the people in Ontario, Quebec and the western portions of the Dominion realize how necessary a link we are? Do they fully recognize that without the Provinces-by-the-Sea there would be no exit or entrance for passengers, mails or freight for six months of the year, excepting through the ports of another country? And that the tremendous development in the foreign trade that is being looked forward to with such promise, would be impossible? If not fully cognizant of the importance of the Maritime Provinces to the full development of the country at large, should this not be made clear?

"There is no need for us to be over-modest in pressing our case. Let us make every effort to impress upon the governing powers that no time should be lost in having our ports and approaches so equipped that the expressed policy of Sir Wilfrid Laurier—that he would not rest satisfied until every pound of Canadian freight should be carried through Canadian ports and over Canadian railways—can be carried into effect.

The All-Red Route.

"Before leaving the question of transportation, brief reference should be made to the All-Red Route, the practical outcome of the agitation for years for a fast mail service on the Atlantic. The great importance of this movement to the country is recognized and the question is absorbing much attention throughout the Empire, and if carried out, as proposed, will be fraught with great results.

"The question of better communication between Prince Edward Island and the mainland, a matter of great moment to those on both shores, appears on the agenda, and will, with the above, be discussed in due course.

Agricultural Development.

"Our agricultural development is not what we should expect or what the opportunities demand. We are not producing nearly enough for our own needs. We are every year bringing in from Ontario and Quebec large quantities of oats, beans, bacon, poultry, beef, onions, cheese, butter, lard, etc.; whereas we should not only be producing all we need for home consumption, but in all these lines and many more we should have a large surplus for export for which we are so admirably situated.

"A glance at the list of Canadian exports or a visit during the winter to the export warehouses in St. John and Halifax, will serve to show the vast quantities of these goods being exported by Ontario and Quebec.

"Why is it that our farmers are so little alive to the opportunities offered them through the excellent market right at their doors, as well as to the still greater market abroad for our surplus products, which, geographically, we are in such an unequalled position to cater to?

"The only Maritime Province that is to-day producing a surplus of food products is Prince Edward Island, mainly owing to the fact that farming there is carried on on a more scientific basis, which proves what could and should be done in the other Provinces as well.

"It is worthy of note in this connection that the counties of Carleton, Victoria and Madawaska, in New Brunswick, are developing a large trade in potatoes with Ontario, where, it is stated, they cannot be grown to so good advantage and where the quality is not so good.

Future of Dairying.

"Another reference in passing might also be made to the development in dairying that is going on in Sussex, where last year several hundred thousand dollars were paid out for milk and cream by one concern alone whose further extension is solely a matter of increased capital. A great deal more might be said to emphasize the need and the opportunities, but this is not necessary as the facts are patent to us all. It would seem that a more aggressive agricultural policy would in a large measure gradually remedy this and by infusing more life into the farmers through agricultural societies and farmers' institutes, awaken a more lively and intelligent interest in their vocation, which is the chief and most important of all and the real basis of our greatest prosperity and progress.

"Much is said and written deploring the tendency of the young people to leave the farms and drift towards the cities because of their apparently greater attractions. With the spread of that practical education referred to elsewhere and the awakened intelligence and interest aroused in farming pursuits in consequence, as well as the better financial returns which will follow, life on the farm will be made much more attractive.

"In addition to this, the extension of the telephone into the rural districts, the advent of better roads, with improved and more frequent mail service, will do much to remove that sense of isolation which is now an important factor in deterring young people from remaining on the farms.

Immigration.

"In the past we have been rich in men, have done our share in providing men of great breadth of thought, who have been important factors in creating, cementing and developing this Dominion. Of this same material we may feel sure the supply will not fail. For many years, unfortunately, quite too many of our young men found it necessary to seek their fortunes in the adjoining republic; and go where you will, you will find they are rendering a good account of themselves. For some time past the current has changed and those of our young men who seek other fields and what appears to them greater opportunities, are finding their way to our own West. While we regret the fact that so many leave the Maritime Provinces, it is a great satisfaction to know that they are helping to build up and develop our own country.

"What can we do to fill up the vacancies thus made and thus maintain a population sufficient to develop our own resources? How can we better do this than by making a greater effort to secure our share of that stream of immigrants who are flocking to our shores, seeking that freedom and opportunity which they fail to find at home? Very few of these know anything of the Maritime Provinces or of the openings that exist for them here, mainly owing to the fact that in the advertising matter issued in the past by the Federal Government little or no reference is made or information given about this section, their effort having been concentrated in exploiting the West.

"Our board has been urging for years that this be remedied and it is gratifying to note that the Dominion immigration authorities have at last awakened to the need, and at the present time have a staff collecting illustrations which it is understood will, with other necessary information, be used in the advertising matter to be distributed abroad in the future. If this is done, and the local government of the Lower Provinces co-operate in offering inducements and in making known what we have to offer, good results must follow.

Co-operating With the Army.

"Within the past year the several provincial governments, co-operating with the Salvation Army, have secured some very good citizens from among those whom this excellent body have been instrumental in bringing across the ocean; but much more remains to be done.

"We have not been alone among the eastern provinces in feeling the drain and consequent scarcity of labor, owing to this outflow to the far west, as in Ontario, also, this became so serious that some years ago the government there, assisted by the Dominion Immigration Department, started a special colonization bureau to procure settlers who would replace those drifting westward, with the result that last year more than 30,000 over-sea emigrants were secured and are now chiefly settled in the agricultural sections of that province. This year they expect a large increase over last; and, as we have as much to offer as Ontario, why should we not take a leaf out of their book?

"There are many thousands among the sturdy, industrious people of Northern Europe whose conditions of life are hard, and who would be glad to come to these provinces and make their home among us if the proper effort were made to show them what the opportunities are and to give them such encouragement as they need to induce them to come; and when here, to get them started on the right lines.

Technical Education.

"Technical education is a subject of great moment and one that is absorbing much attention. What has been done for years in Germany, Great Britain, and the United States, and what is now beginning to be done in Canada, shows the deepening interest taken by educationalists the world over in providing that practical education which teaches not only the value of dignity of labor, but that theory and practice must go hand in hand in securing that all-round training so needed to develop the best in man. To train boys and girls in merely literary accomplishments to the total exclusion of industrial, manual and technical training, tends to unfit them for industrial work, and in real life most work is industrial. The calling of the skilled tiller of the soil or of the skilled mechanic should alike be recognized, just as emphatically as the calling of the lawyer, the banker or the merchant. The abandonment of the old apprentice system has resulted in a great scarcity of trained mechanics and this lack must be made up by the trade school if we are to hold our own in industrial pursuits.

"Nova Scotia is moving in the right direction through the Government Agricultural Institute already established and the Technical College about to be. Sir Wm. Macdonald has done much for the country in establishing in each province consolidated schools with excellent teachers and departments devoted to manual training, domestic science and school gardens, supplemented by careful indoor studies in agricultural and other lines. We also are aware of the great work he has done at McGill and at Guelph, and is now doing at St. Anne de Belliveau. These are movements that will prove a great boon in fitting the coming generation to grapple with the problems confronting them as they take their places in the ranks of the workers.

Industrial Situation.

"With raw material such as coal, iron, lumber, etc., in abundance; with a climate in which men can work in comfort at all seasons; with an intelligent people available as operatives; with good facilities for gathering together the necessary raw materials at many points; with good local markets for many lines now imported; with excellent steamship services to the largest markets abroad; with a rapidly growing market in our own West, the opportunities for industrial development are not lacking.

"At the present time more than one hundred manufacturing firms in these Provinces are shipping goods to western points and the number of these is constantly growing. Ten years ago the number of our manufacturers doing business outside the limits of the Maritime Provinces could almost be counted on one's fingers. It means effort and the expenditure of labor and capital to obtain a foothold there; but when we consider the prospect, is it not worth while?

"What is the prospect? The present population of Canada is estimated at six millions in round numbers. The immigration this year will likely exceed 300,000. At all events, we are safe in figuring that during the next ten years it will at least average that number; thus making a total increase during that time from immigration alone of three millions or more. Add to that as a conservative estimate the natural increase after all allowances at an average of one hundred thousand yearly, and we will have a total population in Canada ten years hence of not less than ten millions.

Immigration as Wealth.

"It is estimated that every immigrant is worth to the country not less than one thousand dollars, which—if correct—means an increase in our natural wealth from immigration alone, of three hundred millions yearly.

GEO. E. HUGHES
of Charlottetown P.E.I. the newly chosen 1st Vice-President of Maritime Board of Trade.

"Think what this vast accession to our population and wealth means in increased demand for every conceivable class of goods, especially when it is remembered that practically all this new population is made up of adult workers who require everything from the ground up. Is it not, therefore, well worth our determining to secure a share of the trade in manufactured goods that this growing market will afford?

"Referring briefly to the industrial growth around us, it is only necessary to mention one or two examples of what is possible. Amherst, where we met last year, and where is seen the result of the work of a few progressive spirits, is again in the van this year in being the first section to carry into effect Edison's scheme of producing electric power at the pit mouth and transmitting it by wire direct to the factory; the Sydneys, where two of the greatest iron and steel industries in Canada have been developed in the past ten years. These might be enlarged upon or others given, but are sufficient as examples of what can be done.

"It is surely very clear that in this, as in the exploitation of all our opportunities, we have touched only the very fringe, and that it only needs men with the requisite knowledge and possessed of the right parts to take hold and secure results.

"For this the money should be easily available. We know from experience that when some well spoken gentleman from a safe distance appears with an alluring promise of high returns and waves his magic wand, the wherewithal is not lacking. If some of this surplus wealth which during the past has been lost in experiments of this kind, had been invested at home, the results both to the country and to the individual would have been much better.

Federal Insolvency Law.

"Before concluding, I wish to draw attention to the need that exists for a general insolvency law that will apply to the whole Dominion, looking to the proper protection of creditors and to the equitable distribution of insolvent estates. This is becoming more urgent as trade between the different sections of the country develops. Those firms who are doing business in every Province find the present system very troublesome, and the number of these is increasing rapidly. The present laws are inadequate. The merchant wants no uncertainty as to his rights and at present there is such, owing to the variety of laws in existence in the different provinces.

Summing Up.

"To sum up, how shall we secure that measure of progress and prosperity which should be ours?

"First, through the development of our rich natural resources; viz.: agriculture in all its branches; mining, lumbering, fishing.

"Second, through the development of our manufacturing industries.

"Third, through being on the highway of the all-the-world-round traffic, and by holding the key to the position as providing the only access from the interior to the Atlantic on Canadian territory for six months in the year.

"Where in the whole of Canada or elsewhere can be found a pleasanter country in which to live, a happier and healthier people, more real comfort and fewer drawbacks, than in the Maritime Provinces? As with men, so with communities; it is the strong, hopeful one that wins. Let us, therefore, strike a more hopeful note. Let us more fully realize the advantages we possess. Let us abandon any inclination towards pessimism, and with that courage born of faith and common sense, co-operate in securing our share of the progress and prosperity with which this country is being so abundantly blessed."

Mr. Fisher was cheered and applauded as he sat down.

Maritime Union.

The president called upon C. A. Duff-Miller, agent-general of New Brunswick, in London, and Hance J. Logan, M.P. for Cumberland County, N.S., to address the meeting.

Mr. Miller had left the meeting, but Mr. Logan made a few very interesting remarks. He thought there was more business on the list than could be gone through with properly, and said he had noticed that oftentimes matters of importance were rushed through without thought. The Maritime Provinces should be bonded more closely together and not pull so much for the separate Provinces. He was always ready to work in the interests of the Maritime Provinces, not Nova Scotia alone. He referred to the new electrical power plant at Chignecto Mines by which the Amherst industries were supplied with power.

The list of subjects submitted by the committee was then taken up.

The All-Red Line.

The first subject discussed was the All-Red Line project of a fast line to the Orient through Canada. G. M. Campbell, of the Halifax board, introduced the subject. He said he thought it was generally realized that such a line would be a great help to Canada, and in particular to the Maritime Provinces, as the terminus would be down here.

At present nearly all notable people and bodies of men coming to Canada do not see the Maritime Provinces at all. This line would make us better known. In Halifax they had been making strenuous efforts the past few days to induce the British journalists, now touring Canada, to come here for a day, and they had made satisfactory arrangements to that effect. If the proposed line were inaugurated, all such bodies would pass through the Maritime Provinces. He said he wouldn't ask them to endorse Halifax, as the western terminus, but asked that the resolution be adopted in some shape, so that the feeling of the Maritime Provinces might be known.

The resolution was seconded by M. G. De Wolfe, who said he thought the matter a most important one, not only to the seaboard towns, but inland as well.

A. M. Bell, of Halifax, also spoke on the subject, referring to the remarks made by Earl Grey at Halifax, that England had made a colossal blunder in subsidizing the Cunard steamers at the enormous amount that they had given them. He thought the proposed scheme would be a great benefit to the Maritime Provinces and Canada.

Geo. Robertson also spoke on the subject, and the resolution was unanimously adopted.

It reads as follows:

"This convention of the Maritime Board of Trade, believing that fast high-class mail, passenger and express freight service between the British possessions would be of immense material and political advantage to Canada and all ports of the empire, hereby heartily endorses the 'All Red Line' project proposed by the Premier of Canada, and unanimously adopted, at the recent Imperial Conference held in London, and urge the Canadian Government to take all possible steps to hasten the inauguration of the scheme."

The second subject—a national banking system, introduced by the Chatham board, was laid over for consideration later on.

Cold Storage in Steamers.

A resolution favoring the installation of cold storage facilities in steamers plying between ports in the Maritime Provinces and Newfoundland was presented by Hon. Geo. E. Hughes, of Charlottetown, and seconded by Hector McDougall, of Sydney. It was discussed at length by a large number of the members present including Rev. A. E. Burke, Hon. Wm. Pugsley, Hance J. Logan, M.P.; J. E. Birch, J. E. De Wolfe, Joseph Likely and others. Several changes and amendments were suggested, and the sense of the resolution as finally adopted was that the Federal Government should see that subsidized freight steamers plying between the ports of Prince Edward Island and Sydney and of Newfoundland be equipped with cold storage facilities.

There was a lengthy discussion on the question of some changes in the Canadian coinage, but the matter was allowed to lie on the table.

WEDNESDAY AFTERNOON.

The Railways and Express Rates—Pulpwood Export.

At the opening of the afternoon session, W. B. Snowball, of Chatham, announced that the member from their board who was to have submitted a resolution on "A National Banking System" was not present, and he thought that subject had better lie over.

E. H. Armstrong, of Yarmouth, brought up the subject of the advisability of the Dominion Government taking over the railways of western Nova Scotia and making them a part of the Intercolonial system. He referred to the great passenger traffic at Yarmouth by steamer and also by rail.

If the I. C. R. was to be a government road, he thought the roads in western Nova Scotia should be taken into the system. There were two railways in Nova Scotia, the D. A. R., owned by English parties, and the Halifax and Southwestern owned by McKenzie & Mann. He said that the eastern part of the province and New Brunswick had received more from the government in the I.C.R. than had the western part of Nova Scotia. The I.C.R., he said, was built to bind together the four provinces, Quebec, Ontario, New Brunswick and Nova Scotia. The western part of Nova Scotia had contributed more than its share in the expense of building this road, as it was not served as were the other districts. The I. C. R. had been extended to various towns of the eastern part of the province, and if it was good for the east it was good for the west.

He quoted Hon. H. R. Emmerson as advocating the extension of the I. C. R. by taking in branch lines, and claimed that the pro rata cost of the western railways was below that of any other railway in Canada. If this was so, the I. C. R. could to advantage take over these railways. He then read a resolution, which was seconded by H. B Short, of Digby. Mr. Short pointed out that only three freight trains a week passed between Digby and Yarmouth. He heartily endorsed the arguments of Mr. Armstrong.

W. Frank Hatheway also cordially endorsed the proposal that the I. C. R. should take over these lines. He hoped the resolution would be carried unanimously.

All the Branch Lines.

H. J. Logan, M.P., thought that the resolution should include all the branch railways in the Maritime Provinces. He could think of several in New Brunswick and one or two in Nova Scotia that he did not think the people along the lines would object to having the I. C. R. take over. Many of the roads were unsafe. He was strongly in favor of all these small roads being taken over by the I. C. R.

The resolution was as follows: "That in the opinion of this Maritime Board of Trade, it would be greatly in the public interest and to the advantage of the Intercolonial Railway, if the Federal Government were to acquire by purchase or lease the railways of western Nova Scotia, and also the branch lines connecting with the Intercolonial Railway in Nova Scotia and New Brunswick, and to operate them as part of the Intercolonial system, and that the government be urged to take the necessary steps to acquire such railways, provided, the same can be purchased or leased on reasonable terms, based not merely on the original cost, but also on the earning power of the lines."

F. A. Dykeman also spoke in favor of the small roads being taken over by the government. The resolution, he said, should be changed to take in all branch lines. W. B. Snowball spoke in support of the resolution being enlarged to take in all small railways. He referred to improved conditions on small lines that were taken over by the I. C. R. Frank Black, of Sackville, thought the resolution was too local and should be broadened to take in all lines. He knew, for instance, that people along the line of the Cape Tourmentine Railway would not object to the line being taken over by the I. C. R. Mr. Armstrong, the mover of the resolution, said he was agreeable to changing it to embrace all branch lines if the members wished it. It was then decided that the matter should lie over for a time to see if it could be drafted in more suitable form.

Express Rates.

Hon. Geo. E. Hughes presented a resolution asking for an improvement in the express rates, as now in force in Prince Edward Island. The Canadian Express Company, he said, was now the only company doing business there and it had a monopoly. Rev. Father Burke seconded the resolution and cited instances of excessive charges for sending away shipments of fresh fish. The result of these exorbitant rates was to kill the trade in fresh fish as far as Prince Edward Island was concerned.

Mr. Wilson, of Halifax, who was interested in the fish trade, spoke of difficulties in getting fair express rates when only one company was doing business in the territory. He cited a number of instances of unfair charges. W. B. Snowball advocated that the resolution be so enlarged that the I. C. R. should grant running rights to all express companies.

H. B. Short, of Digby, spoke of the difficulties experienced at Digby, and urged that the resolution should take in all subsidized steamers as well as railways. A. J. Logan, Amherst; J. E. DeWolfe and G. S. Campbell, Halifax, also spoke on the question.

Position Reaffirmed.

Mr. DeWolfe read a resolution on the subject which was passed at the meeting in Moncton in 1904, and on motion it was reaffirmed. The resolution read as follows :

"Whereas, under the Railway Act all express companies may have same privileges over all railways in Canada, except over the Intercolonial Railway, and

"Whereas, under the contract between the government and the Canadian Express Company, paragraph 15, stipulates that this agreement is subject to the condition that equal facilities and equal terms shall and may be granted to any or all express companies which may contract with his majesty for the conduct of an express business over said railway, or any part thereof, and

"Whereas, over the I. C. R. between Montreal and Moncton, and from Truro eastward in Cape Breton and in Prince Edward Island, there is only one express company permitted to do business, and

"Whereas, the shippers of fish and other merchandise from eastern Nova Scotia and Prince Edward Island, and shippers of fruit from western Canada to points on the I. C. R. east of Truro are handicapped by being shut out from a competing express service and compelled to pay extra rates of carriage, and

"Whereas, numerous resolutions have from time to time been addressed to the government asking that the Dominion Express Company be granted same privileges over government railways as are granted to the Canadian Express Company ;

"Therefore, resolved, that this Maritime Board of Trade endorse said resolutions and urge upon the government the early granting of said privileges to other companies than that one now having the monopoly of the express business ;

"Further, resolved, that copies of this resolution be forwarded to the Minister of Railways and Canals, Minister of Finance and maritime members at Ottawa."

I. C. R. and Branch Lines.

Mr. Armstrong, of Yarmouth, at this point presented his resolution on the subject of the I. C. R. taking over branch railways, in which he had made several changes as suggested in the discussion previously.

Hon. Mr. Pugsley thought the board should not pass the resolution without inserting a clause that the government should only take over the branch lines on paying a "reasonable" price for them.

Mr. Whitman, of Annapolis, claimed that the board was only offering a principle, and of course the government would use its best judgment in whatever action they took. Mr. Armstrong also spoke along the same lines and urged the adoption of the resolution.

Alderman Baxter thought the McKenzie & Mann roads would not generally be considered branch roads. There was a probability that McKenzie & Mann would build a line down the St. John valley ; it would not be a branch road, but probably part of a transcontinental system, and it wouldn't be expected that the I. C. R. should take it over. He thought the resolution should be worded differently.

Hon. Mr. Pugsley said he was strongly in favor of the government taking over branch lines, but he thought that they should put in the resolution the words "at a reasonable cost."

H. J. Logan, M.P., thought there was but little difference between the suggestions, and it might be well if the mover of the resolution and those who desired it changed somewhat, should get together and draft it up in suitable form. On motion of Ald. Baxter, it was decided to allow the matter to stand over until Messrs. Armstrong, Pugsley, Short and Logan will present a new resolution to the meeting, embodying the changes suggested.

Pulpwood Export.

The next matter taken up was "the export duty on rossed pulpwood," presented by W. B. Snowball, of Chatham. The resolution was as follows :

"Whereas, it has been the policy of the Federal Government to encourage and promote manufacture within the bounds of the Dominion by duties sufficiently high to keep out foreign manufacture, and by bounties to encourage the use of home raw material, and

"Whereas, the Federal Government has expended large amounts to promote agriculture and to boom the western country, and to encourage emigration from the motherland and foreign countries ; and

"Whereas, we have within the bounds of our eastern Canada as a natural product, spruce pulpwood, capable of providing employment for a large number of our present people, and for many who may come to find homes here at profitable wages, instead of going west and thus populating our eastern provinces ; and

"Whereas, this wood is being raised in New Brunswick, Nova Scotia and Quebec, and shipped in that state to the United States to keep their pulp mills running.

"Therefore, resolved, that this Maritime Board of Trade is of the opinion that the exportation of pulpwood should be prohibited by the Federal Government."

Mr. Snowball advocated that a prohibitive export duty be put on pulpwood, so that manufacturers would be induced to come here instead of taking the wood out of the country and manufacturing it elsewhere.

In Chatham, there was a rossing plant that provided more pulpwood than could be used by the mill there now, and other mills should be built there. He thought also that paper should be manufactured here as well as the pulp. No raw materials should be allowed to cross the border.

Mr. Ward, of Kentville, in seconding the resolution, gave some interesting information on the scarcity of spruce in Nova Scotia, which is used for pulp. He thought no rossed pulpwood should be exported, as the forests would soon be depleted.

Pulpwood Silk.

Mr. Andrews, of Sackville, showed to the members a piece of black silk made from pulpwood. He thought there was a possibility, some time in the future, of Canada manufacturing goods of that kind.

James Beveridge, of the pulp mill at Chatham, also spoke on the subject of pulpwood and the manufacture of pulp. He thought this board should support the resolution so ably put forth by Mr. Snowball.

Geo. Robertson, M.P.P., spoke interestingly of the uses to which the product of pulp could be put, and he thought it would be only a matter of time when Canada would be manufacturing clothing, shoes and other products from pulp.

Should Prohibit Export.

H. B. Schofield, of the St. John board, said he hoped the time would come when the exporting of pulp would be prohibited altogether, and all the paper and other materials made from the pulp be manufactured in Canada. J. C. Anderson, of Kentville, also spoke on the subject, and Mr. Dawson, of Bridgewater, urged that the resolution should deal with all pulpwood, not only with the "rossed," i.e. that from which the bark has been removed.

F. C. Whitman, of Annapolis, and Mr. Pineo, of Kentville, also discussed the subject, and it was pointed out that the resolution would prevent individuals from selling their woods as they might wish.

Mr. Snowball said he had no objection to striking out the word "rossed." He did not know that any but rossed pulpwood was exported. Regarding the contention that private owners would be prevented from selling their wood as they wished, he said it was a matter of national interest and they should deal with it in a broad manner. They should not let their private feelings come in a matter of this kind.

J. H. Whitman, of the Halifax board, submitted a resolution to the effect that a fisheries board should be appointed along the lines of the fishing board of Scotland, to act under the department of marine and fisheries for the regulation and control of the Canadian Atlantic fisheries. He spoke at some length on the subject and cited many instances to show that an inspection of fish and packages for shipping them was needed.

Mr. Eakin, of the Yarmouth board, in seconding the resolution, presented another dealing with the lobster fisheries and their preservation.

Some, he said, advocated having a size limit on lobsters of 10½ inches, others advocated closing the fishing for a period of five years. His own view was that pounds might be provided near the canning factories, where a government inspector could measure the lobsters, and all those under size could be put in the pound. For a diet they could be fed on pickled dogfish, and this would help to do away with this dogfish pest. He strongly favored the resolution submitted by Mr. Whitman. He read a paper dealing exhaustively with the lobster question which was heartily applauded.

Mr. Wilson, of the Halifax board, also spoke at length on the subject, and quoted statistics to show that a fisheries board would be a benefit. Geo. Robertson, M.P.P., M. G. DeWolf, of Kentville, and F. C. Whitman, of Annapolis, also spoke on the subject.

As it was then six o'clock, it was decided to defer further discussion until the evening session.

WEDNESDAY EVENING.

The Lobster Fisheries Question—Maritime Provinces as a Home.

At the evening session the debate on the resolution presented by Mr. Whitman, of Halifax, on the fisheries question, was resumed. H. B. Short, of Digby, strongly supported the resolution. He thought steps should be taken at once to protect the lobster fisheries.

W. S. Loggie, M.P., of Chatham, said he could not agree with the gentlemen who had spoken on the subject. He thought the interest in lobster fishing was not waning, and that the government was taking every reasonable precaution to protect the fisheries. He thought a greater danger was the pressure that was being brought to bear by people in Nova Scotia to increase the number of factories. There were factories enough, but they wanted more hatcheries. However, if it was felt that a fishery board would be an improvement to the fisheries he would give it his hearty support.

The president said a letter had been received to-day from E. C. Bowers, a member of the fisheries commission, which had been handed to the gentlemen who had this matter in hand. F. C. Witman then read some extracts from the letter, in which Mr. Bowers advocated the establishment of a fishery board, free from political influence.

Mr. Wilson, of Halifax, read some extracts from a lobster canner in Cape Breton, showing that the supply was falling off. John Sealy, of St. John, followed in a few remarks, in which he advocated the forming of a fishery board to regulate the fishing of this section. The board should be free from political influence.

W. F. Hatheway also spoke on the subject, and among other things said he thought a plant for turning the dogfish into fertilizer might be established between this city and New River. Mr. Whitman, of Halifax, gave some further information on the matter of forming the board as referred to in the resolution.

The motion was then put and carried as follows: "That this meeting of the Maritime Board of Trade endorses the movement for the appointment of a fisheries board to act under the department of marine and fisheries for the regulation and control of the Canadian Atlantic fisheries, and that steps be taken to have this important matter acted upon at the next session of the Dominion Parliament."

The Modus Vivendi.

Mr. Wilson, of Halifax, brought up the matter of the abrogation of the modus vivendi and submitted a report reaffirming the resolution of last year which was referred to a committee to report. This report was read by Mr. Wilson and a letter from Mr. Whitman, of Canso, another member of the committee, was read. Mr. Wilson also read extracts from the speech of A. K. McLean in parliament, giving reasons why

A. M. BELL, HALIFAX.
New President of the Maritime Board of Trade.

the act should be abolished. Mr. Wilson moved the adoption of the report, which was seconded by Mr. Stewart, of Chatham, and carried.

The report was as follows: "Your special committee on the modus vivendi beg leave to report that it is highly in the interests of the Canadian Atlantic fisheries that the Dominion Government abrogate the same, until such time as the Government of the United States is disposed to negotiate a new treaty.

S. Y. WILSON
J. W. STEWART
F. C. WHITMAN."

George Robertson M.P.P., in introducing a resolution embracing a number of the subjects from various boards of trade, spoke of the natural resources of the Maritime Provinces and reviewed conditions of many years ago up to the present time. He deplored the fact that so many of our young men were leaving the country and going to the west. He had many industries but not enough of them apparently to keep the young men here.

Maritime Provinces Neglected.

"We must," he said, "take a grip on this enormous depletion of our Maritime Provinces." The tide of immigration, he said, was setting altogether to the west and the Maritime Provinces were being given the go-by. What the Maritime Provinces wanted was a good class of immigrants from the British Isles. He rose to heights of eloquence in referring to the Maritime Provinces as a home for settlers, and his remarks elicited frequent applause.

F. L. Potts, of St. John, spoke of the flowery eloquence of previous speakers and referred to the fact that the expense of living here had increased fifty per cent. of late years. He was going on to tell that what was needed in this province was more manufacturing industries when he was reminded by the president that the resolution submitted by Mr. Robertson had not been seconded and was, therefore, not yet debatable, and also that only delegates were supposed to speak on the subjects before them.

Mr. Potts apologized for his action and took his seat. The motion was then seconded by Prof. Andrews, of Sackville, and carried.

Maritime Advantages.

The resolution was as follows:

"Resolved, That the Maritime Provinces, with their rich and varied resources, afford in farming, lumbering, fishing, mining and manufacturing, and in commercial life and other spheres of activity exceptional opportunities for the obtaining of a comfortable livelihood, and the making of pleasant homes with all the surrounding essentials to the highest degree of happiness. That there are here great opportunities for development and progress, and that all reasonable efforts should be put forth by the various provincial governments and municipalities as well as by the federal administration, to induce our young men to remain at home, and also to divert to the Maritime Provinces a portion of the ever increasing stream of immigration which is now flowing from the motherland and other European countries to the Canadian west."

Technical Education.

Prof. Andrews, of Sackville, introduced a resolution on technical education, the public support of schools—common, consolidated and technical colleges, the relation of industries of N.S. and N.B. to this question, and the value of such education for provincial revenue. Among other remarks he spoke of the probability of a chair of forestry being established in the U.N.B. Some arrangement, he thought, should be made whereby the three provinces could work together in the matter of technical education.

Frank Black, of Sackville, seconded the resolution, endorsing the remarks of Prof. Andrews.

A. H. Wetmore, of the St. John board, spoke in favor of the resolution, and thought that New Brunswick should do its part in forming a technical college for the three provinces. He was in favor of making the matter national in its character. He referred to the system of technical education in Norway and of the colleges in the United States.

G. S. Campbell said he thought the resolution placed Nova Scotia in an awkward light. Their position in N.S. was now this, that they were about to get what they wanted in a technological college. This resolution would have the effect of asking the legislature of N.S. to discontinue their work on the institution now to be built in Halifax. He suggested that N.B. and P.E.I. should erect schools or that the scholars from these provinces should attend the school at Halifax.

Prof. Andrews said it was not the intention of the resolution to stop the work in N.S., but to get the three provinces interested in having a central college somewhere. He did not want to touch at all on the location.

E. B. Elderkin, of Amherst, spoke in favor of the resolution and thought the matter of location should be left to the authorities of the three provinces. E. H. Armstrong, of Yarmouth, said he would have to take the same objections to the resolution as had Mr. Campbell, of Halifax. Nova Scotia had already provided for the building of a college for a higher technical education. New Brunswick and P.E.I. should also establish colleges of a similar type.

A. M. Bell, of Halifax, said the Halifax board had agitated for the establishment of the technical college because they had been hounded to it by the Mining Association. He thought it would be inconsistent for the board to vote to have consolidated colleges, after they had asked last year that the governments separately should establish the institutions.

Unite the Provinces.

Geo. E. Corbett, of Annapolis, said he would strongly support the resolution of Prof. Andrews, and would give $300 to help carry it out. They wanted to bind the three provinces closer together. "Bury your dirty old politics," he said, "and vote for the resolution."

Mr. Bell, of Halifax, again argued that the Nova Scotia members could not consistently support the resolution, nor did he think the Maritime board should do so. J. L. Stewart, of Charlottetown, said he could see no objection to the resolution going through as read. It would not interfere with what was being done by Nova Scotia.

Mr. Armstrong—"What is New Brunswick going to do?"

Mr. Stewart—"I don't think your remark is apropos and I don't feel called upon to answer it." He went on to say that he thought the board should adopt the resolution and that much good would result from the establishment of such institutions.

Mr. Birch, of Alberton (P.E.I.), thought the arguments being used in favor of giving the young men of the provinces a technical education were directly in opposition to the resolution moved by Mr. Robertson that had just been adopted.

As soon as the young men learned the mysteries of some trade, they left their farms and went to the west where they got positions on the railways and so on. They wanted to keep the young men home. "Chickens at a distance," he said, "have very fair plumage, and they want to follow that plumage. It don't look quite so good when they get close to it. Oftentimes when the farmer wakes up in the morning now, he sees his young men disappearing in the dawn on a train for the west." (Applause and laughter.)

He thought manual training or technical education made young men feel that they wanted to go out and find a soft spot on which to lie down. Mr. Birch's remarks were greeted with hearty laughter and applause and cries of encore.

W. L. LOGGIE,
Member Executive Committee of the Maritime Board of Trade.

Prof. Andrews said that manual training helps to keep the young men here, but anyway, they should give them the best education possible.

President Fisher told of his recent trip to California, where he had visited several technical education colleges and had seen the men at work at forges and other forms of labor. He saw over 500 men at work in these institutions. Such education as this he thought would keep the young men at home, as they would be needed.

The motion was adopted as follows:

"Resolved, That for the sake of greater practical results, we urge the governments of the three Maritime Provinces to confer or appoint a commission to confer for the purpose of devising some plan for consolidating the work of technical education and industrial scientific research in Maritime Canada."

E. B. Elderkin of Amherst moved that a committee of two from each province be appointed to approach the governments of New Brunswick, Nova Scotia and Prince Edward Island, to see what could be done along the lines of the resolution. Carried.

THURSDAY MORNING.

Favor Extension of Government Control of Railways.

When the meeting was called to order on Thursday morning, E. H. Armstrong again brought forward his motion regarding the acquirement by the Government of the railways in Nova Scotia, and it was carried.

Dr. Pugsley supported Mr. Armstrong's resolution. Mr. Loggie, M.P., referred to the fact that the Dominion Government had done a wise thing to take over the Canada Eastern Railway. The great difficulty was that no percentage of dividend had ever been paid upon the millions invested in the I.C.R. Who was paying for this loss? The people, of course, were the ones who paid. He knew that at the present time the people of Canada were unitedly opposed to any extension of the I.C.R., but Dr. Pugsley's suggestion was a good one.

A. M. Bell, of Halifax, thought that the commercializing of the I.C.R. was a most desirable one. He suggested that a committee should be appointed to wait upon the various Provincial Governments with a view to fulfilling Dr. Pugsley's resolution. He read a short resolution to this effect, and it was passed.

Mails at Sydney.

H. F. McDougall, of Sydney, then brought up the question of the landing of mails at Sydney. He referred to the success of the various trials of landing the mails at Sydney, and the speedy distribution of these mails. Sydney, therefore, looked for the support unaminously of the three Maritime Provinces in reference to the use of Sydney as the mail port of call, during the open navigation season.

Running Rights for C. P. R.

He then presented the following resolution:

Whereas, Several experiments have been made in the transfer of the English mails to and from Canada at Sydney harbor;

And whereas, These experiments have demonstrated beyond doubt that by the Sydney route the English and European mails can be delivered to all points in Canada earlier than by any other route;

And whereas, It is the general belief that the Canadian Pacific Railway Company, as well as the Allan Company, are opposed to the landing and embarking of mails and passengers at Sydney, and prefer, during the navigation of the St. Lawrence, landing mails and passengers at Quebec;

And whereas, It is inconsistent with reason that Canada should continue to subsidize a mail and passenger service to run parallel with the Government system of railway over a distance of some 800 miles, and thereby delay for many hours the delivery of mails;

And whereas, The Maritime Provinces are deeply interested in the landing of

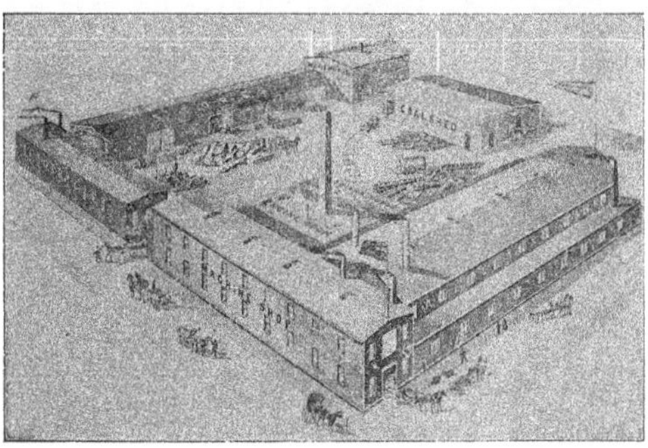

mails and passengers at Sydney during the season of St. Lawrence navigation and until such time as a more permanent and definite service is established;

And whereas, The C.P.R. system is more closely interested and connected with this traffic than any other in Canada, the Sydney Board of Trade, after due consideration, express the opinion that the privilege of running rights over the Intercolonial from St. John to Sydney, to the C.P.R., would greatly facilitate the adoption of the Sydney route.

Therefore, resolved, That this Maritime Board of Trade recommend the granting of such running rights for mails and passengers to the C.P.R. from St. John to Sydney;

Further resolved, That this Board reaffirm the resolution passed at the annual meeting last year at Amherst recommending the Government of Canada to insist on all mails to and from Canada by the St. Lawrence, to be transferred to Sydney.

The resolution was seconded by M. G. De Wolfe.

Mr. McDougall continued by stating that the resolution had no thought of interfering with any movement regarding the landing of mails at Halifax or St. John during winter. He spoke at considerable length upon the resolution, claiming that Sydney was the only point of call uniquely advantageous for distributing mails promptly and satisfactorily. Sydney would, no doubt, do their utmost to give certain concessions to the C.P.R.; for instance, running rights might be pointed out, and when any other trans-continental road could offer anything like a parallel claim, it should be considered.

Mr. De Wolfe asked if this were the first effort made by Sydney to get recognition.

Mr. McDougall replied that it was the first.

W. F. Hatheway, St. John, referred to a principle established by resolution a year previously, at Amherst, regarding Sydney. He was glad to know that the mail venture at Sydney had proven a success. He suggested that one or two words should be added, such as after "running rights," the words "for mails and passengers."

Mr. McDougall acknowledged that those words should be added.

Mr. Hatheway pointed out that the term "running rights" was vague and to the railway men it was peculiarly broad.

Mr. Loggie urged that the subject matter of the resolution be separated. All would concur regarding the advantage of Sydney as a port of call, but the granting of running rights to railways was another matter. He would oppose the portion of the resolution which would grant running rights to any outside railway.

Mr. McDougall was willing to eliminate the clause re running rights.

Mr. De Wolfe referred to the trouble they had last year over this same question. Halifax Board of Trade and St. John Board had an interest in these mail questions. There was a "nigger in the fence" somewhere, and Halifax and St. John men should come out and say what they thought.

Mr. Campbell said he agreed with Mr. McDougall re the concessions to the C.P.R. The C.P.R. had the fastest steamers on the route of Canadian call, and, no doubt, the hitch was in the possibility of the C.P.R. not getting running rights.

Treated Scandalously.

He thought the Maritime Provinces had been treated scandulously by the mail service passing their doors and then being sent back after twenty-four hours or more. In the matter of passengers, they should be given the privilege of landing at the nearest port. He would urge strongly the resolution being maintained as it stood.

H. J. Logan, M.P., agreed with the resolution in the main, and spoke strongly regarding the carrying of mails away up the St. Lawrence and then returning them to the provinces. It was folly to fear the running rights of any railway. The more trains run over our tracks meant more business. Every railway company should have running rights over the I.C.R. He endorsed the resolution.

The resolution was put and carried.

Mr. Duff-Miller was then called upon. He spoke of his pleasure at meeting the prominent business men of the provinces.

M. G. DE WOLFE, KENTVILLE.

Auditor Maritime Board of Trade, who has not missed a meeting in a dozen years.

The agents-general in all ports required more co-operation from districts, so that information supplied would not be lacking in detail. He illustrated it by a printed form for the labor demand, giving details exhaustively as to kind of work available, how many men were wanted, also the class of men, young or old, married or single. Then such information as farms being for sale, other business chances, etc.

The information regarding the Maritime Provinces was vague. He instanced it, as time after time, immigrants went west, instead of east, in Canada.

The Government should be appealed to to get a special grant for the immigration to the Maritime Provinces. He was anxious to have the literature of the province re immigration, more concentrated.

The Eleanor Home was mentioned in passing, and he believed several others similar in nature would be established at different ports of the provinces.

He mentioned the Anglo-French exhibition to be held in London, Eng., next year, where the provinces should be elaborately represented. The Duke of Argyll had suggested a panoramic view of Annapolis valley, the St. John river and others.

A. M. Bell then moved that a vote of thanks be accorded Mr. Duff-Miller. Mr. De Wolfe seconded it and spoke strongly in approval of Mr. Duff-Miller's work.

West Indian Trade.

The betterment of trade with the West Indies was then brought forward by the St. John Board of Trade.

H. J. Schofield, who had visited the Indies on a trip, accorded by Pickford & Black last winter, on behalf of the St. John board, spoke. He was anxious to find out what we were able to export ta the West Indies that we were not selling at present. He thought there was a large market for manufactured goods, but Canadians were so occupied with the work of filling orders at home, that he believed it would be years before any large attention could be given to the Indies. The Americans had caught the market by pleasing the customers.

The following resolution, moved by Mr. Schofield, and seconded by A. M. Bell, was adopted.:

Resolved, That the interchange of natural products between Canada and the British West Indies with British Guiana should be encouraged in every possible way by preferential tariff in both countries, by the establishment at the seaboard of jobbing houses to handle export and import traffic, and by a more frequent steamship connection as soon as possible.

It was also resolved that the Maritime Board of Trade approves of the exchange of visits between Boards of Trade in Canada and Chambers of Commerce and Agricultural Societies in the West Indies and British Guiana, and extends a hearty invitation to these bodies in the various communities of the West India islands and British Guiana to visit Canada at the earliest opportunity.

Mr. Fisher asked that a vote of thanks should be tendered the delegates of the St. John and Halifax boards, who had, at great expense of time, if nothing else, visited the Indies and had taken such excellent care of the subjects in hand. This was put in the form of a resolution and carried, as follows:

Resolved, That the very hearty thanks of the Maritime Board of Trade be tendered to the gentlemen who visited the islands of the British West Indies and British Guianas in the interest of reciprocal trade between Canada and the West Indies, and also to Messrs. Pickford and Black, whose enterprise and courtesy in suggesting the idea and providing free passages made the scheme possible.

The Three Short-Haul System.

Transportation facilities in Prince Edward Island then came up, being presented by H. A. Hughes, of Charlottetown. The following resolution was adopted:

Whereas, The trade of Prince Edward Island has been seriously handicapped for many years past owing to the difficulties of navigation and the excessive freight rates demanded by reason of three short-haul freight rates between points on Prince Edward Island and points on the mainland;

And whereas, The freight rates prevailing to and from Prince Edward Island are altogether disproportionate with those prevailing from Montreal eastward, as will be seen from the following :

From Montreal to St. John, 9c per 100 lbs., plus 1½c for elevation; distance, 735 miles.

From Montreal to Halifax, 9c per 100 lbs., plus 1½c ; distance, 832 miles.

From Montreal to Sydney, 16½c per 100 lbs.; distance, 977 miles.

Regular rates on grain in carloads from Tignish to St. John, 26c per 100 lbs.; distance, 428 miles.

From Tignish to Halifax, 23c per 100 lbs.; distance, 309 miles.

From Tignish to Sydney, 26c per 100 lbs.; distance, 380 miles.

And whereas, The Province of Prince Edward Island, being a part of the Domioion of Canada, has to contribute its full share of the cost of maintenance of all Government-owned and subsidized railways and steamers, should, therefore, be entitled to equal privileges, mile per mile, with the upper provinces, over all such railways and steamers, due allowance being made for extra handling of freight from cars to and from steamers ;

Resolved, That in the opinion of this board, the Government be and is hereby requested to so equalize the three short-hauls system of transportation of P. E. Island with the continuous haul of the other provinces, with which it is in competition ;

And further resolved, That Prince Edward Island be placed on the same basis as to import and export rates to and from ports outside of Canada as are accorded to the other provinces of the Dominion.

Inadequate Transportation.

The following resolution was also passed :

That whereas, Steamboat passenger rates to and from the Province of Prince Edward Island are excessive ;

Therefore, resolved, That this Maritime Board of Trade respectfully urge upon the Government the necessity of adjusting such steamship rates, whether on Government boats or those subsidized by the Government, so as to permit to the Island Province the minimum rates, mile per mile, charged by the Dominion Systems of Railways.

Mr. Hughes said that they were competing with Ontario and evidently transportation facilities were against them. Mr. Birch, of Alberton, also spoke. The Ontario man, he said, could market his stuff in the provinces at one-third less than the man in Prince Edward Island.

Mr. Logan spoke of the lack of cars on the I.C.R. at certain seasons, and referred to it as a vital issue. The trouble existed all over, but he offered the following resolution, which was seconded by Mr. Elderkin, and carried :

Whereas, The lack of sufficient rolling stock on the I.C.R. is the cause of great inconvenience and loss to patrons of that road, and seriously impairs its earning power ;

Therefore, resolved, That this board strongly urges the Railway Department to purchase as soon as possible sufficient additional rolling stock to meet the rapidly-increasing traffic along the line.

Mr. Logan then continued by referring to the Grand Trunk Pacific. He spoke of the unsuitable selection of the route of the I.C.R., so that the grades of the road in certain points were almost impracticable. He took up the various grades of the I.C.R. to show that the road was unfit to form a part of the Grand Trunk Pacific, and offered the following resolution, which was seconded by A. M. Bell, and adopted :

Whereas, The portions of the Intercolonial Railway, between Moncton and Halifax, and between Moncton and St. John, are under the N.T. Ry. contract to form part of a transcontinental railway system ;

And whereas, The curves and grades on these portions of the line, owing to the location of the road, in many cases, are severe and excessive ;

And whereas, The increasing business will cause a congestion of traffic on a single track ;

Therefore, resolved, That in the opinion of this board steps should be taken at once by changing route where necessary, by reducing curves and grades and by double-tracking to put these portions of the I.C.R. in a proper condition to become part of a great transcontinental railway system.

CHAS. M. CREED, HALIFAX.
Permanent Secretary-Treasurer of the Maritime Board of Trade.

A Tunnel for Prince Edward Islanders.

Rev. A. E. Burke then took up his time-honored proposal of the tunnel. He cited carefully the sufferings of the islanders who were sometimes shut up there without any communication and often at great peril, not only to the lives of the citizens, but especially to the trade of the province. He pointed out that there was a strong request from P.E.I. to have direct and continuous communication with the other provinces. He felt that the justice of the tunnel was growing every year.

Continuous communication with P.E. I. was a part and parcel of the compact of Confederation, and it certainly had not been fulfilled.

Rev. A. E. Burke quoted a poem, which he had printed on a picture post card, which, when held to the light, revealed the tunnel as it existed in his prophetic mind. This caused great amusement. Mr. Logan, M.P., followed in support of the distinguished reverend delegate's proposal. He believed that the Government should keep its promise to P. E. Island, and he believed they would keep it.

The following resolution was then adopted:

Resolved, that this Board does hereby reaffirm its resolution passed unanimously at Yarmouth, in 1905, and reaffirmed at Amherst in 1906, asking for the speedy construction of a tunnel between Prince Edward Island and the mainland.

Rebating on Scheduled Rates.

A. E. McMahon, of Kings County, spoke in regard to the rebating on freight rates in the matter of shipping apples and other farm produce from the Annapolis Valley. He said it was costing about 20 cents a barrel more than it should cost. This was caused by the commission merchants making outside arrangements with the steamship and railway companies. He presented the following resolution which was seconded by Mr. Peel, and adopted:

Whereas, certain steamship companies are receiving subsidies from the Dominion Government to aid them in conducting a legitimate carrying trade and whereas it is known that said steamship companies have made and are still making private arrangements with speculators, foreign apple receivers, or their agents at home or abroad, the same being detrimental to the interest of the growers, and legitimate dealers in the Maritime Provinces.

Therefore resolved that all steamship companies receiving subsidies from the Dominion Government shall be obliged by the Government to advertise rates for carriage of goods to foreign ports and that said rate shall be subject to and under control of the Railway Commission, and that any steamship company departing from said advertised rate shall be deprived of said subsidy by the Government.

Mr. Kinsman then brought forward a resolution in the interests of the growers in Kings County, which was worded as follows:

Whereas, arsenic of lead is taking the place of paris green in the extermination of pests, it is therefore resolved that the attention of the Government be called to this matter, and that the Government be requested to have the duty on arsenic of lead abolished.

This was seconded by M. G. De Wolfe, who mentioned one tree which yielded 27 barrels of Gravensteins in one year, and this was owned by Mr. Kinsman, to whom he offered his compliments.

Grants to Shipbuilders.

Shipbuilding and a bonus was then brought up by the Halifax Board. J. E. De Wolfe, of Halifax, presented the following resolution, which was seconded by Geo. Robertson, and adopted:

"Whereas, the substitution of steel for wood in the construction of ships has reduced Canada from her position as one of the four greatest shipbuilding and shipowning countries of the world to a place amongst the most unimportant;

"And, whereas, her coast line on two oceans of over two thousand (2,000) miles, with enormous lake and river navigation leading to the interior of the continent, has produced not only a great coastwise and foreign trade, but a large population skilled in maritime affairs;

"And, whereas, Canada possesses all the natural facilities and materials for the production of steel vessels, but is prevented from utilizing them by reason of the skill and capital employed in British yards, which have established England's supremacy in this industry, the products of which are admitted free to our coastwise as well as foreign trade;

"And, whereas, the above conditions prove that Canada can never regain her position as a maritime country unless the Government offers inducements by bonus or otherwise to shipbuilding companies to establish that industry;

"And, whereas, several municipalities and Boards of Trade in the Maritime Provinces have memorialized the Dominion Government to grant aid to the industry by means of bonus or otherwise;

"And, whereas, a similar memorial from parties interested in this industry in Canada has also been presented to the Government;

"Therefore, resolved, that the Maritime Boards of Trade hereby endorse and approve the principle of a bonus to steel shipbuilding as offering a new and profitable field for the utilization of our steel products, the employment of skilled labor, and, more than all, in the creation of a new mercantile marine, which was formerly a great source of wealth to our country;

"And, further resolved, that this Board urges the Government to pass necessary legislation at the next session of Parliament, granting a bonus to steel shipbuilding within the Dominion."

The Dog Nuisance.

The question of the sheep industry in certain sections being ruined by dog nuisances, was next taken up by D. C. Crosby, Berwick. He backed up his statement by quoting from Mr. Chipman, agricultural secretary of Nova Scotia. Large numbers of farmers had given up their occupation owing to this dog nuisance. He offered the following resolution, which was adopted:

That whereas our country is now overrun by thousands of dogs which are of no benefit to it, but are the cause of much damage in many lines, especially in that of sheep products;

"Therefore, resolved, that the Maritime Board of Trade recommend that the local Governments of the Maritime Provinces at the next session of Parliament, do enact such laws as may be effective in controlling the dog nuisance and ridding the country of this pest.

Prohibitive Postage Rates.

J. L. Stewart, of Chatham, brought in the following resolution, which was seconded by J. P. Maloney, and adopted:

Whereas, although the Postmaster-General reports a surplus of receipts over expenditures in his department, the postage on miscellaneous printed matter has been doubled, and,

Whereas, the postage on papers sent by publishers to subscribers in the United States has been increased from one-half a cent to four cents a pound, and,

Whereas, this oppressive and unreasonable tax has caused Canadian newspapers to lose the greater portion of their circulation in the United States, thereby severing the strongest tie between Canada and our expatriated countrymen; therefore be it

Resolved that, in the opinion of this Maritime Board of Trade, these increased postage rates should be reduced to the rates prevailing before the present increase was adopted.

The Steel and Coal Dispute.

H. F. McDougall then brought up the Iron and Steel Company trouble with the Coal Company, and offered the following resolution, which was adopted:

W. S. FISHER, ST. JOHN.
Retiring President Maritime Board of Trade.

Whereas, the Maritime Board of Trade assembled at St John, N.B., on this the 22nd day of August, 1907, feel a deep interest in the successful development of all industries that have for their object the upbuilding and general advancement of Canada, and more particularly that which concerns the Provinces of Nova Scotia, New Brunswick and Prince Edward Island.

And, whereas, this Board is pleased with the progress of development of our coal and iron industries, and the visible and undoubted prospects for their great future advancement;

And, whereas, this Board learns with deep regret that a dispute has arisen between the two largest corporations engaged in those industries in Canada, namely, the Dominion Coal Co., engaged in the development of our largest coal fields, and the Dominion Iron & Steel Co., engaged in the manufacture of iron and steel—both in Cape Breton—and that such dispute is in respect to the supplying of coal for use in the making of iron and steel, and that such dispute has assumed the form of expensive litigation in the courts, and that such dispute and litigation is now and has from its inception hampered and most seriously injured the trade and commerce, as well as the general financial standing of the communities immediately concerned in the ssuccessful development of these two very important industries;

And, whereas, the facts relating to this dispute are now before the public, taken in sworn evidence before a judge of the Supreme Court of Nova Scotia, at Sydney;

And, whereas, a continuation of this dispute before the courts threatens one or the other, or possibly both, of those corporations with serious loss and injury, and thereby further hamper and retard the advancement of the industries in question and the progress of the business interests of the country generally;

And, whereas, both these industries are in the enjoyment of important franchises from the public through the Parliament of Canada and the Legislature of Nova Scotia in the form of bounties, duties, mining rights and royalties, and are also enjoying municipal assistance in respect to taxation, etc.;

Be it therefore resolved, that this Board regard it the duty of the Federal Government representing Parliament and the people to call upon the two great corporations involved in this unfortunate dispute and insist upon an immediate friendly settlement, and in so doing have the co-operation and assistance of the Provincial Government of Nove Scotia.

And further resolved that in the event of such immediate settlement being delayed for any unreasonable time the disputing corporations be advised that Parliament and the Legislature of Nova Scotia may be called on to consider the advisability of withdrawing in whole or in part the valuable concessions granted to those corporations for the proper development of those important industries.

The Nominations.

The report of the nominating committee was recited and A. M. Bell, the new president, was asked to take the chair, which he did, amid much applause. The committee reported as follows:

For President, A. M. Bell, Halifax; First Vice-President, Hon. Geo. E. Hughes, Charlottetown; Second Vice-President, W. B. Snowball, Chatham; Secretary, E. A. Saunders, Halifax; Permanent Secretary-Treasurer, Chas. M. Creed, Halifax; Auditor, M. G. De Wolfe, Kentville. Committee—A. E. Burke (chairman), W. L. Loggie, E. H. Armstrong, J. E. De Wolfe.

It was recommended that $25 be added to the Permanent Secretary's stipend for present year.

Rev. Father Burke paid a graceful tribute to W. S. Fisher, the retiring President, which was cordially received by a vote of thanks. The whole body rose and sang "He's a Jolly Good Fellow." Three cheers were given for Mr. Fisher, who responded suitably.

M. G. De Wolfe moved a vote of thanks to the Press. and especially to the MacLean Publishing Co., which was carried.

It was decided to meet in Halifax next year.

A special vote of thanks was tendered the St. John Board of Trade for their cordial entertainment of delegates.

The following financial statement was submitted by Mr. Creed and adopted:

St. John, N.B., Aug. 22, 1907.

The Moncton Board of Trade, to Charles M. Creed, Secretary-Treasurer:

Dr.

General expenses		$137.72
Secretary's salary		150.00
		$287.72

Cr.

By balance on hand ....$ 71.48		
Salaries paid	265.36	
do., unpaid	57.40	$394.24
Balance		$106.52

Examined and found correct.
M. G. DE WOLFE,
Auditor.

A Pleasant Excursion.

The convention concluded by a well arranged excursion up the St. John River to Evandale, about thirty miles from the city.

The visiting delegates, and their lady friends were the guests of the St. John Board of Trade. The weather was perfect, the sail and scenery most enjoyable, and the whole event unique from every standpoint. About three hundred guests attended, and dinner was served to part of these on the steamer, and another portion at John O. Vanwart's dining-room, at Evandale. The return trip was made by moonlight, and the scene will not be forgotten easily.

Music was furnished by a first-class orchestra, and songs and speeches filled in the programme. On the way down the majority of the passengers assembled in the saloon, where the witty island clergyman, Rev. A. E. Burke, eulogized the retiring president of the Board, W. S. Fisher. In closing he conferred upon Mr. Fisher "The Order of the Sun Flower," by pinning a handsome life-sized speciman of the flower on his breast. Cheer after cheer, greeted Mr. Fisher, as he rose to respond, "For He's a Jolly Good Fellow" was sung heartily, and Mr. Fisher responded suitably. Among the others who spoke were Mr. Logan, M.P., Hon. Wm. Pugsley. Mayor Sears, of St. John; Geo. Robertson. M.P.P., J. E. Birch, of Prince Edward Island. A. M. Bell, the new President of the Maritime Boards of Trade, was also called upon, but it was said that he was either too modest, or too busy entertaining the ladies. "We do not see him neither do we hear his tongue," said Rev. Mr. Burke, amid laughter.

Rev. Burke also said that he had been asked by the Prince Edward Island and the Nova Scotia Boards to express their hearty thanks for "the magnificent hospitality" of the St. John Board of Trade. J. E. De Wolfe at this point extended a cordial invitation to the Maritime Boards to attend the meeting at Halifax next year, when an effort would be made to reciprocate to some extent the kindness of the St. John Board.

"Auld Lang Syne" and the National Anthem brought to a close an excursion that will leave behind memories as unfading as the sky. The committee in charge of the arrangements were T. H. Estabrooks, H. B. Schofield and Ernest Barbour, who were ably assisted by the other members of the Board.

On the excursion a special committee meeting was held, in connection with the resolution passed at the convention, referring to the purchase of railway branches by the Intercolonial Railway. The following committee was selected to interview the Governments of their respective Provinces. on this subject: New Brunswick—T. H. Estabrooks, J. H. McRobbie, St. John; W. B. Snowball, Chatham; John Morrisey, M.P.P., Newcastle; J. T. Hawke, Moncton. Nova Scotia—A. M. Bell, Geo. E. Faulkner, Halifax; F. P. Whitman, Annapolis; P. C. Moore, Sidney; E. H. Armstrong, M.P.P., Yarmouth.

Notes.

Never were the funds of the convention so satisfactory. Charlie Creed won great favor by his excellent financial report. The old secretary seems to get younger and more popular every year.

A number of the old guard were noticeably absent. Although there were four past presidents at the convention, there were such time honored faces missing as those of the Hon. George J. Troop, really the founder of the Maritime Board; also W. M. Jarvis, one of the first presidents.

It was stated that when Rev. Mr. Burke gets that famous tunnel constructed the first passenger train will carry only members of the Maritime Board. "More power to the reverend gentleman" was the general wish.

M. G. De Wolfe. the much liked delegate from Kentville, was as usual busy making others happy. He is looking younger every year, and no man does more to keep up the general enthusiasm than this same gentleman. During the life of the Board, Mr. De Wolfe has the honor of not having missed a meeting either of the Board or of the Council.

The Kentville Board sent six delegates, which is a pretty nice showing from a farming community. Each delegate from this section pays his own expenses, which is unique in itself, and thus tends towards the best of representation.

Mr. Anderson, Secretary of the St. John Board of Trade, did spdendid work as an entertainer, and is an ideal official.

The weather for this convention was of the usual brand, "fine," and it is a notable fact that the Board have never had any bad weather during their thirteen years.

"Thirteen" is said to be an unlucky number, but this idea was thoroughly dissipated into thin air by the finest convention in the Board's history. W. S. Fisher will be long remembered for his excellent work. He is a master of the craft of diplomacy.

The influence of the Board is growing every year as was evidenced by the presence of the many Federal and Provincial members of Parliament.

The excellent management of the President and delegates enabled the convention to get through in two days what ordinarily took three and gave practically a whole day for pleasure and sight-seeing.

The Halifax men will have their work cut out for them to equal the St. John convention. Halifax sent a fine delegation and the probabilities are that the Haligonians will acquit themselves nobly.

H. J. Logan, the sturdy M.P. for Cumberland made things lively, as did also Mr. Loggie, the well-balanced member from Chatham.

The St. John ladies took a great interest in the entertainment part and the greatest praise is due them. Mayor Sears is a cordial official, and his well-poised speech on the steamer impressed the visitors.

QUEBEC PERSONALS.

L. C. Rugen, general manager of the Standard Paint Co., Montreal, has been spending a few days at his home in New York city.

J. Berube, of the sales department of Lewis Bros., Montreal, has severed his connection with this firm to go into business for himself in Montreal.

W. H. Evans, of the Canada Paint Co., Montreal, is due to arrive with his wife from England next week. They are passengers on S.S. Southwark.

E. Dowsley, manager of A. Ramsay & Son, paint manufacturers, Montreal, has returned from a two weeks' vacation, spent in Ontario and the Muskoka lakes.

Mr. McNamara, of McNamara & Jones, Bedford; C. B. Vanantwerp, Frelighsburg; Mr. James, of James & Reid, Perth; Mr. Lapointe, of Lapointe & Lapointe, Quebec city; J. Loughrin, Mattawa; Mr. Corriveau, of Corriveau & Frere, St. Sebastien, were callers in Montreal last week.

Geo. Brown, of Caverhill Learmont & Co., Montreal, has just returned, after spending a few holidays in the United States.

BEAUTIES OF THE MARITIME PROVINCES

A natural scenic panorama whose beauty is always fresh, and whose attraction never tires.

The charm of the cool summer climate, the magnificent scenery, the endless variety of tourist routes, the unequalled sporting wealth of river and forest, the many places of historic interest are yearly attracting to Eastern Canada an ever-increasing number of summer tourists, and of these, a goodly portion wend their way to Quebec and the Maritime Provinces. The natural Canadian gateways for travel to this portion of the Dominion are Montreal, Quebec, St. John, Halifax and the Sydneys, but to single out one beauty spot for special mention above its many hundred competitors would be a thankless, not to say hopeless, task.

Perhaps with more justice than in the case of any other of Canada's Provinces may the term picturesque be applied to Nova Scotia. It is a land where nature seems to have handed out in most abundance her varied assortment of physical pecularities and glories. Nova Scotia, with all but a few miles of her borders sea-washed, is a land of hill and vale, of lake and river, of waterfall and sunshine; a country of changes, of surprises, of delights. Longfellow has immortalized the glories of old Acadia, the Minas Basin section; but only the Nova Scotia lover who has seen this land from end to end, and revelled in its glories, has any adequate idea of its picturesque beauties.

For the sportsman, for the fisherman, especially, Nova Scotia is indeed the Arcadian land. Visiting anglers are welcomed by the people, and strangers who have enjoyed an outing in this charming country are emphatic in its praise. Some of the best fishing in the Province is within easy reach of Halifax. Tuna fishing, which has made Santa Catalina, California, so famous, may be indulged in at Mira Bay and other places in Cape Breton. July and August are said to be the best months, and a well known sportsman tells of a tuna that he played for over seven hours last season in Mira Bay. A delightful trip through the famous Bras d'Or Lakes may be combined with the fishing by strangers who have the time.

The section from Truro to Halifax includes a magnificent farming district, well watered, and abundantly fertile, a very garden of productivity. West of this lies, too, the famous Annapolis Valley, the land of Evangeline.

Grand Falls of the Nipisiguit.

The beauties of Halifax—the Arm, the Basin, the magnificent harbor, with its islands and its fleets, the distant hills and forests, and the city itself, with the old historic citadel and suggestive breastworks overlooking the busy streets and delightful public gardens—cannot be forgotten by anybody who has once seen them.

God has been good to Nova Scotia. Though it has seen changes of Governments and peoples which have involved untold suffering and which have made history, it is a land of peace and plenty, a land whose natural features have conduced to the rearing of a sturdy race who are proud of their heritage.

New Brunswick.

St. John, the capital of New Brunswick, has a history which extends back to the days when the land was Acadia and the banner of France waved from the forts of the harbor and river. But the founding of the city dates from the landing of the loyalists in 1783. The latter, and those who took their places, labored faithfully and well to build a city, and thus they continued to laboi for nearly a century, when the fire of 1877 came and the greater portion of the city was swept out of existence in a few hours. The destruction was swift and complete. With a surprising energy, however, considering the far-reaching effects of the calamity, the people began their work anew, and the city of to-day is far more substantial and beautiful than the city of former years.

A sail to Fredericton by the St. John River is indeed one continuous panorama of beautiful scenery. Leaving the city, the lower portion reveals some bold scenery with high hills on either side of the noble river. At Grand Bay the reaches widen and on the right is seen Kennebecasis Bay, where the river of that name enters. This bay is a beautiful stretch of water on which a yacht may sail for twenty miles without starting a sheet. Indeed, the whole river between St. John and Fredericton will delight the heart of a yachtsman. Gagetown, Sheffield, Maugerville and Oromocto are among the many attractive places along the river and they are in a glorious farming country. There are commodious summer hotels at the Cedars and Evandale.

The situation of Fredericton, at the head of river navigation for the larger steamers, is most picturesque. Water sports are a feature of the summer pleasures of the Fredericton people, and yachting and canoeing are freely indulged in.

The Garden of the Gulf.

The Garden of the Gulf, as Prince Edward Island has been so appropriately termed, is reached in summer, either from Point du Chene, N.B., or Pictou, N.S., on the fast steamers of the Charlottetown Steam Navigation Company. Going by the first-named route the landing is made at Summerside, and Charlottetown is reached by a journey of forty-nine miles on the Prince Edward Island Railway, which stretches from Souris in the east to Tignish in the west. Leaving Pictou, the trip is direct to Charlottetown. There is a daily service on each route.

The run across the Strait of Northumberland on a fine day is a most enjoyable trip. There are times when the water is as calm as that of a placid lake. As the island shore is approached the red of the earth and the bright green of the verdure show a most picturesque effect as a background to the smooth

stretch of water, in which is mirrored the glory of the sunlight from the western sky. Under such conditions the first impressions of Prince Edward Island must always be such as will long be remembered.

Charlottetown, the capital, and the commercial centre of the island, has a population of over 12,000. It has a fine harbor, and when seen from the water the city makes an especially fine appearance. Of all the magnificent scenery to be seen in the Maritime Provinces, Prince Edward Island, perhaps, contains more than her fair share. Visitors who have spent a summer or part of a summer, or even a month or a week, in the enjoyment of the remarkable beauty with which nature has endowed the island, return another year with a feeling of intimate affection and admiration as sincere as it is deserved.

HALIFAX

A city of expanding commercial importance.
One of the finest harbors of the world and the Dominion's Atlantic gateway.
Some of the more important business houses.

For many years Halifax has been known the world over as one of the important British military and naval stations. This knowledge has largely passed into history, and to-day, Halifax is garrisoned by Canadian troops and is merging into a city of commercial rather than of military importance. It is admirably located for commerce, being founded on a rock and situated on a peninsula. Its time-honored title, "the gateway to the Atlantic," is well-chosen. The Atlantic Ocean appears on the east and west of the city like a tranquil lake, and there forms a harbor unsurpassed in the world. This harbor is open for navigation the year round, and there is friendly rivalry between Halifax and St. John as to which shall be the winter port of the Dominion. At the Maritime Board of Trade Convention this was a live subject, and there is a feeling among the Halifax delegates that their city will yet win out with the government with regard to the mail contract. Halifax is the capital of the province and is provided with excellent railway facilities, including that of the Intercolonial, the Dominion Atlantic and the Canadian Pacific. No guide book of any of these transportation companies can be read without seeing a glowing appreciation of Halifax harbor. This harbor is protected by eleven forts and batteries. At the top of the hill at the base of which the city is built is the citadel. This is 250 feet above the level of the harbor. On it are fortifications begun by the Duke of Kent 100 years ago, and notwithstanding the various improvements made in order to keep pace with the advances in the science of warfare, most of them are now regarded as obsolete. Modern systems of fortification, however, are in vogue all about the city, and a large armament of quick-firing and disappearing guns form part of the equipment.

Near Antigonishe, on the I.C.R.

In Rockwood Park, St. John.

Nearly every business man in Halifax is wealthy, and his business appears to be more of a pleasure than an aggressive campaign for money. Large manufacturing industries are carried on, including iron castings, machinery, nails, paint, gunpowder, sugar, leather, cordage, boots and shoes, soap and candles, also, woodenware and cotton and woolen goods. Local distilleries and breweries are also in evidence. It was only during the second week in August that Halifax celebrated its founding by Lord Cornwallis in 1749. To be accurate, the Hon. Edward Cornwallis, the Earl of Halifax, was the founder of the name "Halifax." In 1750 it was chosen the capital of the province. It was in 1817 declared a free port and in 1842 incorporated as a city.

There has never been a time when trade in Halifax was so active, and the merchants have awakened to the greater possibilities of the city and its surroundings; in fact, a well-known member of the Board of Trade stated the other day, that the Maritime Provinces should be the New England of Canada, and with the great west opening up as a market for manufactured goods there is no reason why this prophecy should not be fulfilled. Firms like the Dominion Molasses Co., William Stairs, Son & Morrow, The Starr Manufacturing Co., of Dartmouth, and many others, do business all over Canada. Large business is done with the West Indies, Great Britain, the leading American cities and other parts of the world. The exports are chiefly lumber, fish and agricultural products, and the imports include sugar, rum, molasses and other sub-tropical products. The best evidence of the growth of the city is the steady increase of exports and imports.

During holiday time, such as the carnival week referred to, Halifax business men vie with each other to make social life the dominant note, and the cordial entertainment accorded all visitors would be hard to surpass. Everywhere one sees the avoidance of extravagance in action and word, as well as the harmony of behaviour which is so noticeable in older countries. The celebration of the landing of Cornwallis and his caravel at the North West Arm was carried out in a thrilling and creditable manner. It is the home of sound amateur sport, and the regattas of Halifax are notable for honest competition and spirited rivalry, which are a feature of every event. The streets are well paved; many new buildings are in course of construction, and Halifax seems to have taken on new life and to be sharing the general prosperity of the Dominion.

William Stairs, Son & Morrow.

Sterling integrity marks this fine old hardware house, and for long over a century the name Stairs & Morrow have been identified with the growth of Halifax. Off Chebucto Bay, about a hundred years ago, stood the premises of Kitson, a Glasgow iron merchant. In 1854, William Stairs bought this business, and shortly afterwards Mr. Morrow joined the firm. From small beginnings the business has grown to one of the most influential and important shelf and heavy hardware stores in the

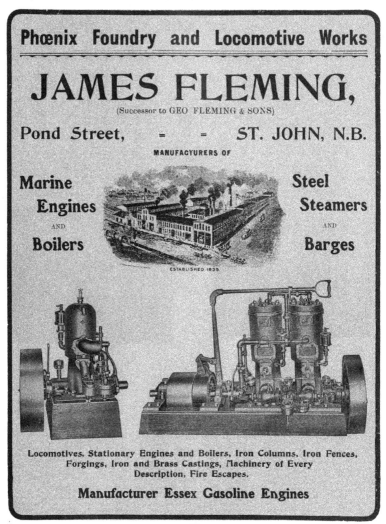

Dominion. It is said they are the largest and oldest hardware merchants in the Maritime Provinces. A stock of rare diversity is carried, and the atmosphere pervading the establishment, both in character of goods and method of doing business, is of the highest order. Everything is conducted upon a distinct Old Country tone, and Stairs, Son & Morrow are in every way a credit to their city.

The Halifax Hotel.

No one could visit the Maritime Provinces without remembering well the "Halifax Hotel." It is, in fact, the most popular house east of the Chateau Frontenac. By "popular," is meant that it is frequently the home of Canada's leading politicians and most prominent citizens, while celebrities from all parts of the world stop here from time to time. Since 1861, it has been the leading hotel of the Maritime Provinces, and especially of Nova Scotia. There is an air of homelike comfort about the house from basement to attic and aside from the excellent furnishings and accessories of beauty, a great deal of the attraction is due to good management, and for this the name of E. L. MacDonald, who is in charge, must be mentioned. Every guest at the "Halifax" may be assured of all the attention due him, because the organization in every department is complete. His idea is to give his customer something for his money, such as supplying clean, airy sleeping rooms, the best of cuisine, cozy parlors and reading rooms for ladies and gentlemen, the best of baggage and train attendants, so that in all no combination of hotel accommodation could be better than that supplied by the "Halifax." This means that the house is always busy, yet never is there a time when so many guests are taken that any one of them can complain of neglect. What more, then, could a traveler wish for? The house is enviably located, overlooking the harbor in the rear, and with a frontage on one of the most attractive streets, within easy reach of all the chief attractions of the city.

ST. JOHN

A city that is coming to its own after many vicissitudes.

Some of the leading business houses.

Earl Grey, in a recent speech at Halifax, said: "That he believed that the Maritime Provinces were the front door of America, and the natural entry for all postal matter from Europe to this continent. He hoped to see all mail subsidies given by the British Government paid to vessels heading for Canadian ports. To pay subsidies on steamers plying to New York was a colossal blunder. The laws of nature were on the side of those who recommended that the trans-Atlantic mails from Great Britain should be via Canada and not New York." In the money poured out so freely for harbor improvements, the deepening of the channel, the supplying of magnificent new wharfage and increased facilities for the handling of the passenger, and freight traffic that must naturally flow into St. John, its citizens have been anticipating the march of Nature.

Fate and Fortune are not over gentle in their dealings with some cities, as with some men. Both the cities and the men seem compelled to climb the most rugged paths, constantly being retarded in the upward climb by the very hand outstretched to help some favored rival. So when by dogged determination, persistent striving, and unlimited enthusiasm, any city has overcome its obstacles, and has reached that position of commercial power and prosperity towards which it so patiently moved, we feel like saying, "Hats off, gentlemen."

The history of St. John has been one of continued loyalty and determined striving. From the days of LaTour, as he fought his life struggle with the wily Charnisey, simply for the mere love and loyalty he felt towards his dear St. John, and that he might have the privilege of being buried in what was to him the dearest spot on earth, until the present day, the story is the same. Patient plodding has been finally successful, and St. John is coming into its own. All honor to its brave, determined and loyal citizens.

St. John means to have what it ought to have, and its citizens are going to get what they have worked for. Of the city itself, volumes could be written; its natural beauty, its business energy; its splendid strategic position in relation to the lines of travel, both by land and water; its grand harbor; its delightful climate; all combining to give the impression, indelible as true, that St. John is destined to become the Liverpool of America. And why not?

The natural beauty in and around St. John is unsurpassed anywhere in Canada. Its location at the mouth of the St. John river, at the head of a large and growing inland traffic, gives it a position, commercially speaking, of great advantage. Its winter port business this year bids fair to far surpass all previous records. And in every industry in the city the cry is "More capital" to keep pace with the growth of the trade in all lines. Direct connection with every Atlantic trade centre; raw material right at her very doors, such as iron, coal, lime and lumber; splendid facilities from a distributive standpoint; with business men thoroughly alive to the interests of the city, and yet liberal in the extreme towards their fellows; an intelligent and industrious population; St. John to-day calls a halt! not to the natural western march, but certainly to the unnatural western stampede. Look before passing, then possibly there may be no passing.

Halifax from the Citadel.

Fort Mulgrave.

W. H. Thorne & Co.

Foremost among St. John's leading houses is that of W. H. Thorne & Co, wholesale and retail hardware merchants, whose handsome premises on Market

Square are, of necessity, among the memories retained in the minds of all visitors to St. John. The business was established in 1867 by the present president of the company, W. H. Thorne, and later, in 1895, turned into a joint stock company. Their large and growing trade is distributed over the whole of eastern Canada. Some sixty-eight thousand four hundred feet of floor space, every foot of which is found absolutely necessary to accommodate the tremendous stock that this concern carries, would give some idea of the extent of the trade done annually by the company.

Emerson & Fisher.

The sky scraper of St. John is the magnificent store and warehouse of Emerson & Fisher, wholesale and retail hardware merchats. Founded in 1878, the progress has been rapid and continuous, until to-day the excellent standing of this joint stock company, composed of R. B. Emerson, W. S. Fisher, Fred. R. Murray and Stanley L. Emerson, places it in the front rank of wholesale firms in Canada. Politeness and geniality is in evidence in every department of the concern, from the heads of the firm down.

The company is interested in the Enterprise foundry at Sackville, the product from which foundry has made a great name for itself in the east. Enterprise "Monarch" stoves having proved to be good sellers. A complete line of all hardware goods in the wholesale and retail line is, of course, carried in stock, the system and order prevailing in handling the trade being worthy of special mention.

Josiah Fowler & Co.

A high standard of excellence is what has built up the trade of this firm, a trade extending over the whole Dominion from Vancouver to Halifax. An experience of over forty years goes into every axe turned out from their plant at St. John, and in these days that fact alone is a recommendation of vast weight. Personal touch with both ends of the trade is the secret of the success that has attended the progress of this firm. That the factory cannot keep up to the demand, is the only thing that is troubling Mr. Fowler to-day, although the plant is going at its full capacity. This is the case not only with their celebrated axes, but also in the other branches of their trade, such as railway springs, chisels and heavy hammers.

James Pender & Co.

Established in 1892, taking over the plant of the James Pender Company, a business of twenty years' standing, Jas. Pender & Company, St. John, are in a position to-day second to none in the country, in the matter of equipment. A unique feature of the plant is the wire cleaning apparatus, this being the only one of its kind in Canada. It has an automatic mechanism for controlling an hydraulic crane and operating a washing device as complete as practical experience can make it.

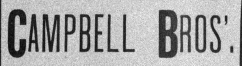

The company have a wire straightening machine, which, while fulfilling the original purpose for which such a machine is used, has also incorporated in its make-up a stretching device, this feature increasing the value of the product enormously, and, together with its large productive power, makes the machine the best for this purpose in operation in Canada to-day. In wire nail machinery this firm have a larger variety of makes than any other domestic manufacturer. James Pender & Co. were the first manufacturers of the "Special Bull Dog" brand of nail, used in the making of boxes, crates, etc., and are specialists in the manufacture of coated nails, which have an adhesive power over one hundred per cent. greater than the bright wire nails. The firm have recently gone into the production of the Acme galvanized wire nail, which, on account of its brightness and heavy coating, commends itself to the trade.

In the manufacturing of horse-shoe nails, the company have lately installed and perfected a new process, which, on account of its automatic action, uniform heating, and gradual reduction, makes possible the production of two and one-half times the quantity, makes a very much stronger article, and insures a uniformity, impossible under the old conditions. Their plant is also the home of the standard pattern toe calks, which brand, on account of its merit, being sold through the ordinary channels of trade, has secured more than half the business of the United States in this ably covered by Mr. Smith, who had been with the company for some time prior to the change of hands.

The Maritime Nail Co.

This company has as perfect a plant of its kind as can be found the world over, it being a veritable "beehive of industry." Every day is a record-breaker, and every man in the concern from Manager Elkins, down to the last laborer in the mill, keeps at it from morning to night.

In some concerns, the management never thinks of mixing with the inside workings of the plant. Here, however, every department is directly under the practiced eye of the indefatigable manager, who knows the workings of the concern from floor to roof. Modern methods, the latest ideas of value, indeed, any new good thing, is at once adopted to be put into practice at the earliest possible opportunity.

Experts operate the very best of machinery, and use only the very finest metal in the world. that imported from Sweden, so that not a single shipment can be below a certain standard of excellence. Nothing but the absolutely perfect production is allowed to go through the sorters' hands, to be passed out to the trade. The Maritime Nail Co. is in for quality, first, last and always.

The capacity of the plant has been lately doubled, and yet they are rushed to the very limit, filling the orders. The horse nail business is the large feature of this company's business, but they are also manufacturers of the Monarch brand of wire nails, tacks and rivets, well known throughout the country. The orders for these last would keep a large sized plant at full working tension the year through. Any communication with the firm has the direct attention of the management, and receives the most careful consideration.

Union Foundry and Machine Works.

In 1902, Geo. H. Warring, William Brackhoff, James Manchester, and A. H. Hannington, all well-known business men of St. John, took over the old-established works of W. H. Allan, general foundrymen; incorporating themselves under the firm name of the Union Foundry and Machine Works, Limited, for the purpose of carrying on a machine repair, and general foundry business.

A very extensive trade is carried on with the various steamship companies that make St. John their headquarters in the winter, and with the natural growth in this line, a growth annually assuming tremendous increased proportions, together with the extensive general foundry business carried on; this firm has a future ahead of it that will loom up large in the annals of St. John's business development, in this most important branch of its trade.

The Union Foundry employs some fifty hands, and has a splendidly situated and equipped plant.

Halifax and Harbor From the Citidel.

line, in less than thirteen years, in competition with plants of twenty-five years' standing.

The company has splendid facilities in the transportation line right at the factory door, with ample provision made for future enlargement. Their commodious storehouse is carefully systematized, and holds a large stock of carefully graded nails. The company generate their own light and has unlimited power from three immense engines.

Robertson, Foster & Smith.

Formerly known as Kerr & Robertson, this company has one of the finest warehouses in the city, situated in the direct route of all traffic in St. John. They have splendid facilities for handling the large trade that comes their way. The hardware trade of the Maritime Provinces is large and increasing every year, with Robertson, Foster & Smith getting their fair share of the trade.

The present management of the concern rests in the competent hands of F. A. Foster, while the outside field is

T. McAvity & Sons.

Possibly one of the oldest houses in the hardware line in the Dominion, is that of T. McAvity & Sons, St. John, brass and iron founders, and wholesale and retail hardware merchants. For some hundreds of feet along the wharf the incoming voyager sees nothing else but the sign of T. McAvity & Sons. Then as he journeys up-town to his hotel, on the principal street of the city, again he strikes it, in a well-appointed and commodious retail store, and when in a drive round the city later, he finds that the large plant in the south end of the town, the Vulcan Iron Works, also belongs to this firm, he is inclined to ask in all good faith, "Does this firm own the whole town."

With its large warehouses on Water street, backing right on to the wharf, its extensive retail and wholesale hardware trade on King street, and the splendidly equipped plant, of the Vulcan Iron Works on Broad street, this firm occupies a position in St. John's business life that is unique, and is, perhaps, the most widely known local concern in Dominion trade circles.

Employing hundreds of hands, the year through, and giving to the trade an article that can be depended on at all times, thus giving it a reputation second to none in the country, St. John ought to feel proud of the family whose energy and brains have made possible an industry that would be a credit to any city.

The goods turned out by this firm are to-day in use in all the biggest plants in Canada, in street railways, in the large steel plants, in machine works, in fact, everywhere the Dominion through.

P. Campbell & Co.

Messrs. P. Campbell & Co., manufacturers of "Scientific" acetylene gas generators, and general lighting and heating engineers, contractors, etc., are a St. John firm of many years' standing, and well known throughout the Maritime Provinces and eastern Canada. Their "Scientific" generator is found in many of the largest hotels, finest private residences, and best equipped plants in the country. Accepted by the boards of underwriters in all the provinces, it is yearly proving its claim of being the cheapest and most practical illuminant on the market.

In their general plumbing and heating business this firm annually do a very extensive trade, and they believe that carefully selected material and the very best of workmanship, have enabled them to achieve the standing that they have to-day in the trade.

Phoenix Foundry & Locomotive Works.

Established in 1835, by the grandfather of the present proprietors, the Phoenix Foundry, St. John, has continued successfully through all the intervening years, and has to-day a well-equipped plant under a progressive and aggressive management.

Last year the Phoenix people started the manufacturing of the Essex Gasoline Engine.

General foundry work, of course, is the main production of the Phoenix Foundry, and years of satisfactory handling of the large business in this line that has come their way, have given this firm a reputation second to none in the country.

Guy H. Humphrey.

Guy H. Humphrey has been in the coffee business but a little over a year, and has a trade that has already assumed large proportions, and one that is rapidly growing. Guy H. Humphrey, while one of the youngest of St. John's merchants, has the necessary qualifications; determination, business training, and ambition, to assure success.

After some five years' training, right in the centre of this trade, Mr. Humphrey felt that he had the necessary practical knowledge that would enable him to start in business for himself. As a coffee importer and roaster, Mr. Humphrey is a success, which fact is proven by his continued progress, and the general favor in which his brands are held. Starting in the coffee business right in what might be called the Dominion tea centre, required pluck, but a fight won where odds are heavily against, is something to be proud of.

New Brunswick Wire Fence.

No industry in the East has taken on more special impetus than that of the New Brunswick Wire Fence Co., Moncton, N.B. Since its foundation in 1899, the business has steadily grown and the management have kept pace with the times. During the past year the well equipped factory was rejuvenated by elaborate new machinery for modern work. This has placed the establishment on a par with the largest in the country as far as the facility for turning out high grade fencing is concerned. The system of erecting fences by contract at so much per rod has proven very successful and has been largely carried on in the Provinces during the year. It is a unique convenience for the farmer and the country dweller to be able to leave his order for so much fencing, the company doing the rest, even to the unpleasant duty of digging post holes, etc.

All the much desired elements of staying power and capacity for withstanding the weather are embodied in the fences made by this company. The corrosive effect of the atmosphere in the seaside provinces is completely offset by the company's system of galvanizing. Each fence is done according to its needs; namely, that special galvanizing is necessary for certain fences to be used in certain parts of the country. High-priced carbon wires are even supplied to meet special cases. Their ornamental fences are the pride of the smiling gardens and homes of the Maritime Provinces for the notable beauty they add to houses and all premises is proverbial. In these no wood at all is used; they are made from steel posts, steel wire and steel clamps. The posts are set in stone and hence its durability may readily be appreciated. C. A. Murray is the president of this company and A. C. Chapman the secretary and manager. To Mr. Chapman great credit is due for promoting the industry to its present high state of excellence.

George E. Smith & Co.

Reference to Halifax business men would be incomplete without mentioning the firm of George E. Smith & Co. Its history dates as far back as 1875, when it was carried on under the name of Irish & Smith. Shortly afterwards Mr. Irish retired and the present firm name was taken, comprising as partners, Messrs. George G. and A. W. Smith. A. W. Smith is the man who deals with the public, and he is certainly a bright and capable representative. Their premises are admirably located, fronting on Cheapside and Bedford Row, and are considered one of the most valuable sites in Halifax. The firm have many fine agencies and do a leading business in the city.

A. M. Bell & Co.

Thirty-three years ago, in 1874, Andrew M. Bell started in the hardware business in Halifax, and in 1896, Arthur B. Wiswell was taken into partnership, having entered the employ of Mr. Bell in 1879. By their enterprise and energy these gentlemen have built up one of the most successful and largest business houses in the Maritime Provinces to-day,

In 1903, the steady growth of the business made new premises imperative and one of the best business stands in Halifax was purchased, and the present fine concrete building was erected and occupied in April, 1904. The floor space is about 30,000 square feet and there are 7,000 square feet of glass in this up-to-date building.

The firm's policy throughout has been to handle "good goods," looking for value rather than cheapness. Their growing business has proven the wisdom of this course, so that the saying, "If you want it good get it from Bell's, has become a truism.

Amherst Foundry Co.

Since last year's convention of the Maritime Board of Trade, the Amherst Foundry Co. suffered a severe set-back by the destruction by fire of a large portion of their plant. Characteristic of the enterprise of the management, the plant has been rebuilt and will be in running order inside of a few days. The foundry was established in 1904, when they put in operation the second plant in the Dominion for the manufacture of porcelain, cast iron enamel baths, lavatories, sinks and kindred goods for the plumbing trade. The business has grown steadily, and the company to-day ranks among the leaders for the manufacture of these lines in the Dominion. They have a large and growing trade in the great west, and their facilities for accommodating the trade have grown accordingly. The destruction of their premises, which took place on the first of May has served the purpose of causing the new building to be larger and better equipped than ever had been thought of previously.

In 1890, J. A. Crossman and J. A. Laws, started in a small way to manufacture stoves, hot air furnaces and sinks. William Knight and J. A. Black were at the same time carrying on a retail business in stoves, furnaces and tinware. On January 1, 1895, the four gentlemen, together with C. A. Lusby, formed a partnership under the firm name of the Amherst Foundry & Heating Co., for the purpose of continuing the business of Crossman & Laws and Knight & Black. This business rapidly grew into prominence and its influence spread all over the provinces. Handsome new premises were erected of brick and stone, and the plant to-day is unsurpassed in Canada.

About five years ago, a new company was formed known as the "Amherst Foundry Co., Limited," leaving out the "heating" term. They now manufacture an extensive line of stoves, furnaces, baths, sinks, etc. Durability, sanitary, accuracy and beauty, are the striking elements of attraction in their goods. The "Beaver Brand" of enamel goods stands alone, and is sold widely in all parts of the Dominion. Their special lavatory is called "The Eastern Beauty," and it well deserves the name.

C. A. Lusby, the secretary-treasurer, is president of the Amherst Board of Trade, and recently was brought into prominence by his identification with the establishment of electric power, generated at the mouth of the Chignecto mines and utilized for industrial work. On this occasion, a special telegram of congratulation was received from Thomas Edison, the famous electrical wizard of Menlo Park, who stated that Amherst was the first town on the American continent to install such an equipment.

I. Matheson & Son.

This firm was established in 1867, by I. Matheson, at New Glasgow, and steadily grew in prominence, manufacturing marine boilers and mining machinery for most of the leading firms of the Maritime Provinces, including also Newfoundland. About fifteen years ago, after the death of Mr. Matheson, senior, a stock company was formed, including Charles M. Crockett, W. G. Matheson, and J. C. McGregor, the latter being appointed president. Their plant covers a large area and is well-equipped with modern machinery and all the accessories to up-to-date work. Coal mining machinery is a specialty, a staff of from 175 to 200 hands being kept constantly busy.

A unique and important part of their business is shipbuilding and charter work. They have now in the course of construction, the first steel sailing vessel ever built in the provinces. It will be a three masted schooner rigged vessel of 350 tons. The steel used in the construction, is largely supplied by the Nova Scotia Steel & Coal Co. Mr. Charles Crockett is widely known in New Glasgow for his proverbial hospitality and social qualities. He is a keen business man, and is devoted to the up-building of the industries of New Glasgow and the provinces.

Cragg Bros.

A smart, up-to-date hardware store is that owned by Cragg Bros., located on the corner of two busy streets in Halifax. For nearly a quarter of a century they have been known as the leading cutlery house of the east. There is a style and "go" about the place which suggests an American store, yet everywhere are characteristics of high-class, dignified business methods.

Nova Scotia Steel & Coal Co.

One of Canada's leading and largest industries is that of the Nova Scotia Steel & Coal Co., Limited, with head office and works at New Glasgow, also at Sydney Mines and North Sydney, and at Wabana, Newfoundland. This company began work almost with the initiation of the Dominion, and it has kept pace with it ever since. They employ a staff of over 4,000 hands, distributed as follows: At the collieries, Sydney Mines, 1,800; Iron & Steel Works, 750; Rolling Mills Forge, 1,000; Wabana, 600. They pay out $180,000 per month in wages, so that some idea may be gathered of the immensity of this enterprise.

Their market is practically the world, and they ship iron ore to points such as Philadelphia, Rotterdam, Middleboro, as well as supplying Canada's leading manufacturing and railway companies. They have ore properties in Newfoundland and coal in Cape Breton, where they expect to mine and ship 650,000 tons of coal, about one-third of which is used by the company and the balance in the Provinces of Nova Scotia, New Brunswick, Prince Edward Island and Quebec, including as far west as Montreal. About 200,000 tons of coal from this company finds its way up the St. Lawrence during the season.

They have now under course of construction, a new forge building for the manufacture of car axles, which will be in addition to their present plant at New Glasgow. This will enable them to increase their output at a decided decrease in cost. As the new building has been constructed over the site of part of the old building, the present manufacturing work has not been interfered with.

The general manager, Thomas Cantlie, has been in charge during the past four years, and has showed marked capability and energy in all departments He has been with the company almost since its inception and has won the good-will and respect of all who know him.

J. S. Neill, Fredericton.

It is something to be able to accredit to one family the brains, push and energy sufficient to have carried on for over fifty years, a progressive, successful, and increasing business.

In the year 1851, John Neill (father of the present James S. Neill, proprietor and manager of the concern to-day), started in the hardware business with a small store, a limited stock, and a somewhat confined field, but with unlimited energy and enterprise.

Young James grew up with the business until in 1870, he was in a position to relieve his father, buying out the business, succeeding to the good-will, and establishing the present house of J. S. Neill, wholesale and retail hardware merchant, developing and progressing, until to-day the firm can say that they do more business in two weeks than was done in the first year, after the business was taken over by the present management.

A more complete line of general hardware would be hard to find, the firm carrying full lines of the following: Bar iron, steel, mill supplies, steamfitters' and plumbers' supplies, paints and oils, glass, cement, fire brick, fire clay, etc, and a very complete and full line of stoves and ranges.

Fredericton is a large and growing centre of a large jobbing business. The firm keep two men constantly on the road, and find that as a distributing point Fredericton is in no sense handicapped, that they have no trouble in competing with any house in the larger centres. Direct importations are made from Great Britain, Germany, and the United States, goods being landed in Fredericton about as cheaply as in Montreal, St. John and Halifax. The firm have a large warehouse on a spur of the I.C.R. to facilitate the handling of goods, with a large warehouse in town for storage purposes. The store itself on Queen St. has a frontage of 78 feet and a total floor space of 32,410 feet.

Mr. Neill has the control of New Brunswick of the celebrated "Gillingham" brand Portland Cement manufactured by the Associated Portland Cement Manufacturers, London, England, and has made some very nice contracts.

He sold last year upwards of 9,000 casks of this celebrated cement,—6,000 casks for the Maine & New Brunswick Electrical Power Company for damming the Aroostook Falls, as well as the cement used in the Gibson Water Works and Fredericton Sewerage. This year he has contracts for upwards of 14,000 casks:—6,000 casks for the C.P.R. Bridge at Grand Falls, 5,500 casks for the C. P. R. Bridge at Aroostock Junction, 600 casks for the Fredericton Bridge and also for cement used for important works in the different parts of the province.

HARDWARE TRADE GOSSIP

Ontario.

John Johnson, harness dealer, Clinton, Ont., has sold his business to A. R. McBrien.

A new hardware store will be opened on Bathurst street, Toronto, by W. A. Stephenson.

Jeremiah Tangney, harness dealer, Mount Forest, Ont., has assigned to M. O. MacGregor.

M. A. Wigle, of the hardware firm of M. A. Wigle & Co., Amherstburg, Ont., spent Tuesday in Toronto.

The death occurred recently of W. H. Bragg, late of the hardware firm of W. H. Bragg & Co., Chatham.

Harry Moore, Oakville, and J. J. Coffee, Barrie, Ont., were hardware men noticed in Toronto on Wednesday.

Mr. McDonald, of the hardware firm of McDonald & Hay, North Bay, Ont., was a caller in Toronto this week.

GEORGE KELLY
With Lewis Bros., Montreal, Single Senior Canoe Champion of Canada.

Burglars recently broke into the hardware store of McLenan & Co., Lindsay, Ont., and carried away a considerable amount of goods.

Wm. O. Greenway, manufacturers' agent, Montreal, is in Toronto this week calling on the paint manufacturers and others who are his customers.

S. A. Archibald, of Chesney & Archibald, Seaforth was one of the exhibition visitors who called on Hardware and Metal in Toronto this week.

Cameron Browne and John Tait, Galt, Ont., have bought out the hardware business of Theron Buchanan of the same town. The new proprietors will take possession on September 3rd.

The hardware store of the John Armstrong Co., Brigden, Ont., was recently entered by burglars, but the latter were frightened away by A. D. Armstrong and Purvis Dawson, before any of the stock had been appropriated.

The heating and ventilating contract for the new wing, operating room, and alterations to the General Hospital, Orillia, has been awarded to the Orillia Hardware Co., Orillia, Ont., at $1,625. The plumbing contract was awarded to the same firm at $1,025.

W. H. Rumhall, formerly in the hardware business at South Woodslee, Ont., has returned from a trip to the west and has decided to establish a hardware, furniture and undertaking business at Victoria Harbor. He is now building a cement block store, 50x80 feet with two flats and basement, occupying the new premises in October.

Harry Wilson, traveler for H. S. Howland, Sons & Co., wholesale hardware merchants, Toronto, has left for a few weeks' rest at his home in Whitby. For several weeks previous to his departure, Mr. Wilson had been confined to a private hospital in the Queen City, suffering from an attack of typhoid fever. While Mr. Wilson is recuperating, Alex. Lowry will take his place on the road.

The Walkerville Hardware Company, Walkerville, have moved into a fine new store in the up-town residential district and will close their down-town business as soon as quarters are ready for their plumbing department. The new store is decidedly attractive, being almost square, 48x49 feet, with two stories and basement. Photos of the fine windows and novel interior arrangement have been promised Hardware and Metal for publication in an early issue when a full description will be given.

Quebec.

E. H. Bazin, of the Bazin Mfg. Co., Quebec city, was in Montreal last week.

Mr. Shillington, purchasing agent for the Canadian Copper Co. Copper Cliff, Ont., was in Montreal last week. His company is completing a new warehouse, which, when finished, will have a large storage capacity.

Western Canada.

A. J. Falconer, Deloraine, Man., president of the Western Ratail Hardware Association, was in Winnipeg last week on business.

Graham & Rolston, Winnipeg, are selling out their business, near the corner of Portage and Main streets. For some years this firm have had one of the most desirable locations in town, but the premises have recently been acquired by the Winnipeg Street Railway Co. In conversation with Hardware and Metal Mr. Graham said he had made no plans for the future.

METAL MERCHANT'S EARLY DEATH.

A. H. Campbell, of A. C. Leslie & Co., metal merchants, Montreal, died at his home in that city on Tuesday of this week after a prolonged illness.

Albert Havelock Campbell was born in 1867 at Vankleek Hill, Ont., being the son of the late Donald P. Campbell, M. D., who died while his son was a young boy. He lived with his widowed mother in Montreal, and was educated at the Royal Arthur and high schools, graduating from the latter in 1883. He then entered the employ of Frothingham & Workman, wholesale hardware merchants where he rose to a responsible position, only resigning at the end of 1896 to enter into partnership with William S. Leslie, in the firm of A. C. Leslie & Company, and on the incorporation of this firm at the beginning of 1907, he became vice-president of the company.

In April of this year he had a serious attack of pleurisy, and when apparently convalescent, was taken down with a lingering attack of typhoid fever, being confined to his bed for ten weeks, and his constitution, which was weakened by his previous illness, was unable to stand the strain.

He leaves a widow, the daughter of City Clerk Henderson, of Ottawa, but no family. His mother, who had always lived with him, also survives. Mr. Campbell was a member and faithful worker in the American Presbyterian church. He was a member of the Board of Trade, Dominion Commercial Travelers' Association, Canada Club, M.A.A.A., Y.M.C.A., and St. Andrew's Society, but devoted his time outside of business hours to his home and church.

Among all who knew him personally or in business connections, he was very

THE LATE A. H. CAMPBELL
Vice-President of A. C. Leslie Co., Montreal.

highly esteemed as a true friend, a manly and consistent Christian and a business man of high character, determination and undoubted ability. Hardware and Metal unites in an expression of condolence, both to the widow of the deceased and to the company with whom he was connected.

CANADA'S CHAMPION CANOEIST.

George Kelly, employed in the financial department of Lewis Bros., wholesale hardwaremen, Montreal whose picture is reproduced herewith from a photo kindly loaned Hardware and Metal by Capt. Stranger, is the hardwareman who recently won first honors in the senior single championship race at Montreal, winning both his trial heat and the final race. Mr. Kelly had previously made his mark in Rugby football, and this latest achievement adds to his reputation for athletic prowess. Mr. Kelly is a member of the Grand Trunk Boating Club, Montreal, and his friends predict a bright career for him, both as an athlete and as a business man.

HARDWARE AND METAL

Established 1888

The MacLean Publishing Co.
Limited

JOHN BAYNE MACLEAN . President

Publishers of Trade Newspapers which circulate in the Provinces of British Columbia, Alberta, Saskatchewan, Manitoba, Ontario, Quebec, Nova Scotia, New Brunswick, P E. Island and Newfoundland.

OFFICES:

MONTREAL, ... 232 McGill Street
Telephone Main 1255
TORONTO . . . . . 10 Front Street East
Telephones Main 2701 and 2702
WINNIPEG, . . . 511 Union Bank Building
Telephone 3726
LONDON, ENG. . . . . 88 Fleet Street, E.C.
J. Meredith McKim
Telephone, Central 12960

BRANCHES:

CHICAGO, ILL. . . . . 1001 Teutonic Bldg
J. Roland Kay
ST. JOHN, N.B. . . . No. 7 Market Wharf
VANCOUVER, B.C. . . . Geo. S. B. Perry
PARIS, FRANCE - Agence Havas, 8 Place de la Bourse
MANCHESTER, ENG. . . . 92 Market Street
ZURICH, SWITZERLAND . . . Louis Wolf
Orell Fussli & Co.
Subscription, Canada and United States, $2.00
Great Britain, 8s. 6d., elsewhere . 12s

Published every Saturday.

Cable Address { Adscript, London / Adscript, Canada

MARITIME BOARD OF TRADE.

The annual meeting of the Maritime Board of Trade is one of the important gatherings of businessmen in Canada. Representing every section and every commercial interest of the three seaboard provinces, these men assemble and discuss, as the president says, "a wide range of interesting and live topics." The venerable secretary, Chas. W. Creed, was able to report that last year practically all their resolutions and recommendations had received consideration in Parliament. While this was satisfactory it was no more than right. The board is enthused by public spirit and ambition for the welfare of the provinces. It is made up of the ablest business men, men whose success has demonstrated their fitness to offer wise counsel in regard to matters of paramount public interest. Their deliberations and conclusions are worthy of the most serious consideration of every man in public life.

The people of the Maritime Provinces have felt that their part of the Dominion has not shared to the extent it should the wave of prosperity that has swept the West. They feel that the possibilities for success in the East are as large as in the West. They want to keep their own people at home and secure a share of the many immigrants crowding into Canada from Europe. It is to help solve such problems, to gather up and focus public opinion and business counsel, that the board holds its annual meeting.

Through the efforts of the board we hope some day to see a union of the three provinces that will far more than combine their present influence in directing the destiny of the great Dominion.

NOVA SCOTIA COAL MARKET.

Amherst, Nova Scotia, bulks large this summer upon the world's horizon. The transmission of electrical energy from coal mine to factory by wire is now achieved and Amherst, always progressive, may blaze the way to great industrial achievements.

The Maritime Provinces can only forge ahead by building up manufactures and it is the irony of fate, or the fate of misgovernment, that Nova Scotia has been unable to manufacture largely because coal costs too much. Nova Scotia coal is not expensive, except in Nova Scotia. It sells f.o.b. in Sweden for about half what it sells for in Halifax, and the poor man who goes to the pit mouth in Sydney and loads the coal on his own cart pays for that coal considerably more than the citizen in Montreal or in Boston pays for it put in his cellar.

Figures from the U.S. Foreign Commerce report show that the existing duties on soft coal imposed by both the Canadian and U.S. Governments are not only a hardship upon the many manufacturers of Ontario, but also upon the users of coal in the Maritime Provinces.

For the year ending June, 1907, Canada exported $3,089,254 worth of bituminous coal to the States, this being half a million less than in 1906, and about the same as in 1905. The figures show a slump and, of course, include the exports of the Vancouver Island mines. In addition to this, the U.S. imported about $1,000,000 worth from other countries. Canada, therefore, supplied seventy-five per cent. of the imports and if the tariff wall were down would undoubtedly have sent more to the New England States, with every likelihood of lower prices ruling in the Maritime market.

The hardship on Ontario manufacturers is indicated by the failure of the Nova Scotia mine owners to bring regular cargoes farther west than the St. Lawrence river points, in spite of the fact that for years they have been paying a heavy impost on every ton of U.S. bituminous coal used, in order to encourage the Nova Scotia producers. The imports are steadily on the increase, figures for the year ending June, 1905, showing exports from the States to Canada of $11,381,381 worth of anthracite, and $11,690,784 worth of bituminous. For the same period up to June, 1907, the figures were $12,007,843 worth of anthracite, and $15,014,627 worth of bituminous. Canada is the only buyer of U.S. hard coal, and also takes over two-thirds of the soft coal exported.

A study of the figures seems to prove conclusively that the duty on soft coal imports serves no good purpose to the people of the Maritime Provinces, while they, as well as Ontario consumers, would stand to gain by the abrogation of the duty.

INTERESTING LEGAL DECISION.

The evil results of installing heating plants, which are inadequate for the purposes for which they were intended, is emphasized in a recent decision of the supreme court of Wisconsin. According to the Engineering Review, a school board, having in charge the erection of a new high school building, entered into an agreement with a furnace company, in which the furnace company agreed to "furnish and set in position in a neat and workmanlike manner," a heating plant and to superintend the setting of the apparatus. By the terms of the contract the furnace company agreed that, with good care, the apparatus would warm the rooms to an average temperature of 70 degrees during the coldest weather, and that, at the same time, good ventilation would be secured.

The plant was put in, but upon the arrival of cold weather it was found that it did not sufficiently heat the building and it became necessary to close the school. The company was notified of this by wire and sent men to make repairs, but even then the plant would not bring the temperature up to the desired degree. The school board had paid $2,800 of the contract price of $3,500, but refused the final installment of $700 on the ground that the apparatus was a failure. The company sued for the balance claimed due, and the school board entrenched itself behind a clause in the contract, which provided that if the plant did not heat the rooms to 70 degrees during the coldest weather, and at the same time permit proper ventilation, it would be made to do so or the money paid would be refunded. A judgment was given in favor of the defendant for the sum of $3,088.57, representing the amount paid by the school board with interest, in addition to which sum the furnace company had to meet the expenses of the action.

It doesn't do much good to grasp an opportunity unless you know what to do with it.

Markets and Correspondence

(For detailed prices see Current Market Quotations, page 90.)

MARKETS IN BRIEF.

Montreal.

Antimony—Weak.
Copper—Strengthening.
Linseed Oil—Dull.
Pig Iron—Firm.
Sporting Goods—Active.
Tin—Firmer.
Turpentine—Dull.

Toronto.

Antimony—Steadier.
Copper—Stronger.
Linseed Oil—Steadier.
Turpentine—Firmer.
Tin—Firmer.
Old Materials—Still declining.

MONTREAL HARDWARE MARKETS.

Montreal, Aug. 30.—The usual early fall activity characterizes the hardware trade at present. Well established brands of goods are experiencing steady and lively call and manufacturers are exceedingly busy, not only in keeping pace with current demands, but in the endeavor, in some cases almost futile, to replete their supplies for future rushes.

Transportation facilities still constitute a somewhat annoying problem. Although they are in an eminently more satisfactory condition than that of a month or so ago, there is yet large room for improvement. The only possible solution, one which would ensure satisfaction to all parties interested, would be an increase by the railroads in the number of freight cars. While it is true they are steadily turning out large numbers of cars, the numbers are not commensurate with the call for them.

Screws—The situation is steadily improving. The factories are turning out increasing supplies. Prices remain firm and unchanged.

Bolts, Nuts and Rivets—Supplies are being steadily repleted and jobbers are now in a fair position to handle the trade. No changes in prices are noted.

Sporting Goods—The volume of business transacted this fall in all lines of sporting goods will very probably be a record-breaker. An increased number of sportsmen, both native and tourist, is expected to explore the wilds of the north, and the demand for guns and ammunition will, therefore, be greatly increased.

Poultry Netting—An unseasonably strong demand exists for this. The factories have laid in a heavy stock of raw material and the possibilities for another acute shortage are very slight. The discounts for the coming season are 50 and 5 off.

Wire Goods—A large volume of business is being done this season and the amount of shipments, which will be made in the next two months to the Northwest will, without, doubt, be unparalleled. All makers are looking forward to a bumper season.

Builders' Hardware and Mechanics' Tools—The demand is steady and strong. Supplies are adequate, prices are firm and unchanged.

Building Paper—Business is well maintained. The situation is unchanged. Prices are firm.

TORONTO HARDWARE MARKETS.

Toronto, Aug. 30.—For the past two or three weeks business has been continually improving in local hardware circles and this steady gain in the volume of business is more than ever in evidence this week. With September practically here, fall goods are beginning to be in increased demand and local jobbers are kept busy attending to outside enquiries which are daily arriving from those retailers who have not yet placed their orders for fall lines. The great national exposition was formally opened here on Tuesday and thousands of visitors will arrive in Toronto daily from now till the 9th of September, and among these will be many retail hardware merchants from all over the country. As many of these retailers will make it a point to call on the wholesale hardware houses while in the city, jobbers will have an excellent opportunity to get personally acquainted with many outside merchants, and there is not a doubt that such personal acquaintance means a resultant increase in outside business.

Screws—The situation is slowly improving, but the scarcity in most sizes is still very marked and manufacturers are far from being with the trade. Since the recent advance in price, the demand has not been too strong, but it is bound to set in again before long, as supplies in the hands of retailers are known to be very low.

Nails—A good demand exists and supplies are fairly well up with the trade. A few jobbers still experience a little difficulty in obtaining adequate supplies of the best selling sizes, but on the whole, the manufacturers have the situation well in hand and there is no reasonable cause for complaint. Prices remain firm at figures last quoted.

Wire—Trade is dull in all kinds of wire, though jobbers are expecting an increased demand for hay-baling wire during September. The makers are determined to prevent a recurrence of the recent scarcity next year and the present quietness is, therefore, being utilized in getting adequate supplies on hand to meet the strong demand that invariably sets in with the commencement of January.

Building Material and Supplies—The strong demand which has prevailed all season for builders' hardware still continues and will certainly last all through the coming fall. Cement is moving much faster for, with the rush of harvest over, farmers now have more time to plan out improvements and repairs, many of which call for cement in their construction.

Sporting Goods—Guns and ammunition are now selling well and displays of firearms and ammunition are quite in evidence around local hardware showrooms. A lively trade is being done in all easily carried fall sporting requisites just at present on account of the immense number of people in the city attending the national exhibition.

Fall Goods—The autumn trade is already beginning to set in and indications point to a large volume of business being done this fall. Increased enquiries are coming in for axes, saws, horse blankets and sleigh bells and travelers of local houses continue to book large orders for these lines throughout the province.

MONTREAL METAL MARKETS.

Montreal, Aug. 30.—"When things are at their worst they begin to mend." The serious depression which has ruled over the metal market for the past two months is beginning to ameliorate, and it is possible that in another month the trade will be in a very satisfactory condition.

The English market is firm. Although supplies are not over-abundant they are adequate for the present demand. The sources of supply during the past fortnight or month have been rather unsteady, as many furnaces were idle and others working were not in a very satisfactory condition.

Absolute stagnation characterizes the American situation. Little or no buying is being done and the buying public have no confidence whatever in the present state of the market.

In Ontario, especially the western districts, there is a tendency amongst consumers of iron to place orders with nearby American furnaces, taking advantage of the prices ruling there which are considerably better than the English. Locally, an active demand for most of the products exists.

Pig Iron—English and Scotch iron, especially the best grades, are hardly available. The prices there are firm and are likely to remain so for three or four months hence. Heavy enquiries have been made this week in the local market. Prices, locally, are firm and unchanged: Middlesboro, No. 1, $21.50; No. 2, $20.50; Summerlee, $25.50.

Antimony—Continues dull. Prices are weak. Cookson's, 15c; Hallett's, 14½c.

Ingot Tin—The English market is a little firmer. In the United States it is dull. Little call is experienced in the local market. Lamb & Flag and Straits, $44.

Copper—Is a little firmer. More activity is experienced. Prices are firm and unchanged.

Old Materials—The situation is unchanged. Little or no demand is experienced, and a general weakness will prevail for another ten days or fortnight. Heavy copper and wire, 15½c; light copper, 13½c; heavy red brass, 13c; heavy yellow brass, 10c; light brass, 7c.

TORONTO METAL MARKETS.

Toronto, Aug. 30.—A marked improvement is noticeable in the Toronto metal markets this week and many consumers have evidently come into the market in order to get the advantage of present low prices. While no unusually large sales are reported in any of the metals it is, nevertheless, quite evident that the summer quietness is now a thing of the past and jobbers are confidently looking forward to an increasingly large business throughout the fall months.

Antimony—The market is much steadier, locally, and sales have largely increased during the week. The American market is dull, with little or nothing doing. Cookson's is still quoted locally at 15c, and Hallett's at 14½c.

Pig Iron—Local business is much better and an increased number of buyers were in evidence throughout the week. In regard to the American market the manager of a large syndicate which controls the output of fifteen mills, stated as late as yesterday that the demand for certain grades of cast iron was becoming so great that the mills could hardly supply it. Local prices are firm at figures last quoted.

Tin—The demand is much heavier this week and prices are beginning to stiffen in response to advances in the primary markets of supply. The London tin market yesterday scored a very sharp advance of £3, which had the effect of advancing the New York prices in sympathy, without, however, creating any life in business. The large arrivals of the present week undoubtedly have had a tendency to convince American consumers that a stable range of higher price is not a probability. The reason for the advance in the London market is not known, and it seems to be a purely speculative movement.

Lead—During the early part of the week lead weakened considerably on the local market, but with the increased demand at the end of the week, lead prices have more than regained their former strength. Imported pig lead is now firmly held at $5.35.

Copper—The demand is much stronger locally, and a number of good sales have been made during the week. Casting ingot is still firm at 22c and many buyers are being attracted at that figure. As for the American situation the copper market yesterday was dull to the point of stagnation. The talk continues of a price adjustment during the present week which will bring the larger consumers into the market, but even the adoption of a new selling basis by the big producers might not meet the views of the consumers under existing conditions.

Old Materials—The local market is still exceptionally quiet and prices are still declining as a consequence. The British market is extremely dull and the local situation is but a reflection of the conditions there. Heavy copper and wire has declined ½c and is now quoted at 14½c; light copper is ½c easier and is now 12½c; heavy red brass is 1c cheaper and now brings 12c; yellow brass has dropped from 10½c to 10c; scrap zinc from 3¼c to 3½c, and No. 2 wrought iron from 6c to 5c.

U. S. METAL MARKETS.

Cleveland, O., August 29.—The Iron Trade Review to-day says: All centres report a further weakening of prices, and the midsummer lull has only been broken by purchases of a few odd lots to cover consumers' immediate needs.

Southern iron, No. 2, has declined to $18.50, Birmingham, for delivery the remainder of the year, and one producer has named $17 for the first six months of 1908. The purchase of 6,000 tons of Bessemer by the Jones & Laughlin Steel Company at $22 valley for September shipment, records a decline of fifty cents on this grade. The blowing out of several eastern stacks is scheduled in the next few weeks, and this curtailment of production is expected to have a stimulating effect. For the first time in several years, the Illinois Steel Co. has a surplus iron producing capacity, and from 5,000 to 10,000 tons a month are being shipped to other plants of the steel corporation.

Several independent tinplate mills have temporarily suspended operations, owing to the scarcity and high cost of raw material. Specifications for sheets and other finished lines are heavy, although new business is light. Orders for cast iron pipe, include 5,000 tons for Kansas City, 2,500 tons for Columbus, while New York closes for 6,000 tons to-day.

LARGEST CAVE IN THE WEST.

Two gold prospectors recently discovered in the Santa Susanna Mountains, about fifty miles from Los Angeles, Cal., the largest and most remarkable cave in western America. While looking for indications of gold, they found an opening which they entered. The opening led to a great cavern consisting of many passages, some of them wide, but most of them narrow and lofty. The passages lead into great halls, some containing an acre studded with stalagmites and stalactites, in some cases so thickly that it is difficult to get through. The walls of one of these halls are covered with rude drawings, some almost obliterated, but others still clear. The drawings represent incidents of the chase, showing Indians on foot pursuing bear, deer and other animals. One wall-painting shows the bear pursuing the hunter. The work is done with a soft red stone much used by the Indians for that purpose.

CORRESPONDENCE

Travelers, hardware merchants and clerks are requested to forward correspondence regarding the doings of the trade and the industrial gossip of their town and district. Addressed envelopes, stationery, etc., will be supplied to regular correspondents on request. Write the Editor for information.

HALIFAX, N.S.

Halifax, N.S., Aug. 26.—There is nothing of special note in hardware prices which now seem to be pretty steady all round. Business is unusually quiet, but this dull period is not unexpected at this season of the year. Collections on the whole are satisfactory.

Wire nails are quoted at $2.60 base, and cut at $2.70 base. Sheet lead is worth $6.50 per 100 pounds in rolls, and $7.50 in small quantities. There is very little doing in tin plates, the quotations for which are: 1X charcoal, $5.25 to $5.50; 1C charcoal, $4.25 t o$4.75; 1C coke, $4 to $4.25. Turpentine is quoted at 90 cents in barrel lots and 95 cents for smaller quantities. Raw oil, in barrels, is quoted at 73 cents, and smaller quantities at 78 cents. Boiled, in barrels, is worth 76 cents, and in smaller quantities, 95 cents. Axes are worth $6.50 to $6.75 for single bitt and $13 for double bitt.

* * *

On the report of the city engineer, the Halifax Works Department has recommended the purchase of two thousand half-inch, and one hundred three-fourth-inch Trident meters from the Neptune Meter Company of New York, and nine hundred half-inch Lambert meters. The half-inch tridents cost $13.26 and the three-quarter-inch Tridents $19.89. The Lambert half-inch meters cost $11.20, each with connections.

* * *

Some changes are to be made in the Windsor Foundry and Machine Company at Windsor, N.S. The business since the reorganization has not been as successful as its promoters might wish, arising probably from two reasons, lack of larger capital and a thoroughly up-to-date practical man in charge of the practical side of the enterprise. It is now proposed to borrow $20,000 on mortgage to put the company in a good position to compete for outside business, as well as to engage a competent and expert manager. Failing to obtain from the shareholders the authority to do this, the property and assets of the company will be sold.

* * *

The chair of civil engineering in Dalhousie University, Halifax, recently made vacant by the resignation of Prof. Brydone Jack, has been filled by the appointment of A. E. Stone, M.E.M. Canadian Society Civil Engineers, structural engineer, of the Canada Foundry Company (Canadian General Electrical

Company), Toronto. Mr. Stone has a brilliant career as a student and for the past sixteen years has been constantly engaged in construction work of importance. He was with the C.P.R. for four years, and for two years he was engineer and chief draughtsman for the Montreal Steel Works. In 1900-1902 he was engineer for the Canadian Electro Chemical Company, and in that capacity designed and constructed their plant at Sault Ste. Marie. Since 1902 he has been structural engineer of the Canada Foundry Company, superintending the design of all kinds of structural steel work. He is a native of Prince Edward Island.

ST. JOHN, N.B.

St. John, N.B., Aug. 27.—Business during the past week has been very fair for the season. Building operations seem to be on the increase and as a result, there is quite a demand for hardware used in this connection.

* * *

Work on the wharves and warehouses on the west side is going ahead well; a new foundation is being placed under one of the east side warehouses; contractors are also busy on the big cold storage plant, the Y.M.C.A. building, the Royal Bank block and a flour shed for the I.C.R., as well as a dozen or more smaller contracts.

* * *

A fire, which broke out in the wood in conveyor of the Partington Pulp and Paper Company's mill at Union Point, on Friday last, did about $3,000 damage. It will not interfere with the work to any great extent

* * *

A number of Fredericton capitalists have organized a company to manufacture acetylene gas generators. It will be known as the Monitor Manufacturing Co., and will be capitalized at $24,000. John Kilburn, T. B. Kidner, Dr. W. H. Irvine, W. T. Chestnut and J. H. Barry, are among those interested in the venture.

* * *

Geo. W. McKeon, W. E. Golding, Samuel G. Kilpatrick, of St. John; Hamilton Benn and Edward G. Price, of London, England, are seeking incorporation as The George McKeon Company, Limited. The object is to take over and carry on the business of George McKeon, lumber merchant. The proposed capital stock is to be $100,000, and St. John is to be the chief place of business.

* * *

C. A. Duff-Miller, agent-general for New Brunswick in London, England, was in the city last week, and in speaking of the attitude of England toward Canada said that England and the continent now recognized that Canada was a safe and lucrative field for the investment of capital and it would be only a short time before a large amount of capital would flow in here to develop the oil fields of Westmorland county.

QUEBEC, QUE.

Quebec, Aug. 26.—The inauguration of the first section of 20 miles of the new Atlantic, Quebec & Western Railway, between New Carlisle, county seat of Bonaventure, and Port Daniel, took place Sunday and was a great success. This means an extension to the Atlantic & Lake Superior Railway, and brings the railway within eighty miles of Gaspe, and one hundred and twenty miles from Metapedia, the junction with the Intercolonial. It is hoped that the terminus port would be reached within two years. No doubt that this country will become, in future, a great centre for industries.

* * *

Gingras & Nadeau, plumbers, of Levis, have been awarded important contracts for installing heating by hot water at Indian Lorette and St. Alphonse churches, at Thedford Mines.

* * *

This is the beginning of the season for much trade in supplies for camping and hunting parties, as well as for prospecting and fishing camp trade. Business is, therefore, fairly brisk, with sporting goods business.

* * *

It is announced here that the Department of Marine will establish a regular winter mail service with Anticosti for the coming winter, placing the Montcalm, ice-breaker, on the route.

* * *

The stringency in the money market is still affecting Quebec, as well as other business centres. The various banking houses here have raised the discount rate, and it is thought the rate may go still higher.

* * *

The steamer Starmount, from Sydney, has just completed discharging, at the Quebec and Lake St. John dock, 34,000 steel rails, to be used for the Transcontinental, which will be re-shipped via the Q. & L. St. J. Ry.

* * *

After an almost unanimous request of taxpayers and the consent of the Municipal Council, it was resolved by the corporation of St. Romauld d'Etchemin to offer to any manufacturer who wishes to established any industry in its limits, an amount of two and one-half per cent. yearly of the salaries paid out, besides ten thousand dollars towards the construction of the building and an exemption of taxes during twenty-five years. Further details would be given on applying to the Mayor of the municipality.

* * *

Business conditions in hardware trade are quiet and much about the same as last week. The greatest part of the travelers are returned from well-earned vacations. They are already preparing their autumn samples. Plumbers and tinsmiths report good business. Builders' material trade is quiet. Woodworking machinery has a good demand.

HAMILTON, ONT.

Hamilton, Aug. 27.—The local plumbers are indignant over certain insinuations which have been made in this city within the past week relative to the existence of a plumbers' combine. The statement that such an organization existed is said to have originated with one of the aldermen, as the result of the attitude of the plumbers toward the contract which was let some time ago for a new police patrol station. In response to the call for tenders only one plumber submitted figures, and the aldermen, after opening the tender, refused to accept it, as it was too high. Another request was made for tenders, and this time another plumber offered to do the work for about $100 less than his former confreres. This caused one of the aldermen to say that the reason that more tenders were not received was because the plumbers had an understanding among themselves by which they were not to compete against each other on contracts.

Your correspondent had an interview with a prominent plumber and former member of the Plumbers' Association, and he flatly denied that there was any combine in Hamilton. There was a Master Plumbers' Association in Hamilton until a year ago, when the chief officers of the organization were charged with conspiracy, and heavily fined in the courts, but since that time it has disbanded and no attempt has been made to revive it. The plumber in question explained the plumbers' attitude regarding the patrol station contract by the fact that the plumbers were so busy that they did not consider it worth while to tender on a city job, since there is so much red tape about such matters. At any rate, he considered it grossly unfair to the local plumbers to insinuate as to their business methods and he assured your correspondent that there was absolutely nothing in the combine charge.

* * *

Nearly every week now there is a new industry of greater or lesser importance locating in Hamilton, and the latest arrival is the Berlin Machine Tool Co., of Beloit, Wis. This will, doubtless, prove to be one of the biggest concerns which the city has secured for some time, and already it has started on the erection of its first building, which will cost over $150,000. The company has a large plant in the States, and manufactures a high-class of tools, especially for planing mills. Its location here will mean the employment of a vast number of skilled mechanics and will, doubtless, prove a great boon to the city.

* * *

"This is the busiest season in our history," said a prominent plumber to your correspondent a few days ago, when discussing the condition of the plumbing trade. Although last year's business was a marvel to those engaged in this line, this year's work has surpassed even the most sanguine expectations. Perhaps in no other city in Canada is there proportionately more building being done than in Hamilton, and as a result of this the plumbers find themselves hardly able to meet the demand. Apart from the great amount of work being done is the fact that it is mostly all of a high-class character. The number of new office buildings which are springing up in the business section are in themselves supplying enough work to keep some plumbers constantly engaged, while the numerous factories, houses, churches and other buildings being built have almost trebled the plumbing work throughout the city. However, with peace and harmony among the employers and the unions, everyone is reaping the fruits of the harvest.

GALT, ONTARIO.

Galt, Ont., Aug. 28.—While the hardware and allied trades in Galt are very lively there is very little of interest to chronicle. The trade has kept up better during the past few months than in any previous year. The increased activity is probably due to the many operations at present under way.

* * *

The works of the Canadian Brass Manufacturing Co. are completed and the machinery is being installed. It is expected that active operations will be commenced the latter portion of this week.

* * *

The building trades unions in Galt are in a state of unrest. Owing to trouble over the employment of non-union workmen there have been several partial strikes and there is a possibility of a general strikes in all building trades lines. The plumbers are affected to a certain extent, but as yet no men are on strike.

* * *

The Old Boys' Reunion held here recently was most successful, and there will be a handsome surplus to be divided among local charities. The local hardwaremen did themselves proud on the matter of decoration.

* * *

J. W. Murtry, of the local hardware firm, is spending the week in St. Thomas, looking after his hardware business in that town. Mr. McMurtry is a progressive merchant, and his trade in Galt and vicinity is to be envied.

* * *

Cameron Brown and John Tait, have taken over the hardware business for some years conducted by Theron Buchanan, jr., and are actively preparing for a prosperous career. Both are men of exceptional commercial capacity. Cameron Brown, who is the youngest son of A. S. Brown, manager of the Galt branch of the United Empire Bank, has been in the business for four years, while his partner, a well known local young man, has had three years' experience in the same establishment

CHATHAM, ONT.

Chatham, Ont., Aug. 27.—The organization meeting of the shareholders of the Canadian Wolverine Brass Co was held in this city last week, L. A. Cornelius being elected president, and H. C. Cornelius vice-president, for the ensuing year. Claud Cornelius, nephew of Cornelius Bros., who hold the controlling interests in the company, will be manager and general superintendent of the Chatham concern. The building operations are progressing rapidly.

The Chatham Fireproof Construction Co. have been awarded the contract for a new reinforced concrete factory on King street, to be occupied, when completed, by the Western Bridge & Equipment Co. This concern, which recently located here, will employ about twenty-five hands and will handle all branches of steel construction.

* * *

Dr. Hodgetts, chief provincial health inspector, was in the city last week. Incidentally, he gave to the press an interview in which he criticized some of the local plumbing, and expressed the view that there was great need for the appointment of a plumbing inspector. He qualified his remarks however, by adding that he did not blame the master plumbers for any defects he had found. One of the latter, commenting on the interview, expressed a perfect willingness to have a plumbing inspector appointed, but did not see that there was any great need for one. Practically all work here now has to undergo the smoke test.

* * *

One of the leading hardware men of the city passed away Sunday morning in the person of William Henry Bragg, senior member of the firm of Bragg & Sons. Mr. Bragg was 55 years and 2 months of age, and had been ill some seven weeks. A native of Port Hope, he came to this section while still young, settling first in Harwich, later in Dover and finally in Chatham, where he had resided for 26 years. Shortly after coming to this city, he took a position in the hardware store of Geo. Stephens & Co., with which firm he was identified for 18 years. Six years ago he set up in business for himself, in partnership with M. O'Neil, the firm name being Bragg & O'Neil. Later, Mr. O'Neil retired, and Mr. Bragg's sons entered the partnership. Mr. Bragg leaves a widow, two sons and two daughters.

The wide popularity of the late Mr. Bragg was attested by the large attendance at the funeral, which took place on Tuesday afternoon to Maple Leaf cemetery.

* * *

This is a quiet time in the hardware trade For the next three or four weeks there will be little doing—that is, compared with the past few months, which are reported to have been exceedingly busy.

"Now is a good time to talk organization," suggested one hardwareman to your correspondent. "Chatham hasn't any hardware organization. It ought to have one."

The suggestion is timely. Merchants in othes lines all have trade organizations. Some of them are rather somnolent, 'tis true; but at least they exist, and, in case of emergency, they can do good work. The hardwaremen of Chatham are a superior class of men, and it is rather surprising that they have not long since seen the benefits of organization, and proceeded to grasp them. The past season can furnish a passable excuse. The hardwaremen have been pretty busy. But, now that a comparatively quiet time has come, a good opportunity offers to take the matter up and bring it to a head."

LONDON, ONTARIO.

London, Aug. 28.—Hardware business has been very good, both wholesale and retail, during the past week.

The little village of Clandeboye, on the L. H. & B. railway, has recently been the scene of rapid business changes. It was announced in this correspondence last week that S. C. Chown, of that place, had purchased the hardware store of his neighbor, Kestler, the supposition being that, with the combined stocks, he would be the only hardware dealer in the village, and would, therefore, have a clear field. Now it is announced that he has disposed of the entire business to Samuel Lamport, of Devizes.

* * *

Gillean McLean, the Talbot St. hardware dealer, has been compelled, owing to increasing business, to erect an addition of thirty feet to the rear of his store.

* * *

Ed. McIntyre, traveler for the Hobbs Hardware Company, north of Toronto, is at his home in Hamilton, suffering from a fractured kneecap, the result of a recent accident. He will be laid up for several weeks, and meantime, his duties are being looked after by Harvey Baynon.

* * *

The big gas producer plant for the McClary Manufacturing Co. has arrived in the city from Kansas City, and is now on Grand Trunk cars awaiting removal to the company's power-house at the new works in the south-eastern part of the city. The new plant is an immense affair, weighing 20 tons, and its installation and operation will be watched with a great deal of interest and concern by manufacturers in general throughout the city, it being the first one to installed here. Col. Gartshore, the vice-president and manager, is anxious to get the plant installed as early as possible, so that it can be tested and the exact cost of producing power by this means arrived at. It will be remembered that the McClary Co. did not give the hydro-electric power commission much encouragement when they canvassed the city looking for manufacturers who would contract for power, as they then had the gas producer plant under consideration and figured that they could obtain their energy at a cheaper rate by this method than by contracting for Niagara power.

* * *

The Southwestern Traction Company has ordered four new cars to take the place of those that were burned in the recent disastrous fire at their barns. The manager expects they will arrive here in time to be running regularly by Saturday next. The laying of rails to Port Stanley has been completed.

* * *

The display of hardware and metal goods at the Western Fair promises to be unusually good, several local firms having entered exhibits.

MANITOBA HARDWARE AND METAL MARKETS

Corrected by telegraph up to 12 noon Friday, Aug. 30. Room 511, Union Bank Bldg. Winnipeg, Man.

Wholesale houses report a satisfactory business for this season of the year, with no changes in price, and stocks well in hand for the expected activity looked for as soon as the results of the harvest can be sized up by the dealers throughout the country. While a smaller crop is likely, indications are that higher prices will prevail, thus evening matters up.

Rope—Sisal, 11c per lb., and pure manila, 15¾c.

Lanterns—Cold blast, per dozen, $7; coppered, $9; dash, $9.

Wire—Barbed wire, 100 lbs., $3.22½; plain galvanized, 6, 7 and 8, $3.70; No. 9, $3.25; No. 10, $3.70; No. 11, $3.80; No. 12, $3.45; No. 13, $3.55; No. 14, $4; No. 15, $4.25; No. 16, $4.40; plain twist, $3.45; staples, $3.50; oiled annealed wire, No. 10, $2.90; No. 11, $2.96; No. 12, $3.04; No. 13, $3.14; No. 14, $3.24; No. 15, $3.39; annealed wires (unoiled), 10c less; soft copper wire, base, 36c; brass spring wire, base, 30c.

Poultry Netting—The discount is now 47½ per cent. from list price, instead of 50 and 5 as formerly.

Horseshoes—Iron, No. 0 to No. 1, $5.65; No. 2 and larger, $4.40; snowshoes, No. 0 to No. 1, $4.90; No. 2 and larger, $4.65; steel, No. 0 to No. 1, $5; No. 2 and larger, $4.75.

Horsenails—No. 10 and larger, 22c; No. 9, 24c; No. 8, 24c; No. 7, 26c; No. 6, 28c; No. 5, 30c; No. 4, 36c per lb. Discounts: "C" brand, 40, 10, 10 and 7½ p.c.; "M.R.M" cold forged process, 50 and 5 p.c. Add 15c per box. Capewell brand, quotations on application.

Wire Nails—$3 f.o.b. Winnipeg and $2.55 f.o.b. Fort William.

Cut Nails—Now $3.20 per keg.

Pressed Spikes—¼ x 5 and 6, $4.75; 5-6 x 5, 6 and 7, $4.40; ⅜ x 6, 7 and 8, $4.25; 7-16 x 7 and 9, $4.15; ½ x 8, 9, 10 and 12, 4.05; ⅝ x 10 and 12, $3.90. All other lengths 25c extra, net.

Screws—Flat head, iron, bright, 80, 10, 10 and 10; round head, iron, 80; flat head, brass, 75; round head, brass, 70; coach, 70.

Nuts and Bolts—Bolts, carriage, ⅜ or smaller, 60 p.c.; bolts, carriage, 7-16 and up, 50; bolts, machine, ⅜ and under, 50 and 5; bolts, machine, 7-16 and over, 50; bolts, tire, 65; bolt ends, 55; sleigh shoe bolts, 65 and 10; machine screws, 70; plough bolts, 55; square nuts, cases, 3; square nuts, small lots, 2½; hex nuts, cases, 3; hex nuts, small lots, 2½ p.c. Stove bolts, 70 and 10 p.c.

Rivets—Iron, 60 and 10 p.c.; copper, No. 7, 43c; No. 8, 42½c; No. 9, 45½c; copper, No. 10, 47c; copper, No. 12, 50½c; assorted, No. 8, 44½c, and No. 10, 48c.

Coil Chain—¼-in., $7.25; 5-16, $5.75; ⅜, $5.25; 7-16, $5; 9-16, $4.70; ⅝, $4.65; ¾, $4.65.

Shovels—List has advanced $1 per dozen on all spades, shovels and scoops.

Harvest Tools—60 and 5 p.c.

Axe Handles—Turned, s.g. hickory, $3.15; No. 1, $1.90; No. 2, $1.60; octagon extra, $2.30; No. 1, $1.60.

Axes—Bench axes, 40; broad axes, 25 p.c. discount off list; Royal Oak, per doz, $6.25; Maple Leaf, $8.25; Model, $8.50; Black Prince, $7.25; Black Diamond, $9.25; Standard flint edge, $8.75; Copper King, $8.25; Columbian, $9.50; handled axes, North Star, $7.75; Black Prince, $9.25; Standard flint edge, $10.75; Copper King, $11 per dozen.

Churns—45 and 5; list as follows: No. 0, $9; No. 1, $9; No. 2, $10; No. 3, $11; No. 4, $13; No. 5, $16.

Auger Bits—"Irwin" bits, 47½ per cent. and other lines 70 per cent.

Blocks—Steel blocks, 35; wood, 55.

Fittings—Wrought couplings, 60; nipples, 65 and 10K Ts and elbows, 10; malleable bushings, 50; malleable unions, 55 p.c.

Hinges—Light "T" and strap, 65.

Hooks—Brush hooks, heavy, per doz., $8.75; grass hooks, $1.70.

Stove Pipes—6-in., per 100 feet length, $9; 7-in., $9.75.

Tinware, Etc.—Pressed, retinned, 70 and 10; pressed, plain, 75 and 2½; pieced, 30; japanned ware, 37½; enamelled ware, Famous, 50; Imperial, 50 and 10; Imperial, one coat, 60; Premier, 50; Colonial, 50 and 10; Royal, 60; Victoria, 45; White, 45; Diamond, 50; Granite, 60 p.c.

Galvanized Ware—Pails, 37½ per cent.; other galvanized lines, 30 per cent.

Cordage—Rope sisal, 7-16 and larger, basis, $11.25; manila, 7-16 and larger, basis, $16.25; lathyarn, $11.25; cotton rope, per lb., 21c.

Solder—Quoted at 27c per pound. Block tin is quoted at 45c per pound.

Wringers—Royal Canadian, $36; B.B., $40.75 per dozen.

Files—Arcade, 75; Black Diamond, 60; Nicholson's, 62½ p.c.

Locks—Peterboro and Gurney 40 per cent.

Building Paper—Anchor, plain, 66c; tarred, 69c; Victoria, plain, 71c; tarred, 84c; No. 1 Cyclone, tarred, 84c; No. 1 Cyclone, plain, 66c; No. 2, Joliette, tarred, 69c; No. 2 Joliette plain, 51c; No. 2 Sunrise, plain, 56c.

Ammunition, etc.—Cartridges, rim fire, 50 and 5; central fire, 33⅓ p.c.; military, 10 p.c. advance. Loaded shells: 12 gauge, black, $16.50; chilled, 12 gauge, $17.50; soft, 10 gauge, $19.50; chilled, 10 gauge, $20.50. Shot: ordinary, per 100 lbs., $7.75; chilled, $8.10. Powder: F.F., keg, Hamilton, $4.75; F.F.G., Dupont's, $5.

Revolvers—The Iver Johnson revolvers have been advanced in price, the basis for revolver with hammer being $5.30 and for the hammerless $5.95.

Iron and Steel—Bar iron basis, $2.70. Swedish iron basis, $4.95; sleigh shoe steel, $2.75; spring steel, $3.25; machinery steel, $3.50; tool steel, Black Diamond, 100 lbs., $9.50; Jessop, $13.

Sheet Zinc—$8.50 for cask lots, and $9 for broken lots.

Corrugated Iron and Roofing, etc.—

Corrugated iron, 28 gauge, painted, $3; galvanized, $4.10; 26 gauge, $3.35 and $4.35. Pressed standing seamed roofing, 28 gauge, $3.45 and $4.45. Crimped roofing, 28 gauge, painted, $3.20; galvanized, $4.30; 26 gauge, $3.55 and $4.55.

Pig Lead—Average price is $6.

Copper—Planished copper, 44c per lb.; plain, 39c.

Iron Pipe and Fittings—Black pipe, ¼-in., $2.65; ⅜, $2.80; ½, $3.50; ¾, $4.40; 1, $6.35; 1¼, $8.65; 1½, $10.40; 2, $13.85; 2½, $19; 3, $25. Galvanized iron pipe, ⅜-in., $3.75; ½, $4.35; ¾, $6.65; 1, $8.10: 1¼, $11; 1½, $13.25; 2-inch, $17.65. Nipples, 70 and 10 per cent.; unions, couplings, bushings and plugs, 60 per cent.

Galvanized Iron—Apollo, 16-gauge, $4.15; 18 and 20, $4.40; 22 and 24, $4.65; 26, $4.65; 28, $4.50; 30-gauge or 10¾-oz., $5.20; Queen's Head, 20, $4.60; 24 and 26, $4.90; 28, $5.15.

Lead Pipe—Market is firm at $7.80.

Tin Plates—IC. charcoal, 20x28, box, $10; IX charcoal, 20x28, $12; XXI charcoal, 20x28, $14.

Terne Plates—Quoted at $9.50.

Canada Plates—18x21, 18x24, $3.50; 20x28, $3.80; full polished, $4.30.

Lubricating Oils—600W, cylinders, 80c; capital cylinders, 55c and 50c; solar red engine, 30c; Atlantic red engine, 29c; heavy castor, 28c; medium castor, 27c; ready harvester, 28c; standard hand separator oil, 35c; standard gas engine oil, 35c per gal on.

Petroleum and Gasolene—Silver Star, in bbls., per gal., 20c; Sunlight, in bbls., per gal., 22c; per case, $2.35; Eocene, in bbls., per gal., 24c; per case, $2.50; Pennoline, in bbls., per gal., 24c; Crystal Spray, 23c; Silver Light, 21c. Engine gasoline, in barrels, gal., 27c; f.o.b. Winnipeg, in cases, $2.75.

Paints and Oils—White lead, pure, $8.50 to $7.50, according to brand; bladder putty, in bbls., 2½c; in kegs, 2¾c; turpentine, barrel lots, Winnipeg, 90c; Calgary, 97c; Lethbridge, 97c; Edmonton, 98c. Less than barrel lots, 5c per gallon advance. Linseed oil, raw, Winnipeg, 72c; Calgary, 79c; Lethbridge, 79c; Edmonton, 80c; boiled oil, 3c per gallon advance on these prices.

Window Glass—16-oz., O. G., single, in 50-ft. boxes—16 to 25 united inches, $2.25; 26 to 40, $2.40; 16-oz. O.G. single, in 100-ft. cases—16 to 25 united inches, $4; 26 to 40, $4.52; 41 to 50, $4.75; 50 to 60, $5.25; 61 to 70, $5.75; 21-oz. C.S., double, in 100-ft. cases, 26 to 40 united inches, $7.35; 41 to 50, $8.40; 51 to 60, $9.45; 61 to 70, $10.50; 71 to 80, $11.55; 81 to 90, $17.30.

Small gasoline and kerosene stoves for campers' use are good things to feature during the summer, as they are mighty convenient things, much better than the romantic camp fire.

The purposes of an invention by O. Reppert, J. F. Fruchte and F. L. Litterer, Decatur, Ind., is to provide details of construction for a gate hinge, especially well adapted as an adjunct for a farm gate, and that will release and permit the lateral opening movement of such a gate when the hinge is rocked on its support in either direction.

Heating and Housefurnishings

SUCCESSFUL HEATING BY HOT AIR FURNACE.

In considering the heating of a modern building by means of a hot air furnace, the first and most important consideration will be the location of the furnace itself, and of the registers, says an exchange. Air, like water, will always flow in the direction of least resistance; therefore, it naturally follows that in placing registers in a room great care should be taken to favor that location where the least resistance will be met with from the incoming flow of air. As cold air is denser and heavier than warm, it follows that the proper location for a register in a room should be the warmest place in that room, i.e., on that side farthest from outside influences.

Having first located the registers, place the furnace, keeping three facts in mind. First, remember that the greater the elevation of a warm air pipe the more rapid the flow of air; second, that the air will flow more rapidly toward the point of least resistance; third, that the velocity of the air is dependent on the height of the outlet above the furnace, and on the amount of frictional resistance in the pipe, in other words, on the length of the run and the pressure resistance in the room in which the register is placed. Therefore, rooms having the greatest exposure in the direction of the prevailing winds, on the first floor, naturally should be nearest to the furnace, and should have a larger pipe and register. Rooms which are remote from the furnace, necessitating a long horizontal run of pipe, should have larger pipes.

Aim to minimize the frictional resistance in all pipes by avoiding all square turns or abrupt angles. Insist on having at least one inch rise to the running foot of pipe from the furnace to the register. Long runs of pipe, especially when going through cold rooms, should be wrapped with asbestos paper; pipes going through stone or brick walls should have thimbles one inch larger diameter than the pipe.

In the adjustment of the pipe work, bear in mind that the pressure of the air is equal on all pipes at the furnace. If, therefore, some of the pipes do not flow as freely as others, the cause of that trouble may be looked for either in the frictional resistance in the pipes, pressure resistance in the rooms into which these pipes lead, or on the pressure of an adjacent pipe, having the advantage of elevation, and taking more than its proportion of the heated air. Should the trouble be caused by frictional resistance, look for obstructions to the free and natural flow of air such as abrupt angles, etc., and remove them. If this does not furnish the remedy, then increase the size of the pipe. If the trouble is caused by pressure resistance in the room itself, this resistance is caused by air pressure in the room, and some outlet must be provided before satisfactory results can be obtained.

A very satisfactory solution to this difficulty can be had by cutting an opening in the base-board of an inside partition between two studs, and utilizing the space between two studs and plaster walls for the vent duct. The plates on top of the studs must be cut, and the duct be unobstructed to the attic. Generally speaking, the air will find its way out of the attic, but in case it does not do so, an opening can be made in some unused chimney, or some other means employed to overcome the difficulty. When inside air is used all doors must be left open and chimneys or fireplaces closed.

—

TIN PLATE FOR FIRE PROTECTION

A number of rather serious conflagrations have been reported in the daily press in the vicinity of New York recently, probably the fire covering the largest territory being that at Coney Island, where some 35 acres are said to have been burned over. In this instance, when the fire had reached a brick hotel and a large frame structure covered with tin, used for advertising purposes, its extension in that direction was stopped. This, points out the Metal Worker, is only another evidence of the fire protecting qualities of tin plate.

During the past twelve months, manufacturers and master sheet metal workers have been devoting energy to showing that buildings covered with metal roofs withstand fire much more successfully than those covered with a roofing material of the same or lower cost. The desirability of tin plate for a roof covering is fully realized and the endeavor to extend its field so as to compete with materials of much less cost has resulted in the use of low grades of tin plate which naturally rendered poor service and brought tin roofing under adverse criticism. This not only caused the use of other roofing materials, and thereby robbed the owners of the buildings of the fire protection that would have been provided had tin plate been used, but it gave business to other tradesmen which by a little courage and judgment could have been retained by the tin roofer.

Reports from various sources are to the effect that there has been a marked difference in the character of tin plates for roofing purposes, which have been demanded by the roofers during the past year. Those who have had the wisdom and judgment to refuse to use cheaper plates in the past have stated again and again that they have had little or no trouble with their tin roofing. It is probable, from the trend in tin plate buying, that a larger number of roofers will be added to this corps of supporters of the tin roof, and the effect of tin plate in stopping a 35-acre fire should be brought to the attention of all who build, particularly those who build homes. The Coney Island experience is too valuable an addition to the testimony of the safeguarding against fire provided by the tin roof to be allowed to pass unnoticed, and roofers will do well to preserve a copy of the papers making special note of this example of fire resistance.

—

BRITISH COLUMBIA ZINC INDUSTRY.

The zinc industry is yet in its infancy in British Columbia, according to a recent report of a Government commission investigating the subject. While several of the mines now being worked are essentially zinc mines, there are some silver-lead mines in which zinc blend is found in considerable quantities. Until recently the zinc in the silver-lead ores was regarded as an impurity to be gotten rid of by the easiest and cheapest means possible. The increased demand for zinc of late has made valuable as a by-product this zinc blend, which was formerly a troublesome impurity. The commission regards 15,000 tons of zinc ore of 50 per cent. grade as a liberal estimate of the present annual production of the Slocan district, and it is calculated that the Ainsworth district can produce 100 tons daily of 50 per cent. ore. There are many mines and prospects in other portions of the province which are supposed to carry zinc in paying quantities. A zinc smelter has been established at Frank, Alberta, the plant being close to a coal mine and on the route from the mines to the market. The large amount of fuel required in reducing zinc ore renders it necessary that coal shall be near the smelter. Coal can be delivered at the Frank smelter at less than $1 per ton.

—

SETTING UP STOVES.

No stove or range ever made has of itself what is called a draft; that must be furnished by the chimney or flue, and even when the draft in chimney or flue is perfect, that is not all that is wanted to insure good work or good bread or biscuits; the other things necessary being proper setting up, good fuel, good material—and last, but not least, a good cook.

In making a sale, do not make any unreasonable guarantees or promises; do not say this stove or range will do good work set up to any kind of an old chimney and with any kind of fuel; because you know it will not, no matter what may be its name or who may be its maker. To obtain the best results there is needed a good flue or chimney, proper setting up with pipe full size of collar on the stove or range and good fuel, and as stated before, a good cook to run it.

When possible, the dealer should set up every stove or range he sells and the man doing this work should not be the cub or the poorest workman about the place or shop. He should know enough about the business to know if the flue he is going to use has a good draft or not, and if not, should so tell the party buying the stove or range and if possible correct the trouble before leaving the job. If this were always done there would be fewer kicks afterwards.

When the stove or range goes into the country where a man cannot well be sent to set it up, instruct them as far as possible how to set it up and what faults to avoid. Many of the manufacturers have booklets or circulars on this subject that they would be glad to furnish if they knew they would be used.

A good chimney should be 8x8 inches inside, says Hardware Trade, and the top of it should be a little higher than the highest part of the comb of the house, and should not have a tree overhanging the chimney, as is sometimes the case, and flue should end from 4 to 6 inches below the opening into which the stove pipe runs, and should not run down to the floor or into the cellar, where, perhaps, there is an opening into said flue without a cover or stopper of any kind; and remember putting up two or three stoves to one flue is like hitching up two or three wagons to a single team just large enough to handle one wagon easily.

COMPARATIVE COST OF COOKING.

Tests made by cooking meat for one hour gave results for five different methods, as shown in the accompanying table:

	Cents.
Electricity, per hour	4.128
Coal, per hour	3.675
Gas, per hour	2.000
Gasoline, per hour	1.248
Kerosene, per hour	1.092

An investigation of the table shows that electricity would cost twice as much as gas. In most localities, however, the unit cost of electricity would be higher. Also the prices of gasoline and kerosene are very low for a good many places. It is usually considered that cooking by coal is cheaper than by gas.

The advantages of electric cooking are that there are no ashes, smoke, or soot, very little heat and no danger of explosion.

ENAMELED WARE MENDER.

For mending holes in enameled ware James S. Longhurst, Jr., Lynbrook, L. I., N.Y., is now putting up a material called Agatite in a round, turned wood box 1 5-16 in. in diameter, that retails for 15 cents. This mending material is also sold in larger sizes at 25 and 50 cents each. A catch phrase the manufacturer uses is that "Agatite makes Agate tight," mending all kinds of enameled wares, tin, sheet iron, glass, crockery, etc., but designed primarily for repairing instantly at slight cost house furnishings and cooking utensils. The substance is a black paste, a little of which is pressed through and around both sides of a hole, then heated and lastly covered with cold water, when the

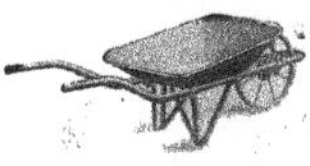
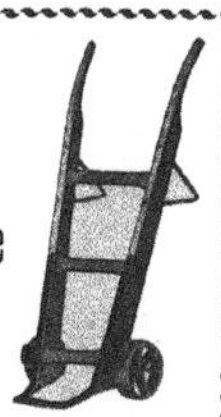

repair is made, later heatings improving rather than deteriorating the repair. It is absolutely harmless, we are informed, and is made in black only. Early in the coming fall the maker expects to have on the market the same material in stick form for use on steam, water and gas pipe, which, when applied will withstand 40 pounds pressure.

EASILY MADE PIPE COVERING.

A home-made pipe covering can be made from asbestos paper and links of wood, it is stated, as follows. First wrap the pipe with asbestos paper and then lay strips of wood lengthwise along the pipe, using five to ten of these, according to the size of the pipe, and binding them with wire or cord. Finally, wrap roofing paper around the strips. The spaces between the strips form a dead air space, which, of course is a nonconductor of heat. At flanges room enough may be left to give access to the bolts and the space afterward filled up with hair felt. Tarred paper may be used for the outside wrapping where the paper is exposed to the weather.

STOVES ON INSTALLMENTS.

Since some of the large city house-furnishing establishments have been so successful in selling stoves and everything else on the installment plan, many dealers in the smaller towns have taken it up with marked success. It is a whole lot easier for the young married couple to pay four or five dollars down and a dollar or so a month than it is for them to plank down the whole price of the stove in one big lump, and if the business is carefully managed and payments are made when they are due, it is about as satisfactory to the dealer. Stoves can be sold in this way to people who haven't the ready cash and whom the dealer would not care to trust indefinitely.

ELECTRIC VENTILATING FANS.

To be successful, ventilation must be positive. Natural means may serve at some seasons and under some conditions, but mechanical means are always to be relied upon. Even on the high seas where the breezes blow free, the old time ventilator funnel is giving way to the compact and adaptable fan blower. A wonderfully complete installation was made on the recently completed U. S. battleship New Hampshire, which is equipped with no less than twenty-five electrically-driven Sturtevant fans of varied types and sizes. These are scattered all over the ship, being applied for boiler and engine room ventilation, for renewing the air in cabins, mess rooms and holds; in fact they have been placed with the utmost ease just where they were wanted.

The small fans with cast-iron casings are driven by bi-polar motors, while the larger fans of steel-plate construction have four-pole or eight-pole motors. All of the fans are so made that they may be adjusted to discharge in any direction by slacking the nuts and turning the casings. For fire-room service, the motors are of the enclosed type, to prevent damage from dust in the air; for general ventilation in other parts of the ship they are of the semi-enclosed, or of the open type. The motors are so designed that there is an allowance of 20 per cent. field variation below the maximum speed.

What is done on shipboard is only suggestive of the much more extended usefulness of the electric fan on land. It can be placed anywhere and arranged to force fresh air into a room, or withdraw the foul air from it. By means of suitable hoods the ventilation may be rendered absolutely local. Odors, vapor and heated air may be drawn directly away from the kitchen range; the dye-house vat, the paper machine, and the steaming kettle may be in like manner rendered inoffensive; from the emery wheel and the tumbling barrel the dust may be completely removed and the planing mill kept clear of shavings. In a word, with a fan the air is made to move regardless of wind and weather, and with an electric fan the operation is simplicity itself.

THE CHIMNEY FLUE.

The first essential in setting up a range or stove is to examine the chimney and see that it has sufficient flue space and is of proper height. The flue should be at least eight inches in diameter and as high as any wall near it. If not high enough it should be extended with brick, if possible, but if a sheet-iron cap is found necessary the pipe should be at least eight inches in diameter and should fit the chimney tightly. If any air is admitted around the top of the chimney it will decrease the draft. It should be straight and should never have two openings exactly opposite each other. The flue that is used for the stove should not be used for ventilating the cellar and all openings above or below the stove should be tightly closed.

THE WORLD'S COAL.

The world still has a considerable supply of coal. Germany is credited with 280,000,000,000 tons, sufficient to last two thousand years at the present rate of consumption; Great Britain and Ireland claim 193,000,000,000 tons, with an annual consumption about double that of Germany; Belgium has 23,000,000,000 tons; France, 19,000,000,000; Austria, 17,000,000,000 and Russia, 40,000,000,000. North America is believed to have 681,000,000,000,000 tons, more than the total of the other countries named. It is the tremendous increase in the use of coal that justifies alarm, for, while the supply of the United States would last four thousand years at the rate of consumption in 1905, it will be exhausted within a century if the rate of increase of the last ninety years continues. No estimates of the coal of other parts of the world can be made, but Asia is known to have an enormous store.

MANUFACTURING AMERICAN FLAGS.

Japan is manufacturing the American flag and exporting it to the United States under a tariff tax of fifty per cent.

CATALOGUES AND BOOKLETS.

When sending catalogues for review, manufacturers would confer a favor by pointing out the new articles that they contain. It would assist the editor in writing the review.

By mentioning HARDWARE AND METAL to show that the writer is in the trade, a copy of these catalogues or other printed matter will be sent by the firms whose addresses are given.

Ranges and Stoves.

The firm of A. Belanger, Montmagny, Quebec, have issued to the trade an artistically made catalogue of 65 pages, illustrating and describing their line of ranges, stoves, harrows and plows. The booklet throughout is printed both in English and French, the information is arranged in an interesting manner, and the cuts illustrating their various commodities are very clear.

At the back of the booklet are also illustrations of their sink pumps and washing utensils. All interested in these lines and not having yet received this booklet will do well to procure a copy.

MARKET FOR OLD TINFOIL.

Save and sell your tinfoil. The recent rise in the price of tin has led to a curious development in this and other countries. Several of the best-known chocolate manufacturers on the continent have issued the following notice: "Do not throw away the tinfoil in which the chocolate is enveloped. It is composed of pure metal, a metal which is dear. Keep it, and before long it will be called for by our agents, who will pay for it at its market value. The chocolate industry in Europe spends nearly $4,000,000 per annum in tinfoil, and these $4,000,000 are generally thrown to the winds." It is further explained, that the present high price of tin is due to the action of English and Dutch speculators, who have forced it far beyond its actual value. What seems to give some color to the alleged preciousness of the paper wrapped around the chocolate, is the story told by a socialist journal of Hamburg, to the effect that a group of workmen were able to procure a part of their common library, by collecting and selling these fugitive sheets of tinfoil.

FRESH WATER IN OCEAN.

Shallow ponds of fresh water on the surface of the ocean are occasionally encountered. This segregation is believed to be due principally to the melting of icebergs, with a subsequent lack of winds and currents to cause the usual intimate mixture. A still more curious feature is that these strata of fresh water offer considerably more resistance to the passage of a ship through them than does salt water. This appears to be because the vessel in passing produces two sets of waves in the two strata, causing relative movement between them, with resultant friction and retardation of motion. Experiments have shown the plausibility of both the separate wave and the loss of headway theories to be based upon correct reasoning, for such phenomena may be readily reproduced and their effects measured.

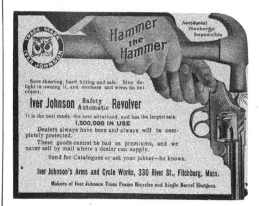

BUILDING AND INDUSTRIAL NEWS

For additional items see the correspondence pages. The Editor solicits information from any authoritative source regarding building and industrial news of an- sort, the formation or incorporation of companies, establishment or enlargement of mills, factories or foundries, railway or mining news.

Industrial Development.

A biscuit factory will be erected in Toronto to cost $10,000.

A large modern laundry will soon be erected at Vancouver, B C.

Fire recently did considerable damage to the power house at Stratford, Ont.

Fisher Bros., Toronto, will erect a galvanized iron warehouse to cost $3,-000.

Wells, Kaine & Co., Grand Bay, Que., will start an Excelsior factory in that place.

Work is progressing on the new high pressure plant for the Winnipeg waterworks.

The Cataract Power Co. have the contract for supplying power to Brantford, Ont.

Fire did damage to a mill and elevator at Russell, Man , to the extent of $25,000.

The Eagle Knitting Co , Hamilton, are going to erect a new factory in that city.

The Ontario Power Co's. transformer house near Welland, Ont., was burned out recently.

The Bell Telephone Co. have installed a metallic line from Tara to Owen Sound, Ont.

The machinery is being installed in the plant of the Western Canneries at Medicine Hat.

The new plant for the Hespeler Hoisting Machinery Co., Hespeler, will soon be in operation.

Plans are under way for a big iron and steel plant for Vancouver, B.C , to cost $15,000,000.

The official tests of the new 800 kilowatt plant at Morrisburg, Ont., were completed recently.

The new power plant at Woodstock was recently put in operation, and is giving entire satisfaction.

It is said that the power capable of being developed at Kakabeka Falls, Ont., is 100,000 horse power.

The Stoney Lake Navigation Co., Peterborough, Ont., will build a new boat for service on the lakes.

The electric plant on the Wabi river will soon be in operation, supplying power to New Liskeard, Ont.

The Canadian Pneumatic Tool Co., Ottawa, have replaced their producer gas plant with electric motors.

The Northumberland & Durham Power Co. are contemplating the transmission of power to Kingston, Ont.

The Canadian Independent Telephone Co , Toronto, is organized, and will erect works in Hamilton or Toronto.

The capital stock of the Canadian General Electric Co., Peterborough, Ont., has been increased to $8,000,000.

The new firm of Thackeray & Sproule, Ottawa, will erect a factory for the manufacture of office supplies and fixtures.

The British Columbia Electric Railway Co., will supply Ladner, B.C., with light. The power is generated at Lake Buntzen.

T. Gray & Co., New Westminster, B.C., will erect a saw mill on the Pitt River. The machinery for the mill has been purchased.

A company has been formed at Minnedosa, Man , for the development of power there A hydro-electric plant will be installed.

A large power plant is contemplated on the Yukon River about 50 miles below Dawson. The initial capacity will be 1,000 horse power.

The Steel-Coal trial which has been going on recently in Sydney, N.S., has been brought to a close. The decision will be given shortly.

The Moleton Truck and Storage Co's. building, Toledo, Ohio, occupied by the International Harvester Co., was destroyed by fire recently.

The Graham Co., Vancouver, will erect a large saw mill at the Queen Charlotte Islands. It will have a capacity of 200,000 feet per day.

The puddling furnaces and rolling recently established at Winnipeg make the fourteenth iron and steel plant now in operation in Canada.

The Copeland-Chatterson Co., Toronto, will take over the entire business of the Elliott Fisher Billing Machine Co., for the Dominion.

E. H. Keating and Wm. H. Breithaupt, Toronto, have formed an engineering partnership, with offices at Aberdeen Chambers, Toronto.

United States capitalists are considering the erection of a large dry dock, capable of holding the large lake vessels, near St. Catharines, Ont.

The contract has been awarded to the Canadian Iron & Foundry Co , Montreal, for 21 hydrants and 140 tons of pipe by the town of St. Mary's, N B.

The new plant of the Canadian Brass Manufacturing Co., at Galt, is now in full operation, and will manufacture plumbing supplies and brass castings.

The Saraguay Electric Light & Power Co., Montreal, are erecting a new 11,-500 volt, three phase transmission line to Notre Dame de Grace, a distance of nine miles.

The second brick plant for Rosthern, Sask., will be erected soon. A company has been formed and a plant with a daily capacity of 60,000 bricks will be erected.

The new blast furnace of the Atikokan Iron Co., Port Arthur, Ont., will soon be running at its full capacity. The company is considering the doubling of its plant at Port Arthur.

Plans are being made for the erection of an enormous plant in the Kootenay district for the manufacture of steel rails It is hoped that the plant will be able to fill orders in two years.

The British Columbia Electric Railway Co , Vancouver, B C., are pushing a scheme which supplies electric power to portable saw mills in the forests. It is claimed to reduce the cost of cutting timber.

The Excelsior Factory, operated at Chilliwack, B.C., by Messrs. Kipp & Sons, has been purchased by the Barbara Mattress Company, of Vancouver, and the machinery has been removed to that city.

A telephone company has been formed in Princeton, Ont., to be known as the Princeton, Drumbo, Rural Telephone Co. A line will be run connecting Princeton, Drumbo, Eastwood and Gobles F. J. Daniel, Princeton, is president.

Companies Incorporated.

Monterey Plumbing Co., Toronto; capital, $50,000 To carry on a plumbing business. Incorporators include R. Ruel, G. F. Macdonnell and A. J. Mitchell, all of Toronto.

Record Stove & Furnace Co., Winnipeg; capital, $40,000. To deal in and manufacture stoves and furnaces. Incorporators: J. Peters, A. E. Peters, D. I. Welsh, all of Moncton, N.B.

The Ozone Sterilization Co., Haileybury, Ont.; capital, $100,000. To acquire and operate any processes and inventions for purifying liquids, ores, etc. Provisional directors: W. A. Gordon, R. O. Morrow, F. A. Day, all of Haileybury.

The Canadian Lash Steel Process Co., Toronto; capital, $100,000. To manufacture and sell iron, steel and other metals and their alloys. Provisional directors: J. A. Macintosh, B. W. Essery, Eleanor J. Potts, J. G. Adair, and John Adair, all of Toronto.

The Algoma Co-operative Co., Sault Ste. Marie, Ont., capital $40,000, to sell, manufacture and deal in general wares, hardware, paints, oils, etc. Provisional directors: W. Stringer, J. Cleland, D. Robertson, D. Dewar and D. Donald, all of Sault Ste. Marie.

Building Notes.

A. Peters will erect a hotel in New Westminster to cost $10,000.

A new Anglican Church will be erected at Elmwood, Man., to cost $5,-000.

The Anglican Parish of St. Albans, Burnaby, B.C., will erect a church to cost $2,000.

The Bank of Commerce will build a branch on Yonge street, opposite College street, Toronto.

Mining News.

A large ore crusher will be added to the plant of the Granby Co., at the Boundary mines.

The Salt Lake Smelting & Refining Co., will commence the erection of their smelter at North Cobalt soon.

During the first month of the last half of the year the Boundary smelters, B.C., reduced 150,000 tons of copper ore.

The plant of the Algoma Steel Co., Sault Ste. Marie, has shut down on account of the lack of ore and need of repairs.

Owing to the shortage of coke in the west, it is feared that the furnaces at

CONDENSED OR "WANT" ADVERTISEMENTS.

Advertisements under this heading 2c. a word first insertion; 1c. a word each subsequent insertion.

Contractions count as one word, but five figures (as $1,000) are allowed as one word.

Cash remittances to cover cost **must** accompany all advertisements. **In no case** can this rule be overlooked. Advertisements received without remittance cannot be acknowledged.

Where replies come to our care to be forwarded, five cents must be added to cost to cover postage, etc.

AGENTS WANTED.

AGENT wanted to push an advertised line of Welsh tinplates; write at first to "B.B.," care HARDWARE AND METAL, 88 Fleet St., E.C., London, Eng. [tf]

ONE who has sufficient organization to travel Canada east and west, fine line of Sheffield cutlery, brands already well known in Canada; none but applications from experienced and well-established men can be considered. Box 650, HARDWARE AND METAL, Toronto. [38]

BUSINESS CHANCES.

FOR SALE — Well established hardware, tinshop, implement and undertaking business, also good lumber yard, well fenced, with lumber and lime sheds in good condition; we will sell above altogether, or divide same to suit purchaser; proprietors are retiring from business in Manitoba, and therefore wish for immediate sale. Apply to Eakins & Griffin, Shoal Lake, Man. [33]

HARDWARE, tinware, stove and plumbing business in manufacturing town in the Niagara Peninsula; no competition; $250,000 factory and watermarks will be completed this summer; stock about $3,000; death of owner reason for selling. Box 85, Thorold, Ont. [tf]

HARDWARE Business, in good position for sale, and store with modern brick dwelling to rent. Proprietor retiring; Stock $2,000; Rent $35; Business more than doubled in the last two years; Good investment for young business man. Box 651 HARDWARE & METAL, Toronto. (36)

HARDWARE and Tin Business for Sale in good Western Ontario town of 3,000: stock about $3,500; good reasons for selling. Address, Box 647 HARDWARE AND METAL, Toronto. (34)

HARDWARE, Tinware, Stove and Furnace Business for sale. in live Eastern Ontario Village; first class chance for a practical man; English speaking community; stock can be reduced to suit purchaser; can give possession September 15th. 1907; premises for Sale or Rent, Apply to D. Courville, Maxville, Ont. (35)

HARDWARE Business and Tinshop for sale in Saskatchewan; population 1500; stock carried about $14,000 turnover, $45,000 practically all cash business; cash required, $8,000 would rent building; Do not answer without you have the money and mean business; it will pay to investigate this. Box 648 Hardware & Metal, Toronto. (41)

WELL established tinsmith business in thriving Ontario village, including Eavetroughing, metal roofing, furnaces, pumps, pipefitting, hardware, paints, glass and stoves. Present stock about $1500; could be reduced. Shop and house combined, can be rented, including tools, horse and wagon, for $14.00 per month; possession given at once. Box 682, HARDWARE AND METAL. (33)

FOR SALE.

FOR SALE — First-class set of tinsmith's tools second-hand but almost as good as new; includes an 8-foot iron brick almost new. Apply Pease Waldon Co., Winnipeg. [tf]

FOR SALE—A quantity of galvanized plain twist wire. Apply to C. B. Miner, Cobden, Ont. (34)

TEN gross of "Antisplash" Tap Filters. Made to fit on all ordinary kitchen taps, acts as a filter and prevents water splashing. Supplied with showcards to hold 1 dozen. Price per dozen $1.00; price per gross $11.00. Sample on request. G. T. Cole, Box 460, Owen Sound, Ont. [36]

SITUATIONS VACANT.

TINSMITHS WANTED — First-class tinsmiths wanted for points west of Winnipeg; must be good mechanics capable of taking charge of a metal department; thorough knowledge of furnace work necessary. Pease Waldon Co., Winnipeg, Man. [tf]

WANTED general hardware clerk, experienced, sober, good stock-keeper and salesman; steady job for right man; state experience and salary expected. C. Richardson & Son, Box 325, Harrow, Ont. [36]

WANTED—Hardware Clerk, for retail in Hamilton, with 4 years' experience. Address stating experience, also salary expected. Box 652 HARDWARE & METAL (35)

SITUATIONS WANTED.

COMMERCIAL gentleman with nine years' trade connection with ironmongers, architects and public institutions in Great Britain desires position as representative of a Canadian manufacturing firm. Box X, HARDWARE AND METAL, Montreal.

HARDWARE Salesman wants position; can furnish first class references; capable of managing city or town business. Box 653 HARDWARE & METAL, Toronto. (39)

WANTED.

WANTED—Two store ladders, "Myers" preferred, complete with fixtures. Send full description and price to Ilsey Bros., Red Deer, Alta. (36)

THE WANT AD.

The want ad. has grown from a little used force in business life, into one of the great necessities of the present day.

Business men nowadays turn to the "want ad" as a matter of course for a hundred small services.

The want ad. gets work for workers and workers for work.

It gets clerks for employers and finds employers for clerks. It brings together buyer and seller, and enables them to do business, though they may be thousands of miles apart.

The "want ad." is the great force in the small affairs and incidents of daily life.

the Trail smelter, near Rossland, will have to shut down.

It has been found possible to turn out pig iron capable of being turned into the best quality of Bessemer steel, at the Atikokan furnace, Port Arthur, Ont.

The successful experiments made by the Dominion Government with the Heroult electric smelter, at Heroult, Cal., have been corroborated. The process is found satisfactory and much cheaper than usual methods.

Railroad Construction.

Work has commenced on the construction of the new C.P.R. depot at Saskatoon, Sask.

Work is progressing rapidly on the construction of the Brandon-Regina line of the C.N.R.

The G.T.R. will proceed without delay to construct the line from Kingston to Ottawa, Ont.

The new Guelph to Goderich line of the C.P.R. was opened recently. Trains are now running regularly.

There is a rumor that the proposed Hamilton-Galt electric line will be pushed through without delay.

The first steel section of the M.C.R. tunnel under the Detroit river at Detroit, will be laid in a few days.

Active operations are in full swing on the construction of the new addition to the G.T.R. shops at Stratford, Ont.

Construction will begin on the Kootenay Central branch of the C.P.R. from Golden, B.C., to the Crow's Nest Pass.

The Intercolonial Railway have recently given to the Rhodes, Curry Co., Amherst, N.S., the contract for 260 flat cars, 400 box cars, 25 refrigerator cars and four conductor's vans.

The Canadian Railway Commission have issued new regulations regarding the watching of trestles and bridges, fire protection for same, watching of tracks and fire protection for cars.

The C.P.R. will build two large steamers of the Princess Victoria type for service on the Pacific. The builders are from the Fairfield Shipbuilding & Engineering Co., Govan, Scotland.

The British Columbia Electric Railway have ordered 26 cars for city service. In order to turn these out, the number of men employed at the car shops will be doubled and two shifts organized to work night and day. The cars cost $6,000 each.

Municipal Undertakings.

Public buildings will be erected at Neepawa, Man., and Selkirk, Man.

Moose Jaw, Sask., will raise $90,000 for a municipal lighting system.

The town of Gracefield, Ont., will instal a municipal waterworks system.

The foundations have been laid for the new civic buildings at Saskatoon, Sask.

Surveys are being made for the new government telephone system for Winnipeg.

The ratepayers of Cornwallis, Man., are petitioning for a municipal telephone system.

The new light plant for Claresholm, Man., has been completed. It replaces the one destroyed by fire recently.

A Provincial asylum will be erected by the government of British Columbia at Coquitlam, B.C. It will cost $200,000.

Bids will be asked by the Winnipeg Board of Control for a $40,000 bridge in connection with the Point du Bois power plant.

The city of Revelstoke, B.C., is calling for tenders for a 500 horsepower electric plant, to be run by a producer gas engine.

The new building of the government telephone system at Winnipeg will shortly be commenced. It will be completed by the first of the year.

The council of Ottawa, Ont., are considering the erection of a municipal lighting plant, to be put in operation when the Hydo-electric contract expires.

The ratepayers of the village of Tetreauville, Que., will vote on a by-law for issuing $15,000 of debentures and establishing a municipal waterworks system.

The ratepayers of Aylmer, Ont., decided to grant a loan of $10,000 to the Canadian Condensed Milk Co., as well as a free site and exemption from taxes for ten years. They will establish a plant there.

WORLD'S SUPPLY OF MANGANESE ORE.

It is reported that so far as the United States is concerned manganese ore development in recent years has amounted to very little. The condition of the United States in respect to true manganese is very much like that in regard to tin—a practically complete dependence upon outside sources. The latest figures show that out of the world's production of 1,100,000 tons, Russia produced 426,813 gross tons in 1905; Brazil, 233,950 tons; India, 150,297 tons; Germany, in the previous year, 52,866 tons; Turkey, 49,100 tons; Spain, 26-985 tons; Chile, 17,110 tons in 1903; Austria, 13,788 tons; Hungary, 11,527 tons in 1904; with France, Japan, Greece and Cuba following with lesser supplies.

LUBRICATING AND BEARINGS.

With a babbitted bearing, any kind of grease or oil may be used, it has been stated, but when the former has been softened down with lead or with any alloy having zinc and lead in its composition, care should be exercised in the selection of the oil used to lubricate it, as such selection forms a considerable factor in the life of the bearing. It is often a question in designing the commoner class of machinery, such as agricultural implements or coal elevating outfits, whether to put in high priced bearings, with which any kind of cheap lubricant can be used; or to employ bearings of less cost, which, to be equally as efficient and to wear as long, would necessitate the use of a lubricant costing two or three times as much as in the former case. In such a case as this, it is largely a question of balancing the greater cost of the better lubricant against the interest and depreciation on the more expensive bearing, and choosing from the basis of such a commercial comparison.

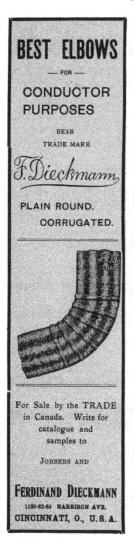

News of the Paint Trade

PAINTS AT THE CONTRACTORS' SHOW.

Amongst the very effective exhibits at the Contractors' and Builders' Show held at Victoria Rink, Montreal, last week, were two representing the products of two prominent paint manufacturers, Martin-Senour Co., and P. D Dods & Co. This exhibition, at which were shown products of the various manufacturers of building and decorative supplies, afforded a splendid opportunity to paint makers to display their goods effectively and to a class of spectators, the majority of whom were consumers or probable consumers.

The exhibit of Martin-Senour Co's. 100 per cent. pure paints was highly deserving of praise. The arrangement of their various commodities was very attractive, especially in the evenings, when the electric lights, which were of varied colors, were turned on. Not only was the arrangement of their commodities alone deserving of praise, but the kind attention paid to all who came to the booth interested in the display by J. A. Naud, who was in charge of the exhibit. Informing and interesting literature was distributed by Mr. Naud. One of the booklets distributed at the booth is entitled, "Cheerfulness Producers and Money-Makers for Thrifty Folks," a booklet full of information and interest, and one which all interested in paint will do well to procure.

P. D. Dods & Co. had a very attractive and practical display of their commodities and what their products would do. The booth was under the able supervision of J. H. Weber, who took great pains to give demonstrations of the finishes capable of being produced by paint and varnishes of the "Island City" Paint Co. Many contractors and builders were deeply interested in a peculiar and effective finish effected on a piece of maple by the application of stain and varnish respectively. Valuable information was given out by Mr. Weber verbally and through pamphlets.

The Standard Paint Company, Montreal, devoted their space chiefly to a display of Ruberoid Roofing.

PAINTING CEMENT SURFACE.

After a cement surface has stood exposed to the weather for a year it is safe enough to paint it over. But it will not do to paint a cement surface under much less time than that, unless surface has first been sized with acid water, to kill the alkali, and even then there is some danger of bad results. Here is a somewhat tedious method for prepairing and painting such a surface, but it has the sanction of some of our best painters, says Hardware Trade. Slack one-half bushel of fresh stone lime in a barrel, and add in all 25 gallons of water; when slacked, and cold, add six gallons of the best cider vinegar and five pounds of best dry venetian red. Now mix well and then strain through a fine wire strainer. Use it when about the consistency of thin cream. Give the cement surface a coat of this, and alter standing a day or so apply a coat of red lead and linseed oil

paint. After this has dried you may paint the surface any color you wish. Some jobs require two coats of paint over the red lead paint. In this case make the second coat of paint serve as filler and paint both. This second coat may be made with plaster of paris and oil, of the consistency of buttermilk. Then break up some white lead and oil to make a paint the same consistency as the plaster paint. Now take equal parts of each of the two mixtures and "box" them together, and thin to a working consistency with turpentine. This second coat should be applied as heavy as possible, or as heavy as you can spread it well. After this coat is dry apply your next and finishing coat of paint, which should be quite glossy, or about as you would for the last coat on woodwork outside. The object in giving it this plaster paint is to prevent the running and wrinkling of the paint where considerable paint is to be applied to the surface. And it must be made to dry quickly, so that you will not likely give the finishing coat before the second coat is dry enough, for if you do that there will be blistering or cracking. Observe particularly that no plaster is to go in the last or finishing coat.

PAINT AND OIL MARKETS

MONTREAL.

Montreal, August 30.—Manufacturers are busy on specialties, but, in general, trade is characterized by the usual midsummer quietness. Local prices are firm and unchanged.

Owing to the uncertainty existing in Western grain circles considerable apprehension is expressed by local dealers regarding the prospects for Western paint trade during the coming season. The scarcity of money to lift the crops this fall has incited the banking institutions throughout the country to adopt a sternly conservative policy and to hold out little encouragement to real estate men. There is little cause for any excitement amongst paint makers regarding the Western situation, as the consumption of building and decorative supplies is bound to be maintained by the steady industrial progress throughout Canada, especially in the North-West districts.

Linseed Oil.—The local market at present is quiet; consumers are calling for moderate supplies, and we continue to quote: Raw, 1 to 4 barrels, 60c.; 5 to 9, 59c.; boiled, 1 to 4 barrels, 63c.; 5 to 9, 62c.

Turpentine.—Dullness characterizes the turpentine market, both local and southern, conditions quite in keeping with what is commonly known as the "tween" season. Little buying is being done. Prices are firm and unchanged: Single barrels, 80c.

Ground White Lead.—Business is moderate. Local corroding work is active. Prices are firm: Government standard, $7.50; No. 1, $7; No. 2, $6.75; No. 3, $6.35.

Dry White Zinc.—The situation is unchanged. Prices are firm and business moderate. We quote: V.M. Red Seal, 7½c.; Red Seal, 7c.; French V.M., 6c; Lehigh, 5c.

White Zinc Ground in Oil.—We continue to quote: Pure, 8½c.; No. 1, 7c.; No. 2, 5⅝c.

Red Lead.—Trade at present is moderate with probabilities of enlivenment in about a month. Prices are unchanged: Genuine red lead, in casks, $6.25; in 100-lb. kegs, $6 50; in less quantities at $7.25 per 100 lbs. No 1 red lead, casks, $6; kegs, $6.25, and smaller quantities, $7.

Gum Shellac.—Demand is quiet. We continue to quote: Fine orange, 60c per lb., medium orange, 55c. per lb., white (bleached), 65c. per lb.

Shellac Varnish.—Seasonable business is being done. Pure white bleached shellac, $2.80; pure orange, $2.60; No. 1 orange, $2.40.

Putty.—An active demand exists and makers are busy endeavoring to supply the current demand and lay up stocks for future rushes. Prices are firm: Pure linseed oil, $1.85 in bulk; in barrels, $1.60; in 25-lb. irons, $1.90; in tins, $2; bladder putty, in barrels, $1.85.

Paris Green.—The season is almost concluded and makers are eminently satisfied with this season's business, which although of short duration, was large in volume.

TORONTO.

Toronto, August 28.—Business continues rather quiet in local paint and oil circles, but jobbers are not complaining, for they do not figure on being particularly rushed at the close of August. A fair amount of orders, however, continue to arrive for paints and oils in sorting quantities, and business in glass and putty is steadily increasing with the approach of fall. Yesterday the Toronto Exhibition was formally opened, and from now till the 9th of September managers of local supply houses will be kept busy attending to outside customers, many of whom make it a point to call during the fair.

White Lead.—Trade in this commodity is good for the closing week in August, and quite a number of sorting orders continue to arrive. Prices are unchanged and remain firm as under: Genuine pure white lead is quoted at $7.65, and No. 1 is held at $7.25.

Red Lead.—A few enquiries for red lead in sorting quantities continue to arrive, but otherwise there is little doing in the market. Prices show no disposition to change, and remain firm at the following figures: Genuine, in casks of 500 lbs., $6.25; ditto, in kegs of 100 lbs., $6.75; No. 1 in casks of 500 lbs., $5; ditto, in kegs of 100 lbs., $5.50.

Glass and Putty.—Fall trade is beginning to open up, and many enquiries are coming in from retailers whose orders were not booked in advance. There is a strong demand for putty, and the makers are kept busy supplying the needs of the trade. Prices on both glass and putty are unchanged.

Petroleum.—The demand is steadily increasing owing to the lengthening nights, and in a month at most the

autumn trade will be in full swing. Prices continue unchanged as follows: Prime white, 15c., water white, 14½c. Pratt's astral, 18c.

Shellac.—Trade is dull in shellac as is usual at the close of August; jobbers have had a good season in shellac, and even yet a few repeat orders are being filled. Present prices are: Pure orange, in barrels, $2.70; white, $2.82½ per barrel; No. 1 (orange), $2.50.

Linseed Oil.—The local market has assumed a stronger tone since last week, though the market has not stiffened enough to warrant an advance in price. The British market is slightly firmer at present, although large shipments of Indian seed have arrived within the last few days, and larger consignments are said to be on the ocean en route for the British Isles. The local demand for oil is very weak, and few sales are reported for the week. Prices continue as follows: Raw, 1 to 3 barrels, 63c; 4 barrels and over, 62c Add 3c. to this price for boiled oil f.o.b. Toronto, Hamilton, London, Guelph and Montreal, 30 days.

Turpentine. — Prices have gained strength locally since last week, but as the demand is only fair, no advance has been made. The following figures still hold good: Single barrels, 79c.; two barrels and upwards, 78c. f.o.b. point of shipment, net 30 days; less than barrels, 5c. advance. Terms, 2 per cent., 30 days.

For additional prices see current market quotations at the back of the paper.

ORIGIN OF PETROLEUM.

A. E. McKercher, of Humble, Texas, has a theory of the origin of petroleum. In a letter to the Oil Investor's Journal he says:

"Oil is in different sands, because it originated at different depths, and this brings me to its origin. I am forced to believe that crude petroleum is the refined pitch of bituminous coal, because numerous veins that I have traced to their extreme end invariably trickle out from under beds of that commodity, and there are just as many different kinds and grades of coal as there are different kinds of hydro-carbons. Take, for instance, elaterite, gilsonite and ozokerite; they are just as dissimilar as are the paraffine and asphalt oils, and yet they show a kindred origin and were at one time a liquid.

"This leads me to a much discussed subject, viz.: salt water. Every trend or vein of oil must be accompanied by or in close proximity to a flow of salt water. Why? To keep the even temperature that sustains its life; otherwise it would become dead and solidify. I can show you evidences of this in the States of Colorado, Utah and Wyoming, where the oil and salt water flowing up against a solid wall, the water has evaporated, leaving a bed of salt, and the oils in consequence have turned to elaterite, gilsonite or asphaltum, owing to the difference in the oil as the case might have been. Hence, salt water is a necessary evil in the oil fields, but it is never found in the same sands with the oil (except in porous rock where the oil has eaten its way through with the assistance of the water), unless let in through the carelessness of some operator. And I have known it to travel a distance of from ten to twenty miles, driving the oil before it, and ruining fields at such a distance that they were puzzled to know whence it came. There ought to be a law in this State—if not one already such—as exists in many other States, making it an offense and imposing a heavy fine for failure to plug a salt well just as soon as it is found to be so.

"I can show you trends of oil and gas in this coast country that are from two to five miles wide, that can be traced without having to develop them, for a distance of over 100 miles. Why are they so wide and pronounced in the south? For the simple reason that the smaller streams to the north gradually getting into the same sands augment their forces until they form the mighty river that sweeps on and on until it finds its gravity basis as stated heretofore. It is not unlike the waters of our great rivers as seen upon the surface. Oil, like water, casts off or picks up sediment, and is purified or befouled by the stratification through which it passes, which accounts for its gravity, in a measure."

CRACKING OF VARNISH.

One common cause of cracking of varnish is the adding of terebine to a varnish in order to make it harden quickly, especially when exposed to the sunlight. These cracks at first give the varnish a silky appearance, due to their hairlike fineness and great numbers. Subsequently many of the cracks open out wider under atmospheric variations.

The application of any hard, quick-drying coat of paint or varnish on a soft undercoat is liable to cause cracking, and would affect any second coat in the same way.

Another cause of cracking is the application of a coat of size upon a hard, non-porous ground prior to varnishing, such as sometimes occurs when re-varnishing old work in cheap jobs, if the size be fairly strong, the cracks caused being generally of polygon shape and the edges having a tendency to curl outwards.

To avoid tendency to cracking, there is no better course than to take care that every coat prior to varnishing be thin and allowed to dry hard before applying the following coat. It is important also that no quick-drying medium, such as gold-size or terebine, be used in painting over a coat mixed with ordinary linseed or boiled oil, though the reverse order may be employed without danger. A hard varnish may be used as an undercoat, and an elastic finishing over-varnish over that.

"FALCONITE."

A special white enamel very largely used in various Government departments and by leading decorators throughout the world for high-class decorative work inside and out, is being supplied to the trade by Wilkinson, Heywood & Clark, London, Eng. "Falconite" dries very hard with a brilliant gloss, and is very durable. The makers claim for it a great flowing power, elasticity, and easy application. F. C. Reynolds, Montreal, is the Canadian representative of the manufacturers.

ASBESTOS
STOVE LINING

Mr. Dealer! It's up to you to handle the best Floor Lining on the market. Our Asbestos Lining has proved itself a more durable and econ-nomical lining for Cooking and Heating Stoves, Grates, Ranges, Hot Air Furnaces, than Fire Brick, and is especially adapted for Grates, Ranges, etc., where natural Gas is used. Fits any stove; can be applied by anyone in a few minutes.

Put up in 6-lb Packages and Bulk.

ASBESTOS FIREPROOF FURNACE CEMENT

is used for mounting and repairing Furnaces, Ranges, Heaters, Stoves, etc., making them gas-tight. Always ready for use; adheres firmly to the castings and sets in a few hours as hard as iron. Odorless, and cannot be injured by acids or chemicals. A thoroughly-satisfying Furnace Cement and a profitable line.

Benjamin Moore & Co., Limited

(Formerly The Francis-Frost Co., Limited)

New Offices and Works: Toronto Junction, Canada

New York, Chicago, Cleveland, Carteret, N. J.

AGENTS:

W. PORTEOUS JACK, 1256 Robson, Vancouver.
GEO. PEATT & CO., P. O. Box 466, Winnipeg, Man.
BAXTER, PATERSON & CO., 102 St. Antoine St., Montreal.

Specify the Following Products

MANUFACTURED BY

American Sheet and Tin Plate Company

PITTSBURG, PA.

IF YOU WANT THE BEST:

**"Apollo Best Bloom" Galvanized Sheets,
Blue Annealed and American Bessemer
Black Sheets,
Wellsville Polish, Range Steel,
Wood's "A" Planished Iron,
Black Steel specialties for all purposes,
Tinplates, Coke and Charcoal, all grades;
Terne Plates, Old Style and New
Method Processes.**

PLACE YOUR ORDER WITH

B. & S. H. THOMPSON & CO., Limited

Dominion of Canada Sales Agents, MONTREAL, CANADA

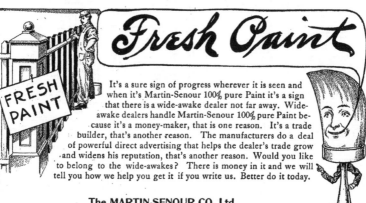

CURRENT MARKET QUOTATIONS.

August 29, 1907

These prices are for such qualities and quantities as are usually ordered by retail dealers on the usual terms of credit, the lowest figures being for larger quantities and prompt pay. Large cash buyers can frequently make purchases at better prices. The Editor is anxious to be informed at once of any apparent errors in this list, as the desire is to make it perfectly accurate.

METALS.

ANTIMONY.

Cookson's ....per lb....	0 15
Hallett's	0 14½

BOILER PLATES AND TUBES.

Plates, ¼ to ½ inch, per 100 lb.	2 40	2 50
Heads, per 100 lb	2 65	2 75
Tank plates 3-16 inch	2 60	2 70
Tubes per 100 feet, 1½ inch ..	8 25	8 50
" " 2 " ..	9 00	9 10
" " 2½ " ..	10 50	11 00
" " 3 " ..	12 00	12 50
" " 3½ " ..	15 00	16 00
" " 4 " ..	19 25	20 00

BOILER AND T.K. PITTS

Plain tinned and Spun, 25 per cent. off list.

BABBIT METAL.

Canada Metal Company—Imperial genuine 60c; Imperial Tough, 56c; White Brass, 50c.; Metallic, 35c; Harris Heavy Pressure, 25c.; Hercules, 25c; White Bronze, 15c; Star Frictionless, 14c; Alluminoid, 10c; No. 4, 9c per lb.

James Robertson Co.—Extra and genuine Monarch, 60c; Crown Monarch, 50c; No. 1 Monarch, 40c; King, 30c.; Fleur-de-lis, 20c; Thurber, 15c; Philadelphia, 12c; Canadian, 10c; hardware, No 1, 15c.; No. 2, 12c.; No. 3, 10c. per lb.

BRASS.

Rod and Sheet, 14 to 30 gauge, 25 p c. advance.

Sheets, 12 to 14 in.	0 30
Tubing, base, per lb 5-16 to 2 in	0 33
Tubing, ⅜ to 3-inch, iron pipe size..	0 31
" 1 to 3-inch, seamless..	0 36

Copper tubing, 6 cents extra

COPPER. Per 100 lb

Casting ingot........... ...	23 00	23 50
Cut lengths, round, bars, ½ to 2 in...		35 00
Plain sheets, 14 oz......		36 00
Plain, 16 oz., 14x48 and 14x60		35 00
Tinned copper sheet, base		38 00
Planished base..................		43 00
Braziers' (in sheets). 4x6 ft., 25 to 30 lb. each, per lb., base..	0 34	0 35

BLACK SHEETS.

	Montreal.	Toronto
10 gauge	2 70	2 75
12 gauge.....................	2 70	2 75
14 "	2 50	2 60
17 "	2 50	2 60
18 "	2 50	2 60
20 "	2 50	2 60
22 "	2 55	2 65
24 "	2 55	2 70
26 "	2 65	2 85
28 "	2 70	3 00

CANADA PLATES.

Ordinary, 52 sheets	2 75	3 05
All bright "	3 75	4 05
Galvanized—	Dom.Crown.	Ordinary.
18x24x52..................	4 45	4 35
60..................	4 70	4 60
20x28x80	8 90	8 70
....	9 40	9 20

GALVANIZED SHEETS.

B.W. gauge	Queen's Head	Fleur-de-Lis	Gordon Crown	Colborne-Crown Gorbal's Best
16-20 ..	3 95	3 80	3 95	....
22-24 ..	4 20	4 05	4 00	4 05
26........	4 45	4 30	4 40	4 30
28........	4 70	4 55	4 60	4 55

Less than case lots 10 to 25c. extra.

Apollo Brand.

24 guage, American	3 85
26 " "	4 10
28 " "	4 55
10¾ oz.	4 85

25c. less for 1,000 lb. lots.

IRON AND STEEL.

	Montreal.	Toronto.
Middlesboro, No. 1 pig iron..	21 50	24 50
Middlesboro, No. 3 pig iron	20 50	23 50
Summerlee, " ..	25 50	26 50
" special " .	24 50	....
" soft " ..	24 00	....
Carron...	26 00	....
Carron Special	24 50	....
Carron Soft	24 00	....
Clarence, No. 3	21 50	23 50
Glengarnock, No. 1		27 00
Midland, Londonderry and Hamilton, off the market but quoted nominally at		26 00
Radnor, charcoal iron.........	32 00	34 00
Common bar, per 100 lb......	2 20	2 30
Lowmoor iron......................		6 50
Angles...		2 50
Forged iron "	2 45	..
Refined " "	2 60	2 70
Horseshoe iron "	2 60	2 70
Band iron, No. 10 x ⅜ in.		3 60
Sleigh shoe steel "	2 25	2 30
Iron finish steel		2 40
Reeled machinery steel....		3 50
Tire steel	2 40	2 50
Best sheet steel......		0 12
Mining cast steel....		0 08
Warranted cast steel		0 14
Annealed cast steel....		0 15
High speed...........		0 60
B.P.L. tool steel	10½	0 11

INGOT TIN.

Lamb and Flag and Straits—56 and 28-lb. ingots, 100 lb. $43 00 $43 50

TINPLATES.

Charcoal Plates—Bright

M.L.S., Famous (equal Bradley)		Per box
I C, 14 x 20 base	....	$6 50
IX, 14 x 20 "	....	8 00
IXX, 14 x 20 base	....	9 50
Raven and Vulture Grades—		
I C, 14 x 20 base	....	5 00
I X "	....	6 00
I X X "	....	7 00
I X X X "	....	8 00

'Dominion Crown Best"—Double Coated, Tissued.		Per box.
I C, 14 x 20 base..........	5 50	5 75
I X, 14 x 20 "	6 50	6 75
IXX x 20 "	7 50	7 75
"Allaway's Best"—Standard Quality.		
I C, 14 x 20 base..........	4 65	5 00
I X, 14 x 20 "	5 40	5 75
I XX, 14 x 20 "	6 15	6 50
Bright Cokes. Bessemer Steel—		
I C., 14 x 20 base	4 25	4 35
20x28, double box	8 50	8 70
Charcoal Plates—Terne. Dean or J. G. Grade—		
I C., 20x28, 112 sheets	7 25	8 00
IX., Terne Tin	...	9 50

Charcoal Tin Boiler Plates.

Cookley Grade—
X X, 14x56, 50 sheet bxs. } 7 50
" 14x60, " }
" 14x65, " }

Tinned Sheets.

72x30 up to 24 gauge...... ...	....	8 50
" " 26 "	....	9 00

LEAD.

Imported Pig, per 100 lb......	5 25	5 35
Bar, "	5 75	6 00
Sheets, 2½ lb. sq. ft., by roll ..	6 75	6 50
Sheets, 3 to 6 lb. "	6 50	6 25

Cut sheets ½c. per lb., extra.

SHEET ZINC.

5-cwt. casks	7 75	8 00
Part casks....................	8 00	8 25

ZINC SPELTER.

Foreign, per 100 lb	6 75	7 00
Domestic	6 50	6 75

COLD ROLLED SHAFTING.

9-16 to 11-16 inch	....	0 06
¾ to 1 7-16 "	....	0 05½
1 7-16 to 3 "	....	0 05

30 per cent.

OLD MATERIAL.

Dealers buying prices:

	Montreal	Toronto
Heavy copper and wire, lb.	0 15½	0 15
Light copper...............	0 13½	0 13
Heavy red brass....	0 13	0 13
" yellow brass	0 10	0 10½
Light brass...............	0 07	0 07½
Tea lead	0 03½	0 03½
Heavy lead...............	0 04	0 04
Scrap zinc	0 03½	0 03½
No. 1 wrought iron	15 50	11 50
" " " "	6 00	6 00
Machinery cast scrap ...	17 00	16 50
Stove plate...............	13 00	12 00
Malleable and steel	8 00	8 00
Old rubbers..	0 10½	0 10
Country mixed rags, 100 lbs.	1 00	1 25

PLUMBING AND HEATING

BRASS GOODS, VALVES, ETC.

Standard Compression work, 57½ per cent.
Cushion work, discount 40 per cent.
Fuller work, 70 per cent.
Flatway stop and stop and waste cocks, 60 per cent; roundway, 55 per cent.
J.M.T. Globe, Angle and Check Valves, 45; Standard, 55 per cent
Kerr standard globes, angles and checks, special, 42½ per cent.; standard, 47½ p c.
Kerr Jenkins' disc, copper-alloy disc and heavy standard valves, 40 per cent.
Kerr steam radiator valves, 60 p c., and quick-opening hot-water radiator valves, 60 p c.
Kerr brass, Weber's straightway valves, 40; straightway valves, I B. H. M., 60.
J. M. T. Radiator Valves 50; Standard, 60; Patent Quick-Opening Valves, 65 p c.
Jenkins' Valves—Quotations on application to Jenkins' Bros, Montreal.

No. 1 compression bath cock......net		2 00
No. 4 " "	"	1 90
No 7 Fuller's......	"	2 25
No. 4½, "	"	2 35
Patent Compression Cushion, basin cock, hot and cold, per doz., ...		$16.20
Patent Compression Cushion, bath cock, No. 2208		2 25

Square head brass cocks, 50; iron, 60 p.c.
Thompson Smoke-test Machine 25.00

BOILERS—COPPER RANGE.

Copper, 30 gallon, $33, 15 per cent.

BOILERS—GALVANIZED IRON RANGE.

30-gallon, Standard, $5; Extra heavy, $7.75

BATH TUBS.

Steel clad copper lined, 15 per cent.

CAST IRON SINKS.

16x24, $1; 16x30, $1; 18x36, $1.30.

ENAMELED BATHS, ETC.

List issued by the Standard Ideal Company Jan. 3, 1907, shows an advance of 10 per cent. over previous quotations.

ENAMELED CLOSETS AND URINALS

Discount 15 per cent.

HEATING APPARATUS.

Stoves and Ranges—40 to 70 per cent.
Furnaces—45 per cent.
Registers—70 per cent.
Hot Water Boilers—50 per cent.
Hot Water Radiators—50 to 55 p.c
Steam Radiators—50 to 55 per cent.
Wall Radiators and specials—50 to 55 p.c.

LEAD PIPE

Lead Pipe, 5 p.c. off
Lead waste, 5 p c. off.
Caulking lead, 6¼c. per pound.
Traps and bends, 40 per cent.

IRON PIPE.

Size (per 100 ft.)	Black.		Galvanized
¼ inch........	2 35	¼ inch........	3 20
⅜ "	2 35	⅜ "	3 20
½ "	2 90	½ "	3 75
¾ "	3 90	¾ "	5 00
1 "	5 40	1 "	7 25
1¼ "	7 65	1¼ "	9 90
1½ "	9 20	1½ "	11 90
2 "	12 25	2 "	15 80
2½ "	20 10	2½ "	26 00
3 "	26 75	3 "	34 00
3½ "	34 25	3½ "	42 75
4 "	39 00	4 "	48 60

Malleable Fittings—Canadian discount 30 per cent.; American discount 25 per cent.
Cast Iron Fittings 57½; Standard bushings 57½; headers, 57½; flanged unions 57½, malleable bushings 55; nipples, 70 and 10; malleable lipped unions, 65 and 5 p.c.

SOIL PIPE AND FITTINGS

Medium and Extra heavy pipe and fittings, up to 6 inch, 60 and 10 to 70 per cent.
7 and 8-in. pipe, 40 and 5 per cent.
Light pipe, 50 p.c.; fittings, 50 and 10 p c.

OAKUM.

Plumbers ....per 100 lb... .. 4 50 5 00

STOCKS AND DIES.

American discount 25 per cent.

SOLDERING IRONS.

½-lb. to 1½	per lb.	0 45½	0 48
2-lb. or over	"	0 42½	0 46

SOLDER. Per lb.

	Montreal	Toronto
Bar, half-and-half, guaranteed	0 25	0 26
Wiping........................	0 22	0 23

PAINTS, OILS AND GLASS.

BRUSHES

Paint and household, 70 per cent.

CHEMICALS.

	In casks per lb.
Sulphate of copper (bluestone or blue vitrol)	0 09
Litharge, ground	0 06
" flaked	0 06½
Green copperas (green vitrol)	0 01
Sugar of lead........	0 08
Lump olive..........	0 01½

COLORS IN OIL.

Venetian red, 1-lb. tins	pure.	....	0 08
Chrome yellow	"	....	0 15
Golden ochre	"	....	0 10
French "	"	....	0 08
Marine black	"	....	0 04½
Chrome green	"	....	0 09
French permanent green	"	....	0 13
Signwriters' black	"	....	0 15

GLUE.

Domestic sheet	0 10	0 10½
French medal	0 12	0 12½

PARIS GREEN.

	Berger's	Canadian
600-lb. cask	....	0 27½
250-lb. drums	0 28	0 27½
100-lb. "	0 29½	...
50-lb. "	0 29½	0 28½
1-lb. pkgs, 100 in box	0 30½	0 29½
½-lb. "	0 32½	0 31½
1-lb. tins, 100 in box	0 31½	0 30½
½-lb. tins	0 33½	0 32

PARIS WHITE.

In bbls	0 90

PIGMENTS.

Orange mineral, casks	....	0 08
" " 100-lb. kegs	....	0 08½

PREPARED PAINTS.

Barn (in bbls.)	0 65	0
Sherwin-Williams paints,	....	1
Canada Paint Co.'s pure	....	1
Standard P. & V. Co.'s "New Era."		1
Benj. Moore Co.'s "Ark" B'd	....	1
" British Navy deck	....	1
Brandram-Henderson's "English"		1
Ramsay's paints, Pure, per gal.	....	1
" Thistle, "	....	1
Martin-Senour's 100 p.c. pure,	....	1
Senour's Floor Paints	....	1
Jamieson's "Crown and Anchor"		1
Jamieson's floor enamel	....	1
"Island City" paint	....	1
Sanderson Pearcy's, pure	....	1 90
Robertson's pure paints	....	1 [illegible]

PUTTY.

Bulk in bbls	1 60
Bladders in bbls	1 85
25-lb. tins	1 90
Bladders in bulk or tins less than 100 lb	2 00
Bulk in 100-lb. irons	1 80

SHINGLE STAINS.

In 5 gallon lots	0 85	0 90

SHELLAC.

White, bleached	....	2 65
Fine orange	....	2 60
Medium orange	....	2 55

TURPENTINE AND OIL.

Prime white petroleum	....	0 13
Water white "	....	0 14½
Pratt's astral "	....	0 18
Castor oil	0 08	0 10
Gasoline	...	0 22½
Benzine, per gal	0 17	0 20
Turpentine, single barrels	0 78	0 79
Linseed Oil, " raw	0 59	0 63
" " boiled	0 62	0 67

WHITE LEAD GROUND IN OIL. Per 100 lbs

Canadian pure	7 15	7
No. 1 Canadian	6 80	7
Munro's Select Flake White	....	7
Elephant and Decorators' Pure	....	7
Monarch	....	7
Standard Decorator's	....	7
Essex Genuine	....	6
Brandram's B. B. Genuine	....	8
"Anchor," pure	....	7 50
Ramsay's Pure Lead	....	7 [illegible]
Ramsay's Exterior	....	6
"Crown and Anchor," pure	....	7
Sanderson Pearcy's	....	7 65
Robertson's C.P., lead	....	7 [illegible]

RED DRY LEAD.

Genuine, 560 lb. casks, per cwt	....	6 25
Genuine, 100 lb. kegs, "	....	6 50
No. 1, 560 lb. casks, per cwt	....	6 00
No. 1, 100 lb. kegs, per cwt	....	6 25

WINDOW GLASS

Size United inches.	Star	Double Diamond
Under 26	$4 25	$6 25
26 to 40	4 65	6 75
41 to 50	5 10	7 50
51 to 60	5 35	8 50
61 to 70	5 75	9 75
71 to 80	6 25	11 00
81 to 85	7 00	12 50
86 to 90	....	15 00
91 to 95	....	17 50
96 to 100	....	20 50
101 to 105	....	24 00
106 to 110	....	27 50

Discount—16-oz., 25 per cent.; 21-oz. 30 per cent. per 100 feet. Broken boxes 50 per cent.

WHITING.

Plain, in bbls	....	0 70
Gilders bolted in bands	....	0 90

WHITE DRY ZINC.

Extra Red Seal, V.M.	0 07½	0 08

WHITE ZINC IN OIL.

Pure, in 25-lb. irons	....	0 08½
No. 1, " "	....	0 07
No. 2, " "	....	0 05½

VARNISHES.

		Per gal cans
Carriage, No. 1	....	1
Pale durable body	....	3
" hard rubbing	....	3
Finest elastic gearing	....	3
Elastic oak	....	1
Furniture, polishing	....	2
Furniture, extra	....	1
" " No. 1	....	0
" " union	....	0
Light oil finish	....	1
Gold size japan	....	1
Brown japan	...	0
No 1 brown japan	....	0
Baking black japan	....	1
No. 1 black japan	....	0
Benzine black japan	....	0
Crystal Damar	....	2
No. 1 "	....	2
Pure asphaltum	....	1
Oilcloth	....	1
Lightning dryer	....	0
Elastilite varnish, 1 gal. can, each		2
Granitine floor varnish, per gal	....	2
Maple Leaf coach enamels; size 1,		1
Sherwin-Williams' kopal varnish, gal.,		2
Canada Paint Co's sun varnish	....	2
"Kyanize" Interior Finish	....	2
"Flint-Lac," coach	....	1
B.H. Co's "Gold Medal," in cases		2 50
Jamieson's Copaline, per gal.	....	2 [illegible]

BUILDERS' HARDWARE.

BELLS.

Brass hand bells, 60 per cent.
Nickel, 55 per cent.

Gongs, Sargeant's door bells	5 50	8 00
American, house bells, per lb.	0 35	0 40

Peterboro' door bells, 37½ and 10 off new list.

BUILDING PAPER, ETC.

Tarred Felt, per 100 lb.		2 25
Ready roofing, 2-ply, not under 45 lb. per roll		1 00
Ready roofing, 3-ply, not under 65 lb., per roll		1 25
Carpet Felt .... per ton		60 00
Heavy Straw Sheathing .... per ton		40 00
Dry Surprise		0 45
Dry Sheathing .... per roll, 400 sq. ft.		0 40
Tar " " 400 "		0 50
Dry Fibre " 400 "		0 55
Tarred Fibre " 400 "		0 65
O. K. & I. X. L. " 400 "		0 70
Resin-sized " 400 "		0 45
Oiled Sheathing " 600 "		1 00
Oiled " " 400 "		0 70
Roof Coating, in barrels .... per gal.		0 17
Roof " small packages "		0 25
Refined Tar .... per barrel		5 00
Coal Tar "		4 00
Coal Tar, less than barrels .... per gal.		0 15
Roofing Pitch .... per 100 lb.	0 80	0 90
Slater's felt .... per roll		0 70
Heavy Straw Sheathing f. o. b. St. John and Halifax		42 50

BUTTS.

Wrought Brass, net revised list.
Wrought Iron, 70 per cent.
Cast Iron Loose Pin, 60 per cent.
Wrought Steel Fast Joint; and Loose Pin, 70 per cent.

CEMENT AND FIREBRICK.

Canadian Portland	2 00	2 10
Belgium	1 60	1 90
White Bros. English	1 80	2 05
"Lafarge" cement in wood	....	3 40
"Lehigh" cement, in wood	....	2 54
"Lehigh" cement, cotton sacks	....	2 39
"Lehigh" cement, paper sacks	....	2 31
Fire brick, Scotch, per 1,000	27 00	30 00
" English	17 00	21 00
" American, low	23 00	25 00
" " high	27 50	35 00
Fire clay (Scotch), net ton	....	4 95

Paving Blocks per 1,000.

Blue metallic, 9"x4½"x3", ex wharf	35 00
Stable pavers, 12"x6"x2", ex wharf	50 00
Stable pavers, 9"x4½"x3", ex wharf	36 00

DOOR SETS.

Peterboro, 37½ and 10 per cent.

DOOR SPRINGS.

Torrey's Rod .... per doz.	....	1 75
Coil, 9 to 11 in. "	0 95	1 65
English "	2 00	4 00

Chicago and Reliance Coil 25 per cent.

ESCUTCHEONS.

Discount 50 and 10 per cent., new list
Peterboro, 37½ and 10 per cent.

ESCUTCHEON PINS.

Iron, discount 40 per cent.
Brass, 45 per cent.

HINGES.

Blind, discount 50 per cent.

Heavy T and strap, 4-in., per lb. net		06
" " 5-in., "		05¾
" " 6-in., "		05½
" " 8-in., "		0 05¼
" " 10-in. and larger		0 05

Light T and strap, discount 65 p.c.

Screw hook and hinge—

under 12 in. .... per 100 lb.	....	4 65
over 12 in. "	....	3 65

Crate hinges and back flaps, 65 and 5 p. c.
Hinge hasps, 65 per cent.

SPRING HINGES.

Spring, per gro., No. 5, $17 50 No. 10, $18; No. 20, $10.80; No. 120, $20; No. 51, $10; No. 50, $27.50
Chicago Spring Butts and Blanks 12½ per cent.
Triple End Spring Butts, 30 and 10 per cent.
Chicago Floor Hinges, 37½ and 5 off.
Garden City Fire House Hinges, 12½ p.c.
"Chief" floor hinge, 50 p c.

CAST IRON HOOKS.

Bird cage .... per doz.	0 50	1
Clothes line, No. 61 "	0 00	0
Harness "	0 60	12
Hat and coat .... per gro.	1 10	10 10
Chandelier .... per doz.	0 50	1 [illegible]

Wrought hooks and staples—

¼ x 5 .... per gross	....	2 65
5-16 x 5 "	....	3 30

Bright wire hooks, 60 p c.
Bright steel gate hooks and staples, 40 p.c.
Cresent hat and coat wire, 60 per cent.
Screw, bright wire, 65 per cent.

KNOBS.

Door, japanned and N.P., doz	1 50	2 50
Bronze, Berlin .... per doz.	2 75	3 25
Bronze, Genuine .... "	6 00	9 00
Shutter, porcelain, F. & L. screw .... per gross	1 30	2 00
White door knobs .... per doz.	....	2 00

Peterboro knobs, 37½ and 10 per cent.
Porcelain, mineral and jet knobs, net list.

KEYS.

Lock, Canadian 40 to 40 and 10 per cent.

LOCKS.

Peterboro, 37½ and 10 per cent
Russell & Erwin, steel rim $2 50 per doz
Eagle cabinet locks, discount 30 per cent
American padlocks, all steel, 10 to 15 per cent.; all brass or bronze, 10 to 25 per cent.

SAND AND EMERY PAPER.

B. & A. sand, discount, 35 per cent
Emery, discount 35 per cent.
Garnet (Burton's). 5 to 10 per cent. advance

SASH WEIGHTS.

Sectional .... per 100 lb.	2 00	2 25
Solid "	1 50	1 75

SASH CORD.

Per lb.	....	0 31

BLIND AND BED STAPLES.

All sizes per lb.	0 07½	0 10

WROUGHT STAPLES.

Galvanized	....	2 75
Plain	....	2 50

Coopers', discount 45 per cent.
Poultry netting staples, discount 40 per cent.
Bright spear point, 75 per cent. discount.

TOOLS AND HANDLES.

ADZES.

Discount 22½ per cent.

AUGERS.

Gilmour's, discount 60 per cent. off list.

AXES.

Single bit, per doz.	6 00	9 00
Double bit, "	10 00	11 00
Bench Axes, 40 per cent.		
Broad Axes. 25 per cent.		
Hunters' Axes	5 50	6 00
Boys' Axes	6 25	7 00
Splitting Axes	7 00	12 00
Handled Axes	7 00	9 00
Red Ridge, boys', handled	....	5 75
" hunters	....	5 25

BITS.

Irwin's auger, discount 47½ per cent.
Gilmour's auger, discount 60 per cent.
Rockford auger, discount 50 and 10 per cent.
Jennings' Gen. auger, net list.
Gilmour's car, 47½ per cent.
Clark's expansive, 40 per cent.

Clark's gimlet, per doz	....	0 65
Diamond, Shell, per doz.	....	1 00
Nail and Spike, per gross	....	2 25

BUTCHERS CLEAVERS

German .... per doz.	7 00	9 00
American "	12 00	18 00

CHALK.

Carpenters' Colored, per gross	0 45	0 75
White lump .... per cwt.	0 60	0 65

CHISELS.

Warnock's, discount 70 and 5 per cent.
P. S. & W. Extra, discount, 70 per cent.

CROSSCUT SAW HANDLES.

S. & D., No. 3 .... per pair	0 13
S. & D., " 5 "	0 11½
S. & D., " 6 "	0 18
Boynton pattern "	0 20

CROWBARS.

3½c. to 4c. per lb.

DRAW KNIVES.

Coach and Wagon, discount 75 and 5 per cent.
Carpenters' discount 75 per cent.

DRILLS.

Millar's Falls, hand and breast. net list.
North Bros., each set, 50c.

DRILL BITS.

Morse, discount 37½ to 40 per cent.
Standard, discount 50 and 5 to 55 per cent.

FILES AND RASPS.

Great Western	75 per cent.
Arcade	75 "
Kearney & Foot	75 "
Disston's	75 "
American	75 "
J. Barton Smith	75 "
McClellan	75 "
Eagle	75 "
Nicholson	62½ "
Globe	75 "

Black Diamond, 60, 10 and 5 p.c.
Jowitt's, English list, 27½ per cent.

GAUGES.

Stanley's discount 50 to 60 per cent.

Winn's, Nos. 26 to 33 .... each	1 65	2 40

HANDLES.

Second growth ash fork, hoe, rake and shovel handles, 40 p c.
Extra ash fork, hoe, rake and shovel handles, 45 p c.
No. 1 and 2 ash fork, hoe, rake and shovel handles, 50 p c.
White ash whiffletrees and neckyokes, 35 p c.
All other ash goods, 40 p c.
All hickory, maple and oak goods, excepting carriage and express whiffletrees, 40 p c.
Hickory, maple, oak carriage and express whiffletrees, 45 p c.

HAMMERS.

Maydole's, discount 5 to 10 per cent.
Canadian, discount 25 to 27½ per cent.

Magnetic tack .... per doz.	1 10	1 20
Canadian sledge .... per lb.	0 07	0 08½
Canadian ball pean, per lb.	0 22	0 25

HATCHETS.

Canadian, discount 40 to 42½ per cent.
Shingle, Red Ridge 1, per doz. 4 40
" " 2, " 4 80
Barrel Underhill 5 05

HOES

Mortar, 50 and 10 per cent.

MALLETS.

Tinsmiths' per doz. 1 25 1 50
Carpenters', hickory, " 1 25 3 75
Lignum Vitae " 3 85 5 00
Caulking, each 0 60 2 00

MATTOCKS.

Canadian per doz. 5 50 6 00

MEAT CUTTERS.

German, 15 per cent.
American discount, 33⅓ per cent.

PICKS.

Per dozen 6 00 9 00

PLANES.

Wood bench, Canadian, 40, American, 25.
Wood, fancy, 37½ to 40 per cent.
Stanley planes, $1 55 to $3 60, net list prices.

PLANE IRONS.

English per doz. 2 00 5 00
Stanley, 2⅝ inch, single 24c., double 39c.

PLIERS AND NIPPERS.

Button's genuine, 37½ to 40 per cent.
Button's imitation per doz. 5 00 9 00
Berg's wire fencing 1 72 5 50

PUNCHES.

Saddlers per doz. 1 00 1
Conductor's " 3 00 15
Tinners', solid per set 0 85
" hollow per inch 1 00

RIVET SETS.

Canadian, discount 35 to 37½ per cent.

RULES.

Boxwood, discount 70 per cent.
Ivory, discount 20 to 25 per cent.

SAWS.

Atkins, hand and crosscut, 25 per cent.
Disston's Hand, discount 12½ per cent
Disston's Crosscut per foot 0 35 0 55
Hack, complete each 0 75 2 75
" frame only each 0 50 1 25

S. & D. solid tooth circular shingle, concave and band, 50 per cent; mill and ice, drag, 30 per cent.; cross-cut, 35 per cent.; hand saws, butcher, 35 per cent.; buck, New Century, $6 25, buck No. 1 Maple Leaf, $5 25; buck, Happy Medium $4 25; buck, Watch Spring, $4.25; buck, common frame, $4.00.

Spear & Jackson's saws—Hand or rip, 26 in., $12 75; 24 in., $11 25; panel 18 in., $8 25; 20 in., $9; tenon, 10 in., $9 90; 12 in., $10.90; 14 in., $11.50

SAW SETS.

Lincoln and Whiting 4 75
Hand Sets, Perfect 4 00
X-Cut Sets, " 7 50
Maple Leaf and Premiums saw sets, 40 off.
S. & D. saw swages, 40 off.

SCREW DRIVERS.

Sargent's per doz. 0 65 1 00
North Bros., No. 30 per doz. 16 80

SHOVELS AND SPADES.

Canadian, discount 45 per cent.

SQUARES.

Iron, discount 20 per cent.
Steel, discount 65 and 10 per cent.
Try and Bevel, discount 50 to 52½ per cent.

TAPE LINES.

English, ass skin per doz. 2 75 5 00
English, Patent Leather 5 50 9 75
Chesterman's each 0 90 2 85
" steel each 0 80 8 00
Berg's, each 0 75 2 50

TROWELS.

Disston's discount 10 per cent.
S. & D. discount 35 per cent.
Berg's, brick, 924x11 4 00
" pointing, 924x5 2 10

FARM AND GARDEN GOODS

BELLS.

American cow bells, 63⅓ per cent.
Canadian, discount 45 and 50 per cent.
American, farm bells, each 1 35 3 00

BULL RINGS.

Copper, $2.00 for 2½-inch

CATTLE LEADERS

Nos. 32 and 33 per gross 7 50 8 50

BARN DOOR HANGERS

		doz. pairs
Stearns wood track	4 50	6 00
Zenith	....	9 00
Atlas, steel covered	5 00	6 00
Perfect	8 00	11 00
New Milo, flexible	....	6 50
Steel, track, 1 x 3-16 in(100 ft)	....	3 25
" " 1½ x 3-16 in(100 ft)	....	4 75
Double strap hangers, doz. sets	....	6 40
Standard jointed hangers, "	....	6 40
Steel King hangers "	....	6 25
Storm King and safety hangers	....	7 00
" " rail	....	4 25

Chicago Friction, Oscillating and Big Twin Hangers, 5 per cent.

HARVEST TOOLS

50 and 10 per cent.
S. & D. lawn rakes, Dunn's, 40 off
" sidewalk and stable scrapers, 40 off.

HAY KNIVES.

Net list.

HEAD HALTERS.

Jute Rope, ⅜-inch per gross 9 00
" " ½ " " 10 00
" " ⅝ " " 12 00
Leather, 1-inch per doz. 4 00
Leather, 1¼ " " 5 20
Web " 2 45

HOES.

Garden, 50 and 10 per cent.
Planter per doz. 4 00 4 50

LAWN MOWERS

Low wheel 12, 14 and 16-inch $2 30
9-inch wheel, 12-inch 2 85
" " 14 " 3 10
" " 16 " 3 12½
High wheel, 12 " 4 05
" " 14 " 4 50
" " 16 " 4 75

SCYTHES.

Per doz. net 6 25 9 25

SCYTHE SNATHS.

Canadian, discount 40 per cent.

SNAPS.

Harness, German, discount 25 per cent.
Lock, Andrews' 4 50 11 00

STABLE FITTINGS.

Warden King, 35 per cent
Dennis Wire & Iron Co., 33⅓ p.c.

WOOD HAY RAKES.

40 and 10 per cent.

HEAVY GOODS, NAILS, ETC.

ANVILS.

Wright's, 80-lb. and over 0 10½
Hay Budden, 80-lb. and over 0 09½
Brook's, 80-lb. and over 0 11½
Taylor-Forbes, prospectors 0 05
Columbia Hardware Co., per lb. 0 09½

VISES.

Wright's 0 13½
Berg's, per lb. 0 12½
Brook's 0 12½
Pipe Vise, Hinge, No. 1 .... 3 50
" " " No. 2 .... 5 50
Saw Vise 4 50 5 00
Blacksmiths' (discount) 60 per cent.
" parallel (discount) 45 per cent.

BOLTS AND NUTS

	Per cent.
Carriage Bolts, common ($1 list)	
" " ⅜ and smaller	60, 10 and 10
" " 7-16 and up	55 and 5
" " Norway Iron ($3 list)	50
Machine Bolts, ⅜ and less	60 and 10
Machine Bolts, 7-16 and up	55 and 5
Plough Bolts	55 and 10
Blank Bolts	55
Bolt Ends	55
Sleigh Shoe Bolts, ⅜ and less	60 and 10
" " " 7-16 and larger	50 and 5
Coach Screws, conepoint	70 and 5

Nuts, square, all sizes, 4c. per cent. off
Nuts, hexagon, all sizes, 4½c. per cent off
Stove Rods per lb., 5½ to 6c.
Stove Bolts, 75 per cent.

CHAIN.

Proof coil, per 100 lb. ¼ in., $6 00; 5-16 in., $4.85; ⅜ in., $4 25; 7-16 in., $4.00; ½ in., $3 75; 9-16 in. $3 70; ⅝ in., $3 65; ¾ in., $3 60; ⅞ in., $3 45; 1 in. $3.40.

Halter, kennel and post chains, 40 to 40 and 5 per cent; Cow ties, 40 per cent.; Tie out chains 65 per cent; Stall fixtures, 35 per cent; Trace chain, 45 per cent.; Jack chain iron, 50 per cent.; Jack chain, brass, 50 per cent.

HORSE NAILS

M R M. cold forged process, list May 15, 1907, 50 and 5 per cent.
"C" brand, 57½ per cent. off list.
Capewell brand, quotations on application

HORSESHOES.

M R.M. brand: iron, light and medium No. 1 and smaller, $3 90; No 2 and larger $3 65; snow pattern No 1 and smaller $4.15, No 2 and larger, $3 90; "X L" new light steel, No 1 and smaller, $4 75; No 2 and larger, $4; "X L" featherweight steel No. 0 to 4, $5 60; toe-weight, all sizes, $5 85 F o b Montreal. Extras for packing

Belleville brand: No. 0 and 1, light and medium iron $3 90; snow, $4 15; light steel, $4 25; No 2 and larger, light and medium iron, $3 65; snow $3.90; light steel, $4. F o b. Belleville. Two per cent., 30 days

Toecalks—Standard No 1 and smaller, $1 50; No. 2 and larger, $1.25 Blunt No. 1 and smaller, $1 75; No 2 and larger, $1 50 per box Sharp. Put up in 2, lb. bxs

HORSE WEIGHTS.

Taylor-Forbes, 4½c. per lb.

NAILS.	Cut	Wire.
2d	4 00	3 50
3d	3 15	3 20
4 and 5d	2 90	2 90
6 and 7d	2 80	2 80
8 and 9d	2 65	2 65
10 and 12d	2 60	2 60
16 and 20d	2 55	2 55
30, 40, 50 and 60d (base)	2 50	2 50

F.o b. Montreal. Cut nails, Toronto 20c. higher.
Miscellaneous wire nails, discount 75 per cent
Coopers' nails, discount 40 per cent.

PRESSED SPIKES

Pressed spikes, ⅝ diameter, per 100 lbs $3.15

RIVETS AND BURRS.

Iron Rivets, black and tinned, 60, 10 and 10.
Iron Burrs, discount 60 and 10 and 10 p.c.
Copper Rivets, usual proportion burrs, 15 p c
Copper Burrs only, net list.
Extras on Coppered Rivets, ½-lb. packages 1c. per lb.; ¼-lb. packages 2c. lb.
Tinned Rivets, net extra, 4c. per lb.

SCREWS.

Wood, F. H., bright and steel, 65 and 10 p c.
" R. H., bright, 80 and 10 per cent
" F. H., brass, 75 and 10 per cent.
" R. H., " 70 and 10 per cent.
" F. H., bronze, 70 and 10 per cent.
" R. H., " 65 and 10 per cent.
Drive Screws, dis. 87½ per cent.
Bench, wood per doz. 3 25 4 00
" iron " 4 25 5 00
Set, case hardened, dis. 60 per cent.
Square Cap, dis. 50 and 5 per cent.
Hexagon Cap, dis. 45 per cent.

MACHINE SCREWS.

Flat head, iron and brass, 35 per cent.
Felister head, iron, discount 30 per cent.
" " brass, discount 25 per cent.

TACKS, BRADS, ETC.

Carpet tacks, blued 75 p.c.; tinned, 80 and 10, (in kegs), 40; cut tacks, blued, in dozens only, 75; ½ weights, 60; Swedes cut tacks, blued and tinned, bulk 75 dozens, 75; Swedes, upholsterers', bulk, 85 and 12½; brush, blued and tinned, bulk, 70; Swedes, gimp, blued, tinned and japanned, 75 and 12½; zinc tacks, 35; leather carpet tacks, 40; copper tacks, 25; copper nails 30; trunk nails, black, 65; trunk nails, tinned and blued, 65, clout nails, blued and tinned 65; chair nails, 35; patent brads, 40; fine finishing, 40; lining tacks, in papers, 10; lining tacks, in bulk, 15; lining tacks, solid heads, in bulk, 75; saddle nails, in papers, 10; saddle nails, in bulk, 15; tufting buttons, 22 line in dozens only, 60; zinc glaziers' points, 5; double pointed tacks, papers, 90 and 10; double pointed tacks, bulk, 40; clinch and duplex rivets, 45; cheese box tacks, 85 and 5 trunk tacks, 80 and 10.

WROUGHT IRON WASHERS.

Canadian make, discount 40 per cent.

SPORTING GOODS.

CARTRIDGES.

"Dominion" Rim Fire Cartridges and C.B. caps, 50 and 7½ per cent; Rim Fire B B. Round Caps 60 and 2½ per cent; Centre Fire, Pistol and Rifle Cartridges 30 p c; Centre Fire Sporting and Military Cartridges, 2½ and 5 p c; Rim Fire, Shot Cartridges, 5 and 7½ p c; Centre Fire, Shot Cartridges, 30 p.c; Primers, 25 p c.

LOADED SHELLS

"Crown" Black Powder, 15 and 5 p.c.; "Sovereign" Empire Bulk Smokeless Powder, 30 and 5 p.c., "Regal" Ballistite Dense smokeless Powder, 30 and 5 p.c; "Imperial" Empire or Ballistite Powder, 30 and 10 p.c.

EMPTY SHELLS

Paper Shells, 25 and 5; Brass Shells, 55 and 5 p.c.

Wads.	per lb.
Best thick brown or grey felt wads, in ½-lb. bags	$0 70
Best thick white card wads, in boxes of 500 each, 12 and smaller gauges	29
Best thick white card wads, in boxes of 500 each, 10 gauge	0 35
Thin card wads, in boxes of 1,000 each, 12 and smaller gauges	0 20
Thin card wads, in boxes of 1,000 each, 10 gauge	0 25
Chemically prepared black edge grey cloth wads, in boxes of 250 each—	Per M
11 and smaller gauge	0 60
9 and 10 gauges	0 70
7 and 8 "	0 90
5 and 6 "	1 10
Superior chemically prepared pink edge, best white cloth wads in boxes of 250 each—	
11 and smaller gauge	1 15
9 and 10 gauges	1 40
7 and 8 "	1 65
5 and 9 "	1 90

SHOT.

Ordinary drop shot, AAA to dust $7.50 per 100 lbs. Discount 5 per cent; cash discount 2 per cent, 30 days; net extras as follows subject to cash discount only; Chilled, 40 c; buck and seal, 80c.; no 28 ball, $1 20 per 100 lbs.; bags less than 25 lbs., ½c. per lb.; F.O.B. Montreal, Toronto, Hamilton, London, St. John and Halifax, and freight equalized thereon.

TRAPS (steel)

Game, Newhouse, discount 30 and 10 per cent.
Game, Hawley & Norton, 50, 10 & 5 per cent.
Game, Victor, 70 per cent.
Game, Oneida Jump (B. & L.) 40 & 2½ p.
Game, steel, 60 and 5 per cent.

SKATES.

Skates, discount 37½ per cent.
Empire hockey sticks, per doz .. 3 00

CUTLERY AND SILVER-WARE.

RAZORS. per doz.

Elliot's 4 00 18 00
Boker's 7 50 11 00
" King Cutter 13 50 18 50
Wade & Butcher's.......... 3 60 10 00
Lewis Bros.' "Klean Kutter 8 50 10 50
Henckel's 7 50 20 00
Berg's.......... 7 50 20 00
Clauss Razors and Strops, 50 and 10 per cent

KNIVES.

Farriers-Stacey Bros., doz 3 50

PLATED GOODS

Hollowware, 40 per cent. discount.
Flatware, staples, 40 and 10, fancy, 40 and 5.
Hutton's "Cross Arrow" flatware, 47½; "Singalese" and "Alaska" Nevada silver flatware, 42 p c.

SHEARS.

Clauss, nickel, discount 60 per cent.
Clauss, Japan, discount 67½ per cent.
Clauss, tailors, discount 40 per cent.
Seymour's, discount 50 and 10 per cent.
Berg's.......... 6 00 12 00

HOUSE FURNISHINGS.

APPLE PARERS.

Hudson, per doz., net 5 75

BIRD CAGES.

Brass and Japanned, 40 and 10 p. c.

COPPER AND NICKEL WARE.

Copper boilers, kettles, teapots, etc. 30 p.c.
Copper pitts, 30 per cent.

KITCHEN ENAMELED WARE.

White ware, 75 per cent.
London and Princess, 50 per cent.
Canada, Diamond, Premier, 50 and 10 p.c.
Pearl, Imperial, Crescent and granite steel, 50 and 10 per cent.
Premier steel ware, 40 per cent.
Star decorated steel and white, 25 per cent.
Japanned ware, discount 45 per cent.
Hollow ware, tinned cast, 35 per cent. off.

KITCHEN SUNDRIES.

Can openers, per doz.......... 0 40 0 75
Mincing knives per doz 0 50 0 80
Duplex mouse traps, per doz.... .... 0 65
Potato mashers, wire, per doz... 0 60 0 70
" wood " .. 0 50 0 60
Vegetable slicers, per doz 2 25
Universal meat chopper No. 1.. .... 1 15
Enterprise chopper, each 1 30
Spiders and fry pans, 50 per cent.
Star Al chopper 5 to 32 .... 1 35 4 10
" " 100 to 103.......... 1 35 2 00
Kitchen hooks, bright 0 60

LAMP WICKS.

Discount, 60 per cent.

LEMON SQUEEZERS.

Porcelain lined...... per doz. 2 20 5 60
Galvanized.......... " 1 87 3 85
King, wood.......... " 2 75 2 90
King, glass.......... " 4 00 4 50
All glass .... " 0 50 0 90

METAL POLISH.

Tandem metal polish paste 6 00

PICTURE NAILS.

Porcelain head ......per gross 1 35 1 50
Brass head.......... 0 40 1 00
Tin and gilt, picture wire, 75 per cent.

SAD IRONS.

Mrs. Potts, No. 55, polished....per set 0 90
" No 50, nickle-plated, " 0 95
" handles, japaned, per gross 9 25
" " nickled, " 9 75
Common, plain.......... 4 25
" plated 5 50
Asbestos, per set.......... .... 1 50

TINWARE.

CONDUCTOR PIPE.

2-in. plain or corrugated., per 100 feet, $3.30; 3 in., $4 40; 4 in., $5.80; 5 in., $7.45; 6 in., $9.90.

FAUCETS.

Common, cork-lined, discount 35 per cent.

EAVETROUGHS.

10-inch .... per 100 ft. 3 30

FACTORY MILK CANS.

Discount off revised list, 35 per cent
Milk can trimmings, discount 25 per cent.
Creamery Cans, 45 per cent

LANTERNS.

No 2 or 4 Plain Cold Blast....per doz. 6 50
Lift Tubular and Hinge Plain, " 4 75
No. 0, safety.......... " 4 00
Better quality at higher prices.
Japanning, 50c. per doz. extra.
Prism globes, per doz., $1.20.

OILERS.

Kemp's Tornado and McClary's Model galvanized oil can, with pump, 5 gallon, per dozen 10 92
Davidson oilers, discount 40 per cent.
Zinc and tin, discount 50 per cent.
Coppered oilers, 20 per cent off.
Brass oilers, 50 per cent. off.
Malleable, discount 25 per cent

PAILS (GALVANIZED).

Dufferin pattern pails, 45 per cent.
Flaring pattern, discount 45 per cent.
Galvanized washtubs 40 per cent.

PIECED WARE.

Discount 35 per cent off list, June, 1899.
10-qt. flaring sap buckets, discount 35 per cent.
6, 10 and 14-qt. flaring pails dis. 35 per cent.
Copper bottom tea kettles and boilers, 30 p.c.
Coal hods, 40 per cent.

STAMPED WARE.

Plain, 75 and 12½ per cent. off revised list.
Retinned, 72½ per cent. revised list.

SAP SPOUTS.

Bronzed iron with hooks ....per 1,000
Eureka tinned steel, hooks 8 00

STOVEPIPES.

5 and 6 inch, per 100 lengths 7 64 7 91
7 inch.......... " " 8 18
Nestable, discount 40 per cent.

STOVEPIPE ELBOWS

5 and 6-inch, common..........per doz. 1 32
7-inch.......... " 1 48
Polished, 15c. per dozen extra.

THERMOMETERS.

Tin case and dairy, 75 to 75 and 10 per cent.

TINNERS' SNIPS.

Per doz.......... 3 00 15
Clauss, discount 35 per cent.

TINNERS' TRIMMINGS.

Discount, 45 per cent.

WIRE.

ANNEALED CUT HAY BAILING WIRE.

No. 12 and 13, $4; No 13½, $4 10; No. 14, $4 25; No 15, $4 50; in lengths 6' to 11', 25 per cent; other lengths 20c. per 100 lbs extra; if eye or loop on end add 25c. per 100 lbs. to the above.

BRIGHT WIRE GOODS

Discount 60 per cent.

CLOTHES LINE WIRE.

7 wire solid line, No. 17, $4.90; No. 18, $3.90; No. 19, $2.70; 6 wire solid line, No. 17, $4.45; No. 18, $3.10, No. 19, $2 80. All prices per 1000 ft. measure; 6 strand, No. 18, $2.60; No. 19, $2 90. F.o.b. Hamilton, Toronto, Montreal.

COILED SPRING WIRE

High Carbon, No. 9, $2.95; No. 11, $3.50; No. 12, $3.20.

COPPER AND BRASS WIRE.

Discount 37½ per cent.

FINE STEEL WIRE.

Discount 25 per cent List of extras In 100-lb. lots: No. 17, $5 — No. 18, $5.50 — No. 19, $6 — No. 20, $6.65 — No. 21 $7 — No. 22, $7 30 — No. 23, $7,65 — No 24, $8 — No. 25, $9—No. 26, $9 50—No. 27, $10—No. 28, $11—No. 29, $12—No. 30, $13—No. 31, $14—No. 32, $15—No. 33, $16—No. 34, $17. Extras net—tinned wire, Nos. 17-25 $2—Nos. 26-31, $4—Nos. 32-34, $6 Coppered, 75c.—oiling, 10c.—in 25-lb. bundles, 15c.—in 5 and 10-lb. bundles, 25c.—in 1-lb. hanks, 25c. —in ½-lb. hanks, 38c.—in ¼-lb. hanks, 50c. packed in casks or cases, 15c.—bagging or papering, 10c

FENCE STAPLES.

Bright. .... 2 80 Galvanized.... 32

HAY WIRE IN COILS.

No 13, $2.70; No. 14, $2.80; No. 15, $2.95; f.o.b., Montreal.

GALVANIZED WIRE.

Per 100 lb.—Nos. 4 and 5, $3 90 — Nos. 6, 7, 8, $3.35 — No. 9, $2.85 — No. 10, $3 40 — No. 11, $3.45—No. 12, $3.00 —No. 13, $3 10—No. 14, $3.95—No. 15, $4.30 —No. 16, $4.30 from stock. Base sizes, Nos. 6 to 9, $2.35 f.o.b. Cleveland. Extras for cutting.

LIGHT STRAIGHTENED WIRE.

Gauge No.	Over 20 in. per 100 lbs.	10 to 20 in.	5 to 10 in.
0 to 5	$0 50	$0.75	$1 25
6 to 9	0.75	1 25	2 00
10 to 11	1 00	1 75	2.50
12 to 14	1 50	2.25	3 50
15 to 16	2.00	3.00	4.50

SMOOTH STEEL WIRE.

No. 0-9 gauge, $2.40; No. 10 gauge, 6c extra; No. 11 gauge, 12c extra; No. 12 gauge, 20c. extra; No. 13 gauge, 30c. extra; No 14 gauge, 40c. extra; No. 15 gauge, 55c. extra; No. 16 gauge, 70c. extra. Add 60c. for coppering and $2 for tinning.

Extra net per 100 lb.—Oiled wire 10c., spring wire $1.25, bright soft drawn 15c., charcoal (extra quality) $1 25, packed in casks or cases 15c., bagging and papering 10c., 50 and 100-lb. bundles 10c., in 25-lb. bundles 15c., in 5 and 10-lb. bundles 25c., in 1-lb hanks, 50c., in ½-lb. hanks 75c., in ¼-lb. hanks $1.

POULTRY NETTING.

2-in. mesh, 19 w. g., 50 and 5 p.c. off. Other sizes, 50 and 5 p.c. off.

WIRE CLOTH.

Painted Screen, in 100-ft. rolls, $1.72½, per 100 sq. ft.; in 50-ft. rolls, $1.77½, per 100 sq. ft.

WIRE FENCING.

Galvanized barb.......... .... 2 95
Galvanized, plain twist 3 30
Galvanized barb, f.o.b. Cleveland, $2.70 for small lots and $2 60 for carlots

WIRE ROPE

Galvanized, 1st grade, 6 strands, 24 wires, ⅜, $5; 1 inch $16 80
Black, 1st grade, 6 strands, 19 wires, ⅜, $5; 1 inch $15.10. Per 100 feet f o.b. Toronto.

WOODENWARE.

CHURNS.

No. 0, $9; No. 1, $9; No. 2, $10; No. 3, $11; No. 4, $13; No. 5, $16; f.o.b. Toronto Hamilton, London and St. Marys, 30 and 30 per cent.; f.o.b. Ottawa, Kingston and Montreal, 40 and 15 per cent. discount.

CLOTHES REELS.

Davis Clothes Reels, dis. 40 per cent.

FIBRE WARE.

Star pails, per doz.......... $ 3 00
0 Tubs, " 14 00
1 " " 12 00
2 " " 10 00
3 " " 8 50

LADDERS, EXTENSION.

3 to 6 feet, 12c. per foot; 7 to 10 ft., 13c.
Waggoner Extension Ladders, dis 40 per cent.

MOPS AND IRONING BOARDS.

"Best" mops.......... .... 1 25
"900" mops.......... .... 1 25
Folding ironing boards..........12 00 16 50

REFRIGERATORS

Discount, 40 per cent.

SCREEN DOORS.

Common doors, 2 or 3 panel, walnut stained, 4-in. style..........per doz. 7 25
Common doors, 2 or 3 panel, grained only, 4-in., style ....per doz. 7 55
Common doors, 2 or 3 panel, light stair per doz.......... 9 55

WASHING MACHINES.

Round, re-acting per doz. 60 00
Square " 63 00
Eclipse, per doz 54 00
Dowswell " 39 00
New Century, per doz 75 00
Daisy.......... 54 00
Stephenson.......... 74 00

WRINGERS.

Royal Canadian, 11 in., per doz. 35 00
Royal American, 11 in. 35 00
Eze 10 in., per doz 36 75

MISCELLANEOUS

AXLE GREASE.

Ordinary, per gross.......... 6 00 7 00
Best quality 10 00 12 00

BELTING.

Extra, 60 per cent.
Standard, 60 and 10 per cent.
No. 1, not wider than 6 in., 60, 10 and 10 p.c.
Agricultural, not wider than 4 in., 75 per cent.
Lace leather, per side, 75c.; cut laces, 80c.

BOOT CALKS.

Small and medium, ball ....per M 2 25
Small heel 4.50

CARPET STRETCHERS.

American ....per doz. 1 00 1 50
Bullard's.......... " 6 50

CASTORS.

Bed, new list, discount 55 to 57½ per cent.
Plate, discount 52½ to 57½ per cent.

PINE TAR.

½ pint in tins ....per gross ... 7 80
" " 9 60

PULLEYS.

Hothouse ....per doz. 0 55 1 00
Axle.......... " 0 22 0 33
Screw .... " 0 22 1 00
Awning .... " 0 35 2 50

PUMPS.

Canadian cistern 1 40 2 00
Canadian pitcher spout 1 80 3 18
Berg's wing pump, 75 per cent.

ROPE AND TWINE.

Sisal.......... 0 10½
Pure Manilla 0 15
"British" Manilla.......... 0 12
Cotton, 3-16 inch and larger..........0 21 0 23
" 5-32 inch 0 25 0 27
" ¼ inch 0 25 0 28
Russia Deep Sea 0 16
Jute.......... 0 09
Lath Yarn, single 0 10
" double 0 10½
Sisal bed cord, 48 feet..........per doz. 0 65
" " 60 feet.......... " 0 80
" " 72 feet.......... " 0 95

Twine.

Bag, Russian twine, per lb. 0 27
Wrapping, cotton, 3-ply 0 25
" " 4-ply 0 29
Mattress twine per lb. 0 33 0 45
Staging " " 0 27 0 35

BINDER TWINE.

500 feet, sisal 0 09½
500 " standard 0 09½
550 " " manilla 0 10½
600 " " " 0 12½
650 " " " 0 13½
Car lots, ¼c. less; 5-ton lots, ⅛c. less. Central delivery.

SCALES.

Gurney Standard, 35; Champion, 45 p.c.
Burrow, Stewart & Milne — Imperial Standard, 35; Weigh Beams, 35; Champion Scales, 45.
Fairbanks Standard, 30; Dominion, 50 Richelieu, 50.
Warren new Standard, 35; Champion, 45 Weigh Beams, 30.

STONES—OIL AND SCYTHE.

Washita..........per lb. 0 25 0
Hindostan .... " 0 06 0 10
" slip .... " 0 18 0 20
" Axe.......... " 0 10
Deer Creek.......... " 0 10
Deerlick .... " 0 25
" Axe .... " 0 15
Lily white .... " 0 42
Arkansas.......... " 1 50
Water-of-Ayr .... " 0 10
Scythe ....per gross 3 50 5 00
Grind, 40 to 200 lb., per ton.... 20 00 22 00
" under 40 lb., " 24 00
" 200 lb. and over 28 00

INDEX TO ADVERTISERS.

CLASSIFIED LIST OF ADVERTISEMENTS.

Alabastine.
Alabastine Co., Limited, Paris, Ont.

Auditors.
Davenport, Pickup & Co., Winnipeg.

Automobile Accessories.
Canada Cycle & Motor Co., Toronto Junction.
Carriage Mountings Co., Ltd.. Toronto.

Awnings.
Tobin Tent and Awning Co., Ottawa

Babbitt Metal.
Canada Metal Co., Toronto.
Canadian Fairbanks Co., Montreal.
Frothingham & Workman, Montreal.
Robertson, Jas. Co., Montreal.

Bar Urns.
Buffalo Mfg. Co., Buffalo, N.Y.

Bath Room Fittings.
Buffalo Mfg. Co., Buffalo, N.Y.
Carriage Mountings Co., Ltd , Toronto.
Forsyth Mfg. Co., Buffalo, N.Y.
Ontario Metal Novelty Co., Toronto

Belting, Hose, etc.
Gutta Percha and Rubber Mfg. Toronto.
Sadler & Haworth, Toronto.

Bicycles and Accessories.
Johnson's, Iver, Arms and Cycle Works Fitchburg, Mass

Binder Twine.
Consumers Cordage Co., Montreal.

Bolts.
Toronto & Belleville Rolling Mills, Belleville.

Box Strap.
J. N. Warminton, Montreal.

Brass Goods.
Frothingham & Workman, Montreal.
Glauber Brass Mfg. Co., Cleveland, Ohio.
Kerr Engine Co., Walkerville, Ont.
Lewis, Rice, & Son, Toronto.
Morrison, Jas., Brass Mfg. Co., Toronto.
Mueller Mfg. Co., Decatur, Ill.
Penberthy Injector Co., Windsor, Ont.
Taylor-Forbes Co., Guelph, Ont.

Bronze Powders.
Canadian Bronze Powder Works, Montreal.

Brushes.
United Factories, Toronto.

Cans.
Acme Can Works, Montreal.

Builders' Tools and Supplies.
Covert Mfg. Co., West Troy, N.Y.
Frothingham & Workman Co., Montreal.
Howland, H. S. Sons & Co., Toronto.
Hyde, F., & Co., Montreal.
Lewis Bros. & Co., Montreal.
Lewis, Rice, & Son, Toronto.
Lockerby & McComb, Montreal.
Lufkin Rule Co., Saginaw, Mich.
Newman & Sons, Birmingham.
North Bros. Mfg. Co., Philadelphia, Pa.
Stanley Rule & Level Co., New Britain.
Stanley Works, New Britain, Conn.
Stephens, G. F., Winnipeg.
Taylor-Forbes Co., Guelph, Ont.

Carriage Accessories.
Carriage Mountings Co , Ltd., Toronto.
Covert Mfg. Co., West Troy, N.Y.

Carriage Springs and Axles.
Guelph Spring and Axle Co., Guelph.

Carpet Beateas.
Ontario Metal Novelty Co., Toronto.

Cartridges.
Dominion Cartridge Co., Montreal.

Cattle and Trace Chains.
Greening, B., Wire Co., Hamilton.

Chafing Dishes.
Buffalo Mfg. Co., Buffalo, N.Y.

Churns.
Dowswell Mfg. Co., Hamilton.

Clippers—All Kinds.
American Shearer Mfg. Co., Nashua, N.H

Clothes Reels and Lines.
Hamilton Cotton Co., Hamilton,

Clutch Nails.
J. N. Warminton, Montreal.

Congo Roofing
Buchanan Foster Co., Philadelphia, Pa.

Cordage.
Consumers' Cordage Co., Montreal.
Hamilton Cotton Co., Hamilton.

Cork Screws.
Erie Specialty Co., Erie, Pa.

Cow Ties
Greening, B., Wire Co., Hamilton

Cuspidors.
Buffalo Mfg. Co., Buffalo, N.Y.

Cut Glass.
Phillips, Geo., & Co., Montreal.

Cutlery—Razors, Scissors, etc.
Birkett, Thos., & Son Co., Ottawa.
Clauss Shear Co., Toronto
Dorken Bros. & Co., Montreal.
Frothingham & Workman, Montreal.
Heinisch's, R., Sons Co., Newark, N.J.
Howland, H. S. Sons & Co., Toronto.
Hutton, Wm., & Sons, Ltd., London, Eng.
Lamplough, F. W., & Co., Montreal.
Phillips, Geo., & Co., Montreal.
Round, John, & Son, Montreal.

Electric Fixtures.
Canadian General Electric Co., Toronto.
Morrison James, Mfg. Co., Toronto.
Munderloh & Co., Montreal.

Electro Cabinets.
Cameron & Campbell Toronto.

Euameled Ware
Kemp Mfg. Co., Toronto.

Engines, Supplies, etc.
Kerr Engine Co., Walkerville, Ont.

Eavetroughs
Wheeler & Bain, Toronto

Fencing—Woven Wire.
Dominion Wire Mfg. Co., Montreal.
Owen Sound Wire Fence Co., Owen Sound.
Banwell Hoxie Wire Fence Co., Hamilton.

Files and Rasps.
Barnett Co., G. & H., Philadelphia, Pa.
Nicholson File Co., Port Hope

Firearms and Ammunition.
Hamilton Rifle Co., Plymouth, Mich.
Harrington & Richardson Arms Co., Worcester, Mass.
Johnson's, Iver, Arms and Cycle Works Fitchburg, Mass.

Fishing Tackle.
Enterprise Mfg. Co., Akron, Ohio

Food Choppers.
Enterprise Mfg. Co., Philadelphia, Pa.
Lamplough, F. W., & Co., Montreal.
Shirreff Mfg. Co., Brockville, Ont.

Furnaces.
Pease Foundry Co , Toronto.

Galvanizing.
Canada Metal Co., Toronto.
Dominion Wire Mfg. Co., Montreal.
Montreal Rolling Mills Co., Montreal.
Ontario Wind Engine & Pump Co., Toronto.

Glass Ornamental
Hobbs Mfg. Co., London
Consolidated Plate Glass Co., Toronto

Glaziers' Diamonds.
Gibsone, J. B., Montreal.
Pelton, Godfrey S.
Sharratt & Newth, London, Eng.
Shaw, A., & Son, London, Eng.

Handles.
Still, J. H., Mfg. Co.

Harvest Tools.
Maple Leaf Harvest Tool Co., Tillsonburg, Ont.

Hockey Sticks
Still, J. H. Mfg. Co., St. Thomas.

Hoop Iron.
Frothingham & Workman Montreal.
Montreal Rolling Mills Co , Montreal.
J. N. Warminton, Montreal.

Horse Blankets.
Heney, E. N., & Co., Montreal.

Horseshoes and Nails.
Canada Horse Nail Co., Montreal.
Montreal Rolling Mills, Montreal.
Capewell Horse Nail Co., Toronto
Toronto & Belleville Rolling Mills Belleville.

Hot Water Boilers and Radiators.
Cluff, R. J., & Co Toronto.
Pease Foundry Co. Toronto.
Taylor-Forbes Co., Guelph.

Ice Cream Freezers.
Dana Mfg. Co., Cincinnati, Ohio.
North Bros. Mfg. Co., Philadelphia, Pa.

Ice Cutting Tools.
Erie Specialty Co., Erie, Pa.
North Bros. Mfg. Co., Philadelphia, Pa.

Injectors—Automatic.
Morrison, Jas., Brass Mfg. Co., Toronto.
Penberthy Injector Co., Windsor, Ont.

Iron Pipe.
Montreal Rolling Mills, Montreal.

Iron Pumps.
Lamplough, F. W., & Co., Montreal.
McDougall, R., Co., Galt, Ont.

Ladders.
The Waggoner Ladder Co., London, Ont.

Lanterns.
Kemp Mfg. Co., Toronto.
Ontario Lantern Co., Hamilton, Ont.
Wright, E. T., & Co., Hamilton.

Lawn Mowers.
Birkett, Thos., & Son Co., Ottawa.
Frothingham & Workman, Montreal.
Maxwell, D., & Sons, St. Mary's, Ont.
Taylor, Forbes Co., Guelph.

Lawn Mower Grinders
Root Bros. & Co., Plymouth, Ohio.

Ledgers—Loose Leaf.
Business Systems, Toronto.
Copeland-Chatterson Co , Toronto.
Crain, Rolla L., Co., Ottawa.
Universal Systems, Toronto.

Lithographing.
London Printing & Lithographing Co., London, Ont.

Locks, Knobs, Escutcheons, etc.
Peterborough Lock Mfg. Co., Peterborough, Ont.
National Hardware Co., Orillia, Ont.

Lumbermen's Supplies.
Pink, Thos., & Co., Pembroke Ont.

Lye
Gillett, E. W., & Co., Toronto

Manufacturers' Agents.
Fox, C. H., Vancouver
McIntosh, H. F., & Co , Toronto.
Gibb, Alexander, Montreal.
Scott, Bathgate & Co , Winnipeg.

Metals.
Canada Iron Furnace Co., Midland, Ont.
Canada Metal Co., Toronto.
Eadie, H. G., Montreal.
Frothingham & Workman, Montreal.
Gibb, Alexander, Montreal.
Kemp Mfg. Co., Toronto
Leslie, A. C., & Co., Montreal.
Lysaght, John, Bristol, Eng.
Nova Scotia Steel and Coal Co., New Glasgow, N.S.
Robertson, Jas , Co., Montreal.
Roper, J. H., Montreal.
Samuel, Benjamin & Co., Toronto.
Stairs, Son & Morrow, Halifax, N.S.
Thompson, B. & S. H. & Co. Montreal.

Metal Lath.
Galt Art Metal Co., Galt.
Metallic Roofing Co., Toronto.
Metal Shingle & Siding Co., Preston, Ont.

Metal Polish, Emery Cloth, etc.
Oakey, John, & Sons, London, Eng.

Nails Wire
Dominion Wire Mfg. Co., Montreal.

Oil Tanks
Bowser, S. F., & Co., Toronto.

Ornamental Iron and Wire.
Dennis Wire & Iron Co , London, Ont.

Packing.
Gutta Percha & Rubber Co Toronto

Paints, Oils, Varnishes, Glass.
Blanchite Process Paint Co., Toronto.
Brandram-Henderson, Montreal
Canada Paint Co., Montreal.
Canadian Oil Co. Toronto.
Consolidated Plate Glass Co., Toronto.
Dods, P. D., & Co., Montreal
Imperial Varnish and Color Co., Toronto.
Jamieson, R. C., & Co., Montreal.
Lucas John & Co., New York
McArthur, Corneille & Co., Montreal.
McCaskill, Dougall & Co., Montreal.
Moore, Benjamin, & Co. Toronto.
Ottawa Paint Works, Ottawa
Queen City Oil Co., Toronto.
Ramsay & Son, Montreal.
Sanderson Pearcy & Co., Toronto
Sherwin-Williams Co., Montreal.
Standard Paint Co., Montreal
Standard Paint and Varnish Works Windsor, Ont.
Stephens & Co., Winnipeg.
Martin-Senour Co., Montreal
Winnipeg Paint & Glass Co., Winnipeg

Perforated Sheet Metals.
Greening, B., Wire Co., Hamilton.

Plumbers' Tools and Supplies.
Canadian Fairbanks Co., Montreal.
Cluff, R. J., & Co., Toronto
Frothingham & Workman, Montreal.
Glauber Brass Co., Cleveland, Ohio.
Jardine, A. B., & Co , Hespeler, Ont.
Jenkins Bros., Boston. Mass.
Kerr Engine Co , Walkerville, Ont.
Lewis, Rice, & Son, Toronto.
Merrell Mfg. Co., Toledo, Ohio.
Montreal Rolling Mills Montreal.
Morrison, Jas., Brass Mfg. Co., Toronto.
Mueller, H., Mfg. Co., Decatur, Ill.
Oshawa Steam & Gas Fitting Co., Oshaw
Robertson Jas., Co. Montreal.
Robertson, Jas , Co , Limited, Toronto
Somerville, Limited, Toronto
Stairs, Son & Morrow, Halifax, N.S.
Standard Ideal Sanitary Co., Port Hope,
Standard Sanitary Co., Pittsburg.
Stephens, G. F , & Co., Winnipeg, Man.
Turner Brass Works, Chicago.
Vickery, Orlando, Toronto.

Polishes.
Majestic Polishes, Toronto

Portland Cement.
International Portland Cement Co. Ottawa, Ont.
Hanover Portland Cement Co., Hanover, Ont.
Hyde, F., & Co., Montreal.
Thompson B. & S. H. & Co., Montreal.

Poultry Netting.
Greening, B., Wire Co., Hamilton, Ont.

Printing.
London Printing & Lithographing Co., London, Ont.

Razors.
Clauss Shear Co., Toronto.

Refrigerators.
Fabien, C. P., Montreal.

Registers.
Pease Foundry Co., Toronto.

Roofing Supplies.
Brantford Roofing Co., Brantford.
Barrett Mfg. Co., New York.
F. W. Bird, East Walpole, Mass.
Buchanan Foster Co., Philadelphia, Pa.
McArthur, Alex., & Co., Montreal.
Metal Shingle & Siding Co., Preston, Ont.
Metallic Roofing Co., Toronto.
Paterson Mfg. Co., Toronto & Montreal.
Wheeler and Bain, Toronto

Saws.
Atkins, E. C., & Co., Indianapolis, Ind
Shurly & Dietrich, Galt, Ont.
Spear & Jackson, Sheffield, Eng.

Scales.
Canadian Fairbanks Co., Montreal.
Frothingham & Workman, Montreal.

Screw Cabinets.
Cameron & Campbell, Toronto.

Screws, Nuts, Bolts.
Dominion Wire Mfg Co , Montreal.
Montreal Rolling Mills Co., Montreal.

Soil Pipe
McFarlane, Walter, Glasgow

Sewer Pipes.
Canadian Sewer Pipe Co., Hamilton
Hyde, F., & Co., Montreal.

Shelf Boxes.
Cameron & Campbell, Toronto.

Shears, Scissors.
Clauss Shear Co., Toronto.

Shovels and Spades
Eclipse Mfg Co , Ottawa
Frothingham & Workman, Montreal.
Peterboro Shovel & Tool Co., Peterboro.

Silverware.
Hutton, Wm , & Sons, Ltd., London, Eng
McGlashan, Clarke Co., Niagara Falls, Ont.
Phillips, Geo., & Co., Montreal
Round, John, & Son, Sheffield, Eng.

Skates.
Canada Cycle & Motor Co., Toronto.
McFarlane, Walter, Glasgow.

Sprayers
CaVers Bros., Galt

Spring Hinges, etc.
Chicago Spring Butt Co., Chicago, Ill.

Stable Fittings
Dennis Wire & Iron Co , London

Steel Rails.
Nova Scotia Steel & Coal Co., New Glasgow, N.S.

Stove Pipe.
Chown, Edwin, and Son, Kingston

Stoves, Tinware, Furnaces
Canadian Heating & Ventilating Co. Owen Sound.
Copp, W J., Son & Co., Fort William
DaVidson, Thos , Mfg. Co., Montreal
Down Draft Furnace Co., Galt
Guelph Stove Co., Guelph.
Gurney Foundry Co., Toronto.
Harris, J. W., Co., Montreal.
Howard, Wm , Toronto
Kemp Mnfg. Co. Toronto.
McClary Mfg. Co. London.
Merrick Anderson, Winnipeg
Pease Foundry Co., Toronto.
Smart, James, Mfg Co., Brockville
Stewart, Jas., Mfg. Co., Woodstock, Ont.
Taylor-Forbes Co., Guelph, Ont.
Wright, E. T., & Co., Hamilton.

Tacks.
Montreal Rolling Mills Co., Montreal.
Ontario Tack Co., Hamilton.

Tents.
Tobin Tent and Awning Co., Ottawa

Tin Plate.
American Sheet & Tin Plate Co , Pittsburg, Pa.
Baglan Bay Tin Plate Co , Briton Ferry South Wales
Ly agut, John, Bristol, Newport and Montreal

Turpentine
Defiance Mfg. Co., Toronto.

Ventilators.
Harris, J. W., Co , Montreal.
Pearson, Geo. D., Montreal.

Wall Paper
Staunton Limited, Toronto.

Wall Paper Cleaner.
Gilbert, Frank U. S., Cleveland

Washing Machines, etc
Dowswell Mfg Co , Hamilton, Ont.
The Shultz Bros. Co , Brantford.
Taylor-Forbes Co., Guelph, Ont.

Water Filters.
Buffalo Mfg. Co., Buffalo, N.Y.

Wheelbarrows
London Foundry Co , London Ont.
Schultz Bros. Co., Ltd., The Brantford.

Wholesale Hardware
Birkett, Thos., & Sons Co., Ottawa.
Caverhill, Learmont & Co., Montreal.
Frothingham & Workman, Montreal.
Hobbs Hardware Co., London
Howland, H. S., Sons & Co., Toronto.
Lamplough, F. W., & Co., Montreal.
Lewis Bros. & Co., Montreal.
Lewis, Rice, & Son, Toronto.

Window and Sidewalk Prisms.
Hobbs Mfg. Co., London, Ont.

Wire, Wire Rope, Cow Ties, Fencing Tools, etc
Banwell-Hoxie Fence Co., Hamilton
Dennis Wire and Iron Co., London, Ont.
Dominion Wire Mnfg. Co., Montreal
Greening, B., Wire Co., Hamilton.
Owen Sound Wire Fence Co., Owen Sound
Montreal Rolling Mills Co., Montreal.
Western Wire & Nail Co , London, Ont.

Wrapping Papers.
Canada Paper Co., Toronto.
McArthur, Alex., & Co , Montreal.
Stairs, Son & Morrow, Halifax, N.S.

Wringers
Connor, J. H. & Son, O awa, Ont

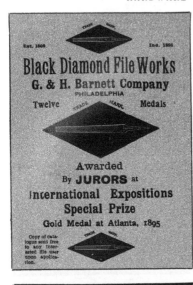

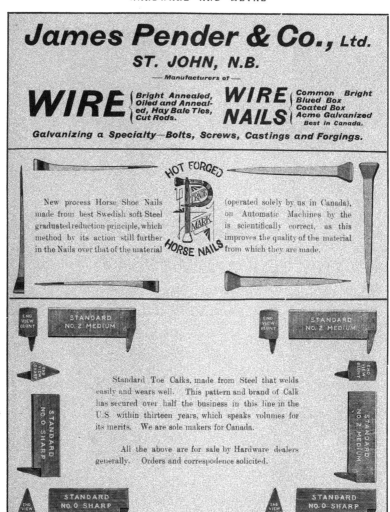

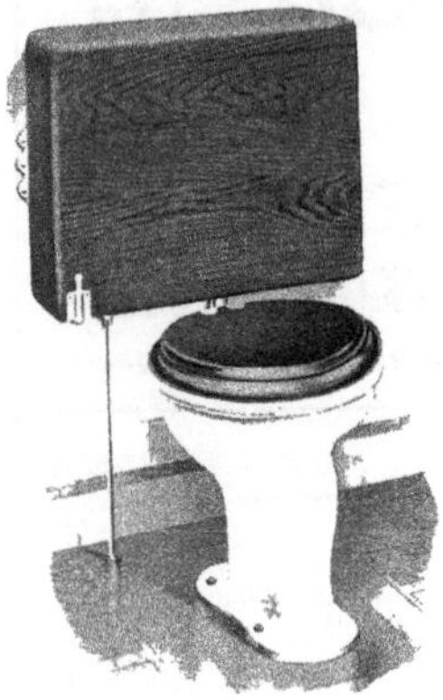

O. K. "ROYAL"

ARCHITECTS
SPECIFY
THEM

No. 08 "ROYAL" (AUTOMATIC)

"ROYAL"

LOW-TANK COMBINATIONS

ARE FIT FOR ANY BATHROOM

No. 05 "ROYAL"

THE BEST
PLUMBERS
INSTALL THEM

No. 10 "ROYAL"

MADE BY

ST. JOHN, N. B.

CIRCULATES EVERYWHERE IN CANADA
Also in Great Britain, United States, West Indies, South Africa and Australia.

HARDWARE AND METAL

A Weekly Newspaper Devoted to the Hardware, Metal, Heating and Plumbing Trades in Canada.

Office of Publication, 10 Front Street East, Toronto.

VOL. XIX. MONTREAL, TORONTO, WINNIPEG, SEPTEMBER 7, 1907 **NO. 36.**

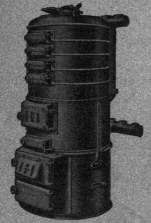

Read "Want Ads." on Page 53

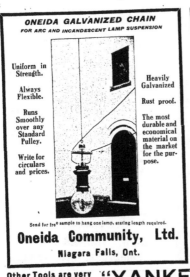

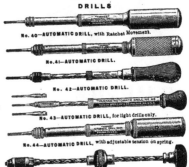

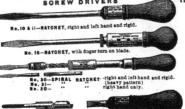

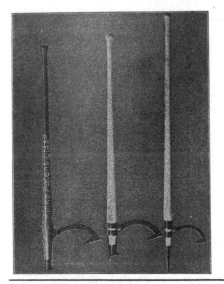

BUILDERS' HARDWARE

ALBY-DESIGN

No. O. E. 802 A. N.

Are you going to add a new design to your stock of Lock Sets? If so, ask our salesman to show you a pattern of Alby-Design. This is the newest and most up-to-date pattern of a medium-priced set that we have seen. Antique brass,b eautifully finished. This finish is going to have quite as big a sale as antique copper.

No. O. E. 316 A. N.

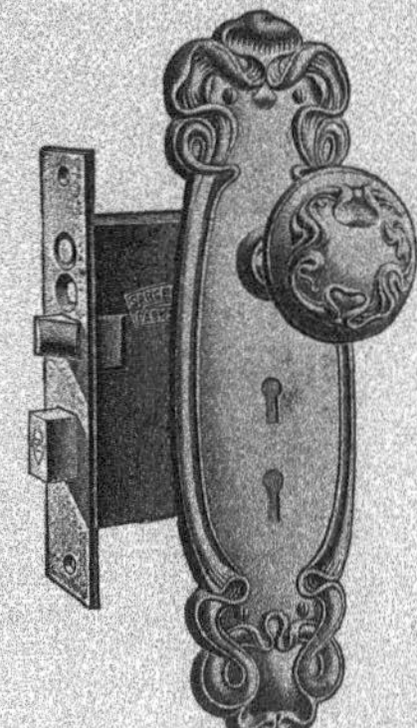

Nos. O. E. 663 A. N. and O. E. 652 A. N.

In addition to the lines illustrated above we stock front door sets in both steel and bronze metal, and we also have the following lines of the same finish to match this set:

Bronze Metal Push Button for Electric Bells
Wrought Steel Flush Sash Lights
Sliding Door Latches, both astragal & flat front
Rotary Door Bells, Bronze Metal Turn Plate
Wrought Steel Push Plates
Triple End Butts
Letter Box Plates
Cupboard Catches
Drawer Pulls
Sash Fasteners
Sash Lifts
Hat and Coat Hooks
Plated Butts

AT THE PRICE WE SELL THE ABOVE LINE YOU CAN MAKE A PROFIT OF 100 PER CENT IF YOU BUY NOW BEFORE OTHER MANUFACTURERS COPY THIS DESIGN

EXCLUSIVE CANADIAN DISTRIBUTORS

LEWIS BROTHERS, LIMITED

SHELF AND HEAVY HARDWARE

Office and Warerooms: 14 to 28 Bleury St., Montreal, and at Toronto, Ottawa, Vancouver, Calgary, Winnipeg

BUSINESS SUCCESS

Mr. Dealer, if you could take out a policy assuring commercial success in your line you certainly would do so. The next best thing is to place your orders for

Carpenters' Tools and Builders Hardware

with a specialty house—a house which guarantees quality and prompt deliveries. We handle

Stanley's, Union Hardware Co. and Sarjent's Iron and Wood Bottom Planes.
Disston's, Atkins', and Shurly & Dietrich—High Grade Hand, Rip and Panel Saws.
Sarjent's Front, Store Door and Inside Locks.
Stanley's Butt Hinges—all kinds, all finishes.
Also headquarters for Wood Sash Pegs, sizes 7/8, 1 1/4, 1 5/8-in. in one-bushel bags.

Remember our prices are right and we deliver like lightning

THOS. BIRKETT & SON CO., Ottawa, Canada

Patented 1900. Improved 1905.

Davidson's "Empire" Stove Pipe

Improved Patent

Neatly packed in crates of 25.
Secures lowest possible freight rate.

Deep Swage Prevents Telescoping.
Coated to Prevent Rust.

Sold as low as ordinary Stove Pipes.

SIMPLE and EASY to put together.
No Profane Language Necessary.
The only tools required are a pair of hands.
Pipes being cut out with dies ensure absolute fit and conformity.

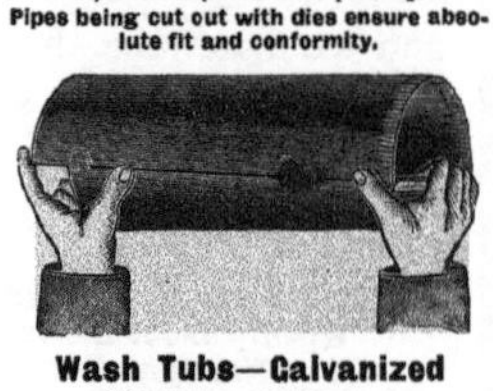

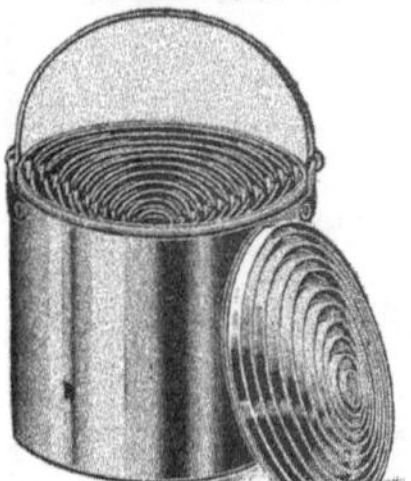

Miners' Camp Kettles

Strong and substantially made in extra heavy tin.

Nineteen sizes, from 1/8 quart to 29 quarts.

Wash Tubs—Galvanized

With Wringer Attachment

Nos.	1	2	3
Top measurement, inches	21 1/2	22 3/4	25
Bottom " "	17 7/8	18 1/4	20
Height " "	9 1/2	10 1/2	11

The THOS. DAVIDSON MFG. CO., Ltd., Montreal and Winnipeg

H. S HOWLAND, SONS & CO. LIMITED

HARDWARE MERCHANTS

Only Wholesale — 138-140 WEST FRONT STREET, TORONTO. — Wholesale Only

Registers, etc.

"Diamond" Fire Fronts

Adjust in height, 4 to 6 inches
" " length, 11 to 21 "

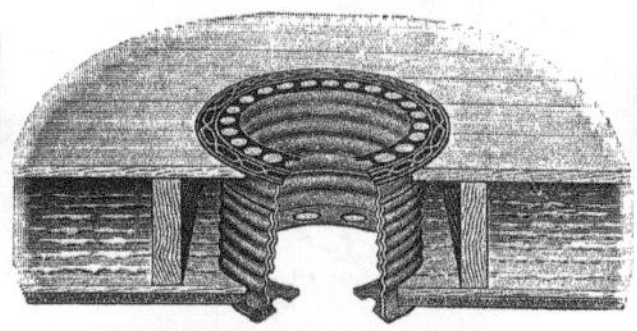

"O.K." Stove Pipe Thimble—Adjustable

CLOSED

"Diamond" Fire Backs

Adjust in height, 4½ to 7 inches
" " length, 15 to 25 "

Side Wall Registers

Japanned Black
Sizes, 7 x 10 and 12 inches

Adams' Stove Pipe Registers

Japanned Floor Plates, White Ceiling Plates

Cast Round Registers—Complete

Sizes, 7 and 8 inches

Borders

Stamped Steel

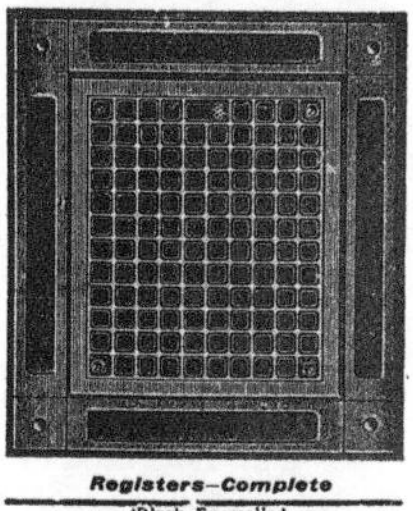

Registers—Complete

Black Enamelled
Stock all sizes

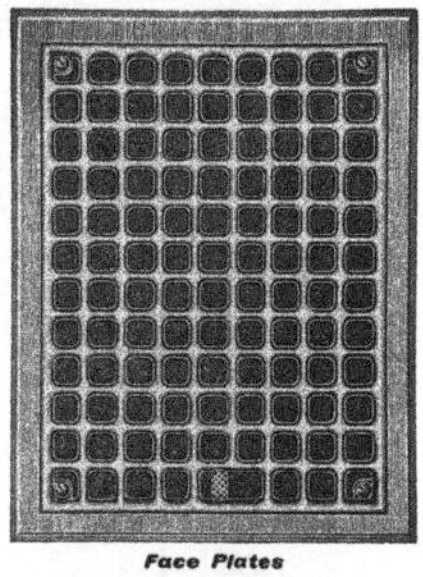

Face Plates

Stamped Steel

H. S. HOWLAND, SONS & CO., LIMITED

Opposite Union Station

GRAHAM WIRE NAILS ARE THE BEST

Factory—Dufferin Street

Our Prices are Right — Are you receiving our monthly illustrated circular? IF NOT, WRITE FOR IT. — We Ship Promptly

One of the best selling specialties we have ever taken up is the line of Asbestos Sad Irons.

You know their construction, the solid iron core and asbestos-lined hood and handle. What we wish to persuade you to do now is to handle them or add to your line. We want you to handle them because they will sell.

They will sell :

Because they supply the want of a better iron.

Because they are well advertised in the papers women read.

Because their makers supply you with stands and advertising matter to assist you.

Because the line has variety. It includes an iron for every purpose, and helps you to meet every demand.

Ask our travellers for prices and more detailed information or send to us for a catalogue.

Caverhill Learmont & Co

MONTREAL

Winnipeg **Ottawa** **Quebec** **Fraserville**

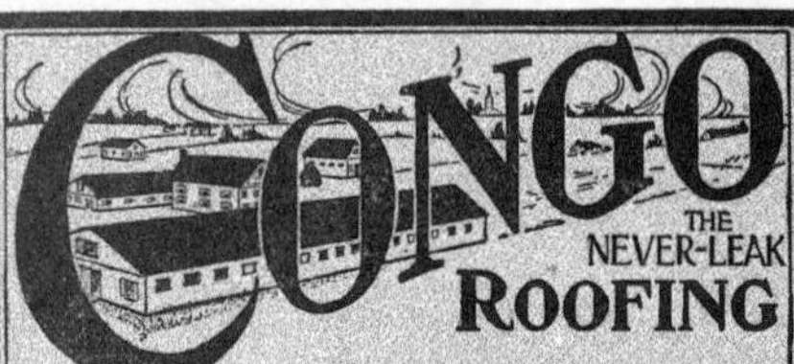

Congo can't leak, or rust, or rot, or melt in the sun, or freeze in the cold, or dry up, or get brittle, or crack, or rip off in a wind.

Congo can fit snugly on a roof, keep out the rain, the snow, the damp and the cold for many years, with a minimum of care and attention.

Its strongest point is its durability. Send for a free sample.

Buchanan-Foster Co., Philadelphia Pa.
S. Hayward Company, St John, N.B.
The Starke Hardware Co., Montreal
Geo. E. Smith & Co., Halifax.
Mechanics' Supply Co., Quebec.
H. S. Howland, Sons & Co., Ltd. Toronto.

REGISTERED TRADE MARK
GRANTED 1682

World's Best Cutlery

You make no mistake in specifying goods made by

JOSEPH RODGERS & SONS, Limited

when ordering cutlery. A reputation of over 200 years at stake, insures the quality of every article.

Write for catalogues from Sole Canadian Agents

JAMES HUTTON & COMPANY, MONTREAL

NO "CLOUDY" SURFACES

come from using "Majestic" furniture polishes. When applied according to directions the result is a brilliant lustre—none of those smeared surfaces so common to many so-called polishes.

Free from grit and dirt, "Majestic" furniture polish is all a furniture polish should be.

Write for samples and prices. 72

MAJESTIC POLISHES, Limited
575 Yonge Street, Toronto, Canada

PAGE WHITE FENCES

The Acme style you see above costs only from 16 cents a linear foot. Handsome and durable for lawns, parks, cemeteries, etc. Any height or length. Gates to match from $2.25. Write for catalog, or inspect this ideal fence. From us direct or any Page dealer.

The PAGE Wire Fence Co., Limited Walkerville Toronto Montreal St. John Winnipeg 812

Persons addressing advertisers kindly mention having seen their advertisement in Hardware and Metal.

TINPLATES

All made out of the

BEST SIEMEN'S MARTIN STEEL

Brand
Best Coke or Charcoal "Rhyd"
Good Charcoal "Edith"
Second Charcoal "Sawel"
Best Charcoal "Dryw"

We are ready to offer the Canadian Trade very best value and finest quality in Tinplates.

Baglan Bay Tinplate Co., Limited
Briton Ferry, South Wales

Please mention above brands in your indents and inquiries.

Forwell Foundry Co.

BERLIN, ONT.
Manufacturers of
SOIL PIPE, FITTINGS, and CAST IRON SINKS
Ask Jobbers for "F. F. CO." Brand

ALEXANDER GIBB

Manufacturers' Agent and Metal Broker,
13 St. John Street, Montreal

Representing Canadian, British and American Manufacturers. Correspondence invited from firms wishing to be represented.

Expert Bookkeeping

Investigation and auditing of all classes of accounts undertaken. Trading, profit and loss accounts and balance sheets prepared. Sets of books opened and office systems devised in an up-to-date manner. Collection of accounts a specialty.

Davenport, Pickup & Co.
622 McIntyre Block and 422 Ashdown Block
WINNIPEG, MAN.,
AND AT BRANDON, MAN.

Square your Accounts

CHARLES H. FOX

Vancouver, Canada
MANUFACTURERS' AGENT
Selling to the Hardware Jobbers Only.
Representing Canadian, British and American Manufacturers, correspondence invited. Cable Address, Chasfox Vancouver, P.O.Box 1105.

H. W. MITCHELL

WINNIPEG, MAN.

Correspondence is solicited from manufacturers wishing a representative in Winnipeg.
Travellers call regularly on the wholesale and retail trade in Western Canada.
Highest references and financial responsibility.

ONE CENT

Is it worth one cent to know well not only the paints and brushes you are selling but also those your competitor is selling? A Post Card sent to our advertisers will bring you full information about their goods.

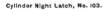

Are you interested in any of the lines that are advertised?
A Post Card will bring you price list and full information.
Don't forget to mention Hardware and Metal.

To Manufacturers' Agents

THE CANADIAN GROCER has enquiries from time to time from manufacturers and others wanting representatives in the leading business centres here and abroad.

Firms or individuals open for agencies in Canada or abroad may have their names and addresses placed on a Special list kept for the information of enquirers in our various offices throughout Canada and n Great Britain without charge.

Address, BUSINESS MANAGER,
HARDWARE AND METAL
Montreal and Toronto.

"Good Cheer" Furnaces

No. 100 SERIES

For Anthracite or Butuminous Coal, Wood and Coke

The class of Furnace which means every satisfaction to the dealer in selling and installing and to the householder in operating

The radiating qualities of this furnace are so greatly enhanced by the adoption of the cast iron tube principle therein as to place it in a class entirely by itself, and, in combination with the solidity of construction and air-blast feature, make it a warm air generator of such capacity, durability and economical fuel-consuming qualities as to place it beyond comparison with any other of the moderately-priced furnaces in the market.

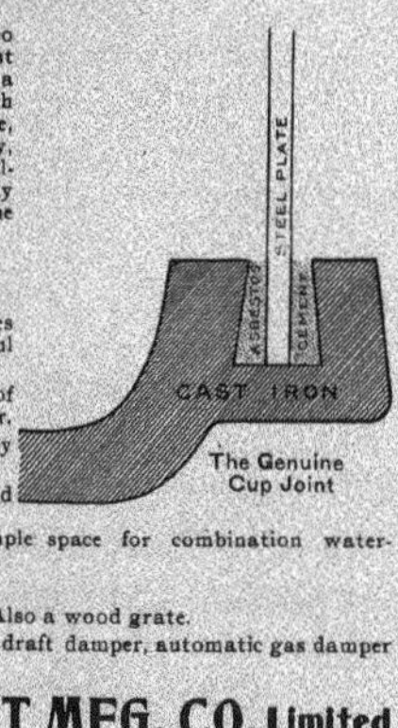

The Genuine Cup Joint

A Few of its Many Good Points

Cast iron tube radiating principle. Tubes surround fire chambers and form most powerful heat producers.

No sheet steel exposed to direct action of fire, being used only in outside jacket of radiator.

All steel set in genuine cup joints, positively gas and dust tight.

Commen-sense, positively unblockable and most easily cleaned flues.

Large combustion chamber, affording ample space for combination water-heating attachment if required.

Large double-feed door with smoke shield.

Roller grates for either hard or soft coal. Also a wood grate.

Roomy ashpit, waterpan, dust flue, direct draft damper, automatic gas damper and AIR BLAST FUEL ECONOMIZER.

The JAMES STEWART MFG. CO. Limited
WOODSTOCK, ONT.
Western Branch: Foot of James St., WINNIPEG, MAN.

Buy the Best.

HERCULES

Sash Cord.
Star Brand Cotton Rope
Star Brand Cotton Clothes Lines
Star Brand Cotton Twine

For Sale by all Wholesale Dealers.

CARRIAGE SPRINGS & AXLES

ANCHOR BRAND

GUELPH AXLE WORKS

THE GUELPH SPRING & AXLE CO. LIMITED
GUELPH, ONT.

Hardware Merchants

who undertake

Plumbing and Heating Contracts

cannot do better than subscribe for the

Plumber and Steamfitter of Canada

$1.00 per year—Samples Free

MacLean Publishing Co.,
MONTREAL. 10 Front St. E., TORONTO... WINNIPEG

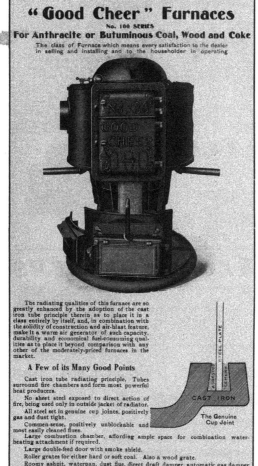

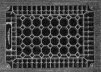

I WILL TALK

to practically every Hardware merchant in Canada from the Atlantic to the Pacific. I cannot do it all in one day, but during the first twenty-four hours I will deliver your message to every Hardware merchant in Ontario. I travel all day Sunday and on Monday morning there will not be a village within the limits of Halifax in the East and Brandon in the West, into which I will not have penetrated.

I cannot go any further East, so I now devote all my energies to the West, and so many new towns are springing up here each week that I haven't as much time as I used to have to enjoy the scenery. But I like talking to hardwaremen, clerks, travellers and manufacturers, especially as they are always glad to see me and hear the news I have to tell them. Tuesday noon I am at Calgary, Wednesday noon at Kamloops, and by Thursday morning I reach Vancouver, having been in all the mining towns and all through the fruit districts of British Columbia.

I have been eighteen years on the road and I have a pretty good connection. I never intrude when a man is busy, but just bide my time, because I know men pay far more attention to what you have to say if you catch them when they have a few moments to spare. So I often creep into their pocket when they are going home at night, and when supper is over Mr. Hardwareman usually finds me. He must be glad to see me, because he listens to what I have to say for an hour or more.

I try to always tell the truth, and men put such confidence in what I say that I would feel very sorry to deceive them even inadvertently. Probably some other week I will tell you about the different classes of people I meet. In the meantime if you want a message delivered to HARDWAREMEN, PLUMBERS, CLERKS, MANUFACTURERS or TRAVELLERS—and want it delivered quickly—I'm your man.

THE WANT AD MAN

Condensed Advertisements in Hardware and Metal cost 2c. per word for first insertion, 1c. per word for subsequent insertions. Box number 5c. extra. Send money with advertisement. Write or phone our nearest office

Hardware and Metal

MONTREAL **TORONTO** **WINNIPEG**

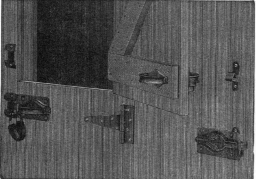

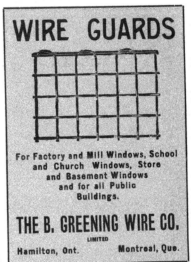

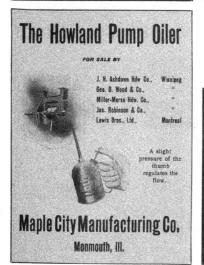

Retail Merchant and His Window

By H. L. Hall, Circulation Manager Business Man's Magazine.

The properly used show window is the best advertising medium within the reach of the retail merchant, and yet, curiously enough, it seems to be the least appreciated of all the means he uses. His show window is his best medium, because it will sell goods for him at a less percentage of cost than any other means at his command. I say that it is the least appreciated because of the fact that it is so very generally neglected. Most retail merchants have not yet learned its true value. Many of them seem to think that it is merely a space which must be filled up with something—it does not really matter what or how. All this is wrong.

Your department store manager appreciates his window space and makes good use of it. If you do not believe that he considers window space valuable, go to him and try to rent one of those he is using. You will soon get a larger idea of its value. The big store even goes to the length of employing an artist who spends all his time and thought in getting up and executing attractive window displays. And these window displays sell goods enough to make the window dresser and his big salary a good investment. We called these window trimmers artists, and many of them are nothing less, as an inspection will soon show, but it is not enough that a display shall please the eye. It must sell goods or it is not successful. The big store demands that there shall be a material increase in the sale of a displayed article while it is in the window, and if the increase does not come the fact is chalked up against the window dresser in the records of the manager.

In show windows the acme of achievement is to be found in the big stores on Broadway and State Sts., New York, and there are not lacking those who will whisper that the latter is in the lead. The other extreme is to be found in the window of the little store in a side street, where the sole decorations consist of a smoky lamp or a flaring gas jet and a choice collection of fly-specks. And the latter has just as great a relative value as the former if the merchant did but know it. The advertising agent of your local opera house knows the value of show windows, for he is willing to exchange seats worth money for the privilege of hanging his lithographs in your window, and I cannot conceive of him doing it unless your window has a real value to him—and if to him, why not to you?

The real potential value of a show window lies in the number of people who will pass it within a given space of time. Its value may be computed on the same basis as that used for the computation of any other means of publicity. Magazines charge so much a line per thousand of circulation. If you buy space in the pages of a magazine you pay for the privilege of exhibiting your announcement in a place where a given number of people will pass. It is up to you to make your announcement in such a manner that these people, or a goodly number of them, will stop to see what you have to say. It is just the same with your show window. No matter what your location, there will be about so many people pass your window each day, and it is your task to make that window catch and hold their attention to such an extent that some of them will feel a want of what you there offer.

Of course the average retail merchant cannot afford to pay a big salary to an exclusive window trimmer, but the chances are that there is some one within reach who can do many times better than he is now doing with the means at his command. It may be his clerk or porter. It may be his wife or daughter. The proper thing to do is to experiment till he finds the right one, and then let that one do his best. Window trimming is an art, but like most other arts it can be studied and acquired. Get out on the street and study other windows. Analyze the ones which appeal to you as being above the average. Learn what it is which makes them better than most of the others. Pick out the good points and emulate them. Pick out the faults and avoid them. I do not mean to advise copying, but we can learn from the mistakes and successes of others without copying.

It is at night when the outside world is dark that your window will look the most attractive. Hence the best time for window display is in the fall and winter, when the evenings are longest. For this reason, too, it follows that one of the first things to be seen to is that the window must be well illuminated. Nothing so surely kills off a window display as poor lights. And at the same time it may be stated that there is no other investment which will pay a merchant so well as good lights throughout the store as well as in the window, but if the lights must be cut down anywhere, let it not be in the windows.

The chief fault of the ordinary window display is crowding. Don't try to put your entire stock in your windows, but leave room for an effective arrangement of what you do put there. Too much stuff will defeat your purpose, which is to call attention to the items displayed with enough force to make the gazer want to buy. The handsomest and most effective window the writer has seen in a long time was at the same time the simplest. At the season of beautiful window displays—Christmas—this one was easily the best. It was in the window of a shoe store. The back and sides of the window were finished in plain wood of a dark shade, and in the window were three pedestals about fourteen inches high. Over these pedestals were draped three skins, one white, one red and one blue. On the top of each was a single slipper, matching in color the skin under it. In the centre of the window, on a white mat, was placed a single shoe, one of a new pattern. This shoe was marked "Our Marquise Shoe—$3.50." I'll warrant that the display sold that brand of shoes and sold plenty of them.

There is one more point. Do not expect a window display to sell goods indefinitely. Those who pass your place will get tired seeing the same thing day after day. Give them something new to look at once in a while. Let them get into the habit of looking to see what you are going to offer them next. Sooner or later you will catch the fancy of the regular gazer and sell him, or her, something. If your offerings are made on the basis of attractive prices, make the price a part of the display. In most cases it is well to do this anyway, as the combination of the article and the price together sometimes make an irresistible combination to the vagrant fancy of the window-gazer.

One word more and I am done. Wash your windows. Is this an unnecessary suggestion? Not so much so as it might be, as you will see if you will take a walk along any street you like, for I will warrant that you will find at least half the windows you pass would be all the better for a little attention from the porter. It is a big job to wash windows frequently when other work presses, but do not allow yourself to fall into the habit of letting it go till a more convenient season. Even if you do not do it for the benefit of the window display, do it anyway for the sake of the akpearance of the store generally. If nothing better is possible, cover the floor of your window with clean paper of a light color It looks much better than stained and blotched boards. A little attention to your windows will pay well for all the time and trouble it will cost you, as you will soon find out.

HARDWAREMAN IN COMMAND.

Amongst the latest militia orders issued at Ottawa are noticed the promotions of several Montreal officers. One of these is Major Robert Starke, of the 3rd Regiment. Victoria Rifles of Canada, who has been promoted to the command of that regiment.

Col. Starke is prominent in mercantile circles in Montreal, and has for many years been connected with the hardware trade. He was born in 1865, and in 1880 he entered the pioneer hardware firm of Benny, MacPherson & Co., with

COL. ROBERT STARKE, MONTREAL.
The New Commander of the 3rd. Regiment Victoria Rifles.

whom he was connected for four years. Two or three years were then spent in the employ of the Dominion Transport Co., and in 1888 Mr. Starke became identified with Peck, Benny & Co , latterly known as the Peck Rolling Mills, with whom he stayed for nine years.

On the death of the late Jas. G. Howden, Mr. Starke joined his brother, William Starke, as partner in the firm of Howden, Starke & Co., later on consolidated into the firm of Starke Hardware Co., Ltd., which, on Januaey 1, 1907, was amalgamated with the Seybold Co., under the name of Starke, Seybold, Ltd. William Starke is president of the new company ; Robert Starke, vice-president, and Gordon Seybold, secretary-treasurer.

It is rather a remarkable coincidence that the two brothers, William and Robert, should have served similar terms in the Victoria Regiment.

The Cost of Doing Business

Montreal merchant inVites a discussion on this subject and submits a series of questions to be answered by merchants in an effort to help the retail dealer in figuring out what it costs him to conduct his business.

Just as the store of Amiot, Lecours & Lariviere, St. Lawrence Boulevard, Montreal, stands in a class far above the average of Canadian retail hardware stores, so does the proprietor of the business, Fred. C. Lariviere, stand forward as one of the most liberal-minded, progressive and studious hardware merchants in all Canada. On several occasions readers of Hardware and Metal have been favored with articles on trade subjects from his pen or reports of practical talks made by him before his fellow-retailers, and on every such occasion Mr. Lariviere's arguments have been followed with keen interest. His article on profit sharing with employes, published a few months ago, attracted the attention of some of the largest hardware jobbers in the United States, and Mr. Lariviere is now in communication with them regarding their experiences. He has also established a system of profit-sharing in his own business, and may have more to say on this subject later.

Another matter which has been occupying Mr. Lariviere's attention recently is the cost of doing business, and in a recent letter to the editor he submitted the accompanying form and plan to secure information on which to base definite calculations. He writes:

> Every dealer claims to be in business to acquire an honest, well-acquired wealth to provide for his family and secure an income for old age.
>
> This being admitted, how can we explain why merchants, as a rule, are selling goods at such a small percentage over cost, and still be expecting to make a living? An answer that could apply to many jobbers and retail merchants is, they do not know the cost of doing business, and are too much inclined to seek a large volume of unprofitable goods rather than a small amount, bringing better returns.
>
> You might print in Hardware and Metal the enclosed questions, leaving sufficient space for answers, asking those who wish to take advantage of our joint offer to fill in the blanks without signature or address, using some nom de plume, such as "B2306," for the dealer's guidance. Replies could be sent to Hardware and Metal, in Toronto, and we would undertake to demonstrate to every retailer what it costs to do business, both on cost and selling prices, giving the answers through your valuable paper free of charge.

The suggestion is a practical one, and readers are invited to take advantage of the offer made to comment on all replies sent in. Dealers can fill in the accompanying blank columns and forward to the editor in a plain envelope over some fictitious signature understood only by the sender.

The receipt of a fair number of replies will be of considerable assistance in reaching a solution of this question, at present too little understood. The series of questions follows:

	Answer	Amount	% on cost prices of yearly sales	% on selling prices of yearly sales
What was your turnover last year?....				
What were your gross profits?.........				
The difference in the amount of goods sold at cost prices?.................				
Have goods returned been deducted from the amount of your turnover?..				
If not, what is the amount?				
Are the charges you have paid for boxing, packing, freights and cartage to your store added to your cost price?.				
If not, what is the amount?				
What was the amount of your general and administrative expenses? (in a lump sum or detailed as below)				
For Advertising				
Insurance				
Store heating and cleaning				
Licenses, business taxes of all kinds, except those imposed on real estate				
Donations				
Lighting				
Stationery for office, wrapping paper and twine for store				
Postage stamps				

	Answer	Amount	% on cost prices of yearly sales	% on selling prices of yearly sales
Telephone services				
Store rent (or if you own your premises, 6 per cent. per annum on the value, plus taxes, insurance and wear and tear of building and ordinary repairs to which tenants are generally obliged)				
Salary of all employes or as detailed hereafter:				
Salary of travelers..............				
Traveling expenses of travelers ...				
Salary of indoor salesmen				
Salary of general store help				
Salary of office help				
Your own salary, if not included, and also that of your partner or partners, if you have any				
Sundry expenses not heretofore accounted for, such as car fares, entertaining, trade papers, legal advice, legal costs, commercial agencies, telegrams, etc.				
Expenses for horses and carriages, etc., in a lump sum or detailed as below:				
Feed of horses				
Veterinary surgeon				
Horse shoeing				
Repairs to harness				
Repairs to vehicles				
Rent of stables or value of same if your property				
Salary of drivers and stable men..				
Depreciation of horses, vehicles, etc.				
If you are not the owner of horses and delivery wagons, what have you paid for that service?........				
What was the amount of the allowance you gave your customers outside of cash discount?				
What was amount of your bad debts?.				
What was the amount of interest you paid to your bankers for advances on your customers' or personal notes, also to the firms you buy from, for overtime requested and the cash discount allowed to your customers?.				
What is your capital in business?				
Is, in this amount, included any real estate used for business? If so, what is the amount?...				
Outside of capital above named and not included in, have you any real estate used for your trade? If so, what is the value of same?				
Did you allow anything for wear and tear of store and office fixtures and furniture? If not, what do you estimate it to be?......................				
Did you allow anything for breakage of merchandise, deterioration of stock for goods delivered and omitted to be charged or taken from your store without your consent or knowledge? If not, what do you allow for this?..				
How many traveling salesmen have you?				
Total of their annual sales?				
Their gross profits, the amount of their salary and expenses?........				
Can you detail the operations of each, as follows?:				
Traveler No. 1				
Annual sales				
Gross profits				
Salary				
Traveling expenses				

WESTERN TRADE GOSSIP.

Graham & Rolston, hardware merchants, Winnipeg, Man., are retiring from business.

Malcolm Bros., Dauphin, Man., have sold their hardware business to McDonald & Wright.

J. R. Chisholm, hardware and lumber merchant, North Battleford, has compromised with his creditors.

The Summers & Kelley Co., plumbers, Regina, Alta., are being succeeded by the Kelley Plumbing, Heating & Lighting Co.

Jackson Bros., hardware merchants, Pincher Creek, Alta., will move into the Hunter block as soon as the latter is completed.

John Blair, Hartney, Man., has purchased the Sturdy Hardware Co.'s establishment, and, assisted by his son, William, will, in future, carry on the business.

Theo. Miles, Kamsack, Sask., is a progressive hardware merchant, who has put his business on a strictly cash basis. F. A. Healey, Lloydminster, has done the same.

Chas. H. Fox, manufacturers' agent, Vancouver, B.C., was in his old home, Portage la Prairie, last week, renewing acquaintances. He has since returned to Vancouver.

The London Fence Co., whose factory in Portage la Prairie was destroyed by fire this spring, have decided to rebuild on a new site. The old site is to be expropriated by the C.N.R. for railway purposes.

H. V. Lawlor, Brandram-Henderson paint specialist for the Marshall-Wells Co., Winnipeg, has returned to Winnipeg from an extended business trip through the West. He reports that the crops in Alberta and Northern Saskatchewan will beat all records this year. The South is not so good.

ANOTHER SAW FACTORY.

C. Atkins & Co., who have been looking over the Canadian field for some time, have definitely decided to establish a branch factory in Canada. E. C. Atkins, head of the company, was in Toronto last week conferring with Mr. Ten Eyck, the company's Canadian representative, and it is now announced that a factory building has been secured at Hamilton at an expenditure of $50,000, and machinery will be installed as rapidly as possible.

SAD-IRON HEATER.

An improved sad-iron heater has been invented by N. A. Westerlund, Superior, Wis. The object of the improvement is to provide an arrangement whereby a number of sad-irons may be heated and moved conveniently from place to place. More specifically, the invention contemplates an arrangement whereby the same handle may be used for moving the heater or for applying the irons to the ironing board.

HARDWARE AND METAL

Established 1888

The MacLean Publishing Co.
Limited

JOHN BAYNE MACLEAN - President

Publishers of Trade Newspapers which circulate in the Provinces of British Columbia, Alberta, Saskatchewan, Manitoba, Ontario, Quebec, Nova Scotia, New Brunswick, P.E. Island and Newfoundland.

OFFICES:

MONTREAL, ... 232 McGill Street, Telephone Main 1255
TORONTO . . . . . 10 Front Street East, Telephones Main 2701 and 2702
WINNIPEG, . . . 511 Union Bank Building, Telephone 3726
LONDON, ENG. . . . . 88 Fleet Street, E.C., J. Meredith McKim, Telephone, Central 12960

BRANCHES:

CHICAGO, ILL. . . . . 1001 Teutonic Bldg, J. Roland Kay
ST. JOHN, N.B. . . . No. 7 Market Wharf
VANCOUVER, B.C. . . . Geo. S. B. Perry
PARIS, FRANCE . Agence Havas, 8 Place de la Bourse
MANCHESTER, ENG. . . . 92 Market Street, Louis Wolt
ZURICH, SWITZERLAND . . . Orell Fussli & Co.

Subscription, Canada and United States, $2.00
Great Britain, 8s. 6d., elsewhere - 12s

Published every Saturday.

Cable Address { Adscript, London / Adscript, Canada

ANOTHER ADVANCE IN STOVES.

The five per cent. advance to be made on stoves and ranges on October 1 is, in the opinion of many prominent stove men who have visited Toronto during the Exhibition, an altogether necessary step and, in fact, one which should have been made a year ago, when the last change was made. The five per cent jump made at that time did not cover the advance in cost of manufacture and as both labor and material have continued to increase, the second five must now be added. Both association members and non-members, it is understood, will advance prices as announced.

Like last year, the present advance will not affect the major portion of the current year's business, as few dealers will leave off buying till after the rise is made. All sorting orders will cost the extra five, however, and as goods in hand on Oct. 1 cannot be replaced at the old discount, dealers should not hesitate to take advantage of the increase to mark up their retail prices in accordance with the higher cost rate.

Speaking to Hardware and Metal, a prominent Orillia hardwareman said that the dry season had affected crops unfavorably in his district and farmers are likely to hug their bank books between the present and next harvest. In his opinion, however, no difficulty should be experienced in marking up prices, although many dealers would neglect to re-mark their stoves and continue to sell at present rates.

As dealers stand to lose if caught in a drop in the market, they should not hesitate to take advantage of a rise when circumstances make it possible for them to mark up selling prices in keeping with advanced cost prices.

THOSE MAIL ORDERS.

A man went into a hardware store to buy an axe. Being shown the article and informed that the price was $1.15, he said: "Why, I can get that same axe from a mail order house for 90 cents."

"Very well," said the hardwareman, "I will give it to you for the same price, provided that you will do the same with me as you would with them."

"All right," replied the customer, as he handed over a dollar bill, the merchant giving him back 10 cents in change.

"Now," said the hardwareman, "I want 25 cents more to pay express charges," which the purchaser gave him. "How much did your axe cost you?"

"One dollar and fifteen cents," the man answered.

"Very good; now give me the five cents for money order fees and postage," which the purchaser had to hand over. "Now, how much did your axe cost you?"

"One dollar and twenty cents," said the customer.

"Not so cheap after all," said the merchant, whereupon he picked up the axe, tossed it back on the shelf and told the customer to call for it in ten days, as that would be as soon as he could get it from the mail order house.

THE ENDORSING HABIT.

There are few if any business customs sanctioned by law and usage so utterly hostile to the very foundations of business principle as the endorsing habit. As an American exchange says: "Value received for value given," is one of the unalterable fundamentals of honest business; one that both custom and law make themselves parties in violating whenever a disinterested third party is called into the deal.

However complicated the business transaction, it resolves itself into two parties, each represented by one or more individuals, a buyer and a seller. Both are supposedly benefited in some way by the trade; "value received for value given." Each party is paid for his original possession, his work, his property, his money, in an equivalent of work, property or money at some time, past, present or future. If goods are sold, cash or its equivalent is given for them, the seller receiving as his price for making the trade a certain advance over what they cost him, or, are worth to him to keep his profits. The buyer receives his price or profit in the use the articles will be to him above their cost, either to barter or to keep.

If money is borrowed, the lender sells the work of his money at so much per year, the same as a livery keeper hires a horse out to a customer. The price paid in each case represents the profit. On the other hand, the borrower expects to get enough service, work, out of the money to warrant him in paying for its services. All business transactions conducted on an honest business basis only involve two parties, the buyer and the seller. Any attempt to drag in a third is rank robbery, whether it be done by the merchant in the name of friendship or by the law in the shape of bail bonds, etc.

The system is so well established that it looks like blowing in wind to controvert it; still it is a bit of commercial brigandage that ought to be frowned upon by every honest man that all the others, including the court and lawyers, would have to drop it from their book of legal traditions.

What is the remedy? There are several in the hands of every retailer. First of all, and this is the oldest of all, too, never endorse a note or any form of commercial paper for any one. Let those of your friends whom you think worthy use you as a reference, in other words, occasionally permit the loan of your own good opinion. Give all men to understand that when it comes to borrowing your money or your credit there is but one business way, that is, to pay for the accommodation, same as they would pay a bank.

Another remedy, harder to accept, but quite as important as the first; never permit any business man to tempt you into dragging one of your friends in as an endorser. You have no right to ask him to pay for your goods in his own credit any more than in cash. If you must go to him, ask him to lend you the money and pay him for the accommodation, not some banker. Be honest with him, anyway, and give the interest to the man who really assumes the risk.

Last, but not least, never ask a creditor to get an endorser. Deal with him on the even exchange of value basis, or else don't deal. You have no more right to tempt him into taking something from his friends he has no right to take than you have to ask him to steal for you; no better business right, that is. The legal right is quite another thing.

GOSSIP OF THE TRADE IN WESTERN CANADA

August has been a comparatively quiet month with the Winnipeg retail hardwaremen. Of course, it is ordinarily a quiet month, but this year there has been less business than usual in building supplies and mechanics' tools, owing to the temporary falling off in building, and this has served to emphasize the dullness of a dull season. Nevertheless, several stores complain of the difficulty in getting good clerks and the proprietors say that in an active season they would be in considerable trouble, owing to the smallness of their selling staffs. There are many inexperienced men wanting positions as hardware clerks, but it is becoming increasingly difficult to find experienced clerks. There are several good openings in Winnipeg for experienced hardware clerks. So many clerks have started stores of their own in the new towns in the west, that there are openings, not only in Winnipeg, but throughout the west, for hardware clerks with some experience.

* * *

The Winnipeg hardwaremen continue to accord their association loyal and hearty support. It was in Winnipeg that the Hardware Association movement in the West had its birth, and the hardware dealers in the Manitoba capital are still true to the traditions of the past. It would be hard to find anywhere in Canada, a body of business men who are such keen competitors, but who are, nevertheless, able to work together so well in all matters of common interest. In many of the western towns the dealers seem to have lost interest in their association, but it is not so in Winnipeg. Meetings are still held regularly, and the Winnipeg hardwaremen have learned to have confidence in the good faith of their business rivals.

* * *

There is considerable grumbling among Winnipeg retailers at the action of the city council, in calling for the business tax in August. Ordinarily this tax is collected in December, but owing to the stringency of the money market, the city authorities were in need of ready money and selected this way of getting it. There have been some changes in the basis of this particular tax, and in some instances, it would seem that grave injustice has been done some individuals. But the complaint heard most frequently, has reference to the unusual time of collecting the tax. It is a heavy item and coming in the dull month of August, has to some men seemed like the last straw. Nevertheless, the tax has been met promptly, a very large proportion of the Winnipeg business men taking advantage of the discount offered for payment before the middle of the month.

In Winnipeg, as elsewhere, there is bitter complaint among the business men over the curtailment of credit by banks. At the meeting of the Western Retail Hardware Association last month, a strong resolution was introduced, condemning the action of the banks in placing so much money in call loans outside Canada, and calling upon the government to compel Canadian banks to invest only in Canadian securities. This resolution was considered too drastic and it was not carried, but the discussion showed how keen was the feeling among the business men of the west.

"Let the banks shut down on speculative accounts if they want to," said a leading Winnipeg hardwareman to Hardware and Metal one day last week. "It's quite right and peoper that they should do so, for they are not supposed to lend on real estate accounts. But if they are not prepared to discount notes and advance money on good security for their customers in established legitimate lines of trade, I want to know why we give them their charters."

"A couple of months ago," he continued, "I went to my bank with a note in my favor for $350 and wanted to get it discounted. The bank wouldn't look at it."

"Anything wrong with the security, I asked?"

"No, the security is good. The man who gave you the note is good for it and we know that you are, but we can't advance the money. We are not doing it now."

"To show you how good the security actually was," concluded Hardware and Metal's informant, the note was paid me in full when it fell due."

ANOTHER PRIZE OFFERED.

It's not too early to lay plans for the Christmas holiday trade. Dealers have already bought some lines, of course, and in these days of slow deliveries other orders should be placed as soon as opportunity offers. "Goods well bought are half sold" is a time-worn truism in mercantile life.

But it doesn't do to feel too secure on the buying part until the selling has been done. Plans for increasing holiday sales must be worked out months ahead. Window displays must be figured out and a series of rough sketches planned. A series of ads. for the local papers should also be prepared and, where deemed advisable, a neat booklet gotten up, to be mailed to all probable customers in the surrounding district.

It is in these selling plans that Hardware and Metal can be helpful to its readers. Let all partake of the spirit of co-operation and exchange ideas through these columns. To help the suggestion along the editor invites any reader to join the friendly contest and hangs up a prize of $10 cash for the best answer to the following questions:—

How can the hardware merchant increase his sales of holiday goods next December? What special lines should he stock? What selling plans should be adopted? What special advertising should be done? What novel window displays can be suggested? Should souvenirs (calendars, knives, trays, etc.) be given to customers?

The prize of $10 will be awarded to the writer of the most practical and original letter of from 500 to 1,000 words received by the editor before October 15, 1907, and the best half-dozen replies will be published in Hardware and Metal.

GOOD TRADE LOGIC.

In the course of a conversation with a traveling man, writes F. A. Parker, the salesman said: "Whenever I go into a town and find a man there who has been a customer of one of our competitors for years, I at once lay plans to land him as a customer of my house, because I know he is a good customer for any house. He will, in nearly every instance, tell me that there is no use talking to him, as he has been buying all his goods of that particular house and has no reason for changing. I generally tell him that he is exactly the kind of a customer I want, for, if I can show him it is to his advantage to give me his business, instead of to the customary house, he will then be as loyal to my house as he now is to a competitor.

"I may not get my man for a year or so, but I eventually get him, and when I do, I know he will be with us for years, and we will get all his business, either by mail or when I go to his town, for he is a sticker."

Too many good business men are overlooking this point to-day. This traveler strikes right at the centre of big business building when he goes after the customer who is an old customer of a good competitor, for his trade is well worth much effort to land; while the dealer who is easily persuaded to give you an order on short acquaintance is just as easily persuaded to give the next man an order, and as a result you never feel that you can depend upon his regular business.

This idea should be just as valuable to the retailer as to the manufacturer or jobber, for the retailer will also find that the "sticker" is a good customer.

Statistical Boarder: "Have you the remotest idea, for instance, what the world's supply of honey is?"

Sentimental Boarder: "Yes, sir. The world's supply of honey weighs exactly 116 pounds, and her name is—well, I'm not going to mention it in this crowd."

Markets and Correspondence

(For detailed prices see Current Market Quotations, page 62.)

MARKETS IN BRIEF.

Toronto.

Antimony—Steadier.
Copper—Weak.
Lead—Firmer.
Linseed Oil—Advance of one cent.
Turpentine—Easier.
Pig Iron—In better demand.
Tin—Fluctuating.
Old Materials—Weak.

Montreal.

Antimony—Easy.
Copper—Weak.
Lead—Firming.
Linseed Oil—Steady.
Old Materials—Weak.
Pig Iron—Firm, with active call.
Tin—Firming.
Turpentine—Steady.

MONTREAL HARDWARE MARKETS

Montreal, September 6.—During the past few weeks the hardware trade has been subject to many fluctuations and, it might almost be said, regular. One week things have been active and encouraging, and the next week shows a slackening in the demand. Why there should be fluctuations at regular intervals it is difficult to explain. But there is a good reason for the slackening in demand which has occurred this week. It seems that the buying public are hesitating to place big or many orders because of a lack of confidence in the stability of the money market and also the uncertainty of this year's crops. The frequent rains which have fallen during the past fortnight, have done something to make people doubt big crops this fall.

No changes have occurred in the prices of any lines. A general firmness is reported.

Screws—Factories still find it exceedingly difficult to cope with the current demand. The difficulty lies now not so much in scarcity of raw material as lack of capacity on part of factories. It is expected that it will be some time yet before conditions are satisfactory.

Sporting Goods—Trade is rapidly opening up and jobbers are very hopeful as to the outlook for business this season. The increased number of tourists will be largely responsible for the large volume of trade expected this fall.

Poultry Netting—A strong demand still prevails, with firm and unchanged prices. The factories are making strong endeavors to place the trade in this by working for reserve stocks to cope with any contingency.

Builders' Hardware—A heavy business is being done in all lines, owing to the exceptionally large amount of building being done. Prices are firm.

Building Paper—Business is steady without change in prices. The factories are busy endeavoring to lay up reserve stock.

Wire Goods—Are moving freely at present, especially hay wire. The hay crop in the Province of Quebec this year is heavy and a strong demand exists for baling material. Strong demand prevails for wire nails, of which there is not much surplus stock.

Bolts, Rivets and Nuts—Are still scarce. The demand is moderate and prices firm.

TORONTO HARDWARE MARKETS.

Toronto, Sept. 6.—Trade in all seasonable lines of hardware continues to increase, and local jobbers are quite pleased at the way September business has opened up. While most of the orders for fall goods were booked in advance, still there is always a considerable amount of trade which is impossible to procure until the season for retailing is close at hand; with the approach of fall, however, travelers find it comparatively easy to land this business. and fall orders from late buyers all over the Province are now daily arriving to swell the volume of September trade. Hundreds of retail hardware merchants have called on local jobbers this week, while in the city visiting the National Exhibition, and a large amount of business has been booked as a result.

Screws—The scarcity in all good selling sizes is still very marked, though the manufacturers are putting forth their best efforts to catch up with the trade. As the demand is very strong, it looks as if considerable time must yet elapse before supplies will be sufficient to relieve the situation to any great extent.

Nails—A strong demand still prevails for all the best selling sizes, but supplies are now fairly adequate to meet the call. Prices are well maintained at figures last quoted.

Wire—A few inquiries concerning hay-baling wire are arriving, but, on the whole, trade in wire is dull, as it usually is at this time of the year. An increased business is looked for by jobbers before the end of this month, but the real wire trade will not set in again in earnest until the beginning of the new year. Meanwhile manufacturers are getting a supply on hand in order to prevent a repetition of the serious scarcity which prevailed all through this season.

Building Material and Supplies—Builders' hardware of all kinds is in strong demand. Glass and putty are now selling fast, and the sale of both promises to be exceptionally heavy all fall. Cement is moving out fast, as farmers are now over the rush of harvest, and are giving their attention to repairs. Strawboard and tarred building paper are in good demand, with supplies quite adequate for present needs.

Sporting Goods—As the shooting season for ducks and partridges opens on Sept. 15, guns and ammunition are more than ever in demand. Sporting goods dealers are receiving a great deal of trade from Exhibition visitors, and attractive displays of ammunition and firearms are everywhere in evidence.

Fall Goods—September business has opened up well and large orders continue to arrive for the various lines of fall goods. Book orders for the West are not quite as heavy as might be expected, but, as far as Ontario is concerned, business is away ahead of last year.

MONTREAL METAL MARKETS.

Montreal, September 6.—A general quietness prevails in local circles, with the exception of the trade in pig iron, which, for some time past, has been very firm and continues so at the present. The buying public are at present suspicious, and hesitant in placing orders. They strongly suspect declines in the prices of certain commodities and are therefore hesitating to buy.

Latest cable advices from England report a slight advance in lead and tin. Throughout the summer the English market has continued remarkably strong considering the weakness of surrounding markets.

Pig Iron—Scotch and English, first-class grades, are scarce and hard to procure, owing to the fact that many of the furnaces have not been working satisfactorily and have been producing little high-grade material. Consumers in some districts, especially, have not come into the market very actively. It is expected that when the new furnaces at Hamilton commence operations that the orders which have been placed by western Ontario firms will be diverted from neighboring American furnaces. Locally, a strong demand for high-grade products prevails, large enquiries having been received by various firms this week. Prices are firm and unchanged: Middlesboro, No. 1, $21.50; No. 2, $20.50; Summerlee, $25.50.

Antimony—Continues easy. Prices are weak and unchanged. It is reported that in the American market the continuous decline of the past weeks has at last stopped. Cookson's, 15c; Hallett's, 14½c.

Ingot Tin—Recent messages from England report an advance there. The American market is unusually quiet, consumers being unwilling to buy at present prices. Local prices are firmer, although no serious advance has been made in prices yet, Lamb and Flag and Straits are still quoted at $44.

Copper—In consonance with conditions as they exist in the United States, very little activity is noticeable. Much uncertainty is expressed by local manufacturers, who will not buy, as they are expecting a decline in prices. Prices at present are unchanged and easy.

Lead—An advance is reported in Great British. The local market is unusually strong, and a steady call for this commodity prevails. Imported pig, $5.25; bar, $5.75.

Old Materials—Scrap iron in local circles is quite plentiful, and consequently prices are weak. A slight demand at present prevails. Little change is noted in other lines. Heavy copper and wire,

15c; light copper, 13½c; heavy red brass, 13c; heavy yellow brass, 10c, light brass, 7c; machinery cast scrap, $17; heavy machinery scrap (imported), 50c higher.

Black Sheets—Are very firm, and a large number of contracts are on hand. Prices are firm and unchanged.

TORONTO METAL MARKETS.

Toronto, September 6.—While no exceptionally heavy transactions are reported in metals locally this week, a steady business is, however, being done, and jobbers have no reason to complain with the volume of trade coming in for the first week in September; indeed, considering the depressed state of the foreign markets, it is surprising that local business shows as much life as it does. The metal markets, generally, are not nearly as bearish as they were two or three weeks ago, and many consumers who have been holding back their orders, or merely purchasing from hand to mouth, are now showing evidences of coming into the market in order to take advantage of present low prices.

Antimony—There is a steady demand locally, but no large transactions have taken place during the week. The American market continues dull and little buying is being done. Cookson's is still quoted locally at 15c, and Hallett's at 14½c.

Pig Iron—The increase in pig iron business, which began last week, is well maintained and additional buyers are in evidence this week. Many consumers who have been playing the waiting game have now come into the market, and more prompt buying is now being done.

Tin—Last week tin developed a stronger tone locally, but the strength has not been maintained, and at present the market is a trifle easier. The foreign markets, however, have been alternately advancing and declining of late, and the local market is simply fluctuating in response to foreign influences. Prices are unchanged locally.

Lead—There has been a slightly increased demand during the week, and local prices are a little stiffer in response to recent advances, which have taken place in the British market. The local price of imported pig lead is very firm at $5.35.

Copper—Local sales of copper have been rather small during the week, and consumers seem to show a disposition to hang back and watch the market. The London market continues dull, with little doing. The leading American producers and selling agents announced on Tuesday that they have established their prices for electrolytic copper at eighteen cents a pound, and it is hoped there that by this action the deadlock between buyers and sellers will be broken. It is reported from reliable sources that considerable business has already been done there at this figure, and it looks as if the first real buying movement in several months has at last started. Locally, casting ingot copper is still quoted at 22c.

Old Materials—Prices are still declining and there is little or no life to the local situation, which continues to reflect the dullness of the over-sea market. The price of heavy copper and wire has declined one cent, and is now 14c; light copper and heavy red brass have each declined one cent, and are now 12c. The market for scrap iron continues very dull though prices are locally unchanged.

LONDON METAL MARKETS.

London, Sept 3.—Cleveland warrants are quoted at 56s, and Glasgow standards at 55s 3d, making prices, as compared with a fortnight ago, on Cleveland warrants 1s 3d lower, and on Glasgow standards 1s 3d lower.

Tin—Spot tin opened irregular at £166 10s, futures at £166 10s, and after sales of 200 tons of spot, and 300 tons of futures, closed easy at £166 for spot and £166 for futures, making price as compared with two weeks ago 12s 6d lower on spot, and on futures unchanged.

Copper—Spot copper opened weak at £75 5s, futures at £74 10s, and after sales of 200 tons of spot and 1,600 tons of futures, closed weak at £74 for spot, and £74 for futures. making price, as compared with two weeks ago, £5 lower on spot, and £2 15s lower on futures.

Spelter—The market closed at £21 12s 6d, making price as compared with two weeks ago 7s 6d lower.

Lead—The market closed at £19 5s, making price as compared with two weeks ago 5s higher.

U. S. METAL MARKETS.

Cleveland, O., Sept. 5.—The Iron Age to-day says:

The general tone of the market conditions during the week has been slightly better than during the previous week. Pig iron is just as quiet as it has been, and no sales of Bessemer, except in very small lots, are reported. Foundry iron is moving in tonnages ranging from carloads to 100-ton lots, with prices unchanged. It is in finished lines that the improvement is noted.

The rail market has shown a decided improvement, and the leading interest during the week sold 60,000 tons of standard sections. The Pennsylvania, New York Central and Wabash contracts are expected within a short time, or as soon as the committee having the 1908 specification under advisement makes its report. Contracts already placed will be given rails according to the decision of the committee.

Since the dissolution of the shafting association, prices have declined sharply. The demand for merchant pipe is active, and mills are having difficulty in making shipments fast enough to satisfy buyers. New business in tin plate is light, but shows some improvement.

Owing to the long continued heavy demand for wire products an advance of $1 per ton is announced. The prosperity of the farmers and the favorable prospects for crops have been an important factor in the prosperity of the wire industry. The demand for structural material is strong. although not quite as active as last month, when the American Bridge Co. booked the largest tonnage of any month in its history.

Machine tool manufacturers report improvement in business, with an active demand for all they can turn out. Orders from foreign countries are coming in at an encouraging rate.

The coke market is firm and the demand active. Old material is weak but prices are not as low as might be expected. considering all the unfavorable conditions.

U. S. IRON AND METAL MARKETS.

New York, September 5.—The Iron Age to-day says: There are indications that the long continued deadlock in the Foundry Pig Iron markets is soon to be broken, although those buyers who are testing the market do not reach far into the future in covering their requirements. Still, it is a fact that the prices to which the markets have worked down, notably in the east, are proving attractive to some melters. The movement, however, is at best only in its

The leading southern pig iron interests preliminary stages.

are still behind in their deliveries and are maintaining a conservative attitude toward the market. Smaller interests, however, are willing to make concessions, and $18 to $18.50 is being quoted for delivery during the last quarter. There is a feeling in the southern iron trade that shortage of lake ores may, in the winter and spring, divert pig iron tonnage to the southern producers. It is a fact that normal conditions have not yet been restored on the Mesaba range.

The rolling mills and steel works are running at the highest speed, and it is probable that before the close of October some records will be broken.6There is still a shortage of steel, which has been emphasized in the central west in sheet bars, by the shut down of the Shenango works of the Steel Corporation and the breakdown of the sheet bar mill of the Youngstown Sheet & Tube Company.

While specifications against old orders continue heavy and are crowding the mills, still new business is coming in to some of the large interests in considerably reduced volume, foreshadowing a lighter business in the winter and spring. This is not unwelcome, since it will permit of weeding out of the poor material in the labor crews, and restore to some extent the greatly impaired efficiency of labor.

Conferences are now in progress between the representatives of the railroads and of the rail mills, over the specifications and the sections, and an early settlement is hoped for. In the meantime, there is a little movement in the rail trade.

The American Steel & Wire Company has just announced an advance of $1 per ton on wire products. The same company has established new discounts on shafting, and has announced that the former practice of base territory quotations has been discontinued.

The leading copper selling agencies have opened their books for September and October deliveries of electrolytic at 18c, and the lower quotations made recently by smaller sellers have been withdrawn. Large buyers state that they will purchase only for early requirements, not being thoroughly satisfied that the situation is sound on the new basis. Evidence is accumulating that rather more copper was shipped upon the market during the past month under official prices than the trade was generally cognizant of, quite apart from liberal sales for shipment abroad.

Many a young man has passed into the old bachelor class because of his intimate acquaintance with married men.

HARDWARE TRADE GOSSIP

Quebec.

A. J. Petit, plumber, Levis, Quebec, has made an assignment.

Mr. Stewart, of Hamilton Cotton Company, Hamilton, was in Montreal last week.

T. Fortye, of the Peterborough Lock Company, Peterborough, was in Montreal this week.

The assets of the hardware firm of Dubois, Lapierre & Bourbonnais, Montreal, were recently sold.

Captain Strange, sales manager of Lewis Bros., Montreal, is spending his vacation at Portland, Me.

Mr. Cameron, of the Winchester Repeating Arms Co., New Haven, Conn., was in Montreal this week on business.

Mr. Stanley, representing Landers, Frary & Clarke, New Britain, Conn., is calling on the Montreal trade this week.

P. D. Dods & Company, Montreal, have an exhibit of their products at the Toronto Exhibition under the direction of J. H. Weber.

H. W. Benedict, sales manager of the Standard Paint Company, Montreal, has charge of his company's exhibit at the Toronto Exhibition.

Gordon Seybold, secretary-treasurer of Starke-Seybold, Limited, Montreal, is confined to his home with a light attack of typhoid fever.

The Montreal Rolling Mills Company intend having an exhibit at the Sherbrooke Fair. The exhibit will consist of horseshoes, horse nails, bolts and nuts.

The clerks in the employ of Starke-Seybold, Limited, Montreal, have organized a hockey team for the coming winter. Officers were elected at a meeting held this week.

Jean Baptiste Drapeau and Alphonse Champagne, have been registered to carry on a plumbing and heating business in Montreal under the firm name of Drapeau & Champagne.

E. R. Denis, managing director of the Denis Wire & Iron Co., London, Ontario, was in Montreal a few days last week inspecting the important constructions now in progress, in which ornamental iron work manufactured by his company is being used.

The plumbing firm of F. X. Bissonnette & Company have dissolved partnership, and Mrs. Joseph G. Duquette and Francois Xavier Bissonnette have been registered to carry on the business.

Ontario.

E. F. Earle, of Milton, was a visitor at Preston last week.

J. F. Smythe, Toronto, city traveler for United Factories, is absent from his work, owing to illness.

Mr. Cove, of the hardware firm of Christenson & Cove, MacLennan, Ont., spent last week visiting in Toronto.

The death occurred recently of W. H. Bragg, late of the hardware and implement firm of Bragg & Sons, Chatham, Ont.

Charles D. Chown, of Montreal, who has been preparing a catalogue for the Kennedy Hardware Company, Toronto, returned to Montreal on Tuesday.

B. Stephenson, of Chesney & Archibald, hardware merchants, Seaforth, was in Preston a few days ago calling on some manufacturers of hardware lines in that town.

Charles Bailey, manager of the Toronto branch of Wm. Jessop & Son, Sheffield, is said to aspire to represent Ward Three in the Toronto city council in 1908.

Clare Bros. & Company, stove and furnace manufacturers, Preston, are now comfortably settled in their fine, new office, which immediately adjoins their foundry.

C. Dolph, president and general manager of the Metal Shingle & Siding Co., Preston, has returned from a ten-days' holiday in Quebec and other St. Lawrence points.

D. A. McNab, Orillia, and Mr. Harrison, of Bloor & Harrison, Parkhill, were hardware merchants who called on the Toronto office of Hardware and Metal on Thursday.

J. M. Robertson, Tweed, Ont., was a visitor at the Toronto office of Hardware and Metal on Tuesday. Owing to ill-health, Mr. Roberston is advertising his business for sale.

Mrs. J. J. Turner, sen., wife of the head of the firm of J. J. Turner & Sons, tent manufacturers, Peterboro, died last Saturday, and was buried at Port Hope on Monday. Deceased was 51 years of age.

Mr. Humphries, hardware merchant, Parkhill, Ont., an ex-president of the Ontario Retail Hardware Merchants' Association, was a caller at the Toronto office of Hardware and Metal on Wednesday.

M. W. Howell, formerly engaged in the hardware business at Goderich, Ont., was a caller at the Toronto office of Hardware and Metal on Tuesday. Mr. Howell has just returned from an extended business trip to the Pacific coast.

Wm. P. Miller, sales manager of the James Smart Manufacturing Company, Brockville, has returned from a trip to western Canada, and was in Toronto on Thursday. Owing to ill-health, Mr. Miller expects to retire and move west shortly.

W. A. Hillhouse, formerly a hardware merchant at Shelburne, and more recently a traveler for Findlay Bros., of Carleton Place, was one of those injured in the Caledon mountain wreck, on the Owen Sound branch of the C.P.R., on Tuesday.

J. R. Hambly, hardware merchant, Barrie, and vice-president of the Ontario Retail Hardware Merchants' Association, called on the Toronto office of Hardware and Metal early this week. Mr. Hambly spent Sunday visiting friends in Lockport, Vt.

W. L. Clucas, manager of the W. L. Clucas Plumbing & Heating Co., St. Louis, Mo., spent a few days in Toronto last week. Mr. Clucas reports business good in the building lines throughout the southwestern States, especially in the State of Missouri, and particularly in the city of St. Louis.

D. Darrach, proprietor of the firm of Reuben, Tuplin & Co., Kensington, P. E. I., was in Toronto during the Exhibition. He reports business good in the island, but deplores the unsatisfactory conditions which too often prevail in transportation facilities between the island and the mainland during the winter months. He is not pessimistic, however, and believes that it is only a question of time before more satisfactory conditions will obtain.

Among the many hardware merchants in the city visiting the National exhibition, Hardware and Metal noticed the following : J. W. Hartman, Thornbury; W. Glassford, of Glassford & Co., Beaverton ; J. W. McMurtry, Galt ; Roy McPherson, of Dreany Bros., Englehart; Ralph Taylor, of Geo. Taylor Hardware Co., Cobalt ; D. McConnell, Hillsburg ; James Clarke, Breckin ; G. A. Binns, Newmarket ; Wm. Kriesel, New Dundee; Louis Mehew, Penetang ; Geo. Peaker, Brampton ; Wilson brothers, Ingersoll ; H. J. Morden, of H. J. Morden Co., Colborne ; William Ough, of Aurora ; Carnegie brothers, Port Perry ; J. W. Rork, Norwood ; E. Pollard, of E. S. Pollard & Co., Petrolea ; Walker Holborne, Sutton West ; A. E. Bottum, Bobcaygeon ; W. B. Clifton, Alliston ; G. Bohlander, Elmwood ; W. H. Stratford, Deseronto ; J. A. Coleman, Winger ; F. H. Dole, Hillsdale ; D. W. McDonald, Brooklyn ; J. A. Hartman, Thornbury ; S. W Moore, Churchill ; A. G. Snider, Vandorf ; S. E. Eakins, Millbrook ; Thos. McDonald, Canfield ; E. L. Gerex, Little Britain ; S. Purvis, Orton ; L. Richardson, Maple ; V. N. Chapman, Pickering ; W. Glassford, Beaverton ; W. H. Johns, Southampton; J. B. White, Port Hope ; Chas. Reynolds, Bethany ; W. H. Creeper, Hayden; W. W. Baker, Granton ; R. H. Carson, Gorrie ; B. Mutrie, Rockwood ; Thos. Lawrence, Lucknow ; A. Cain, Grand Valley ; V. Hunt, Tweed ; G. H. Baker, Gormley ; T. Hill, Craighurst ; J. Heatherington, Sharbot ; G. McLean, Otterville ; L. Wittlaufer, Brighton ; C. W. Carey, Connor ; Chas. Sargent, Claremont ; W. D. Stinson, Omemee ; J. J. McGill, Caronville ; F. A. Hoar, Barrie; A. Bowrie, Argyle ; A. E. Nichols, Fergus ; John Ritter, Millbank ; S. B. McClung, Trenton ; Carter brothers, Picton ; J. Hamilton, Haysville ; Ramsay & Co., Englehart ; J. W. Franks, Woodbridge ; H. Child, of Aaron Child & Son, Gravenhurst.

Maritime Provinces.

H. S. McDowell, harness dealer, Parrsboro, N.S. has made an assignment for the benefit of his creditors.

W. Wylie Rockwell, Kentville, N.S., is opening up a plumbing, heating and sheet metal establishment, and the new business will be under the management of Lewis G. Ells.

Western Canada.

A. F. Grady, hardware merchant, Coleman, Man., has been succeeded by the Coleman Hardware Co.

The harness firm of McLean & McKinnon, Lumsden, Man., have dissolved.

CORRESPONDENCE

Travelers, hardware merchants and clerks are requested to forward correspondence regarding the doings of the trade and the industrial gossip of their town and district. Addressed envelopes, stationery, etc., will be supplied to regular correspondents on request. Write the Editor for information.

HALIFAX, N.S.

Halifax, N.S., Sept. 2.—The town of Windsor, N.S., suffered a severe blow last week, when at a meeting of the directors and shareholders of the Windsor Foundry Company, it was decided to close out the business, as it could not be carried on successfully under the existing circumstances. The stockholders authorized the directors to sell, either at public auction, or at private sale, the foundry, building, plant, etc. There has been a foundry on the present site for half a century, under various managements, and the loss of this industry to the town will be very keenly felt.

* * *

Some forty years ago a rich bed of copper was worked by a United States company on Black Brook, Waughs River, Colchester County, N.S. This lead, from other metamorphic rocks having been tilted cross them etal vein, was lost, and though strenuous and almost untiring efforts have been made at different times in all these years to again pick up this lead, the results have been fruitless, until a few days ago, when Abraham Currie, of Waughs River, a practical miner, and Wellwood Currie, came across the long-looked-for ore streak. The find is thought to be a very valuable one, and the specimens of the ore taken out show sixty per cent. copper.

* * *

A report from Belle Island says that the Dominion Iron and Steel Co. will make a record year in their ore shipments this season in spite of the fire that stopped shipping. There are now 180,000 tons of ore ready to be conveyed to the pier. It is expected that within a few days General Manager Jones will announce that a new record has been established for the production of the pig iron at the blast furnaces.

* * *

The new foundry at Lunenburg, N.S., which is now under construction is nearing completion. It is expected that the plant will soon be in operation. This industry will mean much for the town.

* * *

F. J. Cragg, of the firm of Cragg Bros., retail dealers in hardware, left last week on a business trip to Boston and other American cities.

* * *

Spencer Bros. and Turner, Truro, have completed their contract to supply 70,000 feet of building material for the new cold storage plant to be erected at Port Hawkesbury, by the Halifax Cold Storage Company.

* * *

Major Maddocks, treasurer of the Sydney Cement Company, who has been on a visit to England, returned to Sydney last week.

* * *

James Crosby, manager of the Halifax Tram Company, has left on a visit to Boston and New York, and before he returns he will attend the electrical convention in Toronto.

ST. JOHN, N.B.

St. John, N.B., Aug. 26.—Your correspondent was shown through the spacious seven-storey building occupied by Emerson & Fisher, a few days ago. A number of changes and improvements have been made, and the various departments are all in first-class shape. The ground floor is utilized for general shelf hardware, cutlery, enamelware, woodenware, garden utensils, etc., and in the rear the offices have their place. The first floor is taken up with showrooms for mantels of all kinds, and an endless variety of stoves and ranges. On the floors above are the tinware manufacturing department, wholesale stock rooms, packing rooms, etc. A half-storey between the ground floor and first floor, just over the driveway that runs under one side of the big structure, is used for storing paints, oils and varnishes of the Sherwin-Williams manufacture, for which this firm has the agency. The cellar, which runs under the entire building, is used for storing stove fittings, and as a shipping department, the exit at the rear being on a level with the ground. Two large elevators, one on each side of the building, facilitate the work greatly. In the rear of the building are a number of sheds for storing tar paper and other heavy materials. This firm are also the proprietors of the Enterprise Foundry Co., of Sackville. They have just turned out a new steel range, a perfect beauty, known as the "Enterprise Regal." It is very highly finished, the entire surface, excepting the cooking holes, being completely nickeled.

* * *

The matter of rates to be charged by the city for winterport business is now being considered by the steamship companies. The city has offered to lease berths 1, 2, 3 and 4 to the C.P.R., at a rental of $45,000 a year, and a proposition was put forward to charge on steamers loading a flat rate of fifteen cents a ton on all freight. The city expects to derive enough revenue from the wharves to pay the interest and sinking fund.

St. John, N.B., Sept. 2.—There appears to be but little doing in hardware circles. The retailers report that business is rather quiet while the jobbers are by no means rushed with orders. This state of affairs is, however, expected at this time of year. No doubt there will be a change very shortly, as with cooler weather thoughts will turn to stoves and ranges and other cold weather necessities.

To-day, St. John is thronged with visitors who are drawn here by the big firemen's tournament, which is being held to-day and to-morrow. On the morrow merchants in all lines of trade are looking for good sales from the buyers who have taken advantage of the cheap excursions on railway and steamship to come to the New Brunswick metropolis and combine business with pleasure.

* * *

Good progress is being made with the wharf building operations on the West Side, and it is now certain that at least one additional berth and warehouse will be provided more than last year, to take care of the immense import and export trade that is expected this winter. From present indications the coming winter season will be the greatest in the history of St. John. It is thought that the claims of St. John for assistance in providing more wharves and better shipping facilities to meet the rapidly growing trade of the port will receive more attention than heretofore, now that Hon. Wm. Pugsley has been appointed minister of public works.

* * *

Another Cushing case is now before the courts. Geo. S. Cushing has taken action against the liquidators of the Cushing Sulphite Fibre Co. for over $15,000. He claims this amount as compensation for fees in promoting the company and time expended in the interests of the company, and also for moneys paid to George Robertson, M.P.P., $2,000 for promotion, for which Mr. Cushing claims he should be recouped.

* * *

Joseph Rowley's carriage factory and blacksmith shop, 94 Brussels St., was completely destroyed by fire last Thursday night. A barn and several houses were also considerably damaged. Mr. Rowley's loss amounted to several thousand dollars. He had no insurance.

* * *

The "city fathers" have passed an order that no work costing $50 or over shall be done except by tender. The aldermen claim to have an offer from a syndicate for a lease of certain city property in the vicinity of Fairville. The purpose of the applicants is, it is stated, to establish a cement industry on a large scale. There is plenty of limestone on the property.

QUEBEC CITY.

Quebec, Sept. 3.—Retailers' business in the hardware line is good, and specially for sporting goods. The stove trade is very active. All the houses report a very satisfactory volume of business, although the increase on stoves show an average of 12½ per cent. on last year. The wholesale movement of hardware at this centre is fairly good for the season of the year, and collections are such as to occasion now unfavorable comment; indeed, the reports of credit men are that payments are satisfactory. In all the metals the demand is considerably heavier than usual and local stocks are not sufficient for filling all the orders.

* * *

A concrete and brick machine shop and boiler house will be constructed by the Federal Government at Riviere-on-Larp, P.Q. Tenders will be received up to and including September 14th.

* * *

The St. Maurice Lumber Company have purchased the timber limits of the Pentecost Lumber Company, and took possession of the property on the first of September. The Pentecost Lumber Company owns valuable limits at River Pentecost, North Shore, and also have a mill and a small settlement there. A considerable amount of pulpwood is harvested and shipped to the United States.

* * *

Fowler Ross, vice-president of Floyd Silver Mines Co., was in our city Thursday. Prominent Quebec people are largely interested in this company, which possesses a property located in the township of Bucke, at the end of Sharpe Lake, and near Cobalt itself. Samples of the ore are reported to have been recently taken from the property and analyzed by J. W. Evans, chemist, of Cobalt. They ran $681 to the ton, and other made by P. H. Walsh, Magog, ran $2,987 to the ton, which shows conclusively that this property very shortly will be a dividend-payer.

* * *

The Quebec Railway, Light and Power Company are about to erect a magnificent new fireproof car barn on St. John street, the work on which will be begun this fall and pushed through to completion with energy. The new building will be a thoroughly modern and up-to-date structure, with steel girders and concrete roof, and its construction will mean a large saving in insurance for the company.

* * *

The municipal electors of the parish of Notre Dame de Quebec voted, Monday, on the by-law to authorize a loan for the introduction of water and drainage into the municipality. Every vote was given in favor of the by-law, so that the residents of that parish will, in a short time, be supplied with a water system, which will be taken from two lakes situated on the heights of Charlesbourg.

* * *

Something unique in navigation aids will be the telephone marine signal system, which is now being established in the St. Lawrence, between Quebec and Montreal, and which will be in use at the beginning of September. This service is to provide for a system of communication between ships and shore, and will enable the owners and agents to know of the progress of ships between these cities It will also enable the captains of vessels to learn of the condition of affairs in the river and to avoid any unusual or unexpected danger. There will be eleven stations which will be situated at the following points: Montreal, Longue Point, Vercheres, Sorel, Three Rivers, Batiscan, Cape a la Roche, Portneuf, St. Nicholas, Cape Rouge and Quebec. At each station will be erected a mast sixty-five feet in height, with a cross-spar twenty-five feet in length, and placed about twenty feet from the top of the mast. From the cross-spar signals will be displayed, one one side being for downstream; the other for upstream. When signals are being displayed, an ensign and pennant will be flown from the masthead. At night a white light at the masthead will call attention to signals.

* * *

A deal in large tracts of timber lands and mill property in Cookshire, Que., was effected last week, when the Pejepscot Pulp and Paper Co., of Skinner, Me., became the owners of the limits of the R. H. Pope. The company is a large and prosperous concern and have been supplied with pulpwood by the Cookshire Mill Co., and its successors. The price is reported as between $200,000 and $300,000.

* * *

The only changes we note this week in hardware quotations are for metals, which decline from 5c to 10c. We presently quote: Copper, ingot, 19c to 20c per pound; leaf, 30c to 35c; tin, ingot, 42c to 45c per pound; bars, 45c to 47c. Also ordinary iron bars lose 5c, and actually are sold at $2.30 to $2.15 per 100 pounds.

KINGSTON.

Business has been very fair here the past few weeks for this time of the year; a good trade has been done and is still being done in all lines of builders' supplies. Building operations are still going on rapidly. Some of the stores are at present a little short handed, owing to a number of the clerks and employees being away on their well-earned summer vacations. The sporting trade is gradually quieting down, as the various sports are settling down to work once more after the summer holidays.

* * *

Work has been commenced on the new parsonage to be erected for the minister of the Princess street Methodist Church.

* * *

What might have resulted in a very serious accident happened last Thurday, when William Neilson, tinsmith for Lemmon & Sons, met with a very painful injury. The young man was engaged at the Kingston Milling Company putting on a new tin roof, and was working on a scaffold forty feet above the ground. In some manner the board turned on the worker, hurling him head foremost to the ground. His fellow-workers and some men employed by the mill hurried to his assistance, and an ambulance was summoned to remove him to the hospital. Upon examination, it was found that his left elbow was very badly dislocated, but not broken, as at first thought to be. Further examination brought forth no signs of any internal injuries and the young man is doing nicely now.

* * *

The new roller skating rink, which opened last spring for a short time, but closed later, for the summer months, reopened again on Labor Holiday and large crowds attended all day. During the summer a gang of men were employed in getting the rink ready, and many improvements have been made. They have placed a beautiful polished hardwood floor for the skaters, and they have one of the largest and best-equipped roller rinks in Canada. The workmen made a fine job of the new floor, and the skaters say that they never skated on a smoother or better surface than offered at the Kingston rink, which speaks well for the men that laid it.

* * *

W. B. Dalton, hardware merchant, has returned to the city, after having spent his summer vacation at Old Orchard Beach.

* * *

Among visitors to the city last week was Joseph A. Taylor, a former Kingstonian, who has been absent from the city for a number of years. He was visiting his sister here on Brock street. Mr. Taylor is located in New York, where he is engaged in the manufacture of all kinds of machinery, gasoline and other electrical engines.

* * *

Time after time there has been spasmodic attempts made to have the shops close earlier on Saturday night. If the merchants would combine and say that they would close their stores at nine o'clock, the customers would rapidly fall in line. Saturday night is generally used as visiting night, and not a few merchants spend prolonged hours in business places talking with other dealers and with friends who drop in. It is a night for relaxation, and possibly the exchange of views is both helpful and stimulating. This question of early closing has just been brought up fresh again, and there has been a good deal of discussion over the matter, and the public generally are wondering if the merchants will act now or not, and release their clerks at any hour on Saturday they all agree upon.

* * *

Knox Co. opened their new 5c., 10c. and 15c. store here on Saturday, and large crowds were present all day. Quite a bit of discussion was going on among the merchants, owing to the cheap prices offered, as they sell all lines of goods, and nothing sold is over 15c. This company have now eighty-six stores, which undoubtedly enables them to sell their goods at very small profit.

* * *

It is stated that the Grand Trunk Railway Company will proceed without delay to the construction of a new line of railway between Kingston and Ottawa as soon as its route map, now before the department, is approved. In certain places the line projected, which is known as the Kingston, Smith's Falls & Ottawa Railway, is almost identical with the proposed Canadian Northern line, and there has been so much conflict about it that it is not yet straightened out. The assurance has been given that as soon as the approval of routes is secured, the work will be commenced, the Grand Trunk being the real backers of the scheme. It will substantially shorten the line between Toronto and Ottawa.

* * *

All the public schools reopened here to-day, after the summer vacation.

During the summer months they have gone under all necessary repairs, and are now all in good shape for the re-opening day.

* * *

On Saturday night, about ten o'clock, a small fire started in the engine room of the binder twine department, at the penitentiary. An alert watchman discovered it in time, however, and the fire was quickly subdued, with little damage beyond scorching the woodwork and blackening the walls. Spontaneous combustion among the oil and waste about the engine is thought to have been the cause ; no fire had been in the building for four or five days, as the department had been closed down for that time, owing to a shortage of raw material. The promptness of the watchman averted what might have been a serious conflagration.

* * *

The sub-committee of the light, heat and power committee met a few days ago to award tenders for a supply of oil that is needed at the city lighting plant. The contract has been awarded to the Queen City Oil Company, at $5.27 per barrel, this being the lowest tender. Last year, the McCourt Oil Company secured the contract.

* * *

D. E. Harding, Waltham, Mass., arrived the latter part of last week with six men to construct the new telescope gas holder ; work has now been commenced and will be completed in about six weeks. Six bricklayers were engaged here for the work. Mr. Harding came here from Quincy, Mass., where he erected a large tank. Another car load of machinery for the city lighting plant has just arrived from the west and will be installed immediately

PETERBORO, ONT.

Peterboro, Aug. 4.—MacGregor & Reid, of this city, have the contract for making some important changes in the heating and plumbing of the British American Bank Note Co.'s building at Ottawa. Mr. McGregor has been in the capital for some time superintending the work.

* * *

F. R. J. MacPherson & Co. have completed the contract for renovating the lighting system of the Peterboro Customs house. The work has been done in a highly satisfactory manner.

* * *

Ray Best, of the hardware firm of Best Bros., of this city, made a somewhat remarkable catch while fishing in Chemong Lake, near Peterboro, on Labor Day. He was fishing for minnows with a No. 1 minnow hook, and a small line. He hooked a five-pound 'lunge and succeeded in landing him after an exciting struggle. Many people will hardly believe that a 'lunge was caught under these peculiar circumstances, but the story is true, nevertheless.

* * *

The hardware stores of this city do a big business in fishing tackle and other requisites of the angler. Of course, trade in this line is confined largely to the spring and summer months, and during this period the different stores make a big bid for this trade. Peterboro is situated in close proximity to some of the best fishing waters in Canada, which accounts for the big demand for angling goods.

* * *

Lewis Dobson, of the hardware firm of Adamson & Dobson, caught his first big fish on Friday of last week. While traveling with a friend at Rice Lake he hooked a twelve-pound 'lunge and succeeded in landing him after an exciting struggle.

* * *

Vincent Eastwood, formerly manager of the Peterboro Hardware Co., but now manager of the Peterboro branch of the Royal Bank, is on a trip to the Old Country. He is accompanied by Mrs. Eastwood, and has been absent for several weeks. He is expected home within the next fortnight.

HAMILTON, ONT.

Hamilton. Sept. 3.—Building operations continue on the increase in this city, with the incidental increase in business to the handware and other merchants. The latest firm to accede to the demands of its growing business is the Gartshore, Thompson pipe foundry, which has taken out a permit for a $10,000 addition to its already extensive plant in the northwestern portion of the city. The expansion of this firm's business has been gradual and healthy, Owing to the great amount of outside contracts which it has been receiving of late years, added to its already extensive city business, it is now one of the most thriving of the city's industries. When the new addition is added it will be one of the city's largest.

* * *

The announcement that the franchise of the Dominion Heating and Power Co. has elapsed and that the company will likely abandon the idea of erecting a central heating system in the city will be received with regret by the various manufacturers and the hardware and metal trade. This company was to have erected a large plant in the central portion of the city and to have undertaken the heating of many of the big buildings in the central portion of the city. The installation of this plant would have meant a large sale of heating apparatus, piping, etc.; in fact, it would have proven one of the most extensive orders of the kind ever placed on the Hamilton market. However, there is a possibility of the plan being carried out after all, and if this course is pursued, it will be pleasing news, not only to those merchants who might reap monetary benefit therefrom, but by the people of Hamilton generally.

LONDON, ONT.

London, September 4.—Hardware jobbers report continued activity in trade, there being no sign of let up to the briskness that has prevailed throughout the season.

* * *

The Purdom-Gillespie Hardware Company are displaying in the windows of their store a novelty in the shape of Russian hammered brass ornamental ware, which is attracting not a little attention.

* * *

Features of the Western Fair next week will be a big display of stoves, ranges and enameled ware, by the McClary Manufacturing Company, and a lead glazing plant, operated by the Hobbs Manufacturing Company.

* * *

A number of local capitalists are considering the establishment here of a big metallic ceiling and siding manufactory. There is felt to be room in this city for such an industry, the possibilities of which are said to be almost limitless.

* * *

The recent visit to this city, St. Thomas and Port Stanley, of a party of American railway and coal magnates, has given rise to the report that while ostensibly on a trip of inspection of the city's own railroad—the L. & P. S. R. —which is under lease to the Pere Marquette Railway Company, the party were really sizing up Port Stanley as a site for a new steel plant, which they intend to locate in Canada. According to the report, the capitalists, who are interested in the big coal mines of Pennsylvania, as well as in different railways of New York, Ohio, Pennsylvania and other states, have a scheme which will promote the interests of the coal companies and the railways. If a steel plant were established on this side of the line, it would secure all its coal and supplies through the coal companies, and the heavy duties imposed by the Canadian Government would be avoided. There is such a demand for steel in this country, that it is felt a large plant at Port Stanley would pay, and pay well. Port Stanley is looked upon as a very good site, from the fact that there is plenty of water there, and steel plants, it is said, use more water than either the north or south branch of the Thames could supply, at certain seasons of the year. When they were in London the Americans gave no intimation of any intention of establishing a steel plant here. It is understood that the consummation of the steel plant project depends upon whether the city of London will extend the present lease of the L. & P. S. R. to the Pere Marquette. The party consisted of J. P. O'Dell, president Marquette and Bessemer Navigation Company, New York ; E. H. Butley, general manager Bessemer and Lake Erie Railway, Pittsburg; D. W. Bigomey, treasurer of Erie Railroad, New York ; R. W. Elliott, Pittsburg ; John Stack, president Fidelity Trust Company, Pittsburg ; A. D. Starr, general superintendent of the Pennsylvania Railroad, Pittsburg ; H. I. Poter, chief engineer of B. & L. E. Railway, Pittsburg, and also A. Leslie, general manager of the Lake Erie Coal Company, Walkerville.

* * *

Plans have been completed for the erection on the site of the wrecked Crystal Hall of a six-storey building, and it is expected work on the structure will be begun without delay. W. J. Reid & Co. will occupy a large portion of the new building.

SASKATOON, SASK.

Saskatoon, Aug. 24.—The recent heavy rains have been a great blessing to this district, coming at a critical time as they did. Rain was badly needed, and a good crop is now assured, where before doubts were entertained of any crop at all. Speaking generally, farmers are more optimistic now than they were all the season, and they expect that if only they escape frost this week, the crops will be practically safe, and they will reap as good a harvest as in former years. Some farmers are well into their barley and oats, but wheat cutting will not be general until the first of September, which will be considerably later than last year.

* * *

Although farmers were somewhat slow in placing early orders for binder twine, they are now coming forward, as they realize that they will need twine after all, notwithstanding bad appearances, and wagons may be seen daily wending their way countrywards loaded with implements and twine.

The implement houses report good sales, one warehouse alone having sent out as many as three hundred binders within a few weeks. The immigrant end of the implement business has been conspicuous by its absence this year, but this has been somewhat recouped by the settled farmers having more acreage under crops.

Some good samples of wheat and oats are to be seen round town, and, judging from their appearance, as well as reports from country districts, there is every reason to believe that there will be a good harvest.

* * *

Work at the C.P.R. bridge is going ahead, but there is no hope of its being completed this year. The company, however, ordered a temporary bridge to be built at once, and this is now well under way. It is their intention to have this line in operation this fall, in order to haul the grain from the surrounding districts, thus saving the farmers many miles of teaming.

MAKING WAR ON CATALOGUE HOUSES

Half a Million Retailers Combining for Offensive and Defensive Action.

War has been declared on the great catalogue houses of Chicago by the 500,000 retail merchants of the western states. The struggle between these two great interests has been in progress for many years. The advance made by the catalogue houses has been such as to threaten the very existence of thousands of smaller traders throughout the west, and they are now fighting, as they say, for their lives. In the struggle, jobbers and manufacturers are involved with the retail merchants.

The retail merchants are being guided at the present juncture in the struggle by an organization known as the Home Trade League of America, which has existed for several months as a publicity bureau, with headquarters in the Monadnock building, in the city of Chicago, but which is shortly to incorporate and institute a more systematic and aggressive campaign against the catalogue houses, through a board of twenty-one managers, with a vice-president and a committee of ten members in each state. This Chicago league has, as its support, commercial associations in the states of Illinois, Wisconsin, Minnesota, Iowa, Michigan and Kansas. The connection between these scattered associations and the Home Trade League of Chicago, is very close, the officers of the associations being in constant correspondence with the league, the new officers of which will be chosen from the body of active business men in all the leading towns and cities of the entire west.

Plan of Campaign.

The steps which are being taken by the Home Trade League to aid the retail merchant in the competition with the catalogue houses is indicated by the press of Chicago, as follows.

First.—Bringing pressure on manufacturers, jobbers and wholesalers generally, to give to local merchants the same rates as are granted to mail order companies. Under present conditions, retail merchants are, it is claimed, compelled to sell certain lines of goods at higher prices than are charged by the catalogue houses, owing to the fact that the wholesalers grant a lower price to the catalogue house than will be granted to the small retailer.

Second.—Pressure is brought to bear on the consumer, in all possible ways, to persuade him to give the local merchant a chance to figure on the order which is to be sent to the catalogue house, before it is sent, in order that the retail merchant may have the opportunity to show the consumer that he can compete with the catalogue house.

Third.—Merchants are pledging themselves to buy no goods from any wholesaler who sells any goods of any kind to any catalogue house. The results of the operation of this method of warfare are already apparent in Chicago, where a number of the large wholesale mercantile institutions are said no longer to sell to catalogue houses. The League is using all means to enlarge the number of merchants who refuse to buy from concerns which sell to catalogue houses.

Fourth.—Appeals are made to local pride and the loyalty of every community to support its local institutions.

Fifth.—Local merchants in many of the larger towns and cities have organized excursions to their towns, for the purpose of persuading buyers that they can compete with the large catalogue houses of Chicago and elsewhere.

Sixth.—Farmers who deal with catalogue houses are refused all accommodation at the retail stores of the country towns and cities.

New Method of Attack.

In the existing struggle, injury has been done to the large catalogue houses, owing to the fact that their competitors have organized a system through the operation of which the catalogue houses have received thousands of bogus letters requesting that catalogues be furnished them. Catalogues cost about one dollar each, together with the cost of mailing. Cases are now pending in the courts of the United States, in which catalogue houses are taking action against those who have occasioned them losses in this way. This is true, however, apparently not of the houses dealing in all classes of goods, but of the houses which have been selling lumber by mail order. This method of competition does not meet with the approval of the Chicago league. The business done by the mail order houses in the western states is now said to amount to hundreds of millions of dollars annually, and if it should continue to increase in the future as it has in the past, it is feared that hundreds of villages of the country will actually disappear, since the entire reason for their existence is to furnish the needed supplies to the farmers of their vicinity. Those which do not disappear, will steadily decrease in size and importance. The jobbers and wholesalers are interested, since their fortunes have been created through the business done by them with the small retail dealers of the country. It is claimed that the jobbers of the western states lost, approximately, sixty million dollars in 1905, owing to the great advance made in the business of the mail order houses. As much business was done from Chicago through the mail order houses in that year as all the jobbers of the city combined. Not many years ago, all of this business was done by the jobbers. That is to say, the jobbers and the country merchants have already lost one-half of the total trade of the country.

Manager Clark, of the Chicago league, stated, in conversation with reference to the work of the institution, that all that the retail merchants of the country wanted was a square deal. If they could get the same terms from the jobbers and wholesalers that are granted to the catalogue houses they would be satisfied to take their chances with consumers. They were in direct contact and touch with them, and, with the same advantages, could more than hold their own. He added that the majority of the jobbers and wholesale dealers of Chicago had taken but little interest in the work of the league, owing to the fact that Chicago was the home of the great mail order houses, whose accounts had been of great value to these wholesalers and jobbers. There were, however, notable exceptions, certain houses giving up accounts, amounting to hundreds of thousands of dollars per year, rather than run the risk of the entire destruction of the business of the country merchants, whose continued existence was of so great importance to the houses referred to.

CATALOGUES AND BOOKLETS.

When sending catalogues for review, manufacturers would confer a favor by pointing out the new articles that they contain. It would assist the editor in writing the review.

By mentioning HARDWARE AND METAL to show that the writer is in the trade, a copy of these catalogues or other printed matter will be sent by the firms whose addresses are given.

Eye Shields and Goggles.

The Chicago Eye Shield Co., 112 Randolph street, Chicago, Ill., have just issued a neat and fully illustrated 5x8-inch booklet, setting forth the excellencies of the various styles of eye shields, shades and goggles manufactured by that firm. Railwaymen, motorists, students, and all others who need protection for the eyes, and are wise enough to provide it, will find much to interest them in this convenient little booklet, which may be had by applying to the company mentioning this notice.

The Howland Pump Oiler.

We are in receipt of a neat 5x7½-inch catalogue describing the merits of Howland patented oilers, manufactured by the Maple City Manufacturing Co., Monmouth, Ill. The catalogue is replete with engravings of the various patented oilers, made by this firm, and the reading matter supplies convincing facts to show that the Holland Pump Oiler is the latest and most effective oiler of its kind in the market. The manufacturers have spent a great deal of time and money in securing a most perfect mechanism and beautiful design. The mechanism is very simple, indeed, but at the same time absolutely positive. It is made with a force pump having two valves working automatically, which force the oil to the bearings quickly and in just the quantity desired. It is claimed to be the only oiler ever presented having the union spout connection, thus making all spouts detachable and interchangeable. The position of the plunger is the most convenient possible, being directly under the thumb and working with a downward pressure. These oilers are made in four sizes, half-pint, pint, pint and one-half, and two pint, and are nicely finished in tin, copperized and solid brass. This convenient pocket catalogue is available to the trade by writing the above company mentioning this paper.

REVISED DUTY ON BRICK.

The Board of Customs has declared the rates of duties on the undermentioned articles, as follows:

Fire brick (9-in.x4½-in.x2½-in.) valued at $10 and upwards per thousand at place of export, free, under tariff item No. 281.

Pressed brick and other brick for building purposes, including fire brick valued at less than $10 per thousand at place of export—and stove linings manufactured from fire clay, are subject to duty under tariff item No. 282.

More frequently than others some dealers realize a good price may be obtained by simply asking for it.

MANITOBA HARDWARE AND METAL MARKETS

Corrected by telegraph up to 12 noon Friday, Sept 6 Room 511, Union Bank Bldg, Winnipeg, Man.

Business continues fairly active and there is a more confident feeling among the trade than at any time for the last three months. Crop reports are reassuring, and another year of prosperity is confidently expected.

Rope—Sisal, 11c per lb., and pure manila, 15¾c.

Lanterns—Cold blast, per dozen, $7; coppered, $9; dash, $9.

Wire—Barbed wire, 100 lbs., $3.22½; plain galvanized, 6, 7 and 8, $3.70; No. 9, $3.25; No. 10, $3.70; No. 11, $3.80; No. 12, $3.45; No. 13, $3.55; No. 14, $4; No. 15, $4.25; No. 16, $4.40; plain twist, $3.45; staples, $3.50; oiled annealed wire, No. 10, $2.90; No. 11, $2.96; No. 12, $3.04; No. 13, $3.14; No. 14, $3.24; No. 15, $3.39; annealed wires (unoiled), 10c less; soft copper wire, base, 36c; brass spring wire, base, 30c.

Poultry Netting—The discount is now 47½ per cent. from list price, instead of 50 and 5 as formerly.

Horseshoes—Iron, No. 0 to No. 1, $5.65; No. 2 and larger, $4.40; snowshoes, No. 0 to No. 1, $4.90; No. 2 and larger, $4.65; steel, No. 0 to No. 1, $5; No. 2 and larger, $4.75.

Horsenails—No. 10 and larger, 22c; No. 9, 24c; No. 8, 24c; No. 7, 26c; No. 6, 28c; No. 5, 30c; No. 4, 36c per lb. Discounts: "C" brand, 40, 10, 10 and 7½ p.c.; "M.R.M" cold forged process, 50 and 5 p.c. Add 15c per box. Capewell brand, quotations on application.

Wire Nails—$3 f.o.b. Winnipeg and $2.55 f.o.b. Fort William.

Cut Nails—Now $3.20 per keg.

Pressed Spikes—¼ x 5 and 6, $4.75; 5-6 x 5, 6 and 7, $4.40; ⅜ x 6, 7 and 8, $4.25; 7-16 x 7 and 9, $4.15; ½ x 8, 9, 10 and 12, 4.05; ⅝ x 10 and 12, $3.90. All other lengths 25c extra, net.

Screws—Flat head, iron, bright, 80, 10, 10 and 10; round head, iron, 80; flat head, brass, 75; round head, brass, 70; coach, 70.

Nuts and Bolts—Bolts, carriage, ⅜ or smaller, 60 p.c.; bolts, carriage, 7-16 and up, 50; bolts, machine, ⅜ and under, 50 and 5; bolts, machine, 7-16 and over, 50; bolts, tire, 65; bolt ends, 55; sleigh shoe bolts, 65 and 10; machine screws, 70; plough bolts, 55; square nuts, cases, 3; square nuts, small lots, 2½; hex nuts, cases, 3; hex nuts, small lots, 2½ p.c. Stove bolts, 70 and 10 p.c.

Rivets—Iron, 60 and 10 p.c.; copper, No. 7, 43c; No. 8, 42½c; No. 9, 45½c; copper, No. 10, 47c; copper, No. 12, 50½c; assorted, No. 8, 44½c, and No. 10, 48c.

Coil Chain—¼-in., $7.25; 5-16, $5.75; ⅜, $5.25; 7-16, $5; 9-16, $4.70; ⅝, $4.65; ¾, $4.65.

Shovels—List has advanced $1 per dozen on all spades, shovels and scoops.

Harvest Tools—60 and 5 p.c.

Axe Handles—Turned, s.g. hickory, $3.15; No. 1, $1.90; No. 2, $1.60; octagon extra, $2.30; No. 1, $1.60.

Axes—Bench axes, 40; broad axes, 25 p.c. discount off list; Royal Oak, per doz, $6.25; Maple Leaf, $8.25; Model, $8.50; Black Prince, $7.25; Black Diamond, $9.25; Standard flint edge, $8.75; Copper King, $8.25; Columbian, $9.50; handled axes, North Star, $7.75; Black Prince, $9.25; Standard flint edge, $10.75; Copper King, $11 per dozen.

Churns—45 and 5; list as follows: No. 0, $9; No. 1, $9; No. 2, $10; No. 3, $11; No. 4, $13; No. 5, $16.

Auger Bits—"Irwin" bits, 47½ per cent. and other lines 70 per cent.

Blocks—Steel blocks, 35; wood, 55.

Fittings—Wrought couplings, 60; nipples, 65 and 10K Ts and elbows, 10; malleable bushings, 50; malleable unions, 55 p.c.

Hinges—Light "T" and strap, 65.

Hooks—Brush hooks, heavy, per doz., $8.75; grass hooks, $1.70.

Stove Pipes—6-in., per 100 feet length, $9; 7-in., $9.75.

Tinware, Etc.—Pressed, retinned, 70 and 10; pressed, plain, 75 and 2½; pieced, 30; japanned ware, 37½; enamelled ware, Famous, 50; Imperial, 50 and 10; Imperial, one coat, 60; Premier, 50; Colonial, 50 and 10; Royal, 60; Victoria, 45; White, 45; Diamond, 50; Granite, 60 p.c.

Galvanized Ware—Pails, 37½ per cent.; other galvanized lines, 30 per cent.

Cordage—Rope sisal, 7-16 and larger, basis, $11.25; manila, 7-16 and larger, basis, $16.25; lathyarn, $11.25; cotton rope, per lb., 21c.

Solder—Quoted at 27c per pound. Block tin is quoted at 45c per pound.

Wringers—Royal Canadian, $36; B.B., $40.75 per dozen.

Files—Arcade, 75; Black Diamond, 60; Nicholson's, 62½ p.c.

Locks—Peterboro and Gurney 40 per cent.

Building Paper—Anchor, plain, 66c; tarred, 69c; Victoria, plain, 71c; tarred, 84c; No. 1 Cyclone, tarred, 84c; No. 1 Cyclone, plain, 66c; No. 2, Joliette, tarred, 69c; No. 2 Joliette plain, 51c; No. 2 Sunrise, plain, 56c.

Ammunition, etc.—Cartridges, rim fire, 50 and 5; central fire, 33⅓ p.c.; military, 10 p.c. advance. Loaded shells: 12 gauge, black, $16.50; chilled, 12 gauge, $17.50; soft, 10 gauge, $19.50; chilled, 10 gauge, $20.50. Shot: ordinary, per 100 lbs., $7.75; chilled, $8.10. Powder: F.F., keg, Hamilton, $4.75; F.F.G., Dupont's, $5.

Revolvers—The Iver Johnson revolvers have been advanced in price, the basis for revolver with hammer being $5.30 and for the hammerless $5.95.

Iron and Steel—Bar iron basis, $2.70. Swedish iron basis, $4.95; sleigh shoe steel, $2.75; spring steel, $3.25; machinery steel, $3.50; tool steel, Black Diamond, 100 lbs., $9.50; Jessop, $13.

Sheet Zinc—$8.50 for cask lots, and $9 for broken lots.

Corrugated Iron and Roofing, etc.—Corrugated iron, 28 gauge, painted, $3; galvanized, $4.10; 26 gauge, $3.35 and $4.35. Pressed standing seamed roofing, 28 gauge, $3.45 and $4.45. Crimped roofing, 28 gauge, painted, $3.20; galvanized, $4.30; 26 gauge, $3.55 and $4.55.

Pig Lead—Average price is $6.

Copper—Planished copper, 44c per lb.; plain, 39c.

Iron Pipe and Fittings—Black pipe, 1/4-in., $2.70 ; 3/8, $2.85 ; 1/2, $3.75 ; 3/4, $4.75 ; 1, $6.75 ; 1 1/4, $9.25 ; 1 1/2, $11.00 ; 2, $14.80 ; 2 1/2, $24.60 ; 3, $32.30 ; 3 1/2, $40.50 ; 4. $46.00 ; 4 1/2, $54.00. Galvanized : 1/4 in., $3.65 ; 3/8, $3.80 ; 1/2, $4.50 ; 3/4, $5.80 ; 1, $8.40 ; 1 1/4, $11.40 ; 1 1/2, $13.80 ; 2, $18.40. Nipples, 70 per cent.; unions, couples, bushings and plugs, 50 per cent.; malleables, 20 per cent.

Iron Pipe and Fittings—Black pipe, 1/4-in., $2.65; 3/8, $2.80; 1/2, $3.50; 3/4, $4.40; 1, $6.35; 1 1/4, $8.65; 1 1/2, $10.40; 2, $13.85; 2 1/2, $19; 3, $25. Galvanized iron pipe, 3/8-in., $3.75; 1/2, $4.35; 3/4, $6.65; 1, $8.10; 1 1/4, $11; 1 1/2, $13.25; 2-inch, $17.65. Nipples, 70 and 10 per cent.; unions, couplings, bushings and plugs, 60 per cent.

Galvanized Iron—Apollo, 16-gauge, $4.15 ; 18 and 20, $4.40 ; 22 and 24, $4.65 ; 26, $4.65 ; 28, $4.50 ; 30-gauge or 10 3/4-oz., $5.20 ; Queen's Head, 20, $4.60; 24 and 26, $4.90 ; 28, $5.15.

Lead Pipe—Market is firm at $7.80.

Tin Plates—IC. charcoal, 20x28, box, $10 ; IX charcoal, 20x28, $12 ; XXI charcoal, 20x28, $14.

Terne Plates—Quoted at $9.50.

Canada Plates—18x21, 18x24, $3.50 ; 20x28, $3.80; full polished, $4.30.

Lubricating Oils—600W, cylinders, 80c; capital cylinders, 55c and 50c; solar red engine, 30c; Atlantic red engine, 29c ; heavy castor, 28c; medium castor, 27c; ready harvester, 28c; standard hand separator oil, 35c; standard gas engine oil, 35c per gal on.

Petroleum and Gasolene—Silver Star, in bbls., per gal., 20c; Sunlight, in bbls., per gal., 22c; per case, $2.35; Eocene, in bbls., per gal., 24c; per case, $2.50; Pennoline, in bbls., per gal., 24c; Crystal Spray, 23c; Silver Light, 21c. Engine gasoline, in barrels, gal., 27c; f.o.b. Winnipeg, in cases, $2.75.

Paints and Oils—White lead, pure, $6.50 to $7.50, according to brand; bladder putty, in bbls., 2 1/2c; in kegs, 2 3/4c ; turpentine, barrel lots, Winnipeg, 90c; Calgary, 97c; Lethbridge, 97c; Edmonton, 98c. Less than barrel lots, 5c per gallon advance. Linseed oil, raw, Winnipeg, 72c; Calgary, 79c; Lethbridge, 79c; Edmonton, 80c; boiled oil, 3c per gallon advance on these prices.

Window Glass—16-oz., O. G., single, in 50-ft. boxes—16 to 25 united inches, $2.25 · 26 to 40, $2.40 ; 16-oz. O.G. single, in 100-ft. cases—16 to 25 united inches, $4 ; 26 to 40, $4.52 ; 41 to 50, $4.75 ; 50 to 60, $5.25 ; 61 to 70, $5.75 ; 21-oz. C.S., double, in 100-ft. cases, 26 to 40 united inches, $7.35 ; 41 to 50, $8.40 ; 51 to 60, $9.45 ; 61 to 70, $10.50 ; 71 to 80, $11.55 ; 81 to 90, $17.30.

NEW ELECTRIC RAILWAY.

The first of the electric line between New Westminster and Chilliwack, B.C., was turned yesterday by the B.C. Electric Railway Company. The line will give connection to all parts of the lower mainland; will be 62 miles long, and will cost two and a half million dollars. Construction will take a couple of years.

Practical Talks on Warm Air Heating

The Third of a Series of Articles by E. H. Roberts.

Heating a Modern Eight-roomed House.

One of the first essentials of a successful warm air heating plant is simplicity and simplicity is the essence and the significant feature of the plant here shown.

Side wall registers are indicated for all the first floor rooms, and these are intended to connect with the wall stacks leading to rooms on the second floor, so that only four basement pipes are necessary to supply warm air to the eight rooms. There are deflecting valves in these warm air registers that can be set to proportion the heat between the two rooms in the manner described in the last article.

Unlike this other plant, however, is the method of securing an inside cold air circulation; here we have a cold air face 16x20 near outside door and where it can take the air from both parlor and front bedroom, if the door between these two rooms is left open.

This face connects with a 16-inch galvanized cold air pipe run overhead in basement to the point indicated by circle where it drops vertically and runs through laundry partition to connect with cold air shoe at bottom of casings.

The other cold air face is set in baseboard just in front of china closet in dining-room and is intended to circulate the cold air coming down stairway from second floor as well as from the dining-room. The bottom of the china closet should be raised one foot, so that the air can pass through from the cold air face to the opening in bottom of closet floor, which is indicated by circle on first floor plan. Here a top flange collar connects with an 18-inch galvanized elbow and an 18-inch cold air pipe runs overhead across furnace room of basement to a point near furnace, then turns and drops obliquely to connect with cold air shoe just above basering.

These two cold air returns give an aggregate air supply of 450 square inches, which is within two inches of the total areas of the front warm air pipes. While this method of securing inside cold air circulation is quite different from that described in our previous chapter, the results are practically the same and the best method for taking the cold air supply in any building can only be determined after a careful study of the arrangement of the rooms and the general conditions.

In this house it is practically impossible to reach the bathroom with a warm air pipe, so we have indicated a hot water radiator with ¾-inch wrought iron pipe connection to furnace. If the furnace is made with an offset in side of feed chute, it is, of course, a very simple matter to put in a coil; but, if there is no provision of this kind it will be necessary to drill either the door frame or the body of the furnace and then care should be taken to prevent a leakage of gas into furnace chamber.

The return hot water pipe should be connected with water supply without a shut off so that as the water expands from the heat it can force itself back into the supply pipes, thus doing away with the necessity for an expansion tank and overflow.

No matter how well a warm air plant is installed it is impossible to get the best results unless the furnace itself has ample capacity to do the work. Economy of fuel can only be obtained with a slow fire and any furnace that requires constant forcing will necessarily waste fuel and cause clinkers to form on the grates.

Another reason for selecting a furnace of ample size, is the fact that a furnace which is never overheated will last two or three times as long as one that is sometimes crowded to the limit. The quality of air delivered from a furnace not overheated is better also—at least it is pleasanter to breathe, although chemists tell us that air does not undergo any chemical change from being heated to a high temperature.

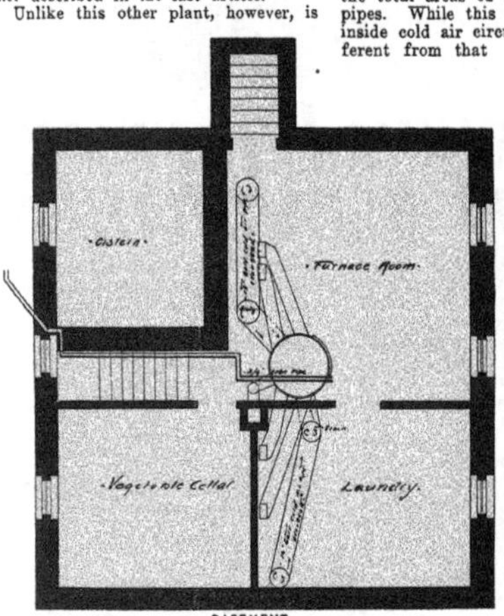

BASEMENT.

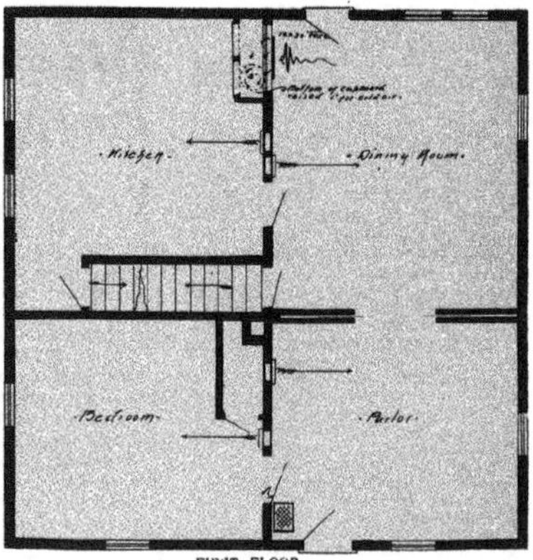

FIRST FLOOR.

Stick to one good line of stoves and ranges and thus prove to your trade that you know that line and have confidence in it. But be sure to select a good, dependable line, one that possesses real merit and quality.

CHIMNEY-COWL.

An invention relating to chimneys has recently been patented by E. A. Gerrard, Monroe, Neb. In this improvement the action of the wind on the uppermost blades operates to draw up smoke and the like from a revolving

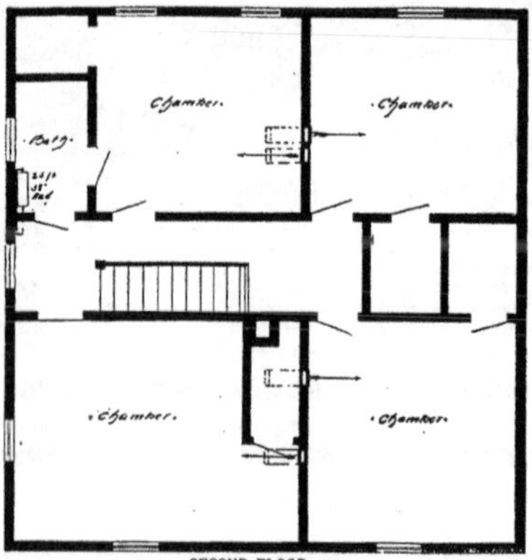

SECOND FLOOR

tube and thus increases or improves the draft and clearance at the top of the chimney as desired. The cover is formed from a flat plate of metal approximately square and bent to bring its diagonally opposite corners downward to form the side wings. The wheel is mounted to turn loosely on its shaft, the latter being bushed to operate without unnecessary noise. The vane is preferably made of sheet metal.

LARGE CHIMNEYS.

A trouble found mostly in the suburbs and the country is that of chimneys that are extra large or much larger than the stove pipe connected with them. Some of these old chimneys have been used in connection with open fires and are as much as 1½x3 feet.

Such a flue is not likely to draw, says Hardware Trade, even if the chimney is a high one, owing to the small amount of heated air discharged into it from the stove. The heated gases will lose their heat by becoming diluted with so much cold air in the chimney, and will fail to rise to the top of the chimney and pass out.

Sometimes such a trouble can be remedied by placing on the top of the chimney a close sheet iron cover, with a 10-in. round opening in the centre. This will prevent cold air from falling down the chimney at the corners when there is an attempt of the hot air to escape through the centre. It will also have the effect of retarding the outflow until the chimney has been warmed sufficiently to draw as it should. A stove burning properly can only give off a certain amount of heat to the chimney, and if this is not sufficient to raise the temperature of the chimney properly the draft will not be established, which is necessary for the operation of the stove.

The more the sides of the chimney are protected from the outside cold air the better the draft. A chimney placed on the outside of a wooden house, as is often seen in the country, will be surrounded on all sides by cold air, and the cooling effect will often be sufficient to spoil the draft.

This is the season of the year when the steam cookers look especially good to the housewife who does her own work. It doesn't take much talking to sell one.

COMBINATION BOILER AND FURNACE.

N. Frost, Bloomington, Ill., has invented a combination boiler and furnace. The invention relates to heating apparatus, having a boiler arranged within a furnace and both located within a warm air chamber, thus utilizing the generated heat for hot air heating and ventilating purposes as well as for direct or indirect steam heating purposes. The object is to utilize the heat generated by the burning fluid to the fullest advantage.

A GURNEY-TILDEN CHANGE.

C. H. Fox, Vancouver, has taken over the agency and warehouse of the Tilden-Gurney Co., in that city. We are informed that this will not conflict in any way with the agencies which he is at present holding.

GASOLINE AND OIL STORAGE.

On account of the frequent requests for suggestions, plans and specifications received by S. F. Bowser & Co., that company has recently issued a book entitled "Plan Drawings of Model Oil Storage Systems," which they are placing in the hands of those interested. This book contains a full set of plans to meet every phase of the oil storage question, giving sizes of oil tanks, dimensions of oil houses, proposed layouts for retail oil stores and wholesale oil houses and all other necessary information for the equipment of garages, factories and stores with oil handling devices.

The plan drawing shown on this page is a reduction of plan drawn up for a retail store equipment and shows position of buried tanks, dimensions of hole

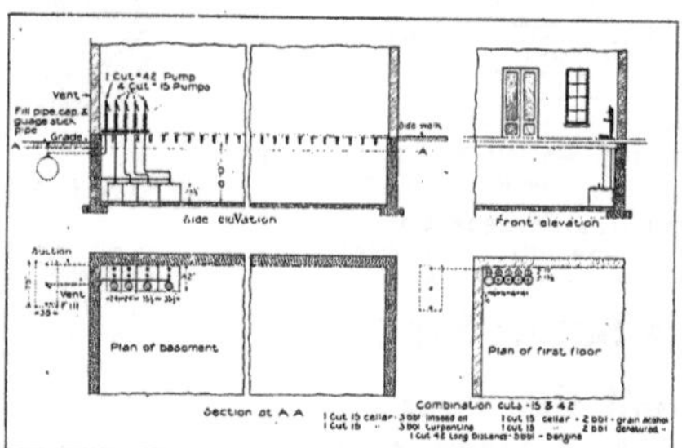

Plan Drawing, No. 15.—Oil Storage Plans.

to be dug and space occupied by pump for both gasoline and lubricating oils, other drawings show different equipments for both private and public garages.

The complete set of plan drawings, all of which are compiled from actual installations made by this firm, will be sent upon application to Messrs. Bowser & Co.

CONDENSED OR "WANT" ADVERTISEMENTS.

Advertisements under this heading 2c. a word first insertion; 1c. a word each subsequent insertion.

Contractions count as one word, but five figures (as $1,000) are allowed as one word.

Cash remittances to cover cost **must** accompany all advertisements. **In no case** can this rule be overlooked. Advertisements received without remittance cannot be acknowledged.

Where replies come to our care to be forwarded, five cents must be added to cost to cover postage, etc.

AGENTS WANTED.

AGENT wanted to push an advertised line of Welsh tinplates; write at first to "B.B.," care HARDWARE AND METAL, 88 Fleet St., E.C., London, Eng. [tf]

ONE who has sufficient organization to travel Canada east and west, fine line of Sheffield cutlery, brands already well known in Canada; none but applications from experienced and well-established men can be considered. Box 650, HARDWARE AND METAL, Toronto. [38]

BUSINESS CHANCES.

AN old-established house of brass founders, making a wide range of brass foundry furnishings, such as brass cast tubing, stair rods, cornice poles, art metal work, electric light fittings, ecclesiastical metal work, are anxious to secure a thoroughly reliable and trustworthy agent or traveller for Canada, to sell goods on commission. An excellent opportunity for any firm covering Canada thoroughly. Address all enquiries to Box 654, HARDWARE AND METAL. [37]

HARDWARE, tinware, stove and plumbing business in manufacturing town in the Niagara Peninsula; no competition; $250,000 factory and watermarks will be completed this summer; stock about $3,000; death of owner reason for selling. Box 85, Thorold, Ont. [tf]

HARDWARE Business, in good position for sale, and store with modern brick dwelling to rent. Proprietor retiring; Stock $2,000; Rent $35; Business more than doubled in the last two years; Good investment for young business man. Box 651 HARDWARE & METAL, Toronto. (36)

HARDWARE, Tinware, Stove and Furnace Business for sale, in live Eastern Ontario Village; first class chance for a practical man; English speaking community; stock can be reduced to suit purchaser; can give possession September 15th, 1907; premises for Sale or Rent, Apply to D. Courville, Maxville, Ont. (35)

HARDWARE Business and Tinshop for sale in Saskatchewan; population 1500; stock carried about $14,000 turnover, $45,000 practically all cash business; cash required, $8,000 would rent building; Do not answer without you have the money and mean business; it will pay to investigate this. Box 648 Hardware & Metal, Toronto. (41)

HARDWARE Business for Sale in Tweed, a thriving Eastern Ontario town. Best business in town. Must sell owing to ill-health. Liberal terms to satisfactory purchaser. Write for particulars. J. M. Robertson, Tweed. [40]

WELL established tinsmith business in thriving Ontario Village, including Eavetroughing, metal roofing, furnaces, pumps, pipefitting, hardware, paints, glass and stoves. Present stock about $1500; could be reduced. Shop and house combined, can be rented, including tools, horse and wagon, for $14.00 per month; possession given at once. Box 682, HARDWARE AND METAL. (33)

FOR SALE.

FOR SALE — First-class set of tinsmith's tools second-hand but almost as good as new; includes an 8-foot iron brick almost new. Apply Pease Waldon Co., Winnipeg. [tf]

FOR SALE—A quantity of galvanized plain twist wire. Apply to C. B. Miner, Cobden, Ont. (34)

TEN gross of "Antisplash" Tap Filters. Made to fit on all ordinary kitchen taps, acts as a filter and prevents water splashing. Supplied with showcards to hold 1 dozen. Price per dozen $1.00; price per gross $11.00. Sample on request. G. T. Cole, Box 480, Owen Sound, Ont. [36]

LIVE AGENT WANTED FOR CUTLERY

Large and Old Established Sheffield Firm

manufacturing cutlery in all its branches,

Table and Dessert Knives and Forks, Carvers, Painters' and Glaziers' Goods, Butcher Knives, Steels, Pen Knives, Pocket Knives, Daggers, Shears, Scissors, Razors, Case Goods, Spoons and Forks, Electroplate in all its Branches

Is anxious to correspond with a First Class Man or Firm, NOT HANDLING GERMAN GOODS, who would be prepared to take up the agency, and push same thoroughly in Canada from coast to coast.

None but persons thoroughly acquainted with the business need apply. Highest references given and required.

Liberal commission to the right party.

Correspondence, in first instance, to "CUTLER," care of Hardware and Metal 10 Front St. E., Toronto

SITUATIONS VACANT.

TINSMITHS WANTED — First-class tinsmiths wanted for points west of Winnipeg; must be good mechanics capable of taking charge of a metal department; thorough knowledge of furnace work necessary. Pease Waldon Co., Winnipeg, Man. [tf]

TINSMITH for general country work. Used to furnace work. Also Helper with 2 or 3 years' experience. Apply to G. A. Binns, Newmarket. [37]

WANTED general hardware clerk, experienced, sober, good stock-keeper and salesman; steady job for right man; state experience and salary expected. C. Richardson & Son, Box 325, Harrow, Ont. [36]

WANTED—First-class tinsmith in New Ontario town. Must be capable of managing work-shop and estimating. Good position, good salary, and no lost time for the right man. Address, Box 655, HARDWARE AND METAL, Toronto. [40]

SITUATIONS WANTED.

COMMERCIAL gentleman with nine years' trade connection with ironmongers, architects and public institutions in Great Britain desires position as representative of a Canadian manufacturing firm. Box X, HARDWARE AND METAL, Montreal.

HARDWARE Salesman wants position; can furnish first class references; capable of managing city or town business. Box 653 HARDWARE & METAL, Toronto. (39)

WANTED.

WANTED—Two store ladders. "Myers" preferred, complete with fixtures. Send full description and price to Ilsey Bros., Red Deer, Alta. (36)

THE WANT AD.

The want ad. has grown from a little used force in business life, into one of the great necessities of the present day.

Business men nowadays turn to the "want ad" as a matter of course for a hundred small services.

The want ad. gets work for workers and workers for work.

It gets clerks for employers and finds employers for clerks. It brings together buyer and seller, and enables them to do business, though they may be thousands of miles apart.

The "want ad." is the great force in the small affairs and incidents of daily life.

BUILDING AND INDUSTRIAL NEWS

For additional items see the correspondence pages. The Editor solicits information from any authoritative source regarding building and industrial news of any sort, the formation or incorporation of companies, establishment or enlargement of mills, factories or foundries, railway or mining news.

Industrial Development.

The capital of the new company is $250,000.

The Canadian Bag Co. are building a factory at Winnipeg.

Peers Bros., Vancouver, will erect a sawmill at Port Moody, B.C.

Alterations to G. S. Britnell's factory, Toronto, will cost $12,000.

R. Bigley, stove manufacturer, Toronto, will build a new stove foundry in that city.

A two-storey biscuit factory will be built by Symons & Rae, Toronto, at a cost of $10,000.

Work has commenced on the erection of the steel plant of the Coughlan Co., at Vancouver, B.C.

The Brantford Linen Manufacturing Co. are going to transfer their factory to Tillsonburg, Ont.

The Canadian Ornamental Iron Co., Toronto, will build a one-storey brick factory, to cost $5,000.

The Miramichi Lumber Co. are negotiating for the purchasing of the Clark Spool Mill, Newcastle, N.B.

The Canadian Pin Co., a new concern, will erect a factory at Woodstock, Ont., for the manufacture of pins.

The Ladysmith Lumber Co. will erect a saw mill near Nanaimo, B.C., to cut 35,000 feet of lumber per day.

The warehouse of Shim & Son, Saskatoon, was destroyed by fire, and $15,000 worth of goods was destroyed.

Fire did damage in Janesville, near Ottawa, to the extent of $9,000. The paint shop of A. Proulx suffered.

A telephone signal service is being established on the St. Lawrence River, between Montreal and Quebec.

Fire recently destroyed the Maritime Mfg. Co.'s plant at Pugwash, N.S. Loss $20,000, with insurance of $10,000.

The New Liskeard Concrete Co., New Liskeard, Ont., are installing a new mixer, and a concrete block machine.

Hamilton purchasers have acquired the timber limits, mill and machinery of McManus & McKelvie, New Liskeard, Ont.

A big addition will be made to the coal and furnace docks of the Canadian Northern Coal Co., at Port Arthur, Ont.

J. Davidson, Millbrook, Ont., has purchased the lighting plant and has been awarded the contract for lighting the town.

The Dominion Iron & Steel Co., Sydney, N.S., have received an order from an American concern for one million tons of ore.

Messrs. Mooney, Toronto, are considering the establishment of a large plant in that city for the manufacture of Portland cement.

The Railway Paint Co., Edmonton, Alta., will build a factory in that city at a cost of $100 000. The company's capital is $250,000.

The power situation in Moose Jaw is reaching an alarming condition. The present plant is incapable of keeping up with the demand.

The Canadian branch of the Royal mint will cost, when completed, about $500,000. The salaries will run as high as $80,000 per annum.

The Wapella Roller Mills, Wapella, Sask., owned by R. J. Lund, were destroyed by fire last week. Loss, $22,000, partly covered by insurance.

R. Forbes Co., Hespeler, Ont., has secured a Dominion charter and will manufacture textile goods. The capital of the company is $1,000,000.

A large plant for the manufacture of glass, tiles and pipes will be established at Morinville, twenty-four miles north of Edmonton, by a San Francisco syndicate.

It is reported that a large English firm will compete with the Standard Oil Co. for Western Canada business. The new company will build a large oil refinery at Vancouver, B.C., with a capacity of one thousand barrels a day.

It may be some days yet before Judge Longley delivers his decision in the Steel-Coal case. After leaving Sydney, the judge went to Spa Springs, in Chamberlain county, N.S., where he usually spends the summer with his family.

Fort William, Ont., has made definite arrangements with the International Snow Plow Mfg. Co., whereby the company will receive a free site and ten years' tax exemption, in return for the establishment of steel car shops in that city.

D. Hall and Alex. Robertson have gone into the lumber business, and will build a large lumber and shingle mill at Chilliwack, B.C. They also intend putting in machinery for the equipment of an up-to-date box factory, and will manufacture fruit boxes of all sizes desired right at home.

Crowland township last week passed a by-law granting a fixed assessment of $20,000 per year to the Bemis Bag Co., of Boston, who will build a million and a half-dollar factory close to Welland, Ont., and agree to employ fifteen hundred hands.

The Crow's Nest Coal Co. are said to have exported large quantities of coke to American smelters. As there is a great shortage at home, Canadian industries are suffering for want of coke and the Yale and Kootenay districts, B.C., are calling for enforcement of the penalty clause in the company's charter.

A syndicate of wealthy men, headed by A. D. McRae, of Winnipeg, and Peter Jansen, of Jansen, Neb., have purchased a controlling interest in the Fraser River Saw Mills, Millside, B.C. Large timber holdings have been bought and the total investment made by the McRae-Jansen syndicate is said to exceed $2,500,000.

The report of the Toronto City Architect for the month of August shows a falling off in the value of buildings erected during the month, when compared with the same month last year, but there was an increase in the number of permits issued. The number of buildings this August was 497, compared with 470 last August, and the value, $1,201,410, against $1,271,620 last year. The total value from January to September 1, was $11,440,740 this year, as compared with $8,663,525 in 1906, an increase of $2,800,000 for the eight months.

Tenders for steel rails have been accepted by the Government from the Dominion Iron and Steel Company, of Sydney, and the Algoma Steel Company, of the Soo. The price of the Algoma Company rails, which is given free on board at Fort William, is about thirty-four dollars a ton. The Sydney company undertakes to deliver its rails at Quebec for about fifty cents a ton less. The tenders of the Canadian companies, when duty is taken into consideration, were much lower than the prices quoted by the American and British firms which tendered. There are about thirty-six thousand tons of rails in the order, about half going to each company. The rails from Sydnew will be laid on the National Transcontinental to east and west of Quebec, and the rails from the Soo will be laid on the Superior Junction-Winnipeg section of the road.

Companies Incorporated.

Keystone Lorrain Mining Co., Haileybury, Ont.; capital, $1,000,000. To prospect for and deal in minerals. Incorporators: E. J. DuMee, C. J. Suplee, G. T. Armitage, George G. Thompson, all of Philadelphia, Pa., and G. M. Davis, Wilmington, Del.

The Dominion Oil Co., Hamilton; capital, $100,000. To prospect for and deal in oil, gas, salt and minerals. Incorporators: William Melville McClement, H. Harry Bicknell, D. P. Kappele, Florence Austin and Lizzie Eldon Anderson, all of Hamilton, Ont.

The Jenks Dresesr Company, Sarnia, Ont.; capital, $50,000. To carry on the general manufacture and sale of iron and steel work. Provisional directors: William G. Jenks, Andrew A. Dresser, Roy M. Norton, all of Port Huron, Mich.; Merton Fuller, Richmond, Mich., and Hiram Manning, Sarnia, Ont.

Building Notes.

The new collegiate school at Ottawa will cost $225,000.

Rogers and McKay, Vancouver, will build a residence at a cost of $12,000.

E. Hobson, Vancouver, B.C., will build an apartment house at a cost of $12,000.

H. Dorenwend, Toronto, will erect an apartment house in that city to cost $20,000.

P. Roach, Toronto, will build three attached 2-storey brick dwellings at a cost of $10,000.

A new business block will be erected in Vancouver by W. Hepburn, at a cost of $22,000.

W. O. McTaggart, Toronto, will build three 3-storey brick stores and dwellings at a cost of $12,000.

The Nelson Theatre Co., Nelson, B.C., will erect a large theatre in that city. C. W. Busk is one of the directors.

L. C. Sheppard, Toronto, will build seven pairs semi-detached 2½-storey brick dwellings at a total cost of $25,000.

The J. Y. Griffin Company will erect a large office block and storage house in Edmonton, Alta., to cost in the neighborhood of $100,000.

Municipal Undertakings.

The Government will build a large concrete dam near Trenton in connection with the Trent Canal.

Vancouver, B.C., will shortly install a new creosoting plant for the treatment of wood blocks to be used in street paving.

A new court house and a new registry office will be built at Sudbury, Ont. The contract for the court house has been awarded to M. Healey, Toronto. The registry office will be built by the O'Boyle Construction Company, North Bay.

London, Ont., has decided to submit the Komoka water scheme to the people in two by-laws, one to provide $293,500 for extending the domestic water supply, and one for $182,000, to provide for a hydraulic power plant and reservoir.

Railroad Construction.

Fifteen powerful locomotives have just been completed for the C.P.R.

A new electric line is proposed between Wallaceburg and Petrolia, Ont.

Construction has commenced on the electric line from Eburne, B.C., to New Westminster.

The Michigan Central will construct a line from Charing Cross, Ont., to Chatham, Ont.

The C.P.R. are installing telegraphones along the line between Fort William, Brandon and Winnipeg.

The contract for the new roundhouse for the C.N.R., at Brandon, has been let to the Sharp Construction Co., Winnipeg.

News of the Paint Trade

TINTING A CEMENT HOUSE.

It not infrequently happens that in building a concrete house, the cement will dry out in several colors or shades, and it is found desirable to tint the entire surface of a uniform color, that shall not be paint, but practically a part of the house itself. This result may be secured, says Edward Hurst Brown, in an article on "Painting," in the American Carpenter and Builder, by washing the whole house with cement, but there is a trick in doing this properly that is not always understood. The cement wash is made by mixing two parts of Portland cement and one part of marble dust with enough water to reduce it to about the same consistency of whitewash, and is applied with a whitewash brush The wall must be thoroughly wet with water for several hours before the wash is applied, and kept constantly wet during the application, and for at least a day afterward. The important thing to remember is, that the wash must not be applied to a dry wall, as it will not adhere. This work will be worth at least a dollar a square yard, or more, according to the price of labor, but the result will fully justify the cost.

CHALK IN PAINTS.

In reply to an inquiry as to the use of chalk in paints, the Oil and Color Trades Journal says: Chalk is, of course, used as an ordinary ingredient paints for ships, and of other kinds of paints up to as much as 50 per cent. of dry, solid matter. We never heard of a pure chalk paint, though if it could be laid on thick enough to hide the under surface it might be invisible, but we do not think it has enough body without something to tint it. It could, of course, be used in emergency as a sort of whitewash, or possibly rubbed dry on to the iron, or applied as blacklead is to the grate. In oil it would be so transparent as not to hide the under surface. Graphite, i.e., blacklead, and zinc oxide would make a better combination. It takes a lot to burn graphite. A little graphite would tint a lot of chalk, and a combination of the two might answer the purpose. But, again, it would take a lot of chalk in oil to kill the black tint of the graphite, and it is difficult to see how chalk can be got to adhere without oil The waves would wash it off, and chalk in oil has no body. French chalk (talc, soapstone, agalite, etc.) and graphite might be got to adhere mechanically; but, again, it would take much French chalk to kill the graphite so as to get an invisible grey.

HARDWARE FINISH.

It is safer to rub hardwood finish that has not stood many days with water and pumice stone and not with oil, as the oil acts as a solvent on varnish that has not become quite hard, and the result will be seen in the rubbing through of the varnish in spots, spoiling the work. Water, on the other hand, tends to harden the varnish

---

PAINT AND OIL MARKETS

MONTREAL.

Montreal, Sept. 6.—Trade conditions continue satisfactory, considering the unfavorable season of the year. Linseed oil, turpentine, and all staples in white lead and color departments are without change. Customary to this season of the year, and owing to the fact that a large number of travelers are taking their yearly rest, business is slightly quieter. Factories, however, are running full time, and it is anticipated that heavy autumn business will be coming forward very shortly. Prices in all lines are firm and unchanged.

Shipments of raw material, especially from the United States, are rather tardy in arrival. Transportation conditions on the whole are very much improved.

Linseed Oil—No change has occurred in the local market. Moderate calls arrive and supplies are adequate. Raw, 1 to 4 barrels, 60c; 5 to 9, 59c; boiled, 1 to 4 barrels, 63c; 5 to 9, 62c.

Turpentine—Local market conditions are quiet, there being little demand at present, as inventory-taking and vacations are the order of the day. Single barrels, 80 cents.

Ground White Lead—Trade is quiet. Prices are firm and unchanged. Government standard, \$7.50; No. 1, \$7; No. 2, \$6.75; No. 3, \$6.35.

Dry White Zinc—The situation continues unchanged. We continue to quote: V.M. Red Seal, 7½c; Red Seal, 7c; French V.M., 6c; Lehigh, 5c.

White Zinc Ground in Oil—No change has occurred. Prices are firm: Pure, 8½c; No. 1, 7c; No. 2, 5⅜c.

Red Lead—Demand at present is moderate, with bright prospects for heavy fall business. Prices are unchanged:—

Genuine red lead, in casks, \$6.25; in 100-lb. kegs, \$6.50; in less quantities at \$7.25 per 100 lbs. No. 1 red lead, casks, \$6; kegs, \$6.25, and smaller quantities, \$7.

Gum Shellac—Prices are firm and unchanged: Fine orange, 60c per lb., medium orange, 55c per lb., white (bleached), 65c per lb.

Shellac Varnish—We continue to quote: Pure white bleached shellac, \$2.80; pure orange, \$2.60; No. 1 orange, \$2.40.

Putty—The demand at present is strong and manufacturers are working long and hard to lay up large stocks for fall orders. Prices are steady: Pure linseed oil, \$1.85 in bulk; in barrels, \$1.[illegible]; in 25-lb. irons, \$1.90; in tins, \$2; bladder putty, in barrels, \$1.85.

Paris Green—A large volume of business was transacted during the season, short though it was. Makers are highly pleased with the trade done.

TORONTO.

Toronto, Sept. 6.—No special feature marks the paint and oil markets this week, and with the exception of turpentine, which has declined one cent, prices remain the same as last quoted. Generally speaking, the paint and oil markets here are a trifle quiet at present, but jobbers do not expect a heavy trade at this time of the year, and express themselves as being well satisfied with the amount of business coming in for the first week in September. While in the city visiting the National Exhibition, many outside paint dealers have made it a point to call on local jobbers, and, as a result, the latter have become better acquainted with their customers, and, incidentally, have landed a considerable amount of business.

White Lead—Business for the first week in September is quite up to expectations, and a good sorting order trade is being done. Prices are unchanged and are quoted as follows: Genuine pure white lead is quoted at \$7.65, and No. 1 is held at \$7.25.

Red Lead—The market is quiet at present, but business this week is in excess of the corresponding period of last year. Prices are well maintained at the following figures: Genuine, in casks of 500 lbs., \$6.25; ditto, in kegs of 100 lbs., \$6.75; No. 1 casks of 500 lbs., \$6; ditto, in kegs of 100 lbs., \$5.50.

Glass and Putty—Trade in glass and putty is daily increasing and present evidences point to a heavy fall trade in these lines. The immense amount of building which has been going on all over the country this season makes it certain that a hustling fall business in glass will set in toward the end of September, and there is no doubt that a splendid trade in glass will be done right along till the snow flies. Prices on glass and putty are unchanged.

Petroleum—As the nights grow longer, the demand for petroleum is steadily increasing. Quotations are still unchanged and present prices are: Prime white, 15c; water white, 14½c; Pratt's astral, 18c.

Shellac—Trade is quiet in shellac, though a few repeat orders continue to arrive. The following prices show no disposition to change: Pure orange, in barrels, \$2.70; white, \$2.82½ per barrel; No. 1 (orange), \$2.50.

Linseed Oil—During the past week there has been a further firmness in the English price, both for present delivery and futures. English oil delivered in Toronto now costs fully three cents per gallon more than it did ten days ago, and, as a consequence, Toronto jobbers have advanced the local prices one cent. We now quote: Raw, 1 to 3 barrels, 64c; 4 barrels and over, 63c. Add 3c to this price for boiled oil, f.o.b., Toronto, Hamilton, London, Guelph and Montreal, 30 days.

Turpentine—There has been a slight decline in turpentine in the Southern States, but as the export demand from there to Great Britain is expected to improve now, it is not anticipated that it will go much further in this direction. The local price of turpentine is much lower than it was at this time last year, and jobbers are hoping that many retailers, attracted by present prices, will now lay in their stocks for fall and winter trade. The holiday season for the past month has tended to decrease the local demand, but as travelers are now going out for fall orders, business should show considerable improvement during the next couple of weeks. We still quote: Single barrels, 79c; two barrels and upwards, 78c f.o.b. point of shipment, net 30 days; less than barrels, 5c advance. Terms, 2 per cent., 30 days.

For additional prices see current market quotations at the back of the paper.

PROMOTING PAINT SALES.

The Paint Manufacturers' Association of the United States, has formed a bureau for the promotion and development of the use and sale of prepared paint. A letter recently sent to the retailers of paint by this bureau, contains the following:

The manufacture of prepared paint was started only forty years ago. The intrinsic merit and popularity of the idea to prepare paint ready for use, is proved by the fact, that the sale of prepared paint in the United States has grown from nothing in 1866 to over eighty million gallons in 1906. Still this industry may be said to be yet in its infancy. Its possibilities in the future are practically limitless. But its further growth depends not only upon the efforts of the manufacturers to constantly increase the excellence of their product, but also upon the efforts of the retail dealers in educating the public to the great advantages of using prepared paint.

Push the Trade.

One of the most essential factors in this campaign of education, is, that the retail dealers properly display their advertising matter, make an attractive show upon their shelves of their stock of prepared paint and paint specialties, and push their sale by every legitimate method.

Paints offer the dealer a field for the establishment and steady growth of a

most remunerative business. They should not be sidetracked and simply "handled" indifferently, but sold by vigorous and progressive methods. They will respond to proper business treatment more promptly than most lines. There are many merchants in the United States to-day, who, with a very small investment, are doing a large volume of paint business on a thoroughly satisfactory margin of profit There is no reason why every dealer should not be doing the same thing. We earnestly urge any dealer who is not, to give this matter serious thought. Someone is getting the business and the corresponding profit. Investigation shows that most dealers realize a margin of profit on their paint sales, at least equal to the margin derived from the sale of other staples. In fact, many merchants secure a better margin from their paint business than other lines.

Increased Cost.

The accompanying chart indicates the increased cost of representative items in the principal classes of crude material used by makers of good paint.

Linseed oil, zinc oxide, lead carbonate (dry white lead), tin cans, and packing boxes are distinct classes. Turpentine and Japan dryer represent volatile liquids. Barium sulphate and silica represent the class "reinforcing pigments" (inert materials).

The same increase in cost, as shown on this chart, will apply to practically everything affecting the cost of production, including labor, taxes, insurance, etc., etc.

A proper knowledge of paints and paint ingredients is the greatest selling force that can be incorporated in any paint business. With this thought before us, we are arranging to place in the hands of paint dealers a line of practical paint facts and selling suggestions, in the hope that the ideas advanced may be incorporated into tangible sales arguments by every paint dealer and his clerks in the country. This matter will be sent from time to time with a series of paint letters during the next few months. Beginning Sept. 1st, we will send to a list of sixty thousand paint dealers in the United States a valuable paint book, or paint "Catechism," compiled by Dr. G. B. Heckel, a well-known authority on paint subjects. This "catechism" contains a vast amount of practical information useful in selling paints, and which has never before been accessible to the dealer in concise and comprehensive form. It tells all about paints and paint pigments, how they are used and why they are used. It enables the reader to realize the great progress made in the manufacture of scientific paint by machinery, and to better understand the growth of a great industry, which has developed within a few years from practically nothing, to an output that is recognized and utilized as an established commodity in every home.

We trust that when this "catechism" is received, it will be carefully read and studied, not only by the dealer himself, but by every clerk having anything to do with selling paint.

It is also our earnest hope, that the general ideas advanced in this letter may be given careful consideration by the dealer personally, as it is a matter of great importance to him from the standpoint of present profit and future volume of business. The paint manufacturers must look to the dealer for an increased business from year to year, and they fully realize that this increase cannot be obtained without the dealer's interest and his hearty co-operation.

PAINT

COMPARISON OF COSTS—CRUDE MATERIALS.

SOME OF THE ADVANCES AFFECTING COST OF PRODUCTION.

1897 - 1907.

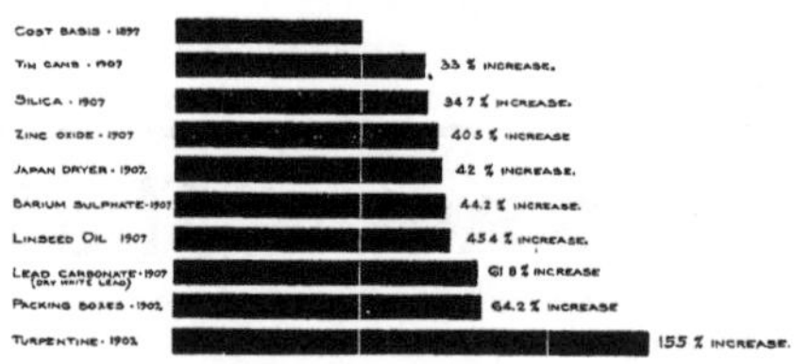

ALL OTHER CRUDE MATERIALS, PACKAGES, LABOR, INSURANCE AND TAXES HAVE ADVANCED DURING THE SAME PERIOD FROM TEN PERCENT TO TWO HUNDRED PERCENT.

VARNISH OIL.

A German inventor has patented a process for preventing the formation of coagulum, due to the impurities in oil when boiled for varnish. A small proportion of an alkali, preferably quick or fresh slacked lime in powder, though other alkalies or their carbonates may be used, is added to the oil, the amount ranging from one-eighth to one-half per cent. of lime, and that of the other caustic alkalies or carbonates varying in accordance with their molecular weights. Under this treatment oil can be heated to temperatures of 250 degrees C. to 300 degrees C., and will furnish a clear varnish, free from turbidity or coagulum.

CURRENT MARKET QUOTATIONS.

September 6, 1907

These prices are for such qualities and quantities as are usually ordered by retail dealers on the usual terms of credit, the lowest figures being for larger quantities and prompt pay. Large cash buyers can frequently make purchases at better prices. The Editor is anxious to be informed at once of any apparent errors in this list, as the desire is to make it perfectly accurate.

METALS.

ANTIMONY.

Cookson's ....per lb.	0 15
Hallett's	0 14½

BOILER PLATES AND TUBES.

Plates, ¼ to ½ inch, per 100 lb..	2 40	2 50
Heads, per 100 lb	2 65	2 75
Tank plates, 3-16 inch	2 60	2 70
Tubes per 100 feet, 1½ inch ..	8 25	8 50
" " 2 " ..	9 00	9 10
" " 2½ " ..	10 50	11 00
" " 3 " ..	12 00	12 50
" " 3½ " ..	15 00	16 00
" " 4 " ..	19 25	20 00

BOILER AND T.K. PITTS.

Plain tinned and Spun, 25 per cent. off list.

BABBIT METAL.

Canada Metal Company—Imperial, genuine 60c; Imperial Tough, 60c; White Brass, 50c.; Metallic, 35c.; Harris Heavy Pressure, 25c.; Hercules, 25c; White Bronze, 15c.; Star Frictionless, 14c; Alluminoid, 10c.; No 4, 9c. per lb

James Robertson Co—Extra and genuine Monarch, 60c.; Crown Monarch, 50c.; No. 1 Monarch, 40c; King, 30c; Fleur-de-lis, 20c; Thurber, 15c; Philadelphia, 12c; Canadian, 10c.; hardware, No. 1, 15c, No. 2, 12c.; No. 3, 10c. per lb.

BRASS.

Rod and Sheet, 14 to 30 gauge, 25 p c. advance.

Sheets, 12 to 14 in.	0 30
Tubing, base, per lb 5-16 to 2 in	0 33
Tubing, ⅜ to 3-inch, iron pipe size..	0 31
" ⅜ to 3-inch, seamless..	0 36

Copper tubing, 6 cents extra.

COPPER.

	Per 100 lb.	
Casting ingot........... ...	23 00	23 50
Cut lengths, round, bars, ½ to 2 in...		35 00
Plain sheets, 14 oz.................		36 00
Plain, 16 oz., 14x48 and 14x60		35 00
Tinned copper sheet, base		38 00
Planished base...............		43 00
Braziers' (in sheets), 4x6 ft., 25 to 30 lb each, per lb., base .	0 34	0 35

BLACK SHEETS.

	Montreal.	Toronto
10 gauge	2 70	2 75
12 gauge......................	2 70	2 75
14 "	2 50	2 60
17 "	2 50	2 60
18 "	2 50	2 60
20 "	2 50	2 60
22 "	2 55	2 65
24 "	2 55	2 70
26 "	2 65	2 85
28 "	2 70	3 00

CANADA PLATES

Ordinary, 52 sheets	2 75	3 05
All bright "	3 75	4 05
Galvanized—	Dom. Crown.	Ordinary
18x24x52	4 45	4 35
60................	4 70	4 60
20x28x80	8 90	8 70
.....	9 40	9 20

B.W GALVANIZED SHEETS.

gauge	Queen's Head	Fleur-de-Lis	Gordon Crown	Colborne Crown Gorbal's Best
16.-20				
..	3 95	3 80	3 95	....
22-24 ..	4 20	4 05	4 00	4 05
26........	4 45	4 30	4 40	4 30
28........	4 70	4 55	4 60	4 55

Less than case lots 10 to 25c. extra.

Apollo Brand.

24 guage, American	3 85
26 " "	4 10
28 " "	4 55
10¾ oz.	4 85

25c. less for 1,000 lb. lots.

IRON AND STEEL.

	Montreal.	Toronto.
Middlesboro, No. 1 pig iron..	21 50	24 50
Middlesboro, No. 3 pig iron..	20 50	23 50
Summerlee, " ..	25 50	26 50
" special " .	24 50	....
" soft " ..	24 00	....
Carron	26 00	....
Carron Special	24 50	....
Carron Soft	24 00	....
Clarence, No. 3	21 50	23 50
Glengarnock, No. 1		27 00
Midland, Londonderry and Hamilton, off the market but quoted nominally at		26 00
Radnor, charcoal iron	32 00	34 00
Common bar, per 100 lb......	2 20	2 30
Lowmoor iron.................... ...		6 50
Angles		2 50
Forged iron "	2 45	..
Refined " "	2 60	2 70
Horseshoe iron "	2 60	2 70
Band iron, No. 10 x ⅜ in.		3 60
Sleigh shoe steel "	2 25	2 30
Iron finish steel		2 40
Reeled machinery steel		3 50
Tire steel	2 40	2 50
Best sheet steel.		0 12
Mining cast steel....		0 08
Warranted cast steel....		0 14
Annealed cast steel....		0 15
High speed		0 60
B.P.L. tool steel	10½	0 11

INGOT TIN.

Lamb and Flag and Straits—

56 and 28-lb. ingots, 100 lb.	$43 00	$43 50

TINPLATES.

Charcoal Plates—Bright

M.L.S., Famous (equal Bradley) Per box.

I C, 14 x 20 base		$6 50
IX, 14 x 20 "		8 00
IXX, 14 x 20 base		9 50

Raven and Vulture Grades—

I C, 14 x 20 base		5 00
I X "		6 00
I X X "		7 00
I X X X "		8 00

"Dominion Crown Best"—Double Coated, Tissued. Per box.

I C, 14 x 20 base..........	5 50	5 75
I X, 14 x 20 "	6 50	6 75
I X X x 20 "	7 50	7 75

"Allaway's Best"—Standard Quality.

I C, 14 x 20 base..........	4 65	5 00
I X, 14 x 20 "	5 40	5 75
I XX, 14 x 20 "	6 15	6 50

Bright Cokes.

Bessemer Steel—

I C., 14 x 20 base	4 25	4 35
20x28, double box	8 50	8 70

Charcoal Plates—Terne

Dean or J G Grade—

I.C., 20x28, 112 sheets	7 25	8 00
IX., Terne Tin	...	9 50

Charcoal Tin Boiler Plates.

Cookley Grade—

X X, 14x56, 50 sheet bxs. } " 14x60, " } " 14x65, " }	....	7 50

Tinned Sheets.

72x30 up to 24 gauge...... ...	....	8 50
" " 26 "	....	9 00

LEAD.

Imported Pig, per 100 lb......	5 25	5 35
Bar, ...	5 75	6 00
Sheets, 2½ lb. sq. ft., by roll ..	6 75	6 50
Sheets, 3 to 6 lb. "	6 50	6 25

Cut sheets ½c. per lb., extra.

SHEET ZINC.

5-cwt. casks	7 75	8 00
Part casks....................	8 00	8 25

ZINC SPELTER.

Foreign, per 100 lb	6 75	7 00
Domestic	6 50	6 75

COLD ROLLED SHAFTING.

9-16 to 11-16 inch	0 06
¾ to 1 7-16 "	0 05½
1 7-16 to 3 "	0 05

30 per cent.

OLD MATERIAL.

Dealers buying prices:

	Montreal	Toronto
Heavy copper and wire, lb.	0 15	0 14
Light copper..............	0 13½	0 12
Heavy red brass....	0 13	0 12
" yellow brass	0 10	0 10½
Light brass................	0 07	0 07½
Tea lead	0 03¾	0 03½
Heavy lead................	0 04	0 04
Scrap zinc	0 03¾	0 03¾
No 1 wrought iron	15 50	11 50
" 2 " "	6 00	6 00
Machinery cast scrap ...	17 00	16 50
Stove plate................	13 00	12 00
Malleable and steel	8 00	10 00
Old rubbers	0 10½	0 10
Country mixed rags, 100 lbs.	1 00	1 25

PLUMBING AND HEATING

BRASS GOODS, VALVES, ETC.

Standard Compression work, 57½ per cent.
Cushion work, discount 40 per cent.
Fuller work, 70 per cent.
Flatway stop and stop and waste cocks, 60 per cent.; roundway, 55 per cent.
J.M.T. Globe, Angle and Check Valves, 45; Standard, 55 per cent.
Kerr standard globes, angles and checks, special, 42½ per cent; standard, 47½ p.c.
Kerr Jenkins' disc, copper-alloy disc and heavy standard valves, 40 per cent.
Kerr steam radiator valves, 60 p.c., and quick-opening hot-water radiator valves, 60 p.c.
Kerr brass, Weber's straightway valves, 40; straightway valves, I.B.B.M., 60.
J. M. T. Radiator Valves 50; Standard, 60; Patent Quick-Opening Valves, 65 p.c.
Jenkins' Valves—Quotations on application to Jenkins' Bros., Montreal.

No. 1 compression bath cock......net	2 00
No. 4 " " ...... "	1 90
No 7 Fuller's .... "	2 25
No. 4½, " .... "	2 35
Patent Compression Cushion, basin cock, hot and cold, per dcz., ...	$16.20
Patent Compression Cushion, bath cock, No. 2208	2 25

Square head brass cocks, 50; iron, 60 p.c.
Thompson Smoke-test Machine $25.00

BOILERS—COPPER RANGE.

Copper, 30 gallon, $33, 15 per cent.

BOILERS—GALVANIZED IRON RANGE.

30-gallon, Standard, $5; Extra heavy, $7.75

BATH TUBS.

Steel clad copper lined, 15 per cent.

CAST IRON SINKS.

16x24, $1; 18x30, $1; 18x36, $1.30.

ENAMELED BATHS, ETC.

List issued by the Standard Ideal Company Jan. 3, 1907, shows an advance of 10 per cent. over previous quotations.

ENAMELED CLOSETS AND URINALS

Discount 15 per cent.

HEATING APPARATUS.

Stoves and Ranges—40 to 70 per cent.
Furnaces—45 per cent.
Registers—70 per cent.
Hot Water Boilers—50 per cent.
Hot Water Radiators—50 to 55 p.c.
Steam Radiators—50 to 55 per cent.
Wall Radiators and specials—50 to 55 p.c.

LEAD PIPE

Lead Pipe, 5 p.c. off
Lead waste, 5 p.c. off.
Caulking lead, 5½c. per pound.
Traps and bends, 40 per cent.

IRON PIPE.

Size (per 100 ft.)	Black.		Galvanized
⅛ inch........	2 35	⅛ inch........	3 20
¼ "	2 35	¼ "	3 20
⅜ "	2 90	⅜ "	3 75
½ "	3 90	½ "	5 00
1 "	5 80	1 "	7 25
1¼ "	7 65	1¼ "	9 90
1½ "	9 20	1½ "	11 90
2 "	12 25	2 "	15 80
2½ "	20 10	2½ "	26 00
3 "	26 75	3 "	34 00
3½ "	34 25	3½ "	42 75
4 "	39 00	4 "	48 60

Malleable Fittings—Canadian discount 30 per cent.; American discount 25 per cent.
Cast Iron Fittings 57½; Standard bushings 57½; headers, 57½; flanged unions 57½, malleable bushings 55; nipples, 70 and 10; malleable lipped unions, 55 and 5 p.c.

SOIL PIPE AND FITTINGS

Medium and Extra heavy pipe and fittings, up to 6 inch, 60 and 10 to 70 per cent.
7 and 8-in. pipe, 40 and 5 per cent.
Light pipe, 50 p.c.; fittings, 50 and 10 p.c.

OAKUM.

Plumbers ....per 100 lb......	4 50	5 00

STOCKS AND DIES.

American discount 25 per cent.

SOLDERING IRONS.

½-lb. to 1½ .... per lb.	0 45½	0 48
2-lb. or over .... "	0 42½	0 46

SOLDER.

	Per lb. Montreal	Toronto
Bar, half-and-half, guaranteed	0 25	0 26
Wiping........................	0 22	0 23

PAINTS, OILS AND GLASS.

BRUSHES

Paint and household, 70 per cent..

CHEMICALS.

	In casks per lb.
Sulphate of copper (bluestone or blue vitrol)	0 09
Litharge, ground	0 06
" flaked	0 06½
Green copperas (green vitrol)	0 01
Sugar of lead..........	0 08
Lump olive..........	0 01½

COLORS IN OIL.

Venetian red, 1-lb. tins pure.	0 08
Chrome yellow "	0 15
Golden ochre "	0 10
French " "	0 08
Marine black "	0 04½
Chrome green "	0 09
French permanent green "	0 13
Signwriters' black "	0 15

GLUE.

Domestic sheet	0 10	0 10½
French medal	0 12	0 12½

PARIS GREEN.

	Berger's	Canadian
600-lb. cask	....	0 27½
250-lb. drums	0 28	0 27¾
100-lb. "	0 29¼	....
50-lb. "	0 29¼	0 28¼
1-lb. pkgs, 100 in box	0 30¼	0 29¼
½-lb. "	0 32¼	0 31¼
1-lb. tins, 100 in box	0 31¼	0 30¼
½-lb. tins	0 33¼	0 32

PARIS WHITE.

In bbls	0 90

PIGMENTS.

Orange mineral, casks	....	0 08
" " 100-lb. kegs	....	0 08½

PREPARED PAINTS.

Barn (in bbls.)	0 65	0 90
Sherwin-Williams paints,	....	1 55
Canada Paint Co.'s pure	....	1 40
Standard P. & V. Co.'s "New Era."		1 30
Benj Moore Co.'s "Ark" B'd	....	1 25
" British Navy deck	...	1 50
Brandram-Henderson's "English"		1 45
Ramsay's paints, Pure, per gal.	....	1 30
" Thistle, "	....	1 10
Martin-Senour's 100 p c. pure,	....	1 55
Senour's Floor Paints	....	1 35
Jamieson's "Crown and Anchor"		1 20
Jamieson's floor enamel	....	1 50
"Island City" paint	....	1 25
Sanderson Pearcy's, pure	....	1 20
Robertson's pure paints	....	1 20

PUTTY.

Bulk in bbls	1 60
Bladders in bbls	1 85
25-lb. tins	1 90
Bladders in bulk or tins less than 100 lb.	2 00
Bulk in 100-lb. irons	1 80

SHINGLE STAINS.

In 5 gallon lots	0 85	0 90

SHELLAC.

White, bleached	....	2 65
Fine orange	....	2 60
Medium orange	....	2 55

TURPENTINE AND OIL.

Prime white petroleum	....	0 13
Water white "	....	0 14½
Pratt's astral "	....	0 18
Castor oil	0 08	0 10
Gasoline	....	0 22½
Benzine, per gal	0 17	0 20
Turpentine, single barrels	0 78	0 79
Linseed Oil, " raw	0 60	0 64
" " boiled	0 63	0 67

WHITE LEAD GROUND IN OIL. Per 100-lbs

Canadian pure	7 15	7 50
No. 1 Canadian	6 80	7 15
Munro's Select Flake White	....	7 65
Elephant and Decorators' Pure	....	7 65
Monarch	....	7 40
Standard Decorator's	....	7 15
Essex Genuine	....	6 80
Brandram's B. B. Genuine	....	8 70
"Anchor," pure	....	7 40
Ramsay's Pure Lead	....	7 25
Ramsay's Exterior	....	6 65
"Crown and Anchor," pure	....	7 25
Sanderson Pearcy's	....	7 40
Robertson's C.P., lead	....	7 20

RED DRY LEAD.

Genuine, 560 lb. casks, per cwt	....	6 25
Genuine, 100 lb. kegs, "	....	6 50
No. 1, 560 lb. casks, per cwt	....	6 00
No. 1, 100 lb. kegs, per cwt	....	6 25

WINDOW GLASS

Size United inches.	Star	Double Diamond
Under 26	$4 25	$6 25
26 to 40	4 65	6 75
41 to 50	5 10	7 50
51 to 60	5 35	8 50
61 to 70	5 75	9 75
71 to 80	6 25	11 00
81 to 85	7 00	12 50
86 to 90	....	15 00
91 to 95	....	17 50
96 to 100	....	20 50
101 to 105	....	24 00
106 to 110	....	27 50

Discount—16-oz., 25 per cent.; 21-oz. 30 per cent. per 100 feet. Broken boxes 50 per cent.

WHITING.

Plain, in bbls	....	0 70
Gilders bolted in bands	....	0 90

WHITE DRY ZINC.

Extra Red Seal, V.M.	0 07½	0 08

WHITE ZINC IN OIL.

Pure, in 25-lb. irons	....	0 08½
No. 1, " "	....	0 07
No. 2, " "	....	0 05½

VARNISHES.

		Per gal. cans.
Carriage, No. 1	....	1 50
Pale durable body	....	3 50
" hard rubbing	....	3 00
Finest elastic gearing	....	3 00
Elastic oak	....	1 50
Furniture, polishing	....	2 00
Furniture, extra	....	1 15
" " No. 1	....	0 90
" " union	....	0 80
Light oil finish	....	1 40
Gold size japan	....	1 80
Brown japan	...	0 95
No. 1 brown japan	....	0 95
Baking black japan	....	1 35
No. 1 black japan	....	0 90
Benzine black japan	....	0 70
Crystal Damar	....	2 80
No. 1 "	....	2 50
Pure asphaltum	....	1 40
Oilcloth	....	1 50
Lightning dryer	....	0 70
Elastilite varnish, 1 gal. can, each		2 00
Granitine floor varnish, per gal	....	2 50
Maple Leaf coach enamels; size 1,		1 20
Sherwin-Williams' kopal varnish, gal.,		2 50
Canada Paint Co's sun varnish	....	2 00
"Kyanize" Interior Finish	....	2 40
"Flint-Lac," coach	....	1 80
B.H. Co's "Gold Medal," in cases		2 00
Jamieson's Copaline, per gal.	....	2 00

BUILDERS' HARDWARE.

BELLS.

Brass hand bells, 60 per cent.

Nickel, 55 per cent.

Gongs, Sargeant's door bells	5 50	8 00
American, house bells, per lb.	0 35	0 40

Peterboro' door bells, 37½ and 10 off new list.

BUILDING PAPER, ETC.

Tarred Felt, per 100 lb.		2 25
Ready roofing, 2-ply, not under 45 lb. per roll		1 00
Ready roofing, 3-ply, not under 65 lb., per roll		1 25
Carpet Felt ......per ton		60 00
Heavy Straw Sheathing ......per ton		40 00
Dry Surprise		0 45
Dry Sheathing ......per roll, 400 sq. ft.		0 40
Tar " " 400 "		0 50
Dry Fibre " 400 "		0 55
Tarred Fibre " 400 "		0 65
O. K. & I. X. L " 400 "		0 70
Resin-sized " 400 "		0 45
Oiled Sheathing " 600 "		1 00
Oiled " " 400 "		0 70
Roof Coating, in barrels ......per gal.		0 17
Roof " small packages "		0 25
Refined Tar ......per barrel		5 00
Coal Tar "		4 00
Coal Tar, less than barrels ......per gal.		0 15
Roofing Pitch ......per 100 lb.	0 80	0 90
Slater's felt ......per roll		0 70
Heavy Straw Sheathing f. o. b. St. John and Halifax		42 50

BUTTS.

Wrought Brass, net revised list.

Wrought Iron, 70 per cent.

Cast Iron Loose Pin, 60 per cent.

Wrought Steel Fast Joint, and Loose Pin, 70 per cent.

CEMENT AND FIREBRICK.

Canadian Portland	2 00	2 10
Belgium	1 60	1 90
White Bros. English	1 80	2 05
"Lafarge" cement in wood	....	3 40
"Lehigh" cement, in wood	....	2 54
"Lehigh" cement, cotton sacks	....	2 39
"Lehigh" cement, paper sacks	....	2 31
Fire brick, Scotch, per 1,000	27 00	30 00
" English	17 00	21 00
" American, low	23 00	25 00
" " high	27 50	35 00
Fire clay (Scotch), net ton	....	4 95

Paving Blocks per 1,000.

Blue metallic, 9"x4½"x3", ex wharf	35 00
Stable pavers, 12"x6"x2", ex wharf	50 00
Stable pavers, 9"x4½"x3", ex wharf	36 00

DOOR SETS.

Peterboro, 37½ and 10 per cent.

DOOR SPRINGS.

Torrey's Rod ......per doz.	....	1 75
Coil, 9 to 11 in. "	0 95	1 65
English "	2 00	4 00

Chicago and Reliance Coil 25 per cent.

ESCUTCHEONS.

Discount 50 and 10 per cent., new list

Peterboro, 37½ and 10 per cent.

ESCUTCHEON PINS.

Iron, discount 40 per cent.

Brass, 45 per cent.

HINGES.

Blind, discount 50 per cent.

Heavy T and strap, 4-in., per lb. net	0 06
" " 5-in., "	0 05¾
" " 6-in., "	0 05½
" " 8-in., "	0 05¼
" " 10-in. and larger	0 05

Light T and strap, discount 65 p.c.

Screw hook and hinge—

under 12 in. ... per 100 lb.	4 65
over 12 in. "	3 65

Crate hinges and back flaps, 65 and 5 p. c.

Hinge hasps, 65 per cent.

SPRING HINGES.

Spring, per gro., No. 5, $17.50 No. 10, $18; No. 20, $10 80; No. 120, $20; No. 51, $10; No. 50, $27.50

Chicago Spring Butts and Blanks 12½ per cent.

Triple End Spring Butts, 30 and 10 per cent.

Chicago Floor Hinges, 37½ and 5 off.

Garden City Fire House Hinges, 12½ p.c.

"Chief" floor hinge, 50 p.c.

CAST IRON HOOKS.

Bird cage ......per doz.	0 50	1 10
Clothes line, No. 61 "	0 00	0 70
Harness "	0 60	12 00
Hat and coat ......per gro.	1 10	10 00
Chandelier ......per doz.	0 50	1 00

Wrought hooks and staples—

¼ x 5 ......per gross	....	2 65
5-16 x 5	....	3 30

Bright wire hooks, 60 p c.

Bright steel gate hooks and staples, 40 p.c.

Cresent hat and coat wire, 60 per cent.

Screw, bright wire, 65 per cent.

KNOBS.

Door, japanned and N.P., doz	1 50	2
Bronze, Berlin ......per doz.	2 75	3 50
Bronze, Genuine "	6 00	9 00
Shutter, porcelain, F. & L. screw ......per gross	1 30	2 00
White door knobs ......per doz.	....	2 00

Peterboro knobs, 37½ and 10 per cent.

Porcelain, mineral and jet knobs, net list.

KEYS.

Lock, Canadian 40 to 40 and 10 per cent.

LOCKS.

Peterboro, 37½ and 10 per cent

Russell & Erwin, steel rim $2.50 per doz

Eagle cabinet locks, discount 30 per cent

American padlocks, all steel, 10 to 15 per cent.; all brass or bronze, 10 to 25 per cent.

SAND AND EMERY PAPER.

B. & A. sand, discount, 35 per cent

Emery, discount 35 per cent.

Garnet (Rurton's), 5 to 10 per cent. advance

SASH WEIGHTS.

Sectional ......per 100 lb.	2 00	2 25
Solid "	1 50	1 75

SASH CORD.

Per lb.	....	0 31

BLIND AND BED STAPLES.

All sizes per lb.	0 07½	10

WROUGHT STAPLES.

Galvanized	....	2 75
Plain	....	2 50

Coopers', discount 40 per cent.

Poultry netting staples, discount 40 per cent.

Bright spear point, 75 per cent. discount.

TOOLS AND HANDLES.

ADZES.

Discount 22½ per cent.

AUGERS.

Gilmour's, discount 60 per cent. off list.

AXES.

Single bit, per doz.	6 00	9 00
Double bit, "	10 00	11 00

Bench Axes, 40 per cent.

Broad Axes, 25 per cent.

Hunters' Axes	5 50	6 00
Boys' Axes	6 25	7 00
Splitting Axes	7 00	12 00
Handled Axes	7 00	9 00
Red Ridge, boys', handled	....	5 75
" hunters	....	5 25

BITS.

Irwin's auger, discount 47½ per cent.

Gilmour's auger, discount 60 per cent.

Rockford auger, discount 50 and 10 per cent.

Jennings' Gen. auger, net list.

Gilmour's car, 47½ per cent.

Clark's expansive, 40 per cent.

Clark's gimlet, per doz	....	0 65
Diamond, Shell, per doz.	....	1 00
Nail and Spike, per gross	....	2 25

BUTCHERS CLEAVERS

German ......per doz.	7 00	9 00
American "	12 00	18 00

CHALK.

Carpenters' Colored, per gross	0 45	0 75
White lump ......per cwt.	0 60	0 65

CHISELS.

Warnock's, discount 70 and 5 per cent.

P. S. & W. Extra, discount, 70 per cent.

CROSSCUT SAW HANDLES.

S. & D., No. 3 ......per pair	0 13
S. & D., " 5 "	0 11½
S. & D., " 6 "	0 18
Boynton pattern "	0 20

CROWBARS.

3½c. to 4c. per lb.

DRAW KNIVES.

Coach and Wagon, discount 75 and 5 per cent.

Carpenters' discount 75 per cent.

DRILLS.

Millar's Falls, hand and breast. net list.

North Bros., each set, 50c.

DRILL BITS.

Morse, discount 37½ to 40 per cent.

Standard, discount 50 and 5 to 55 per cent.

FILES AND RASPS.

Great Western	75	per cent.
Arcade	75	"
Kearney & Foot	75	"
Disston's	75	"
American	75	"
J. Barton Smith	75	"
McClellan	75	"
Eagle	75	"
Nicholson	66⅔	"
Globe	75	"

Black Diamond, 60, 10 and 5 p.c.

Jowitt's, English list, 27½ per cent.

GAUGES.

Stanley's discount 50 to 60 per cent.

Winn's, Nos. 26 to 33 ....each	1 65	2 40

HANDLES.

Second growth ash fork, hoe, rake and shovel handles, 40 p c.

Extra ash fork, hoe, rake and shovel handles, 45 p c.

No. 1 and 2 ash fork, hoe, rake and shovel handles, 50 p c.

White ash whiffletrees and neckyokes, 35 p c.

All other ash goods, 40 p.c.

All hickory, maple and oak goods, excepting carriage and express whiffletrees, 40 p c.

Hickory, maple, oak carriage and express whiffletrees, 45 p c.

HAMMERS.

Maydole's, discount 5 to 10 per cent.

Canadian, discount 25 to 27½ per cent.

Magnetic tack ......per doz.	1 10	1 20
Canadian sledge .... per lb.	0 07	0 08½
Canadian ball pean, per lb.	0 22	0 25

HATCHETS.

Canadian, discount 40 to 42½ per cent.
Shingle, Red Ridge 1, per doz. 4 40
" " 2, " 4 80
Barrel Underhill 5 05

HOES.

Mortar, 50 and 10 per cent.

MALLETS.

Tinsmiths' ....per doz. 1 25 1 50
Carpenters', hickory. " 1 25 3 75
Lignum Vitae........ " 3 85 5 00
Caulking, each 0 60 2 00

MATTOCKS.

Canadian............ per doz. 5 50 6 00

MEAT CUTTERS.

German, 15 per cent.
American discount, 33⅓ per cent.

PICKS.

Per dozen 6 00 9 00

PLANES.

Wood bench, Canadian, 40, American, 25.
Wood, fancy, 37½ to 40 per cent.
Stanley planes, $1.55 to $3 60, net list prices.

PLANE IRONS.

English ....per doz. 2 00 5 00
Stanley, 2⅝ inch, single 24c , double 39c.

PLIERS AND NIPPERS.

Button's genuine, 37½ to 40 per cent.
Button's imitation....per doz. 5 00 9 00
Berg's wire fencing.... ...... 1 72 5 50

PUNCHES.

Saddler's ....per doz. 1 00 1 85
Conductor's.......... " 3 00 15 00
Tinners', solid..........per set 0 72
" hollow......per inch 1 00

RIVET SETS.

Canadian, discount 35 to 37½ per cent.

RULES.

Boxwood, discount 70 per cent.
Ivory, discount 20 to 25 per cent.

SAWS.

Atkins, hand and crosscut, 25 per cent.
Disston's Hand, discount 12½ per cent
Disston's Crosscut ....per foot 0 35 0 55
Hack, complete..........each 0 75 2 75
" frame only........each 0 50 1 25

S. & D solid tooth circular shingle, concave and band, 50 per cent; mill and ice, drag, 30 per cent; cross-cut, 35 per cent.; hand saws, butcher, 35 per cent; buck, New Century, $6.25; buck, No 1 Maple Leaf, $5 25; buck, Happy Medium, $4.25; buck, Watch Spring, $4.25; buck, common frame, $4.00

Spear & Jackson's saws—Hand or rip, 26 in, $12 75; 28 in., $14.25, panel, 18 in., $8.25; 20 in, $9; tenon, 10 in., $9 90; 12 in., $10 90; 14 in., $11 50.

SAW SETS.

Lincoln and Whiting 4 75
Hand Sets, Perfect.................. 4 00
X-Cut Sets, " 7 50
Maple Leaf and Premiums saw sets, 40 off.
S. & D. saw swages, 40 off.

SCREW DRIVERS.

Sargent's ....per doz 0 65 1 00
North Bros, No. 30 . per doz. 16 80

SHOVELS AND SPADES.

Canadian, discount 45 per cent.

SQUARES.

Iron, discount 20 per cent.
Steel, discount 65 and 10 per cent.
Try and Bevel, discount 50 to 52½ per cent.

TAPE LINES.

English, ass skin .....per doz. 2 75 5 00
English, Patent Leather 5 50 9 75
Chesterman's ....each 0 90 2 85
" steel.........each 0 80 8 00
Berg's, each........ .... 0 75 2 50

TROWELS.

Disston's, discount 10 per cent.
S & D., discount 35 per cent
Berg's, brick, 924x11 .... 4 00
" pointing, 924x5.......... .... 2 10

FARM AND GARDEN GOODS

BELLS.

American cow bells, 63⅓ per cent.
Canadian, discount 45 and 50 per cent.
American, farm bells, each .. 1 35 3 00

BULL RINGS.

Copper, $2.00 for 2½-inch

CATTLE LEADERS

Nos. 32 and 33 .. ...per gross 7 50 8 50

BARN DOOR HANGERS.

doz. pairs.
Stearns wood track 4 50 6 00
Zenith........................ .... 9 00
Atlas, steel covered 5 00 6 00
Perfect 8 00 11 00
New Milo, flexible 6 50
Steel, track, 1 x 3-16 in(100 ft) 3 25
" " 1¼ x 3-16 in(100 ft) 4 75
Double strap hangers, doz. sets.... 6 40
Standard jointed hangers, " 6 40
Steel King hangers " 6 25
Storm King and safety hangers 7 00
" " rail................ .... 4 25
Chicago Friction, Oscillating and Big Twin Hangers, 5 per cent.

HARVEST TOOLS.

50 and 10 per cent.
S. & D. lawn rakes, Dunn's, 40 off.
" sidewalk and stable scrapers, 40 off.

HAY KNIVES.

Net list.

HEAD HALTERS.

Jute Rope, ⅜-inch....per gross 9 00
" " ½ " .... " 10 00
" " ⅝ " .... " 12 00
Leather, 1-inch ......per doz. 4 00
Leather, 1¼ " ...... " 5 20
Web................. " 2 45

HOES.

Garden, 50 and 10 per cent.
Planter.............. per doz. 4 00 4 50

LAWN MOWERS

Low wheel, 12, 14 and 16-inch $2 30
9-inch wheel, 12-inch 2 85
" " 14 " 3 00
" " 16 " 3 12½
High wheel, 12 " 4 05
" " 14 " 4 50
" " 16 " 4 75

SCYTHES.

Per doz. net.................... 6 25 9 25

SCYTHE SNATHS.

Canadian, discount 40 per cent.

SNAPS.

Harness, German, discount 25 per cent.
Lock, Andrews' 4 50 11 00

STABLE FITTINGS

Warden King, 35 per cent
Dennis Wire & Iron Co., 33⅓ p c.

WOOD HAY RAKES.

40 and 10 per cent.

HEAVY GOODS, NAILS, ETC.

ANVILS.

Wright's, 80-lb. and over.............. 0 10½
Hay Budden, 80-lb. and over 0 09½
Brook's, 80-lb. and over 0 11¾
Taylor-Forbes, prospectors 0 05
Columbia Hardware Co , per lb. 0 09½

VISES.

Wright's...................... .... 0 13½
Berg's, per lb........ .... 0 12½
Brook's...................... .... 0 12½
Pipe Vise, Hinge, No. 1 .... 3 50
" " " No. 2...... .... 5 50
Saw Vise......................... 4 50 5 00
Blacksmiths' (discount) 60 per cent.
" parallel (discount) 45 per cent.

BOLTS AND NUTS

Carriage Bolts, common ($1 list Per cent.
" " ⅜ and smaller.. 60, 10 and 10
" " 7-16 and up .... 55 and 5
" " Norway Iron ($3 list) 50
Machine Bolts, ⅜ and less .... 60 and 10
Machine Bolts, 7-16 and up.... 55 and 5
Plough Bolts .... 55 and 10
Blank Bolts.................... 55
Bolt Ends...................... 55
Sleigh Shoe Bolts, ⅜ and less .. 60 and 10
" " " 7-16 and larger 50 and 5
Coach Screws, conepoint...... 70 and 5
Nuts, square, all sizes, 4c. per cent. off
Nuts, hexagon, all sizes, 4½c. per cent. off.
Stove Rods per lb., 5½ to 6c.
Stove Bolts, 75 per cent.

CHAIN.

Proof coil, per 100 lb ¼ in , $6 00; 5-16 in., $4.85; ⅜ in., $4 25; 7-16 in , $4.00; ½ in., $3 75; 9-16 in. $3 70; ⅝ in., $3 65; ¾ in., $3 60; ⅞ in., $3.45; 1 in.. $3.40.

Halter, kennel and post chains, 40 to 40 and 5 per cent ; Cow ties, 40 per cent. ; Tie out chains 65 per cent. ; Stall fixtures, 35 per cent ; Trace chain, 45 per cent. ; Jack chain iron, 50 per cent. ; Jack chain, brass, 50 per cent.

HORSE NAILS.

M R.M. cold forged process, list May 15, 1907, 50 and 5 per cent.
"C" brand, 57½ per cent. off list.
Capewell brand, quotations on application.

HORSESHOES.

M.R.M brand: iron, light and medium No. 1 and smaller, $3 90; No. 2 and larger $3 65; snow pattern No. 1 and smaller $4.15, No. 2 and larger, $3 90; "X.L." new light steel, No 1 and smaller, $4 25; No 2 and larger, $4; "X.L" featherweight steel, No. 0 to 4, $5 60, toe-weight, all sizes, $6.85. F o b. Montreal. Extras for packing

Belleville brand: No. 0 and 1, light and medium iron, $3 90; snow, $4 15; light steel, $4 25; No. 2 and larger, light and medium iron, $3 65; snow, $3 90; light steel, $4. F.o b. Belleville. Two per cent , 30 days.

Toecalks —Standard No 1 and smaller, $1 50; No. 2 and larger, $1 25 Blunt No. 1 and smaller, $1.75; No. 2 and larger, $1.50 per box Sharp Put up in 25 lb. bxs

HORSE WEIGHTS.

Taylor-Forbes, 4½c. per lb.

NAILS.	Cut.	Wire.
2d	4 00	3 50
3d	3 15	3 20
4 and 5d	2 90	2 90
6 and 7d	2 80	2 80
8 and 9d	2 65	2 65
10 and 12d	2 60	2 60
16 and 20d	2 55	2 55
30, 40, 50 and 60d (base)	2 50	2 50

F.o.b. Montreal. Cut nails, Toronto, 20c. higher
Miscellaneous wire nails, discount 75 per cent
Coopers' nails, discount 40 per cent.

PRESSED SPIKES.

Pressed spikes, ⅝ diameter, per 100 lbs $3.15

RIVETS AND BURRS.

Iron Rivets, black and tinned, 60, 10 and 10.
Iron Burrs, discount 60 and 10 and 10 p.c.
Copper Rivets, usual proportion burrs, 15 p.c.
Copper Burrs only, net list.
Extras on Coppered Rivets, ½-lb. packages 1c. per lb.; ¼-lb. packages 2c. lb.
Tinned Rivets, net extra, 4c. per lb.

SCREWS.

Wood, F. H., bright and steel, 85 and 10 p c.
" R. H., bright, 80 and 10 per cent.
" F. H., brass, 75 and 10 per cent.
" R. H., " 70 and 10 per cent.
" F. H , bronze, 70 and 10 per cent.
" R. H., " 65 and 10 per cent.
Drive Screws, dis. 87½ per cent.
Bench, wood ....per doz. 3 25 4 00
" iron .... " 4 25 5 00
Set, case hardened, dis. 60 per cent.
Square Cap, dis. 50 and 5 per cent.
Hexagon Cap, dis. 45 per cent.

MACHINE SCREWS.

Flat head, iron and brass, 35 per cent.
Felister head, iron, discount 30 per cent.
" " brass, discount 25 per cent.

TACKS, BRADS, ETC.

Carpet tacks, blued 75 p c.; tinned, 80 and 10; (in kegs), 40; cut tacks, blued, in dozens only, 75; ½ weigh's, 60; Swedes cut tacks, blued and tinned, bulk, 75 dozens, 75; Swedes, upholsterers', bulk, 85 and 12½; brush, blued and tinned, bulk, 70; Swedes, gimp, blued, tinned and japanned, 75 and 12½; zinc tacks, 35; leather carpet tacks, 40; copper tacks, 25; copper nails 30; trunk nails, black, 65; trunk nails, tinned and blued, 65; clout nails, blued and tinned 65; chair nails, 35; patent brads, 40; fine finishing, 40; lining tacks, in papers, 10; lining tacks, in bulk, 15; lining tacks, solid heads, in bulk, 75; saddle nails, in papers, 10; saddle nails, in bulk, 15; tufting buttons, 22 line in dozens only, 60; zinc glaziers' points, 5, double pointed tacks, papers, 90 and 10; double pointed tacks, bulk, 40; clinch and duck rivets, 45; cheese box tacks, 85 and 5 trunk tacks, 80 and 10.

WROUGHT IRON WASHERS.

Canadian make, discount 40 per cent.

SPORTING GOODS.

CARTRIDGES.

"Dominion" Rim Fire Cartridges and C.B. caps, 50 and 7½ per cent; Rim Fire B B. Round Caps, 60 and 2½ per cent.; Centre Fire, Pistol and Rifle Cartridges, 30 p c; Centre Fire Sporting and Military Cartridges, 20 and 5 p c.; Rim Fire, Shot Cartridges, 50 and 7½ p.c.; Centre Fire, Shot Cartridges, 30 p.c; Primers, 25 p.c.

LOADED SHELLS.

"Crown" Black Powder, 15 and 5 p c.; "Sovereign" Empire Bulk Smokeless Powder, 30 and 5 p.c.; "Regal" Ballistite Dense smokeless Powder, 30 and 5 p c; "Imperial" Empire or Ballistite Powder, 30 and 10 p.c.

EMPTY SHELLS

Paper Shells, 25 and 5; Brass Shells, 55 and 5 p.c.

Wads. per lb.
Best thick brown or grey felt wads, in ½-lb. bags $0 70
Best thick white card wads, in boxes of 500 each, 12 and smaller gauges 29
Best thick white card wads, in boxes of 500 each, 10 gauge........... 0 35
Thin card wads, in boxes of 1,000 each, 12 and smaller gauges 0 20
Thin card wads, in boxes of 1,000 each, 10 gauge.................. 0 25
Chemically prepared black edge grey cloth wads, in boxes of 250 each— Per M
11 and smaller gauge 0 60
9 and 10 gauges.................. 0 70
7 and 8 " 0 90
5 and 6 " 1 10
Superior chemically prepared pink edge, best white cloth wads in boxes of 250 each—
11 and smaller gauge 1 15
9 and 10 gauges 1 40
7 and 8 " 1 65
5 and 9 " 1 90

SHOT.

Ordinary drop shot, AAA to dust $7 50 per 100 lbs. Discount 5 per cent; cash discount, 2 per cent, 30 days; net extras as follows subject to cash discount only; Chilled, 40 c; buck and seal, 80c.; no. 28 ball, $1 20 per 100 lbs; bags less than 25 lbs., ½c. per lb.; F O.B. Montreal, Toronto, Hamilton, London, St. John and Halifax, and freight equalized thereon.

TRAPS (steel.)

Game, Newhouse, discount 30 and 10 per cent.
Game, Hawley & Norton, 50, 10 & 5 per cent.
Game, Victor, 70 per cent.
Game, Oneida Jump (B. & L.) 40 & 2½ p.
Game, steel, 60 and 5 per cent.

SKATES.

Skates, discount 37½ per cent.
Empire hockey sticks, per doz .. 3 00

CUTLERY AND SILVERWARE.

RAZORS.

	per doz.	
Elliot's	4 00	18 00
Boker's	7 50	11 00
" King Cutter	13 50	18 50
Wade & Butcher's	3 60	10 00
Lewis Bros. "Klean Kutter"	8 50	10 50
Henckel's	7 50	20 00
Berg's	7 50	20 00

Clauss Razors and Strops, 50 and 10 per cent

KNIVES.

Farriers-Stacey Bros., doz 3 50

PLATED GOODS

Hollowware, 40 per cent. discount.
Flatware, staples, 40 and 10, fancy, 40 and 5.
Hutton's "Cross Arrow" flatware, 47½; "Singalese" and "Alaska" Nevada silver flatware, 42 p.c.

SHEARS.

Clauss, nickel, discount 60 per cent.
Clauss, Japan, discount 67½ per cent.
Clauss, tailors, discount 40 per cent.
Seymour's, discount 50 and 10 per cent.
Berg's 6 00 12 00

HOUSE FURNISHINGS.

APPLE PARERS.

Hudson, per doz., net 5 75

BIRD CAGES.

Brass and Japanned, 40 and 10 p. c.

COPPER AND NICKEL WARE.

Copper boilers, kettles, teapots, etc. 30 p.c.
Copper pitts, 20 per cent.

KITCHEN ENAMELED WARE.

White ware, 75 per cent.
London and Princess, 50 per cent.
Canada, Diamond, Premier, 50 and 10 p.c.
Pearl, Imperial, Crescent and granite steel, 50 and 10 per cent.
Premier steel ware, 40 per cent.
Star decorated steel and white, 25 per cent.
Japanned ware, discount 45 per cent.
Hollow ware, tinned cast, 35 per cent. off.

KITCHEN SUNDRIES.

Can openers, per doz	0 40	0 75
Mincing knives per doz	0 50	0 80
Duplex mouse traps, per doz	....	0 65
Potato mashers, wire, per doz	0 60	0 70
" wood "	0 50	0 60
Vegetable slicers, per doz	....	2 25
Universal meat chopper No. 1	....	1 15
Enterprise chopper, each	...	1 30

Spiders and fry pans, 50 per cent.

Star Al chopper 5 to 32	1 35	4 10
" " 100 to 103	1 35	2 00
Kitchen hooks, bright	....	0 60

LAMP WICKS.

Discount, 60 per cent.

LEMON SQUEEZERS.

Porcelain lined	per doz.	2 20	5 60
Galvanized	"	1 87	3 85
King, wood	"	2 75	2 90
King, glass	"	4 00	4 50
All glass	"	0 50	0 90

METAL POLISH.

Tandem metal polish paste 6 00

PICTURE NAILS.

Porcelain head ......per gross	1 35	1 50
Brass head	0 40	1 00

Tin and gilt, picture wire, 75 per cent.

SAD IRONS.

Mrs. Potts, No. 55, polished....per set	0 90
" No. 50, nickle-plated, "	0 95
" handles, japaned, per gross	9 25
" " nickled, "	9 75
Common, plain	4 25
" plated	5 50
Asbestos, per set	1 50

TINWARE.

CONDUCTOR PIPE.

2-in. plain or corrugated., per 100 feet, $3 30; 3 in., $4 40; 4 in., $5.80; 5 in., $7.45; 6 in., $9.90.

FAUCETS.

Common, cork-lined, discount 35 per cent.

EAVETROUGHS.

10-inch ....per 100 ft. 3 30

FACTORY MILK CANS.

Discount off revised list, 35 per cent
Milk can trimmings, discount 25 per cent.
Creamery Cans, 45 per cent

LANTERNS.

No. 2 or 4 Plain Cold Blast....per doz.		6 50
Lift Tubular and Hinge Plain,	"	4 75
No. 0, safety	"	4 00

Better quality at higher prices.
Japanning, 50c. per doz. extra.
Prism globes, per doz., $1.20.

OILERS.

Kemp's Tornado and McClary's Model galvanized oil can, with pump, 5 gallon, per dozen 10 92
Davidson oilers, discount 40 per cent.
Zinc and tin, discount 50 per cent.
Coppered oilers, 20 per cent. off.
Brass oilers, 50 per cent, off.
Malleable, discount 25 per cent

PAILS (GALVANIZED).

Dufferin pattern pails, 45 per cent.
Flaring pattern, discount 45 per cent.
Galvanized washtubs 40 per cent.

PIECED WARE.

Discount 35 per cent off list, June, 1899.
10-qt. flaring sap buckets, discount 35 per cent.
6, 10 and 14-qt. flaring pails dis. 35 per cent.
Copper bottom tea kettles and boilers, 30 p.c.
Coal hods, 40 per cent.

STAMPED WARE.

Plain, 75 and 12½ per cent. off revised list.
Retinned, 72½ per cent. revised list.

SAP SPOUTS.

Bronzed iron with hooks ....per 1,000
Eureka tinned steel, hooks 8 00

STOVEPIPES.

5 and 6 inch, per 100 lengths	7 64	7 91
7 inch " "	....	8 18

Nestable, discount 40 per cent.

STOVEPIPE ELBOWS

5 and 6-inch, common........per doz.	1 32
7-inch "	1 48

Polished, 15c. per dozen extra.

THERMOMETERS.

Tin case and dairy, 75 to 75 and 10 per cent.

TINNERS' SNIPS.

Per doz 3 00 15
Clauss, discount 35 per cent.

TINNERS' TRIMMINGS.

Discount, 45 per cent.

WIRE.

ANNEALED CUT HAY BAILING WIRE.

No. 12 and 13, $4; No. 13½, $4 10; No. 14, $4 25; No. 15, $4.50; in lengths 6' to 11', 25 per cent; other lengths 20c. per 100 lbs. extra; if eye or loop on end add 25c. per 100 lbs. to the above.

BRIGHT WIRE GOODS

Discount 60 per cent.

CLOTHES LINE WIRE.

7 wire solid line, No. 17, $4.90; No. 18, $3.00; No. 19, $2.70; 6 wire solid line, No. 17, $4.45; No. 18, $3.10; No. 19, $2 80. All prices per 1000 ft. measure; 6 strand, No. 18, $2 60; No. 19, $2 90. F.o.b. Hamilton, Toronto, Montreal.

COILED SPRING WIRE.

High Carbon, No. 9, $2.95; No. 11, $3.50; No. 12, $3.20.

COPPER AND BRASS WIRE.

Discount 37½ per cent.

FINE STEEL WIRE.

Discount 25 per cent. List of extras In 100-lb. lots: No. 17, $5 — No. 18, $5.50 — No. 19, $6 — No. 20, $6.65 — No. 21 $7 — No. 22, $7.30 — No. 23, $7.65 — No 24, $8 — No. 25, $9—No. 26, $9.50—No. 27, $10—No. 28, $11—No. 29, $12—No. 30, $13—No. 31, $14—No. 32, $15—No. 33, $16—No. 34, $17. Extras net—tinned wire, Nos. 17-25 $2—Nos. 26-31, $4—Nos. 32-34, $6. Coppered, 75c.—oiling, 10c.—in 25-lb. bundles, 15c.—in 5 and 10-lb. bundles, 25c.—in 1-lb. hanks, 25c. —in ½-lb. hanks, 38c.—in ¼-lb. hanks, 50c. packed in casks or cases, 15c.—bagging or papering, 10c

FENCE STAPLES.

Bright. .... 2 80 Galvanized.... 32

HAY WIRE IN COILS.

No 13, $2 70; No. 14, $2 80; No. 15, $2.95; f.o.b., Montreal.

GALVANIZED WIRE.

Per 100 lb.—Nos. 4 and 5, $3.90 — Nos. 6, 7, 8, $3.35 — No. 9, $2.85 — No. 10, $3.40 — No. 11, $3.45—No. 12, $3 00 —No. 13, $3.10—No. 14, $3.95—No. 15, $4.30 —No. 16, $4.30 from stock. Base sizes, Nos. 6 to 9, $2.35 f.o.b. Cleveland. Extras for cutting.

LIGHT STRAIGHTENED WIRE.

Gauge No.	Over 20 in. per 100 lbs.	10 to 20 in.	5 to 10 in.
0 to 5	$0 50	$0.75	$1.25
6 to 9	0 75	1.25	2 00
10 to 11	1 00	1.75	2.50
12 to 14	1.50	2.25	3.50
15 to 16	2.00	3.00	4.50

SMOOTH STEEL WIRE.

No. 0-9 gauge, $2.40; No. 10 gauge, 6c extra; No. 11 gauge, 12c extra; No. 12 gauge, 20c. extra; No. 13 gauge, 30c. extra; No 14 gauge, 40c. extra; No. 15 gauge, 55c. extra; No. 16 gauge, 70c. extra. Add 60c. for coppering and $2 for tinning.

Extra net per 100 lb.—Oiled wire 10c., spring wire $1.25, bright soft drawn 15c., charcoal (extra quality) $1.25, packed in casks or cases 15c., bagging and papering 10c., 50 and 100-lb. bundles 10c., in 25-lb. bundles 15c., in 5 and 10-lb. bundles 25c., in 1-lb hanks, 50c., in ½-lb. hanks 75c., in ¼-lb. hanks $1.

POULTRY NETTING.

2-in. mesh, 19 w. g., 50 and 5 p.c. off. Other sizes, 50 and 5 p.c. off.

WIRE CLOTH.

Painted Screen, in 100-ft rolls, $1.72½, per 100 sq. ft; in 50-ft. rolls, $1.77½, per 100 sq ft.

WIRE FENCING.

Galvanized barb	2 95
Galvanized, plain twist	3 30

Galvanized barb, f.o.b. Cleveland, $2.70 for small lots and $2.60 for carlots

WIRE ROPE

Galvanized, 1st grade, 6 strands, 24 wires, ⅜, $5; 1 inch $16 80.
Black, 1st grade, 6 strands, 19 wires, ⅜, $5; 1 inch $15 10. Per 100 feet f o b Toronto.

WOODENWARE.

CHURNS.

No. 0, $9; No. 1, $9; No. 2, $10; No 3, $11; No. 4, $13; No. 5, $16; f o b. Toronto Hamilton, London and St. Marys, 30 and 30 per cent.; f o b. Ottawa, Kingston and Montreal, 40 and 15 per cent. discount,

CLOTHES REELS.

Davis Clothes Reels, dis. 40 per cent.

FIBRE WARE.

Star pails, per doz	$ 3 00
0 Tubs, "	14 00
1 " "	12 00
2 " "	10 00
3 " "	8 50

LADDERS, EXTENSION.

3 to 6 feet, 12c. per foot; 7 to 10 ft., 13c.
Waggoner Extension Ladders, dis 40 per cent.

MOPS AND IRONING BOARDS.

"Best" mops	....	1 25
"900" mops	....	1 25
Folding ironing boards	12 00	16 50

REFRIGERATORS

Discount, 40 per cent

SCREEN DOORS.

Common doors, 2 or 3 panel, walnut stained, 4-in. style..........per doz. 7 25
Common doors, 2 or 3 panel, grained only, 4-in., style ....per doz. 7 55
Common doors, 2 or 3 panel, light stair per doz. 9 55

WASHING MACHINES.

Round, re-acting per doz.	60 00
Square "	63 00
Eclipse, per doz	54 00
Dowswell "	39 00
New Century, per doz	75 00
Daisy	54 00
Stephenson	74 00

WRINGERS.

Royal Canadian, 11 in., per doz.	35 00
Royal American, 11 in.	35 00
Eze 10 in., per doz	36 75

MISCELLANEOUS

AXLE GREASE.

Ordinary, per gross	6 00	7 00
Best quality	10 00	12 00

BELTING.

Extra, 60 per cent.
Standard, 60 and 10 per cent.
No. 1, not wider than 6 in., 60, 10 and 10 p.c.
Agricultural, not wider than 4 in., 75 per cent.
Lace leather, per side, 75c.; cut laces, 80c.

BOOT CALKS.

Small and medium, ball ....per M	2 25
Small heel	4 50

CARPET STRETCHERS.

American ....per doz.	1 00	1 50
Bullard's "	....	6 50

CASTORS.

Bed, new list, discount 55 to 57½ per cent.
Plate, discount 52½ to 57½ per cent.

PINE TAR.

½ pint in tins ....per gross	...	7 80
" " "	....	9 60

PULLEYS.

Hothouse	per doz.	0 55	1 00
Axle	"	0 22	0 33
Screw	"	0 22	1 00
Awning	"	0 35	2 50

PUMPS.

Canadian cistern	1 40	2 00
Canadian pitcher spout	1 80	3 16

Berg's wing pump, 75 per cent.

ROPE AND TWINE.

Sisal		0 10½
Pure Manilla		0 15
"British" Manilla		0 12
Cotton, 3-16 inch and larger	0 21	0 23
" 5-32 inch	0 25	0 27
" ⅛ inch	0 25	0 28
Russia Deep Sea		0 15
Jute		0 09
Lath Yarn, single		0 10
" double		0 10½
Sisal bed cord, 48 feet.......per doz.		0 65
" " 60 feet "		0 80
" " 72 feet "		0 95

Twine.

Bag, Russian twine, per lb.	....	0 27
Wrapping, cotton, 3-ply	....	0 25
" " 4-ply	....	0 29
Mattress twine per lb.	0 33	0 45
Staging " "	0 27	0 35

BINDER TWINE.

500 feet, sisal	0 09½
500 " standard	0 09½
550 " " manilla	0 10½
600 " " "	0 12½
650 " " "	0 13½

Car lots, ¼c. less; 5-ton lots, ½c. less. Central delivery.

SCALES.

Gurney Standard, 35; Champion, 45 p c.
Burrow, Stewart & Milne — Imperial Standard, 35; Weigh Beams, 35; Champion Scales, 45.
Fairbanks Standard, 30; Dominion, 50 Richelieu, 50.
Warren new Standard, 35; Champion, 45 Weigh Beams, 30.

STONES—OIL AND SCYTHE.

Washita	per lb.	0 25	0
Hindostan	"	0 06	0 10
" slip	"	0 18	0 20
" Axe	"	....	0 10
Deer Creek	"	....	0 10
Deerlick	"	....	0 25
" Axe	"	....	0 15
Lily white	"	....	0 42
Arkansas	"	....	1 50
Water-of-Ayr	"	....	0 10
Scythe	per gross	3 50	5 00
Grind, 40 to 200 lb., per ton		20 00	22 00
" under 40 lb., "		....	24 00
" 200 lb. and over		....	26 00

INDEX TO ADVERTISERS.

CLASSIFIED LIST OF ADVERTISEMENTS.

Alabastine.
Alabastine Co., Limited, Paris, Ont.

Auditors.
Davenport, Pickup & Co., Winnipeg.

Automobile Accessories.
Canada Cycle & Motor Co, Toronto Junction.
Carriage Mountings Co., Ltd., Toronto.

Awnings.
Tobin Tent and Awning Co., Ottawa

Babbitt Metal.
Canada Metal Co., Toronto.
Canadian Fairbanks Co., Montreal.
Frothingham & Workman, Montreal.
Robertson, Jas. Co., Montreal.

Bar Urns.
Buffalo Mfg. Co., Buffalo, N.Y.

Bath Room Fittings.
Buffalo Mfg. Co., Buffalo, N.Y.
Carriage Mountings Co., Ltd., Toronto.
Forsyth Mfg. Co., Buffalo, N.Y.
Ontario Metal Novelty Co., Toronto

Belting, Hose, etc.
Gutta Percha and Rubber Mfg. Toronto.
Sadler & Haworth, Toronto.

Bicycles and Accessories.
Johnson's, Iver, Arms and Cycle Works Fitchburg, Mass

Binder Twine.
Consumers Cordage Co, Montreal.

Bolts.
Toronto & Belleville Rolling Mills, Belleville

Box Strap.
J. N. Warminton, Montreal.

Brass Goods.
Frothingham & Workman, Montreal.
Glauber Brass Mfg. Co., Cleveland, Ohio.
Kerr Engine Co., Walkerville, Ont.
Lewis, Rice, & Son., Toronto.
Morrison, Jas., Brass Mfg. Co., Toronto.
Mueller Mfg Co, Decatur, Ill.
Penberthy Injector Co., Windsor, Ont.
Taylor-Forbes Co., Guelph, Ont.

Bronze Powders.
Canadian Bronze Powder Works, Montreal.

Brushes.
United Factories, Toronto.

Cans.
Acme Can Works, Montreal

Builders' Tools and Supplies.
Covert Mfg. Co., West Troy, N.Y.
Frothingham & Workman Co., Montreal.
Howland, H. S. Sons & Co., Toronto.
Hyde, F., & Co., Montreal.
Lewis Bros. & Co., Montreal.
Lewis, Rice, & Son, Toronto.
Lockerby & McComb, Montreal.
Lufkin Rule Co., Saginaw, Mich.
Newman & Sons, Birmingham.
North Bros. Mfg. Co., Philadelphia, Pa.
Stanley Rule & Level Co., New Britain.
Stanley Works, New Britain, Conn.
Stephens, G. F., Winnipeg.
Taylor-Forbes Co., Guelph, Ont.

Carriage Accessories.
Carriage Mountings Co, Ltd., Toronto.
Covert Mfg. Co., West Troy, N.Y.

Carriage Springs and Axles.
Guelph Spring and Axle Co., Guelph.

Carpet Beateas.
Ontario Metal Novelty Co., Toronto.

Cartridges.
Dominion Cartridge Co, Montreal.

Cattle and Trace Chains.
Greening, B., Wire Co., Hamilton.

Chafing Dishes.
Buffalo Mfg. Co., Buffalo, N.Y.

Churns.
Dowswell Mfg. Co., Hamilton.

Clippers—All Kinds.
American Shearer Mfg. Co., Nashua, N.H.

Clothes Reels and Lines.
Hamilton Cotton Co., Hamilton,

Clutch Nails.
J. N. Warminton, Montreal.

Congo Roofing
Buchanan Foster Co., Philadelphia, Pa.

Cordage.
Consumers' Cordage Co., Montreal.
Hamilton Cotton Co., Hamilton.

Cork Screws.
Erie Specialty Co., Erie, Pa.

Cow Ties
Greening, B., Wire Co., Hamilton

Cuspidors.
Buffalo Mfg. Co., Buffalo, N.Y.

Cut Glass.
Phillips, Geo., & Co., Montreal.

Cutlery—Razors, Scissors, etc.
Birkett, Thos., & Son Co., Ottawa.
Clauss Shear Co., Toronto
Dorken Bros. & Co., Montreal.
Frothingham & Workman, Montreal.
Heinisch's, R., Sons Co., Newark, N.J.
Howland, H. S. Sons & Co., Toronto.
Hutton, Wm., & Sons, Ltd., London, Eng.
Lamplough, F. W., & Co, Montreal.
Phillips, Geo., & Co., Montreal.
Round, John, & Son, Montreal.

Electric Fixtures.
Canadian General Electric Co., Toronto.
Morrison James, Mfg. Co., Toronto.
Munderloh & Co., Montreal.

Electro Cabinets.
Cameron & Campbell Toronto.

Enameled Ware
Kemp Mfg. Co., Toronto.

Engines, Supplies, etc.
Kerr Engine Co., Walkerville, Ont.

Eavetroughs
Wheeler & Bain, Toronto

Fencing—Woven Wire.
Dominion Wire Mfg. Co., Montreal.
Owen Sound Wire Fence Co., Owen Sound.
Banwell Hoxie Wire Fence Co., Hamilton.

Files and Rasps.
Barnett Co., G. & H., Philadelphia, Pa.
Nicholson File Co., Port Hope

Firearms and Ammunition.
Hamilton Rifle Co., Plymouth, Mich.
Harrington & Richardson Arms Co., Worcester, Mass.
Johnson's, Iver, Arms and Cycle Works Fitchburg, Mass.

Fishing Tackle.
Enterprise Mfg. Co., Akron, Ohio

Food Choppers.
Enterprise Mfg. Co., Philadelphia, Pa.
Lamplough, F. W., & Co., Montreal.
Shirreff Mfg. Co., Brockville, Ont.

Furnaces.
Pease Foundry Co., Toronto.

Galvanizing.
Canada Metal Co., Toronto.
Dominion Wire Mfg. Co., Montreal.
Montreal Rolling Mills Co., Montreal.
Ontario Wind Engine & Pump Co., Toronto.

Glass Ornamental
Hobbs Mfg. Co., London
Consolidated Plate Glass Co., Toronto

Glaziers' Diamonds.
Gibsone, J. B., Montreal.
Pelton, Godfrey S.
Sharratt & Newth, London, Eng.
Shaw, A., & Son, London, Eng.

Handles.
Still, J. H., Mfg. Co.

Harvest Tools.
Maple Leaf Harvest Tool Co., Tillsonburg, Ont.

Hockey Sticks
Still, J. H. Mfg. Co, St. Thomas.

Hoop Iron.
Frothingham & Workman Montreal.
Montreal Rolling Mills Co., Montreal.
J. N. Warminton, Montreal.

Horse Blankets.
Heney, E. N., & Co., Montreal.

Horseshoes and Nails.
Canada Horse Nail Co., Montreal.
Montreal Rolling Mills, Montreal.
Capewell Horse Nail Co., Toronto
Toronto & Belleville Rolling Mills Belleville.

Hot Water Boilers and Radiators.
Cluff, R. J., & Co Toronto.
Pease Foundry Co. Toronto.
Taylor-Forbes Co., Guelph.

Ice Cream Freezers.
Dana Mfg. Co., Cincinnati, Ohio.
North Bros. Mfg. Co., Philadelphia, Pa.

Ice Cutting Tools.
Erie Specialty Co., Erie, Pa.
North Bros. Mfg. Co., Philadelphia, Pa.

Injectors—Automatic.
Morrison, Jas., Brass Mfg Co., Toronto.
Penberthy Injector Co., Windsor, Ont.

Iron Pipe.
Montreal Rolling Mills, Montreal.

Iron Pumps.
Lamplough, F. W., & Co., Montreal.
McDougall, R., Co., Galt, Ont.

Ladders.
The Waggoner Ladder Co., London, Ont.

Lanterns.
Kemp Mfg. Co., Toronto.
Ontario Lantern Co., Hamilton, Ont.
Wright, E. T., & Co., Hamilton.

Lawn Mowers.
Birkett, Thos., & Son Co., Ottawa.
Frothingham & Workman, Montreal.
Maxwell, D., & Sons, St. Mary's, Ont.
Taylor, Forbes Co., Guelph.

Lawn Mower Grinders
Root Bros. & Co., Plymouth, Ohio.

Ledgers—Loose Leaf.
Business Systems, Toronto.
Copeland-Chatterson Co., Toronto.
Crain, Rolla L., Co., Ottawa.
Universal Systems, Toronto.

Lithographing.
London Printing & Lithographing Co., London, Ont.

Locks, Knobs, Escutcheons, etc.
Peterborough Lock Mfg. Co., Peterborough, Ont.
National Hardware Co., Orillia, Ont.

Lumbermen's Supplies.
Pink, Thos., & Co., Pembroke Ont.

Lye
Gillett, E. W., & Co., Toronto

Manufacturers' Agents.
Fox, C. H., Vancouver.
McIntosh, H. F., & Co., Toronto.
Gibb, Alexander, Montreal.
Scott, Bathgate & Co., Winnipeg.

Metals.
Canada Iron Furnace Co., Midland, Ont.
Canada Metal Co., Toronto.
Eadie, H. G., Montreal.
Frothingham & Workman, Montreal.
Gibb, Alexander, Montreal.
Kemp Mfg. Co., Toronto
Leslie, A. C., & Co., Montreal.
Lysaght, John, Bristol, Eng.
Nova Scotia Steel and Coal Co., New Glasgow, N S.
Robertson, Jas., Co., Montreal.
Roper, J. H., Montreal.
Samuel, Benjamin & Co., Toronto.
Stairs, Son & Morrow, Halifax, N S.
Thompson, B. & S. H. & Co. Montreal.

Metal Lath.
Galt Art Metal Co., Galt.
Metallic Roofing Co., Toronto.
Metal Shingle & Siding Co., Preston, Ont.

Metal Polish, Emery Cloth, etc.
Oakey, John, & Sons, London, Eng.

Nails Wire
Dominion Wire Mfg. Co., Montreal.

Oil Tanks
Bowser, S. F., & Co., Toronto.

Ornamental Iron and Wire.
Dennis Wire & Iron Co, London, Ont.

Packing.
Gutta Percha & Rubber Co Toronto

Paints, Oils, Varnishes, Glass.
Blanchite Process Paint Co., Toronto.
Brandram-Henderson, Montreal
Canada Paint Co., Montreal.
Canadian Oil Co., Toronto.
Consolidated Plate Glass Co., Toronto.
Dods, P. D., & Co., Montreal
Imperial Varnish and Color Co., Toronto.
Jamieson, R. C., & Co, Montreal.
Lucas John & Co., New York
McArthur, Corneille & Co., Montreal.
McCaskill, Dougall & Co., Montreal.
Moore, Benjamin, & Co. Toronto.
Ottawa Paint Works, Ottawa
Queen City Oil Co., Toronto.
Ramsay & Son, Montreal.
Sanderson Pearcy & Co., Toronto
Sherwin-Williams Co., Montreal.
Standard Paint Co., Montreal
Standard Paint and Varnish Works Windsor, Ont.
Stephens & Co., Winnipeg.
Martin-Senour Co., Montreal
Winnipeg Paint & Glass Co., Winnipeg

Perforated Sheet Metals.
Greening, B., Wire Co., Hamilton.

Plumbers' Tools and Supplies.
Canadian Fairbanks Co., Montreal.
Cluff, R. J., & Co., Toronto
Frothingham & Workman, Montreal.
Glauber Brass Co., Cleveland, Ohio.
Jardine, A. B., & Co, Hespeler, Ont.
Jenkins Bros., Boston. Mass.
Kerr Engine Co., Walkerville, Ont.
Lewis, Rice, & Son, Toronto.
Merrell Mfg. Co., Toledo, Ohio.
Montreal Rolling Mills, Montreal.
Morrison, Jas., Brass Mfg. Co., Toronto.
Mueller, H., Mfg. Co., Decatur, Ill
Oshawa Steam & Gas Fitting Co., Oshaw
Robertson Jas., Co. Montreal.
Robertson, Jas.; Co., Limited, Toronto
Somerville, Limited, Toronto
Stairs, Son & Morrow, Halifax, N.S.
Standard Ideal Sanitary Co., Port Hope,
Standard Sanitary Co., Pittsburg.
Stephens, G. F., & Co., Winnipeg, Man.
Turner Brass Works, Chicago.
Vickery, Orlando, Toronto.

Polishes.
Majestic Polishes, Toronto

Portland Cement.
International Portland Cement Co. Ottawa, Ont.
Hanover Portland Cement Co., Hanover, Ont.
Hyde, F., & Co., Montreal.
Thompson B. & S. H. & Co., Montreal.

Poultry Netting.
Greening, B., Wire Co., Hamilton, Ont.

Printing.
London Printing & Lithographing Co., London, Ont.

Razors.
Clauss Shear Co., Toronto.

Refrigerators.
Fabien, C. P., Montreal.

Registers.
Pease Foundry Co., Toronto.

Roofing Supplies.
Brantford Roofing Co., Brantford.
Barrett Mfg. Co., New York.
F. W. Bird, East Walpole, Mass.
Buchanan Foster Co., Philadelphia, Pa.
McArthur, Alex., & Co., Montreal.
Metal Shingle & Siding Co., Preston, Ont.
Metallic Roofing Co., Toronto.
Paterson Mfg. Co., Toronto & Montreal.
Wheeler and Bain, Toronto

Saws.
Atkins, E. C., & Co., Indianapolis, Ind
Shurly & Dietrich, Galt, Ont.
Spear & Jackson, Sheffield, Eng.

Scales.
Canadian Fairbanks Co., Montreal.
Frothingham & Workman, Montreal.

Screw Cabinets.
Cameron & Campbell, Toronto.

Screws, Nuts, Bolts.
Dominion Wire Mfg. Co, Montreal.
Montreal Rolling Mills Co., Montreal.

Soil Pipe
McFarlane, Walter, Glasgow

Sewer Pipes.
Canadian Sewer Pipe Co., Hamilton
Hyde, F., & Co., Montreal.

Shelf Boxes.
Cameron & Campbell, Toronto.

Shears, Scissors.
Clauss Shear Co., Toronto.

Shovels and Spades.
Eclipse Mfg. Co, Ottawa
Frothingham & Workman, Montreal
Peterboro Shovel & Tool Co., Peterboro.

Silverware.
Hutton, Wm., & Sons, Ltd., London, Eng.
McGlashan, Clarke Co., Niagara Fal's, Ont.
Phillips, Geo., & Co., Montreal.
Round, John, & Son, Sheffield, Eng.

Skates.
Canada Cycle & Motor Co., Toronto.
McFarlane, Walter, Glasgow.

Sprayers
Cavers Bros., Galt

Spring Hinges, etc.
Chicago Spring Butt Co., Chicago, Ill.

Stable Fittings
Dennis Wire & Iron Co, London

Steel Rails.
Nova Scotia Steel & Coal Co., New Glasgow, N.S.

Stove Pipe.
Chown, Edwin, and Son, Kingston

Stoves, Tinware, Furnaces
Canadian Heating & Ventilating Co. Owen Sound
Copp, W. J., Son & Co., Fort William
Davidson, Thos., Mfg Co., Montreal.
Down Draft Furnace Co., Galt
Guelph Stove Co., Guelph.
Gurney Foundry Co., Toronto.
Harris, J. W., Co., Montreal.
Howard, Wm, Toronto
Kemp Mnfg. Co. Toronto.
McClary Mfg. Co. London.
Merrick Anderson, Winnipeg
Pease Foundry Co., Toronto.
Smart, James, Mfg. Co., Brockville
Stewart, Jas., Mfg. Co., Woodstock, Ont.
Taylor-Forbes Co., Guelph, Ont.
Wright, E. T., & Co., Hamilton.

Tacks.
Montreal Rolling Mills Co., Montreal.
Ontario Tack Co., Hamilton.

Tents.
Tobin Tent and Awning Co., Ottawa

Tin Plate.
American Sheet & Tin Plate Co, Pittsburg, Pa.
Baglan Bay Tin Plate Co., Briton Ferry South Wales
Lysaght, John, Bristol, Newport and Montreal

Turpentine
Defiance Mfg. Co., Toronto.

Ventilators.
Harris, J. W., Co., Montreal.
Pearson, Geo. D., Montreal.

Wall Paper
Staunton Limited, Toronto.

Wall Paper Cleaner.
Gilbert, Frank U. S., Cleveland

Washing Machines, etc
Dowswell Mfg. Co., Hamilton, Ont.
The Shultz Bros. Co, Brantford.
Taylor-Forbes Co., Guelph, Ont.

Water Filters.
Buffalo Mfg. Co., Buffalo, N.Y.

Wheelbarrows
London Foundry Co., London Ont
Schultz Bros. Co, Ltd., The Brantford.

Wholesale Hardware.
Birkett, Thos., & Sons Co., Ottawa.
Caverhill, Learmont & Co., Montreal.
Frothingham & Workman, Montreal.
Hobbs Hardware Co., London.
Howland, H. S., Sons & Co., Toronto.
Lamplough, F. W., & Co., Montreal.
Lewis Bros. & Co., Montreal.
Lewis, Rice, & Son, Toronto.

Window and Sidewalk Prisms
Hobbs Mfg. Co., London, Ont.

Wire, Wire Rope, Cow Ties, Fencing Tools, etc.
Banwell-Hoxie Fence Co., Hamilton
Dennis Wire and Iron Co., London, Ont.
Dominion Wire Mnfg. Co., Montreal
Greening, B., Wire Co., Hamilton.
Owen Sound Wire Fence Co., Owen Sound
Montreal Rolling Mills Co., Montreal.
Western Wire & Nail Co., London, Ont.

Wrapping Papers.
Canada Paper Co., Toronto.
McArthur, Alex., & Co., Montreal.
Stairs, Son & Morrow, Halifax, N.S.

Wringers
Connor, J. H. & Son, Ottawa, Ont

"Seeing is Believing"

¶ As the Ross Sporting Rifle represents the best in modern rifle manufacture, the highest grade materials and minute accuracy in workmanship, it is to your interest to learn fully about this Canadian Rifle—

¶ The Ross Straight Pull 303 Sporting Model M is one of our leaders to retail at $26. The barrel is the celebrated Ross high pressure, non-fouling, gun-barrel steel — The breech action is dependable.

Our representatives are now demonstrating the merits of this rifle—It places you under no obligation to drop us a line to make sure you will be on their calling list.

Our advertising policy is co-operative.

THE ROSS RIFLE CO., - QUEBEC

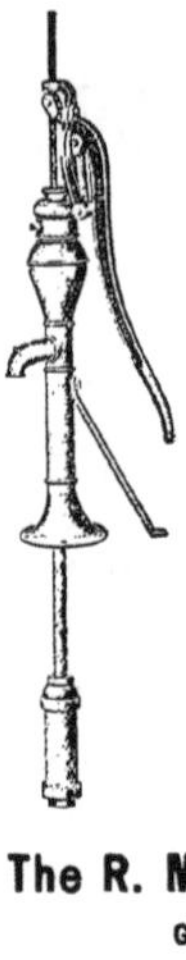

For twelve weeks we have been working on the strain.

McDougall Pumps Are Standard

Many letters come to us addressed "Standard." They find us all right. Customers find us all right, too. Do not forget: Quality first and price not out of the way. In fact, but—Well, could we not do a lot of explaining in response to an enquiry?

Our Permanent Address:

The R. McDougall Co., Limited

GALT - CANADA

That Stuffy Room

is more deadly than disease itself. Bad air is the curse of crowded living. In all public places such as asylums, clubs, big offices, schools and theatres, PURE AIR CAN BE OBTAINED ONLY BY OUR SYSTEM OF VENTILATION.

It has stood the tests of scientific experts

It is in use in the finest buildings in Canada.

Its price is reasonable and its benefits lasting.

Testimony

Montreal, June 7th, 1901

Messrs J. W. Harris Co., Limited
Montreal

Dear Sirs,

I have been using your "AEOLIAN" Ventilators on several houses in this city with the best results. The tenants of these houses declared that your Ventilators are of great advantage to them. I also installed two (2) 18" on my stable and I can say that your Ventilator cannot be surpassed by any to take out the sweating or the bad smell in any room.

F. D. MONK

Ask for booklet re Aeolian and Zephyr Ventilators, also booklet on Filter, etc.

THE J. W. HARRIS CO., LIMITED

General Contractors - MONTREAL

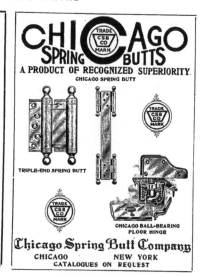

CIRCULATES EVERYWHERE IN CANADA
Also in Great Britain, United States, West Indies, South Africa and Australia.

HARDWARE AND METAL

A Weekly Newspaper Devoted to the Hardware, Metal, Heating and Plumbing Trades in Canada.

Office of Publication, 10 Front Street East, Toronto.

VOL. XIX. MONTREAL, TORONTO, WINNIPEG, SEPTEMBER 14, 1907 NO. 37.

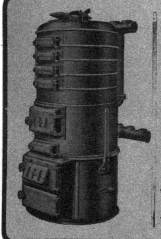

Read "Want Ads." on Page 66

CUTLERY

Pen Knives and Scissors

Cases of Carvers

Razors and Pocket Knives

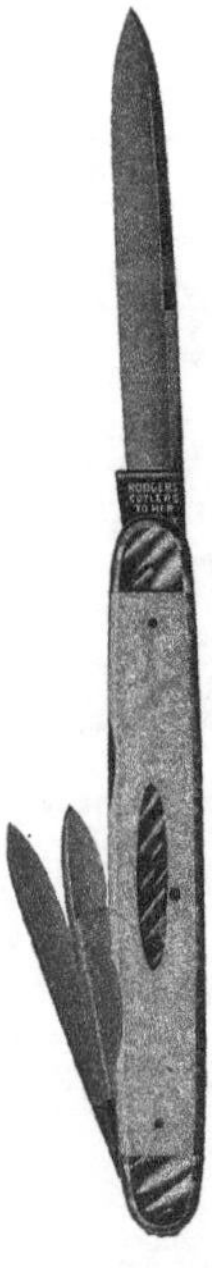

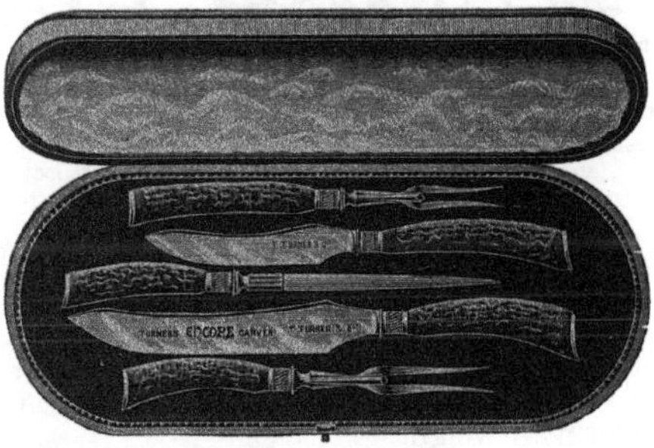

5 Piece Case Carvers

The new lines of Cutlery Samples shown by our travellers this season are far superior to anything heretofore put on the market, and it will be interesting to you, as well as profitable, to examine them.

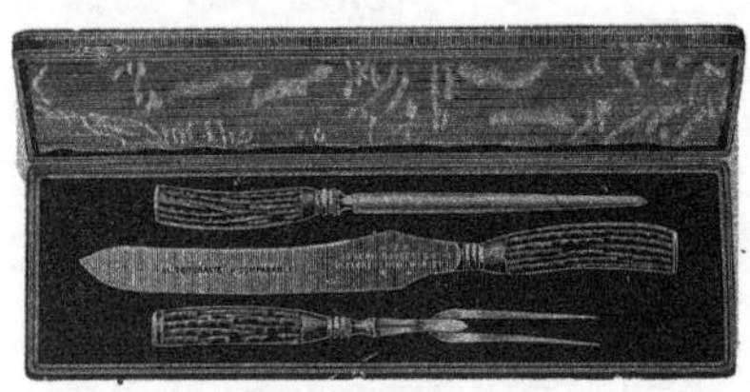

Pocket Knives

3 Piece Case Carvers

Pocket Knives

RICE LEWIS & SON

LIMITED

TORONTO.

FURS ARE VALUABLE

Don't allow your catch to escape because caught in a poor trap. **A GENUINE NEWHOUSE** trap will hold the game and earn its extra cost several times in a season.

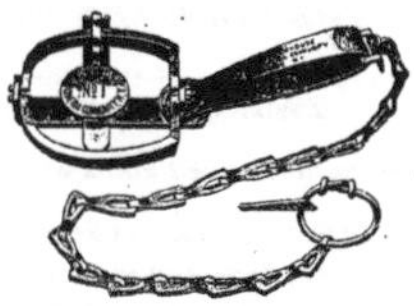

Newhouse Steel Traps

ARE ABSOLUTELY GUARANTEED

Made since 1848 by

ONEIDA COMMUNITY, Limited

Write for Catalogue NIAGARA FALLS, ONT.

CHICAGO SPRING BUTTS

A PRODUCT OF RECOGNIZED SUPERIORITY

CHICAGO SPRING BUTT

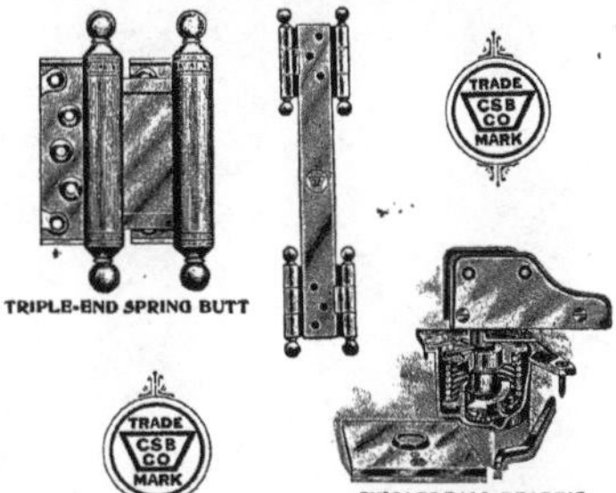

TRIPLE-END SPRING BUTT

CHICAGO BALL-BEARING FLOOR HINGE

Chicago Spring Butt Company

CHICAGO NEW YORK

CATALOGUES ON REQUEST

Freezes two flavors of Ice Cream or an Ice or Sherbet and Ice Cream at one and same time, in one freezer.

SOMETHING ENTIRELY NEW —NEVER DONE BEFORE.

THE can is divided in two parts by a vertical partition. One flavor is placed in one, the other in opposite side of can, in each of which the scrapers, etc., operate. The can cannot be turned as in old style freezers, because of the partition. It is therefore swung to and fro by a rocking motion of the crank. This is a more comfortable and much easier movement. It is very much less tiresome than turning a crank. One can sit back in a chair with freezer alongside and rock the crank to and fro without discomfort or undue exertion and even read at same time. Although two flavors are frozen at once and with less exertion, no more time is required than in old style freezers.

The **Pails have electric welded wire hoops,** which are guaranteed not to break or fall off; Cans are of heavy tinplate with **drawn steel bottoms** that are guaranteed not to fall out or break and do not leak, the strongest and most durable Freezer Can made; the **Automatic Twin Scrapers** by their positive action insure perfect scraping of frozen particles from side of can. All inside parts are heavily coated with pure block tin, and outside parts all thoroughly galvanized.

SOLD BY LEADING JOBBERS

ICE CREAM FREEZERS THAT SELL

THEY ARE	THEY EXCEL IN
Well Advertised	Easy Running
In Demand	Quick Freezing
Easily Sold	Economy
Satisfactory in Use	Convenience
Of Known Reputation	Practical Results

Send for Catalog.

North Bros. Mfg. Co.

Philadelphia, Pa.

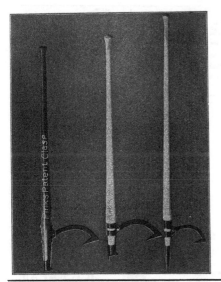

Auger Bits

Save Muscle by Easy Boring **Save Time by Rapid Boring**

You know that every article that leaves your store is some kind of an advertisement of the store.

If it is good, it is a good advertisement; if it is not, it is a bad one.

This applies as much to tools as it does to everything else you sell.

Modern retail hardware methods have reached such a high place that no modern dealer can afford to run any risks.

He can afford to sell nothing but the best goods to his customers.

The artisan, mechanic, the man who uses tools, is an important consideration with you.

These men are critical judges of tools. The tools they use are their daily bread.

We submit for your attention a new line of Auger Bits illustrated on this page. These are Canadian make, and equal in quality and finish to any solid centre bit that is made in the United States.

If you have not bought them we would recommend that you buy a sample lot and satisfy yourselves that they are better than the line you have been handling. And then look at the price. We are satisfied that you will stock this line in future to the exclusion of all other makes of the same pattern.

We carry all sizes from 3/- to 20/16. We also stock sets in wooden boxes and rolls.

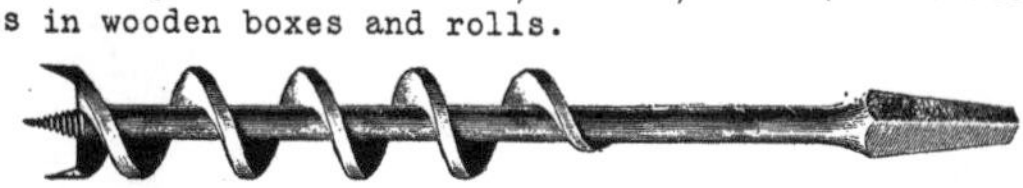

IF INTERESTED, LET US NAME YOU A PRICE

Lewis Bros., Limited - Montreal

OTTAWA
TORONTO

WINNIPEG

VANCOUVER
CALGARY

Transmission Rope

The transmission of power by Rope Drives is increasing daily, on account of the durability and low cost. Rope Drives were first manufactured and introduced into the United States by Sewall, Day & Co., of Boston. A few years later we manufactured and introduced them into this country.

If you contemplate putting in a Rope Drive, we will be pleased to furnish you with any information you may require. As we have had years of experience, we are naturally in a position to turn out the most up-to-date product.

Our mills are now thoroughly equipped for manufacturing our famous Tallow Laid Transmission Rope, three or four strand, and in all sizes and lengths, and we always carry a large stock of the most select Manila Fibre for the manufacture of this Rope. We carry a fully assorted stock of all the standard sizes, and can make special orders promptly.

CONSUMERS' CORDAGE COMPANY, Limited

Mills: MONTREAL, QUE., and HALIFAX, N.S.

BRANCHES:

F. H. ANDREWS & SON, Quebec. CONSUMERS CORDAGE CO., Limited, St. John, N.B. MacGOWAN & CO., Vancouver
TEES & PERSSE, Limited, Winnipeg, Calgary and Edmonton. GEO. WOOD, London, England.
MERRICK, ANDERSON & CO., Winnipeg, Binder Twine Distributors for Northwest.

Patented 1900. Improved 1905.

Davidson's "Empire" Stove Pipe

Improved Patent

Neatly packed in crates of 25.
Secures lowest possible freight rate.

Deep Swage Prevents Telescoping.
Coated to Prevent Rust.

Sold as low as ordinary Stove Pipes.

SIMPLE and EASY to put together.
No Profane Language Necessary.
The only tools required are a pair of hands.
Pipes being cut out with dies ensure absolute fit and conformity.

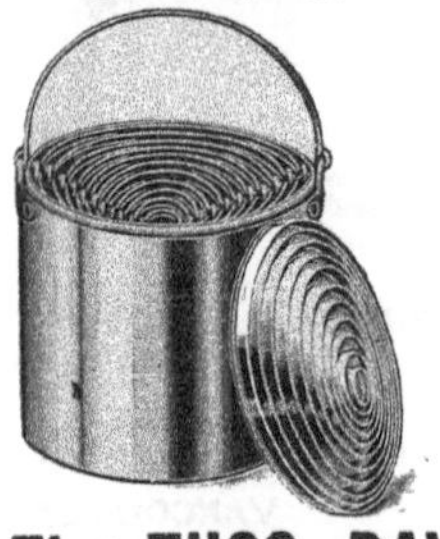

Miners' Camp Kettles

Strong and substantially made in extra heavy tin.

Nineteen sizes, from ⅛ quart to 29 quarts.

Wash Tubs—Galvanized

With Wringer Attachment

Nos.	1	2	3
Top measurement, inches ...	21½	22¾	25
Bottom " "	17⅞	18¼	20
Height " "	9½	10½	11

The THOS. DAVIDSON MFG. CO., Ltd., Montreal and Winnipeg

H. S. HOWLAND, SONS & CO. LIMITED

HARDWARE MERCHANTS

Only Wholesale — 138-140 WEST FRONT STREET, TORONTO. — Wholesale Only

Single and Double Barrel Shot Guns

Single Barrel Shot Guns
Hopkins and Allen. Iver Johnson. Stevens Arms Co.

Double Barrel Shot Guns
Belgium. Neumann Bros. Stevens Arms Co.

Hammerless Repeating Shot Guns
Winchester Model 1897----12 Gauge

American Dead Shot Powders
Black and Smokeless

Dominion and Eley's Loaded Shells
Black or Smokeless Powders

For full description see our catalogue.

H. S. HOWLAND, SON & CO., LIMITED

Opposite Union Station

GRAHAM WIRE NAILS ARE THE BEST

Our Prices are Right. — Are you receiving our monthly illustrated circular? IF NOT, WRITE FOR IT. — We Ship Promptly

HOUSEHOLD NECESSITIES

The **Universal Brand** guarantees mechanical perfection and unquestionable superiority in operation.

UNIVERSAL

The "Universal" Bread Maker

mixes and kneads Bread in three minutes. The hands do not touch the dough.

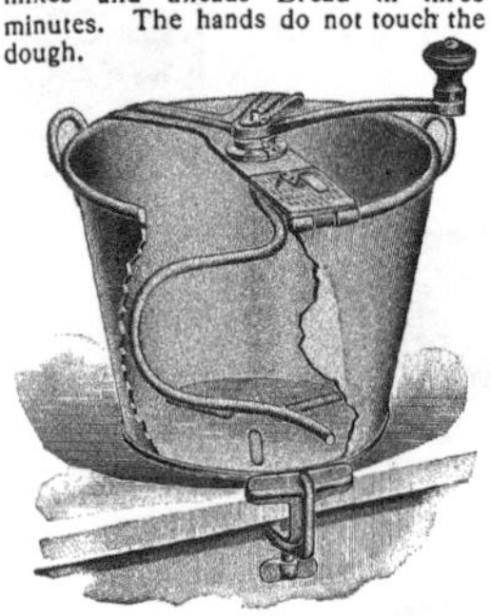

The "Universal" Cake Maker

mixes the batter for all kinds of cake rapidly and easily. Simple to operate.

The "Universal" Food Chopper

chops all kinds of food, whether meat or vegetables, rapidly and easily, either coarse or fine as required.

Does away with drudgery of the chopping knife and bowl altogether.

The "Universal" Coffee Percolator

makes Perfect Coffee. A healthful, appetizing beverage, clear as wine, though no eggs are used.

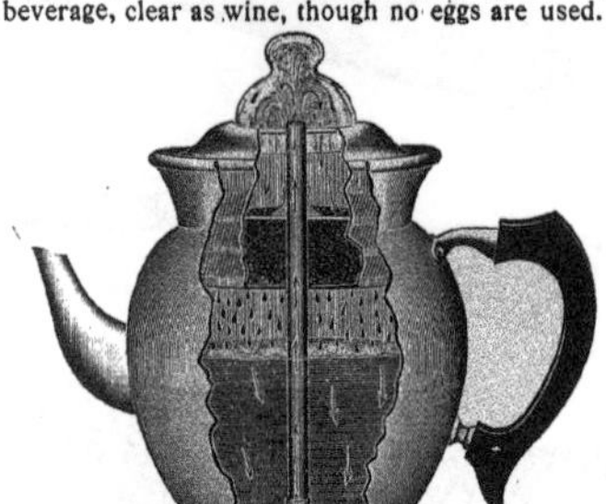

Any one can make Perfect Coffee in the Universal. Uniform in results on any kind of a stove.

ELECTROS AND DESCRIPTIVE BOOKLETS SUPPLIED FREE OF CHARGE.

Quotations Cheerfully Given on Request.

Frothingham & Workman, Limited

Montreal, - Canada

F&W

One of the best selling specialties we have ever taken up is the line of Asbestos Sad Irons.

You know their construction, the solid iron core and asbestos-lined hood and handle. What we wish to persuade you to do now is to handle them or add to your line. We want you to handle them because they will sell.

They will sell :

Because they supply the want of a better iron.

Because they are well advertised in the papers women read.

Because their makers supply you with stands and advertising matter to assist you.

Because the line has variety. It includes an iron for every purpose, and helps you to meet every demand.

Ask our travellers for prices and more detailed information or send to us for a catalogue.

Caverhill Learmont & Co

MONTREAL

Winnipeg — Ottawa — Quebec — Fraserville

Wheelbarrows for all purposes

Drag and Wheel Road Scrapers

Contractors' Dump Carts

Store, Warehouse, Steamboat, Railway and Packing House

Trucks of every size and description

BEST QUALITY — NEWEST PATTERNS

LOWEST PRICES — PROMPT SHIPMENTS

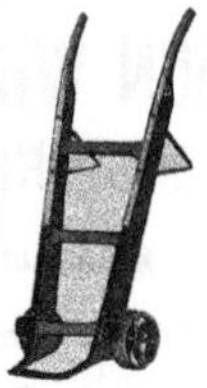

THE MEAFORD WHEELBARROW CO. Limited

MEAFORD, - ONTARIO

Eastern Agent—Alexander Gibb, Montreal — Western Agent—Bassett & Leucks, Winnipeg

Queen City Water White Oil

GIVES PERFECT LIGHT

The most economical high-grade oil ever sold in Canada.

FOR SALE BY ALL DEALERS

NO ACID OR GRIT

will be found in "Majestic" polishes. The brilliant and lasting lustre imparted by "Majestic" polishes is not susceptible to atmospheric conditions and weather changes.

"Majestic" polishes are unequalled for brass, copper, nickel and all finished metallic surfaces.

Write for sample and prices. 71

MAJESTIC POLISHES, Limited

575 Yonge Street - - Toronto, Canada

Can. Feb. 27, 1889

FACTS

about the best

CONE ROTARY VENTILATOR

on the market

Known everywhere as PEARSON'S

This Ventilator is set in motion by the least current of air and will give a perfect draught at all times.

Substantial in make, Ornamental in appearance.

For balky chimneys and Ventilating Churches, Schools, Halls, Factories, Steamers, Public and Private Buildings, Stables, Water-Closets, &c.

This Ventilator is in use throughout the Dominion, and has proved a great success.

Remember every Ventilator is guaranteed to work satisfactorily.
They are made in sizes from 2½ to 48 inches.
All orders by mail promptly attended to Testimonials on application.

Write to-day for particulars to the owners:

Geo. D. Pearson & Co., Office and Works: 17 Pontiac Street, MONTREAL
Bell Telephone East 2106

Persons addressing advertisers kindly mention having seen their advertisement in Hardware and Metal.

Past Due Accounts.

We have just issued an eight-page booklet regarding our Collection Department. If you have not receivd one write us. It will nterest you.

The Beardwood Agency

Claims, Collections and Commercial Reports

313, New York Life Bldg., MONTREAL

H. W. MITCHELL

WINNIPEG, MAN.

Correspondence is solicited from manufacturers wishing a representative in Winnipeg.

Travellers call regularly on the wholesale and retail trade in Western Canada.

Highest references and financial responsibility.

ALEXANDER GIBB

Manufacturers' Agent and Metal Broker,
13 St. John Street, Montreal

Representing Canadian, British and American Manufacturers. Correspondence invited from firms wishing to be represented.

Expert Bookkeeping

Investigation and auditing of all classes of accounts undertaken. Trading, profit and loss accounts and balance sheets prepared. Sets of books opened and office systems devised in an up-to-date manner. Collection of accounts a specialty.

Davenport, Pickup & Co.

622 McIntyre Block and 422 Ashdown Block

WINNIPEG, MAN.,
AND AT BRANDON, MAN.

Square your Accounts

CHARLES H. FOX

Vancouver, Canada

MANUFACTURERS' AGENT

Selling to the Hardware Jobbers Only.

Representing Canadian, British and American Manufacturers, correspondence invited. Cable Address, Chasfox Vancouver, P.O. Box 1105.

Covert New Patent Breast Chain

Just what every teamster is looking for.

Adjustable to any length.

Harness Snaps, Chains, Wagon Jacks, Rope Halters, etc.

Sold by all leading Jobbers.

Sizes 28, 32, 36, 40, 44 and 48 inches.

Manufactured by

COVERT MFG. CO., - TROY, N.Y.

Simonds
Crescent-Ground Cross-Cut Saws

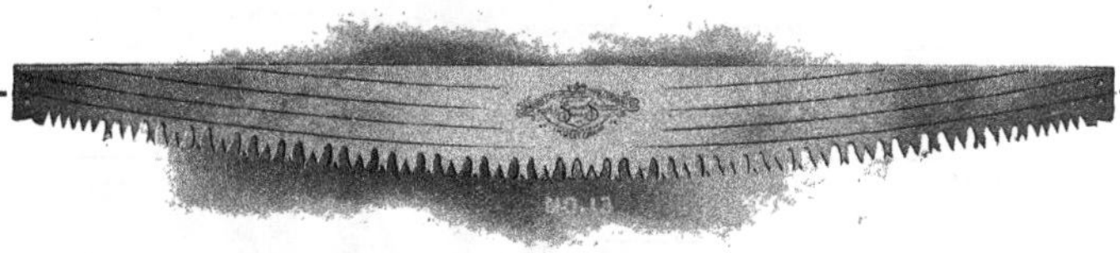

For logging camps where work must be fast and economical there is no other saw made that can give results equal to the Crescent-Ground Saw. **Made of Simonds Steel.** It cuts easy, runs fast and gives good results. This cross-cut saw is sold by most of the leading jobbers in Canada.

Simonds Canada Saw Co., Limited

TORONTO, ONT. **MONTREAL, QUE.** **ST. JOHN, N.B.**

Pratt and Whitney Small Tools and Gauges

MADE IN CANADA

have an established reputation for quality and accuracy. They are used exclusively by those who require the best. These tools of merit yield a profit to the dealer.

They win friends. They win sales. They win distinction.

Stock **PRATT AND WHITNEY SMALL TOOLS** and add to your prestige as a dealer carrying the best.

SPECIAL TERMS TO DEALERS

The Canadian Fairbanks Company, Limited,

Exclusive Canadian Selling Agents

MONTREAL - TORONTO - WINNIPEG - VANCOUVER

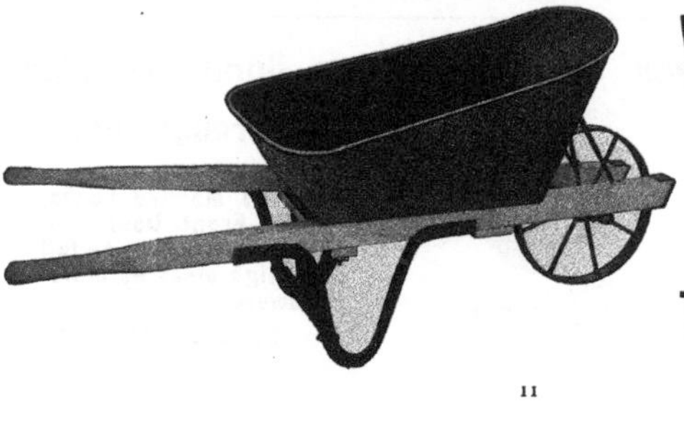

Wheelbarrows

All kinds and sizes. The cut reproduced here is just of one of the many, but the kind every contractor should use. The substantial, satisfactory, steel tray Contractor's Barrow.

The London Foundry Co.

LONDON, CAN.

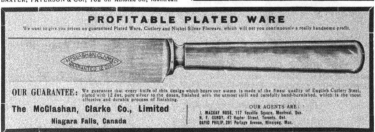

"Good Cheer" Furnaces

No. 100 SERIES

For Anthracite or Butuminous Coal, Wood and Coke

The class of Furnace which means every satisfaction to the dealer in selling and installing and to the householder in operating

The radiating qualities of this furnace are so greatly enhanced by the adoption of the cast iron tube principle therein as to place it in a class entirely by itself, and, in combination with the solidity of construction and air-blast feature, make it a warm air generator of such capacity, durability and economical fuel-consuming qualities as to place it beyond comparison with any other of the moderately-priced furnaces in the market.

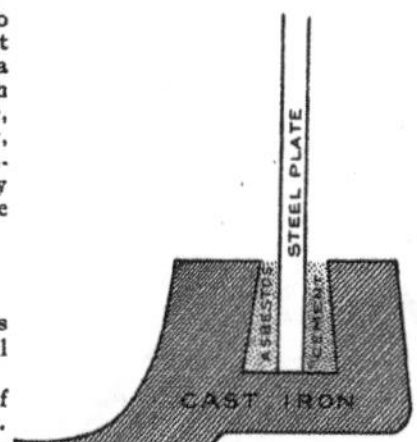

The Genuine Cup Joint

A Few of its Many Good Points

Cast iron tube radiating principle. Tubes surround fire chambers and form most powerful heat producers.

No sheet steel exposed to direct action of fire, being used only in outside jacket of radiator.

All steel set in genuine cup joints, positively gas and dust tight.

Commen-sense, positively unblockable and most easily cleaned flues.

Large combustion chamber, affording ample space for combination water-heating attachment if required.

Large double-feed door with smoke shield.

Roller grates for either hard or soft coal. Also a wood grate.

Roomy ashpit, waterpan, dust flue, direct draft damper, automatic gas damper and **AIR BLAST FUEL ECONOMIZER.**

The JAMES STEWART MFG. CO. Limited

WOODSTOCK, ONT.

Western Branch : Foot of James St., WINNIPEG, MAN.

Varnishes

For All Purposes.

OILS

Linseed Oil, Raw, Boiled and Double Boiled
Cotton, Rape, Colza, Sanctuary, Lubricating Oils, etc., etc.

Special Turpentines

Finely Blended Substitutes and Deodorized Russian Turpentine.

Any Color or Shade Matched if Sample Forwarded.

AGENT :

W. BRODIE

226 Board of Trade Bldg.

Tel. Main 15 MONTREAL

You May Be a Good Salesman

But, is everything in your favor? Are you selling the new Imperial Oxford? A continuous stream of repeat orders tells the story to us.

Removable Nickel
Removable Fire Linings
An Oval Firebox
The "Divided Draft"

All contained by the best looking stove in the market.

It gives a good salesman a chance to spread himself.

Drop us a card, or, better still, tell us the size range you want on your floor. If you get this agency you're lucky.

The Gurney Foundry Company, Limited

Toronto Winnipeg Vancouver Hamilton and London

THE GURNEY-MASSEY CO., LIMITED, MONTREAL, Que.
THE GURNEY STANDARD METAL CO., LIMITED, CALGARY, Alta., and EDMONTON, Alta.

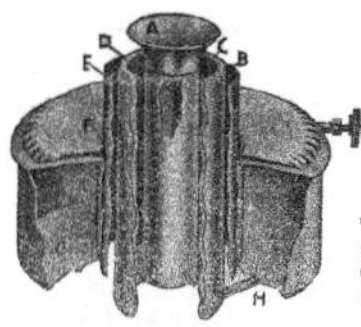

This Burner

A—Flame Spreader; B—Air Space outside of Wick; C—Air Space inside of Wick Tube D—Wick; E—Outside Casing to Burner; F—Air Space between Fount and Outside Casing, G—Fount for oil, entirely separate from Burner; H—Pipe Feed carrying oil from Fount to Burner.

combined with the many other superior features, found in no other make, has **SOLD OVER 1,000,000**

"Alumino" and "Electric" OIL HEATERS

and the demand this year is greater than ever before. Every sale has been a seller. People have only to know the "Alumino" and "Electric" to want 'em.

The burner is the very heart, the successful "innerds," of any, every and all Oil Heaters, and because our "heart" is right, because it eliminates odor and danger and smoke and trouble, common with all other Oil Heaters, our "Safety Burner" has made "Alumino" and "Electric" the only standard Oil Heaters in the world.

Sell "Alumino" or "Electric" and you sell satisfaction. Must be so or the demand wouldn't almost double every year.

Let us have your orders without delay, because somebody is liable to be disappointed.

Lewis Bros., Ltd.	**Emerson & Fisher, Ltd.**
Montreal Toronto Ottawa Winnipeg Vancouver Calgary	ST. JOHN, N.B.

H&R GUNS AND REVOLVERS

are profitable articles for you to handle, Mr. Dealer, because of the absolute satisfaction they have given to discriminating purchasers for over 35 years.

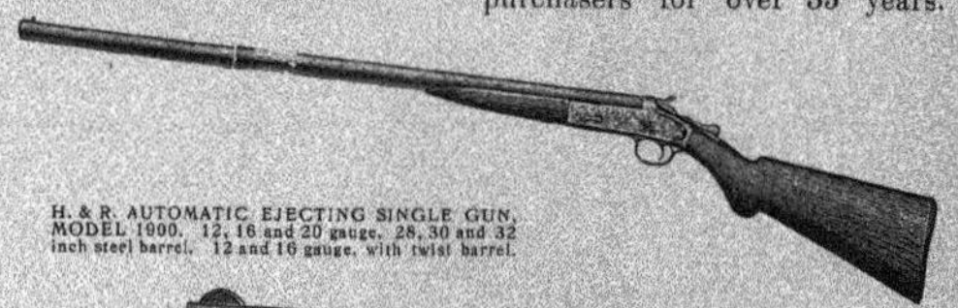

H. & R. AUTOMATIC EJECTING SINGLE GUN, MODEL 1900. 12, 16 and 20 gauge. 28, 30 and 32 inch steel barrel. 12 and 16 gauge, with twist barrel.

We are constantly impressing upon the mind of the public—your customers—the superiority of H. & R. Guns and Revolvers; the simplicity of construction, accurate and reliable shooting qualities, perfect balance and beautiful finish.

The great care and watchfulness exercised in the various processes of manufacture have gained the confidence of the public, which simply means an increased demand and larger profits to you.

Our Catalog tells about our full line and gives valuable selling points.

Write for it.

H. & R. HAMMERLESS in 32 and 38 calibres, five shot, with independent cylinder stop and automatic ejector. Weight of 32 calibre 3-inch barrel, 13 ounces; 38 calibre, 3½-inch barrel, 17 ounces. Also made with 4, 5 and 6 inch barrels. Finished in nickel or blue.

H & R ARMS CO.

HARRINGTON & RICHARDSON ARMS CO.

415 PARK AVE., WORCESTER, MASS., U.S.A.

The Howland Pump Oiler

FOR SALE BY

J. H. Ashdown Hdw Co.,	Winnipeg
Geo. D. Wood & Co.,	"
Miller-Morse Hdw. Co.,	"
Jas. Robinson & Co.,	"
Lewis Bros., Ltd.,	Montreal

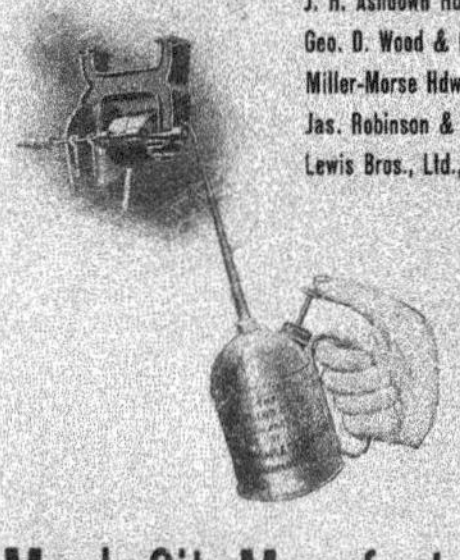

A slight pressure of the thumb regulates the flow.

Maple City Manufacturing Co.

Monmouth, Ill.

AXE HANDLES

are just as important as axe blades. If you sell a man an axe handle which breaks at the critical moment, it's pretty safe betting that your name will be the centre of a cyclone of profanity. You may lose that man's trade. This will never occur if you sell only our high-grade hickory handles

Which will not Break

upon the slightest hard use. They are made to be used. They will stand a lot of abuse. The very **cleanest, toughest hickory** is what they are made of.

FREE

We want to send you our list of Goods, Prices and Discounts, free.

J. H. Still Manufacturing Co.
Limited

ST. THOMAS, ONTARIO

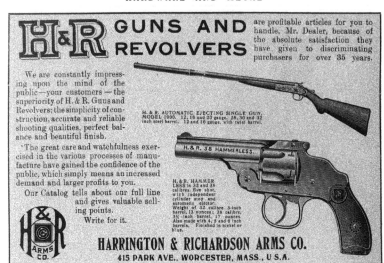

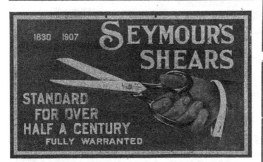

Kemp's Seamless Savory Roaster

Self-Basting *Seamless*

Self-Browning *Sanitary*

MADE IN SHEET IRON AND ENAMEL WARE.

			Per Doz
No. 200	Finest Siemens-Martin Blued Steel,	List Price,	$16.00
No. 400	Finest Siemens-Martin Blued Steel,	List Price,	33.60
No. 200	Pearl Enameled Ware,	List Price,	40.00

Perfectly smooth and seamless, without a corner or an angle, or a corrugation or fold of any kind—just a plain, smooth surface, easy to clean. This is what makes the **SAVORY** an easy seller to the women.

MANUFACTURED BY

The Kemp Manufacturing Co.

Cor. Gerrard and River Sts. Toronto, Can.

The Kemp Mfg. & Metal Co., Limited
111 Lombard St., WINNIPEG, MAN.

The Kemp Mfg. Co. of Montreal
39 St. Antoine St., MONTREAL, QUE.

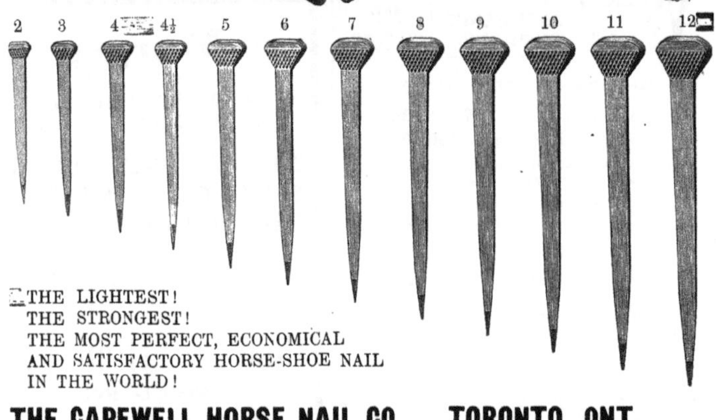

THE LIGHTEST!
THE STRONGEST!
THE MOST PERFECT, ECONOMICAL
AND SATISFACTORY HORSE-SHOE NAIL
IN THE WORLD!

THE CAPEWELL HORSE NAIL CO., TORONTO, ONT.

Branch Offices and Warehouses: WINNIPEG AND VANCOUVER

Matters of Mutual Interest

The Exhibition supplement on heavy coated paper included in this issue will help many merchants who were unable to visit Toronto during the Exhibition to note the progress which has been made, both in the event of itself and in the exhibits of particular interest to the hardware trade. Much expense has been gone to by manufacturers in displaying their newest improvements and Hardware and Metal, through the photographs and written descriptions, has endeavored to give absent merchants as complete an idea as possible, so that the different lines can be compared and the merchant thereby enabled to select what he considers the best selling lines, containing the most serviceable improvements appealing to customers.

* * *

Several special numbers have been published during the past three months. The Western Boards of Trade Convention the Western Retail Hardware Convention, Winnipeg Exhibition, Maritime Provinces Boards of Trade Convention, and the Toronto Exhibition have all been graphically reported, in order that readers in every part of Canada will become familiar with the progress and development of other sections as well as getting a grasp of the problems interesting and affecting merchants throughout Canada.

The summer holiday and convention season being over, more attention will be given in future issues to the departments of the paper which have been receiving minor attention during the hot summer months.

* * *

Attention is drawn to the advertisement on another page offering a prize of ten dollars for the best article suggesting means whereby the hardware merchant can increase his volume of holiday business during the Christmas season.

The offer should encourage many merchants, travelers and clerks to give thought to the problems which will be commanding their attention a couple of months hence and the editor hopes that the offer of this prize (and the spirit of friendly competition) will result in a large number of short letters containing practical ideas on holiday merchandising advertising and window display being received for publication

Let none hang back through indifference or modesty. Anything done to help the trade as a whole will help the doer.

* * *

Incidents are constantly occurring indicating the valuable use the condensed advertising page in Hardware and Metal has been put to by men in different branches of the trade.

The department proved its usefulness years ago and it has now been suggested that its scope be widened by adding an "exchange" heading to those already established. Under the new heading dealers could offer to make exchanges of left over or slow-selling stock for other goods or of store fixtures not now in use for other articles.

For instance, a dealer might advertise that he would exchange a set of tinsmith's tools for some store fixtures, a silent salesman for a set of store ladders, or some cut nails for other more salable goods. Not long ago a Northern Ontario firm spent half a dollar on a small ad. offering to buy a quantity of cut nails and they received about seventy replies. Possibly other stores in the northern district could also take cut nails and give other articles in exchange.

SUGGESTIONS FOR FAIR SEASON.

The time of year has arrived when country fairs, harvest home festivals and carnivals of various character, are in order, and it is important that the merchant should perceive to the full the excellent opportunities which they offer for advertising himself and his business, as well as "boosting" his home town and the surrounding country. There is no occasion during the entire year, the Hardware Trade truly points out, when conditions are more favorable for some resultful advertising on the part of the merchant and he should embrace the opportunity.

In the first place, these celebrations give the merchant a chance to show his fellow townsmen that he is a "live one." He should be prominent in the work of planning and conducting the fair or other celebration and making it a success, and should spare neither himself not his time, for every moment spent in work of this character will bring rich returns. It will teach the people of his town and the adjacent territory that he is a wide-awake, progressive man, full of enterprise and hustle. If he is all that in the public affairs of the town he must be the same in his personal business.

A Golden Opportunity.

During the country fair or the harvest festival people come in from far and near, with money in their pockets and time at their disposal. They are in a happy frame of mind, out to have a good time, and there is no more auspicious time for the merchant to get acquainted and to make himself popular with all. He should be careful not to let anyone get the idea that he regards it as a golden opportunity to lure people into his store and get their money away from them and he should be careful that he is not working along this principle. It is not the right one.

He should devote himself to increasing his personal popularity and that of his store. He should greet the farmers by name, talk to them about their crops and their farms, evince a genuine personal interest in the things in which they are interested, chat with the wives and children, and in general be a "good fellow." He can do this without indulging in any cheap "jolly" or coarse work of any sort.

Store Must Be Attractive.

The store should be made as comfortable and attractive to visitors as possible. A resting place and all other conveniences should be provided for the women and their babies, and everyone should be made to feel at home. There should be plenty of good cold water on tap and it would not be amiss to have a dainty little maid serving free lemonade at a neat little booth in the store.

If the fair grounds are out some distance from the town it might be well to arrange with the local liveryman to run a buss line or carriage service from the front of the store. If it is mentioned in the store's ads. and a large sign is erected in front of the store the liveryman will probably be glad to do this and it will be an excellent advertisement for the store.

NEW WIRE FACTORY AT HAMILTON.

William Holmes has taken into partnership Alexander Donald, late with the B. Greening Wire Co., Hamilton, the new firm style being the Canada Wire Goods Manufacturing Company, which will occupy the premises known as the Young Bros.' brass foundry, 162-168 King William street, Hamilton. The principal lines of manufacture will be wire cloth and ornamental wire work of every description.

On leaving the Greening Company's employ last Saturday Mr. McDonald was presented with a handsome mantel clock by the office staff and foreman of the Greening plant, the presentation being made by D. F. Griffith. Mr. Donald was with the Greening Company for nearly sixteen years, for the past eight having supervision of the office.

PROPOSED BY DEALERS.

By a law which has been proposed in Springfield, O., by dealers in firearms and ammunition, persons who buy such goods in the future will be required to sign their names on a record at the place of purchase, the same as is now required when one buys poisonous drugs at a drug store. It is believed by the dealers and authorities of Springfield that if such a law goes into effect that the careless buying by men and boys of cartridges and revolvers will be lessened, as unresponsible parties will not be pleased with the idea of signing their names and addresses when making the purchase. This will be a protection to the dealers in the way of accusations being made of careless selling of such goods.

HARDWARE AND METAL

Established - - - - - - 1888

The MacLean Publishing Co.
Limited

JOHN BAYNE MACLEAN - President

Publishers of Trade Newspapers which circulate in the Provinces of British Columbia, Alberta, Saskatchewan, Manitoba, Ontario, Quebec, Nova Scotia, New Brunswick, P.E. Island and Newfoundland.

OFFICES:

MONTREAL, - - - - - - 232 McGill Street
Telephone Main 1255
TORONTO - - - - - - 10 Front Street East
Telephones Main 2701 and 2702
WINNIPEG, - - - - 511 Union Bank Building
Telephone 3726
LONDON, ENG. - - - - - 88 Fleet Street, E.C.
J. Meredith McKim
Telephone, Central 12960

BRANCHES:

CHICAGO, ILL. - - - - - 1001 Teutonic Bldg
J. Roland Kay
ST. JOHN, N.B. - - - - No. 7 Market Wharf
VANCOUVER, B.C. - - - - Geo. S. B. Perry
PARIS, FRANCE - Agence Havas, 8 Place de la Bourse
MANCHESTER, ENG. - - - - 92 Market Street
ZURICH, SWITZERLAND - - - - Louis Wolf
Orell Fussli & Co.

Subscription, Canada and United States, $2.00
Great Britain, 8s. 6d., elsewhere - 12s

Published every Saturday.

BUSINESS CONDITIONS.

Our Winnipeg editor reports the outlook for the western crop to be very favorable, as while the quantity will show a decrease the price will be higher. It is also likely that the season of 1907 will see more grain shipped by January 1st than in any previous year. The railways are on the alert and are getting everything in readiness to move the crop quickly. The farmers need the money and will sell as soon as the grain is fit to market. The season is late, but the growth of straw is not rank and threshing will be fast.

There is little danger of drugging the market by rushing the wheat in at an early date, as the world demand can stand all of Canada's wheat and still cry for more. Just what the price will be it is hard to say, but the Canadian Thresherman predicts that it will not fall below 80 cents to the farmers.

In the east, crops have also been affected by the late spring and dry summer, and farmers may have their purse strings tightened somewhat, but reports from the retailers throughout Ontario and the east indicate that the outlook for business is bright and buying will be as heavy as a year ago.

In the larger cities, the tightness of money will affect building operations and incidentally the sale of builders' hardware. With no loans procurable under seven per cent, building cannot be as active as it was when money was available at from five to six per cent. August showed a falling off in Toronto compared with a year ago, but the year as a whole is yet millions ahead of 1906, and the demand for houses for newly arrived immigrants and workmen for new factories is such as to continue building operations at a fairly active rate in spite of the dearness of money.

Chicago reports a falling off of $1,000,000 in building operations in August and is $7,000,000 behind the record for 1906 for the months from January to September. This indicates that something of a depression exists across the border. Canada, however, is in a favorable position to escape the ill effects of a financial crisis, it being the safety valve of the American continent to-day, its vast undeveloped resources attracting hundreds of thousands of immigrants yearly and incidentally creating a demand for the products of Canadian factories and business for merchants both in the vicinity of the factories and in the newly developed country.

TORONTO'S GREATEST ENTERPRISE.

Acting as a magnet to draw visitors from every portion of the civilized world, Toronto's annual fall fair, now known as the Canadian National Exhibition, stands forth as the greatest feature of Toronto's civic and business life and also as an enduring monument to the enterprise and public spirit of the business men who, with few exceptions, have given their time and labor without remuneration.

From being a very ordinary provincial fair the Exhibition has developed into what is admittedly the greatest annual event of its kind in the world. It is greater than many so-called world's fairs and should be a valuable forerunner and training school for the world's fair which will be held in Toronto some time in the not distant future, possibly in 1912.

Originally an agricultural show with live stock as an additional attraction, the Exhibition to-day is as wide in its scope as our national life, industry in all its various phases being represented. Retail merchants, therefore, find as much to interest them in the exhibits as do the farmers who are the real backbone of the Exhibition. The cheap rates on the railways create practically a suspension of business over a wide area and both farmers and merchants enjoy an annual holiday with pleasure and profit to themselves. Wholesalers use the opportunity to get better acquainted with their customers and retailers take advantage of their visit to more carefully inspect goods and get in closer touch with jobbers than is possible at any other time. The Exhibition, therefore, has become a fixture in the lives of all classes of the community.

Being held in a city with a large urban, rural and tourist population to draw from, the directors of the fair are fortunate in being able to year after year show a balance on the right side of the ledger. Rain or shine, a large attendance is assured and, the Exhibition being held practically under municipal auspices, the treasury of the city is always available in emergencies requiring the erection of new buildings or the holding of a world's fair.

Toronto is fortunate in possessing such an attraction as the Canadian National Exhibition and the directors can be trusted to continue the work so ably carried on up to the present, reconstructing old and adding new buildings, creating a new entrance and making other improvements made necessary by the continued development of the Exhibition.

THE EVILS OF SUBSTITUTION.

An interesting campaign has been inaugurated by the large magazines against the substitution of cheap or unknown articles for more widely known and advertised products put on the market by enterprising concerns.

The magazines believe that the advertisers who stand behind their goods and stake their capital and reputation upon the good quality of their goods, are entitled to protection against the imitators and oftimes unscrupulous concerns who market the "just-as-good" lines.

Plumbers, of all business men, cannot afford to encourage substitution, as it is upon the quality of their work and the fixture they install that the future of their business depends. The plumber who makes a practice of installing only the best-known and highest quality of goods, backed by the most careful workmanship, is the one who will win the esteem of the architects and contractors who have the placing of the best business. They, too, are the ones who will secure the best prices and reap the greatest financial reward in the long run.

Substitution has a twin evil—the appropriation of patterns of successful lines made by established manufacturers. This evil should also be frowned down upon by the trade and business given only to reliable manufacturers and supply houses, who show their confidence in the good quality of their goods by advertising them in the trade press.

Canada's National Exhibition

The Exhibition An Important Factor in the Industrial Development of the Dominion—New Records Established for 1907—Many Improvements Over Last Year—Average Attendance More Than 85,000 Daily.

The Canadian National Exhibition, designated by impartial critics as the greatest annual fair in the world, ended its thirteen-day course for 1907 on Saturday, Sept. 7. In those thirteen days it broke all existing records. It established new figures for attendance, and a new high-water mark for cash receipts. It has to its credit one brand new building and some important additions to existing structures. To avoid going too minutely into detail, it may be asserted briefly, without fear of contradiction, that the Exhibition of 1907 surpassed that of any previous year in practically every respect. To regard it from no other standpoint, the Exhibition possesses a value incalculable in the opportunity it affords—first, to the manufacturer, of openly demonstrating to the retailer the methods adopted to produce his goods, and in this way of forming and strengthening a personal bond of confidence between himself and his customers; second, to the retailer, of increasing his knowledge of the goods he handles daily and thus enabling him to represent their quality to his customers with a greater measure of appreciation and intelligence; and, thirdly, to the public, of becoming more intimately acquainted with articles of daily use or consumption which they had hitherto been in the habit of taking pretty much for granted as one accepts the assistance of the sun, moon and stars without worrying particularly as to their origin.

The Process Building, in which these demonstrations took place, ranked in popular favor about equally with that of the Manufacturers', which was crowded with people every afternoon and evening. In the former building, about twenty different articles, ranging from bicycles to bread, were shown in process of manufacture. In the building set apart for manufacturers, the number of exhibits was legion. One hundred and fifty would-be exhibitors were unable to obtain space for their booths, and the committee were obliged to refuse their applications for this reason. There were more than 120 different exhibits in the Manufacturers' and Liberal Arts Building alone. Counting those in the Implement, Transportation, Dairy, Automobile, Process, and Stove buildings, and in Machinery Hall and Manufacturers' Annex, as well as some forty-odd, scattered about the grounds, there must have been close upon 400.

Every feature of Canadian life, agricultural, industrial, educational or artistic, was covered by the Exhibition, which, though owing its origin and development solely to Toronto, is, of course, in scope and representative character, broadly Canadian. Its influence as a factor in Canada's prosperity was expressed very succinctly by Earl Grey in his opening remarks. He said: "The object of Canadian manufacturers and producers through their processes of manufacture and the methods they employ for marketing their goods, must be to associate the name of Canada with high quality. The Exhibition, in the success which has attended it, is an excellent illustration of the advantages a country derives where the Government and the people pull at the same end of the rope and work together for the common good. If the people and the Government work together and so secure to buyers the knowledge that the goods they buy are of a high quality, I am sure there is no limit to the prosperity which awaits the people of Canada. Let our inspections be thorough, and let our condemnation of inferior articles be absolutely merciless. No man is such a blackleg, in my opinion, as the man who endeavors to pass as sound, unsound articles, and secures an inferior reputation in the markets of the world."

The newly-erected Agricultural Building, one of the largest and finest on the grounds, containing a wonderfully comprehensive selection of fruits and vegetables from all parts of Canada, was a feature of this year's exhibition. Another much needed improvement, and one which was rendered necessary by last year's fire, is the erection of a new grand stand, which has a seating capacity of upwards of 5,000. It was filled to overflowing many times during the course of the Exhibition. Two improvements which will probably be instituted by next year, are a new transportation building, to replace the one burned last year, and a more extended street car service. The cars were unable to handle the enormous crowds that thronged the Exhibition, and next year it is likely that an eastern extension will be arranged. President W. K. George, during the course of some remarks at the concluding luncheon, said that the present entrance to the grounds appeared too small and cheap by comparison with the splendor that lay within its gates, but that next year they hoped to have this remedied by the erection of a new and more suitable entrance at a different part of the grounds. Mr. George also recorded the interesting fact that the increased receipts this year were attributable entirely to the enlarged grand stand accommodation and other receipts within the grounds, the gate receipts being actually less than last year.

The Process Building from the North—The Stove Building, the "Mecca" of all Hardware Merchants is the right hand corner section shown in this view.

The Display of "Treasures."

The D. Moore Company, of Hamilton, probably the oldest established stove manufacturing house in Canada, the business being commenced in Hamilton in 1827, had a display in the Stove Building which was undoubtedly one of the finest permanent exhibits on the grounds.

The beautiful Art Treasure, is, without exception, the most beautiful heating stove manufactured to-day, being equipped with all the latest improvements, and having as a chief feature, the taking of cold air from the floor and distributing to the apartments above. This feature equally appeals to customers, who see the advantage of saving of fuel accomplished by this plan. The stove is a double heater and self-ventilator, having a screw draft in the front, which is so adjusted that it can be regulated to a nicety. The shaking ring and duplex grate is something of the latest design and is to be seen to be appreciated. In the Art Treasure there is no complicated mechanism to get out of order. The Empire Treasure differs in design from the Art Treasure, but possesses all the features of the Art so far as improvements and operation is concerned, it also being a double heating stove. The Crown Treasure is of the same class as the Art and Empire, is a trifle smaller, but the large sales of this stove indicate the merits it contains and the popularity it has won throughout the Dominion It is claimed that more Crown Treasures are in use in Canada than any other heating stove The smallest self-feeding stove manufactured in Canada is the Ruby Treasure, made with and without ovens.

The Treasure Heater is the largest, strongest and handsomest heater made, absolutely air tight, fitted with duplex grate and shaking ring, double register draw centre grate for wood, the steel leg base that never breaks (used in this line only), large double feed door (which admits using large pieces of wood or soft coal). Perfect control of the fire is ensured by this use of their patent circular draft register. The hot blast consumer effects perfect combustion of gases arising from the coal, and gives the most intense heat from the cheapest and smallest amount of fuel.

Among the cheaper lines of heaters shown by the D. Moore Company was the Oak Treasure, which this season is fitted with a shaking ring and duplex grate.

In ranges, the D. Moore Company made a large display, among which was noticed their latest addition to the steel line, The Sovereign Treasure, which is admitted by many to be the finest steel range manufactured in Canada to-day. The body is composed of two linings of steel with an asbestos inter-lining. One of the latest features it contains is the Top Lift, patented by the Moore Company, which enables the operator to elevate the front part of the key top to any angle, which is convenient for toasting, broiling, or adding fuel. The fire box is of special construction, and so arranged that the duplex coal grate can be drawn out at the end of the fire box and rock wood grate substituted in one minute's time. Each range is supplied with coal and wood grates. The oven is large and commodious, and each range is supplied with nickel legs or a cast base, as desired. The Sovereign Treasure is furnished with screw draft in the end ash pit door, and also has a large draft damper in front of the range, which insures even burning of fuel in the fire box from one end to the other. It was also exhibited with an oven thermometer and all the latest devices. The Premier Steel Range has a cast-end reservoir and is made similarly to the Sovereign, all the body being lined with asbestos between the steel sheets. Within the last two years the Premier has made itself a favorite in many Ontario homes. The Western Treasure is similar in construction to the Premier, but plainer in its dress. It has an equally large oven and will burn anything in the fuel line. Also on exhibition was the Domestic Treasure Steel Range, nickeled, and possessing all new features of the Sovereign and Premier ranges. It is claimed to be the cheapest steel range of its class manufactured in this country. The Sunset Steel Range, so well and favorably known, was also included.

Something new was noticed in Natural Gas Heaters, the Charm Treasure, made in two sizes, being a feature of the display. This stove has a very attractive appearance, it being placed on the market only last Fall in the gas burning districts. The large orders booked from dealers speak well for its future.

The exhibit was in charge of Mr. J. W. Parrish, eastern traveling representative, ably assisted by Archie Denny, western traveling representative, and by Colin Munro, who covers the northern part of Ontario for the company.

Canadian National Exhibition—The D. Moore Company's Display of "Treasures."

Peninsular Stoves and Ranges.

A great advance over any previous year was made by Clare Bros. & Co., of Preston, they having secured one of the finest locations in the Stove Building, immediately to the right of the main entrance. In keeping with the progress made, a most artistically arranged exhibit was enclosed in the beautiful new booth finished in Antwerp with parallel columns with gold lettering. The company exhibited several new lines this season, as well as their established ranges and heaters.

The Grand Peninsular, which has reduced as much as possible in order to combine beauty with plainness, is removed and differs from the former efforts of the majority of firms attempting this style of ornamentation, in that it is always found to fit perfectly. The base is of a massive design, perfectly free of embellishment, and has been most admired of all for its artistic lines. These stoves are sold with or without thermometers.

The Peerless Peninsular promises to duplicate its success of last year when it makes its initial bow to the stove trade of Canada. The Peerless contains won a high place in the estimation of the trade. In heating stoves the Elegant Base Burner occupied a prominent place on the floor, and found many admirers among hardwaremen and others who inspected the different lines. They contain all the features which commend themselves to dealers and customers, and their popularity is constantly growing.

At the end of the booth several Hecla Furnaces were also included. The Hecla was placed on the market four or five years ago, and the claim that it was superior to anything in the line of hot

Canadian National Exhibition—The "Peninsular" Display in Clare Bros. & Co.

been credited with being the most finished product of the stove foundry turned out in the last ten years, is the result of Canadian brains from the time the designs were sketched until the finished article was placed on the market. This range is fitted with a very heavy iron box, easily removed, has two shaker bars, the latter being removable through the side of ash door. The oven is of steel construction, fitted with a slide oven rack, fitting neatly into retaining slides, easily permitting the raising or lowering of shelf. All the nickeling, which, by the way, has been all the good features of a steel range, viz., drip oven door, key plate top, etc., but still occupies a foremost position among cast iron ranges. During 1906 it proved to be a very popular stove for city trade, and is unrivalled as to working ability.

The Home Peninsular, the stove making its initial bow to the public at this year's Fair, also deservedly attracted considerable attention.

The Royal Steel Range, introduced a year or so ago, was another feature of the exhibit, which commanded much attention, it having proved its worth and air heaters, well constructed, still holds good, experience having proven that the claim has facts and truth behind it.

Hecla Furnaces have several features which cannot be used by any others. The cast iron dome and steel radiator is put together with a patent fused joint which does away with all bolts, rivets and cement, thereby making it absolutely gas and dust proof.

Clare Bros. & Co. invite the trade to communicate with them regarding the agency for either stoves, ranges or furnaces manufactured at their recently enlarged foundry plant at Preston.

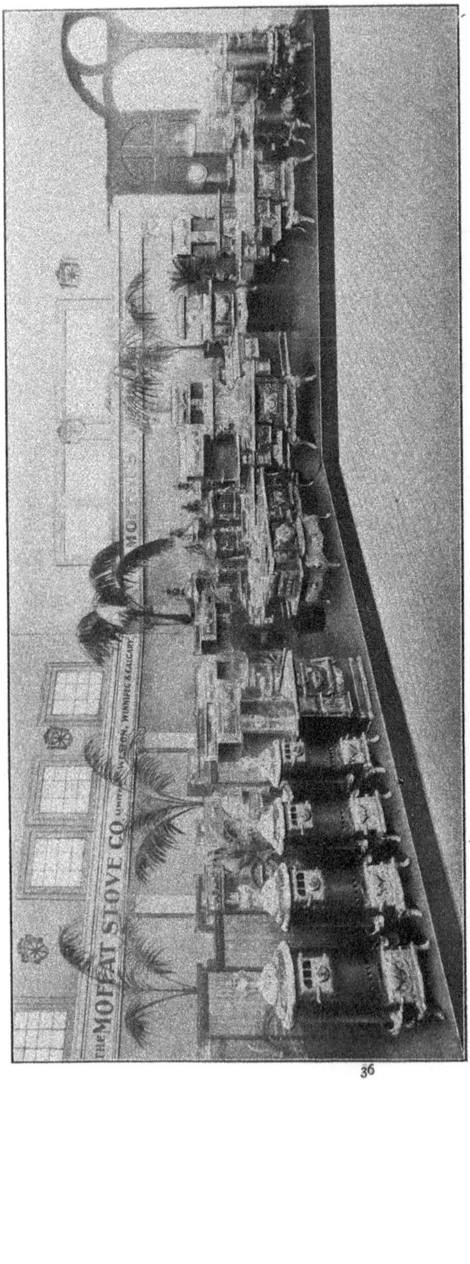

Canadian National Exhibition—The Exclusively Stove Display Made by the Moffat Stove Company.

Moffat's National Stoves.

Located in the accustomed position just inside the main entrance to the Stove Building, was the Moffat display, bright and neat in its arrangement, even more so than in previous years. Situated in the corner where visitors most naturally pass first on entering the building, the display requires to be kept at a high standard, in order to attract and hold attention, and it can truly be said that this year's exhibit "made good" in this respect. The beautiful nickel ornamentation on the ranges and stoves, lit up by the afternoon sun or the evening's artificial electric lights, made a showing worthy of the high standard of the "National" line.

Every year the Moffat Exhibit is the gathering place of the numerous dealers who are selling the famous National Stoves. For the last sixteen years, one or two of the members of this firm have made a practice of being on hand every day at the National Exhibition, for the purpose of meeting old friends and customers and having a talk on stove matters in general. Their popular and able traveling representatives were ever ready to welcome and see that every one had a good time, not forgetting of course, the important matter of orders.

There is no better place in Canada than Toronto Exhibition to see the different styles of stoves and ranges and make comparisons. The leading feature of the Moffat Exhibit was their "Canada B" Steel Range, with the new steel top and other patented and registered improvements. The first steel top on a Canadian make of coal range was exhibited by the Moffats at the 1906 Fair. Since then some others have appeared on the market, but Moffats are the specialists of the Canadian stove trade and may be expected to continue to lead, making nothing but one line of stoves and ranges, and keeping closely in touch with everything pertaining to stoves all over the world.

For a full description of the "Canada B" and other specialities, send for catalogue No. 11, a copy of which will be mailed any dealer who means business.

It has been the custom when a Canadian stove manufacturer wished to get a "move on" and have a new dress on his stoves, to take a run over to the United States, pick out a set of patterns and put it on the Canadian market, without a thought as to its adaptibility for Canadian conditions.

The Moffats have given up this method in the "Canada B" Steel Range and other specialities, and have produced a steel range, the original drawings, design, clay models, dies, etc., of which were made in the Weston factory, by and under the personal eye of some member of the firm, so they are in a position to offer to the stove trade, an all Canadian product, designed for Canadian homes, climate and conditions.

Already this year, some of their enterprising customers have disposed of two and three car loads of this leading Canadian range, and one firm writes, stating they expect to sell six more car loads between now and Christmas.

How is this for business these days, with money tight and uncertain markets? If you want something to help your stove trade this year and give you a reasonable profit, write the Moffat or the branches at Winnipeg or Calgary. Stove Company, Limited, Weston, Ont.,

Canadian National Exhibition—The Gurney, Tilden Display of Stoves and Furnaces.

Souvenir Stoves and Ranges.

The Gurney-Tilden Company, Limited, Hamilton, had a magnificent display of modern stove construction, in their permanent booth in the Stove Building. This celebrated line of stoves are known for their great conformity of design, elegance in finish and appearance, and their wonderful working qualities, both as regards heating and cooking. Every stove in the exhibit was a sample from stock, and just the same in finish as those shipped from their warehouse to the trade. Many were fitted with a nickel-plated base, the company supplying this on some of their makes, without extra charge.

In the cast iron range series, the Champion and Domestic Souvenirs, 4 and 6 hole ranges, were displayed, with patented aerated ovens, the latest and most useful invention made in range construction in recent years. The 1907 series are fitted with full nickel legs or base with draw-out grates and automatic oven shelf, which draws out the pan of biscuits or roast beef, if so desired when opening the oven door. They can be fitted with handsome steel high shelves or warming closets as well as with hot water reservoir or waterfront for attaching to kitchen range boiler. The "Model" Souvenir four-hole range is a handsome medium priced range, possessing all improvements which go with the other Souvenir Ranges but being a little plainer in finish.

In the steel range class, the "Royal" and "Supreme" six-hole ranges indicate their class by their names, having every essential point known in good steel range construction which the trade can safely recommend to their customers. The reservoir is constructed on an unique plan, which allows its removal in an instant.

The Art Souvenir Base Burners are made in three sizes, the middle having an oven attachment.

In "Oak" heaters, the Gurney-Tilden Company make three different sizes, viz: "F," "G," and "H," which enables the dealer handling these goods to supply his trade with any style and price of stove that they want. Special mention may be made regarding the "H" series. This is an entirely new production, and was appreciated by all visitors who saw the series at the exhibit. It has the combination revolving and duplex grates, deep ash pot, foot rails, nickel wings, nickel bands, etc., which go to make a very handsome heater.

The "New Idea" coal burning furnaces shown at the exhibit are the "acme" of furnace construction, the one-piece flange fire-box combining the heating power of two fire pots in one. Other features contained in the "New Idea" are the combination revolving and duplex grates, extra large radiating surface and warm air chambers. The "New Idea" has placed itself in the front rank with the heating trade of Canada, and being fitted with the only grate that will take the ashes away from the fire pot without disturbing the centre, has met with an exceedingly large sale throughout Canada

Bowser Oil Storage System.

Probably no display at the Toronto Exhibition attracted more attention from hardwaremen this year than the exhibit of Bowser Oil Tanks in the Process Building. Several expert salesmen were present at all times, and they had few idle moments, large numbers of hardwaremen, grocers, automobile owners and factory users, being constantly present, studying the advantages to be derived from the use of Bowser Tanks, which have now become such a standard article in the oil and gasoline trade throughout Canada. The large new factory established by the Bowser Company last October, in the vicinity of the Toronto Exhibition grounds, was also visited by many dealers who desired to get in closer touch with the men who are providing the trade with such an economical, clean and money-saving system of handling all classes of oils. The factory since its establishment less than a year ago has been kept in operation to its fullest capacity, in an endeavor to keep pace with the enormous demand for Bowser Tanks.

A representative of this paper spent an interesting hour at the Bowser display, and learned much regarding the Bowser system, and was greatly impressed by the interest manifested by retailers of oil, who already knew much, but desired to learn more regarding oil storage by the Bowser system, and spent considerable of their time while at the Exhibition in gathering additional information. W. R. Hance, the manager, kindly outlined the remarkable development of the business, which was established about twenty-two years ago. In less than a quarter of a century the business has grown until, to-day, Bowser Oil Tanks can be found in all parts of the world wherever oil is used by enterprising men, who desire to handle it in the most economical manner. Competition to-day forces every business man to adopt every device which reduces cost of fire risk, or tends to save time, labor, or material. The Bowser oil equipments accomplish this, and are so complete in detail, and so simple and comprehensive in construction, that they win the approval of every handler of oil who is sufficiently progressive to adopt modern methods of doing business.

Mr. Hance also informed our representative that his company had recently issued a book entitled "Plan Drawings of Model Storage Systems," which they are placing in the hands of everyone interested in handling oil economically.

There is absolutely no waste in using a Bowser outfit, and the saving made in consequence, is sufficient in a short time to fully cover the original expense of installing the system. The purchase of the Bowser outfit is, therefore, one of the most profitable investments that can be made in the way of store fixtures, this being borne out by the experience of thousands of merchants.

The Bowser Oil Storage outfit is the best system for handling oil out of the store and away from it beyond the distance required by the insurance companies. The company places outside, at a suitable distance from the building, a tank holding any desired amount, generally about twenty barrels. The pump is placed at the most convenient point on the store floor, and connected with the tank by a $1\frac{1}{4}$-inch galvanized iron pipe. The pump is a combination suction and force pump, built entirely of metal, is equipped with all the latest improvements, and measures an accurate gallon, quart or pint, at a stroke.

The handling of paint oils, varnishes, and other non-lubricants, has always been more or less costly, because of the large amount of oil wasted by the dripping faucets and by using measures and funnels, causing besides the waste of oil, time and labor, a dirty and disagreeable oil room. Bowser Measuring Outfits handle these oils in an economical, clean and convenient manner, and are guaranteed to handle all classes of heavy, gummy oils and liquids easily and satisfactorily. They positively will not "gum" up. The pumps are so constructed that all working parts are always submerged in the oil, and hence, never being exposed to the air, never become gummed and always work easily.

The company has a corps of able salesmen covering every part of the country, who have been kept busy during the past year in installing several thousand new outfits made by the Toronto plant, hardware dealers and grocers everywhere finding it to their advantage to install a system of handling oil which not only saves time and money, but has the unqualified approval of all fire insurance companies.

Canadian National Exhibition—Display of S. F. Bowser & Company, Toronto.

J.M.T. Brass Goods.

A brilliant display of electrical fixtures was made by the James Morrison Brass Manufacturing Co., Toronto, in their old position to the right of the central entrance of the Machinery Hall. Hundreds of beautiful white and colored lights, in artistic brass fixtures, for use in library, drawing-room or hall, made a beautiful picture, and attracted the attention of every visitor. The crystal drawing-room fixtures in their novel and artistic designs, are worthy of special mention, the most beautiful of all, however, being the special "dragon" design of lighting fixture, for hanging from the ceiling over a dining-room table. Its massiveness and artistic beauty, made it stand out from amongst the rest, and innumerable expressions of surprise were heard from these who studied the display.

The second feature of the exhibit, was that of plumbers' supplies, a full line of bath-room necessities being shown, amongst which were noticed a beautiful bird's eye maple low down closet; an English porcelain lavatory, with a very large basin; Acme instantaneous water-heaters with and without shower attachments; a massive French bath, with a unique waste drain in the centre; porcelain handle stop valves, and handsomely decorated throughout; a marble lavatory designed especially for the Morrison Company, and in the opinion of many, the handsomest ever made in Canada. Another feature of the display was a low-down closet, having a glass front in the tank, the closet being in operation with a patent high-up ballcock with reversible seat, and shut-off cock inside tank, to take the place of shut-off valve on supply tank. This patent is a new feature which makes repairs easy, and renders unnecessary the tearing down of the whole tank. Two closets were also fitted up with the "Nethery" flush-valve, working under direct and tank pressure.

In connection with the plumbing display, a full line of bath-room fittings were also shown, including such lines as soap holders, towel racks, sponge and tooth brush holders, and other toilet fittings.

The third feature of the exhibit, was the display of J.M.T. globe and angle valves, steam gauges, locomotive mountings, and a full line of marine work. Included in this display were also samples of Hancock inspirators and J.M.T. locomotive injectors. The Morrison Company have for many years been engaged in making high-class plumbers' and engineers' brass goods, and the steady growth of their business indicates that the quality of their manufactures are of a very high standing.

In addition to the names mentioned by Hardware and Metal last week, the following hardware merchants were noticed in the city toward the close of the National Exhibition: D. McConnell, Hillsburg; G. F. Johnston, Norwood; F. W. Lee, Enniskillen; R. H. Carson, Gorrie; H. Gilles, St. Jacobs; W. W. Leonard, Lakefield; J. Nasmith, Lotus; J. W. Hambly, Picton; R. C. Fair, Bancroft; J. G. James, Earnbray; J. McArthur, Priceville; J. R. Lougheed, Gore Bay; Wilson Bros., Balsom; D. Shiller, Cooksville; E. Pillow, Chapleau; Bean & McKay, Otterville; A. W. Crosby, Goodwood; S. L. McCabe, Lotus; W. H. McDougall, White River; A. Scott, Mono Centre; W. M. Deverell, Rathburn; G. H. McColl, Victoria; F. Killog, Carmel; F. Richardson, Maple; D. Bell, Bellview; J. A. Henderson, Brampton; J. Webb, Allandale; James Dandy, Streetsville; W. J. Scott, Mount Forest; A. C. Clemens, New Dundee.

Canadian National Exhibition—The James Morrison Co.'s Display of Plumbers' Supplies, Lighting Fixtures and Engineers Brass Work.

Canadian National Exhibition—The Pease Foundry Company's Beautiful Display of Heating Apparatus.

Pease Economy Furnaces and Boilers.

The annual recurrence of Canada's National Exhibition is an unfailing sign that summer is nearly over and that one and all must begin preparations for cold weather. For this purpose the exhibit of the Pease Foundry Co., Limited, was peculiarly well fitted and their comprehensive display of Economy warm air furnaces, Economy combination heaters, and Economy steam and hot water boilers, together with several lines of registers and radiators, was one calculated to set people thinking.

A glance at the complete line of "300 Series Economy" warm air furnaces ranging from their No. 308, which has capacity for warming small residences, etc., to their No. 318—the largest warm air heater manufactured—which is adapted for use in churches, schools and large buildings, was sufficient to convince the most sceptical that "Pease goods" are made to last a life time and that only the best materials are used in their construction. The large double doors, special gas burner, automatic gas damper, two-piece fire pot, long fire travel and ample air space around and between extensive heating surfaces called furth favorable comment from all comers.

The Economy hot water and steam boilers attracted considerable attention. They have many special features, not contained in boilers of other construction—the fire pots are extra deep, corrugated on the inner sides and have radial arms, which increases the fire surface one-third—the sections are connected with cast iron push nipples, doing away entirely with the use of rubber packing, the central water columns make rapid circulation certain—the oscilating grates are connected to shakers so arranged as to permit of shaking without stopping. There are also many other important details in these boilers, which are worthy of consideration. They are made in twenty-four sizes for use in warming residences, stores, etc., and range in capacity from 275 to 2,700 square feet of water radiation and from 200 to 1,650 square feet of steam radiation

An Economy hot water and warm air combination furnace, which cannot be excelled for use in building with certain rooms hard to reach with warm air only, a "Victor 100 Series" warm air furnace, a high-class heated capable of holding its own with the ordinary furnace on the market and selling at a small price, complete line of registers and faces for both sidewall and floor finished in black Japan, oxidized or nickle-plated, and an assortment of radiators in different heights and designs completed a display of heating goods.

The Pease Company maintain a drafting department for the benefit of customers and are always prepared to furnish plans of heating plants for use when installing systems that are a little different from the ordinary run of heating plants. A complete set of blue prints of a heating and ventilating apparatus now doing good work in the Northwest Provinces, was an interesting study.

Plumbers' Supplies.

The James Robertson Company, Toronto, had a decidedly neat exhibit near the middle entrance to the Process Building, the same location as the exhibit made by the company last year. Improvements had been made, however, by the addition of an artistic roof to the two booths, and the display, therefore, was enhanced in beauty and general appearance.

In the bath room were samples of the most modern fixtures, a highly decorated Premier bath on a solid base being the centre of attraction, and fitted above it was one of the latest patterns of shower baths, while modern fittings were used throughout. A magnificent Copley lavatory was the second feature in the room, the lines of its design being very artistic, while combining all the features of efficiency in operation. In place of the Sitz bath displayed a year ago, a serviceable foot bath was shown, the fixtures being rounded up by the addition of a beautiful Naturo closet with flushometer connections. The room was well lighted and contained every requirement for the modern bath room.

The main display in the corner exhibit had as its chief feature many different designs of beautiful lavatories, including several marble lavatories in artistic new designs produced in the Robertson Marble Works. These contained permanent fixtures, such as shelves and brackets for various articles used in the lavatory. The different types were all greatly admired, but none more than two pedestal vitreous white china lavatories also on display.

On the floor were also beautiful samples of highly decorated Yale, Occident and Gladstone enameled baths and nickeled fittings. The outside decorations attracted much attention from visitors to the exhibit. One of the latest models of kitchen sinks was also shown, with both enameled iron and wooden drain boards fitted with rubber mats. This feature of the display attracted the eye of many a housewife who has older and less sanitary fixtures now in use.

The closets on exhibition included the Naturo, Acme, Acme Centripetal and Fleur de lis. all being fitted with Robertson's centre bush fittings and top supply ball cocks. With this modern apparatus for flushing closets, nothing can get out of order. The operation of the apparatus is exceedingly simple. As will be seen from the picture, the push button is located in the centre and near the top of the tank.

The Marble Bathroom in the James Robertson Exhibit.

Canadian National Exhibition. Exhibit of James Robertson Co. Ltd.

Gillett's Lye Eats Dirt.

There was no dearth of artistic exhibits in the Manufacturers' Building, and a booth which comes under that heading alone failed in the purpose for which it was intended. More preferable would be a display designed entirely from a practical standpoint, and devoid of all features pleasing to the eye of the general public. It would at least have a certain distinction that would catch the attention of interested parties. The aspiration of every exhibitor who has studied the matter is to combine these artistic and concrete qualities. E. W. Gillett Company have come very near to attaining the point of perfection in this respect, with the result that their showing ranks high among the best in the building.

Canadian Eational Exhibition—The E. W. Gillett Co.'s Display of Baking Powder and Lye.

The accompanying illustration gives a much clearer idea of the arrangement than could possibly be written. There is no superfluous decoration—everything is an advertisement for Gillett's Lye, Magic Baking Powder, etc., made doubly effective by being well displayed, and the booth brilliantly illuminated.

Now-a-days, when there is an ever-increasing demand for a good cleaner in every household and in every trade, it behooves the hardware dealer to be on the lookout for that article. Lye, when not adulterated, is the best all-round cleaning agency that is known, and is one of the best paying lines that the modern hardware dealer can handle. When one thinks of its many uses it practically appeals to all classes. It is as much used in the farmer's orchard for spraying his trees as it is in the home for removing old paint from floors, doors, walls, etc., in fact the housewife finds many uses for this valuable cleaner. Machinists, foundrymen, engineers, nut and bolt makers, brewers, and bottlers, also find it valuable as a cleaner.

The following is a complete list of E. W. Gillett Company's lines: Gillett's Perfumed Lye, Gillett's Washing Crystal, Gillett's Cream Tartar, Gillett's Caustic Soda, Royal Yeast Cakes, Magic Baking Soda, Magic Baking Powder, Imperial Baking Powder and Cream Yeast Cakes.

The Gillett factory building (to which a $20,000 addition is now being made) has a frontage of 80 feet by a depth of 300 feet, facing on three streets. On the fifth floor is located the baking powder room, with its large revolving mixer, taking a ton of material at a time, and also the paper box department, where in the neighborhood of 100,000 complete packages are turned out daily. On the fourth floor is the machinery for making Royal Yeast, including the machines, which each turn out 1,460 yeast cakes every minute. Drying rooms, with a capacity of over 15,000,000 yeast cakes, also take up part of the fourth floor. It has been calculated that at the lowest estimate possible, each batch of yeast which the company turn out will make 2,600,000 loaves of bread, and from four to six batches are turned out per week the year round. A striking feature is the fine arrangement of every detail looking toward the careful and expeditious preparation of the goods.

Here are a few of the many things Gillett's lye is useful for:

For family soap making.

For washing dishes.

For softening water.

For disinfecting sinks, closets, drains, etc.

For cleaning and sweetening milk cans, pans and cheese utensils.

For photographers' and machinists' uses, foundrymen, bolt and nut makers.

For engineers as a boiler cleaner and anti-incrustator.

For brewers and bottlers, for washing barrels, bottles, etc.

For painters to remove old paint.

For washing trees, etc.

For use instead of Sal Soda.

For killing roaches, vermin, etc.

For house-cleaning, scrubbing, cleaning chambers, cuspidors, bath tubs, tile floors, etc.

Used extensively for scrubbing floors of theatres, churches, street and railroad cars, hotels, decks of boats, steamships, etc.

Canada Metal Company's Display.

Hardwaremen and plumbers found much to interest them in the exhibit of the Canada Metal Company, Toronto. in Machinery Hall, where, under the management of W. G. Harris, an extensive display of metals in ingot and manufactured form was atractive.

In the background a good showing was made of lead traps and bends, and lead and waste pipe and babbitt metals, samples of practically all their many lines handled and manufactured being shown. A specialty was made of their Imperial Genuine Babbitt Metal, as well as one of the latest additions to their product, that of lead washers, used largely by tinsmiths.

The wide extent of the company's business, covering so many varied lines, caused many to express surprise, few realizing the wonderful expansion of Mr. Harris' business since its establishment a few years ago.

In order to give the trade a complete idea of the enormous volume of business transacted, and the rapid growth of the company, the following partial list of metals they manufacture and supply is given:

For Plumbers: Lead pipe and lead waste; Hydraulic Drawn Traps; Non-Siphon Centrifugal Cast Trap; Strictly Bar Solder; Star Extra Wiping; Acme Wiping; Brass Ferrules, tinned; Iron and Lead Combination Ferrule Bends or Spun End Test, and Sheet Lead.

Machinery Metals: Imperial Genuine; Imperial Tough; Armature Special; White Brass; Grayburn Street Car; Metallic Genuine; Harris Heavy Pressure; Hercules Genuine; White Bronze; Star Frictionless; Aluminoid; No. 0, 1. 2, 3, 4 Babbitt, Cotton Waste and Sutton Boiler Compound.

For Brass Founders: Phosphor Tin; Phosphor Bronze; Zinc Spelter; Pig Lead; Ingot Copper and Antimony Ingot Tin.

For Electricians: Fuse Wire; Round or Flat, any size; Battery Zincs; Pencils, Crowfoot or any Design, and Wire solder.

For Tinsmiths: Guaranteed Half and Half (best made), Strictly Brand Half and Half (most used), Commercial Half and Half, and Lead Washers (this is a new specialty).

For Builders: Lead Sash Weights.

The company makes a specialty of Rolling Britannia Metal; Coffin Plate Metal; Pure Sheet Block Tin and Sheet Lead; also Galvanizing Cast, Wrought and Steel Iron; Canada Plate or Corrugated Iron. Nails, Chain and Tinning Copper, Brass and Iron. They also make castings from patterns in brass, copper, bronze, aluminum and lead, and always carry in stock Pig Lead; Antimony; Sheet Lead; Brass Brazing Spelter; Aluminum and Copper in Crucible shape; Pig Tin; Ingot Copper, and Bismuth.

Canadian National Exhibition—Canada Metal Co.'s Display of Babbitt, Solder, and Other Metals.

Splendid Silverware Show.

Hardwaremen who handle silverware found much to interest them in the Standard Silver Co.'s exhibit in the centre of the Manufacturers' Building where is located the beautiful reception room maintained by the company, the head of which is W. K. George, also President of the Canadian National Exhibition. This reception room is a convenient meeting place, where hardwaremen desirous of gathering ideas regarding holiday silverware can do so to good advantage.

The reception room contains little in the line of display, although the windows are neatly dressed with attractive wares. Resting on a table inside, however, was a magnificent epergne, having three arms fitted with detachable gold-lined bowls for receiving fruit, with a large vase rising up from the centre for a bouquet of flowers.

In one window were a large number of prize cups, this being one of the special lines to which the Standard Silver Co. devote particular attention, and dealers who cater for business of sporting clubs had much to study in this window display. In the other window was a large variety of table ware, including cake baskets, candlesticks, teapots, sugar bowls, pickle dishes, card receivers, fruit bowls, butter dishes, and other similar articles. Two articles which attracted particular attention, however, were two nut bowls, which, on inquiry, were found to have been made by hydraulic pressure only, the flower art work being finished in gray with the centre burnished bright.

Booklets, handsomely illustrated, are supplied by the company to their customers, and these are very helpful in developing business, when sent by dealers to their probable customers before the rush of the Christmas season arrives.

Canadian National Exhibition—Display of Standard Silverware.

Maximum Prismatic Glass.

The Hobbs Manufacturing Company, of London and Toronto made a very fine exhibit of Prismatic Art Glass, in the Process Building. The booth was undoubtedly one of the most beautiful in the building, on the floor being many samples of the beautiful work turned out of their London factory.

Hardwaremen and retailers generally who had in mind the erection of new store fronts, were particularly interested in the display of Maximum Prismatic Glass, the utility of which has been proven by years of successful use. Samples were shown in small square form as well as in large sheets in artistic designs, for decorative store fronts.

Utility and beauty were combined in the rest of the exhibit. From the ceiling hung innumerable samples of beautiful leaded art glass in Art Neveau designs, which had an individuality all its own, and every art lover could find much to enthuse over in this new class of work designed and introduced by the Hobbs Company.

In bevelled plate glass there were also some decidedly attractive designs which attracted the attention of many visitors, the designs being in keeping with the other high-class work shown in the exhibit. There were also some fine samples on exhibit of V-cut work, which were worthy of particular mention. The samples of British Mirror Plate, were also in keeping with the high standard maintained in their lines.

For originailty in designs, the Hobbs Company have a wide reputation, and all who saw their exhibit this year went away with the idea that this branch of Canadian industry is on a par with that of any other part of the world, young, though, the industry is in this country. The Hobbs Manufacturing Company are sole agents for the Maximum Prismatic Glass, which is undoubtedly the best Prismatic glass on the market, being guaranteed to furnish 25 per cent. more light than any other kind. Dealers are invited to write for samples before purchasing, and hardwaremen are urged to pay particular attention to securing every order placed in their district for Prismatic or decorative glass work.

Canadian National Exhibition—Hobbs Mfg. Co.'s Display of "Maximum" Prismatic and Leaded Art Glass.

Maple Leaf Binder Twine.

The binder twine display made by the Brantford Cordage Company, in the Process Building, was in the form of a pyramid of bundles of twine, as will be seen by the accompanying illustration. The distinguishing features of first-class binder twines are very evident in the Brantford Cordage Company's four Maple Leaf brands: "Gilt Edge" is a 650-ft. pure Manilla twine, and in quality it is all that its name signifies. Thousands of farmers have certified to this after years of experience with it. "Gold Leaf" brand is a 600-ft. Manilla twine, smooth, clean, even, long and strong, and claimed to be the best twine of its length on the market. "Silver Leaf" brand is a standard Manilla, 550 feet in length, being unsurpassed as a satisfactory general purpose twine, capable of being used on any binder properly adjusted, in wheat, rye, barley or oats. "Maple Leaf" is the shortest brand manufactured being a standard 500-foot twine. For old binders, etc., the "Maple Leaf" gives full measure of satisfaction, and is the cheapest to buy.

Hardware dealers in districts not already represented by Brantford Cordage Company's agents, are urged to communicate with the company, and arrange to handle their twines in future. The high reputation of the company makes certain that dealers will have satisfied customers if Brantford cordage is sold by them.

Canadian National Exhibition—The Brantford Cordage Company's Display of Binder Twine.

Berlin Woodenware.

The display of ladders made by the Berlin Woodenware Company, near the centre of the Process Building, interested every hardwareman who visited that part of the Exhibition. As will be seen by the illustration, dozens of ladders in all manner of shapes and sizes, were shown, some extending to the roof, while others were merely of the stepladder variety.

The exhibit was in charge of D. B. Betzner, manager of the company, and he was kept busy describing the advantages of the celebrated "double truss, high carbon steel wire extension" on their ladders, a feature of which is the steel wire on each rail attached to the ladder, to prevent side swaying, and materially strengthening the whole ladder. Norway and Georgia pine is used for the woodwork and all attachments are of steel.

Canadian National Exhibition—Berlin Woodenware Company's Display of Ladders.

Canadian National Exposition—Display of Spooner's Copperine."

Spooner's "Copperine."

The photo of this exhibit at the Toronto Fair, which appears in this issue, is worthy of more than passing note, because it represents strictly the Hardware Merchants' Metal, which is made for them and handled only by the wholesale and retail men of Canada.

Alonzo W. Spooner is the man who first introduced Babbitt Metal to the store trade of Canada many years ago. Before that the foundries made composition metal, and people had to order from them. Consequently, this Copperine, for this and many other reasons, has a moral claim to the support of the trade. Its patrons have always been fairly and honestly treated, and there are no travelers out selling to merchants and their customers, and appointing everybody agents.

Many improvements have since been made in bearing metals. "Copperine" was got on earth by Mr. Spooner, and it bears the reputation of being the best metal extant. No one questions this fact, who knows or has any practical experience. It having been so long before the competent Canadian engineers, machinists and machinery owners, there can be no doubt about it. The important use of this metal should impress the minds of dealers that they cannot substitute any other metal. None can possibly take the place of "Copperine." "A boy cannot do a man's work," applies to this matter.

The hardware dealers who have been faithful to Copperine have the best trade and make the most money. Considering the quality, Copperine is the cheapest metal to use. Considering nothing but price, the lowest grade price of Copperine, has other Babbitts skinned to death in quality.

Majestic Polishes.

In a tastefully decorated booth on the west aisle of the Manufacturers' Annex, the exhibit of Majestic Polishes attracted a great deal of attention. This was one of the displays that appealed to everyone, no matter what their business or profession.

There are two kinds of polishes on the market, that is, ordinary polishes. One requires an inordinate amount of labor to bring satisfactory results. The other kind, while eliminating this feature, are almost invariably injurious either eating the metal or scratching its surface. Majestic Polishes were evolved, after a great deal of study, to avoid the undesirable features of the general run of polishes, and they have proved remarkably successful.

In Majestic liquid metal polish, only the highest quality of refined oils are used, which are entirely free from all acid and grit. Majestic paste metal polishes are made in three colors—white, red and black—of the same high-grade ingredients as the liquid polish. Housewives generally seem to be prejudiced against a liquid silver polish, and to provide for this, the company have a silver polishing powder.

Majestic furniture polish is another special product, manufactured particularly for high-grade polished wood surfaces. Take a piano, for instance. Most people would hesitate a moment before using an ordinary furniture polish to restore its lustre, but there need be no hesitancy in using Majestic Furniture Polish on this or any other article of the finest furniture.

Canadian National Exhibition—Display of Majestic Polishes.

Canadian National Exhibition—J. J. Turner & Son's Display of Tents.

Turner's Tents.

On the grounds, in front of the new Agricultural Hall, the exhibit of J. J. Turner & Sons, Peterborough, was found of interest by many hardwaremen interested in the display of tents, adjoining "Society Row." The Turner Company claim to be the largest and most up-to-date tent and awning makers in Canada, and their exhibit indicated that they are also the most enterprising, they having a large display of various sizes of tents, as well as the largest "Union Jack" ever made in the Dominion, being 20x40 feet in size, and having 240 yards of bunting used in its construction.

Harriston Stove Company.

The Harriston Stove Company, Harriston, Ont., formerly known as the Canada Stove Works, had one of the neatest displays of up-to-date stoves and ranges, in the stove building. The high standard of excellence as maintained by this firm, is building up a splendid trade for them, and they are steadily going ahead and expanding their business. John E. Cave, the manager of the company, was in personal charge of the exhibit, and stated that his only trouble is that in spite of increased factory facilities, the plant is unable to keep pace with the growing demand for the products of the Harriston Stove Foundry.

No line of stoves on the market show such marked improvement during the past year as the Royal stoves and ranges.

A feature of the exhibit this year was the new range for city trade, being displayed with nickle base and nickle high shelf, square with water front. The new cook stove added to the "Royal" line should prove a very popular seller, it having quality as well as the advantage of cheapness It is made with the top, bottom, front frame, and doors of cast iron, and has a duplex grate with an 18x20-inch oven. Dealers who are looking for something which can be sold for $18 retail, besides allowing a good margin of profit, should look into the advantages of handling this stove.

In the heater line, the Harriston Company have something in a cast-lined, down-draft heater, with swing, top duplex grate, and mica door. This heater is a wonderful fuel saver, and is built to burn corn-cobs, soft coal, hard coal, or any old thing, and will produce surprising results.

Included in the exhibit were samples of most of the various stoves and ranges made in Harriston this line including the "Royal Corona" range, which holds a remarkable degree of popularity with the public for its appearance and good working qualities; the "Royal Champion," a popular coal or wood cook, beautiful in appearance, and economical in fuel; the "Royal Consort," a modern coal cook stove, with large ventilated oven; the "Royal Cook," fitted with pouch feed; the "Royal Palace," a modern wood cook stove, with large fire box; the "Royal Princess," another wood cook stove; the "Royal Pluto," a perfect operating hot blast stove; and the "Gem Oak," a splendid cold weather heater.

Canadian National Exhibition—Harriston Stove Co.'s Exhibit of Stoves.

Empire Stoves and Ranges.

The Canadian Heating & Ventilating Company, Owen Sound, had an attractive exhibit of stoves, ranges, furnaces and registers, which attracted unusual

favorable impression on the women visitors, as the figured range has seen its day, and the busy matron is now looking for a stove or range which will retain its attractiveness without the per-

Range and Walker Success Furnace, received a great deal of attention by the general public, and especially those in-

As fuel economy is the most important question in the heating problem,

Canadian National Exposition—Display of "Empire" Stoves, Ranges and Furnaces.

attention, and kept their demonstrators busy from the opening to the close of the Exhibition each day, in showing to the throng of interested visitors the many points of merit in the "Empire"

range shown this year, especially depetual scouring necessary in the more elaborately decorated and less useful lines. The smooth finish of the Empire line, eliminates all this unnecessary la-

strict attention and years of experience terested in the trade. The Walker Pilot Range is a combination of cast iron and steel, cast iron for those parts subject to fire and wear, and steel for those

Canadian National Exhibition—Display of Walker Stoves and Ranges and Success Furnaces.

line. Robert G. Christie, the Eastern sales representative, and James B. Brownlee, Western Ontario representative, were in charge of the display and made many good sales.

The plain but attractive finish of the Empire Stoves and Ranges made a very

bor. The Empire "Excelda" was a new signed for city trade, or for people desiring a small range, and it was favorably commented upon

Walker's Stoves, Ranges and Furnaces.

As manifested by their excellent exhibit at the Fair, the Walker Pilot

parts which are to radiate the heat. have been given in getting out the Walker Success Furnace.

Write for circulars and prices to the Walker Steel Range Company, Grimsby, Ont.

Ruberoid Roofing.

The exhibit of the Standard Paint Company of Canada, Limited, was a centre of attraction, and a great deal of interest was evidenced by the crowds which thronged their space in the model factory illustrating the application of ruberoid roofing, and the model cottage which clearly showed the adaptability of ruberoid roofing for such purposes.

Ruberoid Roofing maintains the prestige which it has established in all parts of the world. The factory of the Standard Paint Company of Canada, which is located adjacent to Montreal, is a model of perfectness in every respect.

A fact which should appeal to dealers is that Ruberoid is the standard by which roofing quality is judged. It is the only composition roofing for which a satisfactory service of fifteen years can be proven.

Dealers would profit by writing to the manufacturers for full data.

Canadian National Exhibition—Standard Paint Co.'s Display of Ruberoid Roofing.

Island City Paints.

The display of P. D. Dods & Company, proprietors of the Island City Paint & Varnish Works, Montreal, Toronto, Winnipeg and Vancouver, won the admiration of visitors for the attractive arrangement of the various products and raw materials shown by the company. The illustration speaks for itself, and it will be admitted that few exhibits had a more attractive appearance.

Canadian National Exhibition—P. D. Dods & Co.'s Display of "Island City" Paints and Varnishes.

Metallic Roofing Co.'s Exhibit.

The accompanying photo shows what is probably the finest exhibit of sheet metal building material that has ever been made. The workmanship could not have been excelled. This very attractive exhibit clearly shows that sheet metal pressed zinc ornaments. Haye patent metallic lath, ventilators, conductor pipe, eave trough, spun zinc balusters, metallic ceiling, centre pieces, cores, moldings, etc., fireproof wired glass windows with hollow sheet metal sash and frames, "Eastlake" and "Empire" tistic and durable articles, and point out that theirs is the only exhibit of sheet metal building materials that has ever been considered worthy of a gold medal, and this high honor has been given no less than three years in succession. What more need be said?

Canadian National Exhibition—The Metallic Roofing Co., Toronto.

can be successfully used for nearly every portion of a building and produces a handsome, durable, fire and lightning proof finish.

The exhibit included cornices, skylights, finials, embossed fireproof doors, metallic shingles, corrugated iron, both galvanized and painted, straight or curved. Some of these articles are most beautifully decorated in colors and only need to be seen to be appreciated.

The company makes only reliable, ar-

The company's products are exported to nearly all foreign countries and their Canadian trade has increased so greatly that they find it necessary to operate their factories both day and night.

Artistic "Metallic" Building Materials

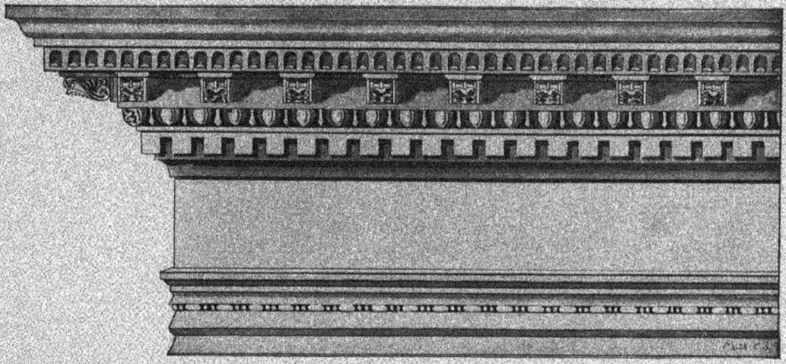

We can produce any style of ***"Cornice"*** to suit your requirements. We know how to do it right. No trouble to erect our cornices because our workmanship is so exact that the parts we send you will fit easily and accurately.

Tympanium Panels are what add a finish to every fine residence. Every builder wants them.

Centre Pieces
that will not drop off like plaster.

Pressed Zinc Ornaments
for all purposes. Hundreds of designs.

Capitals
For either interior or exterior columns. Every size and design that you can think of.

We also make Corrugated Iron, "Metallic" Shingles, Siding, Ceiling, Lath, Doors, Windows, Skylights, Finials, Cresting, Conductor Pipe, and in fact, all kinds of reliable Sheet Building Materials.

HAVE YOU OUR CATALOGUE?

In competition with the world we received:

The Highest Award at Paris, France, 1900.
Gold Medal, Jamaica, B.W.I., 1891.
Silver Medal, Toronto, 1901.
Gold Medal, Toronto, 1902.
Silver Medal, Winnipeg, 1906.
Gold Medal, Toronto, 1903.
Gold Medal, Toronto, 1904.
Bronze Medal, London, 1896.
Silver Medal, London, 1890

22 Years' Experience

The Metallic Roofing Company of Canada, Limited

MANUFACTURERS AND EXPORTERS

Toronto and Winnipeg, Canada

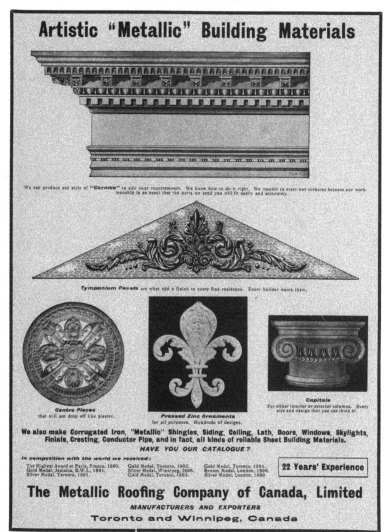

Automobile and Cycle Skates.

The Canada Cycle & Motor Company had several exhibits in the Process Building, the latest and best in automobiles, bicycles and skates being shown by them. The most interesting of all to hardwaremen, however, was the display of "Automobile" and "Cycle" skates, under the direction of Mr. Lambert. Since the introduction of these lines about three years ago, they have come into such prominence with hockey players and skaters, that the hardware dealer who does not carry them in stock is standing in his own light. We understand, however, that few representative hardwaremen in the various provinces where skating is a popular pastime, fail to carry this line in their stock of sporting goods.

The "Automobile" skate, for lightness, strength and beauty of design, cannot be equalled. The tops are made of aluminum, which gives lightness to the skate, strength being added by the runners being made of nickel steel, the highest grade of material for this purpose. The workmanship and finish of the "Automobile" skate, is of the very best, making it a perfect skate, which is guaranteed throughout, and to which has been applied the well-earned title of "Canada's Famous Skate."

The same careful attention has been given to the "Cycle" skate, it being made solely of selected material and under carefully inspected workmanship. These skates will be in great demand during the coming winter season, and dealers not already handling them will do well to communicate with the Canada Cycle & Automobile Company, Limited, Toronto Junction.

Canadian National Exhibition—Display of Automobile and Cycle Skates.

Joy Malleable Ranges.

An interesting exhibit in the Stove Building, and one which won much commendation from other exhibitors, was that of the Joy Mfg. Company, Toronto.

Speaking to Hardware & Metal, Mr. Joy stated that many dealers hold a wrong impression of the wants of their customers the argument of cheapness being used entirely too often. Mr. Joy spent many years of his life in the retail hardware business, and knows whereof he speaks when he emphasizes the fact that dealers should talk quality in stoves, as in other lines of hardware. "Many dealers," said Mr. Joy "have the idea that their customers will not pay the price for a good stove, and, consequently, refuse to handle the highest grade of goods, and sell cheaper lines which allow only a small profit of $5 or $6. They might just as well sell a high-grade Malleable range, giving their customers a stove which will last a lifetime, and making on the sale a profit of two or three times the margin made on the cheaper lines. That the public are willing to buy high-priced malleable ranges, is proven by the large business done by the stove peddlers throughout the country, who have little difficulty in selling ranges at prices which the retail hardwareman imagines it impossible to procure."

Canadian National Exhibition—Display of Ranges, Unbreakable from Fire, Accident, Use or Abuse.

HARDWARE TRADE GOSSIP.

Ontario.

M. W. Howell Goderich, formerly organizer of the Ontario Retail Hardware Association, has returned from a business trip to Western Canada, and is now on a trip to Montreal.

J. A. Page, formerly connected with the tinshop of the McClary Manufacturing Company, has opened a new store at 807 Dundas street, London, and will carry a large line of stoves and furnaces.

Rice, Lewis & Son, Toronto, have taken over the selling agency for the Pittsburg Fence Company, Hamilton, and the traveling staff have been straddling the fence proposition during the past month.

Austin Eaton, manager of the Gurney Oxford Stove Company's stores on Yonge and College streets, Toronto, has resigned and gone to Winnipeg to go into the piano business. J. H. Garbutt has assumed the management of the two stores.

Quebec.

J. McKay, representing the Corbin Lock Co., New Britain, Conn., was in Montreal this week.

D. W. Clark, representing B. K. Morton & Co., Sheffield, Eng., passed through Montreal last week on his way to Newfoundland.

E. C. Dingman, advertising manager for Caverhill, Learmont & Co., Montreal, is holidaying in Western and Northern Ontario.

Robert Munro, managing director of the Canada Paint Co., Montreal, who was paying a short visit to Winnipeg, has returned to Montreal.

George Boyd, one of the sales managers at the Montreal Rolling Mills, Montreal, is spending a two weeks' holiday at Lac L'Achigan, Que.

W. H. Smith, of Robertson, Foster & Smith, the well-known hardware firm of St. John, N.B., accompanied by Miss Smith, was in Montreal this week.

At a special meeting of the Board of Directors of A. C. Leslie & Co., Montreal, held at the offices in Montreal on Sept. 10, for the purpose of filling the vacancy on the Board caused by the death of A. H. Campbell, late vice-president of the company, T. H. Jordan was elected vice-president, and E. H. Copland, director.

Maritime.

Wm. Stairs, the founder of the firm of Wm. Stairs, Son & Morrow, Halifax, commenced business on his own account in 1810, and when the Kidstons left for Glasgow, in 1825, he bought them out and succeeded them. W. J. Stairs, his son, was made a partner in 1841, while Robert Morrow, his son-in-law, joined the firm in 1854, the business having since been carried on under the name of Wm. Stairs, Son & Morrow. In 1900 the firm was incorporated, Edward Stairs being the only surviving incorporator, the others being John F. Stairs, James W. Stairs, George Morrow and Wm. J. Stairs, jr. This data is given correcting a recent article in which some dates in the firm's history were wrongly given.

TEN YEARS OF DEVELOPMENT.

In this age of active opposition in all business, ten years is not a long period in which to establish a sound success. As manufacturers' selling agent and metal broker, Alexander Gibb of Montreal stands out as an especial example in this connection, having just completed ten years of steady development. Mr. Gibb had the advantage of a splendid commercial training in Scotland, his native land, where thoroughness in everything is a characteristic. At an early age he acquired shorthand and his sterling education, coupled with concentration and energy at whatever he was set to do, in a large railway office, soon attracted attention both in Edinburgh and Glasgow. Detail and accuracy became second nature to him.

He received agreeable promotions, but his ambitions turned towards the Dominion, over a quarter of a century ago, so he came to Montreal, under engagement with the G. T. R. Shortly afterwards he took the position of secretary to James Crathen, a leading hardware merchant of Montreal. This place he held with satisfaction for fifteen years. It is not surprising that he learned well the ins and outs of the hardware and iron business, with such an opportunity at his hand. In addition, he was possessed of that disposition that took advantage of every opportunity to learn. So, then, when the Gilbertson Galvanized Sheet Company, of Pontardawe, Wales, sought an agent in Canada ten years ago, Alex. Gibb applied and was highly recommended. He secured this agency, with others, and at once started out for himself. With keen judgment and hard work he built up a fine business which to-day is a credit to all concerned. He has added many other agencies in different lines, including that of J. Beardshaw & Son, Limited, of Sheffield, for "Conqueror" brand high speed steel and high speed drills, which business is now assuming large proportions.

ALEX GIBB, MONTREAL
A Leading Metal Merchant and Manufacturers Agent.

He also sells American chain, representing the Standard Chain Co., now about to erect a factory in Canada, and of which company Mr. Gibb is a director and vice-president. He is also a director of the Meaford Wheelbarrow Co.

He sells copper, brass, wood accessories of all descriptions, English hardware, iron bars, steel, etc., shovels, wheelbarrows, wrought iron and scrap, also dry colors, window glass, music spring wire, etc. These various goods are mentioned to give an idea of the manifold character of a one-man business, developed from crudest beginnings into a property which is earning well for its owner to-day. His business is entirely wholesale and is carried on like clockwork by the judicious use of trade paper advertising, booklets, the mails, travelers, and strong individual effort and connection. Not one branch of the business is left to anyone but Mr. Gibb. He supervises a steel test in the Angus shops one day, attends a special meeting of some board of directors another day, and during almost any day will have called upon personally many of the heads of the largest manufacturing and jobbing houses in the city. His correspondence is tremendous and his stenographers have little time to whistle the latest tunes or to waste in idleness. Work is the watchword of the Gibb offices and warerooms and no customer or probable buyer is left unattended.

There are greater things looming up for Mr. Gibb, as his judgment and organization ability have identified him with large industrial ventures. His motto has been: "Equity and fair dealing, without a single adroit measure," so that both his principals and himself enjoy an almost endless chain of satisfaction all over the Dominion. Every man who orders goods from Alex. Gibb, either by mail or over the 'phone, knows that promptness and rigid fairness will rule throughout. Personally Mr. Gibb allows the cares of life to rest easily upon him. One of his fads is his fondness for well bred dogs, while another is gardening. He likes manly sport and drives a good horse, and, like most Old Countrymen, takes his pleasure seriously. The accompanying picture of Mr. Gibb suggests the well dressed, well taken care of British gentleman; easy to see and pleasant in all his dealings.

ELECTRICAL EXHIBITION.

The first annual show to be held in Canada by the Canadian Electrical Exhibition Co., under the direction of R. S. Kelsch, Montreal, opened in the Drill Hall in Montreal on Sept. 2, and will last until Sept. 14. The show is the first of its kind in the Dominion in seventeen years, and it is claimed that it surpasses anything previously attempted on this continent. The variety of exhibits presents the latest of everything electrical from all over the world.

The Exhibition throughout is exceedingly attractive and very educative. It is open every afternoon and evening, and large crowds daily are in attendance. It is providing an excellent opportunity for the various manufacturers of electrical supplies to demonstrate the value of their products to the buying public.

Markets and Correspondence

(For detailed prices see Current Market Quotations, page 78.)

MARKETS IN BRIEF.

Montreal.

Antimony—Strengthening.
Copper—Unsteady.
Lead—Very firm.
Linseed Oil—Firm.
Old Materials—Weak.
Tin—Variable.
Turpentine—Steady.
White Lead—Unsteady.

Toronto.

Antimony—Decline of 2 cents.
Copper—Decline of 2 cents.
Lead—Firm.
Linseed Oil—Very firm.
Old Materials—General decline in prices.
Tin—Decline of 2 cents.
Wire Hat and Coat Hooks—60 per cent. off instead of 65.

MONTREAL HARDWARE MARKETS

Montreal, Sept. 13.—Trade has again enlivened, and the usual fall activity characterizes conditions at present. The large number of fairs being held at this particular time throughout the country, has done much to deter business transactions. Pessimistic jobbers say there is more harm done to the year's trade by these fairs than anything else. The buying public are yet somewhat hesitant in placing orders owing to some uncertainty as to financial and agricultural conditions. There has been no changes in prices, which continue firm.

Screws—Although making some headway, factories still find it rather difficult to supply the demand for certain sizes. It will be a few months yet before the factories will be able to handle the trade with ease and despatch. Difficulty in securing labor and raw material are the chief reasons for the delays in shipments. Prices are firm and unchanged.

Sporting Goods—The prospect for this season's business are very bright and encouraging. A heavy demand for all lines of guns and ammunition exist, and jobbers find it very interesting getting prompt and complete shipments.

Poultry Netting—A good demand prevails, although it is somewhat weaker than previously. The season is about over, and factories are making strong endeavors to lay up a reserve stock for future calls.

Cement—A very heavy demand exists, heavier than the producers can supply. An enormous volume of business in this line has been transacted this summer, and is still being transacted. The only possible limits to the present output is lack of capacity. Producers are experiencing great difficulty in effecting prompt shipments, owing to shortage of cars. As an instance, one carload has reached its destination, after covering a distance of 100 miles, in three months.

Building Paper—Business is quiet. Supplies are adequate and a moderate demand exists.

Builders' Hardware—Trade is still very brisk, an active demand for all lines existing. Supplies are adequate and prices firm.

Wire Goods—Hay wire is moving rapidly. The heavy hay crop throughout Quebec has created a strong demand. A good demand also prevails for wire nails, and there is no surplus stock.

TORONTO HARDWARE MARKETS.

Toronto, Sept. 13.—As September advances, business in local hardware circles continues to steadily improve and fall trade is opening up in a way that is quite gratifying to local jobbers. The rush and excitement of the National Exposition is now over, and city retailers, who, for the past fortnight have been busy entertaining visitors or personally attending the Fair, have now more time to scrutinize their stocks, and jobbers are experiencing an increased city business as a result. As far as outside business is concerned, the volume of Eastern trade for the second week in September is much ahead of last year, although Western business is a trifle lighter, owing to the stringency of the money market, which has caused Western buyers to temporarily adopt a more conservative policy in buying. Fall fairs, which are now in full swing all over the Province, cause travelers to lose an occasional order in small places where the proprietor happens to be away at his own or a neighboring fair, but otherwise Ontario business is in a flourishing condition.

Screws—Supplies are slowly increasing, but at a rate which is scarcely noticeable in face of the strong demand which exists. The fact is that manufacturers are so far behind in their orders that they cannot possibly get abreast of the trade for some weeks, at least. Prices remain firm at figures last quoted.

Nails—The demand for the best selling sizes is exceedingly strong at present, but manufacturers are now in a fair position to satisfy consumers' needs. Prices are firm and unchanged.

Wire—The demand for hay-baling wire is fairly good, but outside of that there is little life to the market. While business may revive a trifle during the next few weeks, jobbers do not expect any great demand till after the close of the year. Mindful of the lamentable scarcity which characterized the wire markets throughout the past summer, manufacturers are devoting their spare energy to amassing supplies in order to avoid a repetition of the shortage when the rush of spring trade sets in.

Building Materials—The immense number of buildings in course of construction maintains the exceptionally strong demand for all kinds of builders' hardware and supplies. The fall trade in glass and putty is opening up strong and the season's sale of both promises to be exceptionally heavy. Cement sales are steadily increasing with the approach of autumn, when farmers have more time to give to repairs. The increasing popularity of cement for stable floors, water-troughs, wells, etc., is evidenced by the greatly increased rural trade which is being done in this material. Strawboard and tarred building papers are in good demand, and supplies are well up with the trade.

Sporting Goods—Firearms and ammunition are in heavy demand, and indications point to a heavy season's business. The shooting season for duck and partridge opens on Monday, and, consequently, guns and cartridges are moving fast. Just here it may be remarked that there is considerable confusion among retailers in regard to the new law on Automatic Shotguns. Section 23 of the Ontario Game and Fisheries Act prohibits the use of automatic shotguns for the killing of game, but their sale is not forbidden in any way, and they may still be legitimately used for trap shooting, etc. Retailers will do well to bear in mind that the new Act does not apply in any way to automatic rifles.

Wire Coat and Hat Hooks—There has been a slight advance on these goods and the discount is now 60 per cent. off list, instead of 65 per cent., as formerly. The new discount does not apply to 3-inch coppered hooks, on which jobbers have individually adopted special net prices.

MONTREAL METAL MARKET.

Montreal, Sept. 13.—A little more activity is noticeable in local metal circles, due probably to the approach of the fall season when consumers are forced to procure a certain amount of surplus stock for fall and winter consumption. Quite a large volume of business is being transacted by pig iron dealers at the present time, as consumers are anxious to stock up before the 1s. 6d. extra freight charge comes into effect (Oct. 1). Some suspicion still exists, especially amongst buyers of copper, tin, and antimony. They are anticipating a decline in prices and are, therefore, holding off. Some have held off too long, especially consumers of zinc, spelter, and copper, and will soon be face to face with the impossibility of securing freight room.

English market conditions are very satisfactory, a remarkable firmness having been maintained throughout the summer months. The American market is still very uncertain, due to speculative variations.

Pig Iron—Strictly No. 1 grades of English and Scotch iron are exceedingly scarce, and the maximum tonnage offered by the English dealers on ore shipment is 500 tons. Although supplies in Great Britain are short, yet prices are remarkably low. One explanation of this peculiar situation might be that the furnace owners are anxious to have a large-sized tonnage contracted for when the furnaces are again working satisfactorily, and producing sufficient high-grade material. Furnaces at Londonderry, N.S., which have been idle for some time, for re-lining, have blown in again. Eastern consumers in Canada are filling up for fall and winter requirements very rapidly. Prices are firm and unchanged: Middlesboro, No. 1, $21.50; No. 2, $20.50; Summerlee, $25.50.

Antimony—Is somewhat stronger; American market conditions are still very unsatisfactory. Prices are firm and unchanged. Recent despatches from Great Britain report an advance of £3 a ton for Cookson's. Cookson's, 15c; Hallett's, 14½c.

Copper—There is still very little activity in local circles, consumers refusing to do much buying until they are assured of a settlement amongst American dealers. The basis of 18c, recently established, failed to produce the buying activity anticipated, and there is a slight shading of prices, although this is denied by leading producers. Local prices are unchanged.

Ingot Tin—The market at present is variable, wide fluctuations being prevalent. Much uncertainty exists. Prices are unchanged. Lamb, and Flag, and Straits, $44.

Lead—The local market continues very strong, a heavy demand existing. Imported pig, $5.25; bar, $5.75.

Zinc Spelter—Continues weak. Little buying is being done owing to the uncertainty in prices. It will, perhaps, be too late for local dealers to secure sufficient storage room.

Old Materials—Scrap iron continues weak, there being little demand. Supplies, however, are abundant. A general decline has occurred in other lines: Heavy copper and wire, 14c; light copper, 12c; heavy red brass, 11c; heavy yellow brass, 9c; light brass, 7c; No. 1 wrought iron, $14.50; stove plate, $12.

TORONTO METAL MARKETS.

Toronto, Sept. 13.—The week furnishes a marked improvement in business through the local metal market, and heavy buying in the various metals is decidedly more in evidence than it has been for the past six weeks. With the exception of lead and pig iron, prices generally continue to decline. The low prices which rule this week's market have succeeded in attracting many buyers, and are directly responsible for much of the increase in the week's business. Jobbers are quite satisfied at the way September business has opened up, and, notwithstanding the depressed state of foreign markets, are quite optimistic concerning the prospects for a heavy autumn trade.

Antimony—The demand has increased during the week, and consumers are now showing a disposition to buy in larger quantities. The markets abroad continue dull and inactive, and the local price has been reduced two cents during the week. Cookson's is now quoted locally at 13c, and Hallett's at 12½c.

Pig Iron—Business in old contracts continues good and a considerable amount of new business has come in during the week. Evidently consumers are being brought into the market by the fact that freight rates on imported pig iron will advance 1s. 6d. per ton on the first of October.

Tin—The local price of tin continues to vary, in sympathy with fluctuations in the primary markets of supply. During the last couple of days, tin has taken a sharp drop in the British Isles, and as a consequence, local quotations have been reduced $2 during the week. Lamb & Flag and Straits ingot tin is now quoted from $41 to $42.

Lead—The increase of business which was reported last week is more than maintained, and more prompt buying is being done this week than for some time. The price of imported pig lead has stiffened a trifle and is firm at $5.35.

Copper—Some small sales are reported for the week, but apparently the majority of local consumers are still content to play the waiting game. The London market is dull and inactive, with little buying being done. The establishment last week of the price of electrolytic copper in the United States at eighteen cents a pound, has failed to produce the buying movement that the producers expected, and little confidence is felt by consumers in the market there. A further decline has taken place in the American market this week, and at present it is impossible to accurately determine the actual price of copper in the United States, owing to the fact that the large producers there are attaching as much secrecy as possible to their movements. Locally, copper has declined 2c per pound and casting ingot is now quoted at 21c.

Old Materials—The market continues extremely dull and prices are still declining. Heavy copper and wire, light copper, heavy red brass and heavy yellow brass have all declined ½c., and are now quoted at 13½c, 11½c, 11¼c and 10c, respectively; heavy lead and scrap zinc have each dropped ¼c, and are now held at 3¾c and 3½c, respectively; No. 2 wrought iron has dropped from $6 to $5, machinery cast scrap from $16.50 to $16, and stove plate from $12 to $11. While scrap iron is very dull at present, it is expected that domestic cast iron scrap will be in strong demand this coming winter, owing to the fact that the heavy business of the ocean transportation companies will prevent them from devoting much carrying space to imported scrap.

LONDON, ENG., METAL MARKETS.

London, Sept. 10.—Cleveland warrants are quoted at 54s 9d and Glasgow standards at 53s 9d, making prices as compared with last week, on Clevelend warrants 1s 3d lower, and on Glasgow standards 1s 6d lower.

Tin—Spot tin opened easy at £167 15, futures at £167, after sales of 130 tons spot and 600 tons futures, closed steady at £167 15s for spot and £167 for futures, making price as compared with last week £1 15s higher on spot and £1 higher on futures.

Copper—Spot copper opened weak at £71, futures at £71 10s, after sales of 400 tons spot and 1,400 tons futures, closed weak at £70 10s for spot and £70 17s 6d for futures, making price as compared with last week £3 10s lower on spot and £3 2s 6d lower on futures.

Spelter—The market closed at £20 15s, mak'ng price as compared with last week 17s 6d lower.

Lead—The market closed at £19 15s, making price as compared with last week unchanged.

U. S. IRON AND METAL MARKETS

New York, Sept. 12.—The Iron Age today says: The statistics of the production of pig iron collected by The Iron Age, show that the output in August fell about 5,000 tons below that of July, the production of anthracite and coke iron in August having been 2,250,410 gross tons, as compared with 2,255,660 tons in July, both being months of 31 days.

There is more cheerful feeling in eastern pig iron trade, due to the fact that there has been increased activity, some good concerns having bought not only for delivery during the last quarter, but also during the first quarter of next year. The movement has been most marked in basic pig, but has been observable also in forge and foundry irons. The earlier sales were made at the lowest prices close to $18.50 delivered for basic, but there was not enough available at that figure and buyers have had to pay $19 for additional purchases. It is estimated that the purchases in the eastern markets aggregate about 50,000 tons. In the west and south buyers and sellers are still apart.

Reports from Cleveland ore shippers vary widely as to the threatened shortage of ore. Some of them have notified their customers that they will be forced to cut down their shipments, and in the case of some, it is particularly the Bessemer grades which are affected. However, it must be remembered that the ore purchases were based originally on the expectation of a demand at the rate of the summer. How much the winter and spring will cut this down is a question. That a smaller tonnage will be used than was originally estimated is averred by some leading steel interests.

The copper market is utterly demoralized. The effort to induce buying, by naming 18c for electrolytic copper has been a flat failure, and the conviction is growing in the trade that there will be no halt until 15c is reached. Electrolytic copper has been sold for shipment abroad at the equivalent of 16⅝c, and is now being offered at 16½c without takers.

CORRESPONDENCE

Travelers, hardware merchants and clerks are requested to forward correspondence regarding the doings of the trade and the industrial gossip of their town and district. Addressed envelopes, stationery, etc., will be supplied to regular correspondents on request. Write the Editor for information.

HALIFAX HARDWAREMAN MARRIED.

Halifax, N.S., Sept. 9.—The present is what is known by the hardware trade as the off-season, and, as a result, business is very quiet. Only a few lines are moving. Some sales of hay wire, and lanterns are being made, and axes ordered early in the summer are now being delivered. Wire nails are steady at $2.60 base. There has been an advance in raw and boiled linseed oil. Raw is quoted at 68 cents, and boiled at 71. Turpentine is down, and is now quoted at 90 cents.

* * *

A large and rich deposit of iron ore has been discovered at Lakeville, Antigonish county. The property is owned by American capitalists. Near the new discovery is an iron property owned by North Sydney parties, and it is reported that the American syndicate is negotiating for the purchase of the same. Samples of the ore taken from the mine have been sent to Montreal, and the assay is high.

* * *

Some enterprising young businessmen of Stewiacke, are talking of the feasibility of an electric railway from Stewiacke to Musquodoboit, by way of the Stewiacke Valley. The idea is based on both a passenger and freight service. It is thought sufficient water power can be obtained in the vicinity of where the line would be located to operate it. If sufficient power cannot be obtained at any one point, it is proposed to put in auxiliary stations at different points.

* * *

Wiliam Hutton, a popular member of the staff of the hardware firm of William Stairs, Son & Morrow, was married last week to Miss Laura Boutilier. The young couple received many valuable wedding gifts, including one from his fellow employes.

* * *

A large number of delegates from Provincial points left here to attend the Canadian Electrical Association convention in Montreal this week. The delegates from Halifax are: C. C. Starr, local manager of the Canadian Westinghouse Co.; J. W. Crosby, manager Halifax Electric Tram Co.; P. A. Truman, chief electrician Halifax Electric Tram Co.; P. G. Colpitt, city electrician; J. T. Murphy, Canadian General Electric Co.; James Farquhar, of Farquhar Bros., dealers in electrical supplies; G. L. Burritt, manager of the Canadian Rand Drill Co.

From other Maritime points the delegates are: G. G. Chambers, Truro; J. A. Davis, Amherst; D. H. Robb, Amherst; W. Pickley, N. S. Steel & Coal Co.; L. H. McLeod, Sydney; N. Lodge, Moncton; A. E. Morrison, Charlottetown, P.E.I.; A. M. McKay, Chatham, N.B.

* * *

The directors of the Dominion Iron & Steel Co., who were in session for two days at Sydney last week, are enthusiastic over the conditions existing at the plant. The blast furnaces are now in full swing, and the rod mill is turning out miles of good material. Senator Forget says that it has been definitely decided to undertake as soon as convenient the further development of the iron ore at Wabana, not only for the purpose of supplying the Sydney plant, but for sale and export abroad. The directors appear confident that the properties at Wabana are practically inexhaustible and see in this departure a new source of revenue.

Harney Graham, who has just returned from Rio Janeiro and South American ports, where he was inspecting some iron ore properties for the Steel Co., was in North Sydney last week on business.

NATURAL GAS AT QUEBEC.

Quebec, Aug. 9.—Business conditions are somewhat strengthened this week. The ammunition and gun trade is opening in splendid style. Supplies are rather short, especially for European goods. Prices remain unchanged. The demand for builders' materials is still good. The trade throughout the year has been splendidly maintained. Retailers are satisfied with the present state of business. Trade is beginning to revive again for wholesalers. All their travelers are on the road and good orders are yet received.

* * *

John S. Shaw, tug and lighter owner, has been named a Harbor Commissioner in succession to Harold Kennedy.

* * *

R. H. Scougall has sold to an American syndicate, composed of Hon. A. Cochran, Hon. J. F. Hatche and C. S. Sowle, all of Vermont, U.S., his timber limits and mill at Marsoni's. The property sold consists of an up-to-date saw mill, store, ten horses, lumbering camp, etc., six hundred acres of freehold land, and about seventy thousand acres of timber limits. These limits are known as the Townships of Christie and Duchesnay, Gaspe, and are estimated to cut about five hundred million feet of spruce and pine. It is stated that the purchasers intend to operate the limits at once.

* * *

Although there was a decrease of 10 per cent. this week in stoves, business continues to be good. Some dealers that your correspondent met told him that the present state of prices will not last long. They expect a new increase on account of the high cost of iron.

* * *

Work is still being pushed ahead on the artesian well at the Quebec Railway Light & Power Co.'s power house on Queen Street. It has now reached a depth of six hundred and eighty feet. Natural gas has been struck to a limited extent, and a small flow of water, but not enough to satisfy the company.

* * *

The construction work of the Indian River Railway, which will connect Lake Megantic with the United States, should be commenced within a year; such was the decision of the company at a meeting held on Wednesday, in our city. The election of officers resulted as follows: President, P. Davis; vice-president and general manager, W. Tettengill, Rumford Falls; directors, W. P. Davis, L. A. Cannon, Geo. Parent; treasurer, G. W. York, Portland; solicitor, L. A. Cannon.

PAINTING KINGSTON HARDWARE STORE.

Kingston, Sept. 10.—Business continues here about the same as last week. The fine weather assists greatly with the contractors and builders, who seem to be kept very busy still, and as a result, there is quite a demand for all lines of hardware used in this connection.

* * *

Several workmen have been engaged by the city all summer in laying new walks and it is estimated that when the work of laying the walks concludes for the year, about November 1, this city will have nearly thirty miles of permanent walks, most of which are of concrete. There are fifty-two miles of walks in Kingston, so that after one or two more years only the outskirts of the city will be able to boast of having wooden walks. The city engineer is now agitating to have the board of works try an experiment with a piece of road pavement since the walks have been such a success and made such an improvement. The roads are in a very bad condition and there is little doubt but that the city council will submit a by-law to the ratepayers next January for the raising of money for debentures to rebuild the roads.

* * *

Elliott Bros. have the contract for the plumbing and heating work in the new Hotel Quinte, in Belleville. Alderman R. F. Elliott and seven workmen left yesterday for that place to begin the work. It will be some time before the contract is completed, but when completed will be a fine piece of work, as all the latest improvements are being put in.

* * *

A. Chown & Co., hardware dealers, are having their four-storey stone building on Bagot St. painted a nice grey shade. This makes a great improvement on the appearance of the building, as well as advertising their paints.

* * *

Flies of late have become a great nuisance in these parts, especially to the horses standing outside the stores awaiting deliveries. One of the local dealers noticed this and to show the value of "Cow Ease," of which he sells quite a quantity, took down a can and rubbed the horse with it, which proved what is good for the cow is good for the horse as well, this, the poor animal appreciated very much.

* * *

W. A. Mitchell has returned to the city after spending a few days in Toronto.

* * *

Lemmon & Sons, hardware merchants, have received another cargo of cement from Belleville, which arrived here on the schooner "Maggie L," containing 550 barrels, which they are at present busy unloading.

New government offices are being erected on Clarence St. for the customs examining department, which has occupied offices on Ontario St. for a number of years. The new quarters are to be ready for occupation by the first of November. Workmen have been employed at the work for some time and building operations are now nearing completion.

The work on the new gas tank is progressing rapidly, a number of tiers of sheet iron have already been placed in position. Mr. Harding, of Waltham, Mass., who is superintending the work, is thoroughly acquainted with all parts of the work, and expects to have his tank up in a few weeks' time. Six riveters are at work every day putting in the plates.

* * *

The walls of the new laboratory building at Queen's were completed on Friday by contractor William McCartney, who has rushed the construction. The roof is now being put on and it is expected that the building will be ready for occupation by the end of the year.

The firm of Moore & Company, machinists, have removed from their old premises on Montreal St., to new quarters on Wellington St., and are ready to attend to the needs of all their customers. They have now an up-to-date establishment, and one that reflects great credit upon the management. They have an automobile garage, twenty-five feet by seventy-five feet, with a large cement wash basin for cleaning automobiles, and there is everything that goes to form a quick repair shop. In addition, there is a well-equipped machine shop, and special attention is given to all orders. The firm is one that should receive a liberal share of patronage.

HAMILTON GETS ATKINS' PLANT.

Hamilton, Sept. 10.—One of the most important items of interest to the local hardware trade this week is the announcement that the E. C. Atkins Co., of Indianapolis, Ind., one of the largest saw manufacturing concerns on the continent, has decided to erect a large branch here for the Canadian trade. Assessment Commissioner Macleod has landed the new concern, which has already bought the site of the old Hoepfner Refining Co., on North Sherman avenue. The company is capitalized at $1,500,000, and it will spend about $150,000 in the erection and fitting up of its plant here. The firm will manufacture all kinds of saws and will employ for a beginning about 100 men. The local hardwaremen look upon the new industry as an important thing for them, as they will be able to have a more ready supply of saws and will, therefore, be better qualified to supply the local as well as the outside trade at good figures.

For some time past the local plumbing trade has been agitating for the appointment of a plumbing inspector by the city. The matter has been dealt with several times, but the City Fathers have not yet become sufficiently enthused over the matter to do as they are asked. However, the question has been referred to the Board of Health, and, as the members of that body look with favor upon the proposition, it is likely that an appointment will be made if at all possible, this year. Both the Plumbers' Union and the master plumbers appear to be in favor of having a plumbing inspector, which would not only result in more plumbing work being ordered for sanitary purposes, but would also result in the work being well done.

BUSINESS BRISK AT INGERSOLL.

Ingersoll, Sept. 11.—"Business was never better," this was the cheerful response of one of Ingersoll's leading hardware merchants to a query by your correspondent as to how he had found trade during the summer. The other dealers of the town are equally emphatic in declaring that this has been one of the most successful years that they have known. Contrary to expectations early in the season when the weather was backward and the outlook in general anything but bright, there has been much building in both town and country and all the dealers have enjoyed a large demand for building supplies. The hardware merchants who also carry on a plumbing business have had a busy season in this branch and the prospects are bright for another season. A number of new houses have been built this summer and a number are in contemplation for next year. It is gratifying to note that the plumbers and their employers know nothing of strikes and disaffection which is so characteristic of life in the larger centres. Everything goes along here with a merry swing that is satisfactory to all concerned.

* * *

Autumn has once more rolled around and as the leaves begin to take on their variegated hues, imparted by Jack Frost, much attention is directed to the purchasing of stoves and other heating appliances. The local hardware dealers do a large stove and furnace trade and they are all busy preparing for the approaching season.

* * *

Not a few Canadian towns and cities looked with covetous eyes on the announcement recently made at Hamilton that negotiations had been completed with the E. C. Atkins Co., of Indianapolis, the largest manufacturers of saws in the United States, to locate a branch factory there. Several places, Ingersoll included, were anxious that they should secure this valuable industry. Representatives of the company visited Ingersoll and looked over proposed sites and for a time there was a strong feeling that the company might locate here, as there was much that was in the town's favor. However, just when little or nothing was being heard, Hamilton came out with the announcement that they had landed the industry.

WESTERN FAIR AT LONDON.

London, Ont., Sept. 11.—This is a very busy week with the wholesale hardware people. The Western Fair has brought a great influx of visitors to the city, among them being retail dealers from all over Western Ontario, to whom the jobbing houses are proving even a greater attraction than the Fair itself. The travelers are all in this week, and assist the warehouse staffs in making things pleasant for their patrons and selling them goods. It is a big week for the wholesale men, but the retailers would stand a little more activity of a business kind.

* * *

The displays of stoves, hardware, carriage stuff, glass and binder twine take up a good deal of room at the Fair. Many of the local firms are represented by exhibits, which, taken as a whole, are most creditable.

* * *

Cameron Brown, son-in-law of ex-Premier Ross, is head of a firm which proposes to start in this city an establishment for the manufacture of gas lights. The Manufacturers' Committee has under consideration an application from the firm for an inducement in the shape of free water, and if the amount required is not too great it will probably be granted. The committee is also in correspondence with several firms from other towns, who are willing to locate here if the Council will make it worth their while.

* * *

John Hayman, a local builder, has been awarded the contract for the erection of a $50,000 addition to the Grand Trunk car shops, in this city. The work will commence immediately, and will mean that the capacity of the shops will be nearly doubled. Mr. Hayman has also been given the contract for the erection of a new roundhouse for the Grand Trunk at Durand, Mich., upon which work has already been started. The contract is worth about $75,000, and the new building will cover an area of six acres, and have a capacity for forty locomotives.

PEDDLERS' TROUBLE AT CHATHAM.

Chatham, Sept. 10.—Officials of the C.W. & L.E. electric road describe as premature the report that that line had come to an arrangement with the M.C.R. for a crossing at Essex, and also for the making of direct connections between the M.C.R. and Chatham. The matter of a crossing will probably go before the Railway Commission for adjudication. The M.C.R. have a watchman steadily on duty to prevent the electric line rushing their tracks across. Construction work on the C.W. & L.E. is, however, progressing briskly.

* * *

Representatives of a large 5c, 10c and 15c store were in the city last week with a view to locating a branch in Chatham. They were unable, however, to secure a vacant store large enough for the purpose; so the scheme has been abandoned, at least, for the present.

* * *

C. H. Mills, the energetic secretary of the Civic Board of Trade, intends shortly to sever his connection with the firm of Thos. Stone & Sons. He is returning to his old home in Berlin, having accepted the management of the Berlin Leatherette Co., a new concern.

* * *

Chatham will not have free mail delivery for some time to come. The matter has been agitated off and on for a couple of years, but a recent letter from R. M. Coulter, Deputy Postmaster-General, states that Chatham, though possessing more than the required $20,000 annual revenue, falls short of the necessary 12,000 population, and is, consequently, ineligible.

* * *

The old trouble over peddlers and transients has broken out anew, and, as a result, the City Council on Monday was the scene of some caustic comments on the stand taken by the local police magistrate. This time grocers and fruit dealers are particularly interested, though the matter is being keenly watched by all the retailers, since, sooner or later, every line feels the effects of competition from this source.

* * *

Complaint was made to the police regarding an Italian named Favata, who, having no license, undertook to sell goods from a pushcart on King street. When warned by P. C. Dezalia, Favata declined to quit, assuring the policeman that he had a good lawyer who told him he didn't need a license, and that the magistrate wouldn't do anything to him, anyway. The magistrate verified this statement by refusing to take an information against Favata. He alleged that this could not very well be done, pending a decision on the appeal in the Brodie case, which the magistrate dismissed some time ago on the ground that a $100 license fee was prohibitive.

The Property Committee were instructed on Monday night to ascertain if an information could not be laid against Favata before two justices of the peace. Comments on the magistrate's refusal to take an information were distinctly of the sizzling type, the aldermen declaring that this case and the Brodie case had nothing in common.

SEPTEMBER SILLINESS.

Davisville, September 10.—Hay-fever is prevalent this month. Hay-fever is something to be sneezed at. If you sneeze you lose your teeth, if they're the loose kind. People who have hay-fever would rather feel a sneeze going than a sneeze coming. It is said the mountains will cure hay-fever. Dose—One mountain after each meal. A man without a handkerchief has no business to have hay-fever.

* * *

The farmer picks his apples this month. He puts them in barrels, the rusty ones first and on top a few good ones. This kind of a barrel of apples is called a gold brick. A farmer does not always buy gold bricks—he sometimes sells them. Nothing beats dried apples for internal upholstering. Dried apples and water constitute a swell diet. Adam was the first man who bit into an apple. He liked it, he said. "There ain't going to be any core!" Apple pie was Adam's favorite dish.

TRAVELING SALESMAN.

PREPARE FOR HOLIDAY BUSINESS

It's not too early to lay plans for the Christmas holiday trade. Dealers have already bought some lines, of course, and in these days of slow deliveries other orders should be placed as soon as opportunity offers. "Goods well bought are half sold" is a time-worn truism in mercantile life.

But it doesn't do to feel too secure on the buying part until the selling has been done. Plans for increasing holiday sales must be worked out months ahead. Window displays must be figured out and a series of rough sketches planned. A series of ads. for the local papers should also be prepared and, where deemed advisable, a neat booklet gotten up, to be mailed to all probable customers in the surrounding district.

It is in these selling plans that Hardware and Metal can be helpful to its readers. Let all partake of the spirit of co-operation and exchange ideas through these columns. To help the suggestion along the editor invites any reader to join the friendly contest and hangs up a prize of $10 cash for the best answer to the following questions:—

How can the hardware merchant increase his sales of holiday goods next December? What special lines should he stock? What selling plans should be adopted? What special advertising should be done? What novel window displays can be suggested? Should Souvenirs (calendars, knives, trays, etc.) be given to customers?

The prize of $10 will be awarded to the writer of the most practical and original letter of from 500 to 1,000 words received by the editor before October 15, 1907, and the best half-dozen replies will be published in Hardware and Metal.

Address the Editor

Hardware and Metal, 10 Front St. E., Toronto

Modern Conveniences for Farm Homes

The Fourth of a Series of Articles Intended to Help Canadian Plumbers in Educating Residents in Country Districts to the Necessity of Better Sanitary Arrangements.

By Elmina T. Wilson, C.E.

Installation of Bathroom.

The bathroom should be a light, well ventilated room with every facility for cleanliness. Floors and wainscoting of tile or composite material are most desirable, but painted walls are much less expensive and give excellent results. Tile is undoubtedly the most satisfactory material which can be used for the covering of the floors and walls where it can be afforded. Tile floor with covered base and walls finished with cement or hard plaster, painted with enamel paint, are much cheaper. When a tile floor cannot be had, linoleum is an excellent substitute, as it is practically impervious to water. It should be laid before the fixtures are set, in order that there may be no joints. Cement mixed with small chips of marble well rubbed down after setting makes an excellent floor, one that washes as clean as a porcelain plate and has no cracks to harbor dirt; the cost is only about twice that of a double wood floor, or 50 cents per square foot, including the necessary cement bed on which it is laid.

When it is desired to lay a cement, composition or tile floor upon wooden floor joints, proceed as follows: Nail a 2x4 to the side of each of the floor joists flush with the bottom. Upon the top of these stretch wire lath, after the joists have first been covered with tarred paper to prevent them absorbing moisture; and upon this lay cinder concrete, made of 1 part Portland cement, 3 parts loose sand, 6 to 8 parts crushed and screened furnace clinkers; filling in to a level at least 2 inches above the tops of the joists. Upon this is placed the floor finishing. Cinder concrete is used because it is so much lighter than that made of stone. When a tile or cement wainscot is too expensive the walls should be painted. Wall paper is not desirable in a bathroom, nor is wood paneling.

A porcelain lined or enameled iron bathtub is the best medium priced tub. For supplying the tub with water a combination cock is best, allowing hot or cold water to enter the tub separately or the temperature to be regulated to allow of water being drawn into pitchers. The best lavatories are those of porcelain or enameled iron, with back and overflow all formed as integral parts of the fixture. The basin cocks through which the hot and cold water come are of various shapes, the simplest being the best.

The Closet.

The water closet is the most important plumbing fixture in the house, and should be selected and put up with particular care. A good closet should be simple, neat and strong, of a smooth material, with ample water in the bowl. Among the modern closets there is none more satisfactory than the flushing rim, siphon jet closet, which can be had, including the trap, in a single piece of porcelain. Porcelain is used because no other material can be kept so clean and sanitary. But even this is an imperfect protection from dirt and disease unless the bowl is flushed so as to clean

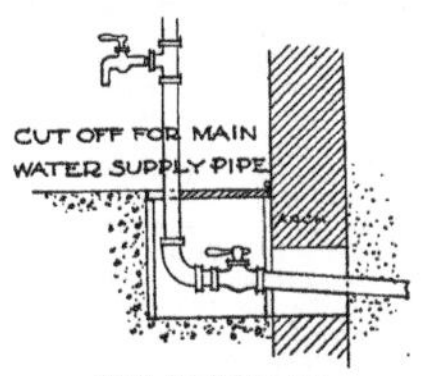

Fig. 3. Cut Off for Main.

it completely and absolutely. The water should be poured from the rim of the bowl, so that every part of it is perfectly cleaned. The washdown and washout closets are similar in make, but are not so thorough in their action. In the washout closet the basin acts as a receiver, a small quantity of water being retained in it, and into this the deposit is made, to be washed out afterward into the trap by the flush. The water in the basin is prevented from leaking into the trap by a raised ridge which is apt to break the force of the flush, so that its whole force is not directed into the trap, which is objectionable. The washdown closet receives the deposit directly into the water held in the bowl by the trap. It has a straight back and a much smaller fouling surface. There is no open vent. The outlet is entirely covered with water, so that the water does not throw the soil against the side. The only advantage the siphon closet has over it is the greater force of discharge given by the siphon.

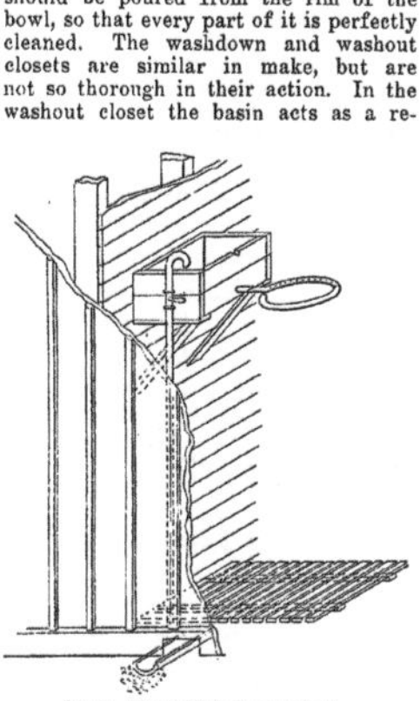

Fig. 4.—Home Made Shower Bath.

The siphon closet, like the washdown closet, retains a certain amount of water into which filth is discharged. In addition there is a siphon trap provided with a long ascending arm, so that the water in the trap is at a lower level than the water in the bowl. The water from the flushing cistern is directed not only into the bowl, but downward into the trap itself. As a result of this discharge into the trap a siphon action is produced whereby the contents of the bowl are sucked through the trap into the soil pipe without soiling the bowl. The seal—that is, the body of water which prevents the sewer gas from escaping into the house—is deep, broad, and always in plain sight.

The flushing cistern or tank for a water closet is always distinct from the main water supply. As a rule, a plain hardwood box, copper-lined, is supported by brackets from the wall about 7 feet above and communicating with the closet by a pipe. This pipe is usually about 1½ inches in diameter and should have as few bends and angles about it as possible. The cistern should hold 2 or 3 gallons of water, all of which should be discharged at one time into the closet. The flush of the closet should be quick, powerful, and noiseless, thoroughly scouring all parts exposed to fouling.

The flow into the cistern is regulated by a float valve which allows the tank to fill, the float rising with the water; when it reaches the proper level the float is entirely raised and the supply shut off. When the tank is emptied by opening the flush valve, which is lifted by pulling a chain attached to it, the process is repeated. The cistern is usually provided with an overflow connected with the flush pipe, so that if the ball-cock fails to act properly in shutting off the water the surplus will escape through the water closet to the drain instead of overflowing.

Soil Pipe Connections.

The best closets are provided with a brass screw soil pipe connection, calked with lead and cemented into the base of the closet. The corresponding threaded brass coupling is soldered into the end of the bend which connects with the soil pipe. The closet is then screwed into the threaded coupling until the base rests on the floor. The closet may be removed at any time by simply unscrewing it. No bolts are necessary through the base flanges. In setting a water closet a neater finish can be obtained if a porcelain floor slab is put with the finished floor.

The important need of the work is simplicity, not only in detail, but in general scheme. Construct the water closet to be used as a urinal and slop sink and arrange to draw water through the bath cocks placed at the top of the tub. It not only saves cost, but is a great advantage to have the fewest possible points requiring inspection and care and to secure the most frequent possible use of every inlet into the drainage system. Great care must be taken not to throw into the water closet hair, matches, strips of cloth, or anything which is insoluble and liable to clog the trap and soil pipe. A burnt match seems small in itself, but if lodged in the trap it will collect other things and cause serious obstruction of the outlet. Tissue toilet paper should be used. Its cost would be exceeded many times if a part of the system needed to be taken out to free it from newspaper obstruction. It is often found more convenient to have the water closet with a separate entrance from the hall and entirely independent from the bathroom.

Traps and Vents.

Every plumbing fixture must have a trap to prevent the foul air from coming back from the drain through the waste pipe. In its simplest form a trap is a downward bend in a pipe, so deep that the upper wall of the pipe dips into the water held in the bend, the extent to which it dips being known as the depth of the seal. With slight modifications this is the trap most commonly used for wash basins, laundry tubs, etc. Its greatest fault is the danger from siphonage—that is, the water seal may be carried out of the trap into the soil pipe by the rush of the water when the fitting itself is emptied, by the flow of water from another fixture on the same branch waste pipe or by the discharge of water from a fixture higher up, but connected to the same soil pipe. This danger is much lessened by the introduction of a system of ventilation pipes extending upward either from the trap itself or from the outlet near the trap. To avoid this extra expense of a third system of pipes it is better to supply each fixture with one of the patent non-siphonage traps, which should also be self-cleansing. There are several good ones on the market. It is a good habit, after emptying the wash basin, bathtub or kitchen sink, to allow some clean water from the faucet to run into the fixture in order to have clean water in the trap. All traps should be provided with trap screws, placed below the water line, and arranged so as to be accessible for cleaning.

Nothing short of continuous use will prevent the evaporation of the water in the trap. One with a large dip is best, but at the same time the trap must be so formed that at each use of the fixture all the filth that is delivered shall be carried away, the trap being immediately refilled with fresh water. Hair and fibers from cloth sometimes carry the water out of the traps by capillary attraction, and care should be taken not to allow such things to enter the pipes.

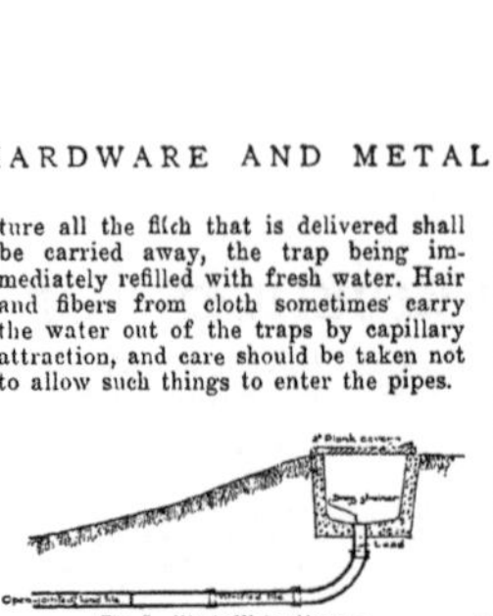

Fig. 5.—Waste Water Hopper.

The Soil Pipe.

The soil pipe should extend from cellar to roof in a straight line, if possible, as each offset or bend forms an obstruction to its proper flushing with both water and air. Use only "extra heavy" soil pipe of uniform thickness throughout, as the hubs stand the calking better.

Avoid if possible plumbing fixtures in the cellar if the drain must go under the floor. If it is necessary to make connections with a fixture in the cellar it is better that the main channel should run under the floor to or near the location of such fixtures, that all or nearly all of its length should constitute a part of the main drain, thoroughly flushed and ventilated like the rest of the system. The pipe should be laid in an open trench and so thoroughly calked that under a pressure equal to one storey in height not a drop of water should escape at any point, and then it should be closed in good concrete, after which the trench should be filled. The soil pipe should pass through the foundation by means of an arch, and the cast iron pipe should extend at least 5 feet outside the foundation; from there on a carefully laid and rigidly inspected vitrified pipe drain is to be preferred. The joint between the iron pipe and the vitrified sewer pipe should be made with neat Portland cement motar.

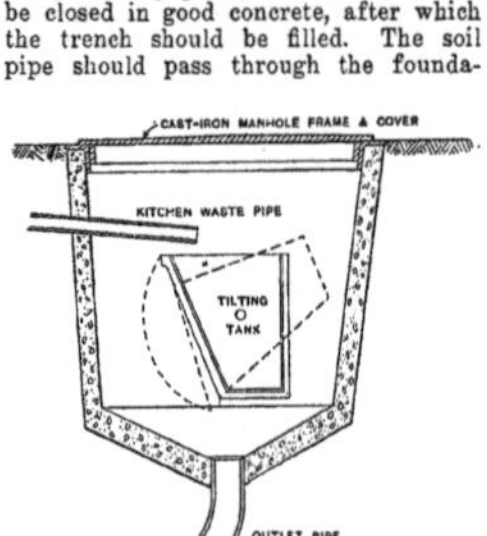

Fig. 6 —Tilting Tank in Hopper.

If there are no fixtures in the cellar carry the drain in full sight along the face of the cellar, or suspended from the floor beams, so the joints may be inspected. At the point where it is to turn up as a vertical soil pipe support it by a post or a brick pier. Use no short turns in the soil pipe, like tees and quarter bends. Two one-eight bends or a Y-branch and a single one-eighth bend give a more gradual and therefore a better change of direction. Water closets should connect to the soil pipe with a Y-branch. The soil pipe should be secured along its entire length at distances not over 5 feet with hangers and clamps or hooks, so that it will be rigidly held in position. The joints in the cast iron soil pipe should be made by first inserting a little picked oakum into the socket, allowing none to enter the pipe; it is better formed into a sort of rope. The oakum prevents the lead from running into the pipe to form an obstruction to the flow. Enough molten lead is then poured into the hub to fill it. After the lead has cooled it is carefully hammered with a special calking tool until the space between the spigot and hub is perfectly gas and water tight. Every joint should be made with a view to being tested with hydraulic pressure.

Joints Must Be Tight.

In making this test the simplest way is to close all openings into the pipe with wooden plugs or disks of India rubber compressed between two plates of iron forced together with a screw. There is no especial advantage in applying a great head of water, for if a joint is not tight it will leak under a head of a few inches. It is generally most convenient to test the vertical pipe storey by storey, the plugs being inserted through the water-closet branches. There is probably no occasion to fear that work once made tight will develop leaks for many years. The tendency to rust after a time, even with tar-coated or enameled pipe, being rather to close such slight leaks as may exist.

Four inches in diameter is sufficient for soil pipe, and the best results are obtained by running it full size straight above the roof and covering the top with a wire basket, such as is used to keep leaves out of gutters.

There should always be a trap between the house and the sewage disposal plant, and there must also be on the house side of it an inlet for fresh air. There can be no real ventilation of the system if it is open only at the top, but a generous inlet for fresh air on the drain outside the house, in connection with the opening at the top of the soil pipe, will insure a free movement throughout the whole system. The fresh air inlet must be guarded from obstruction. It may be brought out close to the foundation walls, but not too near windows and doors. If the trap is formed by the submerging of the inlet pipe in the settling chamber of the disposal system the fresh air in it should be placed close to this.

For all minor waste pipes lead pipe is used, as it may be bent and cut to suit all possible positions and requires but few joints. Only "heavy" lead pipe should be used. As lead is quite a soft and connected with a hopper, as shown in one of the cuts, can be used. This hopper can be made of wood or galvanized iron, and is provided with a strainer and solid cover. That part of the pipe which is more than 10 in. below the surface should be of vitrified pipe and laid with cemented joints. The waste water when poured into such a hopper a pailful at a time will be distributed the length of the tile.

For a more elaborate system, a tilting or tumbling tank, also shown, can be built with pipe connections to the kitchen sink and bathroom, this tank to collect the irregular flow and empty, when full, into the drain.

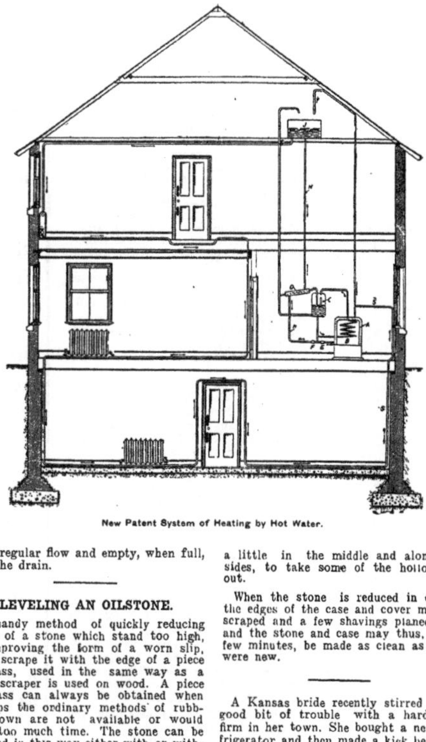

New Patent System of Heating by Hot Water.

LEVELING AN OILSTONE.

A handy method of quickly reducing parts of a stone which stand too high, or improving the form of a worn slip, is to scrape it with the edge of a piece of glass, used in the same way as a steel scraper is used on wood. A piece of glass can always be obtained when perhaps the ordinary methods of rubbing down are not available or would take too much time. The stone can be scraped in this way either with or without water. Without water is perhaps the best, as it is then easier to see how much is being removed. If one end or one corner of the stone stands higher than the rest, it is easier to reduce to a general level in this way than by the ordinary methods, which make the surface flat, but cannot easily remove a slope to one end or one side, says a writer in a recent issue of The Patternmaker. A slight inclination in any direction causes the oil to run off the stone, and it is advisable, therefore, always to leave the stone slightly hollow so that the oil will tend to run to the middle when it is left standing.

The greatest wear occurs not in the middle of a stone, but near the ends, at the places where the movement of the tool is reversed. It is, therefore, chiefly a small area at the extreme ends which requires scraping down, and sometimes a little in the middle and along the sides, to take some of the hollowness out.

When the stone is reduced in depth, the edges of the case and cover may be scraped and a few shavings planed off, and the stone and case may thus, in a few minutes, be made as clean as if it were new.

A Kansas bride recently stirred up a good bit of trouble with a hardware firm in her town. She bought a new refrigerator and then made a kick because milk soured in it. The dealer she bought it of made an investigation and found the dear thing had put no ice in it. She said she would not have bought it if she had supposed that was necessary.

MANITOBA HARDWARE AND METAL MARKETS

Corrected by telegraph up to 12 noon Friday Sept 13 Room 511, Union Bank Bldg, Winnipeg, Man.

Business is fairly active. Indications now point to a fairly good crop in most sections of the west and as prices are nearly thirty cents higher than last year, the outlook is bright. Given anything like favorable weather during September, the prosperity of the west is assured for another year. There is a marked revival of confidence among the trade.

Rope—Sisal, 11c per lb., and pure manila, 15¾c.

Lanterns—Cold blast, per dozen, $7; coppered, $9; dash, $9.

Wire—Barbed wire, 100 lbs., $3.22½; plain galvanized, 6, 7 and 8, $3.70; No. 9, $3.25; No. 10, $3.70; No. 11, $3.80; No. 12, $3.45; No. 13, $3.55; No. 14, $4; No. 15, $4.25; No. 16, $4.40; plain twist, $3.45; staples, $3.50; oiled annealed wire, No. 10, $2.90; No. 11, $2.96; No. 12, $3.04; No. 13, $3.14; No. 14, $3.24; No. 15, $3.39; annealed wires (unoiled), 10c less; soft copper wire, base, 30c; brass spring wire, base, 30c.

Poultry Netting—The discount is now 47½ per cent. from list price, instead of 50 and 5 as formerly.

Horseshoes—Iron, No. 0 to No. 1, $5.65; No. 2 and larger, $4.40; snowshoes, No. 0 to No. 1, $4.90; No. 2 and larger, $4.65; steel, No. 0 to No. 1, $5; No. 2 and larger, $4.75.

Horsenails—No. 10 and larger, 22c; No. 9, 24c; No. 8, 24c; No. 7, 26c; No. 6, 28c; No. 5, 30c; No. 4, 36c per lb. Discounts: "C" brand, 40, 10, 10 and 7½ p.c.; "M.R.M" cold forged process, 50 and 5 p.c. Add 15c per box. Capewell brand, quotations on application.

Wire Nails—$3 f.o.b. Winnipeg and $2.55 f.o.b. Fort William.

Cut Nails—Now $3.20 per keg.

Pressed Spikes—¼ x 5 and 6, $4.75; 5-6 x 5, 6 and 7, $4.40; ⅜ x 6, 7 and 8, $4.25; 7-16 x 7 and 9, $4.15; ½ x 8, 9, 10 and 12, 4.05; ⅝ x 10 and 12, $3.90. All other lengths 25c extra, net.

Screws—Flat head, iron, bright, 80, 10, 10 and 10; round head, iron, 80; flat head, brass, 75; round head, brass, 70; coach, 70.

Nuts and Bolts—Bolts, carriage, ⅜ or smaller, 60 p.c.; bolts, carriage, 7-16 and up, 50; bolts, machine, ⅜ and under, 50 and 5; bolts, machine, 7-16 and over, 50; bolts, tire, 65; bolt ends, 55; sleigh shoe bolts, 65 and 10; machine screws, 70; plough bolts, 55; square nuts, cases, 3; square nuts, small lots, 2½; hex nuts, cases, 3; hex nuts, small lots, 2½ p.c. Stove bolts, 70 and 10 p.c.

Rivets—Iron, 60 and 10 p.c.; copper, No. 7, 43c; No. 8, 42½c; No. 9, 45½c; copper, No. 10, 47c; copper, No. 12, 50½c, assorted, No. 8, 44½c, and No. 10, 48c.

Coil Chain—¼-in., $7.25; 5-16, $5.75; ⅜, $5.25; 7-16, $5; 9-16, $4.70; ⅝, $4.65; ¾, $4.65.

Shovels—List has advanced $1 per dozen on all spades, shovels and scoops.

Harvest Tools—60 and 5 p.c.

Axe Handles—Turned, s.g. hickory, $3.15; No. 1, $1.90; No. 2, $1.60; octagon extra, $2.30; No. 1, $1.60.

Axes—Bench axes, 40; broad axes, 25 p.c. discount off list; Royal Oak, per doz., $6.25; Maple Leaf, $8.25; Model, $8.50; Black Prince, $7.25; Black Diamond, $9.25; Standard flint edge, $8.75; Copper King, $8.25; Columbian, $9.50; handled axes, North Star, $7.75; Black Prince, $9.25; Standard flint edge, $10.75; Copper King, $11 per dozen.

Churns—45 and 5; list as follows: No. 0, $9; No. 1, $9; No. 2, $10; No. 3, $11; No. 4, $13; No. 5, $16.

Auger Bits—"Irwin" bits, 47½ per cent. and other lines 70 per cent.

Blocks—Steel blocks, 35; wood, 55.

Fittings—Wrought couplings, 60; nipples, 65 and 10K Ts and elbows, 10; malleable bushings, 50; malleable unions, 55 p.c.

Hinges—Light "T" and strap, 65.

Hooks—Brush hooks, heavy, per doz., $8.75; grass hooks, $1.70.

Stove Pipes—6-in., per 100 feet length, $9; 7-in., $9.75.

Tinware, Etc.—Pressed, retinned, 70 and 10; pressed, plain, 75 and 2½; pieced, 30; japanned ware, 37½; enamelled ware, Famous, 50; Imperial, 50 and 10; Imperial, one coat, 60; Premier, 50; Colonial, 50 and 10; Royal, 60; Victoria, 45; White, 45; Diamond, 50; Granite, 60 p.c.

Galvanized Ware—Pails, 37½ per cent.; other galvanized lines, 30 per cent.

Cordage—Rope sisal, 7-16 and larger, basis, $11.25; manila, 7-16 and larger, basis, $16.25; lathyarn, $11.25; cotton rope, per lb., 21c.

Solder—Quoted at 27c per pound. Block tin is quoted at 45c per pound.

Wringers—Royal Canadian, $36; B.B., $40.75 per dozen.

Files—Arcade, 75; Black Diamond, 60; Nicholson's, 62½ p.c.

Locks—Peterboro and Gurney 40 per cent.

Building Paper—Anchor, plain, 66c; tarred, 69c; Victoria, plain, 71c; tarred, 84c; No. 1 Cyclone, tarred, 84c; No. 1 Cyclone, plain, 66c; No. 2, Joliette, tarred, 69c; No. 2 Joliette plain, 51c; No. 2 Sunrise, plain, 56c.

Ammunition, etc.—Cartridges, rim fire, 50 and 5; central fire, 33⅓ p.c.; military, 10 p.c. advance. Loaded shells: 12 gauge, black, $16.50; chilled, 12 gauge, $17.50; soft, 10 gauge, $19.50; chilled, 10 gauge, $20.50. Shot: ordinary, per 100 lbs., $7.75; chilled, $8.10. Powder: F.F., keg, Hamilton, $4.75; F.F.G., Dupont's, $5.

Revolvers—The Iver Johnson revolvers have been advanced in price, the basis for revolver with hammer being $5.30 and for the hammerless $5.95.

Iron and Steel—Bar iron basis, $2.70. Swedish iron basis, $4.95; sleigh shoe steel, $2.75; spring steel, $3.25; machinery steel, $3.50; tool steel, Black Diamond, 100 lbs., $9.50; Jessop, $13.

Sheet Zinc—$8.50 for cask lots, and $9 for broken lots.

Corrugated Iron and Roofing, etc.—Corrugated iron, 28 gauge, painted, $3; galvanized, $4.10; 26 gauge, $3.35 and $4.35. Pressed standing seamed roofing, 28 gauge, $3.45 and $4.45. Crimped roofing, 28 gauge, painted, $3.20; galvanized, $4.30; 26 gauge, $3.55 and $4.55.

Pig Lead—Average price is $6.

Copper—Planished copper, 44c per lb.; plain, 39c.

Iron Pipe and Fittings—Black pipe, ¼-in., $2.70; ⅜, $2.85; ½, $3.75; ¾, $4.75; 1, $6.75; 1¼, $9.25; 1½, $11.00; 2, $14.80; 2½, $24.60; 3, $32.30; 3½, $40.50; 4. $46.00; 4½, $54.00. Galvanized: ¼ in., $3.65; ⅜, $3.80; ½, $4.50: ¾, $5.80; 1, $8.40; 1¼, $11.40; 1½, $13.80; 2, $18.40. Nipples, 70 per cent.; unions, couples, bushings and plugs, 50 per cent.; malleables, 20 per cent.

Iron Pipe and Fittings—Black pipe, ¼-in., $2.65; ⅜, $2.80; ½, $3.50; ¾, $4.40; 1, $6.35; 1¼, $8.65; 1½. $10.40; 2, $13.85; 2½, $19; 3, $25. Galvanized iron pipe, ⅜-in., $3.75; ½, $4.35; ¾, $6.65; 1, $8.10; 1¼, $11; 1½, $13.25; 2-inch, $17.65. Nipples, 70 and 10 per cent.; unions, couplings, bushings and plugs, 60 per cent.

Galvanized Iron—Apollo, 16-gauge, $4.15; 18 and 20, $4.40; 22 and 24, $4.65; 26, $4.65; 28, $4.50; 30-gauge or 10¾-oz., $5.20; Queen's Head, 20, $4.60; 24 and 26, $4.90; 28, $5.15.

Lead Pipe—Market is firm at $7.80.

Tin Plates—IC. charcoal, 20x28, box, $10; IX charcoal, 20x28, $12; XXI charcoal, 20x28, $14.

Terne Plates—Quoted at $9.50.

Canada Plates—18x21, 18x24, $3.50; 20x28, $3.80; full polished, $4.30.

Lubricating Oils—600W, cylinders, 80c; capital cylinders, 55c and 50c; solar red engine, 30c; Atlantic red engine, 29c; heavy castor, 28c; medium castor, 27c; ready harvester, 28c; standard har separator oil, 35c; standard gas engi oil, 35c per gal on.

Petroleum and Gasolene—Silver Sta in bbls., per gal., 20c; Sunlight, in bbls per gal., 22c; per case, $2.35; Eocene, bbls., per gal., 24c; per case, $2.50; Pe noline, in bbls., per gal., 24c; Cryst Spray, 23c; Silver Light, 21c. Engi gasoline, in barrels, gal., 27c; f.o.b. Wi nipeg, in cases, $2.75.

Paints and Oils—White lead, pure, $6 50 to $7.50, according to brand; bladd putty, in bbls., 2½c; in kegs, 2¾c; tu pentine, barrel lots, Winnipeg, 90c; Ca gary, 97c; Lethbridge, 97c; Edmonto 98c. Less than barrel lots, 5c per ga lon advance. Linseed oil, raw, Winn peg, 72c; Calgary, 79c; Lethbridge, 79 Edmonton, 80c; boiled oil, 3c per gall advance on these prices.

WIRE NAILS
TACKS
WIRE

Prompt Shipment

The ONTARIO TACK CO.
Limited.
HAMILTON ONT.

INGOT

ANTIMONY
COPPER
LEAD
ZINC
TIN

IN STOCK

A. C. LESLIE & CO.
Limited
MONTREAL

"MIDLAND"

BRAND.

Foundry Pig Iron.

Made from carefully selected Lake Superior Ores, with Connellsville Coke as Fuel, "Midland" will rival in quality and grading the very best of the imported brands.

Write for Price to Sales Agents

Drummond, McCall & Co.
MONTREAL, QUE
or to
Canada Iron Furnace Co.
MIDLAND, ONT. Limited

Nova Scotia Steel & Coal Co., Limited

NEW GLASGOW, N.S.

Manufacturers of

Ferrona Pig Iron

And SIEMENS-MARTIN

OPEN HEARTH STEEL

OVER A MILLION AND A HALF

IVER JOHNSON

SAFETY AUTOMATIC REVOLVERS

ALREADY SOLD

Three Things Have Done it:

The superiority of the goods in every regard.

The immense popularity of "HAMMER ***THE HAMMER***" advertising.

The complete protection of the dealer. These goods cannot be used as premiums, and we never sell by mail where dealers can supply.

Send for Dealers' Catalogues, or ask your jobber—he knows

TRADE MARK IVER JOHNSON

Iver Johnson's Arms and Cycle Works, 330 River Street, Fitchburg, Mass.

New York: 99 Chambers Street. Pacific Coast: P. B. Bekeart Co., Alameda, Cal. Europe: Pickhuben 4, Hamburg, Germany.

Makers of Iver Johnson Single Barrel Shot Guns and Iver Johnson Truss Frame Bicycles

---

OAKEY'S The original and only Genuine Preparation for Cleaning Cutlery, 6d. and 1s. Canisters

'WELLINGTON'

KNIFE POLISH

JOHN OAKEY & SONS, Limited
Manufacturers of
Emery, Black Lead, Emery, Glass and Flint Cloths and Papers, etc.

Wellington Mills, London, England

Agent:
JOHN FORMAN, - 644 Craig Street
MONTREAL.

TINPLATES

All made out of the

BEST SIEMEN'S MARTIN STEEL

Brand

Best Coke or Charcoal	"Rhyd"
Good Charcoal	"Edith"
Second Charcoal	"Sawel"
Best Charcoal	"Dryw"

We are ready to offer the Canadian Trade very best value and finest quality in Tinplates.

Baglan Bay Tinplate Co., Limited

Briton Ferry, South Wales

Please mention above brands in your indents and inquiries.

NICKEL

The Canadian Copper Co.

NICKEL FOR NICKEL STEEL

The Orford Copper Co.

WRITE US FOR
PARTICULARS AND PRICES

General Offices

43 EXCHANGE PLACE

NEW YORK CITY

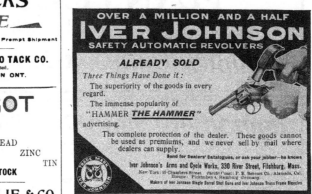

Window Glass—16-oz., O. G., single, in 50-ft. boxes—16 to 25 united inches, $2-25; 26 to 40, $2.40; 16-ox. O.G. single, in 100-ft. cases—16 to 25 united inches, $4; 26 to 40, $4.52; 41 to 50, $4.75; 50 to 60, $5.25; 61 to 70, $5.75; 21-oz. C.S., double, in 100-ft. cases, 26 to 40 united inches, $7.35; 41 to 50, $8.40: 51 to 60, $9.45; 61 to 70, $10.50; 71 to 80, $11.55; 81 to 90, $17.30.

CALGARY DEPARTMENT STORE.

A Calgary paper stated that a departmental store will be established in that city. Work on the building, which is to be five storeys high and of reinforced concrete, will be started about December 1. It is expected that it will be ready for business by April or May. The concern will be known as the Dominion Stores, Limited. The prospectus states that Calgary offers the best location in Canada for a business of this kind. The store will make a feature of its mail order department. The names of the promoters of this enterprise have not as yet been made public. It seems not unlikely that one of the big Toronto stores may be interested in it. The building to be erected will represent an outlay of a million dollars.

NEW WINNIPEG GALVANIZING PLANT.

The new galvanizing plant of the Winnipeg Galvanizing and Manufacturing Co. was installed last week and the company have made an important addition to the list of Winnipeg industries. Sample lots of all classes of galvanized hollowware have been prepared and it is evident that the "Made in Winnipeg" product will stand comparison with other goods on the market. The company are offering special prices to introduce their goods to the trade.

A CALGARY BOOKLET.

A book dealing with Calgary and the surrounding country has just been issued jointly by the City Council and Board of Trade of that city. It is attractively printed and illustrated and contains much of interest. The last report of the present Board of Trade is printed in full. There are also articles dealing with the industrial and climatic conditions of that part of the Dominion. The compilers of the book are to be congratulated.

ENLARGING CLAUSS WORKS.

The officials of the Clauss Shear Co., Fremont, Ohio, are planning improvements which will mean the increase in the number of employes in the large factory by 100 or more men. They will elect a new three-storey brick building, 90 by 40 feet, for the manufacture of safety razors. A building similar in size will also be erected on a new lot. This will also be three storeys high and will be connected with the present shear department. It is the intention to use the ground floor of the new building for office purposes and this will be connected with the present office and will, when complated, give the company an office space on the ground floor 150 feet long and 36 feet wide.

CONDENSED OR "WANT" ADVERTISEMENTS.

RATES.

Two cents per word first insertion; one cent per word subsequent insertions.

Five cents additional each insertion where box number is desired.

Contractions count as one word, but five figures (as $1,000) are allowed as one word

Cash remittances to cover cost **must** accompany all advertisements. **In no case** can this rule be overlooked. Advertisements received without remittance cannot be acknowledged

RULES FOR COPY.

Replies addressed to HARDWARE AND METAL boxes are re-mailed to advertisers every Monday, Wednesday and Friday.

Requests for classification will be followed where they do not conflict with established classified rules

Orders should always clearly specify the number of times the advertisement is to run.

All "Want" advertisements are payable in advance.

AGENTS WANTED.

THIS is the problem of many manufacturers who are anxious to get a foothold in Western Canada. It will be easily solved if HARDWARE AND METAL is given the opportunity to solve it.

AGENT wanted to push an advertised line of Welsh tinplates; write at first to "B.B.," care HARDWARE AND METAL, 88 Fleet St., E.C., London, Eng. [tf]

ONE who has sufficient organization to travel Canada east and west, fine line of Sheffield cutlery, brands already well known in Canada; none but applications from experienced and well-established men can be considered. Box 650, HARDWARE AND METAL, Toronto. [38]

AN old-established house of brass founders, making a wide range of brass foundry furnishings, such as brass cast tubing, stair rods, cornice poles, art metal work, electric light fittings, ecclesiastical metal work, are anxious to secure a thoroughly reliable and trustworthy agent or traveller for Canada, to sell goods on commission. An excellent opportunity for any firm covering Canada thoroughly. Address all enquiries to Box 654, HARDWARE AND METAL. [37]

BUSINESSES FOR SALE.

SOMEWHERE in Canada is a man who is looking for just such a proposition as you have to offer. Our "For Sale" department brings together buyer and seller, and enables them to do business although they may be thousands of miles apart.

HARDWARE, tinware, stove and plumbing business in manufacturing town in the Niagara Peninsula; no competition; $250,000 factory and waterworks will be completed this summer; stock about $3,000; death of owner reason for selling. Box 85, Thorold, Ont. [tf]

HARDWARE Business and Tinshop for sale in Saskatchewan; population 1,500; stock carried, about $14,000; turnover, $45,000; practically all-cash business; cash required, $8,000; would rent building; do not answer without you have the money and mean business; it will pay to investigate this. Box 648, HARDWARE AND METAL, Toronto. (41)

HARDWARE Business for Sale in Tweed, a thriving Eastern Ontario town. Best business in town. Must sell owing to ill-health. Liberal terms to satisfactory purchaser. Write for particulars. J. M. Robertson, Tweed. [40]

FOR SALE — First-class set of tinsmith's tools; second-hand, but almost as good as new; includes an 8-foot iron brick almost new. Apply Pease Waldon Co., Winnipeg. [tf]

GENERAL hardware and sporting goods business for sale, "Kawartha Lakes" District, Trent Valley Canal; open to offers up to Oct 1st. A snap. Box 656, HARDWARE AND METAL, Toronto. (37)

SOLE Agency for Canada. Prominent American firm is now in position to receive bids from reliable wholesale hardware firms or others for the sole right to handle their Aluminum Solder, which will unite aluminum with aluminum, or aluminum with other metals. Only bids from established houses of high standing will be considered. For particulars write under N. Y. 1012 to John M. Munchenberg, 1161-75 Broadway, New York, U.S.A. [38]

LIVE
AGENT WANTED
FOR CUTLERY

Large and Old-Established Sheffield Firm

manufacturing cutlery in all its branches,

Table and Dessert Knives and Forks, Carvers, Painters' and Glaziers' Goods, Butcher Knives, Steels, Pen Knives, Pocket Knives, Daggers, Shears, Scissors, Razors, Case Goods, Spoons and Forks, Electroplate in all its Branches

Is anxious to correspond with a First Class Man or Firm, NOT HANDLING GERMAN GOODS, who would be prepared to take up the agency, and push same thoroughly in Canada from coast to coast.

None but persons thoroughly acquainted with the business need apply. Highest references given and required.

Liberal commission to the right party.

Correspondence, in first instance, to "CUTLER," care of Hardware and Metal 10 Front St. E., Toronto

SITUATIONS VACANT.

YOU can secure a "five thousand a year" manager, or a "five hundred a year" clerk, by stating your wants under "Situations Vacant."

TINSMITHS WANTED — First-class tinsmiths wanted for points west of Winnipeg; must be good mechanics capable of taking charge of a metal department; thorough knowledge of furnace work necessary. Pease Waldon Co., Winnipeg, Man. [tf]

TINSMITH for general country work. Used to furnace work. Also Helper with 2 or 3 years' experience. Apply to G. A. Binns, Newmarket. [37]

WANTED—First-class tinsmith in New Ontario town. Must be capable of managing workshop and estimating. Good position, good salary, and no lost time for the right man. Address, Box 655, HARDWARE AND METAL, Toronto. [40]

ARTICLES WANTED.

IF you cannot afford to buy a new counter, show case, screw cabinet, store ladder, or some other fixture which you could use to advantage, try a "Want Ad." under "Articles Wanted," and you may get what you want at a bargain price.

WANTED—Two store ladders, "Myers" preferred, complete with fixtures. Send full description and price to Ilsey Bros., Red Deer, Alta. (36)

ARTICLES FOR SALE.

DON'T keep any fixtures or tools around your store for which you have no further use. They will be worth more to-day than they will a year hence. Don't keep money tied up which you could use to secure discounts from your wholesaler.

CUT Nails, 20 kegs 5-inch at $2.00 per keg. Cash with order. New stock. Taylor Bros., Limited, Carleton Place, Ont. [37]

GOOD silent salesman for sale. Reason for selling, advertiser wishes to purchase larger one. A bargain for immediate sale. Box 658, HARDWARE AND METAL, Toronto

FOR Sale cheap, latest improved National Cash Register, four drawers, used only eighteen months. The Wilson Grocery Co., Gananoque, Ont. [37]

SITUATIONS WANTED.

TRAVELLERS, clerks, or tinsmiths, wishing to secure new positions, can engage the attention of the largest number of employers in their respective lines, by giving full particulars as to qualifications, etc., under "Situations Vacant."

COMMERCIAL gentleman with nine years' trade connection with ironmongers, architects and public institutions in Great Britain desires position as representative of a Canadian manufacturing firm. Box X, HARDWARE AND METAL, Montreal.

HARDWARE Salesman wants position; can furnish first class references; capable of managing city or town business. Box 653, HARDWARE & METAL, Toronto. (39)

HARDWARE CLERK, Irishman, 27, long experience of trade in best houses Old Country, one year with large Canadian wholesale firm, desires change, would go west. Unexceptionable references. Photo if required. Apply box 657, HARDWARE AND METAL, Toronto. (39)

THE WANT AD.

The want ad. gets work for workers and workers for work.

It gets clerks for employers and finds employers for clerks. It brings together buyer and seller, and enables them to do business, though they may be thousands of miles apart.

The "want ad." is the great force in the small affairs and incidents of daily life.

satisfied ? ' If those questions can be answered for the prospective salesman affirmatively, he is the man for the place.

WICKEDNESS OF STREET LIGHTING.

The wickedness of street lighting was the subject of an article printed in March, 1819, by a newspaper of Cologne, Germany. It was brought to light by the Journal fur Gasbeleuchtung at a celebration some time ago of the eithtieth anniversary of the date when gas lamps were first employed in Berlin. Objections to street lamps were taken by the writer on several different grounds. He declared the illumination of streets was wrong theoretically, for it was :

1. An attempt to evade the law of God, Who appointed darkness to follow light and Who ordained the moon to shine at certain times only. Mortals had no right to criticize or interfere with the plans of the Deity by turning night into day.

2. It was wrong legally, for the cost of the lamps was an indirect tax falling on people who did not require any light, and, indeed, found it an inconvenience.

3. It was wrong hygienically, for the fumes of oil or gas would poison the air and affect the health of delicate persons, while by encouraging others to be abroad after dark the light would assist in the spread of colds and similar affections.

4. It was wrong morally and philosophically, for it would lower the standard of morality and remove that fear of darkness which prevents many weakminded individuals from plunging into crime. It would increase drunkenness, enabling the toper to remain in the public house until late at night, and it would assist lovers.

5. It was wrong from a police point of view, for the light would make horses shy and robbers bold.

6. It was wrong politically, for the money spent on the light would go abroad, decreasing the national wealth.

7. It was wrong nationally, public fetes illuminated by artificial light being potent factors in the growth of patriotism, whereas if the streets were lighted up every night specially illuminated functions would lose their effect upon the imagination of the populace.

BUILDING AND INDUSTRIAL NEWS

For additional items see the correspondence pages. The Editor solicits information from any authoritative source regarding building and industrial news of any sort, the formation or incorporation of companies, establishment or enlargement of mills, factories or foundries, railway or mining news.

Industrial Development.

Fire recently did $15,000 damage to the planing mills of S. Hill & Son, Saskatoon, Sask.

A local telephone company has been formed at Vancouver, B.C., and will immediately apply for a charter.

A large force of men are engaged in laying underground cables for the Bell Telephone Company, at Peterborough, Ont.

Walsh & Arnaud will install an up-to-date brick plant with a capacity of 50,000 bricks a day, near New Westminster, B.C.

The Consumers' Gas Company, Toronto, will build a one-storey brick boiler house, condenser house and smokestack, at a cost of $60,000.

The Canadian Machine Telephone Co., intend erecting a factory which will employ two hundred men. Brantford, Hamilton and Toronto are all after the new industry.

O. & C. Clarke, Montreal, propose to establish a factory in that city for the manufacturer of brass goods providing the city will grant them a free site, with exemptions and other privileges.

The installation of a lighting plant at the new factory of the Gillson Company, Guelph, Ont., is being carried out by A. C. Lyons & Company, Brantford, and will be completed by the end of this week.

The Sechelt Brick & Tile Company, will install a $50,000 plant at Sechelt, Vancouver, B.C. An immense deposit of suitable clay has just been discovered at Sechelt, the quality of which is pronounced by experts to be equal to the best on the continent.

A new company, the Dominion Smoke Consuming Company, is being organized, capitalization $50,000, with head offices in Toronto. The company has applied for an Ontario charter and has purchased the Canadian patents from the National Smoke-Consuming Company, of Buffalo, N.Y.

The largest fire that has happened in New Glasgow, N.S., for years occurred last week, when the magnificent building, plant and machinery of the Standard Brick & Tile Company were burned to the ground, and the great industrial establishment is now a mass of ruins. The loss is $300,000.

R. D. Isaacs, St. John, N.B., has submitted to the common council a request for a free site and exemption from taxation for twenty years for a proposed car works to be erected there. He says a company with $250,000 capital, which will be increased to a million, is behind the scheme. The council referred the matter to a committee.

The Canadian Machine Telephone Co., Brantford, have filed plans for the laying of two miles of underground telephone cable. The cable will be laid in ducts, some eight miles of which will ultimately be constructed. The manholes will all be constructed of concrete. The plans of the company involve considerable expenditure.

The Utah Railway Paint Co., Kansas City, Mo., capitalized at over a quarter of a million of dollars, proposes to establish an immense industry of a similar nature at Edmonton, Alta. The company are said to have located in the vicinity of Edmonton immense beds of a mineral clay, called Kaolin, from which paint can be manufactured. The proposed plant will cost $100,000.

The Chicago and St. Lawrence Steam Navigation Company's new steamer the E. B. Osler, was launched at Bridgeburg, Ont., on Saturday. The Osler has the distinction of being the largest ever constructed in Canada. It has a capacity of 9,000 tons. It is 510 feet long and 56 feet wide, and is built throughout after the most approved modern plans. The launching was witnessed by quite a party of Toronto people.

Building Notes.

H. Gleiser, Oxbow, Sask., will built a new hotel.

Robt. Shaw, Oxbow, Sask., will erect a new business block.

Seymons & Roe, Toronto, will erect a brick dwelling at a cost of $10,000.

S. J. Castleman, Vancouver, will erect a six-storey business block to cost approximately $100,000.

A new home to be erected for the Seamen's Mission, St. John, N.B., will cost $15,000.

The congregation of Paisley St. Methodist Church, Guelph, will erect a new edifice at a cost of $10,000.

C. R. S. Dinnick, Toronto, will erect six pair semi-detached brick residences at a total cost of $30,000.

T. Gingras, Toronto, will build four pair semi-detached houses at an aggregate cost of $14,500.

R. Cassidy, Vancouver, B.C., will erect a six-storey building in that city at an approximate cost of $250,000.

Harry Weineberg, Toronto, will erect a large store on the corner of Yonge and Edward Sts., at a cost of $30,000.

The J. Y. Griffin Co., Edmonton, Alta., will erect a large office block and cold storage warehouse at an estimated cost of $100,000.

The Manitoba Government has let the contract to James M. and John J. Kelley, Winnipeg, for a telephone exchange building to cost $97,172.

The barns of the Macdonald Agricultural College, St. Anne de Bellevue, Que., which were recently destroyed at a loss of $35,000, will be rebuilt.

The value of Vancouver building permits for the month of August was $708,775, being an advance of 131 per cent. over the $306,925 recorded for the corresponding month last year.

Municipal Undertakings.

A new school building will be erected at New Westminster, B.C.

Hull, Que., will erect three new fire stations at an aggregate cost of $25,000.

The town of Orillia, Ont., will make a profit this year of nine thousand dollars from its light and power plant.

The contract for the new city hall to be erected at Calgary, Alta., calls for an expenditure of $142,124, exclusive of equipment.

Railroad Construction.

A new flour shed is about to be erected by the I. C. R. at St John, N.B., at an expenditure of $20,000.

The Ottawa Electric Railway Company propose to extend their line to the Experimental Farm, near that city.

The contract for the G.T.P. bridge over the Kaministikwia river, Fort William, has been awarded to the Canadian Bridge Company.

Contracts for the erection of a $50,000 addition to the car shops at London, have been let to Mr. John Hayman, and work will be commenced immediately.

The C.P.R. are fitting up a plant at Vancouver, B.C., where they intend to generate Pintsch gas, to be utilized in illuminating passenger coaches on their western lines.

The rebuilding of the railway bridge across Rice Lake, near Port Hope, Ont., and the rehabilitation of the old Cobourg and Peterborough Railway, is being seriously discussed.

The Southwestern Traction Company's line, after being tied up for three weeks, on account of the recent fire in the car barns, is now in full operation between London and St. Thomas.

The C.P.R. are contemplating building a million-ton coal dock at Fort William. The entire yards of the company will be remodelled at great expense, and the plans when carried out will make Fort William the finest inland steamship and railway terminal in the world.

A contract, which will involve the expenditure of over $1,000,000 was recently awarded by the C.P.R. to Macdonnell & Gzowski, engineers and contractors, Vancouver, B.C. The contract calls for the alteration of the grade of the main line of the C.P.R. at Field, B.C., the driving of a mile and a half of tunnels on each side of the Kicking Horse River and the construction of two bridges. It is said that the work cannot be completed in less than eighteen months, and that operations will commence as soon as men can be placed on the ground. This is one of the largest pieces of construction work ever undertaken on the C.P.R.

Mining News.

The Canadian Smelting and Refining Co., Sault Ste. Marie, will erect a large smelter and expect to have the plant in operation within six months. $1,000,000 is behind the enterprise.

A large and enormously rich deposit of iron ore has been discovered at Lakevale, Antigonish, N.S. The property is said to be one of the best yet found and is owned by American capitalists.

The directors of the Golden Peak Mining Company, Larder Lake, Ont., have decided to install a mill of 20 stamps on their property. The mill will be put in as soon as the road from Boston to Larder Lake is completed. This will be in about a month.

It is stated authoritatively that a company called the Muggley Concentrators, Limited, will put in a hundred ton concentrating mill shortly at Cobalt. The mill will use twenty screens and crushing rolls to increase the capacity. Owing to the delay in getting in machinery it is probable that the plant will not be running before January.

Eight new engines of the "consolidated" type have been received at the London, Ont., roundhouse of the Grand Trunk, and are now being operated on that division. As a result, the situation has been greatly relieved. The engines are of the heaviest type in Canada. They are able to haul 90 loaded cars and have four drive wheels. They are much heavier than the locomotives of the "900" and "800" class and will be a great addition to the freight service.

Companies Incorporated.

Canadian Gypsum Co., Toronto; capital, $20,000. To manufacture gypsum, gypsum plaster and kindred substances. Incorporators: J. S. Lovell, H. Chambers, Robert Gowans, S. G. Crowell, Walter Gow, all of Toronto.

The Tyrell Cooler & Filter Co., Ottawa; capital, $100,000. To manufacture coolers, filters, and similar articles. Provisional directors: P. D. Herbert, H. W. Tyrell, D. T. Smith, James Herbert and J. R. Osborne, all of Ottawa.

Union Brass Goods Co., Toronto; capital, $150,000. To carry on business as brass and iron founders. Provisional directors: Morley Punshon Vander Voort, Francis Joseph Stanley, and William Andrew Smiley all of Toronto.

The Dickson Bridge Works Co., Campbellford, Ont.; capital, $40,000. To carry on the business of bridge builders and structural steel manufacturers. Provisional directors: James H. Caskey, Campbellford; William Coates Macann, F. C. Downey, Toronto.

Standard Sanitary Manufacturing Co., of Pittsburg, Montreal; capital, $250,000. To manufacture plumbers' enamelware, brass goods, and supplies. Incorporators: T. C. Collins, J. N. Collins, F. J. M. Collins, P. M. Robertson, H. J. Trihey, all of Montreal.

G. A. Rudd & Co., Toronto; capital, $100,000. To manufacture and deal in harness, sadlery, leather and kindred articles. Provisional directors: George A. Rudd, James McGregor Young, John E. Elliott, Gordon C. Rudd, and Gregory S. Hodgson, all of Toronto.

Hamilton Steel & Iron Co., Hamilton, Ont.; capital $5,000 000. To carry on a general rolling mills and smelting business. Incorporators: Robert Hobson, William Southam, John Milne, Albert E. Carpenter, Charles E. Doolittle, George L. Staunton, all of Hamilton.

Ingersoll Sergeant of Canada, Limited, Montreal; capital, $20,000. To manufacture air compressors, rock drills, pumps and pneumatic tools. Incorporators: H. D. Lawrence, William Morris, A. F. Plant, R. F. Morris, Sherbrooke, Que.: W. E. McIver, Richmond, Que.

Anthes Foundry, Limited, Toronto; capital, $100,000. To manufacture castings, iron pipe, soil pipe and fittings, and plumbers' and steamfitters' supplies. Provisional directors: Lawrence L. Anthes, William Wallbridge Vickers, and Herbert C. Sparling, all of Toronto.

Monterey Plumbing Co., Toronto; capital, $50,000. To carry on a general plumbing, steamfitting, heating and lighting business, and to deal in general hardware. Incorporators: Gerard Ruel, G. F. Macdonnell, A. J. Mitchell, F. C. Annesley, Robert P. Ormsby, all of Toronto.

National Oxide Paint & Color Co., Hamilton, Ont.; capital, $50,000. To manufacture paint pigments, colors, oils, paints, varnishes, and japans. Provisional directors: George Strond, G. F. Webb, Helena O'Sullivan, T. J. O'Sullivan, and Alfred Strond, all of Hamilton, Ont.

The Dominion Nickel Copper Co., Toronto; capital, $10 000,000. To carry on the operations of a mining, milling, reduction, and development company. Incorporators: James Houston Spence, Ada May Duncan, Lillian M. Heal, Charles E. Freeman, Gertrude E. Jamieson, all of Toronto.

L. H. Hebert & Co., Montreal; capital, $350,000. To carry on the business of ironmongers and merchants in hardware, paints, oils and chemical compounds. Incorporators: Louis H. Hebert, Alfred Jeannotte, Eugene Poitevin, Gustave Busseau, Montreal; Joseph E. Theriault, Joliette, Que.

The New Liskeard Concrete Co., New Liskeard, Ont.; capital, $40,000. To manufacture concrete cement, artificial stone and kindred materials. Provisional directors: S. Jewell, V. E. Taplin, W. H. Carruthers, W. V. Cragg, A. McKelvie, J. E. Whyte, and F. L. Smiley, all of New Liskeard, Ont.

Ideal Foundry Co., Toronto; capital, $100,000. To engage in the business of general foundrymen and to manufacture cast or wrought iron, steel, brass, aluminum or composition specialties. Provisional directors: Henry Edward Pearce, William Henry Smith, Arthur Gate, Matthew Irving, and William Baggs, all of Toronto.

Simplex Gas Co., Toronto; capital, $40,000. To manufacture under Colwell's patents the Simplex Gas Machine; to manufacture gas burners, stoves, heaters ranges and hardware, and mechanical specialties. Provisional directors: Walter Ernest Colwell, Harry Herbert Colwell, Oakville, Ont, and Henry H. York, Toronto.

CATALOGUES AND BOOKLETS.

When sending catalogues for review, manufacturers would confer a favor by pointing out the new articles that they contain. It would assist the editor in writing the review

By mentioning HARDWARE AND METAL to show that the writer is in the trade, a copy of these catalogues or other printed matter will be sent by the firms whose addresses are given.

Mop Wringer.

The Art of Mop Wringing is the title of an artistic 5½x8½ catalogue issued by the White Mop Wringer Co., Fultonville, N.Y. The various kinds of mop wringers illustrated and described through this booklet are the result of sixteen years of study and experience, and are claimed to be the highest development in the art of mop wringing. They work easily and efficiently, fit any size of pail, do not pull the mops to pieces and are so simple in construction that there is nothing delicate to get out of repair. With White patent mops the housewife can wash the floor and wring the mop as often as desired without wetting or soiling her hands in any way. An attractive display stand is loaned with White wringers and acts as a silent salesman for the retailer. Those desiring catalogues may obtain them by communicating with the company, mentioning this notice.

Electric Supplies.

The Canadian General Electric Co., Toronto, have issued to the trade an artistic and comprehensive catalogue containing 109 8x 10½ pages. The catalogue is devoted to the many lines of electric construction materials, electricians' tools, and special electric supplies manufactured by that firm. The catalogue is copiously illustrated with engravings of the different articles dealt with, and price lists and full descriptions follow each engraving. A full index at the back of the book makes it easy for the reader to get at the particular line of goods in which he is interested without losing any of his valuable time. This comprehensive and conveniently compiled catalogue is available to the trade by applying to the company, mentioning this paper.

Seeing Canada in a Russell.

Little Trail Stories in a Big Land, is the title of a neat little 5x6½ booklet issued by the Canada Cycle and Motor Co., Toronto Junction, Ont. The reading matter contains graphic descriptions of trips made in different parts of Canada in a Russell car. First of all, the trip through the File Hills, Sask., by the Governor-General of Canada, Earl Grey, and his party, is described in a very interesting manner. Next, the thrilling incidents of a run through the heart of the Muskoka District by a party of four are narrated, and in another part of the booklet an interesting deseirption is given of a winter run from Guelph to Toronto in the midst of deep snow. The latter portion of the book is devoted to an explanation of the manufacture of the Russell, every part of which is made in Canada at the company's large plant at Toronto Junction. Copies of this interesting booklet can be had by applying to the company, mentioning this paper.

Facts About Francis.

The Francis, Sask., Board of Trade has just issued an interesting 5x6 booklet descriptive of that growing western town and its immediate vicinity. The booklet is replete with interesting engravings and the compilers have succeeded in giving, concisely, the early history of the town and district, its struggling infancy, a short review of the business houses and their proprietors, a description of the wonderful growth and development of the town and surrounding country, and finally, the reason of such growth and the possibilities of the future. Copies of this instructive booklet can be obtained by writing C. R. Gough, secretary-treasuer Francis Board of Trade, mentioning this notice.

Kraeuter's Tools.

Kraeuter & Co., Newark, N.J., have issued a fifty-page 6x9 catalogue describing the many lines of high-grade mechanics' tools manufactured by that firm. The tools shown in the engravings and described throughout the book include ticket punches, callipers, dividers, gas pliers, combination pliers, rivet sets, round and oval punches, spring punches, belt punches, washer cutters, cold and cape chisels, buttonhole cutters, pinking irons, wood-turners' sizers, milliners' pliers and, in fact, every similar small tool that one could imagine. Any one interested may obtain a copy of this catalogue by writing the company, mentioning this paper.

Weiss Cutlery.

J. Weiss, Sons & Co., Newark, N.J., have just issued an attractive and well-printed 6x9 catalogue, setting forth the excellencies of Weiss shears, scissors and razors. The catalogue is bound in an attractive red cover on which the subject of the catalogue and the firm's name is embossed in heavy black letters which can be read by the busy merchant at a glance. Between the covers, Weiss shears, scissors, snips and razors each in turn receive attention and the various facts concerning the manufacture and wearing qualities of each are marshalled before the reader in a convincing and interesting manner. Weiss cutlery combines best material, highest skilled labor, most careful inspection and over half a century's practical experience in its manufacture. Every detail of manufacture has the company's personal attention and a written guarantee accompanies their output. This guarantee states that every article manufactured by the firm is free from flaws and imperfections; is absolutely perfect in material and workmanship, and that any Weiss dealer is authorized to exchange any unsatisfactory article bearing the company's trade mark, on application to the company.

Vigorous Vancouver.

The twentieth annual report of the Vancouver, B.C., Board of Trade has just been issued to the public in the form of an attractive 6x9 booklet, containing 144 pages, exclusive of a full index at the back of the book. The reading matter from cover to cover is full of valuable and interesting information concerning the commercial, industrial, educational, religious and social life of Vancouver. A full history of Vancouver is given from its birth in 1885 up to the present time, when its population numbers over 60,000 and its bank clearings stand fourth in the official list covering all the cities in Canada. Artistic engravings of the harbor and representative scenes around Vancouver enliven the pages of the report and sustain the interest of the reader. Copies may be obtained by applying to Wm. Skene, secretary Vancouver Board of Trade, mentioning this paper.

HELP THE BOY ALONG.

Don't "fire" the boy! Keep him and make a better boy of him! If you do he will be a better man. Boys are all right if you understand them. In every one of them rightly handled there is a germ of manhood, and possibilities of mighty success in the future. Grown up under kindly influences, the excess energy that made them enjoy their boyish escapades will be directed to the accomplishment of great things. So don't "fire" the boy. Talk to him. Get him interested in his work. Tell him of the things before him in life. Teach him thrift and industry. Remember, he is just a little raw material, out of which you can fashion a better man than you are, no matter how good you are.

CEMENT AS A METAL PRESERVATIVE.

The preservative qualities of cement have been well demonstrated in maritime work during the past few months. The larger vessels on the Pacific Coast have a coating of fine cement on the inside steel plates of the hull. Lately a number of these steel-hulled ships have been docked for repairs and the plates torn away. It was found in every instance that where a plate had received a cement coating it was as sound and bright as when first placed in position.

On the other hand, where a vessel had not been coated, the steel was pitted and corroded, and in some cases this honey-combing extended clear through the plate. The effect was further noticed on vessels which were only lined on the inside. The outside half of the steel was thoroughly "rotted," while the inside, protected by cement, was as bright and sound as when new.

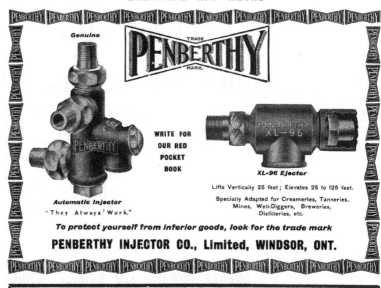

News of the Paint Trade

ONTARIO OIL AND GAS FIELDS.

The production of petroleum and natural gas in Ontario, and the conditions and recent developments of the fields, are the subjects of a paper by Eugene Coste in the report of the Ontario Bureau of Mines, just issued. The production of both oil and gas showed a marked increase in 1906, recent fields becoming considerably enlarged and more productive. In the case of the oil the Leamington field of Essex county, and the Moore field of Lambton county, account for most of the increase, while the gain in natural gas production is due to the opening of new wells in Welland and Haldimand counties, from which several cities are now supplied. From the figures supplied by a number of big logs of deep wells, which are given, Mr. Coste draws the conclusion that, from Osprey township in Grey county, to Petrolea, in Lambton county, the strata dip more or less uniformly in a southwest direction; but continuing further in the same direction, they again rise more or less gradually. It is therefore demonstrated that the Lambton county oilfields are really in the bottom of a broad, deep, syncline, instead of being on the Cincinnati anticline, as is often contended. New records of the wells in the Niagara Peninsula adduced, as well as others previously published, show that the strata of that peninsula have a more or less constant dip to the south-southeast. As the gas fields of Welland and Haldimand counties are on the flank of that long slope, which continues to the south across Lake Erie into New York and Pennsylvania and also to the north across Lake Ontario, it is to be seen that the famous anticlinial theory of oil and gas production is far from being supported by the facts either in the gasfield of the Niagara Peninsula or in the oilfields of Lambton county. Disturbances and faults exist in these fields, sometimes bringing up the strata locally in blocks, terraces, or sharp folds, but the more or less broad anticlines which the supporters of the theory of the organic origin of oil and gas regard as necessary to the large accumulation of those products are conspicuously absent. Another result of the drilling recorded in the logs is that both oil and gas have been found in the lower part of the Trenton and right on the top of the Archean formation, which cannot be explained by any other view of the origin of oil and gas than that of volcanic emanations from below, as held by the French school of geology. The practical bearing of this conclusion upon the oil and gas industry is important. In planning future developments in Ontario it is of consequence to know that oil and gas may be looked for in large quantities just as well under the Petrolea oil rock—the coniferous limestone—as in or above it, and, therefore, drilling along the fissured oil belts which traverse Ontario, in exactly the same manner as they traverse Pennsylvania, Ohio and Indiana, will develop similar fields of oil and gas. In support of his conclusion Mr. Coste adduces the facts that in the Leamington oilfield the oil comes from the Guelph limestone, a stratum 1,500 feet below the coniferous and that the gasfields of Welland and Haldimand derive their supplies which are still lower strata. Latest advices state that the oil production from the Tilbury field is holding up, notwithstanding that a number of drilling and pumping rigs had to shut down on account of a scarcity of water. In the Shallow field only two wells are running owing to the water famine. During the month of July about 38,000 barrels of oil were shipped from the field to the refinery at Sarnia. Two gas lines are being laid to the Shallow field, and when these are completed, which is only a matter of days, operations will be resumed on a large scale there. The Kennedy Oil and Gas Co. brought in a great well on the Crosbury Farm, Tilbury East, recently. It is making over 200 barrels of oil a day.

E. Lossing, painter and paper hanger, Toronto, has assigned to A. E. J. Blackman. A meeting of the creditors was held on Wednesday.

FURNITURE CLEANSER.

Splashes of dirt on polished furniture are removed with soap and water, and the wood is well rubbed with a mixture of equal parts of spirit and oil or spirit and turpentine, applied with a woolen rag. The mixture has both a cleansing and polishing action, the polish being retained for a long time if well rubbed in and the surplus wiped off. Another good preparation for the same purpose is a solution of stearine in oil of turpentine and a little spirit, care being taken not to use so much stearine that white streaks are produced in the mass. When the turpentine and spirit have evaporated, the wood is well rubbed with a woolen rag. This gives an excellent polish that can be renewed by rubbing when dimmed. Furniture with a matt finish can be renovated with a thin solution of white wax in oil of turpentine or by rubbing it over with linseed oil.

Paint and Oil Markets

TORONTO.

Toronto, Sept. 13.—While trade in the paint and oil markets is not as brisk as local jobbers would like it to be, nevertheless, this week's transactions show a considerable increase over those of last week, and, everything considered, there is no just cause for complaint. Strictly local orders are certainly much more in evidence this week and the increase in the volume of local trade can easily be explained by the fact that, with the close of the big Exposition here, local dealers have again settled down to serious business and have now more time to look over their stocks. Outside business is somewhat hampered by the fact that country fairs are now in full swing all over the province; consequently, travelers often find the proprietor of the small retail store away attending the village fair, or that of a neighboring village, and in many cases are forced to pass on their way without receiving his order.

White Lead—Orders are coming in very satisfactorily for the time of the year and many enquiries for white lead in sorting quantities have been received during the week. Prices continue firm as under: Genuine pure white lead is quoted at $7.65, and No. 1 is held at $7.25.

Red Lead—Business shows a slight increase over last week and, in sympathy with white lead, prices are firm. Present quotations are: Genuine, in casks of 500 lbs., $6.25; ditto, in kegs of 100 lbs., $6.75; No. 1 casks of 500 lbs., $6; ditto, in kegs of 100 lbs., $5.50.

Glass and Putty—The fall trade in glass and putty is opening out well and transactions in both are daily increasing. Present indications are that this fall's business in both commodities will be extremely large. Prices on both glass and putty are unchaged.

Petroleum—The demand is steadily in- creasing as the autumn approaches. Prices are unchanged: Prime white, 15c; water white, 14½c; Pratt's astral, 18c.

Shellac—While sales for painting purposes are not very large at present, there is, however, a strong demand for shellac on account of the many and varied industrial uses to which gum shellac is now being put. Prices are well maintained, as follows: Pure orange, in barrels, $2.70; white, $2.82½ per barrel; No. 1 (orange), $2.50.

Linseed Oil—In linseed oil the prices in Great Britain have again advanced, so that the import price for cargo lots has now very nearly reached the selling price here for 4-barrel lots. This state of the market cannot continue very long and it looks as if prices must be advanced in the very near future. As the amount of oil to be produced in Canada from home-grown seed is not yet sufficient to supply the market, the outside price must rule and it is expected by many that for next week the prices will be advanced. We quote: Raw, 1 to 3 barrels, 64c; 4 barrels and over, 63c. Add 3c to this price for boiled oil, f.o.b., Toronto, Hamilton, London, Guelph and Montreal, 30 days.

Turpentine—The demand for turpentine is not very pronounced at the present time and prices remain fairly steady, as under: Single barrels, 79c; two barrels and upwards, 78c; f.o.b. point of shipment, net 30 days; less than barrels, 5c advance. Terms, 2 per cent., 30 days.

For additional prices see current market quotations at the back of the paper.

MONTREAL.

Montreal, Sept. 13.—Owing to the numerous fairs now being held throughout the country, orders being received by paint makers are still somewhat small. Reports from the different districts are varied, some reporting very cheerfully as to present conditions and future prospects, and others reporting that there is some anxiety as to the Fall trade and an apparent unwillingness to stock up liberally for the present.

The factories are all busy, and supplies of raw material are coming along freely. Paint makers are all well satisfied with this year's trade, which has without a doubt greatly exceeded all previous years in volume.

Linseed Oil—Is showing a firmness. Orders are not very heavy and quotations remain firm and unchanged: Raw, 1 to 4 barrels, 60c, 5 to 9, 59c; boiled, 1 to 4 barrels, 63c; 5 to 9, 62c.

Turpentine—There has been a slight easing off in the southern market. A decline of 1c was noted at Savannah. Local prices are unchanged and firm. Single barrels, 80c.

Ground White Lead—The output at present is somewhat limited. Quotations cannot be characterized as strong. We continue to quote: Government standard, $7.50; No. 1, $7; No. 2, $6.75; No. 3, $6.35.

Dry White Zinc—A fair amount of trade is being done Prices are unchanged. V.M. Red Seal, 7½c; Red Seal, 7c; French V.M., 6c; Lehigh, 5c.

White Zinc Ground in Oil—Quotations continue very firm and this commodity seems to be increasing in popularity. Pure, 8½c; No. 1, 7c; No. 2, 5½c.

Red Lead—Shipments at present are spasmodic and largely governed by the call for structural work in some localities. Prices are unchanged Genuine red lead, in casks, $6.20; in 100-lb. kegs, $6 50; in less quantities at $7.25 per 100 lbs. No. 1 red lead, casks, $6; kegs, $6.25, and smaller quantities, $7.

Gum Shellac—A scarcity is reported of white gum shellac, and makers are experiencing some difficulty in keeping up stocks, owing to increasing demand. Some large parcels of fine orange have been received in the local market. Conditions, generally, are strong. Prices are unchanged. Fine orange, 60c per lb.; medium orange, 55c per lb.; white (bleached), 65c per lb.

Shellac Varnish—Together with other grades of varnishes there is a steady output, with quotations unchanged.

Pure white bleached shellac, $2.80, pure orange, $2.60, No. 1 orange, $2.40.

Putty—A good demand exists and stocks are cleared as fast as they come from the mills. Prices are firm. Pure linseed oil, $1.85 in bulk, in barrels, $1.00, in 25-lb. irons, $1.90, in tins, $2, bladder putty, in barrels, $1.85.

COLORS FOR HOUSE PAINTING.

The effect of a house may be improved or marred by the choice of the colors for painting it. There is no doubt that the old New England houses, with their white paint and green blinds, presented a very charming appearance, when seen amid the beautiful elms of the quiet village streets, and, indeed, a white house almost always looks well when surrounded by considerable foliage. On the other hand, a white house becomes very glaring, when it stands out in the open, with the bright sky as a background. This shows at once, says Edward Hurst Brown in an article on "Painting," in the American Carpenter and Builder, the necessity for making the colors harmonize or fit in with the surroundings, hence no general rule for color selection can be given. The idea that one color or one class of colors should be chosen because it is fashionable or the prevailing mode is for this reason rather an absurd one. In general, it may be said that dark colors should be selected only for a large house that stands out in the open, while lighter colors or tints should be chosen for houses that stand among trees or which are not over large. A house looks larger when painted in light colors than when dark colors are used. This is a point that should always be remembered, for most persons experience a keen feeling of disappointment if their house appears small, while if it can be made to look larger than it is, they are more than satisfied.

LINSEED OIL AS FOOD.

It may surprise many of our readers to learn that linseed oil is extensively used as a food by the peasants in Hungary, Poland, Russia and other parts of eastern Europe. The demand for table purposes in Germany is exceedingly limited, although even in that country it is regularly consumed by a small number of people among the poorer classes and its use is slowly extending—possibly owing to the great influx of foreign laborers in recent years, principally of Polish origin.

TO PAINT GALVANIZED IRON.

To get paint to adhere to galvanized iron seems to be a difficult problem and various experiments have been made in order to find a way of getting around the difficulty. The United States Government has adopted a mode of procedure which seems to give satisfactory results. Specifications call for the use of vinegar in washing the surface before painting. This roughens or corrodes the surface and gives the paint much better adhesion.

SHELLAC AND VARNISH.

A fair estimate of the covering powers of shellac and varnish is as follows: One gallon of shellac varnish will cover 400 square feet of white pine, first coat; it will cover 500 square feet on second and succeeding coats. Interior varnish will cover from 350 to 400 square feet to the gallon, first coat, and nearly 600 square feet for succeeding coats. On hard wood, filled with paste filler, interior varnish will cover from 50 to 75 square feet more of surface than on the unfilled wood.

COLORING PUTTY.

Ordinary painters' colors in oil, preferably transparent colors such as burnt and raw sienna, burnt and raw umber and lampblack, are best for coloring putty. Make the putty several shades darker than the wood, as all wood grows darker with age.

A COLOR PUTTY.

When it is desired to make a color putty to match a hardwood finish use dry white lead, and not whiting, for whiting does not give clear tints of color, but lead does. A little whiting would do no special harm, but it is not necessary. To make this putty, use boiled oil, and mix the whiting and oil to a stiff mass, and then add some coloring pigment in oil. Make the putty about the general color effect of the wood, not matching any particular part.

DIPPING SHINGLES.

A very good method of treating shingles is to dip the part that goes to the weather in a solution of persulphate of iron, making 2 to 2½ Baum. After they have been on the roof a while apply a coat of hot linseed oil. The hot oil alone is also very good. The creosote stains are better still, as they not only preserve the wood, but their colors are very durable, and pleasing also from an artistic standpoint.

AMERICAN OIL IN CHINA.

Last year 1,088,800 gallons of American oil were entered for import at Nankin, China, being twenty-five times the amount entered in 1903. The present famine aids the sale of this article, as the bean oil, commonly used, is dearer than the imported kerosene. No part of the native city can be visited without seeing evidences of the success of the Standard Oil Company. Many shanties are erected from the wooden boxes, and the tin cans are everywhere used as buckets, coal scuttles, dustpans, etc. The lamps and oil stoves imported by this company greatly increase the sale of the oil. Sumatra furnished only 31,000 gallons and Russia 5,000 gallons last year.

The death occurred recently of W. J. Townley, of Townley & London, painters and decorators, Toronto.

DON'T SELL PAINT THAT IS CHEAP AND INFERIOR. The **Falcon Brand** for paints and varnishes, etc., means that they are not only **good**, but that they are the very **highest grade** that up-to-date paint making can produce.

There's good money in it for you.

Send us a sample order and we will prove our statement.

PHASANTITE
an ideal water paint.

FALCONITE
White Enamel of high-grade.

LACKERITE
Our new Varnish Stains in 12 colors.

Wilkinson, Heywood & Clark, Limited

EST. 1796

300 Notre Dame St., MONTREAL

Branches and agencies all over the world.

FRANCIS C. REYNOLDS, Can. Manager.

Any manufacturer can make a green paint. But there is only ONE

IMPERIAL FRENCH GREEN

It lasts years longer than any other green. Does not lose its brilliancy. Be sure the label has the name of

L U C A S

PHILADELPHIA NEW YORK CHICAGO BOSTON SAN FRANCISCO

Fresh Paint

It's a sure sign of progress wherever it is seen and when it's Martin-Senour 100% pure Paint it's a sign that there is a wide-awake dealer not far away. Wide-awake dealers handle Martin-Senour 100% pure Paint because it's a money-maker, that is one reason. It's a trade builder, that's another reason. The manufacturers do a deal of powerful direct advertising that helps the dealer's trade grow and widens his reputation, that's another reason. Would you like to belong to the wide-awakes? There is money in it and we will tell you how we help you get it if you write us. Better do it today.

The MARTIN-SENOUR CO. Ltd.,

"Pioneers of Pure Paints"

142-144 Inspector Street, Montreal 2514-2522 Quarry St. and Archer Ave., Chicago (Est. 1878)

The WINNIPEG PAINT & GLASS CO. Ltd., Winnipeg. The MARITIME PAINT & VARNISH CO. Ltd., HALIFAX, N. S.

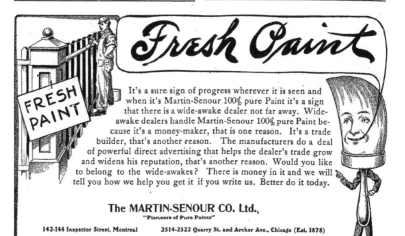

CURRENT MARKET QUOTATIONS.

September 13, 1907

These prices are for such qualities and quantities as are usually ordered by retail dealers on the usual terms of credit, the lowest figures being for larger quantities and prompt pay. Large cash buyers can frequently make purchases at better prices. The Editor is anxious to be informed at once of any apparent errors in this list, as the desire is to make it perfectly accurate.

METALS.

ANTIMONY.

Cookson's ....per lb...		0 15
Hallett's		0 14½

BOILER PLATES AND TUBES.

Plates, ⅜ to ½ inch, per 100 lb.	2 40	2 50
Heads, per 100 lb	2 65	2 75
Tank plates 3-16 inch	2 60	2 70
Tubes per 100 feet, 1½ inch ..	8 25	8 50
" " 2 " ..	9 00	9 10
" " 2½ " ..	10 50	11 00
" " 3 " ..	12 00	12 50
" " 3½ " ..	15 00	16 00
" " 4 " ..	19 25	20 00

BOILER AND T.K. PITTS.

Plain tinned and Spun, 25 per cent. off list.

BABBIT METAL.

Canada Metal Company—Imperial, genuine 60c.; Imperial Tough, 60c; White Brass, 50c. Metallic, 35c; Harris Heavy Pressure, 25c, Hercules, 25c; White Bronze, 15c; Sta Frictionless, 14c.; Alluminoid, 10c.; No. 4 9c per lb.

James Robertson Co—Extra and genuine Monarch, 60c.; Crown Monarch, 50c; No 1 Monarch, 40c.; King, 30c; Fleur-de-lis, 20c; Thurber, 15c; Philadelphia, 12c.; Canadian, 10c.; hardware, No 1, 15c; No, 2, 12c; No. 3, 10c. per lb.

BRASS.

Rod and Sheet, 14 to 30 gauge, 25 p c. advance.

Sheets, 12 to 14 in.	0 30
Tubing, base, per lb 5-16 to 2 in ...	0 33
Tubing, ⅜ to 3-inch, iron pipe size..	0 31
" ⅜ to 3-inch, seamless..	0 36

Copper tubing, 6 cents extra.

COPPER.

	Per 100 lb	
Casting ingot............ ...	22 00	22 50
Cut lengths, round, bars, ½ to 2 in...		35 00
Plain sheets, 14 oz......		36 00
Plain, 16 oz., 14x48 and 14x60		35 00
Tinned copper sheet, base		38 00
Planished base...............		43 00
Braziers' (in sheets). 4x6 ft, 25 to 30 lb. each, per lb, base..	0 34	0 35

BLACK SHEETS.

	Montreal.	Toronto
No 10 gauge	2 70	2 75
12 gauge....................	2 70	2 75
14 "	2 50	2 60
17 "	2 50	2 60
18 "	2 50	2 60
20 "	2 50	2 60
22 "	2 55	2 65
24 "	2 55	2 70
26 "	2 65	2 85
28 "	2 70	3 00

CANADA PLATES.

Ordinary, 52 sheets	2 75	3 05
All bright "	3 75	4 05

Galvanized—	Dom. Crown.	Ordinary.
18x24x52	4 45	4 35
60..................	4 70	4 60
20x28x80	8 90	8 70
. ....	9 40	9 20

B W GALVANIZED SHEETS.

gauge	Queen's Head	Fleur-de-Lis	Gordon Crown	Colborne Crown Gorbal's Best
16.-20 ..	3 95	3 80	3 95	....
22 - 24 ..	4 20	4 05	4 00	4 05
26........	4 45	4 30	4 40	4 30
28...... ..	4 70	4 55	4 60	4 55

Less than case lots 10 to 25c. extra.

Apollo Brand.

24 guage, American	3 85
26 " "	4 10
28 " "	4 55
10¾ oz.	4 85

25c. less for 1,000 lb. lots.

IRON AND STEEL.

	Montreal.	Toronto.
Middlesboro, No 1 pig iron..	21 50	24 50
Middlesboro, No 3 pig iron	20 50	23 50
Summerlee, "	25 50	26 50
" special "	24 50	....
" soft "	24 00	....
Carron	26 00	....
Carron Special	24 50	....
Carron Soft..................	24 00	....
Clarence, No. 3	21 50	23 50
Glengarnock, No. 1		27 00
Midland, Londonderry and Hamilton, off the market but quoted nominally at		26 00
Radnor, charcoal iron	32 00	34 00
Lowmoor iron..................		6 50
Angles		2 50
Common bar, per 100 lb......	2 20	2 30
Forged iron "	2 45	..
Refined " "	2 60	2 70
Horseshoe iron "	2 60	2 70
Band iron, No. 10 x ⅜ in.		3 60
Sleigh shoe steel "	2 25	2 30
Iron finish steel		2 40
Reeled machinery steel		3 50
Tire steel	2 40	2 50
Best sheet steel......		0 12
Mining cast steel....		0 08
Warranted cast steel....		0 14
Annealed cast steel....		0 15
High speed		0 60
B.P.L. tool steel	10½	0 11

INGOT TIN.

Lamb and Flag and Straits—56 and 28-lb. ingots, 100 lb.	$43 00	$43 50

TINPLATES.

Charcoal Plates—Bright

M.L.S., Famous (equal Bradley)		Per box
I C,14 x 20 base	...	$6 50
IX, 14 x 20 "	....	8 00
IXX, 14 x 20 base	....	9 50

Raven and Vulture Grades—		
I C, 14 x 20 base	....	5 00
I X "	....	6 00
I X X "	....	7 00
I X X X "	....	8 00

"Dominion Crown Best"—Double Coated, Tissued.

		Per box.
I C, 14 x 20 base..........	5 50	5 75
I X, 14 x 20 "	6 50	6 75
IXX x 20 "	7 50	7 75

"Allaway's Best"—Standard Quality.

I C, 14 x 20 base..........	4 65	5 00
I X, 14 x 20 "	5 40	5 75
I XX, 14 x 20 "	6 15	6 50

Bright Cokes.

Bessemer Steel—		
I.C., 14 x 20 base	4 25	4 35
20x28, double box	8 50	8 70

Charcoal Plates—Terne

Dean or J. G. Grade—		
I C., 20x28, 112 sheets	7 25	8 00
IX., Terne Tin	...	9 50

Charcoal Tin Boiler Plates.

Cookley Grade—
X X, 14x56, 50 sheet bxs. }
" 14x60, " } ... 7 50
" 14x65, " }

Tinned Sheets.

72x30 up to 24 gauge...... ...	....	8 50
" " 26 "	....	9 00

LEAD

Imported Pig, per 100 lb......	5 25	5 35
Bar, "	5 75	6 00
Sheets, 2½ lb. sq. ft., by roll ..	6 75	6 50
Sheets, 3 to 6 lb. "	6 50	6 25

Cut sheets ½c. per lb., extra.

SHEET ZINC.

5-cwt. casks	7 50	7 75
Part casks..................	7 75	8 00

ZINC SPELTER.

Foreign, per 100 lb	6 50	6 75
Domestic	6 50	6 75

COLD ROLLED SHAFTING.

9-16 to 11-16 inch	....	0 06
¾ to 1 7-16 "	....	0 05½
1 7-16 to 3 "	....	0 05

30 per cent.

OLD MATERIAL.

Dealers buying prices:

	Montreal	Toronto
Heavy copper and wire, lb.	0 14	0 14
Light copper..............	0 12	0 12
Heavy red brass....	0 11	0 12
" yellow brass	0 09	0 10½
Light brass....	0 07	0 07½
Tea lead	0 03½	0 03½
Heavy lead....	0 04	0 04
Scrap zinc	0 03¾	0 03¾
No. 1 wrought iron	14 50	11 50
" 2 " "	6 00	6 00
Machinery cast scrap ...	17 00	16 50
Stove plate..............	12 00	12 00
Malleable and steel	8 00	10 00
Old rubbers	0 10½	0 10
Country mixed rags, 100 lbs.	1 00	1 25

PLUMBING AND HEATING

BRASS GOODS, VALVES, ETC.

Standard Compression work, 57½ per cent.
Cushion work, discount 40 per cent.
Fuller work, 70 per cent.
Flatway stop and stop and waste cocks, 60 per cent; roundway, 55 per cent.
J.M.T. Globe, Angle and Check Valves, 45; Standard, 55 per cent
Kerr standard globes, angles and checks, special, 42½ per cent.; standard, 47½ p.c
Kerr Jenkins' disc, copper-alloy disc and heavy standard valves, 40 per cent.
Kerr steam radiator valves, 60 p c., and quick-opening hot-water radiator valves, 60 p c.
Kerr brass, Weber's straightway valves, 40; straightway valves, I.B.B.M., 60.
J. M. T. Radiator Valves 50; Standard, 60; Patent Quick-Opening Valves, 65 p c.
Jenkins' Valves—Quotations on application to Jenkins' Bros, Montreal.

No. 1 compression bath cock......net		2 00
No. 4 " "	"	1 90
No 7 Fuller's......	"	2 25
No. 4½, "	"	2 35

Patent Compression Cushion, basin cock, hot and cold, per doz., ... $16.20
Patent Compression Cushion, bath cock, No. 2208 25
Square head brass cocks, 50; iron, 60 p.c.
Thompson Smoke-test Machine 25.00

BOILERS—COPPER RANGE.

Copper, 30 gallon, $33, 15 per cent

BOILERS—GALVANIZED IRON RANGE.

30-gallon, Standard, $5; Extra heavy, $7.75

BATH TUBS.

Steel clad copper lined, 15 per cent.

CAST IRON SINKS.

16x24, $1; 18x30, $1; 18x36, $1.30.

ENAMELED BATHS, ETC.

List issued by the Standard Ideal Company Jan. 3, 1907, shows an advance of 10 per cent. over previous quotations.

ENAMELED CLOSETS AND URINALS

Discount 15 per cent.

HEATING APPARATUS.

Stoves and Ranges—40 to 70 per cent.
Furnaces—45 per cent.
Registers—70 per cent.
Hot Water Boilers—50 per cent.
Hot Water Radiators—50 to 55 p.c
Steam Radiators—50 to 55 per cent.
Wall Radiators and specials—50 to 55 p.c.

LEAD PIPE

Lead Pipe, 5 p.c. off.
Lead waste, 5 p c. off
Caulking lead, 6½c. per pound.
Traps and bends, 40 per cent.

IRON PIPE.

Size (per 100 ft.)	Black.		Galvanized
[illegible] inch........	2 35	[illegible] inch........	3 20
[illegible] "	2 35	[illegible] "	3 20
[illegible] "	2 90	[illegible] "	3 75
[illegible] "	3 90	[illegible] "	5 00
1 "	5 60	1 "	7 25
1¼ "	7 65	1¼ "	9 90
1½ "	9 20	1½ "	11 90
2 "	12 25	2 "	15 80
2½ "	20 1[illegible]	2½ "	26 00
3 "	26 75	3 "	34 00
3½ "	34 25	3½ "	42 75
4 "	39 00	4 "	48 60

Malleable Fittings—Canadian discount 30 per cent.; American discount 25 per cent.
Cast Iron Fittings 57½; Standard bushings 57½; headers, 57½; flanged unions 57½, malleable bushings 55; nipples, 70 and 10; malleable lipped unions, 55 and 5 p.c.

SOIL PIPE AND FITTINGS

Medium and Extra heavy pipe and fittings, up to 6 inch, 60 and 10 to 70 per cent.
7 and 8-in. pipe, 40 and 5 per cent.
Light pipe, 50 p c.; fittings, 50 and 10 p.c.

OAKUM.

Plumbers ....per 100 lb......	4 50	5 00

STOCKS AND DIES.

American discount 25 per cent.

SOLDERING IRONS

½-lb. to 1½ .... per lb.	0 45½	0 43
2-lb. or over .... "	0 42½	0 46

SOLDER.

Per lb.	Montreal	Toronto
Bar, half-and-half, guaranteed	0 25	0 26
Wiping.......................	0 22	0 23

PAINTS, OILS AND GLASS.

BRUSHES

Paint and household, 70 per cent.

CHEMICALS.

	In casks per lb.
Sulphate of copper (bluestone or blue vitrol)	0 09
Litharge, ground	0 06
" flaked	0 06½
Green copperas (green vitrol)	0 01
Sugar of lead.......	0 08
Lump olive..........	0 01½

COLORS IN OIL.

Venetian red, 1-lb. tins pure.		....	0 08
Chrome yellow	"	....	0 15
Golden ochre	"	....	0 11
French "	"	....	0 08
Marine black	"	....	0 14½
Chrome green	"	....	0 09
French permanent green	"	....	0 13
Signwriters' black	"	....	0 15

GLUE.

:ic sheet	0 10	0 10½
medal	0 12	0 12½

PARIS GREEN.

	Berger's	Canadian
cask	....	0 27½
drums	0 28	0 27½
"	0 2?½	
"	0 29½	0 28½
pkgs, 100 in box	0 30½	0 29½
"	0 32½	0 31½
tins, 100 in box	0 31½	0 30½
tins	0 33½	0 32

PARIS WHITE.

s	0 90

PIGMENTS.

mineral, casks	....	0 08
" 100-lb. kegs	....	0 08½

PREPARED PAINTS.

n bbls.)	0 65	0 90
n-Williams paints,	....	1 55
Paint Co.'s pure	....	1 40
rd P. & V. Co.'s "New Era."		1 30
oore Co.'s "Ark" B'd	....	1 25
British Navy deck	....	1 50
am-Henderson's "English"		1 45
y's paints, Pure, per gal.	....	1 30
" Thistle, "	....	1 10
n-Senour 100 p.c. pure,	....	1 55
r's Floor Paints	....	1 35
son's "Crown and Anchor"		1 20
son's floor enamel	....	1 50
Dods & Co's "Island City"		1 25
rson Pearcy's, pure	....	1 20
tson's pure paints	....	1 20

PUTTY.

n bbls	1 60
ers in bbls	1 85
tins	1 90
ers in bulk or tins less than 100 lb.	2 00
n 100-lb. irons	1 80

SHINGLE STAINS.

allon lots	0 85	0 90

SHELLAC.

, bleached	....	2 65
orange	....	2 60
m orange	....	2 55

TURPENTINE AND OIL.

white petroleum per gal	....	0 13½
: white "	....	0 15
s astral "	....	0 18½
r oil	0 08	0 10
Gasoline per gal, single		
ls	0 21	0 23½
ne, per gal single bbls	0 17	0 17½
ntine, single barrels	0 78	0 79
ed Oil, " raw	0 60	0 64
" boiled	0 63	0 67

K LEAD GROUND IN OIL. Per 100 lbs

lian pure	7 15	7 50
Canadian	6 80	7 15
o's Select Flake White	....	7 65
ant and Decorators' Pure	....	7 65
rch	....	7 40
ard Decorator's	....	7 15
Genuine	....	6 80
tram's B. B. Genuine	....	8 70
hor," pure	....	7 40
ay's Pure Lead	....	7 25
ay's Exterior	....	6 65
wn and Anchor," pure	....	7 25
Dods & Co's "Island City" pure lead	....	7 25
erson Pearcy's	....	7 40
rtson's C.P., lead	....	7 20

RED DRY LEAD.

ine, 560 lb. casks, per cwt	....	6 25
ine, 100 lb. kegs, "	....	6 50
, 560 lb. casks, per cwt	....	6 00
, 100 lb. kegs, per cwt	....	6 25

WINDOW GLASS

United inches.	Star	Double Diamond
der 26	$4 25	$6 25
o 40	4 65	6 75
o 50	5 10	7 50
o 60	5 35	8 50
o 70	5 75	9 75
o 80	6 25	11 00
o 85	7 00	12 50
o 90	....	15 00
o 95	....	17 50

96 to 100	20 50
101 to 105	24 00
106 to 110	27 50

Discount—16-oz., 25 per cent; 21 oz. 30 per cent. per 100 feet. Broken boxes 50 per cent.

WHITING.

Plain, in bbls	0 70
Gilders bolted in bands	0 90

WHITE DRY ZINC.

Extra Red Seal, V.M.	0 07½	0 08

WHITE ZINC IN OIL.

Pure, in 25-lb. irons	0 08½
No. 1, " "	0 07
No. 2, " "	0 05½

VARNISHES.

	Per gal. cans
Carriage, No. 1	1 50
Pale durable body	3 50
" hard rubbing	3 00
Finest elastic gearing	3 00
Elastic oak	1 50
Furniture, polishing	2 00
Furniture, extra	1 15
" " No 1	0 90
" " union	0 80
Light oil finish	1 40
Gold size japan	1 80
Brown japan	0 95
No 1 brown japan	0 95
Baking black japan	1 35
No. 1 black japan	0 90
Benzine black japan	0 70
Crystal Damar	2 80
No. 1 "	2 50
Pure asphaltum	1 40
Oilcloth	1 50
Lightning dryer	0 70
Elastilite varnish, 1 gal. can, each	2 00
Granitine floor varnish, per gal	2 50
Maple Leaf coach enamels; size 1,	1 20
Sherwin-Williams' kopal varnish, gal.,	2 50
Canada Paint Co's sun varnish	2 00
"Kyanize" Interior Finish	2 40
"Flint-Lac," coach	1 80
B H. Co's "Gold Medal," in cases	2 00
Jamieson's Copaline, per gal.	2 00

BUILDERS' HARDWARE.

BELLS.

Brass hand bells, 60 per cent.
Nickel, 55 per cent.

Gongs, Sargeant's door bells	5 50	8 00
American, house bells, per lb.	0 35	0 40

Peterboro' door bells, 37½ and 10 off new list.

BUILDING PAPER, ETC.

Tarred Felt, per 100 lb.		2 25
Ready roofing, 2-ply, not under 45 lb. per roll		1 00
Ready roofing, 3-ply, not under 65 lb., per roll		1 25
Carpet Felt, per ton		60 00
Heavy Straw Sheathing, per ton		40 00
Dry Surprise		0 45
Dry Sheathing, per roll, 400 sq. ft.		0 40
Tar " " 400 "		0 50
Dry Fibre " 400 "		0 55
Tarred Fibre " 400 "		0 65
O. K. & I. X. L. " 400 "		0 70
Resin-sized " 400 "		0 45
Oiled Sheathing " 600 "		1 00
Oiled " " 400 "		0 70
Roof Coating, in barrels, per gal.		0 17
Roof " small packages "		0 25
Refined Tar, per barrel		5 00
Coal Tar "		4 00
Coal Tar, less than barrels, per gal.		0 15
Roofing Pitch, per 100 lb.	0 80	0 90
Slater's felt, per roll		0 70
Heavy Straw Sheathing f. o. b. St. John and Halifax		42 50

BUTTS.

Wrought Brass, net revised list.
Wrought Iron, 70 per cent.
Cast Iron Loose Pin, 60 per cent.
Wrought Steel Fast Joint and Loose Pin, 70 per cent.

CEMENT AND FIREBRICK.

Canadian Portland	2 00	2
Belgium	1 60	1
White Bros. English	1 80	2
"Lafarge" cement in wood		3 10
"Lehigh" cement, in wood		2 [illegible]
"Lehigh" cement, cotton sacks		2 39
"Lehigh" cement, paper sacks		2 31
Fire brick, Scotch, per 1,000	27 00	30 00
" English	17 00	21 00
" American, low	23 0	25 00
" " high	27 50	35 00
Fire clay (Scotch), net ton		4 95

Paving Blocks per 1,000.

Blue metallic, 9"x4½"x3", ex wharf	35 00
Stable pavers, 12"x6"x2", ex wharf	50 00
Stable pavers, 9"x4½"x3", ex wharf	36 00

DOOR SETS.

Peterboro, 37½ and 10 per cent.

DOOR SPRINGS.

Torrey's Rod, per doz.	....	1 75
Coil, 9 to 11 in. "	0 95	1 65
English "	2 00	4 00

Chicago and Reliance Coil 25 per cent.

ESCUTCHEONS.

Discount 50 and 10 per cent., new list
Peterboro, 37½ and 10 per cent.

ESCUTCHEON PINS.

Iron, discount 40 per cent.
Brass, 45 per cent.

HINGES.

Blind, discount 50 per cent.

Heavy T and strap, 4-in., per lb. net	0 06
" " 5-in., "	0 05¾
" " 6-in., "	0 05½
" " 8-in., "	0 05¼
" " 10-in. and larger	0 05

Light T and strap, discount 65 p.c.

Screw hook and hinge—

under 12 in., per 100 lb.	4 65
over 12 in. "	3 65

Crate hinges and back flaps, 65 and 5 p. c.
Hinge hasps, 65 per cent.

SPRING HINGES.

Spring, per gro., No. 5, $17.50 No. 10, $18; No. 20, $10.80; No. 120, $20; No. 51, $10; No. 50, $27 50
Chicago Spring Butts and Blanks 12½ per cent.
Triple End Spring Butts, 30 and 10 per cent.
Chicago Floor Hinges, 37½ and 5 off.
Garden City Fire House Hinges, 12½ p.c.
"Chief" floor hinge, 50 p.c.

CAST IRON HOOKS.

Bird cage, per doz.	0 50	1 10
Clothes line, No. 61 "	0 00	0 70
Harness "	0 60	12 00
Hat and coat, per gro.	1 10	10 00
Chandelier, per doz.	0 50	1 00
Wrought hooks and staples—		
¼ x 5, per gross	....	2 65
5-16 x 5 "	....	3 30

Bright wire hooks, 60 p c.
Bright steel gate hooks and staples, 40 p c.
Cresent hat and coat wire, 60 per cent.
Screw, bright wire, 65 per cent.

KNOBS.

Door, japanned and N.P., doz	1 50	2 50
Bronze, Berlin, per doz.	2 75	3 25
Bronze, Genuine "	6 00	9 00
Shutter, porcelain, F. & L. screw, per gross	1 30	2 00
White door knobs, per doz.	....	2 00

Peterboro knobs, 37½ and 10 per cent.
Porcelain, mineral and jet knobs, net list.

KEYS.

Lock, Canadian 40 to 40 and 10 per cent.

LOCKS.

Peterboro, 37½ and 10 per cent
Russell & Erwin, steel rim $2.50 per doz
Eagle cabinet locks, discount 30 per cent
American padlocks, all steel, 10 to 15 per cent.; all brass or bronze, 10 to 25 per cent.

SAND AND EMERY PAPER.

B. & A. sand, discount, 35 per cent
Emery, discount 35 per cent.
Garnet (Rurton's), 5 to 10 per cent. advance

SASH WEIGHTS.

Sectional, per 100 lb.	2 00	2 25
Solid "	1 50	1 75

SASH CORD.

Per lb.	0 32

BLIND AND BED STAPLES.

All sizes per lb.	0 07½	10

WROUGHT STAPLES.

Galvanized	2 75
Plain	2 50

Coopers', discount 45 per cent.
Poultry netting staples, discount 40 per cent.
Bright spear point, 75 per cent. discount.

TOOLS AND HANDLES.

ADZES.

Discount 22½ per cent.

AUGERS.

Gilmour's, discount 60 per cent. off list.

AXES.

Single bit, per doz.	6 00	9 0
Double bit, "	10 00	11 00
Bench Axes, 40 per cent.		
Broad Axes, 25 per cent.		
Hunters' Axes	5 50	6 00
Boys' Axes	6 25	7 00
Splitting Axes	7 00	12 00
Handled Axes	7 00	9 00
Red Ridge, boys', handled	....	5 75
" hunters	....	5 25

BITS.

Irwin's auger, discount 47½ per cent.
Gilmour's auger, discount 60 per cent.
Rockford auger, discount 50 and 10 per cent.
Jennings' Gen. auger, net list.
Gilmour's car, 47½ per cent.
Clark's expansive, 40 per cent.

Clark's gimlet, per doz	0 65
Diamond, Shell, per doz.	1 00
Nail and Spike, per gross	2 25

BUTCHERS CLEAVERS.

German, per doz.	7 00	9 00
American "	12 00	18 00

CHALK.

Carpenters' Colored, per gross	0 45	0 75
White lump, per cwt.	0 60	0 65

CHISELS.

Warnock's, discount 70 and 5 per cent.
P. S. & W. Extra, discount, 70 per cent.

CROSSCUT SAW HANDLES.

S. & D., No. 3, per pair	0 13
S. & D., " 5 "	0 11½
S. & D., " 6 "	0 18
Boynton pattern "	0 20

CROWBARS.

3½c. to 4c. per lb.

DRAW KNIVES.

Coach and Wagon, discount 75 and 5 per cent.
Carpenters' discount 75 per cent.

DRILLS.

Millar's Falls, hand and breast, net list.
North Bros., each set, 50c.

DRILL BITS.

Morse, discount 37½ to 40 per cent.
Standard, discount 50 and 5 to 55 per cent.

FILES AND RASPS.

Great Western	75 per cent.
Arcade	75 "
Kearney & Foot	75 "
Disston's	75 "
American	75 "
J. Barton Smith	75 "
McClellan	75 "
Eagle	75 "
Nicholson	66⅔ "
Globe	75 "

Black Diamond, 60, 10 and 5 p.c.
Jowitt's, English list, 27½ per cent.

GAUGES.

Stanley's discount 50 to 60 per cent.

Winn's, Nos. 26 to 33, each	1 65	2 40

HANDLES.

Second growth ash fork, hoe, rake and shovel handles, 40 p c.
Extra ash fork, hoe, rake and shovel handles, 45 p c.
No. 1 and 2 ash fork, hoe, rake and shovel handles, 50 p c
White ash whiffletrees and neckyokes, 35 p c.
All other ash goods, 40 p c.
All hickory, maple and oak goods, excepting carriage and express whiffletrees, 40 p.c.
Hickory, maple, oak carriage and express whiffletrees, 45 p c.

HAMMERS.

Maydole's, discount 5 to 10 per cent.
Canadian, discount 25 to 27½ per cent.

Magnetic tack, per doz	1 10	1 20
Canadian sledge, per lb.	0 07	0 08½
Canadian ball pean, per lb.	0 22	0 25

HATCHETS.

Canadian, discount 40 to 42½ per cent.
Shingle, Red Ridge 1, per doz. 4 40
" " 2, ... 4 80
Barrel Underhill 5 05

HOES

Mortar, 50 and 10 per cent

MALLETS.

Tinsmiths' .... per doz. 1 25 1 50
Carpenters', hickory, " 1 25 3 75
Lignum Vitae 3 85 5 00
Caulking, each 0 60 2 00

MATTOCKS.

Canadian .... per doz. 5 50 6 00

MEAT CUTTERS.

German, 15 per cent
American discount, 33⅓ per cent.

PICKS.

Per dozen 6 00 9 00

PLANES.

Wood bench, Canadian, 40, American, 25
Wood, fancy, 37½ to 40 per cent.
Stanley planes, $1.55 to $3 60, net list prices.

PLANE IRONS.

English .... per doz. 2 00 5 00
Stanley, 2⅝ inch, single 24c., double 39c.

PLIERS AND NIPPERS.

Button's genuine, 37½ to 40 per cent.
Button's imitation .... per doz. 5 00 9 00
Berg's wire fencing 1 72 5 50

PUNCHES.

Saddlers .. .. per doz. 1 00 1 85
Conductor's .... " 3 00 15 00
Tinners', solid .... per set 0 72
" hollow ...... per inch 1 00

RIVET SETS.

Canadian, discount 35 to 37½ per cent.

RULES.

Boxwood, discount 70 per cent.
Ivory, discount 20 to 25 per cent.

SAWS.

Atkins, hand and crosscut, 25 per cent.
Disston's Hand, discount 12½ per cent
Disston's Crosscut .... per foot 0 35 0 55
Hack, complete .... each 0 75 2 75
" frame only .... each 0 50 1 25

S. & D. solid tooth circular shingle, concave and band, 50 per cent; mill and ice, drag, 30 per cent; cross-cut, 35 per cent.; hand saws, butcher, 35 per cent.; buck, New Century $6 25; buck, No. 1 Maple Leaf, $·2·; buck, Happy Medium, $4.25; buck, Watch spring, $1.2·; buck, common frame, $1 00

Spear & Jackson's saws—Hand or rip, 26 in., $12 75; 2· in., $1· 25; panel 18 in., $8 25; 20 in., $·, tenon, 10 in., $9 90; 12 in., $10.90; 1· in., $11 50

SAW SETS.

Lincoln and Whiting 4 75
Hand Sets, Perfect 4 00
X-Cut Sets " 7 50
Maple Leaf and Premiums saw sets, 40 off.
S. & D saw swages, 40 off.

SCREW DRIVERS.

Sargent's .... per doz. 0 65 1 00
North Bros., No. 30 . per doz. 16 80

SHOVELS AND SPADES.

Canadian, discount 45 per cent.

SQUARES.

Iron, discount 20 per cent.
Steel, discount 65 and 10 per cent.
Try and Bevel, discount 50 to 52½ per cent.

TAPE LINES.

English, ass skin ..... per doz. 2 75 5 00
English, Patent Leather 5 50 9 75
Chesterman's .... each 0 90 2 85
" steel .... each 0 80 8 00
Berg's, each 0 75 2 50

TROWELS.

Disston's, discount 10 per cent.
S. & D., discount 35 per cent
Berg's, brick, 924x11 .... 4 00
" pointing, 924x5 .... 2 10

FARM AND GARDEN GOODS

BELLS

American cow bells, 63⅓ per cent.
Canadian, discount 45 and 50 per cent.
American, farm bells, each .. 1 35 3 00

BULL RINGS.

Copper, $2.00 for 2½-inch

CATTLE LEADERS

Nos. 32 and 33 .. ... per gross 7 50 8 50

BARN DOOR HANGERS

doz. pairs
Stearns wood track 4 50 6 00
Zenith 9 00
Atlas, steel covered 5 00 6 00
Perfect 8 00 11 00
New Milo, flexible 6 50
Steel, track, 1 x 3-16 in (100 ft) 3 25
" " 1½ x 3-16 in (100 ft) 4 75
Double strap hangers, doz. sets 6 40
Standard jointed hangers, " 6 40
Steel King hangers " 6 25
Storm King and safety hangers 7 00
" " rail 4 25
Chicago Friction, Oscillating and Big Twin Hangers, 5 per cent

HARVEST TOOLS.

50 and 10 per cent.
S. & D. lawn rakes, Dunn's, 40 off.
" sidewalk and stable scrapers, 40 off.

HAY KNIVES.

Net list.

HEAD HALTERS.

Jute Rope, ⅝-inch .... per gross 9 00
" " ¾ " .... " 10 00
" " ⅞ " .... " 12 00
Leather, 1-inch ...... per doz. 4 00
Leather, 1¼ " ...... " 5 20
Web .... " 2 45

HOES.

Garden, 50 and 10 per cent
Planter .... per doz. 4 00 4 50

LAWN MOWERS

Low wheel 12, 14 and 16-inch $2 30
9-inch wheel, 12-inch 2 85
" " 14 " 3 00
" " 16 " 3 12½
High wheel, 12 " 4 05
" " 14 " 4 50
" " 16 " 4 75

SCYTHES.

Per doz. net 6 25 9 25

SCYTHE SNATHS.

Canadian, discount 40 per cent.

SNAPS.

Harness, German, discount 25 per cent.
Lock, Andrews' 4 50 11 00

STABLE FITTINGS.

Warden King, 35 per cent
Dennis Wire & Iron Co., 33⅓ p.c.

WOOD HAY RAKES.

40 and 10 per cent.

HEAVY GOODS, NAILS, ETC.

ANVILS.

Wright's, 80-lb. and over 0 10½
Hay Budden, 80-lb. and over 0 09½
Brook's, 80-lb. and over 0 11¼
Taylor-Forbes, prospectors 0 05
Columbia Hardware Co., per lb. 0 09½

VISES.

Wright's 0 13½
Berg's, per lb. 0 12½
Brook's 0 12¾
Pipe Vise, Hinge, No. 1 .... 3 50
" " " No. 2 .... 5 50
Saw Vise 4 50 5 00
Blacksmiths' (discount) 60 per cent.
" parallel (discount) 45 per cent.

BOLTS AND NUTS

Carriage Bolts, common ($1 list) Per cent.
" " ⅜ and smaller .. 60, 10 and 10
" " 7-16 and up .... 55 and 5
" " Norway Iron ($3 list) 50
Machine Bolts, ⅜ and less .... 60 and 10
Machine Bolts, 7-16 and up .... 55 and 5
Plough Bolts .... 55 and 10
Blank Bolts 55
Bolt Ends 55
Sleigh Shoe Bolts, ⅜ and less .. 60 and 10
" " " 7-16 and larger 50 and 5
Coach Screws, conepoint ...... 70 and 5
Nuts, square, all sizes, 4c. per cent. off
Nuts, hexagon, all sizes, 4½c. per cent. off
Stove Rods per lb., 5½ to 6c.
Stove Bolts, 75 per cent.

CHAIN.

Proof coil, per 100 lb. ¼ in., $6 00; 5-16 in., $4.85; ⅜ in., $4 25; 7-16 in., $4 00, ½ in., $3 75; 9-16 in., $3 70; ⅝ in., $3 65; ¾ in., $3.60; ⅞ in., $3 45; 1 in., $3.40.

Halter, kennel and post chains, 40 to 40 and 5 per cent; Cow ties, 40 per cent; Tie out chains 65 per cent.; Stall fixtures, 35 per cent; Trace chain, 45 per cent.; Jack chain iron, 50 per cent; Jack chain, brass, 50 per cent.

HORSE NAILS

M.R.M. cold forged process, list May 15, 1907, 50 and 5 per cent.
"C" brand, 57½ per cent. off list.
Capewell brand, quotations on application

HORSESHOES.

M.R.M. brand: iron, light and medium No. 1 and smaller, $3 90; No. 2 and larger $3 65; snow pattern No. 1 and smaller $4 15, No. 2 and larger, $3 90; "X.L." new light steel, No. 1 and smaller, $4 25; No. 2 and larger, $4; "X.L." featherweight steel, No. 0 to 4, $5 60; Special countersunk steel No. 0 to 4, $6 85 pkg; toe-weight, all sizes, $6.85. F.o.b. Montreal. Extras for packing.

Belleville brand: No. 0 and 1, light and medium iron, $3 90; snow, $4 15; light steel, $4.25; No. 2 and larger, light and medium iron, $3 65; snow, $3 90; light steel, $4. F.o.b. Belleville. Two per cent., 30 days.

Toecalks—Standard No. 1 and smaller, $1 50; No. 2 and larger, $1 25. Blunt No. 1 and smaller, $1.75; No. 2 and larger, $1.50 per box. Sharp. Put up in 25 lb. bxs.

HORSE WEIGHTS.

Taylor-Forbes, 4½c. per lb.

NAILS. Cut. Wire.
2d 4 00 3 50
3d 3 15 3 20
4 and 5d 2 90 2 90
6 and 7d 2 80 2 80
8 and 9d 2 65 2 65
10 and 12d 2 60 2 60
16 and 20d 2 55 2 55
30, 40, 50 and 60d (base) 2 50 2 50

F.o.b. Montreal. Cut nails, Toronto 20c. higher
Miscellaneous wire nails, discount 75 per cent
Coopers' nails, discount 40 per cent.

PRESSED SPIKES.

Pressed spikes, ⅝ diameter, per 100 lbs $3.15

RIVETS AND BURRS.

Iron Rivets, black and tinned, 60, 10 and 10.
Iron Burrs, discount 60 and 10 and 10 p.c.
Copper Rivets, usual proportion burrs, 15 p.c.
Copper Burrs only, net list
Extras on Coppered Rivets, ½-lb. packages 1c. per lb.; ¼-lb. packages 2c. lb
Tinned Rivets, net extra, 4c. per lb.

SCREWS.

Wood, F. H., bright and steel, 85 and 10 p.c.
" R. H., bright, 80 and 10 per cent.
" F. H., brass, 75 and 10 per cent.
" R. H., " 70 and 10 per cent.
" F. H., bronze, 70 and 10 per cent.
" R. H., " 65 and 10 per cent.
Drive Screws, dis. 87½ per cent.
Bench, wood .... per doz. 3 25 4 00
" iron .... " 4 25 5 00
Set, case hardened, dis. 60 per cent.
Square Cap, dis. 50 and 5 per cent.
Hexagon Cap, dis. 45 per cent.

MACHINE SCREWS.

Flat head, iron and brass, 35 per cent.
Felister head, iron, discount 30 per cent
" " brass, discount 25 per cent.

TACKS, BRADS, ETC.

Carpet tacks, blued 75 p.c.; tinned, 80 and 10; (in kegs), 40; cut tacks, blued, in dozens only, 75; ½ weights, 60; Swedes cut tacks, blued and tinned, bulk, 75 dozens, 75; Swedes, upholsterers', bulk, 85 and 12½; brush, blued and tinned, bulk, 70; Swedes, gimp, blued, tinned and japanned, 75 and 12½; zinc tacks, 35; leather carpet tacks, 40; copper tacks, 25; copper nails 30; trunk nails, black, 65; trunk nails, tinned and blued, 65; clout nails, blued and tinned 65, chair nails, 35; patent brads, 40; fine finishing, 40; lining tacks, in papers, 10; lining tacks, in bulk, 15; lining tacks, solid heads, in bulk, 75; saddle nails, in papers, 10; saddle nails, in bulk, 15; tufting buttons, 22 line in dozens only, 60; zinc glaziers' points, 5, double pointed tacks, papers, 90 and 10; double pointed tacks, bulk 40; clinch and duck rivets, 45; cheese box tacks, 85 and 5 trunk tacks, 80 and 10

WROUGHT IRON WASHERS.

Canadian make, discount 40 per cent

SPORTING GOODS.

CARTRIDGES

"Dominion" Rim Fire Cartridges and C.B. caps, 50 and 7½ per cent.; Rim Fire R.B. Round Caps 60 and 2½ per cent; Centre Fire, Pistol and Rifle Cartridges 30 p.c; Centre Fire Sporting and Military Cartridges, 2· and 5 p.c.; Rim Fire, Shot Cartridges, 5· and 7½ p.c; Centre Fire, Shot Cartridges, 30 p.c; Primers, 25 p.c.

LOADED SHELLS

"Crown" Black Powder, 15 and 5 p.c.; "Sovereign" Empire Bulk Smokeless Powder, 30 and 5 p.c.; "Regal" Ballistite Dense smokeless Powder, 30 and 5 p.c.; "Imperial" Empire or Ballistite Powder, 30 and 10 p.c.

EMPTY SHELLS.

Paper Shells, 25 and 5; Brass Shells 55 and 5 p.c.

Wads. per lb.
Best thick brown or grey felt wads, in ½-lb. bags $0 70
Best thick white card wads, in boxes of 500 each, 12 and smaller gauges 29
Best thick white card wads, in boxes of 500 each, 10 gauge 0 35
Thin card wads, in boxes of 1,000 each, 12 and smaller gauges 0 20
Thin card wads, in boxes of 1,000 each, 10 gauge 0 25
Chemically prepared black edge grey cloth wads, in boxes of 250 each— Per M
11 and smaller gauge 0 60
9 and 10 gauges 0 70
and 8 " 0 90
5 and 6 " .... 1 1·
Superior chemically prepared pink edge, best white cloth wads in boxes of 250 each—
11 and smaller gauge .... 1 1·
9 and 10 gauges .... 1 4·
7 and 8 " .... 1 6·
5 and 9 " 1 90

SHOT.

Ordinary drop shot AAA to dust $7 50 per 100 lbs. Discount 5 per cent; cash discount 2 per cent, 30 days; net extras as follows subject to cash discount only; Chilled, 40 c.; buck and seal, 80c.; no 28 ball, $1 20 per 100 lbs; bags less than 25 lbs., ½c per lb; F.O.B. Montreal, Toronto, Hamilton, London, St John and Halifax, and freight equalized thereon.

TRAPS (steel)

Game, Newhouse, discount 30 and 10 per cent.
Game, Hawley & Norton, 50, 10 & 5 per cent
Game, Victor, 70 per cent.
Game, Oneida Jump (B. & L.) 40 & 2½ p.
Game, steel, 60 and 5 per cent.

SKATES.

Skates, discount 37½ per cent.
Empire hockey sticks, per doz .. 3 00

CUTLERY AND SILVERWARE.

RAZORS.

	per doz.	
iot's	4 00	18 00
ker's	7 50	11 00
" King Cutter	13 50	18 50
ade & Butcher's	3 60	10 00
wis Bros.' "Klean Kutter"	8 50	10 50
nckel's	7 50	20 00
rg's	7 50	20 00

uss Razors and Strops, 50 and 10 per cent

KNIVES.

rriers-Stacey Bros., doz 3 50

PLATED GOODS

olloware, 40 per cent. discount.
atware, staples, 40 and 10, fancy, 40 and 5.
utton's "Cross Arrow" flatware, 47½; "Singalese" and "Alaska" Nevada silver flatware, 42 p.c.

SHEARS.

auss, nickel, discount 60 per cent.
auss, Japan, discount 67½ per cent.
auss, tailors, discount 40 per cent.
ymour's, discount 50 and 10 per cent.
rg's 6 00 12 00

HOUSE FURNISHINGS.

APPLE PARERS.

idson, per doz., net 5 75

BIRD CAGES.

ass and Japanned, 40 and 10 p. c.

COPPER AND NICKEL WARE.

pper boilers, kettles, teapots, etc. 30 p.c
pper pitts, 20 per cent.

KITCHEN ENAMELED WARE.

hite ware, 75 per cent.
ndon and Princess, 50 per cent.
nada, Diamond, Premier, 50 and 10 p c.
arl, Imperial, Crescent and granite steel, 50 and 10 per cent.
emier steel ware, 40 per cent.
ar decorated steel and white, 25 per cent.
panned ware, discount 45 per cent.
ollow ware, tinned cast, 35 per cent. off.

KITCHEN SUNDRIES.

n openers, per doz.	0 40	0 75
incing knives per doz	0 50	0 80
uplex mouse traps, per doz.	....	0 65
tato mashers, wire, per doz.	0 60	0 70
" wood "	0 50	0 60
getable slicers, per doz	....	2 25
niversal meat chopper No. 1	....	1 15
nterprise chopper, each	...	1 30
iders and fry pans, 50 per cent.		
ar Al chopper 5 to 32	1 35	4 10
" " 100 to 103	1 35	2 00
itchen hooks, bright	....	0 60

LAMP WICKS

scount, 60 per cent.

LEMON SQUEEZERS.

rcelain lined	per doz.	2 20	5 60
alvanized	"	1 87	3 85
ing, wood	"	2 75	2 90
ing, glass	"	4 00	4 50
l glass	"	0 50	0 90

METAL POLISH.

andem metal polish paste 6 00

PICTURE NAILS

rcelain head ......per gross	1 35	1 50
ass head	0 40	1 00

n and gilt, picture wire, 75 per cent.

SAD IRONS.

rs, Potts, No 55, polished ....per set	0 90
" No 50 nickle-plated. "	0 95
" handles, japaned, per gross	9 25
" " nickled, "	9 75
mmon, plain	4 25
" plated	5 50
bestos, per set	1 50

TINWARE.

CONDUCTOR PIPE.

2-in, plain or corrugated., per 100 feet, .30; 3 in., $4.40; 4 in, $5.80; 5 in., $7.45; n., $9 90.

FAUCETS.

mmon, cork-lined, discount 35 per cent.

EAVETROUGHS.

uch ....per 100 ft. 3 30

FACTORY MILK CANS.

scount off revised list, 35 per cent
lk can trimmings, discount 25 per cent.
eamery Cans, 45 per cent

LANTERNS.

No 2 or 4 Plain Cold Blast....per doz.	6 50
Lift Tubular and Hinge Plain, "	4 75
No. 0, safety "	4 00

Better quality at higher prices.
Japanning, 50c. per doz. extra.
Prism globes, per doz., $1.20.

OILERS.

Kemp's Tornado and McClary's Model galvanized oil can, with pump, 5 gallon, per dozen 10 92
Davidson oilers, discount 40 per cent.
Zinc and tin, discount 50 per cent
Coppered oilers, 20 per cent. off.
Brass oilers, 50 per cent. off.
Malleable, discount 25 per cent

PAILS (GALVANIZED).

Dufferin pattern pails, 45 per cent.
Flaring pattern, discount 45 per cent.
Galvanized washtubs 40 per cent.

PIECED WARE.

Discount 35 per cent off list, June, 1899.
10-qt. flaring sap buckets, discount 35 per cent.
6, 10 and 14-qt. flaring pails dis. 35 per cent.
Copper bottom tea kettles and boilers, 30 p.c.
Coal hods, 40 per cent.

STAMPED WARE.

Plain, 75 and 12½ per cent. off revised list.
Retinned, 72½ per cent revised list.

SAP SPOUTS

Bronzed iron with hooks ....per 1,000
Eureka tinned steel, hooks 8 00

STOVEPIPES.

5 and 6 inch, per 100 lengths	7 64	7 91
7 inch " "	...	8 18

Nestable, discount 40 per cent.

STOVEPIPE ELBOWS

5 and 6-inch, common........per doz.	1 32
7-inch "	1 48

Polished, 15c. per dozen extra.

THERMOMETERS.

Tin case and dairy, 75 to 75 and 10 per cent.

TINNERS' SNIPS.

Per doz. 3 00 15
Clauss, discount 35 per cent.

TINNERS' TRIMMINGS.

Discount, 45 per cent

WIRE.

ANNEALED CUT HAY BAILING WIRE.

No. 12 and 13, $4; No. 13½, $4 10; No 14, $4 25; No. 15, $4.50; in lengths 6' to 11', 25 per cent.; other lengths 20c per 100 lbs extra; if eye or loop on end add 25c. per 100 lbs. to the above.

BRIGHT WIRE GOODS

Discount 60 per cent.

CLOTHES LINE WIRE.

7 wire solid line, No. 17, $4.90; No. 18, $3.00; No. 19, $2.70; 6 wire solid line, No. 17, $4.45; No. 18, $3.10; No. 19, $2 80. All prices per 1000 ft. measure; 6 strand, No. 18, $2 60; No. 19, $2 90. F.o.b. Hamilton, Toronto, Montreal

COILED SPRING WIRE

High Carbon, No. 9, $2 95; No. 11, $3.50; No. 12, $3.20.

COPPER AND BRASS WIRE.

Discount 37½ per cent.

FINE STEEL WIRE.

Discount 25 per cent. List of extras In 100-lb. lots: No. 17, $5 — No. 18, $5.50 — No. 19, $6 — No. 20, $6 65 — No. 21 $7 — No. 22, $7.30 — No 23, $7.65 — No 24, $8 — No. 25, $9—No. 26, $9.50—No. 27, $10—No. 28, $11—No. 29, $12—No. 30, $13—No. 31, $14—No. 32, $15—No. 33, $16—No. 34, $17. Extras net—tinned wire, Nos. 17-25 $2—Nos. 26-31, $4—Nos. 32-34, $6. Coppered, 75c.—oiling, 10c.—in 25-lb. bundles, 15c.—in 5 and 10-lb. bundles, 25c.—in 1-lb. hanks, 25c. —in ½-lb. hanks, 38c.—in ¼-lb. hanks, 50c. packed in casks or cases, 15c.—bagging or papering, 10c

FENCE STAPLES.

Bright. ..... 2 80 Galvanized.... 32

HAY WIRE IN COILS.

No 13, $2.70; No. 14, $2 80; No. 15, $2.95; f.o.b., Montreal.

GALVANIZED WIRE.

Per 100 lb.—Nos. 4 and 5, $3.90 — Nos. 6, 7, 8, $3.35 — No. 9, $2.85 — No. 10, $3 40 — No. 11, $3.45—No. 12, $3 00 —No. 13, $3 10—No. 14, $3.95—No 15, $4.30 —No. 16, $4.30 from stock. Base sizes, Nos. 6 to 9, $2.35 f.o.b. Cleveland. Extras for cutting.

LIGHT STRAIGHTENED WIRE.

Gauge No.	Over 20 in. per 100 lbs.	10 to 20 in.	5 to 10 in.
0 to 5	$0 50	$0.75	$1.25
6 to 9	0.75	1 25	2 00
10 to 11	1.00	1 75	2.50
12 to 14	1.50	2.25	3.50
15 to 16	2.00	3.00	4.50

SMOOTH STEEL WIRE.

No. 0-9 gauge, $2.40; No. 10 gauge, 6c extra; No. 11 gauge, 12c extra; No. 12 gauge, 20c. extra; No. 13 gauge, 30c. extra; No 14 gauge, 40c. extra; No. 15 gauge, 55c. extra; No 16 gauge, 70c extra. Add 60c. for coppering and $2 for tinning.

Extra net per 100 lb.—Oiled wire 10c., spring wire $1.25, bright soft drawn 15c., charcoal (extra quality) $1.25, packed in casks or cases 15c., bagging and papering 10c., 50 and 100-lb. bundles 10c., in 25-lb. bundles 15c., in 5 and 10-lb. bundles 25c., in 1-lb hanks, 50c., in ½-lb. hanks 75c., in ¼-lb. hanks $1.

POULTRY NETTING.

2-in. mesh, 19 w g., 50 and 5 p.c. off. Other sizes, 50 and 5 p c. off.

WIRE CLOTH.

Painted Screen, in 100-ft rolls, $1.72½, per 100 sq. ft.; in 50-ft. rolls, $1.77½, per 100 sq ft.

WIRE FENCING.

Galvanized barb 2 95
Galvanized, plain twist 3 30
Galvanized barb, f.o.b. Cleveland, $2.70 for small lots and $2.60 for carlots

WIRE ROPE

Galvanized 1st grade, 6 strands, 24 wires, ¼, $5; 1 inch $16 80.
Black, 1st grade, 6 strands, 19 wires, ¼, $5; 1 inch $15.10. Per 100 feet f.o.b. Toronto.

WOODENWARE.

CHURNS.

No. 0, $9; No. 1, $9; No. 2, $10; No. 3, $11; No 4, $13; No. 5, $16.; f.o b. Toronto Hamilton, London and St. Marys, 30 and 30 per cent; f.o b. Ottawa, Kingston and Montreal, 40 and 15 per cent. discount,

CLOTHES REELS

Davis Clothes Reels, dis., 40 per cent.

FIBRE WARE.

Star pails, per doz		$ 3 00
0 Tubs,	"	14 00
1 "	"	12 00
2 "	"	10 00
3 "	"	8 50

LADDERS, EXTENSION.

3 to 6 feet, 12c. per foot; 7 to 10 ft, 13c.
Waggoner Extension Ladders, dis. 40 per cent.

MOPS AND IRONING BOARDS.

"Best" mops		1 25
"900" mops		1 25
Folding ironing boards	12 00	16 50

REFRIGERATORS

Discount, 40 per cent.

SCREEN DOORS.

Common doors, 2 or 3 panel, walnut stained, 4-in. style..........per doz. 7 25
Common doors, 2 or 3 panel, grained only, 4-in., style ....per doz. 7 55
Common doors, 2 or 3 panel, light stair per doz. 9 55

WASHING MACHINES.

Round, re-acting per doz.	60 00
Square "	63 00
Eclipse, per doz	54 00
Dowswell "	39 00
New Century, per doz	75 00
Daisy	54 00
Stephenson	74 00

WRINGERS.

Royal Canadian, 11 in., per doz.	35 00
Royal American, 11 in.	35 00
Eze 10 in., per doz	36 75

MISCELLANEOUS

AXLE GREASE.

Ordinary, per gross	6 00	7 00
Best quality	10 00	12 00

BELTING.

Extra, 60 per cent.
Standard, 60 and 10 per cent.
No. 1, not wider than 6 in., 60, 10 and 10 p.c
Agricultural, not wider than 4 in., 75 per cent.
Lace leather, per side, 75c.; cut laces, 80c.

BOOT CALKS.

Small and medium, ball ....per M 2 25
Small heel 4 50

CARPET STRETCHERS.

American ....per doz.	1 00	1 50
Bullard's "	....	6 50

CASTORS.

Bed, new list, discount 55 to 57½ per cent.
Plate, discount 52½ to 57½ per cent.

PINE TAR.

½ pint in tins ....per gross	...	7 80
" " "	....	9 60

PULLEYS.

Hothouse ....per doz.	0 55	1 00
Axle "	0 22	0 33
Screw "	0 22	1 00
Awning "	0 35	2 50

PUMPS.

Canadian cistern	1 40	2 00
Canadian pitcher spout	1 80	3 16

Berg's wing pump, 75 per cent.

ROPE AND TWINE.

Sisal		0 10½
Pure Manilla		0 15
"British" Manilla		0 12
Cotton, 3-16 inch and larger	0 21	0 23
" 5-32 inch	0 25	0 27
" ⅛ inch	0 25	0 28
Russia Deep Sea		0 16
Jute		0 09
Lath Yarn, single		0 10
" double		0 10½
Sisal bed cord, 48 feet........per doz.		0 65
" " 60 feet "		0 80
" " 72 feet "		0 95

Twine.

Bag, Russian twine, per lb.	....	0 27
Wrapping, cotton, 3-ply	....	0 25
" " 4-ply	....	0 29
Mattress twine per lb.	0 33	0 45
Staging " "	0 27	0 35

BINDER TWINE.

500 feet, sisal	0 09¼
500 " standard	0 09¾
550 " " manilla	0 10¾
600 " " "	0 12¼
650 " " "	0 13¾

Car lots, ¼c. less; 5-ton lots, ⅛c. less. Central delivery.

SCALES.

Gurney Standard, 35; Champion, 45 p c

Burrow, Stewart & Milne — Imperial Standard, 35; Weigh Beams, 35; Champion Scales, 45

Fairbanks Standard, 30; Dominion, 50 Richelieu, 50.

Warren new Standard, 35; Champion, 45 Weigh Beams, 30.

STONES—OIL AND SCYTHE.

Washita	per lb.	0 25	0
Hindostan	"	0 06	0 10
" slip	"	0 18	0 20
" Axe	"	....	0 10
Deer Creek	"	....	0 10
Deerlick	"	....	0 25
" Axe	"	....	0 15
Lily white	"	....	0 42
Arkansas	"	....	1 50
Water-of-Ayr	"	....	0 10
Scythe	per gross	3 50	5 00
Grind, 40 to 200 lb.,	per ton	20 00	22 00
" under 40 lb.,	"	....	24 00
" 200 lb. and over		....	26 00

INDEX TO ADVERTISERS.

CLASSIFIED LIST OF ADVERTISEMENTS.

Alabastine.
Alabastine Co., Limited, Paris, Ont.

Auditors.
Davenport, Pickup & Co., Winnipeg.

Automobile Accessories.
Canada Cycle & Motor Co, Toronto Junction
Carriage Mountings Co., Ltd., Toronto.

Awnings.
Tobin Tent and Awning Co., Ottawa

Babbitt Metal.
Canada Metal Co, Toronto.
Canadian Fairbanks Co., Montreal.
Frothingham & Workman, Montreal.
Robertson, Jas. Co., Montreal.

Bar Urns.
Buffalo Mfg. Co., Buffalo, N.Y.

Bath Room Fittings.
Buffalo Mfg Co., Buffalo, N.Y.
Carriage Mountings Co., Ltd, Toronto.
Forsyth Mfg. Co., Buffalo, N.Y.
Ontario Metal Novelty Co, Toronto

Belting, Hose, etc.
Gutta Percha and Rubber Mfg. Toronto.
Sadler & Haworth, Toronto.

Bicycles and Accessories.
Johnson's, Iver, Arms and Cycle Works Fitchburg, Mass

Binder Twine
Consumers Cordage Co, Montreal.

Bolts.
Toronto & Belleville Rolling Mills, Belleville.

Box Strap.
J. N. Warminton, Montreal.

Brass Goods.
Frothingham & Workman, Montreal.
Glauber Brass Mfg Co, Cleveland, Ohio.
Kerr Engine Co, Walkerville, Ont.
Lewis, Rice, & Son., Toronto.
Morrison, Jas., Brass Mfg Co., Toronto.
Mueller Mfg. Co., Decatur, Ill.
Penberthy Injector Co., Windsor, Ont.
Taylor-Forbes Co., Guelph, Ont.

Bronze Powders.
Canadian Bronze Powder Works, Montreal.

Brushes.
United Factories, Toronto.

Cans
Acme Can Works, Montreal.

Builders' Tools and Supplies.
Covert Mfg. Co, West Troy, N.Y.
Frothingham & Workman Co., Montreal.
Howland, H. S. Sons & Co., Toronto.
Hyde, F., & Co., Montreal.
Lewis Bros. & Co., Montreal.
Lewis, Rice, & Son, Toronto.
Lockerby & McComb, Montreal.
Lufkin Rule Co., Saginaw, Mich.
Newman & Sons, Birmingham.
North Bros. Mfg. Co., Philadelphia, Pa.
Stanley Rule & Level Co., New Britain.
Stanley Works, New Britain, Conn.
Stephens, G. F., Winnipeg.
Taylor-Forbes Co., Guelph, Ont.

Carriage Accessories.
Carriage Mountings Co., Ltd., Toronto.
Covert Mfg. Co., West Troy, N.Y.

Carriage Springs and Axles.
Guelph Spring and Axle Co., Guelph.

Carpet Beateas.
Ontario Metal Novelty Co., Toronto.

Cartridges.
Dominion Cartridge Co., Montreal.

Cattle and Trace Chains.
Greening, B., Wire Co., Hamilton.

Chafing Dishes.
Buffalo Mfg. Co., Buffalo, N.Y.

Churns.
Dowswell Mfg. Co., Hamilton.

Clippers—All Kinds.
American Shearer Mfg. Co., Nashua, N.H

Clothes Reels and Lines.
Hamilton Cotton Co., Hamilton.

Clutch Nails.
J. N. Warminton, Montreal.

Congo Roofing
Buchanan Foster Co., Philadelphia, Pa.

Cordage.
Consumers' Cordage Co., Montreal.
Hamilton Cotton Co., Hamilton.

Cork Screws.
Erie Specialty Co., Erie, Pa.

Cow Ties
Greening, B, Wire Co., Hamilton

Cuspidors.
Buffalo Mfg Co., Buffalo, N.Y.

Cut Glass.
Phillips, Geo., & Co., Montreal.

Cutlery—Razors, Scissors, etc.
Birkett, Thos., & Son Co., Ottawa.
Clauss Shear Co., Toronto
Dorken Bros. & Co., Montreal.
Frothingham & Workman, Montreal.
Heinisch's, R., Sons Co., Newark, N.J.
Howland, H. S. Sons & Co, Toronto.
Hutton, Wm., & Sons, Ltd., London, Eng.
Lamplough, F. W., & Co., Montreal.
Phillips, Geo., & Co., Montreal.
Round, John, & Son, Montreal.

Electric Fixtures.
Canadian General Electric Co., Toronto.
Morrison James, Mfg. Co., Toronto.
Munderloh & Co., Montreal.

Electro Cabinets.
Cameron & Campbell Toronto.

Enameled Ware
Kemp Mfg. Co., Toronto.

Engines, Supplies, etc.
Kerr Engine Co., Walkerville, Ont.

Eavetroughs
Wheeler & Bain, Toronto

Fencing—Woven Wire.
Dominion Wire Mfg. Co., Montreal.
Owen Sound Wire Fence Co., Owen Sound.
Banwell Hoxie Wire Fence Co., Hamilton.

Files and Rasps.
Barnett Co., G. & H., Philadelphia, Pa.
Nicholson File Co., Port Hope

Firearms and Ammunition.
Hamilton Rifle Co, Plymouth, Mich.
Harrington & Richardson Arms Co., Worcester, Mass.
Johnson's, Iver, Arms and Cycle Works Fitchburg, Mass.

Fishing Tackle.
Enterprise Mfg. Co., Akron, Ohio

Food Choppers
Enterprise Mfg. Co., Philadelphia, Pa.
Lamplough, F. W., & Co., Montreal.
Shirreff Mfg Co., Brockville, Ont.

Furnaces.
Pease Foundry Co, Toronto.

Galvanizing.
Canada Metal Co., Toronto.
Dominion Wire Mfg. Co., Montreal.
Montreal Rolling Mills Co., Montreal.
Ontario Wind Engine & Pump Co., Toronto.

Glass Ornamental
Hobbs Mfg. Co., London
Consolidated Plate Glass Co., Toronto

Glaziers' Diamonds.
Gibsone, J. B., Montreal.
Pelton, Godfrey S.
Sharratt & Newth, London, Eng.
Shaw, A., & Son, London, Eng.

Handles.
Still, J. H., Mfg. Co.

Harvest Tools.
Maple Leaf Harvest Tool Co., Tillsonburg, Ont.

Hockey Sticks
Still, J. H. Mfg. Co, St. Thomas.

Hoop Iron.
Frothingham & Workman Montreal.
Montreal Rolling Mills Co., Montreal.
J. N. Warminton, Montreal.

Horse Blankets.
Heney, E. N., & Co., Montreal.

Horseshoes and Nails.
Canada Horse Nail Co., Montreal.
Montreal Rolling Mills, Montreal.
Capewell Horse Nail Co., Toronto
Toronto & Belleville Rolling Mill Belleville.

Hot Water Boilers and Radiators.
Cluff, R. J., & Co Toronto.
Pease Foundry Co. Toronto.
Taylor-Forbes Co., Guelph.

Ice Cream Freezers.
Dana Mfg. Co., Cincinnati, Ohio.
North Bros. Mfg. Co., Philadelphia, Pa.

Ice Cutting Tools.
Erie Specialty Co., Erie, Pa.
North Bros. Mfg. Co., Philadelphia, Pa.

Injectors—Automatic.
Morrison, Jas., Brass Mfg. Co., Toront
Penberthy Injector Co., Windsor, Ont.

Iron Pipe.
Montreal Rolling Mills, Montreal.

Iron Pumps.
Lamplough, F. W., & Co., Montreal.
McDougall, R., Co., Galt, Ont.

Ladders.
The Waggoner Ladder Co., London, On

Lanterns.
Kemp Mfg. Co., Toronto.
Ontario Lantern Co., Hamilton, Ont.
Wright, E. T., & Co., Hamilton.

Lawn Mowers.
Birkett, Thos., & Son Co., Ottawa.
Frothingham & Workman, Montreal.
Maxwell, D., & Sons, St. Mary's, Ont.
Taylor, Forbes Co., Guelph.

Lawn Mower Grinders
Root Bros. & Co., Plymouth, Ohio.

Ledgers—Loose Leaf.
Business Systems, Toronto.
Copeland-Chatterson Co., Toronto.
Crain, Rolla L., Co., Ottawa.
Universal Systems, Toronto.

Lithographing.
London Printing & Lithographing C London, Ont.

Locks, Knobs, Escutcheons, et
Peterborough Lock Mfg. Co., Peterborough, Ont.
National Hardware Co., Orillia, Ont.

Lumbermen's Supplies.
Pink, Thos., & Co., Pembroke Ont.

Lye
Gillett, E. W., & Co., Toronto

Manufacturers' Agents.
Fox, C. H., Vancouver.
McIntosh, H F., & Co , Toronto.
Gibb, Alexander, Montreal.
Scott, Bathgate & Co , Winnipeg.

Metals.
Canada Iron Furnace Co , Midland, Ont.
Canada Metal Co., Toronto
Eadie. H. G., Montreal.
Frothingham & Workman, Montreal
Gibb, Alexander, Montreal.
Kemp Mfg. Co., Toronto
Leslie, A. C., & Co , Montreal.
Lysaght, John, Bristol, Eng.
Nova Scotia Steel and Coal Co., New Glasgow, N S.
Robertson, Jas., Co., Montreal.
Roper, J. H., Montreal.
Samuel, Benjamin & Co., Toronto.
Stairs, Son & Morrow, Halifax, N.S.
Thompson, B. & S. H. & Co. Montreal.

Metal Lath.
Galt Art Metal Co., Galt.
Metallic Roofing Co., Toronto.
Metal Shingle & Siding Co., Preston, Ont.

Metal Polish, Emery Cloth, etc.
Oakey, John, & Sons, London, Eng.

Nails Wire
Dominion Wire Mfg. Co , Montreal.

Oilers
Maple City Mfg Co., Monmouth, Ill.

Oil Tanks.
Bowser, S. F., & Co., Toronto.

Ornamental Iron and Wire.
Dennis Wire & Iron Co , London, Ont.

Packing.
Gutta Percha & Rubber Co Toronto

Paints, Oils, Varnishes, Glass.
Blanchite Process Paint Co , Toronto.
Brandram-Henderson, Montreal
Canada Paint Co., Montreal.
Canadian Oil Co., Toronto
Consolidated Plate Glass Co., Toronto.
Dods, P. D., & Co , Montreal
Imperial Varnish and Color Co., Toronto
Jamieson, R. C., & Co., Montreal.
Lucas John & Co., New York
McArthur, Corneille & Co , Montreal.
McCaskill, Dougall & Co., Montreal.
Moore. Benjamin, & Co. Toronto.
Ottawa Paint Works, Ottawa
Queen City Oil Co., Toronto.
Ramsay & Son, Montreal.
Sanderson Pearcy & Co., Toronto
Sherwin-Williams Co., Montreal.
Standard Paint Co., Montreal
Standard Paint and Varnish Works Windsor, Ont.
Stephens & Co., Winnipeg.
Martin-Senour Co., Montreal
Winnipeg Paint & Glass Co., Winnipeg

Perforated Sheet Metals.
Greening, B., Wire Co., Hamilton.

Plumbers' Tools and Supplies
Canadian Fairbanks Co., Montreal.
Cluff, R. J , & Co., Toronto
Frothingham & Workman, Montreal.
Glauber Brass Co., Cleveland, Ohio.
Jardine, A. B., & Co , Hespeler, Ont.
Jenkins Bros., Boston. Mass.
Kerr Engine Co , Walkerville, Ont.
Lewis, Rice, & Son, Toronto.
Merrell Mfg. Co., Toledo, Ohio.
Montreal Rolling Mills. Montreal.
Morrison, Jas , Brass Mfg. Co., Toronto.
Mueller, H., Mfg. Co., Decatur, Ill.
Oshawa Steam & Gas Fitting Co., Oshaw
Robertson Jas., Co. Montreal.
Robertson, Jas , Co., Limited, Toronto
Somerville, Limited, Toronto
Stairs, Son & Morrow, Halifax, N.S.
Standard Ideal Sanitary Co., Port Hope,
Standard Sanitary Co., Pittsburg.
Stephens, G. F., & Co., Winnipeg, Man.
Turner Brass Works, Chicago.
Vickery, Orlando, Toronto.

Polishes.
Majestic Polishes, Toronto

Portland Cement.
International Portland Cement Co. Ottawa, Ont.
Hanover Portland Cement Co., Hanover, Ont.
Hyde, F., & Co., Montreal
Thompson B. & S. H. & Co., Montreal.

Poultry Netting.
Greening, B., Wire Co., Hamilton, Ont.

Printing.
London Printing & Lithographing Co., London, Ont.

Razors.
Clauss Shear Co., Toronto.

Refrigerators.
Fabien, C. P., Montreal.

Registers.
Pease Foundry Co., Toronto.

Roofing Supplies.
Brantford Roofing Co., Brantford.
Barrett Mfg Co., New York
F. W. Bird, East Walpole, Mass.
Buchanan Foster Co., Philadelphia, Pa.
McArthur, Alex., & Co., Montreal.
Metal Shingle & Siding Co., Preston, Ont.
Metallic Roofing Co., Toronto.
Paterson Mfg. Co , Toronto & Montreal.
Standard Paint Co., Montreal.
Wheeler and Bain, Toronto

Saws.
Atkins, E. C., & Co., Indianapolis, Ind
Shurly & Dietrich, Galt, Ont.
Spear & Jackson, Sheffield, Eng.

Scales.
Canadian Fairbanks Co., Montreal.
Frothingham & Workman, Montreal.

Screw Cabinets.
Cameron & Campbell. Toronto.

Screws, Nuts, Bolts.
Dominion Wire Mfg. Co., Montreal
Montreal Rolling Mills Co., Montreal.

Soil Pipe
McFarlane, Walter, Glasgow

Sewer Pipes
Canadian Sewer Pipe Co., Hamilton
Hyde, F., & Co., Montreal.

Shelf Boxes.
Cameron & Campbell, Toronto.

Shears, Scissors.
Clauss Shear Co., Toronto.

Shovels and Spades.
Eclipse Mfg. Co., Ottawa
Frothingham & Workman, Montreal.
Peterboro Shovel & Tool Co., Peterboro.

Silverware.
Hutton, Wm., & Sons, Ltd., London, Eng
McGlashan, Clarke Co., Niagara Fal s, Ont.
Phillips, Geo., & Co., Montreal.
Round, John, & Son, Sheffield, Eng.

Skates.
Canada Cycle & Motor Co., Toronto.
McFarlane, Walter, Glasgow.

Sprayers
Cavers Bros., Galt

Spring Hinges, etc.
Chicago Spring Butt Co., Chicago, Ill.

Stable Fittings
Dennis Wire & Iron Co., London

Steel Rails.
Nova Scotia Steel & Coal Co., New Glasgow, N.S.

Stove Pipe.
Chown, Edwin, and Son, Kingston

Stoves, Tinware, Furnaces
Canadian Heating & Ventilating Co. Owen Sound.
Copp, W. J., Son & Co., Fort William
Davidson, Thos., Mfg. Co., Montreal.
Down Draft Furnace Co , Galt
Guelph Stove Co., Guelph.
Gurney Foundry Co., Toronto.
Harris, J. W., Co., Montreal.
Howard, Wm., Toronto
Kemp Mnfg. Co. Toronto.
McClary Mfg. Co. London.
Merrick Anderson, Winnipeg
Pease Foundry Co., Toronto.
Smart, James, Mfg. Co., Brockville
Stewart, Jas., Mfg. Co., Woodstock, Ont.
Taylor-Forbes Co., Guelph. Ont.
Wright, E. T., & Co., Hamilton.

Tacks.
Montreal Rolling Mills Co., Montreal.
Ontario Tack Co., Hamilton.

Tents.
Tobin Tent and Awning Co., Ottawa

Tin Plate.
American Sheet & Tin Plate Co., Pittsburg, Pa.
Baglan Bay Tin Plate Co., Briton Ferry South Wales
Lysaght, John, Bristol, Newport and Montreal

Turpentine
Defiance Mfg. Co , Toronto.

Ventilators.
Harris, J. W., Co., Montreal.
Pearson, Geo. D., Montreal.

Wall Paper
Staunton Limited, Toronto.

Wall Paper Cleaner.
Gilbert, Frank U. S., Cleveland

Washing Machines, etc
Dowswell Mfg. Co., Hamilton, Ont.
The Shultz Bros. Co., Brantford.
Taylor-Forbes Co., Guelph, Ont.

Water Filters.
Buffalo Mfg. Co., Buffalo, N.Y.

Wheelbarrows
London Foundry Co., London Ont.
Schultz Bros. Co., Ltd., The Brantford.

Wholesale Hardware.
Birkett, Thos., & Sons Co., Ottawa.
Caverhill, Learmont & Co., Montreal.
Frothingham & Workman, Montreal.
Hobbs Hardware Co., London.
Howland, H. S., Sons & Co., Toronto.
Lamplough, F. W., & Co., Montreal.
Lewis Bros. & Co., Montreal.
Lewis, Rice, & Son, Toronto.

Window and Sidewalk Prisms.
Hobbs Mfg. Co., London, Ont.

Wire, Wire Rope, Cow Ties, Fencing Tools, etc.
Banwell-Hoxie Fence Co., Hamilton
Dennis Wire and Iron Co., London, Ont.
Dominion Wire Mnfg. Co., Montreal
Greening, B., Wire Co., Hamilton.
Owen Sound Wire Fence Co., Owen Sound
Montreal Rolling Mills Co., Montreal.
Western Wire & Nail Co., London, Ont.

Wrapping Papers.
Canada Paper Co., Toronto.
McArthur, Alex., & Co , Montreal.
Stairs, Son & Morrow, Halifax, N.S.

Wringers
Connor, J. H.&Son, Ottawa, Ont

"Seeing is Believing"

¶ As the Ross Sporting Rifle represents the best in modern rifle manufacture, the highest grade materials and minute accuracy in workmanship, it is to your interest to learn fully about this Canadian Rifle—

¶ The Ross Straight Pull 303 Sporting Model M is one of our leaders to retail at $26. The barrel is the celebrated Ross high pressure, non-fouling, gun-barrel steel — The breech action is dependable.

Our representatives are now demonstrating the merits of this rifle—It places you under no obligation to drop us a line to make sure you will be on their calling list.

Our advertising policy is co-operative.

THE ROSS RIFLE CO., - QUEBEC

Hardwareman and Plumber—

We have pumps suitable to any demand made upon you by most exacting customers.

Our pumps will do anything that a pump can do, and we will be pleased to give the benefit of many years' experience to any one having pump troubles.

McDougall Pumps are Standard Pumps, and with us it is quality first, last and always, while we ask just as little money as we can get along with.

We have a few catalogues for distribution.

The R. McDougall Co., Limited

GALT, - CANADA

It Saves Lives

A Building Necessity

These are the names that have been made synonymous with the

AEOLIAN VENTILATOR

We are the manufacturers of it and control the patents.

We have placed it on some of the finest buildings in the Dominion.

It is a scientific fact that no modern building can afford to neglect sanitation.

We are contractors and sanitary experts. Our ventilator is guaranteed and is placed in position to suit the building as to quality and size.

One Testimony:

Montreal, June 12th, 1903

Canadian Pacific Railway Company.

Messrs. J. W. Harris Co., Limited,
7 St. Elizabeth St., Montreal, Que.

Gentlemen,—Answering yours of June 1st, I am not personally familiar with the merits of the "Aeolian" Ventilators but our Assistant Engineer, Mr. F. Crossley, reports that the Ventilators of this type manufactured by you and used by this Company at the Place Viger Hotel, Chateau Frontenac, Windsor Street Station and elsewhere, have proved satisfactory. Yours truly,

E. H. McHENRY, Chief Engineer.

The J. W. HARRIS CO., Limited

General Contractors and Manufacturers

MONTREAL

CIRCULATES EVERYWHERE IN CANADA

Also in Great Britain, United States, West Indies, South Africa and Australia.

HARDWARE AND METAL

A Weekly Newspaper Devoted to the Hardware, Metal, Heating and Plumbing Trades in Canada.

Office of Publication, 10 Front Street East, Toronto.

VOL. XIX. MONTREAL, TORONTO, WINNIPEG, SEPTEMBER 21, 1907 ***NO. 38.***

Read "Want Ads." on Page 49

GEM FOOD CHOPPER

A modern household utensil that is needed in every kitchen. Carefully made, nicely tinned, self-cleaning, self-sharpening. It has steel cutters, and is made in four sizes.

No. 16	**capacity**	**1 lb.**	**per minute**
No. 20	"	**2 lbs.**	"
No. 22	"	**2½ lbs.**	"
No. 24	"	**3 lbs.**	"

Gem Food Choppers chop all kinds of Raw or Cooked Meat, Fish, Clams, Oysters, Vegetables of all kinds, Fruit, Bread, Crackers or Cheese, and many other articles used in making substantial dishes and dainty desserts. It is more useful than a chopping bowl and a great deal handier. It may be easily clamped to the kitchen table and is ready for use whenever needed.

RICE LEWIS & SON

LIMITED

TORONTO.

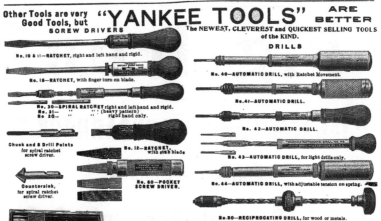

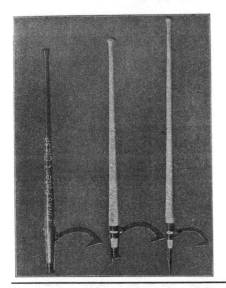

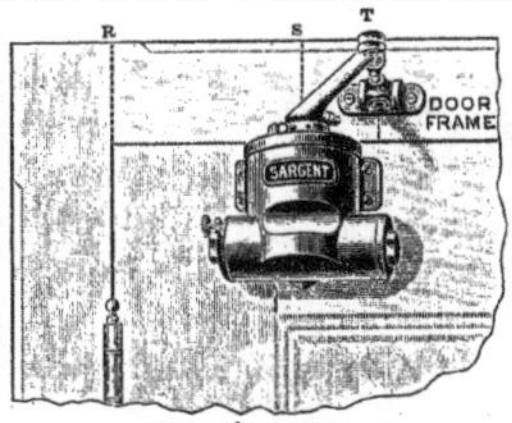

Check attached to Door

Shut the Door!

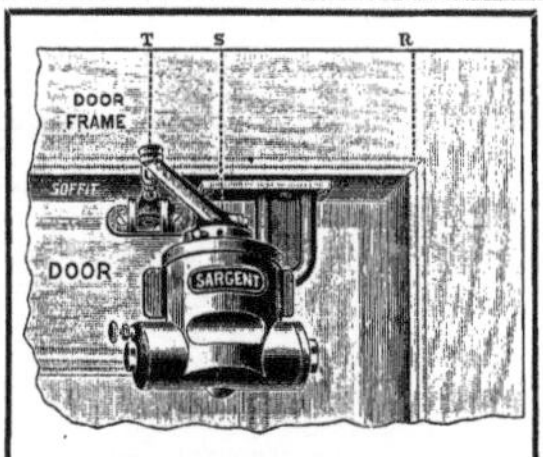

Check attached with Soffit Bracket

That is not a nice way for a storekeeper to greet his customers, and yet there are many merchants, neighbors of yours, who have such a notice attached to their front doors. The up-to-date plan, as used by progressive business men, is to have applied to the door

Sargent's Liquid Door Check and Spring

so that the door will close itself, quickly and noiselessly.

Get the Business

It is in your town if you try for it, and there is money in it for you. Convince these neighboring merchants that they need Sargent's Liquid Door Check and Spring and that now is the time to have it applied.

Begin Early.

Many buildings in your town will need this Check during the coming winter. Look around and with a little effort you can increase your profits by supplying the schools, churches, public buildings, office buildings, hospitals, libraries, fire houses, hotels, stores and dwellings.

Sargent's Liquid Door Check and Spring

Simple, Strong, Durable. **Suitable for all Doors.**

Easy to Apply, Regulate and Keep in Order.

Compact and Powerful Spring. **Prompt and Effective in its Action.**

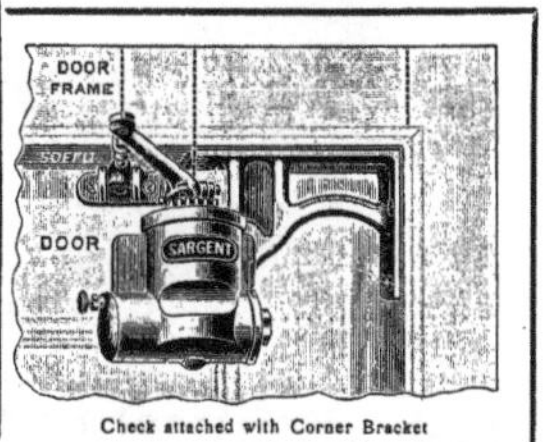

Check attached with Corner Bracket

Lewis Bros.

LIMITED

Montreal

Ottawa Toronto

Winnipeg

Vancouver Calgary

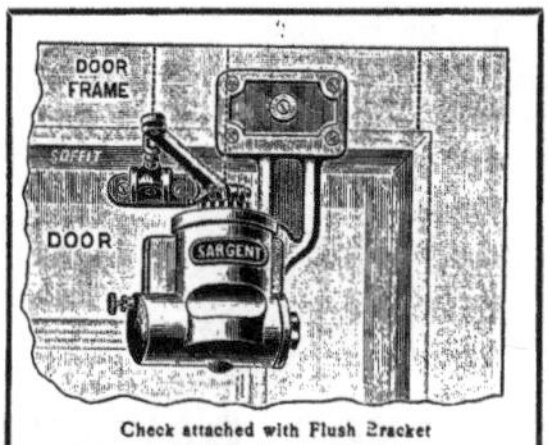

Check attached with Flush Bracket

Transmission Rope

The transmission of power by Rope Drives is increasing daily, on account of the durability and low cost. Rope Drives were first manufactured and introduced into the United States by Sewall, Day & Co., of Boston. A few years later we manufactured and introduced them into this country.

If you contemplate putting in a Rope Drive, we will be pleased to furnish you with any information you may require. As we have had years of experience, we are naturally in a position to turn out the most up-to-date product.

Our mills are now thoroughly equipped for manufacturing our famous Tallow Laid Transmission Rope, three or four strand, and in all sizes and lengths, and we always carry a large stock of the most select Manila Fibre for the manufacture of this Rope. We carry a fully assorted stock of all the standard sizes, and can make special orders promptly.

CONSUMERS' CORDAGE COMPANY, Limited

Mills: MONTREAL, QUE., and HALIFAX, N.S.

BRANCHES:

F. H. ANDREWS & SON, Quebec. CONSUMERS CORDAGE CO., Limited, St. John, N.B. MacGOWAN & CO., Vancouver
TEES & PERSSE, Limited, Winnipeg, Calgary and Edmonton. GEO. WOOD, London, England.
MERRICK, ANDERSON & CO., Winnipeg, Binder Twine Distributors for Northwest.

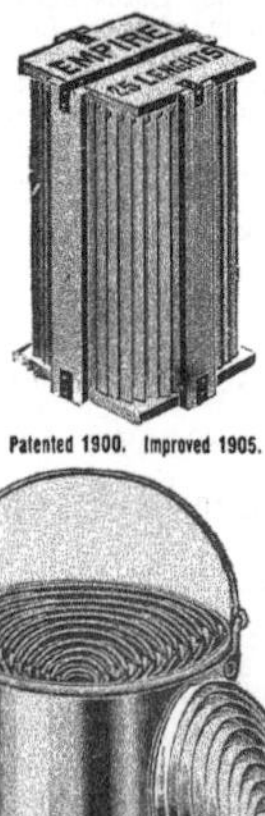

Patented 1900. Improved 1905.

Davidson's "Empire" Stove Pipe

Improved Patent

Neatly packed in crates of 25.
Secures lowest possible freight rate.

Deep Swage Prevents Telescoping.
Coated to Prevent Rust.

Sold as low as ordinary Stove Pipes.

SIMPLE and EASY to put together.
No Profane Language Necessary.
The only tools required are a pair of hands.
Pipes being cut out with dies ensure absolute fit and conformity.

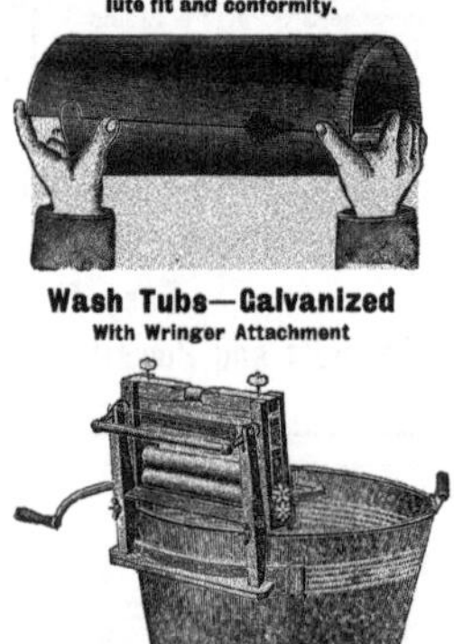

Miners' Camp Kettles

Strong and substantially made in extra heavy tin.

Nineteen sizes, from ⅛ quart to 29 quarts.

Wash Tubs—Galvanized

With Wringer Attachment

Nos.	1	2	3
Top measurement, inches ...	21½	22¾	25
Bottom " "	17⅞	18¼	20
Height " "	9½	10½	11

The THOS. DAVIDSON MFG. CO., Ltd., Montreal and Winnipeg

FURS ARE VALUABLE

No. 91.—Double jaws make it impossible for an animal to wrench itself free from the jaws. Made in two sizes.

Do not allow your catch to escape because of a poor trap. Particular care is taken in the design of these jaws, so that they will hold the game securely, and yet not crush or break the bones.

These traps are thoroughly inspected and tested, and the important parts are hand-made and hand-fitted. The springs are made from the best crucible steel, and they will hold their temper even under severe strain. They are quick and sure in action.

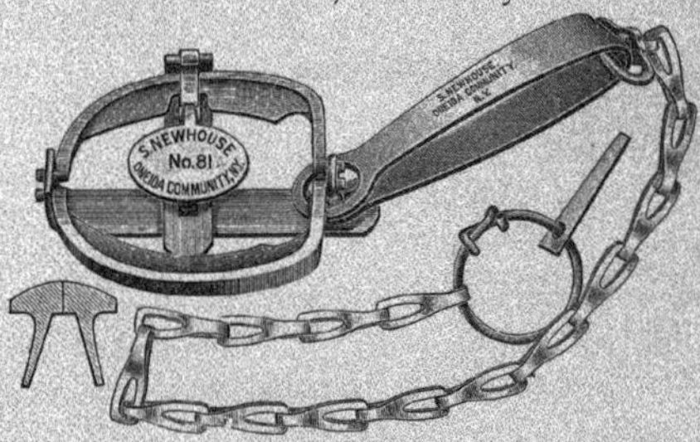

No. 81.—The webbed jaws prevent an animal from gnawing off its leg just below the jaws, where the flesh is numb from pressure.

These lines, as well as many others, are fully illustrated on pages 739 and 740 of our general catalogue.

Bear Trap. With offset jaws. ½-inch space between.

Victor Traps. With or without chain. Six Sizes.

If you ask our prices you will buy from us

Frothingham & Workman, Limited

Wholesale Hardware and Iron Merchants

Montreal, - Canada

F & W.

DOMINION WIRE MANUFACTURING CO.

MONTREAL TORONTO LIMITED

Manufacturers of

PLAIN WIRE *All Sizes and for all Purposes.*

GALVANIZED WIRE *For Fencing, Telegraph and Telephone Purposes.*

Copper and Brass Wire

WIRE NAILS
WOOD SCREWS

Bright Wire Goods

Staples

Steel and Brass Jack Chains

Spring Cotters, Strand Guy Wire, Steel Wire Barrel Hoops

BUSINESS SUCCESS

Mr. Dealer, if you could take out a policy assuring commercial success in your line you certainly would do so. The next best thing is to place your orders for

Carpenters' Tools and Builders Hardware

with a specialty house—a house which guarantees quality and prompt deliveries. We handle

Stanley's, Union Hardware Co. and Sarjent's Iron and Wood Bottom Planes.
Disston's, Atkins', and Shurly & Dietrich—High Grade Hand, Rip and Panel Saws.
Sarjent's Front, Store Door and Inside Locks.
Stanley's Butt Hinges—all kinds, all finishes.
Also headquarters for Wood Sash Pegs, sizes ⅞, 1¼, 1⅝-in. in one-bushel bags.

Remember our prices are right and we deliver like lightning

THOS. BIRKETT & SON CO., Ottawa, Canada

Simonds
Crescent-Ground Cross-Cut Saws

For logging camps where work must be fast and economical there is no other saw made that can give results equal to the Crescent-Ground Saw. **Made of Simonds Steel.** It cuts easy, runs fast and gives good results. This cross-cut saw is sold by most of the leading jobbers in Canada.

Simonds Canada Saw Co., Limited

TORONTO, ONT. MONTREAL, QUE. ST. JOHN, N.B.

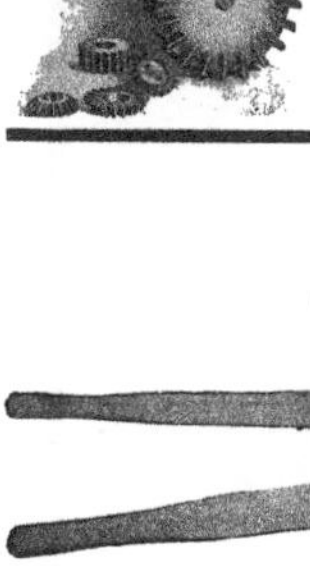

The dealer who carries the best carries

PRATT AND WHITNEY SMALL TOOLS

Made in Canada. The standard everywhere.

THE CANADIAN FAIRBANKS CO., LIMITED

Exclusive Canadian Selling Agents

MONTREAL TORONTO WINNIPEG VANCOUVER

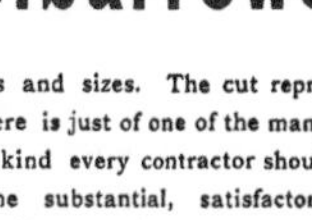

Wheelbarrows

All kinds and sizes. The cut reproduced here is just of one of the many, but the kind every contractor should use. The substantial, satisfactory, steel tray Contractor's Barrow.

The London Foundry Co.

LONDON, CAN.

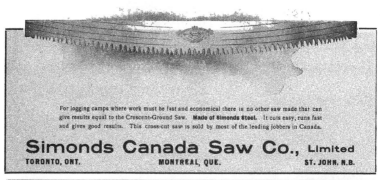

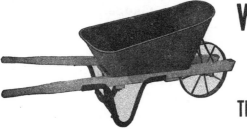

FOR BEST **QUALITY**

IN

ENAMELLED AND GALVANIZED WARE

Buy goods manufactured by

ONTARIO STEEL WARE, LIMITED

115-121 Brook Ave. and 79-91 Florence St.

TORONTO, ONT.

20th Century

WHITE FLAME BURNERS

Light and Heat combined, at the cost of light only

A revolutionizing invention in gas combustion, by which excessive heat is generated from brilliantly illuminating white flame.

Burns either manufactured or natural gas. Fits any gas fixture.

NO DIRT NO ODOR NO TROUBLE
NO RISKY RUBBER TUBE CONNECTION

Orders solicited from the jobbing trade. Write for prices.

MANUFACTURED IN CANADA BY

ONTARIO LANTERN AND LAMP CO.
HAMILTON, ONT. LIMITED

The Buffalo Manufacturing Co.

Buffalo, N.Y.

When you get our goods you know you get THE BEST.

We manufacture
Water Filters
Water Coolers
Chafing Dishes
Table Kettles and Stands
Coffee Extractors
Wine Coolers
Nursery Chests
Baking Dishes
Crumb Trays and Scrapers
Tea and Bar Urns
Bathroom Fixtures
Coal Vases and Hods
Candlesticks
Cuspidors
Match Safes, Etc.

All High Grade and exceedingly presentable.

REPRESENTED BY

H. F. McINTOSH & CO.

34 Yonge Street

Write for Catalogue **TORONTO, ONT.**

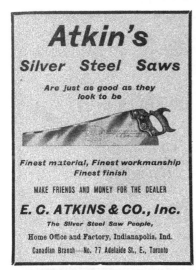

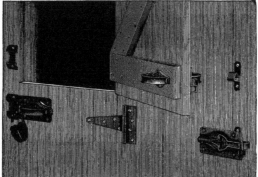

DO YOU KNOW

what the profits are in selling our rifles? Why not find out? It only costs you a two-cent stamp and two minutes of your time. Let us tell you all about the good points—the profit points—of

HAMILTON RIFLES

Hamilton rifles are absolutely safe. The only way to discharge a Hamilton Rifle is to pull the trigger. Hamilton Rifles are made right. They never explode accidentally. They can't do it. The carefully-selected and thoroughly-tested rolled steel and the expert workmanship make bursting impossible.

No. 15 sells at $2.00
No. 19 sells at $2.50
No. 23 sells at $3.50

THE HAMILTON RIFLE CO.

Dept. 71 ***Plymouth, Mich.***

NICKEL

The Canadian Copper Co.

NICKEL FOR NICKEL STEEL

The Orford Copper Co.

WRITE US FOR PARTICULARS AND PRICES

General Offices

43 EXCHANGE PLACE

NEW YORK CITY

Shear making "up-to-date" is found in

HEINISCH

TAILORS' SHEARS, TRIMMERS, SCISSORS, TINNERS' SNIPS, ETC.

"Best in the World"

Since 1825, and a steady increase in sales due to their superior cutting quality.

R. HEINISCH'S SONS CO.
NEWARK, N.J., U.S.A.
New York Office and Salesroom, 155 Chambers St.

OAKEY'S

The original and only Genuine Preparation for Cleaning Cutlery, 6d. and 1s. Canisters

'WELLINGTON'

KNIFE POLISH

JOHN OAKEY & SONS, Limited

Manufacturers of
Emery, Black Lead, Emery, Glass and Flint Cloths and Papers, etc.

Wellington Mills, London, England

Agent:
JOHN FORMAN, - 644 Craig Street
MONTREAL.

REGISTERED TRADE MARK
GRANTED 1682

World's Best Cutlery

You make no mistake in specifying goods made by **JOSEPH RODGERS & SONS, Limited** when ordering cutlery. A reputation of over 200 years at stake, insures the quality of every article.

Write for catalogues from Sole Canadian Agents

JAMES HUTTON & COMPANY, MONTREAL

TINPLATES

All made out of the

BEST SIEMEN'S MARTIN STEEL

Brand

Best Coke or Charcoal	"Rhyd"
Good Charcoal	"Edith"
Second Charcoal	"Sawel"
Best Charcoal	"Dryw"

We are ready to offer the Canadian Trade very best value and finest quality in Tinplates.

Baglan Bay Tinplate Co., Limited

Briton Ferry, South Wales

Please mention above brands in your indents and inquiries.

Are you interested in any of the lines that are advertised? A Post Card will bring you price list and full information. Don't forget to mention Hardware and Metal.

Stauntons

PROFIT-EARNING

WALL PAPERS

WE HAVE A PAPER FOR EVERY WANT AND A WANT FOR EVERY PAPER.

PRICES ARE IN YOUR FAVOR
SEE OUR SALESMAN'S SAMPLES

ALL GOODS 16 YARDS TO DOUBLE ROLL

Stauntons Limited

MANUFACTURERS OF SUPERIOR WALL PAPERS

TORONTO

OIL STOVES

FOR SPRING IMPORTATION

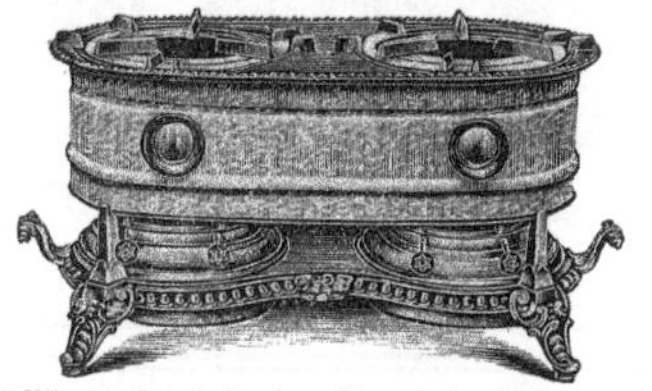

THE up-to-date dealer places his order for this line now so as to get the lowest prices and ensure delivery in good time. There is always a rush when the season comes nearer, and the makers are overwhelmed with orders already now. We have a fine and extensive variety for selection and **our line** of goods is known as being **superior to any other** on the market.

Why not get the best for your money? Drop us a post-card to-day and we shall send you our quotation by return mail along with an attractive booklet showing the most saleable lines and giving full particulars.

F. W. Lamplough & Co.
9 de Bresoles Street, Montreal

PENBERTHY

TRADE MARK.

Genuine

Plain Engine Lubricator
Made with or without Drain and Tube
Iron or Wood Wheel
Strong and Durable

Automatic Injector
"THEY ALWAYS WORK"

Plain Engine Lubricator
Made with or without Drain and Tube
Iron or Wood Wheel
Strong and Durable

PENBERTHY INJECTOR CO., Limited, WINDSOR, ONT.

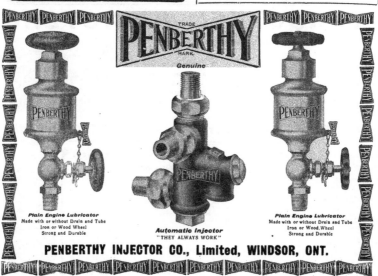

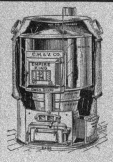

A Well-made Range
A Well-liked Range
A Well-advertised Range

PANDORA

Features:

Oven—Scientifically ventilated.

Flues—Heat forced twice around oven.

Fire-box — Proportioned to size of oven.

Reservoir — Enamelled and easy to clean.

Thermometer--Registers exact heat.

Coal-grates — Acme of strength.

Wood-grate — Will fit when coal-grate is in place.

Fire-box Linings--Heavy cast-iron.

Broiler Door — Takes largest broiler or toaster.

Emery Rod—The handy knife sharpener.

"PANDORA" RANGE

is the one to recommend to your Customers

The McClary Mfg. Co.

London Toronto Montreal Winnipeg

Vancouver St. John, N.B. Hamilton Calgary

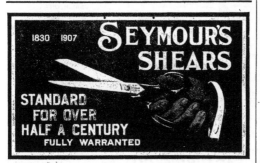

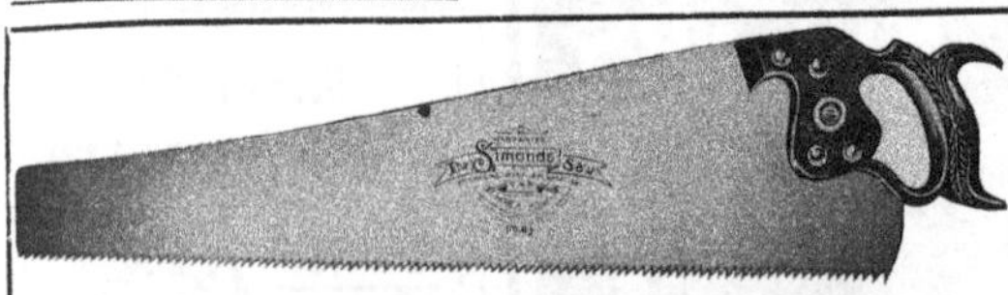

No. 100. Laundry Set.

Consists of one Hood, two oval six pound Cores, one 6½ pound Polishing Core, and one Stand. Packed in a single-set box.

No. 130. Laundry Set.

Consists of one Hood Complete, two oval six-pound Cores, one nine-pound Pressing Core and one Stand. Packed in a single-set box.

No. 140. Laundry Set.

Consists of one Hood complete, one oval six-pound Core, one Polishing Core, one Pressing Core, and one Stand. Packed in a single-set box.

No. 50. Flounce Iron.

A four-pound Iron for ironing flounces, ruffles, etc. Wanted by every housewife who sees it.

No. 40. Sleeve Iron.

A three-pound Iron for light work. Adapted for places hard to get at with a large Iron.

No. 10. Tourist Iron.

A 1½ pound Iron for very light work. Sold in large quantities as a toy. A great seller during the holidays.

Make an "Asbestos" Window.

¶ We are placing a fine lot of Window Display Matter in Asbestos Sad Iron Assortment No. 6. Canadian hardware dealers know that Asbestos Sad Irons yield a good profit. We know that window displays will move the Irons —to the dealer's profit.

The following pieces of Printed Matter are included in the Assortment :

Decalcomania Sign
18 inches wide by 10 inches high. Lithographed in seven colors. To be transferred upon the glass of your door or window. Cannot be washed off. An effective and permanent ad. Will head customers your way.

Hanging Sign
16 x 7 inches. Printed in three colors and gold, with silk cord hanger. A handsome sign.

Transparencies
Ten pieces. Printed in green and red. To be pasted on the window glass.

Card
22 x 11 inches. A chart from which to demonstrate Asbestos Sad Irons.

Booklet
Entitled "Asbestos Sad Irons—Their Story." Tells all about the Irons, with ample illustrations. Twenty-five included in each Assortment.

Leaflet
A piece of salesmanship on paper. 500 included in each Assortment.

The Assortment contains the following Irons :

¼ doz. No. 100 Laundry Sets
1½ " No. 130 Laundry Sets
1¼ " No. 140 Laundry Sets
½ " No. 50 Flounce Irons
¼ " No. 40 Sleeve Irons
¼ " No. 10 Tourist Irons

ORDER A NO. 6 ASSORTMENT FROM YOUR JOBBER

The Dover Manufacturing Co.

Canal Dover, Ohio

RUBEROID

A ROOFING WITH AN EARNED REPUTATION

There is only one prepared roofing for which a service of 15 years can be claimed and proved—that is **RUBEROID.**

It was the first smooth-faced, elastic, weather-proof roofing made. Its record is not and cannot be equalled by any other roofing.

Being absolutely free from tar paper, **RUBEROID** will not melt, rot or corrode. Neither will it rust like tin or metal, nor crack from vibration like asphalt or slate.

On account of its elasticity, it contracts and expands to the requirements of heat or cold.

Send to-day for booklet and samples of **RUBEROID.**

THE STANDARD PAINT CO. OF CANADA,
LIMITED.

Board of Trade Building, - MONTREAL, P.Q.

FACTORIES AT LACHINE CANAL, HIGHLANDS, P.Q.

LOCKERBY & McCOMB, Selling Agents for Ontario and Quebec.

The Howland Pump Oiler

FOR SALE BY

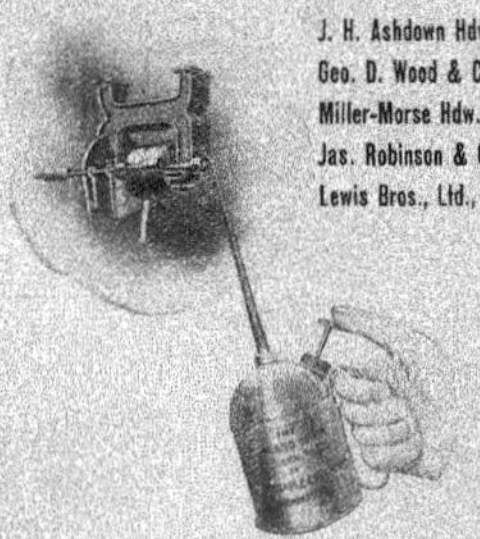

J. H. Ashdown Hdw Co.,	Winnipeg
Geo. D. Wood & Co.,	"
Miller-Morse Hdw. Co.,	"
Jas. Robinson & Co.,	"
Lewis Bros., Ltd.,	Montreal

A slight pressure of the thumb regulates the flow.

Maple City Manufacturing Co.

Monmouth, Ill.

Really Tough

A tough human being generally lands in the coop. Toughness in human beings is very rightly frowned down upon. But as far as

AXE HANDLES

are concerned, the tougher they are the better. ***Still's handles*** are known as the very toughest, wherever axes are wielded by the arms of ***lusty Canucks.*** It takes a lot of abuse to break our handles, because they are made of the ***cleanest and toughest hickory*** we can get hold of.

Let us mail you our lists and discounts. A one-cent post card will bring them.

J. H. Still Manufacturing Co.
Limited

ST. THOMAS, ONTARIO

J. M. T. Injectors

J. M. T. Valves

The Standard for "valve excellence" Made in Brass and Iron.

J. M. T. Valves are made in all the standard designs "Globe" Angle and Check Patterns, also Extra Heavy Pattern.

We make the Genuine Hancock Inspirator and Ejector.

J.M.T.

We Stand Back of our Productions.

Ready to make good any article that may prove unsatisfactory.

Our

Steam Supplies

are made for ***Durability*** and ***Efficiency*** which results are attained by combining the best that is procurable in material and workmanship, to which is added correctness in design properly proportioned.

THE

Jas. Morrison Brass Mfg. Co.

Limited

93-97 Adelaide St. West

TORONTO

EASY MONEY

seldom comes to the Bicycle dealer. It is just possible that our Metal Half Chain Guard

has cashed in more easy money for Hardware men than almost anything else on the market.

Many of these guards will be sold in your town this season. Light, strong, dainty, popular —these four words tell our story. There's a Profit Story attached. Do you know it ?

Forsyth Manufacturing Co.

BUFFALO, N.Y.

W. F. Canavan, 13 St. John Street, Montreal, Representative

H. BOKER & CO.

Adjustable Razor Guard

Retails at 25c. each

Makes the ordinary Razor into a Safety Razor by sliding the guard on to the blade. Fits every Boker Razor.

McGILL CUTLERY CO., *140 McGill Street,* ***Montreal***

Cooper's Improved Patent One-Piece Stove-Pipe Elbows

The Heaviest and Best Elbow on the market. Measures 23 inches long at the back.
Send in for sample dozen.
Polished or Common.

E. T. WRIGHT & CO.

Sole Manufacturers and Patentees

HAMILTON and MONTREAL

Will Hold up A Shelf

That's what a shelf brackets for.

For this purpose there can be NOTHING BETTER, NOTHING CHEAPER than the BRADLEY STEEL BRACKET. It is well Japanned, Strong and Light. The saving on freight is a good profit aside from the lower price at which the goods are sold. Order direct or through your jobbers

ATLAS MFG. CO., New Haven.

"Quality Unsurpassed"

Belleville B Brand

HORSE SHOES

Bar Iron
Machine Bolts
Nuts $\frac{5}{8}''$ and Larger
U. S. STANDARD

Toronto and Belleville Rolling Mills

LIMITED

BELLEVILLE, ONT.

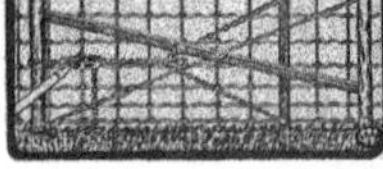

HOW TO BUILD A GOOD FENCE

Everyone intending fence building should send for our folder on Erecting Fences. It's full of valuable information on fence building, tells how to erect woven wire fencing quickly and substantially, describes the manufacture of fence wire and has an article quoted from bulletin of U. S. Dept. of Agriculture on concrete post making, showing how these durable posts can be economically made at home. Don't fail to write for a copy. It's free.

THE BANWELL HOXIE WIRE FENCE CO., Ltd.
Dept. J Hamilton, Ontario. Winnipeg, Manitoba.

THE FENCE WITH THE PEERLESS LOCK
WILL TURN ANY KIND OF STOCK

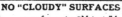

THE MERIT OF

Standard Ideal Porcelain Enameled Ware

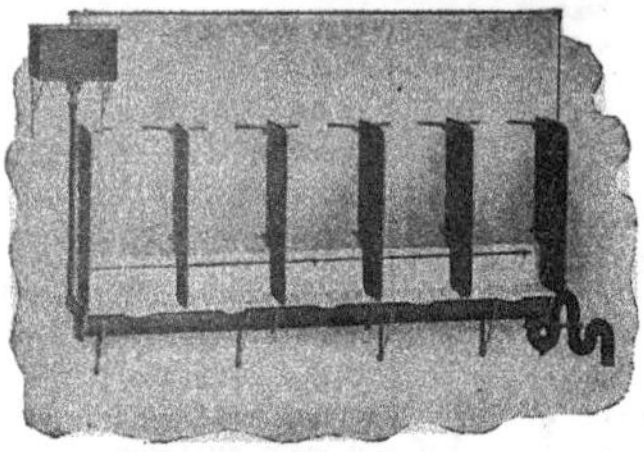

is recognized by the leading dealers in Plumbing Equipment. **Standard Ideal Ware** is well constructed, durable in character and of designs calculated to meet the requirements of the Canadian Trade.

We manufacture Bath Tubs, Sitz Baths, Shower Baths, Lavatories, Urinal Ranges, Slop Hoppers, Laundry Tubs, Sinks, Closet Ranges, Steamship Supplies, Railway Car Supplies, Hospital Appliances.

WRITE FOR CATALOGUE

The Standard Ideal Co., Limited

Head Offices and Factories, - PORT HOPE, Ont.

Sales Offices and Sample Rooms, 50 Colborne Street, TORONTO

128 West Craig Street, MONTREAL, and 24 Telfer Block, WINNIPEG

WIRE
I
R
E

Do You Handle Wire?

We manufacture and sell all kinds of WIRE.

Hay Baling Wire
Oiled and Annealed Wire
Plain Galvanized (Fence) Wire
Coppered Steel Spring Wire

Cut and straightened to length.

WRITE FOR PRICES

THE B. GREENING WIRE CO.

LIMITED

Hamilton, Ont. Montreal, Que.

Make Your Customers Lasting Friends

Hundreds of Dealers from the Atlantic to the Pacific have cornered the paying trade of their town by recommending the

"General" Food Chopper

The "General" differs materially from every Food Chopper ever marketed, excepting the Shirreff. All parts interchangeable and **both of these Choppers have steel knives.**

Shirreff Mfg. Co., Ltd.

Brockville, - Ont.

Tinsmith's Tools
Sheet Metal Working Tools
Presses and Dies
Special Machinery

We specialize on the above lines. **Every Tool is guaranteed.** Let us figure on your requirements.

THE BROWN, BOGGS CO.

LIMITED

HAMILTON, CANADA

Wheelbarrows for all purposes

Drag and Wheel Road Scrapers

Contractors' Dump Carts

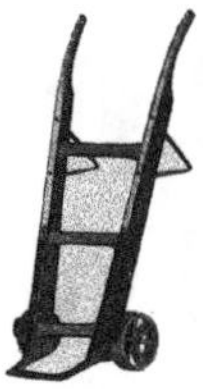

Store, Warehouse, Steamboat, Railway and Packing House

Trucks of every size and description

BEST QUALITY — NEWEST PATTERNS

LOWEST PRICES — PROMPT SHIPMENTS

THE MEAFORD WHEELBARROW CO. Limited

MEAFORD, - ONTARIO

Eastern Agent—Alexander Gibb, Montreal — Western Agent—Bassett & Leucke, Winnipeg

Our Stable Fixtures

are a profitable line for the hardwareman at this season. Many new barns have been built this year, and farmers are now preparing to equip their stables. You can make money by pushing our

" Acorn " Steel Stanchions

Beith's Litter Carriers

The Cowbowl

Have you a supply of our Catalogues 16 and A109? If not, send for them. They will help you to make sales.

"WE WORK WITH THE TRADE"

The Metal Shingle and Siding Co., Limited

Preston

Montreal— Cor. St. Catherine and Delorimer

Toronto— 100 Esplanade Street East

SALES AGENTS:

ST. JOHN, N.B. Emerson & Fisher, Limited — QUEBEC J. A. Bernard — WINNIPEG Clare & Brockest

CALGARY Ellis & Grogan — VANCOUVER McLennan, McFeely & Co., Limited

BEAVER POST-HOLE DIGGER

will please your customer. A pleased customer is an asset to any business. No wood to rot, check or break.

Special Discounts to Trade

CANADA FOUNDRY COMPANY

LIMITED

HEAD OFFICE and WORKS, TORONTO, ONT.

District Offices: Montreal, Halifax, Ottawa, Winnipeg, Vancouver, Rossland

CHARLES H. FOX

Vancouver, Canada

MANUFACTURERS' AGENT

Selling to the Hardware Jobbers Only.

Representing Canadian, British and American Manufacturers, correspondence invited. Cable Address, Chasfox Vancouver, P.O.Box 1105.

H. W. MITCHELL

WINNIPEG, MAN.

Correspondence is solicited from manufacturers wishing a representative in Winnipeg.

Travellers call regularly on the wholesale and retail trade in Western Canada.

Highest references and financial responsibility.

ALEXANDER GIBB

Manufacturers' Agent and Metal Broker, 13 St. John Street, Montreal

Representing Canadian, British and American Manufacturers. Correspondence invited from firms wishing to be represented.

Expert Bookkeeping

Investigation and auditing of all classes of accounts undertaken. Trading, profit and loss accounts and balance sheets prepared. Sets of books opened and office systems devised in an up-to-date manner. Collection of accounts a specialty.

Davenport, Pickup & Co.

622 McIntyre Block and 422 Ashdown Block

WINNIPEG, MAN.,

AND AT BRANDON, MAN.

Square your Accounts

INCANDESCENT

LIGHTING WITH COAL OIL

The most economic modern lighting. The "Ideal" lamp burner, screws on any ordinary coal oil lamp. This simple apparatus produces its own gas, giving a light more brilliant than electricity or gas. Costs only ¼ of a cent per hour. Guaranteed exactly as represented or money refunded. The "Ideal" lamp for the country home. Agents wanted everywhere. Write for free booklet describing everything.

ADDRESS

ST. ARNAUD FRERES

Cor. St. Sulpice and Commissioners Sts.

Montreal, P.Q.

Persons addressing advertisers will kindly mention having seen their advertisement in Hardware and Metal.

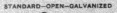

Retail Hardware Association News

Official news of the Ontario and Western Canada Associations will be published in this department. All correspondence regarding Association matters should be sent to the secretaries. If for publication send to the Editor of Hardware and Metal, Toronto.

TWENTY-FOUR DOLLARS AHEAD.

D. A. MacNab, of MacNab Bros., Orillia, called on the secretary of the Ontario association the other day and expressed himself as delighted with the results he had secured from the collection form letters through the association.

He invested $1 in a supply a couple of months ago and picked out fifteen of his worst accounts, ones which he had given up as uncollectable. All were small, from $1 to $5 each, but in each case he added a few cents for interest. Within two weeks he had collected $24 on the accounts, and enough extra in interest to cover his expenditure for postage and for the supply of form letters. MacNab Bros are ahead $24, enough to pay their membership in the association for a few years to come, and they still have a bunch of the form letters on hand. Needless to say, they believe in the Retail Hardware Association.

FARMERS AND MERCHANTS.

The Canadian Society of Equity is a farmers' organization which is rapidly gaining a foothold in western Canada, it being a branch of a similar body having tens of thousands of members in the United States. The object of the organization is to encourage the farmers to combine together to secure better prices for the goods they have to sell, or in other words, for the product of their labor. This they have as much right to do as merchants have to organize to overcome the evils of competition and price-cutting.

The new idea the farmers have, as taught by the Society of Equity, is to pay more attention to getting fair prices for what they sell, and less attention to the bargain-hunting method of Jewing down every person they buy articles from. In this idea merchants can agree, their business experience having taught them the soundness of the position. In the States the new society has encouraged the farmers to oppose the large catalogue houses in extending the parcels post system and Canadian merchants should, therefore, meet the Canadian members half way if the same broad-minded stand is adopted here.

In order that a better idea of the aims of the new organization may be secured, the following address, delivered by Charles Webber, an organizer of the Society of Equity, who was a fraternal delegate to the Michigan Retail Hardware Association, is reproduced:

"I don't think the hardwaremen need any talk about organization. They have the most perfect organization in the country to-day, and have accomplished the most of any of them, but what I want to do is to point out a few things to allay the prejudice which seems to have existed for a long time towards our society. Some people think that it is a great trust and that it is going to attempt to place high prices on farm products, but that is not its object, but it is to regulate the marketing of farm products. Farmers ought to get the best prices they can for their products, and this can only be done through organization.

"Your interests are closely identified with theirs. If farmers can have prosperity, it means prosperity to you hardwaremen. I would refer to our work towards overcoming the tendency to patronize the catalogue house, and I know that in your conventions you have brought out points that we have been working on in our campaign. I know as a young association we felt it was a task for any organization to solve the catalogue house problem satisfactorily, and we decided in our country organization to conduct a campaign of education with the men who were buying from these houses, and we found that the best education we can give them is to get a better feeling, a better relation between the business man of the small towns and the farmer.

"This is what the American Society of Equity stands for, and as an organizer I have been very much disappointed to find that the merchants are prejudiced against us, when we were the very thing that was making money for them. I believe that when you understand the American Society of Equity more, that you will be willing to help us. We need your assistance and your assistance will be of benefit to you. With your co-operation the catalogue house evil can be remedied and the parcels post measure can be defeated. On this platform we can unite both the merchant and the farmer. As I said before, this is the policy of the society, that we are against the catalogue house, and it will go farther than that; it will so closely unite the farmer and the merchant that the farmer will not patronize those dealers who are giving with their legitimate line, something from some other line as a premium."

LEGAL FORM OF STOVE NOTE.

Special provisions are sometimes made for dealers selling articles of necessity on the installment plan, as contrasted with dealers selling luxuries. In some states the statutes are loosely drawn, but in New York State there is a well-defined path marked out and a special statute applied to protect sellers of stoves and some other few articles of necessity. The law regarding stoves is somewhat complex, as the stove is considered an absolute requirement for the welfare of the family and can be retained under the bankruptcy act. It is undoubtedly true that many dealers are familiar with the ordinary form of stove note, which is given by the purchaser of the stove, and which permits the title to be retained in the dealer's name. This is not legal and will not hold in court, as the law expressly states that a note must be given and one received. A copy of the form used by one dealer and reproduced in the Metal Worker, is as follows:

"It is expressly agreed that the title of the stove for which this note is given shall remain in Cox Brothers until full payment of the purchase price thereof. In default of payment hereof, or upon any attempt to sell the said stove, or if it shall be seized under a writ of attachment or execution against me, or if Cox Brothers, with reasonable cause, deem themselves unsecured they may take possession of said property before maturity hereof and sell it at public or private sale, without any notice whatever, applying the net proceeds hereon and in consideration of the use of the said stove, agree to pay any balance unpaid for such application."

This statement is subscribed to by both the dealer and the purchaser on separate notes, the purchaser receiving the dealer's note and the dealer the purchaser's note. This form of contract differs materially from other forms, such as mortgages and deeds, but the reason therefor is that other contracts are matters of record in certain prescribed offices, and there is no prescribed place of record for a contract such as this. When payments are made upon the bill indorsements are made on the back of each note showing the amount. If payments are not made at stated intervals it is perfectly legal for the dealer or his authorized representative to seize the stove, but in practice it has usually been found desirable to have a constable or some other officer accompany the workman, so that in case any dispute should arise the testimony of an outside witness is available.

AN ABSURD PRACTICE.

In Collingwood, nearly all of the stores remain open until eight, nine or ten o'clock, ordinarily, and on Saturday night till eleven or twelve o'clock. One firm had the courage and good sense over a year ago to close at six-fifteen ordinarily, and ten o'clock Saturdays. It has continued to do so, and states that, from the first, business did not suffer in the slightest degree. None of the others has followed its good example, although we believe the results would be equally as satisfactory. Doubtless all of them would like to, but some are too cross-grained to co-operate with their competitors in bringing about this much-needed change, or to admit that it could be made a success.

If the majority of retailers are favorable to a fixed closing hour, the town council might be persuaded to pass a by-law providing that no stores, of certain classes, be kept open after a reasonable, specified time. Collingwood is altogether too big a town to longer tolerate an absurd practice which should have been dropped when it stepped out of its village clothes.

Good Window Dressing Pays

Practical Demonstrations in Toronto during Exhibition Weeks—Cutlery Displays increase Sales—Will keep Windows Lighted at Night in Future.

Two very fine examples of window display advertising were noticed in Toronto by Hardware and Metal during the Industrial Exhibition fortnight. Inquiry was occasioned by the number of people seen standing before the windows of the National Cash Register Company, in the Rossin House Block, and the displays of Rice Lewis & Son, on their rounded corner, opposite the King Edward Hotel.

The Cash Register window consisted solely of a tank of running water, two taps emptying into two large glass tumblers resting on a small stand in a large, flat basin. One tumbler was running over with the waste and emptying into the tray. The other tumbler, while receiving an equal force of water, failed to overflow, the tray below being perfectly dry. Thousands studied the display, wondering where the water went. They saw it enter the glass, they could see it above, below and all round the glass, there being no mirror deception in the display, and they were puzzled. The display did its work in impressing each looker-on, and a simple card asking the question: "Is there a hidden leak in your business?" suggested the moral that to prevent such leakages a cash register should be used.

The secret of the display was the presence of a glass tube inside the tap, the tube being invisible in the centre of the stream of rushing water. It was capable of drawing out of the tumbler by suction the exact quantity of water emptied into the glass, the bubbling water obscuring the end of the tube about a quarter of an inch below the surface of the water. To make the display more effective, a similar tube was fitted into the second tap, but as the suction was not applied, the tumbler overflowed with the inrush of water.

Lighting Windows at Night.

Rice Lewis & Son, Toronto, have a reputation of long standing for high class displays in their windows, and their cutlery display, shown during the Exhibition, and reproduced in the accompanying engraving, sustained their past record. The work was done by W. C. Knight, in charge of the retail cutlery department, and that the display was effective is attested by the fact that the sales of cutlery and razors averaged an increase of $25 per day during the time the display was in the window.

This result was partly due to the lighting of the windows at night. In the past the blinds have been drawn when the doors were closed in the evening, but this year it was decided that much valuable advertising effort was wasted by this practice. The change was a good one, as every evening large numbers of young fellows, with money to spend, were to be seen studying the display in their idle moments.

The pricing of the goods also had its effect, numerous direct sales being traced by customers entering the store and asking to see a certain-priced article, which, after slight inspection, was purchased. Many of these customers would never have entered the store if they had not known that a certain article could be secured for a price within their reach. Another advantage gained by marking the prices plainly was that the time of the clerks was saved by merely having to show one article, instead of submitting half a dozen, from which the buyer could make a selection.

So satisfactory were the results from the experiment of lighting the windows at night that the firm have decided to follow the practice right along.

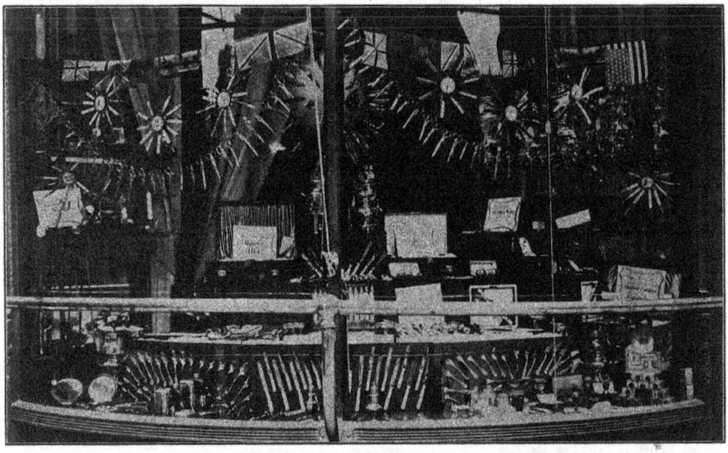

A Cutlery and Silverware Window that made Sales for Rice Lewis & Son, Toronto.

Arrangement of Cutlery Display.

As will be seen from the illustration, the cutlery display included a full line of table flatware, shears, pocket knives, shaving goods, and similar lines. The arrangement followed the semi-circular construction of the window, the top being set off by a display of flags, below which were strings of shears wired together, in the centre being a large ball of cork, used for displaying hunting knives. Also suspended by wires were eight hanging displays of pocket knives and razors, inserted in bung corks on the front of which price tags were attached. In the background were some fine pieces of silverware and a large assortment of cabinet cutlery and other case goods, below this being knives, forks and spoons, displayed on sample boards, and held there by double headed tacks. In the bottom corners were a full line of shaving goods, safety razors, brushes, packages of soap, mirrors, etc., as well as a few carving sets, flasks and other lines usually stocked in the cutlery department.

A Carborundum Window.

Adjoining the cutlery window was the fine exhibit of carborundum articles, including razor hones, which naturally worked in well with the cutlery, especially as a demonstrator was located inside showing visitors how well and how easily knives and razors could be "put on edge" by using carborundum stones as an abrasive. Numerous sales of razors were traced directly to the carborundum demonstration and display.

This window speaks for itself and needs little explanation. The work was done very artistically, the designs being such as can be easily copied by other retailers. As large dealers in carborundum, Rice Lewis & Son were able to devote a whole window to this one line during the demonstration, but the ordinary dealer cannot hope to do so. By grouping his carborundum with a display of tools, however, he will find both the display and the results satisfactory.

Carborundum is now known to the trade as the next hardest material to the diamond, and as an abrasive of the highest class. It is made from coke, sand, salt and sawdust, by electricity, requiring 1,000 horse power of electrical energy and a heat of 7,000 degrees for 36 hours, in the making. In the display besides razor hones, were oil stones, slip stones, knife sharpeners, strops, abrasive wheels, axe and scythe stones, etc., besides samples of their sample cabinet, given free to dealers. The Indian head trade mark and the picture of Niagara Falls, where the Carborundum Company's plant is located, were also features of the display.

An Elaborate Carborundum Display in the window of Rice Lewis & Son, Toronto.

TIMELY WINDOW SUGGESTIONS.

A sporting goods window display, suggested by "Hardware," of New York, is made up of goods ordinarily carried in any gun and ammunition department, with the exception of the cedar or pine boughs used to bank the tent, and the deer's head. The tent is set up in the side wall corner of the window, the sides being rolled back under and pinned or sewed. The evergreen boughs will effectually screen the points where the tent is attached to the wall. The camp cot shown in the tent may be a "dummy," or if the real article is used it must be carried out through the back wall into the store. Ammunition boxes, blanket rolls, etc., are distributed under and on the cot. Ammunition boxes are placed in the front of the window and shells and cartridges arranged on the back wall display. This window offers a simple design on which any trimmer may work up many combinations and striking effects.

A Suggested Gun or Camp Supply Window.

DISPLAYS MAKE SALES.

Few hardware stores in Canada have windows so uniformly well dressed as those of the J. H. Ashdown Hardware Co., Winnipeg. W. J. Illsey, who has charge of the window dressing and advertising departments in this store, is an expert window dresser and an enthusiastic believer in the selling powers of attractive windows. On several occasions he has contributed interesting articles for Hardware and Metal, and he has had prize-winning windows.

A Hardware and Metal representative had an interesting chat with Mr. Illsey the other day. One window was attractively dressed at the time, with coat and trouser hangers, and it could easily be seen that sales were resulting.

"Yes, nearly every hardware store carries that line of goods," said Mr. Illsey, "and no doubt most people know that coat and trouser hangers can be bought in hardware stores. But the difficulty is to make people buy them. If we didn't display them in the window as we have done, they would probably lie on our shelves for years. As it is, you can see for yourself that they are moving satisfactorily."

Mr. Illsey goes west to Red Deer before the year is out, in partnership with his brother, who started last spring the business of Illsey Bros. in that town.

Retailers Must Be Progressive

Ideas of an Experienced Dealer on Bookkeeping—Letter Writing an Art—A New Reply Post Card.

SATISFACTORY SYSTEM OF BOOK-KEEPING

To a great many people engaged in business the keeping of books is an undeniable drudgery, said W. P. Bogardas, in an address before the recent convention of the Michigan Retail Hardware Association. I confess, however, to have never looked upon the work from that standpoint. It has always seemed to me to be the key to any man's business, for the books contain the records of the business.

The great problem for me has been, how to get the complete record with the least amount of work. To sit at a desk and copy items from a blotter to a day-book, and then to a journal, and from there to a ledger, has appealed to me as a useless and profitless method. To have to charge the same item three different times opens a way for mistakes, and does not add anything to the security of the account.

How to change the old methods, and still retain safety and security and correctness of the accounts, has been the problem that the advocates of advanced methods in book-keeping have had to solve. That to a certain extent the problem has been solved, is true. In a modern business I have found that a cash register in connection with a loose-leaf day-book and ledger combined has given me the best results with the smallest amount of work.

Record of Cash Sales.

One inflexible rule must be adopted in connection with the use of a register. A record of every transaction in the possession and use of a register. The possession and use of a register is not simply to have some place to put the money received, but there is a larger purpose in the expenditure of so much money. Our register gives us a record of the total amount of cash sales, with a detailed slip to verify amounts.

At night, when we are ready to close, we have the following information: Total cash sales, total paid out, total received on account, total amount of charges and total business for the day, together with the total amount of business each person in the store who has waited on customers has done. With this information I am enabled to make a record for each day, and I can judge the value my clerks are to me and how much work they are doing.

The knowledge of the amount of work each clerk is doing is known to the other clerks, and is, I think, a spur to increase their efficiency. As each clerk has a drawer that is used by him exclusively, and as he makes up his own record of his day's work, which must agree with the report I take from the register; the mistakes made during the day—and we all make them—are placed where they belong. There is no dodging us, and as a result, there is greater care used by each clerk in the work he does for the store. To make men honest is a difficult matter. To make men careful is often a matter of education. The register furnishes to the book-keeper every night certain records: Slips for all goods sent out of the store and charged; slips of all money received on account and slips of all money paid out.

I have found that a loose-leaf journal ledger is best. That is, a loose-leaf book that is ruled so that the items of an account can be written in on the debit side, and when it becomes necessary to make out an account there is but one book to refer to. The slips from the register are, of course, preserved, as they are the original entries, and should be used to check up all the bills before they are presented to the customer. With the checks and counter checks of this system and the division of the responsibility each one is charged with only the responsibility that justly belongs to him. It seems to me we have a condition of things that will give the greatest amount of capacity for work the store can get.

NOVELTY IN RETURN CARDS.

A Chicago man has submitted a new form of postal to the United States Government, which, it is believed, will be adopted for the convenience of certain classes of business houses. It is called the "reply postal card," and is meant to serve the economy of those sending out postal cards to boom trade.

This postal, it is proposed, shall contain a two-cent due stamp engraved in one corner, and if the inventor's idea is accepted, it is meant to be issued at little more than the cost of printing by the Government, and is to be redeemed for two cents when returned in course of mails by the postman to the firm that sent it out. In other words, this postal does not have its postage paid until the party receiving it has made use of it—if it does at all—by returning it to the writer; and when the Government delivers it to the person or firm that sent it out, he pays the Government two cents for the service.

This plan to pay double postage is designed to save the extravagant waste of regular Government postals sent out in the mails soliciting replies, for it is well known that about 95 per cent. or more of these are never returned to the houses that send them out, and as the senders' names are generally printed on the front of them, it is safe to say that more than three-fourths of them are wasted.

To show the benefit of the new card, let us illustrate:—One thousand ordinary postals, costing $10, are sent out with letters or other printed matter. Of these probably 25 are returned, the remaining 975 being wasted, and causing a loss of $9.75. If the newly proposed two-cent due postal had been used, the 25 coming back would have cost the sender but 50 cents! This would reduce the waste enormously. Such a card would be restricted to the use of business firms, clubs, etc., using them in quantities.

This postal has the warm endorsement of big commercial interests, which expect that it will be adopted. The United States Government is showing every activity at this time in modifying postal usages to suit the growing demands of business, one of the late reforms being the ruling that special delivery postage now will be paid by affixing ten cents' worth of ordinary stamps, instead of requiring a special stamp, as formerly.

LETTERS THAT PULL.

Here is some good advice given by Sherwin Cody, the authority on letter writing:

Letter writing is a distinct art, built principally on applied psychology. A good letter makes a sharp impression at the right place and at the right time. A bad letter lessens the impression that may have been created by a first and stronger one. Two weak letters following one strong one will make no impression whatever.

This is what Mr. Cody says:

"Write a long letter to

"A farmer,

"A woman,

"A customer who has asked a question,

"A customer who is angry and needs quieting down and will be made only more angry if you seem to slight him,

"A man who is interested but must be convinced before he will buy your goods.

"Write a short letter to

"A business man,

"An indifferent man upon whom you want to make a sharp impression,

"A person who has written you about a trivial matter for which he cares little,

"A person who only needs the slightest reminder of something he has forgotten, or, of something he may have overlooked."

THANKSGIVING DAY ON MONDAY.

The several commercial travelers' associations are asking the Dominion Government to change Thanksgiving Day from a Thursday to a Monday. The travelers' plea is that a Thanksgiving Day in the middle of the week interferes with business, whilst its observance on a Monday would enable many to take part in family reunions who are not otherwise able to do so. The chances are that a day will be selected either in the third or fourth week of October.

HARDWARE TRADE GOSSIP

Ontario.

H. R. Manders & Co., Owen Sound, have closed out their hardware business.

E. H. Jeeves, Port Rowan, Ont., is advertising his hardware business for sale.

J. Chapelle, of the C. W. Clarke Co., Burke's Falls, visited Toronto last week.

Alfred Gaudreau, tinsmith, Vankleek Hill, Ont., has assigned to A. Hagar, L'Orignal.

D. E. McPhee, paint dealer, Ottawa, has offered to compromise with his creditors at 50 cents on the dollar.

Mr. Nealey, of the Victoria Mercantile Co., Victoria Harbor, Ont., was a visitor in Toronto last week.

Congdon & Marshall, Dunnville, are establishing a branch store at Port Colborne, a town that is showing very rapid growth.

J. C. Northcott, of the wholesale hardware firm of Rice Lewis & Son, Toronto, has returned from spending a few days in Buffalo.

Fred E. Seigner, hardware merchant, Durham, Ont., has assigned to Wm. Laidlaw. A meeting of the creditors was held on Sept. 13.

H. Henderson, manager of the Henderson Hardware Company, Vancouver, and formerly Toronto city traveler for the Canada Paint Co., was a visitor in Toronto during the week.

G. B. Ion, of Wood, Vallance & Co., Hamilton, was a caller at the Toronto office of Hardware and Metal on Thursday. Mr. Ion has just completed the work of getting out a magnificent 1,000-page catalogue, comprising the full line carried by this old established firm.

The employes of the Toronto branch of the McClary Manufacturing Co. recently made a presentation to N. R. Turner, their eastern representative, on the occasion of his intended marriage to Miss Keeler, of Prescott. The gift was a handsome suit case, and was presented by a brother traveler, Wm. Jeffrey, who in a very appropriate address expressed the sincerest congratulations and well wishes of his colleagues.

Quebec.

Needham & Showers, plumbers, Montreal, have dissolved partnership.

The St. Lawrence Sporting Goods Co. have commenced business in Montreal.

Philip Charland, hardware merchant, St. Cyrille De Wendover, Que., has assigned.

C. Birmingham, general manager of the Canadian Locomotive Co., Kingston, was in Montreal last week.

Amongst those seen in Montreal this week were : C. M. Cutts, of C. M. Cutts & Co., Toronto Junction, and H. Coursall, Oka, Que.

R. B. Cherry, of Sargent & Co., New Haven, Conn., and Mr. North, of O. B. North & Co., New Haven, Conn., were in Montreal this week.

G. T. Davie, a director of the Chinic Hardware Company, Quebec, and a shipbuilder of wide reputation, died a fortnight ago, aged 79 years.

A fire broke out in the pattern shop of A. Gravel, Montreal. Little damage, however, was done to the property, and the business is being conducted as usual.

J. E. Millen, of John Millen & Son, Montreal, is paying a visit this week to their Toronto branch. Owing to a steadily expanding trade in Toronto, the firm have found it imperative to double the capacity of that branch.

B. Welbourn, contract manager of the British Insulated & Helsby Cables Co., Prescott, Eng., has been making an extended tour through Canada, looking up trade conditions. He attended the Canadian Electrical Convention in Montreal.

Western Canada.

Taylor's hardware store at Griswold, Man., was destroyed by fire on August 29.

L. H. King, hardware merchant, Redvers, Sask., has sold his business to A. J. Silcox.

Tees & Persse, manufacturers' agents, Winnipeg, suffered a small loss from fire last week.

W. B. Marshall & Co., Medicine Hat, have moved into a new store, 80 x 28 feet in size.

E. B. Horsman & Son, hardware merchants, Moosomin, Man , are discontinuing the business.

Fraser Bros., hardware merchants, Carstairs, Alta., have disposed of their business to Hammill Bros.

MacLean & Sterling and McRoric & Co., hardware merchants, Souris, Man., have combined their businesses, and the new firm will be known under the name of the Souris Hardware Co.

HOME FROM ENGLAND.

After a three-months' holiday in Great Britain and Scandinavia, T. H. Newman, of Caverhill, Learmont & Co., Montreal, has returned home greatly improved in his general health. Mr. Newman, while in England, visited the large manufacturing centres, giving special attention to Sheffield and Birmingham. In the former city he found the cutlery and other hardware industries very busy, but the iron and steel industries in Birmingham were not so brisk.

Mr. Newman speaks very enthusuastically as to Canada's industrial outlook in reference to the Mother Land. "Without an exception," he said, "British manufacturers with whom I came in contact are looking forward with great anxiety to the time when they shall have established closer commercial relations with Canada, either through a confederation of the colonies in trade policy or in any other way."

One fact which would appear extraordinary in the eyes of a stranger, impressed Mr. Newman. England is exporting a large portion of her coal output to the continent of Europe. Coal being one of the greatest auxiliaries to industrial activity, it seems very strange that so much of it is being exported from England.

Mr. Newman also paid a short visit to Norway and Sweden. The staple industries of those countries are the production of iron, lumber and fish. Mr. Newman ventured the statement that there are more Swedes and Norwegians in America than in their home countries. Scandinavia's most valuable product is bar iron, the quality of which is unsurpassed anywhere. It is used to a great extent in America in the manufacture of nails and other wire goods.

U. S. IRON AND METAL MARKETS.

New York, Sept. 19.—"The Iron Age" says : There is increasing interest in the pig iron markets, particularly in the east. Some of the large iron makers have taken the lead in meeting the new conditions and have reached a basis upon which business is being done. This level is close to $18.75, delivered, for basis pig, and $20, delivered, for No. 2 foundry. It has induced conservative buying for the last quarter, and it is clear that for the time being the market has reached its level. Some low figures are reported to have been accepted by New England business.

While it is acknowledged quite generally in the finished trade that a diminished tonnage must be expected during the winter months, it looks as though prices will be kept steady at close to the present levels by the large interests. The keynote is to be co-operation without any close associations. Offerings will be kept within the requirements of the markets, and the salutary influences which prevented an undue rise during the boom will be exerted to counteract any tendency toward demoralization. It must be kept thoroughly in mind that prices have not at any time during the period of the extraordinary demand risen hysterically, and are not now on a basis which could at all discourage consumption.

Some of the plate mills are seeking work rather more eagerly. The skelp mills, for the first time in a long while, are in a position to make prompt deliveries.

Copper has sold for export on the basis of 15½c for electrolytic, but little appears to have been done by domestic buyers. There is little doubt that 15c would be eagerly accepted by some dealers.

LONDON, ENG., METAL MARKETS.

London, Sept. 18.—Cleveland warrants are quoted at 54s 7½d, and Glasgow standards at 53s 9d, making prices as compared with last week, on Cleveland warrants 1½d lower, and on Glasgow standards unchanged.

Tin—Spot tin opened strong at £168, futures at £165 10s, and after sales of 280 tons spot and 550 tons futures, closed steady at £168 for spot and £165 10s for futures, making price as compared with last week 5s higher on spot, and £1 10 lower on futures.

Copper—Spot copper opened strong at £67 15s, futures at £67 7s 6d, and after sales of 200 tons of spot and 1,500 tons of futures, closed easy at £66 15s for spot, £66 15s for futures, making price as compared with last week, £3 15s lower on spot, and £4 2s 6d lower on futures.

Spelter—The market closed at £21 5s, making price as compared with last week £1 higher.

Lead—The market closed at £19 15s, making price as compared with last week unchanged.

HARDWARE AND METAL

Established 1888

The MacLean Publishing Co.
Limited

JOHN BAYNE MACLEAN - President

Publishers of Trade Newspapers which circulate in the Provinces of British Columbia, Alberta, Saskatchewan, Manitoba, Ontario, Quebec, Nova Scotia, New Brunswick, P.E. Island and Newfoundland.

OFFICES:

MONTREAL, ... 232 McGill Street
Telephone Main 1255
TORONTO . . . . . . 10 Front Street East
Telephones Main 2701 and 2702
WINNIPEG, ... 511 Union Bank Building
Telephone 3726
LONDON, ENG. . . . . . 88 Fleet Street, E.C.
J. Meredith McKim
Telephone, Central 12960

BRANCHES:

CHICAGO, ILL. . . . . . 1001 Teutonic Bldg
J. Roland Kay
ST. JOHN, N.B. . . . . No. 7 Market Wharf
VANCOUVER, B.C. . . . . Geo. S. B. Perry
PARIS, FRANCE - Agence Havas, 8 Place de la Bourse
MANCHESTER, ENG. . . . . 92 Market Street
ZURICH, SWITZERLAND . . . . Louis Wolf
Orell Fussli & Co.

Subscription, Canada and United States, $2.00
Great Britain, 8s. 6d., elsewhere - 12s

Published every Saturday.

LAW SUITS COSTLY.

It is announced that the cost of the case before the courts to settle the dispute between the steel and coal companies at Sydney amounted to about $60,000, and the end is not yet, as the decision in favor of the steel company provides for an arbitration award on certain points, and even this settlement may be appealed and the case carried to higher courts. For every hour the court sat the case is said to have already cost $1,000.

The case was purely a business disagreement and in the light of events it would appear that the most sensible settlement would have been arrived at by an arbitration. The cost would have been much less and the results more satisfactory.

THE WESTERN CROPS.

It is never wise to whistle until one is out of the woods, but if present indications are not utterly misleading, it would seem that the pessimists were wrong. The crop killers were determined that Western Canada should have no crop this year, and their noisy predictions of disaster dominated the situation for a long time. For once, the cheery optimism of the West was unable to stand up against the pessimistic forebodings of the faint-hearted ones. But it is now pretty generally understood that in spite of the calamity howlers, the West has a good crop this year—a crop not quite so large as that of 1906 or that of 1905, but still a satisfactory crop. The Northwest Grain Dealers' Association, after a careful inquiry, have estimated that the West will harvest this year at least 82,000,000 bushels of wheat. October wheat has already reached the dollar mark in Winnipeg, and there is every reason to believe that the Western farmer will get about one dollar per bushel this fall for his wheat. In view of the higher prices, there will be no reason to complain if the crop is somewhat smaller than last year. It is, indeed, not at all improbable that, in spite of all the forebodings of the last month, the Western farmer will find the 1907 crop the most profitable he has ever had.

TAX ON TRAVELERS.

If retail merchants would carefully consider the practice of holding up travelers for subscriptions and to buy tickets in aid of local undertakings, church, charity or sport, they would quit it. Fifty dollars a year is a low estimate of a traveling salesman's expenses on account of such demands. The Editor of this paper meets a great many travelers, and he is constantly hearing complaints of the practice, and if the merchants knew what the travelers think of it and of the men and women who hold them up in this way, they would hesitate for that reason alone, to make the levy. Tickets for entertainments, church socials, bazaars, lotteries, are a constant drain upon a traveler's purse, and his good-nature, and a drain, too, upon his respect for his fellow men. Many travelers are asked to buy membership tickets in local clubs, bowling, tennis, curling, and others. which cannot possibly ever be of the slightest use to them. Some good ladies on the Finance Committee of their church society, will regularly ask a subscription of the traveler who calls to sell her husband goods. Many a traveler's dollar has gone to help defray a church debt, and the debt is all he ever knows about.

There are a great many thoroughly valid reasons why a merchant should not make these demands upon the men who come to him to sell goods. Take the case of the church. The traveler has his own church at home to help support. Why should he, a man of very moderate means and modest income, be called upon to contribute to the funds of churches all over the country? A man would have to be a miniature John D. Rockefeller to keep it up. It isn't once in a hundred times a traveler can use a ticket he buys, and some dealers have objected because a traveler gave away a ticket he had bought. Then the practice engenders bad feeling. The traveler feels it is a hold-up; he gets no return for the money, but can't refuse for fear of offending a customer. There are some travelers with the backbone to say "No," and withstand all such unjust demands upon them, and they are usually among the most successful men on the road, but most men submit to it. Another reason is that the wise merchant will not put himself under obligation to the traveler or the wholesale house. It is the boast of many successful merchants: "We never asked a favor of a wholesale house yet." That is the right attitude in business. The practice of holding up the travelers is unfair. It is taking advantage of a man when his living is at stake. Some merchants salve their consciences by the assumption that, "Oh, the house pays it." The other day a man got a dollar from a traveler for a church fund, then turned over a page in his book and showed where he had got ten dollars from the house which the man represented. Sometimes, no doubt, the tax goes down as expenses, but more often, especially with city travelers, it comes out of the traveler's own pocket.

PRIZE FOR BEST LETTER.

Retail clerks who wish to improve their position should act on the suggestion made on another page, where a prize is offered for the most practical short letter outlining plans for increasing sales during the Christmas holiday season. The winning of the prize is a secondary matter. The advantage gained through the study of the plans and the publicity secured through the publication of a sensible letter is worth far more than the prize.

One Ontario retail clerk was in a few months placed in a responsible position as the manager of a large western business as a direct result of the publicity secured through writing one or two practical articles for Hardware and Metal.

Travelers and merchants are also invited to send in contributions for the benefit of the trade as a whole, and for the advantages to be derived from an exchange of ideas. Every dealer has some ideas of his own on the subject of holiday sales. Jot them down and compare them with the plans of other hardware merchants. And "do it now."

TECHNICAL EDUCATION IN QUEBEC.

The work of the Council of Arts and Manufacturers of the Province of Quebec will be of interest to readers of this paper in face of the fact of such deplorable lack of technical education in this country, and a short resume of its objects is worthy of consideration.

The council is composed of leading men of the province, including several members of the Legislative Assembly, and their work is really a part of the Department of Public Works and Labor. The object, as stated in the annual prospectus, is to provide free evening drawing and industrial classes, and to make the instruction as practical as possible, in order that the pupils may profitably apply the knowledge gained in the various trades and branches of industry in which they are engaged.

A subsidy of $15,000 per annum is granted by the Provincial Government, and this constitutes practically the whole source of income. The city council of Montréal grants the free use of rooms over the St. Lawrence Market for some of the classes, free of charge, which is all that is done by the City of Montreal in the cause. By means of this $15,000, educational work is carried on in nine towns throughout the province, Montreal being the headquarters and principal centre of work. Instruction is given in fourteen or fifteen subjects, and last year 2,258 pupils were enrolled, and showed a very high average of attendance. Instruction in plumbing and heating is given at three of the nine centres, viz., Montreal, Quebec and Levis.

Admission to these classes is entirely free to suitable persons above fifteen years of age, the only condition being a deposit of one dollar, which is forfeited if the student is absent on more than four occasions throughout the session. Examinations are held and prizes given at the end of the season's work, and an exhibition held in Montreal for the benefit of the friends of the students. For the plumbing classes, the prizes are found by the Master Plumbers' Association of Montreal, and the interest taken by the association in the progress of the students in this class is very encouraging. The good work done is very evident to any who visited the exhibition which was held in Montreal during June, and certainly is worthy of very much more interest than is accorded to it. On the evening of the 11th of June, the distribution of prizes was held in the Monument National, and great earnestness was shown on that occasion on this question of technical education by the Hon. W. A. Weir, Minister of Public Works and Labor; Archbishop Bruchesi, the Roman Catholic archbishop; the Hon. J. B. Prevost, and others.

Instruction in practical affairs of men given to youths who voluntarily seek it, is of very great value—beyond, in fact, what can be estimated, and here we find this instruction costing the miserable pittance of about $6.50 per head per annum. On all sides we find this question of technical education coming to the fore. We notice other nations, such as Germany, forging ahead in practical affairs, as a direct result of this specialized education; we deplore the fact that this country's need of properly skilled mechanics is dependent upon men trained in other places; we recognize the great demand of this country, and its resources for properly qualified men to realize its wealth, and yet we find the magnificent sum of $15,000 being spent on 2,258 of the youth of our country seeking after knowledge—a truly national asset. We learn of a grant of $5,000 being made by the Legislative Assembly towards the new Technical High School in Montreal, and we are glad because we recognize in it an awakening—if but a very slow awakening—to the need of the country in this regard.

What appears to be needed, though, is a real vision on the part of the people of the country, and of the men in power, of the tremendous need and even larger demand for technical training amongst the youth of this country, if Canada is to grow in the way we are looking for, and in the rightful way to achieve her destiny.

Conditions are different in different localities, which makes it difficult to make comparisons, much as we would like to do so on this question, but a few general figures and facts will form a clear enough basis to draw rightful conclusions. Huddersfield, a manufacturing town of the west riding of Yorkshire, with a population of just under 100,000 inhabitants, and an extremely large area, spends on its technical school the sum of approximately $42,000 per annum, and has a roll of about 1,550 pupils, a cost of $27 per annum. These pupils and expenditure are outside of the ordinary evening classes which are held at the common schools throughout the city, and which are provided for through another channel. Montreal, with a population of 400,000, has 1,475 pupils under the Council of Arts and Manufactures, at a cost of $6.50 per head, and now we learn of $5,000 more being granted to the Technical High School, to carry on evening classes. If Montreal was in proportion to Huddersfield, $168,000 per annum would be spent on technical education amongst the workers, and a student roll of 6,200 would be in evidence. And what of other portions of the province? Under the council there are but 783 students throughout the whole province, at eight centres, truly a pitiable show if only comparisons were made.

Two-thirds of the cost of technical education is borne by the citizens of Huddersfield directly. How much by Quebec, Sherbrooke, Montreal and other towns in this country? Of course the whole case is different, and we do not intend to make exact comparisons, but enough has been written to show how very far behind we are in this matter of importance. Surely it is time for the vision to come, if we are to realize anything of the nationhood that is within us.

WESTERN ENTERPRISE.

A fine series of booklets advertising the progressive towns in Western Canada are being received by Hardware and Metal, and deserve more than passing mention.

Last week reference was made to a comprehensive 144-page report of the work of the Vancouver Board of Trade, attractive booklets being also gotten out by the Tourist Associations of Vancouver, Victoria and New Westminster, to advertise the coast cities.

Beautiful booklets from Calgary and Francis, Sask., were also referred to last week, they being followed up in this issue by a well-printed booklet, descriptive of the possibilities and advantages of the town of Saskatoon. Edmonton, too, issued a magnifient advertising folder a few weeks ago, and Regina, Moose Jaw, Prince Albert, and other growing centres are also getting out printed matter of the same nature. Hardware and Metal has also to thank Mayor A. F. Grady, hardware merchant, Macleod, Alberta, for a sample package of fall wheat, grown in his district.

The spirit of the West is contagious, and the sending out of such a mass of attractive literature is certain to bring about the result desired—the migration of many to the towns in the West, whose future is bright in promise. There are many merchants, farmers and mechanics in Eastern Canada who are restless under present conditions, and have their eyes turned towards the setting sun, as the land offering the brightest possibilities for the future.

The Board of Trades in the Western towns are doing a service by placing such literature before intending investors and settlers and their work should be rewarded by practical results.

Markets and Correspondence

(For detailed prices see Current Market Quotations, page 66.)

MARKETS IN BRIEF.

Montreal.

Antimony—Firming.
Copper—Very weak.
Lead—Steady.
Linseed Oil—Unchanged.
Old Materials—Quiet.
Tin—Variable.
Turpentine—Weak.
White Lead—Very strong.

Toronto.

Linseed Oil—Very firm.
Turpentine—Weaker.
Hinges—Advanced prices quoted.
Butts—Wrought iron, higher.
Copper—Decline in ingots and bars.
Ingot Tin—Lower quotations made.
Zinc Spelter—Foreign spelter declined.

MONTREAL HARDWARE MARKETS

Montreal, Sept. 20.—Both the wholesale and retail trade are at present characterized by a very marked activity. Makers and distributors of all hardware commodities are up against a demand which, in many cases, exceeds the output. Consequent upon this fact are the delays experienced both by retailers and wholesalers in obtaining goods, which delay is often attributed to the railways, and, in a number of cases, justly. Owing to a shortage of cars, the transportation companies find it impossible to guarantee, or, at least, to secure, prompt delivery. Numerous instances are still extant of very serious delays, ranging from periods of ten days or a fortnight to six weeks or two months. There are, then, only two things which can possibly limit this season's volume of business, lack of factory capacity and shortage of cars.

The condition of the crops, which a week or so ago was sufficiently unsatisfactory to excite considerable apprehension amongst hardwaremen, has very much improved, and prospects for the amelioration of financial stringency are much brighter.

Screws—Owing to the difficulty in securing efficient workmen and raw materials, the factories are far behind in their orders. At present they are confining their efforts to filling back orders, a number of which were placed last spring. Prices are firm and unchanged.

Sporting Goods—There is a heavy demand for all lines of guns, rifles, ammunition, and sportsmen's outfits, so heavy that makers find it exceedingly hard to keep up with it. In many cases they are far behind, and, as a consequence, the retail and wholesale trade is considerably retarded. Complete shipments are not frequent occurrences. Prices are steady.

Poultry Netting—The season is nearing a close, and the demand has about ceased. Despite the shortness of supplies this season, a good volume of business has been transacted.

Builders' Hardware—The demand for all lines is very brisk. An unparalleled activity in building circles is the feature of the year in Montreal. Supplies are adequate.

Building Paper—Trade has lapsed into a period of inactivity, as the mills are, in some cases, limited in their output on account of repairing operations. Supplies are equal to the current demand and prices are steady.

Wire Goods—Heavy shipments are being made to the Northwest, previous to the close of navigation. The factories are taxed to their utmost capacity, and there is no surplus of stock. A strong demand for hay-baling wire exists in Quebec. Prices are very firm.

Cement—The demand at present exceeds the output. Owing to the fact that cement is being used to an increasing degree in architecture, and to the fact that great difficulty is being experienced in securing cars to make prompt shipments, the producers find it absolutely impossible with their present limited capacity to keep pace with the demand. Prices are steady.

TORONTO HARDWARE MARKETS.

Toronto, Sept. 20.—Reports from various branches of the hardware trade throughout Ontario indicate that business since the first of the month has been very brisk, and feeling is more optimistic than for some time. While it is realized that trade could not continue at the rate at which is has been moving for the past two or three years, and a slight reaction must come at some time, still it is felt that the much-talked-of weakness in the money market is not going to affect the hardware trade very materially, so far as Ontario is concerned. If the temporary financial stringency has the effect of discouraging speculation in mining stocks and Western lands, and forces merchants to devote their entire attention to their own business, results will be satisfactory to all concerned.

The volume of business during the eight and a half months since the beginning of the year has far exceeded even last year's big record, and, as buying is keeping up actively, it is unlikely that any month this year will show a smaller total than during 1906. At present, sporting goods are moving very actively, and there is also a good trade doing in builders' hardware, tools, etc. There are also many large shipments of heavy goods being forwarded to the west, prior to the close of navigation on the lakes. Prices on spring lines, such as lawn mowers, poultry netting, etc., are expected shortly, and spring bookings will commence actively about the beginning of October.

A new development is the arrangements now being made by manufacturers of wire nails and similar goods to use metal kegs instead of the wooden packages used up to the present time. The growing scarcity of timber for cooperages has steadily increased the cost of wooden kegs, and this has encouraged inventors to produce metal kegs which will be strong enough to stand the strain of shipments of such heavy goods. It is probable that by next spring, the great volume of wire nails shipped to the trade will be contained in metal kegs costing the manufacturers about 8c each, whereas the wooden kegs cost about 13c. One metal keg factory is to be established in Toronto, its entire output being taken by a single wire nail concern. Another factory is to be located at Oshawa, and kegs are to be produced for carrying nails, lead washers, white lead, and putty.

Hinges—We announce this week advanced prices on heavy T and strap hinges, as follows: 4-inch, per 100 lbs., $7.25; 5 inch, $7; 6 inch, $6.75; 8 inch, $6.50, and 10 inch and larger, $6.25. Wrought butts have also advanced from 65c to 70c and 10 per cent.

Bolts and Nuts—Complaints continue to be received from retailers that lists of shortages of bolt orders are almost as extensive as the list of goods contained in the invoices. All manufactured iron products continue hard to procure.

Screws—Stocks are reported to be improving, but a strong demand exists, and manufacturers are still far behind in their orders. Prices unchanged.

Wire Nails—Stocks are reported satisfactory, with large orders in hand, and prices the same as quoted for some time.

Poultry Netting—Some booking has been done for spring delivery, but no active movement in this direction will develop until October. Prices are in an indefinite state as yet, and the retail trade are not inclined to book until matters are in a more definite position. The demand for hay-baling wire continues good. Jobbers are endeavoring to arrange for a plentiful supply of all wire products, sufficient to handle the business during the coming winter months. It will not be their fault if manufacturers are unable to supply the goods when desired, next spring.

Sporting Goods—Heavy orders continue to be received from merchants whose stocks have run down. Sporting goods is a big feature of the retail trade at present, and shipments of loaded shells continue to be called for. Rifles and shotguns are being sold in large quantities by the retail trade, and both shot and fishing tackle are also selling well.

Cutlery—Buying is increasing with merchants paying more attention to orders for the holiday trade. Table cutlery and case goods are reported to be in good demand, and are likely to hold their popularity as holiday gifts.

Silverware, both flat goods and hollow ware, are being included in retailers' orders in increasing quantities, retailers having learned from last year's experience that it is well to place orders early if they desire to get the goods in time to turn them over into cash before the holiday season is over.

Builders' hardware continues to be included in sorting orders, the demand for building paper, and cement, as well as for finishing hardware, being particularly brisk.

Axes and Handles—More attention is being paid to lumbermen's supplies, and from now on shipments of these articles should increase.

MONTREAL METAL MARKETS.

Montreal, Sept. 20.—The activity existing in metal circles during the past week has been almost wholly confined to pig iron. There had been some possibility of a general improvement in local circles, through a corresponding easing in American districts. Any bright hopes cherished by dealers here will have been dashed to pieces by a recession of the market. The actual necessity of consumers in getting supplies for fall and winter requirements has done something towards creating a little activity. A steady and lively business is being done in pig iron. Consumers of tin, copper and antimony, however, are holding off, confidently looking for a decline in prices. Too much hesitancy may prove fatal, and they may be confronted with the impossibility of securing 1907 delivery, at least, warehouse room.

The English market at present is subject to numerous fluctuations due more than anything to speculative agencies. The American market has been characterized simply as "demoralized."

Pig Iron—High grades of Scotch and English iron are very difficult to procure. The furnaces there are producing a limited quantity of high-grade iron and stocks are low. Recent communications from a prominent American firm to a Canadian dealer are calculated to mean that the former party will not commit themselves for 1907 delivery. Locally, a heavy tonnage is being inquired for, and distributed, and consumers, especially in Eastern districts, are stocking up rapidly for fall and winter requirements. The furnace of the Canada Iron Furnace Co., at Midland, has blown out for repairs. A general decline is reasonably anticipated in American circles with the readjustment of prices now in progress. Consumers found their expectations on the saying, "Whatever goes up, must come down." Prices are unchanged: Midlesboro, No. 1, $21.50; No. 2, $20.50; Summerlee, $25.50.

Antimony—Is a little firmer; there is still a general dullness in the United States. English prices fluctuate. Cookson's, 15c; Hallett's, 14½c.

Ingot Tin—Latest despatches from England report an advance of over £1. It remains to be seen, however, whether this advance will be permanently maintained, as there have been many uncertain fluctuations in the English market lately. Little buying is being done in the local market. Prices are easy and unchanged at 44 cents.

Copper—On account of the unsettled condition of the American market very little buying is being done. Consumers are hopeful of a settlement of trade, and are, therefore, hesitant in buying. The market is weak, and prices unchanged.

Spelter—Is a little firmer. Not much buying is being done.

Old Material—Scrap iron is dull. Supplies are adequate, but the demand is weak. Little activity is noted in old metal circles. Heavy copper and wire, 14c; light copper, 12c; heavy red brass, 12c; heavy yellow brass, 9c; light brass, 7c; No. 1, wrought iron, $14.50; stove plate, $12; machinery cast scrap, $17.

TORONTO METAL MARKETS.

Toronto, Sept. 20.—Business is reported to be of a hand to mouth nature, the uncertain condition of the market not encouraging large buying. The volume of business done is satisfactory to the jobbers, but it is all in small lots. Some jobbers speak of a record business being done in shipments from warehouse for present consumption, but this may only be to stand off buying until prices are more favorable. Supplies from mills in the States are slow.

Copper and tin have both been shaded in Toronto this week, following the recent weakness in outside markets. As stocks in hand were purchased at high prices, the decline here has been slow. There is every indication of easier conditions in these two metals for next year.

The strength of the iron market is being maintained and General Manager McMaster, of the Montreal Rolling Mills has, in a newspaper interview, expressed his opinion that "there is nothing to justify the belief that Canada will see reaction in the iron and steel industry. Possibly there will be a little shading in values here and there, but things are on a sound basis, and with the steady growth of the country the voume of business wil be maintained."

Another indication of strength in pig iron is the imminent advance in stoves and ranges, another five per cent. advance following the rise of a year ago. Furnaces will remain unchanged. Stove founders report a very heavy trade this year, notwithstanding a falling off in western buying in August, pending the size up of the crop harvesting. In British Columbia a particularly good business has been done, one firm alone which had a $55,000 increase last year, being already over $100,000 ahead of the first eight months in 1906, and a quarter of a million in advance of their 1906 total to this date, covering the entire country.

Sheet Metals—Trade this year has been heavy, but only the fag end of the season's business is being done at present Stocks are light, and there will be very little carried over. Prices are being well maintained for present delivery, but the prospect is for easier conditions on next season orders. Bookings are being delayed until a more definite size up of the market can be made, one jobber having advised his customers to defer booking until October.

Pig Iron—The market is seasonably active, with prices holding fairly firm. In fact, iron is, next to lead, the most stable metal on the market. Shipments from England during the first eight months of 1907 have been 25 per cent. greater than any previous year, Canada and the States taking a large portion, but Germany being the largest buyer. In the States, the general impression is that there will be little change in prices unless a sharp contraction in consumption develop—an unlooked for condition.

Lead—The market continues very strong, with no change in prices.

Antimony—We quote the same as last week, Cookson's, 13c, and Hallet's, 12½c.

Copper—Prices have been shaded to from $20 to $21 for casting ingots, and $30 on round bars. This is in response to the recent break in the States, the local market not being subject to immediate changes, following speculative doings in New York. While prices have fallen about nine cents on ingots, and sellers are resisting the decline movement, common opinion is that the present halt is but a temporary check, and that at least three, and, possibly, eight cents more, may be chopped off before the bottom is reached. On the other hand, it is contended that consumption has about reached its minimum, and the surplus stocks piled up during the balloon price era will gradually melt and check the drop

Zinc Spelter—A decline in foreign spelter has developed, attributed to the same causes as have been behind the speculative movement in copper during the past year. We now quote from $6 to $6.25.

Ingot Tin—The declining demand for tin has been the reason for the weakness in tin in foreign markets, and with a growing supply available, easier prices seem probable for next year. We are now quoting from $41 to $42, although foreign prices are below this. Speculation is causing fluctuations, but futures are easier than present deliveries.

Old Materials—There is little to say. Stocks of most lines are plentiful and prices show no changes since our last report.

CORRESPONDENCE

Travelers, hardware merchants and clerks are requested to forward correspondence regarding the doings of the trade and the industrial gossip of their town and district. Addressed envelopes, stationery, etc., will be supplied to regular correspondents on request. Write the Editor for information.

HALIFAX BUYING ENGLISH PIPE.

Halifax, N.S., Sept. 16.—There is a general complaint among the hardware trade regarding the slow delivery of goods. This state of affairs has a very bad effect on trade all round, as the dealers are dependent on the delivery of their goods within a specified period at the time of placing their orders. The manufacturers are to blame for this condition of affairs. Some Halifax firms placed their orders with manufacturers last May for goods which have not yet come to hand, and no reason is offered in explanation of the delay. Not only is this complaint against the American concerns, but Canadian as well. The only explanation that the Halifax firms can offer for this delay is that the manufacturers are too busy filling orders for the west, to pay much attention to the east. Promises are made for deliveries within a certain time, but they are never fulfilled.

Iron pipe in small lots has been very scarce here for some time, and the local dealers have had the greatest difficulty in getting a sufficient supply to fill their orders. One of the largest firms in Halifax, being unable to secure pipe in Canada or the United States, placed an order with a British firm, and the pipe was forwarded promptly. It may seem strange, but it is a fact, that goods ordered on the other side of the Atlantic arrive here much quicker than they do from points in Canada or the United States. One time, quite a lot of British pipe was used here, and it is a long time since any was imported, but as a last resort the dealers were obliged to purchase in that market.

* * *

Another article quite new on this market is the tubular wheelbarrow, manufactured by an American concern. The firm handling these goods in Halifax has found a ready sale for them. These are principally used by the paving companies now completing contracts in the province. The delivery of these is also slow. Weeks ago a shipment was ordered, but up to date they have not arrived.

* * *

Heavy sales of cement are being made, both for building and paving purposes. The consumption of cement for paving in Halifax this year will be enormous.

* * *

The destruction by fire last week of the Standard Drain Company's plant will be a serious loss. The works were most complete in every respect, and the plant was in full operation, having large orders booked ahead. The buildings were valued at $50,000, and the machinery about the same. The loss is partially covered by insurance.

George C. Corbitt, of Annapolis, has just returned from New Brunswick, bringing with him some boxes of sample specimens of silver ore, which are said to be very rich. The property has already been bonded by Mr. Corbitt to New York parties.

* * *

The steamer Borneo arrived here last week with a cargo of sisal for Brantford, Welland, and Peterboro. The shipment took seventy cars.

* * *

Joseph M. Hottel, vice-president of the Delta File Works, of Philadelphia, was in the city last week in the interests of his firm. He has been coming to Halifax for the past twelve years.

CEMENT PLANT AT ST. JOHN.

St. John, N.B., Sept. 16.—With the approach of winter preparations are being made by local hardware dealers to provide for the demand for stoves and ranges and already these lines are being brought to the front, ready for inspection. Business generally is reported very good both in wholesale and retail circles.

* * *

With reference to an application made to the city for a lease of certain property for the establishment of a cement manufacturing plant here, of which mention was made in a recent issue of Hardware and Metal, it is learned that B. Mooney & Sons, who own large brickmaking yards near Fairville, are interested in the scheme and that they recently arranged for Dr. Ells, head of the geological work in these provinces, to visit the property and inspect the clay deposits. At his suggestion samples of the material have been sent to Ottawa for analysis. Although not similar to the material usually employed in the manufacture of cement, a test may show that the red clay deposit here is equally serviceable. There is, of course, lime in abundance. The result of the analysis by the department is being awaited with great interest, as it may mean the establishment of a big industry here.

* * *

The Canada Woodenware Company, which has been looking for a site for their factory since they were burned out at Hampton recently, has decided to locate at South Bay, near Fairville, N.B. They will occupy the Sutton mill, owned by James Lowell, M.P.P. A number of local people have subscribed for stock in the enterprise.

* * *

The buildings, plant and machinery of the F. B. Dunn Packing Co., at Fairville, have been advertised to be sold at public auction in this city. The factory has not been operated for four or five years, but is reported to be in good condition.

* * *

J. M. Seeley, W. G. Kennedy, Hugo Von Hagen, of New York; Isaac Purdy, of Purdy Station, N.Y., and Dr. M. F. Keith, of Moncton, seek incorporation as the North Shore Railway Co., with a capital of $190,000, in $100 shares. The object of the company is to take over and operate the Beersville Coal and Railway Co. (in Kent County) and carry on the business of coal mining.

* * *

Application has been made by C. P. Harris, A. E. Harris, W. E. Harris, Geo. L. Harris, of Moncton, and W. L. Harris, of Providence (R.I.), for incorporation as the E. A. Harris Co., Limited; to carry on a mercantile business in Moncton, with a capital of $10,000.

STOVE ADVANCES AT QUEBEC.

Quebec, Sept. 16.—The Quebec hardware trade is actually in a very satisfactory condition and a fair volume of business is passing. Wholesalers, as well as retailers, are preparing for a heavy season. Prices are firm and we note no changes in quotations of hardware this week. Plumbers are very busy in filling the good contracts they have in hand. Business is dull for painters.

* * *

Tenders will be received till September 21 by the secretary-treasurer of the municipality of Notre Dame de Quebec, who will have constructed a part of their system of waterworks. Separate tenders have to be invited for the digging and filling of the trench, and for the pipe laying.

* * *

Business in stoves continues very active. An advance of 10 per cent., as was predicted in our last report, is registered this week. A manufacturer told your correspondent that next month it is more than probable stoves will be 5 per cent. dearer than they are now. Factories are very busy and travelers are reported to be sending in good orders.

* * *

At the last meeting of the directors of the Agricultural Society of the County of Quebec, it was decided to hold an exhibition on the Quebec Exhibition grounds, on September 25.

WEDDING BELLS AT KINGSTON.

Kingston, Sept. 17.—The quiet season in the hardware line is now on in Kingston, but despite that fact, the local hardware merchants are by no means suffering from lack of trade. The dealers report they have not witnessed a better season than this in the past. They have been very busy all summer, and although the business has fallen off considerably, they are being kept well on the move. The chief demand lately with the hardwaremen has been for builders' supplies and sporting goods. Several sportsmen are out with rod and gun and are making things lively along the line of this class of goods. The merchants are taking advantage of this somewhat quiet spell, and are making ready for the busy fall and winter trade, which from the present standpoint looks quite favorable.

* * *

Mr. Cushman, promoter of the lead smelter which was to have been started here this summer, was in the city the latter part of last week, and stated that they had not altogether given up the purpose of establishing a smelter here, but that they were at present busy at the mines in the northern section of Frontenac. He also stated that the quantity of mineral found there exceeded their expectations. There has been great delay in the company's coming here, but Mr. Cushman says the demand in this line is still brisk, and the values keeping up, and that they were just as much disappointed at the delav as the people of Kingston. This is the bylaw the people voted on in May last and have since been looking forward to their coming.

* * *

Wednesday will be Cupid's day in Kingston, when several weddings are to take place, in one of which the son of a local hardware merchant is to take a prominent stand. The young man concerned is Allan Lemmon, of the firm of Lemmon & Sons. Allan is quite a popular young man and a great favorite in social circles throughout the city. His many friends will wish him every success in his new sphere of life.

* * *

The iron framework of the new boiler shop at the Canadian Locomotive Works is fast nearing completion. This work has all been done without interfering in any way with the regular boiler-making going on in the building underneath. When the new building is completed, it will be one of the best and finest equipped in the Dominion. Gangs of men are also busy working on the interior of the new power house. Two gigantic force fans were installed lately No men need be idle in Kingston; more laborers are wanted at the locomotive works, and there are also several sewer extensions now under way, a number of other buildings going up, and laborers are at present in great demand.

* * *

At a meeting of the inspectors of the George Sears estate, an offer of forty cents on the dollar was received for the hardware stock. The offer is being held over until Friday next.

PAINT FACTORY FOR HAMILTON.

Hamilton, Sept. 17.—A new steam fire engine, built by the Waterous Engine Company, of Brantford, has just been delivered to the local fire department. It was built at a cost of $5,000, weighs six tons, and has a capacity of 800 gallons of water per minute. The new apparatus embodies all of the latest improvements for fire engines, and is rubber-tired. It will throw four streams of water at once. The engine was displayed at the Toronto Exhibition, where it attracted much attention.

* * *

The Dominion Paint Company has just been organized to carry on a paint manufacturing industry in this city. It is capitalized at about $75,000, and it is expected to start business here soon.

* * *

The firm of E. T. Wright & Company, lantern manufacturers, is continually making addition to its already extensive plant in this city. The latest move is to secure all of the property adjoining its present factory for a square block. It is the intention of the firm to make extensive additions in the near future.

* * *

A permit has been taken out for the erection of a $50,000 building, to be erected at James and Main streets, for the Landed Banking and Loan Company, of this city. It is intended to make the new building only the basis for a more pretentious building which it is planned to erect at some future time. While the new structure will be only four storeys high for a beginning, it will be built of handsome material and will be a credit to the downtown section.

* * *

The assessment department has about concluded with the valuations of property and population figures, and it is estimated that there has been an increase in population of at least 4,000 this year, and in the assessments of $2,000,000.

* * *

The local merchants and business people generally are eliciting much satisfaction at the passing of the by-law giving the Hamilton, Galt and Guelph Railway permission to enter the city via Dundurn and Harvey parks. Last year the road applied for permission to enter the city, but as it insisted on being given this route, the permission was denied it. However, since an independent engineer has picked out another route through the parks as the only entrance to the city from the west, the council has passed the by-law granting the company an entrance by that route. This road is expected to prove a great boon to Hamilton, since it will open up a territory which heretofore has not had direct railway connection with this city. By the conditions of the by-law the company will have to have the road to Galt completed by December 31, 1909, and the one to Guelph by 1910.

* * *

Plans are being prepared for a new isolation hospital for incurable consumptives, to cost $10,000. The new institution is the free gift of William Southam, of the Hamilton Spectator.

* * *

The International Harvester Company, which has the largest industry in Hamilton, and which has formerly had branch offices in Toronto, Ottawa and London, has recently had the office staffs in each of these cities transferred to this city, and corraled in a handsome head office, which has been located on the seventh floor of the Federal Life building.

* * *

The new terminal station, which is being built by the Hamilton Terminal Company, one of the subsidiary companies of the Dominion Power and Transmission Company, the $25,000,000 corporation, is nearing completion. It is being erected at King and Catharine streets, in front of the new Bennett theatre, and will form a terminus for all of the electric roads running out of this city. It is four storeys high, of handsome exterior and interior finish, and will contain the head office of all of the companies controlled by the big corporation, including the Hamilton Street Railway Company.

The amalgamation of the Ontario Tack Company and the Canada Screw Company has created considerable interest in the hardware and metal trade in this city and throughout the country generally. Locally, it has had the effect of combining two of the city's busiest industries and of resulting in the erection of large buildings on North Wellington street. Both firms, which are well known to the hardware trade of Canada, will benefit by the amalgamation, as will the merchants.

A number of Buffalo and Hamilton capitalists are considering a proposition to build a large industry here for the manufacture of smoke consumers. There is an increasing demand for consumers now.

INGERSOLL MAN HONORED.

Ingersoll, Sept. 17.—Wilson Bros., hardware merchants, have a very striking and effective window display. It consists of a splendid collection of lamps with globes of various hues, which are illuminated by incandescent lights. The lights alternate and as first one and then another of the various colors flash forth they are certain to arrest attention. Of course patience and pains are essential to successfully carry out a scheme of this kind, but it is such with all undertakings when the aim is to interest the public. Window advertising is undoubtedly valuable when the desired object of creating interest is attained, and it surely has been in the present case.

* * *

The marriage of Miss Annie Jones, daughter of W. H. Jones, the well-known hardware merchant, to Mr. Edgar Hargan, superintendent of the John Morrow Screw Co., Limited, was solemnized at the home of the bride's parents on Tuesday evening at 6.30 o'clock. The officiating clergyman was the Rev. J. E. Hughson, pastor of the King Street Methodist Church. The ceremony was witnessed by the immediate relatives and friends of the contracting parties. The bride and groom are two of Ingersoll's most popular young people and they will have the best wishes of hosts of friends.

* * *

An honor was last week conferred upon J. T. Norton, the genial hardware dealer, when he was elected by acclamation to the board of education for the balance of the year as the representative of ward two. The seat which Mr. Norton is filling was forfeited, and when the ratepayers began to cast about them for a suitable candidate the genial "Tom" was spotted at once. His name was first to go on the nomination paper, and the other two nominees retired, giving him the seat by acclamation.

* * *

Are all hardware merchants optimists? This is a question which suggests itself after one has made the round of the various dealers. They are always smiling and cheerful and seldom if ever have any complaint to make in regard to the volume of business. It is a noteworthy fact that with them business is invariably "good" and they always appear to have well founded expectations as to what the future will bring forth.

LONDON FIRM'S ENLARGEMENT.

London, Ont., Sept. 18.—Local wholesale hardwaremen report business during fair week to have been exceedingly heavy. Throughout the week the warehouses were crowded, dealers being here from all over Western Ontario.

* * *

Not only is hardware business brisk at present, but jobbers report the outlook for next spring as unusually bright. An indication of this is afforded by the fact that one firm here has now booked more orders for spring than it had at the close of last year, which itself was a very good one.

* * *

E. S. Field manager of the Hobbs Hardware Company, and F. S. Woods, manager of the same company's glass works, are camping in the wilds around Wiarton for a couple of weeks.

* * *

For a new firm the McMurtry Hardware Company is certainly "going some." Though but a few weeks in business here its trade has grown to such dimensions as to necessitate the taking in of premises in the rear of its store on Dundas street. This is the first London hardware firm to carry stoves and ranges and it is finding it a very profitable line.

* * *

Parson's Fair has opened a second big store in the east end of the city, where, in addition to the usual stock of hardware, five, ten and fifteen cent goods will be carried as specialties.

* * *

The employes of the McClary Manufacturing Company appear to have imbibed much of the sporting enthusiasm of the company's vice-president and manager, Col. Gartshore, who has ever been an ardent patron of athletics. A result of this enthusiasm is shown in the fact that for some years the men have been represented by a team in the city baseball league. Two years ago "the McClarys" won the championship pennant and last year lost it by a narrow margin only. This year two series of ten games each were played for the championship and "the McClarys" won both—in the first capturing six and losing four, and taking every game of the second series. These victories are notable in view of the fact that amateur ball of the highest class is put up by all the teams in the league. One McClary graduate, "Mooney" Gibson, is to-day considered the premier catcher of the National League. This man played for two years with the local club, and developed so fast as to attract notice outside. The consequence was that he secured an engagement with the Buffalo team of the Eastern League. A year or so later found him with Montreal, of the same league, and to-day he is the star backstop of the Pittsburg National League team. At a recent contest at Cincinnati Gibson won the prize as the most accurate long-distance thrower. The employes of the McClary Company are scarcely less proud of the advancement of their former shopmate in his chosen profession than they are of the achievements of their own representatives on the baseball diamond.

SPORTING GOODS SELLING IN WEST.

Saskatoon, Sept. 12.—The duck shooting season opened on the 2nd inst., and at an early hour the surrounding country re-echoed to the sound of shotgun and rifle, and, ducks being plentiful, some good bags were secured. During the week there was a good demand for guns, ammunition and other sporting goods. At this season of the year, all the leading hardware stores usually make an attractive display of shotguns, rifles, and hunting requisites, and this week some neatly trimmed windows drew the attention of the sporting public.

* * *

Fishing in the river is indulged in extensively, owing to the scarcity of meat, as the butchers' shops have been closed down now for two weeks. The butchers consider themselves unable to comply with the new bylaw passed by the city council regarding the slaughter houses.

* * *

D. M. Leyden has just returned from a business trip to Winnipeg. For the past three and a half years Mr. Leyden was manager of the hardware department of the James Clinkskill department store, and has now taken up his duties with S. A. Clark, Ltd., as vice-president and managing director of the company.

* * *

The excavating for the new C.P.R. station and freight sheds is now complete. The steam heating will be installed by Elford & Cornish and Oxford boilers will be used.

* * *

The tinshops are kept busy making up quantities of stove and hot air pipes for the fall rush. The present cold spell has also brought work to the tinsmiths in the shape of repairs to stoves and furnaces.

SIGHT FOR FIREARMS.

A new shooting sight has been invented by G. B. Crandall, Cherry Valley, Ont. Loose and inaccurate adjustment has been the result of attempts to secure a rear tang-sight having a wind-gauge base. This inventor, however, obtains a rear wind-gauge by use of a sight stem having a wind-gauge top no more likely to become loose and shaky than an ordinary tang-sight. An ordinary sight can be immediately converted into a very satisfactory rear wind-gauge by the simple use of the extra wind-gauge stem. It does away with the high and cumbersome front wind-gauge sight and places a wind-gauge at the rear, which all who use an arm, prefer.

LETTER BOX.

Correspondence on matters of interest to the hardware trade is solicited. Manufacturers, jobbers, retailers and clerks are urged to express their opinions on matters under discussion.

Any questions asked will be promptly answered. Do you want to buy anything, want some shelving, a silent salesman, any special line of goods, anything in connection with the hardware trade? Ask us. We'll supply the necessary information.

Cementing Cellar.

Johnson Bros., Minto, Man., write:—"A customer of ours is having considerable trouble with an underlying vein of water in his cellar. He has cemented the cellar with the best of Portland cement, and the water still makes its appearance to the extent of five to eight pails per day. Thinking that probably you could give us a remedy that would keep this water back, have taken the liberty of writing you, and thanking you in anticipation of any trouble you may have in our behalf."

Answer.—It is doubtful if any satisfactory solution can be found other than to locate the springs and drain the water away. Even the smallest crack in any of the cement work will allow water to work its way through, and once through it will continue coming so long as the source of supply lasts.

Ontario Steel Ware, Limited, Toronto, who manufacture cement laundry tubs, suggest that if the water cannot readily be drained off and the customer is willing to experiment that a new floor, etc., be laid as follows: Use a material consisting of one bag of the best Portland cement to 300 pounds of fine, crushed stone, and finish off by covering with plain cement in paste form, using a trowel to secure a smooth surface.—Editor.

Insurance on Gasoline Stoves.

Murdie & Sutherland, Lucknow, write: "Do you know any insurance company in Canada, or any company doing business in Canada, who will accept gasoline stoves, without extra insurance?"

Answer.—We are unable to locate such a company. All assurance companies affiliated with the Board of Underwriters, charge an extra rate of 25 cents per $100 per annum where gasoline stoves are used.—Editor.

IMPROVED LOCK.

A new style of lock has recently been patented by A. N. Wickham, Lincoln, Neb. This lock may be placed in a door with a portion of the casing bearing the latch-bolt arm arranged uppermost. It may be inverted and placed in the door with that portion of the casing bearing latch tumblers and locking tumbler arranged uppermost. So arranged, these tumblers drag over the inner casing and fall into locking engagement with the casing of their own weight, and operate even if certain springs become inoperative. Means aid in holding the tumbler edges against the inner casing, as the end of the latch bolt extends through the face-plate of the main casing.

FORTY-TWO YEARS WITH ONE FIRM

Few can excel the long record of W. R. Ross with Frothingham & Workman, Montreal.

William Robertson Ross has this spring completed the forty-second year of his connection with the old firm of Frothingham & Workman, Montreal, and in July celebrated the sixtieth anniversary of his birth. To see Mr. Ross performing his daily duties in the big St. Paul Street House, one would never believe that he has lived to the end of Dr. Osler's chloroform period, for he has yet the youthful sparkle in his eye, and the flush of vitality in his cheek. He is another of that illustrious group of men who entered the hardware trade as aspiring lads, and have grown old in it, who have been witnesses of the evolution and remarkable growth of the trade, from the early days of small sales and big profits to the present day of big sales and smaller profits.

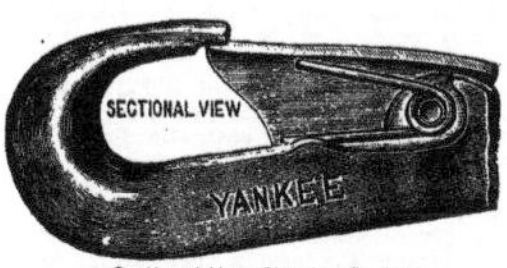

Sectional View Showing Spring.

Mr. Ross was born in the county of Derry, Ireland, in July, 1847. His grandfather was an English Church clergyman in that county, and his father, just previous to coming to America, was in attendance at Dublin College. In the spring of 1853, Mr. Ross' father brought his family to America, landing in the latter part of May at Philadelphia. The early days of the Ross family in America were spent in various towns. A few days after landing at Philadelphia they came to Montreal, and after remaining there a short time, removed to Oswego, N.Y., then to Trenton, Ont., and in 1860, back to Montreal to stay. Mr. Ross' father became a partner in the firm of Ross & Collis, for a number of years in the flour business in that city.

William Robertson Ross attended the public school at Mascouche, a village a short distance from Montreal, in Terrebonne County. After receiving a thorough grounding in the fundamentals of education—reading, writing and arithmetic—he came to Montreal, and, after a year's attendance at a high school, connected himself with the firm of Frothingham & Workman, in the spring of 1865. In those days honesty was considered the essential qualification of anyone entering on a business career. As an illustration of this fact, Mr. Ross, in conversation with a representative of Hardware and Metal, said that one of the principals of the firm asked Mrs. Ross if her son was honest. The implication contained within the query was what Mr. Ross had in mind.

Commencing in the warerooms, he was engaged in all the various departments of the business, excepting the offices. For the last fifteen years he has had full charge of the sporting goods and cutlery departments, and it is said that he knows everything about sporting goods, from BB. shot to machine rifles, and about cutlery, from jackknives to the finest carving sets.

Having been connected with the trade for nearly half a century, Mr. Ross is in a good position to draw interesting comparisons between conditions of trade in the early years and those of the present. For instance, when he was first connected with the business, the spring and fall were the only busy seasons, and it was easy during the summer and winter months to get protracted vacations, as very little business was done. The reason for this was that in those days there were very few commercial travelers to go about regularly each week to solicit and get orders from country dealers, so that the latter were compelled to send larger orders and at longer intervals than they do at the present day. When Mr. Ross became first connected with the firm of Frothingham & Workman, they had only two travelers, whereas at present they have about fifteen. Speaking of the particular lines af goods with which he has now to do, Mr. Ross said that in the early days a large amount of the cutlery, (in fact, all of it), came from England. At the present day German and American goods (especially the cheap lines), are being used, to a great extent.

Hardware and Metal wishes Mr. Ross many more years of usefulness and happiness.

NEW PROCESS IN "GALVANIZING."

The covering of iron and other metals with a thin protective layer of zinc, generally known as "galvanization," although it is not now usually done with the aid of the electric current, may be accomplished by a new and efficient method named "sherardization," after its inventor, Sherard Cowper Coles.

The "zinc grey" used for sherardization, is a by-product of the metallurgy of zinc; it is formed by the condensation of zinc vapor and contains pure zinc associated with its oxide, with traces of cadmium, lead, iron, etc. It is found generally in the form of impalpable particles whose diameter is in the neighborhood of one two-thousandth of a milli-meter (about 1-50,000 inch).

In practice, sherardization is effected as follows: The object to be treated is placed in an iron shell and covered with commercial zinc grey. The shell is closed as hermetically as possible, and if needful is sealed. It is then placed in a furnace and heated to 300 degrees or more. The duration of the operation varies according to the thickness of the layer of zinc that is desired. It is then allowed to cool, and the pieces are removed.

The metallic objects are covered with a fine layer of zinc.

PIPE-CLOSURE.

An improved pipe-closure has been patented by A. A. Fisk, Pomona, Cal. This improvement pertains to irrigation pipes, stand-pipes, and the like, and its object is to provide a closure arranged to permit of conveniently, quickly, and securely closing the end of a pipe, and to allow of opening the same for making connection with another pipe whenever it is desired to do so.

NEW YANKEE SNAPS.

The Covert Mfg. Co., Troy, N.Y., are enjoying exceptional success with their "Yankee" snaps, which have become so popular and have given such good service that they are now made in all the styles and sizes usual to the hardware and harness trade.

The accompanying sectional view of the body and tongue of a snap, demonstrates that unlike other snaps of similar construction, the spring cannot become displaced or jarred out, as the rivet passes through the steel cap or tongue, the spring and the snap, and is riveted by special machinery, thus securely fastening both the cap and spring so they will remain intact and unimpaired, as long as the snap lasts.

The "Yankee" roller snap, also illustrated, shows a favorite pattern of snap with this "Yankee" principle. It is a veritable safety snap, and. like all goods manufactured by this company, is first-class in every particular. It is the only roller snap made with this "Yankee" principle, which easily accounts for the great demand for them.

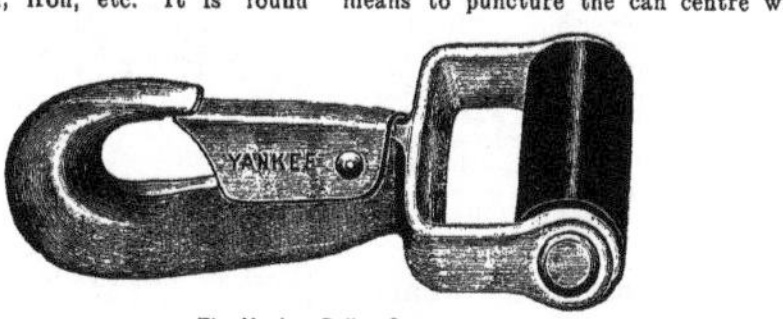

The Yankee Roller Snap.

IMPROVED CAN-OPENER.

J. Codville, Bella Bella, B.C., has invented an improved style of can-opener. A shank member has rigidly fixed to it at its outer end a steel plate having means to puncture the can centre with a slidable blade on the member adapted to co-operate with the puncturing means in cutting out the top of a round can. In connection with the blade is a guiding means to keep the blade at proper distance from the centre. The plate at the end of the shank, in addition to the puncturing means, carries a cutting blade for opening square or irregular cans and a slot for engaging a removable strip as placed on some cans to provide means for their opening.

Sporting Goods a Profitable Side Line

MECHANICAL SHOOTING WINDOW DISPLAY.

In a Chicago sporting goods dealer's window last month was a display showing two riflemen shooting at a target. They were small, automatic figures, each only about a foot and a half in height, but each completely equipped and life-like in appearance and action. The target they were shooting at was tacked to a tree stump over at one side of the window; the regulation target with a bull's-eye and a lot of concentric rings. In the window was a placard with an inscription telling the advantages of doing business by modern methods, and at the feet of one of the riflemen was a little card marked, "The Old Method," and at the other a card marked "The New Method." The two riflemen are standing facing the target, and each with his gun at his shoulder, ready. The Old Method chap, with a powder-horn slung around his shoulder and an old-time gun, bends his head over and down and squints along his gun barrel and fires, and a round spot of light appears on the target, off somewhere on one of its concentric rings. The rifleman raises his head and looks, and then he bends it down once more and fires again—to hit the target this time maybe somewhere off near the rim, further from the centre than before. Then young New Method takes his turn. He bends his head down on his rifle and squints along its barrel, and fires, and bing! the light appears—he has plugged the target square in the bull's-eye! And so these two window riflemen keep it up, the old-time chap scattering his hits all over the target, and the new-time man always hitting the bull's-eye every time. This mechanical contrivance is operated by electricity.

NEW BRUNSWICK TRADE GOOD.

Montreal hardware jobbers report a heavy demand for sporting goods from New Brunswick. There is a heavier call from this province this year than in any previous year—a large business in shotguns, rifles and all kinds of metallic ammunition. Factories are very dilatory in filling orders, not because of any difficulty in securing operators or raw material, it is said, but simply because of lack of capacity.

NEW SPORTING GOODS STORES.

Two more sporting goods stores have opened up in Montreal within the past fortnight. One of these is the St. Lawrence Sporting Goods Co., which commenced business on September 9, on Notre Dame Street. The officers of the company are: President, W. E. Ranger; notary public, J. N. Legault; secretary, J. N. Lemieux; director, E. Ranger; manager, A. D. Leblanc. Mr. Leblanc was formerly connected with Lewis Bros., Montreal. It is the aim of the St. Lawrence Company to develop a high class retail business by carrying a full and up-to-date stock of sporting goods of the latest improved patterns, guns, ammunition, lacrosse supplies, baseball and football outfits, and campers' requisites. The building itself is entirely new and from an architectural standpoint, is very attractive.

The other company is Berube Bros., who commenced business a fortnight ago in the north end of the city. Joseph Berube was formerly with Lewis Bros., Montreal, and his brother partner was formerly with W. J. O'Leary & Co., Montreal.

PRICE NO OBJECT.

It is of no use going hunting without a gun, and the huntsman is bound to get the best money can buy. Big price gets little consideration from him. A good sport cares little how much he pays for his fun, so long as he gets it. If a $50 rifle is necessary to bring down a deer with any degree of certainty, the orthodox sportsman will have one. He is not going to run chances on coming home without some palpable evidence of his prowess because he was foolish enough to set out equipped with "the cheapest gun on the market." What he wants is the surest gun, and, therefore, the best.

ROLLER RINK AT CALGARY.

A new roller rink has been established at Calgary, equipped with 800 pairs of the Richardson ballbearing skate.

RUSH HUNTING GOODS.

The sporting instinct is not yet dead. Men are still hunting, and in steadily increasing numbers. The proportion of hunters to the population is pretty steady, and as the population increases, the number of huntsmen will necessarily be augmented. That there are more men to want and buy guns and ammunition this year than last, sporting goods dealers believe, and wholesalers state that the outlook for this year's trade is very bright and all prospects point to a substantial increase over last year.

AN EDUCATIONAL FACTOR.

We talk of the educational influence of visits to picture galleries, to collections of art and museums in general, but half the people who go to them do so in a perfunctory, careless fashion because they are impelled by a vague sense of duty, and often they come away with very little more than they took with them. But every business thoroughfare of a great city is a delightfully entertaining art gallery, museum and university in one, if you are of a mind to so regard it, and looking in at shop windows is usually due to an eager curiosity that makes it an unconscious benefit.

When you stop to consider what one may see in a leisurely walk of half a mile or so along a street of shops you become aware of the fact that a shopping district is a sort of institute of allied arts, sciences and inventions, and the man who cannot get some ideas or impressions of value out of a judiciously arrested promenade must be a dull observer at the mildest estimate.

I have derived an immense amount of pleasure and benefit from the inspection of shop windows. In fact, it is one of my ways of keeping in touch with the world's aesthetic and practical progress. You can judge of the character of the community much more intelligently from a study of its shop windows than you can from the inspection of its art galleries, public libraries and kindred objects of local pride, for these latter represent the energetic spirit of the cultured few, or may even be the result of one man's generosity. Shop windows, on the contrary, certify the general taste or lack of it; they photograph the people themselves, for the grade and tone of the shop displays almost invariably correspond with the grade and tone of the community that encourages them. I have persuaded myself that I need no other evidence than shop windows afford me of the great improvements steadily going on in this country. I have noted very remarkable changes in the course of the sixteen years that I have been making regular tours among the cities of the United States. On all hands are to be seen increasing evidences of taste, refinement and general educational betterment.—Exchange.

HOME-MADE CATALOGUE CASE.

Many sporting goods dealers have too many catalogues, directories and reference books lying around loose, which makes the general appearance of their office look very unpleasant. This can all be remedied by following out the accompanying instructions: Procure two boards six inches wide and two or three feet long, or just the proper length to fill up some vacant space on the office wall. Square these boards up, commence at one end and with the square draw lines across the boards one inch apart. Then saw slots in these boards one inch apart and one-quarter inch deep. These boards form the top and bottom of a bookcase which will make a valuable addition to any sporting goods store.

A GENTLE REMINDER.

When returning goods to the wholesaler, always enclose your name and address and an itemized list of the goods in the package. Read the conditions on the invoice, as they constitute a part of the sale, and always inspect merchandise as soon as it arrives. Be sure that you do not sign receipts for goods you do not receive. Check the list over carefully before you put your signature at the bottom. Then if there is anything wrong, be prompt in making your claim.

CONDENSED OR "WANT" ADVERTISEMENTS.

RATES.

Two cents per word first insertion; one cent per word subsequent insertions.

Five cents additional each insertion where box number is desired.

Contractions count as one word, but five figures (as $1,000) are allowed as one word.

Cash remittances to cover cost **must** accompany all advertisements. **In no case** can this rule be overlooked. Advertisements received without remittance cannot be acknowledged.

RULES FOR COPY.

Replies addressed to HARDWARE AND METAL boxes are re-mailed to advertisers every Monday, Wednesday and Friday.

Requests for classification will be followed where they do not conflict with established classified rules.

Orders should always clearly specify the number of times the advertisement is to run.

All "Want" advertisements are payable in advance.

AGENTS WANTED.

This is the problem of many manufacturers who are anxious to get a foothold in Western Canada. It will be easily solved if HARDWARE AND METAL is given the opportunity to solve it.

AGENT wanted to push an advertised line of Welsh tinplates; write at first to "B.B.," care HARDWARE AND METAL, 88 Fleet St., E.C., London, Eng. [tf]

ONE who has sufficient organization to travel Canada east and west, fine line of Sheffield cutlery, brands already well known in Canada; none but applications from experienced and well-established men can be considered. Box 650, HARDWARE AND METAL, Toronto. [38]

AN old-established house of brass founders, making a wide range of brass foundry furnishings, such as brass cast tubing, stair rods, cornice poles, art metal work, electric light fittings, ecclesiastical metal work, are anxious to secure a thoroughly reliable and trustworthy agent or traveller for Canada, to sell goods on commission. An excellent opportunity for any firm covering Canada thoroughly. Address all enquiries to Box 654, HARDWARE AND METAL. [37]

BUSINESSES FOR SALE.

Somewhere in Canada is a man who is looking for just such a proposition as you have to offer. Our "For Sale" department brings together buyer and seller, and enables them to do business although they may be thousands of miles apart.

HARDWARE, tinware, stove and plumbing business in manufacturing town in the Niagara Peninsula; no competition; $250,000 factory and waterworks will be completed this summer; stock about $3,000; death of owner reason for selling. Box 85, Thorold, Ont. [tf]

HARDWARE Business and Tinshop for sale in Saskatchewan; population 1,500; stock carried, about $14,000; turnover, $45,000; practically all-cash business; cash required, $8,000; would rent building; do not answer without you have the money and mean business; it will pay to investigate this. Box 648, HARDWARE AND METAL, Toronto. (41)

HARDWARE Business for Sale in Tweed, a thriving Eastern Ontario town. Best business in town. Must sell owing to ill-health. Liberal terms to satisfactory purchaser. Write for particulars. J. M. Robertson, Tweed. [40]

FOR SALE — First-class set of tinsmith's tools; second-hand, but almost as good as new; includes an 8-foot iron brick almost new. Apply Box 664, HARDWARE AND METAL, Toronto. [tf]

GENERAL hardware and sporting goods business for sale, "Kawartha Lakes" District, Trent Valley Canal; open to offers up to Oct. 1st. A snap. Box 656, HARDWARE AND METAL, Toronto. (37)

SOLE Agency for Canada. Prominent American firm is now in position to receive bids from reliable wholesale hardware firms or others for the sole right to handle their Aluminum Solder, which will unite aluminum with aluminum, or aluminum with other metals. Only bids from established houses of high standing will be considered. For particulars write under N. Y. 1012 to John M. Munchenberg, 1161-75 Broadway, New York, U.S.A. [38]

Hits the Mark

Among the thousands of readers of Hardware and Metal it seems reasonable that someone wants what you have to sell, or that someone has to sell what you have tried in vain to buy, that someone is hunting the opportunity you have to offer.

The want ad. columns of Hardware and Metal are the simplest form of simplified advertising. No ad. writer need be employed, no drawing or cut need be made, no knowledge of display type is necessary, for display type will not be used in these columns. All that you have to do is state your wants, state them simply and clearly and send to any of our offices along with remittance to cover cost of advertisement. No accounts opened in this department.

Don't despise the want ad. because of its small size or small cost. Don't forget that practically all the hardware merchants, clerks, manufacturers and travellers read our paper each week.

RATES:—

2c. per word first insertion.
1c. per word subsequent insertions.
5c. additional each insertion for box number.

HARDWARE store and stock for sale, within 25 miles of Toronto. Solid brick double store and dwelling; property worth $4,000. Stock worth about $3,000. Splendid business in stoves and hardware. Good reasons for selling. Full particulars on application. Box 663, HARDWARE AND METAL, Toronto.

SITUATIONS WANTED.

Travellers, clerks, or tinsmiths, wishing to secure new positions, can engage the attention of the largest number of employers in their respective lines, by giving full particulars as to qualifications, etc., under "Situations Vacant."

HARDWARE CLERK, 32, married, sober, steady and reliable, capable of taking charge. Box 10, HARDWARE AND METAL, Winnipeg.

CLERKS WANTED.—Clerks who want to improve their positions are requested to write for information to Box 659, HARDWARE AND METAL, Toronto. 40

HARDWARE Salesman wants position; can furnish first class references; capable of managing city or town business. Box 653, HARDWARE & METAL, Toronto. (39)

HARDWARE CLERK, Irishman, 27, long experience of trade in best houses Old Country, one year with large Canadian wholesale firm, desires change, would go west. Unexceptionable references. Photo if required. Apply box 657, HARDWARE AND METAL, Toronto. (39)

SITUATIONS VACANT.

You can secure a "five thousand a year" manager, or a "five hundred a year" clerk, by stating your wants under "Situations Vacant."

TINSMITH for general country work. Used to furnace work. Also Helper with 2 or 3 years' experience. Apply to G. A. Binns, Newmarket. [37]

WANTED—First-class tinsmith in New Ontario town. Must be capable of managing workshop and estimating. Good position, good salary, and no lost time for the right man. Address, Box 655, HARDWARE AND METAL, Toronto. [40]

THE CANADIAN GROCER wants a Managing Editor. It wants a thoroughly capable man—a man who is live, full of up-to-date ideas, and one who understands the newspaper business from the reglet box to the editorial chair. Furthermore, it wants a man who is thoroughly conversant with the commercial situation in Canada. We realize that this is a big want. Not everyone can fill the bill, but we are willing to pay at the outset $2,500 a year to the man who can do so. The right man can eventually make his place worth $5,000. If you think you are this man we want to hear from you, with your experience and qualifications—by letter only. Address The MacLean Publishing Co., 232 McGill St., Montreal, or 10 Front St. East., Toronto. 38

ARTICLES WANTED.

If you cannot afford to buy a new counter, show case, screw cabinet, store ladder, or some other fixture which you could use to advantage, try a "Want Ad." under "Articles Wanted," and you may get what you want at a bargain price.

WANTED—Two store ladders, "Myers" preferred, complete with fixtures. Send full description and price to Ilsey Bros., Red Deer, Alta. (36)

ARTICLES FOR SALE.

Don't keep any fixtures or tools around your store for which you have no further use. They will be worth more to-day than they will a year hence. Don't keep money tied up which you could use to secure discounts from your wholesaler.

CUT Nails, 20 kegs 5-inch at $2.00 per keg. Cash with order. New stock. Taylor Bros., Limited, Carleton Place, Ont. [38]

TYPEWRITER, No. 2 Sun, in first-class condition. Will sacrifice for $55.00. Box 662, HARDWARE AND METAL, Toronto.

SET of tinsmith's tools, good as new, best American make. $150. Great sacrifice. Box 660, HARDWARE AND METAL, Toronto. tf

TWO hundred dozen rim and mortice knobs, black, white and mineral. 90c. per dozen. Full particulars. Box 661, HARDWARE AND METAL, Toronto.

GOOD silent salesman for sale. Reason for selling, advertiser wishes to purchase larger one. A bargain for immediate sale. Box 658, HARDWARE AND METAL, Toronto. 38

FOR Sale cheap, latest improved National Cash Register, four drawers, used only eighteen months. The Wilson Grocery Co., Gananoque, Ont. [38]

MANITOBA HARDWARE AND METAL MARKETS

Corrected by telegraph up to 12 noon Friday, Sept 20. Room 511, Union Bank Bldg, Winnipeg, Man.

A revival of trade is noticeable and a more optimistic feeling undoubtedly exists Prices of wheat will be higher and with a continuance of fair weather, the west is promised another year of prosperity.

Rope—Sisal, 11c per lb., and pure manila, 15¾c.

Lanterns—Cold blast, per dozen, $7; coppered, $9; dash, $9.

Wire—Barbed wire, 100 lbs., $3.22½; plain galvanized, 6, 7 and 8, $3.70; No. 9, $3.25; No. 10, $3.70; No. 11, $3.80; No. 12, $3.45; No. 13, $3.55; No. 14, $4; No. 15, $4.25; No. 16, $4.40; plain twist, $3.45; staples, $3.50; oiled annealed wire, No. 10, $2.90; No. 11, $2.96; No. 12, $3.04; No. 13, $3.14; No. 14, $3.24; No. 15, $3.39; annealed wires (unoiled), 10c less; soft copper wire, base, 36c; brass spring wire, base, 30c.

Poultry Netting—The discount is now 47½ per cent. from list price, instead of 50 and 5 as formerly.

Horseshoes—Iron, No. 0 to No. 1, $5.65; No. 2 and larger, $4.40; snowshoes, No. 0 to No. 1, $4.90; No. 2 and larger, $4.65; steel, No. 0 to No. 1, $5; No. 2 and larger, $4.75.

Horsenails—No. 10 and larger, 22c; No. 9, 24c; No. 8, 24c; No. 7, 26c; No. 6, 28c; No. 5, 30c; No. 4, 36c per lb. Discounts: "C" brand, 40, 10, 10 and 7½ p.c.; "M.R.M" cold forged process, 50 and 5 p.c. Add 15c per box. Capewell brand, quotations on application.

Wire Nails—$3 f.o.b. Winnipeg and $2.55 f.o.b. Fort William.

Cut Nails—Now $3.20 per keg.

Pressed Spikes—¼ x 5 and 6, $4.75; 5-6 x 5, 6 and 7, $4.40; ⅜ x 6, 7 and 8, $4.25; 7-16 x 7 and 9, $4.15; ½ x 8, 9, 10 and 12, 4.05; ⅝ x 10 and 12, $3.90. All other lengths 25c extra, net.

Screws—Flat head, iron, bright, 80, 10, 10 and 10; round head, iron, 80; flat head, brass, 75; round head, brass, 70; coach, 70.

Nuts and Bolts—Bolts, carriage, ⅜ or smaller, 60 p.c.; bolts, carriage, 7-16 and up, 50; bolts, machine, ⅜ and under, 50 and 5; bolts, machine, 7-16 and over, 50; bolts, tire, 65; bolt ends, 55; sleigh shoe bolts, 65 and 10; machine screws, 70; plough bolts, 55; square nuts, cases, 3; square nuts, small lots, 2½; hex nuts, cases, 3; hex nuts, small lots, 2½ p.c. Stove bolts, 70 and 10 p.c.

Rivets—Iron, 60 and 10 p.c.; copper, No. 7, 43c; No. 8, 42½c; No. 9, 45½c; copper, No. 10, 47c; copper, No. 12, 50½c; assorted, No. 8, 44½c, and No. 10, 48c.

Coil Chain—¼-in., $7.25; 5-16, $5.75; ⅜, $5.25; 7-16, $5; 9-16, $4.70; ⅝, $4.65; ¾, $4.65.

Shovels—List has advanced $1 per dozen on all spades, shovels and scoops.

Harvest Tools—60 and 5 p.c.

Axe Handles—Turned, s.g. hickory, $3.15; No. 1, $1.90; No. 2, $1.60; octagon extra, $2.30; No. 1, $1.60.

Axes—Bench axes, 40; broad axes, 25 p.c. discount off list; Royal Oak, per doz, $6.25; Maple Leaf, $8.25; Model, $8.50; Black Prince, $7.25; Black Diamond, $9.25; Standard flint edge, $8.75; Copper King, $8.25; Columbian, $9.50; handled axes, North Star, $7.75; Black Prince, $9.25; Standard flint edge, $10.75; Copper King, $11 per dozen.

Churns—45 and 5; list as follows: No. 0, $9; No. 1, $9; No. 2, $10; No. 3, $11; No. 4, $13; No. 5, $16.

Auger Bits—"Irwin" bits, 47½ per cent. and other lines 70 per cent.

Blocks—Steel blocks, 35; wood, 55.

Fittings—Wrought couplings, 60; nipples, 65 and 10K Ts and elbows, 10; malleable bushings, 50; malleable unions, 55 p.c.

Hinges—Light "T" and strap, 65.

Hooks—Brush hooks, heavy, per doz., $8.75; grass hooks, $1.70.

Stove Pipes—6-in., per 100 feet length, $9; 7-in., $9.75.

Tinware, Etc.—Pressed, retinned, 70 and 10; pressed, plain, 75 and 2½; pieced, 30; japanned ware, 37½; enamelled ware, Famous, 50; Imperial, 50 and 10; Imperial, one coat, 60; Premier, 50; Colonial, 50 and 10; Royal, 60; Victoria, 45; White, 45; Diamond, 50; Granite, 60 p.c.

Galvanized Ware—Pails, 37½ per cent.; other galvanized lines, 30 per cent.

Cordage—Rope sisal, 7-16 and larger, basis, $11.25; manila, 7-16 and larger, basis, $16.25; lathyarn, $11.25; cotton rope, per lb., 21c.

Solder—Quoted at 27c per pound. Block tin is quoted at 45c per pound.

Wringers—Royal Canadian, $36; B.B., $40.75 per dozen.

Files—Arcade, 75; Black Diamond, 60; Nicholson's, 62½ p.c.

Locks—Peterboro and Gurney 40 per cent.

Building Paper—Anchor, plain, 66c; tarred, 69c; Victoria, plain, 71c; tarred, 84c: No. 1 Cyclone, tarred, 84c; No. 1 Cyclone, plain, 66c; No. 2, Joliette, tarred, 69c: No. 2 Joliette plain, 51c; No. 2 Sunrise, plain, 56c.

Ammunition, etc.—Cartridges, rim fire, 50 and 5; central fire, 33⅓ p.c.; military, 10 p.c. advance. Loaded shells: 12 gauge, black, $16.50; chilled, 12 gauge, $17.50; soft, 10 gauge, $19.50; chilled, 10 gauge, $20.50. Shot: ordinary, per 100 lbs., $7.75; chilled, $8.10. Powder: F.F., keg, Hamilton, $4.75; F.F.G., Dupont's, $5.

Revolvers—The Iver Johnson revolvers have been advanced in price, the basis for revolver with hammer being $5.30 and for the hammerless $5.95.

Iron and Steel—Bar iron basis, $2.70. Swedish iron basis, $4.95; sleigh shoe steel, $2.75; spring steel, $3.25; machinery steel, $3.50; tool steel, Black Diamond, 100 lbs., $9.50; Jessop, $13.

Sheet Zinc—$8.50 for cask lots, and $9 for broken lots.

Corrugated Iron and Roofing, etc.—Corrugated iron, 28 gauge, painted, $3; galvanized, $4.10; 26 gauge, $3.35 and $4.35. Pressed standing seamed roofing, 28 gauge, $3.45 and $4.45. Crimped roofing, 28 gauge, painted, $3.20; galvanized, $4.30; 26 gauge, $3.55 and $4.55.

Pig Lead—Average price is $6.

Copper—Planished copper, 44c per lb.; plain, 39c.

Iron Pipe and Fittings—Black pipe, ¼-in., $2.70; ⅜, $2.85; ½, $3.75; ¾, $4.75; 1, $6.75; 1¼, $9.25; 1½, $11.00; 2, $14.80; 2½, $24.60; 3, $32.30; 3½, $40 50; 4. $46.00; 4½, $54.00. Galvanized: ¼ in., $3.65; ⅜, $3.80; ½, $4.50: ¾, $5.80; 1, $8.40; 1¼, $11.40; 1½, $13.80; 2, $18.40. Nipples, 70 per cent.; unions, couples, bushings and plugs, 50 per cent.; malleables, 20 per cent.

Iron Pipe and Fittings—Black pipe, ¼-in., $2.65; ⅜, $2.80; ½, $3.50; ¾, $4.40; 1, $6.35; 1¼, $8.65; 1½, $10.40; 2, $13.85; 2½, $19: 3, $25. Galvanized iron pipe, ⅜-in., $3.75; ½, $4.35; ¾, $6.65; 1, $8.10: 1¼, $11; 1½, $13.25; 2-inch, $17.65. Nipples, 70 and 10 per cent.; unions, couplings, bushings and plugs, 60 per cent.

Galvanized Iron—Apollo, 16-gauge, $4.15; 18 and 20, $4.40; 22 and 24, $4.-65; 26, $4.65; 28, $4.50; 30-gauge or 10¾-oz., $5.20; Queen's Head, 20, $4.60; 24 and 26, $4.90; 28, $5.15.

Lead Pipe—Market is firm at $7.80.

Tin Plates—IC. charcoal, 20x28, box, $10; IX charcoal, 20x28, $12; XXI charcoal, 20x28, $14.

Terne Plates—Quoted at $9.50.

Canada Plates—18x21, 18x24, $3.50; 20x28, $3.80; full polished, $4.30.

Lubricating Oils—600W, cylinders, 80c; capital cylinders, 55c and 50c; solar red engine, 30c; Atlantic red engine, 29c; heavy castor, 28c; medium castor, 27c: ready harvester, 28c; standard hand separator oil, 35c; standard gas engine oil, 35c per gallon.

Petroleum and Gasolene—Silver Star, in bbls., per gal., 20c; Sunlight, in bbls., per gal., 22c; per case, $2.35; Eocene, in bbls., per gal., 24c; per case, $2.50; Pennoline, in bbls., per gal., 24c; Crystal Spray, 23c; Silver Light, 21c. Engine gasoline, in barrels, gal., 27c; f.o.b. Winnipeg, in cases, $2.75.

Paints and Oils—White lead, pure, $6.-50 to $7.50, according to brand; bladder putty, in bbls., 2½c; in kegs, 2¾c; turpentine, barrel lots, Winnipeg, 90c; Calgary, 97c; Lethbridge, 97c; Edmonton, 98c. Less than barrel lots, 5c per gallon advance. Linseed oil, raw, Winnipeg, 72c; Calgary, 79c; Lethbridge, 79c; Edmonton, 80c; boiled oil, 3c per gallon advance on these prices.

CATALOGUES AND BOOKLETS.

When sending catalogues for review, manufacturers would confer a favor by pointing out the new articles that they contain. It would assist the editor in writing the review.

By mentioning HARDWARE AND METAL to show that the writer is in the trade, a copy of these catalogues or other printed matter will be sent by the firms whose addresses are given.

Tool Catalogue.

Two splendidly-made catalogues are being distributed amongst the trade by Schuchardt & Schutte, New York city. These catalogues contain fully illustrated descriptions of all their lines in tools and appliances. One of the catalogues is bound in grey cloth and is a model of compiling and binding. It contains 87 pages, comprised of one page with information as to terms of transactions, and a full index at the back. The illustrations are vividly reproduced, and the book throughout is of great credit to the compilers. The other of the two catalogues is bound in dark, stiff linen with gold embossed letters on the outside front cover, and contains 851 fully illustrated pages of descriptive and illustrated matter, with full index. These booklets should be procured by all interested in these lines. J. R. Baxter & Co., Montreal, are the Canadian representatives of Schuchardt & Schutte.

Favorite Ranges.

Findlay Bros., Carleton Place, Ont., have in their catalogue No. 47 a thorough and comprehensive work fully describing the whole "Favorite" line of heating and cooking stoves. Findlay Bros. claim to be "makers of the smoothest and best fitted stoves and ranges in Canada," and a study of their new catalogue will give dealers much to think about.

About 50 pages are devoted to steel and cast steel ranges, including the "Universal Favorite," "Crown Favorite," "Colonist," "Gleaner," "Ideal Favorite," "Favorite," and "Pan Favorite." In stoves, the "Ideal Favorite," "Favorite," "Live Acorn," "Mascot," "Marvel," "Royal Favorite," "New Diamond," "Elegant," "Active Favorite," "Favorite Warrior," "New Forest Beauty," and "Eclipse." In base burners and Oaks, the "Favorite," "Brilliant," "Aladdin," "Cheerful Oak," "Winner Oak," and the "Favorite Hot Blast," "Tortoise," "Merit," "Onward," "Patron Parlor Cook," "Woodland," "Economy Todd," "Keynote," "Woodside," "Norway," and "Algoma" heating stoves. A large line of registers, cauldrons and hollow iron ware is also illustrated.

The printing is done very attractively and a feature of the cover is a pocket for the keeping of discount sheets, etc. Dealers mentioning Hardware and Metal in their letter correspondence can secure a copy on request.

Facts About Saskatoon.

The Board of Trade of Saskatoon has issued a useful book of 36 pages, containing an historical sketch, description of the advantages, resources and possibilities of the city of Saskatoon, "the hub of the hard wheat belt." As a railway, farming, dairying, educational and financial centre, the city has great advantages and intending locators in the west should secure a copy of this book.

Heating and Housefurnishings

FRICTION IN FURNACE BUSINESS.

Too little attention is paid by the furnace trade, wholesale and retail, to the effect of friction on furnace construction, furnace sales and furnace heating systems, writes "Anti-friction," in the Metal Worker. The best way for the furnaceman to avoid friction is to spend some time in becoming thoroughly conversant with everything which has an influence on the work of the furnace. When he has acquired this mastery he will be qualified to sell furnaces at a fair and reasonable profit to those who ought to have good work or else he will be sufficiently philosophic to allow their work to go to somebody else without suffering worry. Philosophic furnacemen will put in a great deal of time in telling buyers what they ought to have and explaining why they ought to to have it and then will not be surprised to see them go to the other shop, which will cut his figures before the information he gives sinks deep enough to secure for him the lion's share of the trade. The furnaceman naturally is primarily in business to make money. He cannot make money unless he knows how much to charge for his work.

No man will ever know how much to charge for his work until he has put in one or two seasons of trading and then has carefully and laboriously gone over each job of work which has been completed, and made a record of each piece of material used and every minute of time consumed in completing it. If some men would do this they would find that instead of making a profit on a great deal of their work they have not even made journeymen's wages, but have been working at a day laborer's pay for the privilege of having their name on the sign. I assume that the furnaceman wants to do good work, wants to make a profit and wants to build up a reputation for satisfying his customers, and the best way to accomplish this is to learn how to estimate the cost of all work correctly. To men who have spent their early life in the shop and out on practical work this will be exceedingly irksome and irritating, and they naturally will not give that care to this important work which it is absolutely essential to escape a little later on. A book in which a record is kept of the material used and the time spent in heating the houses of various customers will be invaluable for a reference guide.

Good Work and Good Profit.

The observing man will soon find that there are many grades of furnace work, and that some men do the same kind of work in a fine residence for a high price that they do in a cheap building operation for a cutthroat price. A man of broader gauge and better capabilities will realize that there can be two or three different kinds of work, and that it is good work which pays the best profit. He will gather advice from trade literature, the trade papers and from the various salesmen who come to him until he has a fund of information which will enable him to judge the grade of work and what constitutes really good work.

Good work does not necessarily mean the most expensive character, but that which is designed with the best judgment and which has the proper proportions in all the various parts. A man whose judgment will enable him to design and execute his furnace work so that the competent judge will pronounce it good will be endowed with those qualifications which will enable him also to get a good profit, and it is just as essential to a man's confidence and self-reliance that he should have a good profit as it is that he should do good

Special 25c. Offering

A housewife asked us why we did not offer some specials in kitchen utensils. She was sadly in need of a new supply—and wanted to buy at a bargain. We decided to help her out of her troubles and forthwith set about to gather in a stock of useful and substantial articles which we could place on a special sale at a price low enough to make it possible for hundreds of others to throw away the old and useless and buy new and useful ware for the kitchen. The first real snap to show itself we reached for—and we got it. Now we are offering all that heart and purse can ask for at lowest prices ever asked for like goods. So the housewife's appeal was not in vain.

Housewife's Appeal

GRAY MOTTLED ENAMELEDWARE

Any Two Articles for 25c.

2-qt. deep pudding pans, in size 8½x3
3-qt. pudding pan, 9¾x3½.
2-qt. milk pan, 8½x11.
3-qt. milk pan, 9½x21.
2½-qt. lipped sauce pans, in size 7¼x3½.
Windsor dippers 1 quart.
Lipped sauce pans, 2 qt., [illegible]x3½.
Wash bowls, 11½ inch.
2-qt. round bake pan, riveted handles
3-qt. lipped kettle, 7¾x3½.
3-qt. deep stew pans.
10-inch deep pie plate

Gannon & Campion

work. Nothing so interferes with the judgment and the determination as to be continually anxious as to the financial outcome of any undertaking.

The best furnacemen will occasionally encounter difficulties unforeseen, and unless the contract price includes a sufficient margin of profit they will be unable to finally accomplish satisfactory results without loss and the loss is something which every furnaceman should endeavor strenuously to avoid. A good profit will be sufficient to cover all of the incidental expenses of carrying on a successful business, and a business cannot be considered successful unless it builds up a reputation for good work, pays all of the natural expenses, gives the business man better wages than any of his employes and still has a margin that will take care of some losses which it will be impossible to avoid.

Dealing With the Property Owner.

In too many instances the man who has engaged in the furnace business is too proud of his workmanship and the appearance of his final job and overlooks the need of including everything in the cost, which is a very important part of the game of business. In my opinion he is liable to incur an immense deal of friction of a most disagreeable character if he does not devise some system of making a correct estimate of the cost of work, and keep a record of his expenses. Then he must develop the talent of discovering the integrity of his customers, when they are not men of means, and the degree of promptness to pay on the part of men of property. He must realize that when work is done his bill is due, and that he is not a money lender. When he realizes this he will be able to impress it upon those who trade with him, and when they want a long time in which to pay for work done, the price must include interest on the account, or, if this is objected to, it must be sufficiently large to cover this expense. It will be far better if he can induce those who want time to pay their bills to borrow money from some banker or money lender and then the loan they secure can be repaid or renewed as frequently as may be required.

SAMPLE KITCHEN-UTENSIL AD.

At certain times in the year it is advisable to make a special push of kitchen utensils. In the spring and early summer and again in the fruit-canning season a little energy will increase sales materially. A Toronto racket store is now showing a 15-cent enamelware window, offering the smaller articles at this price, and making a special offer of large preserving kettles, etc., for 15 cents to any purchaser of 50 cents' worth of other goods.

The accompanying advertisement was used by a western hardware firm in a special enamelware sale, which, they claim, brought in particularly good returns. The ad. is somewhat crowded, but occupied a special position, where it stood out by itself and readily caught the eyes of bargain hunters. A cut of a saucepan or preserving kettle would, of course, improve the selling qualities of the advertisement.

Have a memorandum book in which to enter the name of every purchaser, and the name of the stove purchased, with other memoranda.

Do not try to push too many lines of stoves. Find out the style or make of stoves most suitable to your district and in most demand, then push that particular line all you can. Give it a conspicuous place near the front of your store and in the window and talk it more than any other stove to your customers.

Prepare for Holiday Business

It's not too early to lay plans for the Christmas holiday trade. Dealers have already bought some lines, of course, and in these days of slow deliveries other orders should be placed as soon as opportunity offers. "Goods well bought are half sold" is a time-worn truism in mercantile life.

But it doesn't do to feel too secure on the buying part until the selling has been done. Plans for increasing holiday sales must be worked out months ahead. Window displays must be figured out and a series of rough sketches planned. A series of ads. for the local papers should also be prepared, and, where deemed advisable, a neat booklet gotten up to be mailed to all probable customers in the surrounding district.

It is in these selling plans that Hardware and Metal can be helpful to its readers. Let all partake of the spirit of co-operation and exchange ideas through these columns. To help the suggestion along the editor invites any reader to join the friendly contest and hangs up a prize of $10 cash for the best answer to the following questions:

How can the hardware merchant increase his sales of holiday goods next December? What special lines should he stock? What selling plans should be adopted? What special advertising should be done? What novel window displays can be suggested? Should Souvenirs (calendars, knives trays, etc.), be given to customers?

The prize of $10 will be awarded to the writer of the most practical and original letter of from 500 to 1,000 words received by the editor before October 15, 1907, and the best half-dozen replies will be published in Hardware and Metal.

Address the Editor

Hardware and Metal

10 Front St. E., Toronto

BUILDING AND INDUSTRIAL NEWS

For additional items see the correspondence pages. The Editor solicits information from any authoritative source regarding building and industrial news of any sort, the formation or incorporation of companies, establishment or enlargement of mills, factories or foundries, railway or mining news.

Industrial Development.

The power plant for Saskatoon will soon be in operation.

The sawmill of Mohr & Co., Killaloe, Ont., was destroyed by fire.

A seam of coal is said to have been uncovered at Headingly, Man.

McCoskerie's sawmill, at Hartley Bay, B.C., will double its capacity.

The Standard Brick & Tile Company, New Glasgow, N.S., suffered $300,000 loss by fire recently.

The Alberta Portland Cement Co., Calgary, wish to supply power to that city.

The Bell Foundry, St. George, Ont., has been sold to Chapman & Fleury, the price being $25,000.

Development work is being done on coal lands at North Saanich, twenty miles north of Victoria, B.C.

It is expected that No. 1 factory of the new Silliker Car Works, at Halifax, will be opened in a few days.

The Sechelt Brick & Tile Co., a new concern, will erect a brick-making plant at Sechelt, to cost $50,000.

Consignments of harvesting machinery, aggregating three hundred tons, were shipped to Australia recently.

The Cataract Power Co. will run a high potential power transmission line from Hamilton to Brantford, Ont.

The Ottawa Pulp and Paper Co. has been formed, to make paper out of refuse spruce and hemlock. Their capital is $150,000.

The cost of the power plant of the Stave Lake Power Co., Stave Lake, B.C., will be in the neighborhood of $2,500,000.

The Canadian Machine Telephone Co., Brantford, may erect a plant for the manufacture of their telephones for use in Canada.

The entire plant of the Standard Brick and Tile Co., New Glasgow, N.S., was destroyed by fire, entailing a loss of $3,000,000.

The Great West Coal Company has been formed at Port Arthur, to develop 12,000 acres of coal lands recently purchased in Alberta.

The Canadian Woodenware Company, recently burned out at Hampton, N.B., are now talking of erecting a plant at South Bay, N.B.

The Empire Mfg. Co. has been granted a permit to erect a foundry and machine shop on the south shore of False Creek, at Vancouver.

Buchanan & Company, Winnipeg, have been awarded a contract for the erection of a bridge over the Winnipeg river, to cost about $41,000.

It is said that the lumber operators on the branches of the Miramichi river, N.B., will handle 120,000,000 feet of timber this coming winter.

The Dominion Iron & Steel Company have been looking over some iron ore deposits in Gloucester county, New Brunswick, ten miles from the I.C.R.

The New Brunswick Government has decided to grant to the New Brunswick Petroleum Company a lease of 10,000 acres of land under the usual restrictions.

The Development & Finishing Co., New York, will install a plant in St. Catharines, Ont., for the manufacture of chlorine and alkali by the Townsend process.

A syndicate, headed by Walsh & Arnoud, Vancouver, will erect a large brick plant in Barnaby, near New Westminster, which will have a capacity of 50,000 bricks daily.

C. J. Wright, Montreal, is proposing to erect a factory for the manufacture of brass goods in St. Catharines, Ont. A free site is wanted, with exemption and other privileges.

The rail mill of the Superior Corporation was opened recently, after a shutdown of two weeks for repairs. One of the furnaces has been relined and the capacity of the plant increased.

The Silica Brick and Lime Co., Parson's Ridge, B.C., is turning out 15,000 bricks per day, and is filling an order for one million bricks for the Spencer block, being erected at Vancouver.

A. F. Barber & Co., Vancouver, have been awarded the contract for the electric appliances for the new C.P.R. Empress hotel. It is one of the biggest contracts of its kind in western Canada.

Clare Brothers & Co., Preston, are building a new warehouse near their stove foundry. The new building will be 100 feet by 30 feet, and will be used for storage purposes.

The North Atlantic Collieries Co., Port Morien, N.S., are enlarging their plant. They have ordered a Babcock & Wilcox water tube boiler. A pump and hoisting engine will be installed shortly.

Recent improvements have made the Nova Scotia Steel Company's plant at Sydney, N.S., one of the finest and most modernly-equipped institutions on the American continent. The company's rod mill at Sydney has a record of production greater than that of any rod mill of the United States Steel Company.

The American Electric Furnace Company is establishing a plant at St. Catharines, where an 80 h.p. Colby Induction furnace will be installed, and later on another furnace of the R. Jellin type, requiring 200 electric horsepower, will also be installed. This furnace will yield 1,000 pounds of steel at a single heat, and has a capacity of six heatings during twenty-four hours. A large crane of five tons capacity is being built for the plant by the Niagara Falls Machine & Foundry Company.

John Ballantyne & Co., Galt, have invented and are having patented a self-locking device for the movable side pressure and chip-breaker on molding machines. This invention has proved a great time-saver and will greatly enhance the value of their machines. At first used only on their 12-inch molder on base, a massive machine of 7,200 pounds, it proved such an unqualified success that this enterprising firm are placing it on all their molders and have taken the necessary steps to protect their invention.

The Imperial Steel & Wire Company's new plant, at Fort William, is now under construction, the site being alongside the new works of the Canada Iron Foundry Company. The main building is to be 600x70 feet, while a warehouse will also be erected, 60x200 feet. The buildings will be of cement, and will be constructed in time to have machinery installed next spring. It is expected that 250 men will be employed at the works. Alderman R. S. Piper, hardware merchant, presented a handsomely-decorated shovel, which was used by the wife of the general manager, Major J. A. Currie, in turning the first sod.

A summary of the decision handed down in the law suit of the Dominion Iron & Steel Co., against the Dominion Coal Co., Sydney, N.S., is as follows: The Dominion Coal Company committed a breach of contract in furnishing to the Dominion Iron and Steel Company coal that was unfit for the purposes for which it was required. The Steel Company was justified in refusing to accept such coal, and the refusal did not constitute a breach of contract. The contract of October, 1903, is still in operation, and the court has power to appoint a receiver to compel the Coal Company to perform the terms of the agreement. An order, therefore, issues requiring the Coal Company to pay such damages as may be determined by the referee, and to specifically carry out the terms of the contract.

Railroad Construction.

The roadbed of the G.T.P. east of Edmonton is ready for the steel.

The Canadian Northern is considering the building of an air line from Brockville to Ottawa.

The Canadian Northern will build a line from Vancouver to Northern British Columbia to connect with the main line.

The Canadian Pacific Railway are to build a new branch line from Kamloops to Edmonton, via the North Thompson River.

The Canadian Pacific will erect an extensive terminal station at Fort William, Ont., and will also build a large dock.

The Great Northern Railway Company is to erect a new stone and brick station at New Westminster, to cost about $50,000.

The Canadian Pacific Railway Company have awarded the contract for their new $250,000 station at Calgary to A. McDermid, Winnipeg.

The contract for the G.T.P. bridge over the Kaministiqui River, near Fort William, has been awarded to Wylie & Belfour and the Canadian Bridge Co.

The N., St. C. & T. Ry. Company's new power house at Thorold, Ont., is now under construction. It will cost $50,000, and develop 1,00 horse power.

The contract for Section No. 6 of the Trent Valley Canal, near Campbellford, has been let to Brown & Aylmer, about $600,000 being involved in the contract.

The Ontario Municipal and Railway Board have approved the plans for the

completion of the line from Pembroke to Gold Lake, on the Pembroke and Gold Lake R.R.

The C.P.R. will spend $100,000 in additions to their shops at Winnipeg. The contract for the new I.C.R. station at Amherst, N.S., has been let to Mr. LeBabeo, of Moncton, N.B.

The British Columbia Electric Railway Company, Vancouver, are to spend $1,000,000 next year on construction work. A new branch will be built from New Westminster to Chilliwack.

The C.P.R. will spend $1,000,000 in constructing two tunnels through the Rockies near Field, B.C. One will be to the north of the Kicking Horse River and will be 3,400 feet long. The other will be south of the river and will form practically a complete circle through the heart of Cathedral Mountain. It will be 3,800 feet long.

Municipal Undertakings.

Extensions are being made to the waterworks system at Preston, Ont.

The town council of Hull, Que., have decided to erect three new fire halls, to cost $25,000.

A bylaw to raise $40,000 for new electric pumps, is to be voted on by Hamilton ratepayers.

The contract for the new city hall at Calgary, has been assigned, the contract price being $142,000.

The city of Revelstoke, B.C., is calling for tenders for the enlargement of the city's hydro-electric plant.

Revelstoke, B.C., is calling tenders for a 500 horse power electric generator, to be run by a producer gas engine.

Tenders will be asked by the Winnipeg Board of Control, for a building in connection with the waterworks, to cost $20,000.

Aylmer, Que., has passed a bylaw to purchase the waterworks system of that town, and another bylaw to establish a sewerage system.

Ottawa is considering the advisability of purchasing 25,000 water power at Gatineau Falls, and erecting a power works to cost $1,250,000.

The town of Chilliwack, B.C., will have a telephone system which will give service very cheaply, the rates being $1.25 and $2.50 per month.

The town of Kenora, Ont., is contemplating the installation of a new arc lighting system, at a cost of $4,000 and an electric pump for the waterworks.

Vancouver ratepayers will be asked to vote $1,000,000 at the next election, to be spent in the erection of three new bridges connecting the city and its suburbs.

The contract for the erection of the telephone exchange for the Manitoba Government, at Winnipeg, has been awarded to J. J. Kelly, Winnipeg, the figure being $97,172.

The city of Toronto has decided to spend $781,000 in improving its waterworks system. The improvements will include large extensions to the water mains, and considerable new machinery for the pumping plant.

A petition is being circulated, asking the Aylmer, Ont., council to submit a bylaw giving authority to raise $10,000, to be expended in securing an additional supply of water and to furnish a site for the Canadian Condensed Milk Company, who propose to locate there.

Building Notes.

The new city hall for Calgary, Alta., will cost $145,000.

H. Weinberg, Toronto, will erect a store to cost $25,000.

R. Cassidy, Vancouver, will erect a block to cost $250,000.

An Oddfellows' Hall, costing $30,000 will be erected at Strathcona, Alta.

The Imperial Bank will erect a new building at Edmonton, Alta., to cost $90,000.

The congregation of Paisley Street Church, Guelph, Ont., will erect an edifice to cost $10,000.

The J. Y. Griffin Company will erect a cold storage block at Edmonton, Alberta, to cost $100,000.

J. B. Pause & Company, Montreal, have been awarded the contract for building a new jail at Montreal, the contract price being $790,000.

Dr. Alexander Graham Bell has just completed on Lookout Mountain, at his summer residence, near Baddeck, N.S., a lookout tower, built on the tetrahedral principle, said to be the first tower of the kind in the world. A tetrahedron is simply a three-sided pyramid, its three sides and base all being equilateral triangles. Those in the tower are made each of six pieces of galvanized half-inch pipe, four feet long, and four nuts into which the pipes are screwed, and the tower is simply a giant tripod built up of these. Such a structure, Mr. Bell says, is lighter and stronger than any other, more quickly and more cheaply built, and requires no skilled labor. Patents on the principle and on the nuts used at the corners, are being taken out.

Companies Incorporated.

The Peel Oil & Gas Company, Toronto; capital $10,000. To develop oil and natural gas wells. Provisional directors, F. Watts, H. A. Menet, J. L. Galloway, J. C. Colling, and W. E. Sampson, all of Toronto.

ST. JOHN BUILDING RECORD.

Building operations on an unusually extensive scale are being carried on in St. John, N.B. Individually, many of the buildings called for passing notice when construction was commenced, but it is doubtful if it is generally realized that nearly $400,000 has been invested in stone and brick erections during the past year. It is safe to say that were St. John a rising township in the west the increase in value of real estate would be heralded far and wide as direct evidence of the city's progress and enterprise. Buildings for residential, business and philanthropic purposes are all largely represented.

Two Large Structures.

The two buildings which naturally first attract attention on account of their size and importance are the new building owned by the Royal Bank of Canada, in King street, a massive stone structure, and the big brick and concrete cold storage warehouse near the Long wharf, Main street. They each represent an investment of more than $125,000, and will both be completed about the end of the present month. The big, new Y.M.C.A. building, costing $60,000, is also being rapidly put up. Simeon Jones, Ltd., are expending some $14,000 on a brick and concrete addition to their brewery, and James Ready, proprietor of the Fairville Brewery, is having a building 84x45 feet in area erected at the corner of Peel and Chipman streets. It will be of brick, with concrete foundation. In Fairville, the new office building of the Partington Pulp & Paper Co. is nearly finished, at a cost of about $10,000. Brick and stone are the materials used.

A new flour shed is being built by the I.C.R. at York Point wharf, at a cost of about $20,000, and the new Salvation Army citadel, costing about $10,000, is almost finished.

New Foundry and Machine Shop.

McLean & Holt are just completing a new foundry and machine shop in Albion street. It will be equipped to carry on a very extensive business. The cost is not mentioned. With the installation of an up-to-date plant in this building the firm will be in a better position than ever before to turn out "Glenwood" ranges and all lines of stoves and furnaces.

The new home for the Seamen's mission, in Prince William street, will shortly be started, so as to be in readiness for the coming season. The building will cost about $15,000. In addition to those mentioned, there are a score or more big residences and tenement houses being erected, and considerable repair work and remodelling being done, so that hardware dealers, plumbers, masons and carpenters are kept pretty busy.

GERMAN STEEL TO BRITISH MARKETS.

The steel trade in Scotland has been steadily prosperous during the current year, the demand taxing the capacity of the works and prices generally being well maintained. Now, however, there is a sudden and apparently unexpected, revival of German competition, sharply affecting the home trade. In a recent issue, the Edinburgh Scotsman says:

"After a long interval of suspended operations in British markets the competition of the German Steel Syndicate has been renewed in severe form, and now relates to the finished steel, instead of semi-finished products, as formerly. Agents of the Stahlwerk Verband are now offering in the Midlands steel tube, strip, and plates at $4.86 per ton below English prices. A contract for some 2,000 tons of German welding steel strip has just been made by a large Birmingham tube firm having ramifications in Scotland, at less than $36.50 per ton, the Staffordshire and Shropshire quotation being $41.37."

CANADIAN MANUFACTURERS' CONVENTION.

The annual convention of the Canadian Manufacturers' Association will be held at Toronto, commencing September 24. Elaborate arrangements are being made by the Montreal branch of the association to visit the convention. Two special cars have been requisitioned. It is expected that the attendance will exceed all previous years.

News of the Paint Trade

PAINT TRADE IN ENGLAND.

W. H. Evans, of the Canada Paint Co., Montreal, has just returned from spending a short holiday in England. His wanderings through England naturally brought him in contact with a number of manufacturers of paint. In an interview with Hardware and Metal, Mr. Evans stated that organized publicity was not carried on so elaborately and extensively in England as in Canada, and that the English makers were surprised when told of the great amount of literature on paint products that was distributed in Canada. The chief reason for this is that England is very thickly populated and hence where the people live so close together less publicity will do the same number of people than when the people are scattered over a greater territory.

The liquid paint trade in Great Britain is not pushed as it is in Canada. Paint essentials are sold principally in paste form. The reason is that in England there are few wooden houses and very few board barns. Again, the architecture of the metropolitan centres is of more substantial materials than those in Canada. More interior decorative work is done in England and consequently a high standard of workmanship is demanded from painters, who like to adhere to old methods. In this lies the fact also that paint makers do not go in for as much tinting work as Canadians, most of the blending and tinting work being left to the painters themselves.

Not as much attention is paid in England to labels as in Canada, by this being meant that no such elaborate decorations are put on their labels. This, however, does not imply that Englishmen are careless about their marking. Although there is no rigid legislation in England as to the marking of white lead packages, great confidence is reposed in the grinders by the consumers in regard to the faithfulness of the marking. If a lead is marked genuine, it is genuine; if a lead is marked pure, it is pure. Lower grades are marked Nos. 1, 2 and 3.

ADVERTISING WALL PAPER.

It is a well-known maxim among advertising men that in retail advertising price should never be left out. Price, indeed, is the most important point in bringing the customer into the store. But it is a point that, in wall paper advertising, should be used as a sort of final rap. Make it appear that the price is a marvelous one when the quality of the article is considered.

Quality in wall paper is, of course, made up of several things, such as color harmony, exclusiveness of design and durability. In the ad., therefore, some or all of these points should invariably be the theme. First tell your prospective customers what the good points of your wall paper are, then tack on the price. It is doubtful if illustrations are of much value, as it is impossible to give any idea of the color effects of a design in a newspaper ad. As eye-catchers, of course, illustrations are always valuable, and are just as valuable in wall paper, as in any other kind of advertising. It is as well, if possible, to have the eye-catcher appropriate to the subject of the ad.

The high grade of wall paper, both of the cheaper and expensive varieties, turned out by our domestic manufacturers to-day, sells readily. Salesmanship, however, is required in order to sell the grades in which there is most profit for the retailer. The merchant must endeavor to analyze a design, so to speak, and be in a position to explain the excellent points of any design which seems to take the customer's fancy. A conversation upon the subject of wall paper designs, and their appropriateness for this, that, and the other room, or hallway, has a tendency to bolster up a customer's wavering decision to buy. Lots of people are quite unable to decide what they want without some aid. And the retailer who makes the most money is invariably the man who is able to talk interestingly to his customers of their needs in his line of business. A wall paper salesman, to be really successful, must know wall paper, and be able to advise and assist his customers to choose designs which will give them satisfaction.

As a Money Maker

a good line of paint is hard to beat. Things must be kept painted, and people want paint which keeps its freshness for a reasonable length of time. They want a paint like

NEW ERA PAINT

which spreads evenly and covers a lot of space to the pot. It is a significent fact that once a man buys a pot of New Era Paint, he will use no other brand but it.

STANDARD PAINT AND VARNISH WORKS CO., LIMITED

WINDSOR - ONTARIO

Have you ever tried to sell

JAMIESON'S Floor Enamels?

They will please your customers because they possess just those qualities which the ordinary floor paints lack.

We guarantee them to dry hard over night, without cracking, leaving a fine glossy finish.

Floors on which our enamels are used, are never sticky, and never peel off, thus meaning less work to the housewife who desires to keep her floor clean.

Genuine satisfaction awaits the users, and this ensures a continuance of trade to the dealer who handles them.

Manufactured in all desirable colors by

R. C. JAMIESON & CO.

LIMITED

16-30 Nazareth St., MONTREAL

HEADQUARTERS

FOR

Window Glass,
Paints, Oils,
Fine Varnishes

SANDERSON PEARCY & CO., Ltd.

Toronto, Canada

GLAZIERS' DIAMONDS OF PROVED WORTH

Having a Century's Reputation for Reliability.

MADE BY

A. SHAW & SON, London, Eng.

CANADIAN AGENT

GODFREY S. PELTON, - 201 Coristine Building, Montreal

Sharratt & Newth's Glaziers' Diamonds

are unequalled for cutting and wearing qualities.

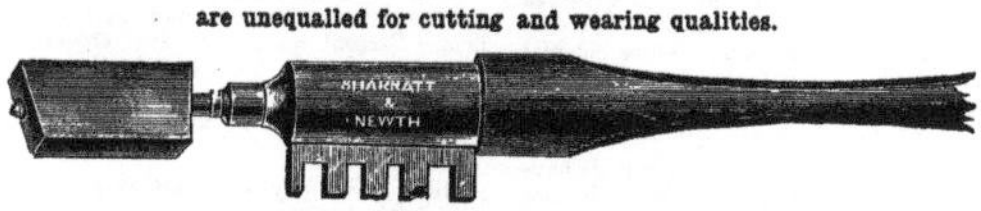

To be obtained from the principal Dealers in Glass, Hardware, and Painters' Supplies.

Contractors to H. M. Government and the principal English Sheet and Plate Glass Works

---

GILBERTSON'S COMET

Gilbertson's Sheets

and the COMET Brand

mean quality and security in whateVer work you may undertake.

GalVanized flat sheets that are unequalled as to flatness; absolute trueness to gauge; admirably suited in softness for any working. Note the brand "COMET." GILBERTSON on eVery sheet means security.

W. GILBERTSON & CO., Limited, Pontardawe, South Wales

ALEX. GIBB, 13 St. John St., Montreal, Agent

PAINT AND OIL MARKETS

MONTREAL.

Montreal, September 20.—Favorable weather has rendered the mind of the buying public easier, and with the receipt of encouraging reports regarding the crops in the Northwest there is more confidence expressed. There has certainly been a marked improvement in the white lead and color trade, both in the size and number of orders being received during the past week. Dealers are reminded of the inevitable rush which comes in the fall months and it will be well if they stock in sorting-up orders without delay.

Linseed Oil—Quotations are slightly higher. There has not been, however, sufficient advance to warrant changes in quotations. Crushers are reporting that stocks are being shipped as rapidly as the oil can be ripened and tanked. Raw, 1 to 4 barrels, 60c; 5 to 9, 59c; boiled, 1 to 4 barrels, 63c; 5 to 9, 62c.

Turpentine—Shows a weakness in primary markets which will probably cause a decline in local quotations of 2 cents a gallon. We continue to quote single barrels at 80 cents.

Ground White Lead—Some slight concessions have been made for round lots of white lead, but later advices indicate that the Old Country corroders are holding very firmly to their quotations. We continue to quote: Government standard, $7.50; No. 1, $7; No. 2, $6.75; No. 3, $6.35.

Red Lead—The output is of a normal character. There are no special features and quotations remain unchanged. Genuine red lead, in casks, $6.26; in 100-lb. kegs, $6.50; in less quantities at $7.25 per 100 lbs. No. 1 red lead, casks, $6; kegs, $6.25, and smaller quantities, $7.

Dry White Zinc—A good demand prevails without any change in prices: V.M. Red Seal, 7½c; Red Seal, 7c; French V.M., 6c; Lehigh, 5c.

White Zinc Ground in Oil—Shipments are very light, but it is expected that there will be a marked increase in shipments very soon. White colonial painting is calling for an increased consumption of zinc. Prices are unchanged: Pure, 8½c; No. 1, 7c; No. 2, 5⅜c.

Gum Shellac—Some large parcels have been coming forward and there is now no difficulty in keeping up with the demand from factories and the hardware dealers. Prices are firm: Fine orange, 60c per lb.; medium orange, 55c per lb.; white (bleached), 65c per lb.

Shellac Varnish—No change has occurred in quotations and all varnishes may be said to be firm with a good turnover: Pure white bleached shellac, $2.80; pure orange, $2.60; No. 1 orange, $2.40.

Putty—It is with difficulty that the grinders fill the heavy orders which are now pouring in, and whereas, some makers have been cutting the price to some extent, others find no difficulty in getting present quotations for their brands: Pure linseed oil, $1.85 in bulk; in barrels, $1; in 25-lb. irons, $1.90; in tins, $2; bladder putty, in barrels, $1.85.

TORONTO.

Toronto, Sept. 20.—Travelers who have been on their territory since the Toronto Exhibition find business good, sorting orders being received in very satisfactory volume for this season. As a result trade has brightened up and Toronto jobbers are looking for a brisk movement of goods from now until well on in November. Mixed paints are, of course, in greatest demand, but a good deal of lead and putty is also being sold. Glass will be a big item from now on, owing to the completion of buildings now in course of erection. All lines are holding firm in price with the exception of turpentine, some quoting this as low as 77 cents.

Turpentine—Being quoted generally at 79c the same as a week ago, but the drop in the South (due, it is said, to a tightness in the money market which prevents the brokers from holding their stocks) has developed a similar weakness here. As low as 77c is being quoted for single barrels in Toronto.

Linseed Oil—Stiffness in the Old Country markets is reflected here as the dominating factor on the Canadian market follows the foreign market closely, leaving only sufficient margin to make it unprofitable to import from England. We still quote: Raw, 1 to 3 barrels, 64c; 4 barrels and over, 63c. Add 3c to this price for boiled oil, f.o.b., Toronto, Hamilton, London, Guelph and Montreal, 30 days.

White Lead—The market is weaker in England, but no similar movement has developed here. Buying is active at unchanged prices, as follows: Genuine pure white lead, $7.65, and No. 1, $7.25.

Red Lead—Business shows a slight increase over last week and, prices are firm. Present quotations are: Genuine, in casks of 500 lbs., $6.25; ditto, in kegs of 100 lbs., $6.75; No. 1 casks of 500 lbs., $6; ditto, in kegs of 100 lbs., $6.50.

Glass and Putty—Indications are that this fall's business in both commodities will be extremely large. Prices are unchanged.

Petroleum—Dealers are stocking up in preparation for winter. Prices are unchanged: Prime white, 13½c; water white, 15c; Pratt's astral, 18½c.

Shellac—There is a strong demand for shellac and prices are well maintained, as follows: Pure orange, in barrels, $2.70; white, $2.82½ per barrel; No. 1 (orange), $2.50.

For additional prices see current market quotations at the back of the paper.

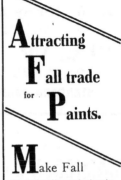

COAL TAR OIL FOR VARNISH MAKING.

The well-known tar oil, obtained as a thick, black mass of disagreeable smell, from the distillation of coal, contains valuable constituents for the production of coloring matters, and, in addition, other constituents that form suitable solvents for resin, and are, therefore, of interest to the varnish industry. These solvents may be recovered from the tar by fractional distillation. At 160 to 170 degrees C. the distillate consists of light coal tar oil, mainly consisting of benzol, toluol, xylol, etc.; and this fraction is succeeded by the heavy tar oil, composed principally of carbolic acid, naphthalene, paraffin, anthracene, etc., leaving as residue coal-tar pitch, which is also of interest for varnish-making. In its crude state the light tar oil is not adapted for the preparation of varnish, but requires further distillation followed by purification. For this purpose the crude oil is distilled in an iron still, fitted with hood, neck, safety valve, thermometer, etc., and gently heated to a temperature not exceeding 110 degrees C. The distillate coming over, principally benzol, is collected, and the temperature is raised to 120 to 125 degrees C. to drive off the light tar oil (sp. gr. 0.940) which is suitable for varnish-making. To eliminate objectionable impurities the oil is treated with potassium chromate, manganese dioxide, and sulphuric acid, whereby the empyreumatic components are destroyed, the impurities settling down as a dark sediment in about twenty-four hours. The clear oil is drawn off, washed several times with water, until all the excess of acid is removed, and is then redistilled. The resulting tar oil is water white, with a density of 0.800—0.885, and forms a useful solvent for resin. It possesses the great advantage of forming very pale varnishes, which dry quickly, and do not darken in color on exposure to air. The unpurified oil is occasionally used for the same purpose, but furnishes only cheap, dark-colored varnishes, which become still darker as time goes on. The pitch left behind in the still is suitable for painting on iron, but should always be mixed with pale resin and Venice turpentine, since it is too brittle when used alone, and easily peels off.—Farben Zeitung.

A PURE PAINT CONVERT.

One of the best demonstrations that it pays to push paints of high quality rather than cheap goods occurred in a northwestern town recently, says the Hardware Trade. One of the retail hardwaremen, who had been selling paints for many years, had always made a specialty of cut prices and cheap paints. He went on the theory that paint was paint. Of course, he claimed much for his brand, as he had been taught to do by the manufacturer, and for its argument he pointed to the price.

But one day this hardwareman saw a light. A new hardware store opened up down the street and the new hardwareman was not much of a "cut price" artist. He also insisted that there was a big difference in paint. He had studied paint enough to know something of its lasting qualities and how it acted in service.

All this information he wove into his selling talks and his educational campaign with the people of that town and the big farming country around it. He had some good advertising ideas and his quality talks were prominent in his advertising. More than this, he charged a good price for his paint.

New Methods Got the Trade.

As time went by, the new hardware store began to get paint business. In fact, the quality talk won so strongly with many paint users in that vicinity that the old hardware store began to see some of its paint trade drift away.

The result was a visit by the old hardwareman to another paint manufacturer, who had a reputation of making absolutely pure paints. He told the paint manufacturer his predicament. The manufacturer of pure paints recommended that he go back, stock up with a line of pure paints and follow about the same plan that the competitor was doing.

He did, and now says his paint department is giving him better results than ever. The profits are better; the customers are better satisfied, and he himself is all done with the cut price idea in paints at least.

One incentive many merchants have for entering a cut price campaign is to meet the prices on the cheap paints put out by the retail-catalogue houses, but the pure paint campaign is putting the retail-catalogue houses in the same boat as all other competition, which makes cheap paints prominent.

In fact, the time is not far away when farmers generally will be educated to fight shy of retail-catalogue-house paints, or the catalogue houses will be forced to sell nothing but pure paints in order to have any paint trade at all.

NEW PAINT WAREHOUSE.

A building permit has been granted the Sherwin-Williams Co. for the erection of a varnish warehouse, on William street, Montreal. The new structure will be built of concrete, and will cost about $20,000.

FLAT WHITE JAPAN.

A "flat white japan" has been brought out in England, for which large claims are made. It is said that it only requires applying with a brush to give a perfectly flat, smooth and non-absorbent surface, and forms the only really perfect ground for enamel, enabling work to be finished the same day as commenced, with the best possible results.

For preparing skirtings, dados, friezes, cornices, and all work with an irregular surface, it is invaluable, as, no rubbing down with sandpaper being required, the coat on the raised parts will be exactly the same as on the flat, the result being really first-class work. The surface produced by the flat japan is so smooth and hard that no rubbing down is necessary; but should it be desired it may be rubbed down to the wood without raising or crumbling.

For ceilings of anaglypta, etc., it is unrivaled, being so thin that it will not injure the finest patterns, whilst it will not discolor with gas fumes, as will white lead paint. When a dull finish is required, it dries with a perfectly flat surface, so hard that it may be washed and scrubbed without deterioration. When an egg-shell gloss is desired, all that is necessary is to give two coats and then rub and felt down the work, when a beautiful semi-gloss will be obtained.

CURRENT MARKET QUOTATIONS.

September 20, 1907

These prices are for such qualities and quantities as are usually ordered by retail dealers on the usual terms of credit, the lowest figures being for larger quantities and prompt pay. Large cash buyers can frequently make purchases at better prices. The Editor is anxious to be informed at once of any apparent errors in this list, as the desire is to make it perfectly accurate.

METALS.

ANTIMONY.

Cookson's .... per lb.	0 13
Hallett's	0 12½

BOILER PLATES AND TUBES.

Plates, ¼ to ½ inch, per 100 lb.	2 40	2 50
Heads, per 100 lb	2 65	2 75
Tank plates 3-16 inch	2 60	2 70
Tubes per 100 feet, 1½ inch	8 25	8 50
" " 2 "	9 00	9 10
" " 2½ "	10 50	11 00
" " 3 "	12 00	12 50
" " 3½ "	15 00	16 00
" " 4 "	19 25	20 00

BOILER AND T.K. PITTS.

Plain tinned and Spun, 25 per cent. off list.

BABBIT METAL.

Canada Metal Company—Imperial genuine 60c.; Imperial Tough, 60c; White Brass, 50c. Metallic, 35c.; Harris Heavy Pressure, 25c, Hercules, 25c; White Bronze, 15c; Sta Frictionless, 14c.; Alluminoid, 10c.; No. 4 9c. per lb.

James Robertson Co.—Extra and genuine Monarch, 60c.; Crown Monarch, 50c; No 1 Monarch, 40c; King, 30c; Fleur-de-lis, 20c; Thurber, 15c.; Philadelphia, 12c.; Canadian, 10c; hardware, No. 1, 15c; No. 2, 12c; No. 3, 10c. per lb.

BRASS.

Rod and Sheet, 14 to 30 gauge, 25 p.c. advance.

Sheets, 12 to 14 in.	0 30
Tubing, base, per lb 5-16 to 2 in	0 33
Tubing, ⅜ to 3-inch, iron pipe size	0 31
" ⅜ to 3-inch, seamless	0 36

Copper tubing, 6 cents extra.

COPPER.

	Per 100 lb	
Casting ingot	20 00	21 00
Cut lengths, round, bars, ½ to 2 in.		30 00
Plain sheets, 14 oz.		36
Plain, 16 oz., 14x48 and 14x60		35
Tinned copper sheet, base		38 00
Planished base		43 00
Braziers' (in sheets), 4x6 ft., 25 to 30 lb. each, per lb., base	0 34	0 35

BLACK SHEETS.

	Montreal.	Toronto
No 10 gauge	2 70	2 75
12 gauge	2 70	2 75
14 "	2 50	2 60
17 "	2 50	2 60
18 "	2 50	2 60
20 "	2 50	2 60
22 "	2 55	2 65
24 "	2 55	2 70
26 "	2 65	2 85
28 "	2 70	3 00

CANADA PLATES.

Ordinary, 52 sheets	2 75	3 05
All bright "	3 75	4 05

Galvanized—	Dom. Crown.	Ordinary.
18x24x52	4 45	4 35
60	4 70	4 60
20x28x80	8 90	8 70
	9 40	9 20

GALVANIZED SHEETS.

B.W. gauge	Queen's Head	Fleur-de-Lis	Gordon Crown	Colborne Crown Gorbal's Best
16-20	3 95	3 80	3 95	....
22-24	4 20	4 05	4 00	4 05
26	4 45	4 30	4 40	4 30
28	4 70	4 55	4 60	4 55

Less than case lots 10 to 25c. extra.

Apollo Brand.

24 guage, American	3 85
26 " "	4 10
28 " "	4 55
10¾ oz.	4 85

25c. less for 1,000 lb. lots.

IRON AND STEEL.

	Montreal.	Toronto.
Middlesboro, No. 1 pig iron	21 50	24 50
Middlesboro, No. 3 pig iron	20 50	23 50
Summerlee, "	25 50	26 50
" special "	24 50	....
" soft "	24 00	....
Carron	26 00	....
Carron Special	24 50	....
Carron Soft	24 00	....
Clarence, No. 3	21 50	23 50
Glengarnock, No. 1	....	27 00
Midland, Londonderry and Hamilton, off the market but quoted nominally at	....	26 00
Radnor, charcoal iron	32 00	34 00
Lowmoor iron		6 50
Angles		2 50
Common bar, per 100 lb.	2 20	2 30
Forged iron "	2 45	..
Refined " "	2 60	2 70
Horseshoe iron "	2 60	2 70
Band iron, No. 10 x ⅜ in.	....	3 60
Sleigh shoe steel "	2 25	2 30
Iron finish steel		2 40
Reeled machinery steel		3 50
Tire steel	2 40	2 50
Best sheet steel	...	0 12
Mining cast steel	....	0 08
Warranted cast steel	....	0 14
Annealed cast steel	....	0 15
High speed		0 60
B.P.L. tool steel	10½	0 11

INGOT TIN.

Lamb and Flag and Straits—56 and 28-lb. ingots, 100 lb.	$41 00	$42 00

TINPLATES.

Charcoal Plates—Bright

M.L.S., Famous (equal Bradley)		Per box.
I C, 14 x 20 base	....	$6 50
IX, 14 x 20 "	....	8 00
IXX, 14 x 20 base	....	9 50

Raven and Vulture Grades—		
I C, 14 x 20 base	....	5 00
I X "	....	6 00
I X X "	....	7 00
I X X X "	....	8 00

"Dominion Crown Best"—Double Coated, Tissued.		Per box.
I C, 14 x 20 base	5 50	5 75
I X, 14 x 20 "	6 50	6 75
IXX x 20 "	7 50	7 75

"Allaway's Best"—Standard Quality.		
I C, 14 x 20 base	4 65	5 00
I X, 14 x 20 "	5 40	5 75
I XX, 14 x 20 "	6 15	6 50

Bright Cokes.

Bessemer Steel—		
I C, 14 x 20 base	4 25	4 35
20x28, double box	8 50	8 70

Charcoal Plates—Terne.

Dean or J. G. Grade—		
I.C., 20x28, 112 sheets	7 25	8 00
IX., Terne Tin	...	9 50

Charcoal Tin Boiler Plates.

Cookley Grade—
X X, 14x56, 50 sheet bxs. }
" 14x60, " }
" 14x65, " } 7 50

Tinned Sheets.

72x30 up to 24 gauge	....	8 50
" " 26 "	....	9 00

LEAD.

Imported Pig, per 100 lb.	5 25	5 35
Bar, "	5 75	6 00
Sheets, 2½ lb. sq. ft., by roll	6 75	6 50
Sheets, 3 to 6 lb. "	6 50	6 25

Cut sheets [illegible] per lb., extra.

SHEET ZINC.

5-cwt. casks	7 50	7 75
Part casks	7 75	8 00

ZINC SPELTER.

Foreign, per 100 lb	6 00	6 25
Domestic	6 50	6 75

COLD ROLLED SHAFTING.

9-16 to 11-16 inch	....	0 06
¾ to 1 7-16 "	....	0 05½
1 7-16 to 3 "	....	0 05

30 per cent.

OLD MATERIAL.

Dealers' buying prices:

	Montreal	Toronto
Heavy copper and wire, lb.	0 14	0 14
Light copper	0 12	0 12
Heavy red brass	0 12	0 12
" yellow brass	0 09	0 10½
Light brass	0 07	0 07½
Tea lead	0 03¾	0 03½
Heavy lead	0 04	0 04
Scrap zinc	0 03¾	0 03¾
No. 1 wrought iron	14 50	11 50
" 2 " "	6 00	6 00
Machinery cast scrap	17 00	16 50
Stove plate	12 00	12 00
Malleable and steel	8 00	10 00
Old rubbers	0 10½	0 10
Country mixed rags, 100 lbs.	1 00	1 25

PLUMBING AND HEATING

BRASS GOODS, VALVES, ETC.

Standard Compression work, 57½ per cent.
Cushion work, discount 40 per cent.
Fuller work, 70 per cent.
Flatway stop and stop and waste cocks, 60 per cent.; roundway, 55 per cent.
J.M.T. Globe, Angle and Check Valves, 45; Standard, 55 per cent.
Kerr standard globes, angles and checks, special, 42½ per cent.; standard, 47½ p.c.
Kerr Jenkins' disc, copper-alloy disc and heavy standard valves, 40 per cent.
Kerr steam radiator valves, 60 p.c., and quick-opening hot-water radiator valves, 60 p.c.
Kerr brass, Weber's straightway valves, 40; straightway valves, I B B.M., 60.
J. M. T. Radiator Valves 50; Standard, 60; Patent Quick-Opening Valves, 65 p.c.
Jenkins' Valves—Quotations on application to Jenkins' Bros., Montreal.

No. 1 compression bath cock......net	2 00
No. 4 " " "	1 90
No 7 Fuller's "	2 25
No. 4½ " "	2 35
Patent Compression Cushion, basin cock, hot and cold, per doz.	$16.20
Patent Compression Cushion, bath cock, No. 2208	2 25

Square head brass cocks, 50; iron, 60 p.c.
Thompson Smoke-test Machine 25.00

BOILERS—COPPER RANGE.

Copper, 30 gallon, $33, 15 per cent.

BOILERS—GALVANIZED IRON RANGE.

30-gallon, Standard, $5; Extra heavy, $7.75

BATH TUBS.

Steel clad copper lined, 15 per cent.

CAST IRON SINKS.

16x24, $1; 18x30, $1; 18x36, $1.30.

ENAMELED BATHS, ETC.

List issued by the Standard Ideal Company Jan. 3, 1907, shows an advance of 10 per cent. over previous quotations.

ENAMELED CLOSETS AND URINALS

Discount 15 per cent.

HEATING APPARATUS.

Stoves and Ranges—45 to 50 per cent.
Furnaces—45 per cent.
Registers—70 per cent.
Hot Water Boilers—50 per cent.
Hot Water Radiators—50 to 55 p.c.
Steam Radiators—50 to 55 per cent.
Wall Radiators and Specials—50 to 55 p.c.

LEAD PIPE

Lead Pipe, 5 p.c. off
Lead waste, 5 p.c. off.
Caulking lead, 5½c. per pound.
Traps and bends, 40 per cent.

IRON PIPE.

Size (per 100 ft.)	Black.	Size	Galvanized
⅛ inch	2 35	⅛ inch	3 20
¼ "	2 35	¼ "	3 20
⅜ "	2 90	⅜ "	3 75
½ "	3 90	½ "	5 00
1 "	5 60	1 "	7 25
1¼ "	7 65	1¼ "	9 90
1½ "	9 20	1½ "	11 90
2 "	12 25	2 "	15 80
2½ "	20 10	2½ "	26 00
3 "	26 75	3 "	34 00
3½ "	34 25	3½ "	42 75
4 "	39 00	4 "	48 60

Malleable Fittings—Canadian discount 30 per cent.; American discount 25 per cent.
Cast Iron Fittings 57½; Standard bushings 57½; headers, 57½; flanged unions 57½, malleable bushings 55; nipples, 70 and 10; malleable lipped unions, 55 and 5 p.c.

SOIL PIPE AND FITTINGS

Medium and Extra heavy pipe and fittings, up to 6 inch, 60 and 10 to 70 per cent.
7 and 8-in. pipe, 40 and 5 per cent.
Light pipe, 50 p.c.; fittings, 50 and 10 p.c.

OAKUM.

Plumbers ....per 100 lb.	4 50	5 00

STOCKS AND DIES.

American discount 25 per cent.

SOLDERING IRONS.

½-lb. to 1½ .... per lb.	0 45½	0 48
2-lb. or over "	0 42½	0 46

SOLDER.

Per lb.	Montreal	Toronto
Bar, half-and-half, guaranteed	0 25	0 26
Wiping	0 22	0 23

PAINTS, OILS AND GLASS.

BRUSHES

Paint and household, 70 per cent.

CHEMICALS.

	In casks per lb.
Sulphate of copper (bluestone or blue vitrol)	0 09
Litharge, ground	0 06
" flaked	0 06½
Green copperas (green vitrol)	0 01
Sugar of lead	0 08
Lump olive	0 01½

COLORS IN OIL.

Venetian red, 1-lb. tins	pure.	0 08
Chrome yellow	"	0 15
Golden ochre	"	0 10
French "	"	0 08
Marine black	"	0 04½
Chrome green	"	0 09
French permanent green	"	0 13
Signwriters' black	"	0 15

GLUE.

Domestic sheet 0 10 0 10½
French medal 0 12 0 12½

PARIS WHITE.

In bbls 0 90

PIGMENTS.

Orange mineral, casks 0 08
" " 100-lb. kegs 0 08½

PREPARED PAINTS.

Barn (in bbls.) 0 65 0 90
Sherwin-Williams paints, ... 1 55
Canada Paint Co.'s pure 1 40
Standard P. & V. Co.'s "New Era." 1 30
Benj. Moore Co.'s "Ark" B'd 1 25
" British Navy deck 1 50
Brandram-Henderson's "English" 1 45
Ramsay's paints, Pure, per gal. 1 30
" Thistle, " 1 10
Martin-Senour 100 p.c. pure, ... 1 55
Senour's Floor Paints 1 35
Jamieson's "Crown and Anchor" 1 20
Jamieson's floor enamel 1 50
P. D. Dods & Co's "Island City" .. 1 25
Sanderson Pearcy's, pure 1 20
Robertson's pure paints 1 20

PUTTY.

Bulk in bbls 1 60
Bladders in bbls 1 85
25-lb. tins 1 90
Bladders in bulk or tins less than 100 lb. 2 00
Bulk in 100-lb. irons 1 80

SHINGLE STAINS.

In 5 gallon lots 0 85 0 90

SHELLAC.

White, bleached 2 65
Fine orange 2 60
Medium orange 2 55

TURPENTINE AND OIL.

Prime white petroleum per gal 0 13½
Water white " 0 15
Pratt's astral " 0 18½
Castor oil 0 08 0 10
Motor Gasoline per gal, single bbls. 0 21 0 23½
Benzine, per gal single bbls 0 17 0 17½
Turpentine, single barrels 0 77 0 79
Linseed Oil, " raw .. 0 60 0 64
" " boiled 0 63 0 67

WHITE LEAD GROUND IN OIL. Per 100 lbs

Canadian pure 7 15
No. 1 Canadian 6 80
Munro's Select Flake White 7 65
Elephant and Decorators' Pure 7 65
Monarch 7 40
Standard Decorator's 7 15
Essex Genuine 6 80
Brandram's B. B. Genuine 8 70
"Anchor," pure 7 40
Ramsay's Pure Lead 7 25
Ramsay's Exterior 6 65
"Crown and Anchor," pure 7 25
P. D. Dods "Island City" pure lead 7 25
Sanderson Pearcy's 7 40
Robertson's C.P., lead 7 20

RED DRY LEAD.

Genuine, 560 lb. casks, per cwt 6 25
Genuine, 100 lb. kegs, " 6 75
No. 1, 560 lb. casks, per cwt 6 00
No. 1, 100 lb. kegs, per cwt 6 50

WINDOW GLASS

Size United Inches.	Star	Double Diamond
Under 26	$4 25	$6 25
26 to 40	4 65	6 75
41 to 50	5 10	7 50
51 to 60	5 35	8 50
61 to 70	5 75	9 75
71 to 80	6 25	11 00
81 to 85	7 00	12 50
86 to 90	....	15 00
91 to 95	....	17 50
96 to 100	....	20 50
101 to 105	....	24 00
106 to 110	....	27 50

Discount—16-oz., 25 per cent.; 21-oz. 30 per cent. per 100 feet. Broken boxes 50 per cent.

WHITING.

Plain, in bbls 0 70
Gilders bolted in bands 0 90

WHITE DRY ZINC.

Extra Red Seal, V.M. 0 07½ 0 08

WHITE ZINC IN OIL.

Pure, in 25-lb. irons 0 08½
No. 1, " " 0 07
No. 2, " " 0 05½

VARNISHES.

Per gal. cans.

Carriage, No. 1 .... 1 50
Pale durable body 3 50
" hard rubbing 3
Finest elastic gearing 3
Elastic oak 1
Furniture, polishing 2
Furniture, extra 1
" " No. 1 0
" " union 0
Light oil finish 1
Gold size japan 1
Brown japan 0
No 1 brown japan 0
Baking black japan 1
No. 1 black japan 0
Benzine black japan 0
Crystal Damar 2
No. 1 " 2
Pure asphaltum 1
Oilcloth 1
Lightning dryer 0
Elastilite varnish, 1 gal. can, each .. 2
Granitine floor varnish, per gal 2
Maple Leaf coach enamels; size 1, 1
Sherwin-Williams' kopal varnish, gal., 2
Canada Paint Co's sun varnish 2
"Kyanize" Interior Finish 2
"Flint-Lac," coach 1
B H. Co's "Gold Medal," in cases 2 00
Jamieson's Copaline, per gal. .... 2 [illegible]

BUILDERS' HARDWARE.

BELLS.

Brass hand bells, 60 per cent.
Nickel, 55 per cent.
Gongs, Sargeant's door bells .. 5 50 8 00
American, house bells, per lb. 0 35 0 40
Peterboro' door bells, 37½ and 10 off new list.

BUILDING PAPER, ETC.

Tarred Felt, per 100 lb. 2 25
Ready roofing, 2-ply, not under 45 lb. per roll 1 00
Ready roofing, 3-ply, not under 65 lb., per roll 1 25
Carpet Felt .... per ton 60 00
Heavy Straw Sheathing .... per ton 40 00
Dry Surprise 0 45
Dry Sheathing .... per roll, 400 sq. ft. 0 40
Tar " .... " 400 " 0 50
Dry Fibre .... " 400 " 0 55
Tarred Fibre .... " 400 " 0 65
O. K. & I. X. L. .... " 400 " 0 70
Resin-sized .... " 400 " 0 45
Oiled Sheathing .... " 600 " 1 00
Oiled " .... " 400 " 0 70
Roof Coating, in barrels .... per gal. 0 17
Roof " small packages " 0 25
Refined Tar .... per barrel 5 00
Coal Tar .... " 4 00
Coal Tar, less than barrels .... per gal. 0 15
Roofing Pitch .... per 100 lb. 0 80 0 90
Slater's felt .... per roll 0 70
Heavy Straw Sheathing f. o. b. St. John and Halifax 42 50

BUTTS.

Wrought Brass, net revised list.
Wrought Iron, 65 and 10 per cent.
Cast Iron Loose Pin, 60 per cent.
Wrought Steel Fast Joint and Loose Pin, 70 per cent.

CEMENT AND FIREBRICK.

Canadian Portland 2 00 2 10
Belgium 1 60 1 90
White Bros. English 1 80 2 05
"Lafarge" cement in wood 3 40
"Lehigh" cement, in wood 2 54
"Lehigh" cement, cotton sacks. 2 39
"Lehigh" cement, paper sacks 2 31
Fire brick, Scotch, per 1,000 .... 27 00 30 00
" English 17 00 21 00
" American, low 23 0 25 00
" " high 27 50 35 00
Fire clay (Scotch), net ton 4 95

Paving Blocks per 1,000.

Blue metallic, 9"x4½"x3", ex wharf .. 35 00
Stable pavers, 12"x6"x2", ex wharf .. 50 00
Stable pavers, 9"x4½"x3", ex wharf .. 36 00

DOOR SETS.

Peterboro, 37½ and 10 per cent.

DOOR SPRINGS.

Torrey's Rod .... per doz. 1 75
Coil, 9 to 11 in. .... " 0 95 1 65
English .... " 2 00 4 00
Chicago and Reliance Coil 25 per cent.

ESCUTCHEONS.

Discount 50 and 10 per cent., new list
Peterboro, 37½ and 10 per cent.

ESCUTCHEON PINS.

Iron, discount 40 per cent.
Brass, 45 per cent.

HINGES.

Blind, discount 50 per cent.
Heavy T and strap, 4-in., 100 lb. net .. 7 25
" " 5-in., " 7 00½
" " 6-in., " 6 75½
" " 8-in., " 6 50½
" " 10-in. and larger .. 6 25
Light T and strap, discount 65 p.c.
Screw hook and hinge—
under 12 in. .... per 100 lb. 4 75
over 12 in. .... " 3 75
Crate hinges and back flaps, 65 and 5 p. c.
Chest hinges and hinge hasps, 65 p.c.

SPRING HINGES.

Spring, per gro., No. 5, $17.50 No. 10, $18; No. 20, $10 80; No. 120, $20; No. 51, $10; No. 50, $27.50
Chicago Spring Butts and Blanks 12½ per cent.
Triple End Spring Butts, 30 and 10 per cent.
Chicago Floor Hinges, 37½ and 5 off.
Garden City Fire House Hinges, 12½ p.c.
"Chief" floor hinge, 50 p.c.

CAST IRON HOOKS.

Bird cage .... per doz. 0 50 1 10
Clothes line, No. 61 .. " 0 00 0 70
Harness .... " 0 60 12 00
Hat and coat .... per gro. 1 10 10 00
Chandelier .... per doz. 0 50 1 00
Wrought hooks and staples—
¼ x 5 .... per gross 2 65
5-16 x 5 .... " 3 30
Bright wire hooks, 60 p c.
Bright steel gate hooks and staples, 40 p.c.
Cresent hat and coat wire, 60 per cent.
Screw, bright wire, 65 per cent.

KNOBS.

Door, japanned and N.P., doz 1 50 2 50
Bronze, Berlin .... per doz. 2 75 3 25
Bronze, Genuine .... " 6 00 9 00
Shutter, porcelain, F. & L. screw .... per gross 1 30 2 00
White door knobs .... per doz. 2 00
Peterboro knobs, 37½ and 10 per cent.
Porcelain, mineral and jet knobs, net list.

KEYS.

Lock, Canadian 40 to 40 and 10 per cent.

LOCKS.

Peterboro, 37½ and 10 per cent.
Russell & Erwin, steel rim $2 50 per doz.
Eagle cabinet locks, discount 30 per cent
American padlocks, all steel, 10 to 15 per cent.; all brass or bronze, 10 to 25 per cent.

SAND AND EMERY PAPER.

B. & A. sand, discount, 35 per cent
Emery, discount 35 per cent.
Garnet (Rurton's). 5 to 10 per cent. advance

SASH WEIGHTS.

Sectional .... per 100 lb. 2 00 2 25
Solid .... " 1 50 1 75

SASH CORD.

Per lb. 0 31

BLIND AND BED STAPLES.

All sizes per lb. 0 07½ 0 10

WROUGHT STAPLES.

Galvanized 2 75
Plain 2 50
Coopers', discount 45 per cent.
Poultry netting staples, discount 40 per cent.
Bright spear point, 75 per cent. discount.

TOOLS AND HANDLES.

ADZES.

Discount 22½ per cent.

AUGERS.

Gilmour's, discount 60 per cent. off list.

AXES.

Single bit, per doz. 6 00 9 00
Double bit, " 10 00 11 00
Bench Axes, 40 per cent.
Broad Axes, 25 per cent.
Hunters' Axes 5 50 6 00
Boys' Axes 6 25 7 00
Splitting Axes 7 00 12 00
Handled Axes 7 00 9 00
Red Ridge, boys', handled 5 75
" hunters 5 25

BITS.

Irwin's auger, discount 47½ per cent.
Gilmour's auger, discount 60 per cent.
Rockford auger, discount 50 and 10 per cent.
Jennings' Gen. auger, net list.
Gilmour's car, 47½ per cent.
Clark's expansive, 40 per cent.
Clark's gimlet, per doz 0 65
Diamond, Shell, per doz. 1 00
Nail and Spike, per gross 2 25

BUTCHERS CLEAVERS

German .... per doz. 7 00 9 00
American .... " 12 00 18 00

CHALK.

Carpenters' Colored, per gross 0 45 0 75
White lump .... per cwt. 0 60 0 65

CHISELS.

Warnock's, discount 70 and 5 per cent.
P. S. & W. Extra, discount, 70 per cent.

CROSSCUT SAW HANDLES.

S. & D., No. 3 .... per pair 0 13
S. & D., " 5 .... " 0 11½
S. & D., " 6 .... " 0 18
Boynton pattern .... " 0 20

CROWBARS.

3½c. to 4c. per lb.

DRAW KNIVES.

Coach and Wagon, discount 75 and 5 per cent.
Carpenters' discount 75 per cent.

DRILLS.

Millar's Falls, hand and breast. net list.
North Bros., each set, 50c.

DRILL BITS.

Morse, discount 37½ to 40 per cent.
Standard, discount 50 and 5 to 55 per cent.

FILES AND RASPS.

Great Western .... 75 per cent.
Arcade .... 75 "
Kearney & Foot .... 75 "
Disston's .... 75 "
American .... 75 "
J. Barton Smith .... 75 "
McClellan .... 75 "
Eagle .... 75 "
Nicholson .... 66⅔ "
Globe .... 75 "
Black Diamond, 60, 10 and 5 p.c.
Jowitt's, English list, 27½ per cent.

GAUGES.

Stanley's discount 50 to 60 per cent.
Winn's, Nos. 26 to 33 .... each 1 65 2 40

HANDLES.

Second growth ash fork, hoe, rake and shovel handles, 40 p c.
Extra ash fork, hoe, rake and shovel handles, 45 p c.
No. 1 and 2 ash fork, hoe, rake and shovel handles, 50 p.c.
White ash whiffletrees and neckyokes, 35 p.c.
All other ash goods, 40 p c.
All hickory, maple and oak goods, excepting carriage and express whiffletrees, 40 p.c.
Hickory, maple, oak carriage and express whiffletrees, 45 p.c.

HAMMERS.

Maydole's, discount 5 to 10 per cent.
Canadian, discount 25 to 27½ per cent.
Magnetic tack .... per doz. 1 10 1 20
Canadian sledge .... per lb. 0 07 0 08½
Canadian ball pean, per lb. .. 0 22 0 25

HATCHETS.

Canadian, discount 40 to 42½ per cent.
Shingle, Red Ridge 1, per doz. 4 40
" " 2, " 4 80
Barrel Underhill 5 05

HOES.

Mortar, 50 and 10 per cent.

MALLETS.

Tinsmiths' ....per doz. 1 25 1 50
Carpenters', hickory, " 1 25 3 75
Lignum Vitae 3 85 5 00
Caulking, each 0 60 2 00

MATTOCKS.

Canadian .... per doz. 5 50 6 00

MEAT CUTTERS.

German, 15 per cent.
American discount, 33⅓ per cent.

PICKS.

Per dozen 6 00 9 00

PLANES.

Wood bench, Canadian, 40, American, 25 p c.
Wood, fancy, 37½ to 40 per cent.
Stanley planes, $1.55 to $3 60, net list prices.

PLANE IRONS.

English .... per doz. 2 00 5 00
Stanley, 2⅝ inch, single 24c., double 39c.

PLIERS AND NIPPERS.

Button's genuine, 37½ to 40 per cent.
Button's imitation....per doz. 5 00 9 00
Berg's wire fencing 1 72 5 50

PUNCHES.

Saddler's ....per doz. 1 00 1 85
Conductor's .... " 3 00 15 00
Tinners', solid..........per set 0 72
" hollow......per inch 1 00

RIVET SETS.

Canadian, discount 35 to 37½ per cent.

RULES.

Boxwood, discount 70 per cent.
Ivory, discount 20 to 25 per cent.

SAWS.

Atkins, hand and crosscut, 25 per cent.
Disston's Hand, discount 12½ per cent
Disston's Crosscut ....per foot 0 35 0 55
Hack, complete..........each 0 75 2 75
" frame only..........each 0 50 1 25

S. & D. solid tooth circular shingle, concave and band, 50 per cent; mill and ice, drag, 30 per cent; cross-cut, 35 per cent; hand saws, butcher, 35 per cent.; buck, New Century. $6 25, buck No. 1 Maple Leaf, $5 25; buck, Happy Medium, $4.25; buck, Watch Spring, $1 25, buck, common frame, $4.00.

Spear & Jackson's saws—Hand or rip, 26 in., $12 75; 28 in., $11 25; panel, 18 in., $8 25; 20 in., $9; tenon, 10 in., $9.90; 12 in., $10 90; 14 in., $11.50.

SAW SETS.

Lincoln and Whiting 4 75
Hand Sets, Perfect 4 00
X-Cut Sets, " 7 50
Maple Leaf and Premiums saw sets, 40 off.
S. & D. saw swages, 40 off.

SCREW DRIVERS

Sargent's ....per doz. 0 65 1 00
North Bros., No. 30 . per doz. 16 80

SHOVELS AND SPADES

Canadian, discount 45 per cent.

SQUARES.

Iron, discount 20 per cent.
Steel, discount 65 and 10 per cent.
Try and Bevel, discount 50 to 52½ per cent

TAPE LINES.

English, ass skin .....per doz. . 2 75 5 00
English, Patent Leather 5 50 9 75
Chesterman's ....each 0 90 2 85
" steel..........each 0 80 8 00
Berg's, each 0 75 2 50

TROWELS.

Disston's, discount 10 per cent.
S. & D., discount 35 per cent.
Berg's, brick, 924x11 .... 4 00
" pointing, 924x5.......... .... 2 10

—

FARM AND GARDEN GOODS

BELLS.

American cow bells, 63⅓ per cent.
Canadian, discount 45 and 50 per cent.
American, farm bells, each .. 1 35 3 00

BULL RINGS.

Copper, $2.00 for 2½-inch

CATTLE LEADERS.

Nos. 32 and 33 .. ...per gross 7 50 8 50

BARN DOOR HANGERS

doz. pairs.

Stearns wood track 4 50 6 00
Zenith 9 00
Atlas, steel covered 5 00 6 00
Perfect 8 00 11 00
New Milo, flexible 6 50
Steel, track, 1 x 3-16 in (100 ft) 3 25
" " 1¼ x 3-16 in (100 ft) 4 75
Double strap hangers, doz. sets 6 40
Standard jointed hangers, " 6 40
Steel King hangers " 6 25
Storm King and safety hangers 7 00
" " rail.......... .... 4 25
Chicago Friction, Oscillating and Big Twin Hangers, 5 per cent.

HARVEST TOOLS

50 and 10 per cent.
S. & D. lawn rakes, Dunn's, 40 off.
" sidewalk and stable scrapers, 40 off.

HAY KNIVES.

Net list.

HEAD HALTERS.

Jute Rope, ⅝-inch....per gross 9 00
" " ⅝ " .... " 10 00
" " ⅝ " .... " 12 00
Leather, 1-inch ......per doz. 4 00
Leather, 1½ " ...... " 5 20
Web.......... " 2 45

HOES.

Garden, 50 and 10 per cent
Planter.......... per doz. 4 00 4 50

LAWN MOWERS

Low wheel, 12, 14 and 16-inch $2 30
9-inch wheel, 12-inch 2 85
" " 14 " .... 3 ℓC
" " 16 " 3 12½
High wheel, 12 " 4 05
" " 14 " 4 50
" " 16 " 4 75

SCYTHES

Per doz. net.......... 8 25 9 25

SCYTHE SNATHS

Canadian, discount 40 per cent.

SNAPS.

Harness, German, discount 25 per cent.
Lock, Andrews' 4 50 11 00

STABLE FITTINGS

Warden King, 35 per cent
Dennis Wire & Iron Co., 33⅓ p c.

WOOD HAY RAKES.

40 and 10 per cent

—

HEAVY GOODS, NAILS, ETC.

ANVILS.

Wright's, 80-lb. and over 0 10½
Hay Budden, 80-lb. and over 0 09½
Brook's, 80-lb. and over 0 11¼
Taylor-Forbes, prospectors 0 05
Columbia Hardware Co., per lb. 0 09½

VISES

Wright's.......... .... 0 13½
Berg's, per lb 0 12½
Brook's.......... .... 0 12½
Pipe Vise, Hinge, No. 1 .... 3 50
" " " No. 2 .... 5 50
Saw Vise 4 50 5 00
Blacksmiths' (discount) 60 per cent.
" parallel (discount) 45 per cent.

BOLTS AND NUTS

Carriage Bolts, common ($1 list Per cent.
" " ⅜ and smaller .. 60, 10 and 10
" " 7-16 and up .... 55 and 5
" " Norway Iron ($3 list) 50
Machine Bolts, ⅜ and less .... 60 and 10
Machine Bolts, 7-16 and up.... 55 and 5
Plough Bolts .... 55 and 10
Blank Bolts 55
Bolt Ends 55
Sleigh Shoe Bolts, ⅜ and less .. 60 and 10
" " " 7-16 and larger 50 and 5
Coach Screws, conepoint ...... 70 and 5
Nuts, square, all sizes, 4c. per cent. off
Nuts, hexagon, all sizes, 4½c. per cent off
Stove Rods per lb., 5½ to 6c.
Stove Bolts, 75 per cent.

CHAIN.

Proof coil, per 100 lb., ¼ in., $6 00; 5-16 in., $4.85; ⅜ in., $4.25; 7-16 in., $4.00; ½ in., $3 75; 9-16 in., $3 70; ⅝ in., $3 65; ¾ in., $3.60; ⅞ in., $3.45; 1 in., $3.40.

Halter, kennel and post chains, 40 to 40 and 5 per cent; Cow ties, 40 per cent.; Tie out chains 65 per cent.; Stall fixtures, 35 per cent; Trace chain, 45 per cent.; Jack chain iron, 50 per cent.; Jack chain, brass, 50 per cent.

HORSE NAILS.

M.R.M. cold forged process, list May 15, 1907, 50 and 5 per cent.
"O" brand, 57½ per cent. off list.
Capewell brand, quotations on application

HORSESHOES.

M R.M brand: iron, light and medium No. 1 and smaller, $3 90; No. 2 and larger $3 65; snow pattern No. 1 and smaller $4 15, No. 2 and larger, $3 90; "X L." new light steel, No 1 and smaller, $4 25; No 2 and larger, $4; "X L" featherweight steel, No. 0 to 4, $5 60; Special countersunk steel No 0 to 4, $6 85 pkg; toe-weight, all sizes, $6.85. F o.b Montreal. Extras for packing.

Belleville brand: No 0 and 1, light and medium iron, $3 90; snow, $4 15; light steel, $4 25; No. 2 and larger, light and medium iron, $3 65; snow, $3 90; light steel, $4. F o b. Belleville. Two per cent., 30 days.

Toecalks—Standard No. 1 and smaller, $1.50; No 2 and larger, $1 25 Blunt No. 1 and smaller, $1.75; No. 2 and larger, $1 50 per box Sharp Put up in 25 lb. bxs.

HORSE WEIGHTS.

Taylor-Forbes, 4½c. per lb.

NAILS.	Cut.	Wire.
2d	4 00	3 50
3d	3 15	3 20
4 and 5d	2 90	2 90
6 and 7d	2 80	2 80
8 and 9d	2 65	2 65
10 and 12d	2 60	2 60
16 and 20d	2 55	2 55
30, 40, 50 and 60d (base)	2 50	2 50

F.o.b. Montreal. Cut nails, Toronto 20c. higher.
Miscellaneous wire nails, discount 75 per cent
Coopers' nails, discount 40 per cent.

PRESSED SPIKES.

Pressed spikes, ⅝ diameter, per 100 lbs $3 15

RIVETS AND BURRS.

Iron Rivets, black and tinned, 60, 10 and 10
Iron Burrs, discount 60 and 10 and 10 p.c.
Copper Rivets, usual proportion burrs, 15 p c.
Copper Burrs only, net list.
Extras on Coppered Rivets, ½-lb. packages 1c. per lb; ¼-lb. packages 2c. lb.
Tinned Rivets, net extra, 4c. per lb.

SCREWS.

Wood, F. H., bright and steel, 85 and 10 p.c.
" R. H., bright, 80 and 10 per cent.
" F. H., brass, 75 and 10 per cent.
" R. H., " 70 and 10 per cent.
" F. H., bronze, 70 and 10 per cent.
" R. H., " 65 and 10 per cent.
Drive Screws, dis. 87½ per cent.
Bench, wood ....per doz. 3 25 4 00
" iron .... " 4 25 5 00
Set, case hardened, dis. 60 per cent.
Square Cap, dis. 50 and 5 per cent.
Hexagon Cap, dis. 45 per cent.

MACHINE SCREWS.

Flat head, iron and brass, 35 per cent.
Felister head, iron, discount 30 per cent
" " brass, discount 25 per cent

TACKS, BRADS, ETC.

Carpet tacks, blued 75 p.c.; tinned, 80 and 10; (in kegs), 40; cut tacks, blued, in dozens only, 75; ½ weights, 60; Swedes cut tacks, blued and tinned, bulk, 75 dozens, 75; Swedes, upholsterers', bulk, 85 and 12½; brush, blued and tinned, bulk, 70; Swedes, gimp, blued, tinned and japanned, 75 and 12½; zinc tacks, 35; leather carpet tacks, 40; copper tacks, 25; copper nails 30; trunk nails, black, 65; trunk nails, tinned and blued, 65; clout nails, blued and tinned 65; chair nails, 35; patent brads, 40; fine finishing, 40; lining tacks, in papers, 10; lining tacks, in bulk, 15; lining tacks, solid heads, in bulk, 75; saddle nails, in papers, 10; saddle nails, in bulk, 15; tufting buttons, 22 line in dozens only, 60; zinc glaziers' points, 5; double pointed tacks, papers, 90 and 10; double pointed tacks, bulk, 40; clinch and duck rivets, 45; cheese box tacks, 85 and 5 trunk tacks, 80 and 10.

WROUGHT IRON WASHERS.

Canadian make, discount 40 per cent.

—

SPORTING GOODS.

CARTRIDGES.

"Dominion" Rim Fire Cartridges and C.B. caps, 50 and 7½ per cent.; Rim Fire B B. Round Caps, 60 and 2½ per cent.; Centre Fire, Pistol and Rifle Cartridges, 30 p.c; Centre Fire Sporting and Military Cartridges, 20 and 5 p.c.; Rim Fire, Shot Cartridges, 50 and 7½ p.c.; Centre Fire, Shot Cartridges, 30 p c.; Primers, 25 p c.

LOADED SHELLS.

"Crown" Black Powder, 15 and 5 p c.; "Sovereign" Empire Bulk Smokeless Powder, 30 and 5 p.c.; "Regal" Ballistite Dense smokeless Powder, 30 and 5 p.c.; "Imperial" Empire or Ballistite Powder, 30 and 10 p.c.

EMPTY SHELLS.

Paper Shells, 25 and 5; Brass Shells, 55 and 5 p.c.

Wads. per lb.

Best thick brown or grey felt wads, in ¼-lb. bags $0 70
Best thick white card wads, in boxes of 500 each, 12 and smaller gauges 29
Best thick white card wads, in boxes of 500 each, 10 gauge 0 35
Thin card wads, in boxes of 1,000 each, 12 and smaller gauges 0 20
Thin card wads, in boxes of 1,000 each, 10 gauge 0 25
Chemically prepared black edge grey cloth wads, in boxes of 250 each— Per M
11 and smaller gauge 0 6C
9 and 10 gauges 0 70
and 8 " 0 90
5 and 6 " 1 1ₗ
Superior chemically prepared pink edge, best white cloth wads in boxes of 250 each—
11 and smaller gauge 1 15
9 and 10 gauges 1 40
7 and 8 " .... 1 6?
5 and 9 " 1 90

SHOT

Ordinary drop shot, AAA to dust $7.50 per 100 lbs. Discount 5 per cent; cash discount, 2 per cent, 30 days; net extras as follows subject to cash discount only; Chilled, 40 c; buck and seal, 80c.; no 28 ball, $1 20 per 100 lbs.; bags less than 25 lbs., ½c. per lb.; F.o.b. Montreal, Toronto, Hamilton, London, St. John and Halifax, and freight equalized thereon.

TRAPS (steel.)

Game, Newhouse, discount 30 and 10 per cent.
Game, Hawley & Norton, 50, 10 & 5 per cent
Game, Victor, 70 per cent.
Game, Oneida Jump (B. & L.) 40 & 2½ p.
Game, steel, 60 and 5 per cent.

SKATES.

Skates, discount 37½ per cent.
Empire hockey sticks, per doz .. 3 00

CUTLERY AND SILVER-WARE.

RAZORS.

	per doz.	
Elliot's	4 00	18 00
Boker's	7 50	11 00
" King Cutter	13 50	18 50
Wade & Butcher's	3 60	10 00
Lewis Bros.' "Klean Kutter"	8 50	10 50
Henckel's	7 50	20 00
Berg's	7 50	20 00

Clauss Razors and Strops, 50 and 10 per cent

KNIVES.

Farriers-Stacey Bros., doz 3 50

PLATED GOODS

Hollowware, 40 per cent. discount.
Flatware, staples, 40 and 10, fancy, 40 and 5.
Hutton's "Cross Arrow" flatware, 47½; "Singalese" and "Alaska" Nevada silver flatware, 42 p c.

SHEARS.

Clauss, nickel, discount 60 per cent.
Clauss, Japan, discount 67½ per cent.
Clauss, tailors, discount 40 per cent.
Seymour's, discount 50 and 10 per cent.
Berg's 6 00 12 00

HOUSE FURNISHINGS.

APPLE PARERS.

Hudson, per doz., net 5 75

BIRD CAGES.

Brass and Japanned, 40 and 10 p. c.

COPPER AND NICKEL WARE.

Copper boilers, kettles, teapots, etc. 30 p c.
Copper pitts, 20 per cent.

KITCHEN ENAMELED WARE.

White ware, 75 per cent.
London and Princess, 50 per cent.
Canada, Diamond, Premier, 50 and 10 p c.
Pearl, Imperial, Crescent and granite steel. 50 and 10 per cent.
Premier steel ware, 40 per cent.
Star decorated steel and white, 25 per cent.
Japanned ware, discount 45 per cent.
Hollow ware, tinned cast, 35 per cent. off.

KITCHEN SUNDRIES.

Can openers, per doz.	0 40	0 75
Mincing knives per doz	0 50	0 80
Duplex mouse traps, per doz.	....	0 65
Potato mashers, wire, per doz.	0 60	0 70
" wood "	0 50	0 60
Vegetable slicers, per doz	....	2 25
Universal meat chopper No. 1	....	1 15
Enterprise chopper, each	...	1 30

Spiders and fry pans, 50 per cent.

Star Al chopper 5 to 32	1 35	4 10
" " 100 to 103	1 35	2 00
Kitchen hooks, bright	....	0 60

LAMP WICKS.

Discount, 60 per cent.

LEMON SQUEEZERS.

Porcelain lined	per doz.	2 20	5 60
Galvanized	"	1 87	3 85
King, wood	"	2 75	2 90
King, glass	"	4 00	4 50
All glass	"	0 50	0 90

METAL POLISH.

Tandem metal polish paste 6 00

PICTURE NAILS.

Porcelain head ......per gross	1 35	1 50
Brass head	0 40	1 00

Tin and gilt, picture wire, 75 per cent.

SAD IRONS.

Mrs. Potts, No. 55, polished....per set	0 90
" No. 50, nickle-plated, "	0 95
" handles, japaned, per gross	9 25
" " nickled, "	9 75
Common, plain	4 25
" plated	8 60
Asbestos, per set	1 50

TINWARE.

CONDUCTOR PIPE.

2-in. plain or corrugated, per 100 feet, $3.30; 3 in., $4 40; 4 in., $5 80; 5 in., $7.45; 6 in., $9.90.

FAUCETS.

Common, cork-lined, discount 35 per cent.

EAVETROUGHS.

10-inch ......per 100 ft. 3 30

FACTORY MILK CANS.

Discount off revised list, 35 per cent.
Milk can trimmings, discount 25 per cent.
Creamery Cans, 45 per cent

LANTERNS.

No. 2 or 4 Plain Cold Blast....per doz.	6 50
Lift Tubular and Hinge Plain, "	4 75
No 0, safety	4 00

Better quality at higher prices.
Japanning, 50c. per doz. extra.
Prism globes, per doz , $1 20.

OILERS.

Kemp's Tornado and McClary's Model galvanized oil can, with pump, 5 gallon, per dozen 10 92
Davidson oilers, discount 40 per cent.
Zinc and tin, discount 50 per cent.
Coppered oilers, 20 per cent. off.
Brass oilers, 50 per cent. off.
Malleable, discount 25 per cent

PAILS (GALVANIZED).

Dufferin pattern pails, 45 per cent.
Flaring pattern, discount 45 per cent.
Galvanized washtubs 40 per cent.

PIECED WARE.

Discount 35 per cent off list, June, 1899.
10-qt. flaring sap buckets, discount 35 per cent.
6, 10 and 14-qt. flaring pails dis. 35 per cent.
Copper bottom tea kettles and boilers, 30 p.c.
Coal hods, 40 per cent.

STAMPED WARE.

Plain, 75 and 12½ per cent. off revised list.
Retinned, 72½ per cent. revised list.

SAP SPOUTS.

Bronzed iron with hooks ....per 1,000
Eureka tinned steel, hooks 8 00

STOVEPIPES.

5 and 6 inch, per 100 lengths	7 64	7 91
7 inch " "	....	8 18

Nestable, discount 40 per cent.

STOVEPIPE ELBOWS

5 and 6-inch, common........per doz.	1 32
7-inch "	1 48

Polished, 15c. per dozen extra.

THERMOMETERS.

Tin case and dairy, 75 to 75 and 10 per cent.

TINNERS' SNIPS.

Per doz. 3 00 15
Clauss, discount 35 per cent.

TINNERS' TRIMMINGS

Discount, 45 per cent.

WIRE.

ANNEALED CUT HAY BAILING WIRE.

No. 12 and 13, $4; No. 13½, $4 10. No. 14, $4.25; No. 15, $4 50; in lengths 6' to 11', 25 per cent.; other lengths 20c. per 100 lbs extra; if eye or loop on end add 25c. per 100 lbs. to the above.

BRIGHT WIRE GOODS

Discount 60 per cent.

CLOTHES LINE WIRE.

No. 7 wire solid line, No. 17. $4.90; No. 18, $3.00; No. 19, $2.70; [illegible] wire solid line, No. 17, $4.45; No. 18, $3.10; No. 19, $2 8[illegible]. All prices per 1000 ft measure; 6 strand, No 18, $2 60; No. 19, $2 90. F.o.b. Hamilton, Toronto, Montreal.

COILED SPRING WIRE

High Carbon, No. 9, $2 95; No. 11, $3 50; No. 12, $3 20.

COPPER AND BRASS WIRE.

Discount 37½ per cent.

FINE STEEL WIRE.

Discount 25 per cent List of extras In 100-lb. lots: No. 17, $5 — No. 18 $5.50 — No. 19, $6 — No. 20, $6.65 — No. 21 $7 — No. 22, $7.30 — No 23, $7.65 — No 24, $8 — No. 25, $9 — No. 26, $9 50 — No. 27, $10 — No. 28, $11 — No. 29, $12 — No. 30, $13 — No. 31, $14 — No. 32, $15 — No. 33, $16 — No. 34, $17. Extras net—tinned wire, Nos. 17-25 $2—Nos. 26-31, $4—Nos. 32-34, $6. Coppered, 75c.—oiling, 10c.—in 25-lb. bundles, 15c.—in 5 and 10-lb. bundles, 25c.—in 1-lb. hanks, 25c. —in ½-lb. hanks, 38c.—in ¼-lb. hanks, 50c. packed in casks or cases, 15c.—bagging or papering, 10c

FENCE STAPLES.

Bright ...... 2 80 Galvanized 32

HAY WIRE IN COILS.

No 13, $2.70; No. 14, $2 80; No. 15, $2.95; f.o.b., Montreal.

GALVANIZED WIRE.

Per 100 lb.—Nos. 4 and 5, $3.90 — Nos. 6, 7, 8, $3.35 — No. 9, $2.85 — No. 10, $3 40 — No. 11, $3.45—No. 12, $3.00 —No. 13, $3 10—No. 14, $3.95—No 15, $4.30 —No. 16, $4.30 from stock. Base sizes, Nos. 6 to 9, $2.35 f.o.b. Cleveland. Extras for cutting.

LIGHT STRAIGHTENED WIRE.

Gauge No.	Over 20 in.	10 to 20 in.	5 to 10 in.
0 to 5	$0 50	$0.75	$1 25
6 to 9	0.75	1.25	2 00
10 to 11	1 00	1.75	2 50
12 to 14	1.50	2.25	3 50
15 to 16	2.00	3.00	4.50

SMOOTH STEEL WIRE.

No. 0.9 gauge, $2.40; No. 10 gauge, 6c extra; No. 11 gauge, 12c extra; No. 12 gauge, 20c. extra; No. 13 gauge, 30c. extra; No 14 gauge, 40c. extra; No. 15 gauge, 55c. extra; No. 16 gauge, 70c. extra. Add 60c. for coppering and $2 for tinning.

Extra net per 100 lb.—Oiled wire 10c., spring wire $1.25, bright soft drawn 15c., charcoal (extra quality) $1.25, packed in casks or cases 15c., bagging and papering 10c., 50 and 100-lb. bundles 10c., in 25-lb. bundles 15c., in 5 and 10-lb. bundles 25c., in 1-lb hanks, 50c., in ½-lb. hanks 75c., in ¼-lb. hanks $1.

POULTRY NETTING.

2-in. mesh, 19 w. g., 50 and 5 p.c. off. Other sizes, 50 and 5 p.c. off.

WIRE CLOTH.

Painted Screen, in 100-ft rolls, $1.72½, per 100 sq. ft.; in 50-ft. rolls, $1.77½, per 100 sq ft.

WIRE FENCING.

Galvanized barb 2 95
Galvanized, plain twist 3 30
Galvanized barb, f.o.b. Cleveland. $2.70 for small lots and $2.60 for carlots

WIRE ROPE

Galvanized, 1st grade, 6 strands, 24 wires, ⅜, $5; 1 inch $16.80
Black, 1st grade, 6 strands, 19 wires, ⅜. $5; 1 inch $15.10. Per 100 feet f o.b Toronto

WOODENWARE.

CHURNS.

No. 0, $9; No. 1, $9; No. 2, $10; No. 3, $11; No. 4, $13; No. 5, $16.; f o b. Toronto Hamilton, London and St. Marys, 30 and 30 per cent.; f o.b. Ottawa, Kingston and Montreal, 40 and 15 per cent. discount.

CLOTHES REELS

Davis Clothes Reels. dis. 40 per cent.

FIBRE WARE.

Star pails, per doz	$ 3 00
0 Tubs, "	14 00
1 " "	12 00
2 " "	10 00
3 " "	8 50

LADDERS, EXTENSION.

3 to 6 feet, 12c. per foot; 7 to 10 ft., 13c.
Waggoner Extension Ladders, dis. 40 per cent.

MOPS AND IRONING BOARDS.

"Best" mops	....	1 25
"900" mops	....	1 25
Folding ironing boards	12 00	16 50

REFRIGERATORS

Discount, 40 per cent.

SCREEN DOORS.

Common doors, 2 or 3 panel, walnut stained, 4-in. style......per doz.	7 25
Common doors, 2 or 3 panel, grained only, 4-in., style ......per doz.	7 55
Common doors, 2 or 3 panel, light stair per doz.	9 55

WASHING MACHINES.

Round, re-acting per doz.	60 00
Square "	63 00
Eclipse, per doz	54 00
Dowswell "	39 00
New Century, per doz	75 00
Daisy	54 00
Stephenson	74 00

WRINGERS.

Royal Canadian, 11 in., per doz.	35 00
Royal American, 11 in.	35 00
Eze- 10 in., per doz	36 75

MISCELLANEOUS

AXLE GREASE.

Ordinary, per gross	6 00	7 00
Best quality	10 00	12 00

BELTING.

Extra, 60 per cent.
Standard, 60 and 10 per cent.
No. 1, not wider than 6 in., 60, 10 and 10 p.c.
Agricultural, not wider than 4 in., 75 per cent.
Lace leather, per side, 75c.; cut laces, 80c.

BOOT CALKS.

Small and medium, ball	per M	2 25
Small heel	"	4 50

CARPET STRETCHERS.

American ......per doz.	1 00	1 50
Bullard's "	....	6 50

CASTORS.

Bed, new list, discount 55 to 57½ per cent.
Plate, discount 52½ to 57½ per cent.

PINE TAR.

½ pint in tins ......per gross	7 80
" "	9 60

PULLEYS.

Hothouse	per doz.	0 55	1 00
Axle	"	0 22	0 33
Screw	"	0 22	1 00
Awning	"	0 35	2 50

PUMPS.

Canadian cistern	1 40	2 00
Canadian pitcher spout	1 80	3 16

Berg's wing pump, 75 per cent.

ROPE AND TWINE.

Sisal		0 10½
Pure Manilla		0 15
"British" Manilla		0 12
Cotton, 3-16 inch and larger	0 21	0 23
" 5-32 inch	0 25	0 27
" ⅛ inch	0 25	0 28
Russia Deep Sea		0 16
Jute		0 09
Lath Yarn, single		0 10
" double		0 10½
Sisal bed cord. 48 feet......per doz.		0 65
" " 60 feet "		0 80
" " 72 feet "		0 95

Twine.

Rag, Russian twine, per lb.	....	0 27
Wrapping, cotton, 3-ply	....	0 25
" " 4-ply	....	0 29
Mattress twine per lb.	0 33	0 45
Staging " "	0 27	0 35

BINDER TWINE.

500 feet, sisal	0 09½
500 " standard	0 09½
550 " " manilla	0 10½
600 " " "	0 12½
650 " " "	0 13½

Car lots. ½c. less; 5-ton lots, ¼c. less. Central delivery.

SCALES.

Gurney Standard, 35; Champion, 45 p.c.
Burrow, Stewart & Milne — Imperial Standard, 35; Weigh Beams, 35; Champion Scales, 45.
Fairbanks Standard, 30; Dominion, 50 Richelieu, 50.
Warren new Standard, 35; Champion, 45 Weigh Beams, 30.

STONES—OIL AND SCYTHE.

Washita	per lb.	0 25	0
Hindostan	"	0 06	0 10
" slip	"	0 18	0 20
" Axe	"	....	0 10
Deer Creek	"	....	0 10
Deerlick	"	....	0 25
" Axe	"	....	0 15
Lily white	"	....	0 42
Arkansas	"	....	1 50
Water-of-Ayr	"	....	0 10
Scythe	per gross	3 50	5 00
Grind, 40 to 200 lb., per ton		20 00	22 00
" under 40 lb., "		....	24 00
" 200 lb. and over		....	28 00

INDEX TO ADVERTISERS.

CLASSIFIED LIST OF ADVERTISEMENTS.

Alabastine.
Alabastine Co, Limited, Paris, Ont

-uditors.
Davenport, Pickup & Co., Winnipeg.

Automobile Accessories.
Canada Cycle & Motor Co., Toronto Junction.
Carriage Mountings Co., Ltd., Toronto.

Awnings.
Tobin Tent and Awning Co., Ottawa

Babbitt Metal.
Canada Metal Co., Toronto.
Canadian Fairbanks Co., Montreal.
Frothingham & Workman, Montreal.
Robertson, Jas. Co., Montreal.

Bar Urns.
Buffalo Mfg. Co., Buffalo, N Y.

Bath Room Fittings.
Buffalo Mfg. Co., Buffalo, N.Y.
Carriage Mountings Co., Ltd, Toronto.
Forsyth Mfg. Co, Buffalo, N Y.
Ontario Metal Novelty Co., Toronto

Belting, Hose, etc.
Gutta Percha and Rubber Mfg. Toronto
Sadler & Haworth, Toronto.

Bicycles and Accessories.
Johnson's, Iver, Arms and Cycle Works Fitchburg, Mass

Binder Twine.
Consumers Cordage Co, Montreal.

Bolts.
Toronto & Belleville Rolling Mills, Belleville

Box Strap.
J. N. Warminton, Montreal.

Brass Goods.
Frothingham & Workman, Montreal.
Glauber Brass Mfg Co., Cleveland, Ohio.
Kerr Engine Co., Walkerville, Ont.
Lewis, Rice, & Son., Toronto.
Morrison, Jas, Brass Mfg. Co., Toronto.
Mueller Mfg Co., Decatur, Ill.
Penberthy Injector Co., Windsor, Ont.
Taylor-Forbes Co., Guelph, Ont.

Bronze Powders.
Canadian Bronze Powder Works, Montreal.

Brushes.
United Factories, Toronto.

Cans
Acme Can Works, Montreal.

Builders' Tools and Supplies.
Covert Mfg. Co., West Troy, N Y.
Frothingham & Workman Co, Montreal.
Howland, H. S. Sons & Co., Toronto.
Hyde, F., & Co., Montreal.
Lewis Bros. & Co., Montreal.
Lewis, Rice, & Son, Toronto.
Lockerby & McComb, Montreal.
Lufkin Rule Co., Saginaw, Mich.
Newman & Sons, Birmingham.
North Bros. Mfg. Co., Philadelphia, Pa.
Stanley Rule & Level Co., New Britain.
Stanley Works, New Britain, Conn.
Stephens, G. F., Winnipeg.
Taylor-Forbes Co., Guelph, Ont.

Carriage Accessories.
Carriage Mountings Co., Ltd., Toronto.
Covert Mfg. Co., West Troy, N.Y.

Carriage Springs and Axles.
Guelph Spring and Axle Co., Guelph.

Carpet Beateas.
Ontario Metal Novelty Co., Toronto.

Cartridges.
Dominion Cartridge Co., Montreal.

Cattle and Trace Chains.
Greening, B., Wire Co., Hamilton.

Chafing Dishes.
Buffalo Mfg. Co., Buffalo, N.Y.

Churns.
Dowswell Mfg. Co., Hamilton.

Clippers—All Kinds.
American Shearer Mfg. Co., Nashua, N.H

Clothes Reels and Lines.
Hamilton Cotton Co., Hamilton,

Clutch Nails.
J. N. Warminton, Montreal.

Congo Roofing
Buchanan Foster Co., Philadelphia, Pa.

Cordage.
Consumers' Cordage Co., Montreal.
Hamilton Cotton Co., Hamilton.

Cork Screws.
Erie Specialty Co., Erie, Pa.

Cow Ties
Greening, B., Wire Co., Hamilton

Cuspidors.
Buffalo Mfg. Co., Buffalo, N.Y.

Cut Glass.
Phillips, Geo., & Co., Montreal.

Cutlery—Razors, Scissors, etc.
Birkett, Thos., & Son Co., Ottawa.
Clauss Shear Co., Toronto
Dorken Bros. & Co., Montreal.
Frothingham & Workman Montreal.
Heinisch's, R., Sons Co., Newark, N.J.
Howland, H. S. Sons & Co., Toronto.
Hutton, Wm., & Sons, Ltd., London, Eng.
Lamplough, F. W., & Co., Montreal.
Phillips, Geo., & Co., Montreal.
Round, John, & Son, Montreal.

Electric Fixtures.
Canadian General Electric Co., Toronto
Morrison James, Mfg. Co., Toronto.
Munderloh & Co., Montreal.

Electro Cabinets.
Cameron & Campbell Toronto.

Enameled Ware
Kemp Mfg. Co., Toronto.

Engines, Supplies, etc.
Kerr Engine Co., Walkerville, Ont.

Eavetroughs
Wheeler & Bain, Toronto

Fencing—Woven Wire.
Dominion Wire Mfg. Co., Montreal.
Owen Sound Wire Fence Co., Owen Sound.
Banwell Hoxie Wire Fence Co., Hamilton.

Files and Rasps.
Barnett Co., G. & H., Philadelphia, Pa.
Nicholson File Co., Port Hope

Firearms and Ammunition.
Hamilton Rifle Co., Plymouth, Mich.
Harrington & Richardson Arms Co., Worcester, Mass.
Johnson's, Iver, Arms and Cycle Works Fitchburg, Mass.

Fishing Tackle.
Enterprise Mfg. Co., Akron, Ohio

Food Choppers
Enterprise Mfg. Co., Philadelphia, Pa.
Lamplough, F. W., & Co., Montreal.
Shirreff Mfg. Co., Brockville, Ont.

Furnaces.
Pease Foundry Co., Toronto.

Galvanizing.
Canada Metal Co., Toronto.
Dominion Wire Mfg. Co., Montreal.
Montreal Rolling Mills Co., Montreal.
Ontario Wind Engine & Pump Co., Toronto.

Glass Ornamental
Hobbs Mfg. Co., London
Consolidated Plate Glass Co., Toronto

Glaziers' Diamonds.
Gibsone, J. B., Montreal.
Pelton, Godfrey S.
Sharratt & Newth, London, Eng.
Shaw, A., & Son, London, Eng.

Handles.
Still, J. H., Mfg. Co.

Harvest Tools.
Maple Leaf Harvest Tool Co., Tillsonburg, Ont.

Hockey Sticks
Still, J. H Mfg. Co., St. Thomas.

Hoop Iron.
Frothingham & Workman Montreal.
Montreal Rolling Mills Co., Montreal.
J. N. Warminton, Montreal.

Horse Blankets.
Heney, E. N., & Co., Montreal.

Horseshoes and Nails.
Canada Horse Nail Co., Montreal.
Montreal Rolling Mills, Montreal.
Capewell Horse Nail Co., Toronto
Toronto & Belleville Rolling Mills Belleville.

Hot Water Boilers and Radiators.
Cluff, R. J., & Co Toronto.
Pease Foundry Co. Toronto.
Taylor-Forbes Co., Guelph.

Ice Cream Freezers.
Dana Mfg. Co., Cincinnati, Ohio.
North Bros. Mfg. Co., Philadelphia, Pa.

Ice Cutting Tools.
Erie Specialty Co., Erie, Pa.
North Bros. Mfg. Co., Philadelphia, Pa.

Injectors—Automatic.
Morrison, Jas., Brass Mfg. Co., Toronto.
Penberthy Injector Co., Windsor, Ont.

Iron Pipe.
Montreal Rolling Mills, Montreal.

Iron Pumps.
Lamplough, F. W., & Co., Montreal.
McDougall, R., Co., Galt, Ont.

Ladders.
The Waggoner Ladder Co, London, Ont.

Lanterns.
Kemp Mfg. Co., Toronto.
Ontario Lantern Co., Hamilton, Ont.
Wright, E. T., & Co., Hamilton.

Lawn Mowers.
Birkett, Thos., & Son Co., Ottawa.
Frothingham & Workman, Montreal.
Maxwell, D., & Sons, St. Mary's, Ont.
Taylor, Forbes Co., Guelph.

Lawn Mower Grinders
Root Bros. & Co., Plymouth, Ohio.

Ledgers—Loose Leaf.
Business Systems, Toronto.
Copeland-Chatterson Co., Toronto.
Crain, Rolla L., Co., Ottawa.
Universal Systems, Toronto.

Lithographing.
London Printing & Lithographing Co., London, Ont.

Locks, Knobs, Escutcheons, etc.
Peterborough Lock Mfg. Co., Peterborough, Ont.
National Hardware Co., Orillia, Ont.

Lumbermen's Supplies.
Pink, Thos., & Co., Pembroke Ont.

Lye
Gillett, E. W., & Co., Toronto

Manufacturers' Agents.
Fox, C. H., Vancouver.
McIntosh, H F., & Co., Toronto.
Gibb, Alexander, Montreal.
Scott, Bathgate & Co., Winnipeg.

Metals.
Canada Iron Furnace Co., Midland, Ont.
Canada Metal Co., Toronto.
Eadie, H. G., Montreal.
Frothingham & Workman, Montreal.
Gibb, Alexander, Montreal.
Kemp Mfg. Co., Toronto
Leslie, A. C., & Co., Montreal.
Lysaght, John, Bristol, Eng.
Nova Scotia Steel and Coal Co., New Glasgow, N.S.
Robertson, Jas., Co., Montreal.
Roper, J. H., Montreal.
Samuel, Benjamin & Co., Toronto.
Stairs, Son & Morrow, Halifax, N.S.
Thompson, B. & S. H. & Co. Montreal.

Metal Lath.
Galt Art Metal Co, Galt.
Metallic Roofing Co., Toronto.
Metal Shingle & Siding Co., Preston, Ont.

Metal Polish, Emery Cloth, etc.
Oakey, John, & Sons, London, Eng.

Nails Wire
Dominion Wire Mfg. Co., Montreal.

Oilers
Maple City Mfg Co., Monmouth, Ill.

Oil Tanks.
Bowser, S. F., & Co., Toronto.

Ornamental Iron and Wire.
Dennis Wire & Iron Co, London, Ont.

Packing.
Gutta Percha & Rubber Co Toronto

Paints, Oils, Varnishes, Glass.
Blanchite Process Paint Co., Toronto.
Brandram-Henderson, Montreal
Canada Paint Co., Montreal.
Canadian Oil Co., Toronto
Consolidated Plate Glass Co., Toronto.
Dods, P. D., & Co., Montreal
Imperial Varnish and Color Co., Toronto
Jamieson, R. C., & Co., Montreal.
Lucas, John & Co, New York
McArthur, Corneille & Co., Montreal.
McCaskill, Dougall & Co., Montreal.
Moore, Benjamin, & Co. Toronto.
Ottawa Paint Works, Ottawa
Queen City Oil Co., Toronto.
Ramsay & Son, Montreal.
Sanderson Pearcy & Co., Toronto
Sherwin-Williams Co., Montreal.
Standard Paint Co., Montreal
Standard Paint and Varnish Works Windsor, Ont.
Stephens & Co., Winnipeg.
Martin-Senour Co., Montreal
Winnipeg Paint & Glass Co., Winnipeg

Perforated Sheet Metals.
Greening, B., Wire Co., Hamilton.

Plumbers' Tools and Supplies
Canadian Fairbanks Co., Montreal.
Cluff, R. J., & Co., Toronto.
Frothingham & Workman, Montreal.
Glauber Brass Co., Cleveland, Ohio.
Jardine, A. B., & Co, Hespeler, Ont.
Jenkins Bros., Boston, Mass.
Kerr Engine Co, Walkerville, Ont.
Lewis, Rice, & Son, Toronto.
Merrell Mfg. Co., Toledo, Ohio.
Montreal Rolling Mills, Montreal.
Morrison, Jas., Brass Mfg. Co., Toronto.
Mueller, H., Mfg. Co., Decatur, Ill.
Oshawa Steam & Gas Fitting Co., Oshaw
Robertson Jas., Co. Montreal.
Robertson, Jas., Co, Limited, Toronto
Somerville, Limited, Toronto
Stairs, Son & Morrow, Halifax, N.S.
Standard Ideal Sanitary Co., Port Hope,
Standard Sanitary Co., Pittsburg.
Stephens, G. F., & Co., Winnipeg, Man.
Turner Brass Works, Chicago.
Vickery, Orlando, Toronto.

Polishes.
Majestic Polishes, Toronto

Portland Cement.
International Portland Cement Co. Ottawa, Ont.
Hanover Portland Cement Co., Hanover, Ont.
Hyde, F., & Co., Montreal.
Thompson B. & S. H. & Co., Montreal.

Poultry Netting.
Greening, B., Wire Co., Hamilton, Ont.

Printing.
London Printing & Lithographing Co., London, Ont.

Razors.
Clauss Shear Co., Toronto.

Refrigerators.
Fabien, C. P., Montreal.

Registers.
Pease Foundry Co., Toronto.

Roofing Supplies.
Brantford Roofing Co., Brantford.
Barrett Mfg. Co., New York.
F. W. Bird, East Walpole, Mass.
Buchanan Foster Co., Philadelphia, Pa.
McArthur, Alex., & Co., Montreal.
Metal Shingle & Siding Co., Preston, Ont.
Metallic Roofing Co., Toronto.
Paterson Mfg. Co., Toronto & Montreal.
Standard Paint Co Montreal.
Wheeler and Bain, Toronto

Saws.
Atkins, E. C., & Co., Indianapolis, Ind
Shurly & Dietrich, Galt, Ont.
Spear & Jackson, Sheffield, Eng.

Scales.
Canadian Fairbanks Co., Montreal.
Frothingham & Workman, Montreal.

Screw Cabinets.
Cameron & Campbell, Toronto.

Screws, Nuts, Bolts.
Dominion Wire Mfg. Co, Montreal.
Montreal Rolling Mills Co., Montreal.

Soil Pipe
McFarlane, Walter, Glasgow

Sewer Pipes.
Canadian Sewer Pipe Co., Hamilton
Hyde, F., & Co., Montreal.

Shelf Boxes.
Cameron & Campbell, Toronto.

Shears, Scissors.
Clauss Shear Co., Toronto.

Shovels and Spades
Eclipse Mfg Co, Ottawa
Frothingham & Workman, Montreal
Peterboro Shovel & Tool Co., Peterboro.

Silverware.
Hutton, Wm, & Sons, Ltd., London, Eng.
McGlashan, Clarke Co, Niagara Falls, Ont.
Phillips, Geo., & Co., Montreal.
Round, John, & Son, Sheffield, Eng.

Skates.
Canada Cycle & Motor Co., Toronto.
McFarlane, Walter, Glasgow.

Sprayers
Cavers Bros., Galt

Spring Hinges, etc.
Chicago Spring Butt Co, Chicago, Ill

Stable Fittings
Dennis Wire & Iron Co., London

Steel Rails.
Nova Scotia Steel & Coal Co., New Glasgow, N.S.

Stove Pipe.
Chown, Edwin, and Son, Kingston

Stoves, Tinware, Furnaces
Canadian Heating & Ventilating Co. Owen Sound.
Copp, W. J., Son & Co., Fort William
Davidson, Thos., Mfg. Co., Montreal
Down Draft Furnace Co, Galt
Guelph Stove Co., Guelph.
Gurney Foundry Co., Toronto.
Harris, J. W., Co., Montreal.
Howard, Wm., Toronto
Kemp Mnfg. Co. Toronto.
McClary Mfg. Co. London.
Merrick Anderson, Winnipeg
Pease Foundry Co., Toronto.
Smart, James, Mfg. Co., Brockville
Stewart, Jas., Mfg. Co., Woodstock, Ont.
Taylor-Forbes Co., Guelph, Ont.
Wright, E. T., & Co., Hamilton.

Tacks.
Montreal Rolling Mills Co., Montreal.
Ontario Tack Co, Hamilton.

Tents.
Tobin Tent and Awning Co, Ottawa

Tin Plate.
American Sheet & Tin Plate Co., Pittsburg, Pa.
Baglan Bay Tin Plate Co., Briton Ferry South Wales
Lysagnt, John, Bristol, Newport and Montreal

Turpentine
Defiance Mfg. Co., Toronto.

Ventilators.
Harris, J. W., Co., Montreal.
Pearson, Geo. D., Montreal.

Wall Paper
Staunton Limited, Toronto.

Wall Paper Cleaner.
Gilbert, Frank U. S., Cleveland

Washing Machines, etc
Dowswell Mfg. Co., Hamilton, Ont.
The Shultz Bros. Co., Brantford.
Taylor-Forbes Co., Guelph, Ont.

Water Filters.
Buffalo Mfg. Co., Buffalo, N.Y.

Wheelbarrows
London Foundry Co., London Ont
Schultz Bros. Co., Ltd., The Brantford.

Wholesale Hardware.
Birkett, Thos., & Sons Co., Ottawa.
Caverhill, Learmont & Co., Montreal.
Frothingham & Workman, Montreal.
Hobbs Hardware Co., London.
Howland, H. S., Sons & Co., Toronto.
Lamplough, F. W., & Co, Montreal.
Lewis Bros. & Co., Montreal.
Lewis, Rice, & Son, Toronto

Window and Sidewalk Prisms
Hobbs Mfg. Co., London, Ont.

Wire, Wire Rope, Cow Ties, Fencing Tools, etc.
Banwell-Hoxie Fence Co., Hamilton
Dennis Wire and Iron Co., London, Ont.
Dominion Wire Mnfg. Co., Montreal
Greening, B., Wire Co., Hamilton.
Owen Sound Wire Fence Co., Owen Sound
Montreal Rolling Mills Co., Montreal.
Western Wire & Nail Co., London, Ont.

Wrapping Papers.
Canada Paper Co., Toronto.
McArthur, Alex., & Co, Montreal.
Stairs, Son & Morrow, Halifax, N.S.

Wringers
Connor, J. H.&Son, Ottawa, Ont

ROBERTSON'S SOLDERS

Bar-King, Strictly Merchant's, Wire (any gauge) Triangular, Oval, Plumbers' and Canners' Solders, Electricians' Solder

ARE STANDARD

Ask Quotations

The James Robertson Co., Limited
144 William St., MONTREAL, QUE.

CHICAGO SPRING BUTTS

A PRODUCT OF RECOGNIZED SUPERIORITY

CHICAGO SPRING BUTT

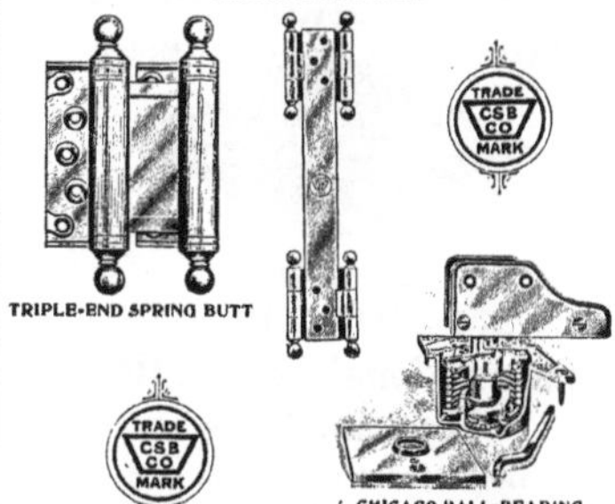

TRIPLE-END SPRING BUTT

CHICAGO BALL-BEARING FLOOR HINGE

Chicago Spring Butt Company
CHICAGO . NEW YORK
CATALOGUES ON REQUEST

WALKER CORK SCREWS

are good holiday souvenirs and live sellers the whole year. Are made well and are made out of good material. That is why we guarantee every one that we put out. When ordering Cork Screws try **Walker's**

Manufactured only by **ERIE SPECIALTY COMPANY,** ERIE, Pa. U.S.A.

LUFKIN MEASURING TAPES

Steel, Metallic, Linen, Pocket, Ass Skin, Pat. Leather, Bend Leather, Etc

ARE THE BEST AND MOST POPULAR TAPES IN THE WORLD.
YOUR STOCK IS NOT COMPLETE WITHOUT THEM.

LUFKIN RULE CO., Saginaw, Mich, U.S.A.

Canadian Factory - - Windsor, Ontario

London Office and Warehouse—24 and 26 Holborn. New York City Branch—280 Broadway.

For sale by ALL PROMINENT CANADIAN HARDWARE JOBBERS.

Est. 1868. Inc. 1895.

Black Diamond File Works

G. & H. Barnett Company

PHILADELPHIA

Twelve TRADE MARK Medals

Awarded
By **JURORS** at
International Expositions
Special Prize
Gold Medal at Atlanta, 1895

Copy of catalogue sent free to any interested file user upon application.

"Redstone"

High Pressure

Sheet Packing

A packing that will hold. For use in highest pressures for steam, hot or cold water and air. Packs equally well for all.

From actual tests, we believe that this packing is the most durable and satisfactory of any on the market. Try a sample lot and see for yourself.

Manufactured Solely by

THE GUTTA PERCHA & RUBBER MFG. CO.

of TORONTO, LIMITED

HEAD OFFICES,
47 Yonge Street, Toronto.
Branches: Montreal, Winnipeg, Vancouver.

Specify the Following Products

Manufactured by

American Sheet and Tin Plate Company

PITTSBURG, PA.

If you want the best.

"Apollo Best Bloom," Galvanized Sheets, Blue Annealed and American Bessemer Black Sheets, Wellsville Polish, Range Steel, Wood's "A" Planished Iron, Black Steel specialties for all purposes, Tinplates, Coke and Charcoal, all grades, Terne Plates, Old Style and New Method Processes.

Place your order with

B. & S. H. THOMPSON & CO.

LIMITED

Dominion of Canada Sales Agents

MONTREAL, CANADA

CIRCULATES EVERYWHERE IN CANADA

Also in Great Britain, United States, West Indies, South Africa and Australia.

HARDWARE AND METAL

A Weekly Newspaper Devoted to the Hardware, Metal, Heating and Plumbing Trades in Canada.

Office of Publication, 10 Front Street East, Toronto.

VOL. XIX. MONTREAL, TORONTO, WINNIPEG, SEPTEMBER 28, 1907 **NO. 39.**

Read "Want Ads." on Page 49

CUTLERY

Pen Knives and Scissors

Razors and Pocket Knives

CASES OF CARVERS

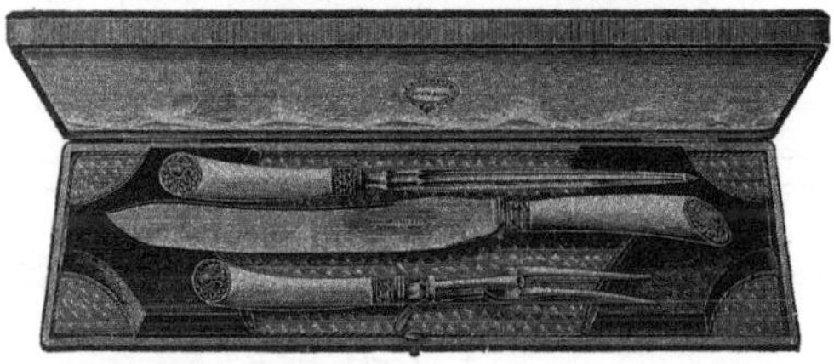

The new lines of Cutlery Samples shown by our travellers this season are far superior to anything heretofore put on the market, and it will be interesting to you, as well as profitable, to examine them.

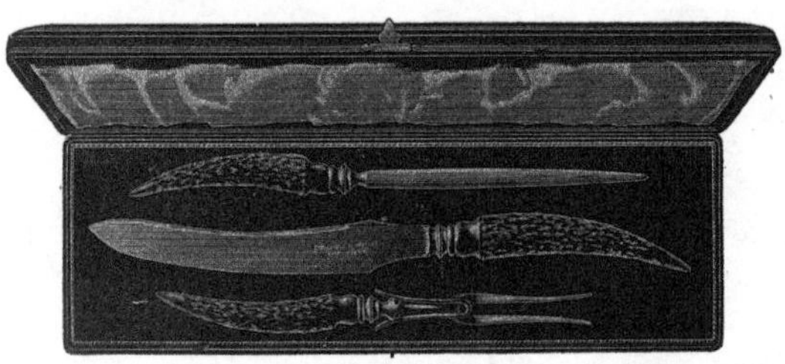

Pocket Knives

Pocket Knives

RICE LEWIS & SON

LIMITED

TORONTO.

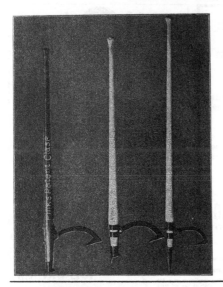

LOADED SHELLS

The name Eley stands for the best quality. This is known to all judges, and a great many people are surprised to learn that their loaded shells are sold at the same price as other makes.

We stock all gauges, 10, 12, 16, 20, both black and smokeless powder, and all popular numbers of shot. Refer to page 74 of our No. 36 catalogue and get full particulars. We have not space to give it to you here.

25

ELEY

"Grand Prix" . .

Cartridges . . . .

12 Gauge.

Central Fire

1 Greaseproof. 1 Felt & 1 Card Wad over Powder.
1 Card Wad over Shot.

Manufactured by

ELEY BROTHERS, Ld.,

Gray's Inn Road, LONDON, England.

If you have never handled this line, we would recommend you to send us a sample order at once. We know your customers will be pleased with the quality and you will be pleased with the price.

WE ALSO STOCK

"CROWN," loaded with Black Powder.	"SOVEREIGN," loaded with Smokeless Powder.	"NEW RIVAL," loaded with Black Powder.

If you have not a copy of our No. 36 Sporting Goods Book, drop us a card, it will pay you to have it for reference.

LEWIS BROS., Limited

Montreal

OTTAWA
TORONTO

WINNIPEG

VANCOUVER
CALGARY

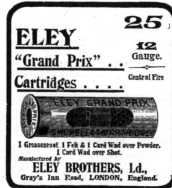

Transmission Rope

The transmission of power by Rope Drives is increasing daily, on account of the durability and low cost. Rope Drives were first manufactured and introduced into the United States by Sewall, Day & Co., of Boston. A few years later we manufactured and introduced them into this country.

If you contemplate putting in a Rope Drive, we will be pleased to furnish you with any information you may require. As we have had years of experience, we are naturally in a position to turn out the most up-to-date product.

Our mills are now thoroughly equipped for manufacturing our famous Tallow Laid Transmission Rope, three or four strand, and in all sizes and lengths, and we always carry a large stock of the most select Manila Fibre for the manufacture of this Rope. We carry a fully assorted stock of all the standard sizes, and can make special orders promptly.

CONSUMERS' CORDAGE COMPANY, Limited

Mills: MONTREAL, QUE., and HALIFAX, N.S.

BRANCHES:

F. H. ANDREWS & SON, Quebec. CONSUMERS CORDAGE CO., Limited, St. John, N.B. MacGOWAN & CO., Vancouver
TEES & PERSSE, Limited, Winnipeg, Calgary and Edmonton. GEO. WOOD, London, England.
MERRICK, ANDERSON & CO., Winnipeg, Binder Twine Distributors for Northwest.

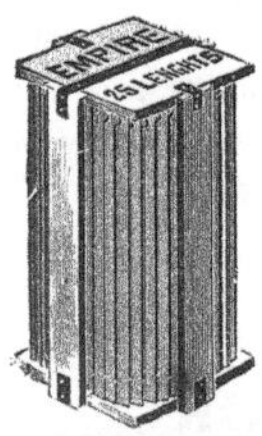

Patented 1900. Improved 1905.

Davidson's "Empire" Stove Pipe

Improved Patent

Neatly packed in crates of 25.
Secures lowest possible freight rate.

Deep Swage Prevents Telescoping.
Coated to Prevent Rust.

Sold as low as ordinary Stove Pipes.

SIMPLE and EASY to put together.
No Profane Language Necessary.
The only tools required are a pair of hands.
Pipes being cut out with dies ensure absolute fit and conformity.

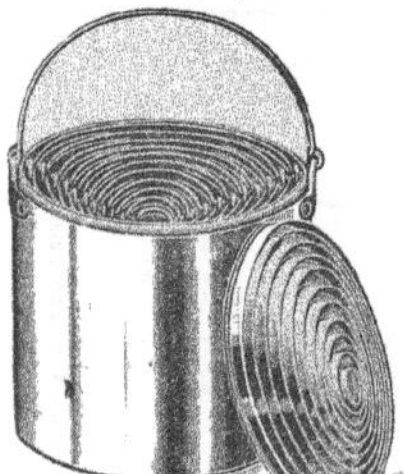

Miners' Camp Kettles

Strong and substantially made in extra heavy tin.

Nineteen sizes, from ⅛ quart to 29 quarts.

Wash Tubs—Galvanized

With Wringer Attachment

Nos.	1	2	3
Top measurement, inches	21½	22¾	25
Bottom " "	17⅞	18¼	20
Height " "	9½	10½	11

The THOS. DAVIDSON MFG. CO., Ltd., Montreal and Winnipeg

H. S. HOWLAND, SONS & CO. LIMITED

HARDWARE MERCHANTS

Only Wholesale — 138-140 WEST FRONT STREET, TORONTO. — Wholesale Only

"Champion" Sweeper
Japanned Trimmings
Nickel "

"Standard" Sweeper
Japanned Trimmings
Nickel "

"Grand Rapids" Sweeper
Japanned Trimmings
Nickel "

Bissell's Sweepers have four Rubber Tired outside wheels, Rubber Furniture Protector, Spring Dumping Device, both sides open with one pressure. Pure Bristle brush.

Parlor Oil Heater
"Buck Eye" 161 Black Trimming
"Puritan" 149 " Aluminum

Carriage Heater
"Standard" B, Carpet covered.

Parlor Oil Heater
"Puritan" 154 N. P. Trimming
" 155 " Ruby Glass

For full discription see our catalogue

H. S. HOWLAND, SON & CO., LIMITED

Opposite Union Station

GRAHAM WIRE NAILS ARE THE BEST

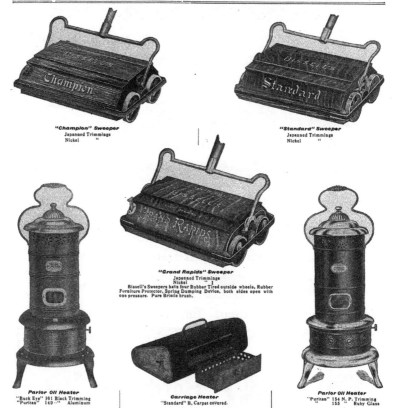

Our Prices are Right — Are you receiving our monthly illustrated circular? IF NOT, WRITE FOR IT. — We Ship Promptly

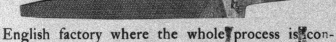

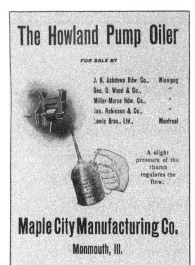

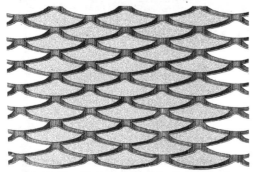

Simonds
Crescent-Ground Cross-Cut Saws

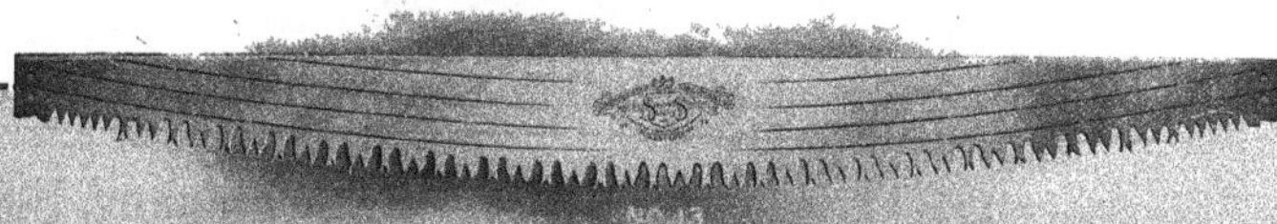

For logging camps where work must be fast and economical there is no other saw made that can give results equal to the Crescent-Ground Saw. **Made of Simonds Steel.** It cuts easy, runs fast and gives good results. This cross-cut saw is sold by most of the leading jobbers in Canada.

Simonds Canada Saw Co., Limited
TORONTO, ONT. MONTREAL, QUE. ST. JOHN, N.B.

Ask for
Pratt & Whitney Small Tools
Made in Canada

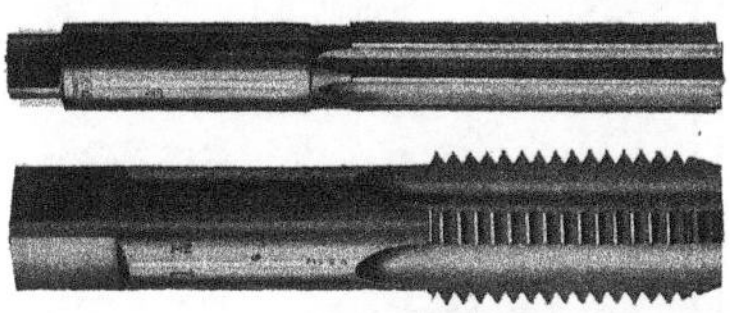

and be sure of getting the best

THE HIGHEST IN QUALITY
THE STANDARD FOR ACCURACY

IF YOUR DEALER DOESN'T CARRY THEM WRITE

THE CANADIAN FAIRBANKS CO., LIMITED
Exclusive Canadian Selling Agents
MONTREAL TORONTO WINNIPEG VANCOUVER

Wheelbarrows

All kinds and sizes. The cut reproduced here is just of one of the many, but the kind every contractor should use. The substantial, satisfactory, steel tray Contractor's Barrow.

The London Foundry Co.
LONDON, CAN.

Do You Handle These Money Makers?

MAXWELL LAWN MOWER

Durability, Finish and Efficiency Unsurpassed by any other mower on earth.

FAVORITE CHURN

By lightening women's labor it adds weight to your bank account.

David Maxwell & Sons, St. Mary's, Ont.

The Peterborough Lock Manufacturing Company, Limited

Cylinder Night Latch, No. 103.

Peterborough, Ont.

Manufacturers of all kinds

Rim and Mortise Locks, Inside, Front Door, and Store Door Sets, also full line High-class Builders' Hardware.

Sold by all Leading Jobbers in the Dominion.

Not How Cheap, but How GOOD

Banner Lamp Burners,
O, A, B and D sizes.

Canada Lamp Burners,
A and B sizes.

Security Lamp Burners,
A and B sizes.

Bing Glass Cone Lamp Burners,
A and B sizes.

Lantern Burners,
No. 1 and No. 2.

Every Burner carefully inspected before leaving the factory, and guaranteed perfect.

Orders solicited through the jobbing trade.

ONTARIO LANTERN AND LAMP CO.
HAMILTON, ONT. LIMITED

The Buffalo Manufacturing Co.

Buffalo, N.Y.

When you get our goods you know you get THE BEST.

We manufacture
Water Filters
Water Coolers
Chafing Dishes
Table Kettles and Stands
Coffee Extractors
Wine Coolers
Nursery Chests
Baking Dishes
Crumb Trays and Scrapers
Tea and Bar Urns
Bathroom Fixtures
Coal Vases and Hods
Candlesticks
Cuspidors
Match Safes, Etc.

All High Grade and exceedingly presentable.

REPRESENTED BY

H. F. McINTOSH & CO.

51 Yonge Street

Write for Catalogue **TORONTO, ONT.**

DOMINION WIRE MANUFACTURING CO.

MONTREAL TORONTO LIMITED

Manufacturers of

PLAIN WIRE *All Sizes and for all Purposes.*

GALVANIZED WIRE *For Fencing, Telegraph and Telephone Purposes.*

Copper and Brass Wire

WIRE NAILS
WOOD SCREWS

Bright Wire Goods

Staples

Steel and Brass Jack Chains

Spring Cotters, Strand Guy Wire, Steel Wire Barrel Hoops

A RECORD BREAKING SELLER

22 CALIBRE
16 SHOT

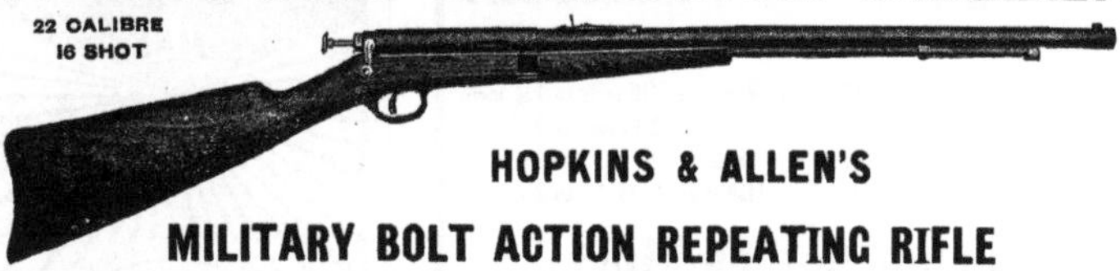

HOPKINS & ALLEN'S

MILITARY BOLT ACTION REPEATING RIFLE

$10.75 LIST There has long been a demand for a good rifle at this price—and the **HOPKINS & ALLEN Junior Repeater** fills the bill. This rifle has been and is being extensively advertised, and a heavy sale for it is created.

SPECIFICATIONS:—22-inch barrel, take-down, 22 calibre. Magazine handles 22 short, 22 long and 22 long rifle cartridges without change of carrier. Shoots 16-22 short or 12-22 long or long rifle cartridges. Action, Military Bolt Pattern with positive safety device. Length over all, 40 inches, weight 5½ pounds.

FOR SALE BY ALL FIRE-ARMS JOBBERS.

PLENTY OF GOOD SUPPLEMENTARY ADVERTISING MATTER—FREE.

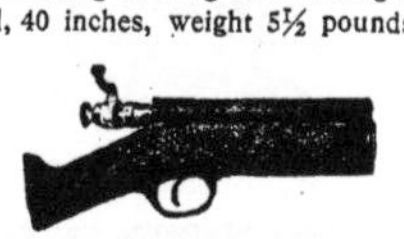

Shows Bolt Drawn Back—Position for Ejecting Shells.

WRITE FOR CATALOGS

THE HOPKINS & ALLEN ARMS CO.

LONDON OFFICE
6 City Road, Finsbury Square, London

Dept. 67, NORWICH, CONN., U.S.A.

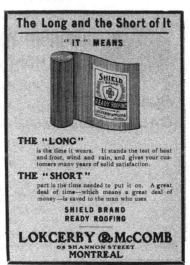

TROJAN HOOP IRON

BLACK and **GALVANIZED**
(Oil Finish) (Bright Finish)

Widths	Gauges
½, ⅝, ¾, 1 in.	26, 27, 29.

½x29 will run 40 feet to the lb.
Packed in bbls. of 500 lbs.

The Only Hoop to Handle.

J. N. WARMINTON,
43 Scott St., Toronto 207 St. James St., Montreal

THE ONLY WAY

you can destroy a

"Dillon" Hinge-Stay Field Fence

is to draw out the posts and cut the wires into pieces. Merely sitting on it, climbing over it, falling against it or otherwise abusing it doesn't hurt the "Dillon." No wonder it is the

best selling fence on the market.

THE OWEN SOUND WIRE FENCE CO., Limited
OWEN SOUND, ONT.

Sold by
Messrs. Caverhill, Learmont & Co., Montreal
Messrs. Christie Bros. Co., Limited, Winnipeg
The Abercrombie Hardware Co., Vancouver

HIGH GRADE ALL THE WAY THROUGH

GUARANTEED
that every piece of Alaska McG., C. brand is the same high grade metal all through.

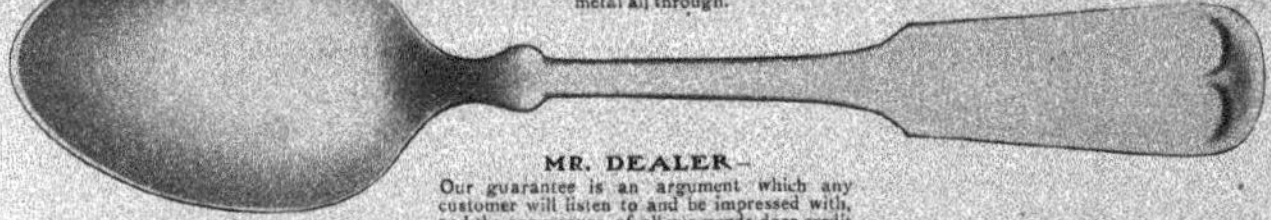

MR. DEALER—
Our guarantee is an argument which any customer will listen to and be impressed with, and the appearance of all our goods does credit to the best hardware store on earth.

THE McCLASHAN, CLARK COMPANY, LIMITED,
NIAGARA FALLS, - CANADA

OUR AGENTS ARE:
J. MACKAY ROSE, 117 d'Youville Square, Montreal, Que.
N. F. GUNDY, 47 Hayter Street, Toronto, Ont.
DAVID PHILIP, 291 Portage Avenue, Winnipeg, Man.

BRANTFORD ROOFING CO. Limited,

MANUFACTURERS OF
"BRANTFORD ROOFING," "Asphalt and Rubber"
"BRANTFORD FLOORINE" (or Linoleum)
"BRANTFORD ELASTIC PAINT" (for Roof Coating)

UNDER OUR OWN "TRADE MARK"

TRADE REGISTERED MARK
BRANTFORD ROOFING CO. LIMITED
"B"
BRANTFORD, CANADA

WRITE US FOR PRICES AND SAMPLES

Office, Factories and Warehouses
Sydenham St., Brantford, CAN.

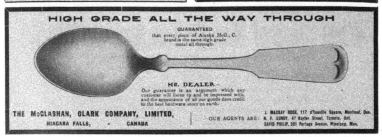

You Might as Well Try

to make bricks without straw as to run a hardware store without a good stock of

BOECKH'S BRUSHES

You've got to ***please the painter*** whether he's a professional or an amateur. If you sell him a brush which moults like a canary in springtime, leaving hairs all over the job he's so proud of, you'll lose his trade sure. But you can win his regard ***and his trade*** by selling him only ***Boeckh's Brushes.***

"Boeckh" means "Brush Protection"

UNITED FACTORIES, Limited

TORONTO, CANADA

TRADE **PENBERTHY** MARK.

Genuine

A Fall Order from your Jobber should include a "Sorting Up" Order for Penberthy Injectors, Oil and Grease Cups, Water Gauges, Gauge Cocks, Whistles, Lubricators, etc.

Automatic Injector

"THEY ALWAYS WORK"

Remember, please, when you say you want "Penberthy" Goods be sure you get them. You have to sell the goods, so

GET THE

GOOD SELLERS.

That's Penberthy.

PENBERTHY INJECTOR CO., Limited, WINDSOR, ONT.

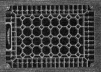

You May Be a Good Salesman

But, is everything in your favor? Are you selling the new Imperial Oxford? A continuous stream of repeat orders tells the story to us.

Removable Nickel
Removable Fire Linings
An Oval Firebox
The "Divided Draft"

All contained by the best looking stove in the market.

It gives a good salesman a chance to spread himself.

Drop us a card, or, better still, tell us the size range you want on your floor. If you get this agency you're lucky.

The Gurney Foundry Company, Limited

Toronto Winnipeg Vancouver Hamilton and London

THE GURNEY-MASSEY CO., LIMITED, MONTREAL, Que.
THE GURNEY STANDARD METAL CO., LIMITED, CALGARY, Alta., and EDMONTON, Alta.

INCREASED BUSINESS

invariably follows an inquiry about our

Asbestos Stove Lining

We know, and we want **you to know,** that this is the best lining ever made for every kind and size of Stove, Range, Grate or Hot-Air Furnace. Better and more economical than fire brick, it fits any stove—even those where natural gas is used and can be applied almost instantly by anyone. For mounting and repairing Stoves, Furnaces, etc., our

ASBESTOS FIREPROOF FURNACE CEMENT

is ahead of everything. It adheres **instantly** to the castings and sets as hard as iron. It is indestructible by fire or chemicals. And then— **there's a lot of money in it for dealers.**

Benjamin Moore & Co., Limited

(Formerly The Francis-Frost Co., Limited)
New Offices and Works: Toronto Junction, Canada
New York, Chicago, Cleveland, Carteret, N. J.

TRADE MARK
FOR PLATE

Get Ready for Xmas

We make many attractive patterns in Fish, Game and Meat Sets, Carvers, Serving Pieces, Cheese Scoops, Nut Crackers, Flasks, etc., that are especially suited to your trade. Orders taken now will be delivered in ample time for the holiday season.

The House of Hutton

Established in England in 1800

Sole Canadian Agent
W. J. GRANT, Lindsay Bldg., MONTREAL

TRADE MARK
FOR CUTLERY

SUNRAY
GOOD CHEER

A Well-made Range
A Well-liked Range
A Well-advertised Range

PANDORA

Features:

Oven—Scientifically ventilated.

Flues—Heat forced twice around oven.

Fire-box — Proportioned to size of oven.

Reservoir — Enamelled and easy to clean.

Thermometer--Registers exact heat.

Coal-grates — Acme of strength.

Wood-grate — Will fit when coal-grate is in place.

Fire-box Linings--Heavy cast iron.

Broiler Door — Takes largest broiler or toaster.

Emery Rod—The handy knife sharpener.

"PANDORA" RANGE

is the one to recommend to your Customers

The McClary Mfg. Co.

London Toronto Montreal Winnipeg
Vancouver St. John, N.B. Hamilton Calgary

Persons addressing advertisers will kindly mention having seen their advertisement in Hardware and Metal.

PLUMBING GOODS

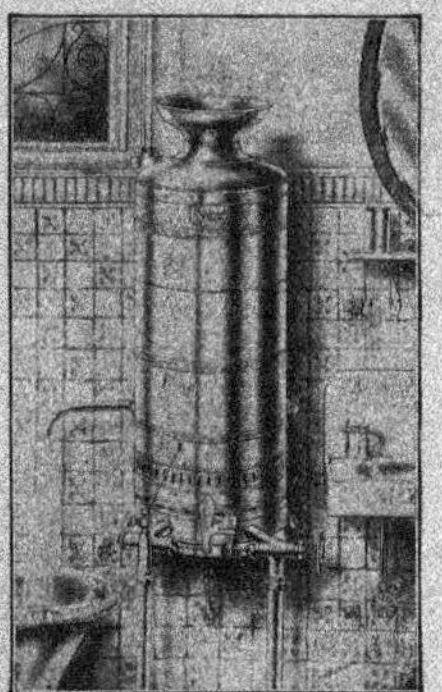

Instantaneous Water Heater—Copper, Nickel-Plated—with Safety Pilot Cock.

We make the largest variety in

PLUMBERS' BRASS and IRON GOODS

Dealers in

Enameled Baths, Lavatories, Sinks and Laundry Tubs, etc.

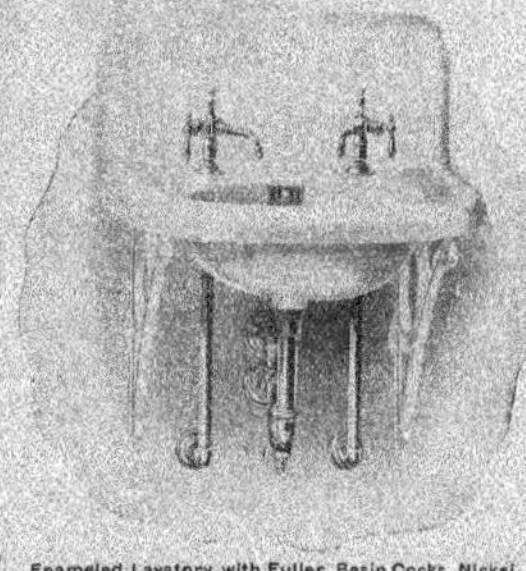

Enameled Lavatory with Fuller Basin Cocks, Nickel-Plated Brass Trap and N. P. Supply Pipes

Cool weather brings a desire for warm water. An instantaneous water heater is a comfort and a great convenience at this season—hot water in a jiffy. Write for Circular.

We handle the best—It's Safe, Reliable, Simple and Attractive.

The James Morrison Brass Mfg. Co., Limited, 95-97 Adelaide St. West, TORONTO

A Better Metal Chain Guard

Just as soon as the trade knew that we were ready to market our improved Metal Chain Guard, our construction department was almost swamped with orders.

Shrewd dealers recognized this Guard as

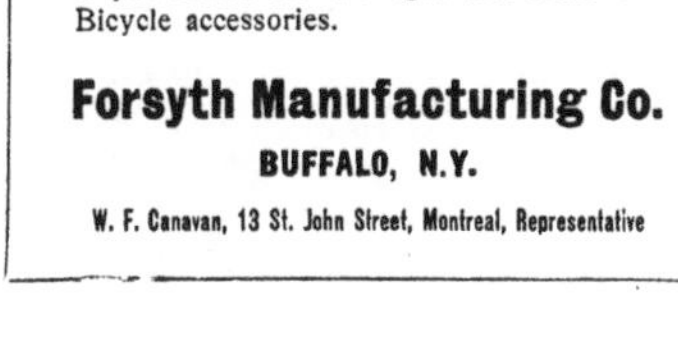

an important constituent of The Wise Buyers' line. The big bicycle boom in Canada should mean lots to you. It will—if you secure our catalogue and stock our Bicycle accessories.

Forsyth Manufacturing Co.

BUFFALO, N.Y.

W. F. Canavan, 13 St. John Street, Montreal, Representative

KERR'S GLOBE AND GATE VALVES

STRICTLY HIGH GRADE. TESTED & PACKED

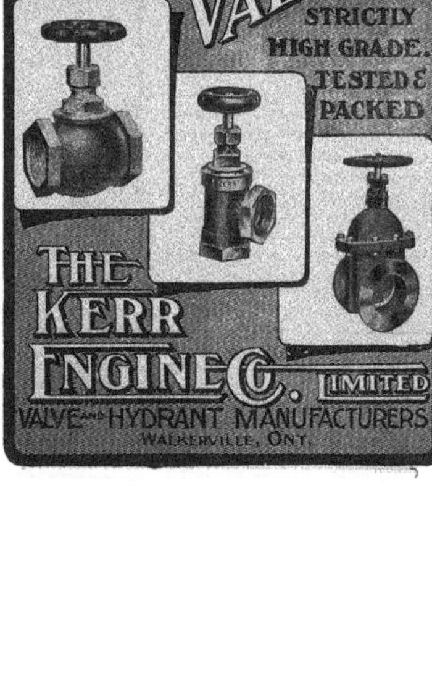

THE KERR ENGINE CO. LIMITED

VALVE AND HYDRANT MANUFACTURERS

WALKERVILLE, ONT.

Persons addressing advertisers will kindly mention this paper.

DIRECTORY OF MANUFACTURERS' PAGE.

Hardware and Metal receives, almost daily, enquiries for the names of manufacturers of various lines. These enquiries come from Wholesalers, Manufacturers and Retail Dealers, who usually intimate they have looked through Hardware and Metal, but cannot find any firm advertising the line in question. In the majority of cases these firms are anxious to secure the information at once. This page enables manufacturers to keep constantly before the trade lines which it would not pay them to advertise in larger space.

ADVERTISING.

MACLEAN PUBLISHING CO.
Montreal. Toronto. Winnipeg.
Hardware Book. Book of ready-made ads. for hardware merchants and stove dealers. Each advertisement to be illustrated with an attractive cut. Price, $1.50.

BRASS GOODS.

THE JAMES MORRISON BRASS MFG. Co., Limited,
89 to 97 Adelaide St. W., TORONTO.
Manufacturers of Brass and Iron Goods for Engineers and Plumbers; Locomotive and Marine Brass Works; Gas and Electric Fixtures. Telephone Main 3836.

PENBERTHY INJECTOR CO., Ltd.,
WINDSOR, ONT.
Manufacturers of "Penberthy" Automatic Injectors, XL-96 Ejectors, Brass Oilers and Lubricators, Water Gauges and Gauge Cocks, Air Cocks, etc.

CEMENT.

The Hanover Portland Cement Co., Limited,
HANOVER, ONTARIO
Manufacturers of the celebrated "Saugeen Brand" OF PORTLAND CEMENT.
Prices on application. Prompt shipment.

CHEESE PRESSES.

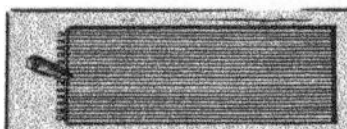

When in the market for GANG CHEESE PRESSES and up-to-date CURD CUTTERS just sit down and write to JAMES & REID, Perth, Ont. For Farmers' Feed Cookers, write JAMES BROS. FOUNDRY CO., Perth, Ont.

CONDENSED ADS.

Agents Wanted, Articles for Sale, Articles Wanted, Business Chances, Businesses for Sale, Situations Wanted, Situations Vacant.
Does not one of these headings suggest some way in which you can use our "want ads." to advantage?

CLIPPESS.

PRIEST'S CLIPPERS

Largest Variety, Toilet, Hand, Electric Power ARE THE BEST.
Highest Quality Grooming and Sheep-Shearing Machines.
WE MAKE THEM.
SEND FOR CATALOGUE TO American Shearer Mfg. Co., Nashua, N.H., USA

Weibusch & Hilger, Limited, special New York representatives, 9-15 Murray Street.

FLOOR SPRINGS.

The Best Door Closer is . . .
NEWMAN'S INVINCIBLE FLOOR SPRING
Will close a door silently against any pressure of wind. Has many working advantages over the ordinary spring, and has twice the wear. In use throughout Great Britain and the Colonies. Gives perfect satisfaction. Made only by
W. NEWMAN & SONS,
Hospital St. - Birmingham

GALVANIZING.

GALVANIZING
Work and Prices Right
ONTARIO WIND ENGINE & PUMP CO., Limited
Toronto, Ont.

GALVANIZING AND TINNING
The CANADA METAL CO.
Toronto Ontario.

LAWN MOWERS.

The people of Canada know that our name on a LAWN MOWER is an unalterable guarantee of quality and Durability.
TAYLOR-FORBES CO., Limited,
Head Office and Works, Guelph, Ont. Toronto—1880 King St. W.; Montreal—122 Craig St. W.; Winnipeg—The Vulcan Iron works, Limited.

Durability Finish
Efficiency
Guaranteed in the
Maxwell Lawn Mower
DAVID MAXWELL & SONS,
St. Mary's, Ont.

SHELF BRACKETS.

Will Hold Up a Shelf
That's what a shelf bracket's for. For this purpose there can be NOTHING BETTER, NOTHING CHEAPER than the BRADLEY STEEL BRACKET. It is well Japanned, Strong and Light. The saving on freight is a good profit aside from the lower price at which the goods are sold. Order direct or through your jobbers
Atlas Mfg. Co., New Haven.

SOIL PIPE.

FORWELL FOUNDRY CO.
BERLIN, ONT.
Manufacturers of
Soil Pipe, Fittings, and Cast Iron Sinks
Ask Jobbers for "F. F. CO." Brand.

CLUFF BROS.,
Toronto.
Can give prompt shipment on SOIL PIPE and FITTINGS

STEEL TROUGHS.

STEEL TROUGHS AND TANKS
We Manufacture
Steel Tanks, Stock Tanks, Steel Cheese Vats, Threshers' Tanks, Hog Troughs, Water Troughs Feed Cookers, Grain Boxes, Coal Chutes, Smokestacks.
AGENTS WANTED.
The STEEL TROUGH and MACHINE CO. Ltd., TWEED, ONT.

TOOLS.

ARMSTRONG CUTTING-OFF TOOLS
are correctly designed and the blades are bevel rolled from special Self Hardening Steel. Straight and Offset shaping 7 sizes each. Write for Catalog
Armstrong Bros. Tool Co.
1⅙ N. Francisco Ave.
CHICAGO, U.S.A.

WANT ADS.

Every Hardware Merchant can use our "want ad." page to advantage. Do not despise the want ad. because it is small, the results are big.

Shear making "up-to-date" is found in

HEINISCH

TAILORS' SHEARS, TRIMMERS, SCISSORS, TINNERS' SNIPS, ETC.

"Best in the World"

Since 1825, and a steady increase in sales due to their superior cutting quality.

R. HEINISCH'S SONS CO.
NEWARK, N.J., U.S.A.
New York Office and Salesroom, 155 Chambers St.

REGISTERED TRADEMARK
GRANTED 1682

World's Best Cutlery

You make no mistake in specifying goods made by **JOSEPH RODGERS & SONS, Limited** when ordering cutlery. A reputation of over 200 years at stake, insures the quality of every article.

Write for catalogues from Sole Canadian Agents

JAMES HUTTON & COMPANY, MONTREAL

Our Stable Fixtures

are a profitable line for the hardwareman at this season. Many new barns have been built this year, and farmers are now preparing to equip their stables. You can make money by pushing our

"Acorn" Steel Stanchions
Beith's Litter Carriers
The Cowbowl

Have you a supply of our Catalogues 16 and A109? If not, send for them. They will help you make sales.

"WE WORK WITH THE TRADE"

The Metal Shingle and Siding Company, - Limited

PRESTON

MONTREAL Cor. St. Catherine and Delorimier
TORONTO 100 Esplanade St. East

SALES AGENTS:

ST. JOHN, N.B. Emerson & Fisher, Limited
QUEBEC J. A. Bernard
WINNIPEG Clare & Brockest
CALGARY Ellis & Grogan
VANCOUVER McLennan, McFeely & Co. Limited

BEAVER POST-HOLE DIGGER

will please your customer. A pleased customer is an asset to any business. No wood to rot, check or break.

Special Discounts to Trade

CANADA FOUNDRY COMPANY
LIMITED

HEAD OFFICE and WORKS, TORONTO, ONT.

District Offices: Montreal, Halifax, Ottawa, Winnipeg, Vancouver, Rossland

ALEXANDER GIBB

Manufacturers' Agent and Metal Broker,
13 St. John Street, Montreal

Representing Canadian, British and American Manufacturers. Correspondence invited from firms wishing to be represented.

Expert Bookkeeping

Investigation and auditing of all classes of accounts undertaken. Trading, profit and loss accounts and balance sheets prepared. Sets of books opened and office systems devised in an up-to-date manner. Collection of accounts a specialty.

Davenport, Pickup & Co.
622 McIntyre Block and 422 Ashdown Block
WINNIPEG, MAN.,
AND AT BRANDON, MAN.

Square your Accounts

INCANDESCENT LIGHTING WITH COAL OIL

The most economic modern lighting. The "Ideal" lamp burner, screws on any ordinary coal oil lamp. This simple apparatus produces its own gas, giving a light more brilliant than electricity or gas. Costs only ¼ of a cent per hour. Guaranteed exactly as represented or money refunded. The "Ideal" lamp for the country home. Agents wanted everywhere. Write for free booklet describing everything.

ADDRESS

ST. ARNAUD FRERES
Cor. St. Sulpice and Commissioners Sts.
Montreal, P.Q.

Past Due Accounts.

We have just issued an eight page booklet regarding our Collection Department. If you have not received one write us. It will nterest you.

The Beardwood Agency

Claims, Collections and Commercial Reports

313, New York Life Bldg., MONTREAL

Where is the British Preference?

War Department of British Government makes contract with American manufacturers—Canadian horse shoe manufacturers given no opportunity to tender on contract—Where were the Canadian Government Trade and Commerce officials?

The recent action of the War Department of the British Government in placing a contract for 100,000 (one report says 216,000) pairs of horse shoes, to be made by manufacturers in United States, has caused a great deal of discussion in the Old Country, and has been brought up in Parliament by labor members who very strongly contend that no Government contracts should be given foreign manufacturers.

British manufacturers of horse shoes claim to have tendered for the contract at a lower rate than they had ever done before, and were confident that in spite of the announced intention of American manufacturers to compete for the contract, that the order would again be placed with British manufacturers. Their expectations, however, were not realized, as Americans, it is stated, quoted 15 per cent. below the old contract figure. By producing in larger quantities the latter will probably be able to still make a good margin of profit, although selling to the British Government at a lower rate than to their home trade.

Canadian Manufacturers Overlooked.

To Canadians the incident is chiefly interesting because of the fact that Canadian manufacturers of horse shoes were given no opportunity to quote prices to the War Department and endeavor to secure the contract.

Whether tenders were openly called for, as is understood, or whether the tenders were only open to certain manufacturers, the Canadian Trade and Commence department officials should have seen to it that manufacturers in Canada were made aware of the business opportunity. This paper has asked the department at Ottawa for an explanation, which will be published when received.

Another phase of the matter, however, is equally serious: Canada has been giving the Mother Country a preferential tariff for a number of years, and less than a year ago increased the preference to such an extent that practically all of Canada's trade in metals is now with Great Britain. Yet, when the British Government has an order for goods which could have been handled by Canadian manufacturers, they are entirely overlooked in favor of United States manufacturers. British workmen are complaining loudly against the contract being given to the Americans, but had the orders been placed with Canadian manufacturers using iron purchased in England, there would not have been so much justice to their criticism of the Government's action.

Definite information as to the style of shoes wanted is not available, but it is certain that had Canadian producers been given an opportunity to tender they would have done so. To Hardware and Metal J. R. Kinghorn, of the Montreal Rolling Mills, stated that his firm had not been notified and they are now making enquiries as to the type of shoe required. Thomas Moore, of the Toronto and Belleville Rolling Mills, also stated that they received no information regarding the contract and regretted it exceedingly. Likewise the Peck Rolling Mills, Montreal, stated that they were given no opportunity to tender for the contract.

Was There a Combine?

In an address at Abernethy on Sept. 7, Hon. R. B. Haldane, secretary of the British War Department, hinted at the existence of a combine amongst the British horseshoe manufacturers and said that a large saving had been made by placing the contracts with foreign firms. In referring to the matter, he said:

"The nation had entrusted his department with the spending of £28,000,000. They had to buy supplies for the army. When they assumed office, they put their heads together and made some investigation, and found that, owing to Protectionist principles having got into their army administration, the nation was paying a great deal too much for a good many things. These were very delicate things to go into, but they made up their minds that there was only one way of handling them, and that was as bold Free Traders.

"Relentless, ruthless, and remorseless were the three R's which an eminent administrator advised him to acquire when he took up office, and he and Mr. Buchanan took them to heart when they started the work of cutting down the cost of the army. They wanted to make the army more efficient, if possible, and at the same time to manage it at a cheaper rate, and they found that Protectionist would not suit them at all.

"His friend and he had been very much assailed about what was a very necessary, but innocent, sort of article, namely, horse shoes. The other day they bought a great quantity of horseshoes for the cavalry and artillery. They put out tenders. They were very careful not to issue tenders to any firms that were known to go in for sweating. What they liked to get was the best article at the most moderate price."

Saved 15 to 17 Per Cent.

"There were such things in this country, he would not call them rings, but understandings among manufacturers when it was the War Office—that was the bird that was to be plucked—that they should take as many of the feathers as they could, consequently they did not tender at very low rates.

"They threw open these contracts and they found a difference of from 15 to 17 per cent. on the prices, and they also found that they could get delivery quicker by taking American shoes. He was in favor of giving work to people at home, not that it mattered so much as some people thought, because if they imported a thing from abroad they always sent something which was manufactured here as a rule. Particularly with army and navy contracts he liked to see them executed at home, but they threw open the contract and bought in the cheapest market the best article. He believed the next time they would find that the British manufacturers would tender at considerably more moderate rates than they did the other day.

"They paid more wages in America than they did here, and yet they got a tender at a very much lower price, with quicker delivery. He hoped that would be a lesson to the home manufacturers, and that for the future they would not have to face contracts in which there seemed to be a general understanding the Government was the purchaser. They that the price should be put up because had got to protect the taxpayer, and they were determined that the taxpayer should not, so long as Mr. Buchanan and he were at the helm, have to pay too much for the goods that were taken on board the ship. These are illustrations in the concrete of the value of a Free Trade policy in a country like Great Britain. We brought in the raw material and the cheap food for the people, and sent out the manufactured goods, and all round it suited our business and our genius to be a Free Trade country."

Now a Political Issue.

The matter has now been taken up by tariff reformers and is being used by them to secure the influence of the workingmen. A despatch from London on Sept. 21 told of a meeting of the employes of the United Horseshoe and Nail Company, at Bubbitt Town, (London), where twenty employes have been laid off, and the balance put on short time because of the loss of the Government contract. The employes gave the speaker, Ben Dent, rousing cheers when he told of the loss of many industries to Britain through the Free Trade policy, mentioning, amongst others, the silk industry, cabinet-making, tanning, wood-turning, and match box making.

Canadians can leave the people of Great Britain to fight out the tariff question, our chief interest being to build up industry in this country. To this end the Canadian Department of Trade and Commerce must answer for its neglect to notify the Canadian manufacturers of horseshoes, and the British Government should be asked to explain why Canadian manufacturers were not given an equal opportunity (if not a preference) when the horseshoe contracts were called for. If we are to have a preferential agreement, let it work both ways.

Manufacturers Want More Protection

Canadian Manufacturers' Association Meets in Toronto—President Cockshutt's Address—Ex-president Ballantyne Honored.

The 36th annual convention of the Canadian Manufacturers' Association, held in Toronto this week, was largely attended by the captains of industry, who direct Canadian manufacturing enterprises. The various sections into which the association is divided got together in groups, the larger assemblage afterwards dealing with all matters of general interest. The Stove Section meeting was attended by most of the large stove founders, all reporting very satisfactory trade conditions, with the exception of the low duty on imported stoves, this tending to prevent the development of the industry.

An interesting feature was the presentation by Mr. Cockshutt on behalf of the association of an illuminated address to C. C. Ballantyne, of Montreal, ex-president of the association, in recognition of his valuable services.

Another feature of the convention was an address by Archibald Blue, chief officer of the Dominion census, illustrative of the growth of Canada in the twentieth century. Capital and products showed large increases in the five years for every Province of the Dominion, except Prince Edward Island, Ontario and Quebec showing the largest development.

President H. Cockshutt of Brantford, in his annual address, made a strong plea for higher protection, as well as for a more vigorous forestry policy, providing for an export duty on pulpwood, so that this industry would be developed in the country to which the raw material belongs. Mr. Cockshutt's address was very favorably received, and is worthy of study by every reader of this paper.

The year just closed has been marked by a steady growth in most branches of Canadian industry said Mr. Cockshutt. Production had materially increased, and, since it was by production that they gauged a nation's prosperity, they might assume that Canada had been fairly prosperous. Though collections from the West had been poor, it was satisfactory to note that there had been fewer commercial failures in 1906 than in any year since 1896. Savings banks deposits on June 30 last showed an increase over 1906 of $53,644,783, while the confidence of their financial institutions in Canadian enterprise was forcibly illustrated by the jump in current loans from $559,336,229 in 1906, to $639,970,696 in 1907. Turning to agriculture, he congratulated the Federal and Provincial Governments on the steps they were taking to foster the growth of that important industry, and on the success which had followed the work of the experimental farms.

Question of Forestry.

It was gratifying, too, to note the awakening of interest in forestry. Per head of population, Canada easily stood first among the countries of the world in forest wealth, with 148 acres of wooded land for every man, woman and child within her borders. Norway and Sweden came next, with about nine acres each, followed by the United States with only seven. Here was surely an estate worthy of their greatest care. And yet, if reports were true, huge tracts were year by year being devastated by fire, involving the loss of millions of dollars. These losses were particularly regrettable in view of the fact that many of them might easily be prevented by the exercise of greater care on the part of railway employes, settlers, prospectors and hunters.

Vigorous Policy Needed.

The whole question was one of such tremendous importance to Canada that it was surely time they were adopting a vigorous forestry policy. It seemed to him that the first step to that end should be a careful stock-taking of their national possessions. They must not lose sight of the fact that in that matter they were simply acting as trustees for posterity. He would favor the immediate creation of more forest reserves; particularly was it important that the height of land where so many of their rivers found their sources should be under perpetual forest. He believed that no reasonable expense should be spared in providing their forest land with fire rangers sufficient in number to properly protect them. He was further of the opinion that legislation should be enacted, making it a criminal offence punishable by imprisonment to expose their forests in any way to danger from fire.

The administration of such an estate was, he believed of sufficient importance to justify the Government in creating a portfolio of forestry. In addition to the duties above referred to, a Minister of the Crown might render valuable service in encouraging the establishment of schools of forestry, where practical instruction would be given on a subject which seemed destined to form the groundwork of one of their foremost professions. He might also extend the scope of the tree-planting operations which the Government had carried on with much success for the past five years.

Export Duty on Pulp.

In that connection he wished to express his strong personal sympathy with the proposal to place an export duty on pulpwood. The United States consumed 2,500,000 cords of pulpwood every year, of which Canada now supplied twenty-five per cent. "By allowing this material to leave out country in its unmanufactured state we are simply contributing to the upbuilding of our greatest industrial rival, whereas the imposition of an export duty would unquestionably compel the investment of United States capital in Canadian pulp mills, would provide employment for thousands of Canadian workmen, and would ultimately open the door for the sale of Canadian pulp across the border. Every year we defer our decision in the the matter we relatively weaken ourselves and strengthen a highly protected competitor whom we already have reason to fear; moreover, we postpone the enjoyment of a prosperity which is well within our reach."

Progress of Manufacturers.

After touching upon the fishing and mining industries, the president dwelt upon the importance of manufacturing, pointing out that, measured by the number of people to whom it gave employment and the amount of money it distributed in wages, manufacturing stood pre-eminent among its fellow industries. Comparison with previous years told a story of progress and expansion of activity in their workshops, of business for their shopkeepers, and of home comforts and more congenial surroundings for their working classes. Conditions more favorable than those for any Canadian producer could scarcely be desired for the competition he had had to meet in the home market from the products of specialized industry and cheap foreign labor had been at its lowest ebb. It was not surprising, therefore, that he should have made considerable progress, and they were quite prepared for the census announcement that during the five-year period the output of their factories had increased from $481,053,371 to $717,118,092.

Standstill in Home Market.

And yet, substantial as that growth might seem to have been, Canadian manufacturers had no more than kept pace with the expansion in other directions. At any rate they had not materially strengthened their hold on the home market. Relatively speaking, they were in the same position to-day that they were five years ago. Their capital had increased, their wage bill had increased, their output had increased, but their share in the Canadian market had remained practically at a standstill. The trade and navigation returns gave further evidence of the fact that they were not progressing as they should. Ever since 1901 their imports had been increasing much more rapidly than their exports, with the result that a balance of trade in their favor that year of $6,072,100 had for the twelve months ending June last been converted into an adverse balance of $104,476,142

More Protection Needed.

Such conditions should not obtain in a young country like Canada, for with the variety and abundance of our natural resources, combined with native enterprise, they should easily be able to produce enough for their own requirements and have a substantial surplus to sell to countries that were more thickly populated and less richly endowed. By the adoption of a policy sufficiently protective, such as their association had always advocated, capital would be irresistibly attracted by the opportunities thus afforded. Their home market would be supplied by home industries, manufacturing would become more specialized, the cost of production would be reduced, and a substantial beginning would be made towards the development of an export trade that would successfully carry them over any ordinary period of domestic depression.

Tariff Inadequate.

"It is deeply to be regretted," he continued, "that the significance of the figures above referred to did not appeal to the Government when making their revision of the tariff, for they clearly indicate that we may expect under a continuance of the present policy, when the inevitable period of depression overtakes us. A tariff which under the most favorable circumstances barely enables our manufacturing establishments to relatively hold their own cannot but prove utterly inadequate to stem the swelling tide of imports when the foreign producer seriously and systematically prepares to unload his surplus products on the Canadian market.

"And this he will do, as surely as the sun will rise on the morrow, the moment that darkening days begin to restrict his trade in other quarters. No hastily improvised tariff will then suffice to secure for Canada a continuance of her present prosperity. Retrenchment, not expansion, will be the policy of capital. Production will be curtailed, workmen will be thrown out of employment, and hard times will once more be found knocking at our doors.

A Word for Preference.

"We believe it to be the part of wisdom to guard against the possibility of such a situation materializing by affording immediate encouragement to the investment of capital in our manufacturing enterprises. It is only by so doing that we can firmly entrench ourselves against foreign competition and build up the industries of our country on a sure and solid foundation.

"We believe in a preference which will give the mother country and our sister colonies the refusal of our trade before passing it on to foreigners. In other words, our motto is 'Keep your money in circulation at home by buying goods made in Canada and when you cannot get what you want at home buy within the British Empire.' "

The Financial Stringency.

Dealing with the question of the present financial stringency, the president indicated what he believed to be some of the causes. None of them pointed to any inherent weakness in their financial institutions. On the contrary, they rather led to the belief that the stringency was due to some slight over-development, and an unfortunate combination of untoward circumstances. He felt that the criticism to which their banks had been subjected was uncalled for by reason of their having increased the amount of their call loans in New York.

The Labor Situation.

A further obstacle to natural expansion of the manufacturing enterprises was the scarcity of skilled labor, and to solve the difficulty they must either import more artisans from other countries or provide educational facilities whereby young Canadians would be able to qualify themselves for the most important positions in their factories. He regretted that the Government had not seen its way clear to meet the wishes of the association in regard to the immigration of skilled mechanics, and had hesitated to act in the matter of technical education. One of the most regrettable features of the labor situation was the steady growth in the number and importance of industrial disputes, and the association would be undertaking a noble work if it would devise means of reducing the number of industrial disputes.

Transportation Question.

Mr. Cockshutt, in concluding emphasized what he regarded as the inadequacy of their transportation facilities, contending that the present equipment of Canadian lines was far from sufficient to meet the demands made upon them by the Canadian shipping public. He quoted figures to show that the companies were making little progress towards taking care of the growing volume of traffic on the older portion of their lines, and regarded the condition of affairs thus created as a national calamity. Railway companies had spent their money in building new lines through undeveloped country when they should have spent it in properly equipping the ones already in operation.

As a result every section had suffered from the shortage of cars. Eastern business houses had been crippled, and western settlers had been left to suffer from cold and hunger. The situation was most serious and he believed Parliament would be acting in the best interests of the community at large if it would refuse to grant any more charters to old roads for the construction of branch lines until such time as the lines they were now operating were properly equipped and in a position satisfactorily to handle all traffic offering.

George F. Clare, for nearly 20 years on the staff of the James Robertson Co., Toronto, and at one time manager of the firm's hardware store at Kenora, was married in Toronto last week, and is now on his honeymoon attending the Jamestown Exhibition.

DELEGATES ATTENDING CONVENTION.

Amongst those connected with the metal and hardware trades seen at the Manufacturers' Association convention were: H. Cockshutt, Cockshutt Plough Co., Brantford; C. B. Frost, Frost Wire Fence Co., Smith's Falls; Robt. Munro, Canada Paint Co., Montreal; F. R. Murray, Emerson & Fisher Co., St. John; Cyrus A. Birge, Canada Screw Co., Hamilton; A. E. Unhler, London Machine Tool Co., London; James A. Straith, Standard Paint & Varnish Works, Windsor; J. J. McGill, Durham Rubber Co., Montreal; A. L. Young, Toronto Brass Mfg. Co., Toronto; C. C. Ballantyne, Sherwin-Williams Co., Montreal; W. K. George, Standard Silver Co., Toronto; W. R. Dunn, International Harvester Co., Hamilton; D. Lorne McGibbon, R. Lloyd Jones, and R. J. Young, Canadian Rubber Co., Montreal; T. W. Kirby, Walker Steel Range Co., Grimsby; David Findlay, Findlay Bros., Carleton Place; Chas. E. Stewart, James Stewart Mfg. Co., Woodstock; C. J. Parker, Wm. Buck Stove Co., Brantford; John Milne, Burrow, Stewart & Milne Co., Hamilton; Edward Gurney, Gurney Foundry Co., Toronto; W. M. Gartshore, McClary Mfg. Co., London; S. D. Robinson, D. Moore Co., Hamilton; J. O. Thorn, Metallic Roofing Co., Toronto; F. Clare, Clare Bros., Preston; Geo. W. Howland, Graham Nail Co., Toronto; Wm. Forwell, Forwell Foundry Co., Berlin; J. T. Sheridan, Pease Foundry Co., Toronto; H. G. Wright, E. T. Wright & Co., Hamilton; Louis A. Payette, Warden, King & Co., Montreal; H. W. Anthes and L. L. Anthes, Toronto Foundry Co., Toronto; T. L. Moffatt and Fred. W. Moffatt, Moffatt Stove Co., Weston; James M. Wilson, Owen Sound Iron Works Co., Owen Sound; A. S. Rogers, Queen City Oil Co., Toronto; R. Scarfe, Scarfe & Co., Brantford; J. A. Coulter, John Morrow Screw Co., Ingersoll; James Maxwell, D. Maxwell Co., St. Mary's; Murton Church, Page Wire Fence Co., Walkerville; J. H. Paterson, Toronto Hardware Mfg. Co., Toronto; H. L. Frost, Frost Wire Fence Co., Hamilton; Geo. B. Meadows, Geo. Meadows Wire, Iron & Brass Co., Toronto; A. E. Kemp, Kemp Mfg. Co., Toronto; C. Dolph, Metal Shingle & Siding Co., Preston; A. H. Chapman, Ontario Wind Engine & Pump Co., Toronto; John H. Tilden, Gurney-Tilden Co., Hamilton; E. C. Boeckh, United Factories, Toronto; W. L. McGregor, McGregor, Banwell Fence Co., Walkerville; Ed. Fairbairn, Ontario Wind Engine & Pump Co., Toronto; H. T. Bush, Standard Ideal Co., Port Hope; W. C. Springer, Belleville Hardware Co., Belleville; B. Fletcher, Fletcher Mfg. Co., Toronto; M. A. Croxall, National Hardware Co., Orillia; Alex. Macpherson and C. N. Candee, Gutta Percha Rubber Co., Toronto; Lloyd Harris and J. B. Detwiler, Brantford Screw Co., Brantford; J. R. Marlow, Canada Cycle & Motor Co., Toronto Junction; H. G. Nicholls, Canada Foundry Co., Toronto.

HARDWARE DISPLAYS AT SHERBROOKE EXHIBITION

In spite of bad weather, the Dominion and Provincial Exhibition, held at Sherbrooke, Que., September 2nd to 14th, by the Eastern Townships Agricultural Association, was a distinct success in every way. In keeping with the added importance this year of the exhibition, the directors made many improvements, which were appreciated by the vast throngs. A new Arts Building, a new Dairy Building, and an enlarged and renovated grand stand greeted visitors. The grounds were improved throughout and large tents accommodated new displays. In point of exhibits and general importance, the exhibition ranks very high among the fairs on the continent.

Vaudeville attractions provided were of a high order, and each week saw a complete change of bill. The trotting and running races attracted high-class fields and good contests resulted.

The first week was interfered with by steady downpours of rain, but the attendance was good. The opening of the second week was attended by good weather and the Governor General's day witnessed the largest paid attendance in the history of the association, over 30,000 passing the turnstiles. The Governor General and suite, the Lieutenant-Governor, and many of the Dominion and Provincial Ministers visited the exhibition.

As usual, the main building containing the various industrial displays, was the centre of attraction.

Canadian Fairbanks' New Factory.

The accompanying illustration shows the United States factory of R. & T. Fairbanks & Co., the manufacturers of Fairbanks' scales, located at St. Johnsbury, Vt.

This firm has recently acquired a block of land in Sherbrooke, P.Q., and are erecting there a large factory to supply the demands of the Canadian trade. They have been manufacturing scales for over three-quarters of a century, and their factory has grown from two small wooden shops in 1830 (from which the finished product was hauled 18 miles by ox teams to the railroad) to an enormous concern employing between 1,200 and 1,400 men, and turning out scales suited to the standards of every country in the world where a weighing machine is used.

Their decision to build in Sherbrooke means another flourishing industry for that place. It is intended to employ about 200 men at first, and it is hoped and expected that in time the Canadian branch of the business will prove a worthy rival to the branch across the border. The building of a Canadian factory merely emphasizes the general tendency of large manufacturing concerns in the States, to recognize that Canada's trade is becoming more and more valuable, and that she is in a position to support her own large industries

The whole output will be handled by the Canadian Fairbanks Company, whose head office is in Montreal, with branch houses in Toronto, Winnipeg and Vancouver.

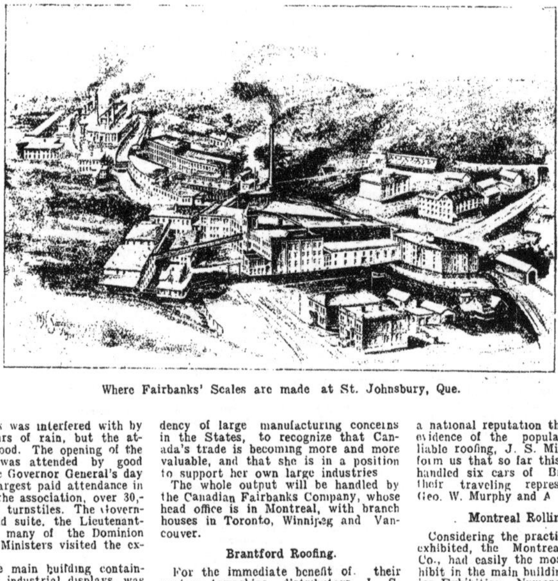

Where Fairbanks' Scales are made at St. Johnsbury, Que.

Brantford Roofing.

For the immediate benefit of their eastern townships distributors, J. S. Mitchell & Co., Sherbrooke, the Brantford Roofing Company made an attractive and instructive exhibit. Explicit literature and samples were distributed daily to visitors.

The exhibit was in the shape of a cottage roofed with the Brantford Roofing Co.'s asphalt and rubber roofing, which is made of Quebec wool felt saturated with pure asphalt, coated each side with sharp-pointed sand worked into the material by machinery, producing durability and insulation. For flat roofs the asphalt is coated each side with rubber, the elasticity of the asphalt protecting the rubber and making it more than ordinarily durable.

Brantford roofing is not a ply roofing, bit is solid wool felt, unaffected by any temperature. Expansion and contraction divide the sections of a ply roofing. It is not cracked by the sun, absorbed by the atmosphere, or affected by acids or gases. It is not applied with tin caps, wood strips, etc., and is practically fireproof.

The company do not give a theoretical guarantee, but a fearless one for ten years, conditional only upon an application, if necessary, within the ten years, of a coating of their genuine asphalt paint, which can be procured from any of their agents at a cost of 25 cents per square. Brantford roofing is not an experiment, having been made in the United States for many years and enjoying a national reputation there. As a local evidence of the popularity of this reliable roofing, J. S. Mitchell & Co. inform us that so far this year they have handled six cars of Brantford roofing, their traveling representatives being Geo. W. Murphy and A. C. Stevens.

Montreal Rolling Mills.

Considering the practical articles they exhibited, the Montreal Rolling Mills Co., had easily the most attractive exhibit in the main building at the Dominion Exhibition. Numerous incandescent lights added to the effect. This was their first display at a Sherbrooke exhibition. and doubtless pleased their jobbers in the section. Fred. Parego and L. Kreiger, practical representatives of the firm, had charge, and explained all details to many interested listeners. Samples of various lines were at the front of the booth, and they were frequently made use of.

Perhaps the outstanding feature of the display was their trade mark formed of all sizes of their horse nails. This may easily be noted in the illustration. Their horseshoe lines occupied the entire space at the back and every description of a horseshoe had a showing. Each horseshoe was briefly described for the advantage of the sightseer. Their representatives told of their true proportion and accuracy in manufacture. All weights and sizes of their large line were on the board. A special feature of the horseshoe exhibit was their countersunk shoe, which was greatly admired by all the visiting farriers.

Although their cold process horse nails are comparatively new, many farriers found using them were free in their praise of their shape, material and driving qualities. At the left from the front their bolts and nuts were fully represented.

The booth was an expensive structure, but results amply proved the advisability of a first-class showing, in keeping with the importance of the concern.

Representatives of all the large stove foundries were in Toronto during the week attending the Canadian Manufacturers' Association convention.

Hyman Millar, of the wholesale hardware firm of Millar, Morse & Co., Winnipeg, spent two or three days in Toronto this week on his return trip from Europe.

The Carriage Mountings Co. have moved their business from Toronto to their new buildings at Niagara Falls, Ont., where they now have an up-to-date plant for the manufacture of automobile and carriage nickel and brass work, and lavatory supplies.

Frederick Southcott has severed his connection with the firm that went so long under the name of Coy Brothers & Southcott, St. Catharines, and Frank E. Coy is now the sole owner of the business. Mr. Southcott has secured some good agencies from manufacturers, and will make Toronto his headquarters.

Montreal Rolling Mills Display at Sherbrooke Exhibition.

HARDWARE TRADE GOSSIP.

Ontario.

J. R. Watson, hardware merchant, Schomburg, Ont., spent Thursday in Toronto.

Mr. Matthews, of the hardware firm of Boxall & Matthews, Lindsay, was in Toronto on Wednesday.

Munroe Bros., Alvinston, Ont., have sold their hardware and tinware business to Johns & Miners.

W. F. Moore, secretary-treasurer of the Ontario Lantern and Lamp Co., died in Belleville on September 10.

Wm. Knowlton has opened a hardware store at the corner of Brunswick avenue and Harbord street, Toronto.

N. L. Stewart, of the Guelph Stove Co., was a visitor at the Toronto office of Hardware and Metal on Wednesday.

Miles Vokes, president of the Vokes Hardware Co., Toronto, will be a candidate for the mayoralty of Toronto next year.

C. R. Banks, who formerly ran a machine shop in Paris and afterwards opened up repair and bicycle shops in Galt and Peterboro, has sold out his business in the former place to George McFarland. Mr. Banks will discontinue his machine repair ship in Peterboro, and devote his time entirely to the bicycle and sporting goods business.

Quebec.

Charles Leblanc, of Joliette, was in Montreal this week.

Mr. Carter, of the Eagle Lock Co., Torrington, Conn., was in Montreal last week.

W. L. Frisby, of E. C. Stearns & Co., Syracuse, N.Y., is calling on the trade in Montreal this week.

A. J. Wood, of Montreal Rolling Mills, has left on a three weeks' business trip to Winnipeg.

A. J. Clymer, representing the Enterprise Mfg. Co., Philadelphia, Pa., called on the trade in Montreal last week.

R. E. Mills, formerly of Page Hersey Iron, Tube & Lead Co., Guelph, and now of New York City, was in Montreal last week.

G. H. Cummings, advertising manager with Frothingham & Workman, Montreal, is spending a few weeks' holiday in the north country with his gun.

Owing to Mr. Paterson having severed his connection with the firm of Baxter, Paterson & Co., Montreal, this business will be carried on at the same address under the name of J. R. Baxter & Co.

Gordon Seybold, secretary-treasurer of Starke-Seybold Limited, Montreal, who has been confined to the hospital with typhoid fever, is improving very favorably, and it is expected he will soon be able to resume his duties at the offices.

The following prominent Montreal manufacturers left this week to attend the Canadian Manufacturers' convention in Toronto: John Watson, P. Hamill, J. A. Christin, S. J. B. Rolland, Col. Burland, J. R. Kinghorn, J. S. N. Dougall.

Capt. C. M Strange, sales manager, Lewis Bros, Montreal, returned last week from a holiday trip to Boston, New York and several Atlantic seaside resorts, ending up at his old home at Kingston, which he still considers the choicest of nature's beauty spots.

J. T. Belaire, formerly city traveler for Starke-Seybold, Limited, Montreal, has bought out the hardware business of Louis Ponton, in Montreal, and has commenced business for himself. Mr. Belaire was well known and respected throughout Montreal, and Hardware and Metal joins in wishing him every success.

Amongst those who visited Montreal this week were: W. Wood, of Douglas & Co., Amherst, N.S.; C. O. Jervas, St. John; Mr. Mathon, of Mathon Freres, Ville Marie; Mr. Southcott, of Coy Bros. & Southcott, St. Catharines; R. J. McKelvey, of McKelvey & Birch, Kingston; H. Desjardin, Terrebonne; S. Bourgeois, Ste. Hyacinthe.

F. R. Murray, director and chief buyer of Emerson & Fisher, Limited, wholesale hardware merchants, St. John, N.B., was in Montreal on Saturday and paid Hardware and Metal a pleasant call. Mr. Murray is combining business with pleasure, and is a delegate to the annual convention of the Canadian Manufacturers' Association, which took place this week in Toronto.

Western Canada.

S. E. Maud, Elstow, Man., has sold his hardware business to S. A. Clark.

W. J. Doyle, hardware merchant, Clanwilliam, Man., has assigned to E. Bailey Fisher, Minnedosa.

W. B. Marshal, of the hardware firm of W. B. Marshal & Co., Medicine Hat, Alta., spent a few days in Toronto last week.

Jas. S. Oliphant, who for some little time has been connected with the Brandon Hardware Co., is now with McDonald & Vogt, of Dauphin.

HARDWARE AND METAL

Established 1888

The MacLean Publishing Co.
Limited

JOHN BAYNE MACLEAN - President

Publishers of Trade Newspapers which circulate in the Provinces of British Columbia, Alberta, Saskatchewan, Manitoba, Ontario, Quebec, Nova Scotia, New Brunswick, P.E. Island and Newfoundland.

OFFICES:

MONTREAL, - - - - - 232 McGill Street, Telephone Main 1255
TORONTO - - - - - - 10 Front Street East, Telephones Main 2701 and 2702
WINNIPEG, - - - - 511 Union Bank Building, Telephone 3726
LONDON, ENG - - - - - 88 Fleet Street, E.C., J. Meredith McKim, Telephone, Central 12960

BRANCHES:

CHICAGO, ILL. - - - - - 1001 Teutonic Bldg, J. Roland Kay
ST. JOHN, N.B. - - - - No. 7 Market Wharf
VANCOUVER, B.C. - - - - Geo. S. B. Perry
PARIS, FRANCE - Agence HaVas, 8 Place de la Bourse
MANCHESTER, ENG. - - - - 92 Market Street
ZURICH, SWITZERLAND - - - - Louis Wolf, Orell Fussli & Co.

Subscription, Canada and United States, $2.00
Great Britain, 8s. 6d., elsewhere - 12s

Published every Saturday

RECIPROCAL DEMURRAGE.

Vexatious delays in the delivery of freight have led to the organization in the Manitoba capital a few months ago of the Winnipeg Jobbers' and Shippers' Association, through which the leading business men are making an earnest effort to find the proper remedy. The remedy which they favor is that now being advocated by boards of trade throughout the United States, and known as the Reciprocal Demurrage Law. Briefly stated, the agitation for this law grew out of a conviction that it is a poor rule that will not work both ways. When shippers are responsible for delay in loading and unloading cars they are forced to pay the company's demurrage charges. This is an absolutely fair and just rule, without which there would be many delays that could easily be avoided. But it is only natural that it should have occurred to someone to suggest that when the railways put shippers to inconvenience by long delays in the delivery of freight, the aggrieved parties should have some speedy and adequate compensation; in short, that there should be a Reciprocal Demurrage Law subjecting the railway companies as well as shippers and consignees to the payment of demurrage rates when delays occur

In the United States there are two main features of the Reciprocal Demurrage Law which the boards of trade desire to have placed on the statute book. In the first place, they desire a provision subjecting the railways to penalties when they are unable to supply cars asked for at given points, and in the second place they ask for penalties when cars are not moved more than a certain number of miles per day when once en route. The former provision is probably unreasonable and it is not being emphasized by the Winnipeg Association. The time may come for such a provision, but it is felt that under existing conditions it would be unreasonable to press for its enactment. However, with the provision penalizing the railways for delays in forwarding freight already shipped, the Winnipeg Jobbers' and Shippers' Association are in hearty and enthusiastic accord, and they have asked the Railway Commission for a clause in all bills of lading providing as follows:

"That all goods shall be transported on an average for the entire journey of at least 100 miles on all main lines and 50 miles on all branch lines, per day of 24 hours; that all goods shall be forwarded and loaded within 24 hours after delivery to the company, and all goods to be unloaded by the company shall be placed for unloading within 24 hours after arrival at destination, Sundays and legal holidays to be excepted. When these conditions are not fulfilled the company shall make the consignee an allowance of $1 per day for all carloads and one cent per 100 lbs.; maximum $1 per day for all less carload shipments for every 24 hours or part thereof until the shipment is unloaded or placed for unloading at destination. Provided that nothing in this clause contained shall be constructed to deprive the owner or consignee of any rights to which he may be entitled under existing law in respect to the subject matter hereof, the foregoing allowances being intended by way of additional remedy."

It is contended that if there were proper handling of freight in terminal points there would be no difficulty in moving freight at the rate provided by the above clause. It is stated that cars are often allowed to accumulate in the railway yards through lack of motive power; that there is little care taken to insure that cars that have been longest delayed shall be the first to be sent on their journey. Usually these cars are not the most easily accessible, and trains are made up of cars that have been in the yards only a few hours, while cars that have been delayed for days or weeks are allowed to be delayed still longer.

It is true that there are regulations covering this point, but in the absence of adequate penalties it is said by the shippers that the rule is honored more in the breach than in the observance. The Winnipeg Jobbers' and Shippers' Association argue that it is not unreasonable to ask the railways to move a car 100 miles per day on the main line or 50 miles per day on the branch lines, but after investigation of the experience of their members for some months, they show that the actual average is only 35 miles per day. They argue that the railway companies need more motive power and more effective supervision of cars in terminals, rather than an increased number of cars. If cars were moved at the rate provided for by the provision which the association desire to have enacted, one car would do the work of three at the present time.

The Winnipeg Jobbers' and Shippers' Association have the co-operation of nearly all the boards of trade in Western Canada in their endeavor to have this provision enacted.

BE A BOOSTER.

Everybody likes the man who goes around with a smile on his face. Some way or other he seems to help you by his good humor. It's an old saying that happiness is contagious, and so is a smile. On the other hand the fellow that turns up at the wrong moment, and other times, with a long face and a tale of woe makes you feel that you have lost something. Therefore look pleasant. Be a booster. If something goes wrong do not tell all the world of your trouble, but quietly go about righting things, and never lose your smile.

People want to be amused, and they like to hear someone talk who always has a good word for somebody or something Be a booster.

MAKE EARLY PREPARATIONS.

Forethought is a great deal more valuable to the dealer than any amount of activity, when the time for action arrives. Some men pride themselves on their skill in handling emergencies, but much of this skill resolves itself into adequate preparation to meet just such emergencies. The holiday trade which is now looming up on the horizon, calls for early and ample preparation. Profit by past experiences and make those arrangements which you deem necessary while there is time to think, and act calmly. Then, when the critical time comes, you can rest assured that you have done everything possible to make the trade a success. There is no reason why an increased business cannot be transacted this year than last year.

WHAT TO DO NEXT MONTH

Inventory days come to men in every phase of activity at almost regular intervals. There is an instinctive desire in men to keep constantly posted as to their latitude and longtitude or, to use a colloquial phrase, to know "where they are at." They must not only be prudent and look forward, but they must also be wise and look backward. They must rid themselves of the obsolete and the "worn-out," and free themselves of all encumbrances to step forward into the future.

* * *

The politician at frequent times in his public career takes an inventory of the various planks in his policy, eliminates one here, adds another there, and thereby preserves that modernity in his policy, a quality which makes all things valuable and attractive. The manufacturer has two things to look after. He must see that his machines are thoroughly up to date, that they are turning out the money-bringing product with the minimum amount of flaw production. He must also see that the goods he is making are modern and keep pace with the constantly changing tastes of men. If any one line of goods is becoming old-fashioned and unseasonable, he must cease producing them, and if there is any superfluous stock he must dispose of it at a minimum loss.

* * *

So it is with the hardware merchant. At regular intervals throughout the commercial year come times when stocks must be inspected and over-hauled, when unseasonable goods must be judiciously selected and disposed of at a minimum loss by a careful advertising campaign in the local papers, by an attractive interior and window display, or by both. The larger the superfluous unseasonable stock, the nearer is the time for re-stocking, the more necessary it is to carry on an aggressive advertising campaign, both by window display and newspaper publicity.

* * *

October is an inventory month, a month for inspecting and over-hauling stock. It is the month of transition from the summer stock to the fall and winter stock, the month when enamelware, tinware, builders' hardware, and agricultural tools must give precedence to sporting supplies and winter goods such as hockey sticks, skates, guns, rifles, ammunition, sleigh bells, cow-ties, horse blankets, curlers' and skaters supplies.

* * *

The first thing to do next month will be to thoroughly overhaul and inspect stocks, not only to see if the supplies of goods for the coming season are calculated to be adequate to the demand, but also to see if there are any goods which will soon be unseasonable, and, therefore, unsalable. If such is unfortunately the case, and it invariably is, these goods should be dragged from their hiding places. Then a profitable campaign of publicity should be considered and devised. If the superfluous stock is sufficiently formidable to warrant not only window displays, but also newspaper publicity, then the various local papers should be communicated with regarding space and rates. As the game is essentially a losing proposition, the prospective advertiser should keep in mind the best-placed and largest space for the least money, if the initial loss is to be at all diminished. The display of unseasonable goods will not in any case warrant the purchase of elaborate publicity. The truth of this is easily discernible. Half a column or a column in a couple of rural weeklies for a week or a fortnight is sufficient publicity to reasonably dispose of a stock rapidly becoming unsalable. Attractive arrangement of the goods in the front windows with price cards attached will aid very materially in disposing of the goods.

* * *

When these goods, which would have become "dead weight," have been gotten rid of, serious attention should then be given to the fall and winter stocks. That the orders for these goods have been placed, and in many cases have been filled, in unnecessary to state. If only a portion have been received, invoices should be looked up and the remainder of the shipment traced up, and, if it has been the fault of the manufacturer, carry on a vigorous correspondence with him. If it is the fault of the railroad or other transportation company, see to it that your goods will soon arrive. Have prompt delivery guaranteed in order not to derange your sales arrangements.

ANOTHER PRIZE OFFERED.

It's not too early to lay plans for the Christmas holiday trade. Dealers have already bought some lines, of course, and in these days of slow deliveries other orders should be placed as soon as opportunity offers. "Goods well bought are half sold" is a time-worn truism in mercantile life.

But it doesn't do to feel too secure on the buying part until the selling has been done. Plans for increasing holiday sales must be worked out months ahead. Window displays must be figured out and a series of rough sketches planned. A series of ads. for the local papers should also be prepared and, where deemed advisable, a neat booklet gotten up, to be mailed to all probable customers in the surrounding district.

It is in these selling plans that Hardware and Metal can be helpful to its readers. Let all partake of the spirit of co-operation and exchange ideas through these columns. To help the suggestion along the editor invites any reader to join the friendly contest and hangs up a prize of $10 cash for the best answer to the following questions:—

How can the hardware merchant increase his sales of holiday goods next December? What special lines should he stock? What selling plans should be adopted? What special advertising should be done? What novel window displays can be suggested? Should souvenirs (calendars, knives, trays, etc.) be given to customers?

The prize of $10 will be awarded to the writer of the most practical and original letter of from 500 to 1,000 words received by the editor before October 15, 1907, and the best half-dozen replies will be published in Hardware and Metal.

* * *

Elaborate plans for advertising your fall and winter goods should then be formulated. That publicity is absolutely essential to successful salesmanship is a commonplace. The goods must be advertised, not only in the windows and on the shelves (by attractive arrangement), but the daily and weekly newspapers also must be brought into active requisition. Even a small regular space in the newspapers filled each day or week by bright, persuading copy, will produce results.

* * *

Sporting goods is the big feature of the fall and winter stock. So far as the ability to sell guns or skates is concerned it is really a "snap." The most durable satisfaction to the salesman is not so much the quantity or number of sales as the quality and the actual money received from the sales. It is obvious that a business in which a sale of $50 was effected with the same ease as ten sales of $5 each in another business are effected, is the better and the more remunerative business.

* * *

Keeping in mind that it is really easy to sell a rifle or a pair of skates, it should be the aim of the merchant to carry the highest class of goods. It not only gives a little better chance of making bigger profits, but it also does much towards enhancing the reputation of the firm. If the merchant is advertising a certain make of rifle in the local paper and he wants to make the ad. apparently attractive by using such superlatives as "The Best," "The Easiest to Manipulate," "The Safest Rifle Made," he must handle the best rifle, the easiest to manipulate, and the safest, so that his words may be substantiated by facts. Profitable advertising must be judicious—must be convincing, without the use of superlative adjectives.

NEW WESTERN WHOLESALE CENTRES.

Portage la Prairie and other western towns and cities having laudable ambitions to become wholesale centres of importance, have recently been urging upon the Railway Commission the unequality of freight rates in the west, contending that there is discrimination in favor of Winnipeg. The Portage la Prairie Board of Trade presented a strong case before the commission last month, asking to be placed on an equality with Winnipeg as a wholesale centre. The secretary of the board and several of the leading business men of Portage la Prairie appeared before the commission, and the whole question was argued at length. The decision of the commission has not yet been announced, but there are well authenticated rumors that the requests of the ambitious western city will be granted almost in their entirety.

Whether this be true or not, there can be no doubt that it is only a question of time until many of the western cities must be given advantageous through and distributing rates which will permit them to become important wholesale centres. Winnipeg has been the first western wholesale centre of importance, but the country is too big to allow it to have a monopoly. In the far west there are several cities of importance already. Calgary, Edmonton and Regina are rapidly growing in importance as wholesale centres and they will have many rivals.

Markets and Correspondence

(For detailed prices see Current Market Quotations, page 66.)

MARKETS IN BRIEF.

Montreal.

Antimony—A little firmer.
Copper—Weak.
Lead—Very firm.
Linseed Oil—Advanced 1 cent all round.
Old Materials—Quiet.
Spelter—Much stronger.
Tin—Unsteady.
Turpentine—Declined 5 cents.
White Lead—Very firm.

Toronto.

Antimony—Advanced half a cent.
Copper—Now quoted at 18c.
Pig Iron—No buying being done.
Old Materials—Several lines decline.
Turpentine—Decline of 3c.
Linseed Oil—Firmer.
Plain and Barbed Wire—5c advance.
Harvest Tools—New discounts out.
Stoves and Ranges—Will advance 5 per cent. on October 1st.

MONTREAL HARDWARE MARKETS

Montreal, September 27.—A large volume of business is being transacted by both the wholesale and retail trade, in a proportionate degree. The usual activity characteristic of the fall and winter season prevails at present. Orders both for immediate requirement and remote delivery are being received in large numbers, and they are not all meagre orders. Supplies from the factories are being forwarded to the jobbers with a little more regularity and promptness than has been usual in the past, and although alarmists persist in publishing discouraging reports as to the western crop conditions, buyers evidence a good deal of confidence in prospective conditions. The reports circulating through the mercantile world regarding the crops are generally of such a contradictory nature that little credence is accorded them. Merchants are very hopeful as to the relief from financial stringency in the near future.

Screws—Jobbers continue to experience considerable difficulty in procuring complete shipments from the factories, and it is thought it will be some time yet before the factories can complete their supplies. Not quite so much activity is noticed at present in buying. Prices are firm and unchanged.

Sporting Goods—A large amount of ammunition and guns has been sold during the past fortnight or three weeks. Retailers are stocking up rapidly in these. Hockey sticks and skates are being enquired for in fairly large quantities and the business in these has not been marked with the same activity as in previous years. The outlook is very encouraging. Prices are firm.

Poultry Netting—The business for this year has closed and until next year's prices are published, a little later, there will be no spring orders booked. A large volume of business has been done this year.

Stoves and ranges—A very heavy demand exists at present, heavier than that of any previous season. This is due probably to the fact that on October 1 this year an advance all round of five per cent. will be made, and dealers are anxious to get in their stock before that time. At present the factories have more orders than they can take care of.

Cutlery and Silverware—During the past year a large amount of business has been done in cutlery. Every line has experienced an active call. At present, however, the buying public, especially in the northwest, are holding off, owing to scarcity of ready money, and the outlook for fall and winter trade is rather uncertain.

Building Paper—Little activity is noticeable at present. The outlook for the fall and winter trade is very bright and encouraging. Prices are steady and unchanged.

Cement—Owing to a serious scarcity of cars, the business in cement is badly crippled. Factories which had been running night and day can now run in the day only. Large orders are coming in, but producers cannot fill them, for the simple reason that because of a great scarcity of cars it is absolutely impossible to get raw material. Prices are very firm, with a strong upward tendency.

Builders' Hardware—The demand is strong and steady and supplies are quite adequate. A large volume of business is being transacted.

Wire Goods—Deliveries in the northwest are being rushed before the close of navigation. A heavy demand prevails and there is no surplus of supplies with the producers. Prices are firm and unchanged.

TORONTO HARDWARE MARKETS.

Toronto, Sept. 27.—An excellent trade is developing in all staple lines of fall and winter hardware and each week the volume of busines scontinues to increase over that of the week before. Western business, which has been light since the present tightness came upon the money markets, is now opening up more favorably and large amounts of fall and winter goods are being rushed west in order to take advantage of lake freight rates before the close of navigation. Ontario business continues in a flourishing state; the heavy demand for builders' hardware and supplies which has characterized the entire season's trade being still decidedly in evidence. As prices are already out for several lines of spring goods, and additional prices are expected within a few days, some heavy booking of spring requirements will undoubtedly commence within the next fortnight. Fall sporting goods, especially guns and ammunition, are in big demand and retailers will do well to have on hand a good supply of rifles, ammunition and shooters' coats before the deer shooting season opens in November.

Wire—Barbed wire has advanced five cents per hundred pounds, and is now quoted at $2.75 f.o.b., Cleveland. Plain galvanized wire has also advanced five cents and numbers 9, 10, 11 and 12 are now quoted from stock at $2.90, $3.45, $3.50 and $3.05, respectively. Base sizes, numbers 6 to 9, are still quoted at $2.35 f.o.b., Cleveland. Extras for cutting.

Harvest Tools—Next year's prices are now out and are slightly lower than present stock prices. The new discount is 50 and 10 off standard list, with a freight allowance of 50 cents to points in Manitoba, British Columbia and the New Provinces. It is expected that heavy booking will commence at once.

Copper and Brass Goods—Jobbers are placing lower prices on cabinet hardware, drawer pulls and all similar brass or copper goods. The reduction is due to the protracted slump in the copper market.

Cutlery and Silverware—A splendid autumn business is being done in both and many large orders are being booked for Xmas requirements. The number of repeat orders coming in to local jobbers is a sure evidence that retailers are receiving a heavy call for hunters' knives and small pocket cutlery.

Nails and Screws—The nail mills are all working to their full capacity and supplies are well up with the trade. Screws are still very scarce, but the demand is not as pronounced as it was a short time ago, and the factories are taking full advantage of every opportunity to catch up with their orders. Prices both on nails and screws are unchanged.

Stoves and Ranges—The factories are working to their fullest capacity in an endeavor to meet the exceptionally heavy demand which prevails. For the last few weeks retailers all over the country have been forwarding their orders to local dealers in order to escape the 5 per cent. advance on stoves and ranges, which will go into effect on Oct. 1, as announced by Hardware and Metal some time ago. Stove dealers who have not placed their orders will do well to figure their winter requirements at once and not be caught napping when the advance comes.

Sporting Goods—An exceptionally heavy business is being done in guns, ammunition, hunting coats and boots, and other fall sporting requisites. The number of enquiries from sportsmen concerning rifles and deer-shooting cartridges would indicate that a heavy business will be done when the season for big game opens in November. Small retailers should be on the alert to secure their share of this profitable trade.

Hinges—Since last week's advance on heavy T and strap hinges there has been no further change, the price remaining firm at the advanced figures.

MONTREAL METAL MARKETS.

Montreal, September 27.—The strength in metal market conditions last week, which was confined almost wholly to the pig iron trade, has become somewhat contagious and found its way into the trade in various of the other commodities. Activity characterizes the metal business generally this week. Dealers throughout the city are becoming better satisfied with the volume of business being transacted.

Most consumers are coming into the field for fall and winter requirements, but few of them show any desire to commit themselves in the way of placing orders for distant delivery, as they have not yet sufficient confidence in the trade, as affected by speculative fluctuations. Prices on all lines are very firm and, in a few cases, display an upward tendency.

Pig Iron—Recent statistics, as published by a prominent Glasgow firm, go to show that there is practically no high-grade Scotch iron available either for immediate or future delivery. The statement shows that in January, 1907, there were 517,386 tons of iron in Scotland, and on September 13 of the same year there were only 166,951 tons—a very large diminution when it is remembered that on September 13, 1906, there were no less than 582,083 tons available. The steadiness of the British pig iron market, according to a recent editorial in the Iron Age, since the middle of the year, is a matter of surprise, especially when it became plainly apparent that the end of demand from the United States was in sight. English iron, by the latest despatches, shows an advance of about 1s 6d. It will be very difficult from now to the end of the season to secure anything like prompt deliveries of English iron, owing to the fact that there is such a heavy demand for English coal on the continent that tramp steamers are nearly all engaging in the transportation of this rather than in carrying iron to Canada, with the uncertainty of return cargos in late November. The American market continues in a demoralized condition. Prices locally are unchanged; Middlesboro, No. 1, $21.50; No. 2, $20.50; Summerlee, $25.50.

Lead—Remains very firm. Prices are very high and the demand is strong. We continue to quote bar at $5.75.

Spelter—Is much stronger. Prices in the United States have been advanced 10c. Curtailment of production is the order of the day.

Copper—There is still considerable hesitancy amongst consumers in buying, as the American market has not yet settled down. The 15-cent basis has been established and opinion pronounces it a reasonable price. Prices are weak and unchanged.

Ingot Tin—Considerable fluctuation is still existant in the English market. Not much business is being transacted. Prices are easy at $44.

Old Materials—There is little doing yet in the old metals market. Prices are very weak. Little demand exists for scrap iron, the supplies of which are large. Heavy copper and wire, 13c; light copper, 11c; heavy red brass, 11c; heavy yellow brass, 8c; light brass, 6c; heavy lead, 4c; No. 1 wrought iron, $14.50; No. 2, $6; stove plate, $12; machinery cast scrap, $17.

TORONTO METAL MARKETS.

Toronto, Sept. 27.—The past week has seen an improvement in metals, jobbers getting prices down to a better buying basis and customers feeling more free to place orders. In London there is a firmer tone, tin being improved by a cent advance. In the States, copper is still the feature, with the price situation about the same as a week ago.

Manufacturers, as a rule, have some unused stocks of metal, bought at higher prices than are now prevailing and they are unwilling to enter the market for additional material until they are compelled to or until the market shows a more firmer future outlook. Manufacturers are in the position of using high-priced material in producing articles which are selling low. Take boilers and radiators for example. The iron being used costs from $1 to $2 more than a year ago, while discounts have been shaded from 50 to 50 and 10, and in some cases to 50 and two tens. On the other hand, iron pipe manufacturers are unable to supply the demand, and some sizes are being imported from the States and sold at a premium. Inch and a quarter, for instance, is listed at $7.65 for black and $9.90 for galvanized, while it is selling for $9.25 and $10.80, respectively. American pipe manufacturers will not promise delivery in less than four months on some sizes.

Pig Iron—At this season buying is usually brisk, but consumers are holding off and waiting for better terms. Prices have been shaded a couple of dollars since spring, but buyers are not yet tempted. There is no scarcity of stock, although the Midland furnace will be closed down until the middle of November. We quote $22 for Middlesboro No. 3.

Copper—Locally the price has been brought down to 18 cents, in keeping with the conditions across the line. Speculation, however, continues as to the future and buying is none too lively. The curtailment in production by the Amalgamated interests is estimated at from 10 to 12 million pounds per month, according to some trade estimates stocks in the hands of producers on the 15th of September were only 186,000,000 pounds, but it should be remembered that this does not include 192,000,000 pounds reported by the Government on the first of the year, only 110,000,000 pounds of which are required to meet consumptive requirements, both domestic and foreign, under present conditions. Adding to surplus we should have stocks of 268,000,000 pounds, which verifies previous estimates, or from 250 to 260 million pounds. Even with the reduction in output now to be expected, the current production will more than equal the current consumption, so that until there is a radical change in the position of manufactured material, small draft, if any, will be made upon the accumulated stocks. It is interesting to recall that in 1901, when surplus stocks had reached about 300,000,000 pounds, copper broke from 17c to 11c, and it was not until 1904 that the surplus was worked off and a gentlemen's agreement entered into by one large lake and two Montana and Arizona producers, which succeeded in turning the market and prices already advanced reached 14c or so; then came a flagging market after more accumulation. At this point the Chinese scheme was conceived, which worked well. After the sea voyage the copper increased in value and was finally melted in Europe and America at enhanced intrinsic value. Then came the rise to 25 and 26 cents under heavy consumption and famine cry until we have to-day another accumulation of 200,000,000 to 275,000,000 lbs. The questions now are, how long will it take to dispose of the surplus, and how low will copper go before the turn comes? The low notch is now set at 12c.

Ingot Tin—The small change that has been evident in the statistical position of tin during the week has been slightly in favor of consumers, but on the other hand the market being so well cornered at London, foreign operators have practically dominated the situation and put up the market 1c per pound on spot tin.

Sheet Metals—Little buying will be done for next year with the market unsettled. Jobbers are hoping to be able to quote satisfactory prices next month.

Lead and Zinc—Both metals keep firm with no change reported.

Antimony—The market is stronger and prices have gained one cent. We now quote 13 to 13½ cents.

Old Material—Stocks are plentiful, but buying is almost nil. Dealers have dropped prices, as their stocks are more than adequate to meet the demand. Copper, brass, zinc and scrap iron have all been shaded half a cent.

LONDON, ENG., METAL MARKETS.

London, Sept. 25.—Cleveland warrants are quoted at 55s 6d, and Glasgow standards at 54s 9d, making prices as compared with last week, on Cleveland warrants, 10½d higher, and on Glasgow standards, 1s higher.

Tin—Spot tin opened easy at £168 15s, futures at £166, and after sales of 130 tons of spot and 500 tons of futures, closed firm at £169 for spot and £166 for futures, making price as compared with last week £1 higher on spot, and 10s higher on futures.

Copper—Spot copper opened irregular at £65 17s 6d, and after sales of 400 tons of spot and 1,300 tons of futures, closed firm at £66 for spot and £66 for futures, making price as compared with last week 15s lower on spot and futures.

Spelter—The market closed at £20 17s 6d, making price as compared with last week 7s 6d lower.

Lead—The market closed at £20 10s, making price as compared with last week 15s higher.

U.S. METAL MARKET.

Cleveland O., Sept. 26.—The Iron Trade Review to-day says: New business in structural material is light in nearly all parts of the country and competition is bringing out lower prices in several finished lines, particularly sheets and iron pipe.

Tin plate business is very light and several plants have shut down several mills. The two eight-mill New Kensington plants of the leading interest are down, presumably for repairs, along the lines of policy of this company to make repairs when specifications are light and the output of the plants is least in demand.

Semi-finished steel for forward deliveries is being offered more freely and at lower prices, although shipments are poor on existing contracts at this time. The third quarter price of sheet bars has been reaffirmed at $31, Pittsburg. Deliveries on contracts placed by the Carnegie Steel Co. with independent producers of billets are about 15,000 tons behind to date.

Prices of southern pig iron are pretty well maintained, but there is much irregularity in northern brands.

WRITE FOR IT.

The Beardwood Agency, claims, collections and commercial reports, Montreal, has issued an attractive little folder advertising their method of doing business. The folder contains several testimonials from representative business men as to the satisfactory manner in which the Beardwood Agency has handled their collections.

Honesty is the best policy. Modesty may also be good policy, but you can't make it sell goods.

Hope is the only elevator of life that is ever an up-lift.

There is plenty of room at the top; that's why the man who reaches it is so lonesome.

CORRESPONDENCE

Travelers, hardware merchants and clerks are requested to forward correspondence regarding the doings of the trade and the industrial gossip of their town and district. Addressed envelopes, stationery, etc., will be supplied to regular correspondents on request. Write the Editor for information.

WINDSOR FOUNDRY SOLD.

Halifax, N.S., Sept. 23.—The hardware business continues very quiet, dullness being reported in nearly all lines. Builders' materials are the most active, general sales being reported. There is a good movement for cement, and the Sydney article is finding large sales in the province, particularly where paving work is being carried on. In Halifax large quantities are being used for concrete sidewalks, and also in the construction of several buildings. The Cement Company is still keeping up its day and night operations to supply the industry's rapidly increasing market. New machinery, in the shape of crushing rollers, is now being installed. The new apparatus is expected to materially increase the output.

* * *

The Windsor Foundry property and plant, situated at Windsor, N.S., was sold at public auction on Saturday and was purchased for $18,000 by George Mounce, of Avondale, N.S., which sum will about pay off the liabilities and leave nothing for the shareholders. A capable manager has been secured, and Mr. Mounce intends to start the foundry again with as little delay as possible. When the concern is in full operation it will prove a valuable addition to the industries of Windsor.

* * *

A coal miners' strike at the Springhill Mines is tying up to a large extent the work at the Torbrook Iron Mines. For want of coal only about half the regular output can be shipped to the blast furnaces at Londonderry. The Londonderry Iron and Mining Company, who recently acquired the properties at Torbrook, are preparing to install an electric system for operating their mines. The power will be drawn from the Nictaux Falls, about a mile from the mines. The mills, which were formerly operated at the falls, have been abandoned to make room for the utilization of the water power for electrical purposes. The estimated cost of the new plant is between $40,000 and $50,000.

* * *

James Pearson, manager of the Brookfield Iron Mines, says that large bodies of new ore have been located and that the prospects for a continued prosperous business are most encouraging. Samuel Archibald, who was for some time with the Nova Steel Company, and is a practical miner of wide experience, has been looking over the property and he states that there are large quantities of first-class ore available, and encourages the full development of the property, having in view the erection of a smelter.

JOHN A. McAVITY ILL.

St. John, N.B., Sept. 23.—There is still considerable activity among the local hardware dealers in the sale of buiders' hardware. Prices in mostly all lines are firm and business generally is reported brisk in both wholesale and retail circles.

* * *

The sheet metal workers last week sent a notice to the various employers that they would expect a new schedule of wages, to go into effect on Saturday, Sept. 21, and in the event of non-compliance with the request, all men employed in the shops would go out on strike. The men have asked that there should be a minimum rate of $12 a week. The matter is still in abeyance.

* * *

Premier C. W. Robinson was in the city last week on his return from a visit to the magnetic iron ore deposits on the Nepisiquit River, in Gloucester county, and he was very favorably impressed with the prospects for successful development. The Premier was accompanied by the Surveyor-General, and their visit was in response to an invitation from T. M. Burns, M.P.P., who is interested in the properties. The deposit is situated about twenty-five miles up the river and two miles from the Grand Falls. They found that the borings made by the Government drill had located large quantities of the ore in one place 400 feet thick. This class of iron is in great demand for smelting purposes.

* * *

The St. John Board of Trade has given its endorsement to the business of the Canada Woodenware Company, which proposes to start operations at South Bay. In an open letter to the newspapers, a committee from the board, after explaining that they have fully examined into the affairs of the concern, recommend it to the favorable consideration of capitalists, merchants and others, who feel like encouraging the promotion of home industries. The company propose to manufacture pails, tubs, etc., for the putting up of candy, jam, lard, etc., and is the only concern of its kind east of Ottawa. The raw materials, viz., spruce and pine wood, is here in abundance. No doubt operations will commence very shortly.

* * *

The Board of Trade has received assurances from the Railway Department at Ottawa that the proposed improvements to be made in the I.C.R. yard here will be completed in time for this winter's traffic. This is welcome news to merchants in all lines of business.

A large new plainer is being placed in position in the Sussex Manufacturing Company's machine shop. The new machine, when in position, will weigh about 20 tons.

* * *

J. K. Flemming's mill at Hartford has just shut down, having had a longer season's run than was expected. Three million laths have been cut and Mr. Fleming has bought 4,000,000 from other mills, the most of which has been shipped to the State of Ohio. The experiment of shipping long lumber, spruce and pine, to Ontario points has been tried and proved a success. Three carloads which Mr. Flemming sent there over the C.P.R. have brought very satisfactory returns and he has orders for twenty carloads more.

* * *

John A. McAvity, of T. McAvity & Sons, who has been in Michigan on a business trip, is now in Montreal, and has been compelled to go to the hospital there because of a severe cold.

* * *

James Pender, of the Pender Nail Works, returned last week from a trip to Upper Canada.

* * *

R. B. Emerson, of Emerson & Fisher, was in Montreal last week, where he attended a meeting of the Canadian Street Railway Association as the representative of the St. John Railway Company.

TRADE GOOD AT QUEBEC.

Quebec, Sept. 23.—Wholesale and retail trading this week has assumed considerable activity, and though prices are unchanged on quotation in hardware, there is a slightly firmer undertone noticeable.

* * *

The Quebec Railway Light and Power Company have commenced the work of erecting a woven wire fence the whole length of their railway, from Quebec to Cape Tourmente. The company are also putting in steel grates at all the farmers' crossings along their track.

* * *

Gingras & Tradeau, plumbers of Levis, have been awarded the contract for heating by hot water the Commercial Academy of Fraserville. Their tender for that installation was $3,680.

* * *

A request is about to be presented to the Quebec Bridge Co. and the Phoenixville Bridge Co. for permission to take a sufficient amount of the wreck and twisted beams of the structure with which to erect a monument to the memory of those who perished in the disaster.

* * *

William Shaw, managing director of the Chinic Hardware Co., and one of the directors of the Union Bank, has just arrived from a trip to Manitoba, Saskatchewan and Alberta. He was delighted with his vacation spent in these countries.

* * *

Many travelers of hardware manufacturers were in our city this week, amongst them we remarked: F. W. Lamplough, of Lamplough & Co.; Thomas Blaikee, of the Dominion Wire Manufacturing Co., Montreal, and F. J. Perego, of the Montreal Rolling Mills.

* * *

The new branch of the Quebec Central Railway between Beauce Junction and St. Joseph is nearing completion, and it is expected that it will be inaugurated next week.

ALLEN LEMMON MARRIED.

Kingston, Sept. 24.—The hardware trade here continues much about the same as last week, if anything a little quieter, the business at present being chiefly along the ammunition line. It is very likely from now on the trade will be brisk as much cooler weather has set in the past week and people are having to prepare for the coming winter, by getting their stoves and furnaces repaired, thus making the jobbers all busy.

* * *

John Lemmon & Sons, hardware merchants, have received another cargo of cement from Belleville, which they are at present unloading.

* * *

The wedding took place on Wednesday last, at the house of the bride's parents, Royal Military College, of Allen Lemmon, eldest son of John Lemmon, of the firm of Lemmon & Sons, hardware merchants, this city, to Miss Martha Haylett. The ceremony was of a quiet nature, only the immediate relatives of the contracting parties being present. Mr. and Mrs. Lemmon left for a bridal tour in the west, and on their return will take up residence on Earl street. They were the recipients of many costly and useful presents, among which was a cabinet of silver from the firm of Wormwith & Co., piano manufacturers, where the bride had been employed, which goes to show the high esteem in which the couple were held in Kingston.

* * *

The extension work is still in progress at the light and power plant. The new gas holder is about half erected, and will likely be ready for operation by the middle of October, quite early enough to supply the big demand for gas this coming winter. Owing to the cheaper rates the people will naturally burn more, and the number of consumers has been increased fully twenty-five per cent. since last winter. Another extra piece of extension is a new oil tank which Selby & Youlden are now erecting. The old tank leaked and was underground, the new one is being placed in the air. It will be large enough to hold nearly the winter's supply of oil, where the old one was far too small. At a special meeting of the City Light Committee on Thursday afternoon, it was decided to supply the required extra light at cost for the "Made in Canada Exhibition," which will be held in October. George R. Thomlinson, electrician, has been awarded the contract for all the electrical fittings. It was also decided at this meeting to put a new floor in the power house, which is badly needed. Mr. Campbell recommended a permanent fireproof one of concrete and steel at a cost of $275. The committee decided to make this expenditure.

* * *

No further word has been received concerning the zinc smelter company, which was to have started here. Money is rather tight at present, and that likely accounts for the inability of the promoters to get the necessary capital. Work of erecting the necessary buildings should have commenced a month ago.

* * *

James Eadie, Toronto, organizer of the Retail Merchants' Association of Canada, was in the city last week, collecting fees and looking up new members. There are now fifty-three members in the city, most of the hardwaremen being members. There is no hardware associations in Kingston, as some of the merchants will not agree to stick to strictly one price, and in this way help one another along.

ENLARGING STEEL GOODS FACTORY.

Hamilton, Sept. 25.—Local hardware men report a slight increase in the amount of business with the first indication of fall. This is partially accounted for by the fact that the stringency in the money market seems to be loosening up with the result that the building trades are making extra efforts to finish up the work on hand, which has, in some cases, been retarded by the condition of the market, before the snow flies. Of late the local hardware business has been comparatively quiet, but a leading hardware merchant stated a few days ago that the outlook was bright for a busy fall.

* * *

On Friday of this week about 200 delegates to the convention of the Canadian Manufacturers' Association in Toronto came to this city as the guests of the local manufacturers and the Board of Trade. They arrived about 9 o'clock in the morning and were met at the Grand Trunk Station by a number of prominent gentlemen and were taken in automobiles and carriages for a trip about the city. They visited several of the large manufacturing concerns, as well as many of the points of interest about the city. At 11.30 they were entertained at luncheon at the Commercial Club, and at 1.30 they left for Niagara Falls.

The Canada Steel Goods Company, the only manufacturers of plated buff hinges in Canada, has been doing such a tremendous business since locating its plant here that plans are out for a large addition to the works. Previous to the establishment of this concern practically all of this class of material was imported from the United States. Mr. Hatch, the manager, is authority for the statement that an addition will have to be made to the plant in the near future in order to afford room to carry on the increasing business.

* * *

Hendrie Legatt, of Wood, Vallance & Legatt, Vancouver, B.C., was a visitor in the city this week. He is here on a combined business and pleasure trip.

* * *

Business in the stove line is particularly brisk just now. Hamilton is a veritable hive of stove manufacturing industries, and they all report a big rush for the fall trade. The D. Moore Company a few weeks ago sent its fifth annual shipment of Treasure stoves to Central China and it is also shipping a large number of stoves out west. The Ontario trade in this line remains steady.

* * *

William Wood, of Wood, Vallance & Company, is at present in the West with his two sons. He is away ostensibly on pleasure bent, but he is also visiting some of the firm's branches in that section of the country.

* * *

The International Harvester Company have just concluded one of the heaviest seasons in several years. This firm have been shipping harvesting machinery to the great West in special train and boat loads for some months past, and now that their shipping season has closed the plant is working to its utmost capacity on next year's output.

* * *

Hon. Adam Beck, of London, who figures prominently in the manufacturing line in the Forest City, will shortly take up his residence in this city. He has recently decided to establish a branch of the box-making industry, in which he is interested in London, in this city. A Brantford box company have also decided to locate here and it is expected that competition in this line will be particularly keen as a result.

* * *

Langsford Robinson has gone on the road for the D. Moore Stove Company, of this city. His territory will comprise that of Northern Ontario.

HARDWARE DISPLAYS AT FAIR.

Ingersoll, Sept. 25.—The annual fall exhibition of the Ingersoll, North and West Oxford Agricultural Society was held here on Tuesday and Wednesday of this week, under the most disagreeable climatic conditions peculiar to the autumn season. The directors, however, had the earnest co-operation of several of the leading hardware merchants, who made extensive and very striking displays in the palace, which did much to enhance the appearance of this building, and at the same time induce additional business their way. Most hardwaremen realize the importance of getting in touch with the public and never before have they made a better attempt to place their articles before the gaze of those who attend the fair.

J. T. Norton had a most attractive exhibit of stoves and bath equipment. The most notable feature of his display being a "bathroom" thoroughly equipped. This gave prospective purchasers a realistic idea and denotes Mr. Norton's enterprise.

Wilson Bros., who do an extensive trade in stoves, were on hand with a large showing of the different styles and makes of stoves which they carry in stock.

Bowman & Co. showed very little in the hardware line, making a specialty of sewing machines, pianos, organs and Edison phonographs. Through the efforts of these hardwaremen the appearance of the palace was greatly improved and they are deserving of congratulations, both on their own enterprise and the able assistance they rendered the fair.

* * *

With the first of September the shooting season opened, and all the dealers are doing a good business in guns and supplies. There are many devotees of the gun in this section, and as they believe in "patronizing home industry," or, in other words, local merchants, there is an extensive trade in this line. The Ingersoll Gun Club keep up the demand for ammunition, as they hold weekly shoots. For the past couple of months they have been shooting for a trophy donated by the Hunter Arms Co., and they will shortly shoot for another, the gift of the Dupont Powder Co.

RETAILER COMMENDS HARDWARE ASSOCIATION.

Galt, Sept. 25.—Business in the hardware and plumbing line in Galt is brisk at present. Several large buildings are under construction, which probably accounts for the brisk movement in plumbers' and painters' supplies. Lockhart & Co. are just completing the installation of a steam heating system in the opera house. The job is an extensive one and has been under way for some months. This firm has also secured the contract for the plumbing and heating of the new Knox Church Sunday School building, to cost about $30,000, operations on which were commenced on Monday.

* * *

The present cold weather has resulted in a sudden demand for gas stoves, which are much in use in Galt, natural gas being available at 35 cents per thousand. It is estimated that the advent of natural gas has decreased the sale of coal in Galt by many hundreds of tons. Many of those having furnaces, both hot air, water and steam, have had them equipped with natural gas burners and obtained excellent results. During the very cold weather of last winter there was a severe shortage of gas, but it is expected that the additional wells opened up this year will do away with the shortage

* * *

W. J. McMurty who is a member of the Retail Hardware Association and a constant reader of Hardware and Metal, in the course of an interview to-day, expressed his appreciation of the efforts being made by that journal to further the organization of retail hardware merchants. "I cannot understand," he said, "why the hardwaremen throughout Ontario do not one and all come into the association. It is of immense benefit to the trade. I have been a member for a couple of years, and would not think of dropping out." As far as your correspondent can ascertain, Mr. McMurty is the only member of the association in Galt.

* * *

A busy man is Jas. Douglas, of the staff of McMurtry's hardware store. In addition to being a member of the city council, Mr. Douglas is secretary of the South Waterloo Agricultural Society, and as the annual exhibition is being held next week, he has little time to spare.

* * *

The Canadian Brass Manufacturing Co., which was given a loan of $10,000 in order that it might be established in Galt, is now in operation. The building erected is a substantial one and a credit to the Manchester of Canada. The plant consists of a foundry 60 feet long by 40 feet wide, an engine room, 40 feet square and the main building 200 feet long by 100 feet wide, consisting of two stories and divided into the following departments, buffing, nickle plating, polishing, pattern, tool, core, assembling, store room and office. The company manufactures plumbers' supplies of every description, in addition to numerous other brass goods. About 25 men are at present employed, and it is expected that this number will be doubled by the end of the year. The manager is J. H. LeFavor.

* * *

The Galt Art Metal Co. last week shipped to the Waterous Engine Co., of Brantford, the first instalment of a large order for metal skylights. The order is the largest ever turned out in Canada. The same company has recently installed a lathing machine for the purpose of making steel lath, and there is already an enormous demand for the new product. The machine is the second of its kind in Canada.

* * *

The Shurly-Dietrich Co.'s saw factory is running over-time, and business is reported to be the best for some years past.

* * *

In short, it would not be over-reaching the mark to say, notwithstanding the stringency of the money market, that the manufacturers of Galt are enjoying a period of prosperity not exceeded in former years, with every indication that it will continue indefinitely.

The Down Draft Furnace Company is experiencing the busiest time in its history. The new addition has been completed and is now occupied.

* * *

The Jas. Warnock Company, manufacturers of edged tools, have recently installed several new machines in order to keep pace with their increased orders.

* * *

The Retail Merchants' Association, of Galt, will hold a big banquet in the near future.

GEORGE O. McCLARY DEAD.

London, Ont., Sept. 24.—The death of Geo. O. McClary, treasurer of the McClary Manufacturing Company, which occurred on Saturday morning last at the residence of his brother, A. E. McClary, 388 Dundas street, proved a shock to his many friends, for he was one of London's most popular business men and most estimable citizens. Early in January last Mr. McClary developed a bad cold, which eventually culminated in an attack of pneumonia. In the course of the summer he made a trip to Muskoka and seemed to be much improved in health. A little over two weeks prior to his death, he returned to the city and had another attack of pneumonia, but seemed to be convalescing nicely until last Friday night, when his condition became alarming. From that time he rapidly grew worse, until the end came. Mr. McClary had been identified with the McClary Manufacturing Company for over twenty years. He was the oldest son of Mr. and Mrs. Oliver McClary, both of whom passed away a number of years ago, and is survived by his brother, A. E. McClary, and a sister, Mrs. John M. Moore, of 478 Waterloo street. Mr. McClary was born on Wellington street, this city, in 1861, and has been a life-long resident of London. The deceased was connected with the Brunswick Club for a number of years. Mr. McClary was a member of the English Church, and attended St. Paul's Cathedral.

* * *

A meeting of local machinists was held a few nights ago to discuss the question of shorter hours. The greater number of the machinists are employed at Leonard's and White's, and all seem to be of the opinion, that shorter hours should be granted by the employers or wages advanced.

* * *

On the invitation of manager Gartshore, a large party of Collegiate Institute students visited the extensive works of the McClary Manufacturing Co. a few days ago, in order that they might view the processes of manufacture. The first place visited was the works in the southeastern part of the city, the visitors being in charge of Mr. Stewart, science master at the institute. They found that arrangements had been made for their reception, Mr. Vallier, one of the McClary staff, personally conducting the trip of the students, and explaining to them the various processes of manufacture. Manager King also gave them a welcome. The students were enabled to witness at close range the various processes of manufacture from the draft on paper to the finished stove or furnace. A very interesting machine was that which extracts the refuse iron from the floor sweepings. The new gas engine was also an object of much curiosity. Subsequently on the initiation of Geo. White, the students were taken through the rolling mills adjoining, and saw here all the intensely interesting processes of developing the finished iron product. The object lessons thus received is not likely to be lost on the students.

WESTMAN BROS. NEW STORE.

Chatham, Sept. 25.—The Brodie transient trader appeal, which was to have come up before Judge Dowlin this week, has been once more adjourned, the parties concerned being at yet unprepared to proceed. The object of the appeal, in a nutshell, is to decide whether or not the magistrate has power to pass upon the validity of the local by-laws.

* * *

Plans are out for a new three-storey brick block on King St., to be erected for Ald. W. H. Westman, of Westman Bros. The plans call for pressed brick and cut stone, and the block, when complete, will cost about $12,000. The site is a vacant lot between the Central Drug Store and the News Office, which hitherto has been the home of a more or less ornamental sign board. With the filling in of this gap, the main part of King St. will be solidly built up.

* * *

The Blonde Lumber & Manufacturing Co. are engaged in removing the stock and fittings of their present hardware store on St. Clair St. to the more commodious premises on the north side of the street, formerly occupied by Alf. Deloge and recently remodeled by the company.

* * *

Jack Blewett, of Edmonton, Alta., traveling for one of the largest hardware firms in the West, was a city visitor last week. He will spend some time with his mother, Mrs. Jean Blewett, in Toronto, prior to returning to Edmonton.

* * *

W. M. Drader, the well-known stave and lumber dealer and contractor, was the victim last week of a couple of the most disastrous fires known here in recent years. Friday morning fire broke out in the Queen St. planing mill, which was pretty will destroyed. The ensuing Sunday morning the stave mill in North Chatham caught fire, and, being a frame building and quite dry, speedily went up in smoke. The fire department fought well on both occasions, but were hampered, on Friday, by a high wind and on Sunday by the necessity of making a long detour, owing to the streets being blocked by paving operations. The total loss is in the neighborhood of $30,000, the buildings being well insured. About 100 men will be thrown out of work. Mr. Drader will probably rebuild.

* * *

Despite unfavorable weather, the Peninsular Fair this year put on one of the best shows in its history. The attendance was very encouraging, and the exhibits were of a high-class, every inch of space in the main building being taken up. The displays by local merchants were most attractive.

CALGARY'S HUNDRED THOUSAND CLUB.

Calgary, Sept. 24.—E. L. Richardson Secretary of the Calgary Board of Trade writes: The building returns for the month of August amounted to $108,200, making a total of $1,023,820 for the eight months.

Customs returns for the month of August, 1907, were $51,075.48, being an increase of $14,280.23 over August, 1906. The total receipts for the five months, April to August, inclusive, 1906, were $162,358.81. The total receipts for a similar period in 1907 were $293,196.63, showing an increase of $130,837.82.

The bank clearing house returns for week ending Sept. 5 were $1,171,378, being an increase of $191,023 over the same week of last year The total receipts for the month of August, 1907, were $5,900,541, being an increase of $1,558,316 over the month of August, 1906.

Six new wholesale warehouses have been started or completed this year.

* * *

The Hundred Thousand Club has instituted a series of business men's excursions into the different parts of the Province tributary to this city. The objects of these excursions is to promote trade and advance Calgary's position as a manufacturing and wholesale distributing centre. The first excursion took place Aug. 6 to 8, to Lethbridge; the club chartered a special train of sleepers and visited all the important points along the line. Nearly 100 of Calgary's leading firms were represented on this excursion. The next excursion will visit the northern towns.

BUILDING IN SASKATOON.

Saskatoon, Sept. 21.—The sewer and water pipes are being extended along Twenty-first street. The following applications for connections have been received by the city council.: Windsor hotel, Butler hotel, Western hotel, City hotel, Iroquois hotel, King Edward hotel, Bank of Commerce, P. C. Collins, W. R. C. Willis, J. C. Wilson, H. Wells, Willoughby & Butler.

The question of a coal supply for Saskatoon, in view of the approaching winter, was brought up at a meeting of the executive of the Board of Trade on Saturday afternoon. It was decided to advise, through the press, the citizens generally to send in orders at once, for it is felt that deliveries are none too prompt, and early orders may serve to avoid disappointment when the cold weather comes.

The Government has reserved as provincial property certain areas in the Eagle Lake district, where surface coal has been found. An inspector was dispatched to make thorough investigations as to the prospects of obtaining a fuel supply for future winters. Reports received show that prospecting has been commenced and a promising amount of lignite coal located. Development of these stores will proceed as quickly as possible in order to supply the urgent needs of the district.

* * *

Building operations have begun on the new court house and work is being rushed. The contractors are the Saskatchewan Building & Construction Co, of Regina.

* * *

Bricklayers are busy this week brick-veneering the King Edward hotel.

* * *

The excavations for the C.N.R. roundhouse have been completed, and a gang of men are laying the cement foundation.

* * *

Stewart & M[illegible], implement agents, will build a warehouse on a site adjacent to their present office.

* * *

Goetz & Flodin Co., Chicago, are installing the engines, machinery and fittings in the new brewery, which is expected to be in operation by the end of the year.

* * *

The following building permits, amounting to $20,492, were taken out in Prince Albert during August: Church of England mission; E. Kirkbright, residence; A. E. Doak, residence; O. B. Manville, residence; M. Hooper, residence; Dr. P. D. Tyerman, warehouse; O. B. Manville, warehouse; R. Stanley, stable; Horkson Larson, residence; D. A. Hopkins, residence; A. Houle, residence.

* * *

The Windsor hotel, Prince Albert, is undergoing some elaborate improvements. O. B. Manville has the contract for putting in baths and lavatories. The public washroom and lavatory, on the ground floor, will be modern in every respect, with tile floor and wainscotting. The contract for heating the hotel with steam was awarded to W. S. Russell.

* * *

W. Harris & Son, Saskatoon, will erect a house and implement warehouse at Asquith, Sask.

* * *

The new office of the Union Bank, Asquith, is now completed. The building is a handsome two-storey structure and is a credit to the town.

BUSINESS CHANCE AT PRINCE ALBERT.

Prince Albert, Sask., Sept. 23.—B. W. Wallace, Secretary of the Board of Trade, says the active building operations of the present season have shown beyond question that the manufacture of building materials has not yet reached anything like adequate proportions. Despite the number of new residences built throughout Saskatchewan and Alberta during the last six months there are still at the end of summer many families living in tents, and many others coming into the country who will find great difficulty in securing suitable houses for the winter. The lack is not so much in carpenters and builders as in materials, such as brick, cement blocks and lumber. Scarcely a building, be it residence or business block, has gone up this season without having been delayed through lack of builders' supplies.

There is a rare opportunity for some shrewd business men to establish at a point where raw material is abundant and shipping facilities good, manufacturing plants for the production of pressed brick and sand lime brick. Mr. Wallace had the privilege of examining, a few days ago, a deposit of clay fifteen feet thick and covering scores of acres in extent. This clay is of the purest possible variety and when properly treated will make the best of pressed brick or pottery. Sand, which is from 95 per cent. to 98 per cent. pure silica, is plentiful in the neighborhood of this clay, and wood in abundance for the firing of kilns grows on the soil topping this enormous clay bed.

The settlement and development of Central Saskatchewan is just beginning. Buildings will be required for many years to come to meet the needs of the rapidly increasing population. The raw materials for supplying these buildings are all here and they call loudly for men of shrewd foresight who will change the clay and sand into cosy homes.

AUSTRALIAN IRON PROTECTION.

It is anticipated a bill to give a bonus to the iron industry in Australia will be introduced by the ministry during the present session of the Commonwealth Parliament. The acting prime minister recently paid a visit to the works of William Sanford, at Lithgow, New South Wales, the firm likely to be most benefited by such a bonus, and after the visit said he "was very much impressed with the works, which were more extensive than he had imagined. A great deal of money had already been expended, but there was still much to be done. Pig iron of various qualities had been produced, but the industry in its higher grades was a question for the future. The promoters of the works said they wanted some assistance. They would prefer a duty on imported iron to a bonus, but, owing to the extent of the other industries which would be affected by a duty on iron, he did not think that would be practicable. Considering the amount of enterprise that had been shown, however, and the extent of the capital embarked, every encouragement should be given to the industry." This iron-making firm has just increased its capital from $750,000 to $1 250,000.

CONDENSED OR "WANT" ADVERTISEMENTS.

RATES.

Two cents per word first insertion; one cent per word subsequent insertions.

Five cents additional each insertion where box number is desired.

Contractions count as one word, but five figures (as $1,000) are allowed as one word.

Cash remittances to cover cost **must** accompany all advertisements. **In no case** can this rule be overlooked. Advertisements received without remittance cannot be acknowledged.

RULES FOR COPY.

Replies addressed to HARDWARE AND METAL boxes are re-mailed to advertisers every Monday, Wednesday and Friday.

Requests for classification will be followed where they do not conflict with established classified rules.

Orders should always clearly specify the number of times the advertisement is to run.

All "Want" advertisements are payable in advance.

AGENTS WANTED.

This is the problem of many manufacturers who are anxious to get a foothold in Western Canada. It will be easily solved if HARDWARE AND METAL is given the opportunity to solve it.

AGENT wanted to push an advertised line of Welsh tinplates; write at first to "B.B.," care HARDWARE AND METAL, 88 Fleet St., E.C., London, Eng. [tf]

AN old-established house of brass founders, making a wide range of brass foundry furnishings, such as brass cast tubing, stair rods, cornice poles, art metal work, electric light fittings, ecclesiastical metal work, are anxious to secure a thoroughly reliable and trustworthy agent or traveller for Canada, to sell goods on commission. An excellent opportunity for any firm covering Canada thoroughly. Address all enquiries to Box 654, HARDWARE AND METAL. [37]

SOLE Agency for Canada. Prominent American firm is now in position to receive bids from reliable wholesale hardware firms or others for the sole right to handle their Aluminum Solder, which will unite aluminum with aluminum, or aluminum with other metals. Only bids from established houses of high standing will be considered. For particulars write under N. Y. 1012 to John M. Munchenberg, 1161-75 Broadway, New York, U.S.A. [38]

ARTICLES FOR SALE.

Don't keep any fixtures or tools around your store for which you have no further use. They will be worth more to-day than they will a year hence. Don't keep money tied up which you could use to secure discounts from your wholesaler.

GOOD silent salesman for sale. Reason for selling, advertiser wishes to purchase larger one. A bargain for immediate sale. Box 658, HARDWARE AND METAL, Toronto. 38

SET of tinsmith's tools, good as new, best American make. $150. Great sacrifice. Box 660, HARDWARE AND METAL, Toronto. tf

TWO hundred dozen rim and mortice knobs and blocks, black, white and mineral. 90c. per dozen. Full particulars. Box 661, HARDWARE AND METAL, Toronto.

TYPEWRITER, No. 2 Sun, in first-class condition. Will sacrifice for $55.00. Box 662, HARDWARE AND METAL, Toronto. tf

ARTICLES WANTED.

If you cannot afford to buy a new counter, show case, screw cabinet, store ladder, or some other fixture which you could use to advantage, try a "Want Ad." under "Articles Wanted," and you may get what you want at a bargain price.

WANTED—Two store ladders, "Myers" preferred, complete with fixtures. Send full description and price to Illsey Bros., Red Deer, Alta. (36)

BUSINESS CHANCES.

MERCANTILE Agency Business, with established offices in Montreal, needs capital for further development; promising outlook; or would entertain active partner with moderate capital; splendid opportunity for energetic business man; highest references given and required; interview by an appointment. Address Box 28, HARDWARE AND METAL, Montreal. [40]

Hits the Mark

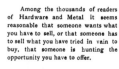

Among the thousands of readers of Hardware and Metal it seems reasonable that someone wants what you have to sell, or that someone has to sell what you have tried in vain to buy, that someone is hunting the opportunity you have to offer.

The want ad. columns of Hardware and Metal are the simplest form of simplified advertising. No ad. writer need be employed, no drawing or cut need be made, no knowledge of display type is necessary, for display type will not be used in these columns. All that you have to do is state your wants, state them simply and clearly and send to any of our offices along with remittance to cover cost of advertisement. No accounts opened in this department.

Don't despise the want ad. because of its small size or small cost. Don't forget that practically all the hardware merchants, clerks, manufacturers and travellers read our paper each week.

RATES:—

2c. per word first insertion.
1c. per word subsequent insertions.
5c. additional each insertion for box number.

BUSINESSES FOR SALE.

Somewhere in Canada is a man who is looking for just such a proposition as you have to offer. Our "For Sale" department brings together buyer and seller, and enables them to do business although they may be thousands of miles apart.

HARDWARE Business and Tinshop for sale in Saskatchewan; population 1,500; stock carried, about $14,000; turnover, $45,000; practically all-cash business; cash required, $8,000; would rent building; do not answer without you have the money and mean business; it will pay to investigate this. Box 648, HARDWARE AND METAL, Toronto. (41)

HARDWARE Business for Sale in Tweed, a thriving Eastern Ontario town. Best business in town. Must sell owing to ill-health. Liberal terms to satisfactory purchaser. Write for particulars. J. M. Robertson, Tweed. [40]

HARDWARE store and stock for sale, within 25 miles of Toronto. Solid brick double store and dwelling; property worth $4,000. Stock worth about $3,000. Splendid business in stoves and hardware. Good reasons for selling. Full particulars on application. Box 663, HARDWARE AND METAL, Toronto. [41]

SITUATIONS VACANT.

You can secure a "five thousan. a year" manager, or a "five hundred a year" clerk, by stating your wants under "Situations Vacant."

WANTED—First-class tinsmith in New Ontario town. Must be capable of managing workshop and estimating. Good position, good salary, and no lost time for the right man. Address, Box 655, HARDWARE AND METAL, Toronto. [40]

WANTED—First class opening for tinsmithing, plumbing and heating. Box 202, Hespeler. [43]

SITUATIONS VACANT.

WANTED—Experienced hardware clerk; must be good salesman and stockkeeper Address, stating age, experience and salary expected, Box 664 HARDWARE AND METAL, Toronto. [40]

THE CANADIAN GROCER wants a *Managing* Editor. It wants a thoroughly capable man—a man who is live, full of up-to-date ideas, and one who understands the newspaper business from the reglet box to the editorial chair. Furthermore, it wants a man who is thoroughly conversant with the commercial situation in Canada. We realize that this is a big want. Not everyone can fill the bill, but we are willing to pay at the outset $2,500 a year to the man who can do so. The right man can eventually make his place worth $5,000. If you think you are this man we want to hear from you, with your experience and qualifications—by letter only. Address The MacLean Publishing Co., 232 McGill St., Montreal, or 10 Front St. East., Toronto. 38

SITUATIONS WANTED.

Travellers, clerks, or tinsmiths, wishing to secure new positions, can engage the attention of the largest number of employers in their respective lines, by giving full particulars as to qualifications, etc., under "Situations Vacant."

CLERKS WANTED.—Clerks who want to improve their positions are requested to write for information to Box 659, HARDWARE AND METAL, Toronto. 40

HARDWARE CLERK, 32, married, sober, steady and reliable, capable of taking charge. Box 10, HARDWARE AND METAL, Winnipeg.

HARDWARE Salesman wants position; can furnish first class references; capable of managing city or town business. Box 653, HARDWARE & METAL, Toronto. (39)

HARDWARE CLERK, Irishman, 27, long experience of trade in best houses Old Country, one year with large Canadian wholesale firm, desires change, would go west. Unexceptionable references. Photo if required. Apply box 657, HARDWARE AND METAL, Toronto. (39)

MANITOBA HARDWARE AND METAL MARKETS

Corrected by telegraph up to 12 noon Friday, Sept 27. Room 511, Union Bank Bldg, Winnipeg, Man.

General business is fairly active. In some districts where crops are poor there is a distinct falling off, but, taking the country through, the wholesalers feel that they have no good reason for complaint. A fairly good crop is now assured, probably 70,000,000 bushels of wheat. Even in districts where the wheat is frozen the price of frozen wheat is about the same as No. 1 hard last year, and the frozen wheat will yield a profit.

A few advances will be noted. Wire nails have been advanced 10 cents; hinges are higher.

Rope—Sisal, 11c per lb., and pure manila, 15¾c.

Lanterns—Cold blast, per dozen, $7; coppered, $9; dash, $9.

Wire—Barbed wire, 100 lbs., $3.22½; plain galvanized, 6, 7 and 8, $3.70; No. 9, $3.25; No. 10, $3.70; No. 11, $3.80; No. 12, $3.45; No. 13, $3.55; No. 14, $4; No. 15, $4.25; No. 16, $4.40; plain twist, $3.45; staples, $3.50; oiled annealed wire, No. 10, $2.90; No. 11, $2.96; No. 12, $3.04; No. 13, $3.14; No. 14, $3.24; No. 15, $3.39; annealed wires (unoiled), 10c less; soft copper wire, base, 33c; brass spring wire, base, 30c.

Poultry Netting—The discount is now 47½ per cent. from list price, instead of 50 and 5 as formerly.

Horseshoes—Iron, No. 0 to No. 1, $5.65; No. 2 and larger, $4.40; snowshoes, No. 0 to No. 1, $4.90; No. 2 and larger, $4.65; steel, No. 0 to No. 1, $5; No. 2 and larger, $4.75.

Horsenails—No. 10 and larger, 22c; No. 9, 24c; No. 8, 24c; No. 7, 26c; No. 6, 28c; No. 5, 30c; No. 4, 36c per lb. Discounts: "C" brand, 40, 10, 10 and 7½ p.c.; "M.R.M" cold forged process, 50 and 5 p.c. Add 15c per box. Capewell brand, quotations on application.

Wire Nails—$3.10 f.o.b. Winnipeg and $2.65 f.o.b. Fort William.

Cut Nails—Now $3.20 per keg.

Pressed Spikes—¼ x 5 and 6, $4.75; 5-6 x 5, 6 and 7, $4.40; ⅜ x 6, 7 and 8, $4.25; 7-16 x 7 and 9, $4.15; ½ x 8, 9, 10 and 12, 4.05; ⅝ x 10 and 12, $3.90. All other lengths 25c extra, net.

Screws—Flat head, iron, bright, 80, 10, 10 and 10; round head, iron, 80; flat head, brass, 75; round head, brass, 70; coach, 70.

Nuts and Bolts—Bolts, carriage, ⅜ or smaller, 60 p.c.; bolts, carriage, 7-16 and up, 50; bolts, machine, ⅜ and under, 50 and 5; bolts, machine, 7-16 and over, 50; bolts, tire, 65; bolt ends, 55; sleigh shoe bolts, 65 and 10; machine screws, 70; plough bolts, 55; square nuts, cases, 3; square nuts, small lots, 2½; hex nuts, cases, 3; hex nuts, small lots, 2½ p.c. Stove bolts, 70 and 10 p.c.

Rivets—Iron, 60 and 10 p.c.; copper, No. 7, 43c; No. 8, 42½c; No. 9, 45½c; copper, No. 10, 47c; copper, No. 12, 50½c; assorted, No. 8, 44½c, and No. 10, 48c.

Coil Chain—¼-in., $7.25; 5-16, $5.75; ⅜, $5.25; 7-16, $5; 9-16, $4.70; ⅝, $4.65; ¾, $4.65.

Shovels—List has advanced $1 per dozen on all spades, shovels and scoops.

Harvest Tools—60 and 5 p.c.

Axe Handles—Turned, s.g. hickory, $3.15; No. 1, $1.90; No. 2, $1.60; octagon extra, $2.30; No. 1, $1.60.

Axes—Bench axes, 40; broad axes, 25 p.c. discount off list; Royal Oak, per doz, $6.25; Maple Leaf, $8.25; Model, $8.50; Black Prince, $7.25; Black Diamond, $9.25; Standard flint edge, $8.75; Copper King, $8.25; Columbian, $9.50; handled axes, North Star, $7.75; Black Prince, $9.25; Standard flint edge, $10.75; Copper King, $11 per dozen.

Churns—45 and 5; list as follows: No. 0, $9; No. 1, $9; No. 2, $10; No. 3, $11; No. 4, $13; No. 5, $16.

Auger Bits—"Irwin" bits, 47½ per cent. and other lines 70 per cent.

Blocks—Steel blocks, 35; wood, 55.

Fittings—Wrought couplings, 60; nipples, 65 and 10K Ts and elbows, 10; malleable bushings, 50; malleable unions, 55 p.c.

Hinges—Light "T" and strap, 65.

Hooks—Brush hooks, heavy, per doz., $8.75; grass hooks, $1.70.

Stove Pipes—6-in., per 100 feet length, $9; 7-in., $9.75.

Tinware, Etc.—Pressed, retinned, 70 and 10; pressed, plain, 75 and 2½; pieced, 30; japanned ware, 37½; enamelled ware, Famous, 50; Imperial, 50 and 10; Imperial, one coat, 60; Premier, 50; Colonial, 50 and 10; Royal, 60; Victoria, 45; White, 45; Diamond, 50; Granite, 60 p.c.

Galvanized Ware—Pails, 37½ per cent.; other galvanized lines, 30 per cent.

Solder—Quoted at 27c per pound. Block tin is quoted at 45c per pound.

Wringers—Royal Canadian, $36; B.B., $40.75 per dozen.

Files—Arcade, 75; Black Diamond, 60; Nicholson's, 62½ p.c.

Locks—Peterboro and Gurney 40 per cent.

Building Paper—Anchor, plain, 66c; tarred, 69c; Victoria, plain, 71c; tarred, 84c; No. 1 Cyclone, tarred, 84c; No. 1 Cyclone, plain, 66c; No. 2, Joliette, tarred, 69c; No. 2 Joliette plain, 51c; No. 2 Sunrise, plain, 56c.

Ammunition, etc.—Cartridges, rim 50 p.c.; pistol sizes 25 p.c.; military, 20 p.c. Primers, $1.55. Loaded shells: 12 gauge, black, $16.50; chilled, 12 gauge, $17.50; soft, 10 gauge, $19.50; chilled, 10 gauge, $20.50. Shot: ordinary, per 100 lbs., $7.75; chilled, $8.10. Powder: F.F., keg, Hamilton, $4.75; F.F.G., Dupont's, $5.

Revolvers—The Iver Johnson revolvers have been advanced in price, the basis for revolver with hammer being $5.30 and for the hammerless $5.95.

Iron and Steel—Bar iron basis, $2.70. Swedish iron basis, $4.95; sleigh shoe steel, $2.75; spring steel, $3.25; machinery steel, $3.50; tool steel, Black Diamond, 100 lbs., $9.50; Jessop, $13.

Sheet Zinc—$8.50 for cask lots, and $9 for broken lots.

Corrugated Iron and Roofing, etc.—Corrugated iron, 28 gauge, painted, $3; galvanized, $4.10; 26 gauge, $3.35 and $4.35. Pressed standing seamed roofing, 28 gauge, $3.45 and $4.45. Crimped roofing, 28 gauge, painted, $3.20; galvanized, $4.30; 26 gauge, $3.55 and $4.55.

Pig Lead—Average price is $6.

Copper—Planished copper, 44c per lb.; plain, 39c.

Iron Pipe and Fittings—Black pipe, ¼-in., $2.70; ⅜, $2.85; ½, $3.75; ¾, $4.75; 1, $6.75; 1¼, $9.25; 1½, $11.00; 2, $14.80; 2½, $24.60; 3, $32.30; 3½, $40.50; 4, $46.00; 4½, $54.00. Galvanized: ¼ in., $3.65; ⅜, $3.80; ½, $4.50; ¾, $5.80; 1, $8.40; 1¼, $11.40; 1½, $13.80; 2, $18.40. Nipples, 70 per cent.; unions, couples, bushings and plugs, 50 per cent.; malleables, 20 per cent.

Galvanized Iron—Apollo, 16-gauge, $4.15; 18 and 20, $4.40; 22 and 24, $4.65; 26, $4.65; 28, $4.50; 30-gauge or 10¾-oz., $5.20; Queen's Head, 20, $4.60; 24 and 26, $4.90; 28, $5.15.

Lead Pipe—Market is firm at $7.80.

Tin Plates—IC. charcoal, 20x28, box, $10; IX charcoal, 20x28, $12; XXI charcoal, 20x28, $14.

Terne Plates—Quoted at $9.50.

Canada Plates—18x21, 18x24, $3.50; 20x28, $3.80; full polished, $4.30.

Lubricating Oils—600W, cylinders, 80c; capital cylinders, 55c and 50c; solar red engine, 30c; Atlantic red engine, 29c; heavy castor, 28c; medium castor, 27c; ready harvester, 28c; standard hand separator oil, 35c; standard gas engine oil, 35c per gallon.

Petroleum and Gasolene—Silver Star, in bbls., per gal., 20c; Sunlight, in bbls., per gal., 22c; per case, $2.35; Eocene, in bbls., per gal., 24c; per case, $2.50; Pennoline, in bbls., per gal., 24c; Crystal Spray, 23c; Silver Light, 21c. Engine gasoline, in barrels, gal., 27c; f.o.b. Winnipeg, in cases, $2.75.

Paints and Oils—White lead, pure, $6.50 to $7.50, according to brand; bladder putty, in bbls., 2½c; in kegs, 2¾c; turpentine, barrel lots, Winnipeg, 90c; Calgary, 97c; Lethbridge, 97c; Edmonton, 98c. Less than barrel lots, 5c per gallon advance. Linseed oil, raw, Winnipeg 70c, Calgary 77c, Lethbridge, 77c, Edmonton 78c; boiled oil, 3c per gallon advance on these prices.

WINNIPEG GOSSIP.

Fall trade in Winnipeg among the retail stores has been fairly satisfactory to date, although somewhat affected by two or three special sales at closing out prices. While the building this year has not been up to the record of 1906, there has, nevertheless, been a large trade in building supplies, and the retailers have no special reason for complaint. Department store competition is felt most in enamelware, and general household supplies, but even in these lines a large business is being done by straight line hardware stores at regular prices. The year 1907 has not been the most prosperous on record; in fact, owing to a combination of circumstances which are well understood and which need not be entered upon here, it has been a year of uncertainty and surprises. Nevertheless, there has been a steady increase in the population of the city, and there has been plenty of trade for practically all classes of stores. A fair crop at good prices is now assured, and the continued prosperity of Winnipeg is not a matter for doubt or conjecture.

Oliver Gilmer, a Main street hardwareman, well-known among the hardware trade of the West, makes every Saturday night a "mechanics' tool night," and in this way has worked up a big trade in this line. Winnipeg stores close at 6 p.m. on every night but Saturday, and hence this is the only night when mechanics are able to buy tools. Mr. Gilmer will not allow his clerks to show rifles or any similar line which requires time to sell on Saturday nights. He says that he has not the staff to handle anything on Saturday nights that requires much time to sell. Mechanics wanting tools have right-of-way.

Graham & Rolston have been selling out their stock, as was noted in a recent issue of this paper. They have been forced to vacate their premises in the Street Railway Building, near the corner of Portage and Main, and, hence, their closing-out sale. They are removing the balance of their stock this week to a new location on Notre Dame avenue, and they will continue their sale there.

Watt & Gordon are conducting a big sale, as they are about to remodel their premises, and, consequently, are obliged to reduce stock.

J. E. Riley has removed from the north side of Market Square to the west side.

NEW CARPET SWEEPERS.

The Shirreff Manufacturing Company, Brockville, are preparing to introduce a nice line of carpet sweepers on the market. The Shirreff line is rapidly growing, both in its variety and in popularity with the trade and their new sweeper should meet with a large advance sale.

JAPAN'S PECULIAR EDGE-TOOLS.

Alexander MacLean, Canadian Commercial Agent to Japan, writes an interesting letter in answer to an inquiry from Canada respecting the market in Japan for Canadian-made axes. After showing imports of foreign axes at the leading Japanese ports to be almost nil, and stating that there is not the slightest probability of the pattern commonly known as the American or Canadian axe coming into general use, he says:

"In the course of personal observation, I have some recollection of having seen a wood-chopper's axe of the standard American pattern in a hardware store somewhere, but only one. I have seen, not often, the small size, hand axe, or hatchet, of the same pattern, in shop windows; and from the value of the axes mentioned as imported, probably these latter were mostly of the hand-axe sort. The customs published returns do not specify in respect of these articles, and the statistics given here are obtained through the courtesy of the Superintendent of Customs.

"The foreign merchants here, especially those interested in American goods, would certainly have a different customs record achieved by this time if it were practicable to introduce the American pattern of axe. The same may be said of edge-tools generally. Some years ago a Sheffield manufacturer prepared a complete set or kit of carpenter and joiner tools of the very best in quality and finish, and put them in the hands of an influential agency, determined to make trade in his line go. After a sufficient lapse of time, and in default of a single order, the attempt was abandoned, and the goods would not realize enough to pay the cost of carrying them out here. The reasons are clear enough, after the event. The Japanese mechanic pulls the plane and the saw, whereas the like English tools were made to be pushed; whether in the pattern of the other tools or in the way of using them, the difference is correspondingly great. That the native should adopt the English or foreign style of tool makes him smile, just as the foreigner would smile if offered the Japanese article. It may be thought by some that in the foreign settlements foreign workmen would be employed, and that they would be using the foreign style of tools. That would be an entire misapprehension. There are no foreign workmen in this country: nor is there any possible opportunity for them.

"In the case of the axe, it may be said the chances of its acceptance by the Japanese woodman are still more remote; not on the ground of scientific merit, assuredly, but from overruling custom. The Japanese woodmen are not choppers. Mostly, they fell the trees and cut them up with the saw. The Japanese axe seems to be used for lopping off brush and slitting. The Canadian concerned in wood-cutting, knows the 'iron wedge' used in splitting cordwood and fence rails. Take a common iron wedge and put a hole through it for the handle, a straight handle, and you have the pattern if not the size of the Japanese axe. No mortal man could chop with such an implement; yet there is certainly no future apparent for the Canadian pattern of axe.

"What has been said of axes, of the plane and the saw, may apply to the miscellaneous others, that go to make up a carpenter and joiner kit, or chest of tools. There may be seen on view, in some hardware stores, a specimen or two, not much more than a sample, of the foreign style of edge tools and other appliances. Possibly some Japanese workmen may have crossed the Pacific and acquired some practice with the foreign hammer, screw-driver, brace and bits, and so on, and may sometimes fancy that sort of thing if only to gratify the curiosity of his untraveled associates. But that does not make Japan a market for edged tools of the foreign pattern."

SECRET OF INFLUENCE.

To have a proper influence over your fellows:
Be absolutely straight.
Have no "side."
Be sympathetic.
Show a sense of humor.
Have the spirit of faith.

BISHOP INGRAM.

COPPER IN JAMAICA.

Copper deposits exists in Jamaica, and numerous attempts have been made to work this metal, beginning as early as 1857, but the mining of copper has not hitherto proved successful commercially. It appears now, however, as the result of the investigations conducted by an expert American mining engineer, that the difficulties which have prevented the profitable working of this metal have been overcome, and that paying copper will be added to the list of staple products of the island. The ore contains gold and silver as well as copper.

Two thousand acres of land have been acquired by the interests which the mining engineer represents, and shafts have been sunk in three places—in one place to a depth of 300 feet with tunnels 125 and 255 feet long. Tests of qualities of ore have been made by the government chemist, and several shipments have been made to a Denver assay works. The results obtained in both cases yielded copper in paying quantities. It is announced that a company with a capital of $250,000 will be organized to work the mines, and that within a short time operations on a large scale will be begun.

The declared exports show that copper ore of the value of 1,671 was shipped from this port to the United States during the last fiscal year.

CATALOGUES AND BOOKLETS.

New "Sovereign" Catalogue.

The Taylor-Forbes Co., Guelph, have in their new "Catalogue G" a very handy and complete book for hot water and steam fitters in which the manufacturers give complete pictorial and printed descriptions of the construction of the various types, embodying the most scientific details known to the heating trade.

In steam boilers the "Sovereign" series now includes the "Canadian," "Western," and "Western Junior"; in hot water boilers, the "Sovereign," "Advance," and "Western Junior," and in tank heaters, the "Little Giant," "Improved Giant," and "Laundry Heater." All of these are shown in sectional and complete forms, the construction of the sections, grates and heaters being thoroughly illustrated and the many advantages peculiar to the "Sovereign" series are clearly brought out. Ratings, dimensions, sizes and price lists are also given in tabulated form, together with many valuable tables added to the back of the book for handy reference.

The book is one which should be in the hands of every person engaged in heating work, and if this paper is mentioned and a business card enclosed, the firm will forward copies to any who have not already received a copy.

Neat Series of Booklets.

The D. Moore Co., Hamilton, Ont., have just issued several very attractive advertising booklets, which take the place of a large catalogue, and being gotten out in small, convenient size and in original style, are exceptionally good for distribution amongst the possible customers of a retail dealer. One booklet of 16 pages and cover, entitled, "Happy Home Ideas," is devoted to the D. Moore line of "Happy Home" Ranges, all the different styles and sizes being illustrated, with additional cuts and interesting descriptive matter, emphasizing the special features of the "Happy Home." The cover is adorned with a winter scene, very appropriately chosen. This booklet has been issued in both French and English, so that hardwaremen having French customers, will find the booklet particularly attractive.

Another booklet of 18 pages and cover, is entitled "Mrs. Tom's Treasure." The reading matter is in the form of a short story, which introduces the various good qualities of the Treasure Ranges, and, in addition, contains many valuable suggestions for the housekeeper. This booklet is one which customers are bound to read and study, as well as preserve for future reference.

Two smaller booklets, illustrating Treasure Steel Ranges, and Treasure Base Burners, have also been issued, each containing features which make the booklets valuable for distribution. As stated before, copies can be had by mentioning this paper when writing the D. Moore Co., Hamilton, Ont.

Practical Talks on Warm Air Heating

The Fourth of a Series of Articles by E. H. Roberts.

A Gravity Warm Air System.

Churches are perhaps more universally heated by warm air furnaces than any other class of buildings. The reasons for this are quite obvious, for the majority of church heating plants are not fired up more than once or twice a week and with hot water or steam there would be danger of freeze-ups if the janitor should neglect to drain the boiler and pipes.

It is also apparent that churches can be heated more quickly with a warm air furnace than with hot water or steam and at a less expense for fuel.

In modern churches the basements are usually utilized for Sunday school or lecture rooms and to heat these as well as the auditorium with a furnace requires a little different arrangement than would otherwise be necessary.

The plan here shown was selected because it illustrates what we consider the best method of heating a church with warm air connection to auditorium and basement.

The church committee for whom these plans were made, wish to heat the auditorium and the basement rooms either independently or together and they stipulate that there shall be no warm air pipes or cold air ducts running overhead in either the school-room or the society room.

Referring to the basement plan, you will notice that in order to get comparatively short warm air connections with the basement rooms, also the small room near entrance on first floor, it is necessary to locate the furnace about half way between the front and rear of building.

Ordinarily a long smoke pipe is to be avoided, but in this case the pipe is not more than eighteen feet, which, with a good chimney flue, is not so long as to be objectionable. There is another chinmey shown on the plans nearer furnace, but it has been decided to leave this out.

The auditorium of the church contains about 38,000 cubic feet, and the basement rooms together about 12,000, making a total of 50,000 cubic feet to heat. To warm this space quickly and properly we would recommend a furnace with firepot 30 inches in diameter, with at least a 60-inch casing. The size of the casings in this instance is quite as important as the size of the firepot, for unless the air has ample room to circulate rapidly through the casing it is impossible to heat the building properly no matter how much heat is generated in the furnace.

Leading from the furnace and connecting with 24x24-inch registers in floor of auditorium are two 22-inch galvanized warm air pipes, there are also 16-inch galvanized warm air pipes connecting with each of the two rooms in basement. All the warm air pipes are provided with dampers so that the heat can be shut off from any of the rooms or proportioned to suit requirements.

In order to heat such a building as this in the least possible time and with a minimum quantity of fuel, it is necessary to circulate the air; in other words the cold air must be taken back to the furnace just as rapidly as the warm air is discharged.

There are four cold air registers shown in floor of auditorium and these are so placed that the air must pass across basement rooms before reaching the under-ground ducts leading to furnace pit. As a result of this arrangement the temperature of the basement rooms will gradually rise as the air in auditorium gets warmer and by the time the church services are over the basement rooms will be almost or quite warm enough for Sunday school. If a little more heat is required in basement, however, the dampers in the basement warm air pipes can be opened, and, if necessary, all the heat from furnace can be thrown into the Sunday school rooms.

When it is desirable to heat the basement rooms without heating auditorium, the cold air registers in the auditorium floor should be closed, also the warm air dampers in the pipes connecting with first floor. Then the warm air is discharged directly into the lower rooms and the cold air taken back through the cold air registers in the basement floor.

The advantages of such a plant as we have described are, first, its efficiency—for every room having warm air connections can be quickly and thoroughly

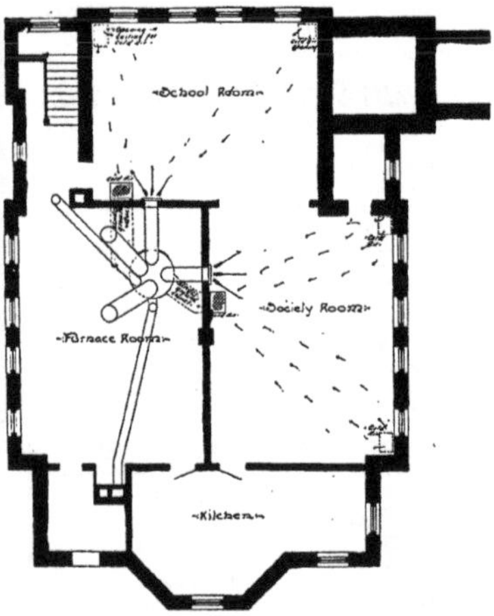

Basement.

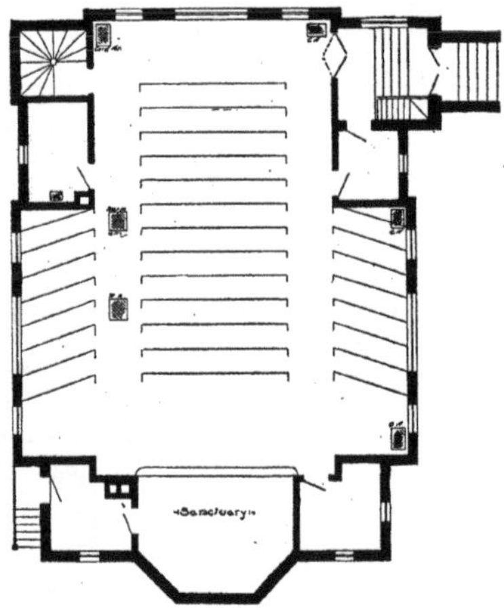

Auditorium.

Prepare for Holiday Business

It's not too early to lay plans for the Christmas holiday trade. Dealers have already bought some lines, of course, and in these days of slow deliveries other orders should be placed as soon as opportunity offers. "Goods well bought are half sold" is a time-worn truism in mercantile life.

But it doesn't do to feel too secure on the buying part until the selling has been done. Plans for increasing holiday sales must be worked out months ahead. Window displays must be figured out and a series of rough sketches planned. A series of ads. for the local papers should also be prepared, and, where deemed advisable, a neat booklet gotten up, to be mailed to all probable customers in the surrounding district.

It is in these selling plans that Hardware and Metal can be helpful to its readers. Let all partake of the spirit of co-operation and exchange ideas through these columns. To help the suggestion along the editor invites any reader to join the friendly contest and hangs up a prize of $10 cash for the best answer to the following questions:

How can the hardware merchant increase his sales of holiday goods next December? What special lines should he stock? What selling plans should be adopted? What special advertising should be done? What novel window displays can be suggested? Should Souvenirs (calendars, knives trays, etc.), be given to customers?

The prize of $10 will be awarded to the writer of the most practical and original letter of from 500 to 1,000 words received by the editor before October 15, 1907, and the best half-dozen replies will be published in Hardware and Metal.

Address the Editor

Hardware and Metal

10 Front St. E., Toronto

heated—and in the second place there are no unsightly cold air ducts to mar the appearance of the basement rooms.

It is also apparent that a plant like this can be installed at a very moderate cost and there is no possibility of the plant failing to heat the building satisfactorily, providing the furnace has sufficient capacity.

DISPLAYS HELP TO SELL.

Nothing so helps making sales and at a good profit as good display and good salesmanship. Five dollars expended in labor for cleaning up the store, doing a little painting, rearranging some platforms and getting ready for the fall trade will go a long way toward changing the general atmosphere and appearance of any stove dealer's establishment. In these days, when stove trucks and casters enable the stoves to be moved out where a customer can see all sides or can be wheeled out to the store front or sidewalk, the dealer who fails to make use of them is neglecting an opportunity.

If the dealer has a good show window he can change it every week prior to the time when stove buying is most active. It is only necessary for him to plan what he will use to make the display. He may alternate a range and a heater, a large range and a small range, a parlor and a store stove. With these he may display kitchen utensils, or the coal hod, shovel, poker, tongs, coal sifters and other equipment which often need to be replaced in the fall. An ingenious man could readily make a display of fire shovels, pokers, sieves, scoops and such things, to leave on the minds of many who pass an impression which later on will bring them to his store for some needed article.

In a large store where there is sufficient help a baking contest may be carried on when there is something special bringing the people to town, like market day or some entertainment. If this is announced in the local weekly paper and by means of circulars in the wagons of farmers, there is a certainty of several farmers' wives being among those who stop in to see it.

It is just such enterprise as this which some people lack, and it is these opportunities which other dealers utilize to get more than their share of the trade.—The Hardware Trade.

HEATING ATTACHMENT FOR GRATES.

F. A. Delph, New Orleans, La., has invented an attachment for use with open grates. It consists of one or more return flues or pipes detachably connected with a front plate adapted to cover the space above the grate basket and projecting therefrom into the room where the grate is located. The plate is hinged to the grate frame so that it may be swung laterally together with the return pipe or flues, when desired, to replenish the grate basket.

FIRING HOT WATER BOILERS

Written for The Plumber and Steamfitter by M. J. Quinn, Toronto

In common with hundreds of other mechanical devices, a hot water heating system often fails to give the maximum of satisfaction because of a lack of knowledge respecting its operation on the part of the owner, and in that way it often happens that a steamfitter obtains an undeserved bad reputation because he has either not realized or has overlooyed the necessity of properly instructing his customer how to operate a heating system after the same has been properly installed, and in order to correct some false impressions existing in the minds of many people respecting the care and method of dealing with their heating appliances, it may be well to explain what has been found in practice to be the best rule to be followed in the care of a hot water boiler.

First, it must be assumed, of course, that the boiler is connected to a suitable flue, i.e., one that is of suitable diameter and height, reasonably smooth and with tight mortar joints, so that there will be no suction of air from any point except through the smoke pipe, and that the boiler sections will be thoroughly swept out once a week, or oftener, according to the character of the fuel and the amount of ashes and soot deposited on the sections, for it must be borne in mind that a layer of ashes, say half an inch thick, on the top of each section will reduce the capacity of the boiler perhaps 30 per cent. by insulating the iron from the passing heat.

The complaint most often heard is that the house is cold in the morning, and that a fair temperature cannot be obtained until about the middle of the day, and this in most cases is due to faulty firing on the part of the owner, because, if a fair temperature may be obtained during any part of the 24 hours, it follows that the system has ample capacity, and that, if a little care is taken, an equally high temperature may be obtained at any other time during the day.

Now, what is the cause of a drop in temperature in the morning? It is due to a combination of two circumstances, viz., first, to the well known fact that the lowest temperature during the 24 hours is about four o'clock a.m., and, secondly, and vastly more important, to the fact that the owner usually adopts a wrong course in preparing a fire for the night just before he goes to bed; for instance, it is a common practice to shake the grates, put in a considerable amount of fresh coal and turn on the drafts for twenty or thirty minutes "to burn the gas off," and herein lies the great mistake, because the fire gets such a good start in this way and there is so much fresh fuel that the hottest fire during the 24 hours is likely to be about two o'clock in the morning, and from that time it gradually dies down so that at the end of the next six or seven hours, or about eight o'clock in the morning, the fire is almost entirely burnt out and the house is cold.

To overcome the resulting difficulty it has been found good practice, first, to shake the fire very little in the morning and put on some fresh coal, to which may be added two or three times during the day enough coal to keep the temperature sufficiently high, so that about ten or eleven o'clock at night there will be three or four inches of good clean fire on the surface and several inches of ashes on the grate.

If the grate then is given a good vigorous shaking and all of the ashes cleaned out, there will be enough room to place in the fire pot a considerable body of fresh coal. The furnace doors then should be entirely closed up, and the key damper in the smoke pipe placed in such a position as experience with each individual chimney will dictate would produce the best results. The damper in the fire dor should be left open in order that sufficient oxygen may be drawn in to mingle with the gas given off by the coal and produce combustion, otherwise, all of the gases will pass up through the chimney without burning, and will be a total loss.

It will be found that the fire will begin to burn up slightly about one o'clock in the morning and will gradually increase in intensity until, say, five or six o'clock, when there will be found to be a good hot bed of coal in the boiler and a very comfortable temperature throughout the house.

The necessity for a vigorous shaking of the fire in the morning is obviated, and the ashes are permitted to accumulate, so that the operation of the night before may be repeated.

Firing the boiler in this way will be found to give the very best of satisfaction, and, if the radiators are not kept hot when there is a good fire on, the plumber might pay special attention to the quality of the coal, which is entirely responsible in many cases where he, and the boiler that he installed, has received unmerited blame.

MALAY'S TIN MINES.

A Straits Settlement newspaper outlines the magnitude of the tin supply of the Malay Peninsula as follows:

A correspondent from the Peninsula states that the projected railway from Hong Kong will be likely to traverse the rich mineral regions of Siamese Malaya. Lang Suan has a tin supply that cannot be exhausted in a hundred years to come while the same may be said of Penang. There are 70 miles in the region of Lang Suan, most of which are worked by natives but the European concessions in the latter place, as well as in Penang, are exceptionally encouraging and already are giving excellent returns. There is no lack of capital, even the natives making themselves better acquainted with modern machinery and bringing it into use.

WIRE NAILS
TACKS
WIRE

Prompt Shipment

The ONTARIO TACK CO.
Limited.
HAMILTON ONT.

MACHINERY STEEL
NORWAY IRON

Large Stock of
All Sizes.

CLOSE PRICES

A. C. LESLIE & CO.
Limited
MONTREAL

"MIDLAND"
BRAND
Foundry Pig Iron.

Made from carefully selected Lake Superior Ores, with Connellsville Coke as Fuel, "Midland" will rival in quality and grading the very best of the imported brands.

Write for Price to Sales Agents
Drummond, McCall & Co.
MONTREAL, QUE.
or to
Canada Iron Furnace Co.
MIDLAND, ONT. Limited

NOVA SCOTIA STEEL & COAL CO., Limited
NEW GLASGOW, N.S.

Manufacturers of
Ferrona Pig Iron
And SIEMENS-MARTIN
OPEN HEARTH STEEL

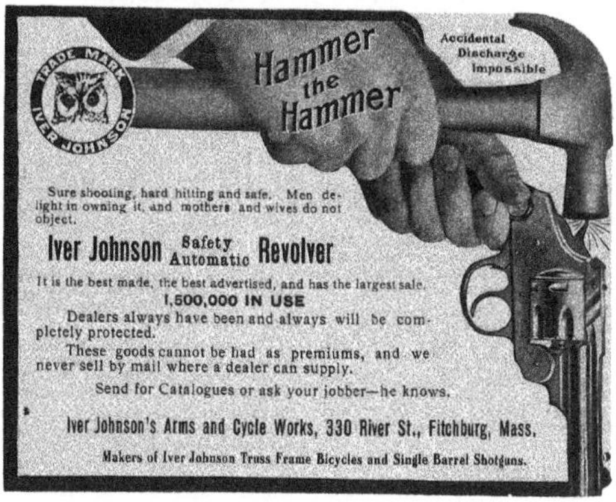

OAKEY'S The original and only Genuine Preparation for Cleaning Cutlery, 6d. and 1s. Canisters
'WELLINGTON'
KNIFE POLISH

JOHN OAKEY & SONS, Limited
Manufacturers of
Emery, Black Lead, Emery, Glass and Flint Cloths and Papers, etc.
Wellington Mills, London, England
Agent:
JOHN FORMAN, - 644 Craig Street
MONTREAL.

TINPLATES

All made out of the
BEST SIEMEN'S MARTIN STEEL

Brand

Best Coke or Charcoal	"Rhyd"
Good Charcoal	"Edith"
Second Charcoal	"Sawel"
Best Charcoal	"Dryw"

We are ready to offer the Canadian Trade very best value and finest quality in Tinplates.

Baglan Bay Tinplate Co., Limited
Briton Ferry, South Wales

Please mention above brands in your indents and inquiries.

NICKEL
The Canadian Copper Co.

NICKEL FOR NICKEL STEEL
The Orford Copper Co.

WRITE US FOR
PARTICULARS AND PRICES

General Offices
43 EXCHANGE PLACE
NEW YORK CITY

BUILDING AND INDUSTRIAL NEWS

For additional items see the correspondence pages. The Editor solicits information from any authoritative source regarding building and industrial news of any sort, the formation or incorporation of companies, establishment or enlargement of mills, factories or foundries, railway or mining news.

Industrial Development.

J. E. Doak, Saskatoon, Alta., will erect a factory for the manufacture of all kinds of house finishings.

The capital stock of the McLaughlin Carriage Co., Oshawa, has been increased from $400,000 to $1,500,000.

Zuelsdorf Bros., Berlin, Ont., will erect a factory in that city for the manufacture of high-grade furniture.

The Grand Bay lumber mill, St. John, N.B., was destroyed by fire on Monday. The loss of $15,000 was covered by insurance to the amount of $11,000.

The Interwest Peat Fuel Company's plant, Lac du Bonnet, Man., was destroyed by fire last week. Loss, $40,-000, partly covered by insurance.

Defective electric wiring is supposed to have been the cause of a fire that destroyed the building and plant of Parker's laundry works, Peterboro, on Tuesday.

The Tacoma Construction Co. are asking New Westminster, B.C., for a free site on Lytton Square, in order to erect a hotel and business block, at a cost of $200,000.

M. L. Aubert, Montreal, a member of a firm of French capitalists, is negotiating with Hull, Que., for the erection of a large biscuit and confectionery factory in that city.

The Porto Rico Co., Nelson, B.C., will erect a lumber mill, with a daily capacity of 45,000 feet. The limits possessed by the company are conservatively estimated at one hundred million feet.

The Albert Sawmills, Barachois, Que., were recently totally destroyed by fire, together with the wharf and all the buildings connected with the mills. Five million feet of lumber also disappeared.

A company with a capital of $50,000 proposes to erect a factory for the manufacture of cement bricks at New Westminster, B.C., providing suitable deposits of sand and clay can be found in the vicinity.

A. C. Flumerfelt and H. N. Galer have acquired a large area of rich coal deposits near Lethbridge, Alta. A company has been formed to develop the deposits and a large plant will be installed at once.

Owing to the closing down of the Rosamond Woolen Mills, Almonte, Ont., about 300 operatives will be thrown out of employment. The factory closed last week for an indefinite period, presumably on account of labor troubles.

The Canadian Steel Specialty Co., of Gravenhurst, Ont., will erect a two-storey factory in that town, together with a boiler house and dry kiln, at a cost of $10,000. The company will manufacture steel furniture, electric fixtures, and novelties. The town has granted a free site and exemption from taxation.

The Peel Oil and Gas Co. has let contracts for the sinking of oil wells on the property of the company at Cooksville, 16 miles west of Toronto. This organization has over 1,000 acres of land under option, and it is expected that deposits of oil or natural gas will be encountered. Some years ago natural gas was obtained in a shaft sunk a few miles west of Mimico, but, owing to a lack of facilities for handling the product, the well was plugged. The above company expects that if the venture proves successful it will be able to supply natural gas at a cheap rate to Toronto.

Railroad Construction.

Trains are now running regularly over the new Guelph to Goderich line of the C.P.R. The roadbed is said to be the smoothest in Canada.

The new double track of the Galt, Preston and Hespeler Railroad is now completed, and the first car was run between Preston and Galt on Monday.

Plans for the new railway station and hotel, which are to be erected at Ottawa on Government property by the G.T.R., were approved. The station is to be for the use of all the companies. The hotel will be located on the southwest corner of Major's Hill Park, with a frontage of 135 feet on Wellington Street.

The second of the C.P.R. new lake boats, the Keewatin, reached Montreal this week from across the Atlantic. She, with her sister ship, the Assiniboia, will be the two largest and best steamers in the C.P.R. lake service. As soon as her cargo is unloaded, the Keewatin will return to Levis, where she will be cut in two. She will then be towed up the St. Lawrence and the lakes as far as Buffalo, where she will be put together again.

The accounts of the Canadian Pacific Railway Company for the year ended June 30, 1907, show the following results :

Gross earnings	$72,217,520.64
Working expenses	46,914,218.83
Net earnings	$25,303,308.81
Surplus	$19,156,033.51
Net surplus for the year...	$9,339,005.32

Companies Incorporated.

Wilbur Iron Ore Company, Toronto; capital, $500.000. To prospect for and deal in ores, metals and minerals. Incorporators : C. L. Dunbar, E. A. Dunbar, and H. C. Scholfield, all of Guelph, Ont.

Building Notes.

Improvements to the Y.W.C.A. building, Simcoe Street, Toronto, will cost $12,000.

The congregation of St. Joseph's Church, Sydney, N.S., will erect a large edifice to replace the one recently destroyed by fire.

The Masons of Alpha Lodge, Toronto, have voted to join with other lodges in the erection of a new Masonic Hall, to be built in the northwestern part of that city, at a cost of $60,000.

NEW FOUNDRY AT ST. JOHN.

An industry that bids fair to hold a foremost position among the manufacturers of the Maritime Provinces is McLean, Holt & Co.'s up-to-date foundry on Albion St., St. John, N.B., which within a short time will be in full running order. The buildings are all but completed, really all that is required being the placing of the machinery.

The two large, two-storey, wooden structures erected, cover an area of 105x135 feet and are splendid architectural specimens for their purpose. The lower storey of the first building, which faces on Albion St., will be devoted to the fitting and mounting departments with a floor space of 130x35 feet. In the upper storey is the nickel-plating department and the buffing room as well as a section devoted to a tin shop and stock room.

Going downstairs again the grinding room is passed through, leading into the moulding department, which covers an area of 130x54 feet. This portion is carried by strong concrete abuttments and sixteen piers with re-enforced iron braces. In the centre of this room, and running from end to end, is a heavy steel track, which carries a truck for distributing iron to different parts of the building The huge steel cupola, fitted only with the most modern attachment, has a capacity of something like fifteen tons, and the iron will be raised to the upper door by means of a steel elevator.

In the yard adjoining the wooden buildings is the power-house, built of brick, 30x20 feet, and containing a boiler-house and engine-room, which is equipped with a Leonard high-speed, automatic engine, of 54 horse-power.

Next spring the McLean, Holt Co. will erect another structure. measuring 130x60 feet. Already a retaining wall, measuring 150 feet, fronting on Courtney Bay is in process of erection. The new buildings will form the new home of the "Glenwood Range," of which a large number are sold in all parts of the Maritime Provinces. The new foundries were made necessary by the company's rapidly growing business, which could not be handled in the limited space now occupied on the City Road.

TRUSTS TRY TO PLEASE PUBLIC.

Judge Elbert H Gary, chairman of the board of directors of the United States Steel Corporation, who has returned from a two months' trip to Europe, says the commission of experts appointed to make estimates of a standard rail that will be strong enough to meet the requirements of the heavier equipment now in use, has not yet handed in its report, but when it does so it will be found that the steel manufacturers and the railroad companies will make every effort to adopt a standard that will insure the safety of the public.

"The time has gone past," said Judge Gary, "for the great corporations to ignore the public and the public interest. The 'Public be Damned' policy, if that phrase was ever actually used, will not go in these days. The heads of our great industrial enterprises can never adopt that attitude, not alone as a matter of policy, but in the interest of their own undertakings. There has been a great change of late years in the attitude of corporations toward the public and their employes.

"When the steel manufacturers and the railroad officials met several months ago to appoint the steel rail commission, I stated that the meeting was an evidence of the spirit of conciliation. Here were rival manufacturers, with varied interests, who came together without airing any differences or displaying the slightest opposition, and they, in turn, met the heads of the country's railroad systems, all of them in a conference which had for its ultimate aim the benefit of the public."

When asked for his views on the condition of the steel industry abroad Judge Gary said: "The steel industry appears rather prosperous in Europe, especially in Germany, where the Government favors the industrial combinations. The Germans are wonderfully progressive and successful. We have no agreement with their steel combination, except such as relates to consultations respecting the interests of the various manufacturers in the different countries. But there is nothing in the nature of a binding contract between us. We claim to excel, of course. The Germans, however, in my opinion, surpass the English in the making of steel."

TAR SANDS OF ATHABASKA.

In a paper read a few days ago Dr. Robert Bell, of Ottawa, said he did not know that in any other part of the world was there so great an accumulation of tar sand, due to the outpouring of petroleum, as in the valley of the Athabaska. He estimated that the sands covered 1,350 square miles, and had an average depth of 150 feet. This would be altogether about 11,000,000 tons of tar substance apart from the sands. He considers the sand would be good material for the manufacture of oil, and for that purpose the resources were practically illimitable. When the railway was built the manufacture of oil would be greatly facilitated.

CANADIAN CEMENT PRODUCTION.

The manufacture of cement in Canada dates from the year 1891, when operations were begun in a very small way at Marlbank and Shallow Lake, both in the Province of Ontario. The first year's output was 2,053 barrels, which is only a little more than the daily production of one of the modern plants. In 1892 the output was 20,247 barrels, and 31,924 barrels in 1893.

The next ten years in the cement industry of Canada witnessed an increase of fifty per cent. each year over the production of the preceeding year. Four plants were in operation in 1901, and now there are nineteen.

In 1906, there were fifteen plants in operation, with a total daily capacity of about 10,500 barrels, and according to the published figures of the Canadian Geological Survey, during the year, 2,152,562 barrels of Canadian cement were made, which represents an increase of 610,994 barrels, or 39.6 per cent. over the production of 1906.

News of the Paint Trade

MANUFACTURERS EDUCATING DEALERS.

The Bureau of Promotion and Development Paint Manufacturers' Associaton of the United States, the headquarters of which are in Chicago, is doing some effective missionary work in promoting a proper knowledge of paints and paint ingredients, claiming that this is the greatest selling force that can be incorporated in any paint business. A forty-four page pamphlet entitled "A Paint Catechism for Paint Men," compiled by Mr. George B. Heckel is being sent free to a list of sixty thousand paint and hardware dealers in the United States. An interesting circular letter accompanies each pamphlet, from which we make the following extracts:

To the Paint Dealer:

How to convey a proper knowledge of paints and paint ingredients, and how to obtain this knowledge have been problems of great importance to the manufacturer and dealer respectively.

Eighty-eight questions, some comparatively simple, others of a more complex nature, are answered in a plain, straightforward, practical way in the enclosed "Paint Catechism for Paint Men."

The information in this "Catechism" represents the best brains in the business, condensed for your convenience and quick reference. Dr. Heckel, in compiling this "Cathechism," has endeavored to answer all of the every-day questions presented to the dealer; he has gone even further by embodying a great many of the more technical features of paint manufacture, all of which are of interest to you as a dealer. This "Catechism" is worth more to you from a sales standpoint than a complete reference library; it is condensed, plainly written, practical. Let it be a text book, a constant reference and guide for yourself and every clerk in your establisment. If you do, it won't be long before you know it by heart and can talk to your trade intelligently and practically on any paint subject.

Your manufacturer or this Bureau will be glad to answer personally any questions regarding the information in this "Catechism," or any questions not covered in this book, at any time. We welcome inquiry on subjects of general paint interest not embodied in this "Catechism" to enable as to make it more complete in the next edition.

Your Silent Salesman.

"Goods well advertised are half sold." Do you realize that advertising is a salesman, working for you twenty-four hours a day and seven days a week? You arrange the affairs of your store so that your salesmen can work to the best advantage; do you do this with your silent salesman—your advertising matter? To put it plainly, every piece of advertising matter sent you by your manufacturer costs something; it is prepared and sent to you to accomplish

NEW ERA PAINT

is made from pure materials. It is made **right,** by men who know how. That's why it always works so satisfactorily. It spreads evenly; it covers a lot of space to the pot; it keeps its freshness for a long time. Naturally enough, it **makes money for the dealer.**

STANDARD PAINT AND VARNISH WORKS CO., LIMITED
WINDSOR - ONTARIO

JAMIESON'S LIQUID HARD FLOOR WAX

The Modern Finish for Waxed Floors

Easiest to apply and keeps its appearance for a longer period than any other floor wax.

Put up in Pints, Quarts, Half Gallons and Gallons.

Manufactured by

R. C. JAMIESON & CO.
LIMITED
16-30 Nazareth St., MONTREAL

HEADQUARTERS

FOR

Window Glass,
Paints, Oils,
Fine Varnishes

SANDERSON PEARCY & CO., Ltd.
Toronto, Canada

GLAZIERS' DIAMONDS OF PROVED WORTH
Having a Century's Reputation for Reliability.
MADE BY
A. SHAW & SON, London, Eng.
CANADIAN AGENT
GODFREY S. PELTON, - 201 Coristine Building, Montreal

Sharratt & Newth's Glaziers' Diamonds

are unequalled for cutting and wearing qualities.

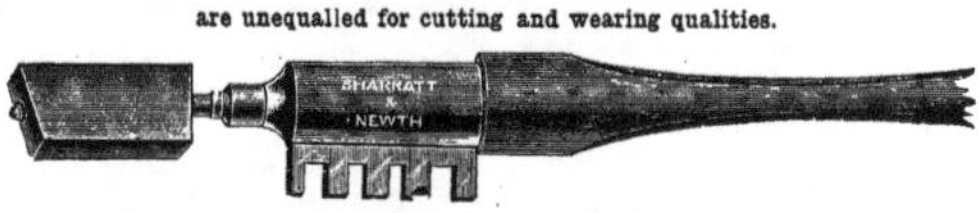

To be obtained from the principal Dealers in Glass, Hardware, and Painters' Supplies.

Contractors to H. M. Government and the principal English Sheet and Plate Glass Works

GILBERTSON'S COMET

Gilbertson's Sheets
and the COMET Brand

mean quality and security in whatever work you may undertake.

Galvanized flat sheets that are unequalled as to flatness; absolute trueness to gauge; admirably suited in softness for any working. Note the brand "COMET." GILBERTSON on every sheet means security.

W. GILBERTSON & CO., Limited, Pontardawe, South Wales
ALEX. GIBB, 13 St. John St., Montreal, Agent

some definite, specific purpose. If you are not using it wisely and to the best advantage, you are causing your manufacturer needless waste, and what is of more consequence, you are wasting an opportunity to increase your own business on profitable lines.

As a demonstration, suppose the day this letter reaches you, you have every piece of paint literature and advertising matter in your store (which is not displayed or placed where it is accessible and working for you) collected and put in one pile where you can see just what it amounts to. You and your manufacturer are to be congratulated if such a demonstration does not unearth a miscellaneous lot of color cards, show cards, practical literature, etc., all a part of your selling organization, which have become soiled and practically worthless, due to being put away carelessly "any old place" when received and then forgotten. You wouldn't do this with a shipment of paint, yet the advertising matter is of more ultimate value to you, as its proper use means doubling, yes, tripling, the sales of the goods it advertises if you use it right.

Doesn't your front need brightening up?

Wouldn't a coat of enamel, stain or paint make your counters, shelves, walls and ceiling inside more inviting?

How does your window look?

Does the passerby get a good or bad impression from it?

Do you change it often and make it interesting?

Does it drive away the "blues" or does it create "that tired feeling" when your friends and customers see it?

Does your window advertise you or does it drive trade to your competitor?

Which of these two classes are you in?

You could not hire the service your show window will give you for the price of two clerks if you use it right.

Make your store appear so "wide awake," so up-to-date, so bright and attractive, that if you were the man you wish to influence you would just naturally consider it "headquarters" for paints. Give it that "know-your-business" air counts with live people.

Let's get together for a better and bigger paint business.

PAINT AND OIL MARKETS

MONTREAL.

Montreal, Sept. 27.—Strong activity characterizes the local market for white lead, varnishes and colors. This week orders, while not being of any great weight, have been extremely numerous, showing the desire on the part of jobbing and retail trade to keep stocks well in hand, at the same time not being too lavish in expressing requirements. Quite a large number of repeat orders are coming in, and a few orders for next spring's requirement are being booked.

Linseed Oil—Shows a decided upward tendency this week. Quotations all round are marked up a cent. We now quote: Raw, 1 to 4 barrels, 61c; 5 to 9, 60c; boiled, 1 to 4 barrels, 64c; 5 to 9, 63c.

Turpentine—Latest despatches from Wilmington, N.C., report a weakness in quotation, and local prices have been affected very materially. We now quote single barrels at 74c; wood distilled turpentine, pure, 68c.

Ground White Lead—One or two of the grinders have given way slightly in quotations on good-sized orders, and, it may be added, without any apparent necessity, because all well established brands of dry white lead are strong, and it would seem that a slight advance in quotations would be more in order than concessions. We continue to quote: Government standard, $7.50; No. 1, $7; No. 2, $6.75; No. 3, $6.35.

Red Lead—Some heavy importations have arrived recently, and, having been bought upon favorable terms, quotations have been eased to the extent of 25 cents per 100 pounds. We quote: Genuine red lead, in casks, $6.25; in 100-lb. kegs, $6.50; in less quantities at $7.25 per 100 lbs. No. 1 red lead, casks, $6; kegs, $6.25, and smaller quantities, $7.

Dry White Zinc—Pursues the even tenor of its way, and there is no special feature worthy of note. Prices are firm and unchanged: V.M. Red Seal, 7½c; Red Seal, 7c; French V.M., 6c; Lehigh, 5c.

White Zinc Ground in Oil—Not many shipments have been made, stocks are ample and there is no special feature. Prices are unchanged: Pure, 8½c; No. 1, 7c; No. 2, 5¾c.

Gum Shellac—Is in good supply, but enquiries are not of a very keen character. There is no change quotations: Fine orange, 60c per lb.; medium orange, 55c per lb.; white (bleached), 65c per lb.

Shellac Varnish—It is noted that more orders are coming in for shellac in small packages being in handy form for retailers. No change in quotations has been made: Pure white bleached shellac, $2.80; pure orange, $2.60; No. 1 orange, $2.40.

Putty—This article is being rolled out with increasing volume, and hardware merchants should look well forward in placing their requirements: Pure linseed oil, $1.85 in bulk; in barrels, $1; in 25-lb. irons, $1.90; in tins, $2; bladder putty, in barrels, $1.85.

TORONTO.

Toronto, Sept. 27.—There has been considerable increase in paint and oil sales during the week. Very few large orders are coming in, but the volume of sorting order business for the last

Window Glass

Come to us for all kinds of glass. Let us quote you on every specification. We have immense stocks in our large new warehouses.

Star Window Glass
Diamond Picture Glass
Double Diamond Glass
Colored Glass
Rolled Glass
Cathedral Glass
Wire Glass
Heavy 26 oz. 32 oz. Glass
and Glass of all kinds.

LET US QUOTE

A. RAMSAY & SON COMPANY

Glass Dealers — MONTREAL — 10 to 22 Inspector St.

A Fire-proof Gasolene Tank

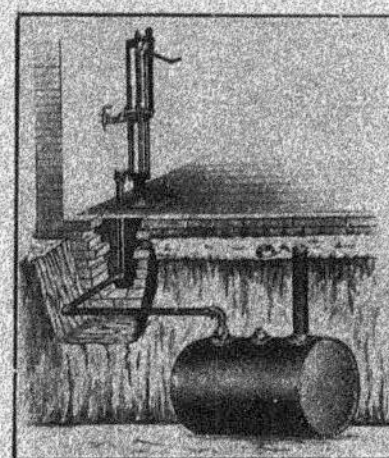

Cut 42—Pump with Underground Tank

That old tin gasolene tank setting out behind your store is dangerous, it's liable to explode and set fire to your store at any time, and because it is so close to the building, you may have trouble collecting your insurance.

You're inviting a blaze

On the other hand, put in a Bowser Gasolene Tank and Self-Measuring Pump and you remove every particle of danger.

The tank is fire-proof, the pump is fire-proof—there's no evaporation, no leakage, no waste. Being self-measuring, there is no spilling or overflow. All the gasoline is in the fire-proof tank or in the customer's can. **It's the safest, surest, cheapest method of handling gasolene.**

Then, too, the Bowser is a money-saver, it's convenient, and it's speedy. Your customers like it and it draws trade.

SEND FOR CATALOG V.

F. S. BOWSER & COMPANY, Limited
66-68 Fraser Ave., TORONTO, ONT.

"If you have an old Bowser and want a new one write us for our liberal exchange offer."

BERRY BROTHERS' VARNISHES

are one grand story of success

BECAUSE THEY ARE THE BEST

They are the safest goods to handle,
and the most reliable goods to use.

Write for Catalogue, Information and Prices.

BERRY BROTHERS, Limited

Varnish Manufacturers
DETROIT
Canadian Factory—WALKERVILLE, ONT.

NEW YORK
BOSTON
PHILADELPHIA
BALTIMORE

CHICAGO
CINCINNATI
ST. LOUIS
SAN FRANCISCO

week in September is away ahead of the amount of business transacted for the corresponding period of last year. As might be expected, the demand for mixed paints is slowly falling off, with the approach of cooler weather, but this loss of trade is more than made up by the daily increase in glass and putty sales. Turpentine has taken a sharp decline, and is now quoted locally at 76c. otherwise prices are unchanged.

Linseed Oil—Foreign prices continue to stiffen and during the week the local market has developed a stronger tone as a result. Hitherto large arrivals of English oil at the port of Montreal have kept the price of the domestic fluid from advancing, but now, with higher prices abroad, the price tendency of domestic oil is decidedly upward. Present prices are: Raw, 1 to 3 barrels, 64c; 4 barrels and over, 63c. Add 3c to this price for boiled oil, f.o.b., Toronto, Hamilton, London, Guelph and Montreal, 30 days.

Turpentine—Very little heavy buying is being done, but the demand is good for this time of the year, with supplies quite adequate to meet it. On account of the stringency of the money markets, many large Southern holders of turpentine have lately eagerly sought to discharge their stocks, and this anxiety to sell has precipitated a sharp drop in the Southern price. In sympathy with this Southern decline, local quotations have been reduced 3c during the week. We now quote: Single barrels. 76c; two barrels and upwards, 75c; f.o.b. point of shipment, net 30 days; less than barrels, 5c advance. Terms, 2 per cent., 30 days.

White Lead—A good seasonable demand exists and local prices are stiffer as a result of the strong tone which characterizes the pig iron markets. Prices are firm, as follows: Genuine pure white lead, $7.65, and No. 1, $7.25.

Red Lead—There is a fair demand and sales are slightly in excess of those of last week. We quote: Genuine, in casks of 500 lbs., $6.25; ditto, in kegs, of 100 lbs., $6.75; No. 1 casks of 500 lbs., $6; ditto, in kegs of 100 lbs., $6.50.

Glass and Putty—Putty is in strong demand and prices are firmly held. Glass sales are steadily increasing each week and indications are that a rushing business will be done in November. Prices on glass are unchanged.

Petroleum—Week by week the demand is growing heavier, and orders are now coming in fast from dealers who are stocking up to meet the winter trade. Prices are firm and unchanged: Prime white. 13½c; water white, 15c; Pratt's astral, 18½c.

Shellac—A strong autumn demand exists, due to the great number of industrial uses to which shellac can now be put. Prices are firm, as under: Pure orange, in barrels, $2.70; white, $2.8[illegible] per barrel; No. 1 (orange), $2.50.

For additional prices see current market quotations at the back of the paper.

The Standard Varnish Co. has bought 12 acres at Chicago Heights, and will build the biggest varnish plant known, to cost $1,000,000.

LINSEED OIL LEGISLATION.

The portion of the North Dakota law relating to linseed oil reads as follows:

2113. Boiled Linseed Oil.—No person, firm or corporation or agent or employe of any person, firm or corporation shall manufacture for sale or offer or expose for sale in this state any flaxseed or linseed oil unless the same answers a chemical test for purity recognized in the United States Pharmacopoeia or any flaxseed or linseed oil as "boiled linseed oil" unless the same shall have been put in its manufacture to a temperature of two hundred and twenty-five degrees Fahrenheit.

2114. Painted, Stamped or Stenciled.—No person, firm or corporation or agent or employe or any person, firm or corporation shall sell, expose or offer for sale any flaxseed or linseed oil unless it is done under its true name and each tank car, tank, barrel, keg or any vessel of such oil has distinctly and durably painted, stamped, stenciled or labeled thereon the true name of such oil and in ordinary bold-face capital letters the words "pure linseed oil, raw," or "pure linseed oil, boiled," and the name and address of the manufacturer thereof and sold under the brand of such manufacturer.

PROHIBITS THE USE OF LEAD.

The municipal council of Vienna, Austria, has passed a resolution prohibiting the use of red and white lead in connection with public works in that city, except where technical reasons render such employment essential, in which event white lead paint must be mixed with an equal quantity of zinc white paint Coating girders and iron work with red lead, except for the purpose of protecting the metal from corrosion in vaulted roofs, must be reduced to a minimum, and omitted entirely when the girders are in wooden roofs, except where the girder ends are embedded in masonry. All water pipes and connections must be coated with anti-corrosive paint, and packed with some authorized innocuous packing. The peculiar feature of this regulation, says a British exchange, is that, while prohibiting lead paints "as a matter of principle," it allows any number of exceptions; while it is difficult to see what beneficial effect the addition of zinc white to white lead paint is expected to produce, the paint being just as poisonous afterwards as before. Either white lead should be prohibited altogether, or else allowed to be used without restriction, any intermediate course being useless.

PAINTING AN OLD BRICK WALL.

The first thing to do is to get every particle of loose stuff from the walls, which may be done with a coarse fibre brush; then dust off clean. If you have a lot of old paint and enough to do the job, thin it down with oil and a little benzine, strain, and apply quite thin to the wall, says Master Painter. Brush this well into the surface, and let it have several days to become hard. The next coat should be a lead paint, of fresh materials, with raw oil and just enough driers to dry it well in reasonable time. A little turps also will be an advantage. This will now give you a good foundation for whatever color of paint you may want to apply.

FOLLOWING IS A LIST OF THE LEADING HARDWARE FIRMS IN CANADA **WHO DO NOT** SELL THE MANUFACTURES OF THE CANADA PAINT COMPANY :—

NAME.	TOWN.
....	....
....	....
....	....
....	....
....	....
....	....
....	....
....	....
....	....
....	....
....	....
....	....
....	....
....	....
....	....
....	....
....	....
....	....
....	....

CURRENT MARKET QUOTATIONS.

September 27, 1907

These prices are for such qualities and quantities as are usually ordered by retail dealers on the usual terms of credit, the lowest figures being for larger quantities and prompt pay. Large cash buyers can frequently make purchases at better prices. The Editor is anxious to be informed at once of any apparent errors in this list, as the desire is to make it perfectly accurate

METALS.

ANTIMONY.

Cookson's ....per lb....	0 13½
Hallett's	0 13

BOILER PLATES AND TUBES

	Montreal.	Toronto
Plates, ¼ to ½ inch, per 100 lb	2 40	2 50
Heads, per 100 lb	2 65	2 75
Tank plates 3-16 inch	2 60	2 65
Tubes per 100 feet, 1½ inch ..	8 25	8 75
" " 2 " ..	9 0	9 10
" " 2½ " ..	10 50	12 00
" " 3 " ..	12 00	13 50
" " 3½ " ..	15 00	16 70
" " 4 " ..	19 25	20 80

BOILER AND T.K. PITTS

Plain tinned and Spun, 25 per cent. off list.

BABBIT METAL.

Canada Metal Company—Imperial genuine 60c; Imperial Tough, 66c; White Brass 50c. Metallic, 35c; Harris Heavy Pressure, 25c, Hercules, 25c; White Bronze, 15c.; Sta Frictionless, 14c; Alluminoid, 10c; No. 4 9c. per lb

James Robertson Co.—Extra and genuine Monarch, 60c.; Crown Monarch, 50c; No. 1 Monarch 40c; King, 30c.; Fleur-de-lis, 20c; Thurber, 15c, Philadelphia, 12c.; Canadian, 10c.; hardware, No. 1, 15c.; No. 2, 12c.; No 3, 10c. per lb.

BRASS.

Rod and Sheet, 14 to 30 gauge, 25 p c. advance.

Sheets, 12 to 14 in.	0 30
Tubing, base, per lb 5-16 to 2 in	0 33
Tubing ½ to 3-inch, iron pipe size..	0 31
" 1 to 3-inch, seamless..	0 36

Copper tubing, 6 cents extra.

COPPER.

		Per 100 lb
Casting ingot..........	....	18 00
Cut lengths, round, bars, ½ to 2 in...		30 00
Plain sheets, 14 oz.....		36 00
Plain, 16 oz., 14x48 and 14x60	....	35 00
Tinned copper sheet, base	....	38 00
Planished base...............	....	43 00
Braziers' (in sheets) 4x6 ft., 25 to 30 lb each, per lb., base..	0 34	0 35

BLACK SHEETS.

No 10 gauge	2 70	2 70
12 gauge.......................	2 70	2 75
14 "	2 50	2 60
17 "	2 50	2 60
18 "	2 50	2 80
20 "	2 50	2 60
22 "	2 55	2 65
24 "	2 55	2 70
26 "	2 65	2 85
28 "	2 70	3 00

CANADA PLATES.

Ordinary, 52 sheets	2 75	3 05
All bright "	3 75	4 05
Galvanized—	Dom.Crown.	Ordinary
18x24x52	4 45	4 35
60................	4 70	4 60
20x28x52	8 90	8 70
60	9 40	9 20

GALVANIZED SHEETS.

B.W gauge	Queen's Head	Fleur-de-Lis	Gordon Crown	Colborne Crown Gorbal's Best
16-20	3 95	3 80	3 95	....
22-24 ..	4 20	4 05	4 00	4 05
26........	4 45	4 30	4 40	4 30
28..... ..	4 70	4 55	4 60	4 55

Less than case lots 10 to 25c. extra.

Apollo Brand.

24 guage, American	3 85
26 " "	4 10
28 " "	4 55
10¾ oz.	4 85

25c. less for 1,000 lb. lots.

IRON AND STEEL.

	Montreal.	Toronto.
Middlesboro, No 1 pig iron..	21 50	23 50
Middlesboro, No 3 pig iron	20 50	22 00
Summerlee, " ..	25 50	25 00
" special " .	24 50	....
" soft " ..	24 00	....
Carron...	26 00	....
Carron Special	24 50	....
Carron Soft	24 00	....
Clarence, No. 3	21 50	22 00
Glengarnock, No. 1		25 50
Midland, Londonderry and Hamilton		23 00
Radnor, charcoal iron........	30 00	32 00
Lowmoor iron......................		6 50
Angles		2 65
Common bar, per 100 lb......	2 20	2 30
Forged iron "	2 45	..
Refined " "	2 60	2 80
Horseshoe iron "	2 60	2 80
Band iron, No. 10 x ½ in.		2 30
Sleigh shoe steel "	2 25	2 30
Iron finish steel		2 30
Reeled machinery steel		3 00
Tire steel	2 40	2 50
Sheet cast steel......		0 12
Mining cast steel		0 08
Warranted cast steel....		0 14
Annealed cast steel....		0 15
High speed	0 60	0 65
B.P.L. tool steel	10½	0 11

INGOT TIN.

Lamb and Flag and Straits—

56 and 28-lb. ingots, 100 lb.	$41 00	$42 00

TINPLATES.

Charcoal Plates—Bright

M.L.S., Famous (equal Bradley)		Per box.
I C,14 x 20 base	....	$6 50
IX, 14x20 "	....	8 00
IXX, 14 x 20 base	....	9 50
Raven and Vulture Grades—		
I C, 14 x 20 base	....	5 00
I X "	....	6 00
I X X "	....	7 00
I X X X "	....	8 00

"Dominion Crown Best"—Double Coated, Tissued.

		Per box.
I C, 14 x 20 base..........	5 50	5 75
I X, 14 x 20 "	6 50	6 75
IXX x 20 "	7 50	7 75

"Allaway's Best"—Standard Quality.

I C, 14 x 20 base..........	4 65	5 00
I X, 14 x 20 "	5 40	5 75
I XX, 14 x 20 "	6 15	6 50

Bright Cokes.

Bessemer Steel—		
I C., 14 x 20 base	4 25	4 35
20x28, double box	8 50	8 70

Charcoal Plates—Terne.

Dean or J. G. Grade—		
I C., 20x28, 112 sheets	7 25	8 00
IX., Terne Tin	...	9 50

Charcoal Tin Boiler Plates.

Cookley Grade—

X X, 14x56, 50 sheet bxs / " 14x60, " / " 14x65, "	....	7 50

Tinned Sheets.

72x30 up to 24 gauge...... ...	....	8 50
" " 26 "	....	9 00

LEAD.

Imported Pig, per 100 lb......	5 35	5 35
Bar, "	5 75	6 00
Sheets, 2½ lb. sq. ft., by roll ..	6 75	6 50
Sheets, 3 to 6 lb. "	6 50	6 25

Cut sheets ½c. per lb., extra.

SHEET ZINC.

5-cwt. casks	7 50	7 75
Part casks.....................	7 75	8 00

ZINC SPELTER.

Foreign, per 100 lb	6 50	6 25
Domestic	6 50	6 75

COLD ROLLED SHAFTING.

9-16 to 11-16 inch	0 06
¾ to 1 7-16 "	0 05½
1 7-16 to 3 "	0 05

30 per cent.

OLD MATERIAL

Dealers' buying prices:

	Montreal	Toronto
Heavy copper and wire, lb	0 14	0 13
Light copper..............	0 12	0 11½
Heavy red brass....	0 12	0 11½
" yellow brass	0 09	0 1½
Light brass	0 07	0 07½
Tea lead	0 03½	0 03½
Heavy lead................	0 04	0 04
Scrap zinc	0 03½	0 03½
No 1 wrought iron	14 50	11 50
" 2 " "	6 00	6 00
Machinery cast scrap ...	17 00	16.00
Stove plate................	12 00	11 70
Malleable and steel	8 00	10 00
Old rubbers	0 10½	0 10
Country mixed rags, 100 lbs.	1 00	1 25

PLUMBING AND HEATING

BRASS GOODS, VALVES, ETC.

Standard Compression work, 57½ per cent.
Cushion work, discount 40 per cent.
Fuller work, 70 per cent.
Flatway stop and stop and waste cocks, 60 per cent.; roundway, 55 per cent.
J.M.T. Globe, Angle and Check Valves, 45; Standard, 55 per cent
Kerr standard globes, angles and checks, special, 42½ per cent.; standard, 47½ p c
Kerr Jenkins' disc, copper-alloy disc and heavy standard valves, 40 per cent.
Kerr steam radiator valves 60 p.c., and quick-opening hot-water radiator valves, 60 p c.
Kerr brass, Weber's straightway valves, 40; straightway valves, I B B M, 60.
J. M. T. Radiator Valves 50; Standard, 60; Patent Quick-Opening Valves, 65 p c
Jenkins' Valves—Quotations on application to Jenkins' Bros., Montreal

No. 1 compression bath cock......net		2 00
No. 4 " "	"	1 90
No 7 Fuller's......	"	2 25
No. 4½, "	"	2 35
Patent Compression Cushion, basin cock, hot and cold, per doz., ...		$16.20
Patent Compression Cushion, bath cock, No. 2208		2 25

Square head brass cocks, 50; iron, 60 p c.
Thompson Smoke-test Machine 25.00

BOILERS—COPPER RANGE.

Copper, 30 gallon, $33, 15 per cent.

BOILERS—GALVANIZED IRON RANGE.

30-gallon, Standard, $5; Extra heavy, $7.75

BATH TUBS.

Steel clad copper lined, 15 per cent.

CAST IRON SINKS.

16x24, $1; 18x30, $1; 18x36, $1.30.

ENAMELED BATHS, ETC.

List issued by the Standard Ideal Company Jan. 3, 1907, shows an advance of 10 per cent over previous quotations.

ENAMELED CLOSETS AND URINALS

Discount 15 per cent.

HEATING APPARATUS

Stoves and Ranges—45 to 50 per cent.
Furnaces—45 per cent.
Registers—70 per cent.
Hot Water Boilers—50 and 10 per cent
Hot Water Radiators—50 to 55 p c
Steam Radiators—50 to 55 per cent
Wall Radiators and specials—50 to 55 p c.

LEAD PIPE

Lead Pipe, 5 p.c. off
Lead waste, 5 p.c. off
Caulking lead, 6½c. per pound.
Traps and bends, 40 per cent.

IRON PIPE.

Size (per 100 ft.)	Black.		Galvanized
¼ inch........	2 35	¼ inch........	3 20
⅜ "	2 35	⅜ "	3 20
½ "	2 90	½ "	3 75
¾ "	3 90	¾ "	5 00
1 "	5 60	1 "	7 25
1¼ " 7 65 to	9 25	1¼ " 9 90 to	10 80
1½ "	9 20	1½ "	11 90
2 "	12 25	2 "	15 60
2½ "	20 1	2½ "	26 00
3 "	26 75	3 "	34 00
3½ "	34 25	3½ "	42 75
4 "	39 00	4 "	48 60

Malleable Fittings—Canadian discount 30 per cent.; American discount 25 per cent.
Cast Iron Fittings 60; Standard bushings 60.; headers, 60; flanged unions 60, malleable bushings 55 and 5; nipples, 70 and 10; malleable lipped unions 55 and 5 p.c.

SOIL PIPE AND FITTINGS

Medium and Extra heavy pipe and fittings, up to 6 inch, 60 and 10 to 70 per cent.
7 and 8-in. pipe, 40 and 5 per cent.
Light pipe, 50 p.c.; fittings, 50 and 10 p.c.

OAKUM.

Plumbers ....per 100 lb.... ..	4 50	5 00

STOCKS AND DIES.

American discount 25 per cent.

SOLDERING IRONS.

½-lb. to 1½	per lb.	0 45½	0 48
2-lb. or over	"	0 42½	0 46

SOLDER.

	Per lb. Montreal	Toronto
Bar, half-and-half, guaranteed	0 25	0 26
Wiping........................	0 22	0 23

PAINTS, OILS AND GLASS.

BRUSHES

Paint and household, 70 per cent.

CHEMICALS.

	In casks per lb.
Sulphate of copper (bluestone or blue vitrol)	0 09
Litharge, ground	0 06
" flaked	0 06½
Green copperas (green vitrol)........	0 01
Sugar of lead..........................	0 08
Lump olive.............................	0 01½

COLORS IN OIL.

Venetian red, 1-lb. tins pure.	....	0 08
Chrome yellow "	....	0 15
Golden ochre "	....	0 10
French " "	....	0 08
Marine black "	....	0 04½
Chrome green "	....	0 09
French permanent green "	....	0 13
Signwriters' black "	....	0 15

GLUE.

Domestic sheet	0 10	0 10½
French medal	0 12	0 12½

PARIS WHITE.

In bbls		0 9)

PIGMENTS.

Orange mineral, casks	....	0 (8
" " 100-lb. kegs	....	0 08½

PREPARED PAINTS.

Barn (in bbls.)	0 65	0 90
Sherwin-Williams paints,	....	1 55
Canada Paint Co.'s pure	....	1 40
Standard P. & V. Co.'s "New Era."		1 30
Benj. Moore Co.'s "Ark" B'd	....	1 25
" British Navy deck	....	1 50
Brandram-Henderson's "English"		1 45
Ramsay's paints, Pure, per gal.	....	1 20
" Thistle, "	....	1 10
Martin-Senour 100 p.c. pure,	....	1 55
Senour's Floor Paints	....	1 35
Jamieson's "Crown and Anchor"		1 20
Jamieson's floor enamel	....	1 50
P. D. Dods & Co's "Island City"		1 25
Sanderson Pearcy's, pure	....	1 20
Robertson's pure paints	....	1 20

PUTTY.

Bulk in bbls		1 67
Bladders in bbls		1 85
25-lb. tins		1 90
Bladders in bulk or tins less than 100 lb.		2 00
Bulk in 100-lb. irons		1 80

SHINGLE STAINS.

In 5 gallon lots	0 85	0 93

SHELLAC.

White, bleached	....	2 65
Fine orange	....	2 60
Medium orange	....	2 55

TURPENTINE AND OIL.

Prime white petroleum per gal	....	0 13½
Water white "	....	0 15
Pratt's astral "	....	0 18½
Castor oil	0 08	0 10
Motor Gasoline per gal, single bbls	0 21	0 23½
Benzine, per gal single bbls	0 17	0 17½
Turpentine, single barrels	0 77	0 79
Linseed Oil, " raw	0 60	0 64
" " boiled	0 63	0 67

WHITE LEAD GROUND IN OIL. Per 100 lbs

Canadian pure	...	7 15
No. 1 Canadian	....	6 80
Munro's select Flake White	....	7 65
Elephant and Decorators' Pure	....	7 65
Monarch	....	7 40
Standard Decorator's	....	7 15
Essex Genuine	....	6 80
Brandram's B. B. Genuine	....	8 70
"Anchor," pure	....	7 40
Ramsay's Pure Lead	....	7 25
Ramsay's Exterior	....	6 65
"Crown and Anchor," pure	....	7 25
P. D. Dods "Island City" pure lead	....	7 25
Sanderson Pearcy's	....	7 40
Robertson's C. P., lead	....	7 20

RED DRY LEAD.

Genuine, 560 lb. casks, per cwt	....	6 25
Genuine, 100 lb. kegs, "	....	6 75
No. 1, 560 lb. casks, per cwt.	....	6 00
No. 1, 100 lb. kegs, per cwt	....	6 50

WINDOW GLASS

Size United inches.	Star	Double Diamond
Under 26	$4 25	$6 25
26 to 40	4 65	6 75
41 to 50	5 10	7 50
51 to 60	5 35	8 50
61 to 70	5 75	9 75
71 to 80	6 25	11 00
81 to 85	7 00	12 50
86 to 90	....	15 00
91 to 95	....	17 50
96 to 100	....	20 50
101 to 105	....	24 00
106 to 110	....	27 50

Discount—16-oz., 25 per cent; 21-oz. 30 per cent. per 100 feet. Broken boxes 50 per cent.

WHITING.

Plain, in bbls	....	0 70
Gilders bolted in bands	....	0 90

WHITE DRY ZINC.

Extra Red Seal, V.M.	0 07½	0 08

WHITE ZINC IN OIL.

Pure, in 25-lb. irons	....	0 08½
No. 1, " "	....	0 07
No. 2, " "	....	0 05½

VARNISHES.

		Per gal. cans.
Carriage, No. 1	....	1 50
Pale durable body	....	3 50
" hard rubbing	....	3 00
Finest elastic gearing	....	3 00
Elastic oak	....	1 50
Furniture, polishing	....	2 00
Furniture, extra	....	1 15
" " No. 1	....	0 90
" " union	....	0 80
Light oil finish	....	1 40
Gold size japan	....	1 80
Brown japan	...	0 95
No 1 brown japan	....	0 95
Baking black japan	....	1 35
No. 1 black japan	....	0 90
Benzine black japan	....	0 70
Crystal Damar	....	2 80
No. 1 "	....	2 50
Pure asphaltum	....	1 40
Oilcloth	....	1 50
Lightning dryer	....	0 70
Elastilite varnish, 1 gal. can, each		2 00
Granitine floor varnish, per gal	....	2 50
Maple Leaf coach enamels; size 1,		1 20
Sherwin-Williams' kopal varnish, gal.,		2 50
Canada Paint Co's sun varnish	....	2 00
"Kyanize" Interior Finish	....	2 40
"Flint-Lac," coach	....	1 80
B H. Co's "Gold Medal," in cases		2 00
Jamieson's Copaline, per gal.	....	2 00

BUILDERS' HARDWARE.

BELLS.

Brass hand bells, 60 per cent.

Nickel, 55 per cent.

Gongs, Sargeant's door bells	5 50	8 00
American, house bells, per lb.	0 35	0 40

Peterboro' door bells, 37½ and 10 off new list.

BUILDING PAPER, ETC.

Tarred Felt, per 100 lb.		2 25
Ready roofing, 2-ply, not under 45 lb. per roll		1 00
Ready roofing, 3-ply, not under 65 lb., per roll		1 25
Carpet Felt ... per ton		60 00
Heavy Straw Sheathing ... per ton		40 00
Dry Surprise		0 45
Dry Sheathing ... per roll, 400 sq. ft.		0 40
Tar " " 400 "		0 50
Dry Fibre " 400 "		0 55
Tarred Fibre " 400 "		0 65
O. K. & I. X. L. " 400 "		0 70
Resin-sized " 400 "		0 45
Oiled Sheathing " 600 "		1 00
Oiled " " 400 "		0 70
Roof Coating, in barrels ... per gal.		0 17
Roof " small packages "		0 25
Refined Tar ... per barrel		5 00
Coal Tar "		4 00
Coal Tar, less than barrels ... per gal.		0 15
Roofing Pitch ... per 100 lb.	0 80	0 90
Slater's felt ... per roll		0 70
Heavy Straw Sheathing f. o. b. St. John and Halifax		42 50

BUTTS.

Wrought Brass, net revised list.

Wrought Iron, 65 and 10 per cent.

Cast Iron Loose Pin, 60 per cent.

Wrought Steel Fast Joint and Loose Pin, 70 per cent.

CEMENT AND FIREBRICK.

Canadian, Portland	2 10	2 10
Belgium	1 60	1 90
White Bros. English	2 00	2 05
"Lafarge" cement in wood		3 40
"Iron Clad" cotton	....	2 10
"Iron Clad" paper	....	2 15
" wood	....	2 25
Fire brick, Scotch, per 1,000	27 00	30 00
" English	17 00	21 00
" American, low	23 00	25 00
" " high	27 50	35 00
Fire clay (Scotch), net ton	....	4 95

Paving Blocks per 1,000.

Blue metallic, 9"x4½"x3", ex wharf	..	35 00
Stable pavers, 12"x6"x2", ex wharf	..	50 00
Stable pavers, 9"x4½"x3", ex wharf	..	36 00

DOOR SETS.

Peterboro, 37½ and 10 per cent.

DOOR SPRINGS.

Torrey's Rod ... per doz.	....	1 75
Coil, 9 to 11 in. "	0 95	1 65
English "	2 00	4 00

Chicago and Reliance Coil 25 per cent.

ESCUTCHEONS.

Discount 50 and 10 per cent., new list

Peterboro, 37½ and 10 per cent.

ESCUTCHEON PINS.

Iron, discount 40 per cent.

Brass, 45 per cent.

HINGES.

Blind, discount 50 per cent.

Heavy T and strap, 4-in., 100 lb. net	7 25
" " 5-in., "	7 00½
" " 6-in., "	6 75½
" " 8-in., "	6 50½
" " 10-in. and larger	6 25

Light T and strap, discount 65 p.c.

Screw hook and hinge—

under 12 in. ... per 100 lb.	....	4 75
over 12 in. "	....	3 75

Crate hinges and back flaps, 65 and 5 p. c.

Chest hinges and hinge hasps, 65 p.c.

SPRING HINGES.

Spring, per gro., No. 5, $17 50 No. 10, $18; No. 20, $10 80; No. 120, $20; No. 51, $10; No. 50, $27.50

Chicago Spring Butts and Blanks 12½ per cent.

Triple End Spring Butts, 30 and 10 per cent.

Chicago Floor Hinges, 37½ and 5 off.

Garden City Fire House Hinges, 12½ p.c.

"Chief" floor hinge, 50 p.c.

CAST IRON HOOKS.

Bird cage ... per doz.	0 50	1 10
Clothes line, No. 61 "	0 00	0 70
Harness "	0 60	12 00
Hat and coat ... per gro.	1 10	10 00
Chandelier ... per doz.	0 50	1 00

Wrought hooks and staples—

¼ x 5 ... per gross	....	2 65
5-16 x 5 "	....	3 30

Bright wire hooks, 60 p c.

Bright steel gate hooks and staples, 40 p.c.

Cresent hat and coat wire, 60 per cent.

Screw, bright wire, 65 per cent.

KNOBS.

Door, japanned and N.P., doz	1 50	2
Bronze, Berlin ... per doz.	2 75	3 50
Bronze, Genuine "	6 00	9 00
Shutter, porcelain, F. & L. screw ... per gross	1 30	2 00
White door knobs ... per doz.	....	2 00

Peterboro knobs, 37½ and 10 per cent.

Porcelain, mineral and jet knobs, net list.

KEYS.

Lock, Canadian 40 to 40 and 10 per cent.

LOCKS.

Peterboro, 37½ and 10 per cent

Russell & Erwin, steel rim $2.50 per doz

Eagle cabinet locks, discount 30 per cent

American padlocks, all steel, 10 to 15 per cent.; all brass or bronze, 10 to 25 per cent.

SAND AND EMERY PAPER.

S. & A. sand, discount, 35 per cent

Emery, discount 35 per cent.

Garnet (Rurton's). 5 to 10 per cent. advance

SASH WEIGHTS.

Sectional ... per 100 lb.	2 00	2 25
Solid "	1 50	1 75

SASH CORD.

Per lb.	....	0 31

BLIND AND BED STAPLES.

All sizes per lb.	0 07½	0 10

WROUGHT STAPLES.

Galvanized	....	2 75
Plain	....	2 50

Coopers', discount 45 per cent.

Poultry netting staples, discount 40 per cent.

Bright spear point, 75 per cent. discount.

TOOLS AND HANDLES.

ADZES.

Discount 22½ per cent.

AUGERS.

Gilmour's, discount 60 per cent. off list.

AXES.

Single bit, per doz.	6 00	9 10
Double bit, "	10 00	11 00

Bench Axes, 40 per cent.

Broad Axes, 25 per cent.

Hunters' Axes	5 50	6 00
Boys' Axes	6 25	7 00
Splitting Axes	7 00	12 00
Handled Axes	7 00	9 00
Red Ridge, boys', handled	....	5 75
" hunters	....	5 25

BITS.

Irwin's auger, discount 47½ per cent.

Gilmour's auger, discount 60 per cent.

Rockford auger, discount 50 and 10 per cent.

Jennings' Gen. auger, net list.

Gilmour's car, 47½ per cent.

Clark's expansive, 40 per cent.

Clark's gimlet, per doz	....	0 65
Diamond, Shell, per doz.	....	1 00
Nail and Spike, per gross	....	2 25

BUTCHERS CLEAVERS

German ... per doz.	7 00	9 00
American "	12 00	18 00

CHALK.

Carpenters' Colored, per gross	0 45	0 75
White lump ... per cwt.	0 60	0 65

CHISELS.

Warnock's, discount 70 and 5 per cent.

P. S. & W. Extra, discount, 70 per cent.

CROSSCUT SAW HANDLES.

S. & D., No. 3 ... per pair		0 13
S. & D., " 5 "		0 11½
S. & D., " 6 "		0 18
Boynton pattern "		0 20

CROWBARS.

3½c. to 4c. per lb.

DRAW KNIVES.

Coach and Wagon, discount 75 and 5 per cent.

Carpenters' discount 75 per cent.

DRILLS.

Millar's Falls, hand and breast. net list.

North Bros., each set, 50c.

DRILL BITS.

Morse, discount 37½ to 40 per cent.

Standard, discount 50 and 5 to 55 per cent.

FILES AND RASPS.

Great Western	75	per cent.
Arcade	75	"
Kearney & Foot	75	"
Disston's	75	"
American	75	"
J. Barton Smith	75	"
McClellan	75	"
Eagle	75	"
Nicholson	66⅔	"
Globe	75	"

Black Diamond, 60, 10 and 5 p.c.

Jowitt's, English list, 27½ per cent.

GAUGES.

Stanley's discount 50 to 60 per cent.

Winn's, Nos. 26 to 33 ... each	1 65	2 40

HANDLES.

Second growth ash fork, hoe, rake and shovel handles, 40 p c.

Extra ash fork, hoe, rake and shovel handles, 45 p.c.

No. 1 and 2 ash fork, hoe, rake and shovel handles, 50 p.c.

White ash whiffletrees and neckyokes, 35 p.c.

All other ash goods, 40 p c.

All hickory, maple and oak goods, excepting carriage and express whiffletrees, 40 p c.

Hickory, maple, oak carriage and express whiffletrees, 45 p c.

HAMMERS.

Maydole's, discount 5 to 10 per cent.

Canadian, discount 25 to 27½ per cent.

Magnetic tack ... per doz.	1 10	1 20
Canadian sledge ... per lb.	0 07	0 08½
Canadian ball pean, per lb.	0 22	0 25

HATCHETS.

Canadian, discount 40 to 42½ per cent.

Shingle, Red Ridge 1, per doz.	4 40
" " 2, "	4 80
Barrel Underhill	5 05

HOES.

Mortar, 50 and 10 per cent.

MALLETS.

Tinsmiths' ... per doz.	1 25	1 50
Carpenters', hickory, "	1 25	3 75
Lignum Vitae ... "	3 85	5 00
Caulking, each	0 60	2 00

MATTOCKS.

Canadian ... per doz.	5 50	6 00

MEAT CUTTERS.

German, 15 per cent.
American discount, 33⅓ per cent.

PICKS.

Per dozen	6 00	9 00

PLANES.

Wood bench, Canadian, 40, American, 25 p c.
Wood, fancy, 37½ to 40 per cent.
Stanley planes, $1.55 to $3 60, net list prices.

PLANE IRONS.

English ... per doz.	2 00	5 00

Stanley, 2⅝ inch, single 24c, double 39c

PLIERS AND NIPPERS.

Button's genuine, 37½ to 40 per cent.

Button's imitation ... per doz.	5 00	9 00
Berg's wire fencing	1 72	5 50

PUNCHES.

Saddlers ... per doz.	1 00	1 85
Conductor's ... "	3 00	15 00
Tinners', solid ... per set	....	0 72
" hollow ... per inch	....	1 00

RIVET SETS.

Canadian, discount 35 to 37½ per cent.

RULES.

Boxwood, discount 70 per cent.
Ivory, discount 20 to 25 per cent.

SAWS.

Atkins, hand and crosscut, 25 per cent.
Disston's Hand, discount 12½ per cent

Disston's Crosscut ... per foot	0 35	0 55
Hack, complete ... each	0 75	2 75
" frame only ... each	0 50	1 25

S. & D solid tooth circular shingle, concave and band, 50 per cent; mill and ice, drag, 30 per cent.; cross-cut, 35 per cent; hand saws, butcher, 35 per cent.; buck, New Century $6 25; buck No. 1 Maple Leaf, $1.25; buck, Happy Medium, $1.25; buck, Watch Spring, $1 25; buck, common frame, $4.00

Spear & Jackson's saws—Hand or rip, 26 in., $12 75; 24 in., $11 25, panel 18 in., $8 25; 20 in, $9; tenon, 10 in., $9 90; 12 in., $10 90; 14 in, $11.50.

SAW SETS.

Lincoln and Whiting	4 75
Hand Sets, Perfect	4 00
X-Cut Sets, "	7 50

Maple Leaf and Premiums saw sets, 40 off.
S. & D. saw swages, 40 off.

SCREW DRIVERS

Sargent's ... per doz.	0 65	1 00
North Bros, No. 30. per doz.	....	16 80

SHOVELS AND SPADES.

Canadian, discount 45 per cent.

SQUARES.

Iron, discount 20 per cent.
Steel, discount 65 and 10 per cent.
Try and Bevel, discount 50 to 52½ per cent.

TAPE LINES.

English, ass skin ... per doz.	2 75	5
English, Patent Leather	5 50	9
Chesterman's ... each	0 90	2
" steel ... each	0 80	8 00
Berg's, each	0 75	2 [illegible]

TROWELS.

Disston's, discount 10 per cent.
S & D., discount 35 per cent.

Berg's, brick, 924x11	4 00
" pointing, 924x5	2 10

FARM AND GARDEN GOODS

BELLS.

American cow bells, 63⅜ per cent.
Canadian, discount 45 and 50 per cent.

American, farm bells, each	1 35	3 00

BULL RINGS.

Copper, $2.00 for 2½-inch

CATTLE LEADERS

Nos. 32 and 33 ... per gross	7 50	8 50

BARN DOOR HANGERS

		doz. pairs.
Stearns wood track	4 50	6 00
Zenith	...	9 00
Atlas, steel covered	5 00	6 00
Perfect	8 00	11 00
New Milo, flexible	....	6 50
Steel, track, 1 x 3-16 in (100 ft)	....	3 25
" " 1¼ x 3-16 in (100 ft)	....	4 75
Double strap hangers, doz. sets	....	6 40
Standard jointed hangers, "	....	6 40
Steel King hangers "	....	6 25
Storm King and safety hangers	....	7 00
" " rail	....	4 25

Chicago Friction, Oscillating and Big Twin Hangers, 5 per cent.

HARVEST TOOLS.

50 and 10 per cent.
S. & D. lawn rakes, Dunn's, 40 off.
" sidewalk and stable scrapers, 40 off.

HAY KNIVES.

Net list.

HEAD HALTERS.

Jute Rope, ⅜-inch ... per gross	....	9 00
" " ½ " ... "	....	10 00
" " ⅝ " ... "	....	12 00
Leather, 1-inch ... per doz.	....	4 00
Leather, 1¼ " ... "	....	5 20
Web ... "	....	2 45

HOES.

Garden, 50 and 10 per cent.

Planter ... per doz.	4 00	4 50

LAWN MOWERS

Low wheel 12, 14 and 16-inch	$2 30
9-inch wheel, 12-inch	2 85
" " 14 "	3 00
" " 16 "	3 12½
High wheel, 12 "	4 05
" " 14 "	4 50
" " 16 "	4 75

SCYTHES.

Per doz. net	6 25	9 25

SCYTHE SNATHS.

Canadian, discount 40 per cent.

SNAPS.

Harness, German, discount 25 per cent.

Lock, Andrews'	4 50	11 00

STABLE FITTINGS.

Warden King, 35 per cent
Dennis Wire & Iron Co., 33⅓ p.c.

WOOD HAY RAKES.

40 and 10 per cent.

HEAVY GOODS, NAILS, ETC.

ANVILS.

Wright's, 80-lb. and over	0 10¼
Hay Budden, 80-lb. and over	0 09¼
Brook's, 80-lb. and over	0 11¼
Taylor-Forbes, prospectors	0 05
Columbia Hardware Co., per lb.	0 09½

VISES.

Wright's	0 13½
Berg's, per lb.	0 12½
Brook's	0 12¼
Pipe Vise, Hinge, No. 1	3 50
" " " No. 2	5 50

Saw Vise	4 50	5 00

Blacksmiths' (discount) 60 per cent.
" parallel (discount) 45 per cent.

BOLTS AND NUTS

	Per cent.
Carriage Bolts, common ($1 list	
" " ⅜ and smaller	60, 10 and 10
" " 7-16 and up	55 and 5
" " Norway Iron ($3 list)	50
Machine Bolts, ⅜ and less	60 and 10
Machine Bolts, 7-16 and up	55 and 5
Plough Bolts	55 and 10
Blank Bolts	55
Bolt Ends	55
Sleigh Shoe Bolts, ⅜ and less	60 and 10
" " " 7-16 and larger	50 and 5
Coach Screws, conepoint	70 and 5

Nuts, square, all sizes, 4c. per cent. off
Nuts, hexagon, all sizes, 4¼c. per cent. off.
Stove Rods per lb., 5½ to 6c.
Stove Bolts, 75 per cent.

CHAIN.

Proof coil, per 100 lb., ¼ in, $6 00; 5-16 in., $4.85; ⅜ in, $4 25; 7-16 in, $4 00; ½ in, $3 75; 9-16 in $3 70; ⅝ in., $3 65; ¾ in., $3 60; ⅞ in., $3.45; 1 in., $3.40.

Halter, kennel and post chains, 40 to 40 and 5 per cent.; Cow ties, 40 per cent.; Tie out chains 65 per cent.; Stall fixtures, 35 per cent; Trace chain, 45 per cent.; Jack chain iron, 50 per cent.; Jack chain, brass, 50 per cent.

HORSE NAILS.

M R.M cold forged process, list May 15, 1907, 50 and 5 per cent.
"U" brand, 57½ per cent off list.
Capewell brand, quotations on application.

HORSESHOES.

M R M brand: iron, light and medium No. 1 and smaller, $3 90; No 2 and larger $3 65; snow pattern No. 1 and smaller $4.15, No. 2 and larger, $3 90; "X L" new light steel, No 1 and smaller, $4 25; No 2 and larger, $4; "X.L" featherweight steel, No. 0 to 4, $5 60, Special countersunk steel No 0 to 4, $6 85 pkg; toe-weight, all sizes, $6.85. F o.b. Montreal Extras for packing.

Belleville brand: No 0 and 1, light and medium iron, $3.90; snow, $4 15; light steel, $4.25; No 2 and larger, light and medium iron, $3 65; snow, $3.90; light steel, $4. F o b Belleville. Two per cent., 30 days.

Toecalks—Standard No. 1 and smaller, $1 50; No 2 and larger, $1.25 Blunt No. 1 and smaller, $1.75; No. 2 and larger, $1.50 per box Sharp Put up in 25 lb. bxs.

HORSE WEIGHTS.

Taylor-Forbes, 4½c. per lb.

NAILS.

	Cut.	Wire.
2d	4 00	3 50
3d	3 15	3 20
4 and 5d	2 90	2 90
6 and 7d	2 80	2 80
8 and 9d	2 65	2 65
10 and 12d	2 60	2 60
16 and 20d	2 55	2 55
30, 40, 50 and 60d (base)	2 50	2 50

F.o b Montreal. Cut nails, Toronto 20c. higher.
Miscellaneous wire nails, discount 75 per cent
Coopers' nails, discount 40 per cent.

PRESSED SPIKES

Pressed spikes, ⅝ diameter, per 100 lbs $3.15

RIVETS AND BURRS.

Iron Rivets, black and tinned, 60, 10 and 10.
Iron Burrs, discount 60 and 10 and 10 p.c.
Copper Rivets, usual proportion burrs, 15 p c.
Copper Burrs only, net list.
Extras on Coppered Rivets, ½-lb. packages 1c. per lb.; ¼-lb. packages 2c. lb
Tinned Rivets, net extra, 4c. per lb.

SCREWS.

Wood, F. H., bright and steel, 85 and 10 p c.
" R. H., bright, 80 and 10 per cent.
" F. H., brass, 75 and 10 per cent.
" R. H., " 70 and 10 per cent.
" F. H., bronze, 70 and 10 per cent.
" R. H., " 65 and 10 per cent.
Drive Screws, dis. 87½ per cent.

Bench, wood ... per doz.	3 25	4 00
" iron ... "	4 25	5 00

Set, case hardened, dis. 60 per cent.
Square Cap, dis. 50 and 5 per cent.
Hexagon Cap, dis. 45 per cent.

MACHINE SCREWS.

Flat head, iron and brass, 35 per cent.
Felister head, iron, discount 30 per cent.
" " brass, discount 25 per cent.

TACKS, BRADS, ETC.

Carpet tacks, blued 75 p c.; tinned, 80 and 10; (in kegs), 40; cut tacks, blued, in dozens only, 75; ½ weights, 60; Swedes cut tacks, blued and tinned, bulk, 75 dozens, 75; Swedes, upholsterers', bulk, 85 and 12½; brush, blued and tinned, bulk, 70; Swedes, gimp, blued, tinned and japanned, 75 and 12½; zinc tacks, 35; leather carpet tacks, 40; copper tacks, 25; copper nails 30; trunk nails, black, 65; trunk nails, tinned and blued, 65; clout nails, blued and tinned 65; chair nails, 35; patent brads, 40; fine finishing, 40; lining tacks, in papers, 10; lining tacks, in bulk, 15; lining tacks, solid heads, in bulk, 75; saddle nails, in papers, 10; saddle nails, in bulk, 15; tufting buttons, 22 line in dozens only, 60; zinc glaziers' points, 5; double pointed tacks, papers, 90 and 10; double pointed tacks, bulk, 40; clinch and duck rivets, 45; cheese box tacks, 85 and 5 trunk tacks, 80 and 10.

WROUGHT IRON WASHERS.

Canadian make, discount 40 per cent.

SPORTING GOODS.

CARTRIDGES.

"Dominion" Rim Fire Cartridges and C B caps, 50 and 7½ per cent.; Rim Fire B B. Round Caps 60 and 2½ per cent; Centre Fire, Pistol and Rifle Cartridges, 30 p c.; Centre Fire Sporting and Military Cartridges, 20 and 5 p c; Rim Fire, Shot Cartridges, 50 and 7½ p c; Centre Fire, Shot Cartridges, 30 p c; Primers, 25 p c.

LOADED SHELLS.

"Crown" Black Powder, 15 and 5 p c; "Sovereign" Empire Bulk Smokeless Powder, 30 and 5 p.c; "Regal" Ballistite Dense smokeless Powder, 30 and 5 p c; "Imperial" Empire or Ballistite Powder, 30 and 10 p.c.

EMPTY SHELLS.

Paper Shells, 25 and 5; Brass Shells 55 and 5 p.c.

Wads.	per lb.
Best thick brown or grey felt wads, in ½-lb. bags	$0 70
Best thick white card wads in boxes of 500 each, 12 and smaller gauges	29
Best thick white card wads, in boxes of 500 each, 10 gauge	0 35
Thin card wads, in boxes of 1,000 each, 12 and smaller gauges	0 20
Thin card wads, in boxes of 1,000 each, 10 gauge	0 25
Chemically prepared black edge grey cloth wads, in boxes of 250 each—	Per M
11 and smaller gauge	0 60
9 and 10 gauges	0 70
and 8 "	0 90
5 and 6 "	1 10
Superior chemically prepared pink edge, best white cloth wads in boxes of 250 each—	
11 and smaller gauge	1 15
9 and 10 gauges	1 40
7 and 8 "	1 65
5 and 6 "	1 90

SHOT.

Ordinary drop shot, AAA to dust $7.50 per 100 lbs. Discount 5 per cent, cash discount 2 per cent, 30 days; net extras as follows subject to cash discount only; Chilled, 40 c; buck and seal, 80c.; no. 28 ball, $1 20 per 100 lbs.; bags less than 25 lbs., ½c. per lb; F.O.B. Montreal, Toronto, Hamilton, London, St. John and Halifax, and freight equalized thereon.

TRAPS (steel.)

Game, Newhouse, discount 30 and 10 per cent.
Game, Hawley & Norton, 50, 10 & 5 per cent
Game, Victor, 70 per cent.
Game, Oneida Jump (B. & L.) 40 & 2½ p.
Game, steel, 60 and 5 per cent.

SKATES.

Skates, discount 37½ per cent.
Empire hockey sticks, per doz .. 3 00

CUTLERY AND SILVERWARE.

RAZORS.

RAZORS.	per doz.	
Elliot's	4 00	18 00
Boker's	7 50	11 00
" King Cutter	13 50	18 50
Wade & Butcher's	3 60	10 00
Lewis Bros.' "Klean Kutter"	8 50	10 50
Henckel's	7 50	20 00
Berg's	7 50	20 00

Clauss Razors and Strops, 50 and 10 per cent

KNIVES.

Farriers-Stacey Bros., doz ... 3 50

PLATED GOODS

Hollowware, 40 per cent. discount.
Flatware, staples, 40 and 10, fancy, 40 and 5.
Hutton's "Cross Arrow" flatware, 47½; "Singalese" and "Alaska" Nevada silver flatware, 42 p.c.

SHEARS.

Clauss, nickel, discount 60 per cent.
Clauss, Japan, discount 67½ per cent.
Clauss, tailors, discount 40 per cent.
Seymour's, discount 50 and 10 per cent.
Berg's ... 6 00 12 00

HOUSE FURNISHINGS.

APPLE PARERS.

Hudson, per doz., net ... 5 75

BIRD CAGES

Brass and Japanned, 40 and 10 p. c.

COPPER AND NICKEL WARE.

Copper boilers, kettles, teapots, etc. 30 p. c.
Copper pitts, 30 per cent.

KITCHEN ENAMELED WARE.

White ware 75 per cent.
London and Princess, 50 per cent.
Canada, Diamond, Premier, 50 and 10 p. c.
Pearl, Imperial, Crescent and granite steel, 50 and 10 per cent
Premier steel ware, 40 per cent.
Star decorated steel and white, 25 per cent.
Japanned ware, discount 45 per cent.
Hollow ware, tinned cast, 35 per cent. off.

KITCHEN SUNDRIES.

Can openers, per doz.	0 40	0 75
Mincing knives per doz	0 50	0 80
Duplex mouse traps, per doz.	....	0 65
Potato mashers, wire, per doz.	0 60	0 70
" wood "	0 50	0 60
Vegetable slicers, per doz	....	2 25
Universal meat chopper No. 1	....	1 15
Enterprise chopper, each	...	1 30
Spiders and fry pans, 50 per cent.		
Star Al chopper 5 to 32	1 35	4 10
" " 100 to 103	1 35	2 00
Kitchen hooks, bright	....	0 60

LAMP WICKS.

Discount, 60 per cent.

LEMON SQUEEZERS.

Porcelain lined	per doz.	2 20	5 60
Galvanized	"	1 87	3 85
King, wood	"	2 75	2 90
King, glass	"	4 00	4 50
All glass	"	0 50	0 90

METAL POLISH.

Tandem metal polish paste ... 6 00

PICTURE NAILS.

Porcelain head ... per gross	1 35	1 50
Brass head	0 40	1 00

Tin and gilt, picture wire, 75 per cent.

SAD IRONS.

Mrs. Potts, No. 55, polished ... per set	0 90
" No. 50, nickle-plated, "	0 95
" handles, japaned, per gross	9 25
" " nickled, "	9 75
Common, plain	4 25
" plated	5 50
Asbestos, per set	1 50

TINWARE.

CONDUCTOR PIPE.

2-in. plain or corrugated., per 100 feet, $3.30; 3 in., $4 40; 4 in., $5.80; 5 in., $7.45; 6 in., $9 90.

FAUCETS.

Common, cork-lined, discount 35 per cent.

EAVETROUGHS.

10-inch ... per 100 ft. 3 30

FACTORY MILK CANS.

Discount off revised list, 35 per cent
Milk can trimmings, discount 25 per cent.
[illegible]

LANTERNS.

No 2 or 4 Plain Cold Blast ...	per doz.	6 50
Lift Tubular and Hinge Plain,	"	4 75
No 0, safety	"	4 00

Better quality at higher prices.
Japanning, 50c. per doz. extra.
Prism globes, per doz., $1.20.

OILERS.

Kemp's Tornado and McClary's Model galvanized oil can, with pump, 5 gallon, per dozen ... 10 92
Davidson oilers, discount 40 per cent.
Zinc and tin, discount 50 per cent
Coppered oilers, 20 per cent off.
Brass oilers, 50 per cent. off.
Malleable, discount 25 per cent

PAILS (GALVANIZED).

Dufferin pattern pails, 45 per cent.
Flaring pattern, discount 45 per cent.
Galvanized washtubs 40 per cent.

PIECED WARE.

Discount 35 per cent off list, June, 1899.
10-qt. flaring sap buckets, discount 35 per cent.
6, 10 and 14-qt. flaring pails dis. 35 per cent
Copper bottom tea kettles and boilers, 30 p.c.
Coal hods, 40 per cent.

STAMPED WARE.

Plain, 75 and 12½ per cent. off revised list.
Retinned, 72½ per cent. revised list.

SAP SPOUTS.

Bronzed iron with hooks ... per 1,000
Eureka tinned steel, hooks ... 8 00

STOVEPIPES.

5 and 6 inch, per 100 lengths	7 64	7 91
7 inch " "	....	8 18

Nestable, discount 40 per cent.

STOVEPIPE ELBOWS

5 and 6-inch, common ... per doz.	1 32
7-inch "	1 48

Polished, 15c. per dozen extra.

THERMOMETERS.

Tin case and dairy, 75 to 75 and 10 per cent.

TINNERS' SNIPS.

Per doz. ... 3 00 15
Clauss, discount 35 per cent.

TINNERS' TRIMMINGS.

Discount, 45 per cent.

WIRE.

ANNEALED CUT HAY BAILING WIRE.

No. 12 and 13, $4; No. 13½, $4 10; No. 14, $4.25; No. 15, $4 50; in lengths 6' to 11', 25 per cent., other lengths 20c per 100 lbs extra; if eye or loop on end add 25c. per 100 lbs. to the above.

BRIGHT WIRE GOODS

Discount 60 per cent.

CLOTHES LINE WIRE.

No. 7 wire solid line, No. 17, $4.90; No. 18, $3.00; No. 19, $2.70; # wire solid line, No. 17, $4.45; No. 18, $3.10; No. 19, $2 80. All prices per 1000 ft measure; 6 strand, No. 18, $2 60; No. 19, $2 90. F.o.b. Hamilton, Toronto, Montreal.

COILED SPRING WIRE

High Carbon, No. 9, $2 95; No. 11, $3.50; No. 12, $3.20.

COPPER AND BRASS WIRE.

Discount 37½ per cent.

FINE STEEL WIRE.

Discount 25 per cent List of extras In 100-lb. lots: No. 17, $5 — No. 18 $5.50 — No. 19, $6 — No. 20, $6.65 — No. 21 $7 — No. 22, $7.30 — No 23, $7,65 — No 24, $8 — No. 25, $9 — No. 26, $9 50 — No. 27, $10 — No. 28, $11 — No 29, $12 — No. 30, $13 — No. 31, $14 — No. 32, $15 — No. 33, $16 — No. 34, $17. Extras net—tinned wire, Nos. 17-25 $2 — Nos. 26-31, $4 — Nos. 32-34, $6. Coppered, 75c.—oiling, 10c.—in 25-lb. bundles, 15c.—in 5 and 10-lb. bundles, 25c.—in 1-lb. hanks, 25c. —in ½-lb. hanks, 38c.—in ¼-lb. hanks, 50c. packed in casks or cases, 15c.—bagging or papering, 10c

FENCE STAPLES.

Bright ... 2 80 Galvanized ... 32

HAY WIRE IN COILS.

No 13, $2 70; No. 14, $2 80; No. 15, $2 95; f.o.b., Montreal

GALVANIZED WIRE.

Per 100 lb.—Nos. 4 and 5, $3.90 — Nos. 6, 7, 8, $3.35 — No. 9, $2 85 — No. 10, $3 40 — No. 11, $3.45 — No. 12, $3 00 — No. 13, $3 10 — No. 14, $3.95 — No 15, $4.30 — No. 16, $4.30 from stock. Base sizes, Nos 6 to 9, $2 35 f.o.b. Cleveland Extras for cutting.

LIGHT STRAIGHTENED WIRE.

Gauge No.	Over 20 in.	10 to 20 in.	5 to 10 in.
0 to 5	$0.50	$0.75	$1 25
6 to 9	0 75	1 25	2 00
10 to 11	1.00	1 75	2 50
12 to 14	1 50	2.25	3 50
15 to 16	2.00	3.00	4.50

SMOOTH STEEL WIRE.

No. 0-9 gauge, $2.40; No. 10 gauge, 6c extra; No. 11 gauge, 12c extra; No. 12 gauge, 20c. extra; No. 13 gauge, 30c. extra; No 14 gauge. 40c. extra; No. 15 gauge, 55c. extra; No. 16 gauge, 70c. extra. Add 60c. for coppering and $2 for tinning.

Extra net per 100 lb.—Oiled wire 10c., spring wire $1.25, bright soft drawn 15c., charcoal (extra quality) $1.25, packed in casks or cases 15c., bagging and papering 10c, 50 and 100-lb. bundles 10c., in 25-lb. bundles 15c., in 5 and 10-lb. bundles 25c., in 1-lb hanks, 50c., in ½-lb. hanks 75c., in ¼-lb. hanks $1.

POULTRY NETTING.

2-in. mesh, 19 w g., 50 and 5 p.c. off. Other sizes, 50 and 5 p c. off.

WIRE CLOTH.

Painted Screen, in 100-ft. rolls, $1.72½, per 100 sq. ft.; in 50-ft. rolls $1 77½, per 100 sq ft.

WIRE FENCING.

Galvanized barb	2 95
Galvanized, plain twist	3 30

Galvanized barb, f.o.b. Cleveland, $2 70 for small lots and $2.60 for carlots

WIRE ROPE

Galvanized, 1st grade, 6 strands, 24 wires, ½, $5; 1 inch $16 80
Black, 1st grade, 6 strands, 19 wires, ½, $5; 1 inch $15.10. Per 100 feet f.o.b. Toronto

WOODENWARE.

CHURNS.

No. 0, $9; No. 1, $9; No. 2, $10; No. 3, $11; No. 4, $13; No. 5, $16; f.o b. Toronto Hamilton, London and St. Marys, 30 and 30 per cent.; f o b Ottawa, Kingston and Montreal, 40 and 15 per cent. discount.

CLOTHES REELS.

Davis Clothes Reels, dis., 40 per cent.

FIBRE WARE.

Star pails, per doz		$ 3 00
8 Tubs,	"	14 00
3 "	"	12 00
2 "	"	10 00
3 "	"	8 50

LADDERS, EXTENSION.

3 to 6 feet, 12c. per foot; 7 to 10 ft., 13c.
Waggoner Extension Ladders, dis. 40 per cent.

MOPS AND IRONING BOARDS.

"Best" mops		1 25
"900" mops		1 25
Folding ironing boards	12 00	16 50

REFRIGERATORS

Discount, 40 per cent.

SCREEN DOORS.

Common doors, 2 or 3 panel, walnut stained, 4-in. style ... per doz. 7 25
Common doors, 2 or 3 panel, grained only, 4-in., style ... per doz. 7 55
Common doors, 2 or 3 panel, light stair per doz. ... 9 55

WASHING MACHINES.

Round, re-acting per doz.	60 00
Square "	63 00
Eclipse, per doz	54 00
Dowswell "	39 00
New Century, per doz	75 00
Daisy	54 00
Stephenson	74 00

WRINGERS.

Royal Canadian, 11 in., per doz.	35 00
Royal American, 11 in.	35 00
Ezo 10 in., per doz	26 75

MISCELLANEOUS

AXLE GREASE.

Ordinary, per gross	6 00	7 00
Best quality	10 00	12 00

BELTING.

Extra, 60 per cent.
Standard, 60 and 10 per cent.
No. 1, not wider than 6 in., 60, 10 and 10 p.c.
Agricultural, not wider than 4 in., 75 per cent
Lace leather, per side, 75c.; cut laces, 80c.

BOOT CALKS.

Small and medium, ball ... per M	2 25
Small heel	4 50

CARPET STRETCHERS.

American ... per doz.	1 00	1 50
Bullard's "	....	6 50

CASTORS.

Bed, new list, discount 55 to 57½ per cent.
Plate, discount 52½ to 57½ per cent.

PINE TAR.

½ pint in tins ... per gross	7 80
1 " "	9 60

PULLEYS.

Hothouse ... per doz.	0 55	1 00
Axle "	0 22	0 33
Screw "	0 22	1 00
Awning "	0 35	2 50

PUMPS.

Canadian cistern	1 40	2 00
Canadian pitcher spout	1 80	3 16

Berg's wing pump, 75 per cent.

ROPE AND TWINE.

Sisal		0 10½
Pure Manilla		0 15
"British" Manilla		0 12
Cotton, 3-16 inch and larger	0 21	0 23
" 5-32 inch	0 25	0 27
" ⅛ inch	0 25	0 28
Russia Deep Sea		0 16
Jute		0 09
Lath Yarn, single		0 10
" double		0 10½
Sisal bed cord. 48 feet ... per doz.		0 65
" " 60 feet "		0 80
" " 72 feet "		0 95

Twine.

Bag, Russian twine, per lb.	....	0 27
Wrapping, cotton, 3-ply	....	0 25
" " 4-ply	....	0 29
Mattress twine per lb.	0 33	0 45
Staging " "	0 27	0 35

BINDER TWINE.

500 feet, sisal	0 09½
500 " standard	0 09½
550 " " manilla	0 10½
600 " " "	0 12½
650 " " "	0 13½

Car lots, ½c. less; 5-ton lots, ¼c. less. Central delivery.

SCALES.

Gurney Standard, 35; Champion, 45 p.c.
Burrow, Stewart & Milne — Imperial Standard, 35; Weigh Beams, 35; Champion Scales, 45
Fairbanks Standard, 30; Dominion, 50 Richelieu, 50.
Warren new Standard, 35; Champion, 45 Weigh Beams, 30.

STONES—OIL AND SCYTHE.

Washita	per lb.	0 25	0
Hindostan	"	0 06	0 10
" slip	"	0 18	0 20
" Axe	"	....	0 10
Deer Creek	"	....	0 10
Deerlick	"	....	0 25
" Axe	"	....	0 15
Lily white	"	....	0 42
Arkansas	"	....	1 50
Water-of-Ayr	"	....	0 10
Scythe	per gross	3 50	5 00
Grind, 40 to 200 lb., per ton		20 00	22 00
" under 40 lb., "		....	24 00
" 200 lb. and over		....	28 00

INDEX TO ADVERTISERS.

CLASSIFIED LIST OF ADVERTISEMENTS.

Alabastine.
Alabastine Co., Limited, Paris, Ont.

Auditors.
Davenport, Pickup & Co., Winnipeg.

Automobile Accessories.
Canada Cycle & Motor Co., Toronto Junction
Carriage Mountings Co., Ltd., Toronto.

Awnings.
Tobin Tent and Awning Co, Ottawa

Babbitt Metal.
Canada Metal Co., Toronto.
Canadian Fairbanks Co., Montreal.
Frothingham & Workman, Montreal
Robertson, Jas. Co., Montreal.

Bar Urns.
Buffalo Mfg. Co., Buffalo, N Y.

Bath Room Fittings.
Buffalo Mfg. Co., Buffalo, N.Y.
Carriage Mountings Co., Ltd, Toronto.
Forsyth Mfg. Co., Buffalo, N Y.
Ontario Metal Novelty Co., Toronto

Belting, Hose, etc.
Gutta Percha and Rubber Mfg. Toronto.
Sadler & Haworth, Toronto.

Bicycles and Accessories.
Johnson's, Iver, Arms and Cycle Works Fitchburg, Mass

Binder Twine.
Consumers Cordage Co., Montreal.

Bolts.
Toronto & Belleville Rolling Mills, Belleville.

Box Strap.
J. N. Warminton, Montreal.

Brass Goods.
Frothingham & Workman, Montreal.
Glauber Brass Mfg. Co., Cleveland, Ohio.
Kerr Engine Co., Walkerville, Ont.
Lewis, Rice, & Son., Toronto.
Morrison, Jas., Brass Mfg. Co., Toronto
Mueller Mfg Co, Decatur, Ill.
Penberthy Injector Co., Windsor, Ont.
Taylor-Forbes Co., Guelph, Ont.

Bronze Powders.
Canadian Bronze Powder Works, Montreal.

Brushes.
United Factories, Toronto.

Cans.
Acme Can Works, Montreal.

Builders' Tools and Supplies.
Covert Mfg. Co., West Troy, N Y.
Frothingham & Workman Co., Montreal.
Howland, H. S. Sons & Co., Toronto
Hyde, F., & Co., Montreal.
Lewis Bros. & Co., Montreal.
Lewis, Rice, & Son, Toronto.
Lockerby & McComb, Montreal.
Lufkin Rule Co., Saginaw, Mich.
Newman & Sons, Birmingham.
North Bros. Mfg. Co., Philadelphia, Pa.
Stanley Rule & Level Co., New Britain.
Stanley Works, New Britain, Conn.
Stephens, G. F., Winnipeg.
Taylor-Forbes Co., Guelph, Ont.

Carriage Accessories.
Carriage Mountings Co, Ltd, Toronto
Covert Mfg. Co., West Troy, N.Y.

Carriage Springs and Axles.
Guelph Spring and Axle Co., Guelph.

Carpet Beateas.
Ontario Metal Novelty Co., Toronto.

Cartridges.
Dominion Cartridge Co., Montreal.

Cattle and Trace Chains.
Greening, B., Wire Co., Hamilton.

Chafing Dishes.
Buffalo Mfg. Co., Buffalo, N.Y.

Churns.
Dowswell Mfg. Co., Hamilton.

Clippers—All Kinds.
American Shearer Mfg. Co., Nashua, N.H

Clothes Reels and Lines.
Hamilton Cotton Co., Hamilton,

Clutch Nails.
J. N. Warminton, Montreal.

Congo Roofing
Buchanan Foster Co., Philadelphia, Pa.

Cordage.
Consumers' Cordage Co., Montreal.
Hamilton Cotton Co., Hamilton.

Cork Screws.
Erie Specialty Co., Erie, Pa.

Cow Ties
Greening, B., Wire Co., Hamilton

Cuspidors.
Buffalo Mfg. Co., Buffalo, N.Y.

Cut Glass.
Phillips, Geo., & Co., Montreal.

Cutlery—Razors, Scissors, etc.
Birkett, Thos., & Son Co., Ottawa.
Clauss Shear Co., Toronto
Dorken Bros. & Co., Montreal.
Frothingham & Workman Montreal.
Heinisch's, R., Sons Co., Newark, N.J.
Howland, H. S. Sons & Co., Toronto.
Hutton, Wm., & Sons, Ltd., London, Eng.
Lamplough, F. W., & Co., Montreal.
Phillips, Geo., & Co., Montreal.
Round, John, & Son, Montreal.

Electric Fixtures.
Canadian General Electric Co., Toronto
Morrison James, Mfg. Co., Toronto.
Munderloh & Co., Montreal.

Electro Cabinets.
Cameron & Campbell Toronto.

Enameled Ware
Kemp Mfg. Co., Toronto.

Engines, Supplies, etc.
Kerr Engine Co., Walkerville, Ont.

Eavetroughs
Wheeler & Bain, Toronto

Fencing—Woven Wire
Dominion Wire Mfg. Co., Montreal
Owen Sound Wire Fence Co., Owen Sound.
Banwell Hoxie Wire Fence Co., Hamilton.

Files and Rasps.
Barnett Co., G. & H., Philadelphia, Pa.
Nicholson File Co., Port Hope

Firearms and Ammunition.
Hamilton Rifle Co., Plymouth, Mich.
Harrington & Richardson Arms Co., Worcester, Mass.
Johnson's, Iver, Arms and Cycle Works Fitchburg, Mass.

Fishing Tackle.
Enterprise Mfg. Co., Akron, Ohio

Food Choppers
Enterprise Mfg. Co., Philadelphia, Pa.
Lamplough, F. W., & Co., Montreal.
Shirreff Mfg. Co, Brockville, Ont.

Furnaces.
Pease Foundry Co, Toronto.

Galvanizing.
Canada Metal Co., Toronto.
Dominion Wire Mfg Co., Montreal.
Montreal Rolling Mills Co., Montreal.
Ontario Wind Engine & Pump Co., Toronto.

Glass Ornamental
Hobbs Mfg. Co., London
Consolidated Plate Glass Co, Toronto

Glaziers' Diamonds.
Gibsone, J B., Montreal.
Pelton, Godfrey S.
Sharratt & Newth, London, Eng.
Shaw, A., & Son, London, Eng.

Handles.
Still, J. H., Mfg. Co.

Harvest Tools.
Maple Leaf Harvest Tool Co., Tillsonburg Ont.

Hockey Sticks
Still, J. H. Mfg. Co, St. Thomas.

Hoop Iron.
Frothingham & Workman Montreal.
Montreal Rolling Mills Co., Montreal.
J. N. Warminton, Montreal.

Horse Blankets.
Heney, E. N., & Co., Montreal.

Horseshoes and Nails.
Canada Horse Nail Co., Montreal.
Montreal Rolling Mills, Montreal.
Capewell Horse Nail Co., Toronto
Toronto & Belleville Rolling Mills Belleville.

Hot Water Boilers and Radiators.
Cluff, R. J., & Co Toronto.
Pease Foundry Co. Toronto.
Taylor-Forbes Co., Guelph.

Ice Cream Freezers.
Dana Mfg. Co., Cincinnati, Ohio.
North Bros. Mfg. Co., Philadelphia, Pa.

Ice Cutting Tools.
Erie Specialty Co., Erie, Pa.
North Bros. Mfg. Co., Philadelphia, Pa.

Injectors—Automatic.
Morrison, Jas., Brass Mfg. Co., Toronto.
Penberthy Injector Co., Windsor, Ont.

Iron Pipe.
Montreal Rolling Mills, Montreal.

Iron Pumps.
Lamplough, F. W., & Co., Montreal.
McDougall, R., Co., Galt, Ont.

Ladders.
The Waggoner Ladder Co., London, Ont.

Lanterns.
Kemp Mfg. Co., Toronto.
Ontario Lantern Co., Hamilton, Ont.
Wright, E. T., & Co., Hamilton.

Lawn Mowers.
Birkett, Thos., & Son Co., Ottawa.
Frothingham & Workman, Montreal.
Maxwell, D., & Sons, St. Mary's, Ont.
Taylor, Forbes Co., Guelph.

Lawn Mower Grinders
Root Bros. & Co., Plymouth, Ohio.

Ledgers—Loose Leaf.
Business Systems, Toronto.
Copeland-Chatterson Co, Toronto.
Crain, Rolla L., Co., Ottawa.
Universal Systems, Toronto.

Lithographing.
London Printing & Lithographing Co., London, Ont.

Locks, Knobs, Escutcheons, etc.
Peterborough Lock Mfg. Co., Peterborough, Ont.
National Hardware Co., Orillia, Ont.

Lumbermen's Supplies.
Pink, Thos., & Co., Pembroke Ont.

Lye
Gillett, E. W., & Co., Toronto

Manufacturers' Agents.

Fox, C. H., Vancouver.
McIntosh, H. F., & Co., Toronto.
Gibb, Alexander, Montreal.
Scott, Bathgate & Co., Winnipeg.

Metals.

Canada Iron Furnace Co., Midland, Ont
Canada Metal Co., Toronto.
Eadie, H. G., Montreal.
Frothingham & Workman, Montreal.
Gibb, Alexander, Montreal.
Kemp Mfg. Co., Toronto
Leslie, A. C., & Co., Montreal.
Lysaght, John, Bristol, Eng.
Nova Scotia Steel and Coal Co., New Glasgow, N.S.
Robertson, Jas., Co., Montreal
Roper, J. H., Montreal.
Samuel, Benjamin & Co., Toronto.
Stairs, Son & Morrow, Halifax, N.S.
Thompson, B. & S. H. & Co. Montreal.

Metal Lath.

Galt Art Metal Co., Galt.
Metallic Roofing Co., Toronto.
Metal Shingle & Siding Co., Preston, Ont.

Metal Polish, Emery Cloth, etc.

Oakey, John, & Sons, London, Eng.

Nails Wire

Dominion Wire Mfg. Co., Montreal.

Oilers

Maple City Mfg Co., Monmouth, Ill.

Oil Tanks

Bowser, S. F., & Co., Torono.

Ornamental Iron and Wire.

Dennis Wire & Iron Co., London, Ont.

Packing.

Gutta Percha & Rubber Co Toronto

Paints, Oils, Varnishes, Glass.

Blanchite Process Paint Co., Toronto.
Brandram-Henderson, Montreal
Canada Paint Co., Montreal.
Canadian Oil Co. Toronto
Consolidated Plate Glass Co., Toronto.
Dods, P. D., & Co., Montreal
Imperial Varnish and Color Co., Toronto
Jamieson, R. C., & Co., Montreal.
Lucas John & Co., New York
McArthur, Corneille & Co., Montreal.
McCaskill, Dougall & Co., Montreal.
Moore Benjamin, & Co. Toronto.
Ottawa Paint Works, Ottawa
Queen City Oil Co., Toronto.
Ramsay & Son, Montreal.
Sanderson Pearcy & Co., Toronto
Sherwin-Williams Co Montreal.
Standard Paint Co., Montreal
Standard Paint and Varnish Works Windsor, Ont
Stephens & Co., Winnipeg.
Martin-Senour Co., Montreal
Winnipeg Paint & Glass Co., Winnipeg

Perforated Sheet Metals.

Greening, B., Wire Co., Hamilton.

Plumbers' Tools and Supplies

Canadian Fairbanks Co., Montreal.
Cluff, R. J., & Co., Toronto
Frothingham & Workman, Montreal.
Glauber Brass Co., Cleveland, Ohio.
Jardine, A. B., & Co., Hespeler, Ont.
Jenkins Bros., Boston, Mass.
Kerr Engine Co., Walkerville, Ont.
Lewis, Rice, & Son, Toronto.
Merrell Mfg. Co., Toledo, Ohio.
Montreal Rolling Mills Montreal.
Murrison, Jas., Brass Mfg. Co., Toronto.
Mueller, H., Mfg. Co., Decatur, Ill.
Oshawa Steam & Gas Fitting Co., Oshaw
Robertson Jas Co. Montreal.
Robertson, Jas., Co., Limited, Toronto
Somerville, Limited, Toronto
Stairs, Son & Morrow, Halifax, N.S.
Standard Ideal Sanitary Co., Port Hope,
Standard Sanitary Co., Pittsburg.
Stephens, G F., & Co., Winnipeg, Man.
Turner Brass Works, Chicago.
Vickery, Orlando, Toronto.

Polishes.

Majestic Polishes, Toronto

Portland Cement

International Portland Cement Co. Ottawa, Ont.
Hanover Portland Cement Co., Hanover, Ont.
Hyde, F., & Co., Montreal.
Thompson B. & S. H. & Co., Montreal.

Poultry Netting.

Greening, B., Wire Co., Hamilton, Ont.

Printing.

London Printing & Lithographing Co., London, Ont.

Razors.

Clauss Shear Co., Toronto.

Refrigerators

Fabien, C. P., Montreal.

Registers.

Pease Foundry Co., Toronto.

Roofing Supplies

Brantford Roofing Co Brantford.
Barrett Mfg. Co., New York.
F. W Bird, East Walpole Mass.
Buchanan Foster Co., Philadelphia, Pa.
McArthur Alex., & Co Montreal
Metal Shingle & Siding Co., Preston, Ont.
Metallic Roofing Co., Toronto
Paterson Mfg. Co., Toronto & Montreal.
Standard Paint Co. Montreal.
Wheeler and Bain, Toronto

Saws

Atkins, E. C., & Co., Indianapolis, Ind
Shurly & Dietrich, Galt, Ont.
Spear & Jackson, Sheffield, Eng.

Scales.

Canadian Fairbanks Co., Montreal.
Frothingham & Workman, Montreal.

Screw Cabinets.

Cameron & Campbell, Toronto.

Screws, Nuts, Bolts.

Dominion Wire Mfg Co., Montreal.
Montreal Rolling Mills Co., Montreal.

Soil Pipe

McFarlane, Walter, Glasgow

Sewer Pipes.

Canadian Sewer Pipe Co., Hamilton
Hyde, F., & Co., Montreal.

Shelf Boxes.

Cameron & Campbell, Toronto.

Shears, Scissors.

Clauss Shear Co. Toronto.

Shovels and Spades

Eclipse Mfg. Co., Ottawa
Frothingham & Workman Montreal.
Peterboro Shovel & Tool Co., Peterboro.

Silverware.

Hutton, Wm., & Sons, Ltd., London, Eng.
McGlashan, Clarke Co., Niagara Falls, Ont.
Phillips, Geo., & Co., Montreal
Round, John, & Son, Sheffield, Eng.

Skates.

Canada Cycle & Motor Co., Toronto.
McFarlane, Walter, Glasgow.

Sprayers

Cavers Bros. Galt

Spring Hinges, etc.

Chicago Spring Butt Co., Chicago, Ill.

Stable Fittings

Dennis Wire & Iron Co., London

Steel Rails.

Nova Scotia Steel & Coal Co., New Glasgow, N.S.

Stove Pipe.

Chown, Edwin and Son, Kingston

Stoves, Tinware, Furnaces

Canadian Heating & Ventilating Co. Owen Sound.
Copp, W. J., Son & Co., Fort William
Davidson, Thos., Mfg Co., Montreal
Down Draft Furnace Co., Galt
Guelph Stove Co., Guelph.
Gurney Foundry Co., Toronto.
Harris, J. W., Co., Montreal.
Howard Wm., Toronto
Kemp Mufg Co Toronto.
McClary Mfg. Co. London.
Merrick Anderson, Winnipeg
Pease Foundry Co., Toronto.
Smart, James, Mfg. Co. Brockville
Stewart, Jas., Mfg. Co., Woodstock, Ont.
Taylor-Forbes Co., Guelph, Ont.
Wright, E. T., & Co., Hamilton.

Tacks.

Montreal Rolling Mills Co., Montreal.
Ontario Tack Co., Hamilton.

Tents.

Tobin Tent and Awning Co., Ottawa

Tin Plate.

American Sheet & Tin Plate Co., Pittsburg, Pa
Baglan Bay Tin Plate Co., Briton Ferry South Wales
Lysaght, John, Bristol, Newport and Montreal

Turpentine

Defiance Mfg. Co., Toronto.

Ventilators.

Harris, J. W., Co., Montreal.
Pearson, Geo. D., Montreal.

Wall Paper

Staunton Limited, Toronto.

Wall Paper Cleaner.

Gilbert, Frank U. S., Cleveland

Washing Machines, etc

Dowswell Mfg. Co., Hamilton, Ont.
The Shultz Bros. Co., Brantford.
Taylor Forbes Co., Guelph, Ont.

Water Filters.

Buffalo Mfg Co., Buffalo, N.Y.

Wheelbarrows

London Foundry Co., London Ont.
Schultz Bros. Co., Ltd., The Brantford.

Wholesale Hardware.

Birkett, Thos., & Sons Co., Ottawa.
Caverhill, Learmont & Co., Montreal.
Frothingham & Workman, Montreal.
Hobbs Hardware Co., London
Howland, H. S., Sons & Co., Toronto.
Lamplough, F. W., & Co., Montreal.
Lewis Bros. & Co., Montreal.
Lewis, Rice, & Son, Toronto

Window and Sidewalk Prisms

Hobbs Mfg. Co., London, Ont.

Wire, Wire Rope, Cow Ties, Fencing Tools, etc

Banwell-Hoxie Fence Co., Hamilton
Dennis Wire and Iron Co., London, Ont.
Dominion Wire Mufg. Co., Montreal
Greening, B., Wire Co., Hamilton.
Owen Sound Wire Fence Co., Owen Sound
Montreal Rolling Mills Co., Montreal.
Western Wire & Nail Co., London, Ont.

Wrapping Papers

Canada Paper Co., Toronto.
McArthur, Alex., & Co., Montreal.
Stairs Son & Morrow, Halifax, N.S.

Wringers

Connor, J. H. & Son, Ottawa, Ont

"Seeing is Believing"

¶ As the Ross Sporting Rifle represents the best in modern rifle manufacture, the highest grade materials and minute accuracy in workmanship, it is to your interest to learn fully about this Canadian Rifle—

¶ The Ross Straight Pull 303 Sporting Model M is one of our leaders to retail at $26. The barrel is the celebrated Ross high pressure, non-fouling, gun-barrel steel — The breech action is dependable.

Our representatives are now demonstrating the merits of this rifle—It places you under no obligation to drop us a line to make sure you will be on their calling list.

Our advertising policy is co-operative.

THE ROSS RIFLE CO., - QUEBEC

When a man asks for a pump he means he wants the best and of course everyone knows that

McDougall Pumps Are Standard Pumps

Do not disappoint your customers, because a second-hand Pump on your hands is about as good stock as a last year's bird's nest.

Remember, here it is Quality first, with price to suit.

A lot of reading matter for the asking.

The R. McDougall Co., Limited

GALT, - CANADA

Chances for Business

In these days of prosperity large public buildings are constantly being planned in all parts of the country.

These are your chances for getting business.

Good air is a necessity in all buildings, but more especially in those built for public use.

The best way to get good air is by the use of our

AEOLIAN VENTILATORS

They have been tried all over Canada, and have never failed to give complete satisfaction.

Montreal, June 15th, 1903

Messrs. J. W. Harris Co., Limited, Montreal.

Dear Sirs,

In answer to your request, we take very much pleasure in saying that we have your system of ventilation installed in our Church for several years and it has given us entire satisfaction.

Fre. MARIE RAYMOND, O.F.M

Write us for terms.

THE J. W. HARRIS COMPANY, LIMITED

Contractors ***Montreal***

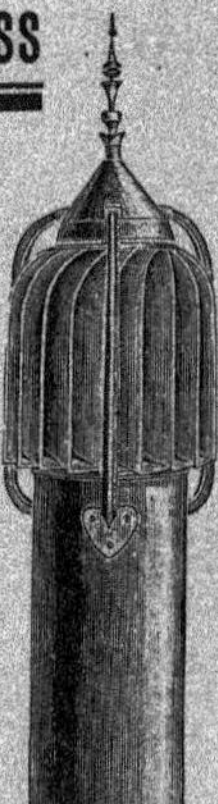

THE AEOLIAN VENTILATOR

(Can be supplied in copper, if so desired.)

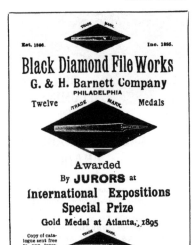

CIRCULATES EVERYWHERE IN CANADA
Also in Great Britain, United States, West Indies, South Africa and Australia.

HARDWARE AND METAL

A Weekly Newspaper Devoted to the Hardware, Metal, Heating and Plumbing Trades in Canada.

Office of Publication, 10 Front Street East, Toronto.

VOL. XIX. MONTREAL, TORONTO, WINNIPEG, OCTOBER 5, 1907 ***NO. 40.***

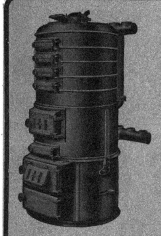

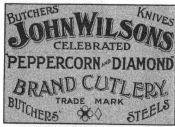

Read "Want Ads." on Page 57

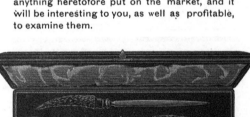

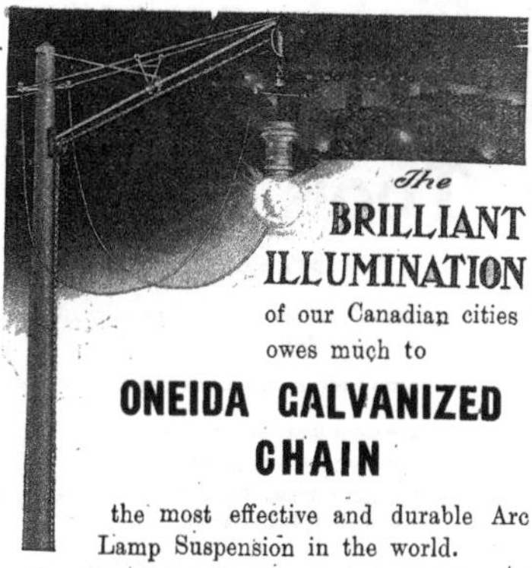

The
BRILLIANT ILLUMINATION

of our Canadian cities owes much to

ONEIDA GALVANIZED CHAIN

the most effective and durable Arc Lamp Suspension in the world.

Heavily galvanized—therefore absolutely rust proof. Impervious to ice and sleet. Uniform in strength. Perfectly flexible. ***Will outwear cord, cable or rope many times over.***

MILLIONS OF FEET IN USE.

ONEIDA COMMUNITY, Ltd, Niagara Falls, Ont.

Your Business at Your Finger Tips

The merchant, if he is wise, no longer keeps his accounts upon tickets, in memorandum books, or in a loose way, but has a system by which, at a glance, he can tell what he is doing.

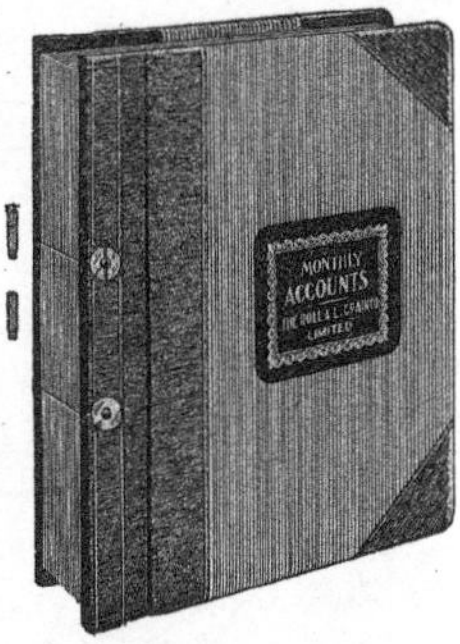

THE CRAIN MONTHLY ACCOUNT SYSTEM

has been in use by successful business men for years. It does the work in one-half the time, and in a way that is much more satisfactory to his customers and more pleasant to himself than plodding along in the old way.

In this system, too, you require a binder, such as the Crain, with an unlimited capacity, which binds one sheet as easily as it does 1,000, not one you must have a certain number of sheets in before it will bind, and then only allow you to extend to a limited number of accounts.

Write us for particulars

The Rolla L. Crain Co., Limited

OTTAWA, CANADA

TORONTO: 18 Toronto Street

MONTREAL: 28 Alliance Bldg., 107 St. James St

WINNIPEG: 11 Nanton Block

Other Tools are very Good Tools, but

"YANKEE TOOLS" ARE BETTER

The NEWEST, CLEVEREST and QUICKEST SELLING TOOLS of the KIND.

SCREW DRIVERS

No. 10 & 11—RATCHET, right and left hand and rigid.

No. 15—RATCHET, with finger turn on blade.

No. 30—SPIRAL RATCHET right and left hand and rigid
No. 31— " " (heavy pattern)
No. 20— " " right hand only.

Chuck and 8 Drill Points for spiral ratchet screw driver.

No. 12—RATCHET, with stub blade

Countersink, for spiral ratchet screw driver.

No. 60—POCKET SCREW DRIVER.

DRILLS

No. 40—AUTOMATIC DRILL, with Ratchet Movement.

No. 41—AUTOMATIC DRILL.

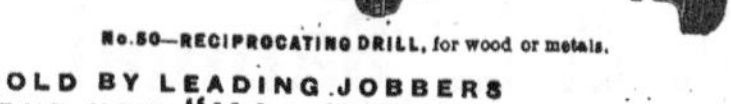

No. 42—AUTOMATIC DRILL.

No. 43—AUTOMATIC DRILL, for light drills only.

No. 44—AUTOMATIC DRILL, with adjustable tension on spring.

No. 50—RECIPROCATING DRILL, for wood or metals.

SOLD BY LEADING JOBBERS

SEND FOR OUR NEW "YANKEE" TOOL BOOK

NORTH BROS. MFG. CO.

PHILADELPHIA, Pa.

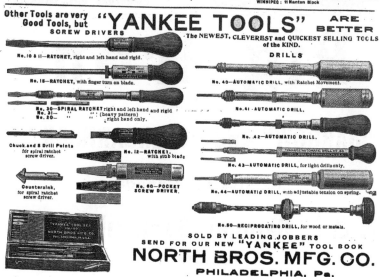

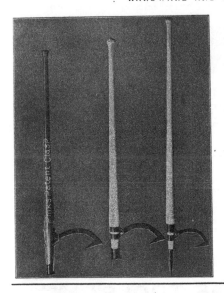

No Woman Should Do Work a Machine Can Do

The Buckeye Power Washing Machines

If there is a water works system in your town send us a sample order before your competitor. This is a line **you can sell** and at a profit.

The Buckeye motor is the only motor constructed on scientific principles, the same as a steam engine, only one valve in its entire construction and that is on the outside of cylinder, and is operated the same as an engine valve. It has no valves inside of the cylinder, as in the construction of other motors. Everything outside, where it can be got at.

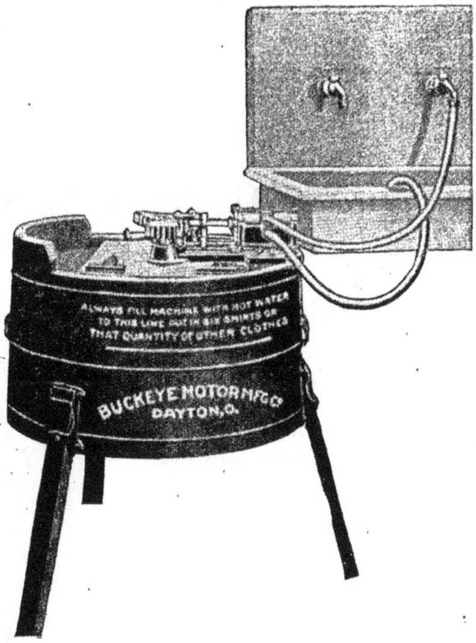

Ours is the largest tub and is made expressly for the Buckeye motor. It will hold more clothes than any other tub, and each one is tested before being sent out. Our motor is larger than any used by other makers. It will pull a heavier load and runs faster on 18 to 25 pounds pressure than any motor washing machine will do on 40 pounds. There are 10,000 of these machines in use in the United States, and these have all been sold in 18 months.

We are many hundreds behind in our orders, but we are receiving them daily and will guarantee to fill your order within ten days. **Order to-day.**

Exclusive Canadian Distributors

LEWIS BROS., Limited

Montreal

OTTAWA
TORONTO

WINNIPEG

VANCOUVER
CALGARY

Up-To-Date Lumbermen's Supplies

AXES

Forest King
Made by Josiah Fowler A lumberman's axe. Hand-made, unequalled in quality and finish.

Blood's Champion
Made from Firth's best cast steel.

Forest Clipper—Black Prince Mikado
Fowler's and Blood's double bit axes.

Atkin's Victor Cross-cut Saws
They saw well. They sell well.

Disston's and Shurly & Dietrich Maple Leafs

Horse Blankets
Every size and kind.

Pink's Lumbering Tools
Full assortment.

WRITE FOR PRICES

Thos. Birkett & Son Co., Limited, Wholesale Hardware **Ottawa, Ont.**

Patented 1900. Improved 1905.

Davidson's "Empire" Stove Pipe

Improved Patent

Neatly packed in crates of 25.
Secures lowest possible freight rate.

Deep Swage Prevents Telescoping.
Coated to Prevent Rust.

Sold as low as ordinary Stove Pipes.

SIMPLE and EASY to put together.
No Profane Language Necessary.
The only tools required are a pair of hands.
Pipes being cut out with dies ensure absolute fit and conformity.

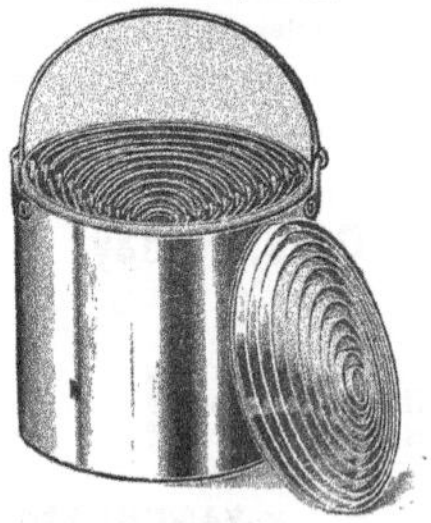

Miners' Lamp Kettles

Strong and substantially made in extra heavy tin.

Nineteen sizes, from ⅛ quart to 29 quarts.

Wash Tubs—Galvanized
With Wringer Attachment

Nos.	1	2	3
Top measurement, inches	21⅛	22¾	25
Bottom " "	17⅞	18¼	20
Height " "	9½	10½	11

The THOS. DAVIDSON MFG. CO., Ltd., Montreal and Winnipeg

H. S. HOWLAND, SONS & CO. LIMITED

HARDWARE MERCHANTS

Only Wholesale — 138-140 WEST FRONT STREET, TORONTO. — Wholesale Only

SAWS

Made by SHURLY & DIETRICH, Galt, Canada

Trade Mark — Trade Mark

Narrow Cross-Cut Saws

"The Lance" Cross-Cut Saws

"The Racer" Cross-Cut Saws

"Premier" Cross-Cut Saws

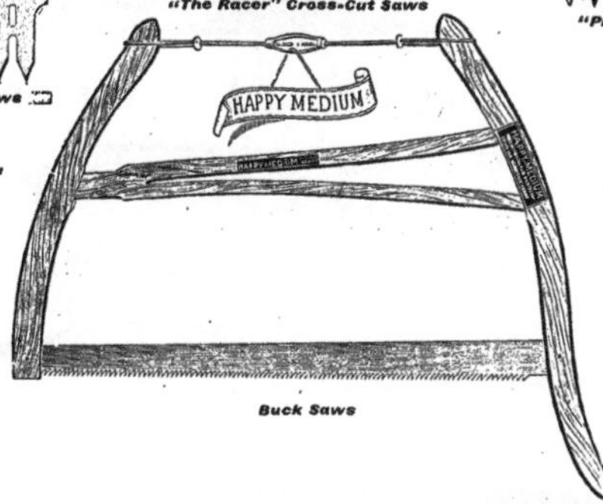

Buck Saws

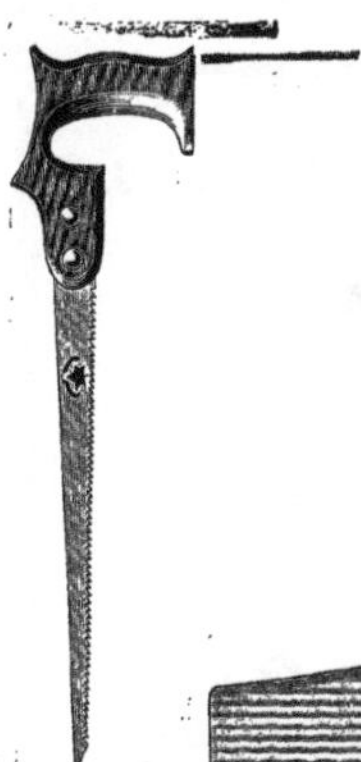

Compass Saws

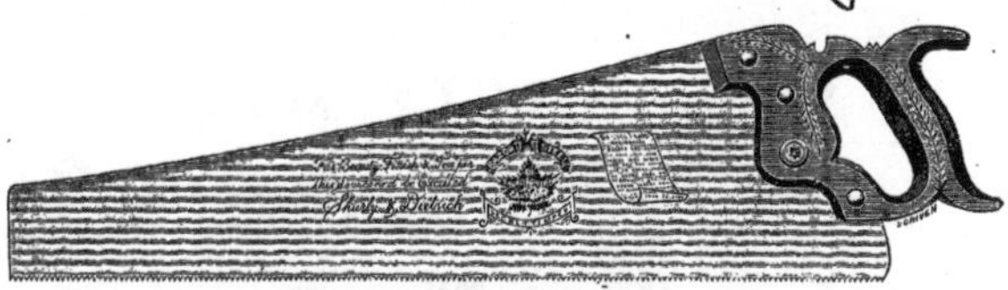

Panel, Hand, Rip and Back Saws, etc.

Buck Saw Blades

FOR OTHER LINES OF SAWS, SEE OUR HARDWARE CATALOGUE

H. S. HOWLAND, SONS & CO., LIMITED

Opposite Union Station

GRAHAM WIRE NAILS ARE THE BEST

Our Prices are Right — If you are not receiving our illustrated monthly circular, WRITE FOR IT. — We Ship Promptly

Canada Leads the World in the Manufacture of

HIGH-CLASS SAWS

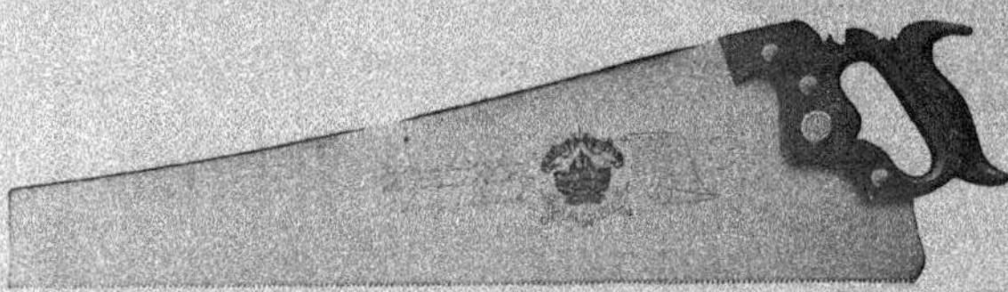

No. 81

The best and cheapest Hand Saw on the market, quality, temper and finish considered. Every saw warranted. Carved applewood handle. Patented, wave finish blade.

No 1 Maple Leaf Lance Tooth Cross-Cut Saw, and the No. 0 Racer are tempered under our secret process.

No. 0 Narrow Racer Cross-Cut Saw

If you want saws that will sell and give your customer satisfaction, you will stock the Maple Leaf Saws.

Manufactured by **SHURLY & DIETRICH, - GALT, Ontario**

First Time Made in Canada

M.R.M. CO.

SPECIAL COUNTERSUNK STEEL

HORSE SHOES

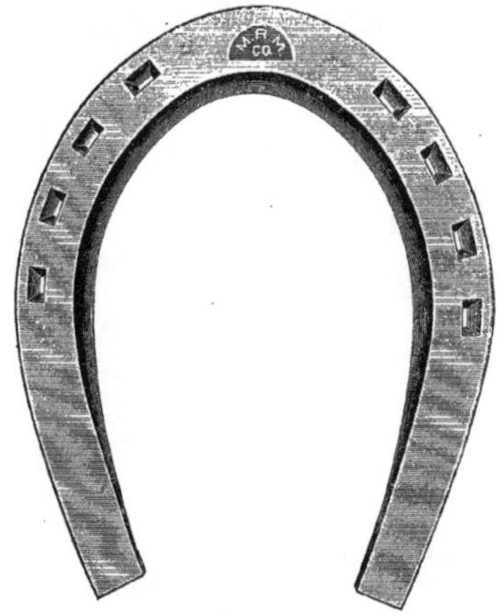

FRONT

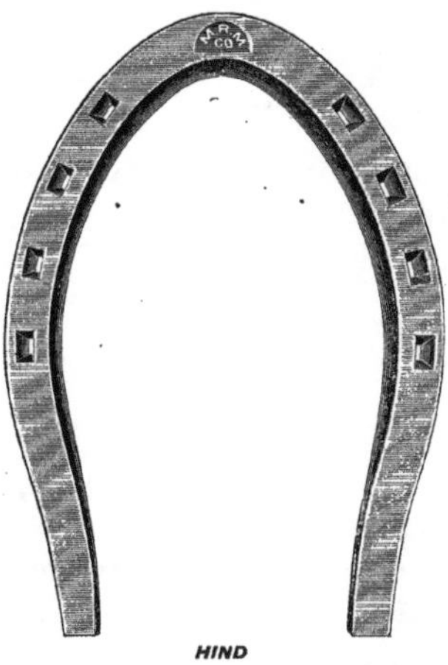

HIND

We are now ready to put on the market our Special Countersunk Steel Shoes, made by us in sizes 0 to 4 inclusive, both front and hind.

This style of Shoe, which has hitherto been imported, is now made in Canada for the first time.

They are made with the inside edge bevelled, similar to a snowshoe, being especially adapted for light driving work in winter, and make a very nice plate shoe for summer wear, and are noted for the ease with which they can be fitted to the horse's foot.

We are the sole manufacturers in Canada of Special Countersunk Steel Shoes.

THE MONTREAL ROLLING MILLS CO.

The Howland Pump Oiler

FOR SALE BY

J. H. Ashdown Hdw Co., Winnipeg
Geo. D. Wood & Co., "
Miller-Morse Hdw. Co., "
Jas. Robinson & Co., "
Lewis Bros., Ltd., Montreal

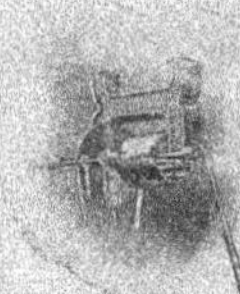

A slight pressure of the thumb regulates the flow.

Maple City Manufacturing Co.

Monmouth, Ill.

ONE WEAK SPOT

in an axe handle is enough to condemn it. That's why we make handles which contain

NO WEAK SPOTS

We use only the cleanest, toughest hickory, because hickory has been proved the best material on earth for

AXE HANDLES

If you don't already handle Still's famous hickory handles you are neglecting a most profitable line.

Let us mail you our lists and discounts.

J. H. Still Manufacturing Co. Limited

ST. THOMAS, ONTARIO

SOLDERS

Guaranteed Half and Half. Strictly Half and Half.
Commercial Half and Half.

Star Wiping. Acme Wiping. Angle or Oval.

Wire, any Size or Gauge.

Segment Ribbon.

CLOSE PRICES. QUICK DELIVERIES.

The CANADA METAL CO., LIMITED

TORONTO

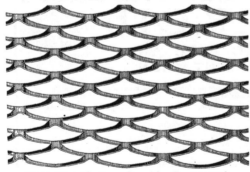

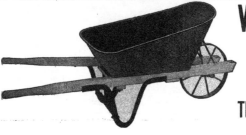

The Peterborough Lock Manufacturing Company, Limited

Cylinder Night Latch, No. 103.

Peterborough, Ont.

Manufacturers of all kinds

Rim and Mortise Locks, Inside, Front Door, and Store Door Sets, also full line High-class Builders' Hardware.

Sold by all Leading Jobbers in the Dominion.

TRADE RAPID MARK

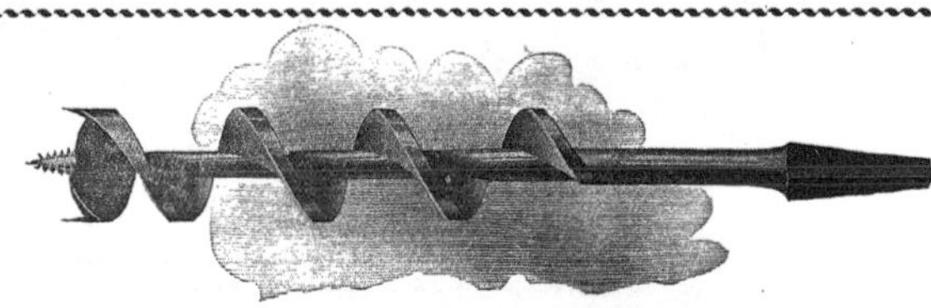

The ***"Rapid"*** Patent Centre Stem Auger Bits are made with double screw and ***guide point.*** This assures a ***perfectly straight bore.*** The "Rapid" cuts cleaner holes, and cuts them quicker, than any other bit on the market.

The Rapid Tool Company, Limited, - Peterborough, Ontario

Not How Cheap, but How GOOD

Banner Lamp Burners,
O, A, B and D sizes.

Canada Lamp Burners,
A and B sizes.

Security Lamp Burners,
A and B sizes.

Bing Glass Cone Lamp Burners,
A and B sizes.

Lantern Burners,
No. 1 and No. 2.

Every Burner carefully inspected before leaving the factory, and guaranteed perfect.

Orders solicited through the jobbing trade.

ONTARIO LANTERN AND LAMP CO.
HAMILTON, ONT. LIMITED

The Buffalo Manufacturing Co.

Buffalo, N.Y.

When you get our goods you know you get THE BEST

We manufacture

Water Filters
Water Coolers
Chafing Dishes
Table Kettles and Stands
Coffee Extractors
Wine Coolers
Nursery Chests
Baking Dishes
Crumb Trays and Scrapers
Tea and Bar Urns
Bathroom Fixtures
Coal Vases and Hods
Candlesticks
Cuspidors
Match Safes, Etc.

All High Grade and exceedingly presentable.

Represented by
H. F. McINTOSH & CO
51 Yonge Street
TORONTO, - - ONT.

rite for Catalogue

DOMINION WIRE MANUFACTURING CO.

MONTREAL TORONTO LIMITED

Manufacturers of

PLAIN WIRE *All Sizes and for all Purposes.*

GALVANIZED WIRE *For Fencing, Telegraph and Telephone Purposes.*

Copper and Brass Wire

WIRE NAILS

WOOD SCREWS

Bright Wire Goods

Staples

Steel and Brass Jack Chains

Spring Cotters, Strand Guy Wire, Steel Wire Barrel Hoops

A RECORD BREAKING SELLER

22 CALIBRE
16 SHOT

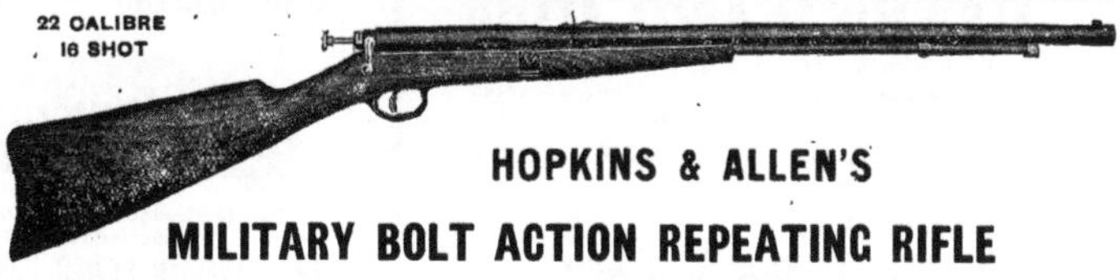

HOPKINS & ALLEN'S

MILITARY BOLT ACTION REPEATING RIFLE

$10.75 LIST There has long been a demand for a good rifle at this price—and the **HOPKINS & ALLEN Junior Repeater** fills the bill. This rifle has been and is being extensively advertised, and a heavy sale for it is created.

SPECIFICATIONS:—22-inch barrel, take-down, 22 calibre. Magazine handles 22 short, 22 long and 22 long rifle cartridges without change of carrier. Shoots 16-22 short or 12-22 long or long rifle cartridges. Action, Military Bolt Pattern with positive safety device. Length over all, 40 inches, weight 5½ pounds.

FOR SALE BY ALL FIRE-ARMS JOBBERS.

PLENTY OF GOOD SUPPLEMENTARY ADVERTISING MATTER—FREE.

WRITE FOR CATALOGS

THE HOPKINS & ALLEN ARMS CO.

LONDON OFFICE
6 City Road, Finsbury Square, London

Dept. 67, NORWICH, CONN., U.S.A.

Shows Bolt Drawn Back—Position for Ejecting Shells.

Persons addressing advertisers will kindly mention having seen their advertisement in Hardware and Metal.

THE WANT AD.

The want ad. gets work for workers and workers for work.

It gets clerks for employers and finds employers for clerks. It brings together buyer and seller, and enables them to do business, though they may be thousands of miles apart.

The "want ad." is the great force in the small affairs and incidents of daily life.

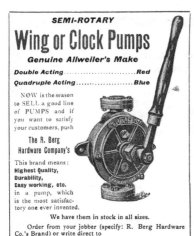

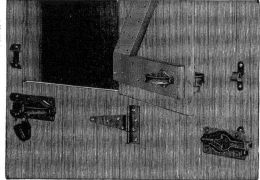

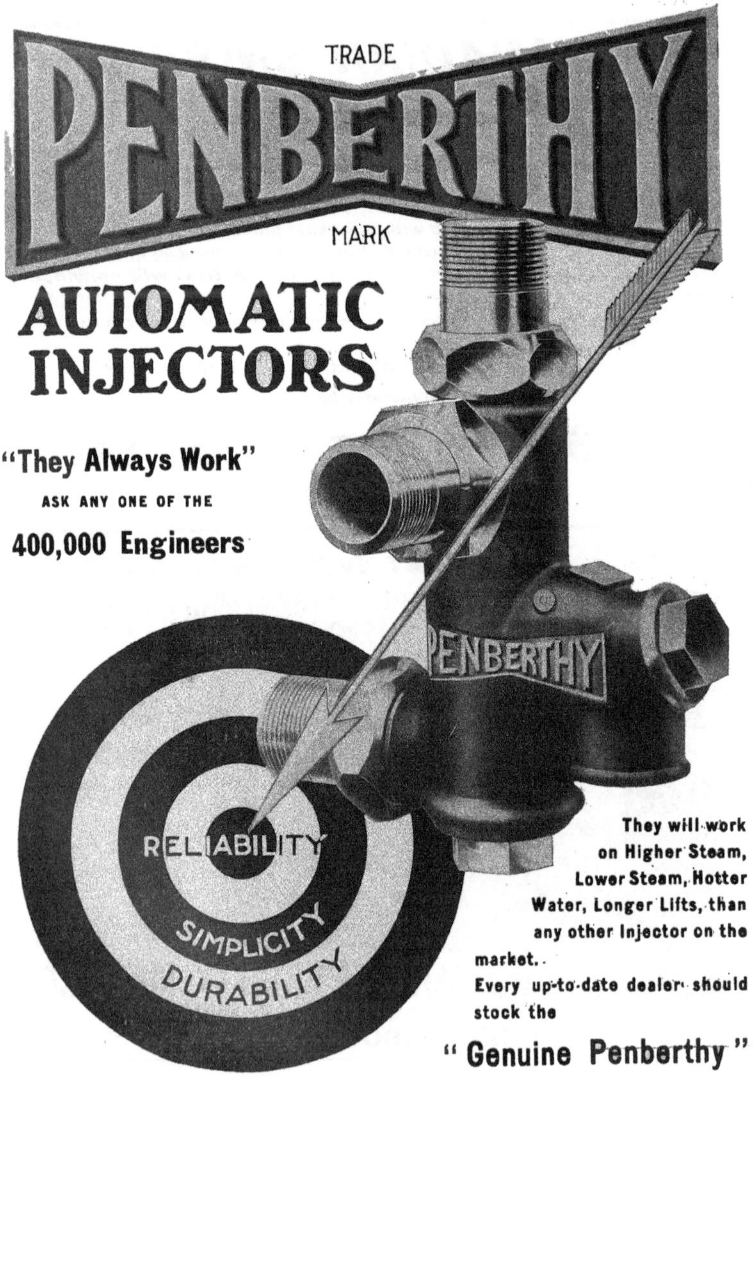
TRADE
PENBERTHY
MARK
AUTOMATIC
INJECTORS
"They Always Work"
ASK ANY ONE OF THE
400,000 Engineers
PENBERTHY
RELIABILITY
SIMPLICITY
DURABILITY
They will work on Higher Steam, Lower Steam, Hotter Water, Longer Lifts, than any other Injector on the market.
Every up-to-date dealer should stock the
"Genuine Penberthy"

EFFICIENCY MEANS MONEY!

There is no doubt about it. Efficiency does mean money. Let the hardwareman sell his customer an article which fulfills all promises made for it, and that customer immediately becomes a free advertising medium for the hardwareman. It so happens that

JONES' SIDE-WALL REGISTERS

give that perfect satisfaction which results in pleased customers. It saves their money, both in installation and in operation.

Let us send you our Plan of Profits.

The Jones Register Co.

732 King St. West Toronto, Canada

Wrought Steel Registers

Standard Sizes, All Finishes
Perfect Operating Device
Unusually Large Air Space
For Sale by Leading Jobbers

HART & COOLEY CO.

New Britain, Conn., U.S.A.

H. & C. No. 100.

FERROSTEEL REGISTERS

The only house in the world manufacturing

- CAST IRON FLOOR
- SEMI STEEL FLOOR
- WROUGHT STEEL FLOOR
- CAST IRON SPECIAL SIDE WALL
- WROUGHT STEEL SPECIAL SIDE WALL
- FLAT OR CONVEX EXTRA SHALLOW SIDE WALLS
- DEFLECTORS, PANCAKES, WAFERS

In all Designs and Finishes.
Mechanism Perfect.
Capacity Greatest.
Finish Unsurpassed.
No reputable line can undersell us.

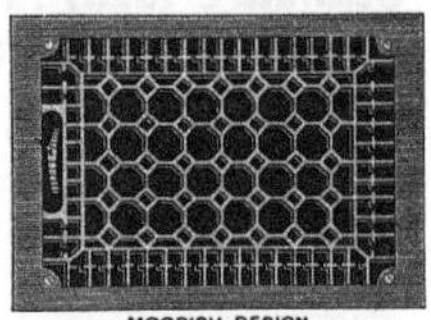

MOORISH DESIGN

FERROSTEEL COMPANY, - - CLEVELAND, OHIO

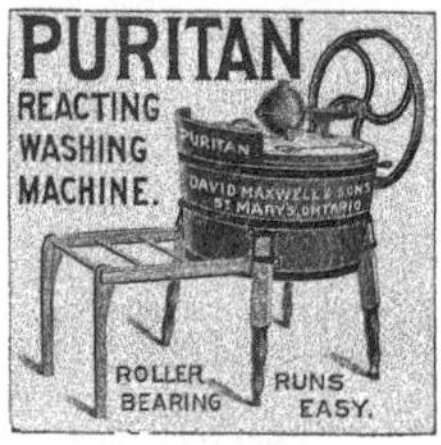

Do You Handle These Money Makers?

MAXWELL LAWN MOWER

Durability, Finish and Efficiency Unsurpassed by any other mower on earth.

FAVORITE CHURN

By lightening women's labor it adds weight to your bank account.

David Maxwell & Sons, St. Mary's, Ont.

You May Be a Good Salesman

But, is everything in your favor? Are you selling the new Imperial Oxford? A continuous stream of repeat orders tells the story to us.

Removable Nickel
Removable Fire Linings
An Oval Firebox
The "Divided Draft"

All contained by the best looking stove in the market.

It gives a good salesman a chance to spread himself.

Drop us a card, or, better still, tell us the size range you want on your floor. If you get this agency you're lucky.

The Gurney Foundry Company, Limited

Toronto Winnipeg Vancouver Hamilton and London

THE GURNEY-MASSEY CO., LIMITED, MONTREAL, Que.
THE GURNEY STANDARD METAL CO., LIMITED, CALGARY, Alta., and EDMONTON, Alta.

FIRE!

An imperfectly-lined stove, range or furnace is liable to cause a lot of damage. Our

Asbestos Stove Lining

beats everything yet discovered for lining every kind and size of stove, range, grate or hot-air furnace. Fits any of them, and can be put in by anyone in a few minutes. For mounting and repairing stoves, furnaces, etc., nothing can excel our

ASBESTOS FIREPROOF FURNACE CEMENT

Easily applied, it adheres instantly and permanently. Neither fire nor chemicals can destroy it. Both these lines are **highly-profitable lines.**

Benjamin Moore & Co., Limited

(Formerly The Francis-Frost Co., Limited)
New Offices and Works: Toronto Junction, Canada
New York, Chicago, Cleveland, Carteret, N. J.

TRADE MARK
FOR PLATE

Xmas Gift Goods

Original and pleasing designs in Fish and Meat Sets, Meat and Game Carvers, Cheese Scoops, Cake Knives, Nut Crackers, Flasks, etc.

Samples and prices on request.

The House of Hutt on

Established in England in 1800

Sole Canadian Agent
W. J. GRANT, Lindsay Bldg., MONTREAL

TRADE MARK
FOR CUTLERY

If you want to do a great deal more Stove business write and ask us about the

PERFECT IDEA RANGE

The ***Perfect Idea*** has more ***selling points*** than any other stove on the market. Not the least of them is the fact that ***all nickel parts are removable.*** This, with the automatically opening oven door, and the patent electric sliding oven tray, has resulted in some ***dealers netting hundreds of dollars'*** worth of new business. ***What about representing the Perfect Idea in your district?*** We have some interesting profit pointers for dealers.

GUELPH STOVE CO., LIMITED, GUELPH, ONT.

There is Profit in

selling high-grade articles at a price within reach of all. We want you to understand that there are no **safer** or more **efficient** sporting rifles made than

HAMILTON RIFLES

With breech and barrel made of **specially-selected** and **tested rolled steel**, and the parts assembled by men of the highest skill, a HAMILTON RIFLE CANNOT BURST and the popular prices make money for the hardwareman.

No. 15 sells at $2.00
No. 19 sells at $2.50
No. 23 sells at $3.50

THE HAMILTON RIFLE CO.

Dept. 71 ***Plymouth, Mich.***

NEW NIAGARA FACTORY FALLS

We've hitched our Factory to the Falls to enable us to make ***better Bath-Room Fixtures*** than can be obtained anywhere else.

NEW ADDRESS:
The Carriage Mountings Co., Ltd., Niagara Falls, Canada

Feed Cookers

Large fire box. Tank directly over fire. Cooks rapidly. Saves fuel. Used anywhere. Safe as a stove. A necessity this fall with feed short. Prices very reasonable. Write us. Get our free catalogue.

THE STEEL TROUGH & MACHINE CO., Ltd.
No. 10 James St., TWEED, ONT.

Are you interested in any of the lines that are advertised? A Post Card will bring you price list and full information. Don't forget to mention Hardware and Metal.

THE WAGGONER EXTENSION LADDER

is the lightest, handiest, strongest and safest ladder made. Re-inforced with steel wire Patent safety lock. Made in all lengths from 20-ft., 24-ft., 28-ft., etc., up to 100 ft. if desired. Every painter, tinner, stacker and farmer needs one

Write for circulars and prices.

The Waggoner Ladder Co., Ltd.
LONDON, - - CANADA
E. H. Briggs Co., Ltd., Western Representative Winnipeg, Man.

Windsor Pump and Foundry Co.

LIMITED

We Make a Specialty of

Brass Cylinders for Wind-Mill and Hand Force Pumps Fitted with Patent Rotating Plunger Valves
Designers of Special Machinery Tools and Dies, Plug Valves, Foot Valves, Brass Tubing, Iron and Brass Castings

100 London Street West, WINDSOR, Ont.

PHONE 369

WRITE FOR PRICES

$ $ $ $ $ $ $ $ $ $

CONTRACTS

mean dollars for the pockets of contractors. The weekly reports in the

CANADIAN CONTRACT RECORD

tell where contracts may be had. $2 per year buys them Address

THE CONTRACT RECORD
TORONTO, MONTREAL, WINNIPEG and VANCOUVER.

Persons addressing advertisers will kindly mention having seen their advertisement in Hardware and Metal.

TRADE WITH ENGLAND

Every Canadian who wishes to trade successfully with the Old Country should read

"Commercial Intelligence"

(The address is 168 Fleet St., London, England.)

The cost is only 6c. per week. (Annual subscription, including postage, $4.80.) Moreover, regular subscribers are allowed to advertise without charge in the paper. See the rules.

Specify the Following Products

Manufactured by

American Sheet and Tin Plate Company

PITTSBURG, PA.

If you want the best.

"Apollo Best Bloom," Galvanized Sheets, Blue Annealed and American Bessemer Black Sheets, Wellsville Polish, Range Steel, Wood's "A" Planished Iron, Black Steel specialties for all purposes, Tinplates, Coke and Charcoal, all grades, Terne Plates, Old Style and New Method Processes.

Place your order with

B. & S. H. THOMPSON & CO.

LIMITED

Dominion of Canada Sales Agents

MONTREAL, CANADA

The Woman Who Kneads

the dough for the family bread supply should know about the

"GENERAL" BREAD MAKER

The **hardwareman** who pushes the "General" makes a host of lasting friends among the **women** of his locality. No wonder, either! The "General" mixes more dough thoroughly in **three minutes** than a woman's aching arms can in **thirty minutes.** Every woman appreciates the merchant who calls her attention to such a labor-saver as the "General."

Then, think of your Profit!

Shirreff Mfg. Co., Ltd.

Brockville, - Ont.

Packers!

The Government insists upon pure food and sanitary factories.

Commence right by using clean, sanitary cans, the kind that is to be depended upon, made by

ACME CAN WORKS

Montreal

You can improve your pack still more by using ***Cans*** made of ***extra coated tin plate.*** Cost is a little greater, but the quality is there, indisputable evidence that it is worth while. In the long run the best pays.

Secure the best. In this case, ***Acme Cans, extra coated tin plate,*** are the best.

Do not experiment.

HAVE YOU SEEN

the new line of

Dandy Brushes

which we are showing?

Solid backs, elegantly finished, steel fastened. Made in a variety of materials and sizes at popular prices.

Write for Samples.

Stevens-Hepner Company

Limited

Port Elgin, Ontario

Persons addressing advertisers will kindly mention having seen their advertisement in Hardware and Metal.

DIRECTORY OF MANUFACTURERS

Hardware and Metal receives, almost daily, enquiries for the names of manufacturers of various lines. These enquiries come from Wholesalers, Manufacturers and Retail Dealers, who usually intimate they have looked through Hardware and Metal, but cannot find any firm advertising the line in question. In the majority of cases these firms are anxious to secure the information at once. This page enables manufacturers to keep constantly before the t ade lines which it would not pay them to advertise in larger space.

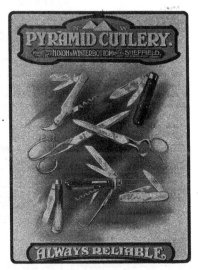

Shear making "up-to-date" is found in

HEINISCH

TAILORS' SHEARS, TRIMMERS, SCISSORS, TINNERS' SNIPS, ETC.

"Best in the World"

Since 1825, and a steady increase in sales due to their superior cutting quality.

R. HEINISCH'S SONS CO.
NEWARK, N.J., U.S.A.
New York Office and Salesroom, 155 Chambers St.

REGISTERED TRADE MARK GRANTED 1682

World's Best Cutlery

You make no mistake in specifying goods made by **JOSEPH RODGERS & SONS, Limited** when ordering cutlery. A reputation of over 200 years at stake, insures the quality of every article.

Write for catalogues from Sole Canadian Agents

JAMES HUTTON & COMPANY, MONTREAL

Metal Ceilings

are in greatest demand at this time of the year. This profitable business naturally falls to the hardwaremen and tinsmiths who boost

Our "Classified" Designs

They are the handsomest ceilings on the market, and delicately harmonize with any surroundings. The plates are guaranteed perfectly square, and fit together closely with **invisible joints. They are the easiest and therefore the most profitable to erect.** They make delighted customers wherever used.

We are always glad to assist our dealers by submitting plans and estimates.

"WE WORK WITH THE TRADE."

The Metal Shingle & Siding Co., Limited

Montreal **PRESTON** **Toronto**

SALES AGENTS:

ST. JOHN, N.B. Emerson & Fisher, Limited — QUEBEC J. A. Bernard — WINNIPEG Clare & Brockest — CALGARY Ellis & Grogan — VANCOUVER McLennan, McFeely & Co. Ltd

THE MOST LIGHT FOR THE LEAST MONEY

C. G. E. Incandescent Lamps

Long Life ***Efficiency*** ***Reliability***

Write for prices and quantity discounts.

Canadian General Electric Co., Limited

Head Office, - - Toronto, Ont.

District Offices: Montreal, Halifax, Ottawa, Winnipeg, Vancouver, Rossland

ALEXANDER GIBB

Manufacturers' Agent and Metal Broker,
13 St. John Street, Montreal

Representing Canadian, British and American Manufacturers. Correspondence invited from firms wishing to be represented.

Expert Bookkeeping

Investigation and auditing of all classes of accounts undertaken. Trading, profit and loss accounts and balance sheets prepared. Sets of books opened and office systems devised in an up-to-date manner. Collection of accounts a specialty.

Davenport, Pickup & Co.
622 McIntyre Block and 422 Ashdown Block
WINNIPEG, MAN.,
AND AT BRANDON, MAN.

Square your Accounts

INCANDESCENT

LIGHTING WITH COAL OIL

The most economic modern lighting. The "Ideal" lamp burner, screws on any ordinary coal oil lamp. This simple apparatus produces its own gas, giving a light more brilliant than electricity or gas. Costs only ¼ of a cent per hour. Guaranteed exactly as represented or money refunded. The "Ideal" lamp for the country home. Agents wanted everywhere. Write for free booklet describing everything.

ADDRESS

ST. ARNAUD FRERES
Cor. St. Sulpice and Commissioners Sts.
Montreal, P.Q.

CARRIAGE SPRINGS & AXLES

ANCHOR BRAND

GUELPH AXLE WORKS

THE GUELPH SPRING & AXLE CO. LIMITED
GUELPH, ONT.

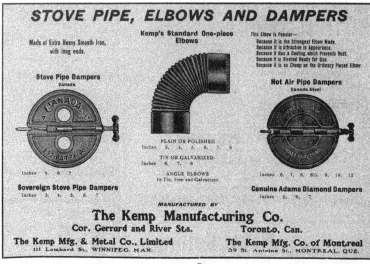

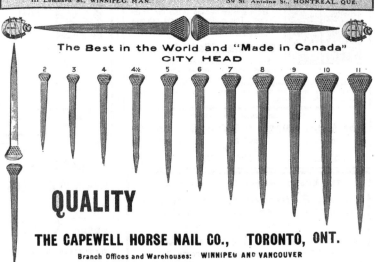

Retail Hardware Association News

Official news of the Ontario and Western Canada Associations will be published in this department. All correspondence regarding Association matters should be sent to the secretaries. If for publication send to the Editor of Hardware and Metal, Toronto.

Officers Retail Hardware and Stove Dealer's Association of Western Canada:

President—A. J. Falconer, Deloraine.
First Vice-President—J B Curran, Brandon.
Second Vice-President—W. M. Gordon, Winnipeg.
Secretary-Treasurer—J. E McRobie, Winnipeg.
Executive—Alberta, A. E. Clemens, Sedgewick; C. F. Comer, Calgary; A. R. Auger, Okotoks
Manitoba—H S. Price, Boissevain; A. P. Macdonald, Winnipeg; O. Gilmer, Winnipeg.
Saskatchewan—G. K. Smith, Moose Jaw; S. A. Clark, Saskatoon; J. R. Fox, Weyburn.
Association offices, 53 Scott building, Main street, Winnipeg.

Officers Ontario Retail Hardware and Stove Dealers' Association:

President—W. G. Scott Mount Forest.
1st Vice-President—J. R. Hambly, Barrie
2nd Vice-President—J. Walton Peart, St. Marys.
Treasurer—John Caslor, Toronto.
Secretary—Weston Wrigley, 10 Front St. East, Toronto.
Executive Committee—The officers and H Becker, Hamburg; G. A Binns, Newmarket; D. Brocklebank, Arthur; A. W Humphries, Parkhill; W. A. Mitchell, Kingston, and Frank Taylor, Carleton Place.
Auditors—J. W. Peacock and C. F. Moorhouse, Toronto

Officers British Columbia Retail Hardware Association.

President—L B. Lusby, New Westminster.
Vice-President—C Snell, Vancouver
Secretary-Treasurer—John Burns, Vancouver.
Executive Committee—W Stearman, Vancouver H. T. Kirk, New Westminster; R A. Ogilvie, Victoria

COLLECT EARLY THIS FALL.

The money shortage this fall is forcing merchants to realize the importance of insisting upon prompt settlement of accounts. Many retailers have been lax in making their collections in recent years, the general expansion of industry and the plenitude of money encouraging them to deal too liberally with credit customers. As a result, most merchants find it hard to meet the demands of the jobbers, who are being urged by the banks to reduce their loans.

If jobbers are forced to pay higher rates for money the burden will be passed on to retailers who are behind in their payments, and to escape this addition, retailers must collect hard and gather in every cent available, and the time to act is before the real squeeze comes during the winter.

The Ontario Retail Hardware Association is helping its members in a practical manner by supplying form letters at a nominal cost to enable dealers to collect hard accounts. These "collection department" letters have been in use by some merchants since last May, and the results have exceeded the most sanguine expectations of any member.

A couple of weeks ago reference was made to the enthusiastic endorsation of the collection letters by McNab Bros, Orillia, and this week another testimonial is published, in which G. A. Binns, a member of the association executive, tells how he gathered in over fifty per cent. of the amount of the accounts rendered to a score of "dead beats" marked off his books as "bad accounts." It was like finding $34 for Mr. Binns, as the form letters used and the cost of postage would amount to less than $1.

Mr. Binns' endorsation of the association collection department should encourage many other members to use the system, the cost to members being only $1 for 50 form letters, 40 follow-up letters and 100 envelopes. It should also encourage many hardwaremen who want to see some immediate advantages before joining the association to send in their application for membership. From September to May membership in the association costs only $2, so that for $3 a firm can become a member for eight months and test the advantages of organization. In Mr. Binns' case, he cleared enough out of his "bad accounts" scrap heap to pay his association dues for ten or eleven years to come. His letter follows:

> Dear Sir,—Replying to your favor of the 18th, asking for my opinion with reference to my experience with the "collection department" letters, I am glad to say I have found them more effective than any plan I have ever tried.
>
> I sent out the forms to twenty dead heads that I had crossed off my books as N.G. The amounts were all small, from $1.50 to $5, amounting in all to about $60. From the first letters I have had $33.89 paid in. I think the business-like heading of the forms carry more weight than the ordinary collection department letter heads.
>
> I feel confident there is nothing that will repay better than a few of these letters sent to dead heads.
>
> G. A. BINNS.
>
> Newmarket, Sept, 25, 1907.

MUST KEEP ON FIGHTING.

The annual report of Sears, Roebuck & Co, Chicago, which has just been published, emphasizes the continued growth of the retail mail order business and the great place it is coming to hold in the distribution of merchandise. It appears while during the calendar year 1905 the company's net sales were $27,614,253, and in 1906 were $37,943,472, that for the year ending June 30 last they were $50,722,839. There is thus an increase in the volume of business for the year covered by the report of more than 80 per cent. over that of the calendar year 1905. These figures enforce the seriousness of the competition which the regular merchants are called upon to meet and the need for unrelaxing efforts and increased enterprise and energy in their conduct of business.

There is also brought up anew to wholesale distributors and manufacturers serious problems in regard to the effect of the mail order business on the regular distribution of merchandise. In this condition of things, which does not promise to become more favorable for the marketing of goods in old channels, the Iron Age points out that it may be that the manufacturers and jobbers will have to see to it that the retail merchants of the country obtain their goods on something like as advantageous terms as do the catalogue houses.

THOUGHT HE WAS A TRAVELER.

At the Michigan Retail Hardware Association convention at Detroit a few weeks ago, Secretary Peck, of the Wisconsin Retail Hardware Mutual Fire Insurance Company, told some of his experiences while on the road selling mutual insurance. He went into one store and asked the man in charge if the proprietor was in.

"No," was the reply of the clerk, who thought he was saving his boss from being interviewed by a strange traveling salesman.

"When will he be in?" asked Mr. Peck.

"I don't know"

"Well, give him this card and tell him he can write me if he wants to," said Mr, Peck, making for the door.

"Oh, you're Mr. Peck. Wait, Mr. Jones is upstairs and he wants to see you I thought you were a traveler."

"Well, so I am, young fellow. But I have been forty-two years in the retail hardware business and I long ago learned not to turn travelers down that way. In future treat travelers decently if you want to hold your job and succeed, as travelers usually have some good things to offer a merchant, more particularly if they are strangers making their first call."

Mr. Peck gave the clerk practical advice and did not forget to also rub it into Mr. Jones, the dealer who was so short-sighted as to encourage his clerks to treat travelers as though they were nuisances, while in reality the man on the road has had a wider experience and is a far better business man than many customers he calls upon for orders. The merchant and the clerk make no mistake in treating a traveler as a gentleman and friend.

How to Figure Profits

An article by J. Malcolm Stewart, in the Furniture Journal, in which he contends a plumber did not get as much profit as he thought he did.

An article by J. Malcolm Stewart, in the Furniture Journal, in which he contends a plumber did not get as much profit as he thought he did.

Various letters have been received by the writer for more information on the subject of figuring profits on merchandise. The information desired is on the method of figuring those profits in the way most likely to avoid error, and unconscious losses, with the accompanying disaster that must overtake those groping in the dark. I contended that the only way to arrive at the proper percentage of profit on any commercial transaction, was to ascertain what proportion of the sale was, or would be, profit, never attempting to arrange the percentage of gain by adding to the cost, a percentage of that cost, equal to the amount of profit desired, or in other words, in adding 25 cents to a dollar, assuming you are making a profit of 25 per cent.

Of course, we all know that it is difficult to convince everyone on any subject, and bring them all to one way of thinking. In the case like the one before me it probably would be the best way out of the difficulty to refer those in search of information to the head of the office in any large wholesale house. A visit to any of these houses would, I think, furnish the seeker after information with plenty of it, all of which would be in favor of my contention. I would like much to see this matter brought before the public at large more forcibly than I ever could hope to do it, for, as one of my correspondents remarks, "It is a matter of the most intense and vital importance to the business world," and I might add, but little understood.

Only a few days ago, an amusing case came under my observation, going to show how poorly armed for business is he who doesn't understand this subject clearly.

Says the Plumber Figured Wrong.

A young business man, a plumber, strange as that may seem, was figuring on a contract. His labor and raw material, he figured, would cost him in round figures $500. Wishing to make a profit of twenty per cent. he proceeded to add twenty per cent. to the cost, making his complete bid $600. When he got through I asked him how much profit he would make at that rate if he did a business of $10,000. He answered with a knowing smile and triumphantly, $2,000. I asked him to apply the mode of reasoning by which he had reached that conclusion to the bid before him and see if it would work, and his surprise when he found it wouldn't was great. He had been doing business all along with the idea that adding twenty per cent. to the cost would give him a profit of twenty per cent. on his business.

It is not necessary to state that figuring one of these on the investment, the other on the amount of business, must lead to error. Don't say you understand that and provide for it. What is the use of doing that; why not figure right in the first instance? Try and find any advantage to be gained by figuring the profit on the cost. There is none that I know of. Ask some merchant who figures his profit on the cost the question I asked the plumber and see what answer you will get. A little study will convince any one with any business experience of the justice of my contention.

How to Figure Correctly.

To figure the profit on any article correctly: Subtract cost from the selling profit, the difference, of course, being the profit. Divide the latter (decimally) by the selling price, and the result will be the true profit; thus: Cost $5, selling price $8, profit $3, and $8 is contained in $3 .375 times, showing a profit of 37½ per cent. For the benefit of those who did not read the former article on this subject, I will repeat the table by which goods can readily be marked at any of the percentages common to business:

To make a profit of 16 2-3 per cent., add 20 per cent. to cost.

To make a profit of 20 per cent., add 25 per cent. to cost.

To make a profit of 25 per cent., add 33 1-3 per cent. to cost.

To make a profit of 33 1-3 per cent., add 50 per cent. to cost.

To make a profit of 50 per cent., add 100 per cent. to cost.

The highest profit that really can be made in business is 100 per cent., and this can be done only when you get something for nothing, and having sold it for something, the entire transaction is profit; in all other cases the sale being 100 per cent. and the cost something, the profit is less than 100 per cent. An article costing 10 cents and selling for $1 pays a profit of 90 per cent. of the selling price, 10 per cent. being the cost and 90 per cent. the profit. Figure your profits according to the above table, then try if you can to figure yourself out of anything.

How Rapid Turnover Counts.

As to the methods of proof submitted by me in these articles from time to time, I have this to say: Captious criticism doesn't amount to anything. The man who is constantly looking for an argument and trying to frame up impossibilities to prove his case only injures himself. I can say for those rules that I recently made an inventory, amounting to over $3,000,000 and consisting of over 40,000 hard extensions, and 1,500 pages of footing. I think I can figure as quickly and correctly as the next man, but I was glad to use those rules, proving any extension or footing I was afraid of by their aid. I was rewarded for my work by a fine present from the general manager, in addition to my regular compensation, he saying at the same time, "I would not believe it possible, but I have had your work carefully examined and have not found a single error." Of course, I was trying to use those rules, not abuse them.

The accompanying table shows the immense advantage of quick sales. It is based on the careers of four $100 bills invested in business for the term of ten years at a profit of 20 per cent., and turned over with their respective profits at various intervals. The difference in earning power is almost incredible, but can be easily verified

$100 turned every two years at 20 per cent., profit for ten years will total $305.18.

$100 turned every year at 20 per cent., profit for ten years will total $931.

$100 turned every six months at 20 per cent., profit for ten years will total $10,844.16.

$100 turned every three months at 20 per cent., profit for ten years will total $891,752.50.

ASBESTOS HORSESHOES.

Visitors to the volcano of Kilauea, on the island of Hawaii, generally ride on horseback, and in crossing what is known as the "pit," the horses suffer much from the great heat. The earth is so hot that the hoofs of the horses are not infrequently scorched. As some protection became necessary, a clever blacksmith in Honolulu has recently devised a very successful method by which asbestos may be used. The idea is to provide the hoofs of the horses with an asbestos covering much after the fashion of the outer shield of iron-studded leather or canvas over the automobile tires. These hoof shields may be put on and removed at pleasure.

NO ONE WOULD BUY.

A farmer, living in the vicinity of Spring Valley, Minn., recently endeavored to market a load of produce in that town. He was known among the merchants as a strong patron of one of the Chicago supply houses. He worked hard until night endeavoring to find a buyer for his produce. No one seemed disposed to buy. Finally, one of the business men suggested that he send it to the Chicago supply house, giving the name of the supply house. The farmer went home mad, but he saw the point.

At Harmony, Minn., the merchants have an intelligent understanding regarding certain farmers who buy nearly everything they use of the retail catalogue houses. Some of the farmers who are in this class have aready been told that the Chicago supply houses should take their butter and eggs and other produce, as long as the supply houses get the most of their money. To make this sound like the real goods, the local merchants have refused to take the produce.

ANNOUNCEMENT!

We have just mailed to the trade our

Supplementary Catalogue (No. 59B)

Showing all recent additions to our line of

"GOOD CHEER"

Stoves, Ranges and Furnaces

If we are not already represented in your town, we shall be pleased to forward you a copy on request, and, as the stove selling season is close at hand, it will be to your advantage to

WRITE US NOW

We carry a full assortment of all lines at our Winnipeg Branch, and can also supply you through our distributing agencies at Calgary, Alta., Nelson and Vancouver, B.C.

The JAMES STEWART MFG. CO.

LIMITED

Western Warehouse:
Foot of James St., WINNIPEG, MAN.

WOODSTOCK, ONT.

"Good Cheer" Ranges

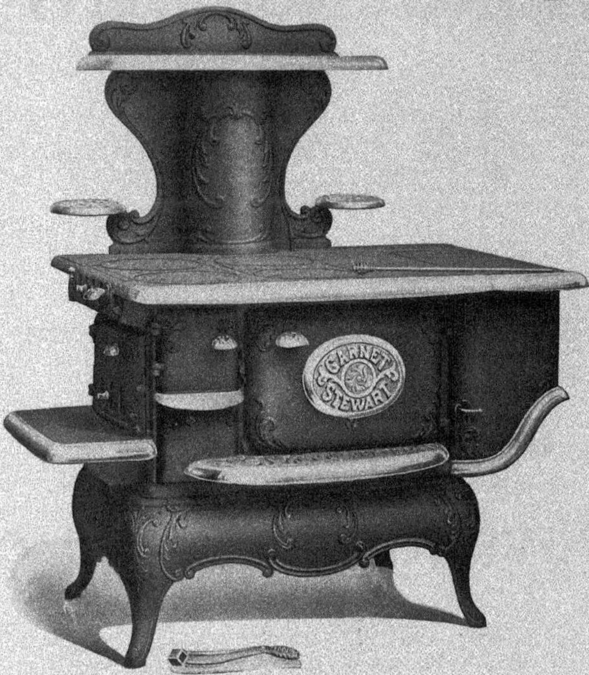

THE "GARNET"

Just the range to immediately catch the eye of the housekeeper. Entire absence of that heavy ornamentation usually found on cast ranges. Nice smooth castings, so easy to keep bright and clean, and the nickel parts all removable for burnishing or when the range is being used on washday, etc. Fine square oven 18 x 18 x 11. Firebox is a model one, with all cast linings including our long-lasting coal back. Duplex grates are removable without disturbing linings or waterfront.

It is a splendid operator and will be found just right in every way.

The JAMES STEWART MFG. CO.,
LIMITED

Western Warehouse:
Foot of James St., WINNIPEG, MAN.

WOODSTOCK, ONT.

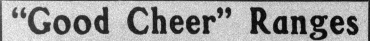

The "Good Cheer" Art

Undoubtedly the most popular Heater in the market.

Beautifully proportioned and the design so artistic as to make a most appropriate furnishing in any room or hall.

Then, as to operating qualities. Did you ever hear of a user of a "Good Cheer" Art being otherwise than delighted with his purchase? And what better advertiser could you wish for than a well satisfied customer?

"Good Cheer" Arts have THE sale throughout the Dominion, and as this well and favorably known line ***costs you no more than the average base-burner of similar size and class,*** now is the time to secure the agency, get in your stock and do the business in the heater line in your locality.

Firepot is removable through side doors without moving a bolt. Ransom duplex shaking ring grate. Good-sized ash-pan, and every stove a double heater.

Two sizes, with or without oven.

No. 44—13-in. firepot.
" 55—14-in. "

THE JAMES STEWART MFG. CO., LIMITED, WOODSTOCK, ONT.

Western Warehouse, James Street, Winnipeg, Man.

"GOOD CHEER" OAKS

FOR COAL OR WOOD

NO. 100 SERIES.

An artistically nickel trimmed line in four sizes.

HEAVY POLISHED STEEL BODY.
GOOD SIZED BASE SECTION.
LARGE ASH PAN.
HEAVY FIREPOT.
DOUBLE FEED DOORS.
SHAKING DRAW CENTRE GRATE.
MAGAZINE, IF REQUIRED, FOR THE THREE LARGER SIZES.

No. 112 Dia. of body 11 inches. No. 116 Dia. of body 15 inches.
No. 114 Dia. of body 13 inches. No. 118 Dia. of body 17 inches.

NO. 50 SERIES.

Single and Double Heaters

Here's a Trade Winner You Should Have!

A handsome Oak with a powerful **Double Heating Feature** which makes the stove sell like hot cakes. Only placed on the market late last season, but wherever sold dealers have placed large orders for this year's supplies.

Body of stove is of heavy polished steel, handsomely nickel trimmed. Large feed door with good sized mica lights, making it a very cheerful stove, while screw dampers in both upper and lower doors make it easy to regulate the fire.

A shaking ring draw centre grate is provided, as also a good sized ash pan.

Made in two sizes. No. 54 with 13½-in. pot and No. 56 with 15½-in. pot, either single or double heater. Feeders for both sizes. The price is right, too.

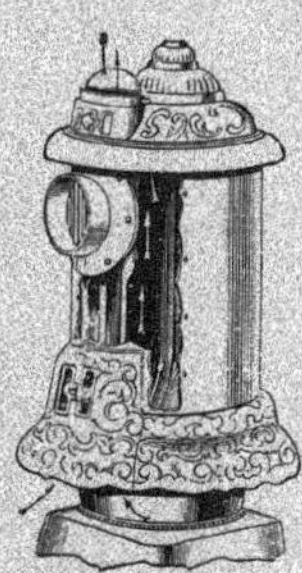

SECTIONAL VIEW

The JAMES STEWART MFG. CO.

LIMITED

Western Branch: Foot of James Street, WINNIPEG, MAN.

WOODSTOCK, ONT.

"Good Cheer" Steel Cooks

FOR COAL OR WOOD

The Bright

No. 8....4 8-in. holes, 14 x 19 x 10 oven.
No. 9....4 9-in. holes, 16 x 19 x 10 oven.

Contact galvanized iron or copper tank.

Flat shaking and dumping grate.

The Pearl

Polished Steel Body.

No. 9 ....4 9-in. holes; 16 x 19 x 12 oven.

Contact galvanized iron or copper tank, either steel-incased or as above. Supplied on feet if required, and with either high shelf, high closet or water-front.

Fitted with Duplex Coal Grates.

The Pride

Polished Steel Body.

No. 9....4 9-in. holes; 16 x 19 x 12 oven.

Contact galvanized iron or copper tank, high shelf or high closet.

Fitted with Duplex Coal Grates.

The JAMES STEWART MFG. CO.
LIMITED

Western Branch:
Foot of James St., WINNIPEG, MAN.

WOODSTOCK, ONT.

"Good Cheer" Furnaces

No. 100 Series

The class of furnace which means every satisfaction to the dealer in selling and installing, and to the householder in operating.

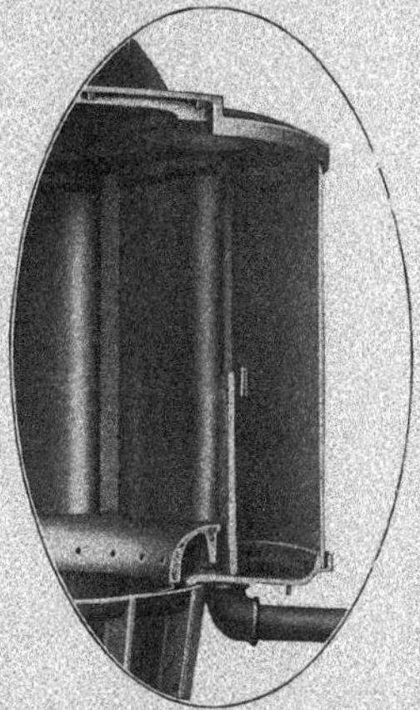

A Few of Its Many Good Points:

Cast iron tube radiating principle. Tubes surround fire chamber and form most powerful heat producers. No sheet steel exposed to direct action of fire, being used only in outside jacket of radiator.

Genuine cup joints, positively gas and dust tight.

Air-blast ring, distributing necessary oxygen to surface of fire and assuring perfect combustion.

Easiest cleaned flues of any furnace made; can be cleaned in five minutes with fire going.

Large feed doors with smoke shield.

Direct and automatic gas dampers.

Sectional cup-jointed fire-pot.

Revolving grates for hard or soft coal. Also grate for wood burning.

A good solid weight furnace of great heating power, for those who want the best.

The James Stewart Mfg. Co., Limited, - Woodstock, Ont.

WESTERN BRANCH: FOOT OF JAMES STREET, WINNIPEG, MAN.

If you have a sign over your door, you are an advertiser. The sign is intended to advertise your goods to passers-by.

An advertisement in Hardware and Metal is just so many thousands of signs spread over the Dominion of Canada.

If you want to secure a clerk or a tinsmith, if you want to buy or sell a business, if you have a counter or a show case for sale, or if you want to buy a set of tinsmith's tools or a second-hand typewriter—in fact, if you have any proposition which you want placed before hardwaremen, stove and tinware dealers in Canada, state your wants briefly on our "want ad." page and your message will be put on a sign board which will be seen and read by practically every dealer and manufacturer in the hardware, stove and tinware lines in Canada.

Pig Iron Industry at Port Arthur

New Blast Furnace, Ore Roasting Kilns and Coking Plant of the Atikokan Iron Company

Another link has been added to the extensive chain of metal industries in Canada with the completion of the works of the Atikokan Iron Company, at Port Arthur, Ont. The past few years have seen a marvellous development in Canadian metal industries, the latest and one of the most important of which is the Atikokan plant, where operations are shortly to commence on an extensive scale for the production of high class pig iron. The market for the output is already assured. The establishment of this industry is timely and the location well chosen.

Atikokan Iron Co., Port Arthur, Ont.— Coke Ovens.

The entire plant of the Atikokan Company consists of coal docks, a blast furnace, ore roasting kilns and coke ovens, which are situated on the water front just west of the Canadian Northern elevators, where they have a water frontage of 1,800 feet. The coal dock and unloader plant represents the best known equipment in this line, while the blast furnace and ore roasters are the most modern of their kind.

For some time past it has been known that enormous deposits of magnetic iron ore existed in Northern Ontario, both north and west of Port Arthur, but the Atikokan Company is the first to take active steps towards the utilization of this enormous natural resource. Property owned by the company is known to contain at least 5,000,000 tons. Some of this ore can be used in the blast furnace without roasting, while that from other veins contains sulphur to such an extent that it must be roasted before being smelted.

Figs. 1 and 2 show the general arrangement of the plant, including the power house, furnaces and beehive coke ovens.

The Roberts-Cooper stoves, 18 feet in diameter by 70 feet in height, constitute the hot blast equipment. The power house equipment includes four water tube boilers of 200 h.p., manufactured by the Canada Foundry Co. The blowing engine is a disconnected cross compound condensing horizontal engine manufactured by the Southwark Foundry & Machine Co., of Philadelphia. The design of the engines permits of operating either the high or low pressure side as an independent engine. Duplicate water circulating and boiler feed pump are provided.

The furnace is designed to produce 100 tons of ore a day, it being proposed to increase the capacity to 200 tons. The construction is the same as that at the Hamilton Steel & Iron Company's furnaces.

Experts on the subject advise the roasting of all magnetic ores, whether they contain sulphur or not, so it was decided by this company to provide roasting kilns capable of handling all the ore supply required by the furnace, using the waste gas of the furnace as fuel. The gases are consumed in a combustion chamber, the products of combustion passing through the ore to the chimney chamber, whence they are drawn by an exhaust fan delivering into a chimney. The use of the exhaust fan for creating the draft permits of varying the supply of air according as the ore to be roasted is coarse or fine, the latter requiring a greater draft than the former. In the design adopted at Port Arthur the ore is elevated to the top of the roaster by means of a skip hoist, discharged into an automatic railroad truck, and thence delivered in bins on top of the roaster. The ore "self feeds" from the bins into the roasting chambers as the roasted ore is drawn out of the roaster at the bottom. Bins are also provided at the bottom of each roasting chamber, the ore being delivered therefrom into electrically operated stock transfer cars, which in turn deliver the roasted ore to the furnace skip hoist. The bottom of the roaster is also designed so as to permit the delivery of roasted ore into ordinary railroad cars for shipment. The roaster and equipment are arranged to permit additional roasting chambers being erected later, the hoisting equipment being of ample capacity to meet possible requirements.

Atikokan Iron Co.—Power Plant, Blast Furnace and Roasting Kilns.

Atikokan Iron Co.—Roasting Kiln and Blast Furnace.

Beehive coke ovens to the number of

100 have been built, each 12 feet 3 inches in diameter. The decision to put in this plant was due to the difficulties and expense of transporting coke the great distance necessary to supply the plant. The coal is brought to the docks by vessels and delivered automatically by electrically operated transfer cars.

The capacity of the stock bins is: Coke, 200 tons; ore, 500 tons; limestone, 175 tons. This includes one large coke bin discharging directly into the furnace skip cars and fuel bins to discharge into an electrically operated transfer weigh car, which delivers the material to the skip hoist, and also transports the roasted ore from the roaster to the skip hoist.

Part of the output of this plant will be used in Fort William by the Canadian Iron & Foundry Co., for the manufacture of car wheels, iron pipe and iron castings.

ELECTRIC POWER FROM MOUTH OF COAL MINE

The new power plant of the Maritime Coal & Railway Co. at Chignecto, Mines, N.S., is unique in being the first adoption of the prediction made some years ago by Thomas A. Edison, when in England, that the ultimate solution of the problem of cheap power was the utilization of the waste products of the coal mine and the situation of the power house at the mouth of the mine, thus eliminating the necessity of expensive transportation of coal over long distances.

In developing power, the first great saving is in the use of fuel which would otherwise go to waste. As there are no vertical shafts at the mines, the coal is hauled up a steep slope in cars, which are pulled by a cable. The cars hold fifteen hundred pounds each, and there are six cars to a train. When the cars reach the surface they continue on their journey up a slope to the top of the bank head. Here a complete system of tracks and switches places each car exactly where it is wanted. Its contents are automatically dumped into a series of screens and hoppers, where the material of commercial value is separated from the waste screenings or "culm." The good coal works its way to the railway cars waiting to receive it. The culm is dumped on an endless conveyer, which carries it to the power-house, 200 feet away. Here it is deposited in a huge bin above the boilers, which has a capacity of 85 tons. From this bin chutes, through which the coal is carried, lead to Jones' Underfeed Stokers, which automatically feed the fires. The boilers are four 200 horse-power return tubular, arranged in banks of two each. They supply steam at a pressure of 150 lbs., and the arrangement is such that any two can supply steam to the engine. A Cochrane open type feed water heater is provided, and the induced draft system, consisting of two blowers, complete, with engines, supplied by the Canadian Buffalo Forge Co., has been installed. This does away with a high smoke stack, one only twenty-five feet high being used.

Water is supplied to the boilers by a pond made by the damming of a small stream. The pond is 1,000 feet away from the power house, and the water is conveyed by gravity to a pit underneath the boiler room, whence feed water and cooling water for the condenser is taken.

The power house is built of brick and is 75 by 50 feet, and is divided into the boiler and engine rooms. The engine room is 30 feet wide and runs parallel to the boiler room.

The engine is a 17-inch and 33-inch x 16-inch 750 h.p. Robb Armstrong centre crank cross compound vertical condensing engine, of the English high-speed type, using the pressure oiling system, which carries oil to all bearings continuously at a pressure of 10 to 15 pounds. This is a new departure for Canadian engine builders, and has been successfully carried out by the Robb Engineering Co., Amherst, N.S., for the past year, a great many installations of this type of engine having been made throughout Canada by the company.

Directly connected to the engine is a 500 k.w. 11,000 volt three-phase alternator, built by the Canadian Westinghouse Co., Hamilton, Ont. On the same shaft as the generator is the small direct current 125 volt exciter, for exciting the generator fields, making a very compact and substantial unit.

Power From the Mine.

Provision has been made for a future unit when the demand for power warrants its installation.

At the end of the engine room the switchboard is placed, which, at present, consists of a generator panel containing an 11,000 volt oil switch, with volt-meter, ammeters and wattmeters. There is also provided a synchronizing outfit to be used when the second unit is installed. The line is also fused with a set of high voltage fuse switches, and the whole equipment is protected by lightning arresters and choke coils of the standard type, manufactured by the Canadian Westinghouse Company.

The transmission line extends to Amherst, a distance of 6½ miles. Several long spans have been necessary in crossing the tidal rivers, varying from 400 to 700 feet. To cross these two poles were set at either end, guyed by steel ropes. The transmission wires are No. 4 B & S gauge hard drawn copper wire supported on 15,000 volt glass insulators. In general, the poles are 125 feet apart. Provision has been made for running a second line, as room for three more wires has been left on the cross arms.

At Amherst, a brick sub-station has been erected two storeys high. The upper portion is occupied by the three Westinghouse oil-cooled, self-contained transformers, which are used to reduce the voltage from 11,000 volts to 2,400 volts. Here also are located the fuses and lightning arresters. The secondary circuits are carried to the ground floor, where an oil switch is installed. From the sub-station the power is distributed locally to the various consumers in Amherst and vicinity at a pressure of 2,400 volts. To date, 400 horse power in motors has been installed, and it is expected that more will be in use in a short time. It is hoped that this will be the means of bringing many more industries to Amherst and vicinity.

Power House — Bank Head — Machine Shop — Miners' Cottages

Panoramic View of the Maritime Coal, Railway & Power Company's Plant at Chignecto, N.S.

HARDWARE TRADE GOSSIP

Quebec.

Lagare, Edouard & Bro., hardware merchants, Roxton Falls, Que., have assigned.

J. A. Marcotte has been appointed curator for Monette, Alphonse & Co., graniteware dealers, Montreal.

Mr. Fisher, representing the Reading Hardware Co., Reading, Pa., is calling on the trade in Montreal this week.

Mr. Stewart, of Well & Emmerson, Port Arthur, called in Montreal this week on his way home from England.

J. H. Plummer, president of the Dominion Iron & Steel Co., returned to Montreal this week from New York City.

Philias Lapierre and Joseph Peloquin, Montreal, have been registered to conduct a hardware business. The new firm will be styled Lapierre & Peloquin.

Mr. McAvitty, of St. John, was in Montreal this week on his way east from Kingston, where he was visiting his son, who is in attendance at the Royal Military College.

C. L. Devitt representative of Lewis Bros., Montreal, is confined to the general hospital in Ottawa, with typhoid fever. J. A. Jackson, of the sample department of Lewis Bros., is taking his place.

Amongst those who called on the trade in Montreal this week were: William Coslett, Fort William; P. Phoenix, Granby; John MacDonald, Williamstown; W. B. Skinner, Kingston; Charles Archibald, New York City.

Ontario.

G. A. McDonald, plumber, College street, Toronto, has assigned.

Fred Hawes, hardware merchant, Listowel, is ill, suffering from an attack of typhoid fever.

G. Membray has established a hardware store on Main street, Dovercourt, a suburb of Toronto.

O. B. Henry, formerly a hardware merchant, of Drayton, was a visitor in Toronto during the week.

The Canadian Wholesale Hardware Association is to hold its annual convention in Montreal in about a fortnight's time.

The Latchford Hardware Co., Latchford, Ont., are seeking a winding-up order for a big mining corporation unable to pay its debts.

C. A. Fletcher, of the A. L. Swett Iron Works, manufacturers of hardware and plumbing specialties, Medina, N.Y., was in Toronto on Thursday.

Simon C. Ray, foreman for D. W. Howden & Co., London, had a narrow escape from suffocation on Tuesday while constructing a cesspool and drain at the rear of his dwelling.

Fred. P. Hall, traveling salesman for Rice Lewis & Son, wholesale hardware merchants, Toronto, has severed his connection with that firm. J. C. Northcote is now on the road covering Mr. Hall's territory in Western Ontario.

John Wise, a workman at McClary's, London, met with a serious accident on Tuesday. Wise backed into the elevator shaft, and fell 25 feet, with the truck on top of him. His arm was split open from the elbow down, and the bones broken, besides which he was otherwise crushed, and received internal injuries.

R. J. Gardiner, Chatham, who has made quite a success of his automobile enterprise in that city, has secured the Canadian manufacturing rights for the Ewer pipe wrench, and is making preparations to establish a plant for the manufacture of the wrench at Chatham. The new wrench is constructed on entirely new lines, and contains features which will undoubtedly make it one of the most popular sellers on the market.

Western Canada.

The hardware firm of Glenwright Bros., Winnipeg, have assigned to C. H. Newton.

John B. Persse, of Tees & Persse, Winnipeg, has been elected president of the Winnipeg Board of Trade.

E. Brown & Co., dealers in builders' supplies, Regina, Alta., have been succeeded by the Central Supply Co.

Great Britain.

F. T. Murray, joint managing director of Messrs. Webley & Scott, London, Eng., has arranged to undertake a lengthy tour in Canada in the interests of his firm. Mr. Murray leaves Liverpool on Oct. 11 for Canada.

United States.

The Stanley Works Company, New Britain, Conn., have received permission to double their capital stock of $1,500,000. Permission has been granted also to transmit electricity for power from their plant on the banks of the Housatonic River, near Kent, by high tension cables to their plant in New Britain.

The Yale & Towne Manufacturing Company, Stamford, Conn., and New York City, has filed a certificate increasing their capital stock from $1,000,000 to $1,500,000. The company were authorized by the recent Connecticut Legislature to increase their capital to $5,000,000, and further issues of new capital may be made from time to time.

Having sold the valuable property, which they have occupied for over a third of a century, at 43 to 45 Chambers street, New York City, the Russell & Erwin Manufacturing Company, will shortly move to a new building now in course of construction at 94-98 Lafayette street. This building will be of steel cage construction, brick and stone, fireproof, modern in every way.

THE HARDWARE REVIEW.

Vol. 1, No. 1, of the Hardware Review has been received, it being a new monthly, published in the interests of the retail hardware trade, by the Pentz Publishing Company, J. W. Pentz and Henry Hopkins, formerly manager and editor of "Hardware," New York, holding the same positions on the new paper. The first issue is gotten out in good style, 50 pages and special cover, the leading feature being an article on "New Britain, the Hardware City." The first issue gives indication of a successful career for the "Review."

WELL-KNOWN MERCHANT DEAD.

The death occured last Thursday of Wesley Bingham, at his home, 472 Manning avenue, Toronto. The late Mr. Bingham was born at Glanford, Ont., in 1835. From 1862 to 1878 he carried on a hardware business in Orillia, and during part of that time he was Orillia's postmaster. Subsequently he was in business in Wallaceburg and Lindsay, and up to the commencement of the illness which resulted in his death he was a director and traveling representative of the Western Foundry Company.

Mr. Bingham was twice married. By his first wife, who pre-deceased him by 28 years, he leaves one son and six daughters. Interment took place at Orillia.

SHORT-LENGTH BINDER TWINE.

Joseph L. Haycock, Dominion inspector of binder twine, has within a short time, collected fines aggregating $3,600 from firms selling twine short measure. In one case, however, an American company, which was fined $1,475, had, in reality, to pay $12,000 for its crooked business methods. The inspector in this case seized 1,180 balls of twine, which contained 500 feet instead of the 600 feet that was stamped on the package, imposing a fine of one dollar a ball. The company bought the twine back at 25 cents each on the conditions that they were re-tagged the correct length. The representative of the firm having admitted that 250 tons of twine had been sold short length, was ordered to notify all dealers to whom it was sold to have each ball re-tagged. This cost $5 per ton, besides reducing the value of the twine by $40 per ton, or $10,000.

BELANGER'S STOVE CATALOGUE.

A catalogue consisting of 62 pages and cover has been issued by A. Belanger, Montmagny, Que., who for forty years has been engaged in manufacturing stoves, ploughs, harrows, ornamental iron work, washing machines, etc. The catalogue is well illustrated, showing all the various lines of "Prince Crawford" ranges and stoves and other products, full descriptions being given in both French and English. Some novel forms of cook stoves with high ovens are shown, as well as several styles of double decked heaters, peculiar to the district in which they are made. Copies of the catalogue can be secured by mentioning Hardware and Metal.

HARDWARE AND METAL

Established 1888

The MacLean Publishing Co.
Limited

JOHN BAYNE MACLEAN - President

Publishers of Trade Newspapers which circulate in the Provinces of British Columbia, Alberta, Saskatchewan, Manitoba, Ontario, Quebec, Nova Scotia, New Brunswick, P.E. Island and Newfoundland.

OFFICES:

MONTREAL, ... 232 McGill Street
Telephone Main 1255
TORONTO . . . . . . 10 Front Street East
Telephones Main 2701 and 2702
WINNIPEG, ... 511 Union Bank Building
Telephone 3726
LONDON, ENG. . . . . . 88 Fleet Street, E.C.
J. Meredith McKim
Telephone, Central 12960

BRANCHES:

CHICAGO, ILL. . . . . . 1001 Teutonic Bldg
J. Roland Kay
ST. JOHN, N.B. . . . . No. 7 Market Wharf
VANCOUVER, B.C. . . . . Geo. S. B. Perry
PARIS, FRANCE - Agence Havas, 8 Place de la Bourse
MANCHESTER, ENG. . . . . 92 Market Street
ZURICH, SWITZERLAND . . . . Louis Wolf
Orell Fussli & Co.

Subscription, Canada and United States, $2.00
Great Britain, 8s. 6d., elsewhere - 12s

Published every Saturday

APPORTIONING AN "OPENING" ORDER.

An agricultural implement firm in a Northwest town of less than 1,000 population, has written Hardware and Metal stating that they are just about to start in the hardware business, and have about $4,000 to invest. They ask the editor for suggestions as to the best apportionment of their capital among the various lines of hardware, admitting that they are "green" at the business.

The question asked is a hard one to answer, as, while it is known that the town has between 500 and 1,000 population, it is uncertain whether the farming population is increasing or whether the district is already built up. If new people are constantly coming in with money to spend for building, it will be readily understood that the apportionment of money spent for stock such as paint, glass, builders' hardware, etc., would be much greater than if the district is already populated. Again, there might be a large demand for household furnishings, harvest tools and similar lines, or there might not, the demand for these lines differing according to the class of people and section of the country.

It will be seen that it is practically impossible for the editor to give satisfactory advice in response to the question asked. Roughly speaking, he would be safe in advising that about 25 per cent. of the firm's capital be spent in such necessary side lines as stoves and paints. To divide the remaining 75 per cent., however, the advice of a traveling salesman for a reputable wholesale house should be taken. Such a man knows the different demands of dealers in the various sections of the country, and his advice as to the proportion to be spent in builders' hardware, harvest tools, house furnishings, cutlery, glass, etc., would be worth considering.

The prospective merchants being new to the business, can also receive valuable advice as to the selling value of the goods, and the proper methods of conducting their business. They will, undoubtedly, find the retail hardware business much different in its operations from an agricultural implement business. The latter is largely a business done on long-term notes, practically a credit business, while a retail hardware business in western Canada should be conducted on as near the cash system as possible.

Whether the present is an opportune time to establish a business, is, of course, a question to be decided by the parties concerned. They know their own financial resources. They have probably studied their field carefully, and see that their town is sufficiently large to warrant the establishment of another business. They must consider, however, the financial situation existing this fall, money being tight, and wholesale houses very cautious in their business transactions. Buying for fall and winter has been done in a very careful manner by western hardwaremen, all feeling that the present is a critical time, a reaction being felt from the over-enthusiasm of the past three or four years. All these points must be carefully considered by the intending investors before putting their money into stock.

The firm in question, however, appears to be in a very satisfactory position, with several thousand dollars to expend, and if their location is well chosen, they can afford to go ahead, regardless of the money stringency being felt throughout the country.

AMERICAN HORSESHOES FOR BRITISH ARMY.

There seems to be some misunderstanding regarding the purchase of American horseshoes by the British War Department. The Trade and Commerce Department of the Canadian Government write that they knew nothing of the order being placed, while the Portland Rolling Mills, Limited, St. John, state that they "had a communication from the Department of Trade and Commerce of the Dominion Government offering us this business a number of months ago."

The latter, in a letter to Hardware and Metal, say:

Editor Hardware and Metal: I have yours of yesterday's date in re contract for horseshoes awarded by the War Department to American manufacturers, and asking if such work as notifying Canadian manufacturers of such work as this does not come within the scope of this department.

In reply I beg to state that this department had no knowledge of any intention on the part of the War Department asking for such supplies. In fact, the War Department has a way of its own of doing its business, and when a question was asked, I believe on the floor of the House, in connection with this identical lot, why they purchased them from American firms instead of from home manufacturers, the curt reply was that they bought their horseshoes wherever they could get them cheapest.

The probability is that their wants were not advertised; but they presumably have a list of contractors and they simply write to them, asking for figures. It is only in two or three instances that tenders have been asked for in Canada for supplies for the War Office.

F. B. Parmelee, Deputy Minister.

Ottawa, Sept. 26.

The "Ironmonger" reports the receipt of three consignments of 417, 808 and 435 kegs of horseshoes from the United States for the British Army, on the total contract of 108,000 pairs, and adds:

"It is said that several British manufacturers tendered for the contract at low prices, but were underbid by the Bryden Horseshoe Co., of Catasaugua, who secured the order. According to J. F. Richardson, a member of the Walsall firm who introduced the pattern of shoe now used in the army, it has hitherto been the rule of the War Office to divide the contract among firms in this country, and his works have been compelled to close the department usually set apart for Government work. In the past the War Office specifications have always called for tenders at per cwt., but on this occasion the time-honored practice has been abandoned in favor of a tender at per pair. Possibly the clue to the transfer of the order to America may be sought in this change in the specification. With reference to the suggestion that the American works are losing money on the order and have accepted it only because they belong to a trust which makes them a compensating allowance out of a pool, the agents for the Bryden Co. deny the assertion completely."

"Mr. Haldane, when cross-examined on the subject in the House of Commons," continues the Ironmonger, "was altogether unrepentant, his answer being that it was his business to administer the army as economically and efficiently as he could, and that the American horsehoes were the best and cheapest he could get for the particular object in view. If this answer correctly represents the facts, there does not seem to be much room for criticism, although it may be argued that it would be worth our while to pay more money if by doing so we could benefit the British employer and workman. Those who take this line of argument, however, should be prepared to say what is the limit of price-difference below which work should be reserved for home factories (unless it be contended that all goods bought for the public service should be of British manufacture no matter how much the price may thereby be enhanced), and secondly, they should be prepared to prove that any extra money the British public may be called upon to pay for indulging in the luxury of excluding foreign goods will be distributed in just proportion between all persons concerned in the manufacture of those goods in this country."

LAW SUITS COSTLY.

It is announced that the cost of the case before the courts to settle the dispute between the steel and coal companies at Sydney amounted to about $60,000, and the end is not yet, as the decision in favor of the Steel Company provides for an arbitration award on certain points, and even this settlement may be appealed and the case carried to higher courts. For every hour the court sat the case is said to have already cost $1,000.

The case was purely a business disagreement, and in the light of events it would appear that the most sensible settlement would have been arrived at by an arbitration. The cost would have been much less and the results more satisfactory.

MERIT OR POLITICAL PULL.

The postmastership of Toronto is vacant, and there is a scramble among politicians for the place. It has hitherto been given as a reward for political service, and under Sir William Mulock and some of the present permanent officials, the post office has been used to a more or less extent as a political machine. The consequence is that the entire service is in a most inefficient state which will tax all the energies of the present Postmaster-General to put in a satisfactory condition.

Business men who have influence with the Government should urge that this appointment should go as a promotion to some of the permanent officials and not to a politician. The United States Postmaster-General has recently set an excellent example in this respect, and he had harder prejudices to overcome than the Postmaster-General of Canada will have. In the States nearly all public appointments are given as the reward for political service, and when a change of Government takes place, it is followed by a complete reorganization of the entire public service. The New York postmastership is one of the most important in the world, has large salary and perquisites, yet it has been given to one of the permanent officials, a man who began in the post office as a postman over thirty years ago.

The efficiency of the service demands that a similar policy should be followed in Toronto. There are, no doubt, many young men among even the postmen who could fill the position with greater success than any politician who can be appointed. The efficiency of the service demands some plan of steady promotion of deserving men, the men who do their work better than their fellows. If a country postmaster or postman show ability, they should be given the first vacancies in the higher ranks. In this way a young man may begin in a country store post office and rise step by step to be postmaster in one of the big cities and end as Deputy or even Postmaster-General. This is the right system, and when it comes we will have a much more efficient service.

A PECULIAR METAL MARKET.

Conditions in the metal market are puzzling. Manufacturers of many lines have been building new plants or enlarging existing buildings in order to supply the demand for goods. This demand still continues, manufacturers of firearms, wire nails, cut nails, iron pipe, screws, bolts and nuts, and many other lines being unable to supply goods in anything like the quantity required for prompt shipment. The iron pipe mills, for instance, have had to shut down for want of skelp, and some Canadian houses have had to import some sizes from the States in order to allow contracts to be filled in specified time. Premiums are freely offered for the scarce sizes by buyers, while manufacturers are willing to pay premiums for raw material to keep their plants in operation.

All of this is very bullish, but there is another side to the story. The drop in the copper market has, of course, affected all finished brass and copper articles, the reductions going gradually into force. The slight weakening in the iron market has encouraged buyers to look for a similar reduction in iron and steel products, a reduction which seems unlikely to materialize. Most manufacturers contracted for iron freely a year or so ago and did not have to pay the long prices asked this spring. On many lines, in consequence, no advances were made and a drop cannot, therefore, be reasonably looked for. The stoppage in building may have an effect on the demand for manufactured articles, but there will have to be a very marked falling off to warrant a reduction in prices on tools and similar articles.

The advance on boilers and radiation this week is another peculiar feature of the market. Notwithstanding the exceedingly brisk demand for these lines during the past year or two prices have been slashed to rock bottom figures. Now when the metal market is considered by many to be on the toboggan slide and building at a standstill, prices are advanced. All of this would be hard to understand if it were not known that the manufacturers of radiation are loaded up with business and may be unable to complete all orders this winter. The same reason which compelled stove founders to advance prices a year ago, and again this fall, has at last forced the heating apparatus men to advance their product to a profitable figure.

As stated in the first paragraph, the metal market has many puzzling features at present, and it is a brave man who essays to act the role of the prophet.

ENGLISH RED TAPE.

As an example of English red tape, the following experience of a Canadian manufacturer would be hard to duplicate. He writes:

"The English methods of getting after the pennies makes me quite certain that they have descended from the lost ten tribes. We have had some very amusing experiences of this in our office, especially when they get after a penny and spend several to get it. We were paying a small item to an English firm, with whom we do a large business, and had asked for a sample of an article, which they sent, along with a bill at the same time for two shillings two pence, which we paid by a two-shilling postal note, but omitted to affix the two odd pennies. The firm sent back a receipt for the two shillings only on a stamped form, which cost a penny, and was unnecessary in this case, and further asked us to remit the two pennies short. We did this, of course, and they sent a receipt, thereby spending two pence to get two pence. In order to save postage, they sent the bill in an open envelope inside a circular."

BAMBOO WATER PIPES.

The Japanese of British Columbia have recently begun the cultivation of bamboo on a large scale. The experiment is proving quite successful. Bamboo roots are being brought in large quantities from Japan and transplanted in British soil.

The cultivation of bamboo is very profitable in Japan, the returns varying from $20 to $90 per acre. With a successful cultivation of the plant in British Columbia, the Japanese hope to work up a big trade in bamboo furniture. Willow is the best that can be secured in this country, but bamboo is far superior to the willow that is now used.

Another industry which the Japanese will endeavor to work up will be the use of bamboo for water pipes. In Japan bamboo is largely used as water pipes, and the growers see no reason why it could not be used in this country.

Markets and Correspondence

(For detailed prices see Current Market Quotations, page 74.)

MARKETS IN BRIEF.

Montreal.

Antimony—Much firmer.
Copper—Unsteady.
Lead—Very firm.
Linseed Oil—Steady.
Old Materials—Weak.
Tin—Declined.
Turpentine—Weak.
White Lead—Unchanged.

Toronto.

Boilers and Radiators—Advanced.
Ingot Tin—One cent lower.
Casting Copper—Now quoted at 17c.
Turpentine—Steadier.
Linseed Oil—Firmer.
White Lead—Firmer.
Copper Rivets and Burrs—Lower discounts.
Lawn Mowers—Next year's prices out.
Nails—Firmer.

MONTREAL HARDWARE MARKETS

Montreal, October 4.—The fall season is now in full swing and orders for all lines of hardware are being received in abundance. Retailers throughout the eastern provinces are showing their desire to be well stocked up in the seasonable lines as early as possible, and thereby preclude annoying delays throughout the season—delays which are annoying to retailers and consumers alike.

To offset what would otherwise be perfectly satisfactory trade conditions there are two defects which, if not becoming worse, are not showing any signs of improvement. Scarcity of cars and scarcity of money are doing much to hamper business transactions. Wholesalers are complaining of bad collections, which cannot be attributed to tardiness on the part of retailers, for they are reliant upon the farmers for cash. The railways are still unable to effect prompt delivery of shipments. Local factories in more than one instance are seriously handicapped in their output because of delay in receiving consignments of raw material.

Screws—There is still a serious shortage of small sizes and jobbers experience a good deal of difficulty in effecting either prompt or complete shipments. It is expected and hoped that by the end of next January the factories will have caught up to the demand. The call at present is somewhat diminished. Prices remain unchanged.

Wire Goods—A strong demand exists there is no surplus supply with the makers, who are working at their utmost capacity, and prices remain firm and unchanged. Deliveries for the Northwest are being rushed before navigation closes.

Poultry Netting—The volume of business transacted this year, keeping in mind the shortness of the season, is considered very large. Next year's list is being awaited.

Sporting Goods—Retailers who have not yet gotten in their supply of hockey skates and sticks are losing no time in doing so now. Guns and ammunition are still moving very freely, and it is expected they will continue in good demand for some time yet. Prices are without change.

Stoves and Heaters—During the past month the number of orders for stoves and heaters, especially, has almost been phenomenal. Dealers have been exceedingly anxious to have their orders booked before October 1, when an advance of 5 per cent. all round was made on prices. The factories now have more orders than they can look after, and it is expected that the number of stoves sold this year will exceed that of any other year.

Building Paper —Prospects for fall and winter business are very encouraging. Little activity exists at present. Conditions remain unchanged.

Builders' Hardware—The amount of business being done is steady and does not vary much. Supplies are adequate and prices unchanged.

Cement—Conditions in the cement industry have experienced little improvement. Cars are very scarce, owing to the fact that almost every car is being requisitioned for Northwest trade, and there being little export freight, few cars are available for local traffic. Prices are very firm, with a limited supply.

Fall Goods—Cow ties, horse blankets, axes, shovels and lumbermen's supplies are experiencing a very active demand. Deliveries of orders are tardy, but supplies are adequate.

Cutlery and Silverware—There is some hesitancy in placing orders, owing to the uncertain conditions of the money market, but prospects for winter business are encouraging. The past year's trade has been very satisfactory.

TORONTO HARDWARE MARKETS.

Toronto, Oct. 4.—A steady autumn trade is being done by local jobbers and the amount of orders received for the first few days in October indicate that the month's business will be well up to jobbers' expectations. Western business continues to show a slight improvement and large quantities of heavy hardware and builders' materials are being pushed westward along the lakes before the close of navigation. In Ontario the bulk of fall and early winter goods have already gone forward. As the majority of retailers delayed placing fall orders later than usual this year—on account of the late spring and summer—sorting orders are not yet coming in to any extent. A splendid business is, however, being done in all seasonable lines and prospects are very bright for a heavy October and November business.

Screws—There is still a great scarcity in the best selling sizes, but a peculiar feature of the situation is that—despite the known scarcity—the demand is very light. As supplies in the hands of consumers are known to be very low, it looks as if the latter were engaging in hand to mouth buying in the hope that there may be a reduction from present prices.

Wire Nails—The supply is none too plentiful and prices this week are considerably firmer, as a result of the heavy demand. Some even go so far as to say that higher prices may be looked for, but the present condition of the iron markets would not seem to bear out such a prediction.

Cut Nails—A heavy demand for cut nails is making itself felt from Northern Ontario and the western provinces. The factories have had considerable difficulty in obtaining raw material and are said to be from two to three months behind the demand. Immediate delivery is out of the question and in face of the scarcity, prices are tending upwards.

Wire—Since the new prices on wire were given out last week, very little booking for future requirements has been done. As the new prices have not been out in former years till the first of the year, buyers are evidently skeptical as to the permanency of present quotations.

Copper Rivets and Burrs—Lower prices are now out on these goods, on account of the low quotations prevailing in the world's copper markets. Copper rivets, with usual proportion of copper burrs, that is to say, weight of copper burrs not exceeding 1-3 of combined weight of copper rivets and copper burrs, are now quoted at 27½ per cent. off list. Copper burrs in excess of 1-3 of combined weight of copper rivets and copper burrs are quoted at 15 per cent., and copper burrs alone are also held at 15 per cent. off list.

Lawn Mowers—Next year's prices have now been given to the trade and are approximately 10 per cent. higher than the old figures. Some heavy booking for spring delivery has already been made.

Sporting Goods—Firearms, ammunition and shooters' coats are selling well and orders are being placed for skates and hockey sticks by retailers who had neglected to place their orders earlier. Sportings goods prices are unchanged.

Fall Goods—Ash sifters, stove pipe wire, dampers, axes, and other seasonable lines are in heavy demand and in some cases repeat orders are already coming in.

Stoves and Ranges—September business was a record breaker, due, no doubt, to the fact that retailers were pre-advised of the five per cent. advance which went into effect on Oct. 1. Manufacturers now have enough business to keep them busy for months.

Copper and Brass Goods—We repeat again this week that lower prices may be expected in all goods into the manufacture of which copper or brass enters largely. Copper rivets and burrs have already fulfilled our last week's prediction and it is only reasonable to suppose that other lines will follow.

Lamps and Lanterns—There is a splendid demand for lamps and heavy October orders have already been received. The demand for lanters is abnormally light this year, due to the fact that all shrewd dealers stocked up well last year when the price was almost half as low as it is now. Prices remain unchanged.

MONTREAL METAL MARKETS.

Montreal, Oct. 4.—More satisfaction with the existent conditions of trade is being expressed by local dealers. Now that the fall months have been entered upon, firm confidence is reposed in the maintenance of increasing activity, and, although there may and will be slumps in the market, caused by speculative influences, there is no reason to apprehend any absolute stagnation for months to come. The volume of business transacted during the past week, though not large, was of sufficient magnitude to encourage more confidence in prices, and a little heavier buying may be looked for this week.

Shipments from Europe are tardy, and few Canadian dealers who have placed orders very recently will have great difficulty in securing delivery this year, as tramp steamers are much averse to coming to Montreal after the second or third week of October. Shipments from the United States also are delayed, and from now to the close of the season, premiums only will be effective in insuring prompt delivery.

Pig Iron—Further statistics published by the prominent Glasgow firm referred to in last week's issue show a still further heavy decrease in the stocks of Scotch iron. On Sept. 13, 1907, there were 166,951 tons available, and a week later, Sept. 19, there were only 157,053 tons, showing a decrease in one week of 9,898 tons. On Sept. 20, 1906, there were 582,272 tons of Scotch iron for sale. There is a strong upward tendency in the English market. English iron has been marked up 6d. this week. A little more activity is noticeable in the United States, consumers showing a willingness to do business on the readjusted basis. In the local market there has been a fair volume of business done. Consumers are pretty well stocked up for fall and winter requirements and as there is no inclination to book for delivery in the remote future, it is not expected that the amount of business being done will increase. Local prices are very firm, with an upward tendency. Prices are unchanged: Middlesboro No. 1, $21.50; No. 2, $20.50; Summerlee, $25.50.

Spelter—Is very firm and shows an upward tendency. Prices at St. Louis, Mo., have been marked up 30 cents a 100 lbs. American producers persist in curtailing their output. A little more buying has been done in the local market. Prices are unchanged.

Ingot Tin—Is still weak. An ominous decline has occurred in the English market. During the past week English prices have been actually marked down £8. Local buying is slight. Prices are easy at $44.

Lead—Has eased off considerably in Great Britain. A mark-down of 10s. occurred this week. It is probable that high premiums will have to be offered to secure deliveries before Nov. 1. Locally, lead is very firm, owing altogether to the steady demand. Prices remain unchanged.

Copper—Is still very weak, and further reductions have been made in the United States. It is said to be down to a figure now when purchases of whatever dimension can be made with comparative safety. Buyers show, however, little inclination to buy, there being a small demand for the metal. Curtailment of production by leading companies has done something towards aggravating the situation. Prices, locally, are unchanged.

Old Materials—A little more activity is noted. The demand for scrap iron is enlivening. Supplies of cast scrap are short. A general decline in prices has occurred this week: Heavy copper and wire, 12½c; light copper, 11½c; heavy red brass, 11c; heavy yellow brass, 8c; light brass, 6c; tea lead, 3½c; heavy lead, 3¾c; scrap zinc, 3½c; malleable and steel, $9.

TORONTO METAL MARKETS.

Toronto, Oct. 4.—Business in metals continues very slow, no large inquiries being received and all buying being in small quantities. At this season it is customary to expect large booking orders for spring business, but none of this has developed yet, confidence in the strength of the market being none too strong.

Antimony is showing a fair amount of strength, the latter being an indication of the result of asking what is considered a reasonable price. A few months ago antimony was at the top of the ladder, but it gradually dropped until a point was reached where buyers were willing to place orders. Then a reaction took place and prices stiffened up again.

Copper—What has happened to antimony may be copied with copper, but the public have not confidence in the reasonableness of the price asked, and buying is still done in limited quantities for immediate use. Locally the price has been shaded again, following outside reductions, we now quoting 17c for casting in Toronto. The only feature of interest in New York during the week was the inquiry through brokers for 200 tons of copper for export to China, still indicating an endeavor to create interest in this speculative position, shipments to be made at the rate of 100 tons per month. In both New York and London the tendency is still downward.

Pig Iron—No inquiries are being made and buying is practically at a standstill. Thresher manufacturers usually buy at this season, but it is understood that production during the coming year may be curtailed. If this condition exists in the implement industry, it is peculiar to that branch of trade, and due either to the smaller crop harvested or to a commendable desire on the part of manufacturers to keep production within safe bounds. In other branches of trade, opposite conditions exist. Manufacturers of stoves, boilers and radiators, for instance, finding it hard to keep pace with the demand in spite of increased factory capacity. On all these lines, advances have been made. The new discounts on radiation will be announced next week, going into effect to-day.

Ingot Tin—Further weakness has developed in the market for pig tin in Europe, especially in futures, and a sympathetic decline occurred in the local market We now quote 40 to 41 cents.

Lead and Spelter—Both of these metals are very firm with prices unchanged locally.

Old Materials—No changes are reported. Dealers' stocks are heavy and they are not buying very freely.

U. S. IRON AND METAL TRADES.

"The Iron Age," of October 3, says:

Reports of the closing down of steel works, apparently started with an object, should be received with reserve. An instance of these false statements has cropped up this week, which dealt with stoppages at Edgar Thomson and Homestead. As a matter of fact, the Edgar Thomson, Homestead and Duquesne works of the Carnegie Steel Company are in full operation.

There has been little movement in the pig iron markets. Bessemer, in the central west, is firm, and is expected to be scarce for the balance of the year, on account of scarcity of suitable ores. Foundry irons, however, are weaker, the furnaces using lake ores finding some relief in that direction. East of the Allegheny mountains the market is fairly steady, but with more than the usual pressure to market misfit iron and off grades. Southern makers, while recognizing the downward tendency, see no motive for cutting prices since they are well booked to the end of the year and the stocks are very light.

Steel is easier in the central west, and Pittsburg now quotes $28.50 for Bessemer, and $30.50 for open hearth billets. Reported sales for export are incorrect. There has, however, some business been done in steel rails for export. A part of the first installment of the new orders for the Manchurian railroads, amounting to 12,000 tons, has been taken by the Steel Corporation; the balance is still under negotiation. In all the new requirements for the Manchurian lines are about 70,000 tons.

The September bookings of fabricated material by the American Bridge Company foot up 33,000 tons, as compared with 53,000 tons in August, outside interests taking a larger proportion. It is estimated that the total tonnage placed in September is between 120,000 and 125,000 tons. During the past week a good run of contracts has been placed, the aggregate being about 15,000 tons, mostly railroad work. It is estimated that there is to be closed, in the Mississippi valley, in the next four months railroad work aggregating 30,000 tons, and that an equal amount is in sight for buildings. On the other hand, it is announced that the contract for 11,000 tons for the new Oliver building, at Pittsburg, has been postponed.

The lake shipbuilders have secured three additional boats, with one more about to be let.

Western thresher manufacturers have contracted for a large quantity of seamless tubes for the season's requirements. It is stated that the quantity was as large as it was last year.

In the wire trade the principal item of news is that the leading company has sold 26,000 tons of wire for shipment to Canada.

Bids have been opened in Cuba for two lots of cast iron water pipe, aggregating 15,000 tons. It is estimated that other municipalities in the island will need about 30,000 tons more.

It is difficult to get at the quantity of business being done in the domestic market in copper. It is known that some sales have been made around 15c for electrolytic. Sales for shipment abroad have taken place at the range of 14⅝c to 14¾c. The exports of copper during September were 17,157 tons, an encouraging exhibit.

U. S. METAL MARKET.

Cleveland, O., Oct. 3.—The Iron Trade Review to-day says: Total obligations on the books of the big iron and steel producers are decreasing from week to week, but they are working around to readjust themselves to the new conditions. The situation is being taken philosophically, and there are no indications of what, with the greatest stretch of imagination, could be construed as a panicky feeling.

To the fact that the railroads are doing very little buying is attributed the lack of interest at this time in new buying, except for immediate requirements on the part of those interests which have not already covered. The pig iron market is without feature, and, aside from a few scattered lots, involving small tonnage, there has been no movement of note.

Structural bookings of September aggregated more than 120,000 tons, of which only about 32,000 tons went to the American Bridge Co. The bulk of the business placed was made up of small jobs on which competition was very brisk and independent interests shared liberally. The McClintic-Marshall Construction Co. took a 1,900-ton bridge order from the Chicago and Northwestern. Grade crossing work is coming out well, Western roads being in the market for more than 5,000 tons. The Carnegie Steel Co. has sold a lot of 2,500 tons of steel bars for concrete reinforcement, and there is a Philippine inquiry for 900 tons.

LONDON, ENG., METAL MARKETS.

London, Oct. 2—Cleveland warrants are quoted at 55s 4½d, and Glasgow standards at 53s 4d, making prices as compared with last week, on Cleveland warrants, 1½d lower, and on Glasgow standard 1s 5d lower.

Tin—Spot tin opened weak at £158 5s, futures at £155 10s, and after sales of 200 tons of spot and 550 tons of futures, closed weak at £156 10s for spot and £154 10s for futures, making price as compared with last week £13 10s lower on spot and £11 10s lower on futures.

Copper—Spot Copper opened easy at £62 15s, futures at £62 15s, and after sales of 300 tons of spot and 1,800 tons of futures, closed weak at £61 15s for spot and £61 15s for futures, making prices as compared with last week £4 5s lower on spot and futures.

Spelter—The market closed at £21, making price as compared with last week 2s 6d higher

Lead—The market closed at £19 15s, making price as compared with last week 15s lower.

BRITISH TARIFF ASSOCIATION.

London, Eng., Oct. 4.—Less than two years ago the electors of the United Kingdom gave the Liberal party, with Sir Henry Campbell-Bannerman at its head, the biggest majority ever known. There are at the present time unmistakable indications that the electors regret the decision of the general election.

The favorite arguments which free importers used in opposition to Mr. Joseph Chamberlain's policy of fiscal reform and closer union with the colonies was that it would materially increase the price of food. There was much talk about the big loaf and the little loaf, which was an appeal to the fears of the working classes. The Radicals promised the big loaf at low price. They told the electors that the duty on corn which was proposed by Mr. Chamberlain would raise the price of bread and that the continuance of the free trade policy would certainly lead to a reduction, since the corn duty of one shilling a quarter was taken off.

Bread has gone up, not down. The price this week has gone higher and according to the corn dealers and millers the price is likely to go still higher.

Workingmen Aroused.

There is a feeling among the working classes that as they had no time to examine the economic questions themselves before the election, the Liberals deceived them. They are now willing listeners to the argument that if one or two countries acquire practical monopolies of the supply of food those countries will make them pay dear for it and that it would be wiser to stimulate production in the colonies and the possessions of the British Empire and make the Empire self-sufficient, as Mr. Chamberlain desired it should be.

Coal is too, as mentioned a week ago, much dearer and this threatens British home industries. The Radicals at the general election frequently stated that as no coal was imported into this country the coal miners had no interest in tariff reform and that even if these industries might suffer free import was at any rate the best policy for the coal miners.

The Radicals have admitted that Mr. Chamberlain's proposal to impose two shillings duty on corn might not be harmful, but they assured the electors that it would not stop there. The people will be told at the meetings and in lectures that the duty would stop exactly where the British workingmen desire it to stop, that electoral power is in the hands of the democracy and that industrial democracy constitutes the greatest majority of the voting strength of the United Kingdom. Therefore they will be assured that the amount of duty would always be absolutely controlled by the workingman's vote in their own interests.

For a long time America has been able to put steel in the British market at four shillings a ton less than it can be made in England. With the present price of coal and the prospective rise during the winter months, America may be able to do it even at a less price. It is known that she has built furnaces and works on the most gigantic scale with the express intention of capturing the British market. She produces nearly three times as much steel as England, and men in the iron and steel trade here will be reminded that steel made in England and sent to America has to pay a heavy duty, while America can sell steel to Great Britain for nothing, thus taking their bread away from them.

Coal is Going Up.

The government has taken off the coal duty, and the result is that most of the English coal is going out of the country. While this lasts the miners will earn good wages, but it isn't likely to go on forever. Then, the working classes are asking, what will happen? Practically no industry can decay without injuring the coal miners, though there is actually no coal dumped into England, but manufactured goods, in the production of which coal has been used, are constantly being dumped on British shores, and when, in consequence of that the industry which is used to produce these goods ceases to exist, there is so much less English coal used and less work and less wages for the English coal miners.

Owing to the high price of coal the iron and steel works in England are practically at a standstill.

The ordering of horseshoes for the army from America is also, according to the officials of the Imperial Tariff Reform Association, helping to bring that question nearer to the thoughts of the people, and believing that the people are already tired of the present government, there is the greatest activity at the Tariff Reform Association's headquarters. Arrangements are being made for holding large demonstrations and meetings in various parts of the country during the autumn and winter. These will be in addition to the small local meetings and lectures which now are so numerous.

Tariff Reform Gaining.

One of the leading men in the tariff reform movement has just stated that no movement in the world of politics is making such headway as tariff reform. The working classes are paying greater attention to it because they feel from what they see around them that it is not a matter to be ignored. The establishment of a federation for the purpose of the propaganda is reported to be working well. The electors in the smallest and remotest villages are being told how they were deceived at the general election by promises of a big and cheap loaf.

There are not only Americans but Germans to be feared. Both can make steel and iron cheaper than England can. One reason of the dearness of steel is that England is exhausting her coal. Coal is much dearer in England than in the United States. Another reason which will be given to the electors is that business men do not care to risk money in providing new machinery and up-to-date appliances when any day a foreigner may ruin their trade by dumping his cheap goods on the English market. Coal rules the destiny of nations; it is the source of Great Britain's commercial prosperity, and the secret of her vaunted naval supremacy, so that this matter of a rise in the price of coal will appeal to the British people just as powerfully as did the dear loaf.

CORRESPONDENCE

Travelers, hardware merchants and clerks are requested to forward correspondence regarding the doings of the trade and the industrial gossip of their town and district. Addressed envelopes, stationery, etc., will be supplied to regular correspondents on request. Write the Editor for information.

HARDWARE DISPLAYS AT HALIFAX.

Halifax, N.S., Sept. 30.—Two Halifax hardware firms have fine displays at the Nova Scotia Exhibition, both of which attract the attention of the patrons of the Fair. This is the first time for many years that the big wholesalers of Halifax have entered the field with exhibits, and to say the least, both have made an excellent start. In the main building is situated the booth of H. H. Fuller & Co., and the display is a glittering one to the eye of the visitor. Mounted on blocks of wood are samples of the brass and metal goods handled by this firm, such as hinges, door plates, knobs, latches, bolts, locks, etc. All are suspended from the sides of the booth, and as they are highly polished and glisten brightly with innumerable lights at night, the display is a most pleasing and attractive one. Henry C. Bennett, the manager for this firm, arranged the exhibit.

Another firm making a fine display is that of William Robertson & Son, whose booth is situated in the Fisheries Building. This exhibit is an extensive and costly one, and as the old saying goes, "almost everything from a needle to an anchor is shown." In addition to making a big display of lines, twines and nets, there is everything in the metal line used in the construction of motor boats. Lanterns, bells and gongs are there, fishing supplies of every description, and ship chandlery in abundance. Everything is most artistically arranged and the whole presents a very striking appearance. On the top of the booth are the letters "W. R. S.," lighted by electricity, and they can be seen all over the building. This exhibit was arranged at a cost of about $600, and it is most creditable to the firm.

* * *

A. M. Bell & Co., the enterprising Halifax hardware firm, is again to the front with a very attractive window display. It is one of the most original and unique window displays ever seen in Halifax. It is a real motion picture, and attracts the attention of all passers-by. The show is a competitive one, the Pike Manufacturing Company having offered prizes for the best window dressing with their manufactures, which include oil stones, sand stones, etc. Mr. Wetmore, the firm's decorator, planned the display, and the arrangement shows the Pike factory, built of oil stones, the bridge built of sand stones, station house, flume, pond, boat, etc., a miniature railway in operation, and stone being hoisted from the quarry by water power. The mill wheel is operated by water power also, and the railway by electric power. The arrangement is, perhaps, the best window display yet seen here. It will be in competition with window displays all over America. Mr. Wetmore is receiving hundreds of congratulations on his successful work as a window decorator.

* * *

All the Halifax hardwaremen have had a very busy time the past week meeting their customers who came to town to take in the Exhibition, thus combining pleasure with business. These annual fairs, as a rule, are very beneficial to the trade.

NEW BRUNSWICK TRADE GOOD.

St. John, N.B., Oct. 1.—September in the hardware trade has been very good. The jobbers report that orders are coming in satisfactorily and retailers have been kept busy dealing out builders' supplies. Since the opening of the big game season, guns and ammunition have been in great demand. Axes and lumbering supplies that have been on order will be shipped shortly, and stoves, ranges, shovels, coal hods, and the like are already being secured by householders. Considering the weather conditions, the sale of summer goods, such as hammocks, tents, refrigerators, ice cream freezers, etc., has been very good.

* * *

One of the biggest lumber deals in the history of the North Shore counties of New Brunswick was consummated the other day, when Sumner & Co. sold their entire property in Gloucester county to American capitalists. The property sold includes a large sawmill at Bathurst, a general store at the same place, and extensive lumber areas throughout the country. The price paid is said to be in the neighborhood of $250,000.

* * *

A new general store has been opened in Dorchester. The proprietors are Payzant & Card. They carry a well-assorted stock of hardware, paints, oils, groceries, dry goods, boots and shoes, clothing, hats and caps, house furnishings, stationery, etc.

* * *

J. A. Hunt, of Mexico City, Mexico, was in the city on Friday last on his way to the southern country, after a visit to his former home in Toronto. Mr. Hunt left Toronto seven years ago and went to Mexico to act as a manufacturers' agent. Since that time he has revisited Canada five times. Mr. Hunt says that the advance made by the Mexican republic during his residence there has been wonderful. Railway and commercial development has been most marked. There are splendid opportunities, he thinks, for developing trade in almost all lines between this country and Mexico. Canadian and other foreign capital has gone into Mexico in large quantities. The electric light and power companies are almost entirely controlled by Canadians.

* * *

Edward A. Lowe, of St. Martins, William E. Golding and Samuel C. Kirkpatrick, of St. John; John W. Lowe, of Aylesford, N.S., and Luther B. Smith, of Central Blissville, have applied for incorporation as the Oromocto Lumber Co., Ltd., with a capital stock of $99,000, divided into shares of $500 each.

* * *

There is quite a building boom at Woodstock. Gibson & Hayden are erecting a new brick block, Wm. M. McConnell has a concrete structure well under way, churches for the Baptists and Methodists are being built, improvements are being made to the Baird Company's building, and a number of private residences are being erected, and improvements and additions made to others.

* * *

The threatened strike among the sheet metal workers, referred to in last week's issue, did not materialize, as the employes decided to grant the increase asked for, of a minimum wage of $12 a week.

* * *

It is a notable fact that the position of president of the Maritime Board of Trade has been held by a hardwareman in three out of the past four years. The first was E. K. Spinney, of Yarmouth, 1904-05; then W. S. Fisher, of St. John, in 1906-7, and now A. M. Bell, of Halifax. Mr. Bell is also president of the Halifax Board of Trade.

SOME LARGE CONTRACTS.

Quebec, Sept. 30.—Raoul Rinfret, of Montreal, has filed plans and specifications for the construction of Notre Dame de Quebec system of waterworks. Tenders will be received up to the tenth day of October for the pipes and work, by L. E. Taschereau, secretary-treasurer of the municipality.

* * *

A building permit was issued this week to Villeneuve & Co. for the erection of the convent of Jacques Cartier. Its dimensions will be 105 feet by 58 feet, four storeys high, and will cost $55,000.

* * *

All the local dealers report a very brisk trade this week. Sporting goods are in very good demand. Plumbers and electricians are still occupied. Business is fairly active for wholesalers, and all hardware is in good demand. There was but little change in the market prices during the past week. A decline of three cents was noted in oils, and we now quote: Raw linseed oil, 52c to 64c; boiled linseed oil, 65c to 67c; turpentine, 87c to 90c. Also we quote a cent lower than last week for ingot copper, now sold at from 18c to 19c per pound.

The syndics of the parish of Baie St. Paul, County of Charlevoix, have awarded a contract for the construction and finishing of the interior of their new church, to cost $115,000. Jos. P. Mellet, of Quebec, is the architect. The same architect is also preparing plans for the erection of a chapel, the property of the Ursuline Nuns of Roberval, which will cost $36,000.

MADE-IN-CANADA EXHIBITION.

Kingston, Ont., Oct. 1.—Among the busiest men in this city at present are the tinsmiths, who are tumbling over each other in their zeal to put up stoves and pipes for the citizens. Very cool weather has set in within the past week, and, as a result, furnaces and stoves are going a month or more ahead of last year.

* * *

It is expected that the new telescope gas tank being erected on Barrack street, will be completed in about two weeks' time. Mr. Hardy, the foreman, has certainly done a good job in putting up the holder in this city, installing it in less than six weeks.

* * *

Tinsmiths, plumbers, electricians and carpenters have all been very busy the past few weeks in order to rush the completion of opening two new five-cent theatres in Kingston. This makes a total of three of these cheap amusement houses here.

* * *

On Sunday morning George R. Tomlinson electrical contractor, was found dead in his bed. The deceased retired on Saturday night in apparently good health, having been at his place of business that evening. Death was caused by a clot of blood around the heart. Mr. Tomlinson was forty-three years of age, and had carried on an electrical business in this city for a number of years.

* * *

Carpenters and workmen are busy at work at present on the curling rinks here. The rink is to be made twice its present size, and will be entirely refitted throughout.

* * *

Next week means a busy time for most of the merchants and people of Kingston, when an exhibition is to be held in the Armories in aid of the Kingston General Hospital. It is called the "Made and Produced in Canada Exhibition," and a number of the merchants are donating their different lines of goods towards helping with this worthy cause. The various merchants are to have booths where they may show the goods they donate.

NEW INDUSTRIES AT LONDON.

London, Ont., Oct. 2.—The fall hardware trade has opened out good, and both wholesalers and retailers are looking forward to a busy time. The local demand for stoves is unusually active. One dealer this morning told your correspondent that last week he sold more stoves and ranges than any previous week since he has been in business.

* * *

London continues to add to her industries. Scarcely a month passes but some new factory is established here. The latest enterprise is known as the National Light and Manufacturing Company, under which title half a dozen Londoners and two Americans have applied for incorporation with a capital of $50,000. The three-storey building at 348 Talbot Street has been secured and practically all the machinery has been installed. They have acquired the patent rights for Canada of the Cody inverted gas lights, and inverted arc lights, and the Keller gas governing devices, and have also become owners of the Cody oil gas patents, including the only known method of generating a non-carbonizing gas from common coal oil. The latter, on account of its absolute safety, freedom from odor, and lower cost, is calculated to entirely replace gasoline and wick-using lamps, for both lighting and cooking, and the firm will handle a full line of stoves and arc lights constructed on this principle. The premises, leased for six years, cover some 10,000 feet of floor space, and a staff of about 25 men will be employed as soon as the work of equipment is completed. The firm is already shipping out gas lights, and will be accepting orders for the oil-gas lights and stoves in a few days.

* * *

Local capitalists purpose organizing a large brick manufacturing company and as soon as a manager has been appointed, he will be sent to England to purchase machinery. The present difficulty in procuring brick of all descriptions caused the floating of the company.

* * *

The probability is that when the contract of the London Electric Company for lighting the city streets expires in November of next year, it will find strong competition. Already a firm has come to the city, and is looking over the ground, preparatory to making a bid for a lighting franchise. The Dominion Steam Heating and Lighting Company, with headquarters in Toronto, has a representative here, and he is making inquiries into the matter. This concern is an American compay that has extensive plants all over the United States. Over 200 towns and cities are heated and lighted by the company, and it is making rapid progress. In Canada it is now getting a strong foothold, having installed plants in several cities. Berlin has now closed a contract with the firm covering a period of ten years. Light and heat is furnished Berlin for eight months of the year. The company is not afraid of competition from Niagara power. With the installation of the double plant for heat and lighting, power for lighting can be furnished cheaper than Niagara power. Berlin, a town in line for Niagara power, has agreed to the new scheme, and has adopted it. In addition to the lighting, the company will heat a large number of business houses and other concerns in the city. The lighting will, it is said, be about 30 per cent. cheaper than under the present schedule. If the company can get a franchise in the city, a large plant will be erected, giving employment to a large number of men.

* * *

There is a possibility of the establishment of another large shoe factory here, the Hamilton Brown Shoe Co., of St Louis, Mo., one of the largest concerns of the kind in the United States, is looking for a site to establish a Canadian branch. A. D. Brown, one of the company, has been interviewing citizens on the project, and the London agent, who is an alderman, has written the company, setting forth the advantages of London as a commercial centre, and asking the firm to consider this city before finally deciding to locate the Canadian branch. A reply is expected shortly.

* * *

Work on the construction of an up-to-date building on the site of the collapsed Crystal Hall, has been commenced.

ROBERT HOBSON HONORED.

Hamilton, Oct. 1.—Local plumbers report a slight falling off in the condition of trade. This is largely due to the fact that the building business has taken a slight drop, and also to the condition of the money market, as stated last week, which has hampered the builders somewhat. However, at the present time there is sufficient doing in this line to keep the jobbers occupied and while the indications are that this winter will not be as heavy a one as last winter in this particular line, it does by no means forecast that business will be slack as last winter was exceptionally busy.

* * *

The death of Frank E. Walker, one of this city's most influential business men, in London, England, on Thursday, Sept. 26, was the cause of deep regret among the business people of the city. Mr. Walker had conducted a furniture business in this city for a great many years, but owing to poor health he took a trip abroad, accompanied by his daughter He was taken with a sudden attack while in London, and died there. His body will be brought to this city for interment. The deceased, in addition to his extensive business connections, was prominent in politics, was an ex-alderman and a member of the city parks board.

* * *

The Dunelm Steamship Company has been incorporated to carry on a general shipping and freight carrying business. It is capitalized at $100,000, and the provisional directors are: R. O. Mackay and J. P. Steadman, Hamilton, and George Somerville, of Toronto. A new boat is now in course of construction for the company.

* * *

The delegates to the Canadian Manufacturers' convention, who came here last Friday, expressed themselves as being thoroughly pleased with what they saw. There were over 200 in the party. The ladies were taken in carriages to the principal points of interest about the city, including the mountain, while the manufacturers themselves were taken in automobiles for a trip through the manufacturing district. The party visited the Canadian Westinghouse works, the International Harvester plant, Imperial Cotton works, Hamilton Bridge works, and several other large industries.

* * *

The election of Robert Hobson, of the Hamilton Steel & Iron Company, to the office of vice-president for Ontario of the Canadian Manufacturers' Association was popular with the local manufacturers. Mr. Hobson is one of the most level-headed business men in the city and as manager of the large blast furnace he has extensive scope for his talents. He is a son of Joseph Hobson, of this city, consulting engineer for the Grand Trunk Railway System.

* * *

The Producers Natural Gas Company has been organized. Frank R. Lalor, of Dunnville, is one of the principal directors, and the others are: William Southam, John Milne, R. S. Lees and Francis A. Magee, of this city. The capital is $100,000, and the company is chartered to carry on all dealings in gas, oils, etc.

NEW STORE AT WALLACEBURG.

Chatham, Oct. 1.—Jas. Watt & Sons have received the contract for the plumbing and heating of the new residences of W. D. and H. A. Stonehouse, Wallaceburg. They are also installing a complete hot water system in Stonehouse Bros.' store.

* * *

J. C. Shaw will open a hardware store in his own block, Wallaceburg, which was lately occupied by Hancock & Co. Mr. Shaw will take into partnership his son William and his nephew, Bailey Shaw.

* * *

David Crombie, who recently severed his connection with the firm of George Stephens & Co., has left his home in London for Montreal, where he has been appointed assistant master of transportation for the G.T.R.

* * *

J. W. Fleming & Son, hardware merchants, of Blenheim, have donated a new ball as a trophy for the next game between the Ridley and South Harwich baseball teams.

* * *

The Chatham Malleable Iron & Steel Works this week shipped a carload of children's hand sleighs and cutters to Winnipeg, and are at present busy getting out other large orders for the winter trade.

* * *

Another brick block is to be added to the building operations for the year. W. E. Merritt is having plans prepared for a new block on Fifth street.

* * *

An enthusiastic meeting of the local Board of Trade was held on the evening of Sept. 27, when the work of the ensuing season was inaugurated. There was a good turnout. It was decided to hold regular meetings every month, at which addresses will be delivered by capable speakers on subjects appropriate to the work of the Board.

* * *

Ald. W. H. Westman, of Westman Bros. returned last week from a two weeks' business trip to Eastern Ontario and Quebec.

* * *

Jas. Watt, Sr., of Jas. Watt & Sons, who has been quite ill for some time past, is reported to be steadily improving.

BUSINESS CHANGES IN WEST.

Saskatoon, Sept. 27.—There is now a good demand for threshers' supplies and engine fittings, belting, packing, lace leather, rubber hose, etc. Heaters have begun to move rapidly within the past week, and, consequently, there is a good sale for stove boards, pipes and dampers. Lamps are being brought forward to be reticketed and polished for the coming season. The chicken shooting season opened on Sept. 15, and customers report some good sport. The demand for ammunition still keeps up, and sporting weather is ideal.

* * *

F. Clark, of Saskatoon, has bought the stock and building of S. E. Maud, of Elstow, Sask.

* * *

J. Clinkskill, accompanied by Mrs. Clinkskill, have left on a trip to Scotland.

* * *

Smith & Fleming have purchased the stock and premises of Haynes Tupper, Lanigan, Sask.

* * *

The hardware department of James Clinkskill's store is now in charge of J. P. Wing. For the past year, Mr. Wing was connected with the Battleford branch of the business.

* * *

The Canadian White Co. are removing their construction equipment from Saskatoon to Clover Bar, near Edmonton, to continue their contract with the G.T.P. Railway.

* * *

The Canadian Bridge Co. have completed their contract on the traffic bridge, and will proceed at once to erect the steel work on the G.T.P. bridge.

* * *

Engineer S. Paget has arrived in town to superintend the steel laying on the Grand Trunk Pacific Railway. The yards are well filled with steel rails, ties, etc., and supplies are arriving daily.

Best quality store fixtures are the least costly in the long run. The cheap lines soon find their way to that pile of has-been in the back room upstairs.

LETTER BOX.

Correspondence on matters of interest to the hardware trade is solicited. Manufacturers, jobbers, retailers and clerks are urged to express their opinions on matters under discussion.

Any questions asked will be promptly answered. Do you want to buy anything, want some shelving, a silent salesman, any special line of goods, anything in connection with the hardware trade? Ask us. We'll supply the necessary information.

Ornamental Iron Work.

Stillman Bros., Keene, Ont., wrote in asking where they could secure iron ends for church pews.

Answer.—The Canadian Ornamental Iron Co., 25 Yonge St. Arcade, Toronto, make a specialty of this class of work. The Canada Foundry also have a department devoted to ornamental iron work.—Editor.

Ferrosteel Registers.

Phillips Bros., Havelock, Ont., asks if the Ferrosteel Register Company, Cleveland, have a Canadian agency.

Answer.—We understand this company are establishing a Canadian factory at Bridgeburg, Ont., but correspondence should be sent to the head office at Cleveland, for the present.—Editor.

How to Make Cellar Dry.

In Hardware and Metal of September 21, Johnson Bros., Minto, Man., enquired how to prevent water coming through a concrete cellar floor. In addition to the suggestions given in response to their enquiry, the following very practical method of overcoming such troubles has been forwarded to Hardware and Metal, by Richardson, Simard & Co., 84 Prince Arthur street, Montreal. This firm are specialists in cement and concrete work and have found the method outlined below very satisfactory where the water pressure is not too strong. The plan they suggest is as follows: Lay six inches of good, strong concrete; over this lay four-ply felt, well stuck with pitch, and a good coat over the same. Then put four inches more of the concrete, and one inch of cement top finish.

BRIDGE DISASTER AFFECTS TRADE.

One of the most serious of the results due to the Quebec bridge disaster will be the delay caused in the completion of several railway connections badly needed for the opening up of the Lake St. John region, and the territory to the north of it.

Various industrial enterprises projected in this belt, enriched by vast forests yet untouched and streams affording water power of priceless value, will have to be deferred for a year or more, as transportation facilities are wanting. The American concern in charge of the construction states that it will take two years to regain the condition that existed before the accident. The Dominion Government might itself undertake the completion of the bridge as soon as the investigation into the causes of the disaster has been completed.

THE BECK-IDEN LAMP.

A lamp which is gaining increasing popularity is the Beck-Iden acetylene lamp. The features of this lamp, which

Beck Iden Acetylene Lamp.

the manufacturers claim to be outstanding, and which they claim make it superior to all ordinary lamps and lighting appliances, are as follows: It provides a house light, having the cleanliness and convenience of gas or electricity without the necessity of piping or wiring. It is lighted and burned from an open burner, as city gas. It burns about nine or ten hours at full force without refilling. Attached to the lamp is an ingenious device to regulate the amount of gas in the jet. We herewith reproduce a cut, illustrating this lamp. This firm has offices in Montreal.

NEW DOOR HINGE.

The Standard Manufacturing Co., of Shelby, Ohio, have recently added to their line the "Champion" double acting spring door hinge, herewith illustrated, on which patents have recently been allowed. They claim for it many points of merit not found in other double acting spring hinges.

The hinge is of the type secured to the lower corner of the door with a floor plate secured to the surface of the floor. This hinge is detachable at the floor plate so that the door can be removed without removing the floor plate from the floor. In this floor plate is also the adjustment for lining up the door to bring it in line with opposite jamb or door when in pairs. This is accomplished by loosening one of the headless screws shown on floor plate and tightening the opposite one which will bring the door to exact alignment without disturbing any screws that secure the floor plate to the floor. The hinge is applied to the door in a very easy and simple manner, it being only necessary to make a straight cut the depth of the hinge and a straight cut the length of the hinge with the short extention at the end let in bush at the bottom of the door. The hinge is secured to the door by one long wood screw near the heel, this screw has machine screw head threaded into top rib of hinge, and in putting hinge on, this wood screw is turned by using the hinge as a handle until the hinge comes up tight at that point. Two wood screws are used at the opposite end to firmly secure the hinge in place. The tension of spring is adjusted to suit the swing of the door by tension nut, shown at end of spring.

The finish plate can be removed by taking out only one screw, as shown in cut, the other end being secured by tongues which enter slots in that end of the finish plate (not shown in cut).

The weight of the door is carried on ball-bearings, directly on top of the pivot post, reducing friction to the minimum and allowing the door to swing very freely and easily. This ball-bearing being located at the top of the post prevents any liability of water or grit getting to and destroying its smooth and easy operation.

Champion Double-acting Door Hinge.

The spring plunger which controls the swing of the door is guided by the inside of the spring, this spring acts as a cushion against the side motion of the plunger at each swing of the door, eliminating to a large degree, the clicking noise found so objectionable in other spring hinges.

NEW HACKSAW MACHINE.

E. C. Atkins & Co. the silver steel saw people of Indianapolis with branches in ten of the largest cities have just put on the market a most unusual invention, in the shape of a power hacksaw machine, a picture of which we reproduce.

The machine is called the Kwik-Kut on account of its time-saving features. The principles are entirely new and revolutionize this, at best, slow operation. It is claimed the machine will do 25 per cent. more work than any other. This is accomplished first by the adjustment of the saw arm, which uses the whole blade from end to end, thus employing the entire cutting surface,

CONDENSED OR "WANT" ADVERTISEMENTS.

RATES.

Two cents per word first insertion; one cent per word subsequent insertions.

Five cents additional each insertion where box number is desired.

Contractions count as one word, but five figures (as $1,000) are allowed as one word.

Cash remittances to cover cost **must** accompany all advertisements. **In no case** can this rule be overlooked. Advertisements received without remittance cannot be acknowledged.

RULES FOR COPY.

In addressing replies care of HARDWARE AND METAL don't fail to give box number.

Replies addressed to HARDWARE AND METAL boxes are re-mailed to advertisers every Monday, Wednesday and Friday.

Requests for classification will be followed where they do not conflict with established classified rules.

Orders should always clearly specify the number of times the advertisement is to run.

All "Want" advertisements are payable in advance.

AGENTS WANTED.

This is the problem of many manufacturers who are anxious to get a foothold in Western Canada. It will be easily solved if HARDWARE AND METAL is given the opportunity to solve it.

AGENT wanted to push an advertised line of Welsh tinplates; write at first to "B.B.," care HARDWARE AND METAL, 88 Fleet St., E.C., London, Eng. [tf]

AN old-established house of brass founders, making a wide range of brass foundry furnishings, such as brass cast tubing, stair rods, cornice poles, art metal work, electric light fittings, ecclesiastical metal work, are anxious to secure a thoroughly reliable and trustworthy agent or traveller for Canada, to sell goods on commission. An excellent opportunity for any firm covering Canada thoroughly. Address all enquiries to Box 654, HARDWARE AND METAL. [37]

AGENCY WANTED

EXPERIENCED and reliable business man, Calgary, could represent good lines of hardware, plumbing and heating goods. Address Box 665, HARDWARE AND METAL, Toronto. [40]

ARTICLES FOR SALE.

Don't keep any fixtures or tools around your store for which you have no further use. They are worth more to-day than they will a year hence. Don't keep money tied up which you could use to secure discounts from your wholesaler.

SET of tinsmith's tools, good as new, best American make. $150. Great sacrifice. Box 660, HARDWARE AND METAL, Toronto. tf

TWO hundred dozen rim and mortice knobs and locks, black, white and mineral. 90c. per dozen. Full particulars. Box 661, HARDWARE AND METAL, Toronto.

TYPEWRITER, No. 2 Sun, in first-class condition. Will sacrifice for $55.00. Box 662, HARDWARE AND METAL, Toronto. tf

GOOD silent salesman for sale. Reason for selling, advertiser wishes to purchase larger one. A bargain for immediate sale. Box 658, HARDWARE AND METAL, Toronto. 38

ARTICLES WANTED.

If you cannot afford to buy a new counter, show case, screw cabinet, store ladder, or some other fixture which you could use to advantage, try a "Want Ad." under "Articles Wanted," and you may get what you want at a bargain price.

WANTED—Two store ladders, "Myers" preferred, complete with fixtures. Send full description and price to Illsey Bros., Red Deer, Alta. (36)

Prepare for Holiday Business

It's not too early to lay plans for the Christmas holiday trade. Dealers have already bought some lines, of course, and in these days of slow deliveries other orders should be placed as soon as opportunity offers. "Goods well bought are half sold" is a time-worn truism in mercantile life.

But it doesn't do to feel too secure on the buying part until the selling has been done. Plans for increasing holiday sales must be worked out months ahead. Window displays must be figured out and a series of rough sketches planned. A series of ads. for the local papers should also be prepared, and, where deemed advisable, a neat booklet gotten up to be mailed to all probable customers in the surrounding district.

It is in these selling plans that Hardware and Metal can be helpful to its readers. Let all partake of the spirit of co-operation and exchange ideas through these columns. To help the suggestion along the editor invites any reader to join the friendly contest and hangs up a prize of $10 cash for the best answer to the following questions:

How can the hardware merchant increase his sales of holiday goods next December? What special lines should he stock? What selling plans should be adopted? What special advertising should be done? What novel window displays can be suggested? Should Souvenirs (calendars, knives, trays, etc.) be given to customers?

The prize of $10 will be awarded to the writer of the most practical and original letter of from 500 to 1,000 words received by the editor before October 15, 1907, and the best half-dozen replies will be published in Hardware and Metal.

Address the Editor

HARDWARE AND METAL

10 Front St. E., Toronto

BUSINESSES FOR SALE.

Somewhere in Canada is a man who is looking for just such a proposition as you have to offer. Our "For Sale" department brings together buyer and seller, and enables them to do business although they may be thousands of miles apart.

HARDWARE Business and Tinshop for sale in Saskatchewan; population 1,500; stock carried, about $14,000; turnover, $45,000; practically all-cash business; cash required, $8,000; would rent building; do not answer without you have the money and mean business; it will pay to investigate this. Box 648, HARDWARE AND METAL, Toronto. (41)

HARDWARE Business for Sale in Tweed, a thriving Eastern Ontario town. Best business in town. Must sell owing to ill-health. Liberal terms to satisfactory purchaser. Write for particulars. J. M. Robertson, Tweed. [40]

HARDWARE store and stock for sale, within 25 miles of Toronto. Solid brick double store and dwelling; property worth $4,000. Stock worth about $3,000. Splendid business in stoves and hardware. Good reasons for selling. Full particulars on application. Box 663, HARDWARE AND METAL, Toronto. [41]

SITUATIONS VACANT.

You can secure a "five-thousand-a-year" manager, or a "five-hundred-a-year" clerk, by stating your wants under "Situations Vacant."

WANTED—First-class tinsmith in New Ontario town. Must be capable of managing workshop and estimating. Good position, good salary, and no lost time for the right man. Address, Box 655, HARDWARE AND METAL, Toronto. [40]

WANTED—First class opening for tinsmithing, plumbing and heating. Box 202, Hespeler. [48]

HARDWARE clerk, with two years' experience; must start at once; state wages to A. H. Gingerich, Woodstock, Ont. [39]

SITUATIONS VACANT

WANTED—Experienced hardware clerk; must be good salesman and stockkeeper. Address, stating age, experience and salary expected, Box 664 HARDWARE AND METAL, Toronto. [40]

THE CANADIAN GROCER wants a Managing Editor. It wants a thoroughly capable man—a man who is live, full of up-to-date ideas, and one who understands the newspaper business from the reglet box to the editorial chair. Furthermore, it wants a man who is thoroughly conversant with the commercial situation in Canada. We realize that this is a big want. Not everyone can fill the bill, but we are willing to pay at the outset $2,500 a year to the man who can do so. The right man can eventually make his place worth $5,000. If you think you are this man we want to hear from you, with your experience and qualifications—by letter only. Address The MacLean Publishing Co., 232 McGill St., Montreal, or 10 Front St. East., Toronto. 38

SITUATIONS WANTED.

Travellers, clerks, or tinsmiths, wishing to secure new positions, can engage the attention of the largest number of employers in their respective lines, by giving full particulars as to qualifications, etc., under "Situations Vacant."

CLERKS WANTED.—Clerks who want to improve their positions are requested to write for information to Box 659, HARDWARE AND METAL, Toronto. 40

HARDWARE CLERK, 32, married, sober, steady and reliable, capable of taking charge. Box 10, HARDWARE AND METAL, Winnipeg.

HARDWARE Salesman wants position; can furnish first class references; capable of managing city or town business. Box 653, HARDWARE & METAL, Toronto. (39)

HARDWARE CLERK, Irishman, 27, long experience of trade in best houses Old Country, one year with large Canadian wholesale firm, desires change, would go west. Unexceptionable references. Photo if required. Apply box 657, HARDWARE AND METAL, Toronto. (39)

MANITOBA HARDWARE AND METAL MARKETS

Corrected by telegraph up to 12 noon Friday, Oct 4. Room 511, Union Bank Bldg, Winnipeg, Man.

Business is fairly active. In some districts there is a falling off, but wholesalers have no reason for complaint.

Rope—Sisal, 11c per lb., and pure manila, 15¾c.

Lanterns—Cold blast, per dozen, $7; coppered, $9; dash, $9.

Wire—Barbed wire, 100 lbs., $3.22½; plain galvanized, 6, 7 and 8, $3.70; No. 9, $3.25; No. 10, $3.70; No. 11, $3.80; No. 12, $3.45; No. 13, $3.55; No. 14, $4; No. 15, $4.25; No. 16, $4.40; plain twist, $3.45; staples, $3.50; oiled annealed wire, No. 10, $2.90; No. 11, $2.96; No. 12, $3.04; No. 13, $3.14; No. 14, $3.24; No. 15, $3.39; annealed wires (unoiled), 10c less; soft copper wire, base, 36c; brass spring wire, base, 30c.

Poultry Netting—The discount is now 47½ per cent. from list price, instead of 50 and 5 as formerly.

Horseshoes—Iron, No. 0 to No. 1, $5.65; No. 2 and larger, $4.40; snowshoes, No. 0 to No. 1, $4.90; No. 2 and larger, $4.65; steel, No. 0 to No. 1, $5; No. 2 and larger, $4.75.

Horsenails—No. 10 and larger, 22c; No. 9, 24c; No. 8, 24c; No. 7, 26c; No. 6, 28c; No. 5, 30c; No. 4, 36c per lb. Discounts: "C" brand, 40, 10, 10 and 7½ p.c.; "M.R.M" cold forged process, 50 and 5 p.c. Add 15c per box. Capewell brand, quotations on application.

Wire Nails—$3.10 f.o.b. Winnipeg and $2.65 f.o.b. Fort William.

Cut Nails—Now $3.20 per keg.

Pressed Spikes—¼ x 5 and 6, $4.75; 5-6 x 5, 6 and 7, $4.40; ⅜ x 6, 7 and 8, $4.25; 7-16 x 7 and 9, $4.15; ½ x 8, 9, 10 and 12, 4.05; ⅝ x 10 and 12, $3.90. All other lengths 25c extra, net.

Screws—Flat head, iron, bright, 80, 10, 10 and 10; round head, iron, 80; flat head, brass, 75; round head, brass, 70; coach, 70.

Nuts and Bolts—Bolts, carriage, ⅜ or smaller, 60 p.c.; bolts, carriage, 7-16 and up, 50; bolts, machine, ⅜ and under, 50 and 5; bolts, machine, 7-16 and over, 50; bolts, tire, 65; bolt ends, 55; sleigh shoe bolts, 65 and 10; machine screws, 70; plough bolts, 55; square nuts, cases, 3; square nuts, small lots, 2½; hex nuts, cases, 3; hex nuts, small lots, 2½ p.c. Stove bolts, 70 and 10 p.c.

Rivets—Iron, 60 and 10 p.c.; copper, No. 7, 43c; No. 8, 42½c; No. 9, 45½c; copper, No. 10, 47c; copper, No. 12, 50½c; assorted, No. 8, 44½c, and No. 10, 48c.

Coil Chain—¼-in., $7.25; 5-16, $5.75; ⅜, $5.25; 7-16, $5; 9-16, $4.70; ⅝, $4.65; ¾, $4.65.

Shovels—List has advanced $1 per dozen on all spades, shovels and scoops.

Harvest Tools—60 and 5 p.c.

Axe Handles—Turned, s.g. hickory, $3.15; No. 1, $1.90; No. 2, $1.60; octagon extra, $2.30; No. 1, $1.60.

Axes—Bench axes, 40; broad axes, 25 p.c. discount off list; Royal Oak, per doz, $6.25; Maple Leaf, $8.25; Model, $8.50; Black Prince, $7.25; Black Diamond, $9.25; Standard flint edge, $8.75; Copper King, $8.25; Columbian, $9.50; handled axes, North Star, $7.75; Black Prince, $9.25; Standard flint edge, $10.75; Copper King, $11 per dozen.

Churns—45 and 5; list as follows: No. 0, $9; No. 1, $9; No. 2, $10; No. 3, $11; No. 4, $13; No. 5, $16.

Auger Bits—"Irwin" bits, 47½ per cent. and other lines 70 per cent.

Blocks—Steel blocks, 35; wood, 55.

Fittings—Wrought couplings, 60; nipples, 65 and 10K Ts and elbows, 10; malleable bushings, 50; malleable unions, 55 p.c.

Hinges—Light "T" and strap, 65.

Hooks—Brush hooks, heavy, per doz., $8.75; grass hooks, $1.70.

Stove Pipes—6-in., per 100 feet length, $9; 7-in., $9.75.

Tinware, Etc.—Pressed, retinned, 70 and 10; pressed, plain, 75 and 2½; pieced, 30; japanned ware, 37½; enamelled ware, Famous, 50; Imperial, 50 and 10; Imperial, one coat, 60; Premier, 50; Colonial, 50 and 10; Royal, 60; Victoria, 45; White, 45; Diamond, 50; Granite, 60 p.c.

Galvanized Ware—Pails, 37½ per cent.; other galvanized lines, 30 per cent.

Solder—Quoted at 27c per pound. Block tin is quoted at 45c per pound.

Wringers—Royal Canadian, $36; B.B., $40.75 per dozen.

Files—Arcade, 75; Black Diamond, 60; Nicholson's, 62½ p.c.

Locks—Peterboro and Gurney 40 per cent.

Building Paper—Anchor, plain, 66c; tarred, 69c; Victoria, plain, 71c; tarred, 84c; No. 1 Cyclone, tarred, 84c; No. 1 Cyclone, plain, 66c; No. 2, Joliette, tarred, 69c; No. 2 Joliette plain, 51c; No. 2 Sunrise, plain, 56c.

Ammunition, etc.—Cartridges, rim 50 p.c.; pistol sizes 25 p.c.; military, 20 p.c. Primers, $1.55. Loaded shells: 12 gauge, black, $16.50; chilled, 12 gauge, $17.50; soft, 10 gauge, $19.50; chilled, 10 gauge, $20.50. Shot: ordinary, per 100 lbs., $7.75; chilled, $8.10. Powder: F.F., keg, Hamilton, $4.75; F.F.G., Dupont's, $5.

Revolvers—The Iver Johnson revolvers have been advanced in price, the basis for revolver with hammer being $5.30 and for the hammerless $5.95.

Iron and Steel—Bar iron basis, $2.70. Swedish iron basis, $4.95; sleigh shoe steel, $2.75; spring steel, $3.25; machinery steel, $3.50; tool steel, Black Diamond, 100 lbs., $9.50; Jessop, $13.

Sheet Zinc—$8.50 for cask lots, and $9 for broken lots.

Corrugated Iron and Roofing, etc.—Corrugated iron, 28 gauge, painted, $3; galvanized, $4.10; 26 gauge, $3.35 and $4.35. Pressed standing seamed roofing, 28 gauge, $3.45 and $4.45. Crimped roofing, 28 gauge, painted, $3.20; galvanized, $4.30; 26 gauge, $3.55 and $4.55.

Pig Lead—Average price is $6.

Copper—Planished copper, 44c per lb.; plain, 39c.

Iron Pipe and Fittings—Black pipe, ¼-in., $2.70; ⅜, $2.85; ½, $3.75; ¾, $4.75; 1, $6.75; 1¼, $9.25; 1½, $11.00; 2,

$14.80; 2½, $24.60; 3, $32.30; 3½, $40.50; 4, $46.00; 4½, $54.00. Galvanized: ¼ in., $3.65; ⅜, $3.80; ½, $4.50; ¾, $5.80; 1, $8.40; 1¼, $11.40; 1½, $13.80; 2, $18.40. Nipples, 70 per cent.; unions, couples, bushings and plugs, 50 per cent.; malleables, 20 per cent.

Galvanized Iron—Apollo, 16-gauge, $4.15; 18 and 20, $4.40; 22 and 24, $4.65; 26, $4.65; 28, $4.50; 30-gauge or 10¾-oz., $5.20; Queen's Head, 20, $4.60; 24 and 26, $4.90; 28, $5.15.

Lead Pipe—Market is firm at $7.80.

Tin Plates—IC. charcoal, 20x28, box, $10; IX charcoal, 20x28, $12; XXI charcoal, 20x28, $14.

Terne Plates—Quoted at $9.50.

Canada Plates—18x21, 18x24, $3.50; 20x28, $3.80; full polished, $4.30.

Lubricating Oils—600W, cylinders, 80c; capital cylinders, 55c and 50c; solar red engine, 30c; Atlantic red engine, 29c; heavy castor, 28c; medium castor, 27c; ready harvester, 28c; standard hand separator oil, 35c; standard gas engine oil, 35c per gallon.

Petroleum and Gasolene—Silver Star, in bbls., per gal., 20c; Sunlight, in bbls., per gal., 22c; per case, $2.35; Eocene, in bbls., per gal., 24c; per case, $2.50; Pennoline, in bbls., per gal., 24c; Crystal Spray, 23c; Silver Light, 21c. Engine gasoline, in barrels, gal., 27c; f.o.b. Winnipeg, in cases, $2.75.

Paints and Oils—White lead, pure, $6.50 to $7.50, according to brand; bladder putty, in bbls., 2½c; in kegs, 2¾c; turpentine, barrel lots, Winnipeg, 90c; Calgary, 97c; Lethbridge, 97c; Edmonton, 98c. Less than barrel lots, 5c per gallon advance. Linseed oil, raw, Winnipeg 70c, Calgary 77c, Lethbridge, 77c, Edmonton 78c; boiled oil, 3c per gallon advance on these prices.

NEW TARIFF DECISION.

Tariff decision No. 293, notice of which has just been sent to the Customs officials, reads as follows: "Articles partly nickle or electro-plated are not to be rated for duty under item 362, as nickle-plated ware or electro-plated ware, unless half the metallic surface thereof is nickle or electro-plated."

Modern Conveniences for Farm Homes

The Fourth of a Series of Articles Intended to Help Canadian Plumbers in Educating Residents in Country Districts to the Necessity of Better Sanitary Arrangements.

By Elmina T. Wilson, C.E.

Earth Closets.

Where there is difficulty in the matter of ultimate disposal without the use of a cesspool, and the consequent and apparently unavoidable risk thereby incurred of contaminating the well water, it would be better to use an earth closet. This is not wholly satisfactory, but is safer and far better than the provision so often found on farms and in villages. The house containing it should be well built and substantial, well lighted and ventilated, with a good roof, and preferably plastered on the inside to insure less exposure in cold weather. A carefully made and dry walk, screened by lattice for protection from the wind and for privacy, should be built to it. The excretia should be received in a galvanized-iron pail, not too large and made to fit close under the seat. This seat can be like that of an ordinary water closet. Each time the closet is used dry earth is added. The pail should be emptied very frequently. With proper attention this closet need not be, and should not be, built far from

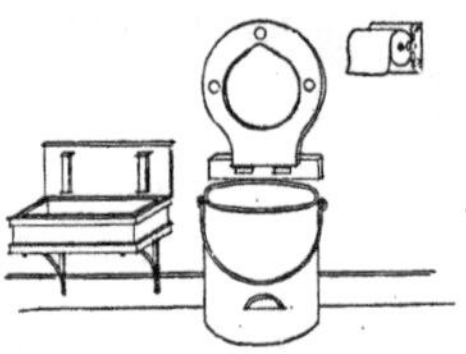

Fig 7. Earth Closet.

the house. It would even be possible to place it in a room built against the house, the room having one door opening from the house and another opening out of doors. This would make it possible to enter from the house in inclement weather, and also to carry out the pail without passing through the house. The room should be well ventilated by a window close to the ceiling, and only tissue paper should be used. (See fig. 7 for arrangement of pail, seat, and dry-earth box).

The earth for use in these places is to be found in nearly every field and garden and should be of rather a loamy nature if possible, and porous. A very sandy soil is next to useless. Large heaps of earth should be collected for the year's use and dried in the summer sun. It is not necessary to use perfectly dry earth, but it is always the best.

After the plumbing has been put into the house, to make it effective, a means must be provided to remove all liquid wastes as soon as possible and to prevent the return of odors.

As has been previously explained, to build a cesspool to foul the ground and air is not a solution; it is only putting the trouble "out of sight." It holds the wastes in a state of putrefaction, and so will give off troublesome gases, which will, if near the house, penetrate to the cellar and then on through the house. Its liquid leachings, if these strike a layer of sand and gravel or a fissure in the rock, will penetrate an unthought-of distance to injure the quality of water of wells and springs. Such a system would never be installed if the householder realized the full effect of the nuisance he was establishing. Fortunately it is no longer necessary to build a cesspool when plumbing is introduced into the house which can not be connected to a public sewer.

Modern Sewage Disposal.

Within the last twenty years many investigations of sanitary methods for the disposal of sewage for isolated houses have been made. The working principle is this: When the air contained in the soil is brought in contact with dead organic matter in a finely divided state, a complete transformation takes place by the natural processes of oxidation and nitrification. This change is brought about by micro-organisms which multiply rapidly under proper conditions and by so doing combine the oxygen of the air with the organic matter. Air is as necessary for this purpose as fuel in a stove that is required to give off heat. It is therefore essential that the waste be deposited on or near the surface of the ground and in such a way that air reaches every particle. If the ground is saturated for a long period of time purification of the liquid will cease. If a foul liquid is thrown on the surface of the ground the water will pass off, but its organic impurities will cling to the particles of earth. Air takes the place of the water and surrounds the waste matter, the bacteria present in all fertile soils effect a combination between the two, converting the dead organic matter to harmless mineral forms needed for vegetable life. This process may be repeated indefinitely if the waste matter be supplied and the air furnished alternately. This is called the "intermittent method" of operating the disposal plant. The process of applying this principle to the disposal of sewage varies, and may be divided into two distinct steps: (1) The collection of the wastes from the house, that these may be applied intermittently; and (2) their application to a natural soil by surface or subsurface irrigation, or to a specially prepared soil, as in a filter bed.

Three Systems Described.

The ground available on one farm can not be, and need not be, exactly the same as that used on another. If the surface is almost level and under cultivation the subsurface system of distribution could be adopted, whatever the method used for collecting the flow from the house. The liquids are retained for a certain length of time and then rapidly discharged into open-jointed tiles, laid near the surface, thus securing a uniform distribution throughout the entire length of the drain. While the tank is refilling time enough is allowed for the water to pass away and the air to enter and accomplish its work. This manner of purifying the liquid wastes of the household where the water-closet is or is not used has been tried with entire satisfaction.

Another method of getting the liquid wastes on the land is by surface irrigation. The most marked feature of the change from the privy to the water carriage system is the great dilution which

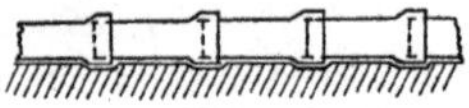

Fig. 8. Method of laying Vitrified Pipe.

the organic matter undergoes. Instead of concentrated organic wastes, unsuited for direct use as a fertilizer, we now have a dilute mixture, which seldom contains more than 1 part in 500 of anything but water. Of course this can not be discharged in a haphazard manner upon the land, but if the intermittent method, as explained above, be followed and the discharge be regulated to avoid saturation of the ground, the sewage can be disposed of without unsightly or offensive results. The area of the ground required is about the same as for subsurface disposal. A sloping surface in grass, or partially wooded, or a cultivated field, say of corn, would be entirely suitable. The area required depends upon the character of the soil and the amount of sewage to be purified. For a family of five, if the soil be reasonably porous, a plat of ground 40 feet by 50 feet should be ample.

The third method of exposing the liquid wastes to the action of the bacteria is used when the available area of the land is limited or if on account of the character of the soil too large an amount of drains would be necessary to handle the sewage to be purified. Under these conditions specially prepared beds of sand, gravel, or screened cinders are used as a filter material. The sizes of these beds vary with the amount and character of the sewage and its previous treatment, but for a family of five under reasonable conditions and a proper preliminary treatment, a bed 25 feet long, 4 feet wide and 4 feet deep should answer, although one 5 feet deep would give a purer effluent.

Collecting Liquid Wastes.

As has been pointed out in the previous paragraphs, intermittent application is the key to success, and in order to apply the liquid waste or sewage to the land intermittently it is necessary to collect the irregular flow from the house in some manner.

The house drain of cast-iron pipe extends at least 5 feet outside of the foundation wall. From this point to the collecting chamber vitrified sewer pipe is used, except in made ground, or in quicksand or where the drain must pass near a well, when cast-iron pipe should be provided. The drain should be given a uniform fall, if possible, and should be at least 3 feet below the surface in cold climates, to avoid freezing. The greater the inclination the less the liability of obstruction. A fall of 1 foot in 40 or 1 foot in 60 feet is desirable; 1 foot in 100 feet is the least that should be used unless special flush tanks are provided. The bottom of the trench should be dug to the exact grade and shaped to fit the lower half of the pipe, with grooves cut for the sockets as shown in figure 8. Inspection and cleaning hand-holes should be located about every 50 feet, and their exact location marked on the plan of the drain, to aid in locating obstructions, if any should occur. Change of direction of the pipe line should be made with special curved pipes of large radius.

Making the Joints.

The joints of the pipe should be well made. The space between spigot and hub should be filled first with a small rope of picked oakum, rammed into place with a hand iron to prevent any cement mortar from entering at the joints. Then fill the remainder of the space with a mortar made of one part Portland cement and one part clean sand. The cement and sand must be thoroughly mixed dry and wetted up only as needed. The bottom of the joint should be made with particular care, using the fingers to press the mortar into place. As soon as the joint is finished the groove in the trench should be filled with earth to prevent the joint being broken before the cement has time to harden. The inside of the pipe must be thoroughly cleaned from projections of oakum or cement. If the drain passes close to trees the joints should be tested by closing the main outlet and filling the pipes with water. Leaky joints are undesirable, both because of the contamination of the soil and because the liquid is needed to carry the solid matters through the pipes.

The volume of the sewage is practieally equal to the water consumption. The waste matter received increases the volume very little, 1 pound in 120 gallons is organic matter, to remove and destroy which is the purpose of the "disposal plant." The garbage or kitchen refuse, of course, is not included. This 1 pound of organic matter with its 120 gallons of water is all carried by the house sewer to a suitable place, the distance from the house varying with the surrounding conditions, and collected in a chamber to which has been given various names, according to the manner of use proposed. We have the "flush tank," the "settling chamber," and the "sseptic tank," and these may be so constructed and used that it is difficult to draw an exact line between them.

Heating and Housefurnishing

FALL STOVE SELLING PLANS.

In order to have a successful season this fall on stoves and make it a profit-maker it is necessary to plan your campaign ahead. Do not wait until the season is on before making a move—get busy early—lay out where and how you can display them to the best advantage, have your advertisements written up ahead, and have your list of prospective buyers made out so that you will be in a position to go right after them as soon as you have your goods on display, says the Hardware Trade.

It is convenient to have stoves on trucks or casters, and ranges and heaters on raised platforms with a neat stove board underneath. They can be neatly arranged in rows, so that they can be immediately drawn out for inspection.

A very necessary adjunct to the stove business is to know the stoves that you are selling, study what the manufacturers say about them, and examine the stove fully. Find out all about it so that you can tell a customer all its meritorious points.

When a buyer comes in don't be afraid to run the stove out where it can be seen. Take it apart, and have him examine its interior, fittings, grates, firebricks, etc. In fact, show every detail about it, no matter how small it may seem, it all helps to pile up evidence that you have a good stove and that you are proud to let it be known.

Live Advertising.

Use printers' ink liberally, but systematically. Have your ads. written weeks ahead, though you may be able to improve them before they are used. Have a cut of each stove you wish to advertise and give your attention to each one in turn, beginning on your leader and following it up with lower priced stoves. Tell all the points about it and give a synopsis of its construction, metal used, grates, size of oven, cooking surface, etc. Be as brief as possible, but to the point, and above all, state the price and terms on which it may be purchased.

Follow this up week after week; have a new ad. each week and something new to say, and you will be agreeably surprised at results, as it makes immediate sales and lays the foundation for future trade.

But while the newspaper is busy talking for you, be busy yourself. Take an hour each day and go out to see your customers and friends. Don't let them forget that you sell stoves. If you have a slack day go out among your farmer friends; get thoroughly acquainted with them; and take an interest in their work and show them by your manner that you would like them to get along well.

Make their interests your interests, study their needs and incidentally tell them of fine lines of goods you carry, and always be pleasant and sociable. It doesn't cost anything to be pleasant and pays big dividends in the way of business. You will find that they will make it a point to call at your store the next time they are in town. They generally stick and make good customers. A little time spent in canvassing your business territory will pay as surely as it does the man who goes about to sell a binder or a cream separator.

Personal Letters.

Another thing that helps to get trade is to keep a list of your customers and

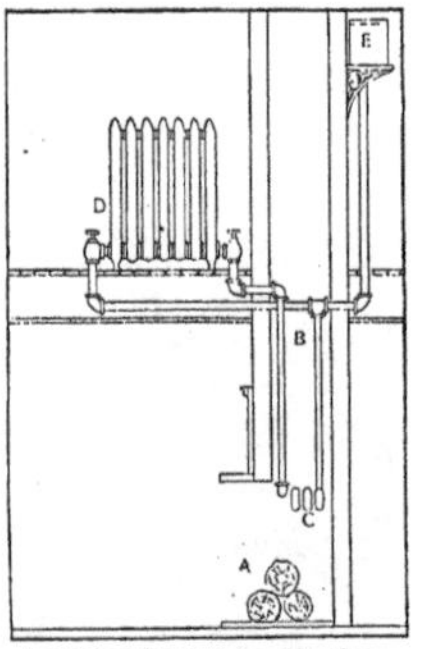

Radiator Heated by Fireplace.

another list of those who are not, but whom you would like to have as c tomers. Write personal letters to them and if you have any snaps going, tell them about them. Tell them candidly about your goods and quote prices on some of the articles. They will certainly be interested, and come to your store to see and perhaps buy some of the goods you mention.

Always try to keep in touch with the people, and at all times be ready to make a deal. You can take second-hand stoves as part pay on a new one. There is money to be made on them and they can generally be handled to advantage by giving them a thorough overhauling and supplying a new grate or fire-brick, by giving the oven and fire-box a coat of whitewash or red oxide, and the interior a liberal coat of black lead.

UTILIZING HEAT FROM CHIMNEY.

A correspondent in the Rural New Yorker tells how the wasted heat in a chimney of a fireplace may be utilized to advantage. Where rooms are directly over the fireplace, he writes, it may be possible to bypass the hot gases in the chimney through an old-fashioned drum, and if the size of the chimney would admit it, it would be a very easy matter to tear out some of the chimney wall and locate a pipe coil right over the fireplace in the chimney so that one or more radiators may be connected with the coil. It will be necessary to use care in placing the coil so as not to obstruct the draft.

The accompanying illustration shows how the hot water system may be installed. A is the fireplace; C the coil; B the conecting pipes; D the radiator, and E, the expansion tank. The pipe connections should be made from one end of the radiator to the top of the coil. Any series of radiators may be connected in the same manner. The expansion tank is for the expansion of the heated water as well as to keep a head of water above the radiators to keep them full.

STOVE EXHIBITS AT FAIRS.

The approach of cold weather makes it imperative that every merchant selling stoves pay particular attention to every possible means of making sales of his stoves and ranges. With prices advancing and money none too plentiful the best judgment will be necessary in order to show satisfactory results at the close of the stove-selling season.

A demonstration of stoves at the fall fair in the dealer's town has proved to be good advertising if conducted in an up-to-date manner. Ideas as to display can be picked up from the large displays at the Toronto Exhibition, the best of which were illustrated in Hardware and Metal of Sept. 14. If desired, arrangements could be made with some grocer to co-operate in a biscuit-baking demonstration. An exhibit, however, should not be undertaken if the dealer is not prepared to put energy into it and take advantage of all the many possibilities for getting in touch with prospective buyers. Circulars and local advertising must be attended to beforehand and some plan adopted of registering the names of visitors who are likely to become buyers.

CANADIAN STOVES IN THE WEST.

At the annual convention of Canadian Manufacturers' Association held in Toronto last week there was considerable discussion regarding the low tariff on stoves. Canadian manufacturers claimed they were making much better stoves than United States manufacturers, but a number of Western dealers were still importing stoves from the United States, owing to the lower price at which they could be procured.

One manufacturer stated that it was the custom of steel range manufacturers in the United States to send over a man to Canada who went through the country with a sledge hammer, demonstrating the fact that their stoves could not be broken. This seemed to have a good deal of weight with many of the farmers in the West, but the dealer should know that the main test of a range is its cooking qualities, and it is the stove that will give satisfaction in this connection that will build up a good stove trade for any dealer.

We reproduce a letter which has just been received by the Moffat Stove Company, Weston, Ont., which seems to bear out the point that Canadian manufacturers are making higher grade stoves than their competitors in the United States.

Seattle, Wash., Sept. 14, '07.

Messrs. Lundy & McLeod,
Edmonton, Alta.

Dear Sirs,—Would you please send me a little information as to where the nearest place, where I can buy a range such as I bought from you last winter?

I forget the name, but it was manufactured by the Moffat Stove Co., and had 20-inch oven.

Have had a great deal of dissatisfaction on this side about ranges, and decided if we can get one like our other one in Vancouver, I would pay the duty and freight.

This will give you an idea as to the quality of your range, when I say, we cannot find anything to near equal it. We were so highly pleased with it, and it did its work so perfectly, that we cannot be satisfied now with anything but a first-class range.

I would order one from you only the freight is too much. Please be kind enough to send name of range, etc., we are not sure of name and take this means of finding out. Send me the manufacturer's name and address so I can write them.

Hoping to hear from you soon, I am,

Yours truly,

D. G. R.

A new flux for soldering has been patented by Oscar J. Lanigan of Chicago, Ill., which it is claimed will not act upon the metal to be soldered unless heated. For this reason, an excess of soldering fluid which may remain on the work after soldering will not injure it. The flux is composed of equal parts of glycerine and ammonium lactate. The ammonium lactate may be made by adding ammonia to lactic acid until the solution is neutral. If desired, the flux may be made as follows: Glycerine, 65 parts; lactic acid, 49 parts, and ammonia, 14 parts.

There are merchants who can get accounts settled by bankable notes when the debtor is unable to pay the money. You can if you go at it right.

CUTTING PATTERNS IN THE TIN SHOP

Readers of Hardware and Metal are requested to make use of this department. Questions regarding patterns will be answered by experts. Discussion is also invited on any matter pertaining to the tin shop.

Patterns for Three-Piece Elbow.

Ross D. Evans, with G. M. Lorimer & Co., Manor, Sask., writes: "I have followed the articles in Hardware and Metal concerning hot air heating and patterns for the tin shop, and would be much obliged if you would send me a

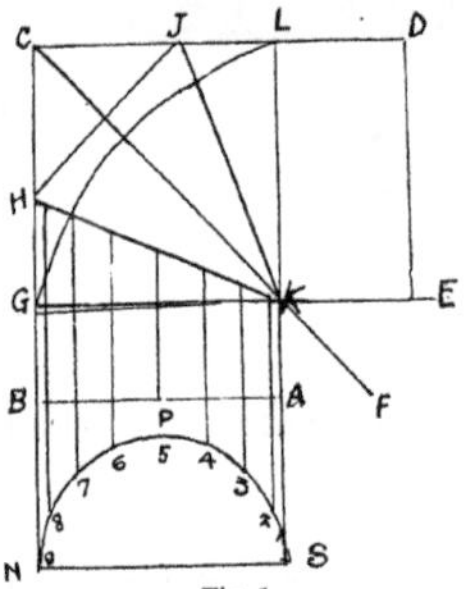

Fig. 1.

description of how to draft a three-piece elbow. Will be willing to compensate you for whatever trouble you are put to in this matter."

This request for information is just what the editor has been looking for. One of our staff answers the question below, without cost to our subscriber, and we are pleased to be able to assist in this way. Hardwaremen should encourage their tinsmiths to send in further questions, as a series of practical talks on the problems met with in the tin shop would be a feature of the paper read with interest in the majority of hardware stores in Canada. Send us additional questions and help to make this department a fixture in the paper.

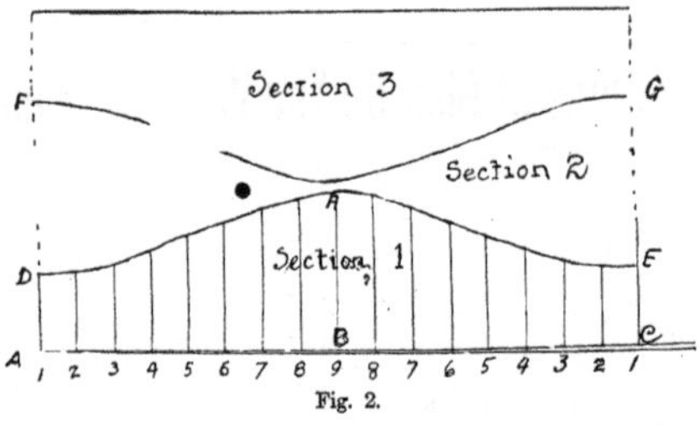

Fig. 2.

The Question Answered.

To draft a three-piece elbow, let A, B, C and D, in Fig. 1, be the given elbow. Draw the line F, C. Make F, K equal to one-half the diameter of the elbow. With F as centre, describe the arc G, L. Divide the arc G, L into four equal parts. Draw the lines H, K and J, K, as well as the line H, J. Then draw the line A, B at right angles to B, C. Describe the semicircle N, P, S, and divide it into any number of equal parts. From the points draw lines parallel to B and H, as 1, 2, 3, 4, etc.

Set off the line A, B, C, as shown in Fig. 2, equal in length to the circumference of the elbow A, B in Fig. 1. Erect the perpendicular lines "A, D," "B, H," "C, E." Set off on each side of B, H the same number of equal distances as in the semi-circle N, P, S. From the points draw lines parallel to B, H, and make B, H, in Fig. 2 equal to B, H in Fig. 1. Make A, D and C, E equal to A, K. Also make each of the parallel lines in Fig. 2 equal to the parallel lines bearing the same number in Fig. 1, as 1 to 1, 2 to 2, etc. Then a line traced through the points will form one of the sections.

Make D, F and G, E in Fig. 2 equal to H, J in Fig. 1. Then reverse the section No. 1 and place D at G, and E at F, and trace a line from G to F. This forms sections 2 and 3. Edges must be allowed.

W. M. M.

NEW STOVE PIPE.

An improvement in stove and other pipe-joints, worked out by J. McGhie and B. Blood, Spokane, Wash., has for an object the provision of a simple construction which can be easily made and which will effectually and safely couple the pipe sections so they may be conveniently detached. In practice the sections may be made with projecting hook-like tongues at one end and the beads at the other end for convenience in fitting the sections together.

Second-hand stoves intended for sale should be thoroughly cleaned, both inside and out, and any requiring new repairs should have the new parts placed in them. Place such stoves in your show room and do not look for too big a profit.

Samples of ranges and heaters in stock should be placed on the floor so that customers cannot help seeing them as they pass in and out of the store. It is a good plan also to have clerks draw the attention of every customer to your display of stoves at every opportunity.

CATALOGUES AND BOOKLETS.

Peterboro Rapid Tools.

Hardware merchants wkho have not received a copy of catalogue No. 1, issued by the Rapid Tool Co., Peterboro, should make it a point to send for one. The book contains only 32 pages, but is devoted to the one line of goods made by the company and it is well worth preserving. All styles of augur bits, car bits and wood-boring tools are illustrated, full descriptions being also given. The company has profited from the experience of others, and claim to have the most modern factory of its kind and size in America. Their "Rapid" boring double-thread screw, patented a year ago, is finding great favor with the trade and amongst mechanics.

* * *

Disston Handbook on Saws.

A new edition of the Lumberman handbook on saws has been brought out by Henry Disston & Sons, saw manufacturerers' Philadelphia, Pa. The book contains over 200 pages, and contains an interesting treatise on the construction of saws, together with information on how they can be kept in order and similar information. The present edition of the book has been made more interesting and attractive than the former edition by the addition of articles on the making of Disston steel saws tools and files, giving the progressive manufacturing steps, numerous illustrations of sections of departments of the great plant, etc. When writing this paper should be mentioned.

* * *

"India" Oil Stones.

The Pike Manufacturing Company, Pike, N.H., are selling agents for the "India" oil stones, made by the Norton Company, Worcester, Mass., and are sending out a booklet descriptive of these goods, with price list, to dealers requesting same. The tiger on the trade mark is characteristic of this line, the "India" oil stones being described as "sharp and quick." "India" oil stones are said also to possess qualities of wonderful durability and uniform grit.—Mention Hardware and Metal.

* * *

"Cheapest on the Job."

Ferdinand Dieckmann, manufacturer of conductor elbows, Cincinnati, Ohio, is very liberal in the matter of samples and catalogues when corresponded with by dealers. The Dieckmann elbow is made of one piece, a method that is not an experiment, but has stood the test of time.

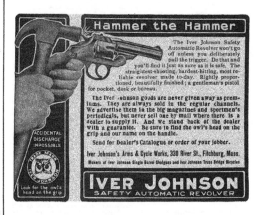

BUILDING AND INDUSTRIAL NEWS

For additional items see the correspondence pages. The Editor solicits information from any authoritative source regarding building and industrial news of any sort, the formation or incorporation of companies, establishment or enlargement of mills, factories or foundries, railway or mining news.

Industrial Development.

The Livingstone flax mill at Brussels, Ont., was destroyed by fire.

The Chatham Gas Co., Chatham, Ont., have completed their plant.

The bridges and locks on the Welland Canal will be operated by electric motors next season.

The Prince Albert Lumber Co., Prince Albert, Sask., suffered loss by fire to the extent of $10,000.

Large deposits of peat have been found near West Zorra, Ont. They will be developed immediately.

The Delaware Seamless Tube Co., Auburn, Pa., will erect a plant at Sarnia, Ont., to cost $200,000.

J. E. Doak, Doaktown, N.B., will erect a factory in Saskatoon, Sask., for the manufacture of house furnishings.

The new factory of the King Radiator Co., Toronto, is nearing completion. The whole plant will cost $250,-000.

The Albert sawmills at Barachois, Que., were destroyed by fire, with all Que., were destroyed by fire, with all lumber.

Walker & Sons, Walkerville, Ont., will erect an addition to their planing mill. Additions will be made to the boiler and engine equipment.

The American Electrical Furnace Co. will erect a Canadian factory at Niagara Falls, Ont. They will manufacture iron and steel castings, the melting being done by electricity.

The strike at the Springhill, N.S., Coal Mines has now been on for two months, and the Government of Nova Scotia will on Wednesday take charge of the work heretofore done by members of the Mechanics' Lodge of the collieries.

Traces of graphite in Haliburton County, Ontario, are being investigated by Eugene McSweeney, president of the United States Graphite Co., Saginaw, Mich. It is said the deposits are rich along the line of the I. B. & O. Railway.

Three industries are looking for a location in Berlin. They will manufacture collapsible go-carts, concrete machinery and washing machines. The outlay will be $100,000 and 300 hands would be employed. Certain conditions will have to be fulfilled.

The organization of a company for the manufacture of railway cars in the Canadian West is being effected in Toronto. Winnipeg is being considered as a location for the factory. The company will have a capital of $2,000,000, and will employ 1,000 hands.

The International Heating and Lighting Co., Edmonton, Alta, will erect a plant for the manufacture of producer gas from straw. The plant will cost $100,000 and mains will be laid to pipe the gas to all parts of the city. Over thirty miles of these mains will be laid, entailing an expenditure of $200,000.

The Dominion Bridge Company, of Lachine, of which James Ross is president, will increase its capital from $1,-000,000 to $1,500,000. This course has been found necessary by the great expansion of business and the large additions to the company's several plants. New shops are being built in Toronto and Winnipeg, and very extensive additions to the plant at Lachine are likewise being undertaken.

London Fence, Ltd., Portage la Prairie, have decided to rebuild their factory, which was destroyed by fire early in the year. For some time there was uncertainty as to the future plans of the company, owing to the dispute with the insurance companies as to the amount of the fire losses. However, the fire losses have been adjusted, and at a meeting of the shareholders, held last week, it was unanimously decided to rebuild.

The Dominion Wrought Iron Wheel Works, Orillia, is enlarging its factory premises by the addition of a 40 x 50 brick building with metal roof to be used as a foundry, and a 40 x 20 metal clad storehouse. The new buildings are west of the present factory, and close to the railway switch, which will be convenient for handling supplies as well as shipping the finished products. The Wheel Works is one of Orillia's rapidly expanding industries.

It has been discovered that the new C.P.R. Assiniboia, which has been cut in two at the graving dock at Levis, Que., for the purpose of taking her through the canals to do service on the lakes, would not float. Her ballast will therefore have to be shifted, and this operation was begun this morning. As soon as this vessel leaves port her sister ship, the Keewatin, will undergo the same operation of being severed in two, and also leave to do service on the lakes.

Municipal Undertakings.

Ottawa will spend $28,000 in sewage construction.

Olds, Alta., has passed a bylaw to raise $12,000 for fire protection.

The ratepayers of Estevan, Sask., will vote on a bylaw to raise $92,000 for waterworks.

Three money bylaws were passed in Vancouver: For new sewers, $300,000; for new roads, $100,000; for schools, $45,000.

The town of Gravenhurst, Ont. are considering the granting of certain concessions to Hess & Co., who will erect a factory to cost $10,000.

Railroad Construction.

The C.P.R. will erect a new station at Listowel, Ont., at a cost of $10,000.

The contract has been let for the construction of the line for the C.P.R. from Regina, Sask., to Bulyea.

English capitalists are contemplating the construction of a line from Edmonton to Dawson, a distance of 1,400 miles.

The C.P.R. will build a line along the shore of Lake Ontario, between Toronto and Montreal, passing through Port Hope.

Mining News.

Large shipments of iron are being made from the Atikokan district.

The Dominion Copper Co. will enlarge its plant at Boundary Falls, B.C.

The smelter of the Montreal Reduction and Refining Co., North Bay, Ont., will be in operation in a year's time.

The Del Nord Co. have located their head office at Sarnia, Ont. They are considering the erection of a large ore reducer in Quebec.

Building Notes.

James Worts, Toronto, will build a residence to cost $18,000.

A. J. Strathy, Toronto, will build a four-storey store at a cost of $12,000.

The number and value of building permits in Toronto for September has decreased considerably in comparison with September, 1906, although the aggregate for the year, so far, is well above that of the same period last year. The total in September was $550,000, but the $200,000 permit for the Alexandra Theatre has been added to this, and thus the total is $750,000. This is $152,803 below the figure for September, 1906, which was $902,803. The permits issued so far this year amount to $12,204,000, as compared with $9,566,-000 for the same period of 1906.

Companies Incorporated.

The Chatham Carriage Co., Chatham, Ont.; capital, $100,000. To manufacture all kinds of vehicles. Provisional directors: Ira Teeter, Arthur Cooke, Frank E. Fisher, all of Chatham, Ont.

Dominion Crown Cork Co., Toronto; capital, $15,000. To manufacture and deal in crown corks, wire, tin and other metals, metal stamping and cork. Provisional directors: W. F. Hayes, Gordon Russel, Violet Waldock, all of Toronto.

The Canadian Smelting & Refining Co., Toronto; capital, $2,500,000. To prospect for and deal in ores, metals and minerals. Provisional directors: J. D. Pringle, D'Arcy Grierson, Victoria Morrison, Alexander G. Roberston, W. Maxime Wallace, all of Toronto.

Haileybury Brick & Tile Co., Haileybury, Ont.; capital, $50,000. To manufacture and sell building material of all kinds. Provisional directors: A. J. Murray, Benson Clothier Beach, Haileybury, Ont.; Chas. F. McArthur, E. E. Wilson, Duncan McArthur, all of Kenmore, Ont.

The Canadian Northern System Terminals, Toronto; capital $2,000,000. To build and operate terminals, stockyards, oil tanks, elevators, smelters and ore furnaces. Incorporators: Gerard Ruel, A. J. Mitchell, J. B. Robertson, R. P. Ormsby, F. C. Annesley, L. W. Mitchell, all of Toronto.

The Calkins Tile & Mosaic Co., Montreal; capital, $20,000. To manufacture and trade in any and every kind of marble, tile, terrasse and mosaic, concrete, granite, sandstones, limestones, clays, slates, plaster, terra cotta, lumber and all classes of structural iron and steel and building supplies. Incorporators: W. J. Henderson, Allan L. Smith, J. W. Hannah, A. C. Calder, John W. Graham, all of Montreal.

The Dominion Tool Company, Peterboro; capital, $100,000. To carry on the manufacture and sale of augurs, bits, gimlets, and all other tools or appliances used for boring or drilling purposes; also to manufacture and sell bolts, rivets, screws, and the necessary machinery for the manufacture of the same. Provisional directors: James Davidson, F. R. J. MacPherson, P. J. Creedon, F. C. Cubitt, F. P. McNulty, all of Peterboro, Ont.

A NOVEL TEST.

A Brooklyn builder has just made a somewhat novel test to get at the relative cost of houses built with different available material. A set of plans was produced and three houses were built from the same plans, one of wood, one of concrete with wood floors, and the third of hollow tile blocks and concrete. The building of wood cost $6,000; the one of concrete $8,900, and the one of tile and concrete $6,500, and the decision was in favor of the latter structure, for obvious reasons—low cost, small future outlay for repairs and immunity from possible destruction by fire. This would seem to relegate lumber to the "also ran" class.

ONTARIO'S RICH MICA MINES.

At Sydenham, Ontario, sixteen miles from Kingston, is located the largest mica mine in the world. The product is mostly amber mica, with some silver amber, the highest quality mined. The mine is one mile from the upper end of Sydenham Lake, and the mica is transported in bulk from the mine by barge to the railroad at Sydenham, where it is shipped to Ottawa for trimming for the market. From Ottawa it is exported to the United States and other points. This valuable mine is owned and worked by Americans, and the output is almost entirely taken by one of the largest American electric companies.

FARMERS TAKE TO PLUMBING.

E. A. Latimer, manager of the plumbing department of the Orillia Hardware Company's business, was a visitor in Toronto a week ago. Mr. Latimer says that of 36 plumbing outfits contracted for this season, three have already been installed in farm houses, and four more are yet to be completed for farmers. Up to this season, none of this work had been done in the rural section near Orillia.

The Orillia Hardware Company has in hand a nice contract for a $3,000 plumbing and heating job on the new Orillia hospital, besides a $1,000 job on a new residence for T. H. Sheppard.

"CHARLEY" WORLD'S CHANGE.

Charley World, for many years with the James Morrison Brass Manufacturing Company, Toronto, has accepted a position with Somerville, Limited, and will introduce the new "Somerville" brass goods to the trade as soon as the new plant is producing goods.

In another month the machinery will be installed and put in motion, and by December a line of high-class brass goods will be offered the trade and prompt shipments made.

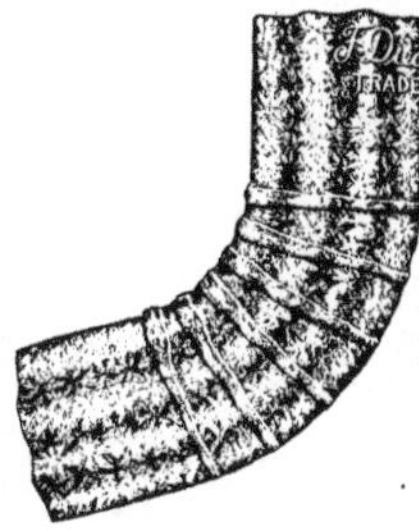

News of the Paint Trade

MURESCO IN MASSEY HALL.

Massey Hall in Toronto is the largest meeting and concert hall in Canada, seating 5,000 persons. The hall caters to the most cultured citizens and the decorative work, therefore, is most important. This fall the hall has been redecorated by the Thornton-Smith Company, the scaffolding for the job costing nearly $2,000 alone.

In decorating such a famous concert hall, the contractors found it necessary to select a material which would combine the qualities of beauty, reasonable cost and durability, and it is a strong commendation of the products of the Benjamin-Moore Company, Toronto Junction, that their "Muresco," so favorably known to the hardware trade throughout Canada, should be selected for the decoration of the vast wall space in the concert hall.

In describing the decorative work the Toronto Globe says: "The hall being principally used at night, the color scheme was devised with a view to obtaining the best effect by electric light consistent with adherence to the Mauresque character of the interior and harmonizing with the more modern style of the main fabric. The dominating note is blue, balanced by a low-tone greenish cream, which, in the vast expanse of the building, takes its place as white, giving that beautiful Moorish combination of blue and white so greatly admired in the Alhambra decorations. The treatment of the south wall, while not intended to be permanent, has removed the bareness which previously existed, and is now in harmony with the decoration of the whole interior."

PAINT TRADE GOSSIP.

L. C. Dumais & Co., Hedleyville, Que., have been registered to carry on business as painters.

The Martin-Senour Company, paint manufacturers, Montreal, have increased their capital stock from $50,000 to $100,000.

Arthur Cornellier and Emile Ayotte, Montreal, have been registered to carry on business as painters. The new firm will be styles Cornellier & Ayotte.

A petition for the winding-up of the Victor Varnish Company has been filed at Osgoode Hall, Toronto, by Albert B. Crosby, president. This action is the result of the decision of the shareholders that this is the only way out of the "hopeless state of the company's finances." The company was incorporated in 1904, to deal in paints, oils and varnishes. Its nominal capital was $40,000 in $50 shares. Two hundred and eighty-six shares were subscribed, and 241 paid-up. The assets are placed at $17,329.21, and there is a deficit of $13,697.66.

HOW TO COLOR OLD WINDOW SHADES.

Mix a sufficient quantity of wall paint, which may be of any color desired, and after placing the shade on a flat surface paint two coats with a brush, putting it on with strokes crosswise. When dry the shade will be like new.

DO YOU WANT TO MAKE MORE MONEY?

This seems to be a needless question. Of course you want to make more money, or you wouldn't stay in business. Well, look over your stock and see if it contains any

NEW ERA PAINT

If it doesn't, and you want to do a more satisfactory paint business, you just send in a sample order. If you haven't got NEW ERA PAINT you are neglecting the BIGGEST MONEY-MAKER in the Canadian paint trade.

STANDARD PAINT AND VARNISH WORKS CO., LIMITED
WINDSOR - ONTARIO

WE KNOW IT'S GOOD

and if we could make it any better, we would do so. We personally test every bit of material that goes into

Jamieson's Floor Enamel

Pigments and all other materials are all the very best we can buy in the markets of the world. The greatest skill and care are used in grinding and mixing. Consequently we are making in our **Floor Enamels** the best possible floor covering that can be used; and yet, they do not cost much more than the ordinary class of floor paints.

These are talking points which move the goods from your shelves, Mr. Dealer !

Be sure to stock up with

Jamieson's Floor Enamel

in all popular colors

Made only by

R. C. JAMIESON & CO.
LIMITED
16-30 Nazareth St., MONTREAL

HEADQUARTERS
FOR
Window Glass,
Paints, Oils,
Fine Varnishes

SANDERSON PEARCY & CO., Ltd.
Toronto, Canada

GLAZIERS' DIAMONDS OF PROVED WORTH
Having a Century's Reputation for Reliability.
MADE BY
A. SHAW & SON, London, Eng.
CANADIAN AGENT
GODFREY S. PELTON, - 201 Coristine Building, Montreal

Sharratt & Newth's Glaziers' Diamonds

are unequalled for cutting and wearing qualities.

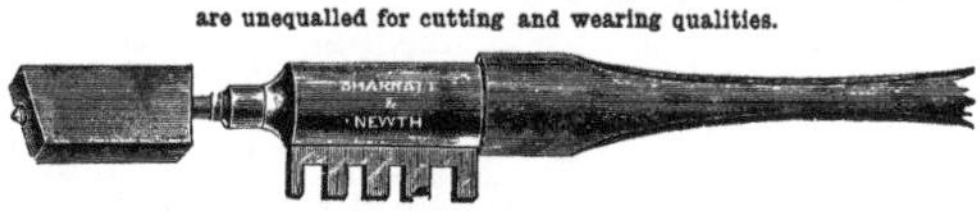

To be obtained from the principal Dealers in Glass, Hardware, and Painters' Supplies.

Contractors to H. M. Government and the principal English Sheet and Plate Glass Works

GILBERTSON'S
COMET

Gilbertson's Sheets
and the COMET Brand

mean quality and security in whatever work you may undertake.

Galvanized flat sheets that are unequalled as to flatness ; absolute trueness to gauge ; admirably suited in softness for any working. Note the brand "COMET." GILBERTSON on every sheet means security.

W. GILBERTSON & CO., Limited, Pontardawe, South Wales
ALEX. GIBB, 13 St. John St., Montreal, Agent

Paint and Oil Markets

TORONTO.

Toronto, Oct. 4.—October business has started in a manner that is quite pleasing to local jobbers, and the number of good orders received for the first four days of the month indicates that the autumn market is assuming a very healthy tone. During the week managers of local supply houses have had an opportunity to sum up September paint and oil business, and they unanimously report the aggregate sales for the month to be far in excess of those for the corresponding month in any previous year. A peculiar feature of the market at the present time is that, instead of declining about this time, as many jobbers expected, prices, in general, have taken on a firmer tone, and, in some commodities, quotations have even been slightly advanced. As Canadian paint and oil quotations are largely dominated by the prices prevailing in the primary markets of supply, only a very minute study of the foreign markets would enable the reader to get at the real cause of the present upward tendency of prices.

Linseed Oil—The market in Great Britain still remains very firm, and recent reports show the amount of seed arrivals for September to be considerably below expectations. As a consequence of this firmness in the British markets, local prices are firmer though there has been no actual advance. We quote: Raw, 1 to 3 barrels, 64c; 4 barrels and over, 63c. Add 3c to this price for boiled oil, f.o.b., Toronto, Hamilton, London, Guelph and Montreal, 30 days.

Turpentine—In spirits of turpentine the advice from Savannah is that the market is firm with a slight advance, and a further advance looked for at an early date. Locally, however, prices are unchanged, though the market is much steadier on account of the Southern advance. We still quote: Single barrels, 76c; two barrels and upwards, 75c; f.o.b. point of shipment, net 30 days; less than barrels, 5c advance. Terms, 2 per cent., 30 days.

White Lead—The large amount of inside painting now being done maintains a good demand for this commodity, and, contrary to expectations, prices are growing firmer. Within the last fortnight English dry and pig lead have each taken a considerable advance, and Canadian quotations have stiffened as a result. Until recently there was a general impression that the Canadian Government would take off the import duty on dry white lead. Such action by the Government is now highly improbable, however, as is evidenced by the fact that a strong company, backed by English capital, will soon have a new white lead corroding factory completed and running in Montreal. Prices are firm, as follows: Genuine pure white lead, $7.65, and No. 1, $7.25.

Red Lead—A good seasonable demand prevails, with prices firmer, in sympathy with the firmer tone of the pig lead markets. We quote: Genuine, in casks of 500 lbs., $6.25; ditto, in kegs, of 100 lbs., $6.75; No. 1 casks of 500 lbs., $6; ditto, in kegs of 100 lbs., $6.50.

Glass and Putty—Both are in heavy demand, and a brisk business is being done for the first week in October. Prices are firm and unchanged.

Petroleum—The fall rush is now on and orders are coming in fast. Prices are firm and unchanged: Prime white, 13½c; water white, 15c; Pratt's astral, 18½c.

Shellac—October enquiries are well up to expectations, and prices are firm, as follows: Pure orange, in barrels, $2.70; white, $2.82½ per barrel; No. 1 (orange), $2.50.

Stove Pipe Varnish—An exceptionally heavy business is being done in this, and orders from small centres are unusually large. Cases of half-pint tins are quoted at $8 per gross.

Bees' Wax—A heavy autumn demand is making itself felt due largely to the popularity of this substance for flattening varnish, in addition to its ordinary domestic uses. At present bees' wax is quoted at 40c per pound.

Pine Tar—This substance is now in heavy demand in rural communities, owing to the variety of uses to which it can be put around a stock farm. Half-pint tins are quoted at 80c per dozen.

Plaster of Paris—As many buildings have already reached the stage where the finishing coat of plaster must be applied, this commodity is now being shipped to retailers by the carload to satisfy the exceptionally heavy demand which the enormous amount of building has created. Plaster of paris is quoted at $2.20 per barrel.

MONTREAL.

Montreal, Oct. 4.—Variable weather conditions during the past week or two have diminished to some extent the production in paint and varnish circles. Orders are coming in rather spasmodically, owing to the unsettled conditions both of the weather and the money market. A fair turnover for October is expected by the makers.

The transportation problem is still a vexing one, some makers having been seriously delayed by tardy arrivals of shipments of raw materials.

Linseed Oil—Shows a firming tendency and some fair shipments are being made. Prices are unchanged: Raw, 1 to 4 barrels, 61c; 5 to 9, 60c; boiled, 1 to 4 barrels, 64c; 5 to 9, 63c.

Turpentine—The Southern market is very firm. Locally it is steady at last week's quotations, and it is not expected that the price for this article will ease off during the balance of the season. We continue to quote: Single barrels at 74c; wood distilled turpentine, pure, 68c.

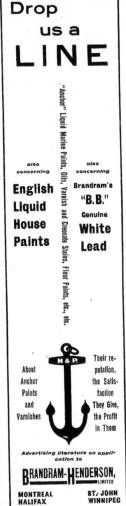

Ground White Lead—There is a fair enquiry for this staple and in localities

where inside painting has been brisk, a good turnover is being experienced. Prices are firm, but some grinders continue to offer slight concessions for big orders. Government standard, $7.50; No. 1, $7; No. 2, $6.75; No. 3, $6.35.

Red Lead—Continues quiet. The slight marking down of quotations last week has not induced any large bookings. Prices, generally speaking, are firm and unchanged. Genuine red lead, in casks, $6.25; in 100-lb. kegs, $6.50; in less quantities at $7.25 per 100 lbs. No. 1 red lead, casks, $6 kegs, $6.25, and smaller quantities, $7.

Dry White Zinc—This is in good request without any change in prices: V.M. Red Seal, 7½c; Red Seal, 7c; French V.M., 6c; Lehigh, 5c.

White Zinc Ground in Oil—Is being shipped a little more freely for decorative work. Prices are firm: Pure, 8½c; No. 1, 7c; No. 2, 5⅜c.

Gum Shellac—Stocks are ample and a fair output is reported. Prices are firm and unchanged: Fine orange, 60c per lb.: medium orange, 55c per lb.; white (bleached), 65c per lb.

Shellac Varnish—As is usual in the fall months, there is a very much better enquiry for this article. Prices are unchanged: Pure white bleached shellac, $2.80; pure orange, $2.60; No. 1 orange, $2.40.

Putty—Is meeting with the usual demand experienced with the approach of winter, and quotations are extremely firm: Pure linseed oil, $1.85 in bulk; in barrels, $1; in 25-lb. irons, $1.90; in tins, $2; bladder putty, in barrels, $1.85.

DAMMAR VARNISHES.

Dammar has been known a long time, and has been largely used for varnishes on account of the ease with which it dissolves in the cold in turpentine and in other hydro-carbons, and on account of the ease with which it can be mixed with linseed oil, also at low temperatures. It is also valued for the very pale varnishes that can be made with it. Nevertheless, dammar varnishes are not altogether so satisfactory in regard to durability, as they are with reference to appearance, and to the ease with which they can be prepared. For out-of-door work, in fact, they are almost useless. Before the invention of enamel colors, which themselves are often made with dammar, they were indispensable for light colored work, and they are still so in cases where a perfectly colorless varnish is wanted, for maps, pictures, etc., which have to be protected indoors. Dammar varnishes, being fairly cheap, are often used to lighten the color and increase the softness of copal varnishes. They form a large constituent of the so-called water-white copal varnishes, and much diluted with turpentine are sold as white driers. Dammar is now less used in varnish-making than formerly, as the rise in price and the substitution for it of Manila copals, soluble in spirit for varnishing paper, has affected it.

Dammar varies, not only in color, but in transparency, in pieces taken from the same chest. Some pieces will be clear and transparent, others opaque, and full of bubbles. The pieces may be sorted out, and the one kind will give a clear solution in oil of turpentine, the other a turbid solution only, which will not clear even after years of standing. The solution can, however, be cleared within a few hours by the addition of a little absolute alcohol, but will retain a yellowish hue. Turbid dammar varnishes cannot be cleared by commercial filtration. Small quantities can be cleared by filter paper, but it takes a long time. As the sorting of the dammar is a long and expensive job, the whole is dissolved in turpentine and cleared by heating, a process discovered by Miller. Another advantage of this heating is that it makes the varnish dry better. Without it the varnish dries very quickly on the top and prevents the lower parts of the coat from drying. The reason is, no doubt, the presence of water in the unheated dammar. The opaque pieces contain water mechanically enclosed. Even when a coat of the unheated varnish does get dry, the contraction and expansion of the enclosed water finally break up the varnish and make it scale off. It is evident that the resin can be dried in a stove, and then dissolved in cold turpentine, with the same good results, but it is found cheaper and more convenient in working on a large scale to heat the dammar with the turpentine.

The varnish is made in the following way: Forty lbs. of unpowdered dammar (attempts to powder it only result in beating it out) is mixed in a pan with 50 lbs. of oil of turpentine and heated. As soon as the mass has lost all its water, it boils quickly, but it froths up until then. The varnish is then put through a sieve, and left to clear. It is a good plan to add 5 per cent. of bleached drying linseed oil to make the varnish more elastic and less brittle. Some makers also add 10—20 per cent. of larch turpentine, and for the commoner varnishes thick turpentine or galipot is often added. Cheap dammar varnishes are also brought down in price by the addition of colophony. In any case, the addition of the bleached oil is very advisable, or of some pale, soft resin. The oil of turpentine can be well replaced by coal-tar distillates, which give a quicker and better drying varnish. Too little has as yet been done in trying combinations of dammar with other resins. It is probable that some very useful results would be achieved in this way. For example, equal parts of hard copal and dammar fused together give a splendid varnish with linseed oil and turpentine. Another is got by dissolving in alcohol containing benzoline or oil of turpentine, a mixture of 30 lbs. of dammar, 10 lbs. of sandarach, 20 lbs. benzoin, and 10 lbs. of climi, or larch turpentine.—English Decorator's Review.

The assets of L. E. McPhee, paint and wall paper dealer, Ottawa, are being sold by tender.

CURRENT MARKET QUOTATIONS.

October 4, 1907

These prices are for such qualities and quantities as are usually ordered by retail dealers on the usual terms of credit, the lowest figures being for larger quantities and prompt pay. Large cash buyers can frequently make purchases at better prices. The Editor is anxious to be informed at once of any apparent errors in this list, as the desire is to make it perfectly accurate.

METALS.

ANTIMONY.

Cookson's ....per lb.	0 13½
Hallett's	0 13

BOILER PLATES AND TUBES.

	Montreal	Toronto
Plates, ¼ to ½ inch, per 100 lb.	2 40	2 50
Heads, per 100 lb	2 65	2 75
Tank plates 3-16 inch	2 60	2 65
Tubes per 100 feet, 1½ inch	8 25	8 75
" " 2 "	9 00	9 10
" " 2½ "	10 50	12 00
" " 3 "	12 00	13 50
" " 3½ "	15 00	16 70
" " 4 "	19 25	20 80

BOILER AND T.K. PITTS.

Plain tinned and Spun, 25 per cent. off list.

BABBIT METAL.

Canada Metal Company—Imperial genuine 60c; Imperial Tough, 60c; White Brass, 50c. Metallic, 35c.; Harris Heavy Pressure, 25c, Hercules, 25c; White Bronze, 15c; Sta Frictionless, 14c.; Alluminoid, 10c.; No. 4 9c, per lb

James Robertson Co.—Extra and genuine Monarch, 60c.; Crown Monarch, 50c; No 1 Monarch, 40c; King, 30c; Fleur-de-lis, 20c; Thurber, 15c.; Philadelphia, 12c; Canadian, 10c; hardware, No 1, 15c.; No. 2, 12c.; No. 3, 10c. per lb.

BRASS.

Rod and Sheet, 14 to 30 gauge, 25 p c. advance.

Sheets, 12 to 14 in.	0 30
Tubing, base, per lb 5-16 to 2 in	0 33
Tubing, ⅜ to 3-inch, iron pipe size	0 31
" 1 to 3-inch, seamless	0 36

Copper tubing, 6 cents extra.

COPPER.

		Per 100 lb
Casting ingot		17 00
Cut lengths, round, bars, ½ to 2 in.		28 00
Plain sheets, 14 oz.		30 00
Plain, 16 oz., 14x48 and 14x60		29 00
Tinned copper sheet, base		32 00
Planished base		30 00
Braziers' (in sheets). 4x6 ft., 25 to 30 lb. each, per lb., base	0 28	0 29

BLACK SHEETS.

No 10 gauge	2 70	2 70
12 gauge	2 70	2 75
14 "	2 50	2 60
17 "	2 50	2 60
18 "	2 50	2 60
20 "	2 50	2 60
22 "	2 55	2 65
24 "	2 55	2 70
26 "	2 65	2 85
28 "	2 70	3 00

CANADA PLATES.

Ordinary, 52 sheets	2 75	3 05
All bright "	3 75	4 05
Galvanized—	Dom Crown	Ordinary.
18x24x52	4 45	4 35
60	4 70	4 60
20x28x80	8 90	8 70
.....	9 40	9 20

GALVANIZED SHEETS.

B.W. gauge	Queen's Head	Fleur-de-Lis	Gordon Crown	Colborne Crown Gorbal's Best
16-20	3 95	3 80	3 95	....
22-24	4 20	4 05	4 00	4 05
26	4 45	4 30	4 40	4 30
28	4 70	4 55	4 60	4 55

Less than case lots 10 to 25c. extra.

Apollo Brand.

24 guage, American	3 85
26 " "	4 10
28 " "	4 55
10¾ oz.	4 85

25c. less for 1,000 lb. lots.

IRON AND STEEL.

	Montreal.	Toronto.
Middlesboro, No. 1 pig iron	21 50	23 50
Middlesboro, No. 3 pig iron	20 50	22 00
Summerlee, "	25 50	25 00
" special "	24 50	....
" soft "	24 00	....
Carron	25 50	....
Carron Special	24 50	....
Carron Soft	24 00	....
Clarence, No. 3	21 50	22 00
Glengarnock, No. 1	....	25 50
Midland, Londonderry and Hamilton	....	23 00
Radnor, charcoal iron	30 00	32 00
Lowmoor iron		6 50
Angles		2 65
Common bar, per 100 lb.	2 20	2 30
Forged iron "	2 45	..
Refined " "	2 60	2 60
Horseshoe iron "	2 60	2 60
Band iron, No. 10 x ⅝ in.	....	2 30
Sleigh shoe steel "	2 25	2 30
Iron finish steel		2 30
Reeled machinery steel		3 00
Tire steel	2 40	2 50
Sheet cast steel	...	0 12
Mining cast steel	....	0 08
Warranted cast steel	...	0 14
Annealed cast steel	...	0 15
High speed	0 60	0 65
B.P.L. tool steel	10½	0 11

INGOT TIN.

Lamb and Flag and Straits—
56 and 28-lb. ingots, 100 lb. $40 00 $41 00

TINPLATES.

Charcoal Plates—Bright

M.L.S., Famous (equal Bradley)		Per box.
I C,14 x 20 base	...	$6 50
IX, 14 x 20 "	....	8 00
IXX, 14 x 20 base	....	9 50

Raven and Vulture Grades—

I C, 14 x 20 base	....	5 00
I X "	....	6 00
I X X "	....	7 00
I X X X "	....	8 00

"Dominion Crown Best"—Double Coated, Tissued.

		Per box.
I C, 14 x 20 base	5 50	5 75
I X, 14 x 20 "	6 50	6 75
IXX x 20 "	7 50	7 75

"Allaway's Best"—Standard Quality.

I C, 14 x 20 base	4 65	5 00
I X, 14 x 20 "	5 40	5 75
I XX, 14 x 20 "	6 15	6 50

Bright Cokes.

Bessemer Steel—

I C., 14 x 20 base	4 25	4 35
20x28, double box	8 50	8 70

Charcoal Plates—Terne

Dean or J. G. Grade—

I.C., 20x28, 112 sheets	7 25	8 00
IX., Terne Tin	...	9 50

Charcoal Tin Boiler Plates.

Cookley Grade—
X X, 14x56, 50 sheet bxs. }
" 14x60, " } 7 50
" 14x65, " }

Tinned Sheets.

72x30 up to 24 gauge	....	8 50
" " 26 "	....	9 00

LEAD.

Imported Pig, per 100 lb.	5 35	5 50
Bar,	5 75	6 00
Sheets, 2½ lb. sq. ft., by roll	6 75	6 50
Sheets, 3 to 6 lb. "	6 50	6 25

Cut sheets ½c. per lb., extra.

SHEET ZINC.

5-cwt. casks	7 50	7 75
Part casks	7 75	8 00

ZINC SPELTER.

Foreign, per 100 lb	6 00	6 25
Domestic	6 50	6 75

COLD ROLLED SHAFTING.

9-16 to 11-16 inch	....	0 06
¾ to 1 7-16 "	....	0 05½
1 7-16 to 3 "	....	0 05

30 per cent.

OLD MATERIAL.

Dealers buying prices:

	Montreal	Toronto
Heavy copper and wire, lb.	0 12½	0 13
Light copper	0 11½	0 11½
Heavy red brass	0 11	0 11½
" yellow brass	0 08	0 10½
Light brass	0 06	0 07½
Tea lead	0 03½	0 03½
Heavy lead	0 03¾	0 04
Scrap zinc	0 03½	0 03½
No. 1 wrought iron	14 50	11 50
" 2 " "	6 00	6 00
Machinery cast scrap	17 00	16 00
Stove plate	12 00	11 70
Malleable and steel	9 00	10 00
Old rubbers	0 10½	0 10
Country mixed rags, 100 lbs.	1 00	1 25

PLUMBING AND HEATING

BRASS GOODS, VALVES, ETC.

Standard Compression work, 57½ per cent.
Cushion work, discount 40 per cent.
Fuller work, 70 per cent.
Flatway stop and stop and waste cocks, 60 per cent.; roundway, 55 per cent.
J.M.T. Globe, Angle and Check Valves, 45; Standard, 55 per cent
Kerr standard globes, angles and checks, special, 42½ per cent.; standard, 47½ p c.
Kerr Jenkins' disc, copper-alloy disc and heavy standard valves, 40 per cent.
Kerr steam radiator valves, 60 p.c., and quick-opening hot-water radiator valves, 60 p.c.
Kerr brass, Weber's straightway valves, 40; straightway valves, I B B M, 60.
J. M. T. Radiator Valves 50; Standard, 60; Patent Quick-Opening Valves, 65 p c.
Jenkins' Valves—Quotations on application to Jenkins' Bros., Montreal.

No. 1 compression bath cock ......net	2 00
No. 4 " " "	1 90
No 7 Fuller's "	2 25
No. 4½ " "	2 35
Patent Compression Cushion, basin cock, hot and cold, per doz.	$16.20
Patent Compression Cushion, bath cock, No. 2208	2 25

Square head brass cocks, 50; iron, 60 p.c.
Thompson Smoke-test Machine 25.00

BOILERS—COPPER RANGE.

Copper, 30 gallon, $33, 15 per cent.

BOILERS—GALVANIZED IRON RANGE.

30-gallon, Standard, $5; Extra heavy, $7.75

BATH TUBS.

Steel clad copper lined, 15 per cent.

CAST IRON SINKS.

16x24, $1; 18x30, $1; 18x36, $1.30.

ENAMELED BATHS, ETC.

List issued by the Standard Ideal Company Jan. 3, 1907, shows an advance of 10 per cent. over previous quotations.

ENAMELED CLOSETS AND URINALS

Discount 15 per cent.

HEATING APPARATUS.

Stoves and Ranges—45 to 50 per cent.
Furnaces—45 per cent.
Registers—70 per cent.
Hot Water Boilers—50 and 10 per cent.
Hot Water Radiators—50 to 55 p.c.
Steam Radiators—50 to 55 per cent.
Wall Radiators and Specials—50 to 55 p.c.

LEAD PIPE

Lead Pipe, 5 p.c. off
Lead waste, 5 p.c. off.
Caulking lead, 6¼c. per pound.
Traps and bends, 40 per cent.

IRON PIPE.

Size (per 100 ft.)	Black.		Galvanized
⅛ inch	2 35	⅛ inch	3 20
¼ "	2 35	¼ "	3 20
⅜ "	2 90	⅜ "	3 75
½ "	3 90	½ "	5 00
1 "	5 60	1 "	7 25
1¼ " 7 65 to	9 25	1¼ " 9 90 to	10 80
1½ "	9 20	1½ "	11 90
2 "	12 25	2 "	15 60
2½ "	20 10	2½ "	26 00
3 "	26 75	3 "	34 00
3½ "	34 25	3½ "	42 75
4 "	39 00	4 "	48 60

Malleable Fittings—Canadian discount 30 per cent.; American discount 25 per cent.
Cast Iron Fittings 60; Standard bushings 60.; headers, 60; flanged unions 60, malleable bushings 55 and 5; nipples, 70 and 10; malleable lipped unions 55 and 5 p.c.

SOIL PIPE AND FITTINGS

Medium and Extra heavy pipe and fittings, up to 6 inch, 60 and 10 to 70 per cent.
7 and 8-in. pipe, 40 and 5 per cent.
Light pipe, 50 p.c.; fittings, 50 and 10 p.c.

OAKUM.

Plumbers ....per 100 lb. 4 50 5 00

STOCKS AND DIES.

American discount 25 per cent.

SOLDERING IRONS.

½-lb. to 1½	per lb.	0 45½	0 45
2-lb. or over	"	0 42½	0 46

SOLDER.

Per lb.	Montreal	Toronto
Bar, half-and-half, guaranteed	0 25	0 26
Wiping	0 22	0 23

PAINTS, OILS AND GLASS.

BRUSHES

Paint and household, 70 per cent..

CHEMICALS.

	In casks per lb.
Sulphate of copper (bluestone or blue vitrol)	0 09
Litharge, ground	0 06
" flaked	0 06½
Green coperas (green vitrol)	0 01
Sugar of lead	0 08
Lump olive	0 01½

COLORS IN OIL.

Venetian red, 1-lb. tins	pure.	....	0 05
Chrome yellow	"	....	0 15
Golden ochre	"	....	0 10
French "	"	....	0 08
Marine black	"	....	0 04½
Chrome green	"	....	0 09
French permanent green	"	....	0 13
Signwriters' black	"	....	0 15

GLUE.

Domestic sheet	0 10	0 10¼
French medal	0 12	0 12½

PARIS WHITE.

In bbls	0 90

PIGMENTS.

Orange mineral, casks	0 08
" " 100-lb. kegs	0 08½

PREPARED PAINTS.

Barn (in bbls.)	0 65	0 90
Sherwin-Williams paints	....	1 55
Canada Paint Co.'s pure	....	1 40
Standard P. & V. Co.'s "New Era."		1 30
Benj. Moore Co.'s "Ark" B'd	....	1 25
" British Navy deck	....	1 50
Brandram-Henderson's "English"		1 45
Ramsay's paints, Pure, per gal.	....	1 30
" Thistle, "	....	1 10
Martin-Senour 100 p.c. pure,	....	1 55
Senour's Floor Paints	....	1 35
Jamieson's "Crown and Anchor"		1 20
Jamieson's floor enamel	....	1 50
P. D. Dods & Co's "Island City"		1 25
Sanderson Pearcy's, pure	....	1 20
Robertson's pure paints	....	1 20

PUTTY.

Bulk in bbls	1 60
Bladders in bbls	1 85
25-lb. tins	1 90
Bladders in bulk or tins less than 100 lb	2 00
Bulk in 100-lb. irons	1 80

SHINGLE STAINS.

In 5 gallon lots	0 80	0 90

SHELLAC.

White, bleached	2 65
Fine orange	2 60
Medium orange	2 55

TURPENTINE AND OIL.

Prime white petroleum per gal	....	0 13½
Water white "	....	0 15
Pratt's astral "	....	0 18½
Castor oil	0 08	0 10
Motor Gasoline per gal, single bbls	0 21	0 23½
Benzine, per gal single bbls	0 17	0 17½
Turpentine, single barrels	0 77	0 79
Linseed Oil, " raw	0 60	0 64
" " boiled	0 63	0 67

WHITE LEAD GROUND IN OIL. Per 100 lbs

Canadian pure	7 15
No. 1 Canadian	6 80
Munro's Select Flake White	7 65
Elephant and Decorators' Pure	7 65
Monarch	7 40
Standard Decorator's	7 15
Essex Genuine	6 80
Brandram's B. B. Genuine	8 70
"Anchor," pure	7 40
Ramsay's Pure Lead	7 25
Ramsay's Exterior	6 65
"Crown and Anchor," pure	7 25
P. D. Dods "Island City" pure lead	7 25
Sanderson Pearcy's	7 40
Robertson's C.P., lead	7 20

RED DRY LEAD.

Genuine, 560 lb. casks, per cwt	6 25
Genuine, 100 lb. kegs, "	6 75
No. 1, 560 lb. casks, per cwt	6 00
No. 1, 100 lb. kegs, per cwt	6 50

WINDOW GLASS

Size United inches	Star	Double Diamond
Under 26	$4 25	$6 25
26 to 40	4 65	6 75
41 to 50	5 10	7 50
51 to 60	5 35	8 50
61 to 70	5 75	9 75
71 to 80	6 25	11 00
81 to 85	7 00	12 50
86 to 90	....	15 00
91 to 95	....	17 50
96 to 100	....	20 50
101 to 105	....	24 00
106 to 110	....	27 50

Discount—16-oz. 25 per cent; 21 oz. 30 per cent. per 100 feet. Broken boxes 50 per cent.

WHITING.

Plain, in bbls	0 70
Gilders bolted in bands	0 90

WHITE DRY ZINC.

Extra Red Seal, V.M.	0 07½	0 08

WHITE ZINC IN OIL.

Pure, in 25-lb. irons	0 08¼
No. 1, " "	0 07
No. 2, " "	0 05½

VARNISHES.

	Per gal cans
Carriage, No. 1	1 50
Pale durable body	3 50
" hard rubbing	3 00
Finest elastic gearing	3 00
Elastic oak	1 50
Furniture, polishing	2 00
Furniture, extra	1 15
" " No. 1	0 90
" " union	0 80
Light oil finish	1 40
Gold size japan	1 80
Brown japan	0 95
No. 1 brown japan	0 95
Baking black japan	1 35
No. 1 black japan	0 90
Benzine black japan	0 70
Crystal Damar	2 80
No. 1 "	2 50
Pure asphaltum	1 40
Oilcloth	1 50
Lightning dryer	0 70
Elastilite varnish, 1 gal. can, each	2 00
Granitine floor varnish, per gal	2 10
Maple Leaf coach enamels; size 1,	1 20
Sherwin-Williams' kopal varnish, gal.,	2 50
Canada Paint Co's sun varnish	2 00
"Kyanize" Interior Finish	2 40
"Flint-Lac," coach	1 80
B H. Co's "Gold Medal," in cases	2 00
Jamieson's Copaline, per gal.	2 00

BUILDERS' HARDWARE.

BELLS.

Brass hand bells, 60 per cent.

Nickel, 55 per cent.

Gongs, Sargeant's door bells	5 50	8 00
American, house bells, per lb.	0 35	0 40

Peterboro' door bells, 37½ and 10 off new list.

BUILDING PAPER, ETC.

Tarred Felt, per 100 lb.	2 25
Ready roofing, 2-ply, not under 45 lb. per roll	1 00
Ready roofing, 3-ply, not under 65 lb., per roll	1 25
Carpet Felt ... per ton	60 00
Heavy Straw Sheathing ... per ton	40 00
Dry Surprise	0 45
Dry Sheathing ... per roll, 400 sq. ft.	0 40
Tar " " 400 "	0 50
Dry Fibre " 400 "	0 55
Tarred Fibre " 400 "	0 65
O. K. & I. X. L. " 400 "	0 70
Resin-sized " 400 "	0 45
Oiled Sheathing " 600 "	1 00
Oiled " " 400 "	0 70
Roof Coating, in barrels ... per gal.	0 17
Roof " small packages "	0 25
Refined Tar ... per barrel	5 00
Coal Tar "	4 00
Coal Tar, less than barrels ... per gal.	0 15
Roofing Pitch ... per 100 lb. 0 80	0 90
Slater's felt ... per roll	0 70
Heavy Straw Sheathing f. o. b. St. John and Halifax	42 50

BUTTS.

Wrought Brass, net revised list

Wrought Iron, 65 and 10 per cent.

Cast Iron Loose Pin, 60 per cent.

Wrought Steel Fast Joint and Loose Pin, 70 per cent.

CEMENT AND FIREBRICK

Canadian Portland	2 10	2 10
Belgium	1 60	1 90
White Bros. English	2 00	2 05
"Lafarge" cement in wood	....	3 40
"Iron Clad" cot on	....	2 10
"Iron Clad" paper	....	2 15
" word	....	2 25
Fire brick, Scotch, per 1,000	27 00	30 00
" English	17 00	21 00
" American, low	23 0	25 00
" " high	27 50	35 00
Fire clay (Scotch), net ton	....	4 95

Paving Blocks per 1,000.

Blue metallic, 9"x4½"x3", ex wharf	35 00
Stable pavers, 12"x6"x2", ex wharf	50 00
Stable pavers, 9"x4½"x3", ex wharf	36 00

DOOR SETS.

Peterboro, 37½ and 10 per cent.

DOOR SPRINGS.

Torrey's Rod ... per doz.	....	1 75
Coil, 9 to 11 in. "	0 95	1 65
English "	2 00	4 00

Chicago and Reliance Coil 25 per cent.

ESCUTCHEONS.

Discount 50 and 10 per cent., new list

Peterboro, 37½ and 10 per cent.

ESCUTCHEON PINS.

Iron, discount 40 per cent.

Brass, 45 per cent.

HINGES.

Blind, discount 50 per cent.

Heavy T and strap, 4-in., 100 lb. net	7 25
" " 5-in., "	7 00
" " 6-in., "	6 75
" " 8-in., "	6 50
" " 10-in. and larger	6 25

Light T and strap, discount 65 p.c.

Screw hook and hinge—

under 12 in. ... per 100 lb.	4 75
over 12 in. "	3 75

Crate hinges and back flaps, 65 and 5 p. c.

Chest hinges and hinge hasps, 65 p.c.

SPRING HINGES.

Spring, per gro., No. 5, $17.50 No. 10, $18; No. 20, $10.80; No. 120, $20; No. 51, $10; No. 50, $27.50

Chicago Spring Butts and Blanks 12½ per cent.

Triple End Spring Butts, 30 and 10 per cent.

Chicago Floor Hinges, 37½ and 5 off.

Garden City Fire House Hinges, 12½ p.c.

"Chief" floor hinge, 50 p.c.

CAST IRON HOOKS.

Bird cage ... per doz.	0 50	1 10
Clothes line, No. 61 "	0 00	0 70
Harness "	0 60	12 00
Hat and coat ... per gro.	1 10	10 00
Chandelier ... per doz.	0 50	1 00

Wrought hooks and staples—

¼ x 5 ... per gross	2 65
5-16 x 5 "	3 30

Bright wire hooks 60 p c.

Bright steel gate hooks and staples, 40 p.c.

Cresent hat and coat wire, 60 per cent.

Screw, bright wire, 65 per cent.

KNOBS.

Door, japanned and N.P., doz	1 50	2
Bronze, Berlin ... per doz.	2 75	3 50
Bronze, Genuine "	6 00	9 00
Shutter, porcelain, F. & L. screw ... per gross	1 30	2 00
White door knobs ... per doz.	....	2 00

Peterboro knobs, 37½ and 10 per cent.

Porcelain, mineral and jet knobs, net list.

KEYS.

Lock, Canadian 40 to 40 and 10 per cent.

LOCKS.

Peterboro 37½ and 10 per cent

Russell & Erwin, steel rim $2.50 per doz

Eagle cabinet locks, discount 30 per cent

American padlocks, all steel, 10 to 15 per cent.; all brass or bronze, 10 to 25 per cent.

SAND AND EMERY PAPER.

R. & A. sand, discount, 35 per cent

Emery, discount 35 per cent.

Garnet (Rurton's) 5 to 10 per cent. advance

SASH WEIGHTS.

Sectional ... per 100 lb.	2 00	2 25
Solid "	1 50	1 75

SASH CORD.

Per lb.	0 31

BLIND AND BED STAPLES.

All sizes per lb.	0 07½	0 10

WROUGHT STAPLES.

Galvanized	2 75
Plain	2 50

Coopers', discount 45 per cent.

Poultry netting staples, discount 40 per cent.

Bright spear point, 75 per cent. discount.

TOOLS AND HANDLES.

ADZES.

Discount 22½ per cent.

AUGERS.

Gilmour's, discount 60 per cent. off list.

AXES.

Single bit, per doz.	6 00	9 00
Double bit, "	10 00	11 00

Bench Axes, 40 per cent.

Broad Axes, 25 per cent.

Hunters' Axes	5 50	6 00
Boys' Axes	6 25	7 00
Splitting Axes	7 00	12 00
Handled Axes	7 00	9 00
Red Ridge, boys', handled	....	5 75
" hunters	....	5 25

BITS.

Irwin's auger, discount 47½ per cent.

Gilmour's auger, discount 60 per cent.

Rockford auger, discount 50 and 10 per cent.

Jennings' Gen. auger, net list.

Gilmour's car, 47½ per cent.

Clark's expansive, 40 per cent.

Clark's gimlet, per doz	0 65
Diamond, Shell, per doz.	1 00
Nail and Spike, per gross	2 25

BUTCHERS CLEAVERS

German ... per doz.	7 00	9 00
American "	12 00	18 00

CHALK.

Carpenters' Colored, per gross	0 45	0 75
White lump ... per cwt.	0 60	0 65

CHISELS.

Warnock's, discount 70 and 5 per cent.

P. S. & W. Extra, discount, 70 per cent.

CROSSCUT SAW HANDLES.

S. & D., No. 3 ... per pair	0 13
S. & D., " 5 "	0 11½
S. & D., " 6 "	0 18
Boynton pattern "	0 20

CROWBARS.

2½c. to 4c. per lb.

DRAW KNIVES.

Coach and Wagon, discount 75 and 5 per cent.

Carpenters' discount 75 per cent.

DRILLS.

Millar's Falls, hand and breast, net list.

North Bros., each set, 50c.

DRILL BITS.

Morse, discount 37½ to 40 per cent.

Standard, discount 50 and 5 to 55 per cent.

FILES AND RASPS.

Great Western	75 per cent.
Arcade	75 "
Kearney & Foot	75 "
Disston's	75 "
American	75 "
J. Barton Smith	75 "
McClellan	75 "
Eagle	75 "
Nicholson	66⅔ "
Globe	75 "

Black Diamond, 60, 10 and 5 p.c.

Jowitt's, English list, 27½ per cent.

GAUGES.

Stanley's discount 50 to 60 per cent.

Winn's, Nos. 26 to 33 ... each	1 65	2 40

HANDLES.

Second growth ash fork, hoe, rake and shovel handles, 40 p c.

Extra ash fork, hoe, rake and shovel handles, 45 p c.

No. 1 and 2 ash fork, hoe, rake and shovel handles, 50 p c.

White ash whiffletrees and neckyokes, 35 p.c.

All other ash goods, 40 p c.

All hickory, maple and oak goods, excepting carriage and express whiffletrees, 40 p c.

Hickory, maple, oak carriage and express whiffletrees, 45 p c.

HAMMERS.

Maydole's, discount 5 to 10 per cent.

Canadian, discount 25 to 27½ per cent.

Magnetic tack ... per doz.	1 10	1 20
Canadian sledge ... per lb.	0 07	0 08½
Canadian ball pean, per lb.	0 22	0 25

HATCHETS.

Canadian, discount 40 to 42½ per cent.
Shingle, Red Ridge 1, per doz.......... 4 40
" " 2, " 4 80
Barrel Underhill...................... 5 05

HOES

Mortar, 50 and 10 per cent

MALLETS.

Tinsmiths'............per doz. 1 25 1 50
Carpenters', hickory, " 1 25 3 75
Lignum Vitae........ " 3 85 5 00
Caulking, each 0 60 2 00

MATTOCKS.

Canadian............ per doz. 5 50 6 00

MEAT CUTTERS.

German, 15 per cent.
American discount, 33⅓ per cent.

PICKS.

Per dozen 6 00 9 00

PLANES.

Wood bench, Canadian, 40, American, 25 p c.
Wood, fancy, 37½ to 40 per cent.
Stanley planes, $1 55 to $3 60, net list prices.

PLANE IRONS.

English ....per doz. 2 00 5 00
Stanley, 2⅝ inch, single 24c., double 33c.

PLIERS AND NIPPERS.

Button's genuine, 37½ to 40 per cent.
Button's imitation....per doz. 5 00 9 00
Berg's wire fencing........... 1 72 5 50

PUNCHES.

Saddler's ....per doz. 1 00 1 85
Conductor's.......... " 3 00 15 00
Tinners', solid...........per set 0 72
" hollow......per inch 1 00

RIVET SETS.

Canadian, discount 35 to 37½ per cent.

RULES.

Boxwood, discount 70 per cent.
Ivory, discount 20 to 25 per cent.

SAWS.

Atkins, hand and crosscut, 25 per cent.
Disston's Hand, discount 12½ per cent
Disston's Crosscut....per foot 0 35 0 55
Hack, complete..........each 0 75 2 75
" frame only........each 0 50 1 25

S. & D. solid tooth circular shingle, concave and band, 50 per cent; mill and ice, drag, 30 per cent.; cross-cut, 35 per cent.; hand saws, butcher, 35 per cent; buck, New Century $6 25; buck No. 1 Maple Leaf, $5.25; buck, Happy Medium, $4.25; buck, Watch Spring, $4.25; buck, common frame, $4 00.

Spear & Jackson's saws—Hand or rip, 26 in., $12.75; 24 in., $11.25; panel 18 in., $8 25; 20 in., $9; tenon, 10 in., $9.90; 12 in., $10 90; 14 in., $11 50.

SAW SETS.

Lincoln and Whiting 4 75
Hand Sets, Perfect 4 00
X-Cut Sets " 7 50
Maple Leaf and Premiums saw sets, 40 off.
S. & D. saw swages, 40 off.

SCREW DRIVERS.

Sargent's ....per doz. 0 65 1 00
North Bros., No. 30 ..per doz. 16 80

SHOVELS AND SPADES.

Canadian, discount 45 per cent.

SQUARES.

Iron, discount 20 per cent.
Steel, discount 65 and 10 per cent.
Try and Bevel, discount 50 to 52½ per cent.

TAPE LINES.

English, ass skin .....per doz. 2 75 5 00
English, Patent Leather 5 50 9 75
Chesterman's ....each 0 90 2 85
" steel ....each 0 80 8 00
Berg's, each................. 0 75 2 50

TROWELS.

Disston's, discount 10 per cent.
S & D. discount 35 per cent.
Berg's, brick, 924x11 .... 4 00
" pointing, 924x5.......... .... 2 10

FARM AND GARDEN GOODS

BELLS.

American cow bells, 63⅓ per cent.
Canadian, discount 45 and 50 per cent.
American, farm bells, each .. 1 35 3 00

BULL RINGS.

Copper, $2.00 for 2½-inch

CATTLE LEADERS.

Nos. 32 and 33 .. ...per gross 7 50 8 50

BARN DOOR HANGERS.

	doz. pairs	
Stearns wood track	4 50	6 00
Zenith	....	9 00
Atlas, steel covered	5 00	6 00
Perfect	8 00	11 00
New Milo, flexible	....	6 50
Steel, track, 1 x 3-16 in(100 ft)	....	3 25
" " 1¼ x 3-16 in(100 ft)	....	4 75
Double strap hangers, doz. sets....		6 40
Standard jointed hangers, "	....	6 40
Steel King hangers "	....	6 25
Storm King and safety hangers	....	7 00
" " rail................	....	4 25

Chicago Friction, Oscillating and Big Twin Hangers, 5 per cent.

HARVEST TOOLS.

50 and 10 per cent.
S. & D. lawn rakes, Dunn's, 40 off.
" sidewalk and stable scrapers, 40 off.

HAY KNIVES.

Net list.

HEAD HALTERS.

Jute Rope, ⅜-inch....per gross 9 00
" " ½ " .... " 10 00
" " ⅝ " .... " 12 00
Leather, 1-inch ......per doz. 4 00
Leather, 1¼ " ...... " 5 20
Web................. " 2 45

HOES.

Garden, 50 and 10 per cent.
Planter.............. per doz. 4 00 4 50

LAWN MOWERS

Low wheel, 12, 14 and 16-inch $2 30
9-inch wheel, 12-inch 2 85
" " 14 " 3 00
" " 16 " 3 12½
High wheel, 12 " 4 05
" " 14 " 4 50
" " 16 " 4 75

SCYTHES.

Per doz. net................... 6 25 9 25

SCYTHE SNATHS.

Canadian, discount 40 per cent.

SNAPS.

Harness, German, discount 25 per cent.
Lock, Andrews' 4 50 11 00

STABLE FITTINGS.

Warden King, 35 per cent
Dennis Wire & Iron Co., 33⅓ p.c.

WOOD HAY RAKES.

40 and 10 per cent.

HEAVY GOODS, NAILS, ETC.

ANVILS.

Wright's, 80-lb. and over............... 0 10½
Hay Budden, 80-lb. and over 0 09½
Brook's, 80-lb. and over 0 11¼
Taylor-Forbes, prospectors 0 05
Columbia Hardware Co., per lb. 0 09½

VISES.

Wright's.......................... .... 0 13½
Berg's, per lb....................... .... 0 12½
Brook's.......................... .... 0 12½
Pipe Vise, Hinge, No. 1...... .... 3 50
" " " No. 2...... .... 5 50
Saw Vise 4 50 5 00
Blacksmiths' (discount) 60 per cent.
" parallel (discount) 45 per cent.

BOLTS AND NUTS

	Per cent.
Carriage Bolts, common ($1 list)	
" " ⅜ and smaller..	60, 10 and 10
" " 7-16 and up....	55 and 5
" " Norway Iron ($3 list)	50
Machine Bolts, ⅜ and less	60 and 10
Machine Bolts, 7-16 and up....	55 and 5
Plough Bolts	55 and 10
Blank Bolts....................	55
Bolt Ends......................	55
Sleigh Shoe Bolts, ⅜ and less ..	60 and 10
" " " 7-16 and larger	50 and 5
Coach Screws, conepoint.......	70 and 5

Nuts, square, all sizes, 4c. per cent. off
Nuts, hexagon, all sizes, 4½c. per cent. off.
Stove Rods per lb., 5½ to 6c.
Stove Bolts, 75 per cent.

CHAIN.

Proof coil, per 100 lb. ¼ in., $6 00; 5-16 in. $4.85; ⅜ in., $4 25; 7-16 in., $4.00; ½ in., $3 75, 9-16 in. $3 70; ⅝ in., $3.65; ¾ in., $3 60; ⅞ in., $3.45, 1 in., $3.40.

Halter, kennel and post chains, 40 to 40 and 5 per cent.; Cow ties, 40 per cent.; Tie out chains 65 per cent.; Stall fixtures, 35 per cent; Trace chain, 45 per cent.; Jack chain iron, 50 per cent.; Jack chain, brass, 50 per cent.

HORSE NAILS

M R M. cold forged process, list May 15, 1907, 50 and 5 per cent.
"C" brand, 57½ per cent off list.
Capewell brand, quotations on application.

HORSESHOES.

M R M. brand: iron, light and medium No. 1 and smaller, $3 90; No. 2 and larger $3 65; snow pattern No 1 and smaller $4.15, No 2 and larger, $3 90; "X L" new light steel, No 1 and smaller, $4 25; No 2 and larger, $4; "X.L" featherweight steel, No. 0 to 4, $5 60; Special countersunk steel No 0 to 4, $6 85 pkg; toe-weight, all sizes, $6.85. F o.b Montreal Extras for packing

Belleville brand: No. 0 and 1, light and medium iron, $3 90; snow, $4 15; light steel, $4 25, No 2 and larger, light and medium iron, $3 65; snow, $3 90; light steel, $4. F o b. Belleville. Two per cent., 30 days.

Toecalks—Standard No. 1 and smaller, $1.50; No 2 and larger, $1 25 Blunt No. 1 and smaller, $1.75; No. 2 and larger, $1.50 per box Sharp Put up in 25 lb. bxs.

HORSE WEIGHTS.

Taylor-Forbes, 4½c. per lb.

NAILS.	Cut.	Wire.
2d	4 00	3 50
3d	3 15	3 20
4 and 5d....	2 90	2 90
6 and 7d	2 80	2 80
8 and 9d	2 65	2 65
10 and 12d	2 60	2 60
16 and 20d....................	2 55	2 55
30, 40, 50 and 60d (base)	2 50	2 50

F.o.b. Montreal. Cut nails, Toronto 20c. higher.
Miscellaneous wire nails, discount 75 per cent
Coopers' nails, discount 40 per cent.

PRESSED SPIKES.

Pressed spikes, ⅝ diameter, per 100 lbs $3.15

RIVETS AND BURRS.

Iron Rivets, black and tinned, 60, 10 and 10.
Iron Burrs, discount 60 and 10 and 10 p.c.
Copper Rivets, usual proportion burrs, 15 p c.
Copper Burrs only, net list.
Extras on Coppered Rivets, ½-lb. packages 1c. per lb; ¼-lb. packages 2c. lb.
Tinned Rivets, net extra, 4c. per lb.

SCREWS.

Wood, F. H., bright and steel, 85 and 10 p c.
" R. H., bright, 80 and 10 per cent.
" F. H., brass, 75 and 10 per cent.
" R. H., " 70 and 10 per cent.
" F. H., bronze, 70 and 10 per cent.
" R. H., " 65 and 10 per cent.
Drive Screws, dis. 87½ per cent.
Bench, wood ....per doz. 3 25 4 00
" iron .... " 4 25 5 00
Set, case hardened, dis. 60 per cent.
Square Cap, dis. 50 and 5 per cent.
Hexagon Cap, dis. 45 per cent.

MACHINE SCREWS.

Flat head, iron and brass, 35 per cent.
Felister head, iron, discount 30 per cent.
" " brass, discount 25 per cent.

TACKS, BRADS, ETC.

Carpet tacks, blued 75 p c.; tinned, 80 and 10; (in kegs), 40; cut tacks, blued, in dozens only, 75; ½ weights, 60; Swedes cut tacks, blued and tinned, bulk, 75 dozens, 75; Swedes, upholsterers', bulk, 85 and 12½; brush, blued and tinned, bulk, 70; Swedes, gimp, blued, tinned and japanned, 75 and 12½; zinc tacks, 35; leather carpet tacks, 40; copper tacks, 25; copper nails 30; trunk nails, black, 65; trunk nails, tinned and blued, 65; clout nails, blued and tinned 65; chair nails, 35; patent brads, 40; fine finishing, 40; lining tacks, in papers, 10; lining tacks, in bulk, 15; lining tacks, solid heads, in bulk, 75; saddle nails, in papers, 10; saddle nails, in bulk, 15; tufting buttons, 22 line in dozens only, 60; zinc glaziers' points, 5; double pointed tacks, papers, 90 and 10; double pointed tacks, bulk, 40; clinch and duck rivets, 45; cheese box tacks, 85 and 5 trunk tacks, 80 and 10.

WROUGHT IRON WASHERS.

Canadian make, discount 40 per cent.

SPORTING GOODS.

CARTRIDGES.

"Dominion" Rim Fire Cartridges and C B. caps, 50 and 7½ per cent.; Rim Fire R B. Round Caps 60 and 2½ per cent; Centre Fire, Pistol and Rifle Cartridges, 30 p c; Centre Fire Sporting and Military Cartridges, 20 and 5 p c; Rim Fire Shot Cartridges, 50 and 7½ p.c; Centre Fire, Shot Cartridges, 30 p c.; Primers, 25 p c.

LOADED SHELLS.

"Crown" Black Powder, 15 and 5 p c; "Sovereign" Empire Bulk Smokeless Powder, 30 and 5 p.c.; "Regal" Ballistite Dense smokeless Powder, 30 and 5 p c; "Imperial" Empire or Ballistite Powder, 30 and 10 p c.

EMPTY SHELLS.

Paper Shells, 25 and 5; Brass Shells 55 and 5 p.c.

Wads.	per lb.
Best thick brown or grey felt wads, in ½-lb. bags	$0 70
Best thick white card wads, in boxes of 500 each, 12 and smaller gauges	29
Best thick white card wads, in boxes of 500 each, 10 gauge.............	0 35
Thin card wads, in boxes of 1,000 each, 12 and smaller gauges	0 20
Thin card wads, in boxes of 1,000 each, 10gauge.........................	0 25
Chemically prepared black edge grey cloth wads, in boxes of 250 each—	Per M
11 and smaller gauge	0 60
9 and 10 gauges...................	0 70
and 8 "	0 90
5 and 6 "	1 10
Superior chemically prepared pink edge, best white cloth wads in boxes of 250 each—	
11 and smaller gauge	1 15
9 and 10 gauges	1 40
7 and 8 "	1 65
5 and 9 "	1 90

SHOT.

Ordinary drop shot, AAA to dust $7.50 per 100 lbs. Discount 5 per cent; cash discount, 2 per cent. 30 days; net extras as follows subject to cash discount only; Chilled, 40 c.; buck and seal, 80c.; no. 28 ball, $1 20 per 100 lbs.; bags less than 25 lbs., ½c. per lb.; F.O.B. Montreal, Toronto, Hamilton, London, St. John and Halifax, and freight equalized thereon.

TRAPS (steel.)

Game, Newhouse, discount 30 and 10 per cent.
Game, Hawley & Norton, 50, 10 & 5 per cent
Game, Victor, 70 per cent.
Game, Oneida Jump (B. & L.) 40 & 2½ p.
Game, steel, 60 and 5 per cent.

SKATES.

Skates, discount 37½ per cent.
Empire hockey sticks, per doz .. 3 00

CUTLERY AND SILVERWARE.

RAZORS.

	per doz.	
Elliot's	4 00	18 00
Boker's	7 50	11 00
" King Cutter	13 50	18 50
Wade & Butcher's	3 60	10 00
Lewis Bros. "Klean Kutter"	8 50	10 50
Henckel's	7 50	20 00
Berg's	7 50	20 00

Clauss Razors and Strops, 50 and 10 per cent

KNIVES.

Farriers-Stacey Bros., doz 3 50

PLATED GOODS

Hollowware, 40 per cent. discount.
Flatware, staples, 40 and 10, fancy, 40 and 5.
Hutton's "Cross Arrow" flatware, 47½; "Singalese" and "Alaska" Nevada silver flatware, 42 p.c.

SHEARS.

Clauss, nickel, discount 60 per cent.
Clauss, Japan, discount 67½ per cent.
Clauss, tailors, discount 40 per cent.
Seymour's, discount 50 and 10 per cent.
Berg's 6 00 12 00

HOUSE FURNISHINGS.

APPLE PARERS.

Hudson, per doz., net 5 75

BIRD CAGES.

Brass and Japanned, 40 and 10 p. c.

COPPER AND NICKEL WARE.

Copper boilers, kettles, teapots, etc. 30 p c.
Copper pitts, 30 per cent.

KITCHEN ENAMELED WARE.

White ware, 75 per cent.
London and Princess, 50 per cent
Canada, Diamond, Premier, 50 and 10 p c.
Pearl, Imperial, Crescent and granite steel, 60 and 10 per cent.
Premier steel ware, 40 per cent.
Star decorated steel and white, 25 per cent.
Japanned ware, discount 45 per cent.
Hollow ware, tinned cast, 35 per cent. off.

KITCHEN SUNDRIES.

Can openers, per doz	0 40	0 75
Mincing knives per doz	0 50	0 80
Duplex mouse traps, per doz	....	0
Potato mashers, wire, per doz	0 60	0
" wood "	0 50	0
Vegetable slicers, per doz	....	2
Universal meat chopper No. 1.	....	1 65
Enterprise chopper, each	...	1 85

Spiders and fry pans, 50 per cent.

Star Al chopper 5 to 32	1 35	4
" " 100 to 103	1 35	2 10
Kitchen hooks, bright	....	0 60

LAMP WICKS.

Discount, 60 per cent.

LEMON SQUEEZERS.

Porcelain lined	per doz.	2 20	5 60
Galvanized	"	1	3 85
King, wood	"	2	2 90
King, glass	"	4	4 50
All glass	"	0 85	0 90

METAL POLISH.

Tandem metal polish paste 6 00

PICTURE NAILS.

Porcelain head .....per gross	1 35	1 50
Brass head	0 40	1 00

Tin and gilt, picture wire, 75 per cent.

SAD IRONS.

Mrs. Potts, No. 55, polished....per set	0 90
" No. 50, nickle-plated, "	0 95
" handles, japaned, per gross	9 25
" " nickled, "	9 75
Common, plain	4 25
" plated	4 50
Asbestos, per set	1 50

TINWARE.

CONDUCTOR PIPE.

2-in. plain or corrugated, per 100 feet, \$3.30; 3 in., \$4.40; 4 in., \$5.80; 5 in., \$7.45; 6 in. \$9 90.

FAUCETS.

Common, cork-lined, discount 35 per cent.

EAVETROUGHS.

10-inch ..... per 100 ft. 3 30

FACTORY MILK CANS.

Discount off revised list, 35 per cent
Milk can trimmings, discount 25 per cent.
[illegible]

LANTERNS.

No 2 or 4 Plain Cold Blast ....per doz.		6 50
Lift Tubular and Hinge Plain,	"	4 75
No. 0, safety	"	4 00

Better quality at higher prices.
Japanning, 50c. per doz. extra.
Prism globes, per doz., \$1.20.

OILERS.

Kemp's Tornado and McClary's Model galvanized oil can, with pump, 5 gallon, per dozen 10 92
Davidson oilers, discount 40 per cent.
Zinc and tin, discount 50 per cent
Coppered oilers, 20 per cent. off.
Brass oilers, 50 per cent, off.
Malleable, discount 25 per cent

PAILS (GALVANIZED).

Dufferin pattern pails, 45 per cent.
Flaring pattern, discount 45 per cent.
Galvanized washtubs 40 per cent.

PIECED WARE.

Discount 35 per cent off list, June, 1899.
10-qt. flaring sap buckets, discount 35 per cent.
6, 10 and 14-qt. flaring pails dis. 35 per cent.
Copper bottom tea kettles and boilers, 30 p.c.
Coal hods, 40 per cent.

STAMPED WARE.

Plain, 75 and 12½ per cent. off revised list.
Retinned, 72½ per cent revised list.

SAP SPOUTS.

Bronzed iron with hooks ....per 1,000
Eureka tinned steel, hooks 8 00

STOVEPIPES.

5 and 6 inch, per 100 lengths	7 64	7 91
7 inch " "	....	8 18

Nestable, discount 40 per cent.

STOVEPIPE ELBOWS

5 and 6-inch, common .....per doz.	1 32
7-inch "	1 48

Polished, 15c. per dozen extra.

THERMOMETERS.

Tin case and dairy, 75 to 75 and 10 per cent.

TINNERS' SNIPS.

Per doz 3 00 15
Clauss, discount 35 per cent.

TINNERS' TRIMMINGS.

Discount, 45 per cent.

WIRE.

ANNEALED CUT HAY BAILING WIRE.

No. 12 and 13, \$4; No. 13½, \$4 10; No 14, \$4.2·; No. 15, \$4 50; in lengths 6' to 11', 25 per cent.; other lengths 20c. per 100 lbs extra; if eye or loop on end add 25c. per 100 lbs to the above.

BRIGHT WIRE GOODS

Discount 60 per cent.

CLOTHES LINE WIRE.

No. 7 wire solid line, No. 17, \$4.90; No. 18, \$3.00; No. 19, \$2 70; a wire solid line, No. 17, \$4 45; No 18, \$3.10; No. 19, \$2 81. All prices per 1000 ft. measure; 6 strand, No. 18, \$2 60; No. 19, \$2 90. F.o.b. Hamilton, Toronto, Montreal

COILED SPRING WIRE

High Carbon, No. 9, \$2 95; No. 11, \$3 50; No. 12, \$3.20.

COPPER AND BRASS WIRE.

Discount 37½ per cent.

FINE STEEL WIRE.

Discount 25 per cent · List of extras In 100-lb. lots: No. 17, \$5 — No. 18, \$5.50 — No. 19, \$6 — No. 20, \$6.65 — No. 21 \$7 — No. 22, \$7.30 — No. 23, \$7.65 — No 24, \$8 — No. 25, \$9—No. 26, \$9.50—No. 27, \$10—No. 28, \$11—No. 29, \$12—No. 30, \$13—No. 31, \$14—No. 32, \$15—No. 33, \$16—No 34, \$17. Extras net—tinned wire, Nos. 17-25 \$2—Nos. 26-31, \$4—Nos. 32-34, \$6. Coppered, 75c.—oiling, 10c.—in 25-lb. bundles, 15c.—in 5 and 10-lb. bundles, 25c.—in 1-lb. hanks, 25c. —in ½-lb. hanks, 38c.—in ¼-lb. hanks, 50c. packed in casks or cases, 15c.—bagging or papering, 10c

FENCE STAPLES.

Bright. ..... 3 80 Galvanized 33

HAY WIRE IN COILS.

No 13, \$2.70; No. 14, \$2 80; No. 15, \$2.95; f.o.b., Montreal.

GALVANIZED WIRE.

Per 100 lb.—Nos. 4 and 5, \$3.95 — Nos. 6, 7, 8, \$3 40 — No. 9, \$2 90 — No. 10, \$3.45 — No. 11, \$3.50 — No. 12, \$3 05 —No. 13, \$3 15—No. 14, \$4 00—No. 15, \$4.35 —No. 16, \$4.35 from stock. Base sizes, Nos. 6 to 9, \$2 35 f.o.b. Cleveland. Extras for cutting.

LIGHT STRAIGHTENED WIRE.

Gauge No.	Over 20 in.	10 to 20 in.	5 to 10 in.
0 to 5	\$0.50	\$0 75	\$1 25
6 to 9	0 75	1 25	2 00
10 to 11	1.00	1.75	2.50
12 to 14	1.50	2.25	3.50
15 to 16	2.00	3.00	4.50

SMOOTH STEEL WIRE.

No. 0-9 gauge, \$2.40; No. 10 gauge, 6c extra; No. 11 gauge, 12c extra; No. 12 gauge, 20c. extra; No. 13 gauge, 30c. extra; No 14 gauge, 40c. extra; No. 15 gauge, 55c. extra; No 16 gauge, 70c. extra. Add 60c. for coppering and \$2 for tinning

Extra net per 100 lb.—Oiled wire 10c., spring wire \$1.25, bright soft drawn 15c., charcoal (extra quality) \$1.25, packed in casks or cases 15c., bagging and papering 10c., 50 and 100-lb. bundles 10c., in 25-lb. bundles 15c., in 5 and 10-lb. bundles 25c., in 1-lb hanks, 50c., in ½-lb. hanks 75c., in ¼-lb. hanks \$1.

POULTRY NETTING.

2-in. mesh, 19 w g., 50 and 5 p.c. off. Other sizes, 50 and 5 p.c. off.

WIRE CLOTH.

Painted Screen, in 100-ft rolls, \$1 72½, per 100 sq. ft.; in 50-ft. rolls, \$1.77½, per 100 sq ft.

WIRE FENCING.

Galvanized barb	2 95
Galvanized, plain twist	3 30

Galvanized barb, f.o.b. Cleveland, \$2.70 for small lots and \$2.60 for carlots

WIRE ROPE

Galvanized, 1st grade, 6 strands, 24 wires, ⅜, \$5; 1 inch \$16 80.
Black, 1st grade, 6 strands, 19 wires, ⅜, \$5; 1 inch \$15.10. Per 100 feet f o.b. Toronto.

WOODENWARE.

CHURNS.

No. 0, \$9; No. 1, \$9; No. 2, \$10; No. 3, \$11; No. 4, \$13; No. 5, \$16; f.o b. Toronto Hamilton, London and St. Marys, 30 and 30 per cent; f o b Ottawa, Kingston and Montreal, 40 and 15 per cent. discount.

CLOTHES REELS.

Davis Clothes Reels, dis. 40 per cent.

FIBRE WARE.

Star pails, per doz	\$ 3 00
0 Tubs, "	14 00
1 " "	12 00
2 " "	10 00
3 " "	8 50

LADDERS, EXTENSION.

3 to 6 feet, 12c. per foot; 7 to 10 ft., 13c.
Waggoner Extension Ladders, dis. 40 per cent.

MOPS AND IRONING BOARDS.

"Best" mops		1 25
"900" mops		1 25
Folding ironing boards	12 00	16 50

REFRIGERATORS

Discount, 40 per cent.

SCREEN DOORS.

Common doors, 2 or 3 panel, walnut stained, 4-in. style.....per doz.	7 25
Common doors, 2 or 3 panel, grained only, 4-in., style .....per doz.	7 55
Common doors, 2 or 3 panel, light stair per doz.	9 55

WASHING MACHINES.

Round, re-acting per doz.	60 00
Square "	63 00
Eclipse, per doz	54 00
Dowswell "	39 00
New Century, per doz	75 00
Daisy	54 00
Stephenson	74 00

WRINGERS.

Royal Canadian, 11 in., per doz.	35 00
Royal American, 11 in. "	35 00
Eze 10 in., per doz	36 75

MISCELLANEOUS.

AXLE GREASE.

Ordinary, per gross	6 00	7 00
Best quality	10 00	12 00

BELTING.

Extra, 60 per cent.
Standard, 60 and 10 per cent.
No. 1, not wider than 6 in., 60, 10 and 10 p.c.
Agricultural, not wider than 4 in., 75 per cent
Lace leather, per side, 75c.; cut laces, 80c.

BOOT CALKS.

Small and medium, ball .....per M	2 25
Small heel	4 50

CARPET STRETCHERS.

American .....per doz.	1 00	1 50
Bullard's "	....	6 50

CASTORS.

Bed, new list, discount 55 to 57½ per cent.
Plate, discount 52½ to 57½ per cent.

PINE TAR.

½ pint in tins .....per gross	...	7 80
" " "	....	9 60

PULLEYS.

Hothouse .....per doz.	0 55	1 00
Axle "	0 22	0 33
Screw "	0 22	1 00
Awning "	0 35	2 50

PUMPS.

Canadian cistern	1 40	2 00
Canadian pitcher spout	1 80	3 16

Berg's wing pump, 75 per cent.

ROPE AND TWINE.

Sisal		0 10½
Pure Manilla		0 15
"British" Manilla		0 12
Cotton, 3-16 inch and larger	0 21	0 23
" 5-32 inch	0 25	0 27
" ¼ inch	0 25	0 28
Russia Deep Sea		0 16
Jute		0 09
Lath Yarn, single		0 10
" double		0 10½
Sisal bed cord, 48 feet.....per doz.		0 65
" " 60 feet "		0 80
" " 72 feet "		0 95

Twine.

Bag, Russian twine, per lb.	....	0 27
Wrapping, cotton, 3-ply	....	0 25
" " 4-ply	....	0 29
Mattress twine per lb.	0 33	0 45
Staging " "	0 27	0 35

BINDER TWINE.

500 feet, sisal	0 09½
500 " standard	0 09½
550 " " manilla	0 10½
600 " " "	0 12½
650 " " "	0 13½

Car lots, ½c. less; 5-ton lots, ¼c. less. Central delivery.

SCALES.

Gurney Standard, 35; Champion, 45 p.c.
Burrow, Stewart & Milne — Imperial Standard, 35; Weigh Beams, 35; Champion Scales, 45
Fairbanks Standard, 30; Dominion, 50 Richelieu, 50.
Warren new Standard, 35; Champion, 45 Weigh Beams, 30.

STONES—OIL AND SCYTHE.

Washita .....per lb.		0 25	0
Hindostan	"	0 06	0 10
" slip	"	0 18	0 20
" Axe	"	....	0 10
Deer Creek	"	....	0 10
Deerlick	"	....	0 25
" Axe	"	....	0 15
Lily white	"	....	0 42
Arkansas	"	....	1 50
Water-of-Ayr	"	....	0 10
Scythe .....per gross		3 50	5 00
Grind, 40 to 200 lb., per ton		20 00	22 00
" under 40 lb., "		....	24 00
" 200 lb. and over		....	28 00

INDEX TO ADVERTISERS.

CLASSIFIED LIST OF ADVERTISEMENTS.

Alabastine.
Alabastine Co., Limited, Paris, Ont.

Auditors.
Davenport, Pickup & Co., Winnipeg.

Automobile Accessories.
Canada Cycle & Motor Co., Toronto Junction.
Carriage Mountings Co., Ltd., Toronto.

Awnings.
Tobin Tent and Awning Co., Ottawa

Babbitt Metal.
Canada Metal Co., Toronto.
Canadian Fairbanks Co., Montreal
Frothingham & Workman, Montreal.
Robertson, Jas. Co., Montreal.

Bar Urns.
Buffalo Mfg. Co., Buffalo, N.Y.

Bath Room Fittings.
Buffalo Mfg. Co., Buffalo, N.Y.
Carriage Mountings Co., Ltd., Toronto.
Forsyth Mfg. Co., Buffalo, N.Y.
Ontario Metal Novelty Co., Toronto

Belting, Hose, etc.
Gutta Percha and Rubber Mfg. Toronto.
Sadler & Haworth, Toronto.

Bicycles and Accessories.
Johnson's, Iver, Arms and Cycle Works Fitchburg, Mass

Binder Twine.
Consumers Cordage Co., Montreal.

Bolts.
Toronto & Belleville Rolling Mills, Belleville.

Box Strap.
J. N. Warminton, Montreal.

Brass Goods.
Frothingham & Workman, Montreal.
Glauber Brass Mfg. Co., Cleveland, Ohio.
Kerr Engine Co., Walkerville, Ont.
Lewis, Rice, & Son., Toronto
Morrison, Jas., Brass Mfg. Co., Toronto.
Mueller Mfg. Co., Decatur, Ill.
Penberthy Injector Co., Windsor, Ont.
Taylor-Forbes Co., Guelph, Ont.

Bronze Powders.
Canadian Bronze Powder Works, Montreal.

Brushes.
United Factories, Toronto.

Cans.
Acme Can Works, Montreal.

Builders' Tools and Supplies.
Covert Mfg. Co., West Troy, N.Y.
Frothingham & Workman Co., Montreal.
Howland, H. S. Sons & Co., Toronto.
Hyde, F., & Co., Montreal.
Lewis Bros. & Co., Montreal.
Lewis, Rice, & Son, Toronto.
Lockerby & McComb, Montreal.
Lufkin Rule Co., Saginaw, Mich.
Newman & Sons, Birmingham.
North Bros. Mfg. Co., Philadelphia, Pa.
Stanley Rule & Level Co., New Britain.
Stanley Works, New Britain, Conn.
Stephens, G. F., Winnipeg.
Taylor-Forbes Co., Guelph, Ont.

Carriage Accessories.
Carriage Mountings Co., Ltd., Toronto.
Covert Mfg. Co., West Troy, N.Y.

Carriage Springs and Axles.
Guelph Spring and Axle Co., Guelph.

Carpet Beateas.
Ontario Metal Novelty Co., Toronto.

Cartridges.
Dominion Cartridge Co., Montreal.

Cattle and Trace Chains.
Greening, B., Wire Co., Hamilton.

Chafing Dishes.
Buffalo Mfg. Co., Buffalo, N.Y.

Churns.
Dowswell Mfg. Co., Hamilton.

Clippers—All Kinds.
American Shearer Mfg. Co., Nashua, N.H

Clothes Reels and Lines.
Hamilton Cotton Co., Hamilton,

Clutch Nails.
J. N. Warminton, Montreal.

Congo Roofing
Buchanan Foster Co., Philadelphia, Pa.

Cordage.
Consumers' Cordage Co., Montreal.
Hamilton Cotton Co., Hamilton.

Cork Screws.
Erie Specialty Co., Erie, Pa.

Cow Ties
Greening, B., Wire Co., Hamilton

Cuspidors.
Buffalo Mfg. Co., Buffalo, N.Y.

Cut Glass.
Phillips, Geo., & Co., Montreal.

Cutlery—Razors, Scissors, etc.
Birkett, Thos., & Son Co., Ottawa.
Clauss Shear Co., Toronto
Dorken Bros. & Co., Montreal.
Frothingham & Workman Montreal.
Heinisch's, R., Sons Co., Newark, N.J.
Howland, H. S. Sons & Co., Toronto.
Hutton, Wm., & Sons, Ltd., London, Eng.
Lamplough, F. W., & Co., Montreal.
Phillips, Geo., & Co., Montreal.
Round, John, & Son, Montreal.

Electric Fixtures.
Canadian General Electric Co., Toronto
Morrison James, Mfg. Co., Toronto.
Munderloh & Co., Montreal.

Electro Cabinets.
Cameron & Campbell Toronto.

Enameled Ware
Kemp Mfg. Co., Toronto.

Engines, Supplies, etc.
Kerr Engine Co., Walkerville, Ont.

Eavetroughs
Wheeler & Bain, Toronto

Fencing—Woven Wire.
Dominion Wire Mfg. Co., Montreal.
Owen Sound Wire Fence Co., Owen Sound.
Banwell Hoxie Wire Fence Co., Hamilton.

Files and Rasps.
Barnett Co., G. & H., Philadelphia, Pa.
Nicholson File Co., Port Hope

Firearms and Ammunition.
Hamilton Rifle Co., Plymouth, Mich.
Harrington & Richardson Arms Co., Worcester, Mass.
Johnson's, Iver, Arms and Cycle Works Fitchburg, Mass.

Fishing Tackle.
Enterprise Mfg. Co., Akron, Ohio

Food Choppers
Enterprise Mfg. Co., Philadelphia, Pa.
Lamplough, F. W., & Co., Montreal.
Shirreff Mfg. Co., Brockville, Ont.

Furnaces.
Pease Foundry Co., Toronto.

Galvanizing.
Canada Metal Co., Toronto.
Dominion Wire Mfg. Co., Montreal.
Montreal Rolling Mills Co., Montreal.
Ontario Wind Engine & Pump Co., Toronto.

Glass Ornamental
Hobbs Mfg. Co., London
Consolidated Plate Glass Co., Toronto

Glaziers' Diamonds.
Gibsone, J. B., Montreal.
Pelton, Godfrey S.
Sharratt & Newth, London, Eng.
Shaw, A., & Son, London, Eng.

Handles.
Still, J. H., Mfg. Co.

Harvest Tools.
Maple Leaf Harvest Tool Co., Tillsonburg, Ont.

Hockey Sticks
Still, J. H. Mfg. Co., St. Thomas.

Hoop Iron.
Frothingham & Workman Montreal.
Montreal Rolling Mills Co., Montreal.
J. N. Warminton, Montreal.

Horse Blankets.
Heney, E. N., & Co., Montreal.

Horseshoes and Nails.
Canada Horse Nail Co., Montreal.
Montreal Rolling Mills, Montreal.
Capewell Horse Nail Co., Toronto
Toronto & Belleville Rolling Mills Belleville.

Hot Water Boilers and Radiators.
Cluff, R. J., & Co Toronto.
Pease Foundry Co. Toronto.
Taylor-Forbes Co., Guelph.

Ice Cream Freezers.
Dana Mfg. Co., Cincinnati, Ohio.
North Bros. Mfg. Co., Philadelphia, Pa.

Ice Cutting Tools.
Erie Specialty Co., Erie, Pa.
North Bros. Mfg. Co., Philadelphia, Pa.

Injectors—Automatic.
Morrison, Jas., Brass Mfg. Co., Toronto.
Penberthy Injector Co., Windsor, Ont.

Iron Pipe.
Montreal Rolling Mills, Montreal.

Iron Pumps.
Lamplough, F. W., & Co., Montreal.
McDougall, R., Co., Galt, Ont.

Ladders.
The Waggoner Ladder Co., London, Ont.

Lanterns.
Kemp Mfg. Co., Toronto.
Ontario Lantern Co., Hamilton, Ont.
Wright, E. T., & Co., Hamilton.

Lawn Mowers.
Birkett, Thos., & Son Co., Ottawa.
Frothingham & Workman, Montreal.
Maxwell, D., & Sons, St. Mary's, Ont.
Taylor, Forbes Co., Guelph.

Lawn Mower Grinders
Root Bros. & Co., Plymouth, Ohio.

Ledgers—Loose Leaf.
Business Systems, Toronto.
Copeland-Chatterson Co., Toronto.
Crain, Rolla L., Co., Ottawa.
Universal Systems, Toronto.

Lithographing.
London Printing & Lithographing Co., London, Ont.

Locks, Knobs, Escutcheons, etc.
Peterborough Lock Mfg. Co., Peterborough, Ont.
National Hardware Co., Orillia, Ont.

Lumbermen's Supplies.
Pink, Thos., & Co., Pembroke Ont.

Lye
Gillett, E. W., & Co., Toronto

Manufacturers' Agents.
Fox, C. H., Vancouver.
McIntosh, H. F., & Co., Toronto.
Gibb, Alexander, Montreal.
Scott, Bathgate & Co., Winnipeg.

Metals.
Canada Iron Furnace Co., Midland, Ont.
Canada Metal Co., Toronto.
Eadie, H. G., Montreal.
Frothingham & Workman, Montreal.
Gibb, Alexander, Montreal.
Kemp Mfg. Co., Toronto
Leslie, A. C., & Co., Montreal.
Lysaght, John, Bristol, Eng.
Nova Scotia Steel and Coal Co., New Glasgow, N.S.
Robertson, Jas., Co., Montreal.
Roper, J. H., Montreal.
Samuel, Benjamin & Co., Toronto.
Stairs, Son & Morrow, Halifax, N.S.
Thompson, B. & S. H. & Co. Montreal.

Metal Lath.
Galt Art Metal Co., Galt.
Metallic Roofing Co., Toronto.
Metal Shingle & Siding Co., Preston, Ont.

Metal Polish, Emery Cloth, etc
Oakey, John, & Sons, London, Eng.

Nails Wire
Dominion Wire Mfg. Co., Montreal.

Oilers
Maple City Mfg Co., Monmouth, Ill.

Oil Tanks.
Bowser, S. F., & Co., Torono.

Ornamental Iron and Wire.
Dennis Wire & Iron Co, London, Ont.

Packing.
Gutta Percha & Rubber Co Toronto

Paints, Oils, Varnishes, Glass.
Blanchite Process Paint Co, Toronto.
Brandram-Henderson, Montreal
Canada Paint Co., Montreal.
Canadian Oil Co., Toronto.
Consolidated Plate Glass Co., Toronto.
Dods, P. D., & Co., Montreal
Imperial Varnish and Color Co., Toronto
Jamieson, R. C., & Co, Montreal.
Lucas John & Co., New York
McArthur, Corneille & Co., Montreal.
McCaskill, Dougall & Co., Montreal.
Moore, Benjamin, & Co. Toronto.
Ottawa Paint Works, Ottawa
Queen City Oil Co., Toronto.
Ramsay & Son, Montreal.
Sanderson Pearcy & Co., Toronto
Sherwin-Williams Co., Montreal.
Standard Paint Co., Montreal
Standard Paint and Varnish Works Windsor, Ont.
Stephens & Co, Winnipeg.
Martin-Senour Co., Montreal
Winnipeg Paint & Glass Co., Winnipeg

Perforated Sheet Metals.
Greening, B., Wire Co., Hamilton.

Plumbers' Tools and Supplies.
Canadian Fairbanks Co, Montreal.
Cluff, R. J., & Co., Toronto
Frothingham & Workman, Montreal.
Glauber Brass Co., Cleveland, Ohio.
Jardine, A. B., & Co, Hespeler, Ont.
Jenkins Bros., Boston, Mass.
Kerr Engine Co, Walkerville, Ont.
Lewis, Rice, & Son, Toronto.
Merrell Mfg. Co., Toledo, Ohio.
Montreal Rolling Mills Montreal.
Morrison, Jas., Brass Mfg. Co., Toronto.
Mueller, H., Mfg. Co., Decatur, Ill.
Oshawa Steam & Gas Fitting Co., Oshaw
Robertson Jas., Co. Montreal.
Robertson, Jas., Co., Limited, Toronto
Somerville, Limited, Toronto
Stairs, Son & Morrow, Halifax, N.S.
Standard Ideal Sanitary Co., Port Hope,
Standard Sanitary Co., Pittsburg.
Stephens, G. F., & Co., Winnipeg, Man.
Turner Brass Works, Chicago.
Vickery, Orlando, Toronto.

Polishes.
Majestic Polishes, Toronto

Portland Cement.
International Portland Cement Co. Ottawa, Ont.
Hanover Portland Cement Co., Hanover, Ont.
Hyde, F., & Co., Montreal.
Thompson B. & S. H. & Co., Montreal.

Poultry Netting.
Greening, B., Wire Co., Hamilton, Ont.

Printing.
London Printing & Lithographing Co., London, Ont.

Razors.
Clauss Shear Co., Toronto.

Refrigerators.
Fabien, C. P., Montreal.

Registers.
Pease Foundry Co., Toronto.

Roofing Supplies.
Brantford Roofing Co., Brantford.
Barrett Mfg. Co., New York.
F. W. Bird, East Walpole, Mass.
Buchanan Foster Co., Philadelphia, Pa.
McArthur, Alex., & Co., Montreal.
Metal Shingle & Siding Co., Preston, Ont.
Metallic Roofing Co., Toronto.
Paterson Mfg. Co., Toronto & Montreal.
Standard Paint Co Montreal.
Wheeler and Bain, Toronto

Saws.
Atkins, E. C., & Co., Indianapolis, Ind
Shurly & Dietrich, Galt, Ont.
Spear & Jackson, Sheffield, Eng.

Scales.
Canadian Fairbanks Co., Montreal.
Frothingham & Workman, Montreal.

Screw Cabinets.
Cameron & Campbell, Toronto.

Screws, Nuts, Bolts.
Dominion Wire Mfg. Co., Montreal
Montreal Rolling Mills Co., Montreal.

Soil Pipe
McFarlane, Walter, Glasgow

Sewer Pipes.
Canadian Sewer Pipe Co., Hamilton
Hyde, F., & Co., Montreal.

Shelf Boxes.
Cameron & Campbell, Toronto.

Shears, Scissors.
Clauss Shear Co., Toronto.

Shovels and Spades
Eclipse Mfg. Co, Ottawa
Frothingham & Workman, Montreal.
Peterboro Shovel & Tool Co., Peterboro.

Silverware.
Hutton, Wm., & Sons, Ltd, London, Eng.
McGlashan, Clarke Co., Niagara Falls, Ont.
Phillips, Geo., & Co., Montreal.
Round, John, & Son, Sheffield, Eng.

Skates.
Canada Cycle & Motor Co., Toronto.
McFarlane, Walter, Glasgow.

Sprayers
Cavers Bros., Galt

Spring Hinges, etc.
Chicago Spring Butt Co., Chicago, Ill.

Stable Fittings
Dennis Wire & Iron Co., London

Steel Rails.
Nova Scotia Steel & Coal Co., New Glasgow, N.S.

Stove Pipe.
Chown, Edwin, and Son, Kingston

Stoves, Tinware, Furnaces
Canadian Heating & Ventilating Co. Owen Sound.
Copp, W. J., Son & Co., Fort William
Davidson, Thos., Mfg. Co., Montreal
Down Draft Furnace Co., Galt
Guelph Stove Co., Guelph.
Gurney Foundry Co., Toronto.
Harris, J. W., Co., Montreal.
Howard, Wm., Toronto
Kemp Mnfg. Co. Toronto.
McClary Mfg. Co. London.
Merrick Anderson, Winnipeg
Pease Foundry Co., Toronto.
Smart, James, Mfg. Co., Brockville
Stewart, Jas., Mfg. Co., Woodstock, Ont.
Taylor-Forbes Co., Guelph, Ont.
Wright, E. T., & Co., Hamilton.

Tacks.
Montreal Rolling Mills Co., Montreal.
Ontario Tack Co., Hamilton.

Tents.
Tobin Tent and Awning Co., Ottawa

Tin Plate.
American Sheet & Tin Plate Co., Pittsburg, Pa.
Baglan Bay Tin Plate Co., Briton Ferry South Wales
Lysaght, John, Bristol, Newport and Montreal

Turpentine
Defiance Mfg. Co., Toronto.

Ventilators.
Harris, J. W., Co., Montreal.
Pearson, Geo. D., Montreal.

Wall Paper
Staunton Limited, Toronto.

Wall Paper Cleaner.
Gilbert, Frank U. S., Cleveland

Washing Machines, etc
Dowswell Mfg. Co., Hamilton, Ont.
The Shultz Bros. Co., Brantford.
Taylor-Forbes Co., Guelph, Ont.

Water Filters.
Buffalo Mfg. Co., Buffalo, N.Y.

Wheelbarrows
London Foundry Co., London Ont.
Schultz Bros. Co., Ltd., The Brantford.

Wholesale Hardware.
Birkett, Thos., & Sons Co., Ottawa.
Caverhill, Learmont & Co., Montreal.
Frothingham & Workman, Montreal.
Hobbs Hardware Co., London.
Howland, H. S., Sons & Co., Toronto.
Lamplough, F. W., & Co., Montreal.
Lewis Bros. & Co., Montreal.
Lewis, Rice, & Son, Toronto.

Window and Sidewalk Prisms
Hobbs Mfg. Co., London, Ont.

Wire, Wire Rope, Cow Ties, Fencing Tools, etc.
Banwell-Hoxie Fence Co., Hamilton
Dennis Wire and Iron Co., London, Ont.
Dominion Wire Mnfg. Co., Montreal
Greening, B., Wire Co., Hamilton.
Owen Sound Wire Fence Co., Owen Sound
Montreal Rolling Mills Co., Montreal.
Western Wire & Nail Co., London, Ont.

Wrapping Papers.
Canada Paper Co., Toronto.
McArthur, Alex., & Co, Montreal.
Stairs, Son & Morrow, Halifax, N.S.

Wringers
Connor, J. H. & Son, Ottawa, Ont

CIRCULATES EVERYWHERE IN CANADA

Also in Great Britain, United States, West Indies, South Africa and Australia.

HARDWARE AND METAL

A Weekly Newspaper Devoted to the Hardware, Metal, Heating and Plumbing Trades in Canada.

Office of Publication, 10 Front Street East, Toronto.

VOL. XIX. MONTREAL, TORONTO, WINNIPEG, OCTOBER 12, 1907 ***NO. 41.***

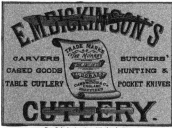

Read "Want Ads." on Page 49

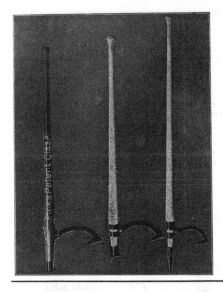

DISSTON BRAND FILES

Every File stamped "DISSTON" is warranted as perfect as a file can be made, and is guaranteed to be made of the best crucible steel. Each file is carefully tested and proved before leaving the factory.

Material — ALL DISSTON BRAND FILES are of the finest crucible steel. This is made in their own steel works and they are thereby enabled to secure absolute uniformity of quality.

Shape — Each blank is carefully forged to the most approved pattern for the class of work for which the file is intended.

Teeth — The teeth are formed to give the greatest strength, with sharp cutting edges, and at the same time allow sufficient clearance to prevent clogging.

Quality — They are hardened under the Disston Special Process, which renders them so tough and strong that the cutting-edge is maintained the longest possible time.

For Cutting and Wearing Qualities the Disston Brand of Files is Unequalled

We carry in stock all regular shapes and sizes, and are prepared to promptly furnish any special patterns and cuts of teeth desired.

Send for Fully Illustrated Catalogue

LEWIS BROS., Limited

Montreal

OTTAWA **TORONTO** **WINNIPEG** **VANCOUVER** **CALGARY**

ARE YOU ALIVE

to every profit-making line of goods in the hardware trade? Do you draw the best trade—**the paying trade** of your town? You will if you handle only goods of unquestioned quality.

EVERY BRUSH OF THE FIRST QUALITY BEARS THE NAME OF

BOECKH

There's a reason why—Boeckh's Brushes are made in the **most complete brush factory** in Canada. The material is **selected by experts** and put together by the most skilled workmen obtainable.

"Boeckh's Brushes Boom Business"

UNITED FACTORIES, LIMITED
TORONTO CANADA

GARBAGE CANS with Galvanized Steel Bodies and Seamless Covers

Will stand any amount of rough handling

Two sizes, with wood slats.
Nos. 400 500
Ins. 17x25½ 18x26
Three sizes, without wood slats.
Nos. 30 40 50
Ins. 14x21 17x25½ 18x26

The Arrow Can No. 1
One size only.
15x24
With or without cover.

The Regulation No. 4
One size only.
14½x20¾
With six wood side protectors.

Three Sizes.
Nos. 1 2 3
12½x13¼ 14x15 15x17
Supplied with bail and back handles.

Strongly made and **heavily galvanized after cans are made up. This is a very important feature,** as it **prevents rust** and **corrosion** from eating out the bottoms.

Descriptive Circular and Prices furnished on application.

The THOS. DAVIDSON MFG. CO., Ltd., Montreal and Winnipeg

A

H. S. HOWLAND, SONS & CO. LIMITED

HARDWARE MERCHANTS

Only Wholesale — 138-140 WEST FRONT STREET, TORONTO. — Wholesale Only

Cow Chains

Special Stall Chains

4/0 Four Chains

13 *Special Stall Chain*

Stall Fixtures Stamped Steel

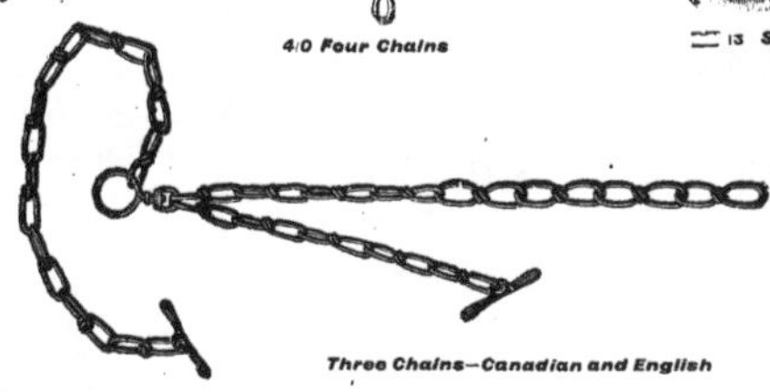

Three Chains—Canadian and English

Closed Ring Canadian and English

Stall Fixtures Round Forged

Open Ring Canadian and English

For other lines of Chains, see our Catalogue

H. S. HOWLAND, SONS & CO., LIMITED.

Opposite Union Station

GRAHAM WIRE NAILS ARE THE BEST

If you are not receiving our illustrated monthly circular, WRITE FOR IT.

Our Prices are Right — We Ship Promptly

NEW LINES

are worth the attention of the up-to-date hardwareman, especially when they bear the trade mark

"UNIVERSAL"

"Universal" goods are articles of merit, made of the best materials on carefully studied principles. They are labor-saving machines which lighten household drudgery, and are well-nigh universally in use.

Nos. 321 and 323 "Universal" Meat Choppers are two new sizes of smaller capacity than the other large choppers. There should be extensive sale for them with hotels, restaurants, butchers and boarding houses.

No. 323

No. 321

"Universal" Coffee Mills Nos. 012 and 0012 are intended for household use. They have a more perfect adjusting device than other mills and will grind coffee as fine or as coarse as wanted. They are simple in construction and adapted for use on the table or attached to the wall.

No. 0012

No. 012

We will send you circulars describing these more fully. Ask for them, and let us quote you prices.

Caverhill Learmont & Co

MONTREAL

Winnipeg Ottawa Quebec Fraserville

HAVE YOU GOT THEM?

The Up-to-date Harvest Tools!

Merchants who appreciate values and aim to give their patrons the best, will stock the **Maple Leaf Harvest Tools,** which are unequalled in Quality, Beauty and Finish.

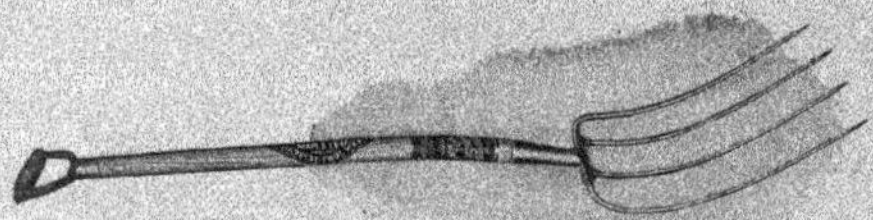

No. 122—Manure Fork

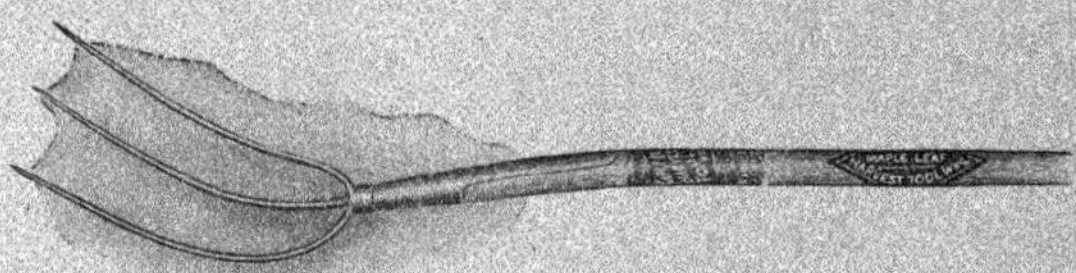

No. 108—Hay Fork

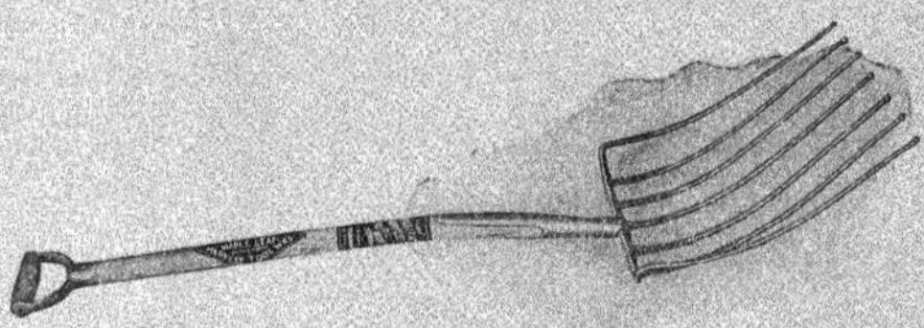

No. 243—Beet Fork

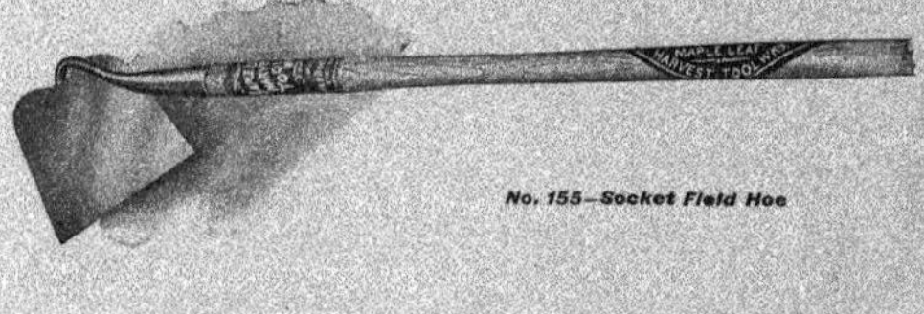

No. 155—Socket Field Hoe

To the Merchants:

The season is now on when your stock of harvest tools will require "sorting" up. Make your orders call for **"Maple Leaf"** Harvest Tools, and if your jobber cannot supply them, send your order to us and it will receive our prompt and careful attention. A trial will convince you of the superiority of the "Maple Leaf" Harvest Tools.

The Maple Leaf Harvest Tool Co., Limited

TILLSONBURG, ONT.

"GOOD CHEER" OAKS

FOR COAL OR WOOD

NO. 100 SERIES.

An artistically nickel trimmed line in four sizes

HEAVY POLISHED STEEL BODY.
GOOD SIZED BASE SECTION.
LARGE ASH PAN.
HEAVY FIREPOT.
DOUBLE FEED DOORS.
SHAKING DRAW CENTRE GRATE.
MAGAZINE, IF REQUIRED, FOR THE THREE LARGER SIZES.

No. 112 Dia. of body 11 inches.
No. 114 Dia. of body 13 inches.
No. 116 Dia. of body 15 inches.
No. 118 Dia. of body 17 inches.

NO. 50 SERIES.

Single and Double Heaters

Here's a Trade Winner You Should Have!

A handsome Oak with a powerful **Double Heating Feature** which makes the stove sell like hot cakes. Only placed on the market late last season, but wherever sold dealers have placed large orders for this year's supplies.

Body of stove is of heavy polished steel, handsomely nickel trimmed. Large feed door with good sized mica lights, make it a very cheerful stove, while screw dampers in both upper and lower doors make it easy to regulate the fire.

A shaking ring draw centre grate is provided, as also a good sized ash pan.

Made in two sizes. No. 54 with 13½-in. pot and No. 56 with 15½-in. pot, either single or double heater. Feeders for both sizes. The price is right, too.

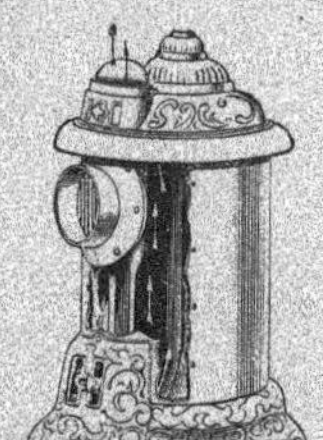

SECTIONAL VIEW

The JAMES STEWART MFG. CO.

LIMITED

Western Branch:
Foot of James Street, WINNIPEG, MAN.

WOODSTOCK, ONT.

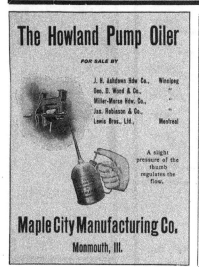

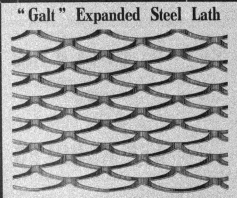

Simonds
Crescent-Ground Cross-Cut Saws

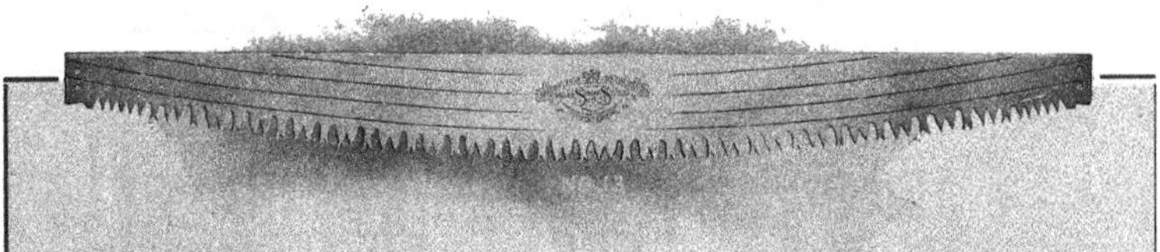

For logging camps where work must be fast and economical there is no other saw made that can give results equal to the Crescent-Ground Saw. **Made of Simonds Steel.** It cuts easy, runs fast and gives good results. This cross-cut saw is sold by most of the leading jobbers in Canada.

Simonds Canada Saw Co., Limited

TORONTO, ONT. **MONTREAL, QUE.** **ST. JOHN, N.B.**

YALE AND TOWNE CHAIN BLOCKS AND TROLLEYS

Fastest and Easiest Lifting Blocks. Frictionless Trolleys.

The TRIPLEX BLOCK will outlift and outlast any other in points of ease and economy. It cannot be equalled as a hand hoist. The view shows the relative weight and height one man can lift with the Triplex and others.

All sizes, up to six tons, in stock

CANADIAN FAIRBANKS CO., LIMITED, MONTREAL

TORONTO ***WINNIPEG*** ***VANCOUVER***

Wheelbarrows

All kinds and sizes. The cut reproduced here is just of one of the many, but the kind every contractor should use. The substantial, satisfactory, steel tray Contractor's Barrow.

The London Foundry Co.

LONDON, CAN.

The Peterborough Lock Manufacturing Company, Limited

Cylinder Night Latch, No. 103.

Peterborough, Ont.

Manufacturers of all kinds

Rim and Mortise Locks, Inside, Front Door, and Store Door Sets, also full line High-class Builders' Hardware.

Sold by all Leading Jobbers in the Dominion.

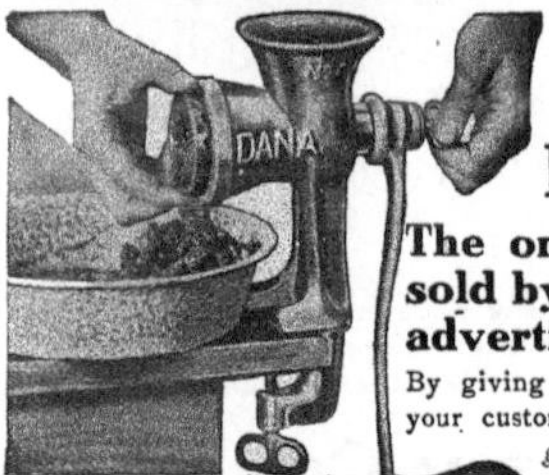

Rotors pull out over dish so that the meat drops into the dish and not on the floor.

Dana
Food Choppers

The only Food Chopper not sold by catalogue houses, and advertised big in home papers.

By giving complete and lasting satisfaction to your customers, Dana Food Choppers help your *general sales.*

ASK YOUR JOBBER

THE DANA MFG. CO. Cincinnati, O.

The Freezer the Women want

Dana Peerless Freezer
Write for catalogue

Dana Mop Wringer takes a woman off her knees.

Not How Cheap, but How GOOD

Banner Lamp Burners,
O, A, B and D sizes.

Canada Lamp Burners,
A and B sizes.

Security Lamp Burners,
A and B sizes.

Bing Glass Cone Lamp Burners,
A and B sizes.

Lantern Burners,
No. 1 and No. 2.

Every Burner carefully inspected before leaving the factory, and guaranteed perfect.

Orders solicited through the jobbing trade.

ONTARIO LANTERN AND LAMP CO.
HAMILTON, ONT. LIMITED

The Buffalo Manufacturing Co,
Buffalo, N.Y.

When you get our goods you know you get THE BEST.

We manufacture
Water Filters
Water Coolers
Chafing Dishes
Table Kettles and Stands
Coffee Extractors
Wine Coolers
Nursery Chests
Baking Dishes
Crumb Trays and Scrapers
Tea and Bar Urns
Bathroom Fixtures
Coal Vases and Hods
Candlesticks
Cuspidors
Match Safes, Etc.

All High Grade and exceedingly presentable.

REPRESENTED BY

H. F. McINTOSH & CO.

51 Yonge Street

Write for Catalogue **TORONTO, ONT.**

DOMINION WIRE MANUFACTURING CO.

MONTREAL TORONTO LIMITED

Manufacturers of

PLAIN WIRE *All Sizes and for all Purposes.*

GALVANIZED WIRE *For Fencing, Telegraph and Telephone Purposes.*

Copper and Brass Wire

WIRE NAILS
WOOD SCREWS

Bright Wire Goods
Staples
Steel and Brass Jack Chains

Spring Cotters, Strand Guy Wire, Steel Wire Barrel Hoops

A RECORD BREAKING SELLER

22 CALIBRE
16 SHOT

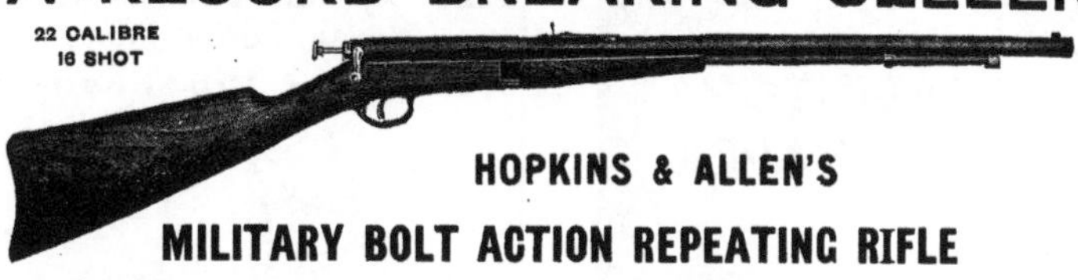

HOPKINS & ALLEN'S

MILITARY BOLT ACTION REPEATING RIFLE

$10.75 LIST

There has long been a demand for a good rifle at this price—and the **HOPKINS & ALLEN Junior Repeater** fills the bill. This rifle has been and is being extensively advertised, and a heavy sale for it is created.

SPECIFICATIONS:—22-inch barrel, take-down, 22 calibre. Magazine handles 22 short, 22 long and 22 long rifle cartridges without change of carrier. Shoots 16-22 short or 12-22 long or long rifle cartridges. Action, Military Bolt Pattern with positive safety device. Length over all, 40 inches, weight 5½ pounds.

FOR SALE BY ALL FIRE-ARMS JOBBERS.

PLENTY OF GOOD SUPPLEMENTARY ADVERTISING MATTER—FREE.

Shows Bolt Drawn Back—Position for Ejecting Shells.

WRITE FOR CATALOGS

THE HOPKINS & ALLEN ARMS CO.

LONDON OFFICE
6 City Road, Finsbury Square, London

Dept. 67, NORWICH, CONN., U.S.A.

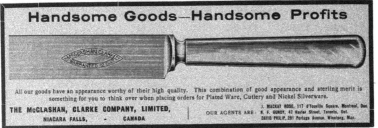

This is *the* Chopper

35 Sizes and Styles for Hand, Steam and Electric Power

The Revolving Steel Blades and Perforated Steel Cutting Plate *cut* the meat clean and clear—don't tear or grind.

No—there's nothing like it on the market.

It is in a distinct class
The "Enterprise" Class.

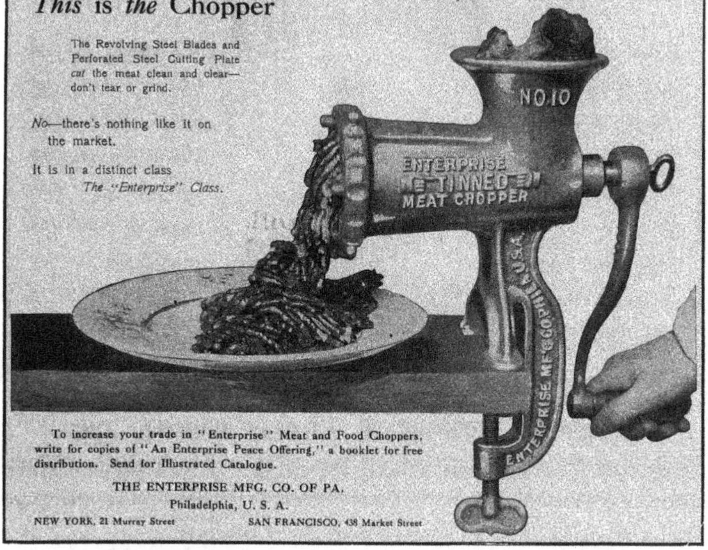

To increase your trade in "Enterprise" Meat and Food Choppers, write for copies of "An Enterprise Peace Offering," a booklet for free distribution. Send for Illustrated Catalogue.

THE ENTERPRISE MFG. CO. OF PA.
Philadelphia, U. S. A.

NEW YORK, 21 Murray Street — SAN FRANCISCO, 438 Market Street

Scrap Iron for Re-working

Better bars for re-working are obtained from selected scrap than from pig iron—have more life and strength.

London Bars are made from selected scrap and come in Flats, Rounds, Squares, Ovals, Half-Ovals, Half-Rounds and Bands.

Large stock always on hand, insuring prompt shipment.

London Rolling Mills
London, Canada.

Windsor Pump and Foundry Co.
LIMITED

We Make a Specialty of

Brass Cylinders for Wind-Mill and Hand Force Pumps Fitted with Patent Rotating Plunger Valves
Designers of Special Machinery Tools and Dies, Plug Valves, Foot Valves, Brass Tubing, Iron and Brass Castings

100 London Street West, WINDSOR, Ont.
PHONE 369
WRITE FOR PRICES

CHARLES H. FOX
Vancouver, Canada
MANUFACTURERS' AGENT

Selling to the Hardware Jobbers Only.
Representing Canadian, British and American Manufacturers, correspondence invited. Cable Address, Chasfox Vancouver, P.O.Box 1105.

Persons addressing advertisers will kindly mention having seen their advertisement in Hardware and Metal.

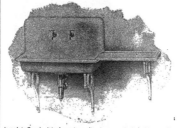

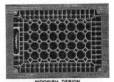

Persons addressing advertisers will kindly mention this paper.

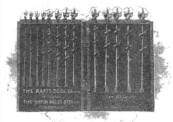

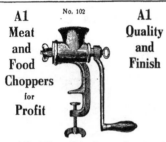

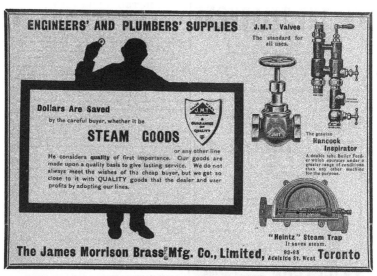

DIRECTORY OF MANUFACTURERS

Hardware and Metal receives, almost daily, enquiries for the names of manufacturers of various lines These enquiries come from Wholesalers, Manufacturers and Retail Dealers, who usually intimate they have looked through Hardware and Metal, but cannot find any firm advertising the line in question. In the majority of cases these firms are anxious to secure the information at once. This page enables manufacturers to keep constantly before the trade lines which it would not pay them to advertise in larger space.

ADVERTISING.

HARDWARE BOOK.
Contains a number of ready-made ads. with suitable illustrations for Hardware Merchants and Stove Dealers. Price, $1.50.
MACLEAN PUBLISHING CO.
Montreal Toronto Winnipeg

BRASS GOODS.

THE JAMES MORRISON BRASS MFG. Co., Limited,
89 to 97 Adelaide St. W., TORONTO.
Manufacturers of Brass and Iron Goods for Engineers and Plumbers; Locomotive and Marine Brass Works; Gas and Electric Fixtures. Telephone Main 3836.

CATALOGUES.

CATALOGUE COMPILING.
Attractive, neat, up-to-date, business-getting, at moderate cost, promptly done.
Before starting a Hardware, Stove or Machinery catalogue, write me for estimate, etc.
CHARLES D. CHOWN,
40 Guilbault St., Montreal, Que.

CEMENT.

The Hanover Portland Cement Co., Limited,
HANOVER, ONTARIO
Manufacturers of the celebrated
"Saugeen Brand"
OF PORTLAND CEMENT.
Prices on application. Prompt shipment.

CONDENSED ADS.

Agents Wanted, Articles for Sale, Articles Wanted, Business Chances, Businesses for Sale, Situations Wanted, Situations Vacant.
Does not one of these headings suggest some way in which you can use our "want ads." to advantage?

CLIPPERS.

PRIEST'S CLIPPERS
BALL BEARINGS

Largest Variety,
Toilet, Hand, Electric Power
ARE THE BEST.
Highest Quality Grooming and Sheep-Shearing Machines.
WE MAKE THEM.
SEND FOR CATALOGUE TO
American Shearer Mfg. Co., Nashua, N.H., USA

Weibusch & Hilger, Limited, special New York representatives, 9-15 Murray Street.

FEED BOILERS.

For Feed Cooking Boilers write the largest Manufacturers in Canada for circulars and prices. You know us.
The
James Bros. Foundry Co.
Perth Ont.

FLOOR SPRINGS.

The Best Door Closer is
NEWMAN'S INVINCIBLE FLOOR SPRING
Will close a door silently against any pressure of wind. Has many working advantages over the ordinary spring, and has twice the wear. In use throughout Great Britain and the Colonies. Gives perfect satisfaction. Made only by
W. NEWMAN & SONS,
Hospital St. - Birmingham

GALVANIZING.

GALVANIZING
Work and Prices Right
ONTARIO WIND ENGINE & PUMP CO., Limited
Toronto, Ont.

GALVANIZING AND TINNING
The CANADA METAL CO.
Toronto, Ontario.

LAWN MOWERS.

The people of Canada know that our name on a LAWN MOWER is an unalterable guarantee of quality and Durability.
TAYLOR-FORBES CO., Limited.
Head Office and Works, Guelph, Ont. Toronto—1880 King St. W.; Montreal—122 Craig St. W.; Winnipeg—The Vulcan Iron works, Limited.

Durability Finish
Efficiency
Guaranteed in the
Maxwell Lawn Mower
DAVID MAXWELL & SONS,
St. Mary's, Ont.

SHELF BRACKETS.

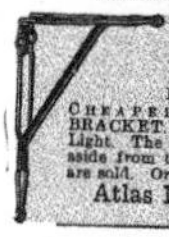
Will Hold Up a Shelf
That's what a shelf bracket's for. For this purpose there can be NOTHING BETTER, NOTHING CHEAPER than the BRADLEY STEEL BRACKET. It is well Japanned, Strong and Light. The saving on freight is a good profit aside from the lower price at which the goods are sold. Order direct or through your jobbers
Atlas Mfg. Co., New Haven.

SOIL PIPE.

FORWELL FOUNDRY CO.
BERLIN, ONT.
Manufacturers of
Soil Pipe, Fittings, and Cast Iron Sinks
Ask Jobbers for "F. F. CO." Brand.

CLUFF BROS.,
Toronto,
Can give prompt shipment on
SOIL PIPE and FITTINGS

STEEL TROUGHS.

STEEL TROUGHS AND TANKS
We Manufacture
Steel Tanks, Stock Tanks, Steel Cheese Vats, Threshers' Tanks, Hog Troughs, Water Troughs. Feed Cookers, Grain Boxes, Coal Chutes, Smokestacks.
AGENTS WANTED.
The STEEL TROUGH and MACHINE CO. Ltd., TWEED, ONT

TOOLS.

ARMSTRONG CUTTING-OFF TOOLS
are correctly designed and the blades are bevel rolled from special Self Hardening Steel. Straight and Offset shaping 7 sizes each. Write for Catalog
Armstrong Bros. Tool Co.
1.6 N. Francisco Ave.
CHICAGO, U.S.A.

WIRE WORK.

Crescent Wire and Iron Works
KINGSTON, ONT.
Manufacturers of every description of Wire Goods, Railings, Fencing, Wire Guards, Coarse Wire Cloth, Heavy Coal, Builders' and Miners' Screens, etc.
PARTRIDGE & SONS - Proprietors

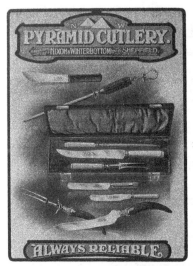

"Shear making "up-to-date" is found in

HEINISCH

TAILORS' SHEARS, TRIMMERS, SCISSORS, TINNERS' SNIPS, ETC.

"Best in the World"

Since 1825, and a steady increase in sales due to their superior cutting quality.

R. HEINISCH'S SONS CO.
NEWARK, N.J., U.S.A.
New York Office and Salesroom, 155 Chambers St.

REGISTERED TRADE MARK
GRANTED 1682

World's Best Cutlery

You make no mistake in specifying goods made by **JOSEPH RODGERS & SONS, Limited** when ordering cutlery. A reputation of over 200 years at stake, insures the quality of every article.

Write for catalogues from Sole Canadian Agents

JAMES HUTTON & COMPANY, MONTREAL

"Acorn Quality" Corrugated Sheets

have captured the market for the roofing and siding of **barns, factories, warehouses, outbuildings,** etc. It has been proven that they make the cheapest and most durable construction, besides being

Positively Fire and Lightning Proof.

Dealers who handle this line are assisted by our widespread advertising and expert salesmen, and protected by our "work-with-the-trade" policy.

We are just now in splendid shape to fill orders for Corrugated Sheets. We have large stocks on hand and **can ship immediately.**

"WE WORK WITH THE TRADE."

The Metal Shingle & Siding Co., Limited

Montreal **PRESTON** **Toronto**

SALES AGENTS:

ST. JOHN, N.B. Emerson & Fisher, Limited — QUEBEC J. A. Bernard — WINNIPEG Clare & Brockest — CALGARY Ellis & Grogan — VANCOUVER McLennan, McFeely & Co. Ltd.

THE MOST LIGHT FOR THE LEAST MONEY

C. G. E. Incandescent Lamps

Long Life ***Efficiency*** ***Reliability***

Write for prices and quantity discounts.

Canadian General Electric Co., Limited

Head Office, - - Toronto, Ont.

District Offices: Montreal, Halifax, Ottawa, Winnipeg, Vancouver, Rossland

---

ALEXANDER GIBB

Manufacturers' Agent and Metal Broker, 13 St. John Street, Montreal

Representing Canadian, British and American Manufacturers. Correspondence invited from firms wishing to be represented.

Expert Bookkeeping

Investigation and auditing of all classes of accounts undertaken. Trading, profit and loss accounts and balance sheets prepared. Sets of books opened and office systems devised in an up-to-date manner. Collection of accounts a specialty.

Davenport, Pickup & Co.
622 McIntyre Block and 422 Ashdown Block
WINNIPEG, MAN.,
AND AT BRANDON, MAN.

Square your Accounts

INCANDESCENT

LIGHTING WITH COAL OIL.

The most economic modern lighting. The "Ideal" lamp burner, screws on any ordinary coal oil lamp. This simple apparatus produces its own gas, giving a light more brilliant than electricity or gas. Costs only ¼ of a cent per hour. Guaranteed exactly as represented or money refunded. The "Ideal" lamp for the country home. Agents wanted everywhere. Write for free booklet describing everything.

ADDRESS

ST. ARNAUD FRERES
Cor. St. Sulpice and Commissioners Sts.
Montreal, P.Q.

DON'T WHINE!

When Collections are Slow

Send them to

The Beardwood Agency

313 New York Life Bldg.
MONTREAL

AND WE WILL GET THE MONEY FOR YOU

Retail Hardware Association News

Official news of the Ontario and Western Canada Associations will be published in this department. All correspondence regarding Association matters should be sent to the secretaries. If for publication send to the Editor of Hardware and Metal, Toronto.

Officers Retail Hardware and Stove Dealers' Association of Western Canada:

President—A. J. Falconer, Deloraine.
First Vice-President—J. B Curran, Brandon.
Second Vice-President—W. M. Gordon, Winnipeg.
Secretary-Treasurer—J. E. McRobie, Winnipeg.
Executive—Alberta, A E Clemens, Sedgewick; C. F. Comer, Calgary; A. R. Auger, Okotoks.
Manitoba—H. S. Price, Boissevain; A. P. Macdonald, Winnipeg; O. Gilmer, Winnipeg.
Saskatchewan—G. K. Smith, Moose Jaw; S A. Clark, Saskatoon; J. R. Fox, Weyburn.
Association offices, 53 Scott building, Main street, Winnipeg.

Officers Ontario Retail Hardware and Stove Dealers' Association:

President—W. G. Scott Mount Forest
1st Vice-President—J. R Hambly, Barrie.
2nd Vice-President—J Walton Peart, St Marys
Treasurer—John Caslor, Toronto.
Secretary—Weston Wrigley, 10 Front St East, Toronto.
Executive Committee—The officers and H Becker, Hamburg; G. A Binns, Newmarket; D Brocklebank, Arthur; A. W. Humphries, Parkhill; W A Mitchell, Kingston, and Frank Taylor, Carleton Place.
Auditors—J. W. Peacock and C. F. Moorhouse, Toronto

Officers British Columbia Retail Hardware Association.

President—L. B. Lusby, New Westminster
Vice-President—C Snell, Vancouver
Secretary-Treasurer—John Burns, Vancouver.
Executive Committee — W. Stearman, Vancouver H T. Kirk, New Westminster; R A. Ogilvie, Victoria

---

SIGNS ON RAILWAY CARS.

Secretary M. L. Corey, of the National Retail Hardware Association announces that practically every railroad in the United States has agreed, through united action, to stop the use of freight cars as advertising bill boards. Drastic instructions were issued June 1 by the American Railway Association to this effect and car inspectors will be instructed to watch for signs and remove them at once. Agents will be notified not to permit shippers to attach signs in any way to cars they are loading. Secretary Corey says:

"This rule will apply to all shippers alike, but it will particularly catch the chief offenders, the catalogue houses, and they are the fellows we are after. They were getting a hundred times more benefit out of this privilege than any other shipper.

"If you see any car with these advertisements on, please report to us the number and the name of the road on which the car belongs."

ASSOCIATION MATTERS BOOMING

The past fortnight has seen quite a revival in interest in Retail Hardware Association matters, nearly every mail bringing in applications for membership and orders for the collection form letters supplied to members by the association.

The offer of an eight-months' membership for $2 is attracting many merchants who will probably attend the next annual convention and get interested in the general work of trade organization. One firm, Christenson & Cove, MacLennan, Ont., in sending in their fee this week, said they would endeavor to make enough out of their "collection" letters to keep on their membership after the present year ends on May 1 next.

Last week a letter, written by G. A. Binns, Newmarket, was reproduced, and to show that other merchants are having equally satisfactory results from the association "collection department," the following letters to the association secretary, are interesting:

Dear Sir,—With reference to the collection forms which we received from the association, we are pleased to say that we received very large returns from the investment.

We had lately transferred our accounts to new ledgers, and before writing a large number of accounts off to "bad debts," sent a number of your collection forms to these parties. In response, several accounts were paid which we considered lost accounts. We trust that your department will prove equally successfully to others.

W. W. CHOWN Co., Limited.
Belleville, Sept, 19th, 1907.

Dear Sir,—In reply to your letter of the 18th, re results of using collecting forms, I might state that I have had excellent returns. In fact, most of the cases that I used them were for accounts that I had considered as lost or no good, as I was anxious to give them a good test, and I have received over $20. This seems like finding money, as I had used several other methods and could not get any results, so that I am convinced that they are the best that I know of, and I intend using them continually.

F. A. HOAR.
Barrie, Sept. 20, 1907.

Dear Sir,—We have your letter of the 18th, making enquiry as to our success with the collection forms of the O.R.H.A., and would say that results so far have been very satisfactory. In all but one instance we have received full settlement with interest, and that one case was a particularly hard one, which will require very vigorous following up.

The first collection made by means of the forms repaid us several times over for their cost, and we have still a good supply on hand for further use.

We think that the more general use of these forms would be greatly to the advantage of the retail trade, and the dealers who use them will find almost immediate benefits, and trust that the collection department may continue the good work with vigor.

A. W. HUMPHRIES & SON.
Parkhill, Sept. 25, 1907.

Money will be high this fall. You will find jobbers and manufacturers collecting hard. If your balance is not paid they will expect good tall rates on past due accounts because they are forced to pay still interest rates themselves. Make up your mind to crowd your collections more than you have ever done. You will need the money this fall in a greater degree than you ever needed it before in good times.

As announced last week, a payment of $3 covers the association membership until May 1, 1908, and a set of the collection letters.

FARMERS SITTING TIGHT.

The Kinburn correspondent to the Arnprior Chronicle last week said: "A business man of this place reports that one day recently he drove all day presenting accounts for collection, yet out of the day's work he did not realize enough to pay horse hire, which amounted to one dollar and fifty cents. Nearly every debtor is said to have a bank account."

If the merchant was a hardwareman, $3 spent in association membership and "collection letters," would have produced enough results to make the merchant feel like spending $1.50 for horse hire for a good time.

MERCHANTS BLAME THE BANKS.

Merchants frequently complain that farmers are not buying as freely as they should, and the banks are blamed for this condition of affairs. All general stores handle a good deal of produce. It was formerly the practice, with most of them, to pay for it in cash, and on certain days a considerable sum had to be on hand. The farmers received their money, and almost invariably spent it liberally. Some time ago the banks requested that produce payments be made by cheque, and, of course, this has been done. Now a farmer takes his eggs, butter, etc., and gets a cheque for the full amount. At the bank he may want to convert it into currency, but is strongly urged to leave it on deposit. In the majority of instances he does so, and then, instead of looking round to see what goods he would like to have, and purchasing freely, he spends as little as possible. As his savings grow he becomes more and more eager to augment them, and, consequently, is likely to cultivate economy.

The banks have adopted this policy all through the country as a means of increasing deposits. One result is that while few farmers had bank accounts some years ago, there are not many who haven't at the present time. General prosperity is benefitted, and, therefore, the condition must be accepted as a good one.

The feature of the situation with which fault can be found is that farmers have comfortable sums in the banks drawing interest, and, at the same time, they expect merchants to sell them goods on long credit. It is this that has given the movement toward cash system such impetus.

A Model Retail System

Business as Conducted by the James Walker Hardware Company, St. James Street, Montreal.

That system minimizes worry and loss and augments efficiency and profit is a truth which every merchant accepts, and, wherever possible, endeavors to actually apply. System prevents or detects laxity on the part of employes and acts as a very powerful auxiliary with the employer. It is an ever active searchlight, playing simultaneously on every part of a business, picking out flaws here, and detecting praiseworthy fidelity there. It never sleeps, but it never obtrudes unpleasantly.

The systematized business, as carried on by the James Walker Hardware Co. in Montreal, is, comparatively, a perfect retail system, and is a great credit to its promoter and originator. That F. M. Hill "knows his business," knows exactly "where he is at," anyone can see when they have occasion to approach him on some business matter. He has had wide experience in business, having been connected with Sargent & Co., New York city, for nine years, Yale & Towne for four years, and with Weed & Co., Buffalo, for eleven years. In Buffalo, Mr. Hill first applied to the business some methods which had been revolving in his brain for some time, and he applied them to the business of the James Walker Hardware Co. in a more perfected form. It is Mr. Hill's aim to extract from system the greatest amount of good possible, and the only way to make it valuable is to apply it rigorously.

Profit Sharing.

It is not always wise to share profits indiscriminately or in a wholesale fashion. Irresponsible employes in firms sometimes do not appreciate to the full either the privilege bestowed or the obligations involved in sharing the profits of their company. If they go so far as to squander the profits they have shared, it is doubly serious. Those profits shared by responsible members of the firm would have been wisely invested in the firm, thus increasing its efficiency.

Profit sharing, as conducted in the James Walker Co., is confined to the oldest employes—those who would have pre-eminently the interests of the company in mind. They are allowed to subscribe to a certain amount of stock, and any indebtedness which they may have contracted may be wiped out by the application of the total earnings, less interest; also a certain number of shares are set aside, pro rata, for certain employes, and a proportion is awarded them as a bonus. Those who have stock are thus encouraged to hold it and increase its value.

Accounting System.

The loose-leaf system of accounting has been adopted by this company, and Mr. Hill considers it the most satisfactory. It is not only a time-saver, but a worry-saver. The charges from day to day are bound together in book form, constituting the Day Book, or Journal, and from either or both of these two the items are carried to the Ledger.

For keeping account of cash sales, a duplicate form book is used. The original sheet is given to the customer, and the carbon sheet is kept by the cashier. A triple record of cash sales is kept by the cashier, the treasurer and Mr. Hill's secretary. Charge sales are recorded by each individual salesman. In making a sale, the item or items are recorded in the charge book, the original slip being given to the customer, and the duplicate slip is checked, according to its consecutive number, and is then detached from the charge book by the office.

F. M. HILL, MONTREAL,

Manager of the James Walker Hardware Company's Fine Retail Store on St. James Street.

Ordering System.

The Order Sheet used in Jas. Walker Co.'s store is very elaborate, and is calculated to preserve a complete record of goods ordered, by statistics, to show, approximately, the demand for certain lines, thereby indicating what lines are becoming obsolete and soon unsalable. A regular stock sheet is used, which covers all transactions for a year, and which contains three extra columns, containing figures for the three preceding years. The remainder of the sheet is divided vertically into twelve columns representing the twelve months of the year, and horizontally into many more columns for the various articles ordered. The squares or rectangles formed by the intersecting lines are divided into four equal spaces. When stock is taken, the number of one kind of article on hand is written in pencil in the lower left hand sub-division. If any of the particular articles are ordered, the quantity is entered in red ink in the upper left hand space. If this quantity is invoiced in the same month as it is ordered, it is entered in the lower right hand space; if only a part is invoiced, the amount is entered as above, showing balance due or unshipped in upper right-hand space in pencil. As was stated before, this system is used to determine accurately how soon an article is obsolete or unsalable.

Back Order Record.

Incomplete shipments of orders are often a source of worry and annoyance to merchants, and it is only by a rigid system of recording partial shipments that worry is done away with. According to Mr. Hill's plan, the regular charge memorandum is made out, the sheet is marked where the goods are to be ordered and turned over to the stenographer. After being ordered the consecutive number notation is affixed to the name of the factory for ready reference. After the goods are ordered, the sheet is placed in the filing case alphabetically arranged for convenience. As soon as the goods are invoiced they are checked on the charge list with date marked awaiting arrival of goods, and placed in the live portion of the filing case. As soon as the goods arrive at the receiver's room, he, being not allowed to see the invoices, and hence not knowing price of goods, as they are kept purposely from him, finds whom the goods are for, gets charge list, makes delivery, and passes charges to office to be billed. The advantage of this system are marked. If a salesman has made a sale he can immediately throw it off his mind, if, on the contrary, he wants to know in what condition the order is, he can refer to the slip in file cases and get exact data.

Handling of Building Hardware Contracts.

There is a special printed form for contracts. It consists of sheets with a notation of character of building, location, name of owner, and name of architects, with proper ruling for items and columns for prices. The sheets are priced in private cipher and put in alphabetical folder to await disposition. If the contract is lost the sheet containing information is removed and placed in correspondence filing case. If, on the contrary, it is secured, the data is copied on to a shipping form with the contract amount and consecutive contract number. The estimate forms are then filed in correspondence case, while the shipping sheets are sent to the stock floors to have goods gotten out and assembled together for delivery. These goods, when delivered, have to be charged against this particular contract on a pad of pink and yellow paper, with carbon between. These sheets show what was actually delivered. The pink sheet is returned to the office and kept on file, and, at the completion of the building, the sheets are priced up to show whether money has been made or lost on the contract.

The yellow sheets, in all cases, accompany the goods as they are delivered at the building, so that the foreman can see what has been delivered, and locate goods more easily by their numbers. By this system, the merchant can readily determine what he has actually gained or lost on the entire contract.

About the Clerk.

By a carefully written and guarded card system, the efficiency of the clerks

Device for Washing the Outside Surfaces of Windows.

is ascertained. A card for each clerk is kept, on which is written his age, name, address, salary, and memoranda regarding his general deportment. In addition, as a method for learning the clerk's efficiency as a salesman, cards are kept on which are written the cash and credit sales made by him every month, with totals for each year carefully recorded.

By all these carefully devised methods, the manager of the store, Mr. Hill, is enabled in a very short time to learn how each department is progressing. Mr. Hill has additional ideas, which he has not yet put in practice. It is his hope to adopt all the devices that he can learn of for increasing the efficiency of his store, and to reduce worry to a minimum.

The way to the Hall of Success may be pointed out, but the door must be opened by each man for himself if he would enter.

DEVICE FOR WASHING WINDOWS

There has long been need for some simple and practical device for washing the outside surfaces of windows. This need has greatly increased in late years with the increased height of modern buildings. The inaccessible exterior surfaces of the windows make the work of cleaning them especially inconvenient and hazardous, so that trained experts are now commonly employed for this service. In the accompanying engraving we illustrate a device with which the exterior surface of a window may be readily cleaned from within the building, and without imperiling the life of the operator. It consists of a cleaning head, which may be projected to the desired point by means of a lazy tongs mechanism that connects the head with the operating handles. The lower legs of the lazy tongs are respectively secured to a pair of dock shafts which are concentrically mounted, one shaft being tubular to receive the other. Each shaft is provided with a handle, and by swinging these handles toward and from each other the lazy tongs may be extended or retracted. A flexible hose connects the head of the washing device with a source of water under compression, so that a flow of water may be had at the desired point. A patent on this window-washing device has been secured by William G. Himrod, of Third and G Sts., N.W., Washington, D.C.

GUTTER-INSERTER.

D. A. Sapp, Towns, Georgia, has invented a substitute for the ordinary hammer commonly used in inserting gutters. According to the old methods, such insertion is a work of considerable difficulty, but the inventor has devised a tool by which it can be effected easily, quickly, and accurately, and also without any danger of injury to the hands of the workmen in operation.

CEMENT SHINGLES.

The use of cement for replacing articles made of wood is increasing every day. Cement has already replaced wood, to a great extent, in building sidewalks, bridges, fence posts, steps, building walls, foundations and many other purposes, and is commanding considerable attention at present in the form of shingles. In the earlier instances of concrete roofing the material was used in the same manner as in laying a floor, but the great strength required in a floor is not necessary in a roof. The cement shingles are only a little heavier than the best wood shingles, and, as they are practically indestructible, they are cheaper in the end than any other material, including tile and slate. These shingles are made in a great variety of designs and are reinforced with metal skeletons, which hold the cement together and terminate in loops at the edges for nailing to the roof. They are practically everlasting, as moisture, the cause of universal decay, is the chemical agent in the process of hardening cement and when properly mixed and tempered the cement shingles become harder and more durable the more they are exposed to the weather.

G. E. DRUMMOND HONORED.

At a meeting of the Directors' Board of the Molsons Bank, held in Montreal this week, G. E. Drummond, of the firm of Drummond & McCall, metal merchants, in Montreal, was elected a member of the Board. Mr. Drummond is a well-known figure in commercial and financial circles, being president of the Drummond Mines, Limited, and the Corrugated Steel Bar Co. He is a director of the Canada Iron Furnace Co., the Canadian Iron & Fundry Co., the Liverpool, London & Globe Assurance Co., the Londonderry Iron & Mining Co., the Montreal Pipe Foundry Co., the Montreal Trust & Deposit Co., and the Radnor Water Co. He is also a

G. E. DRUMMOND, MONTREAL,

Elected a Director of the Molsons Bank.

past president of the Montreal Board of Trade.

THE READY PUMP.

William Barclay Parsons, the famous engineer, is a foe to scamped work, and at a recent dinner in New York he said:

"That man is most unwise who tries to get his work done cheap. Cheap work can always be secured, but the quality of such work is on its face——"

Mr. Parsons, smiling, interrupted himself to tell a story.

"There was a man," he said, "who entered a dairy and asked how much the milk was.

"'Ten cents a quart, sir,' the young woman behind the counter answered.

"The man looked disappointed.

"'Haven't you any for six cents?' he asked.

"'No,' said the young woman, 'but,' she added, 'we can soon make you some.'"

Amongst the Salesmen

KEEPING CLERKS BUSY.

Some of the retailers who are now allowing their assistants to sit around a good part of the day, if there happens to be but few customers, would learn a valuable lesson in how to make money out of hired help if they could see the way their competitors, the big mail-order houses, handle their salesmen, says the Stove Reporter. They have hundreds of people working for them, and none of them is allowed to be idle. The managers of departments realize that idle men eat in on profits faster than almost anything else. A man who is working should be able to show a profit for his firm above his wages, but an idle man shows no profit for either himself or his firm. He is a dead loss. A loss of fifteen minutes' time out of an entire day seems like nothing to the individual, but to the employer of 1,000 men such a loss on each man means 250 hours' loss of time that day, or a good strong month.

For an employe who shirks work there is but one cure—discharge. For the employe who is willing, but does not know what to do next, there is another remedy, and it is the proper kind of management. If all employes were of a first-class variety a manager would hardly be needed, but so many are of the kind who fail to see what to do next that a good manager can keep busy keeping them at work.

Keep a Memorandum Book.

One busy merchant has adopted the following plan. Of course, each employe has certain duties which must be taken care of each day, and which occupy a large proportion of their time, and these duties give very little trouble. It is the extra jobs which are neglected. To avoid neglecting them this man has a little reminder book, which he always carries in his pocket, and everything he notices which needs attention he makes a memorandum of it in this little book, which is divided off into departments under such headings as: "15 Min. Jobs," "30 Min. Jobs," "Hour Jobs," "Half-day Jobs." etc. When he finds an employe with nothing particular to do for an hour, he looks in his book under the "Hour Jobs" department, selects whichever job is there listed which is best suited to the particular talents of this man, and puts him to work on it, at the same time marking right after the memorandum: "Done by ——," and gives the date.

By following up a system of this kind there is never a big accumulation of odd jobs. Everything is kept cleared up in good shape, and the same force of employes will turn out a tenth more work, and at the same time never feel that it is being rushed to death. Every man is kept busy, but not hurried, so no job is ever slighted, and as a result everything in his store looks well taken care of, and with a shorter salary list than most men would use to accomplish the same results. The reminder book is not needed by men who have remarkable memories, but most men will find it useful. There are hundreds of short jobs which should be done every month, but are not started because they are not thought of at the time they could be handled. Most men have a little leisure at different times, and ask themselves what they ought to do, and can think of nothing. The reminder book will suggest a number of jobs which would just fill the time at their disposal, and the job would then be done and out of the way.

CLERKS' BULLETIN BOARD.

In a store where a considerable number of clerks are employed there are obvious advantages in having some convenient way of communicating information to the entire crew at once. This requirement is met by the Leacock Sporting Goods Co., St. Louis, Mo., by establishing a clerks' bulletin board, on which are tacked up notices of interest to the working force. The nature of the information promulgated in this way will be understood from the following characteristic bulletins :

NOTICE
REGARDING SPECIAL ORDERS.

A deposit of 50 per cent. is required on all special orders not usually carried in stock, ordered by people not carrying accounts with us. No exception should be made to this rule except by special permission of the office.

THIS STORE WILL BE OPEN UNTIL 10 P.M., MAY 23.

NOTE NEW PRICES ON BATHING SUITS AND TENNIS GOODS.

Notices are usually typewritten, but short and important bulletins are sometimes boldly lettered with pen and ink or with a brush.

The system can be used by any dealer to an advantage and will doubtless be of much value. The cost of installing it is comparatively small.

A TOO COMMON COMPLAINT.

No more disastrous affliction can come to a man than that commonly know as "swell-head." In business it has ruined thousands of men in the past, and, doubtless, it will continue to add thousands more to the number.

A tendency toward "swell head" is most unfortunate, not only for the employe, who shows signs of the approaching malady, but also for his employer; for, generally, it results in the employer losing what might have been a good and useful man in his business, and in the employe missing a good opportunity to do well for himself.

Generally, the first manifestation of the disease is a marked disposition on the part of the employe to consider himself indispensable to his employer. Gradually the victim assumes that he is the "whole thing," and that the business would wither and die if he should take it upon himself to leave. In the advanced stages of the affliction, such an employe grows impertinent, neglects his duties, sometimes lies, and often deceives his employer. He puts on a lot of airs, treats his employer's views and wishes with disrespect, and even goes to the length of actually antagonizing the man who hires him to carry out his orders. At this stage the wise employer becomes a surgeon. He finds it necessary to cut the afflicted one off from the business, even though he may know that in so doing he is losing a man who has attained to a position of some importance. Large business houses usually make it a definite policy to dismiss the "swell-heads" as fast as they show symptoms of the disease. This should be a warning to those who feel the first approach of the well-known symptoms, so that when they suspect their heads are beginning to swell, even in the slightest, they may follow one of two courses—either hold the swelling in with a band, or hunt another job. Of course if the victim can check the malady with a bandage of his own winding, he will be doing a great service, not only to himself, but to his employer.

THE RIGHT CHANGE.

One of the little things that comes up frequently between customer and clerk is disputing over change. Occasionally the net result is deadly to the store. Many a good customer has been known to leave the store for good because of such a difference.

"I gave you five dollars," said the customer.

"No, ma'am, it was a two-dollar bill," says the clerk.

But the customer knows different and is sure she is right. And the clerk is equally sure.

You know how such differences arise. You also know they are well worth any little care and system that will tend to eliminate such disputes.

One of the best things yet advanced as a solution is to have the clerk always announce the amount of the coin or bill handed over by the customer.

Supposing the customer gives a five-dollar bank note in pay for goods. The clerk says immediately, "You gave me five dollars."

Then if the customer thinks that the bill was larger, the correction can be made before it is too late. This is a simple little method, but it is very effective. Many city stores now insist that this be done in every instance.

HARDWARE TRADE GOSSIP

Ontario.

R. E. Duggan has opened a hardware store at Wallace and Lansdowne avenues, Toronto.

C. G. McLaughlin, Latchford, Ont., is advertising a tinsmithing and plumbing business for sale.

J. Clydsdale, of the hardware firm of Clydsdale & French, Caledonia, Ont., was in Toronto on Tuesday.

The creditors of George A. McDonald, plumber, Toronto, met at the office of Osler Wade on Wednesday and decided to realize on the assets forthwith. The liabilities are $2,080; assets estimated at $1,070.

B. N. Lasher, formerly of the hardware firm of Geo. Lasher & Son, Toronto, was in the Queen City this week. Mr. Lasher is now the proprietor of an up to date hardware store in Leamington Ont.

Chas. F. Smallpeice, warehouse manager for Lewis Bros., wholesale hardware merchants, Montreal, spent Monday in Toronto, and left on Monday night, with a party of friends, for Chicago and St. Louis.

The many friends of S. F. Boyd, traveler for the Paterson Mfg. Co., Toronto, will regret to hear that he is at present laid up in the General Hospital. Toronto, and it will be some weeks before he is again able to resume his regular duties.

George A. Clare, M.P., of Clare Bros, Preston, was in Toronto on Wednesday. Mr. Clare returned from Germany a few days ago, where he spent the summer for the benefit of his health, and friends will be glad to know he is much improved. Mrs. Clare also spent the summer in Germany with her husband.

Palmer Ockenden has established a hardware trade valuation agency at 54 Yorkville avenue, Toronto. Mr. Ockenden's idea is to act as as independent valuator of stocks where hardware stores change hands. He has had several years' experience in Canada, as well as many years' connection with the hardware trade in England.

Oral B. Moore, manager of the Jones Register Co., Toronto, was quietly married Wednesday morning, at the home of Rev. J. B. Kennedy, to Mrs. Sarah Ramsay, formerly of Greenock, Scotland, who is a graduate nurse of one of the large hospitals in Glasgow, and who has practiced her profession with success in the city during the past two years. Both young people have a host of friends whose wishes will go out to them for much of this world's happiness and success. The young couple will spend their honeymoon in Detroit, Battle Creek and Coldwater, Michigan, and they will later reside in Toronto. Mr. Moore has been on the road during the past year, calling on the trade throughout Ontario.

The Parkhill Post of a fortnight ago says: "The christening of the infant daughter of Mr. and Mrs. Ernest A. Humphries by the Rev. F. G. Newton, in St. James' Church on Sunday afternoon was attended by the rather unique fact that the baby wore the dress in which three generations preceding her had been baptized. The garment was brought from India about a hundred years ago by Major Bowen, the present baby's parental great-great-grandfather, who was an officer under George IV., to his daughter in Milford Haven, Wales. The dress is a beautiful example of the hand embroidery of Indian natives, and is still in a splendid state of preservation. Needless to say, it is highly prized by the family in which it has done duty for so many children." Mr. Humphries is the junior member of the well-known retail hardware firm of A. W. Humphries & Son, Parkhill.

Western.

E. Morgan, Mannville, Man., has sold his hardware business.

Glenwright Bros., Winnipeg. have sold their hardware stock.

S. E. Maud, hardware merchant, Denwood, Man., has sold his business to S. A. Clark.

The Sturdy Hardware Co., Hartney, Man., has been succeeded by John Blair & Son.

W. J. Doyle, hardware merchant, Clanwilliam, Man., has assigned to E. Bailey Fisher, Minnedosa.

Quebec.

F. Wilkinson, secretary of B. & S. H. Thompson & Co., Montreal, is spending a few days out of town.

Henry Timmins, formerly of Mattawa, now of LaRose mine, Cobalt, called in Montreal on his way to England.

F. O. Lewis, president of Lewis Bros., Montreal, left on Friday of this week for an extended trip through Europe.

James Kinsman, representing North Bros. Mfg. Co., of Philadelphia, is calling on the trade in Montreal this week.

Mr. Fawcett, representing the Standard Chemical Co., Quebec, is in Montreal this week, purchasing supplies for his company.

R. McOuatt, son of T. McOuatt, Lachute, passed through Montreal this week on his way to eastern cities, accompanied by his bride.

Gravel & Duhamel, carriage and hardware dealers, Montreal, have assigned, with liabilities of $130,000. The Hochelaga Bank holds a note for $70,000.

Hon. William Harty, president of the Canadian Locomotive Co., Kingston, and Mr. McKelvy, senior member of McKelvy & Birch, Kingston, were in Montreal this week on their way home from Caledonia Springs.

N. A Wylie, representing the Kemp Mfg. Co., Toronto, who called in Montreal a short time ago, is now in New Orleans, U.S.A., having made the trip from New York city by steamer.

WHOLESALERS TO MEET.

The Canadian Wholesale Hardware Association will meet in Montreal next Wednesday, Oct. 16. A banquet is to be given by President Newman, at the Canada Club, Wednesday night, at which President W. G. Scott, Mount Forest, will be a guest, representing the Ontario Retail Hardware Association.

ANOTHER EXPATRIATED AMERICAN.

John W. Gates, who began life as a clerk in a retail hardware store, and who is known in the States as the "Bet a million" man, is in England, where he is to make his home in future. Like Richard Croker, Gates will live the life of a country squire and keep a large racing establishment.

More than a year ago Gates announced his determination to retire from active business, and, in fact, left the business practically in the hands of his son and lieutenants, except when he personally took charge of some big deal in Wall Street. Then came the big slump in stocks, in which the Gates firm was hit as heavily as any firm in Wall Street.

Then followed the unexpected announcement of the dissolution of the firm and its retirement from business. This was for the purpose of allowing its debtors time to pay the claims of the Gates firm against them on easy terms, and in the next two or three years large sums of money will be coming to Gates from his clients.

Before his departure, Gates left $40,000 with "Boots" Durnell for the final betting coups of the stable. Durnell lost this within a week in two bets. Gates was so informed when he reached England. His reply was a terse instruction to sell everything at public auction and come on to England.

Gates will lease some large country place in England for his home, but will establish a large training establishment at Newmarket Heath under the direction of Durnell and will purchase a large string of horses in training and young horses in England, hoping to equal Croker's record of winning the Derby.

It is said that Gates will retain his citizenship in the United States, and that his son will probably represent him there, but that he will seldom favor the country with a visit.

IRON FROM PORT ARTHUR.

The first shipment of pig iron to the east from Port Arthur was made on Oct. 9. The steamer Edmonton was chartered and she went to the Atikokan Iron Company, Tuesday, to load the cargo of pig iron for Samuel, Benjamin & Co., Toronto.

Since the blow-in the blast furnace has been constantly operated, and there is a great pile of pig iron in the yards of the Atikokan Company waiting to be shipped.

HARDWARE AND METAL

Established 1888

The MacLean Publishing Co.
Limited

JOHN BAYNE MACLEAN - President

Publishers of Trade Newspapers which circulate in the ProVinces of British Columbia, Alberta, Saskatchewan, Manitoba, Ontario, Quebec, NoVa Scotia, New Brunswick, P.E. Island and Newfoundland.

OFFICES:

MONTREAL, ... 232 McGill Street
Telephone Main 1255
TORONTO . . . . . 10 Front Street East
Telephones Main 2701 and 2702
WINNIPEG, . . . 511 Union Bank Building
Telephone 3728
LONDON, ENG. . . . . 88 Fleet Street, E.C.
J. Meredith McKim
Telephone, Central 12960

BRANCHES:

CHICAGO, ILL. - . . . 1001 Teutonic Bldg
J. Roland Kay
ST. JOHN, N.B. . . . No. 7 Market Wharf
VANCOUVER, B.C. . . . Geo. S. B. Perry
PARIS, FRANCE - Agence HaVas, 8 Place de la Bourse
MANCHESTER, ENG. . . . 92 Market Street
ZURICH, SWITZERLAND . . . Louis Wolf
Orell Fussli & Co.

Subscription, Canada and United States, $2.00
Great Britain, 8s. 6d., elsewhere . 12s

Published eVery Saturday

MISTAKE MANY RETAILERS MAKE

A number of complaints have been made recently by jobbers at the growing laxity on the part of the retail trade in signing railroad receipts. Many a retailer sends his carter to the railroad station for a shipment of goods without impressing on him the necessity of checking the number of pieces and examining the condition of the packages before signing any receipts. The result is that in a number of cases a carter signs the receipt before he obtains the goods. Probably no check is made by the merchant himself for some days. When he finally does do it and discovers an error he immediately writes the jobber to the effect that he is short a number of pieces or that the goods are damaged. When questioned in regard to the receipt given to the Transportation Company he endeavors to be evasive and hints at the withdrawal of his custom unless the claim is settled to his satisfaction.

"A very pretty piece of business," one might exclaim, but it is surprising the number of cases that occur. A retailer would have a very poor opinion of himself or staff if, when a sale of $25 was made, $24 were accepted as a settlement, $4 of which was bad money. Yet, there is an analogy between a merchant who would do that and one who signs for something which he did not receive.

The manufacturers and jobbers in the United States have had so much experience along these lines that they will not recognize any claims of this nature, throwing the onus where it belongs—to the man who signed for the shipment. Nearly every invoice received from the United States has a slip attached bearing this inscription:

> IMPORTANT.
>
> Before accepting and receipting for the material called for by bill of lading herewith, please see that you get the correct number of packages and that they are in good condition.
>
> If the shipment is "short" or the packages broken or damaged so as to expose contents to the weather, do not give the railroad company your receipt until after the station agent notes condition of the shipment on your expense bill.
>
> If you accept shipments from the railroad company "short" or damaged, you do so at your own risk.

Many merchants in Canada are stamping their invoices with a similar notation, and the larger firms refuse to entertain claims for loss or damage unless a proper receipt has been given to the transportation company.

In times like these, when the transportation facilities of the country are overtaxed, it behooves every retail merchant (especially those engaged in the hardware and metal trades) to make a strict check of all their shipments. It is one of the most important details of business.

COMPETITION CLOSES TUESDAY.

A fair number of contributions have been received in answer to the offer to give $10 for the best letter suggesting methods of increasing sales of holiday goods in hardware stores. The number, however, should have been much larger. The editor has made an effort to interest the clerks in these competitions, but so far, very few have availed themselves of the opportunity offered them.

The immediate advantage of winning the prize offered is not everything in the contest. The winner, of course, has the advantage of getting his name known all over Canada as a "live" salesman and, as a result, is likely to get offers of better positions. The advancement of several clerks can be traced directly to the public recognition of the merit of their work in Hardware and Metal.

To sit down and think out a campaign aimed to increase the sales of the store should, however, be sufficient in itself to encourage any clerk to enter the competition. Salesmen cannot rise higher without application and the competitions conducted in Hardware and Metal from time to time tend to encourage clerks to apply themselves to a study of their work. Employers should urge their clerks to send in contributions.

The "holiday sales" competition closes on Oct. 15. If you can do so send along a short letter, whether you are a traveler, merchant or clerk. For particulars see the "want ad." page.

START CHRISTMAS SELLING EARLIER.

The Christmas season tries the facilities of most stores very severely, and the great majority of merchants feel that if they could extend the holiday rush over a little longer period they would do a good deal larger business, and handle it with much more satisfaction to their customers, and, consequently, to themselves. It can be done. In most cases active selling does not commence until the first week in December, or later, and the congestion of trade that follows is well known to everyone. Is there any good reason why you could not attract brisk buying the last week in November? Numbers of other retailers have done so, and you can, too. The Christmas spirit will have permeated the atmosphere by that time, and it rests with you to stir it up. Advertising and display are the chief mediums through which this may be accomplished. Devote your newspaper space to Christmas goods, and emphasize the advantage of buying early, when selections can be made more comfortably and while assortments are unbroken. Make the windows and interior co-operate with the ads. If you can evolve some idea to interest the children in your store it will help very materially.

It is not too early to think of the Christmas trade. Lay your plans now and resolve that your facilities shall be much better this year than ever before.

REDUCTION IN POSTAL RATES.

The Canadian Postal Department has established a new drop letter rate of 1 cent in the cities throughout Canada, while another change has been made in the old rate between British Possessions and Australia and Rhodesea.

The new drop letter rate will be of advantage to all branches of the mercantile community in the large cities where free mail delivery is in force, and the Postal Department is to be congratulated on meeting the long expressed call for this reform, but would it not have been better to retain this rate and spend the money in improving the service outside?

A reduction on the foreign letter rate (2 cents now carrying a letter weighing one ounce, the former rate being 2 cents for half an ounce) will also benefit business houses having a large foreign correspondence as well as the large number of English workmen who have settled in Canada during the past two or three years.

One phase of the matter, however, is worthy of notice. The Toronto Daily Star on Tuesday of last week interviewed a leading post office official regarding the change, the official expressing the opinion that the reduction would be a great boon to business houses handling foreign goods, such as Eaton's and Simpson's, who, he said, send whole sheaves of orders by every mail.

Since the administration of Postmaster-General Mulock, the Postal Department has apparently been in close touch with the managers of the large departmental concerns, being more willing to institute changes in the postal regulations helpful to mail order house interests than to reform abuses in the postal system which effect hundreds of merchants throughout Canada.

The Postal Department is to-day a revenue producing institution, and in the opinion of this paper the officials should centralize their efforts toward improving the mail service, particularly in the large cities and in Western Canada, regarding which so many complaints have been heard during the past few years.

Business life in the West has enough disadvantages to overcome without having a disorganized mail service to put up with, and the improvement of mail conditions in the West should be of far greater importance than the reduction of letter rates or the establishing of a parcel post C.O.D. system, which would benefit chiefly the mail order concerns in the large cities.

THE VALUE OF SYSTEM.

Corporate interests are best maintained only by a thoroughly organized system of operation. Any other means is quite inadequate. It is obvious, too, that the larger the concern to be operated the greater the need for thorough system; the more complex the process the greater need is there for careful oversight. It is needless to say that a large corporation like the United States Steel Corporation or the Canadian Pacific Railway Company will need more systematized operation than John Jones' hardware business, but that is far from saying that John Jones need never trouble himself about devising plans for a more thorough organization of his business. It is simply a difference in degree between his business and the Canadian Pacific Railway's.

Every one who has applied rigid system to his business well knows the value of it. He knows that it does away with worry; that it increases the efficiency of his employes by lightening the load on their minds; that it detects and abrogates laxity on the part of employes; that it lessens the possibility of actual loss of money because of irresponsible trustees; that it makes time by losing none, and he also knows that if he should abandon systematic means his own reputation for shrewdness and executive ability would be lowered in the eyes of his customers.

Systematic operation tends to enhance the reputation of the firm adopting it. The buying public like to have time saved for them and system certainly saves time. To save time, to reduce worry to a minimum, to secure complete return from sales, and to save time both for the employes and the buying public, carry on as thoroughly systematic operations as possible.

RURAL SANITATION.

The "Sanitary Journal," published quarterly by the Provincial Board of Health of Ontario, in its last issue, devotes the bulk of its space to the improvement of the water supplies and sewerage systems for many Ontario towns and cities, notably: Peterboro, Picton, Cobalt, New Liskeard, Bracebridge, Chesley, Bradford, Port Elgin, Gravenhurst, Chapleau and Haileybury.

At the last session of the Ontario Legislature a clause was added to the Public Health Act, giving the Provincial Board power to enforce certain regulations upon municipalities not enjoying modern sewage and water systems. The clause, which stands as subsection 6 of section 30 of the Public Health Act, reads as follows:

(6) No sewage; drainage, domestic or factory refuse, excremental or other polluting matter of any kind whatsoever, which, either by itself or in connection with other matters corrupts or impairs, or may corrupt or impair, the quality of the water of any source of public water supply for domestic use in any city, town, incorporated village or other municipality, or which renders or may render such water injurious to health, shall be placed in or discharged into the waters, or placed or deposited upon the ice of any such source of water supply, near the place from which any municipality shall or may obtain its supply of water for domestic use, nor shall any such sewage, drainage, domestic or factory waste or refuse, excremental or other polluting matter be placed or suffer to remain upon the bank or shore of any such source of water supply near the place from which such municipality shall or may obtain its supply of water for domestic use as aforesaid, nor within such distance thereof as may be considered unsafe by the Provincial Board of Health, after an examination thereof by a member or officer of the said Board, and any person who shall offend against any provision of this section, shall upon summary conviction be liable to a penalty of not more than $100 for each offence, and each week's continuance after notice by the Provincial Board of Health or local Board of Health to abate or remove the same, shall constitute a separate offence.

The first application under this amendment was from Peterboro, the board examining the water supply of that city and making many recommendations towards its improvement. In one case it was found that one hundred sawmill employes were using open closets over the river from which the city water supply was taken. This, along with other similar corrupting influences, was ordered to be discontinued.

In an address before the Ohio State Board of Health, President Josiah Hartzell referred to the comparatively simple sanitary problems in country districts as follows:

"In large cities the immediate and urgent instinct of self-preservation has compelled the intervention of skilled engineers; the establishment of most approved methods. Most acute conditions are often found in towns and villages. For example, in the case of applications for sewers reaching the State Board from such communities, the majority already have in use crude, poorly constructed drains, carrying domestic sewage this way and that, in quest of the nearest watercourse. The abolition of these temporary expedients and the establishment of safe sewers, mixed up as these are with financial troubles, often constitute problems of the most vexatious character.

"Mankind for centuries past has recognized and still recognizes instinctively the powerful agency of the atmosphere and soil in dispelling and absorbing filth. The recognition is so universal that the act of casting the wastes of humanity to the winds and to the earth in the most heedless and haphazard way has become second nature. When the country people group together in villages the careless habit remains, and it is only when the dire results of an over-application of filth to the ground menaces the public health that competent advice is sought.

"As a result of all the study and experimentation of the past century in the field of sanitation, it may be said that these problems, complex as they may appear to be, are usually resolved by just giving the elements of nature a fair chance, instead of hampering their operation. How slow has been our progress in getting back to nature; in learning that even tuberculosis, the dread scourge of civilized man that has persistently mocked at all the empiricisms and puerilities of man, could be constrained to bow its horrid face only in the presence of Nature's pure air!"

Markets and Correspondence

(For detailed prices see Current Market Quotations, page 66.)

MARKETS IN BRIEF.

Montreal.

Antimony—Firmer.
Brass Goods—Declined 10 per cent.
Copper—Weaker.
Lead—Firm.
Linseed Oil—Steady.
Old Material—Firmer.
Tin—Heavy decline in England.
Turpentine—Firmer.
White Lead—Weaker.
Wire Nails—Advanced 5 cents.

Toronto.

Linseed Oil—Some price cutting.
Turpentine—Unsteady.
Wire Nails—Advanced 5c per keg.
Copper Boilers and Kettles—Declined from 30 to 37½ per cent.
Copper and Brass Wire—Reduced from 37½ to 50 per cent.
Radiators—Now 50 and 2½ per cent.
Ingot Tin—Declined 1 cent.
Casting Copper—Down ½ cent.

MONTREAL HARDWARE MARKETS

Montreal, Oct. 11.—The activity characteristic of the hardware business in the fall and winter months is this year very marked. During the past week, with the betterment of weather conditions, trade has experienced an additional impetus. Retailers in the east are very anxious to be well stocked up with autumn and winter goods and orders are coming in to the jobbers in increasing numbers and volume. Despite the disconcerting reports coming from the Northwest, retailers and jobbers in these districts are hopeful and local factories and wholesalers are shipping large orders for the West. The factories are still held back considerably by the delay experienced in receiving raw material. Prices are steady with two or three exceptions.

Screws—The factories are working at high pressure, not only to cope with current demands but also to lay up reserve stocks for future exigencies. It is understood that they are shipping orders direct to the consumers. The present demand is moderate and quotations are firm.

Wire Goods—The business transacted in this line at present is confined mostly to Northwest orders. A moderate demand locally exists. Prices are unchanged.

Stoves and Ranges—As far as order-booking is concerned the business in stoves and ranges for a time is over. Manufacturers report an immense number of orders received during September. From October 1, the ruling prices will be 5 per cent. higher than those previously existing.

Building Paper—In ten days or a fortnight the business in this will be in full swing and the factories at present are working at high pressure endeavoring to get a substantial surplus of stock on hand when the call for it commences.

Builders' Hardware—Business remains remarkably active, as it has during the entire summer and it is not expected that building operations will appreciably diminish with the coming of winter. All lines are moving freely.

Fall Goods—Lumbermen's supplies, farming requisites, and prospectors' tools are in splendid demand and it is expected and hoped by jobbers that a new record of business will be established this season.

Cutlery and Silverware—The public have not yet sufficient confidence in the Western financial situation to commit themselves on large orders, but it is expected that in a fortnight or so the demand will enliven considerably.

Sporting Goods—Are still moving very freely. Shotguns, rifles, ammunition and hockey supplies are in very active demand. Stocks at present are adequate and prices continue unchanged.

Brass Goods—Owing to the continuous decline in the copper market the prices on brass goods have been marked down 10 per cent.

Wire Nails—An advance all round of 5 cents has been made in this article. The reason probably is the limited supplies, the very active demand, and additional cost of raw material.

TORONTO HARDWARE MARKETS.

Toronto, Oct. 11.—Since last week's report sales have increased in almost all departments of the local hardware markets, and a healthy autumn activity now characterizes the trade. It must be admitted, however, that buying for future deliveries is not as active as jobbers would like, but this fact is directly attributable to the prevailing stringency in the money market, which has caused merchants in the West—and those elsewhere in a smaller degree—to make closer estimates of future needs and to adopt a more conservative policy in buying. Jobbers have, however, no fault to find with October business, which, so far, easily exceeds that transacted for the same time last year, and they are satisfied that whatever decrease has taken place in booked orders up to date will be more than offset by the increase in sorting orders, which will be received later on.

Screws—A marked scarcity still exists in all the best selling sizes, and it is evident that the manufacturers will not be able to get abreast of the trade for some weeks. The demand is very strong, and the factories are being run to their fullest capacity in an effort to lessen the scarcity. Prices are firm and unchanged.

Wire Nails—The demand is very strong, and supplies are low in the most popular sizes. The base price to the retail trade has been advanced from $2.50 to $2.55 per keg; with freight equalized on the following points of manufacture: Toronto, Hamilton, London, Brantford and Collingwood. Jobbers claim that the advance is a reasonable one, since so far this season their percentage of profit has been considerably lower than it was a year ago, when nails were sold as low as $1.90.

Copper and Brass Wire—Two weeks ago we warned the trade that lower prices might be expected in brass and copper goods. Our prediction was borne out last week by a reduction in the price of copper burrs and rivets. This week it is further fulfilled by a sharp drop in brass and copper wire, sizes 1 to 26 of which are now quoted at 50 per cent. off list, instead of 37½ off. Now that copper consumers have won their battle against the manipulations of Wall Street, and that copper prices have dropped to somewhere near the real value of the metal, lower prices are expected on many other lines of brass or copper manufacture. Another change made during the week is a decline from 30 to 37½ per cent. on copper boilers and kettles.

Wire—Since the recent advance very few enquiries have been received, and jobbers do not expect much business until the first of the year, when orders should begin to arrive for spring deliveries.

Sporting Goods—A heavy fall trade is being done in firearms and ammunition. The big shoot of the Toronto Junction Gun Club on Thursday—in which eight hundred pigeons were set loose—gave a great stimulus to strictly local trade in shot cartridges, and similar shoots all over the province are daily increasing the outside orders of local jobbers. In order that there may be no more misunderstanding regarding the operation of the new law, which forbids the use of automatic guns for the killing of game, the Government has issued an order-in-council which emphatically states that automatic rifles may still be used for all game, but that automatic guns may only be used for trap shooting.

Fall Goods—Lamps are in splendid demand, and many sorting orders are being received. Ash sifters, stovepipe wire, axes, handles, dampers, and kindred lines are moving out fast. Cutlery and silverware are having a heavy sale and many retailers are placing orders for Christmas requirements in these lines.

Bolts and Nuts—The supply of these has been very limited all summer. While the Canadian manufacturers' prices are low enough to prevent jobbers from buying elsewhere, some retailers are said to have bought a supply from over the border, getting better prices and quicker delivery than could be obtained from Canadian jobbers. As a result, lower prices on some lines are likely to result.

MONTREAL METAL MARKETS.

Montreal, Oct. 11.—For some reason, as yet mysterious to local metal merchants, all metals, with the exception of iron, antimony and lead, have experienced a bad slump, which has almost disheartened those dealers who have stocks of copper, zinc, tin and spelter. Why such a recession is hard to explain, as only during the past fortnight has the metal market assumed this serious aspect. The more optimistic of dealers console themselves by saying that these metals should have been marked down in price at any rate, as they have been boosted. The pessimists are in the dark as to the cause.

Pig Iron—Is maintaining its customary firmness, and although a large number of consumers are well stocked up for fall and winter requirements, considerable business is still being transacted. From now until the close of navigation a large tonnage will be moving, especially from Great Britain. High grades of Scotch and English iron cannot be obtained, as the stocks in Great Britain are practically nil. Recent statistics, published by an English firm, show that at the present time there are less than 150,000 tons available in England at the present time, while at the beginning of this year there were 573,000 tons to be had, and this time last year there was 700,000 tons on hand. Locally, a moderate volume of business is being done, and prices remain unchanged: Middlesboro, No. 1, $21.50; No. 2, $20.50; Summerlee, $25.50.

Spelter—Has experienced a further recession. St. Louis prices have been marked down 10 cents further than last week. The American output is still limited. Locally, little buying is being done.

Lead—Is steadier this week, and a more active call exists. Prices are unchanged. High premiums will be necessary to induce prompt delivery before Nov. 1.

Ingot Tin—For a long time such a decline as tin experienced during the past fortnight has never been seen. Eighteen pounds in two weeks is phenomenal, to say the least. Any who have considerable stock bought at high prices will undoubtedly suffer severe loss. The local price is unchanged.

Copper—The 15-cent basis, considered by all local men as a rational and feasible basis, has proved unsatisfactory to American dealers, and this basis has been lowered by ¼ cent or ½ cent. This decline has affected local prices, especially on brass goods, which have experienced an all round decline of 10 per cent. Very little buying is being done in local circles, as people are still hesitant.

Old Materials—A little more activity is noticed, and prices have firmed a little. Scrap iron is still quiet, with little call. Supplies are limited. Heavy copper and wire, 12½c; light copper, 11½c; heavy red brass, 11c; heavy yellow brass, 8c; light brass, 6c; tea lead, 3½c; heavy lead, 3¾c; scrap zinc, 3c; malleable and steel, $9.

TORONTO METAL MARKET.

Toronto, Oct. 11.—Peculiar conditions still feature the metal market. Copper, with another break on the American market, has gone below the 15-cent basis, at which producers endeavored to hold it, and, in response, another half-cent decline has been made in Toronto.

Ingot tin has also taken on a slump, with prices on Oct. 8 in London £45 per ton lower than the same day a year ago. The local price is practically 10 cents per pound lower than twelve months ago, a cent having been chopped off last week's quotations in response to the big drop in London during the past fortnight.

Pig iron, lead, antimony and zinc spelter all hold firm, although the demoralization of the copper market has its effect on every metal.

Pig Iron—Remarkable conditions exist in iron. Prices are dropping on a market which is bare of stock, and deliveries very hard to get. Dealers are quoting prices to inquirers, but they cannot promise shipments this side of the new year, except in rare instances. Speaking generally, buying is on a much smaller scale than usual at this season. Most buyers have fair stocks in hand, and those who usually come on the market at this time are buying in only partial quantities. Some manufacturers, however, are short, having held off, expecting lower prices. Now its a question with them of getting early shipments, even if a premium must be paid. In one case, the Drummond-McCall firm had to bring 500 tons from the States for a customer, paying about $1 more than our quoted figures, in order to keep their customer supplied. Relief is being looked for by some from the new blast furnace at Port Arthur. A cargo of 2,000 tons will reach Toronto from Port Arthur next week, and this will help to supply the immediate demands of some buyers. Local dealers look for higher prices on iron, although the present condition is such as to make it hard to determine what will develop for next season. We make no changes in our quotations.

Ingot Copper—Another weakening in New York has caused local prices on casting copper to drop from 17 cents to 16½ cents. The sentiment in the copper trade continues to be of a pessimistic character, and every decline in the market only strengthens the buyers' opinion that the situation is by no means healthy, even though prices are 12 cents to 13 cents lower than the early part of the year, and though production is being largely restricted. Manufactured goods, such as copper rivets, copper kettles, etc., have declined locally.

Brass—As dealers were caught with good-sized stocks of sheets and tubing on their hands, they are slow to follow the slump in the copper market. Gradually, however, prices will decline, although no immediate drop is looked for on account of most dealers' stocks being large.

Ingot Tin—Another heavy break has occurred in the markets for pig tin, the drop being attributed to free selling by the Chinese. The outlook is not very encouraging for higher prices. Indeed, a decrease in consumption of about 2,500 tons is expected within the next three months. Prospects are for easier conditions for ingot tin next season.

Sheet Metals—Shipments are large at present, merchants getting in stocks for present use and for working on during the winter. Bookings for spring delivery are very small.

Lead and Spelter—Both lead and zinc keep firm, as also does antimony. No prices have changed.

Old Materials—Business is slow, with dealers offering shaded prices on old copper and brass. Scrap iron and lead are held at the old prices, however.

U. S. IRON AND METAL TRADES.

New York, Oct. 10.—The Iron Age says:

Pig iron production is continuing at the recent gait; in fact, there has been a slight gain in September. The weekly capacity was 511,397 tons on Oct. 1, as compared with 507,768 tons on Sept. 1, two entirely new furnaces, the Lorain and the Inland, having started. The September production of anthracite and coke iron was 2,183,487 gross tons, as compared with 2,250,410 tons in August, a daily rate of 72,783 tons and 72,594 tons, respectively. The steel works product was 1,417,153 tons in September, and 1,445,685 tons in August.

As has been stated again and again, current production and consumption are still proceeding at a very high pressure. As indicative of that fact, we may note that the steel mills of the United States Steel Corporation broke all records of steel production on Oct. 2, the output reaching 48,326 tons of ingots in that one day.

There have been some modest accumulations of stock of pig iron in some sections, but, generally speaking, furnace yards are bare. It is the avowed purpose of the management of the Steel Corporation not to accumulate iron, when the time of lessened finished output comes, but to blow out furnaces as the requirements for pig fall off.

The pig iron markets throughout the country are dull. Western and southern makers are not seriously attempting to place iron for next year, while eastern makers are now in conference in Philadelphia with the object of discussing the situation.

While in some quarters a slight improvement in the volume of orders for finished products has been observed so far this month, as compared with September, others complain that the volume of business for forward delivery is very small. There are some reports of transactions in billets in the Chicago district at lower prices, but in other sections of the country the business in steel is a. a standstill.

In structural steel, a good deal of work is in sight, but there is much delay in its coming out, and the last week has been exceptionally quiet. For two lake boats, 7,000 tons of plates and shapes have been contracted for.

Bids put in by European Cast Iron Pipe Works for the work in Cuba have been a surprise to the American shops, the figures being $7 below our own. Under the circumstances, the prospect of securing any of the coming tonnage from that quarter is slight.

A further decline has taken place in copper and apparently every effort to support the metal has been given up.

U. S. METAL MARKETS.

Cleveland, O., Oct. 9.—The Iron Trade Review to-morrow will say:

While the feeling is better and talk more optimistic, there are sharp lines of demarkation between the strength of the situation, as it relates to finished lines, on one hand, and raw material and semi-finished lines on the other. Taken as a general proposition, prices on finished material are being well maintained, price inducements being offered in isolated cases, and in such a manner that they have no general reflection on the market of the commodities affected. Mill shipments, as has been the case for some time past, are greater than specifications against contracts, and manufacturers are, as a consequence, catching up on their deliveries and promise of shipments are better now than they have been for several months.

Buying of pig iron is confined largely to small lots for prompt shipment, but one large southern interest reports the sale of several round lots at $18.50 for this year's delivery. Other southern producers are, however, selling at somewhat lower figures, and in the east a considerable tonnage has been booked at figures below current quotations. The Bessemer pig iron market is still firm at $22 valley, but is likely soon to be tested by independent interests, changing from the manufacture of foundry iron to Bessemer. One interest has already made this change, and others will do so if sufficient inducements are offered.

New business in structural material is light, but the demand for shipments on accumulated orders was never heavier, and October will see a new production record for the American Bridge Co. The tightness of the money market has interfered with the carrying out of many building operations, but specifications are, nevertheless, heavy. An immense amount of new construction is being planned by the city of Brooklyn, including heavy expenditures for docks, subways, grade crossings and other improvements. Chicago mills are well supplied with plate orders and other interests are busy, but deliveries have much improved.

LONDON, ENG., METAL MARKETS.

London, Oct. 9.—Cleveland warrants are quoted at 54s. 3d., and Glasgow standards at 54s. 3d., making prices as compared with last week, on Cleveland warrants, 1s. 1½d. lower, and on Glasgow standard, 11d. higher.

Tin—Spot tin opened weak, at £149, futures at £146, and after sales of 300 tons spot and 500 tons futures, closed steady at £149 10s. for spot, and £143 5s. for futures, making price as compared with last week £7 lower on spot, and £8 5s. lower on futures.

Copper—Spot copper opened weak at £62, futures at £61 2s. 6d., and after sales of 300 tons spot and 900 tons futures, closed steady at £62 5s. for spot and £61 for futures, making price as compared with last week, 10s. higher on spot and 5s. lower on futures.

Spelter—The market closed at £21 17s. 6d., making price as compared with last week 17s. 6d. higher.

Lead—The market closed at £19 10s., making price as compared with last week 5s. lower.

A GENTLE REMINDER.

What is probably the most interesting circular ever issued from a government department has recently been distributed from the department at Washington of the Internal Revenue Commissioner. It deals in the most emphatic way with the stupidity, misguided zeal and misdirected ambition frequently displayed by revenue agents and other internal revenue employes under the apparently genuine impression that these are the means by which they will soonest attract the notice of their superiors and gain a promotion more nearly according with the self-estimated worth of their services.

The circular is in the form of a personal letter, addressed to an individual agent, and is so phrased as in the most forcible manner to bring before the person to whom it is addressed sundry matters of which he was previously in ignorance.

Here is the gist of its contents:

"A recent case which has been called to my special attention enables me to make my position clear in the matter of what I might term the misguided activity of revenue agents and special employes of this bureau who are operating under collectors or under revenue agents in the field. I fully realize that the arm of the law will be paralyzed in the matter of the enforcement of internal revenue taxes due the government if the officers of this bureau are incapable or corrupt; at the same time I ... and realize that the laws of this government are passed as much for the purpose of enabling honest men to do business under them as for the purpose of detecting and punishing those who violate them.

"There is a disposition, utterly at variance with good service, on the part of many of the internal revenue employes in the field to make what they call 'a record,' and in order to do so subject this bureau to endless annoyance and expense, and to make a perfect fiasco in the courts of their undue and hasty action in seizing distilleries making unfounded assessments, and practically confiscating property of law-abiding citizens who are endeavoring to live under the law in a business—the whisky business in any of its forms—always open to suspicion.

"I have on my desk a record of a case where three revenue agents or their employes hung around a registered distillery, hoping to find some sin of omission on the part of a bookkeeper for three weeks, while during that time there were five blockade (or unlawful) distilleries in full operation within seven miles of the registered distillery. By the operation of these unlawful distilleries the government was losing probably ...00 a day as tax on the whisky they were producing while these revenue agents were sitting about for 'big game'—the registered distillery.

"The men in the field should understand more fully than they seem to that there are two distinct classes of offences—that is, sins of commission and sins of omission. Nine times out of ten when a man in the whisky business is guilty of a sin of commission it is done designedly and for the purpose of defrauding the government; as a usual thing the sins of omission are due to ignorance or negligence.

"In this connection I am astounded to find recommendations for the seizing of distilleries, the upsetting of business enterprises involving investment of hundreds of thousands of dollars, because some clerk has failed to properly keep some blank 'form' prescribed by the regulations. Revenue agents and collectors should proceed upon the proposition guaranteed us by the constitution, that all men are innocent until proven guilty, and when they find minor irregularities they should assume the attitude of advisers and counsellors, and by virtue of their experience in the service, tell the people who know less about it what is expected of them. On the contrary, in too many instances they hastily use these sins of omission as a basis for their own self-glorification, and embarrass this bureau and the entire service by stupid seizures and assessments.

"Recently a revenue agent in the west held up the distillation and manufacture of vinegar for four states because the apparatus they were using could be so misused as to defraud the government. I found upon investigation that the apparatus was identical with that which has been used by vinegar manufacturers for ten years, and that during the entire time but one vinegar manufacturer had been prosecuted for intentional violation of the law. I use these two instances, but there are many others on my desk. I want no let-up whatever in a vigorous search for violations of the law, but I want wisdom on the part of the officers of this bureau to take the place of an activity to make 'a record,' finally at the expense of the government, which is entirely at variance with the letter and spirit of the law."

If this circular does not cause a heart-searching in many departments of our own government, it will have fulfilled only a bare fifty per cent. of its possible value. Its widespread publication throughout Canada might be the means of rewinding many miles of red tape, whose immediate usefulness consists mainly in attracting the nimble wit of neighboring powers to deprecate such a vast expenditure on national bunting by a country that is old enough to know better.

CORRESPONDENCE

Travelers, hardware merchants and clerks are requested to forward correspondence regarding the doings of the trade and the industrial gossip of their town and district. Addressed envelopes, stationery, etc., will be supplied to regular correspondents on request. Write the Editor for information.

NOVA SCOTIA'S IRON INDUSTRIES.

Halifax, N.S., Oct. 7.—With the approach of the cold weather comes a demand for stoves and all the dealers are now rushed with orders. The cold weather has set in earlier this year than usual and some of the dealers were taken unawares. There is a good demand for all classes of stoves, and business in general is reported to be quite brisk. Ranges seem to be even more popular than ever, and the modern housekeeper prefers them to the ordinary stove. In heating stoves, the baseburner still has the call, the Silver Moon, made in Nova Scotia, being very popular. Bedroom stoves of various styles and qualities are finding good sales, the smaller makes being in the greatest demand. All the tinsmiths are working overtime to keep up with the orders, which are unusually heavy for the opening of the season.

* * *

Manager Burchell, of the Sydney Cement Company, who was in the city last week, says everything is booming at the plant. Business this year has been unusually brisk, and the plant has to be kept in operation night and day to fill the orders. The company's business is rapidly expanding. Not only is the local trade good, but orders are coming in from other parts of Canada in large volume.

* * *

Some American capitalists and mineral experts are at present interesting themselves in an iron ore proposition at St. George's, on the west coast of Newfoundland, and are talking of the possible establishment of an electrical steel making plant at Sydney. The iron ore deposits show as high as 60 to 65 per cent. of metallic iron, and the mountain is said to contain thousands of tons. The mountain of ore runs sheer to the waters edge and rises to a height of 200 feet in a practically solid mass of mineral. The face of the cliff runs up in a succession of shelves, and the ore, in place of undergoing the usual mining process, will only have to be quarried away from its present position. Magnetite ore, however, contains a large percentage of titanic acid, which makes this ore practically unfit for steel making with the ordinary blast furnace methods. The new system of purifying this class of mineral, by eliminating the titanium, has lately been perfected in the electric furnace, as reported by the Canadian Commission. St. George's is only a short distance from Sydney, and the cost of transportation would not be very heavy.

Should the company succeed in purchasing the property, they intend erecting several electric furnices at Sydney, for the purpose of turning the ore from these mines into workable steel. The promoters say that Sydney is one of the best centres on the continent in which to run a large plant that might depend largely on electricity for its motive or other power. Mr. Bishop, who owns the psoperty, has refused an offer of $30,000 cash for it.

* * *

Good progress is being made at the plant of the Windsor Foundry Company under the new owners, and the prospects for its future success are very bright.

NEW BRUNSWICK INDUSTRIAL CHANGES.

St. John, N.B., Oct. 8.—Business continues good in both wholesale and retail circles. The jobbers are busy filling and shipping orders for lumbermen's supplies, stoves, stove pipe, snow shovels etc. Retailers report a good demand for all lines, builders' materials especially.

* * *

A destructive fire broke out at St. Martins on Friday afternoon last, which destroyed the saw and grist mill owned by J. Aubrey Vaughan, as well as the handsome residence and barns of J. S. Titus. The windows in the store of S. V. Skillen, opposite the mill, were all broken by the heat, and the paint on the woodwork was destroyed. J. & J. S. Titus' big general store, across the street, also had a narrow escape. Mr. Vaughan's loss will be about $5,000, with no insurance, while Mr. Titus' loss amounts to about $61,000, with only $2,000 insurance.

* * *

Thos. Malcolm, the contractor who is building the Intercolonial Railway from Campbellton to St. Leonard's, was in the city last week. He reports that the work is progressing favorably, and it will probably be completed next fall. It would likely have been finished this year if the weather had been favorable, but the succession of rainstorms has greatly interfered with operations. About 900 men are engaged at the work.

The construction work on the various sections of the G.T.P. Railway in this district is also proceeding very satisfactorily.

* * *

H. N. Coates, D. J. Purdy, M.P.P., Thomas A. Linton, of St. John, and Thomas H. Wilson, of Fairville, are seeking incorporation as the Purity Chemical Co., Ltd., with a capital of $25,000.

The Carman Safety Appliance Co , of St. John, is also making application for incorporation. The applicants are, G. Clawes Carman, Thomas Nagle, Thomas A. Linton, James Christie, and H. A. McKeown. The proposed capital is $50,000.

* * *

A lease for 100 years has been granted the Albertite, Oilite Light & Cannell Coal Company, Limited, in order to facilitate development of the company's shale deposits at Baltimore. Albert county. Work is likely to be commenced on a large scale in the near future. Col. Calhoun and Matthew Lodge, of Moncton, recently returned from England, where they succeeded in obtaining the necessary capital to develop the property, provided an extended lease could be obtained. Until recently, the shale had no commercial value on account of the low price of oil, but a discovery in other places of a method of developing the by-products was now in operation in England, and the Albert company's shale has been found very rich in this particular. In working and testing the shale $30,000 had already been expended. It is proposed to proceed with the development of the by-products and operations will at once be started on a large scale.

* * *

Another big New Brunswick lumber property is passing into the hands of American capitalists. The seller is Ernest Hutchinson, of Chatham, and the purchaser is the International Paper Co. The property consists of forty-five square miles of granted timber land, and 300 square miles of crown lands held under lease from the Government. In andition, there is a general store and a large and well equipped mill. The new company may start a pulp and paper mill in connection. The price paid is said to be in the vicinity of half a million. This is the fifth big property on the north shore to be absorbed by Americans during the past couple of years.

* * *

The Doctor James Walker property, on Germain street, has been sold to Manchester, Robertson, Allison, Limited. This large building was built some 12 years ago, and has been occupied by the Massey-Harris Co. ever since, but, in consequence of this firm having decided to remove their Maritime branch to Moncton, the property becomes vacant in December next. The Walker building adjoins M. R. A.'s Germain street building throughout their whole depth of 200 feet. It is the intention of M. R. A., Limited, to occupy the Walker premises in connection with their wholesale business.

* * *

A proposal by a syndicate of capitalists to establish extensive cement works at Green Head, near Milford, on the St. John river, has recently been made to the council. An appraiser to value the improvements on the properties asked for was appointed at the last meeting of the board, with a view to granting a lease, and should this valuation prove satisfactory to the applicants, it is understood they will pay for the improvements and start operations without delay.

* * *

Henry R. Ross, of Sussex, who has resigned as manager of the Sussex Packing Co., to accept the position of manager of the new cold storage plant in St. John, was dined by citizens of Sussex on Oct. 2, and presented with a handsome gold chain and locket. Mr. Ross is a native of Belleville, Ont., and has been manager of the Sussex concern since it was started six years ago

* * *

Chatham and St. John will get grants of $5,000 each from the Provincial Government for exhibitions to be held next year.

* * *

F. R. Murray, of the Emerson, Fisher Company, returned home last week from Toronto, where he was attending the annual meeting of the Manufacturers' Association.

John A. McAvity, of T. McAvity & Sons, who was reported ill last week, is much improved and expects to resume his duties this week.

PRICE CHANGES AT QUEBEC.

Quebec, Oct. 7.—A marked improvement is noticeable in hardware trade this week. Many consumers have evidently come into the market, and retailers from the country are preparing themselves for winter business. For this reason wholesalers report a brisk business and collections, on the whole, are satisfactory. There is nothing of special note in hardware prices, which seem to be pretty steady all around, although another decrease of one cent is noted for ingot copper, which is quoted at 17c to 18c per pound.

* * *

The Quebec Board of Trade has presented an address to Hon. Messrs. Fielding and Brodeur. The address was one of congratulation to the honorable gentlemen on their successful negotiating of the new commercial treaty with France, the first ever negotiated wholly by Canadian statesmen. The addresses were of a non-political character.

* * *

Lamonde & Linchereau, contractors, have been awarded the contract for the construction of the Brothers' School, on St. Francois and King streets. The cost of the building will be $100,000.

* * *

The stove market is much steadier, and good sales are reported. The demand is much stronger. An advance of 5 per cent. was noted, last week, on stoves. Manufacturers are very busy, and can hardly supply the demand.

HAMILTON FACTORIES ENLARGING.

Hamilton, Oct. 8.—Business in the hardware line is inclined to be slack at the present time. Merchants report a slight falling off in trade, which, however, continues steady.

* * *

William Herbert Wiggin, superintendent of the massive works of the International Harvester Company, in this city, died last week, and in his passing Hamilton lost one of its captains of industry. Mr. Wiggin had under his management the largest concern in Hamilton, in fact, one of the largest in Canada, a concern which estimates it factory space in acres. The late Mr. Wiggin proved himself a general in the administration of the affairs of this company, and he was respected and esteemed by the thousands of employes to an exceptional degree. He was born in Dracut, Wis., 47 years ago. His remains were forwarded to Worcester, Mass., where the interment took place. About 700 of the firm's employes marched in a body to his late residence, and, after viewing the remains, accompanied them to the T., H. & B. Station.

* * *

The Eagle Spinning Company, which was organized not more than a year ago, has let the contracts for a large extension to its premises at the corner of Wilson street and Sanford avenue. The Toronto Construction Company has secured the contract, a feature of which is the stipulation that it has to have the work completed within eight weeks, and that there must be no extras of whatever nature. The extension will cost in the neighborhood of $50,000, and the completion of the contract is being watched with interest.

* * *

Chadwick Brothers, brass manufacturers, have recently acquired a large tract of land on Oak avenue, adjoining their present premises. The business of this firm has been going ahead in leaps and bounds, and it is the intention to make a healthy addition to the plant in the near future.

* * *

The Sawyer-Massey Company has in the course of completion a large extension to its already large plant. The new wing will be used for a paint and carpenter shop, and is 60 feet wide by 320 feet long.

* * *

The company, which it was predicted in Hardware and Metal a short time ago would be organized in Hamilton for the construction of smoke consumers, has at last materialized. It has been incorporated as the Dominion Smoke Consumer Company, with a capital stock of $50,000. The local backers of the concern are A. R. Whyte, customs officer, and Frederick Harold White.

The Regal Shirt Company has also been incorporated. This company is erecting a handsome building at the corner of King and Caroline streets. The local incorporators are W. J. Carroll, Mary Hester Carroll, and J. W. Lameureaux. The capitalization is $200,000.

GALT RETAILERS TO BANQUET.

Galt, Oct. 8.—The approach of the winter season has resulted in the usual increased activity in the stove line. Although the advent of natural gas has affected this line to a certain extent, yet there is no reason why the hardware men should consider the advisability of going out of the stove business, as they state that that business is fairly brisk.

* * *

The new firm, Brown & Tait, successors to Theron Buchanan, is doing remarkably well, and there is every reason to believe that these young men will succeed in their new venture.

* * *

A. Wilson, an employe of the Shurly & Dietrich works, was severely burned on Tuesday. He was lighting one of the big gas fires, when the flame shot out, completely enveloping him. He was removed to his home and his injuries were found to be both serious and painful. It will be some time before he is able to resume his duties.

* * *

The Galt Retail Merchants' Association, which includes a number of local hardwaremen, will hold a monster banquet in the Iroquois Hotel here on Tuesday evening next. Invitations will be sent to members of the association in Berlin, Guelph, Preston, Hespeler, Waterloo, Paris, and surrounding towns. Although the program has not yet been completed, it is understood that prominent speakers will be present, who will speak on subjects of interest to the association.

* * *

A half dozen employes for the Canadian Brass works arrived in town from the United States recently. They are all married men with families, and have taken up their residence in Galt. The works of the company are very busy at present, and there are sufficient orders on hand to keep the plant busy for six months.

* * *

The Down Draft Furnace Co. made a most creditable exhibit of ranges at the fall fair. The display was visited by a large number of citizens, a good many of whom were heard to express their ignorance of the fact that such up-to-date goods were manufactured in Galt. The company is buiser now than at any time in its existence, and is one of Galt's most prominent industries.

* * *

Geo. Clare, M.P. for South Waterloo, and head of the Clare stove factory in Preston, has returned from a six months' tour in Europe, which was necessitated owing to his health breaking down. He returns greatly benefited. The works of which he is the head are very busy, and it is expected that the end of the year will reveal a greatly increased output.

* * *

Louis Semlin, who recently severed his connection with W. J. McMurtry, in order to go into business for himself, is finding a good opening, and his staff of employes has been enlarged.

* * *

In my last correspondence I omitted to mention the fact that the machine for making steel lath and in use by the Galt Art Metal Co. was manufactured in Galt by the Stevens Co. That such an extensive undertaking was successfully accomplished by a local firm is a matter for congratulation.

* * *

Your correspondent was this week invited to inspect the works of the Galt Brass Company, and on doing so found one of the neatest and most compact works that could be found in Ontario. The comppany is making only a modest beginning, but everything is up to date, and the work turned out is the finest quality. It has been found necessary to run two shifts of men night and day, and it is confidently expected that the works will shortly be extended.

* * *

Owing in a large measure to the successful management of Jas. Douglas, of the W. J. McMurtry Co., the annual fall fair of the South Waterloo Agricultural Society was an unprecedented success.

* * *

"The reason for paints is practically over," remarked a local hardware man to-day, "but, without exception, it has been the best in the history of the trade in Galt."

PRICE-CUTTING IN LONDON.

London, Ont., Oct. 9.—The effects of the absence of anything in the nature of organization are continually showing themselves in the local retail hardware trade. It is evident that under existing conditions anything like legitimate prices cannot be maintained. Said one dealer to-day: "People come into my store and ask the price of a certain article. On being told, they reply: 'Oh, so-and-so is selling it at such-and-such a price.' What am I to do but to let the goods go at the figure named, or lose the business? Only to-day a painter came in here and asked the price of putty. I told him. 'Well,' said he, 'I can get it at half a cent less,' and there was nothing left me but to let it go at the lowest price I ever sold it for, and I have been a good many years in the business. And so it is all along the line. I tell you until the retail hardware men of this city get together and come to some agreement, the whole business will go to the dogs."

* * *

"I don't see how it is that so many people are just now predicting the approach of hard times," said a wholesale hardwareman to-day. "If you could see the way and extent orders are coming in to us you would be convinced that the reverse is the case, or that the business is going to be better next year than ever before. You mark my words—next year will see the busiest time ever." And the speaker turned to answer a telephone call.

* * *

The hardware business at Tillsonburg, conducted by the late W. R. Hobbs, has been purchased by Messrs. How and Wilcox. Both are experienced hardwaremen, Mr. How having managed the business since the death of Mr. Hobbs, and Mr. Wilcox having been a clerk in the store. There were a large number of bidders for the business, which is one of the finest in Western Ontario. The new firm take hold on the 24th inst.

* * *

The London Brass Works Co. is an enterprise of which this city is justly proud. The company have had an unusually successful year, having increased their output of steam valves and other mechanical brass goods, over fifty per cent. over the best previous year. The outlook for the fall trade bears a decidedly encouraging aspect. In spite of the recent drop in the price of copper, orders are coming in which quite equal those of any previous year. A few months ago the president of the company, George W. Armstrong purchased a ten-acre lot on which they purpose in the near future erecting an up-to-date factory. The site is located in the south-east end of the city, close to the heart of the manufacturing district, several other manufacturing concerns having also secured lots nearby with the view of erecting premises larger and more convenient than those they now occupy. The difficulty with the London Brass Company is that trade is too brisk at present to give them time to make the anticipated change, but next spring may see the first sod turned.

* * *

Another firm has written to the manufacturers' committee, asking for inducements to open a branch here. The promoter wants the citizens of London to subscribe between $25,000 and $30,000 to float the concern, promising to employ between 75 and 100 skilled laborers. The company will manufacture engines and flash boilers, and such things.

* * *

Papers of incorporation have been sought by local capitalists for the Western Ontario Brickfields, Ltd., capitalized at $100,000. The provisional directors will be Dr. John D. Wilson, S. Frank Glass, Dr. Drake, C. B. Edwards, Chas. H. Ivey, Dr. J. E. Niven and J. Lewis Thomas. The analysis of the clay which is to be used in the manufacture of the brick is reported to be equal to any in the Old Country.

CHATHAM MERCHANTS GET TOGETHER.

Chatham, Oct. 8.—Chatham has added another big industry to its already long list. The new concern is the Chatham Carriage Company, the officers of which are: President and Manager, Ira Teeter; Vice-President, Arthur Cooke, and Secretary-Treasurer and Assistant Manager, F. E. Fisher—all experienced carriage men, and formerly associated with Wm. Gray & Sons, carriage manufacturers, of this city. The new concern, recently chartered, have purchased the property on William street formerly occupied by the McKeough & Trotter launch works, and are equipping it with a view to going in for all classes of high-grade carriage and automobile body and gear work. They expect to start active operations in the course of a few days.

* * *

A variety of matters, chiefly concerning the railways, was dealt with at the last week's meeting of the Council of the Board of Trade. It was decided to communicate with the M.C.R., with reference to securing connections with Chatham, via the C. W. & L. E. Electric Road. The accommodations on the late Wabash trains coming into Chatham were warmly criticized and the free interchange of G.T.R. and Wabash return tickets between here and Detroit was urged. W. R. Landon, manager of the Chatham Wagon Co., was unanimously elected secretary in place of C. H. Mills, recently removed to Berlin. The Executive have secured Hon. Adam Beck to address the Board of Trade, and the citizens in general, on the subject of "Hydro-Electric Power," at a meeting on Monday next.

* * *

At the annual meeting of the Chatham Motor Car Co., recently held here, the reports for the first year of operation were most encouraging. The "Chatham" car attracted considerable attention at the Canadian National Exhibition at Toronto, and elsewhere, where exhibits were put on, and it is felt that the advertising of the past year will count decidedly in the new year's business. Quite a number of improvements will be made in the 1908 car, the most noticeable of which will be an increase in power. To handle the anticipated growth of business, it was decided to increase the company's capital stock to $150,000, the directorate being fixed at seven. Dr. T. K. Holmes was elected president at a subsequent directors' meeting; Ira Teeter, vice-president and manager of the mechanical department, and D. S. McMullen, secretary-treasurer and general manager.

Ald. Westman, of Westman Bros., hardwaremen, has definitely stated that he will not be a candidate for aldermanic honors this year, pleading that he is too busy to give the necessary time to the duties involved. Should he stand by his resolution, next year's Council will lose one of its most aggressive members. As chairman of the Civic Industrial Committee, Ald. Westman has done much toward building up the city, and his place will be hard to fill.

* * *

Hugh Gallagher & Son have been awarded the contract for the new electric light station, at a cost of $2,877.

* * *

J. C. Shaw & Co., the new Wallaceburg hardware firm, announce that they will be open for business about Oct. 18. Their store is now being renovated and painted.

Last week the local Retail Merchants' Association bestirred itself, and energetically inaugurated the work for the coming year, the meeting being the best attended since the local branch was formed last January. Messrs. Pole, Westman and Winterstein were appointed to draft a programme for forthcoming meetings, with a view to rendering them interesting and attractive. It is the intention to secure capable speakers to discuss subjects of interest to merchants. President Cowan also suggested an annual banquet in December—a suggestion which will probably be adopted at the next meeting.

Ald. Westman brought up the matter of a Dominion Day celebration for 1908. A celebration should appeal particularly to merchants, as it tends to keep Chathamites and their money in town for that particular day. Judging by past experiences, the previous autumn is not too early to commence an agitation.

The delay of the Chatham Gas Company in installing purifiers for their natural gas, was also the theme of a resolution.

Last April a resolution was passed "that no member of this association shall advertise on any printed card or programme, nor permit his or their firm name to appear on any subscription list for charitable institutions." The resolution has done a lot of good; but, for some reason or other, the association at the time didn't tackle the ticket nuisance. Chatham has plenty of charitable ladies who are always on the lookout for an excuse to evolve a concert, tea meeting, social or excursion, in aid of some church or charity; and the merchants of King street, being looked upon as millionaires, are invariably the first and most frequent victims. Recently your correspondent chanced to meet a lady with a broad smile making her exit from a place of business on King street. The merchant inside didn't smile. "There ought to be a law against that sort of thing," he said, vehemently. "It's blackmail or highway robbery." The merchant refusing to buy the ticket is threatened with something akin to a boycott. In the absence of any concerted action, the threat has in the past usually worked. The R.M.A., however, dealt with the matter by adding a postscript to last April's resolution, covering the objectionable practice. The words, "Nor purchase tickets for concerts, raffles, socials or excursions," were suggested; but, finally, were rejected in favor of the more comprehensive phrase, "Nor purchase tickets." The resolution, if steadfastly adhered to—and there is no reason why the merchants shouldn't stand pat—will save the average merchant ten times his annual membership fee.

* * *

The association kicked strenuously against unfair competition from the auctioneers on the market. The latter are accustomed to secure new bankrupt stocks and auction them off as second-hand, cutting largely into the Saturday trade. There is scarcely a retail line that isn't injuriously affected; whilst the goods sold are pretty generally declared to be of inferior quality. The R.M.A. claim that the practice is contrary to by-law and should be stopped —or, that if it isn't, the by-laws should be amended. Some months ago the matter was brought before the City Council, but was overlooked. The R.M.A. at their last meeting again brought it up, and it is, as a result, now in the hands of the Civic Property Committee, with instructions to consult the City Solicitor and ascertain what can be done.

INGERSOLL DEALER SELLS 376 STOVES.

Ingersoll, Oct. 9.—This is harvest time with the hardware merchants and the plumbers. The advent of cold weather has set them all on the jump. "This is our busy day," is a sign which could be displayed with no deviation from the truth. And those couple of frosty nights are responsible for all the activity. "Are we busy?" said one hardware merchant to your representative. "Well, I should think we are," was his answer. He then went on to state that calls were constantly coming in from all parts of the town to place stoves in position, furnish pipes or to hurry up and install a furnace. During the warm weather of summer, people overlook the necessity of preparing early for the colder months, and, consequently, when the first cold wave comes along, everyone is in a panic. The orders are sent in almost simultaneously, and, as a consequence, the hardware merchants and the plumbers are now at their wits' end. Of course, there is much fault found with the plumber, but, again, there are times when the plumber feels like taking a hand in the complaining himself, and this is when he is being overwhelmed with work and everyone is clamoring to have their job done first.

* * *

The hardware merchants all report a most successful trade thus far this fall. With the majority of them, last year was a record-breaker, from the standpoint of the volume of business done, and they are looking forward to a great season between now and the new year. "Trade could not be better," was the satisfied manner in which one dealer spoke of the present situation. This same dealer stated that last year he sold 376 stoves and ranges. "Not a bad record," he added, and, in the course of a conversation, he mentioned that he expected that this year's output would be as large, if not larger.

The competition between the members of the Ingersoll Gun Club for the Hunter Arms Co.'s trophy, has been brought to a close, the winner being H. W. Partlo. The trophy was valued at $20, and was well worth striving for. During this contest there was a great demand for the best of ammunition, which is carried in large quantities by local dealers. Mr. Partlo's score was 85 out of 100 blue rocks.

BOOK TO ADVERTISE WOODSTOCK

Woodstock, Oct. 8.—Trade conditions in the city are good, and hardwaremen report that they have no particular reason to complain. Industrial conditions have been better, and some of the big factories are not running their full strength, but the tightness of money, which will probably make itself felt before the winter is over, has not as yet shown itself. However, the people of Woodstock are for the most part industrious and economical, and it is not expected that any particular hardship will be undergone by reason of some of the factory men being laid off.

* * *

A couple of months ago, a communication was received in the city from a firm which desired to establish a plant in western Ontario for the manufacture of pins. Negotiations have been continued between the company and the industrial committee of the Board of Trade, but so far nothing has come of them. It is not known here whether the company has fixed on a place to locate.

* * *

The corner stone of the new $25,000 Y.M.C.A. building was laid last week The building is on a fair way to completion, and will be occupied during the winter some time. The new Carnegie library building is also going up quickly, as is the new hotel. At the municipal elections, the ratepayers will be asked to sanction the construction of a new city hall, at a cost of $60,000. A couple of new schools are also matters of the near future. Being already supplied with very handsome court house and county buildings, a recently remodelled Collegiate, some fine public schools, many fine churches, a new fire hall, and hundreds of beautiful residences, Woodstock will then be splendidly equipped with all that goes to make a city attractive.

* * *

The Board of Trade is entering on an active campaign for the industrial development of Woodstock, and the first step is the preparation and issuing of a beautiful booklet advertising the city and pointing out to prospective manufacturers its many advantages as a place in which to locate. It speaks well for our public-spirited citizens that the whole cost is being borne by business men and others interested in the city's welfare.

CROPS GOOD AT SASKATOON.

Saskatoon, Oct. 5.—Business has been very good this week. There has been a better demand for builders' hardware, paper, nails, etc., and some of the larger buildings are putting on a finished appearance. Shipments of stoves, heaters and fall goods are arriving and a brisk fall trade is anticipated. Wheat cutting is over for another year and threshing is in full swing. The yield so far is very satisfactory and cars of wheat have already been shipped. Good reports come from various districts and samples have been pronounced A1.

* * *

J. McDiarmid Co., the contractors for the post office, have begun operations on the new site.

* * *

Large shipments of steel rails have arrived in the yards of the G.T.P. Railway.

* * *

The walls of the C.P.R. station are in course of construction and the floor of the freight shed is almost completed.

* * *

Water is now turned on in the new mains in the business part of the city, and pipes are being laid in residential parts.

* * *

The Battleford steam laundry is nearing completion and will be ready for business in a week or two. The concrete basement of the Provincial Government registry office, Battleford, is complete, and the steel framework is being placed in position. The building will cost $32,000.

CONDENSED OR "WANT" ADVERTISEMENTS.

RATES.

Two cents per word first insertion; one cent per word subsequent insertions.

Five cents additional each insertion where box number is desired.

Contractions count as one word, but five figures (as $1,000) are allowed as one word.

Cash remittances to cover cost **must** accompany all advertisements. **In no case** can this rule be overlooked. Advertisements received without remittance cannot be acknowledged.

RULES FOR COPY.

In addressing replies care of HARDWARE AND METAL don't fail to give box number.

Replies addressed to HARDWARE AND METAL boxes are re-mailed to advertisers every Monday, Wednesday and Friday.

Requests for classification will be followed where they do not conflict with established classified rules.

Orders should always clearly specify the number of times the advertisement is to run.

All "Want" advertisements are payable in advance.

AGENTS WANTED.

This is the problem of many manufacturers who are anxious to get a foothold in Western Canada. It will be easily solved if HARDWARE AND METAL is given the opportunity to solve it.

AGENT wanted to push an advertised line of Welsh tinplates; write at first to "B.B.," care HARDWARE AND METAL, 88 Fleet St., E.C., London, Eng. [tf]

AN old-established house of brass founders, making a wide range of brass foundry furnishings, such as brass cast tubing, stair rods, cornice poles, art metal work, electric light fittings, ecclesiastical metal work, are anxious to secure a thoroughly reliable and trustworthy agent or traveller for Canada, to sell goods on commission. An excellent opportunity for any firm covering Canada thoroughly. Address all enquiries to Box 654, HARDWARE AND METAL. [37]

AGENCY WANTED

EXPERIENCED and reliable business man, Calgary, could represent good lines of hardware, plumbing and heating goods. Address Box 665, HARDWARE AND METAL, Toronto. [40]

ARTICLES FOR SALE.

Don't keep any fixtures or tools around your store for which you have no further use. They are worth more to-day than they will a year hence. Don't keep money tied up which you could use to secure discounts from your wholesaler.

SET of tinsmith's tools, good as new, best American make. $150. Great sacrifice. Box 660, HARDWARE AND METAL, Toronto. tf

ONE hundred dozen rim locks for sale, 95c. per dozen; also two hundred dozen, rim and mortise knobs, black, white and mineral, 85c. per dozen. Full particulars, Box 661, HARDWARE AND METAL, Toronto.

BUSINESSES FOR SALE.

Somewhere in Canada is a man who is looking for just such a proposition as you have to offer. Our "For Sale" department brings together buyer and seller, and enables them to do business although they may be thousands of miles apart.

HARDWARE Business and Tinshop for sale in Saskatchewan; population 1,500; stock carried, about $14,000; turnover, $45,000; practically all-cash business; cash required, $8,000; would rent building; do not answer without you have the money and mean business; it will pay to investigate this. Box 648, HARDWARE AND METAL, Toronto. (41)

HARDWARE Business for Sale in Tweed, a thriving Eastern Ontario town. Best business in town. Must sell owing to ill-health. Liberal terms to satisfactory purchaser. Write for particulars. J. M. Robertson, Tweed. [40]

Prepare for Holiday Business

It's not too early to lay plans for the Christmas holiday trade. Dealers have already bought some lines, of course, and in these days of slow deliveries other orders should be placed as soon as opportunity offers. "Goods well bought are half sold" is a time-worn truism in mercantile life.

But it doesn't do to feel too secure on the buying part until the selling has been done. Plans for increasing holiday sales must be worked out months ahead. Window displays must be figured out and a series of rough sketches planned. A series of ads. for the local papers should also be prepared, and, where deemed advisable, a neat booklet gotten up. to be mailed to all probable customers in the surrounding district.

It is in these selling plans that Hardware and Metal can be helpful to its readers. Let all partake of the spirit of co-operation and exchange ideas through these columns. To help the suggestion along the editor invites any reader to join the friendly contest and hangs up a prize of $10 cash for the best answer to the following questions:

How can the hardware merchant increase his sales of holiday goods next December? What special lines should he stock? What selling plans should be adopted? What special advertising should be done? What novel window displays can be suggested? Should Souvenirs (calendars, knives, trays, etc.) be given to customers?

The prize of $10 will be awarded to the writer of the most practical and original letter of from 500 to 1,000 words received by the editor before October 15, 1907, and the best half-dozen replies will be published in Hardware and Metal.

Address the Editor

HARDWARE AND METAL

10 Front St. E., Toronto

HARDWARE store and stock for sale, within 25 miles of Toronto. Solid brick double store and dwelling; property worth $4,000. Stock worth about $3,000. Splendid business in stoves and hardware. Good reasons for selling. Full particulars on application. Box 663, HARDWARE AND METAL, Toronto. [41]

SITUATIONS VACANT.

You can secure a "five-thousand-a-year" manager, or a "five-hundred-a-year" clerk, by stating your wants under "Situations Vacant."

WANTED—First-class tinsmith in New Ontario town. Must be capable of managing workshop and estimating. Good position, good salary, and no lost time for the right man. Address, Box 655, HARDWARE AND METAL, Toronto. [40]

WANTED—First class opening for tinsmithing, plumbing and heating. Box 202, Hespeler. [43]

WANTED—Experienced hardware clerk; must be good salesman and stockkeeper Address, stating age, experience and salary expected, Box 664 HARDWARE AND METAL, Toronto. [40]

THE CANADIAN GROCER wants a Managing Editor. It wants a thoroughly capable man—a man who is live, full of up-to-date ideas, and one who understands the newspaper business from the reglet box to the editorial chair. Furthermore, it wants a man who is thoroughly conversant with the commercial situation in Canada. We realize that this is a big want. Not everyone can fill the bill, but we are willing to pay at the outset $2,500 a year to the man who can do so. The right man can eventually make his place worth $5,000. If you think you are this man we want to hear from you, with your experience and qualifications—by letter only. Address The MacLean Publishing Co., 232 McGill St., Montreal, or 10 Front St. East., Toronto. 38

SITUATIONS WANTED.

Travellers, clerks, or tinsmiths, wishing to secure new positions, can engage the attention of the largest number of employers in their respective lines, by giving full particulars as to qualifications, etc., under "Situations Vacant."

CLERKS WANTED.—Clerks who want to improve their positions are requested to write for information to Box 659, HARDWARE AND METAL, Toronto. 40

EXPERIENCED cutlery man, solid connection with best trade throughout Ontario, open for a change of position about 1st January. Box 651, HARDWARE AND METAL, Toronto. [43]

HARDWARE CLERK—32, married, sober, steady and reliable, capable of taking charge. Box 10, HARDWARE AND METAL, Winnipeg.

THE WANT AD.

The want ad. gets work for workers and workers for work.

It gets clerks for employers and finds employers for clerks. It brings together buyer and seller, and enables them to do business, though they may be thousands of miles apart.

The "want ad." is the great force in the small affairs and incidents of daily life.

MANITOBA HARDWARE AND METAL MARKETS

Corrected by telegraph up to 12 noon Friday, Oct 11. Room 511, Union Bank Bldg, Winnipeg, Man.

Business is more active now than for some weeks past, and a further improvement is looked for. The wheat crop is commencing to move and a big improvement in country collections may be looked for.

Rope—Sisal, 11c per lb., and pure manila, 15¾c.

Lanterns—Cold blast, per dozen, $7; coppered, $9; dash, $9.

Wire—Barbed wire, 100 lbs., $3.22½; plain galvanized, 6, 7 and 8, $3.70; No. 9, $3.25; No. 10, $3.70; No. 11, $3.80; No. 12, $3.45; No. 13, $3.55; No. 14, $4; No. 15, $4.25; No. 16, $4.40; plain twist, $3.45; staples, $3.50; oiled annealed wire, No. 10, $2.90; No. 11, $2.96; No. 12, $3.04; No. 13, $3.14; No. 14, $3.24; No. 15, $3.39; annealed wires (unoiled), 10c less; soft copper wire, base, 36c; brass spring wire, base, 30c.

Poultry Netting—The discount is now 47½ per cent. from list price, instead of 50 and 5 as formerly.

Horseshoes—Iron, No. 0 to No. 1, $5.65; No. 2 and larger, $4.40; snowshoes, No. 0 to No. 1, $4.90; No. 2 and larger, $4.65; steel, No. 0 to No. 1, $5; No. 2 and larger, $4.75.

Horsenails—No. 10 and larger, 22c; No. 9, 24c; No. 8, 24c; No. 7, 26c; No. 6, 28c; No. 5, 30c; No. 4, 36c per lb. Discounts: "C" brand, 40, 10, 10 and 7½ p.c.; "M.R.M" cold forged process, 50 and 5 p.c. Add 15c per box. Capewell brand, quotations on application.

Wire Nails—$3.10 f.o.b. Winnipeg and $2.65 f.o.b. Fort William.

Cut Nails—Now $3.20 per keg.

Pressed Spikes—¼ x 5 and 6, $4.75; 5-6 x 5, 6 and 7, $4.40; ⅜ x 6, 7 and 8, $4.25; 7-16 x 7 and 9, $4.15; ½ x 8, 9, 10 and 12, 4.05; ⅝ x 10 and 12, $3.90. All other lengths 25c extra, net.

Screws—Flat head, iron, bright, 80, 10, 10 and 10; round head, iron, 80; flat head, brass, 75; round head, brass, 70; coach, 70.

Nuts and Bolts—Bolts, carriage, ⅜ or smaller, 60 p.c.; bolts, carriage, 7-16 and up, 50; bolts, machine, ⅜ and under, 50 and 5; bolts, machine, 7-16 and over, 50; bolts, tire, 65; bolt ends, 55; sleigh shoe bolts, 65 and 10; machine screws, 70; plough bolts, 55; square nuts, cases, 3; square nuts, small lots, 2½; hex nuts, cases, 3; hex nuts, small lots, 2½ p.c. Stove bolts, 70 and 10 p.c.

Rivets—Iron, 60 and 10 p.c.; copper, No. 7, 43c; No. 8, 42½c; No. 9, 45½c; copper, No. 10, 47c; copper, No. 12, 50½c; assorted, No. 8, 44½c, and No. 10, 48c.

Coil Chain—¼-in., $7.25; 5-16, $5.75; ⅜, $5.25; 7-16, $5; 9-16, $4.70; ⅝, $4.65; ¾, $4.65.

Shovels—List has advanced $1 per dozen on all spades, shovels and scoops.

Harvest Tools—50, 10 and 5 p.c.

Axe Handles—Turned, s.g. hickory, $3.15; No. 1, $1.90; No. 2, $1.60; octagon extra, $2.30; No. 1, $1.60.

Axes—Bench axes, 40; broad axes, 25 p.c. discount off list; Royal Oak, per doz, $6.25; Maple Leaf, $8.25; Model, $8.50; Black Prince, $7.25; Black Diamond, $9.25; Standard flint edge, $8.75; Copper King, $8.25; Columbian, $9.50; handled axes, North Star, $7.75; Black Prince, $9.25; Standard flint edge, $10.75; Copper King, $11 per dozen.

Churns—45 and 5; list as follows: No. 0, $9; No. 1, $9; No. 2, $10; No. 3, $11; No. 4, $13; No. 5, $16.

Auger Bits—"Irwin" bits, 47½ per cent. and other lines 70 per cent.

Blocks—Steel blocks, 35; wood, 55.

Fittings—Wrought couplings, 60; nipples, 65 and 10K Ts and elbows, 10; malleable bushings, 50; malleable unions, 55 p.c.

Hinges—Light "T" and strap, 65.

Hooks—Brush hooks, heavy, per doz., $8.75; grass hooks, $1.70.

Stove Pipes—6-in., per 100 feet length, $9; 7-in., $9.75.

Tinware, Etc.—Pressed, retinned, 70 and 10; pressed, plain, 75 and 2½; pieced, 30; japanned ware, 37½; enamelled ware, Famous, 50; Imperial, 50 and 10; Imperial, one coat, 60; Premier, 50; Colonial, 50 and 10; Royal, 60; Victoria, 45; White, 45; Diamond, 50; Granite, 60 p.c.

Galvanized Ware—Pails, 37½ per cent.; other galvanized lines, 30 per cent.

Solder—Quoted at 27c per pound. Block tin is quoted at 45c per pound.

Wringers—Royal Canadian, $36; B.B., $40.75 per dozen.

Files—Arcade, 75; Black Diamond, 60; Nicholson's, 62½ p.c.

Locks—Peterboro and Gurney 40 per cent.

Building Paper—Anchor, plain, 66c; tarred, 69c; Victoria, plain, 71c; tarred, 84c; No. 1 Cyclone, tarred, 84c; No. 1 Cyclone, plain, 66c; No. 2, Joliette, tarred, 69c; No. 2 Joliette plain, 51c; No. 2 Sunrise, plain, 56c; Jubilee plain, 71c; Jubilee tarred, 84c; Buffalo plain, 66c; Buffalo tarred, 69c.

Ammunition, etc.—Cartridges, rim 50 p.c.; pistol sizes, 25 p.c.; military, 20 p.c. Primers, $1.55. Loaded shells: English and Canadian makes, 12 gauge black soft, $18; 10 gauge, $22.50; 12 gauge smokeless, chilled, English, $24, Canadian, $23; 10 gauge smokeless, chilled, English, $28, Canadian, $27. Shot: Ordinary, per 100 lbs., $7.75; chilled, $8.10. Powder: F.F., keg, Hamilton, $4.75; F.F.G., Dupont's, $5.

Iron and Steel—Bar iron basis, $2.70. Swedish iron basis, $4.95; sleigh shoe steel, $2.75; spring steel, $3.25; machinery steel, $3.50; tool steel, Black Diamond, 100 lbs., $9.50; Jessop, $13.

Sheet Zinc—$8.50 for cask lots, and $9 for broken lots.

Corrugated Iron and Roofing, etc.—Corrugated iron, 28 gauge, painted, $3; galvanized, $4.10; 26 gauge, $3.35 and $4.35. Pressed standing seamed roofing, 28 gauge, $3.45 and $4.45. Crimped roofing, 28 gauge, painted, $3.20; galvanized, $4.30; 26 gauge, $3.55 and $4.55.

Pig Lead—Average price is $6.

Copper—Planished copper, 44c per lb.; plain, 39c.

Iron Pipe and Fittings—Black pipe, ¼-in., $2.70; ⅜, $2.85; ½, $3.75; ¾, $4.75;

1, $6.75; 1¼, $9.25; 1½, $11.00; 2, $14.80; 2½, $24.60; 3, $32.30; 3½, $40.50; 4, $46.00; 4½, $54.00. Galvanized: ¼ in., $3.65; ⅜, $3.80; ½, $4.50; ¾, $5.80; 1, $8.40; 1¼, $11.40; 1½, $13.80; 2, $18.40. Nipples, 70 per cent.; unions, couples, bushings and plugs, 50 per cent.; malleables, 20 per cent.

Galvanized Iron—Apollo, 16-gauge, $4.15; 18 and 20, $4.40; 22 and 24, $4.65; 26, $4.65; 28, $4.50; 30-gauge or 10¾-oz., $5.20; Queen's Head, 20, $4.60; 24 and 26, $4.90; 28, $5.15.

Lead Pipe—Market is firm at $7.80.

Tin Plates—IC. charcoal, 20x28, box, $10; IX charcoal, 20x28, $12; XXI charcoal, 20x28, $14.

Terne Plates—Quoted at $9.50.

Canada Plates—18x21, 18x24, $3.50; 20x28, $3.80; full polished, $4.30.

Lubricating Oils—600W, cylinders, 80c; capital cylinders, 55c and 50c; solar red engine, 30c; Atlantic red engine, 29c; heavy castor, 28c; medium castor, 27c; ready harvester, 28c; standard hand separator oil, 35c; standard gas engine oil, 35c per gallon.

Petroleum and Gasolene—Silver Star, in bbls., per gal., 20c; Sunlight, in bbls., per gal., 22c; per case, $2.35; Eocene, in bbls., per gal., 24c; per case, $2.50; Penoline, in bbls., per gal., 24c; Crystal Spray, 23c; Silver Light, 21c. Engine gasoline, in barrels, gal., 27c; f.o.b. Winnipeg, in cases, $2.75.

Paints and Oils—White lead, pure, $6.50 to $7.50, according to brand; bladder putty, in bbls., 2½c; in kegs, 2¾c; turpentine, barrel lots, Winnipeg, 90c; Calgary, 97c; Lethbridge, 97c; Edmonton, 98c. Less than barrel lots, 5c per gallon advance. Linseed oil, raw, Winnipeg 70c, Calgary 77c, Lethbridge, 77c, Edmonton 78c; boiled oil, 3c per gallon advance on these prices.

LETTER BOX.

Correspondence on matters of interest to the hardware trade is solicited. Manufacturers, jobbers, retailers and clerks are urged to express their opinions on matters under discussion.

Any questions asked will be promptly answered. Do you want to buy anything, want some shelving, a silent salesman, any special line of goods, anything in connection with the hardware trade? Ask us. We'll supply the necessary information.

Brush Manufacturers.

Geo. A. White, hardware merchant, Trenton, writes, asking for the present address of West, Taylor & Bickle, manufacturers of brushes, brooms, woodenware, etc., and Ontario agents for Simms brushes, whom he understands, have moved their Toronto office.

Ans.—This firm on August 14 last had the misfortune of losing their factory at Norwich, Ont., by fire. They are now in temporary premises and have closed their Toronto office, their head office being temporarily at Norwich, where all correspondence should be addressed.—Editor.

Grates for Stove Ovens.

The National Manufacturing Company, Pembroke, write: Can you let us know where pressed steel grates for steel ranges and cook stove ovens can be obtained?

Ans.—We know of no firm making a specialty of this class of work, but the Gurney Foundry Company have equipped their plant with special machinery and can supply outside manufacturers. —Editor.

WATER HEATER.

An improved heater, invented by E. E. Kehnert, is adapted for domestic and shop use. The water is heated by gas, and the volume of the latter admitted to the burner is automatically regulated by the quantity of hot water drawn off. Springs, stuffing-boxes and some of the other usual adjuncts of heaters of this class are dispensed with, and the inventor arranges the gas and water controlling valves, and means for operatively connecting them in one and the same casing, whereby he attains a maximum of simplicity, and efficiency in operation.

Before the electrical industry began to create a demand for copper wire, it was made by rolling copper into sheet, slitting it cold, and then drawing into wire. At that time the manufacture of the wire by hot-rolling copper billets to rod and then drawing was not practiced.

Brass wire is now practically all made by cold rolling cast rods through three-high rolls, transferring to the bull-block, in order to reduce it still further, and then drawing upon a wire-block in the usual manner. The manufacture of wire from slit-sheet has practically been replaced by this method as larger coils are obtained.

Effective Retail Hardware Advertising

A Brief Talk on How to Produce It by T. J. T.

"The advertising man who is not a salesman, like the lawyer shy on law, goes begging for clients."—Rusty Mike's Diary.

* * *

Have you ever been a party to a lawsuit? I have; and equipped with the knowledge that comes of bitter, costly experience, I am resolved never again to enact a similiar role. Litigation is a diversion fit only for a bloated plutocrat, who doesn't have to expend a thought upon the subject of next winter's coal, or new shoes for the baby.

At some time or other you may feel that the only course open to you is an appeal to the courts of justice, perhaps to establish your right to certain properties, or for any one of a thousand other reasons. Don't get blue in the face about it, and rush off at once to set the machinery of the law in motion. In the first place you're liable to bring on an attack of apoplexy, and in the second, it will pay you to reflect that the operation of this machinery is a mighty expensive process. The necessary motive power doesn't come from Niagara Falls, it can only be supplied by real dollars, and you can never figure on how many you will have to burn up before you get through. See first if there is not some way of arriving at an amicable settlement. If you conclude that such is impossible, then consult the best lawyer in your town—the best, remember. Don't be bothered with a second-rate man, even though he would exact a much smaller fee.

That's just where I made my big mistake the time I got mixed up in matters legal—I was not careful to secure the advice and services of a lawyer who knew his business. An acquaintance to whom I confided the particulars of the case, had a friend a member of the profession, and recommended that I call on him. I did. The man didn't impress me very strongly, but I told my story and he listened attentively, from time to time asking sundry questions bearing upon certain features of it. At its conclusion he urged that I permit him to enter action at once, and appeared so enthusiastic over my prospects of winning, that I—innocent idiot that I was—placed myself in his hands. I had a strong case—I was convinced of that—but the verdict went against me because I had a fool lawyer to present my claims to the judge and jury. He had good material to work on, but he marshalled his points with about the same degree of organization as existed in Coxey's Army.

* * *

Some merchants talk to the public through newspaper space, with just about as much skill as this lawyer used in handling my plea in the court-room. Why? Simply because they haven't got down to the root of the matter and absorbed a few of the fundamental principles of advertising. Their publicity is not worthy of the merchandise they carry, and yet they expect it to be effective! Now, what is their reason for buying space in the local papers? Because they believe it will help them to sell goods—any one of them will tell you that. Then we might surely expect to find it filled with matter that could accurately be described as "Printed Salesmanship."

Every ad. should act as a salesman for the store whose goods it represents, and solicit business in every home where it is read.

> **The Hot Weather is with us**
>
> So are the flies and mosquitoes. Let us send up a screen door or two and a few screen windows to keep them out. A nice line of doors at reasonable prices are to be seen at our showrooms. WHITE MOUNTAIN ICE CREAM FREEZERS are the best on earth. CLEVELAND WATER FILTERS insure the best drinking water obtainable. TRY ONE.
>
> **M. ISBISTER & SON**

Many a merchant uses arguments in his ads.—alleged arguments, I should say, for they have never been found guilty of selling goods—that he wouldn't think of employing over his counter. In the latter instance he is face to face with the customer, and knows that he must talk reasonably if he is to make a sale. Ask him, and he cannot but admit that if he adopted the tone of some of his ads. the customer would be apt to conclude that he didn't credit her with ordinary intelligence. How, then, does he figure out that such arguments are going to appeal to the same lady when she reads them in the newspaper at home. He doesn't actually believe that they will, when you get him to reason it out with you. It is just that he has never reasoned it out with himself in the same way. He does not yet appreciate the ability of a good ad. to sell goods.

* * *

There haven't been many hardware ads. sent in for criticism lately, but a glance over some progressive Canadian dailies and weeklies show that hardwaremen, while not advertising as freely as other merchants, use their space to fairly good advantage. Some of the old "never change" ads. are still seen, but a decided change for the better is noticeable, most retailers apparently now realizing that they should buy advertising space (and use it) just as they buy and use any other commodity.

* * *

M. Weichel & Son, Elmira, Ont., known throughout Canada as a live and up-to-date firm, have forwarded copies of local papers containing page ads. devoted to a special "fair week sale" in which special offerings are made, prices being shaded from 25 to 50 per cent. on certain lines, most of them being seasonable. In introducing their sale the firm say:

"This advertisement is on a subject of immediate importance to you, and we feel warranted in drawing your attention to it in this large and lusty manner. Of course you know that we are headquarters for hardware and household goods. We have decided to hold this big money saving sale during fair week commencing Monday, Sept. 23, and ending Saturday, Sept. 28. Come here and buy during this sale, with the absolute guarantee of your money's worth, or money back. We appoint you as the sole judge. A penny saved is two pennies earned. We've changed the old saying somewhat, because it seems to us that the prices below warrant the change. If you want or need anything of the goods mentioned below buy them. You'll never rue it. The profit on these goods is all yours. Every body likes real good bargains, and these are the real thing. Only one week."

* * *

That's a plain matter of fact talk in the merchant's own style—nothing assumed and nothing said that would not be said over the counter. Other adwriters might write it in a different way, but the fact remains that the "talk" is natural; it promises something—not something for nothing—and undoubtedly interested many readers of the local papers, especially when backed up by a full page advertisement set in department store style and illustrated by a reasonable number of engravings. It would be interesting to know what results M. Weichel & Son secured from the ad. Probably they will let our readers know and also send along samples of their regular advertising.

* * *

A. W. Humphries & Son, Parkhill, Ont., have sent three copies of local papers showing their September ads. They use three column space on the editorial page and evidently give the printer well-chosen instructions as to what to display. Catchy, out of the ordinary illustrations are also used to good advantage. One of the ads. in the Parkhill Post of Sept. 18, will illustrate the striking style in which the ads are written. Here are the introductory and closing paragraphs, which were effectively displayed:

Are you prepared for the Fall? With the disappearance of warm weather your thoughts naturally turn to the question of heating. This is a matter which common sense demands should be attended to early, both from the standpoints of health and economy. We are the unquestioned leaders in the stove trade of this district. And are prepared to give you the benefit of long experience and observation. If you want a cheap, effective, useful stove for heating and light housekeeping look at this:

When you come to town make yourself at home at our store. Scores of friends take advantage of our invitation, but we have room for more. On fair day leave your parcels or wraps with us and we will keep them safely for you. You don't need to buy anything.

* * *

M. Iebister & Son, Saskatoon, Sask., use their space effectively, the composition and display of their ads. being commendable. The accompanying sample of their work was used a couple of months ago in a four-inch two-column space, surrounded by a neat border.

* * *

Visitors at the Jamestown Exposition had many stiff necks and flutterings of the heart from gazing skyward at the novel exhibition of E. C. Atkins & Co., Indianapolis. The apparatus consisted of scientific kites from the string of which banners 30x40 feet square were suspended, together with the figures of a man and woman, each hanging from a trapeze. These swaying figures produced a very startling effect upon the spectators upon the ground, and thousands of field glasses were trained upon them to discover if they were real persons. On account of the height above the ground, however (from one-eighth to one-quarter of a mile), this latter was hard to distinguish, and the effect was really startling.

Edmonton's new Y.M.C.A. building will be one of the finest of its kind in Canada. The cost will be $90,000 when completed. Work is progressing rapidly on it, and it will be opened about the new year.

Practical Talks on Warm Air Heating

The Fifth of a Series of Articles by E. H. Roberts.

Heating a Country School-house.

The hardware merchant is continually called upon to estimate the cost and install furnaces in small country school-houses, churches, etc., where there are few, if any, complications, but where it is essential to get the very best results obtainable with the least expense for installation.

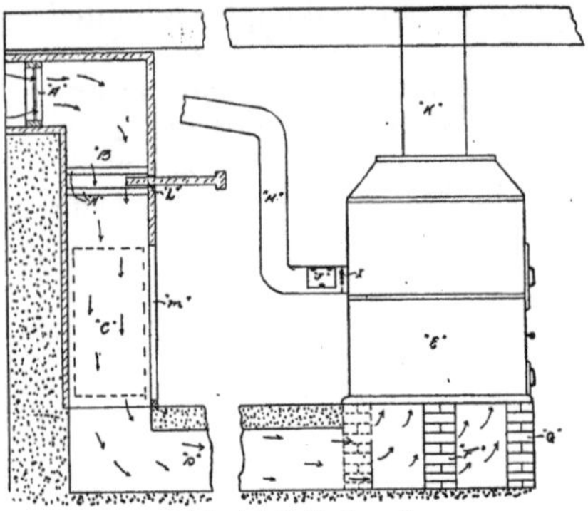

Detail of Cold Air Connection.

A, dust screen ; B, fresh air shaft ; C, dotted lines indicate position of door between fresh air shaft and circulating shaft ; D, underground duct to furnace pit ; E, furnace ; F, centre pier of furnace pit ; G, outer rim of furnace pit ; H, smoke pipe ; I, smoke damper ; J, check draft ; K, warm air pipe ; L, sliding valve in fresh air shaft ; M, door in fresh air shaft ; N, guide or runway for sliding valve.

The plant shown herewith illustrates a type of country school-house which is very common and the heating and ventilating layout indicated can, with very few alterations, be made to apply to almost any one-room school building.

One of the most important features of sanitary school heating is a continuous and ample supply of pure fresh air during school hours and proper provision for removing vitiated or impure air from the building. When school is not in session, however—and this is fully seventeen hours out of the twenty-four—a great saving of fuel can be effected by circulating or revolving the air in the building instead of taking the air supply from outside.

Two Shafts.

Referring to the basement heating plan, you will notice we have shown opposite one of the basement windows, a fresh air shaft leading to underground duct. Next to this fresh air shaft is a circulating shaft separated from the other by a partition of matched stuff in which there is a tight-fitting door that can be opened and closed. Above the circulating shaft is a register face having an area at least equal to the combined areas of the warm air registers and each shaft must have an air carrying capacity equal to, if not larger, than the capacity of the cold air face.

During school hours when outside air is desired the door between circulating shaft and the fresh air shaft should be closed and the sliding valve should be opened so there will be an uninterrupted passageway for the fresh air to the furnace. At other times the sliding valve in the fresh air shaft should be closed and the door into circulating shaft opened so as to establish a circulation from the school room back to the heater.

To Secure Ventilation.

In order to secure ventilation there should be a ventilating flue in chimney and the vent register shown on plan ought to be located as close to floor line as possible. This will carry off the vitiated air, but after school hours this vent register should be closed to prevent the escape of the warm air from the building.

Either a single warm air register directly over the heater can be used or two registers placed as shown on the plans. Generally speaking, however, two registers are more satisfactory than one in a room like this where the children will have to pass over them from time to time and where a single register might create too strong a current of warm air to be pleasant. The single connection has the advantage of being a little cheaper to install and allowing a freer outlet for the heat, but the two short pipes, such as we have indicated, with plenty of elevation, will create very little more friction and the slight loss of heat in transmission will not be noticeable.

10,560 Feet Cubical Capacity.

The school-house for which this heating plant was designed is 24x40 feet with 11-foot ceiling, so the cubical capacity of the building is 10,560 feet. For such a building we would advise a furnace with a 24-inch fire pot and with casings not less than 48 inches. This is a little larger than actually required to heat the building, but there is a great advantage in having considerable reserve power in the furnace for extremely cold weather, and by keeping the radiating surfaces at a low temperature, a pleasanter heat is secured and at the same time the durability of the plant very materially increased.

The construction of the underground duct, furnace pit, etc., are indicated on the section cut which also shows a very satisfactory method of controlling the volume of outside air by means of a sliding valve.

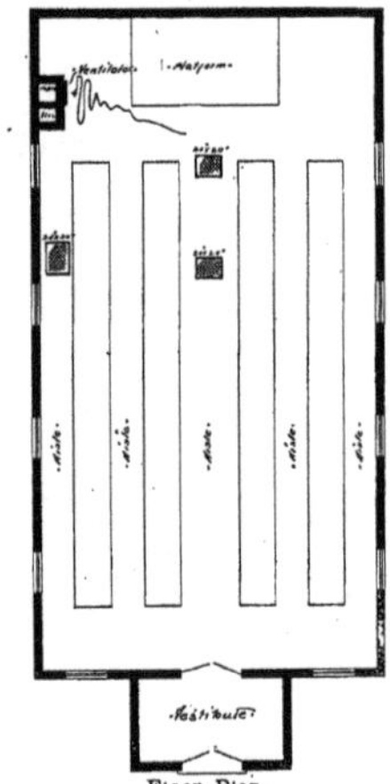

Floor Plan.

The door in fresh air shaft opening into basement is necessary in order to open and close the door between the fresh air shaft and the circulating shaft.

The heating layout illustrated and explained in this article is not always

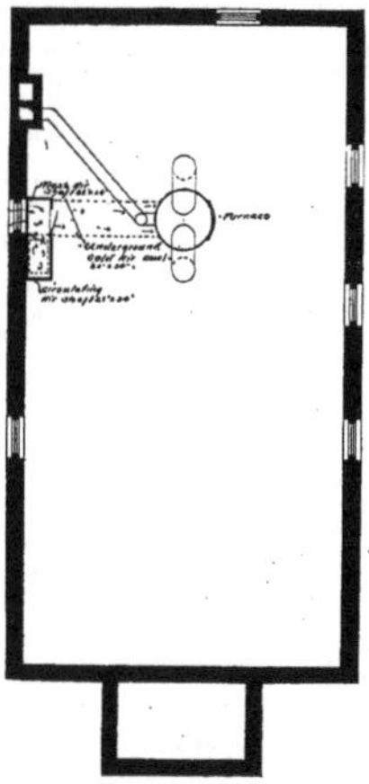

Basement Plan.

practicable because it is sometimes impossible to excavate for furnace pit and underground duct on account of the nature of the soil or fear of striking water. In such cases it is necessary to take the air supply overhead and in a subsequent chapter we will consider the simplest and most economical means of securing an overhead air supply.

KELSEY PATENT INFRINGED UPON

A decision favorable to the Kelsey Heating Company, Syracuse, holding that the furnaces manufactured by the Spear Stove & Heating Company, of Philadelphia, Pa., infringe on the letters patent of the Syracuse Company, has been handed down by Judge R. W. Archbald, of Scranton, in the United States Circuit Court.

The action was brought about two years ago, and a large amount of testimony has been taken. The Kelsey Heating Company manufacture what is known by the trade name of "warm air generator." A furnace claimed to have been copied after it was put out by the Philadelphia manufacturer. Suit was brought, alleging infringement, and the pattern was changed. It was claimed, however, that the principle was the same, and the bill of complaint was amended so as to cover this structure. Judge Archbald finds that both the furnaces made by the Spear Company are infringements on the Kelsey patents. Under the decision, the Spear Company will be prohibited from manufacturing these types of furnaces.

The Kelsey Company regards the decision as very important to its business.

BUILDING AND INDUSTRIAL NEWS

For additional items see the correspondence pages. The Editor solicits information from any authoritative source regarding building and industrial news of any sort, the formation or incorporation of companies, establishment or enlargement of mills, factories or foundries, railway or mining news.

Industrial Development.

A heavy flow of natural gas has been struck near Calgary, Alta.

An electric plant will be installed at Torbrook Mines, N.S., at a cost of $40,000.

A large sawmill, with a daily capacity of 45,000 feet, will be erected at Nelson, B.C.

A Detroit firm, manufacturing malleable iron, wishes to locate in Parry Sound, Ont.

The Lake Huron Lumber Co., Ottawa, have assigned, with liabilities amounting to $45,000.

A system of marine telephones has been installed between seven stations on the St. Lawrence.

Zuelsdorf Bros., will erect a factory in Berlin, Ont. Furniture of all kinds will be manufactured.

Valuable deposits of radium have been found in the Simplon tunnel, between Switzerland and Italy.

The new factory being erected at Hull, Que., by J. W. Woods is nearing completion. It will cost $100,000.

A firm manufacturing engines wishes to locate in London, Ont. They want $25,000 or $30,000 as an inducement.

A new invention makes it possible to reverse steam turbines. This will be appreciated in shipbuilding circles.

A company formed chiefly of Chinamen, and called the Sydney Brick and Tile Co., will erect works at Sydney, B.C.

The new post office building at Edmonton, Alta., will be fireproof throughout. It will be in use early in the spring.

An American company, manufacturing farm implements, garden swings, ladders, etc., is looking for a site for a Canadian factory.

The Fairchild Co., Winnipeg, will sell all interests to the John Deere Plow Co Moline, Ill., under a Dominion charter, with a capitalization of $1,000,000. It is to be known as the John Deere Plow Co of Canada.

The Berlin Machine Works, of Beloit, Wis , has installed a Newton patent cupola, having a capacity of sixteen tons per hour. A cupola of the same make will be installed in their Canadian plant at Hamilton.

The new exchange building for the automatic telephone system now being installed in Edmonton is almost completed, the machinery being installed The Mortimer automatic system is being used. It will be ready by November.

The Canadian Marble & Granite Works have recently been opened in Edmonton, Alta. The machinery for cutting and polishing the marble has been installed. Stone of the best quality is quarried near Nelson, B.C.

For the first eight months of this year the total immigration to Canada was 216,865, an increase of 50,058 as compared with the first eight months of 1906. The total for the eight months is more than the total immigration during the whole six years, 1896 to 1902.

The Canadian Rubber Company, of Montreal, has been appointed sole agents in Canada for "Rainbow" red sheet packing and "Eclipse" gaskets, manufactured by the Peerless Rubber Manufacturing Co., New York. Orders for these goods may be sent direct to any of the company's sales branches.

It is announced from Owen Sound that the Northern Navigation Company are inviting tenders for a new passenger and freight vessel 259 feet in length, to take the place of the wrecked Monarch. Plans and specifications have been prepared by Hugh Calderwood, marine architect. It is intended that the steamer will be ready to go into commission on the opening of navigation next spring.

The new plant of the Standard Chain Company of Canada, now being erected at Sarnia, is progressing very rapidly. The dock has been entirely completed and the main building will shortly be under roof. The blast system is being furnished by Sheldons, Limited, of Galt, Ont. It is anticipated that the plant will be in operation the latter part of October.

If the sale of the iron property at St. George's, Nfld., goes through, it may mean much to Sydney. If the syndicate buy it they intend erecting several electric furnaces at Sydney and will turn out marketable steel. They say Sydney is one of the greatest centres on the continent in which to run a large plant that might depend largely on electricity. The owner of the iron property has been offered $30,000, but wants $50,000.

Insolvencies in the Dominion of Canada for the first nine months of 1907, according to the reports of R. G. Dun & Company, were 870 in number and $8,000,128 in amount of liabilities, compared with 867 failures last year, when the amount involved was $6,826,386. While the comparison is somewhat unfavorable for the last year, it is found by going further back that liabilities in 1907 were smaller than in four of the preceding six years, and the same is true as to the number of failures. The figures are :

	No.	Assets.	Liabilities.
1907	870	$5,919,854	$8,000,128
1906	867	4,751,128	6,826,389

The Dominion Iron and Steel Company's plant at Sydney, N.S., is now in splendid condition, and the company, in order to be in a better position to increase the output, have installed two other batteries of boilers for use under blast furnaces. This means an addition of 1,000 per cent. power to that plant. Altogether, there is now power of nine thousand working to supply steam for the immense plant. The Bessemer plant, which was shut down for a couple of weeks owing to a lack of steam, is now running again full blast, and the product is said to be quite satisfactory. Mr. Jones, general manager, was asked if he had anything to say about the proposed purchase by the Steel Company of the Leapreaux Iron ore areas about eight miles outside of St. John, N.B., and replied : "I have nothing to say." It is thought, however, that the deal will likely go through.

Several prospectors who have recently been up in the Abitibi country, in northern Ontario, bring down word that coal has been found in fair abundance in several sections of that district, and particularly along the line of the new Transcontinental Railway. According to a statement made to several prospectors by a foreman of one railway crew, coal has been found in several places where cuts have been made. The whole country is covered with clay, and in many places the top layer is black earth, which indicates that great growths of timber have fallen there and decayed. Just what the quality of the coal is has not yet been determined, but it is said to lie in huge beds in the places where it has been found. As the season is late for doing any prospecting in the Abitibi country, it is said that no efforts will be made till next spring to ascertain the extent of the discovery.

The McGregor-Banwell Fence Company, manufacturers of Ideal woven wire fence, at Walkerville, Ont., whose factory was destroyed by fire on March 16 last, and who have since been operating in a temporary factory, have almost completed the erection of their new building. This building is of reinforced concrete throughout, the floors being reinforced with Ideal fencing, specially made for that purpose. The factory will have 24,000 square feet of floor space, the ground floor being used for the storage of wire and for the manufacture of woven wire fencing, and the upper storey for the manufacture of gates, lawn fencing and fence supplies. Every up-to-date appliance for the manufacture of these goods will be installed and on completion it will be a model factory in this line. The company will also erect a fireproof storage building, 200x100 feet. It is expected that the looms will be in operation November 15th, and in the meantime the trade is being supplied from the surplus stock made up just before the temporary factory was taken down.

At a meeting of Raven Lake Cement Company shareholders last week, the statement by H. R. Mortin, the assignee, showed a nominal surplus of $350,000, made up as follows : Buildings, etc., $250,000 ; plant, $250,000 ; water power, $35,000 ; liabilities, $179,000, as follows: First mortgage, $50,000 ; interest, $12,000 ; additional cash advanced by mortgage, $3,000 ; second mortgage bonds, National Trust Co. trustees, $50,000 ; interest, $7,000 ; Bank of Montreal personally secured by the directors, $29,000 ; due directors on electrical machinery lien assigned to them, $7,500 ; due unsecured creditors, $20,500. The affairs show a loss of $52,000 in three years, not counting $21,000 charged for depreciation, etc. Of the $52,000, $20,000 was spent for alterations two years ago, another $20,000 a year ago. With these additions the company has been able to run the past year about even. To put it on a paying basis it will be necessary to expend another $60,000. The following plan of reorganization was submitted. The securing of a charter for a $1,000,000 company, half common, half preferred, one share of common to be given for each share of Raven Lake stock to present stockholders. The mortgagee to take first mortgage bonds for his claim, amounting to $62,000. The second mortgage bondholders to take second mortgage bonds in new company, $58,000 ; the directors to get second mortgage bonds in new company for moneys advanced and guaranteed, $37,-

500; the general creditors to get second mortgage bonds of new company, $20,000; second mortgage bonds to be sold for rebuilding and enlarging plant, $72,500, making a grand total of $250,000. After much discussion the shareholders were of the opinion that it was to their interest to raise the money themselves, and that for every hundred-dollar share they held in the present company they should pay in $20 for one share of the new company, as there is about 400,000 shares outstanding.

Mining News.

The Pacific Coal Company, of Banff, Alberta, has let a contract for the erection of 240 coke ovens at the colliery at Hosmer.

A number of American capitalists and mineral experts are proposing to erect electric furnaces at Sydney, C.B., such as are now being used in France and Germany, to develop the iron ore mountain at St. George's, Newfoundland. The mountain rises to a height of 200 feet, and the assay of the ore which is of a black magnetic variety, has shown as high as 60 to 65 per cent. metallic iron.

Municipal Undertakings.

Paris, Ont., will install a duplicate lighting plant.

The new municipal power plant of Saskatoon, Sask., was put in commission recently.

The new municipal lighting plant for the town of Gravenhurst, Ont., was opened recently at South Falls, Ont. The plant cost $45,000, and consists of a 450 k.w. generator, driven by a water turbine of 750 horse power capacity. Power can be supplied to Gravenhurst, at a cost of $7 per horse power. This should be a strong inducement for industries seeking desirable locations.

Railroad Construction.

The bridge for the Halifax & South Western Railway, over the Mersey River, Nova Scotia, has been completed. No accident occurred during its construction.

Building Notes.

The Y.W.C.A., Toronto, will erect an addition to their premises on Simcoe street, to cost $12,000.

Montreal's new post office will cost in the neighborhood of $500,000. It will be completed in May, 1909.

The new building for the Imperial Bank at Edmonton will cost $90,000. It will be completed in the near future.

W. &. D. Dineen, Toronto, will erect a factory at the rear of their premises, for the cutting and making of furs.

Building permits to the value of $385,700 were issued in Stratford, Ont., during September. The new G.T.R. shops, alone, will cost $360,000.

Although more building permits were taken out in Montreal last month than in September, 1906, the value is less. This is said to be due to the scarcity of money. The permits for new buildings amounted to $449,766, and for alterations, $44,979. In 1906, the values were $725,505 and $66,745.

Companies Incorporated.

Shurly & Derrett, Toronto; capital, $75,000; to manufacture and deal in twine, hemp, flax, etc.; provisional directors, C. J. Shurly, T. F. Shurly, Galt, Ont.; G. D. McAllister, Toronto.

Queen City Foundry Co., Toronto; capital, $40,000; to manufacture and sell boilers, hardware, ranges, locks, etc.; provisional directors, E. Gillespie, W. C. Burt, J. W. Clark, J. Brooks and R. J. Smythe, all of Toronto.

The P. Hymmen Company, Berlin, Ont.; capital, $60,000; to take over the hardware and plumbing business of Peter Hymmen, Berlin; provisional directors, P. Hymmen, H. Hymmen and H. S. Hymmen, all of Berlin.

THE NICKEL-COPPER KING.

A New York despatch says that fifteen friends of Col. Robert M. Thompson, financier, retired naval officer and lawyer, are to be his guests on one of the most remarkable yacht cruises on record. The yacht upon which they will voyage round the world, is the 8,000-ton steamer Mineola. The journey will occupy nine months, and the estimated expense of the entertainment to the list is half a million dollars. Among those invited by Col. Thompson to be his guests, are Lord Brassey and Admiral Lord Charles Beresford. A ballroom 100 feet long and as wide as the ship, is to be finished in white and gold, and will cost about $75,000. The plans call for 15 staterooms in styles of Louis XV. and George III. periods, each stateroom to have a sitting-room and bathroom attached. The furnishings alone will cost $150,000.

Col. Thompson, who is said to be worth fifty or sixty millions, is really a Canadian millionaire, as nearly all his great fortune was made from Canadian Nickel and copper investments. He began by smelting the Sudbury nickel ores in New Jersey, and was one of the first to realize the great value of Canada's deposits. With a couple of associates he acquired control at a cost of two or three millions and the stock was floated on a basis of thirty-four millions. Enormous profits have been made, and the property has had a market value of over fifty millions. In addition to this, Col. Thompson has been interested in other properties in that district, and was one of the men to make a big thing out of the Nipissing deal.

GAS GENERATOR.

V. Sepulchre, Paris, France, has invented an improved gas generator. In this patent the invention has for its object a blast gas generator for the production, in a closed receptacle, with combustibles of all kinds, coal, coke, lignite, peat and in particular with the waste of these combustibles, which are but little utilizable, or utilized with difficulty of gases adapted for all purposes.

News of the Paint Trade

TURPENTINE SUBSTITUTES.

The annual consumption of oil of turpentine for the entire world is estimated at 21,400,000 gallons. Practically all of this enormous quantity of spirits is produced in the United States. As the method of extracting turpentine is destructive and the possibility of an extinction of the source of supply appears no longer very remote, there has arisen an increasing demand for turpentine oil substitutes. The value of spirits of turpentine, is largely due to its slow but complete volatilization, allowing the painter time to get an even finish, and it is sufficiently volatile to prevent the paint from running. The petroleum substitutes have been prepared so as to closely simulate the natural product in this respect. Since there is no reaction taking place in a paint which can be credited to the turpentine, it acts as a solvent and thinner, pure and simple, and hence there can be no objection to the use of these substitutes in paint manufacture. The so-called wood turpentine, distilled from stumps and wood, has a very unpleasant odor, due to about 2½ per cent. of extraneous principles, mainly formaldehyde. For this reason it is not looked upon with favor by painters, though for all practical purposes it is as good as the purer article.

SHELLAC AND VARNISH.

A fair estimate of the covering powers of shellac and varnish is as follows: One gallon of shellac varnish will cover 400 square feet of white pine first coat; it will cover 500 square feet on second and succeeding coats. Interior varnish will cover from 350 to 400 square feet to the gallon, first coat, and nearly 600 square feet for succeeding coats. On hardwood, filled with paste filler, interior varnish will cover from 50 to 75 square feet more of surface than on the unfilled wood.

NATURAL RED PAINT DISCOVERED.

The United Paint Co., Dorchester, N. B., recently received a sample barrel of a new and unique variety of red paint which John Ferguson and others have discovered in extensive quantities near Chaplin Island. A barrel of the newly-discovered paint weighs 1,100 pounds, and the discoverer claims it will eclipse any similar paint on the market. The new substance is a reddish mineral, which is 95 per cent. pure oxide, and 5 per cent. silica. It is said that all the preparation it needs is a mixture with oil, when it gives a glossy red coating to whatever it is applied to. Its color cannot be appreciably changed, although at white heat it darkens infinitesimally without losing any weight.

The new plant is not only fireproof, but it has no affinity for turpentine. When placed in a bottle with the latter, the paint precipitates on the sides, while the turpentine remains unmixed in the centre.

The Value of a Pot of Paint

depends upon how far it will go and how long it will keep its freshness. This being the case

NEW ERA PAINT

beats any other brand on the market. It spreads evenly and is a delight to the painter; every pot covers a large space, and it keeps its freshness longer than other paints.

STANDARD PAINT AND VARNISH WORKS CO., LIMITED
WINDSOR - ONTARIO

JAMIESON'S LIQUID HARD FLOOR WAX

The Modern Finish for Waxed Floors

Easiest to apply and keeps its appearance for a longer period than any other floor wax.

Put up in Pints, Quarts, Half Gallons and Gallons.

Manufactured by

R. C. JAMIESON & CO.
LIMITED
16-30 Nazareth St., MONTREAL

HEADQUARTERS

FOR

Window Glass,
Paints, Oils,
Fine Varnishes

SANDERSON PEARCY & CO., Ltd.
Toronto, Canada

GLAZIERS' DIAMONDS OF PROVED WORTH
Having a Century's Reputation for Reliability.

MADE BY

A. SHAW & SON, London, Eng.

CANADIAN AGENT

GODFREY S. PELTON, - 201 Coristine Building, Montreal

Sharratt & Newth's Glaziers' Diamonds

are unequalled for cutting and wearing qualities.

To be obtained from the principal Dealers in Glass, Hardware, and Painters' Supplies.

Contractors to H. M. Government and the principal English Sheet and Plate Glass Works

GILBERTSON'S COMET

Gilbertson's Sheets

and the COMET Brand

mean quality and security in whatever work you may undertake.

Galvanized flat sheets that are unequalled as to flatness; absolute trueness to gauge; admirably suited in softness for any working. Note the brand "COMET." GILBERTSON on every sheet means security.

W. GILBERTSON & CO., Limited, Pontardawe, South Wales
ALEX. GIBB, 13 St. John St., Montreal, Agent

Paint and Oil Markets

MONTREAL.

Montreal, October 11. — Stocks throughout the country appear to be light and those firms who can take their thoughts off stoves or ranges, which are of course seasonable, are now sending in good-sized sorting-up orders.

The factories now seem to be concentrating their energies on the accumulation of winter stocks, and a number of items, including coat colors, car paints, vermilions, aluminum, varnishes, and enamels, have received quite an impetus during the past week. Several firms still experience considerable difficulty in securing sufficient cars to make prompt and complete shipments.

Linseed Oil—This article is strong in demand and no concessions are being offered for large shipments. For ordinary requirements quotations are steady. Raw, 1 to 4 barrels, 61c. 5 to 9, 60c; boiled, 1 to 4 barrels, 64c; 5 to 9, 63c.

Turpentine — Has been affected somewhat by wood distilled turpentine and buyers are not keen to import freely. Southern quotations have been marked up a cent. This, for the present, has not affected the local prices, which are steady: Single barrels at 74c; wood distilled turpentine, pure, 68c.

Ground White Lead—Does not seem to experience the heavy call which existed at this time last year for the same article, and prices for the better grades have been appreciably lowered: Government standard, $7.25; No. 1, $7; No. 2, $6.75; No. 3, $6.35.

Red Lead—Trade is only of a normal character. Prices are weak: Genuine red lead, in casks, $6.25; in 100-lb. kegs, $6.50; in less quantities at $7.25 per 100 lbs. No. 1 red lead, casks, $6; kegs, $6.25, and smaller quantities, $7.

Dry White Zinc—The call is only spasmodic and prices remain unchanged: V.M. Red Seal, 7½c; Red Seal, 7c; French V.M., 6c; Lehigh, 5c.

White Zinc Ground in Oil—A few irons are being shipped in every carload of lead, but the demand is not excessive. Prices are firm: Pure, 8½c; No. 1, 7c; No. 2, 5¾c.

Gum Shellac—There is no feature in the trade in this article, and last week's quotations are well maintained: Fine orange, 60c per lb.; medium orange, 55c per lb.; white (bleached), 65c per lb.

Shellac Varnish—Trade is very brisk and is likely to continue so for the balance of the season. Prices remain firm and unchanged: Pure white bleached shellac, $2.80; pure orange, $2.60; No. 1 orange, $2.40.

Putty—Some heavy bookings have been reported but grinders seem to be able to keep up with the demand and no scarcity is expected. Prices are steady: Pure linseed oil, $1.85 in bulk; in barrels, $1; in 25-lb. irons, $1.90; in tins, $2; bladder putty, in barrels, $1.85.

TORONTO

Toronto, October 11—Paint and oil orders for the last week have not been exceptionally large, but a steady sorting order business is coming in and jobbers have no reason to complain at the volume of trade transacted for the second week in October. The tightness of the money market has reflected itself in the paint and oil markets to the extent that buying for future delivery is practically at a standstill, and hand-to-mouth purchasing has become the order of the day. Prices throughout the market remain the same as last week, but it is quite evident that in some lines—particularly linseed oil and turpentine—one or two large jobbers have not hesitated to slightly shade their prices whenever the order was large enough to be sufficiently tempting.

Linseed Oil—The market is still firm in England and English oil cannot be laid down in Toronto, by the fifty or hundred barrel lots, at a lower price than that which local jobbers are at present quoting to the trade. Canadian crushers, however, are disregarding the English price, and, in face of the present light demand, some cutting has been going on among local jobbers and Toronto linseed oil prices are none too strong. We still quote: Raw, 1 to 3 barrels, 64c; 4 barrels and over, 63c. Add 3c to this price for boiled oil, f.o.b. Toronto, Hamilton, London, Guelph and Montreal, thirty days.

Turpentine—Southern quotations continue to fluctuate from day to day, but the average price at Savannah for last week was one cent higher than the average price of the week preceding. Locally the demand is not very strong, and a little price cutting is going on among local jobbers as a result. Present prices are: Single barrels, 76c; two barrels and upwards, 75c, f.o.b. point of shipment, net 30 days; less than barrels, 5c advance. Terms, 2 per cent., 30 days.

White Lead—A fair autumn demand exists and prices are a trifle weaker than last week, due more to price cutting, however, than any condition in the primary markets which would warrant a reduction in Canadian quotations. English dry lead and pig lead have both recently advanced in price, and, if there is a tendency in some quarters to shade the price in Canada, it is because the money stringency is causing some who have a large stock on hand to convert a part of it into ready cash. White lead is still quoted locally as follows: Genuine pure white lead, $7.65, and No. 1, $7.25.

As the entire supply of dry red lead for use in Canada is imported from Great Britain, the price is much firmer here, in sympathy with the recent advance in English pig lead. There is a good autumn demand and local prices are firm, as under: Genuine, in casks of 500 lbs., $6.25; ditto, in kegs, of 100 lbs., $6.75; No. 1 casks of 500 lbs., $6; ditto, in kegs of 100 lbs., $6.50.

Glass and Putty—Glass is now moving out fast and evidences are that the next six weeks will see a heavy business transacted in this commodity. Putty is in exceptionally heavy demand, with the supply fairly well up to requirements. Prices on both glass and putty are firm and unchanged.

Shellac—A good average October demand exists. Prices are firm, as follows: Pure orange, in barrels, $2.70; white, $2.82½ per barrel; No. 1 (orange), $2.50.

Petroleum—The autumn rush of business continues to increase and the staffs

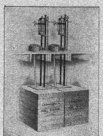

of local supply houses are kept exceptionally busy attending to orders Prices are firm, as follows: Prime white, 13½c; water white, 15c; Pratt's astral, 18½c.

BLACK LETTERING BREAKS PLATE GLASS.

The hazard of black paint on plate glass is, it appears from the insurance press, by no means inconsiderable. Signs painted on plate glass done in solid black paint, are in point of fact extra hazardous, as plate glass insurance companies have discovered to their cost. The practice of painting signs of this kind seems innocent enough, but the experience of conservative casualty companies fully justifies them in refusing to write such windows. The explanation of the hazard under consideration is to be looked for in the property of a black surface to absorb the sun's rays, by means of which, an unequal expansion is produced throughout the plate and under the influence of a sudden gust of cold air or any other sudden change in temperature, a strain is developed which shatters the plate glass bearing the black paint. All black-painted signs on plate glass, no matter with what exposure, must be considered as hazardous risks.

A PAINT CATECHISM.

(By G. B. Heckel.)

1. What is paint? "Any liquid or semi-liquid substance applied to any metallic, wooden or other surface to protect it from corrosion or decay, or to give color or gloss, or all of these qualities, to it," says Wood. More properly speaking, paint is a mixture of opaque or semi-opaque substances (pigments) with liquids, capable of application to surfaces by means of a brush or a painting machine, or by dipping, and of forming an adhesive coating thereon.

2. What is the purpose of paint? To protect or to beautify surfaces, or to perform both offices. Paint is also valuable as a sanitary agent.

3. What is house paint? House paint is paint designed to preserve and to beautify the surfaces of materials used in the construction of buildings.

4. What is the best house paint? That paint which most completely performs the offices named.

5. What are the materials used in manufacturing house paints? Pigments, drying oils, volatile oils or thinners, driers or "Japans" and varnishes.

6. How may these pigments be divided? Into white bases, inert pigments, natural earth colors, chemical colors, pigment lakes, etc.

7. Name the white bases. Oxide of zinc, basic carbonate of lead or corroded white lead, basic oxy-sulphate of lead (or sublimed white lead), zinc-lead, leaded zincs, and lithopone; together with certain inert or subpigments (described later on under their proper titles).

8. What is oxide of zinc? Oxide of zinc is a combination of one atom of metallic zinc with one atom of oxygen. It is produced by two methods, one known as the French and the other as the American process. In the French process metallic zinc is burned in a current of air and the product of combustion, oxide of zinc, is collected in closed chambers. In the American process the ores of zinc mixed with finely powdered anthracite coal are burned in a closed furnace with perforated grate bars, and the resultant oxide of zinc, after passing through a series of cooling flues, is collected in fabric bags.

9. What are the characteristics of oxide of zinc? In texture it is the finest of all white pigments and in color the whitest. A pound of dry oxide of zinc occupies about 50.77 cubic inches, while a pound of corroded white lead in the same condition occupies only 14.69 cubic inches. In consequence of its extreme fineness it requires more oil than any other white pigment to fit it for use with a brush. In one hundred pounds of zinc oxide paint ready for use there are about 46 pounds of oil and 54 pounds of pigment, while the proportions in a corroded white lead paint of similar consistency are about 76 pounds of pigment to 34 pounds of oil. Oxide of zinc is unaffected by any gases present in the atmosphere, has no effect upon any pigment with which it may be mixed, and is non-poisonous.

(To be continued.)

LARGE CHICAGO VARNISH WORKS.

The Standard Varnish Co., Chicago, Ill., have purchased twelve acres of land at Chicago Heights, on which the company will erect what is stated to be the largest varnish factory in the world. The new factory will be as complete as modern equipment and methods can make it, and when placed in full operation, the present output of the Standard Varnish Works will be doubled.

Since its establishment this business has prospered to such an extent that at this time, in addition to their New York and Chicago works, they operate a factory at Toronto, under the name of the International Varnish Co., and have branches in London, Berlin and Brussels. The output of these factories is so large and their reputation so high that Standard varnishes are now to be found in every section of the civilized world, the export of the company being estimated at fully three-fourths of the entire amount of varnishes exported from the United States.

WATERPROOF GLUE.

Dissolve gum shellac, 3 parts, and India rubber, 1 part, by weight, in separate vessels in ether, free from alcohol, subject to a gentle heat. When thoroughly dissolved, mix the two solutions and keep in a tightly sealed vessel. This glue resists the action of hot and cold water, and most acids and alkalies. If thinned with ether and applied to leather along sewn seams it gives a strong watertight joint.

CURRENT MARKET QUOTATIONS.

October 11, 1907

These prices are for such qualities and quantities as are usually ordered by retail dealers on the usual terms of credit, the lowest figures being for larger quantities and prompt pay. Large cash buyers can frequently make purchases at better prices. The Editor is anxious to be informed at once of any apparent errors in this list, as the desire is to make it perfectly accurate.

METALS.

ANTIMONY.

Cookson's ...... per lb.	0 13½
Hallett's	0 13

BOILER PLATES AND TUBES.

	Montreal.	Toronto
Plates, ¼ to ½ inch, per 100 lb.	2 40	2 50
Heads, per 100 lb	2 65	2 75
Tank plates 3-16 inch	2 60	2 65
Tubes per 100 feet, 1½ inch	8 25	8 75
" " 2 "	9 00	9 10
" " 2½ "	10 50	12 00
" " 3 "	12 00	13 50
" " 3½ "	15 00	16 70
" " 4 "	19 25	20 80

BOILER AND T.K. PITTS.

Plain tinned and Spun, 25 per cent. off list.

BABBIT METAL.

Canada Metal Company—Imperial, genuine 60c.; Imperial Tough, 60c.; White Brass, 50c. Metallic, 35c; Harris Heavy Pressure, 25c, Hercules, 25c; White Bronze, 15c.; Sta Frictionless, 14c.; Alluminoid, 10c; No. 4 9c. per lb.

James Robertson Co.—Extra and genuine Monarch, 60c.; Crown Monarch, 50c.; No. 1 Monarch, 40c; King, 30c; Fleur-de-lis, 20c; Thurber, 15c.; Philadelphia, 12c.; Canadian, 10c; hardware, No. 1, 15c.; No, 2, 12c.; No. 3, 10c. per lb.

BRASS.

Rod and Sheet, 14 to 30 gauge, 25 p c. advance.

Sheets, 12 to 14 in.	0 30
Tubing, base, per lb 5-16 to 2 in	0 33
Tubing, ⅜ to 3-inch, iron pipe size	0 31
" 1 to 3-inch, seamless	0 36

Copper tubing, 6 cents extra.

COPPER.

		Per 100 lb
Casting ingot	....	17 00
Cut lengths, round, bars, ½ to 2 in.		28 00
Plain sheets, 14 oz.		30 00
Plain, 16 oz., 14x48 and 14x60	....	29 00
Tinned copper sheet, base	....	32 00
Planished base	....	30 00
Braziers' (in sheets). 4x6 ft., 25 to 30 lb. each, per lb., base	0 28	0 29

BLACK SHEETS.

No 10 gauge	2 70	2 70
12 gauge	2 70	2 75
14 "	2 50	2 60
17 "	2 50	2 60
18 "	2 50	2 60
20 "	2 50	2 60
22 "	2 55	2 65
24 "	2 55	2 70
26 "	2 65	2 85
28 "	2 70	3 00

CANADA PLATES.

Ordinary, 52 sheets	3 75	3 05
All bright "	3 75	4 05
Galvanized—	Dom Crown	Ordinary
18x24x52	4 45	4 35
60	4 70	4 60
20x28x80	8 90	8 70
....	8 40	8 20

GALVANIZED SHEETS

B.W gauge	Queen's Head	Fleur-de-Lis	Gordon Crown	Colborne Crown Gorbal's Best
16-20	3 95	3 80	3 95	....
22-24 ..	4 20	4 05	4 00	4 05
26	4 45	4 30	4 40	4 30
28	4 70	4 55	4 60	4 55

Less than case lots 10 to 25c. extra.

Apollo Brand.

24 guage, American	3 85
26 " "	4 10
28 " "	4 55
10¾ oz.	4 85

25c. less for 1,000 lb. lots.

IRON AND STEEL.

	Montreal.	Toronto.
Middlesboro, No. 1 pig iron	21 50	23 50
Middlesboro, No. 3 pig iron	20 50	22 00
Summerlee, "	25 50	25 50
" special "	24 50	....
" soft "	24 00	....
Carron	25 50	....
Carron Special	24 50	....
Carron Soft	24 00	....
Clarence, No. 3	21 50	22 50
Glengarnock, No. 1	....	25 50
Midland, Londonderry and Hamilton	....	24 00
Radnor, charcoal iron	30 00	32 00
Lowmoor iron		6 50
Angles		2 65
Common bar, per 100 lb.	2 20	2 30
Forged iron "	2 45	..
Refined " "	2 60	2 80
Horseshoe iron "	2 60	2 80
Band iron, No. 10 x ⅜ in.	....	2 30
Sleigh shoe steel "	2 25	2 30
Iron finish steel		2 30
Reeled machinery steel		3 00
Tire steel	2 40	2 50
Sheet cast steel	...	0 12
Mining cast steel	....	0 08
Warranted cast steel	....	0 14
Annealed cast steel	....	0 15
High speed	0 60	0 65
B.P.L. tool steel	10½	0 11

INGOT TIN.

Lamb and Flag and Straits— 56 and 28-lb. ingots, 100 lb.	$40 00	$41 00

TINPLATES.

Charcoal Plates—Bright

M.L.S., Famous (equal Bradley)		Per box
I C, 14 x 20 base	...	$6 50
IX, 14 x 20 "	....	8 00
IXX, 14 x 20 base	....	9 50
Raven and Vulture Grades—		
I C, 14 x 20 base	....	5 00
I X "	....	6 00
I X X "	....	7 00
I X X X "	....	8 00

"Dominion Crown Best"—Double Coated, Tissued.

		Per box.
I C, 14 x 20 base	5 50	5 75
I X, 14 x 20 "	6 50	6 75
IXX 14 x 20 "	7 50	7 75
"Allaway's Best"—Standard Quality.		
I C, 14 x 20 base	4 65	5 00
I X, 14 x 20 "	5 40	5 75
I XX, 14 x 20 "	6 15	6 50

Bright Cokes.

Bessemer Steel—		
I C., 14 x 20 base	4 25	4 35
20x28, double box	8 50	8 70

Charcoal Plates—Terne.

Dean or J. G. Grade—		
I.C., 20x28, 112 sheets	7 25	8 00
IX., Terne Tin	...	9 50

Charcoal Tin Boiler Plates.

Cookley Grade—		
X X, 14x56, 50 sheet bxs. / " 14x60, " / " 14x65, "	....	7 50

Tinned Sheets.

72x30 up to 24 gauge	....	8 50
" " 26 "	....	9 00

LEAD.

Imported Pig, per 100 lb.	5 35	5 50
Bar, "	5 75	6 00
Sheets, 2½ lb. sq. ft., by roll	6 75	6 50
Sheets, 3 to 6 lb. "	6 50	6 25

Cut sheets ½c. per lb., extra.

SHEET ZINC.

5-cwt. casks	7 50	7 75
Part casks	7 75	8 00

ZINC SPELTER.

Foreign, per 100 lb	6 00	6 25
Domestic	6 50	6 75

COLD ROLLED SHAFTING.

9-16 to 11-16 inch	....	0 06
¾ to 1 7-16 "	....	0 05½
1 7-16 to 3 "	....	0 05

30 per cent.

OLD MATERIAL

Dealers' buying prices:

	Montreal	Toronto
Heavy copper and wire, lb.	0 12½	0 13
Light copper	0 11½	0 11
Heavy red brass	0 11	0 11
" yellow brass	0 08	0 09
Light brass	0 06	0 07
Tea lead	0 03½	0 0[illegible]
Heavy lead	0 03½	0 03½
Scrap zinc	0 03	0 03½
No. 1 wrought iron	14 50	11 50
" 2 " "	6 00	6 00
Machinery cast scrap	17 00	16 00
Stove plate	12 00	11 50
Malleable and steel	9 00	10 00
Old rubbers	0 10½	0 10
Country mixed rags, 100 lbs.	1 00	1 25

PLUMBING AND HEATING

BRASS GOODS, VALVES, ETC.

Standard Compression work, 60 per cent.

Cushion work, discount 45 per cent.

Fuller work, 70 per cent.

Flatway stop and stop and waste cocks, 60 per cent.; roundway, 55 per cent.

J.M.T. Globe, Angle and Check Valves, 45; Standard, 55 per cent.

Kerr standard globes, angles and checks, special, 42½ per cent.; standard, 47½ p c.

Kerr Jenkins' disc, copper-alloy disc and heavy standard Valves, 40 per cent.

Kerr steam radiator valves, 60 p.c., and quick-opening hot-water radiator valves, 60 p c.

Kerr brass, Weber's straightway valves, 40; straightway valves, I B B.M., 60.

J. M. T. Radiator Valves 55; Standard, 60; Patent Quick-Opening Valves, 65 p.c.

Jenkins' Valves—Quotations on application to Jenkins' Bros., Montreal.

No. 1 compression bath cock ...... net	2 00
No. 4 " " "	1 90
No 7 Fuller's "	2 25
No. 4½ " "	2 35
Patent Compression Cushion, basin cock, hot and cold, per doz.	$16.20
Patent Compression Cushion, bath cock, No 2208	2 25

Square head brass cocks, 50; iron, 60 p.c.

Thompson Smoke-test Machine 25.00

BOILERS—COPPER RANGE.

Copper, 30 gallon, $33, 15 per cent.

BOILERS—GALVANIZED IRON RANGE

30-gallon, Standard, $5; Extra heavy, $7.75

BATH TUBS.

Steel clad copper lined, 15 per cent.

CAST IRON SINKS.

16x24, $1; 18x30, $1; 18x36, $1.30.

ENAMELED BATHS, ETC.

List issued by the Standard Ideal Company Jan. 3, 1907, shows an advance of 10 per cent. over previous quotations.

ENAMELED CLOSETS AND URINALS

Discount 15 per cent.

HEATING APPARATUS.

Stoves and Ranges—45 to 50 per cent.

Furnaces—45 per cent.

Registers—70 per cent.

Hot Water Boilers—50 and 10 per cent.

Hot Water Radiators—50 to 55 p.c

Steam Radiators—50 to 55 per cent.

Wall Radiators and specials—50 to 55 p.c.

LEAD PIPE

Lead Pipe, 5 p.c. off.

Lead waste, 5 p c. off.

Caulking lead, 6¼c. per pound.

Traps and bends, 40 per cent.

IRON PIPE.

Size (per 100 ft.)	Black.		Galvanized
¼ inch	2 35	¼ inch	3 20
[illegible] "	2 35	[illegible] "	3 20
[illegible] "	2 90	[illegible] "	3 75
[illegible] "	3 90	[illegible] "	5 00
1 "	5 60	1 "	7 25
1¼ "	7 65 to 9 25	1¼ "	9 90 to 10 80
1½ "	9 20	1½ "	11 90
2 "	12 25	2 "	15 50
2½ "	20 1·	2½ "	26 00
3 "	26 75	3 "	34 00
3½ "	34 25	3½ "	42 75
4 "	39 00	4 "	48 60

Malleable Fittings—Canadian discount 30 per cent.; American discount 25 per cent.

Cast Iron Fittings 60; Standard bushings 60.; headers, 60; flanged unions 60, malleable bushings 55 and 5; nipples, 70 and 10; malleable lipped unions 55 and 5 p.c.

SOIL PIPE AND FITTINGS

Medium and Extra heavy pipe and fittings, up to 6 inch, 60 and 10 to 70 per cent.

7 and 8-in. pipe, 40 and 5 per cent.

Light pipe, 50 p c.; fittings, 50 and 10 p.c.

OAKUM.

Plumbers ....per 100 lb.	4 50	5 00

STOCKS AND DIES.

American discount 25 per cent.

SOLDERING IRONS.

½-lb. to 1½	per lb.	0 45½	0 48
2-lb. or over	"	0 42½	0 46

SOLDER.

	Per lb. Montreal	Toronto
Bar, half-and-half, guaranteed	0 25	0 26
Wiping	0 22	0 23

PAINTS, OILS AND GLASS.

BRUSHES

Paint and household, 70 per cent.

CHEMICALS.

	In casks per lb.
Sulphate of copper (bluestone or blue vitrol)	0 09
Litharge, ground	0 06
" flaked	0 06½
Green copperas (green vitrol)	0 01
Sugar of lead	0 08
Lump olive	0 01½

COLORS IN OIL.

Venetian red, 1-lb. tins pure.	....	0 08
Chrome yellow "	....	0 15
Golden ochre "	....	0 1?
French " "	....	0 08
Marine black "	....	0 04½
Chrome green "	....	0 09
French permanent green "	....	0 13
Signwriters' black "	....	0 15

GLUE.

Domestic sheet .... 0 10 — 0 10½
French medal .... 0 12 — 0 12½

PARIS WHITE.

In bbls 0 9)

PIGMENTS.

Orange mineral, casks .... 0 ⌐8
" " 100-lb. kegs 0 08½

PREPARED PAINTS.

Barn (in bbls.) .... 0 85 — 0 90
Sherwin-Williams paints, ... 1 55
Canada Paint Co.'s pure 1 40
Standard P. & V. Co.'s "New Era." 1 30
Benj. Moore Co.'s "Ark" B'd 1 25
" British Navy deck 1 50
Brandram-Henderson's "English" 1 45
Ramsay's paints, Pure, per gal. 1 30
" Thistle, " 1 10
Martin-Senour 100 p.c. pure, ... 1 55
Senour's Floor Paints 1 35
Jamieson's "Crown and Anchor" 1 20
Jamieson's floor enamel 1 50
P. D. Dods & Co's "Island City" .. 1 25
Sanderson Pearcy's, pure 1 20
Robertson's pure paints 1 20

PUTTY.

Bulk in bbls 1 60
Bladders in bbls 1 85
25-lb. tins 1 90
Bladders in bulk or tins less than 100 lb. 2 00
Bulk in 100-lb. irons 1 80

SHINGLE STAINS.

In 5 gallon lots .... 0 81 — 0 90

SHELLAC.

White, bleached 2 65
Fine orange 2 60
Medium orange 2 55

TURPENTINE AND OIL.

Prime white petroleum per gal 0 13½
Water white " 0 15
Pratt's astral " 0 18½
Castor oil .... 0 08 — 0 10
Motor Gasoline per gal, single bbls .... 0 21 — 0 23½
Benzine, per gal single bbls .... 0 17 — 0 17½
Turpentine, single barrels .... 0 75 — 0 76
Linseed Oil, " raw .. 0 60 — 0 63
" " boiled 0 63 — 0 66

WHITE LEAD GROUND IN OIL. Per 100 lbs

Canadian pure 7 15
No. 1 Canadian 6 80
Munro's Select Flake White 7 65
Elephant and Decorators' Pure 7 65
Monarch 7 40
Standard Decorator's 7 15
Essex Genuine 6 80
Brandram's B. B. Genuine 8 70
"Anchor," pure 7 40
Ramsay's Pure Lead 7 25
Ramsay's Exterior 6 65
"Crown and Anchor," pure 7 25
P. D. Dods "Island City" pure lead 7 25
Sanderson Pearcy's 7 40
Robertson's C.P., lead 7 20

RED DRY LEAD.

Genuine, 560 lb. casks, per cwt 6 25
Genuine, 100 lb. kegs, " 6 75
No. 1, 560 lb. casks, per cwt 6 00
No. 1, 100 lb. kegs, per cwt 6 50

WINDOW GLASS

Size United inches.	Star	Double Diamond
Under 26	$4 25	$6 25
26 to 40	4 65	6 75
41 to 50	5 10	7 50
51 to 60	5 35	8 50
61 to 70	5 75	9 75
71 to 80	6 25	11 00
81 to 85	7 00	12 50
86 to 90	....	15 00
91 to 95	....	17 50
96 to 100	....	20 50
101 to 105	....	24 00
100 to 110	....	27 50

Discount—16-oz. 25 per cent.; 21-oz. 30 per cent. per 100 feet. Broken boxes 50 per cent.

WHITING.

Plain, in bbls 0 70
Gilders bolted in bands 0 90

WHITE DRY ZINC.

Extra Red Seal, V.M. .... 0 07½ — 0 08

WHITE ZINC IN OIL.

Pure, in 25-lb. irons 0 08½
No. 1, " " 0 07
No. 2, " " 0 05½

VARNISHES.

Per gal cans.

Carriage, No. 1 .... 1 50
Pale durable body 3 50
" hard rubbing 3 00
Finest elastic gearing 3 00
Elastic oak 1 50
Furniture, polishing 2 00
Furniture, extra 1 15
" " No. 1 .... 0 90
" " union 0 80
Light oil finish 1 40
Gold size japan 1 80
Brown japan 0 95
No 1 brown japan 0 95
Baking black japan 1 35
No. 1 black japan 0 90
Benzine black japan 0 70
Crystal Damar 2 80
No. 1 " 2 50
Pure asphaltum 1 40
Oilcloth 1 50
Lightning dryer 0 70
Elastilite varnish, 1 gal. can, each .. 2 00
Granitine floor varnish, per gal 2 10
Maple Leaf coach enamels; size 1, 1 20
Sherwin-Williams' kopal varnish, gal., 2 50
Canada Paint Co's sun varnish 2 00
"Kyanize" Interior Finish 2 40
"Flint-Lac," coach 1 80
B H. Co's "Gold Medal," in cases 2 00
Jamieson's Copaline, per gal. 2 00

MISCELLANEOUS.

Stovepipe varnish, ½ pints, $8 per gross.
Beeswax, per lb, 4 cents.
Pine tar, half pint tins, 80 cents per doz.
Plaster of Paris, per bbl., $2 25.

BUILDERS' HARDWARE.

BELLS.

Brass hand bells, 60 per cent.
Nickel, 55 per cent.
Gongs, Sargeant's door bells .. 5 50 — 8 00
American, house bells, per lb. 0 35 — 0 40
Peterboro' door bells, 37½ and 10 off new list.

BUILDING PAPER, ETC.

Tarred Felt, per 100 lb. 2 25
Ready roofing, 2-ply, not under 45 lb. per roll 1 00
Ready roofing, 3-ply, not under 65 lb., per roll 1 25
Carpet Felt .... per ton 60 00
Heavy Straw Sheathing .... per ton 40 00
Dry Surprise 0 45
Dry Sheathing .... per roll, 400 sq. ft. 0 40
Tar " .... " 400 " 0 50
Dry Fibre .... " 400 " 0 55
Tarred Fibre .... " 400 " 0 65
O. K. & I. X. L. .... " 400 " 0 70
Resin-sized .... " 400 " 0 45
Oiled Sheathing .... " 600 " 1 00
Oiled " .... " 400 " 0 70
Roof Coating, in barrels .... per gal. 0 17
Roof " small packages " 0 25
Refined Tar .... per barrel 5 00
Coal Tar .... " 4 00
Coal Tar, less than barrels .... per gal. 0 15
Roofing Pitch .... per 100 lb. 0 80 — 0 90
Slater's felt .... per roll 0 70
Heavy Straw Sheathing f. o. b. St. John and Halifax 42 50

BUTTS.

Wrought Brass, net revised list.
Wrought Iron, 65 and 10 per cent.
Cast Iron Loose Pin, 60 per cent.
Wrought Steel Fast Joint and Loose Pin, 70 per cent.

CEMENT AND FIREBRICK.

Canadian Portland .... 2 10 — 2 10
Belgium .... 1 60 — 1 90
White Bros. English .... 2 00 — 2 05
"Lafarge" cement in wood 3 40
"Iron Clad" cotton 2 10
"Iron Clad" paper 2 15
" wood 2 2
Fire brick, Scotch, per 1,000 .... 27 00 — 30 00
" English .... 17 00 — 21 00
" American, low .... 23 0 — 25 00
" " high .... 27 50 — 35 00
Fire clay (Scotch), net ton 4 95

Paving Blocks per 1,000.

Blue metallic, 9"x4½"x3", ex wharf .. 35 00
Stable pavers, 12"x6"x2", ex wharf .. 50 00
Stable pavers, 9"x4½"x3", ex wharf .. 36 00

DOOR SETS.

Peterboro, 37½ and 10 per cent.

DOOR SPRINGS.

Torrey's Rod .... per doz. 1 75
Coil, 9 to 11 in. .... " 0 95 — 1 65
English .... " 2 00 — 4 00
Chicago and Reliance Coil 25 per cent.

ESCUTCHEONS.

Discount 50 and 10 per cent., new list
Peterboro, 37½ and 10 per cent.

ESCUTCHEON PINS.

Iron, discount 40 per cent.
Brass, 45 per cent.

HINGES.

Blind, discount 50 per cent.
Heavy T and strap, 4-in., 100 lb. net .. 7 25
" " 5-in., " 7 00½
" " 6-in., " 6 75½
" " 8-in., " 6 50½
" " 10-in. and larger .. 6 25
Light T and strap, discount 65 p.c.
Screw hook and hinge—
under 12 in .... per 100 lb. 4 75
over 12 in .... " 3 75
Crate hinges and back flaps, 65 and 5 p. c.
Chest hinges and hinge hasps, 65 p.c.

SPRING HINGES.

Spring, per gro., No. 5, $17.50 No. 10, $18; No. 20, $10.80; No. 120, $20; No. 51, $10; No. 50, $27.50.
Chicago Spring Butts and Blanks 12½ per cent.
Triple End Spring Butts, 30 and 10 per cent.
Chicago Floor Hinges, 37½ and 5 off.
Garden City Fire House Hinges, 12½ p c.
"Chief" floor hinge, 50 p.c.

CAST IRON HOOKS.

Bird cage .... per doz. 0 50 — 1 10
Clothes line, No. 61 .. " 0 00 — 0 70
Harness .... " 0 60 — 12 00
Hat and coat .... per gro. 1 10 — 10 00
Chandelier .... per doz. 0 50 — 1 00
Wrought hooks and staples—
¼ x 5 .... per gross 2 65
5-16 x 5 .... " 3 30
Bright wire hooks, 60 p c.
Bright steel gate hooks and staples, 40 p c.
Cresent hat and coat wire, 60 per cent.
Screw, bright wire, 65 per cent.

KNOBS.

Door, japanned and N.P., doz 1 50 — 2 50
Bronze, Berlin .... per doz. 2 75 — 3 25
Bronze, Genuine .... " 6 00 — 9 00
Shutter, porcelain, F. & L. screw .... per gross 1 30 — 2 00
White door knobs .... per doz. 2 00
Peterboro knobs, 37½ and 10 per cent.
Porcelain, mineral and jet knobs, net list.

KEYS.

Lock, Canadian 40 to 40 and 10 per cent.

LOCKS.

Peterboro, 37½ and 10 per cent
Russell & Erwin, steel rim $2 50 per doz.
Eagle cabinet locks, discount 30 per cent
American padlocks, all steel, 10 to 15 per cent.; all brass or bronze, 10 to 25 per cent.

SAND AND EMERY PAPER.

B. & A. sand, discount, 35 per cent
Emery, discount 35 per cent.
Garnet (Rurton's), 5 to 10 per cent. advance

SASH WEIGHTS.

Sectional .... per 100 lb. 2 00 — 2 25
Solid .... " 1 50 — 1 75

SASH CORD.

Per lb. 0 31

BLIND AND BED STAPLES.

All sizes per lb. .... 0 07½ — 0 10

WROUGHT STAPLES.

Galvanized 2 75
Plain 2 50
Coopers', discount 45 per cent.
Poultry netting staples, discount 40 per cent.
Bright spear point, 75 per cent. discount.

TOOLS AND HANDLES.

ADZES.

Discount 22½ per cent.

AUGERS.

Gilmour's, discount 60 per cent. off list.

AXES.

Single bit, per doz. .... 6 00 — 9 10
Double bit, " .... 10 00 — 11 00
Bench Axes, 40 per cent.
Broad Axes, 25 per cent.
Hunters' Axes .... 5 50 — 6 00
Boys' Axes .... 6 25 — 7 00
Splitting Axes .... 7 00 — 12 00
Handled Axes .... 7 00 — 9 00
Red Ridge, boys', handled 5 75
" hunters 5 25

BITS.

Irwin's auger, discount 47½ per cent.
Gilmour's auger, discount 60 per cent.
Rockford auger, discount 50 and 10 per cent.
Jennings' Gen. auger, net list.
Gilmour's car, 47½ per cent.
Clark's expansive, 40 per cent.
Clark's gimlet, per doz 0 65
Diamond, Shell, per doz. 1 00
Nail and Spike, per gross 2 25

BUTCHERS CLEAVERS

German .... per doz. 7 00 — 9 00
American .... " 12 00 — 18 00

CHALK.

Carpenters' Colored, per gross 0 45 — 0 75
White lump .... per cwt. 0 60 — 0 65

CHISELS.

Warnock's, discount 70 and 5 per cent.
P. S. & W. Extra, discount, 70 per cent.

CROSSCUT SAW HANDLES.

S. & D., No. 3 .... per pair 0 13
S. & D., " 5 .... " 0 11½
S. & D., " 6 .... " 0 18
Boynton pattern .... " 0 20

CROWBARS.

3½c. to 4c. per lb.

DRAW KNIVES.

Coach and Wagon, discount 75 and 5 per cent.
Carpenters' discount 75 per cent.

DRILLS.

Millar's Falls, hand and breast, net list.
North Bros., each set, 50c.

DRILL BITS.

Morse, discount 37½ to 40 per cent.
Standard, discount 50 and 5 to 55 per cent.

FILES AND RASPS.

Great Western .... 75 per cent.
Arcade .... 75 "
Kearney & Foot .... 75 "
Disston's .... 75 "
American .... 75 "
J. Barton Smith .... 75 "
McClellan .... 75 "
Eagle .... 75 "
Nicholson .... 66⅔ "
Globe .... 75 "
Black Diamond, 60, 10 and 5 p.c.
Jowitt's, English list, 27½ per cent.

GAUGES.

Stanley's discount 50 to 60 per cent.
Winn's, Nos. 26 to 33 .... each 1 65 — 2 40

HANDLES.

Second growth ash fork, hoe, rake and shovel handles, 40 p c.
Extra ash fork, hoe, rake and shovel handles, 45 p c.
No. 1 and 2 ash fork, hoe, rake and shovel handles, 50 p.c.
White ash whiffletrees and neckyokes, 35 p c.
All other ash goods, 40 p c.
All hickory, maple and oak goods, excepting carriage and express whiffletrees, 40 p c.
Hickory, maple, oak carriage and express whiffletrees, 45 p c.

HAMMERS.

Maydole's, discount 5 to 10 per cent.
Canadian, discount 25 to 27½ per cent.
Magnetic tack .... per doz. 1 10 — 1 20
Canadian sledge .... per lb. 0 07 — 0 08½
Canadian ball pean, per lb. .. 0 22 — 0 25

HATCHETS.

Canadian, discount 40 to 42½ per cent.
Shingle, Red Ridge 1, per doz. 4 40
" " 2, " 4 80
Barrel Underhill 5 05

HOES

Mortar, 50 and 10 per cent

MALLETS.

Tinsmiths' .... per doz. 1 25 1 50
Carpenters', hickory, " 1 25 3 75
Lignum Vitae .... " 3 85 5 00
Caulking, each 0 60 2 00

MATTOCKS.

Canadian .... per doz. 5 50 6 00

MEAT CUTTERS.

German, 15 per cent.
American discount, 33⅓ per cent.

PICKS.

Per dozen 6 00 9 00

PLANES.

Wood bench, Canadian, 40, American, 25 p c
Wood, fancy, 37½ to 40 per cent.
Stanley planes, $1.55 to $3 60, net list prices.

PLANE IRONS.

English .... per doz. 2 00 5 00
Stanley, 2⅝ inch, single 24c., double 39c.

PLIERS AND NIPPERS.

Button's genuine, 37½ to 40 per cent.
Button's imitation .... per doz. 5 00 9 00
Berg's wire fencing 1 72 5 50

PUNCHES.

Saddlers .... per doz. 1 00 1 85
Conductor's .... " 3 00 15 00
Tinners', solid .... per set 0 72
" hollow ...... per inch 1 00

RIVET SETS.

Canadian, discount 35 to 37½ per cent.

RULES.

Boxwood, discount 70 per cent.
Ivory, discount 20 to 25 per cent.

SAWS.

Atkins, hand and crosscut, 25 per cent.
Disston's Hand, discount 12½ per cent
Disston's Crosscut .... per foot 0 35 0 55
Hack, complete .... each 0 75 2 75
" frame only .... each 0 50 1 25

S. & D. solid tooth circular shingle, concave and band, 50 per cent; mill and ice, drag, 30 per cent; cross-cut, 35 per cent.; hand saws, butcher, 35 per cent; buck, New Century, $6.25; buck No. 1 Maple Leaf, $5.25; buck, Happy Medium, $4 25; buck, Watch Spring, $4.25, buck, common frame, $4.00.

Spear & Jackson's saws—Hand or rip, 26 in., $12.75; 28 in., $11.25; panel, 18 in., $8.25; 20 in., $9; tenon, 10 in., $9 90; 12 in., $10 90; 14 in., $11.50

SAW SETS.

Lincoln and Whiting 4 75
Hand Sets, Perfect 4 00
X-Cut Sets, " 7 50
Maple Leaf and Premiums saw sets, 40 off.
S. & D. saw swages, 40 off.

SCREW DRIVERS.

Sargent's .... per doz 0 65 1 00
North Bros., No. 30 . per doz 16 80

SHOVELS AND SPADES.

Canadian, discount 45 per cent.

SQUARES.

Iron, discount 20 per cent.
Steel, discount 65 and 10 per cent.
Try and Bevel, discount 50 to 52½ per cent

TAPE LINES.

English, ass skin ..... per doz. 2 75 5
English, Patent Leather 5 50 9
Chesterman's .... each 0 90 2
" steel .... each 0 80 8 00
Berg's, each 0 75 2 95

TROWELS.

Disston's, discount 10 per cent.
S. & D., discount 35 per cent.
Berg's, brick, 924x11 .... 4 00
" pointing, 924x5 .... 2 10

FARM AND GARDEN GOODS

BELLS.

American cow bells, 63⅔ per cent.
Canadian, discount 45 and 50 per cent.
American, farm bells, each .. 1 35 3 00

BULL RINGS.

Copper, $2.00 for 2½-inch

CATTLE LEADERS.

Nos. 32 and 33 .. ... per gross 7 50 8 50

BARN DOOR HANGERS.

doz. pairs.
Stearns wood track 4 50 6 00
Zenith 9 00
Atlas, steel covered 5 00 6 00
Perfect 8 00 11 00
New Milo, flexible 6 50
Steel, track, 1 x 3-16 in (100 ft) 3 25
" " 1½ x 3-16 in (100 ft) 4 75
Double strap hangers, doz. sets 6 40
Standard jointed hangers, " 6 40
Steel King hangers " 6 25
Storm King and safety hangers 7 00
" " rail 4 25
Chicago Friction, Oscillating and Big Twin Hangers, 5 per cent.

HARVEST TOOLS.

50 and 10 per cent.
S. & D. lawn rakes, Dunn's, 40 off.
" sidewalk and stable scrapers, 40 off.

HAY KNIVES.

Net list.

HEAD HALTERS.

Jute Rope, ⅜-inch .... per gross 9 00
" " ½ " .... " 10 00
" " ⅝ " .... " 12 00
Leather, 1-inch ...... per doz. 4 00
Leather, 1¼ " ...... " 5 20
Web .... " 2 45

HOES.

Garden, 50 and 10 per cent.
Planter .... per doz. 4 00 4 50

LAWN MOWERS

Low wheel, 12, 14 and 16-inch $2 30
9-inch wheel, 12-inch 2 85
" " 14 " 3 00
" " 16 " 3 12½
High wheel, 12 " 4 05
" " 14 " 4 50
" " 16 " 4 75

SCYTHES.

Per doz. net 6 25 9 25

SCYTHE SNATHS.

Canadian, discount 40 per cent.

SNAPS.

Harness, German, discount 25 per cent.
Lock, Andrews' 4 50 11 00

STABLE FITTINGS.

Warden King, 35 per cent
Dennis Wire & Iron Co., 33⅓ p c.

WOOD HAY RAKES.

40 and 10 per cent

HEAVY GOODS, NAILS, ETC.

ANVILS.

Wright's, 80-lb. and over 0 10½
Hay Budden, 80-lb. and over 0 09½
Brook's, 80-lb. and over 0 11½
Taylor-Forbes, prospectors 0 05
Columbia Hardware Co., per lb. 0 09½

VISES.

Wright's 0 13½
Berg's, per lb. 0 12½
Brook's 0 12½
Pipe Vise, Hinge, No. 1 .... 3 50
" " " No. 2 .... 5 50

Saw Vise 4 50 5 00
Blacksmiths' (discount) 60 per cent.
" parallel (discount) 45 per cent.

BOLTS AND NUTS

Carriage Bolts, common ($1 list) — Per cent.
" " ⅜ and smaller .. 60, 10 and 10
" " 7-16 and up .... 55 and 5
" " Norway Iron ($3 list) 50
Machine Bolts, ⅜ and less .... 60 and 10
Machine Bolts, 7-16 and up .... 55 and 5
Plough Bolts .... 55 and 10
Blank Bolts 55
Bolt Ends 55
Sleigh Shoe Bolts, ⅜ and less .. 60 and 10
" " " 7-16 and larger 50 and 5
Coach Screws, conepoint ...... 70 and 5
Nuts, square, all sizes, 4c. per cent. off
Nuts, hexagon, all sizes, 4½c. per cent. off
Stove Rods per lb., 5½ to 6c.
Stove Bolts, 75 per cent.

CHAIN.

Proof coil, per 100 lb., ¼ in., $6.00; 5-16 in., $4.85; ⅜ in., $4 25; 7-16 in., $4.00; ½ in., $3 75; 9-16 in., $3 70; ⅝ in., $3.65; ¾ in., $3 60; ⅞ in., $3 45; 1 in., $3.40.

Halter, kennel and post chains, 40 to 40 and 5 per cent.; Cow ties, 40 per cent.; Tie out chains 65 per cent; Stall fixtures, 35 per cent; Trace chain, 45 per cent.; Jack chain iron, 50 per cent.; Jack chain, brass, 50 per cent.

HORSE NAILS.

M R.M. cold forged process, list May 15, 1907, 50 and 5 per cent.
"C" brand, 57½ per cent. off list.
Capewell brand, quotations on application.

HORSESHOES.

M R M brand: iron, light and medium, No. 1 and smaller, $3 90; No. 2 and larger, $3 65; snow pattern No 1 and smaller $4.15, No. 2 and larger, $3.90; "X L" new light steel, No. 1 and smaller, $4 25; No 2 and larger, $4; "X L" featherweight steel, No. 0 to 4, $5 60; Special countersunk steel No 0 to 4, $6 85 pkg; toe-weight, all sizes, $6.85. F o b. Montreal Extras for packing.

Belleville brand: No. 0 and 1, light and medium iron, $3 90; snow, $4 15; light steel, $4 25; No 2 and larger, light and medium iron, $3 65; snow, $3 90; light steel, $4. F.o b Belleville. Two per cent., 30 days.

Toecalks—Standard No. 1 and smaller, $1 50; No. 2 and larger, $1 25. Blunt No. 1 and smaller, $1.75; No 2 and larger, $1.50 per box. Sharp Put up in 25 lb. bxs.

HORSE WEIGHTS.

Taylor-Forbes, 4½c. per lb.

NAILS.	Cut	Wire.
2d	4 00	3 55
3d	3 15	3 25
4 and 5d	2 90	2 95
6 and 7d	2 80	2 85
8 and 9d	2 65	2 70
10 and 12d	2 60	2 65
16 and 20d	2 55	2 60
30, 40, 50 and 60d (base)	2 50	2 55

F.o.b. Montreal. Cut nails, Toronto 20c. higher.
Miscellaneous wire nails, discount 75 per cent
Coopers' nails, discount 40 per cent.

PRESSED SPIKES.

Pressed spikes, ⅝ diameter, per 100 lbs $3.15

RIVETS AND BURRS.

Iron Rivets, black and tinned, 60, 10 and 10.
Iron Burrs, discount 60 and 10 and 10 p c.
Copper Rivets, usual proportion burrs, 15 p c.
Copper Burrs only, net list.
Extras on Coppered Rivets, ½-lb. packages 1c. per lb.; ¼-lb. packages 2c. lb
Tinned Rivets, net extra, 4c. per lb.

SCREWS.

Wood, F. H., bright and steel, 85 and 10 p.c.
" R. H., bright, 80 and 10 per cent.
" F. H., brass, 75 and 10 per cent.
" R. H., " 70 and 10 per cent.
" F. H., bronze, 70 and 10 per cent
" R. H., " 65 and 10 per cent.
Drive Screws, dis. 87½ per cent.
Bench, wood .... per doz. 3 25 4 00
" iron .... " 4 25 5 00
Set, case hardened, dis. 60 per cent.
Square Cap, dis. 50 and 5 per cent.
Hexagon Cap, dis. 45 per cent.

MACHINE SCREWS.

Flat head, iron and brass, 35 per cent.
Felister head, iron, discount 30 per cent.
" " brass, discount 25 per cent.

TACKS, BRADS, ETC.

Carpet tacks, blued 75 p.c.; tinned, 80 and 10; (in kegs), 40; cut tacks, blued, in dozens only, 75; ½ weigh's, 60; Swedes cut tacks, blued and tinned, bulk, 75 dozens, 75; Swedes, upholsterers', bulk, 85 and 12½; brush, blued and tinned, bulk, 70; Swedes, gimp, blued, tinned and japanned, 75 and 12½; zinc tacks, 35; leather carpet tacks, 40; copper tacks, 25; copper nails 30; trunk nails, black, 65; trunk nails, tinned and blued, 65; clout nails, blued and tinned 65; chair nails, 35; patent brads, 40; fine finishing, 40; lining tacks, in papers, 10; lining tacks, in bulk, 15; lining tacks, solid heads, in bulk, 75; saddle nails, in papers, 10; saddle nails, in bulk, 15; tufting buttons, 22 line in dozens only, 60; zinc glaziers' points, 5; double pointed tacks, papers, 90 and 10; double pointed tacks, bulk, 40; clinch and duck rivets, 45; cheese box tacks, 85 and 5 trunk tacks, 80 and 10.

WROUGHT IRON WASHERS.

Canadian make, discount 40 per cent.

SPORTING GOODS.

CARTRIDGES.

"Dominion" Rim Fire Cartridges and C.B. caps, 50 and 7½ per cent.; Rim Fire B B. Round Caps, 60 and 2½ per cent.; Centre Fire, Pistol and Rifle Cartridges, 30 p c; Centre Fire Sporting and Military Cartridges, 20 and 5 p c; Rim Fire, Shot Cartridges, 50 and 7½ p c.; Centre Fire, Shot Cartridges, 30 p.c.; Primers, 25 p.c.

LOADED SHELLS.

"Crown" Black Powder, 15 and 5 p c; "Sovereign" Empire Bulk Smokeless Powder, 30 and 5 p.c.; "Regal" Ballistite Dense smokeless Powder, 30 and 5 p.c; "Imperial" Empire or Ballistite Powder, 30 and 10 p.c.

EMPTY SHELLS.

Paper Shells, 25 and 5; Brass Shells, 55 and 5 p.c.

Wads. — per lb.
Best thick brown or grey felt wads, in ½-lb. bags $0 70
Best thick white card wads, in boxes of 500 each, 12 and smaller gauges 29
Best thick white card wads, in boxes of 500 each, 10 gauge 0 35
Thin card wads, in boxes of 1,000 each, 12 and smaller gauges 0 20
Thin card wads, in boxes of 1,000 each, 10 gauge 0 25
Chemically prepared black edge grey cloth wads, in boxes of 250 each— Per M
11 and smaller gauge 0 60
9 and 10 gauges 0 70
7 and 8 " 0 90
5 and 6 " 1 10
Superior chemically prepared pink edge, best white cloth wads in boxes of 250 each—
11 and smaller gauge 1 15
9 and 10 gauges 1 40
7 and 8 " 1 65
5 and 6 " 1 90

SHOT.

Ordinary drop shot, AAA to dust $7.50 per 100 lbs. Discount 5 per cent; cash discount, 2 per cent, 30 days; net extras as follows subject to cash discount only; Chilled, 40 c.; buck and seal, 80c; no. 28 ball, $1 20 per 100 lbs.; bags less than 25 lbs., ½c. per lb.; F.O.B. Montreal, Toronto, Hamilton, London, St John and Halifax, and freight equalized thereon.

TRAPS (steel.)

Game, Newhouse, discount 30 and 10 per cent.
Game, Hawley & Norton, 50, 10 & 5 per cent
Game, Victor, 70 per cent.
Game, Oneida Jump (B. & L.) 40 & 2½ p.
Game, steel, 60 and 5 per cent.

SKATES.

Skates, discount 37½ per cent.
Empire hockey sticks, per doz .. 3 00

CUTLERY AND SILVERWARE.

RAZORS. per doz.

Elliot's 4 00 18 00
Boker's 7 50 11 00
" King Cutter 13 50 18 50
Wade & Butcher's.......... 3 60 10 00
Lewis Bros.' "Klean Kutter" 8 50 10 50
Henckel's 7 50 20 00
Berg's 7 50 20 00
Clauss Razors and Strops, 50 and 10 per cent

KNIVES.

Farriers-Stacey Bros., doz 3 50

PLATED GOODS

Hollowware, 40 per cent. discount.
Flatware, staples, 40 and 10, fancy, 40 and 5.
Hutton's "Cross Arrow" flatware, 47½; "Singalese" and "Alaska" Nevada silver flatware, 42 p c.

SHEARS.

Clauss, nickel, discount 60 per cent.
Clauss, Japan, discount 67½ per cent.
Clauss, tailors, discount 40 per cent.
Seymour's, discount 50 and 10 per cent.
Berg's.......... 6 00 12 00

HOUSE FURNISHINGS.

APPLE PARERS.

Hudson, per doz., net 5 75

BIRD CAGES.

Brass and Japanned, 40 and 10 p. c.

COPPER AND NICKEL WARE.

Copper boilers, kettles, teapots, etc. 37½ p.c
Copper pitts, 30 per cent.

KITCHEN ENAMELED WARE.

White ware, 75 per cent.
London and Princess, 50 per cent.
Canada, Diamond, Premier, 50 and 10 p c.
Pearl, Imperial, Crescent and granite steel, 50 and 10 per cent.
Premier steel ware, 40 per cent.
Star decorated steel and white, 25 per cent.
Japanned ware, discount 45 per cent.
Hollow ware, tinned cast, 35 per cent. off.

KITCHEN SUNDRIES.

Can openers, per doz.......... 0 40 0 75
Mincing knives per doz 0 50 0 80
Duplex mouse traps, per doz.... .... 0 65
Potato mashers, wire, per doz... 0 60 0 70
" wood " .. 0 50 0 60
Vegetable slicers, per doz 2 25
Universal meat chopper No. 1.. .... 1 15
Enterprise chopper, each 1 30
Spiders and fry pans, 50 per cent.
Star Al chopper 5 to 32 .... 1 35 4 10
" " 100 to 103.......... 1 35 2 00
Kitchen hooks, bright.......... .... 0 60

LAMP WICKS.

Discount, 60 per cent.

LEMON SQUEEZERS.

Porcelain lined...... per doz. 2 20 5
Galvanized.......... " 1 87 3
King, wood.......... " 2 75 2
King, glass.......... " 4 00 4 60
All glass .... " 0 50 0 90

METAL POLISH.

Tandem metal polish paste 6 00

PICTURE NAILS.

Porcelain head ......per gross 1 35 1
Brass head.......... 0 40 1 60
Tin and gilt, picture wire, 75 per cent.

SAD IRONS.

Mrs. Potts, No. 55, polished....per set 0 90
" No. 50, nickle-plated, " 0 95
" handles, japaned, per gross 9 25
" " nickled, " 9 75
Common, plain.......... 4 25
" plated 5 50
Asbestos, per set.......... .... 1 50

TINWARE.

CONDUCTOR PIPE.

2-in. plain or corrugated., per 100 feet, $3.30; 3 in., $4.40; 4 in., $5.80; 5 in., $7.45; 6 in., $9.90.

FAUCETS.

Common, cork-lined, discount 35 per cent.

EAVETROUGHS.

10-inch .... per 100 ft. 3 30

FACTORY MILK CANS.

Discount off revised list, 35 per cent.
Milk can trimmings, discount 25 per cent.
Creamery Cans, 45 per cent

LANTERNS.

No. 2 or 4 Plain Cold Blast....per doz. 6 50
Lift Tubular and Hinge Plain, " 4 75
No. 0, safety .... " 4 00
Better quality at higher prices.
Japanning, 50c. per doz. extra.
Prism globes, per doz., $1 20.

OILERS.

Kemp's Tornado and McClary's Model galvanized oil can, with pump, 5 gallon, per dozen 10 92
Davidson oilers, discount 40 per cent.
Zinc and tin, discount 50 per cent
Coppered oilers, 20 per cent. off.
Brass oilers, 50 per cent. off.
Malleable, discount 25 per cent

PAILS (GALVANIZED).

Dufferin pattern pails, 45 per cent.
Flaring pattern, discount 45 per cent.
Galvanized washtubs 40 per cent.

PIECED WARE.

Discount 35 per cent off list, June, 1899.
10-qt. flaring sap buckets, discount 35 per cent.
6, 10 and 14-qt. flaring pails dis 35 per cent.
Copper bottom tea kettles and boilers, 30 p.c.
Coal hods, 40 per cent.

STAMPED WARE.

Plain, 75 and 12½ per cent. off revised list.
Retinned, 72½ per cent. revised list.

SAP SPOUTS

Bronzed iron with hooks ....per 1,000
Eureka tinned steel, hooks 8 00

STOVEPIPES.

5 and 6 inch, per 100 lengths 7 64 7 91
7 inch .... " " 8 18
Nestable, discount 40 per cent.

STOVEPIPE ELBOWS

5 and 6-inch, common..........per doz. 1 32
7-inch.......... " 1 48
Polished, 15c. per dozen extra.

THERMOMETERS.

Tin case and dairy, 75 to 75 and 10 per cent.

TINNERS' SNIPS.

Per doz.......... 3 00 15
Clauss, discount 35 per cent.

TINNERS' TRIMMINGS

Discount, 45 per cent.

WIRE.

ANNEALED CUT HAY BAILING WIRE.

No. 12 and 13, $4; No. 13½, $4 10; No. 14, $4.25; No. 15, $4.50; in lengths 6' to 11', 25 per cent; other lengths 20c. per 100 lbs extra; if eye or loop on end add 25c. per 100 lbs. to the above.

BRIGHT WIRE GOODS

Discount 60 per cent.

CLOTHES LINE WIRE.

No. 7 wire solid line, No. 17, $4.90; No. 18, $3.00; No. 19, $2 70; 8 wire solid line, No. 17, $4.45; No. 18, $3.10; No. 19, $2 8[illegible]. All prices per 1000 ft measure; 6 strand, No. 18, $2 60; No. 19, $2 90. F.o.b. Hamilton, Toronto, Montreal.

COILED SPRING WIRE.

High Carbon, No. 9, $2 95; No. 11, $3.50; No. 12, $3.20.

COPPER AND BRASS WIRE.

Discount 50 per cent.

FINE STEEL WIRE.

Discount 25 per cent. List of extras In 100-lb. lots: No. 17, $5 — No. 18, $5.50 — No. 19, $6 — No. 20, $6.65 — No. 21, $7 — No. 22, $7.30 — No. 23, $7.65 — No. 24, $8 — No. 25, $9 — No. 26, $9.50 — No. 27, $10 — No. 28, $11 — No. 29, $12 — No. 30, $13 — No. 31, $14 — No. 32, $15 — No. 33, $16 — No. 34, $17. Extras net—tinned wire, Nos. 17-25 $2 — Nos. 26-31, $4 — Nos. 32-34, $6. Coppered, 75c.—oiling, 10c.—in 25-lb. bundles, 15c.—in 5 and 10-lb. bundles, 25c.—in 1-lb. hanks, 25c. —in ½-lb. hanks, 38c.—in ¼-lb. hanks, 50c. packed in casks or cases, 15c.—bagging or papering, 10c

FENCE STAPLES.

Bright ... 2 80 Galvanized .. 3[illegible]

HAY WIRE IN COILS.

No 13, $2.70; No. 14, $2.80; No. 15, $2.95; f o.b., Montreal.

GALVANIZED WIRE.

Per 100 lb.—Nos. 4 and 5, $3 95 — Nos. 6, 7, 8, $3 40 — No. 9, $2.90 — No. 10, $3.45 — No. 11, $3 50 — No. 12, $3.05 — No. 13, $3 15 — No. 14, $4 00 — No 15, $4.35 — No. 16, $4.35 from stock. Base sizes, Nos. 6 to 9, $2.35 f.o.b. Cleveland. Extras for cutting.

LIGHT STRAIGHTENED WIRE.

Gauge No.	Over 20 in.	10 to 20 in.	5 to 10 in.
0 to 5	$0.50	$0.75	$1.25
6 to 9	0.75	1 25	2 00
10 to 11	1.00	1 75	2.50
12 to 14	1.50	2.25	3 50
15 to 16	2.00	3.00	4.50

SMOOTH STEEL WIRE.

No. 0-9 gauge, $2.40; No. 10 gauge, 6c extra; No. 11 gauge, 12c extra; No. 12 gauge, 20c. extra; No 13 gauge, 30c. extra; No 14 gauge, 40c. extra; No. 15 gauge, 55c. extra; No. 16 gauge, 70c. extra. Add 60c. for coppering and $2 for tinning

Extra net per 100 lb.—Oiled wire 10c., spring wire $1.25, bright soft drawn 15c., charcoal (extra quality) $1.25, packed in casks or cases 15c., bagging and papering 10c., 50 and 100-lb. bundles 10c., in 25-lb. bundles 15c., in 5 and 10-lb. bundles 25c., in 1-lb hanks, 50c., in ½-lb. hanks 75c., in ¼-lb. hanks $1.

POULTRY NETTING.

2-in. mesh, 19 w g., 50 and 5 p.c. off. Other sizes, 50 and 5 p c. off.

WIRE CLOTH.

Painted Screen, in 100-ft. rolls, $1.72½, per 100 sq. ft.; in 50-ft. rolls, $1.77½, per 100 sq ft.

WIRE FENCING.

Galvanized barb...... .... 2 95
Galvanized, plain twist 3 30
Galvanized barb, f.o b. Cleveland, $2.70 for small lots and $2.60 for carlots

WIRE ROPE

Galvanized, 1st grade, 6 strands, 24 wires, ⅜, $5; 1 inch $16 80.
Black, 1st grade, 6 strands, 19 wires, ⅜, $5; 1 inch $15.10. Per 100 feet f o.b Toronto

WOODENWARE.

CHURNS.

No. 0, $9; No 1, $9; No. 2, $10; No. 3, $11; No. 4, $13; No. 5, $16; f.o b. Toronto Hamilton, London and St. Marys, 30 and 30 per cent; f.o b. Ottawa, Kingston and Montreal, 40 and 15 per cent. discount.

CLOTHES REELS.

Davis Clothes Reels, dis. 40 per cent.

FIBRE WARE.

Star pails, per doz .... $ 3 00
0 Tubs, " 14 00
1 " " 12 00
2 " " 10 00
3 " " 8 50

LADDERS, EXTENSION.

3 to 6 feet, 12c. per foot; 7 to 10 ft., 13c.
Waggoner Extension Ladders, dis. 40 per cent.

MOPS AND IRONING BOARDS.

"Best" mops.......... .... 1 25
"900" mops.......... .... 1 25
Folding ironing boards..........12 00 16 50

REFRIGERATORS

Discount, 40 per cent.

SCREEN DOORS.

Common doors, 2 or 3 panel, walnut stained, 4-in. style..........per doz. 7 25
Common doors, 2 or 3 panel, grained only, 4-in., style..........per doz. 7 55
Common doors, 2 or 3 panel, light stair per doz.......... 9 55

WASHING MACHINES.

Round, re-acting per doz. 60 00
Square " 63 00
Eclipse, per doz 54 00
Dowswell " 39 00
New Century, per doz 75 00
Daisy.......... 54 00
Stephenson.......... 74 00

WRINGERS.

Royal Canadian, 11 in., per doz. ... 35 00
Royal American, 11 in. ... 35 00
Eze 10 in., per doz 36 75

MISCELLANEOUS

AXLE GREASE.

Ordinary, per gross.......... 6 00 7 00
Best quality.......... 10 00 12 00

BELTING.

Extra, 60 per cent.
Standard, 60 and 10 per cent.
No. 1, not wider than 6 in., 60, 10 and 10 p.c
Agricultural, not wider than 4 in., 75 per cent.
Lace leather, per side, 75c.; cut laces, 80c.

BOOT CALKS.

Small and medium, ball ....per M 2 25
Small heel 4 50

CARPET STRETCHERS.

American ....per doz. 1 00 1 50
Bullard's.......... " 6 50

CASTORS.

Bed, new list, discount 55 to 57½ per cent.
Plate, discount 52½ to 57½ per cent.

PINE TAR.

½ pint in tins ....per gross ... 7 80
" " .... " 9 60

PULLEYS.

Hothouse ....per doz. 0 55 1 00
Axle.......... " 0 22 0 33
Screw .... " 0 22 1 00
Awning .... " 0 35 2 50

PUMPS.

Canadian cistern 1 40 2 00
Canadian pitcher spout.......... 1 60 3 16
Berg's wing pump, 75 per cent.

ROPE AND TWINE.

Sisal.......... 0 10½
Pure Manilla 0 15
"British" Manilla.......... 0 12
Cotton, 3-16 inch and larger..........0 0 23
" 5-32 inch 0 21 0 27
" ¼ inch ....0 [illegible] 0 28
Russia Deep Sea 0 16
Jute.......... 0 09
Lath Yarn, single 0 10
" double 0 10½
Sisal bed cord, 48 feet..........per doz. 0 65
" " 60 feet.......... " 0 80
" " 72 feet.......... " 0 95

Twine.

Bag, Russian twine, per lb. 0 27
Wrapping, cotton, 3-ply 0 25
" " 4-ply 0 29
Mattress twine per lb. 0 33 0 45
Staging " " 0 27 0 35

BINDER TWINE.

500 feet, sisal .... 0 09[illegible]
500 " standard .... 0 09[illegible]
550 " " manilla .... 0 10[illegible]
600 " " " .... 0 12[illegible]
650 " " " .... 0 13[illegible]
Car lots, ½c. less; 5-ton lots, ¼c. less. Central delivery.

SCALES.

Gurney Standard, 35; Champion, 45 p c.
Burrow, Stewart & Milne — Imperial Standard, 35; Weigh Beams, 35; Champion Scales, 45.
Fairbanks Standard, 30; Dominion, 50 Richelieu, 50.
Warren new Standard, 35; Champion, 45 Weigh Beams, 30.

STONES—OIL AND SCYTHE.

Washita ....per lb. 0 25 0
Hindostan .... " 0 06 0 10
" slip .... " 0 18 0 20
" Axe.......... " 0 16
Deer Creek.......... " 0 10
Deerlick .... " 0 25
" Axe .... " 0 15
Lily white .... " 0 42
Arkansas.......... " 1 50
Water-of-Ayr .... " 0 10
Scythe ....per gross 3 50 5 00
Grind, 40 to 200 lb., per ton.... 20 00 22 00
" under 40 lb., " 24 00
" 200 lb. and over 26 00

INDEX TO ADVERTISERS.

Occasionally advertisements are inserted in the paper after the index has been printed. The insertion of the advertiser's name in this index is not part of contract, but an effort is made to have index as complete as possible.

CLASSIFIED LIST OF ADVERTISEMENTS.

Alabastine.
Alabastine Co., Limited, Paris, Ont.

Auditors.
Davenport, Pickup & Co., Winnipeg.

Automobile Accessories.
Canada Cycle & Motor Co, Toronto Junction.
Carriage Mountings Co, Ltd., Toronto.

Awnings.
Tobin Tent and Awning Co., Ottawa

Babbitt Metal.
Canada Metal Co., Toronto.
Canadian Fairbanks Co., Montreal.
Frothingham & Workman, Montreal.
Robertson, Jas. Co., Montreal.

Bar Urns.
Buffalo Mfg. Co., Buffalo, N.Y.

Bath Room Fittings.
Buffalo Mfg. Co., Buffalo, N.Y.
Carriage Mountings Co., Ltd, Toronto.
Forsyth Mfg. Co., Buffalo, N.Y.
Ontario Metal Novelty Co., Toronto

Belting, Hose, etc.
Gutta Percha and Rubber Mfg. Toronto
Sadler & Haworth, Toronto.

Bicycles and Accessories.
Johnson's, Iver, Arms and Cycle Works Fitchburg, Mass

Binder Twine.
Consumers Cordage Co., Montreal.

Bolts.
Toronto & Belleville Rolling Mills, Belleville

Box Strap.
J. N. Warminton, Montreal.

Brass Goods.
Frothingham & Workman, Montreal.
Glauber Brass Mfg Co., Cleveland, Ohio.
Kerr Engine Co., Walkerville, Ont.
Lewis, Rice, & Son., Toronto.
Morrison, Jas., Brass Mfg. Co., Toronto.
Mueller Mfg Co., Decatur, Ill.
Penberthy Injector Co., Windsor, Ont.
Taylor-Forbes Co., Guelph, Ont.

Bronze Powders.
Canadian Bronze Powder Works, Montreal.

Brushes.
United Factories, Toronto.

Cans
Acme Can Works, Montreal.

Builders' Tools and Supplies.
Covert Mfg. Co., West Troy, N.Y.
Frothingham & Workman Co., Montreal.
Howland, H. S. Sons & Co., Toronto.
Hyde, F., & Co., Montreal.
Lewis Bros. & Co., Montreal.
Lewis, Rice, & Son, Toronto.
Lockerby & McComb, Montreal.
Lufkin Rule Co., Saginaw, Mich.
Newman & Sons, Birmingham.
North Bros. Mfg. Co., Philadelphia, Pa.
Stanley Rule & Level Co., New Britain.
Stanley Works, New Britain, Conn.
Stephens, G. F., Winnipeg.
Taylor-Forbes Co., Guelph, Ont.

Carriage Accessories.
Carriage Mountings Co, Ltd, Toronto.
Covert Mfg. Co., West Troy, N.Y.

Carriage Springs and Axles.
Guelph Spring and Axle Co., Guelph.

Carpet Beateas.
Ontario Metal Novelty Co., Toronto.

Cartridges.
Dominion Cartridge Co., Montreal.

Cattle and Trace Chains.
Greening, B., Wire Co., Hamilton.

Chafing Dishes.
Buffalo Mfg. Co., Buffalo, N.Y.

Churns.
Dowswell Mfg. Co., Hamilton.

Clippers—All Kinds.
American Shearer Mfg. Co., Nashua, N.H

Clothes Reels and Lines.
Hamilton Cotton Co., Hamilton,

Clutch Nails.
J. N. Warminton, Montreal.

Congo Roofing
Buchanan Foster Co., Philadelphia, Pa.

Cordage.
Consumers' Cordage Co., Montreal.
Hamilton Cotton Co., Hamilton.

Cork Screws.
Erie Specialty Co., Erie, Pa.

Cow Ties
Greening, B., Wire Co., Hamilton

Cuspidors.
Buffalo Mfg. Co., Buffalo, N.Y.

Cut Glass.
Phillips, Geo., & Co., Montreal.

Cutlery—Razors, Scissors, etc.
Birkett, Thos., & Son Co., Ottawa.
Clauss Shear Co., Toronto
Dorken Bros. & Co., Montreal.
Frothingham & Workman Montreal.
Heinisch's, R., Sons Co., Newark, N.J.
Howland, H. S. Sons & Co., Toronto.
Hutton, Wm., & Sons, Ltd., London, Eng.
Lamplough, F. W., & Co., Montreal.
Phillips, Geo., & Co., Montreal.
Round, John, & Son, Montreal.

Electric Fixtures.
Canadian General Electric Co., Toronto
Morrison James, Mfg. Co., Toronto.
Munderloh & Co., Montreal.

Electro Cabinets.
Cameron & Campbell Toronto.

Enameled Ware
Kemp Mfg. Co., Toronto.

Engines, Supplies, etc.
Kerr Engine Co., Walkerville, Ont.

Eavetroughs
Wheeler & Bain, Toronto

Fencing—Woven Wire.
Dominion Wire Mfg. Co., Montreal
Owen Sound Wire Fence Co., Owen Sound.
Banwell Hoxie Wire Fence Co., Hamilton.

Files and Rasps.
Barnett Co., G. & H., Philadelphia, Pa.
Nicholson File Co., Port Hope

Firearms and Ammunition.
Hamilton Rifle Co., Plymouth, Mich.
Harrington & Richardson Arms Co, Worcester, Mass.
Johnson's, Iver, Arms and Cycle Works Fitchburg, Mass.

Fishing Tackle.
Enterprise Mfg. Co., Akron, Ohio

Food Choppers
Enterprise Mfg. Co., Philadelphia, Pa.
Lamplough, F. W., & Co., Montreal.
Shirreff Mfg. Co., Brockville, Ont.

Furnaces.
Pease Foundry Co, Toronto.

Galvanizing.
Canada Metal Co., Toronto.
Dominion Wire Mfg. Co., Montreal.
Montreal Rolling Mills Co., Montreal.
Ontario Wind Engine & Pump Co., Toronto.

Glass Ornamental
Hobbs Mfg. Co, London
Consolidated Plate Glass Co., Toronto

Glaziers' Diamonds.
Gibsone, J. B., Montreal.
Pelton, Godfrey S.
Sharratt & Newth, London, Eng.
Shaw, A., & Son, London, Eng.

Handles.
Still, J. H., Mfg. Co.

Harvest Tools.
Maple Leaf Harvest Tool Co., Tillsonburg, Ont.

Hockey Sticks
Still, J. H. Mfg. Co., St. Thomas.

Hoop Iron.
Frothingham & Workman Montreal.
Montreal Rolling Mills Co., Montreal.
J. N. Warminton, Montreal.

Horse Blankets.
Heney, E. N., & Co., Montreal.

Horseshoes and Nails.
Canada Horse Nail Co., Montreal.
Montreal Rolling Mills, Montreal.
Capewell Horse Nail Co., Toronto
Toronto & Belleville Rolling Mills Belleville.

Hot Water Boilers and Radiators.
Cluff, R. J., & Co Toronto.
Pease Foundry Co. Toronto.
Taylor-Forbes Co., Guelph.

Ice Cream Freezers.
Dana Mfg. Co., Cincinnati, Ohio.
North Bros. Mfg. Co., Philadelphia, Pa.

Ice Cutting Tools.
Erie Specialty Co., Erie, Pa.
North Bros. Mfg. Co., Philadelphia, Pa.

Injectors—Automatic.
Morrison, Jas., Brass Mfg. Co., Toronto.
Penberthy Injector Co., Windsor, Ont.

Iron Pipe.
Montreal Rolling Mills, Montreal.

Iron Pumps.
Lamplough, F. W., & Co., Montreal.
McDougall, R., Co., Galt, Ont.

Ladders.
The Waggoner Ladder Co., London, Ont.

Lanterns.
Kemp Mfg. Co., Toronto.
Ontario Lantern Co., Hamilton, Ont.
Wright, E. T., & Co., Hamilton.

Lawn Mowers.
Birkett, Thos., & Son Co., Ottawa.
Frothingham & Workman, Montreal.
Maxwell, D., & Sons, St. Mary's, Ont.
Taylor, Forbes Co., Guelph.

Lawn Mower Grinders
Root Bros. & Co., Plymouth, Ohio.

Ledgers—Loose Leaf.
Business Systems, Toronto.
Copeland-Chatterson Co., Toronto.
Crain, Rolla L., Co., Ottawa.
Universal Systems, Toronto.

Lithographing.
London Printing & Lithographing Co., London, Ont.

Locks, Knobs, Escutcheons, etc.
Peterborough Lock Mfg. Co., Peterborough, Ont.
National Hardware Co., Orillia, Ont.

Lumbermen's Supplies.
Pink, Thos., & Co., Pembroke Ont.

Lye
Gillett, E. W., & Co., Toronto

Manufacturers' Agents.
Fox, C. H., Vancouver.
McIntosh, H F., & Co., Toronto.
Gibb, Alexander, Montreal.
Scott, Bathgate & Co., Winnipeg.

Metals.
Canada Iron Furnace Co., Midland, Ont.
Canada Metal Co., Toronto.
Eadie, H. G., Montreal.
Frothingham & Workman, Montreal.
Gibb, Alexander, Montreal.
Kemp Mfg. Co., Toronto
Leslie, A. C., & Co., Montreal.
Lysaght, John, Bristol, Eng.
Nova Scotia Steel and Coal Co., New Glasgow, N.S.
Robertson, Jas., Co., Montreal.
Roper, J. H., Montreal.
Samuel, Benjamin & Co., Toronto.
Stairs, Son & Morrow, Halifax, N.S.
Thompson, B. & S. H. & Co. Montreal.

Metal Lath.
Galt Art Metal Co., Galt.
Metallic Roofing Co., Toronto.
Metal Shingle & Siding Co., Preston, Ont.

Metal Polish, Emery Cloth, etc.
Oakey, John, & Sons, London, Eng.

Nails Wire
Dominion Wire Mfg. Co., Montreal.

Oilers
Maple City Mfg Co., Monmouth, Ill.

Oil Tanks
Bowser, S. F., & Co., Torono.

Ornamental Iron and Wire.
Dennis Wire & Iron Co, London, Ont.

Packing.
Gutta Percha & Rubber Co Toronto

Paints, Oils, Varnishes, Glass.
Blanchite Process Paint Co., Toronto.
Brandram-Henderson, Montreal
Canada Paint Co., Montreal.
Canadian Oil Co.. Toronto
Consolidated Plate Glass Co., Toronto.
Dods, P. D., & Co., Montreal
Imperial Varnish and Color Co., Toronto
Jamieson, R. C., & Co., Montreal.
Lucas John & Co., New York
McArthur, Corneille & Co., Montreal.
McCaskill, Dougall & Co., Montreal.
Moore, Benjamin, & Co. Toronto.
Ottawa Paint Works, Ottawa
Queen City Oil Co., Toronto.
Ramsay & Son, Montreal
Sanderson Pearcy & Co., Toronto
Sherwin-Williams Co., Montreal.
Standard Paint Co., Montreal
Standard Paint and Varnish Works Windsor, Ont.
Stephens & Co., Winnipeg.
Martin-Senour Co., Montreal
Winnipeg Paint & Glass Co., Winnipeg

Perforated Sheet Metals.
Greening, B., Wire Co., Hamilton

Plumbers' Tools and Supplies.
Canadian Fairbanks Co., Montreal.
Cluff, R. J., & Co., Toronto.
Frothingham & Workman, Montreal.
Glauber Brass Co., Cleveland, Ohio.
Jardine, A. B., & Co, Hespeler, Ont.
Jenkins Bros., Boston, Mass.
Kerr Engine Co, Walkerville, Ont.
Lewis, Rice, & Son, Toronto.
Merrell Mfg. Co., Toledo, Ohio.
Montreal Rolling Mills, Montreal.
Morrison, Jas., Brass Mfg. Co., Toronto.
Mueller, H., Mfg. Co., Decatur, Ill.
Oshawa Steam & Gas Fitting Co., Oshaw
Robertson Jas., Co. Montreal.
Robertson, Jas., Co., Limited, Toronto
Somerville, Limited, Toronto
Stairs, Son & Morrow, Halifax, N.S.
Standard Ideal Sanitary Co., Port Hope,
Standard Sanitary Co., Pittsburg.
Stephens, G. F., & Co., Winnipeg, Man.
Turner Brass Works, Chicago.
Vickery, Orlando, Toronto.

Polishes.
Majestic Polishes, Toronto

Portland Cement.
International Portland Cement Co Ottawa, Ont.
Hanover Portland Cement Co., Hanover, Ont.
Hyde, F., & Co., Montreal.
Thompson B. & S. H. & Co., Montreal.

Poultry Netting.
Greening, B., Wire Co., Hamilton, Ont.

Printing.
London Printing & Lithographing Co., London, Ont.

Razors.
Clauss Shear Co., Toronto.

Refrigerators.
Fabien, C. P., Montreal.

Registers.
Pease Foundry Co., Toronto.

Roofing Supplies.
Brantford Roofing Co. Brantford.
Barrett Mfg. Co., New York.
F. W. Bird, East Walpole, Mass.
Buchanan Foster Co., Philadelphia, Pa.
McArthur, Alex., & Co., Montreal.
Metal Shingle & Siding Co., Preston, Ont.
Metallic Roofing Co., Toronto.
Paterson Mfg. Co., Toronto & Montreal.
Standard Paint Co. Montreal.
Wheeler and Bain, Toronto

Saws.
Atkins, E. C., & Co., Indianapolis, Ind
Shurly & Dietrich, Galt, Ont.
Spear & Jackson, Sheffield, Eng.

Scales.
Canadian Fairbanks Co., Montreal.
Frothingham & Workman, Montreal.

Screw Cabinets.
Cameron & Campbell, Toronto.

Screws, Nuts, Bolts.
Dominion Wire Mfg Co., Montreal.
Montreal Rolling Mills Co., Montreal.

Soil Pipe
McFarlane, Walter, Glasgow

Sewer Pipes
Canadian Sewer Pipe Co., Hamilton
Hyde, F., & Co., Montreal.

Shelf Boxes.
Cameron & Campbell, Toronto.

Shears, Scissors.
Clauss Shear Co., Toronto.

Shovels and Spades.
Eclipse Mfg Co., Ottawa
Frothingham & Workman, Montreal.
Peterboro Shovel & Tool Co., Peterboro.

Silverware.
Hutton, Wm., & Sons, Ltd., London, Eng.
McGlashan, Clarke Co, Niagara Falls, Ont.
Phillips, Geo., & Co., Montreal.
Round, John, & Son, Sheffield, Eng.

Skates.
Canada Cycle & Motor Co., Toronto.
McFarlane, Walter, Glasgow.

Sprayers
Cavers Bros., Galt

Spring Hinges, etc.
Chicago Spring Butt Co., Chicago, Ill.

Stable Fittings
Dennis Wire & Iron Co., London

Steel Rails.
Nova Scotia Steel & Coal Co., New Glasgow, N.S.

Stove Pipe.
Chown, Edwin, and Son, Kingston

Stoves, Tinware, Furnaces
Canadian Heating & Ventilating Co. Owen Sound.
Copp, W. J., Son & Co, Fort William
Davidson, Thos., Mfg. Co., Montreal.
Down Draft Furnace Co., Galt
Guelph Stove Co., Guelph.
Gurney Foundry Co., Toronto.
Harris, J. W., Co., Montreal.
Howard, Wm., Toronto
Kemp Mnfg. Co. Toronto.
McClary Mfg. Co. London.
Merrick Anderson, Winnipeg
Pease Foundry Co., Toronto.
Smart, James, Mfg. Co., Brockville
Stewart, Jas., Mfg. Co., Woodstock, Ont.
Taylor-Forbes Co., Guelph, Ont.
Wright, E. T., & Co., Hamilton.

Tacks.
Montreal Rolling Mills Co., Montreal.
Ontario Tack Co., Hamilton.

Tents.
Tobin Tent and Awning Co., Ottawa

Tin Plate.
American Sheet & Tin Plate Co, Pittsburg, Pa.
Baglan Bay Tin Plate Co., Briton Ferry South Wales
Lysaght, John, Bristol, Newport and Montreal

Turpentine
Defiance Mfg. Co., Toronto.

Ventilators.
Harris, J. W., Co., Montreal.
Pearson, Geo. D., Montreal.

Wall Paper
Staunton Limited, Toronto.

Wall Paper Cleaner.
Gilbert, Frank U. S., Cleveland

Washing Machines, etc
Dowswell Mfg. Co., Hamilton, Ont.
The Shultz Bros. Co., Brantford.
Taylor-Forbes Co., Guelph, Ont.

Water Filters.
Buffalo Mfg. Co., Buffalo, N.Y.

Wheelbarrows
London Foundry Co., London Ont.
Schultz Bros. Co., Ltd., The Brantford.

Wholesale Hardware.
Birkett, Thos., & Sons Co., Ottawa.
Caverhill, Learmont & Co., Montreal.
Frothingham & Workman, Montreal.
Hobbs Hardware Co., London.
Howland, H. S., Sons & Co., Toronto.
Lamplough, F. W., & Co., Montreal.
Lewis Bros. & Co., Montreal.
Lewis, Rice, & Son, Toronto.

Window and Sidewalk Prisms
Hobbs Mfg. Co., London, Ont.

Wire, Wire Rope, Cow Ties, Fencing Tools, etc.
Banwell-Hoxie Fence Co., Hamilton
Dennis Wire and Iron Co., London, Ont.
Dominion Wire Mnfg. Co., Montreal
Greening, B., Wire Co., Hamilton.
Owen Sound Wire Fence Co., Owen Sound
Montreal Rolling Mills Co., Montreal.
Western Wire & Nail Co., London, Ont.

Wrapping Papers.
Canada Paper Co., Toronto.
McArthur, Alex., & Co., Montreal.
Stairs, Son & Morrow, Halifax, N.S.

Wringers
Connor, J. H. & Son, Ottawa, Ont

Lightning Source UK Ltd.
Milton Keynes UK
UKHW031133290119
336362UK00009B/277/P

9 781527 800021